JUPITER

Looking back at Jupiter and Io

JUPITER

Studies of the interior,
atmosphere, magnetosphere
and satellites.

Edited by
TOM GEHRELS

With the assistance of
MILDRED SHAPLEY MATTHEWS

THE UNIVERSITY OF ARIZONA PRESS
TUCSON, ARIZONA

CONTENTS

Part III—ATMOSPHERE AND IONOSPHERE

Part IV — MAGNETOSPHERE AND RADIATION BELTS

PREFACE

We may never do a better book. The timing was opportune as it coincided with the great surge in interest following the first flyby of Jupiter (December 1973). Shortly after that encounter the plan for this book was formulated, and a list of possible chapters was widely circulated, asking for volunteering authors. Our purpose was to make a source and textbook where none existed before. But especially fortunate for the making of this book was the whole-hearted effort and cooperation of the many experts. The overwhelming response, with sometimes four volunteers for the same chapter, caused problems that were partly resolved by asking some authors to write jointly. There was a meeting of most of the authors, but that had little to do with this book of *review* chapters, other than to serve as a catalyst. (The 36 *contributed* papers are in Special Jupiter Issues of Icarus and Journal of Geophysical Research, 1976.) Extensive refereeing has been executed with at least two experts thoroughly inspecting each chapter, and this was followed by partial rewriting by the authors. The *Discussions* were conducted in writing, and they also have been refereed. The effective date of the material in this book is approximately November 1975.

The above method I believe to be superior over others in the making of a comprehensive technical text. The writing by a single author is hardly feasible for the many different aspects of Jupiter. A Compendium-type book with half a dozen or so major chapters would probably be too slow, for lack of excitement and prompting by a conference and by the participation of so many of one's peers; and how would one select the half-dozen authors impartially? Finally, Conference Proceedings — without the special and prior selection of review chapters and their authors, without the strict refereeing and editing, without Introduction and Overview chapters, cross references, Glossary, and Index — generally are inferior volumes of temporary merit. Whether or not to include some contributed papers still was an interesting question; it was done in our previous book (*Planets, Stars, and Nebulae, Studied with Photopolarimetry,* University of Arizona Press, 1974), while

Burns initiated a book of review chapters only (*Planetary Satellites,* University of Arizona Press, 1976). However, a combination in this book would not have been feasible as the topics of Jupiter are too many for a single volume.

A disadvantage of a textbook by the above method, written by many authors, is some overlap of coverage. I actually find it delightful and instructive to see the various approaches. In any case, we encouraged cross referencing in text and footnotes, and we made the Index to facilitate a tying together of parts. The reader may also wish to skim a little before settling down with any one chapter. For instance, Van Allen introduces most of the topics of Part IV, and so do Simpson and McKibben, while their chapters are not the first in Part IV.

The participation by the best scientists in these fields, all very busy people, has been most gratifying. But we owe much to many more. For acknowledgement we can only address the extensive list of advisors, referees and authors found on pp. 1221–25. The University of Arizona Press provided basic publication financing and competent cooperation, especially towards fast publication. Financial backing we also received from NASA Headquarters, the Pioneer Project Office, the National Science Foundation, and the International Astronomical Union.

As for Jupiter, the Introduction and Overview articles (Part I) are aimed at readers who are familiar with the field, providing them with cross references among the various chapters, and are also aimed at readers new in the field by developing some basic concepts. Pictures and description of instrumentation are not presented (there is, however, a diagram of the Pioneer spacecraft on p. 851), nor are discussions of future missions and their experiments. But we do cover all scientific aspects of Jupiter's interior, atmosphere, magnetosphere and satellites.

Tom Gehrels

PART I

Introduction and Overview

ORIGIN AND STRUCTURE OF JUPITER AND ITS SATELLITES

R. SMOLUCHOWSKI
Princeton University

INTRODUCTION

The chemical composition and internal structure of planets cannot be uniquely determined except when they are large and have low density. Only then one can conclude that the planet consists primarily of a few of the lightest elements whose abundance in the solar system is rather well known, and that only a negligible amount of the more volatile constituents may have had a chance to escape its powerful gravitational field. Usually the thermodynamics of these light elements at very high pressures and temperatures is fairly well known so that by using various quantities such as radius, oblateness, heat balance, non-sphericity of the gravitational field as deduced from the motion of satellites or spacecraft, and the presence or absence of a magnetic field one can, at least in principle, arrive at a self-consistent model of the planetary interior. This procedure is indeed feasible for Jupiter and Saturn, and probably also for Uranus and Neptune. Attempts to do the same for the denser terrestrial planets usually fail as is well illustrated by trying to account for many of the various parameters for Earth by assuming that it is all made of iron oxide.

Of fundamental importance for determining the internal structure of Jupiter is the knowledge of the equations of state, that is, the pressure-volume-temperature relations, of its two main components hydrogen and helium. Inserting them into the equations of hydrostatic equilibrium of a rotating self-gravitating body and using an iterative process, one can first obtain a solution for $T = 0$, and then, using suitable corrections, for $T \neq 0$. A typical result for Jupiter is a central core made of so-called rock, that is, SiO_2, MgO, etc., which is comparable in size to Earth, and a mantle or envelope consisting primarily of hydrogen and helium in a mass ratio of ~ 3. Lesser constituents such as H_2O, NH_3 or CH_4 can be assumed to be either in the core or in the mantle or in both. The heat balance of Jupiter indicates that it emits more than twice as much energy as it gets from the sun. This large heat flux leads to very high central temperatures (20,000°K or more) and to the conclusion that the planet is all liquid or fluid, without any solid or liquid surface. Thus, the internal heat transport is dominated by convection rather than by conduction. At 1 bar pressure, which lies close to the outside radius, the temper-

[3]

ature drops to 160–170°K. Of particular interest is the fact that with decreasing radius the pressure rises rapidly from low values in the gaseous atmosphere to about 3 Mbar at 3/4 of the planet's radius where molecular hydrogen transforms into a metallic form with a considerable increase in density. At the center of the planet the pressure is 10–100 Mbar. In addition the motion of the nearest satellites is strongly affected by the radial density gradient in the outside few percent of the planet's radius, and therefore the structure of this region is better known than the deeper layers.

The presence of a metallic liquid interior helps us understand the origin of the powerful Jovian magnetic field, because sufficiently high electrical conductivity and low viscosity permit the operation of the hydromagnetic dynamo mechanism. This mechanism, which is driven by the planetary rotation and by thermal convective motions of the liquid metal, sustains the internal and external magnetic fields. There are several possible ways of explaining the origin of the huge amount of energy generated in the Jovian interior, that differ among themselves in the manner in which the enormous gravitational self-energy of the planet is slowly converted into heat.

While the motion of the four Galilean satellites (Io, Europa, Ganymede and Callisto) of Jupiter has been observed for over 300 years it is only recently that their densities, internal structure and surface properties have been studied in detail. As relatively small and dense bodies, comparable in size to Moon and Mercury, these satellites are very likely made of silicates and some ice because the more volatile constituents could not be gravitationally trapped. Furthermore, the heat of the parent planet removed all of the ice from the nearest Galilean satellite Io but left more of it on the more distant ones. Our knowledge of the constitution of the many outer satellites is highly conjectural, but it is likely that they resemble in chemistry the asteroids or comets.

In practice, the development of satisfactory models of Jupiter and of its satellites encounters a number of problems which have to do with the validity of various necessary approximations and assumptions or with differences in theoretical methods. As a result there is a variety of models each having some particularly useful or carefully analyzed feature. These various aspects are described and compared in the following Overview, which gives the reader perspective and guidance for the study of the major papers dealing with the structure of Jupiter and of its satellites.

OVERVIEW OF THE PERTINENT CHAPTERS

Origin of Jupiter

The problem of the formation of planets from the primitive solar nebula requires taking into account numerous processes such as accretion, various kinds of gravitational instabilities which lead to the formation of the sun or result in the formation of large clumps of matter in the midplane, the removal of the angular momentum of the collapsing cloud by meridional circulation currents, central heating and the radial temperature gradient. The presence or absence of a central object prior to the formation of proto-planets leads to additional questions concerning the influence of solar wind during the solar flare-up phase and the presence of strong tidal effects. A variety of assumptions concerning these phenomena can lead to a variety of theories of the early stages of development of the solar system, and as yet no unique rigorous answer is available. In his theory of the origin of Jupiter, Cameron (see chapter by Cameron and Pollack, p. 65, in this volume for references) assumes that in the midplane of the rotating nebula in the region of formation of Jupiter, the temperature had the relatively low value of $\sim 170°$K so that the nebula consisted of gases and dust. The latter was made of metals, metal oxides, sulfides, silicates and perhaps ice; the former of volatile elements and compounds such as hydrogen, helium, methane, ammonia and water. Through accretion and amalgamation, these several-micron-size grains grew into larger particles (1–10 mm) and eventually formed a midplane layer of small solids that became gravitationally unstable and led to the formation of massive bodies. Some of the asteroids may have been formed in this manner. Through collision the largest of these bodies grew to sufficient size to form a planetary core which could attract its own gaseous cloud co-rotating with it. A hydrodynamic collapse of this cloud released then a large amount of heat which dissociated and ionized H_2 and partially ionized He.

In the course of time this proto-Jupiter attracted more nebular matter, acquiring an atmosphere made primarily of hydrogen and helium with any infalling dust probably evaporating. A subsequent evolution of Jupiter led to an additional slow evolution of heat due to gravitational shrinkage and progressive cooling. The essential feature of this model of Jupiter is that, while the rocky core itself incorporates the various nebular compounds in solar proportions, and the external mantle also retains the characteristic solar ratio of hydrogen to helium (3.5 ± 1 by mass), there is no fixed relation between the two so that the planet as a whole is not solar. In particular, it has a large excess of water in the form of ice.

This origin of Jupiter is to be compared with that analyzed by Bodenheimer (1976) who assumed that the planet was formed by a local self-gravitational condensation of the nebular matter of initial density 10^{-11} g cm^{-3} at $T = 40°$K in a volume with $r = 4600\,R_J$. If the sun was already present, then the initial density had to be ~ 500 times higher to overcome disruptive tidal effects. The subsequent evolutionary contraction first proceeded rapidly in quasi-hydrostatic equilibrium until the hydrodynamic collapse associated with the dissociation of H_2 had occurred and a denser central region had formed. This was followed by a further slow contraction in hydrostatic equilibrium from a radius of about $5\,R_J$ until the present state was reached. Throughout this process the composition was solar and homogeneous.

The evolution following the collapse of the central region has been analyzed in detail by Graboske, Pollack, Grossman and Olness (see Cameron and Pollack p. 71) who have shown that the internal energy transport is primarily convective, while the heat flux to the outside is controlled by radiative transfer in a relatively thin atmospheric surface layer. Thus, the assumptions concerning the opacity of this layer are critical for establishing the rate of the evolution. The early luminosity of Jupiter was $10-10^7$ times higher than it is now, which had an important effect on the Galilean satellites, as discussed further below. At an epoch of about 10^5 years, Jupiter's central temperature reached nearly 50,000°K which may have vaporized the central core. Interestingly enough, much uncertainty concerning the later stages of this process is related to details of the equation of state of the condensed matter that even now have not been completely clarified. In particular, it does not seem possible to give a definite answer to the question whether or not the internal energy built up in the past is sufficient to account for the present Jovian luminosity. This is discussed in more detail below. Figures 1 and 2 in Bodenheimer's paper show how his calculations of the very early history of Jupiter fit into the study of the later stages by Graboske *et al.* It is encouraging that varying — sometimes quite drastically — the initial conditions of the early evolutionary processes has only a minor effect on later evolutionary stages and on the final result. Even the presence or absence of a rocky core, provided it is not too big a fraction of the total mass of the planet, does not affect significantly the evolutionary outcome. Figures 1 to 4 of Cameron and Pollack show the change with time of the various important quantities. From the point of view of models of Jupiter, it is important to note that, whether one takes the Bodenheimer or the Graboske *et al.* or the Cameron evolutionary history, one ends up invariably with a hydrogen-to-helium ratio that is solar. As pointed out by Cameron, a substantial decrease of this ratio, i.e., a major loss of hydrogen, could not be explained even on the scale of several billion years.

Origin and Structure of the Satellites

Cameron (p. 78) and others make the natural assumption that the Galilean satellites, and perhaps also the nearest one Amalthea, were formed by a repetition in the Jovian nebula of the processes which led to the formation of the planet itself in the primeval solar nebula. The temperature in the midplane increased because of infall of gas and dust and very likely also because of direct heating by the proto-planet. The Goldreich-Ward (1973) instability led to the formation of rocky cores which became proto-satellites by further accretion. The existence of ice on the outer satellites and its absence on Io (see below) can be explained if the condensation of water took place after the cores were formed but still in the presence of a strong heat flux from the planet. The more volatile compounds were probably swept away from the Jovian nebula by a strong solar wind before they could condense.

There is also a possibility that there was a strong interaction between the Jovian magnetic field and its nebula which was partially ionized by the planet's radiation. The result was a slowing down of the rotation of the planet and transfer of the angular momentum to the disc which is gradually dissipated.

The more distant, "irregular" satellites of Jupiter have masses comparable to asteroids and are most likely captured bodies, their orbit extending up to the very limit of Jupiter's sphere of influence. They fall into two distinct groups, one prograde and one retrograde; in each group the orbits are quite similar and have high inclinations and eccentricities (see Greenberg p. 123). There are still many unsolved theoretical problems concerning the dynamics of the capture of these bodies such as the need for dissipation of their energy in order to avoid escape. This loss may have been provided by drag in the primitive Jovian disc. Another possibility is collision of larger bodies or comets within the sphere of influence, or break-up upon capture. Nothing is known about the composition of these satellites. The similarity of orbits in each of the groups mentioned above suggests that since each group probably had a similar, if not joint, origin then each should have a similar perhaps asteroid-like or comet-like composition.

As far as the Galilean satellites are concerned (see Morrison and Burns p. 994, and Consolmagno and Lewis p. 1035) we know their size and density and also we can take into account the likely mechanism of their formation from the Jovian nebula. It appears that all of them are essentially a mixture of partially hydrated silicates and water-ice in its various modifications. During the process of their formation the Jovian heat led to an almost complete escape of ice from the nearest, Io, and an almost complete retention of ice on the farthest, Callisto. The stability of water at these distances from the sun is marginal so that small differences in temperature play an important role. The chondritic radioactive content (U, Th, K) of the satellites is expected to have been high enough to melt the interiors and produce differentiation, i.e., a denser silicate core and water mantle with an ice crust wherever present. Depending upon the model, the maximum estimated central temperatures resemble those proposed for the moon, namely 1800–2800°K.

The deep interior of Callisto was undoubtedly molten but it is probable that its thick crust was never differentiated and actually is a primeval mixture of ice and silicates such as serpentine. The low albedo of Callisto is presumably also due to a layer of infalling dust and other matter. Ganymede was probably completely molten and has a dense silicate core. Recent radar data indicate that its surface is very rough, while spectroscopy indicates that it is covered with patches of ice and rocks which account for its dark color. It may possess a very tenuous atmosphere. Europa has also been completely molten and differentiated but the ice mantle is thin. As to Io it is probably made of silicates with a surface covered by evaporite that consists of various salts containing sodium, calcium, ammonia and sulphur, which were leached out from the interior or were left behind when H_2O evaporated and escaped. An important aspect of Io is its very strong interaction with the Jovian plasma which, through ionic bombardment and sputtering, releases from the surface large numbers of ions such as sodium and hydrogen. As a result Io has not only an atmosphere made of these elements, but also an ionosphere. There is also strong electromagnetic interaction of Io with the parent planet which produces the well-known periodic effect on the Jovian decametric radiation. Finally, it should be noted that the Galilean satellites are locked facing the planet. As to Amalthea, it would be very surprising if it were composed of anything other than silicates and perhaps sulfides and oxides. A table of numerical data on the satellites is given in the chapter by Morrison and Burns (see p. 994).

Very interesting commensurabilities exist among the orbital periods of the Galilean satellites so that a certain linear combination of their mean longitudes is constant. This configuration is maintained by the mutual gravitational interactions of the satellites. As pointed out by Greenberg (p. 128), the origin of this resonance is not yet completely understood.

Models of Jupiter

As described briefly in the Introduction, models of Jupiter are obtained using a basically simple procedure: the hydrostatic equilibrium equations for a rotating body are integrated over the volume of the planet using appropriate equations of state of the hydrogen-helium mixture and of the heavier constituents of the core, with the radius and the total mass of the planet providing the necessary boundary conditions. Such a calculation is usually done first at $T = 0$, and then suitable corrections for thermal pressure are introduced. Here the boundary conditions are the observed heat flux, the rate of convection and the estimated temperature T_1 at a pressure of one bar. Essential information about the density profile in the outer layers, i.e., in the molecular H_2 region, is obtained from the gravitational harmonic coefficients J_2 and J_4 (for definition, see Anderson p. 114) which are deduced from the motions of the Galilean satellites and of flyby spacecraft. The higher terms J_6, etc. will probably remain unknown until further data are provided by flyby and orbiting spacecraft. The directly observed oblateness of Jupiter agrees with the oblateness calculated from these coefficients, which indicates that, with good approximation, the planet is in hydrostatic equilibrium. The correction of the equations of state for $T \neq 0$ is very important because it affects the radial temperature profile and thus the thermal expansion and radius of the model.

As a result of a variety of assumptions and theoretical methods which must be used, there exists quite a spectrum of models of Jupiter. Figure 5 in Hubbard and Slattery's chapter (p. 192) shows what a variety of density profiles can satisfy the radius, mass, and both J_2 and J_4 while only one of them corresponds to a realistic equation of state. The major assumptions which affect the outcome of the model calculations concern: 1) the value of the solar hydrogen-helium ratio and of the hydrogen-helium ratio on Jupiter, 2) the presence or absence of a large core made of heavier elements, and 3) uniform or non-uniform distribution of hydrogen and helium in the mantle outside the core and the presence there of an excess of a heavier compound such as H_2O. These assumptions are often interrelated.

According to Cameron (1973) the solar mass ratio of hydrogen to helium is 3.4 ± 0.1; according to Makalkin (see Zharkov and Trubitsyn p. 136) it is 3.62 to 3.94. It should be stressed, however, that other ratios, some of them as low as 2.75, have been called "solar." This is often the case for models based on the assumption of a convectively homogenized hydrogen-helium mixture throughout the planet and the observation that the atmosphere appears to be richer in helium (or in water) than the usual solar composition would suggest.

The problem of a uniform versus layered composition of the hydrogen-helium envelope has two aspects to it; one has to do with the metallic-molecular transition and the other with the incomplete miscibility of helium in metallic and molecular hydrogen. The first follows from the phase rule which requires that in a two-component system a phase change can occur without a change of composition only at a congruent singularity in P-V-T space. It is doubtful that in Jupiter such a singularity would be present now or during a substantial part of the history of the planet. Although no quantitative evaluation of this change in composition has been made, one would expect the solubility of helium to be higher in molecular than in the metallic hydrogen. The lack of complete miscibility of helium in metallic hydrogen, first suspected on a qualitative basis (Smoluchowski 1967) has been confirmed by the exact theory of Stevenson and Salpeter (p. 93). As shown in their Fig. 5, this miscibility gap decreases with increasing pressure and temperature and should be taken into account in all models of the Jovian interior except for those which have a very high central temperature. A similar limited solubility in molecular hydrogen has been found experimentally by Streett (1976) but the theoretical extrapolation to high pressures and temperatures is rather uncertain (Smoluchowski 1973).

Salpeter and Stevenson (1975) have shown that, if a first-order phase transition is accompanied by a large change in density and by a latent heat comparable to the thermal energy, an almost isothermal interface between the two phases would profoundly alter the calculated central temperature of the planet. A negative latent heat could halve the central temperature as compared to the usual isentropic result, while positive latent heat would double it. It follows that if, at the boundary, the specific entropy in the metallic phase were higher than in the molecular phase, then the probability of the existence of a miscibility gap in the metallic region would be high with all the attendant consequences for the mechanism of heat flow and origin of internal energy. In particular, at any such "almost isothermal" interface, whether caused by phase change or miscibility gaps, convection is impeded and heat conduction could play the main role provided the transition is of the first order. Until these thermodynamical aspects of the boundaries, including the composition discontinuities, have been analyzed in more detail, the results of the usual adiabatic calculations of the heat flow and of the radial temperature profile cannot be considered as final.

As discussed further below, it is only recently that the equations of state of hydrogen and helium, of their solutions, and of various compounds are being calculated with sufficient attention to theoretical detail. In particular, the very existence of the miscibility gaps in the hydrogen-helium system indicates that even in this relatively simple system only at very high pressures and temperatures is the usual approximation of volume additivity valid. The situation is somewhat less critical for the constituents of the heavy core because of its relatively small size, but it is very important for H_2O which, at least in the model of Podolak and Cameron (1974), is greatly enhanced and may play an important role in the atmosphere.

The modern approaches to the problem of the internal structure of Jupiter are best illustrated by four categories of models: those of Hubbard and Slattery, those of Stevenson and Salpeter, those of Podolak and Cameron and those of Zharkov and Trubitsyn. They can be briefly summarized as follows.

Hubbard and Slattery (p. 176) base their model of the Jovian interior on a careful analysis of the results of "gravitational sounding" i.e., of the limitations which the most recent observed gravitational coefficients impose on the figure and density profile of a rotating body. As expected, a large variety of density profiles satisfies these boundary conditions and the response coefficients, i.e., the sensitivity of the gravitational coefficients to the ratio of the rotational and gravitational potentials. However, only few of them correspond to realistic equations of state. The interior is described by ignoring the density change at the molecular-metallic boundary and by using either a superposition of a MacLaurin spheroid ($\rho = $ const.) and a point-mass core or a polytrope (i.e., no core) of index of about one. The latter gives a good fit to the data if the density of the hydrogen-helium mantle is increased by 10% which could be due either to extra helium or to another substance such as water; for water its abundance should be about 10 times higher than the solar. These calculations are based on the assumption that the solar mass ratio of hydrogen to helium is 2.85 rather than Cameron's 3.3 to 3.4. The temperature at 1 bar turns out to be 230°K, but in another model in which the density at the center and in the atmosphere is increased, while it is lowered at medium depths, T_1 is about 200°K. This model requires less of the heavy component than the polytrope model.

Stevenson and Salpeter (p. 86) calculate carefully the equation of state of the liquid metallic hydrogen-helium system using the best theoretical procedures presently available. They assume that both hydrogen and helium are fully ionized and that the system can be described as a non-ideal Coulomb plasma. Some of the improvements over other methods stem from the inclusion of a better dielectric function, of quantum mechanical treatment of the ions, of three-body interactions and of a non-linear response of the electron gas to the ions. As a result they are able to show, for the first time, that in this system there is a deviation from volume additivity on the order of several percent and that there is a miscibility gap in the liquid state. As pointed out by the authors, the presence of first-order phase transitions in the liquid part of this system can affect considerably not only the evolutionary cooling of the planet but also its structure, convection and temperature profile. The equation of state of molecular hydrogen is bracketed by the use of three alternative pairs of interaction potentials and of three different assumptions concerning the vibrational and rotational contributions to entropy. Unfortunately, the considerable uncertainty of the helium-molecular hydrogen interaction prevents a reliable theoretical investigation of the upper limit of existence of finite solubility observed in this system in the laboratory (Streett 1976). After a careful analysis of the limitations of the assumption of a fully adiabatic interior, Stevenson and Salpeter arrive at models that contain an isothermal rocky core and a hydrogen-helium-ice mantle in which helium and "ice" (primarily water but also methane and ammonia) occur in solar proportion, and no enrichment is required. Typically for a H-He mass ratio of 3.42 the temperature $T_1 = 150°K$, while for a mass ratio of 2.5, $T_1 = 190°K$. Very importantly it turns out that the uncertainty of the entropy of H_2 plays a bigger role than the uncertainty of the interaction potentials. Inclusion of phase separation in the metallic region raises somewhat the required helium content which, in turn, raises T_1. A typical model which yields the correct gravitational coefficients has $T_1 = 160–170°K$ [which brackets the most likely value as given by Orton (p. 207)], $T_c \sim 20,000°K$, central density 26–27 g cm^{-3}, core radius $\sim 0.3\ R_J$ and the radius of the H-H_2 boundary 0.71–0.75 R_J.

The model of Jupiter of Podolak and Cameron (1974) was the first one directly based on an analysis of the formation of the proto-planet, and it stressed the impossibility of having the ratio of free hydrogen to helium differ from the solar ratio. At the same time the core (made of "rock" SiO_2, MgO, Fe and Ni and "ices" H_2O, CH_4 and NH_3 in solar proportion) did not have to be in solar proportion to the mantle. The equations of state at $T = 0$ of metallic and molecular hydrogen and helium used in this calculation were those of DeMarcus combined with that of Salpeter and Zapolsky for the metallic phase and with that of Trubitsyn for helium. Particular attention was paid to the equation of state for molecular hydrogen. Volume additivity was used to obtain the equation of state of the hydrogen-helium solar mixture. The equations of state of H_2O, NH_3 and CH_4 were calculated from those of Salpeter and Zapolsky for O, N and C and from the equation for metallic hydrogen again assuming additivity of volumes. They were then joined smoothly to the equations valid at low densities. The integration of the equations of hydrostatic equilibrium was then carried up to a radius at which the pressure is one bar. The gravitational coefficients J_2 and J_4 were computed using Peebles' method. A comparison with the observed total mass, radius and the gravitational coefficient J_2 permitted, through an iterative process, to adjust the abundances of various compounds with respect to their solar value. Subsequently J_4 and temperature T_1 at pressure 1 bar were calculated. In their earlier work Podolak and Cameron used a value for J_2 deduced from the occultation of β-Scorpii and then, for a mantle with purely solar composition, the core turned out to have a fractional mass of 0.143, $T_1 = 141°K$ and $T_c = 17,000°K$. For an envelope with a 7.5-fold increase in H_2O content above the solar value, these numbers changed to 0.133, 189°K and nearly 20,000°K. Here the H_2O, rather than excess helium of the early models of Hubbard, provided the extra density of the outer envelope required by the gravitational coefficients. Subsequently it turned out, from Pioneer data, that the actual value of J_2 was nearly 40% higher, which necessitated an even higher density in the outer part of the gaseous envelope and thus an even higher water enrichment. This led to the difficulty that the ratio of the mass of the excess water to the mass of the core became higher than 2, which is the limiting solar value. This is illustrated in Fig. 8 of their second paper. Lowering the H-He ratio to about half of its solar value would bring the water-to-core ratio down to the solar value but this is unacceptable to the authors for cosmogonical reasons mentioned earlier. A change in the pressure-dependent Grüneisen parameter γ from 0.5 to the acceptable higher value 0.6 did not improve the situation and an assumption of higher temperatures T_1 made the result worse as shown in Fig. 4 of the same paper. Another possibility, suggested by the authors but still quantitatively not explored, is that the late-infalling rock material would be suspended in the planetary envelope, thus increasing its density without requiring an abnormal water or helium enrichment.

In the models of Zharkov and Trubitsyn (p. 137) the equations of state at $T = 0$ for hydrogen and helium were those of Trubitsyn and those of Zharkov, Trubitsyn and Tsarevskii. The equations for O, C, N and S are based on quantum-statistical calculations. From these the equations of state of the hydrogen-helium mixtures and of compounds such as H_2O, CH_4, NH_3 and H_2S are obtained using additivity of volumes, with the hydrogen volumes taken from the equation corresponding to its metallic form. In order to adjust these equations to $T \neq 0$, the Debye equation of state is used with the formula of the Grüneisen parameter γ obtained from a generalized Slater formula [in which the pressure-dependent Poisson ratio ν was taken from the Thomas-Fermi calculations of Kopyshev (as used also by Podolak and Cameron), valid at high pressures]. The solar hydrogen-to-helium mass ratio used by Zharkov and Trubitsyn is 3.85 but in some of their models they permit it to vary so as to satisfy the boundary conditions. The transition from the molecular to the metallic state is taken to occur at $p_t = 3$ Mbar which corresponds to $R/R_J = 0.76$. Similarly to the model of Podolak and Cameron, the ratio of gases to ices and rocks which form the core is varied but the relative abundances of constituents in each group are taken as solar. Zharkov and Trubitsyn consider a large number of models all the way from a core consisting of only heavy compounds, with all the more volatile compounds in the gaseous mantle, to a core which includes all solar constituents except hydro-

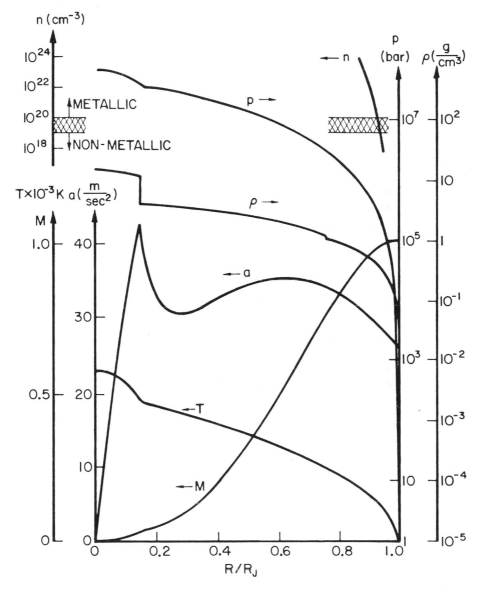

Fig. 1.   Radial profiles of mass M, temperature T, pressure p, density ρ and gravitational acceleration a (Zharkov and Trubitsyn p. 160). Curve n indicates the concentration of conducting electrons in the H_2 layer (from Smoluchowski 1975).

gen and helium, which are in the mantle. It is assumed that all models are fully adiabatic and that $T_1 = 140$ or $250°K$. Model 1 for $T_1 = 140°K$ with a core consisting of a uniform mixture of rock, NH_3 and H_2O and the mantle containing hydrogen, helium and other gases is illustrated in Fig. 1 which shows the density ρ, temperature T, gravitational acceleration a, relative mass M and pressure p as a function of R/R_J. The discontinuities of density at the core boundary and at the metallic-molecular phase boundary are clearly visible. Here the relative mass of the core is 0.04. The value $140°K$ at 1 bar is close to $\sim 160°K$ which is favored by spectroscopic observations and atmospheric models (Orton and Ingersoll p. 212). The hydrogen-to-helium ratio in this model is solar (Makalkin's value around 3.88) but the envelope also contains nearly 10% of core material in gaseous or dispersed form.

Comparison of Models of Jupiter

Until about the end of 1973, most models of the Jovian interior were based on the assumption that all constituents heavier than hydrogen and helium play a negligible role. Their main characteristic was the requirement that, in order to satisfy the gravitational coefficients, the ratio of hydrogen to helium had to be lower than the solar ratio by a factor close to two. The essential features of these models are summarized in the review paper by Hubbard and Smoluchowski (1973). The requirement of a solar hydrogen-to-helium ratio suggests the presence of a sizable core of heavy constituents such as "rock" and "ice" and an enrichment of the envelope in water. A substantial core has now become a common feature in most models mentioned in the previous section.

For a solar hydrogen-to-helium ratio Podolak and Cameron obtain for the relative mass of the core and water the values 0.016 and 0.037 whose ratio is much higher than the solar. Zharkov and Trubitsyn, on the other hand, obtain for a similar model masses 0.10 ± 0.08 and 0.037 which is close to the solar ratio. The main difference between the two calculations seems to be that in the first one $T_1 = 189°K$ while in the second $T_1 = 140°K$. It is known that with increasing T_1 the mass of the excess water increases rapidly while the mass of the core drops rather slowly. Thus, the lower T_1, the better is the approximation of the solar water-to-rock ratio. It should be pointed out that the recent analysis of Orton and Ingersoll (p. 212) indicates $T_1 = 165°K$ while Stevenson and Salpeter (p. 86) conclude that only $140°K < T_1 < 200°K$ are acceptable. It is difficult to say to what extent the difference between the results of Podolak and Cameron on the one hand and those of Zharkov and Trubitsyn on the other are due solely to differences in T_1 and to what extent other factors such as differences in the equation of state play a role.

Hubbard and Slattery (p. 186), in their models (a) and (b), use $T_1 = 230°K$, and a value somewhat lower than $200°K$ in model (c). In all three models the heavy component is present with a relative mass between 0.10 and 0.11 in models (a) and (b) and 0.06 in model (c), but although in the first two it is assumed to be soluble and uniformly distributed in the metallic interior, in the last model it is all concentrated in a distinct central core. Since the heavy component is not further specified, the question of the water-to-rock ratio remains unanswered. The relatively high value of T_1 suggests that if this problem were looked into in the manner outlined by Podolak and Cameron, similar difficulties would arise. It should be kept in mind that in their calculations Hubbard and Slattery assume that the solar hydrogen-to-helium ratio is 2.8 which is nearly 20% lower than the Cameron value and 30% lower than Makalkin's. This low ratio lowers the demand for the additional heavy constituent in the hydrogen-helium envelope as compared to the Podolak and Cameron results. Hubbard and Slattery make a very careful and informative analysis of the importance of precise values of the gravitational coefficients on the structure and composition of the outer layers.

In striking contrast with these models are those of Stevenson and Salpeter which, with a solar hydrogen-to-helium mass ratio of 3.58 or somewhat lower, possess a heavy core of relative mass 0.03 to 0.045 but do not require the presence of water in the hydrogen-helium envelope. As pointed out by these authors, the main difference between their models and all other ones mentioned above is a more accurate equation of state in the metallic region and a more realistic treatment of the uncertainties of the thermodynamics of molecular hydrogen. The improved equations of state are obtained by applying rigorous techniques of the theory of condensed matter and a careful analysis of the validity of the common assumption of the additivity of volumes. The fact that they do obtain a miscibility gap in the hydrogen-helium system in the metallic region, as illustrated in their Fig. 5 (p. 97), may lead to a decrease of the required mass of helium by $\sim 10\%$. In the molecular hydrogen region they use the recent empirical pair potentials and three different sets of assumptions concerning the free energy of the thermal motion of the molecule, dissociation and electronic excitation. The most important conclusion to be drawn from the models of Stevenson and Salpeter is that the remaining uncertainty of the equation of state of the molecular hydrogen includes the possibility that all the observational parameters and boundary conditions may be met without assuming a mantle that has an excess content of heavier constituents such as water. The main effort of theoretical studies should be thus directed at a clarification of this problem including the question of solubility of He in H_2 at pressures and temperatures much higher than those which were investigated experimentally by Streett (1976). Figure 2 illustrates the effect of intermolecular potentials, of pressure at 1 bar and of the addition of helium on the equations of state of hydrogen. Table I provides a rough comparison of some of the typical models developed by each of the four groups of authors. Many other models are described in detail in various

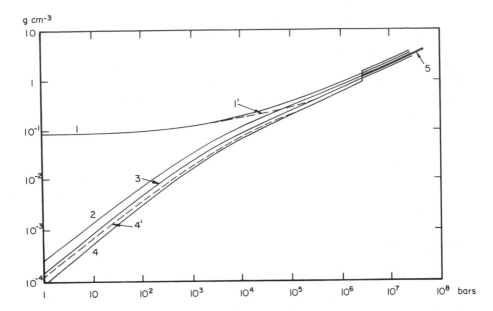

Fig. 2.   Influence of various parameters on the equation of state of hydrogen. Curve $1 - T = 0$
isotherm with a soft intermolecular potential (Zharkov and Trubitsyn p. 137); Curve $1' -$
the same with a hard potential (DeMarcus as used by Podolak and Cameron 1974); Curves 2,
3, 4 – adiabats for $T = 90$, 140 and 250°K respectively (Zharkov and Trubitsyn p. 151);
Curve $4'$ – hydrogen-helium adiabat for mass ratio 2.66 and $T_1 = 250$°K (Hubbard and Slattery p. 187); Curve 5 – hydrogen $T = 6000$°K isotherm in the metallic range (Stevenson and
Salpeter p. 88, and Hubbard and De Witt personal communication), where within the accuracy of this diagram it coincides with the upper part of an adiabat.

chapters in this volume. At the end of their chapter Stevenson and Salpeter
(p. 108) list new data needed for improvement of the Jovian models. Numerical data on Jupiter are given in the Appendix (p. 20); also see p. 784.

Internal Convection

As discussed in detail in this volume by Stevenson and Salpeter and by
Hubbard and Slattery there is little doubt that the interior of Jupiter is liquid.
The metallic solid is expected to have an inherently low (if any) stability, and
in the molecular region the new softer pair-wise interaction potentials indicate a much lower melting point than originally calculated. This conclusion
agrees with the analysis of the gravitational coefficients. Bishop and De-
Marcus (1970) similarly concluded that a purely conductive interior is not
possible or rather, as pointed out by Stevenson and Salpeter, that the predicted thermal conductivity in the metallic phase is an order of magnitude
too low. For these reasons, nearly all models of Jupiter are based on the assumption of a fully adiabatic convective heat flux, $T \sim \rho^\gamma$ with $0.5 < \gamma <$
0.6, and the validity of the mixing-length theory of convection. As discussed

TABLE I

Comparison of Some of the Recent Models of Jupiter

Authors	Relative Mass of Core	Additional Component in the Mantle	T_c (°K)	T(at 1 bar) (°K)	H/He by Mass	Comments
Hubbard and Slattery	0.10 either in a core or dispersed	(He or H_2O)	~ 20,000	200–230	2.85	Homogeneous mantle. Gravitational sounding.
Stevenson and Salpeter	0.03–0.045	Not needed	20,000	160–170	3.6	H-He miscibility gap present if $T < 9000°K$ at the phase boundary.
Podolak and Cameron	0.13–0.15	H_2O	17–20,000	170–190	3.4	Mantle homogeneous
Zharkov and Trubitsyn	0.04 to 0.10 either in a core or dispersed		23,000	140–250	≤ 3.85	Mantle homogeneous

by Hubbard and Smoluchowski (1973) this leads to convective velocities on the order of 1 cm sec^{-1}. Stevenson and Salpeter give a careful analysis of the limitations of the applicability of the mixing-length theory and of full adiabaticity, which stem from the rapid rotation of the planet. This analysis takes into account the coupling to a strong magnetic field and the presence of first-order phase transitions which permit convection within each of the several layers but not across phase boundaries. The latter has been suggested earlier on the basis of qualitative arguments (Smoluchowski 1973). These questions are important for Jupiter's evolution and present temperature profile, but have not yet received sufficiently detailed answers. One should add that within the molecular layer there may be a region where the temperature is low enough so that solid helium-poor H_2 may exist in equilibrium with liquid helium-rich H_2 as suggested by Streett. Such a situation is important for the interpretation of the Great Red Spot as the visible top of a Taylor column based on a mass of solid helium-poor H_2 floating in liquid helium-rich H_2. This theory has recently gained interesting experimental support (Titman *et al.* 1975). Other meteorological models of the GRS are discussed by Ingersoll (1975).

In contrast to the assumption of a highly turbulent convection, which underlies the mixing-length theory used in most models, Busse (1976*a*) discusses convection in the molecular layer of Jupiter in terms of convective cells. Confirming experimentally his basic theoretical model, he concludes that convection cells are of the even type, that is, cylinders (rolls) parallel to the rotation axis extending across the equatorial plane. These long convection cells cannot penetrate into the metallic region and thus, at higher latitudes the convection is described by the usual Bénard cells of the odd type. The fact that the cloud pattern on Jupiter changes just at the latitude separating the even and odd convection cells appears to confirm Busse's model. An interesting consequence of this model is the uniformity of temperature along the even-type convection rolls which might lead to a latitude-dependent radial temperature profile within the molecular layer.

Internal Heat

Many models and theories have been proposed to account for the huge amount of internal heat emitted by Jupiter as deduced from the difference between the calculated and observed effective temperature (see Appendix and Low p. 204). Of these only three need to be further discussed, others having been shown to be too inefficient (Hubbard and Smoluchowski 1973). The three still valid ones are: (1) primordial heat caused by a shrinkage of a homogeneous planet, (2) progressive gravitational shrinkage of a planet with a phase transition, and (3) progressive separation (precipitation) of helium-rich and helium-poor components. The first possibility depends upon fine details of the evolutionary calculations discussed earlier (Bodenheimer 1976; and Graboske *et al.* referred to by Cameron and Pollack, p. 71) and the answer is still uncertain. The second, which is usually interpreted as a progressive outward motion (about one millimeter per year) of the metallic-molecular phase boundary and the associated shrinkage of the planet (Smoluchowski 1967; Flasar 1973), depends upon still unknown details of the change of solubility of helium at the phase boundary. The third one proposed by Salpeter (1973) for the metallic region, on the basis of the expected incomplete hydrogen-helium miscibility, and then extended to the molecular region (Smoluchowski 1973) is very attractive, but its validity depends upon the actual presence of these phase boundaries at pressures and temperatures existing in Jupiter. This is discussed in detail by Stevenson and Salpeter (p. 77). All three theories can account easily for the magnitude of the heat emission ($\sim 10^4$ erg cm^{-2} sec^{-1}) and so it is not possible to decide between them on this basis alone at the present time.

It should be mentioned here that according to Ingersoll (1976) the solar heating of the equatorial region of Jupiter lowers the radial thermal gradient in that part of the planet so that the flux of the internal heat, whatever its origin, is higher towards the poles than towards the equator. This effect would account for the remarkable uniformity of the average surface temperature of the planet, in contrast to the otherwise expected difference of about 20°K between the poles and the equator produced by solar heating of Jupiter whose equatorial inclination is only 3°.

Magnetic Field

Until Pioneer data became available all information about the Jovian magnetic field was based on observations of the decimetric radiation. Ever since it was realized that the metallic interior had to be liquid, the magnetic field was interpreted in terms of a hydromagnetic dynamo (Hide and Stannard p. 777). The necessary, but not sufficient, condition that the magnetic Reynolds number is large enough seems to be amply satisfied. There are at present two sets of data from Pioneer measurements obtained with different instruments: the vector helium magnetometer of Smith *et al.* (p. 789) and the fluxgate magnetometer of Acuña and Ness (p. 830). While the results differ somewhat, the general picture is as follows: the field has opposite orientation from that of the earth; the dipolar magnetic moment is tilted with respect to the axis of rotation and is off center, which is not unlike the situation on Earth. There are, however, strong field components due to quadrupole and octupole magnetic moments, and the field is higher at the North Pole than at the South Pole. The tilt, the displacement from the center and the magnitude of the field can be interpreted in terms of the hydromagnetic dynamo. A comparison of the Jovian and terrestrial magnetic fields is given by Hide and Stannard (p. 777). It will be interesting to see if the slow changes of the earth's field occur also on Jupiter. Problems arise where the magnitudes of the Jovian higher moments are concerned because one would expect the higher field components to fall off rapidly with distance. This appears not to be the case. The answer may lie in the suggestion that the field is generated not in the deep interior of the planet but rather in the outer layers of the metallic region. There are three independent indications that point in this direction: Smoluchowski (1975) has shown that the high pressures and temperatures in the molecular hydrogen layer destroy its insulating properties and that the layer acquires metallic conduction throughout most of its volume. This conclusion is based on the fact that high pressures lower the gap between the valence and conduction bands in H_2 at a known rate, and that this effect accelerates the onset of a huge concentration of thermally excited conduction electrons as illustrated in the top part of Fig. 1. As a result, metallic conduction in the molecular layer extends from the phase boundary at 0.75 R_J to 0.90–0.92 R_J. Secondly, Busse (1976b) concluded on the basis of a detailed analysis that the generation of magnetic flux occurs predominantly in outer regions of the metallic core and that the internal toroidal field is comparable to the poloidal field, in contrast to the usual terrestrial models. Finally, a similar conclusion has been reached by Acuña and Ness (p. 844) on the basis of similarity scaling of planetary dimensions and of multipoles of Earth and Jupiter. Thus it seems that the Jovian magnetic field is generated by a hydromagnetic dynamo mechanism primarily in the layer somewhere between $R/R_J = 0.6$ and 0.9, which is so close to the surface of the planet that the higher multipole field components can be observed. Clearly, much more quantitative work in this area is needed.

APPENDIX
Summary of Data on Jupiter

Mean distance to sun	5.203 A.U.
Present orbital eccentricity	0.048
Sidereal period	11.9 yr
Inclination of equator to orbit	3°07
Mean orbital velocity	13.03 km sec^{-1}
Radius: equatorial	71,398 km
polar	66,770 km
Dynamic oblateness at 1 bar	0.064841
Mass	1.901×10^{30} g
Mean density	1.33 g cm^{-3}
Gravitational acceleration: equatorial	2707 cm sec^{-2}
polar	2322 cm sec^{-2}
J_2	$14{,}750 \pm 50 \times 10^{-6}$
J_3	$10 \pm 40 \times 10^{-6}$
J_4	$-580 \pm 40 \times 10^{-6}$
J_6	$50 \pm 60 \times 10^{-6}$
$|C_{22}|$ and $|S_{22}|$	$< 1 \times 10^{-6}$
Rotation period: IAU System III (1965)	$9^h\ 55^m\ 29\overset{s}{.}71$
decametric	$9^h\ 55^m\ 29\overset{s}{.}68 (\pm 0.02)$
decimetric	$9^h\ 55^m\ 29\overset{s}{.}75 (\pm 0.03)$
Angular velocity ω	1.76×10^{-4} rad sec^{-1}

$q = \omega^2 R_J^3 / GM_J$	0.08885
Bolometric Bond albedo	0.42
Effective temperature: calculated	105°K
terrestrial observations	134 ± 4°K
Pioneer observations	125 ± 2°K
Thermal emission/insolation	1.9 to 2.5
Net heat flux	1.0 to 1.3 × 10⁴ erg cm⁻² sec⁻¹
Magnetic dipole	4.22–4.28 Gauss R_J^3
Magnetic dipole tilt	9°.6–10°.77
Longitude of magnetic dipole tilt	232°–230°.9
Quadrupole/dipole	0.24–0.20
Octupole/dipole	0.21–0.15
North Pole	14 Gauss
South Pole	10.4–11 Gauss
Offset (nearly equatorial)	0.12–0.10 R_J

REFERENCES

Bishop, E. V., and DeMarcus, W. C. 1970. Thermal histories of Jupiter models. *Icarus* 12: 317–330.

Bodenheimer, P. 1976. Contraction models for the evolution of Jupiter. *Icarus* (special Jupiter issue). In press.

Busse, F. H. 1976*a*. A simple model of convection in the Jovian atmosphere. *Icarus* (special Jupiter issue). In press.

———. 1976*b*. Generation of planetary magnetism by convection. In draft.

Cameron, A. G. W. 1973. Abundances of the elements in the solar system. *Space Sci. Rev.* 15: 121–146.

Flasar, F. M. 1973. Gravitational energy sources in Jupiter. *Astrophys. J.* 186:1097–1106.

Goldreich, P., and Ward, W. R. 1973. The formation of planetesimals. *Astrophys. J.* 183:1051–1061.

Hubbard, W. B., and Smoluchowski, R. 1973. Structure of Jupiter and Saturn. *Space Sci. Rev.* 14:599–662.

Ingersoll, A. P. 1975. The atmosphere of Jupiter. *Space Sci. Rev.* 23. In press.

———. 1976. Pioneer 10 and 11 observations and the dynamics of Jupiter's atmosphere. *Icarus* (special Jupiter issue).

Podolak, M., and Cameron, A. G. W. 1974. Models of the giant planets. *Icarus* 22:123–148.

———. 1975. Further investigations of Jupiter models. *Icarus* 25:627–634.

Salpeter, E. E. 1973. On convection and gravitational layering in Jupiter and in stars of low mass. *Astrophys. J.* 181:L83–L86.

Salpeter, E. E., and Stevenson, D. J. 1975. Heat transfer in a stratified two-phase fluid. *Cornell Univ. Ctr. Radiophys. Space Res.* 595:1–24.

Smoluchowski, R. 1967. Internal structure and energy emission of Jupiter. *Nature* 215:691–695.

———. 1973. Dynamics of the Jovian interior. *Astrophys. J.* 185:L95–L99.

———. 1975. Jupiter's molecular hydrogen layer and the magnetic field. *Astrophys. J.* 200: L119–L121.

Streett, W. B. 1976. Phase equilibria in gas mixtures at high pressures. *Icarus* (special Jupiter issue). In press.

Titman, C. W.; Davies, P. A.; and Hilton, P. M. 1975. Taylor columns in a shear flow and Jupiter's Great Red Spot. *Nature* 255:538–539.

ATMOSPHERES AND IONOSPHERES

D. M. HUNTEN
Kitt Peak National Observatory

A general discussion of Jupiter's atmosphere divides naturally into several categories: thermal structure, dynamics, composition, and cloudiness. In addition, the few facts about the atmosphere of Io are listed.

THERMAL STRUCTURE

After years of having to construct theoretical temperature profiles on the basis of a few data points, we are finally close to having reliable measurements for the whole visible atmosphere, and indeed for a significant distance below the cloud tops. These results are summarized in Fig. 1, which also introduces the nomenclature of the different atmospheric regions.

The *troposphere* is rapidly stirred by vertical motions, some combination of free convection and the vertical components of the large-scale circulation. The dominant balance is between the heat flux (the sum of internal and solar heat) and vertical motions, though there is considerable infrared cooling as well in the upper parts. According to Stone[1] the temperature gradient is expected to be very close to the adiabatic value, $-1.9°K$ km^{-1} for Jupiter, except just below the tropopause, which is conventionally defined as the level of minimum temperature. Because of the need to transport the internal heat, the gradient should remain adiabatic to very great depths. With this assumption, the microwave emission spectrum, as discussed by Berge and Gulkis,[2] gives information to depths where the temperature is 400°K or greater, and defines the temperature at 1 bar as 170°K. Both microwave and

---

[1]See p. 598.
[2]See p. 662.

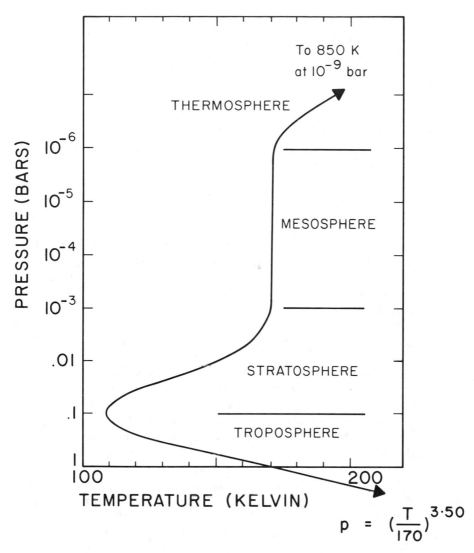

Fig. 1. Adopted temperature profile for Jupiter. For levels deeper than 1 bar, the ideal-gas adiabat is represented by the formula shown. There is some evidence for a temperature maximum ~ 200°K in the mesosphere.

infrared methods exploit the presence of a range of opacities to explore a range of depths, as summarized by Wallace.[3] In addition, a range of zenith angles at a single wavelength can be used, as with the Pioneer data (Ingersoll et al.; Orton and Ingersoll).[4] Various workers have used the classic

[3]See p. 286.
[4]See pp. 200, 206 and 508.

spectrum of Gillett, Low, and Stein (see Ridgway *et al.*)[5] which includes information from CH_4 and H_2 opacities. More recent data at longer wavelengths, again featuring the H_2 opacity, have been exploited by Houck *et al.* (1975). And Lacy *et al.* (1975) have used detailed radiometry of a single ammonia line to the same end.

So far, the radio-occultation experiment has been unable to produce results for the troposphere (Kliore and Woiceshyn; Hunten and Veverka).[6] This method is more and more subject to perturbation as it probes deeper and deeper, and its ultimate potential is still in doubt, though Eshleman (1975) is optimistic that many of the problems can be overcome.

The *stratosphere* and *mesosphere* are regions of radiative balance: the heat sources are planetary and solar radiation, and the sink is infrared cooling. The mesopause is the base of the *thermosphere*, in which the dominant sink is downward conduction. The stratopause on Earth is located by a temperature peak near 50 km, but this peak is probably absent, or not very pronounced, on Jupiter. Because it is so helpful to carry over the terrestrial nomenclature, I propose for now to define the location of the stratopause at the corresponding pressure level, which is 1 millibar.

Computations of radiative balance usually assume a global mean, which seems to work fairly well. The rapid rotation of Jupiter gives good zonal averaging, but one must otherwise depend on atmospheric motions, which may or may not be adequate (Gierasch *et al.* 1970).

There is now plenty of evidence that the upper stratosphere is warm, at least 150°K and perhaps rising to 170–180°K at 1 mb. The presence of a temperature rise is clear from the presence of emission bands of CH_4 and, more recently, other molecules. But there are serious difficulties to making quantitative deductions: the mixing ratios of the emitting constituents are often unknown, and the intensity of the CH_4 band cannot be unambiguously interpreted because it is unusually sensitive to temperature (Wallace).[7] Earthbased center-to-limb measurements of the same band are, however, highly revealing, and Orton and Ingersoll[8] have used them to obtain temperatures up to the 10 mb level. Radio-occultation results cover the same region, and are in general agreement though at present some 10–20° cooler (Kliore and Woiceshyn).[9] If the discrepancy proves to be significant, it could be explained by an underabundance of CH_4 due to photochemical destruction, a question that I shall return to below.

Although the radio occultation experiment gives data to pressures as low as 1 mb, the derived temperatures show a wide spread which is difficult (I suspect impossible) to eliminate. This problem, which is shared by the stellar occultation at pressures below 10 μb, is discussed in detail by Hunten and Veverka.[10]

---

[5]See p. 396.   [8]See p. 208.
[6]See pp. 216 and 274.   [9]See p. 235.
[7]See p. 286.   [10]See p. 279.

The same paper gives a critical discussion of the three sets of data from the 1971 occultation of Beta Scorpii. All observers agree on a temperature of 170°K at 10 μb. They differ on how much deeper into the atmosphere the results can be trusted, with one group (Combes *et al.* 1975) advocating a lower bound of 220°K at 1 mb. Only a very small stretching of their zero-level error would, however, be required for compatibility with an isothermal mesosphere at 170°K. In the absence of any plausible suggestion for an additional heat source concentrated in the lower mesosphere, I prefer to adopt the isothermal profile.

Computations of the heat balance are discussed by Wallace.[11] The overtone bands of methane can be shown to absorb important quantities of solar heat, their weakness being overcome by the stronger flux at shorter infrared wavelengths. Ultraviolet absorption by "smog" particles is important too, but we do not know enough about its vertical distribution. Plausible models can therefore be constructed to fit the observed temperatures, but they are not of much use for prediction.

The height of the mesopause has been calculated by Strobel and Smith (1973). For conditions resembling those adopted here, the pressure is about 1 μb. A higher pressure might be required if the thermospheric heat flux is as large as suggested by the discussion below, or if methane (the principal radiator) is underabundant at these heights.

There is also evidence that the 10-μb temperature is lower at 58° latitude than at 10°, though the variation is not required within the error bars. We have here a strong hint that the mesosphere is not globally symmetric.

The next region for which we have information is the upper ionosphere, which is certainly far hotter than anybody expected; Fjeldbo *et al.*[12] give 850°K for the average of proton and electron temperatures, and the gas should be at the same temperature. The temperature rise from the mesopause is therefore 680°K instead of the 15°K computed by Strobel and Smith (1973). Obviously their heat source (solar ultraviolet forming the ionosphere) is inadequate, and a first estimate of the required factor is 45. This, however, may be too large, because the important quantity is the temperature gradient, not the temperature difference. Adopting it nevertheless, we find that the energy required may be 0.6 erg cm^{-2} sec^{-1} instead of the 0.014 (global average) from the sun.

One plausible source was proposed, and its implications pointed out, even before the Pioneer 10 results had been announced (French and Gierasch 1974). They noted wave-like temperature structure in the stellar occultation results, and concluded that the best explanation was an inertia-gravity wave system carrying an upward flux of 3.4 erg cm^{-2} sec^{-1}. Though the energy is clearly ample, the interpretation is not unique, and moreover we do not know whether the wave energy will be absorbed or reflected by

---

[11]See p. 290.
[12]See p. 242.

the thermosphere. The same difficulties have faced us for years in the corresponding terrestrial problem (see Hunten 1974 for a brief discussion). However, the example of the solar corona is encouraging, since it is thought to be heated by an analogous process. These same arguments are invoked by Atreya and Donahue[13] to justify their assumption of an exospheric temperature around 1100°K. They calculate a model ionosphere that agrees fairly well with the gross features of the observed one.

Nevertheless, it is worthwhile to look for other possible energy sources. There is presumably plenty available in Jupiter's magnetosphere, but the particles that are observed are much too penetrating; the heat has to be deposited at about the same place as ultraviolet photons. Thus, electrons or protons of just a few electron-volts energy would be best.

According to an estimate by Humes,[14] meteoroids may be bringing in as much as 6 ergs cm^{-2} sec^{-1}. Again, this is of the right order, but it seems likely that most of the energy is deposited near the mesopause, where it is very ineffective.

Finally, one could look to the differential rotation of Jupiter's atmosphere and magnetic field. Below 10° latitude the differential velocity is 100 m sec^{-1}, dropping to 3.3 m sec^{-1} at higher latitudes. Ions and electrons will essentially move with the magnetic field; both heat and momentum will be transferred to the neutral gas. Studies of the earth's ionosphere (e.g., Fedder and Banks 1972) suggest that the gas will be dragged along with the ions at great heights, and that a shear layer will exist at some height that may not be far above the mesopause. Since most of the heating is in this layer, it must be located before the corresponding thermal effect can be estimated.

Probably some of these possible energy sources can be ruled out in short order. We may still be left, however, with more than one whose relative importance we do not know.

ATMOSPHERIC MOTIONS

Jupiter's large-scale circulation and other smaller-scale phenomena are reviewed by Stone and by Smith and Hunt.[15] The arrangement of the clouds into enormous bands, and the long lifetime of various features, tell us that we have a regime utterly unlike the familiar one at mid-latitudes on Earth. A remarkable fact observed by Ingersoll et al.[16] is the uniformity of the heat flux between poles and equator. Since the solar input is concentrated at low latitudes, the internal heat must somehow be persuaded to come out with a complementary distribution.

---

[13]See p. 313.
[14]See p. 1064.
[15]See pp. 591 and 564.
[16]See p. 200.

Generally speaking, each region of the Jovian atmosphere rotates at a slightly different rate. These "winds" are summarized by Chapman (1969) and Ingersoll (1975). People continue to be fascinated by the Great Red Spot, and Maxworthy and Redekopp (1976) have suggested that it be identified with their model of a "solitary wave."

Also classifiable under motions is the eddy diffusion coefficient K used by aeronomers to redistribute long-lived constituents. Current ideas are discussed by Prinn and Owen,[17] who also point out that it is time to extend the practice to deeper levels than those usually cultivated by aeronomers. There are several strong hints of minor constituents that should be found at equilibrium only at depths and temperatures too great to observe. Their suggestion is that vertical transport rates may be large enough to overcome the thermochemical destruction of these species. Strobel has found from the photochemical destruction rate of ammonia that K is $\sim 10^4$ cm^2 sec^{-1} just above the cloud tops. For the lower thermosphere, the Lyman-α albedo of Judge and Carlson gives $K = 10^8$ cm^2 sec^{-1}, a large value, resembling that of Mars and Venus. But there is a problem with the model of Wallace and Hunten on which this estimate is based, and which assumed a cold thermosphere. They found the albedo to rise with the temperature of the hydrogen atoms, and a further major increase of K would be required in compensation. Perhaps the situation will be saved by the fact that most of the hydrogen is near the bottom of the thermosphere which is not so hot.

Small-scale turbulence seems to be evident in occultation data, both optical and radio (Elliot and Veverka 1976; Young 1976). Its counterpart in the ionosphere is discussed by Woo and Yang (1976). From time to time, attempts are made to relate such data to the eddy diffusion coefficient. Generally this does not work because the relevant scales are different; the vertical transport is dominated by the larger scales of motion, which carry things the farthest, while observations of turbulence tend to emphasize smaller scales.

COMPOSITION AND CLOUD

The visual appearance of Jupiter is dominated by its cloudy zones, and cloud and haze must be deeply involved in formation of spectral lines. But in both cases it is hard to be quantitative. Composition of the clouds must be obtained from theory. Spectroscopists are usually reduced to subterfuges that avoid the impossible problems of trying to use a realistic model in our present state of ignorance (Prinn and Owen).[18] We know that the cloud must be important, because spectral features do not grow towards the limb as they should in a clear, discrete atmosphere. Several clever theories exist

---

[17]See p. 328; also see p. 435.
[18]See p. 324.

that can be manipulated to reproduce this behavior; the main problem is to define the proper distribution of cloudiness and its spectral properties. Probably the Pioneer photopolarimetry will be very helpful, but the huge mass of data is still being digested (Tomasko).[19] Related Earth-based work is also discussed by Teifel and by Kawata and Hansen.[20]

The Pioneer images almost force us to believe that the clouds are full of fascinating motions, if we were not already convinced from telescopic observations. The available data permit a number of inferences about the various motions, but still represent essentially a set of snapshots. Much more useful would be a movie in which the same face of the planet is repeatedly imaged over a period of months or even years.

Holes in the clouds, which appear blue, and the corresponding hot spots seen at 5 μm, are discussed by Keay et al. (1973). Most of the recent spectroscopic discoveries of trace constituents use this same wavelength band, and therefore refer to an unusually deep atmospheric layer. This fact is well illustrated by the rotational temperature seen for water vapor, near 300°K. This and other aspects of the composition are reviewed by Prinn and Owen and Ridgway et al.[21] There is strong evidence for CO (Beer 1975) and tentative identifications of PH_3, GeH_4, and HCN, none of which would be expected at even the 300°K level. As noted above, there is a possibility that they are produced at greater depths and transported up by atmospheric motions faster than they can reach equilibrium with local conditions.

The question of the coloring material of the clouds, or "chromophores," continues to fascinate people. Current suggestions run all the way from organic material to red phosphorus, but we still seem to lack the means to test them.

Weidenschilling and Lewis (1973) agree with the traditional assessment that the visible clouds should be solid NH_3, located at pressures somewhat below 1 bar. They also predict NH_4SH at about 3 bars, and solid H_2O at 5–7 bars.

The general magnitude of the helium abundance, 10% by number relative to H_2, seems well established, and we are almost ready to argue about the second significant figure. In addition to arguments from the mean density and the infrared spectrum,[22] we have a fairly direct detection of the 584 Å resonance line (Carlson and Judge)[23] which gives a concordant result.

Spectroscopic estimates of the ammonia and methane abundances give values consistent with solar composition. But ammonia is quickly depleted above the cloud tops by photolysis, a fact emphasized by the surprising weakness of ammonia bands in the rocket ultraviolet. What about methane? It too is rapidly destroyed; solar Lyman α alone, if not attenuated, gives a

---

[19]See p. 503.　　　　　　　　　　　　　[22]See p. 206.
[20]See pp. 441 and 516.　　　　　　　　　[23]See p. 435.
[21]See pp. 324 and 396.

lifetime of only 4.7×10^6 sec, or with a diurnal mean, 110 Earth days. Following the procedure used for NH_3 by Strobel (1973), we find that the methane scale height would be only 0.17 that of the atmosphere. But this assumes that Lyman α penetrates all the way to the lower stratosphere. Since the photolysis products of methane are also strongly absorbing, such a penetration is most unlikely. Any attempt at a realistic simulation would have to include the formation of smog particles and their downward settling. While such models can be formulated, they contain too many arbitrary parameters to be of much use. For the present we should at least remember that the distribution of methane in the stratosphere and mesosphere is a complicated question, and the mixing ratio is not necessarily constant.

Ionospheric electron densities are given by Fjeldbo et al.,[24] and further information on irregularities by Woo and Yang (1976). The implications of the topside scale height for the temperature are discussed above. The lower region is extremely complicated, with several distinct layers, some of which, however, may really correspond to horizontal structure. They immediately bring to mind the terrestrial phenomena called "sporadic E," one form of which is due to metallic ions. Bunching into a layer is thought to be due to the magnetic field, interacting with wind shears. It is not at all clear that such a mechanism could produce the multiple layers that seem to exist on Jupiter. Perhaps there is more than one kind of ion, and perhaps one of them is related to the Io sodium cloud (Atreya and Donahue).[25]

SATELLITE ATMOSPHERES

All the work on satellite atmospheres reported in this volume refers to Io. A complete survey of the satellites is given by Morrison and Burns.[26] The initial results on D-line emission from Io by Brown seemed to require an extreme production mechanism to explain the large fluxes.[27] These requirements were removed when Trafton discovered that the sodium cloud is much larger than Io. Resonance scattering now seems adequate, but instead we have the puzzle of how to maintain such a huge cloud. By almost any reasonable estimate of the lifetime of atoms in the cloud, Io must supply a flux of at least 10^7 atoms cm^{-2} sec^{-1}. Sputtering from an evaporite-salt surface is not an implausible mechanism, but it becomes a little hard to accept that if it has been going on continuously for the life of the solar system. The amount of sodium removed is 50 gm cm^{-2} or more, and if sodium is only a few percent of the material, the total quantity is measured in kg cm^{-2} and the depth in meters. Some kind of geological (or Iological?) activity is implied; but is it compatible with the maintenance of an Io-wide salt flat? I have the feeling that an important link may still be missing; can there be

---

[24]See p. 240.   [26]See p. 991.
[25]See p. 315.   [27]See p. 1103.

some way to recycle sodium, or sodium ions, back from the cloud? Can such a mechanism be compatible with the fact that the sodium torus is incomplete?

We also have good evidence for a partial torus of hydrogen atoms from Pioneer 10.[28] They must surely be accompanied by a much larger number of H_2 molecules, although it is dangerous to be dogmatic about such a strange environment. On the Moon H_2 appears to be the only form present, since Lyman α cannot be detected at all. Unfortunately, H_2 is extremely difficult to detect and it may be a long time before we have any real data on it.

The radio-occultation experiment detected an ionosphere on Io with several striking characteristics. The dayside electron density is large, nearly 10^5 cm^{-3}, and so is the scale height; both quantities are much smaller at night. The large diurnal variation implies a short ion lifetime; the large electron density implies a long lifetime or an unusually large production rate. The latter can be obtained with sodium, as shown by McElroy and Yung (1975), but they then had to convect the ions back to the surface at night by condensing the atmosphere. Long lifetime was obtained by Whitten *et al.* (1975) by assuming a thin atmosphere of pure neon, with the dominant loss being at the surface. Such an atmosphere is hard to accept; any impurities will rob the ionization by charge transfer, and we have plenty of evidence for at least one impurity. Moreover, there is no evidence for abundant neon in any other atmosphere. The authors draw an analogy with the Moon, which does indeed have neon, but the absolute amount is very small.

We have a lot to learn about satellite atmospheres; they promise to be a fruitful field, both for observation and for theoretical speculation.

REFERENCES

Beer, R. 1975. Detection of carbon monoxide in Jupiter. *Astrophys. J.* 200:L167–L169.

Chapman, C. R. 1969. Jupiter's zonal winds: variation with latitude. *J. Atmos. Sci.* 26:986–990.

Combes, M.; Vapillon, L.; and Lecacheux, J. 1975. The occultation of β Scorpii by Jupiter. IV. Divergences with other observers in the derived temperature profiles. *Astron. Astrophys.* 45:399–403.

Elliot, J. L., and Veverka, J. 1976. Stellar occultation spikes as probes of atmospheric structure and composition. *Icarus* (special Jupiter issue). In press.

Eshleman, V. R. 1975. Jupiter's atmosphere: problems and potential of radio occultation. *Science* 189:876–878.

Fedder, J. A., and Banks, P. M. 1972. Convection electric fields and polar thermospheric winds. *J. Geophys. Res.* 77:2328-2340.

French, R. G., and Gierasch, P. J. 1974. Waves in the Jovian upper atmosphere. *J. Atmos. Sci.* 31:1707–1712.

Gierasch, P.; Goody, R.; and Stone, P. 1970. The energy balance of planetary atmospheres. *Geophys. Fluid Dyn.* 1:1–18.

Houck, J. R.; Pollack, J. B.; Schaack, D.; Reed, R. A.; and Summers, A. 1975. Jupiter, its infrared spectrum from 16 to 40 micrometers. *Science* 189:720–722.

---

[28]See p. 1079.

Hunten, D. M. 1974. Energetics of the thermospheric energy transport. *J. Geophys. Res.* 79:2533–2534.

Ingersoll, A. P. 1975. The atmosphere of Jupiter. *Space Sci. Rev.* 23:206–242.

Keay, C. S. L.; Low, F. J.; Rieke, G. H.; and Minton, R. B. 1973. High-resolution maps of Jupiter at 5 microns. *Astrophys. J.* 183:1063–1073.

Lacy, J. H.; Larrabee, A. I.; Wollman, E. R.; Geballe, T. R.; Townes, C. H.; Bregman, J. D.; and Rank, D. M. 1975. Observations and analysis of the Jovian spectrum in the 10-micron ν_2 band of NH_3. *Astrophys. J.* 198:L145–L148.

Maxworthy, T., and Redekopp, L. G. 1976. An orbitary wave theory of the Great Red Spot and other observed features in the Jovian atmosphere. *Icarus* (special Jupiter issue). In press.

McElroy, M. B., and Yung, Y. L. 1975. The atmosphere and ionosphere of Io. *Astrophys. J.* 196:227–250.

Strobel, D. F. 1973. The photochemistry of NH_3 in the Jovian atmosphere. *J. Atmos. Sci.* 30:1205–1209.

Strobel, D. F., and Smith, G. R. 1973. On the temperature of the Jovian thermosphere. *J. Atmos. Sci.* 30:718–725.

Weidenschilling, S. J., and Lewis, J. S. 1973. Atmospheric and cloud structures of the Jovian planets. *Icarus* 20:465–476.

Whitten, R. C.; Reynolds, R. T.; and Michelson, P. F. 1975. The ionosphere and atmosphere of Io. *Geophys. Res. Lett.* 2:49–51.

Woo, R., and Yang, F. 1976. Measurements of electron density irregularities in the ionosphere of Jupiter. *J.G.R.* (special Jupiter issue). In press.

Young, A. T. 1976. Scintillations during occultations by planets. *Icarus* (special Jupiter issue). In press.

JUPITER'S MAGNETOSPHERE: PARTICLES AND FIELDS

C. K. GOERTZ
The University of Iowa

I. INTRODUCTION

Even before the Pioneer 10 and 11 encounters with Jupiter there was no doubt that the largest planet in the solar system has an extensive magnetosphere and that *in situ* measurements of Jupiter's magnetic field and energetic particle environment would be of great interest to space physics. A set of rather well-defined models was formulated before encounter [see, e.g., *Proceedings of the Jupiter radiation belt workshop* (Beck 1972)]. These were based on concepts derived from the study of the only magnetosphere to which we have previously had direct access, that of the earth. The *in situ* measurements are therefore not only fascinating in their own right, but also because they provide us with an opportunity to study the applicability and general validity of the theoretical concepts we have devised in order to understand the earth's magnetosphere. Coroniti (1975) shows that the data obtained in the inner magnetosphere are quite compatible with the pre-encounter models. Surely this agreement must be regarded as one of the major successes of space science. However, the outer magnetosphere appears quite different from what had been expected, and even in the inner magnetosphere there exist some numerical and, as we will see, significant discrepancies. Throughout the Jovian magnetosphere and in interplanetary space close to Jupiter, the successful Pioneer missions have yielded some quite unexpected and rather puzzling facts which no pre-encounter model can easily explain. Nevertheless, it seems to me that the pre-encounter models should be extremely useful as a guideline for the analysis of the *in situ* measurements. I shall, therefore, briefly review what I consider to be the main contributions to pre-encounter Jovian magnetospheric physics, list the questions resulting from these, show how some of these have been answered, and how others may be answered by future analyses of the data. Any summary of this kind must necessarily be a personal one and I apologize if the selection I make does not coincide with that which others might make. Much has been written on Jupiter and a thorough discussion even of the most fundamental points is beyond the scope of this overview.

[32]

It is well known that Jupiter is not only an emitter of thermal radiation like any other planet, but that it also emits relatively high-intensity non-thermal radiation in two frequency bands: the decimetric radiation (DIM) with 100 MHz $\lesssim f \lesssim 3000$ MHz and the decametric radiation (DAM) with 0.1 MHz $\lesssim f \lesssim 40$ MHz. Whereas the origin of the decametric radiation has as yet not been explained (R. A. Smith),[1] the decimetric radiation is generally believed to be due to synchrotron radiation from energetic electrons trapped in the inner region of the Jovian magnetosphere ($1.5\ R_J < r < 6\ R_J$). The fact that there is no generally accepted theory of DAM is due to its rich phenomenology and its apparent control by Io, the innermost Galilean satellite. A current review of DAM is given in this volume by Carr and Desch.[2] For a review of the relevant observations and theoretical work on DIM the reader is referred to the excellent review paper by Carr and Gulkis (1969) and the chapter by Berge and Gulkis.[3]

These two forms of radiation provided the only means by which information about the Jovian magnetosphere could be obtained before the Pioneer 10 and 11 encounters. DIM exhibits a broad intensity maximum at 850 MHz. If it is assumed that the radiating electrons all have roughly the same energy E, one finds that the product of $BE^2 \sim 1.8 \times 10^{14}$ G(eV)^2. Warwick (1970) estimated the magnetic field in the radiating region to be $B \sim 2$ G (corresponding to a Jovian magnetic moment of $M = 4 \times 10^{30}$ G cm^3). Thus a typical energy of radiating particles is about 10 MeV. (We now know that the magnetic moment of Jupiter is about one third of this value [$M \approx 1.4 \times 10^{30}$ G cm^3, as given by Smith *et al.*[4] and Acuña and Ness].[5] Hence the energy of the particles should be ~ 17 MeV.)

As these particles radiate synchrotron radiation they lose half of their energy in a time T_M, somewhat less than a year. Although this estimate is very crude it is supported by more rigorous calculations (see, e.g., Birmingham *et al.* 1974). It is precisely this short lifetime of the radiating particles combined with the following observations of DIM that has motivated a number of interesting models for the Jovian magnetosphere. There is almost a complete lack of variation in flux density over periods comparable to T_M (Komesaroff and McCulloch 1976; Klein 1976). Furthermore, there is no or very little correlation between the long-term DIM flux density variation and solar activity (see Klein 1976). In other words, the radiating particles are resupplied to the radiation belts on a time scale of a year and the rate of supply is *not* controlled by the solar wind properties.

Thus two questions emerge: (1) What is the source of particles and how are they accelerated to relativistic energies, and (2) how are they transported from the source to the radiation belts? Two obvious answers to these ques-

---

[1]See p. 1146.   [4]See p. 788.
[2]See p. 693.   [5]See p. 830.
[3]See p. 621.

tions were: The particles are of solar wind origin and they are transported radially inwards by diffusion and/or convection. Both transport processes violate the third adiabatic invariant but conserve the first and second adiabatic invariants. Thus, as the particles are transported inwards towards larger magnetic fields their energies increase (see, e.g., Schulz and Lanzerotti 1974). This model has turned out to be quite successful for the case of the earth. However, Brice and Ioannidis (1970) have ruled out convective transport on the basis of the overwhelming importance of the rapid rotation of Jupiter. Convection is driven by large-scale electric fields generated by the interaction of the solar wind with the magnetosphere. Brice and Ioannidis argue that at Jupiter these fields are very small compared to the large electric field which is due to the rotation of the planet. Recently Kennel and Coroniti (1975) have cast some doubt on this conclusion. Whatever the mode of inward transport may be, one would expect a strong influence of the solar wind properties on the source strength and hence DIM intensity, if the particles are of solar wind origin. That this influence is not observed may be due to any one of three reasons or combinations thereof:

1. The particles may not originate in the solar wind but may be created within the Jovian magnetosphere.
2. The energetic particles may derive their energy in some way or other from the planet's rotational energy.
3. There may be a buffer zone between the outer regions of the Jovian magnetosphere and the inner region where DIM is generated.

Let us briefly review the arguments advanced in favor of each of these possibilities.

1. Van Allen[6] concludes from his measurements that Jupiter may, indeed, be the major source of particles in the Jovian magnetosphere. Simpson and McKibben[7] report periodic bursts of energetic particles ejected from Jupiter. McDonald and Trainor[8] show that many "anomalous" cosmic ray events observed near the earth are of Jovian origin. Although these observations do not exactly locate the source of particles, they strongly suggest that the particles are generated and/or accelerated inside Jupiter's magnetosphere.

This source within the magnetosphere may be inside the radiation region or outside of it. If it were inside, e.g., in Jupiter's ionosphere, one might expect the DIM intensity maximum to be close to the planet and not at $\simeq 2\,R_J$ as observed. ($R_J = 71,372$ km is Jupiter's radius.) It seems more likely that the belts are populated by inward diffusion. Birmingham et al. (1974) and Coroniti (1974) have shown that a radial diffusion model can explain the gross features of DIM. Radial diffusion calculations, however, are quite insensi-

---

[6]See p. 957.
[7]See p. 754.
[8]See p. 978.

tive to the position of the particle source. They only tell us that the source is outside the radiating region.

Possible sources outside the radiating region but within the Jovian magnetosphere are the Jovian satellites. Shawhan (1976) discusses in detail how Io could provide electrons of several 100 keV energy (see also R. A. Smith).[9]

2. If the energetic particles derive their energy from the planet's rotational energy, the arguments of Gold (1976) would quite naturally exclude a strong influence of solar wind properties on DIM even if the particles are of solar wind origin. He argues that in order to tap the rotational energy one needs a recipient for the angular momentum. Three possible recipients exist: (a) the satellites, e.g., as discussed by Goldreich and Lynden-Bell (1969), Goertz and Deift (1973), Piddington and Drake (1968), Carr and Gulkis (1969), and Shawhan (1976); (b) outward flowing plasma may carry away angular momentum as suggested by, e.g., Kennel and Coroniti (1975); or (c) angular momentum can be lost to the solar wind. The first two possibilities do not involve the solar wind directly. Gold investigates the third possibility and shows that even in this case the rate at which energy is made available (presumably to the solar wind particles) is only weakly dependent on solar wind properties. In other words, the energetic particle loading of Jupiter's magnetosphere should be quite insensitive to variations of the solar wind pressure, contrary to the case of the earth where the solar wind has a great influence.

3. The idea of a buffer zone was advanced by Thorne and Coroniti (1972) and extended by Coroniti (1974) and by Kennel and Coroniti (1975). Their model has the following main features: solar wind particles, after being heated as they pass through the bow shock, are injected either through the front-side magnetosphere boundary or through the Jovian magnetotail. They then diffuse or convect inwards. Due to the compression of phase space, the fluxes of energetic particles as well as their energies increase as they are transported towards the planet. In the outer region of the magnetosphere the fluxes should be highly variable owing to solar wind variations. At a critical distance which may vary with solar activity, the energies and fluxes become large enough to excite whistler mode turbulence (Kennel and Petschek 1966). If the turbulence is strong enough, the fluxes of particles with energies above a critical energy $E_c \approx B^2 / 8 \pi N$ will be self-consistently limited to the stable trapping limit. This limit depends only on the thermal plasma density N and the magnetic field and should hence be relatively insensitive to solar wind variations. In the inner region the critical energy for the excitation of whistlers becomes extremely relativistic and the whistlers are stabilized. Based on the density model of Brice and Ioannidis (1970), Coroniti places the boundary of the inner region at 6 R_J. Inside this distance the particles

---

[9]See p. 1158.

are transported inwards by radial diffusion as suggested by Birmingham *et al.*
(1974). Since the flux at 6 R_J is the stable trapping flux and independent of
solar wind properties, the flux inside 6 R_J should also be independent of the
solar wind characteristics. A more detailed discussion of the model is given
by Scarf.[10]

The Jovian satellites may not only accelerate particles but also absorb
them and thus constitute an effective buffer zone. It has been noted by several investigators that, unless the diffusion coefficient is large enough, the
particle fluxes inside the orbit of Io could be very small (Mead 1972; Birmingham *et al.* 1974; Mead and Hess 1973; and for a general discussion see
Mogro-Campero p. 1203).

The foregoing remarks suggest the following basic questions:
1.   What is the source of the energetic particles and where is it located?
2.   What is the mode of transport of particles; i.e., is radial diffusion an
     important process and if so, what is the magnitude of the diffusion coeffi-
     cient? What mechanism could be responsible for it? Or are particles
     transported by convection?
3.   Is pitch angle scattering by whistler turbulence important and if so, in
     what regions of the magnetosphere is it active?
4.   Are the Jovian satellites sources or sinks for particles?
5.   What is the gross shape of the Jovian magnetosphere? Is it open as
     predicted by radial outflow models or is it closed?

Undoubtedly many more questions can be asked, but the advantage of
the set above is that for most of them quantitative answers were proposed
before the encounters. Furthermore, all these basic questions generate a
number of subsequent questions that will undoubtedly stimulate further
research. It will become clear that in many cases these questions can only
be answered if certain assumptions are made. In particular, an accurate
model of the Jovian magnetic field is of prime importance because the energetic particle distributions are frequently displayed and interpreted in terms
of magnetic field coordinates. Because of the inability of any spacecraft
measurement to distinguish between spatial and temporal variations, assumptions must also be made about the time dependence of the phenomena observed. The reader will find that in many instances conclusions are drawn
by tacitly assuming the steady state case. Yet we know that at least the
outer part of the Jovian magnetosphere must be highly variable in time as
shown, for example, by the second magnetosheath encounter at ~ 50 R_J
during the inbound trajectory of Pioneer 10 (see Sec. II.D).

---

[10]See p. 883.

II. EXPERIMENTAL RESULTS AND INTERPRETATIONS

In this section we shall describe the results of 6 experimental packages carried on board Pioneer 10 and of 7 on Pioneer 11. Each spacecraft carried four separate energetic particle detector packages: the University of California at San Diego (UCSD) (Fillius);[11] the Goddard Space Flight Center (GSFC) (McDonald and Trainor);[12] the University of Chicago (Simpson and McKibben);[13] and the University of Iowa package (Van Allen).[14] These instruments were able to measure intensities, angular distributions, and energy spectra of electrons and ions with energies above several hundred keV.

A low-energy plasma detector (Intriligator and Wolfe)[15] was included on both spacecraft. This instrument was designed primarily for measuring the characteristics of the solar wind, but measurements in the bow shock and magnetosheath have been highly successful and significant for our understanding of the Jovian magnetosphere. Recently Frank *et al.* (1976) have used the plasma detector on board Pioneer 10 to infer low-energy proton densities and temperatures (up to 400 eV) in the inner Jovian magnetosphere.

Both Pioneers 10 and 11 carried vector helium magnetometers (Smith *et al.*)[16] which could detect fields from somewhat smaller than $0.01 \gamma = 10^{-7}$ Gauss up to 1.4 Gauss. Pioneer 11 also carried a flux gate magnetometer (Acuña and Ness)[17] capable of detecting fields up to 10 Gauss with a quantization step size of $\pm 600 \gamma$ for field intensities < 2 Gauss.

All instruments have worked extremely successfully and have provided an enormous wealth of data. It is obvious that any summary of these data must be incomplete. The emphasis put on some of the observational results, and the somewhat sketchy treatment of others, reflects my attempt to provide the reader with some suggestions as to how the questions posed above have been or could be answered.

Much of the following discussion will deal with the observation that the particle count rates vary periodically with time. The periodicity is close to 10 hours and presumably identical to the rotation period of Jupiter within the experimental error. Several different rotation periods have been used to describe different recurring features of Jupiter as, for example, surface features, DIM and DAM radiation. The redefinition of the rotation period relating to the radioastronomical observations of DIM and DAM is described by Riddle and Warwick (1976).[18] It is this new rotation period of $9^h 55^m 29\overset{s}{.}71$ instead of the old $9^h 55^m 29\overset{s}{.}37$ which seems most suitable for the description of the energetic particle periodicities. However, instead of writing out this period in full every time, we shall use the rounded value of 10 hours. Thus, whenever we refer to a 10-hour periodicity or 10-hour rotation period, we really mean $9^h 55^m 29\overset{s}{.}71$.

---

[11]See p. 896. [14]See p. 928. [17]See p. 830.
[12]See p. 961. [15]See p. 848. [18]The redefinition is given on p. 826;
[13]See p. 738. [16]See p. 788. also see p. 20.

All experimenters divide the magnetosphere into several distinct regions with only minor discrepancies between the different definitions of these regions. All of them agree that there is an *inner magnetosphere* (roughly inside 10 R_J). Within this inner magnetosphere the magnetic field is approximately dipolar, and energetic particles are relatively well-ordered and most likely time-stationary in a co-rotating frame of reference. Between 10 R_J and 30 R_J there is an *intermediate magnetosphere* where the magnetic field begins to deviate from that of a dipole but where the dipolar component is still significant. Outside 30 R_J the magnetic field has a disc-like configuration presumably due to a thin current sheet. In this *outer magnetosphere*, which is called *middle magnetosphere* by Smith et al.,[19] the dipolar field is small compared with the field due to local current systems (current sheet) and the particle fluxes are variable and irregular. The particle distributions are predominantly isotropic. Smith *et al.* also define an outer magnetosphere which is bounded by the magnetopause on the outside and extends to about 15 R_J inside the magnetopause. Here the magnetic field is very irregular but predominantly southward, indicating that the magnetopause is rather blunt. Particle fluxes are equally irregular in this region. Outside the magnetopause is the *magnetosheath* which contains thermal plasma flow (Intriligator and Wolfe)[20] and in some cases not insignificant magnetic fields. The magnetosheath is separated by a *bow shock* from the solar wind.

A. Inner Magnetosphere (r < 10 R_J)

The Magnetic Field. In the inner magnetosphere the current density is assumed to be zero. The magnetic field can then be expressed as the gradient of a scalar potential and can be modeled as a superposition of fields due to "internal" multipoles, representing a current system flowing within Jupiter, and fields due to "external" multipoles, representing a current system flowing in the intermediate and outer magnetosphere. By fitting various models to the point-by-point measurements along the two fly-by trajectories, Smith *et al.* are able to calculate the first three internal and first two external multipole moments. Acuña and Ness[21] model the field by the first three internal moments.

Acuña and Ness (1975) published a model of the Jovian magnetic field "based upon real time, quick-look data" obtained during the Pioneer 11 encounter. This model differed considerably from that of Smith *et al.* (1975). However, a subsequent analysis of the flux-gate magnetometer data revealed an error in the sensitivity calibration of the instrument. Comparison with charged-particle anisotropies reported by Van Allen[22] and Fillius[23] allowed for a correction of this error. The correction reduced the discrepancy be-

[19]See p. 790.   [22]See p. 954.
[20]See p. 850.   [23]See p. 896.
[21]See p. 830.

tween the flux-gate magnetometer data and those obtained by the vector helium magnetometer considerably. It appears that the latest magnetic field model of Acuña and Ness[24] differs significantly from their original model but is in reasonable agreement with that of Smith *et al.*[25]

In deriving magnetic field models both groups assumed that the field due to internal multipoles is time-stationary in a frame co-rotating with the planet during the time of observation. The external currents which only Smith *et al.* took into account were assumed stationary in a frame fixed with respect to the sun. The internal field can be modeled with reasonable accuracy by the D_2 or D_4 models of Smith *et al.* which approximate the field as that of a dipole of moment M. The northern end of the dipole, which is also the north-seeking pole, is tilted by an angle α towards a longitude (System III, 1957) λ_D. The center of this dipole moment is displaced by r_0 towards a latitude θ_0 and longitude λ_0. The best fit D_4 model (for epoch 1974.9) is obtained from the Pioneer 11 data and yields the following set $(M, \alpha, \lambda_D, r_0, \theta_0, \lambda_0) = (4.225$ G R_J^3, $10°.8$, $230°.9$, 0.101 R_J, $5°.1$, $185°.7$). The D_2 model (epoch 1974.9), which was determined from the Pioneer 10 data, differs only slightly from these $(4.000$ G R_J^3, $10°.6$, $225°$, 0.11 R_J, $15°.9$, $179°.5$). Based on the multipole moments of Acuña and Ness one obtains for the O_4 model (epoch 1974.9) the set $(4.28$ G R_J^3, $9°.6$, $232°$, 0.14 R_J, $4°.1$, $180°.2$).

All models fit the respective data with a relative ratio of rms residuals to rms fieldstrengths of less than 1%. It is thus conceivable that the difference in the strength of the dipole moment between the 1973 Pioneer 10 encounter and the 1974 Pioneer 11 encounter reflects a real change of magnetic fieldstrength within a year. It must, however, be noted that the longitude coverage of Pioneer 10 was very limited due to its prograde trajectory. The longitude coverage of Pioneer 11 was much better and the discrepancy between the O_4, D_4, and D_2 models may be attributable to longitudinal asymmetries. This interpretation is supported by the fact that the multipole expansion yields ratios of the internal quadrupole and octupole moments to the dipole moment that are 20% and 15%, respectively. (These are, however, only slightly larger percentages than the 13 and 9% for the case of the earth.)

Apart from the discrepancy in M, Pioneer 11 models agree fairly well with those of Pioneer 10. Both sets of data are consistent with the magnetic field models derived from DIM and DAM observations (Carr and Gulkis 1969; Mead 1974) except for the magnitude of M. However, the magnitude of M was the least accurately known parameter of the field before the Pioneer 10 encounter because it was estimated from the frequency characteristics of DAM assuming that DAM is emitted at the local gyrofrequency, which may not be the case. The tilt angle of $\alpha = 10° \pm 1°$ as well as the value of $\lambda_D = 230° \pm 5°$ agree well with pre-encounter estimates. The direction

---

[24]See p. 833.
[25]See p. 819.

of the field is southward at the equator (at the earth it is northward), in accordance with pre-encounter models. Finally DIM observations had suggested that the dipole is offset from the planetary center by an amount $\leqslant 0.1\ R_J$ (McCulloch and Komesaroff 1973; Berge 1974) which is in excellent agreement with the offsets reported by Smith et al. (0.101 for D_4 and 0.110 for D_2). Komesaroff and McCulloch (1975) report a longitudinal asymmetry in the DIM polarization characteristics which is compatible with the more sophisticated higher-order multipole models of Smith et al. and of Acuña and Ness.

Outside 10 R_J the models do not fit the data very well. This could be due either to the growing influence of external multipole moments, which may also be time variable, or to the fact that outside 10 R_J the local current density is not negligible anymore. (A multipole expansion of the field is possible only if the local current density is zero.) Frank et al. (1976) observe ring-current-like proton distributions around 10 R_J. Apparently the multipole expansion technique can yield good representations of the magnetic field data only inside 10 R_J. It thus seems advisable to distinguish an intermediate magnetosphere from the inner magnetosphere on the basis of the magnetic field data as well as on the basis of the energetic particle data.

Energetic Particles. In the inner region the particle fluxes display a general increase as the spacecraft approach the planet. The increase is not simply monotonic but shows a "complex, shell-like structure" (Simpson and McKibben).[26] The structure depends on energy and particle species. These particle flux variations are often but not always related to the L shells of the Jovian satellites. Simpson and McKibben suggest that the satellite-unrelated features reflect some anomalous magnetic field structure or property of the magnetospheric plasma density. The Acuña and Ness model contains slightly higher percentages of quadrupole and octupole moments and the particle structures may be reflections of these magnetic "anomalies."

Both electron and proton fluxes show distinct dips or "holes" at the L shells occupied by Io. The decreases are more pronounced for lower-energy particles. Lesser but still marked decreases are observed at the orbits of Europa and Amalthea. Mead (1972), Hess et al. (1973) and Mead and Hess (1973) showed that the Galilean satellites which orbit Jupiter within the Jovian magnetosphere should sweep up energetic particles and deplete the trapped particle density inside their orbits. The variation of particle anisotropies in the vicinity of Io's orbit is consistent with the pre-encounter prediction (Mead and Hess 1973) that particles with small equatorial pitch angles are absorbed preferentially (Simpson et al. 1974b; Trainor et al. 1974; Fillius and McIlwain 1974). These observations have been interpreted by all observers as indicative of absorption of radially inward diffusing particles. Absorption of radially inward convecting particles should not produce holes but merely step-like decreases.

---

[26]See p. 745.

It has been suggested that radial diffusion is either driven by magnetic field disturbances due to compressions of the magnetosphere, by electric field disturbances due to changes of solar wind interaction with the magnetosphere, or by electric field disturbances driven by large-scale turbulence in Jupiter's atmosphere. The diffusion constants for these cases scale as $D = D_0 L^n$ with $n = 10$, $n = 6$, and $n = 2$–3, respectively, the last mechanism leading to an energy dependent diffusion constant (see Mogro-Campero for a review of these).[27] Numerical estimates of the diffusion constant and especially its variation with L are extremely important because they may allow us to identify the dominant physical mechanism responsible for diffusion.

The observed satellite effects have been studied by Simpson et al. (1974b), Fillius et al. (1975), and Thomsen and Goertz (1975) in terms of radial diffusion. That work is reviewed by Mogro-Campero in this book. He indicates that an L-dependent diffusion coefficient of the form $D = 10^{-10} L^4$ $[R_J^2 \text{ s}^{-1}]$ is compatible with the observed satellite effects. D seems to increase with increasing energy. This is to be compared with the best pre-encounter estimates of $D = 2 \times 10^{-9} L^2$ $[R_J^2 \text{ s}^{-1}]$ by Birmingham et al. (1974) obtained from a diffusion solution to DIM intensities and $D = 2 \times 10^{-10} L^3$ $[R_J^2 \text{ s}^{-1}]$ (Coroniti 1974) obtained on purely theoretical grounds. It would seem that these are only marginally consistent with the results reported by Mogro-Campero. Mogro-Campero points out, however, that the quality of the available data is not yet sufficient to obtain reliable estimates of the diffusion coefficient from the satellite effects.

Most estimates for a numerical value of the diffusion coefficient rely in principle on the depth of the observed depletion of the particle fluxes. Mogro-Campero suggests that instead of particle flux variation the phase space density variation should be used. Furthermore, not only the depth of the depletion but also its shape contain information relevant to a numerical estimate of the diffusion coefficient. In addition, most of the methods used so far have assumed azimuthal symmetry which, as shown by Thomsen and Goertz (1975), may not be valid.

A more fundamental point was raised by Fillius and McIlwain (1974) and by Baker and Goertz (1976). The overall phase space density decreases rapidly with decreasing distance from the planet. The decrease is incompatible with loss free diffusion. Local non-adiabatic losses other than those caused by the satellites must occur. It is not clear what causes these losses, but Baker and Goertz suggest that pitch angle scattering due to whistler mode turbulence could account for the losses. If the losses are strong enough they would have to be taken into account in estimating the diffusion coefficient, and the real diffusion coefficient may very well be larger than those quoted above. Thus, although the satellite effects clearly establish radial diffusion as a dominant transport and acceleration process in the inner re-

---

[27]See p. 1200.

gion, reliable estimates of the diffusion coefficients are not available as yet.

The conjecture of losses due to whistler mode turbulence is supported by observations of Van Allen[28] who reports that the variations of omni-directional fluxes of electrons with $E > 21$ MeV along fieldlines, obtained by combining Pioneer 10 and 11 data, are incompatible with an equatorial pancake pitch angle distribution of the form $\sin^m \alpha$ (his Fig. 20). They are, however, in qualitative agreement with a pitch angle distribution one would expect if whistler mode turbulence existed in the inner magnetosphere (Van Allen *et al.* 1975; Scarf;[29] Scarf and Sanders 1976). These observations are the first evidence for the existence of whistler turbulence in the inner Jovian magnetosphere. They suggest that whistler mode turbulence exists on fieldlines with $L \leq 6$.

This is contrary to the pre-encounter predictions that the whistler mode should be stable inside 6 R_J because the critical energy $E_c = B^2/8 \pi N$ becomes too large. That prediction was based on a model for the plasma density N which is due to Brice and co-workers (see Brice and Ioannidis 1970; Ioannidis and Brice 1971; Brice and McDonough 1973; Mendis and Axford 1974). This model predicted that the plasma density inside 7 R_J should decrease with decreasing distance to very small values (10^{-1} cm^{-3} at 4 R_J). Other models for the plasma distribution (Goertz 1973, 1976; Melrose 1967) predicted much larger plasma densities inside 7 R_J and would be compatible with whistler mode turbulence there.

Further evidence for whistler turbulence is provided by Scarf, who compares the predictions of whistler mode limited electron fluxes based on the Brice-Axford-Mendis density model with the UCSD electron fluxes.[30] The agreement is good outside 7 R_J but somewhat poor inside, again indicating that the Brice-Axford-Mendis density model does not correctly describe the plasma density inside 7 R_J. This conclusion is further strengthened by the observation of large plasma densities ($N \gtrsim 40$ cm^{-3}) inside Io's orbit by Frank *et al.* (1976).

The particle data not only show that Io absorbs particles but also that Io injects into the magnetosphere electrons of several hundred keV energy (Van Allen and Fillius).[31] The observations of Frank *et al.* (1976) also indicate the injection of ~ 100 eV protons. This injection tends to confirm the predictions made by Gurnett (1972), Shawhan *et al.* (1973), and Shawhan *et al.* (1975) who have shown that the induced motional *emf* across Io of ~ 540 kV can accelerate particles to energies of this magnitude. A discussion of the salient features of the Gurnett-Shawhan model can be found in the chapter by R. A. Smith.[32]

---

[28] See p. 948.
[29] See p. 885.
[30] See p. 884.

[31] See pp. 953 and 916.
[32] See p. 1160.

Summary. It is clear that most of the pre-encounter ideas about the inner magnetosphere have generally been confirmed by the Pioneer data. In particular, the pre-encounter magnetic field models have turned out to be amazingly close to reality. Radial diffusion has been established as an important mode of transport. There is strong indication for pitch angle scattering due to wave particle interaction. It seems, however, that the observations do not confirm the Brice-Axford-Mendis density model inside, but only outside 7 R_J. It is hoped that further numerical estimates of diffusion coefficients and loss rates based on the satellite effects and the overall phase space density variation may lead to a more decisive test of the possibility of whistler model turbulence and of the various density models. These models may also help to explain the, so far unaccounted for, shell-like structure of particle fluxes.

B. Intermediate Magnetosphere (10 R_J < r < 30 R_J)

The Magnetic Field. Beyond about 8–10 R_J the magnetic field is not dipolar anymore. The fieldlines tend to be more radial than one would expect for a dipole field. The dipole field, however, still represents a significant fraction of the observed field. This can be seen by the fact that the external dipole moment is roughly 4×10^{-3} of the internal dipole moment. The fields due to both are equal at about 15 R_J.

Besides the tendency to become radial, which was observed on all 4 Pioneer passes, the field also tends to deviate from the meridional plane (see Fig. 6 of Smith *et al.*, p. 810). This spiraling, however, is not observed during the high-latitude outbound path of Pioneer 11. Whereas the direction of spiraling is consistently opposite to the direction of Jupiter's rotation for Pioneer 10 outbound, it is extremely variable for Pioneers 10 and 11 inbound. The sense of the magnetic field spiral for Pioneer 10 outbound is consistent with a lagging of the outer parts of the fieldlines with respect to the planet's surface. All these aspects are much better developed in the outer magnetosphere and will be discussed in more detail in the next section.

Energetic Particles. The most striking aspect and the prime motivation for distinguishing an intermediate magnetosphere is the observation of consistent dumbbell distribution functions (i.e., distribution functions that peak at pitch angles of 0° and 180°) near the equator, and field aligned streaming off the equator for both electrons and protons for 12 R_J < r < 25 R_J [Van Allen;[33] McDonald and Trainor;[34] Sentman and Van Allen (1976); Sentman *et al.* (1975)].

The observations in the inner magnetosphere have fairly well established diffusion as the principal mode of transport. If diffusion were also the principal mode in the intermediate magnetosphere one would not expect dumbbell distributions. This is so because diffusion is driven by perturbations with

---

[33]See p. 954.
[34]See p. 971.

frequencies comparable to the azimuthal drift frequency of particles. Drift periods are typically on the order of several hours. Perturbations which occur on this time scale should not destroy the first and second adiabatic invariants. The first adiabatic invariant is $\mu = p_\perp^2/B$ where B is the local magnetic field and $p_\perp$ is the particle momentum perpendicular to the local field. μ will remain invariant with respect to temporal changes of B provided that the time scale of these changes is long compared with the gyroperiod of the particle. The second or longitudinal adiabatic invariant is $J = \oint p_\parallel \, dl$ where $p_\parallel$ is parallel momentum and the integral is evaluated along a complete bounce path of the particle between two mirror points. It is invariant provided that the time scale for changes of B (and possibly μ) is long compared with the bounce period of the particle, i.e., the period of oscillation between the mirror points. Sentman and Van Allen (1976) have shown that particles injected in the outer magnetosphere always drift toward $\alpha = 90°$ pitch angles as they diffuse inward even in the highly distorted magnetic field of the outer magnetosphere. They conclude that an isotropic injection should result in pancake distribution everywhere, interior to injection point, if the first and second adiabatic invariants are conserved and if large-scale electric fields are absent. Apparently such a simple set of assumptions regarding particle diffusion conditions cannot account for the dumbbell distribution found around 20 R_J.

The phase space density variation with L is quite irregular in this region. For the highest energy particles there is some indication for phase space density peaks in this region. These could possibly be due to local injection. The phase space density curves shown by Van Allen[35] are, however, inconclusive in the intermediate and outer magnetosphere.

The dumbbell distribution functions rule out the existence of whistler mode turbulence in the intermediate magnetosphere because a pancake distribution is needed for the amplification of whistlers. Under these circumstances the good agreement between the theoretical prediction by Coroniti (1974) of whistler mode limited flux and the UCSD electron flux outside 12 R_J (as reported by Scarf)[36] must be regarded as fortuitous.

Field aligned streaming of electrons (Van Allen)[37] has been observed at high latitudes (Pioneer 11) on fieldlines which presumably connect to the equatorial intermediate magnetosphere. They seem to be related to the dumbbell distributions observed there. The sense of streaming is *away* from the planet. Van Allen also observes protons at high latitudes streaming *away* from the planet. However, McDonald and Trainor[38] report that at lower latitudes (Pioneer 10) the protons stream *towards* the planet.

In a recent paper by Sentman et al. (1975) the streaming and dumbbell distributions are interpreted in terms of a "recirculation model." Particles diffuse rapidly inwards in the equatorial plane. In the inner magnetosphere

---

[35]See p. 941.   [37]See p. 954.
[36]See p. 883.   [38]See p. 972.

radial diffusion conserving μ and J drives the distribution towards a pancake distribution. Particles are pitch angle scattered by interaction with whistlers and are precipitated towards high latitudes. There they can diffuse across the field due to a diffusion mechanism, proposed by Nishida (1975), which does not conserve the second adiabatic invariant.

Particles which find themselves on fieldlines that connect to the intermediate or outer magnetosphere cannot pass back and forth between mirror points without encountering the strongly curved magnetic field in the equatorial plane. This field is due to a current sheet. As the particles move through the sheet where the scale length of field changes is smaller than their gyroradius they will gain perpendicular energy at the expense of their parallel energy. This scattering-like process raises their mirror point altitudes and increases the first adiabatic invariant of the particles. Thus, the equator acts as a sink for particles injected at high latitudes. The injection provides small pitch angle particles and if the scattering in the current sheet is not too strong, the distribution function of the scattered particles will remain dumbbell-like. Where scattering is strong the distribution will be more isotropic.

The scattered particles are trapped in the outer and intermediate magnetosphere and can diffuse inwards by conventional diffusion, become scattered in the inner magnetosphere, etc. In this way particles are recirculated. This process could account for streaming of both protons and electrons away from the planet. It seems, however, incompatible with streaming *towards* the planet. A further analysis of streaming may resolve the discrepancy or refute the recirculation model in its present crude form.

Summary. The observed pitch angle distributions in the intermediate magnetosphere rule out the simple interpretation of particle injection in the outer magnetosphere and subsequent radial inward diffusion as the principal mode of transport. They may be compatible with the recirculation model recently proposed. Whistler mode turbulence cannot be self-consistently generated by the electrons in the intermediate magnetosphere. These findings are inconsistent with pre-encounter models.

C. Outer Magnetosphere (30 $R_J < r < R_{MP}$)

Magnetic Field. Comparison of Pioneer 10 and 11 data inbound (low latitude, about $10^h\,00^m$ local time) with that of Pioneer 10 outbound (low latitude, $5^h\,00^m$ local time) and Pioneer 11 outbound (high latitude, $12^h\,00^m$ local time) reveals that the characteristics of the outer magnetosphere are dependent on local time. Results from the comparison are the following:

1. The magnetic field becomes radial. This tendency is very pronounced for Pioneer 10 outbound and for Pioneer 11 outbound and inbound. The latitude angle δ ($\delta \to 0$ for radial direction) is more variable and generally more negative (indicating a southward field) for Pioneer 10 inbound than for the other three cases.

2.   The magnetic field magnitude at higher latitudes is consistently larger
     than the extrapolated dipole field. This is true for all four cases, although
     the ratio of observed field to extrapolated dipole field is not quite as
     large for Pioneer 11 outbound as it is for the other three cases.
3.   The magnitude decreases to values comparable to, or less than, the ex-
     trapolated dipole field as the spacecraft approaches the equator. This is
     particularly noticeable for Pioneer 10 outbound.
4.   The field of Pioneer 10 outbound deviates consistently from a meridional
     plane, with the outer parts of a fieldline lagging the inner parts as they
     rotate with the planet. A similar but not as clear tendency is revealed
     by Pioneer 10 inbound and Pioneer 11 inbound data. The Pioneer 11
     outbound fieldlines lie almost precisely in the meridional plane.

During the outbound pass of Pioneer 10 the spacecraft frequently passed
through a region of small magnetic field magnitude. These crossings were
identified as current sheet crossings by Smith *et al.* They correlate well
with maxima in the particle fluxes. Other occasional current sheet crossings
were observed during the Pioneer 11 inbound pass.

An analytic model for all 4 cases is not available and probably will not
be possible because of the local time and latitude effects described above.
Goertz *et al.* (1976) derived a quantitative model valid only for the Pioneer
10 outbound data.

Their model contains a current sheet of $2 R_J$ thickness which lies precisely in the magnetic equatorial plane. This is in agreement with Van Allen's
suggestion (Van Allen *et al.* 1974)[39] of a rigid disk-like magnetosphere symmetric about the magnetic equator, but contradicts the claim of Smith *et al.*
(1974)[40] that the symmetry plane (the current sheet) becomes parallel to the
rotational equator. However, the spacecraft would not pass through the current sheet if it were parallel to the rotational equator as Smith *et al.* themselves point out. They invoke flapping of the disk into the magnetic equatorial plane in order to explain the current sheet crossings which occur *regularly*
with a 10-hour period. A 10-hour periodic flapping of a rotationally aligned
sheet into the magnetic equatorial plane is precisely equivalent to a current
sheet aligned with the magnetic plane and rotating with a 10-hour period.
Thus, it seems almost certain that the pre-dawn magnetosphere contains a
current sheet in the magnetic equatorial plane.

This magnetic field model predicts closed fieldlines near the current sheet
and open fieldlines at higher latitudes, which appears to be consistent with
the observed direction of the fields: southward in the current sheet, radial
or northward at higher latitudes. It seems that these characteristic features
also apply to the other cases: southward fields near the equator, radial or
northward fields at higher latitudes (in particular, for Pioneer 11 outbound).

---

[39]See p. 935.
[40]See p. 813.

The observation of southward fields near the equator shows that low latitude fieldlines are closed. This appears to be in contradiction with radial outflow models (Michel and Sturrock 1974; Eviatar and Ershkovich 1976; Kennel and Coroniti 1975). These models assume that beyond some critical radial distance R_C (usually at 30 R_J) the co-rotation velocity exceeds the Alfvén speed. Then the magnetic field could not enforce co-rotation anymore and the plasma would flow radially outwards very much like the solar wind. The plasma would "blow" open the magnetic field.

It is apparent that these authors must assume a certain model for the plasma density and magnetic field in order to predict the critical distance. Density and field, however, are mutually interdependent and the problem must be solved self-consistently. Apparently this has not been done. Rather, a dipole field was assumed for $r < R_C$ and a radial field $B \propto 1/r^2$ or $B \propto 1/r$ for $r > R_C$. Clearly there must be a transition zone between these two solutions (Mestel 1968). It appears that much of the outer magnetosphere is this transition zone (Northrop et al. 1974) where the fieldlines are greatly extended but closed.

The azimuthal direction of the spiraling field is consistent with a model of radial outflow and the two Pioneer 10 passes have led some authors to conclude that outflow does, indeed, occur. Yet the spiraling is much less obvious on the Pioneer 11 data, particularly the high-latitude outbound data of Pioneer 11 which show no spiraling at all. However, it is precisely at high latitudes where the spiraling should be observable because high-latitude fieldlines can be open and allow for a planetary wind to flow along them. Although outflow models cannot be completely ruled out, the inconsistency of spiraling and the presence of closed fieldlines tend to rule out radial outflow as the reason for the spiraling.

It seems that spiraling which is strong close to the equator and weak at high latitudes, requires a current flow which is field aligned and towards the planet at high latitudes and away from the planet in the equatorial plane. These two currents appear as a part of a meridional current loop. The outward streaming electrons described in the previous section could perhaps carry the necessary inward current. The drift motion of energetic particles in the spiraling field close to the equator is consistent with an outward radial current in the equatorial plane. (Electrons drift radially inwards due to the azimuthal magnetic field.) It remains to be calculated what kind of current density is required to produce the observed spiraling and whether the energetic particle fluxes are large enough to carry that current.

Energetic Particles. Particle flux measurements in the outer magnetosphere also show local time dependence. On all four passes a 10-hour modulation of intensities was observed. This modulation is most obvious for the outbound data of Pioneer 10. There, count rates drop over several orders of magnitude as the spacecraft moves away from the magnetic equator. In fact, for all but the lowest energy channels the count rates drop to interplanetary values. This striking saw-tooth modulation has been interpreted by Goertz *et al.* (1976) as reflecting the motion of Pioneer 10 from closed to open fieldlines and back to closed fieldlines as the current sheet lying in the magnetic equator rotates with the planet and thus sweeps up and down relative to the trajectory plane of the spacecraft. The times of abrupt count rate changes agree, to within 10–30 minutes, with the crossings of the last closed fieldline as predicted by the Goertz model. The fact that the 10-hour modulation is not as pronounced for the inbound data of Pioneer 10 was explained as being due to the possibility that the last closed fieldline extends to higher latitudes on the sunward side. This greater extent or equivalently blunter shape of the front-side magnetosphere could be due to the solar wind compression. Beard and Jackson (1976), for example, show that the Jovian magnetosphere should be almost as blunt as the earth's, even though it contains a current sheet which tends to produce highly distended fieldlines and a thin disk-like magnetic configuration.

A critical test of these ideas was provided by the high latitude outbound pass of Pioneer 11 which occurred at approximately $12^h 00^m$ local time. If the particles are narrowly confined to the magnetic equator, as the Pioneer 10 observations suggested, one would not expect to see large count rates at high magnetic latitudes. Therefore during the Pioneer 11 outbound pass, which occurred at very high latitudes, the count rates should have been small. In fact, they were observed to be equal to or even larger than Pioneer 10 count rates at equal distances from the planet. Thus, unless the overall particle loading of the magnetosphere had increased from 1973 to 1974, the relatively simple model of a magnetodisc and equatorial confinement cannot be valid. A magnetic field model which may reproduce the main features of the Jovian front-side magnetosphere has recently been advanced by Barish and Smith (1975). This model predicts secondary magnetic field minima at high latitudes which could also confine energetic particles. However, no detailed comparisons of that model with the observations are yet available.

The discussion so far has been based on the assumption that variations of particle count rates are due to changes in magnetic latitude of the spacecraft. Simpson *et al.* (1974*a*, 1975) and Simpson and McKibben[41] have presented an alternative interpretation. It has been noted by all experimenters that the count rate maxima in parts of the outer magnetosphere (roughly for $r > 30$ R_J) do not occur when the magnetic latitude λ_m calculated from a purely meridional field (i.e., the field has no azimuthal component) is a minimum. There is a systematic lag between the times when λ_m is a minimum and when the count rate maxima are observed. This lag is very small in the inner and intermediate magnetosphere but quite large in the outer magnetosphere. Apparently the particles are confined to the magnetic equator in the inner

---

[41]See p. 758.

magnetosphere, whereas this may not be true for the outer magnetosphere. The time lag in the outer magnetosphere is a function of radial distance and has been interpreted by Northrop *et al.* (1974) as being due to the finite propagation time of Alfvén waves from the planet's surface to the point of observation. Northrop *et al.* only dealt with Pioneer 10 data and even there had to make extra and perhaps artificial assumptions to explain the Pioneer 10 inbound lags which, in fact, are negative lags (or leads).

Simpson and co-workers show evidence that the variations in count rates are not due to variations in λ_m but that they reflect a regular 10-hour periodic filling and emptying of the whole outer magnetosphere and even the magnetosheath and interplanetary space in the vicinity of Jupiter (see next Sections). This 10-hour modulation is independent of the spacecraft position. It is even more noticeable in the spectral index of the electron energy spectrum. Basically the observations show that the time interval between maxima in the spectral index measured at the beginning of the Pioneer 10 trajectory inside the magnetosphere and at the end of it inside the magnetosphere is an integer multiple of 10 hours. If only λ_m were the determining factor, the interval would be an integer multiple of 10 hours plus an additional 5 hours for the phase change associated with crossing the magnetic equatorial plane near periapsis minus about 2 hours for the local time difference between inbound and outbound. The results of Pioneer 11 show even more clearly that the spectral index maxima do not occur when the spacecraft's magnetic latitude is a maximum but at regular time intervals of 10 hours. Furthermore, even outside the magnetosphere the Jupiter related bursts of particles (Sec. E) show a 10-hour periodicity which is in phase with that inside the magnetosphere. Thus, Simpson and co-workers argue that not only is the outer magnetosphere filled every 10 hours but also, simultaneously, the interplanetary space in the vicinity of Jupiter.

It is, however, not clear that the observations regarding the timing of the Jupiter related bursts in interplanetary space could not be explained equally well by the hypothesis that they are observed only when the magnetic field-line passing through the spacecraft connects with open (and therefore high-latitude) planetary fieldlines. Because the polarity of the interplanetary field would make connection to one hemisphere (either north or south) more likely, the bursts should occur with a 10-hour periodicity (see, e.g., Sentman *et al.* 1975). If Simpson's interpretation turns out to be the only valid one (it is not clear that other observers agree with his basic findings), Jupiter would produce some 10^{28} particles with a typical energy of 10 MeV every 10 hours. This corresponds to an average power of 10^{11} W which is not unreasonable but not immediately plausible either.

Hill and Dessler (1976) have recently developed a model which could explain a periodic filling and emptying of the magnetosphere. The model was based on the highly distorted magnetic field model of Acuña and Ness (1975), which has since been revised. It remains to be seen that the model of Hill and Dessler can work with the less distorted D_2, D_4, or O_4 models. It is also a model which predicts open fieldlines in the outer magnetosphere and is hence subject to the criticism of the previous Section. At the present time the interpretation of Simpson and his co-workers remains one of the most exciting and challenging ideas that have been advanced in response to the data. Whether or not it will hold up depends crucially on the confirmation of their observations by the other observers. We will come back to this point in Sec. E.

Summary. A comprehensive model of the outer magnetosphere is not available as yet due to the obvious local time dependence of this region. It seems that the pre-dawn magnetosphere is similar to the earth's magnetotail. There the region of closed fieldlines and trapping of particles is narrow in latitude and highly distended in the radial direction. The front-side magnetosphere covers a much wider latitude range and no evidence for open fieldlines has been found there. A general 10-hour modulation of particle fluxes as well as spectral index has been found everywhere in the outer magnetosphere. The modulation of fluxes is clearest in the pre-dawn magnetosphere. Two interpretations of the 10-hour modulation have been advanced: (1) that it is due to the 10-hour modulation of the spacecraft's magnetic latitude, or (2) that it is due to an inherent latitude-independent 10-hour filling and emptying of the whole outer magnetosphere. It is not clear which interpretation is correct.

D. Magnetosheath

Plasma Measurements. The inside (magnetopause) and outside (bow shock) boundaries of the magnetosheath are most clearly defined by the results of the plasma analyzers on board Pioneers 10 and 11 (Intriligator and Wolfe).[42] In almost all cases the boundary crossings are also visible in the magnetic field data (see next Section). The data indicate a large number of bow shock as well as magnetopause crossings. This demonstrates that the extent of the Jovian magnetosphere is highly variable.

The first bow shock crossings were observed at 108.9 R_J and 109.7 R_J for Pioneers 10 and 11, respectively. On both occasions the solar wind characteristics outside the bow shock were similar although not quite identical. The kinematic pressure $M_p N_s V_s^2$ was about 1.5 times as large for the Pioneer 10 crossing as for the Pioneer 11 crossing. M_p is the average ion mass, N_s is the number density of ions and V_s^2 is the directed velocity of ions in the solar wind. The first crossings of the magnetopause were encountered at 96.4 R_J and 97.3 R_J, respectively. These distances are considerably larger than those that had been predicted on the basis of balance between solar wind and magnetospheric magnetic pressure $B^2/8\,\pi$. The discrepancy be-

---

[42]See p. 850.

tween observations and predictions indicates that the outer magnetosphere may contain high-beta ($\beta \approx 1$) plasma (β is the ratio of thermal plasma pressure to magnetic pressure).

After crossing the magnetopause at 96.4 R_J, Pioneer 10 remained in the magnetosphere until December 1 when it encountered another magnetopause crossing at 54.3 R_J. It remained in the magnetosheath for about 10 hours and finally moved into the magnetosphere again at 46.5 R_J. Intriligator and Wolfe[43] and Wolfe *et al.* (1974) show that the plasma flow direction in the magnetosheath is consistent with the interpretation that this second magnetopause crossing is due to a sudden compression of the whole magnetosphere from $\sim 100\ R_J$ to $\sim 50\ R_J$. Compressions of this magnitude are very rare for the earth's magnetosphere but may not be surprising at Jupiter because of the slow decrease of magnetic pressure with distance. For the earth $B^2/8\pi \propto r^{-6}$ whereas at Jupiter it varies much more slowly (Smith *et al.*).[44] Thus, Intriligator and Wolfe argue that small changes of solar wind pressure may result in large variations of the magnetopause distance. They claim that the alternative interpretation of the second magnetosheath encounter in terms of a flapping disk (Van Allen *et al.* 1974) cannot be ruled out but is less likely.

It is my personal opinion that the compression of the Jovian magnetosphere by a factor of 2 is not as easily explained by reference to the weakly varying magnetic fieldstrength. The solar wind would not only have to compress B, which should increase at least by a factor of 4 (since $B \propto 1/area$), but also the high-β plasma (if it exists). Even for a magnetospheric field which is independent of r, the solar wind pressure would have to increase by at least an order of magnitude. In fact, the magnetic field inside the magnetosphere is not independent of r and the solar wind pressure increase should be more like two orders of magnitude to produce a compression by a factor of 2. It should also be noted that the observed variation in solar wind pressure by a factor of 1.5 between the Pioneer 10 and 11 encounters did not change the magnetosphere position at all. However, the Jovian magnetopause may not be stable, and calculations based on a static pressure balance may be misleading.

Another plausible scenario for the contraction of the magnetosphere is this: Most of the field in the outer magnetosphere is caused by the currents flowing within the magnetosphere (the current sheet). If this current is interrupted or decreased, the magnetopause should move inwards (E. Smith 1975, personal communication). Also, if the magnetopause moves inwards (due, for example, to an increase in solar wind pressure) the magnetopause would become blunter and the magnetic field gradient and curvature in the outer magnetosphere would be reduced. This in turn would lead to a reduction of

[43]See p. 855.
[44]See p. 815.

gradient and curvature drifts and hence to a reduction of the current sheet density. This would further enhance the tendency towards inward motion of the magnetopause (E. Smith 1975, personal communication) and a very rapid contraction of the whole front-side magnetosphere may result even for small increases of the solar wind pressure.

During Pioneer 10 outbound a large number of both magnetopause and bow shock crossings were also observed. The analysis by Intriligator and Wolfe[45] of some of these crossings shows that the magnetosphere plasma may have β values as large as 2.8. Again, these multiple crossings are explained in terms of an expanding and contracting magnetosphere. Pioneer 11 also encountered several magnetopause and bow shock crossings.

These observations clearly indicate a great variability in the size of the Jovian magnetosphere. It is, however, not certain that this variability is only due to solar wind variations in the manner discussed by Intriligator and Wolfe. It seems equally possible that the Jovian magnetosphere, dominated by the current sheet, is inherently unstable in the manner discussed above.

Magnetic Field. For all magnetopause crossings, which in almost all cases were seen in the magnetic field data, the magnetic field just inside the magnetopause was observed to be quite irregular. The field, however, was on all occasions southward. No systematic deviation from a meridional plane existed. Although in most cases the magnetic field was stronger inside the magnetosphere than outside, the differences were not large and, on at least one occasion (first Pioneer 10 outbound magnetopause crossing), the field outside was even larger than inside. This, again, could indicate the presence of high β plasma in the magnetosphere.

The second Pioneer 10 inbound magnetosheath encounter was not particularly obvious in the magnetic field magnitude data. The fieldstrength during that traversal was about $8\,\gamma$ (Fig. 7 of Smith *et al.*, p. 811) compared with $8\,\gamma$ and $10\,\gamma$ just shortly before and after this encounter. The magnitude of B during the first magnetosheath encounter was typically $2\,\gamma$. The only clearly noticeable effect occurred in the latitude angle. The field during the second encounter was slightly northward pointing (Fig. 5 of Smith *et al.*, p. 809) contrary to the southward direction immediately before and after encountering the sheath. The northward direction is also characteristic of the first magnetosheath traversal. Although it is clear that Pioneer 10 did encounter the magnetosheath on December 1, it is equally clear that the magnetic characteristic of this second magnetosheath crossing is different from others.

Finally, it should be mentioned that Smith *et al.*[46] determine the direction of the normal to the magnetopause. It is "directed radially outward as anticipated for a blunt magnetosphere." This confirms the conclusions of Intrili-

---

[45] See p. 864.
[46] See p. 816.

gator and Wolfe that the magnetosphere is blunt, which were based on the direction of plasma flow in the magnetosheath.

Energetic Particles. Both the bow shock crossings and the magnetopause crossings on the sunlit side (Pioneer 11 inbound, outbound, and Pioneer 10 inbound) are noticeable as changes in the energetic particle count rates. It must be regarded as one of the most striking results of the missions that increased fluxes at high-energy particles ($E_e > 21$ MeV) were found in the magnetosheath and immediately inside the magnetopause.

These are completely incompatible with heating of solar wind particles as they pass through the bow shock. If bow shock heating conserved the first adiabatic invariant, one would expect only a factor of 2 or 3 (ratio of B inside and outside the bow shock) increase in average energy. Even non-adiabatic heating could not easily account for the presence of these particles which have a first adiabatic moment μ of some 10^5 MeV/Gauss. The largest value of μ ever considered before encounter was $\mu \sim 2 \times 10^3$ MeV/Gauss (Birmingham *et al.* 1974). Thorne and Coroniti (1972), for example, predicted that μ should be between 10 MeV/Gauss and 100 MeV/Gauss (solar wind particles with direct access to the magnetosphere). Clearly, any particle with energy above several hundred keV found in the magnetosheath or in the outer regions of the outer magnetosphere must be regarded as unusual.

I believe that the presence of these high-energy magnetosheath particles cannot be explained in terms of solar wind interaction with the magnetosphere alone. They reflect precisely the same problem we encountered in the explanation of DIM: Jupiter's magnetosphere and magnetosheath contain high-energy particles that are apparently not of solar wind origin. The only model which attempts to explain the presence of these particles is the recirculation model of Sentman *et al.* (1975) (see the summary of Sec. B in this chapter).

Summary. The Jovian magnetosphere is blunt on the front side. It is highly variable in size. Very high energy particles are found immediately inside the bow shock with intensities larger than those found in interplanetary space but smaller than those inside the magnetopause. The Jovian magnetosphere possibly contains high β plasma.

E. Interplanetary Space

Energetic Particles. Another unexpected result was provided by the observation of energetic particle bursts several days and even months before the spacecraft crossed into the Jovian magnetosheath (Simpson and McKibben; McDonald and Trainor; and Van Allen).[47] These bursts show a 10-hour periodic variation in intensity as well as in spectral index. Simpson and McKibben (see also Simpson *et al.* 1974*a*) show that the spectral index

---

[47]See pp. 754, 962 and 954.

maxima are in phase with those observed inside the magnetosphere. An amazing aspect is that Simpson *et al.* found that the Pioneer 11 observations are in phase with those of Pioneer 10 made a year earlier. However, this coherence over a one-year period may be fortuitous because if these bursts are of Jovian origin (and most observers believe they are), the models which have been proposed to explain them all imply that the phases of two observed bursts could differ by only integer multiples of 10 hours or 5 hours. There is, of course, no way one can be sure that bursts occurred every 10 hours throughout the time from December 1973 to December 1974. One cannot rule out the possibility that there have been some bursts at 5-hour intervals between the usual 10-hour bursts, even though no 5-hour separations have so far been observed.

These observations establish that Jupiter ejects energetic particles into space every 10 hours. The bursts are observed whenever the interplanetary magnetic fieldline through the spacecraft points towards Jupiter (Smith *et al.* 1976). Simpson and co-workers conclude that these 10-hour variations are most likely explained in terms of 10-hour filling and emptying of the outer magnetosphere, magnetosheath, and the interplanetary region close to Jupiter. The fact that the interplanetary particle variation is observed to be in phase with the variation inside the magnetosphere is inconsistent with the simple idea of filling a closed magnetosphere for 5 hours and emptying it through some opening to interplanetary space for the next 5 hours, because in this case the interplanetary bursts should be in anti-phase with the magnetospheric variations. The observed phase coherence implies that the same source is responsible for the 10-hour variation inside and outside the magnetosphere. One might think of a periodic injection of energetic particles onto fieldlines connecting to the distorted field of the magnetosphere beyond $30\ R_J$, the magnetosheath, and interplanetary space. These fieldlines should be very close to each other near Jupiter and the process of high-latitude cross-L diffusion as suggested by Sentman *et al.* (1975) could possibly account for an injection onto all these fieldlines simultaneously. Yet it is not clear why this injection process should be periodic rather than continuous; or, in other words, why injection (simultaneously at all local times) occurs only when a certain Jovian longitude [about $\lambda_{III} \simeq 270°$ according to Vasyliunas (1975)] faces the sun. No plausible periodic mechanism has been suggested so far.

Clearly the questions raised by Simpson and his co-workers have not been answered and should be pursued further. It is therefore essential that other observers either confirm or refute their basic observations regarding the 10-hour periodicity of the spectral index in the outer magnetosphere, magnetosheath, and interplanetary space.

III. SUMMARY

The Pioneer 10 and 11 missions have significantly increased our knowledge of the Jovian magnetosphere. Although some of the pre-encounter ideas have been roughly confirmed, in particular those relating to the structure of the planetary magnetic field, we now know that Jupiter's magnetosphere is much more complex than was previously believed. Jupiter's magnetosphere is, for example, much larger than anticipated. If one could see it from Earth, it would appear of the same angular size as the moon. The whole Jovian magnetosphere is filled with high fluxes of extremely energetic electrons ($E > 30$ MeV). These particles must be due to acceleration somewhere inside the Jovian magnetosphere. Jupiter also emits energetic particles which can be detected as far away as at the earth. Some other findings are:

1. Radial diffusion is an important transport mechanism in the inner Jovian magnetosphere.
2. Good evidence for pitch angle diffusion, presumably caused by whistler mode turbulence, has been presented.
3. The Jovian satellites which orbit the planet inside the magnetosphere absorb particles.
4. The innermost Galilean satellite, Io, not only absorbs particles but also emits or accelerates them, confirming the important role Io must play in the dynamics of the Jovian magnetosphere. This had been previously established by Io's effective control over the Jovian DAM.
5. The size of Jupiter's magnetosphere, although always extremely large, is highly variable. Contractions by a factor of 2 seem to be common.
6. The Jovian magnetosphere may contain high-β plasma everywhere. No clear evidence for an Earth-like plasmapause has been found.

Many questions remain. To quote from a recent paper by J. G. Roederer (1976):

> "We just cannot wait until the next Mariner-Jupiter-Saturn flybys and, most important, until the proposed Jupiter orbiter mission is carried out. Does Jupiter have a magnetic tail? What is the mechanism for relativistic particle release (from the magnetosphere)? Are there substorms and auroras on Jupiter? Is there plasma convection besides corotation? How do ionosphere and magnetosphere interact? How do magnetosphere and interplanetary field interact?"

To those questions I would add: What is the thermal plasma density? Is whistler mode turbulence present in the inner and intermediate magnetosphere? What is the mechanism driving the radial diffusion of particles? What, finally, is the acceleration mechanism for those enormously energetic particles found everywhere within the Jovian magnetosphere? It is obvious that the answers to these questions will not come quickly. More observations and detailed analyses of data already available are needed. Yet, it is already clear that many of the concepts derived from the study of our own magnetosphere have been very useful and successful in studying the Jovian magnetosphere.

Acknowledgements. I have benefited greatly from numerous discussions with J. A. Van Allen, S. D. Shawhan, M. F. Thomsen, D. A. Baker, D. D. Sentman, B. A. Randall, and R. A. Smith.

REFERENCES

Acuña, M. H., and Ness, N. F. 1975. Jupiter's main magnetic field measured by Pioneer 11. *Nature* 253:327–328.
Baker, D. N., and Goertz, C. K. 1976. Diffusion in Jupiter's magnetosphere. Submitted to *J. Geophys. Res.*
Barish, F. D., and Smith, R. A. 1975. An analytical model of the Jovian magnetosphere. *Geophys. Res. Lett.* 2:269–272.
Beard, D. B., and Jackson, D. L. 1976. The Jovian magnetic field and the magnetosphere shape. *J.G.R.* (special Jupiter issue). In press.
Beck, A. J., ed. 1972. *Proceedings of the Jupiter radiation belt workshop.* Tech. Mem. 33–543. Pasadena, California: Jet Propulsion Laboratory.
Berge, G. L. 1974. Position and Stokes parameters of integrated 21-cm radio emission of Jupiter and their variation with epoch and central meridian longitude. *Astrophys. J.* 191:775–784.
Birmingham, T.; Hess, W.; Northrop, T.; Baxter, R.; and Lojko, M. 1974. The electron diffusion coefficient in Jupiter's magnetosphere. *J. Geophys. Res.* 79:87–97.
Brice, N. M., and Ioannidis, G. A. 1970. The magnetospheres of Jupiter and Earth. *Icarus* 13:173–183.
Brice, N., and McDonough, T. R. 1973. Jupiter's radiation belts. *Icarus* 18:206–219.
Carr, T. D., and Gulkis, S. 1969. The magnetosphere of Jupiter. *Ann. Rev. Astron. Astrophys.* 7:577–618.
Coroniti, F. V. 1974. Energetic electrons in Jupiter's magnetosphere. *Astrophys. J. Suppl.* 27:261–281.
———. 1975. Denouement of Jovian radiation belt theory. *The magnetospheres of the Earth and Jupiter.* (V. Formisano, ed.) pp. 391–410. Dordrecht, Holland: D. Reidel Publ. Co.
Eviatar, A., and Ershkovich, A. I. 1976. The Jovian magnetopause and outer magnetosphere. In preparation.
Fillius, R. W., and McIlwain, C. E. 1974. Measurements of the Jovian radiation belts. *J. Geophys. Res.* 79:3589–3599.
Fillius, R. W.; McIlwain, C. E.; and Mogro-Campero, A. 1975. Radiation belts of Jupiter: a second look. *EOS Trans. Amer. Geophys. Union* 56:428.
Frank, L. A.; Ackerson, K. L.; Wolfe, J. H.; and Mihalov, J. D. 1976. Observations of plasmas in the Jovian magnetosphere. *J. Geophys. Res.* In press.
Goertz, C. K. 1973. Jupiter's ionosphere and magnetosphere. *Planet. Space Sci.* 21:1389–1398.
———. 1976. Plasma in the Jovian magnetosphere. *J. Geophys. Res.* In press.
Goertz, C. K., and Deift, P. A. 1973. Io's interaction with the magnetosphere. *Planet. Space Sci.* 21:1399–1415.
Goertz, C. K.; Jones, D. E.; Randall, B. A.; Smith, E. J.; and Thomsen, M. F. 1976. Evidence for open field lines in Jupiter's magnetosphere. *J. G. R.* (special Jupiter issue). In press.

Gold, T. 1976. Magnetosphere of Jupiter. *J. G. R.* (special Jupiter issue). In press.

Goldreich, P., and Lynden-Bell, D. 1969. Io, a Jovian unipolar inductor. *Astrophys. J.* 156: 59–78.

Gurnett, D. A. 1972. Sheath effects and related charge-particle acceleration by Jupiter's satellite Io. *Astrophys. J.* 175:525–533.

Hess, W. N.; Birmingham, T. J.; and Mead, G. D. 1973. Jupiter's radiation belts: can Pioneer 10 survive? *Science* 182:1021–1022.

Hill, T. W., and Dessler, A. J. 1976. Longitudinal asymmetry of the Jovian magnetosphere and the periodic escape of energetic particles. *J. G. R.* (special Jupiter issue). In press.

Ioannidis, G. A., and Brice, N. M. 1971. Plasma densities in the Jovian magnetosphere: plasma slingshot or Maxwell demon? *Icarus* 14:360–373.

Kennel, C. F., and Coroniti, F. V. 1975. Is Jupiter's magnetosphere like a pulsar's or Earth's? *The magnetospheres of Earth and Jupiter.* (V. Formisano, ed.) pp. 451–477. Dordrecht, Holland: D. Reidel Publ. Co.

Kennel, C. F., and Petschek, H. E. 1966. Limit on stably trapped particle fluxes. *J. Geophys. Res.* 71:1–28.

Klein, M. J. 1976. The variability of the total flux density and polarization of Jupiter's decimetric radio emission. *J. G. R.* (special Jupiter issue). In press.

Komesaroff, M. M., and McCulloch, P. M. 1975. Asymmetries of Jupiter's magnetosphere. *Mon. Not. Roy. Astron. Soc.* 172:91–95.

———. 1976. Evidence for an unexpected time-stable symmetry of the Jovian magnetosphere. *J. G. R.* (special Jupiter issue). In press.

McCulloch, P. M., and Komesaroff, M. M. 1973. Location of the Jovian magnetic dipole. *Icarus* 19:83–86.

Mead, G. D. 1972. The effect of Jupiter's satellites on the diffusion of protons. *Proceedings of the Jupiter radiation belt workshop.* (A. J. Beck, ed.) Tech. Mem. 33–543:271–276. Pasadena, California: Jet Propulsion Laboratory.

———. 1974. Magnetic coordinates for the Pioneer 10 Jupiter encounter. *J. Geophys. Res.* 79:3514–3521.

Mead, G. D., and Hess, W. N. 1973. Jupiter's radiation belts and the sweeping effect of its satellites. *J. Geophys. Res.* 78:2793–2811.

Melrose, D. B. 1967. Rotational effects on the distribution of thermal plasma in the magnetosphere of Jupiter. *Planet. Space Sci.* 15:381–393.

Mendis, D. A., and Axford, W. I. 1974. Satellites and magnetospheres of the outer planets. *Ann. Rev. of Earth and Planet. Sci.* 2:419–474.

Mestel, J. J. 1968. Magnetic braking by a stellar wind. I. *Mon. Not. Roy. Astron. Soc.* 138:359–391.

Michel, F. C., and Sturrock, P. A. 1974. Centrifugal instability of the Jovian magnetosphere and its interaction with the solar wind. *Planet. Space Sci.* 22:1501–1510.

Nishida, A. 1975. Outward diffusion of energetic particles from the Jovian radiation belt. *Inst. of Space and Aero. Science,* University of Tokyo. Preprint.

Northrop, T. G.; Goertz, C. K.; and Thomsen, M. F. 1974. The magnetosphere of Jupiter as observed with Pioneer 10. II. Nonrigid rotation of the magnetodisc. *J. Geophys. Res.* 79:3579–3582.

Piddington, J. H., and Drake, J. F. 1968. Electrodynamic effects of Jupiter's satellite Io. *Nature* 217:935–937.

Riddle, A. C., and Warwick, J. W. 1976. Redefinition of System III longitude. *Icarus* (special Jupiter issue). In press.

Roederer, J. G. 1976. Planetary plasmas and fields. *EOS Trans. Amer. Geophys. Union* 57:53–62.

Scarf, F. L., and Sanders, N. L. 1976. Some comments on the whistler mode instability at Jupiter. *J. Geophys. Res.* In press.

Schulz, M., and Lanzerotti, L. J. 1974. *Particle diffusion in the radiation belts.* New York: Springer-Verlag.

Sentman, D. D., and Van Allen, J. A. 1976. Angular distributions of electrons of energy E_e > 0.06 MeV in the Jovian magnetosphere. *J. Geophys. Res.* In press.

Sentman, D. D.; Van Allen, J. A.; and Goertz, C. K. 1975. Recirculation of energetic particles in Jupiter's magnetosphere. *Geophys. Res. Lett.* 2:465–468.

Shawhan, S. D. 1976. Io accelerated electrons and ions. *J.G.R.* (special Jupiter issue). In press.

Shawhan, S. D.; Goertz, C. K.; Hubbard, R. F.; Gurnett, D. A.; and Joyce, G. 1975. Io-accelerated electrons and ions. *The magnetospheres of Earth and Jupiter.* (V. Formisano, ed.) pp. 375–389. Dordrecht, Holland: D. Reidel Publ. Co.

Shawhan, S. D.; Hubbard, R. F.; Joyce, G.; and Gurnett, D. A. 1973. Sheath acceleration of photoelectrons by Jupiter's moon Io. *Photon and particle interactions with surfaces in space.* (R. Grard, ed.) pp. 405–413. Dordrecht, Holland: D. Reidel Publ. Co.

Simpson, J. A.; Hamilton, D.; Lentz, G.; McKibben, R. B.; Mogro-Campero, A.; Perkins, M.; Pyle, K. R.; Tuzzolino, A. J.; and O'Gallagher, J. J. 1974a. Protons and electrons in Jupiter's magnetic field: results from the University of Chicago experiment on Pioneer 10. *Science* 183:306–309.

Simpson, J. A.; Hamilton, D. C.; Lentz, G. A.; McKibben, R. B.; Perkins, M.; Pyle, K. R.; Tuzzolino, A. J.; and O'Gallagher, J. J. 1975. Jupiter revisited: first results from the University of Chicago charged particle experiment on Pioneer 11. *Science* 188:455–459.

Simpson, J. A.; Hamilton, D. C.; McKibben, R. B.; Mogro-Campero, A.; Pyle, K. R.; and Tuzzolino, A. J. 1974b. The protons and electrons trapped in the Jovian dipole magnetic field region and their interaction with Io. *J. Geophys. Res.* 79:3522.

Smith, E. J.; Davis, Jr., L.; Jones, D. E.; Coleman, Jr., P. J.; Colburn, D. S.; Dyal, P.; and Sonett, C. P. 1975. Jupiter's magnetic field, magnetosphere, and interaction with the solar wind: Pioneer 11. *Science* 188:451–455.

Smith, E. J.; Davis, Jr., L.; Jones, D. E.; Coleman, Jr., P. J.; Colburn, D. S.; Dyal, P.; Sonett, C. P.; and Frandsen, A. M. A. 1974. The planetary magnetic field and magnetosphere of Jupiter: Pioneer 10. *J. Geophys. Res.* 79:3501–3513.

Smith, E. J.; Tsurutani, B. T.; Chenette, D. L.; Conlon, T. F.; and Simpson, J. A. 1976. Jovian electron bursts: correlation with the interplanetary field direction and hydromagnetic waves. *J. Geophys. Res.* 81:65–72.

Thomsen, M. F., and Goertz, C. K. 1975. Satellite sweep-up effects at Jupiter. *EOS Trans. Amer. Geophys. Union* 56:428.

Thorne, R. M., and Coroniti, F. V. 1972. A self-consistent model for Jupiter's radiation belt. *Proceedings of the Jupiter radiation belt workshop.* (A. J. Beck, ed.) Tech. Mem. 33–543:363–372. Pasadena, California: Jet Propulsion Laboratory.

Trainor, J. H.; McDonald, F. B.; Teegarden, B. J.; Webber, W. R.; and Roelof, E. C. 1974. Energetic particles in the Jovian magnetosphere. *J. Geophys. Res.* 79:3600–3613.

Van Allen, J. A.; Baker, D. N.; Randall, B. A.; and Sentman, D. D. 1974. The magnetosphere of Jupiter as observed with Pioneer 10. Part 1. Instrument and principal findings. *J. Geophys. Res.* 79:3559–3577.

Van Allen, J. A.; Randall, B. A.; Baker, D. N.; Goertz, C. K.; Sentman, D. D.; Thomsen, M. F.; and Flindt, H. R. 1975. Pioneer 11 observations of energetic particles in the Jovian magnetosphere. *Science* 188:459–462.

Vasyliunas, V. M. 1975. Modulation of Jovian interplanetary electrons and the longitude variation of decametric emissions. *Geophys. Res. Lett.* 2:87–88.

Warwick, J. W. 1970. *Particles and fields near Jupiter* CR-1685. Washington, D. C.: National Aeronautics and Space Administration.

Wolfe, J. H.; Mihalov, J. D.; Collard, H. R.; McKibbin, D. D.; Frank, L. A.; and Intriligator, D. S. 1974. Pioneer 10 observations of the solar wind interaction with Jupiter. *J. Geophys. Res.* 79:3489–3500.

PART II

Origin and Interior

ON THE ORIGIN OF THE SOLAR SYSTEM
AND OF JUPITER AND ITS SATELLITES

A. G. W. CAMERON
Harvard College Observatory

and

J. B. POLLACK
NASA Ames Research Center

With the assumption (according to the Jupiter models of Podolak and Cameron) that Jupiter contains an excess of chemically condensable material, the formation of Jupiter is discussed within the context of a model of the primitive solar nebula containing gas and dust at relatively low temperature. After much of the dust has clumped into larger particles and has settled toward the midplane of the nebula, the Goldreich-Ward instability mechanism leads to the gravitational concentration of the condensed material into objects of at least asteroidal size. It is postulated that further accumulation occurs as a result of mutual collisions, leading to a proto-Jovian body composed of condensed materials which is substantially more massive than the earth. The gas in the primitive solar nebula becomes gravitationally concentrated toward such an object, and when it is massive enough and the gas cool enough, an instability sets in leading to hydrodynamic collapse of the gas toward what has become the planetary core. Following the hydro-dynamical collapse phase, Jupiter attains a hydrostatic configuration whose dimensions are much larger than its present size. It then undergoes a slow gravitational contraction to its present size that is characterized by an early epoch of high luminosity. Its current excess luminosity is due mostly to a loss of internal energy generated during the initially more rapid contraction period and to a lesser degree to its current release of gravitational energy. Conservation of angular momentum in the infalling gas from the solar nebula will cause some of it to form a flattened gaseous disk in orbit about proto-Jupiter. The thermodynamic conditions in the disk will be governed by heating through the initial compression of the gas and the continued infall of gas onto the surface of the disk, to-gether with heating (particularly important near the inner part of the disk) by radiation flowing out from proto-Jupiter. The luminous output from proto-Jupiter is quite high in the early stages, gradually declining with the passage of time. The regular Jovian satel-lites will collect together through the operation of the Goldreich-Ward instability mech-anism, together with mutual collisions of the resulting planetesimals, with the composition depending upon the local thermodynamic conditions, thus assuring that the mean density of the satellites will decrease outwards in the disk. Comparing theoretical evolutionary tracks with observed mean density of the Galilean satellites, we estimate that the addi-tion of water ice to them was completed between about 1×10^5 and 4×10^6 years after the formation of Jupiter.

[61]

Serious thinking about the origin of the solar system goes back more than three centuries, and the literature on the subject is correspondingly immense (ter Haar and Cameron 1963). This is not an appropriate place to deal with the history of the subject; consequently we shall deal only with the range of schemes currently under discussion, and in detail only with those which make quantitative predictions concerning the formation of Jupiter and its satellites.

This leads to a conceptual primitive solar nebula containing gas and dust. We shall attempt to trace the evolution of the mid-portion of this nebula through the following stages: (a) a gravitational instability which clumps the condensed material into bodies of appreciable size; (b) the formation of a larger body as a result of mutual collisions; (c) the gravitational concentration of the gas in the primitive solar nebula; and subsequent hydrodynamical collapse onto the heavy element, planetary core; (d) the attainment of a highly distended hydrostatic configuration and a subsequent slow gravitational contraction; (e) the formation of a gaseous disk surrounding proto-Jupiter; and (f) the formation of the regular satellites within this disk.

I. THE PRIMITIVE SOLAR NEBULA

Even today there are many grossly different theories of the origin of the solar system. This diversity arises from the differences in the assumptions made by the various authors about the conditions leading to the formation of the solar system. As we learn more about the process of star formation, some of the cosmogonic assumptions made about this process will become increasingly difficult to defend.

Modern observations of our own and other spiral galaxies indicate that star formation takes place in their spiral arms when interstellar gas flows into the spiral arms and increases in density there. Simple virial-theorem considerations show that masses of gas on the order of hundreds to thousands of solar masses must form into dense interstellar clouds before instability toward gravitational collapse occurs (Cameron 1963). The collapse process takes place at very low temperatures, and when the density becomes high enough, it is expected that the lines of force of the interstellar magnetic field will no longer be compressed with the collapsing gas. As the collapse progresses, the cloud will fragment into separate subsystems centered about density fluctuations. This will result in the formation of a system containing many stars. The more massive of these will ionize the residual gas, causing it to escape from the system, and rendering the system unstable against expansion, thus forming an expanding association.

A critical issue arises in connection with the angular momentum contained within the gas. Can the fragments of the collapsing cloud form directly into stars, or must they first flatten into gaseous disks? In general, even if the stars are rotating sufficiently fast to be rotationally unstable at their

equators, their angular momentum is orders of magnitude less than the expected angular momentum in a fragment containing a stellar mass; this results from a combination of the initial spin of the collapsing gas cloud together
with a random angular momentum component in each fragment arising from
the turbulent motions within the gas (Cameron 1973a). If this expected
angular momentum is distributed throughout the gas in a reasonable way,
then the collapse of the gas would form a rotating disk having dimensions
comparable to those of the planetary system (Cameron and Pine 1973).

Some theories of the origin of the solar system assume that the sun can be
formed by direct collapse of gas from an interstellar cloud fragment. Such
theories are incomplete in the sense that they fail to discover a mechanism
by means of which the large amount of angular momentum expected to be
contained in the collapsing cloud fragment can be removed before the gas
reaches stellar densities. This issue of the angular momentum distribution
in the infalling gas is perhaps the most critical of those which determine the
plausibility of various postulated initial conditions for the formation of the
solar system.

We may start our discussion of current theories of the formation of the
solar system at one end of the broad range of postulated initial conditions,
with the theory of Alfvén and Arrhenius (1970a,b, 1973, 1974). This theory
postulates that the sun was formed by an independent process, and that it
had a very strong initial magnetic dipole field. This postulate ignores the
observational difficulty that no main-sequence stars of spectral type G have
been observed to have such a strong magnetic field, and the theoretical difficulty that even such a strong field would be greatly modified by the outflowing solar wind, which is likely to be greatly enhanced above the current
flow rate when the sun was young and passing through its T Tauri phase. The
authors postulate that plasma clouds fall in toward the sun, and various
constituents of the clouds are trapped at different distances from the sun
when the atoms are ionized in the strong magnetic field. It is not explained
how this infall occurs in the presence of the solar wind. The theory is mainly
concerned with an attempted explanation of the formation of satellite systems
within the solar system. The authors recognize that their theories must be
supplemented by gas hydrodynamic processes to form Jupiter, but they do
not attempt to analyze this. Furthermore, they do not attempt to account for
the decrease of the mean density of the regular Jovian satellites with increasing distance from the central planet. Therefore we shall not discuss the
formation of the Jovian system within the context of the Alfvén-Arrhenius
theory.

Less extreme assumptions are involved in the class of conceptual solar
nebula models which can be called "minimum solar nebula theories." Probably the majority of current working models of the solar nebula can be assigned to this class, although no really quantitative model of such a solar
nebula exists. The rationale leading to this concept has been clearly ex-

plained by Hoyle (1960), who makes the unstated assumption that the accumulation of the planetary bodies makes very efficient use of the available chemically-condensed constituents within the primitive solar nebula. Thus, one can attempt to reconstruct the density distribution within the primitive solar nebula by adding to the masses of the planets the masses of the missing volatile (hydrogen and helium) and icy (water, methane, and ammonia) constituents. One should multiply the masses of the inner terrestrial planets by a factor of ~ 300 for these missing constituents. Hoyle assumes that Jupiter and Saturn are of solar composition and thus represents a nearly perfect collection of all of the constituents from the middle region of the solar nebula. Since Uranus and Neptune appear to be mainly composed of rocky and icy constituents, one can multiply their masses by a factor of ~ 50 to find the total mass of the solar nebula associated with the outer part of that nebula. The total mass thus derived is on the order of 10^{-2} solar masses, and hence this general concept for the initial primitive solar nebula requires that the sun was formed by direct collapse of the interstellar cloud fragment, possibly at the same time as the minimum solar nebula was itself formed.

This type of theory is subject to the difficulty that no quantitative analysis is given for a mechanism for removal of angular momentum from the collapsing cloud fragment; magnetic fields are insufficient (Cameron 1963). It ignores the possibility that much of the condensed mass within the solar nebula may be in finely divided form which will follow the motions of the gas during its dissipation, thus leading to the possibility that the accumulation of the planets may have made rather inefficient use of the chemically-condensed materials.

Going now to the other extreme in the range of solar nebula models, we come to the "massive solar nebula model" which has been numerically constructed by Cameron and Pine (1973). This model contains the expected angular momentum of a collapsing cloud fragment; it is therefore spread out over a distance of about 50 A.U. and contains no star at the central spin axis. The sun must form from such a model as a result of gaseous dissipation processes. Cameron and Pine have found angular momentum transport by meridional circulation currents to be the most important of these. With allowance for the loss of several tenths of a solar mass in the T Tauri solar wind, and for the inefficiencies in forming the sun from the primitive solar nebula, the total mass of the model is taken to be two solar masses. Taking into account the compressional heating of the gas as it forms the solar nebula, the internal heat content of the model is sufficient to cause the semithickness of the nebula to exceed one A.U. over most of the radial range.

Two kinds of dissipation time have been estimated for this model. One characteristic time is the angular momentum transport time, the time required to transport the bulk of the inner angular momentum beyond some radial distance by the circulation currents mentioned above. This time was found to range from a few years near the central spin axis to a few thousand years in the general region of planet formation. The other characteristic time is that of radiative cooling, the time required to radiate the bulk of the internal energy from the upper and lower photospheric surfaces of the nebula. This is also found to range from a few years near the central spin axis to a few hundreds or thousands of years in the region of planet formation.

This model has been subjected to a number of criticisms. Its internal energy content seems barely, if at all, sufficient to stabilize the disk against major global deformation, particularly the bar-like deformation (Ostriker and Peebles 1973). It would therefore seem likely that such a deformation would occur after a relatively short period of cooling. There is also the concern that only on the order of one percent of the chemically condensable materials can be retained in the planets, and this may be too low an efficiency for the process. Perhaps most important of all, the two characteristic times mentioned above seem definitely to be shorter than the formation time of the primitive solar nebula, which can be estimated to lie between 10^4 and 10^5 years, comparable to the free-fall time from an initial gas density at which one solar mass can become gravitationally isolated. This renders invalid one principal assumption used in the construction of the model, namely that the formation time of the solar nebula is short compared to the characteristic dissipation times.

It thus appears necessary that a proper evolutionary study of such a massive primitive solar nebula should take into account dissipation during the formation process. Such a study is currently being carried out by Cameron. It is necessary to start with a small amount of material in the solar nebula, and to study its evolutionary behavior in the presence of a continuing infall of gas which is accreting from the collapse of the cloud fragment. One feature of such an evolutionary process is that angular momentum is continually transported from the vicinity of the spin axis towards the outer positions of the growing solar nebula during the infall period; this leads to the formation of the sun while the infall is taking place. This has a stabilizing effect against the bar-like deformation of the solar nebula. It will also produce surface densities of the gas in the solar nebula intermediate between the minimum and the massive solar nebula models described above. Furthermore, the infalling gas passes through an accretion shock at the surface of the solar nebula; the accretion infall energy is promptly radiated away, thus establishing a surface temperature of 10^2 to 10^3 °K over the nebula, sufficiently high to prevent the interior of the nebula from becoming too cool. Additional heating of the nebula near midplane occurs as a result of the gas compression while the surface density of the nebula is growing, thus producing interior temperatures in excess of the shock-produced surface temperatures. It appears likely that the evolutionary study will produce a model of the primitive solar nebula which will be less vulnerable to the above criticisms of the massive solar nebula model of Cameron and Pine.

One concern with solar nebula models is that the sun should have a chance to form before the T Tauri-phase solar wind commences. It must be expected that this early solar wind will be responsible for the removal of residual gas from the primitive solar nebula. Indeed, since main-sequence stars much less massive than the sun can be expected to exhibit such stellar winds, it may appear strange that the internal dissipation of the solar nebula to form the sun should not be terminated by the T Tauri-phase solar wind at a much smaller mass than the sun attained. To address this problem we should first note that the solar wind arises as a consequence of the structure of the sun, in which there is an outer convection zone that heats the outer solar atmosphere through various types of wave generation. We must, therefore, ask when the sun can be expected to acquire something approaching its present structure, with a deep outer convection zone. A potential answer to this problem has been given by Perri and Cameron (1973), who note that the material at the center of the nebula model of Cameron and Pine lies on a low adiabat, which is likely to be rendered even lower as a result of cooling during the dissipation to form the sun. This adiabat is too low to allow the ignition of hydrogen thermonuclear reactions at the center, leading to the expectation that the sun cannot achieve its present structure until the central density becomes so high that pycnonuclear ignition of the hydrogen reactions can take place. This may not occur until very late in the dissipative evolution of the primitive solar nebula.

In our following considerations we shall assume that the primitive solar nebula consisted of a mixture of gas and dust at a relatively low temperature in the region of the formation of Jupiter. The surface density of the nebula will be taken to be intermediate between a minimum value obtained by spreading out the masses of Jupiter and Saturn, and the much larger values in the primitive solar nebula models of Cameron and Pine.

II. JUPITER MODELS

Since Jupiter is composed predominantly of hydrogen and helium, there has long been a tendency to assume that the planet is basically of solar composition. Similar assumptions have often been made about Saturn, but it has been recognized that a solar composition assumption would not work for Uranus and Neptune, for which the mean densities are much too large. Thus, the tendency among Jupiter model builders has been to assume that the interior composition is just hydrogen and helium, the heavier constituents being of such small abundance that they would produce only a minor perturbation of the results. An excellent review of this approach has been given by Hubbard and Smoluchowski (1973).

Hubbard's best models of Jupiter contain nearly twice the solar helium-hydrogen ratio. If this result were true, it would pose enormous cosmogonic problems, as has been discussed by Cameron (1973b). It is extremely difficult to imagine any process which could produce a diffusive separation between hydrogen and helium on such a large scale in any reasonable length of time, to be followed by the physical removal of such large amounts of hydrogen from a deep gravitational potential well.

On the other hand, it is also very unlikely that Jupiter could have formed as a result of a pure gravitational instability in the primitive solar nebula. A local instability would require a fairly large surface density in the nebula, but before this could be attained by any hydrodynamic or evolutionary process, much larger-scale global instabilities would occur (Cameron 1973b). The most rapidly growing of these would be a bar-like instability, which should have produced something much more massive than Jupiter, more like a close binary pair of stars such as is commonly found in the galaxy.

This led Podolak and Cameron (1974a) to suspect that the fairly large excess mass of helium in Hubbard's Jupiter models should really be interpreted as a large excess of heavier elements. Hubbard's models for Saturn contain an even larger helium-hydrogen ratio and he was unable to get a good fit to the gravitational moments of that planet. Furthermore, hydrogen and helium form only a minor mass fraction in Uranus and Neptune, the main constituents of those planets being presumably icy materials and rock. This would lead to a general picture in which all of the giant planets contain large amounts of icy materials and rock, with varying amounts of added hydrogen and helium, the latter gases forming the bulk of the mass in the case of Jupiter and Saturn.

Podolak and Cameron therefore have constructed models of all of the giant planets using the assumption that the helium-hydrogen ratio was the solar value. During the course of this exercise, improved gravitational moments of Jupiter became available following the Pioneer 10 flyby (Anderson et al. 1974), and these moments were confirmed following the one of Pioneer 11 (Null et al. 1975). Typical Jovian models are assumed to contain a simple rocky core, surrounded by a hydrogen-helium envelope into which fairly large amounts of excess water vapor can be admixed. The results indicate that the central rocky core should contain about ten Earth masses of material, and the excess mass, assumed to be water vapor, in the envelope should be about 50 Earth masses. A wide variety of models have been constructed in which fairly extreme variations are tried in the equation of state, but the above results change very little among these models (Podolak and Cameron 1975).

These results pose some very interesting cosmogonic problems. The mass equivalent ratio of water to rock in solar composition is approximately two. The ratio of the excess mass in the envelope of Jupiter to the mass in the core is much greater than this value. It would seem to be very strange if Jupiter were to be assembled with a smaller efficiency for the collection of very refractory material than for the collection of moderately volatile material.

The answer to this dilemma probably lies in the difficulty for accreting solid materials to settle into the central core of Jupiter. For example, even when solid bodies of considerable size fall into the present-day Jupiter, they will swiftly be volatilized. Only extremely massive bodies may penetrate to the core largely intact. The chemical form of the expected rocky vapors should approach thermodynamic equilibrium in the presence of the large amount of hydrogen. As pointed out by Lewis (1969), silicon will form silane. Metal atoms will remain metal atoms or oxides at the temperatures involved, but the oxygen in silicates is likely to form water. These substances probably do not form chemically condensed compounds except when convected close to the Jovian surface. Thus, probably much of the excess mass in the envelopes of the Jovian models of Podolak and Cameron will be in the form of such compounds, rather than just the water vapor which has been assumed in the calculations. It is unlikely that the change in the composition assumed for the excess mass in the envelope would affect Jovian models very much.

III. THE FORMATION OF JUPITER

The picture thus arises that all of the giant planets have considerable excess masses of icy and rocky material. This should provide a clue to the formation mechanisms for these planets.

Let us start our consideration of the formation of planetary bodies by considering the ones that are predominately composed of chemically condensed materials. In an intermediate-mass solar nebula, as described above, both the surface density and the internal temperature will decrease with increasing radial distance from the spin axis. Only very close to the spin axis, within perhaps about one A.U., is it likely that the internal temperature will ever have been high enough to completely vaporize solid materials, except as described below in the vicinity of the accreting giant planets. Therefore, throughout most of the primitive solar nebula, the interstellar grains, which accompany the infall of the interstellar gas to form the solar nebula, will not have been completely vaporized. However, the more volatile substances will evaporate away from the interstellar grains, so that their composition as a function of temperature and pressure within the solar nebula, will be the same as though the corresponding parts of the solar nebula had cooled from a very high temperature. Therefore the grain compositions should be described by the thermodynamic equilibrium calculations of Grossman (1972) and Lewis (1972). The most abundant constituent of the grains, under conditions of thermodynamic equilibrium, will be ice, and this should remain in condensed form beyond a distance of a few astronomical units.

The behavior of the interstellar grains during the infall to form the primitive solar nebula has been examined by Cameron (1975a). The infalling interstellar gas should be highly turbulent, and hence the gas drag effects accelerate the grains in a random fashion and promote collisions between them. The efficiency for the grains to stick together is not known, but for even moderate efficiency it turns out that at least a moderate mass fraction of the grains will have accumulated into particles in the millimeter to centimeter size by the time they reach the primitive solar nebula. Such grain clumps can be expected to retain their entities in the outer portion of the primitive solar nebula where their main icy component will remain intact. However, in the inner portion of the primitive solar nebula, where the ice will evaporate, it is reasonable to expect that an extensive fragmentation of the grain clumps will occur.

This means that in the outer part of the solar system a reasonable part of the condensed mass will settle toward the midplane of the nebula within a time on the order of a few centuries. In the inner part of the solar nebula, a considerably smaller part of the condensed mass will settle toward midplane on a comparable time scale. However, since all of the gas and accompanying solids which form the sun must pass through the solar nebula and be dissipated inwards, only a small part of the condensed solids are large enough to collect at midplane in the available dissipation time. Therefore the first qualitative expectation is that there will be a considerably higher efficiency for utilizing the condensed materials in the outer part of the solar nebula in formation of planetary bodies than in the inner part. The second expectation is that the processes of planetary accumulation begin as soon as the solar nebula starts to form.

When a fairly thin layer of condensed solids has been formed at the midplane of the solar nebula, the Goldreich-Ward (1973) instability process will set in. The thin sheet of condensed solids will become unstable against gravitational contraction into larger solid bodies. Only very local regions of the sheet can contract all the way to form a solid body, however, and the conservation of angular momentum will force much of the material to start orbiting about common gravitating centers. Thus we can expect the formation of large clusters of solid bodies, each body having dimensions of the order of a kilometer (Ward 1975).

On a fairly short time scale these clusters of bodies will start colliding with one another, leading to violent dynamical gravitational relaxation of the merged structures, forming larger clusters with condensed central cores and somewhat expanded "halos," and shedding some of the bodies to take up independent orbits within the solar nebula. Gas drag effects will produce a continued shrinkage in the dimensions of these clusters. By this time the clusters will be large enough to exert considerable dynamical perturbations upon their mutual orbits, thus leading to significant relative velocities and amalgamative collisions among the cores. Since gravitational focusing effects are larger for larger masses, it can be expected that a mass distribution of the amalgamated bodies will be set up, with the larger masses growing much more rapidly than the smaller ones. Thus there will be a growth of massive planetary cores within the primitive solar nebula, and these cores will grow more rapidly in the outer part of the solar nebula than in the inner part, owing to the more efficient use of condensed materials there. Perri and Cameron (1974) have examined the behavior of such planetary cores embedded in the gas of the primitive solar nebula. As the cores grow in size, a significant concentration of the gas from the nebula toward the cores occurs because of the gravitational field of the cores. As the mass of the cores increases by accumulation, the amount of gas concentrated towards them increases even more rapidly. Finally, when the mass of the concentrated gas becomes comparable to the mass of the core, a hydrodynamic instability sets in, which causes the gas to collapse onto the core, the released gravitational energy of the gas going into the dissociation of hydrogen molecules and ionization of hydrogen and helium atoms.

There is thus a critical core mass at which the hydrodynamic instability occurs. Perri and Cameron found that the critical core mass was quite insensitive to the position of the core within the primitive solar nebula and to the background pressure of the gas in the solar nebula, but it was very sensitive to the adiabat of the gas in the primitive solar nebula. The calculations were carried out by assuming that the compression of the gas as it concentrated toward the planetary core was isentropic. For the two trial adiabats assumed by Perri and Cameron in their calculations, their critical core masses were 115 and 65 Earth masses of material. As the adiabat of the gas is lowered, the critical core mass is also lowered. Perri and Cameron therefore concluded that Jupiter and Saturn would be subjected to such a hydrodynamic collapse process during their formation, after the gas in the primitive solar nebula models of Cameron and Pine had undergone a great deal of cooling. It is not yet clear what the adiabats will be in the revised solar nebula models being calculated by Cameron. For the case of Uranus and Neptune, it appeared likely that the critical core masses were never achieved, so that only a residual amount of the solar nebula gas, concentrated toward these planetary bodies, was retained after the T Tauri-phase solar wind removed the remaining gas from the primitive solar nebula. At the same time, it is likely that the lowering of the internal adiabat of the gas due to the rapid outflow of heat would have reduced the entropy of the gas below that of the surrounding solar nebula, thereby inhibiting the mixing between the concentrated gas and the surrounding solar nebula, and preventing the wholesale removal of the steam which was mixed into the concentrated gas.

Podolak and Cameron (1974*b*) have suggested that this physical environment may have played a role in the formation of chondrules and inclusions which ultimately were incorporated into meteoritic material.

The mass of proto-Jupiter, following accumulation of hydrogen and helium in this collapse process, was probably on the order of 60 Earth masses. Perri and Cameron showed that the collapse configuration should act as a gravitational sink for the remaining gas in the Jovian vicinity within the primitive solar nebula. Therefore the bulk of the accretion of Jupiter would result from the later infall of gas. Accompanying this process would be an infall of condensed solids, but as argued above, most of the condensed materials are likely to be vaporized, to form new chemical compounds, and to remain suspended in the convecting Jovian envelope.

This picture leads to an initial Jupiter which would be very hot due to the release of gravitational potential energy. So far, the evolution of such a hot initial model of Jupiter has been followed only for hydrogen-helium models, but the general character of this evolution probably would not be altered by the presence of a rocky core and substantial amounts of heavy elements within the convecting envelope.

IV. EVOLUTION OF JUPITER

As discussed above, the later stages in the formation of Jupiter are characterized by a hydrodynamical collapse of the solar nebula onto the central core. At the end of this phase, Jupiter has attained its current mass and the planet relaxes into a quasi-hydrostatic configuration, whose dimensions far exceed those of current-day Jupiter. Jupiter then undergoes a slow contraction to its present size. In this section, we discuss this gravitational contraction phase and point out its relevance for explaining the observed excess luminosity of Jupiter and for the formation of the Galilean satellites.

Grossman et al. (1972) carried out an initial set of calculations of Jupiter's gravitational contraction history and these results have been refined considerably by Graboske et al. (1975). Below we summarize the results of the latter paper. Graboske et al. used a sophisticated set of thermodynamics and model atmospheres. A standard stellar evolution code was used to determine the time history of Jupiter's radius and luminosity. They assumed that contraction occurred slowly enough so that hydrodynamical effects could be ignored. Thus, at any given time Jupiter was assumed to be in hydrostatic equilibrium. It evolved from one equilibrium configuration to the next as a result of the difference between the amount of energy it radiated to space and the amount it absorbed from the sun.

Several approximations were made. The effects of rotation were neglected. Also, Jupiter was assumed to be homogeneous throughout its interior. Thus, no allowance was made for either the presence of a small core or the separation of hydrogen and helium due to their immiscibility at low temperatures. Finally, the temperature lapse rate in the interior was not calculated. Rather, it was assumed that the interior was convectively mixed and therefore an adiabatic lapse rate was employed. Inclusion of rotation and a core were not expected to greatly alter the results. The effect of altering the other approximations will be commented upon below.

In most of these calculations, Graboske et al. (1975) began with a Jupiter having a radius 16 times larger than the current value. As is typical of most stellar evolution calculations, "memory" of this initial condition is quickly lost, i.e., both the time scale and the properties of later models are very insensitive to the value chosen for the radius of the initial model. An example of this insensitivity will be given below. Bodenheimer (1974) has performed hydrodynamical calculations that provide an estimate of the radius Jupiter would attain following the hydrodynamical collapse phase. This calculation, which allowed for hydrodynamical effects, was made for a homogeneous Jupiter that began as a very diffuse object — 4,600 times its current dimensions. Bodenheimer assumed that Jupiter originated as a result of a gravitational instability within the solar nebula and not as a result of the condensation and collapse processes discussed above. However, the last phases of his calculations were characterized by the hydrodynamical collapse of the outer portion of the proto-planet onto a more condensed central

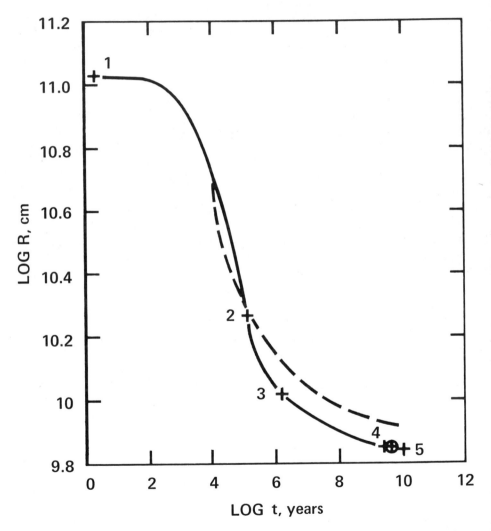

Fig. 1.   Radius of Jupiter R as a function of time t during its hydrostatic contraction phase. The crosses with numbers next to them refer to particular times along the evolutionary track. The solid line refers to calculations for a solar mixture of gases; the dashed line to ones for a pure hydrogen model. The circle enclosing a cross is Jupiter's current mean radius for a time of evolution equal to the age of the solar system. From Graboske *et al.* (1975).

region. At this point his model has some resemblance to the model of Perri and Cameron. Bodenheimer found that, when the hydrodynamical phase was completed and Jupiter had attained a hydrostatic configuration, the radius of Jupiter was 4 to 5 times larger than its current size.

Figure 1 shows the variation of Jupiter's radius R as a function of time t as given by the calculations of Graboske *et al.* The solid line represents the results for a model Jupiter that has a solar mixture of elements, while the dashed line is for a pure hydrogen planet. The numbers accompanied by crosses on the graph represent models at discrete times along the evolutionary track. The circle enclosing a cross in Fig. 1 is the measured value of the mean radius of present-day Jupiter. We see that at the beginning of the evolution Jupiter's radius decreases very rapidly. Thus, the time scale of its subsequent evolution is affected only in a very minor way by the choice of initial radius. Since the track of the solar-mix model lies very close to the observed dimension of Jupiter at 4.6 billion years, the age of the solar system, we conclude that the mass of any hypothetical central core must be much less than Jupiter's total mass. This result is in accord with the calculations of Podolak and Cameron. The agreement between the evolutionary track and Jupiter's current radius provides a valuable check on the overall reliability of the calculations.

The variation of the central temperature T_c and central density ρ_c along the evolutionary tracks is illustrated in Fig. 2. During the first 10^5 years of evolution from Model (1) to Model (2) of the solar-mix case, the central temperature steadily increases to a maximum value of 50,000°K, as a result of the gravitational energy released by Jupiter's contraction. This early phase is analogous to the pre-main sequence phase of stars of low mass. The maximum value of the central temperature changed by only 300°K when calculations were performed for an initial radius of 3.5 times the present value rather than 16 times this size, as used for the models of all the figures in this chapter. This miniscule difference illustrates the great insensitivity of the results to the initial conditions once the earliest stages of evolution are completed.

If Jupiter had a core when it first formed, as assumed earlier in this paper, will the subsequent phases of evolution affect it? The high, peak central temperature of 50,000°K may partially or totally vaporize a core, but will not necessarily mix it into the envelope. This matter warrants further study.

An obvious difference between Jupiter and more massive stars is that the central temperature never becomes high enough for nuclear reactions to occur. The peak central temperature shown in Fig. 2 is about a factor of ten less than the value required for the ignition of deuterium, which is the nuclear fuel with the lowest ignition point.

When Jupiter reaches point 2 the central density has become high enough for electron degeneracy effects to become important at the center. As a result, the rate of contraction in the central portion of the model greatly

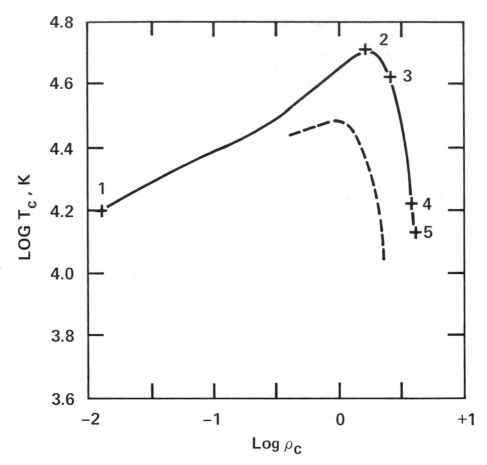

Fig. 2.   Central temperature T_c as a function of the central density ρ_c in g cm^{-3} for Jupiter during its hydrostatic contraction phase. All the symbols have the same meaning as in Fig. 1. From Graboske *et al*. (1975).

slows down, although the outer regions continue to contract rapidly. It is for this reason that the central temperatures begin to decrease; the rate of gravitational energy change is no longer rapid enough to supply the luminosity radiated to space. Thus Jupiter must use some of its internal energy, which has been built up by its prior rapid contraction, to balance its energy budget. During Jupiter's evolution from Model (2) to Model (5) the central zone of partial degeneracy grows and the central temperature steadily declines. This phase is analogous to the cooling phase of white dwarf stars.

At the current epoch, the central temperature has a value of about 15,000°K. This figure is far in excess of the melting point of hydrogen (1,000 to 1,500°K). A similar result holds for all portions of the interior of Jupiter. Thus, at present, just as in the past, the interior of Jupiter is warm enough for

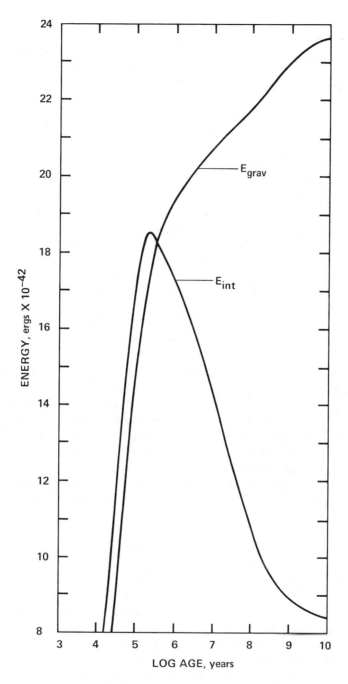

Fig. 3.   Internal energy E_{int} and gravitational potential energy E_{grav} as a function of time for a solar-mix Jupiter during its hydrostatic contraction phase. After Graboske *et al.* (1975).

the solar mixture of gases to be in a fluid rather than a solid state. Furthermore, the pressures for much of the interior are large enough ($\gtrsim$ few megabars) for hydrogen to exist in its metallic phase. The occurrence of a fluid with a high electrical conductivity is consistent with the requirements of a dynamo mechanism for the origin of Jupiter's large magnetic field (Hubbard 1968).

The change with time of Jupiter's internal energy E_{int} and gravitational energy E_{grav} is illustrated in Fig. 3. We can readily discern the two main stages in the evolutionary history as discussed above. During the early rapid contraction phase E_{int} steadily increases, while at later times, once degeneracy begins to set in, E_{int} monotonically decreases. However, even close to the current epoch Jupiter is contracting fast enough so that the change in E_{grav} is not much less than the change in E_{int}.

Finally, the time track of Jupiter's luminosity is shown in Fig. 4. During its early history Jupiter's luminosity is many orders of magnitude larger than its current value. The implications of this result for the formation of the Galilean satellites is discussed in the next section.

If a subadiabatic lapse rate is employed for the interior structure rather than an adiabatic lapse rate, the models evolve more rapidly and have a smaller luminosity at present.

The observed luminosity of Jupiter according to Aumann et al. (1969) is shown by the circle enclosing a cross in Fig. 4. This measurement indicates that Jupiter radiates into space about 2.7 times the amount of energy it absorbs from the sun. The difference between these two energy figures, which we term "excess luminosity," is due to Jupiter's internal energy sources. The value of the internal component predicted by the evolutionary model after 4.6 billion years of evolution falls somewhat below the value obtained by Aumann et al. However, analysis of the measurements obtained by the Pioneer 11 infrared radiometer experiment yields an amount of infrared radiation that is only about 1.9 times the amount of absorbed sunlight (Ingersoll et al.[1]). This new result lies very close to the evolutionary tracks shown in Fig. 4.

The above discussion enables us to make a preliminary assessment of the origin of Jupiter's excess luminosity. The only plausible source is gravitational energy. Alternative sources such as the decay of long-lived radioactive isotopes can readily be shown to be insufficient by orders of magnitude (Graboske et al. 1975). We can distinguish three categories of gravitational energy sources. First, Jupiter could be contracting fast enough *at present* for its current rate of gravitational energy release to meet the luminosity requirements. Second, the excess luminosity could be due to the loss of internal energy, which was built up by a rapid contraction in the *past*. Finally, Salpeter (1973) has proposed that the interior of Jupiter, while in a fluid

---

[1]See pp. 202 and 204.

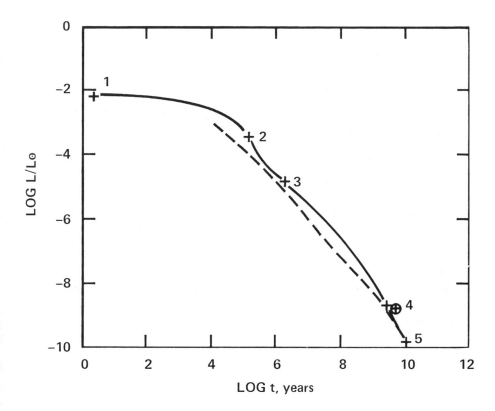

Fig. 4. Internal luminosity in units of the solar luminosity $L/L_\odot$ as a function of time t for Jupiter during its hydrostatic contraction phase. All symbols have the same meaning as in Fig. 1 except for the circle around the cross, which is the Aumann et al. (1969) value for the observed excess luminosity for a time of evolution equal to the age of the solar system. Ingersoll et al. (see p. 202) have derived a smaller value for the excess luminosity, which lies much closer to the evolutionary tracks. From Graboske et al. (1975).

phase, is currently cold enough for helium to become immiscible in hydrogen. As a result, the helium sinks toward the center and produces a release of gravitational energy. The above calculations permit a direct determination of the importance of the first two of these mechanisms. An indirect measure of the importance of the third mechanism is provided by the degree to which the first two fail to supply enough luminosity at present. If the Pioneer 11 results are accepted, then the contraction history of a homogeneous Jupiter is adequate to explain the observations and hence by implication little separation of helium from hydrogen has occurred. According to Fig. 3, most of the excess luminosity is supplied at the expense of the internal energy; the current rate of contraction makes a smaller but still significant contribution to the current power budget.

V. FORMATION OF THE REGULAR SATELLITES

The four large satellites of Jupiter, Io, Europa, Ganymede, and Callisto, form a regular system of bodies, coplanar in the equatorial plane of the planet and with circular orbits of regularly increasing orbital radius forming a kind of "Bode's Law." The small innermost satellite, Amalthea, may also be considered to form a part of this sequence, although very little is known about it.[2] The mean densities of Io, Europa, Ganymede, and Callisto are 3.50, 3.42, 1.95, and 1.65 g cm^{-3}, respectively (Null et al. 1974; Morrison and Burns[3]). Thus the mean densities decrease in a regular fashion with increasing distance from the planet. The inner two of these satellites must be composed predominantly of rocky materials, whereas the outer two must have a substantial component of ice.

The circularity and regularly increasing spacing of these satellite orbits argues for their formation from a gaseous disk emplaced about Jupiter. It is not at all surprising that such a disk should have been formed. During the hydrodynamic collapse of the gas which contributed much of the mass of the early proto-Jupiter, it can be expected that conservation of angular momentum should have placed some of this mass into a gaseous disk orbiting around proto-Jupiter.

The situation thereafter is likely to have been somewhat complicated. Gaseous dissipation within the disk probably causes a net mass flow inwards towards the surface of proto-Jupiter, but with some of the gas moving outwards as it takes up the angular momentum released from the innermost material. The continued accretion of Jupiter will deepen the gravitational potential well in which the disk resides, thereby absorbing the inner part of it into proto-Jupiter. Gas accretion will also occur onto the surfaces of the disk, thereby producing an accretion shock and helping to regulate the internal temperature of the disk in a manner similar to that which occurs through the accretion of interstellar gas onto the primitive solar nebula itself.

If the rocky component of the regular satellites is augmented by the icy and the volatile constituents that would be present in solar proportions, and the resulting total mass is then spread out over the orbital area of the regular satellite system, a characteristic surface density of the expected disk is found to be on the order of 10^6 g cm^{-2} column; the actual surface density may have been greater than this if the process of collecting condensable solids into the satellites was inefficient. Such a disk has a high internal optical depth and therefore it should be modeled by techniques similar to those used in modeling the primitive solar nebula itself.

However, if all of the condensable material has been vaporized and recondensed into particles of appreciable size, they will not be spread uniformly throughout the disk, but rather will be concentrated in a very thin layer near

[2]See p. 127.
[3]See p. 994.

its midplane. Sample calculations with formulae given by Goldreich and Ward (1973) indicate that viscous drag forces cause a particle to settle to the midplane within a fraction of a year after its birth. In this case, the opacity of the disk may not have been very large. Detailed calculations are required before a definite conclusion can be reached on this matter.

No attempt has yet been made to construct a model of such a gaseous disk, but certain features can be anticipated for such a model. The accretion shock due to the infalling matter will maintain a moderately large surface boundary temperature so long as the infall of matter continues. The central region of the disk near the midplane will be heated by compression of the material and the internal thermal structure of the disk will be regulated by the resulting energy transport processes. It would be premature to estimate temperatures for the midplane region, but it would not be surprising if these were high enough so that only rocky constituents could exist in condensed form, and it may well have been true that at one time the temperature was so high that all condensed solids vaporized. The nearby proto-Jupiter would have been very hot, and it seems very likely that there was a continuous fluid medium between the proto-planet itself and the gas in the disk. This suggests the probability that there was a substantial flow of heat from the proto-planet in the radial direction in the midplane of the disk both by convection and radiative transfer. Pollack and Reynolds (1974) have pointed out that if one neglects the internal opacity of the gaseous disk, then the heat output from proto-Jupiter alone would be sufficient to prevent ice from forming a substantial component of the innermost Jovian satellites.

We may expect that the larger solid particles contained within the gaseous disk would settle towards the midplane of the disk fairly rapidly. This would be especially true if the gaseous disk was at one time hot enough to have completely vaporized condensed solids, since particles growing from a supersaturated vapor would settle toward the midplane once their radii became comparable to a centimeter (Goldreich and Ward 1973). Once a thin sheet of condensed solids has formed at midplane, then the Goldreich-Ward instability mechanism would once again operate, causing gravitational instability and the formation of bodies of several kilometers in size (Ward 1975). Thereafter one has a picture similar to that for the formation of planetary bodies within the solar nebula; clusters of condensed bodies would collide and amalgamate, velocity differences would be established by mutual gravitational perturbations, and the larger amalgamations of material would grow more rapidly than their neighbors. One can expect this to lead to the formation of the regular satellites on a relatively short time scale.

An interesting question in this connection is when the icy constituent was added to the outer two regular satellites. The rocky cores of these satellites may well have collected before the temperature in the gaseous disk had fallen low enough to allow the condensation of water as ice. Pollack and Reynolds have estimated that for an optically thin gaseous disk the temperature would not have fallen far enough to allow the condensation of ice for a few million years after the formation of proto-Jupiter. However, for an optically thick gaseous disk, the inner parts of the disk may well provide sufficient shielding from the luminosity of proto-Jupiter to allow the outer parts of the disk to cool enough for the condensation of ice at a much earlier epoch. An important question related to this issue is the lifetime of the gaseous disk, and whether it is short compared to a few million years in the presence of gaseous dissipation processes similar to those operating in the primitive solar nebula.

By considering two extreme models, we can obtain estimates of the epoch at which the addition of water ice to the satellites was completed. The first model is analogous to that of Pollack and Reynolds in which the near-infrared opacity of the disk is neglected and the temperature of a particle that is just forming is assumed to be determined by a balance between the energy it radiates to space and the amount of solar and Jovian energy it absorbs. For the second model we assume that the disk has a high opacity and that the temperature at a given distance from Jupiter is determined by convective energy transport within the disk. Using the primitive solar nebula as a rough analog, we assume for this second case that the temperature of the disk varies inversely with distance from Jupiter (Lewis 1974; Consolmagno[4]). We fix the constant of proportionality by using the evolutionary tracks of Graboske et al. (1975) to specify the photospheric temperature and radius of Jupiter as a function of time.

For each of the two models specified above, we obtain a lower limit to the completion time for satellite formation by determining the time at which water ice can start to condense at the orbit of Ganymede and an approximate upper bound by finding the analogous time for material at the orbit of Europa. As discussed above, Ganymede appears to contain a significant quantity of ices, while Europa does not. According to the first model, the low opacity model, ice condensation ceased between about 1 and 2 million years after the birth of Jupiter. These values are lower than the numbers given by Pollack and Reynolds (1974). This difference stems from their use of a preliminary set of evolutionary tracks from Graboske et al. and our use of a later version of these tracks together with a higher condensation temperature of about 240°K in our assumed disk, which lowers the condensation times by a factor of about three. This temperature results from the assumption that the disk has solar composition. The corresponding limits for the high opacity model are 0.1 and 0.3 million years. There are obviously scenarios in which these latter times could be lowered further. Presumably the cessation of condensation occurred because the planetary disk was dissipated. This may have been brought about by the occurrence of a very strong solar wind which swept gases and small particles out of the solar system.

---

[4]See p. 1045.

In order for water ice to condense out of the disk, the temperature must fall below $\sim 240°K$. The next type of ice to condense out is ammonium hydrate ($NH_3 \cdot H_2O$), which requires temperatures nearly a factor of two lower. As we have seen above, if the condensation of ice in the region of formation of the Galilean satellites was marginal, it is unlikely that more volatile substances would condense. Hence, as Pollack and Reynolds have pointed out, the Galilean satellites probably contain little or no methane or ammonia that was derived from ices made during their formation period.

Another consequence of the above model of formation is that the Galilean satellites probably had a compositional zonation at the end of their formation period. Since rocky materials are stable against vaporization at a much higher temperature than water ice, rocky particles were present at a much earlier epoch than the icy particles, and rocky minerals could have formed proto-satellites before ice condensation was permitted. Thus, we might expect that at the end of the formation period, the rocky material was concentrated toward the center of the satellites and the ices toward their outer portions. Such a zonation is of course most relevant for the outer two Galilean satellites, Ganymede and Callisto, which contain a substantial amount of water ice. Subsequently, the decay of long-lived radioactive isotopes contained in the rocky fraction may have led to a melting of the interior of the satellites and a redistribution of their compositional components.

Jupiter's early evolution may have also profoundly affected the formation of Amalthea, the innermost satellite. Immediately after its formation, Jupiter extended far beyond Amalthea's current position and, according to Fig. 1, did not become smaller until several times 10^4 years later. Hence, the formation of Amalthea began much later than the starting time for the other regular satellites. At the end of the ice formation period for the Galilean satellites, as estimated above, the temperature of the nebula at Amalthea's position had a value between ~ 500 to $600°K$ for the low opacity model and 1000 to $1400°K$ for the high opacity model. These figures imply that Amalthea may be composed principally of highly refractory, rocky material in contrast to the rocky component of the Galilean satellites, which may be much closer to a carbonaceous chondritic composition. A determination of Amalthea's surface composition or mean density may provide a critical test of the two opacity models, as they predict very different temperature conditions when Amalthea formed. The late start of Amalthea's formation may be responsible in part for its being much smaller than the Galilean satellites.

The question of the disk lifetime is related to the present rotational period of Jupiter. In the picture outlined above, it seems highly likely that there was a continuous fluid material connecting proto-Jupiter with the gas in the disk. But this is only possible if the initial proto-Jupiter was rotating faster than at the present time, so that it was rotationally unstable in the equatorial plane. This would have been possible only if a mechanism existed which allowed a substantial loss of angular momentum by proto-Jupiter.

Such a mechanism would almost certainly have involved a primordial strong Jovian magnetic field, which may have involved a different generating mechanism than that responsible for the present field. Such an earlier magnetosphere would exert an accelerating force upon the ionized components of the inner portion of the surrounding gas disk, thus transferring angular momentum to the disk and not only slowing down the rotation rate of Jupiter, but also pushing the disk out to a larger orbital radius. Not all of the loss of early angular momentum by Jupiter need have occurred in this manner; some of it may have happened through loss of plasma from the magnetosphere, as seems to be taking place at a slow rate even today (Cameron 1975b). But as long as the early proto-Jupiter was hot enough to produce some ionization in the inner portion of the disk, it is likely that the resulting angular momentum transferred to the disk would contribute to its dissipation and perhaps severely limit its lifetime.

If the early T Tauri-phase solar wind was much stronger than the present one, then the Jovian magnetosphere would be confined to a much smaller volume than it presently occupies. Under these circumstances the primitive solar wind may also have played a significant role in the removal of gas from the disk surrounding Jupiter.

It will be evident from this discussion that there are many fascinating theoretical questions associated with the formation of the regular Galilean satellites that will form the basis for some interesting investigations within the next few years.

Acknowledgment. This research has been supported in part by grants from the National Aeronautics and Space Administration.

DISCUSSION

W. M. Kaula: I find it difficult to imagine a medium (cool enough for condensation) which simultaneously is quiescent and/or dense enough to grow mm- or cm-sized objects, but violent and/or sparse enough to keep half the mass in micron-sized or smaller objects. Such a collision-interacting population normally would have a coagulation/comminution number density of object sizes, $dN(m)/dm \propto m^{-g}$. The exponent g is always ≤ 1.85 because of purely material properties—the energy density for fragmentation, etc.— and hence half the mass is in objects of more than ~ 0.01 the mass, or ~ 0.2 the radius, of the largest object.

The main point is that once conditions in some region allow *some* condensed matter to collect together sufficiently to attain some property (such as moving independent of gas, immunity to solar influence, etc.), then it is probable that *most* condensed matter in the region will have the same property at any given time.

A. G. W. Cameron: I have been referring to conditions enormously far from thermodynamic equilibrium and nowhere am I assuming condensation.

The condensates are there to begin with: the interstellar grains. The whole issue is the extent to which these can clump together by collision, in competition with clump fragmentation, in a collapsing interstellar cloud (see Cameron 1975). The results indicate that you can have significant mass fractions in both micron and millimeter size ranges.

D. M. Hunten: Do you see any chance of major progress on the question of how the gas and particles were dissipated after formation of the planets? So far, all I have heard is a muttered incantation "T Tauri stellar wind."

A. G. W. Cameron: This is a problem that should be studied but has not been. Observations indicate that the T Tauri stellar wind typically involves mass fluxes $\sim 10^6$ times the present solar wind and that probably several tenths of a solar mass may thus be ejected. The wind presumably carries a magnetic field and will carry away the gas of the solar nebula as it can be ionized by solar ultraviolet radiation and by charge exchange. Rates for these have not been examined. With regard to solids, if these are not collected in planets, there are a variety of radiation processes which can remove them. These include: radiative pressure ejection, Poynting-Robertson effect, radiative torques leading to rotational bursting, and the Yarkovsky effect leading to orbital changes of larger bodies until resonances with planets occur.

D. H. Harris: What about the magnetic field and angular momentum transfer in the early solar nebula.

A. G. W. Cameron: During the collapse phase of an interstellar cloud the ionization level falls very low, and I have estimated that the trapped interstellar magnetic field can slip out once the central density becomes $10^6 - 10^8$ atoms per cm^3. Hence magnetic fields should not play a major role in affecting the hydrodynamics of the formation of the solar system. Angular momentum transfer should take place by dissipative fluid processes.

W. H. McCrea: If you dispose of the light gases by having them fall into the sun, do you not then have an even more difficult problem in disposing of the then disembodied angular momentum originally possessed by such material?

A. G. W. Cameron: In the solar nebula, gas can flow toward the spin axis to form the sun only if angular momentum flows away from the spin axis to draw in material near the outer edge of the disk, thus expanding the disk radius. Viscous dissipative processes in the solar nebula will cause these mass and angular momentum flows to occur.

REFERENCES

Alfvén, H., and Arrhenius, G. 1970a Structure and evolutionary history of the solar system. 1. *Astrophys. Space Sci.* 8:338–421.

———. 1970b. Structure and evolution of the solar system. II. *Astrophys. Space Sci.* 9:3–33.

———. 1973. Structure and evolutionary history of the solar system. III. *Astrophys. Space Sci.* 21:117–176.

———. 1974. Structure and evolutionary history of the solar system. IV. *Astrophys. Space Sci.* 29:63–159.

Anderson, J. D.; Null, G. W.; and Wong, S. K. 1974. Gravitational parameters of the Jupiter system from the tracking of Pioneer 10. *Science* 183:322–323.

Aumann, H. H.; Gillespie, C. M., Jr.; and Low, F. J. 1969. The internal powers and effective temperatures of Jupiter and Saturn. *Astrophys. J.* 157:L69–L72.

Bodenheimer, P. 1974. Calculations of the early evolution of Jupiter. *Icarus* 23:319–325.

Cameron, A. G. W. 1963. Formation of the sun and planets. *Icarus* 1:13–69.

———. 1973a. Accumulation processes in the primitive solar nebula. *Icarus* 18:407–450.

———. 1973b. Formation of the outer planets. *Space Sci. Rev.* 14:383–391.

———. 1975a. Clumping of interstellar grains during formation of the primitive solar nebula. *Icarus* 24:128–133.

———. 1975b. Cosmogonical considerations regarding Uranus. *Icarus* 24:280–284.

Cameron, A. G. W., and Pine, M. R. 1973. Numerical models of the primitive solar nebula. *Icarus* 18:377–406.

Goldreich, P., and Ward, W. R. 1973. The formation of planetesimals. *Astrophys. J.* 183:1051–1061.

Graboske, H. C.; Pollack, J. B.; Grossman, A. S.; and Olness, R. J. 1975. The structure and evolution of Jupiter: the fluid contraction stage. *Astrophys. J.* 199:265–281.

Grossman, A. S.; Graboske, H.; Pollack, J.; Reynolds, R.; and Summers, A. 1972. An evolutionary calculation of Jupiter. *Phys. Earth Planet. Interiors,* 6:91–98.

Grossman, L. 1972. Equilibrium condensation in the primitive solar nebula. *Geochim. Cosmochim. Acta* 36:597–619.

Hoyle, F. 1960. On the origin of the solar nebula. *Quart. J. Roy. Astron. Soc.* 1:28–55.

Hubbard, W. B. 1968. Thermal structure of Jupiter. *Astrophys. J.* 152:745–754.

Hubbard, W. B., and Smoluchowski, R. 1973. Structure of Jupiter and Saturn. *Space Sci. Rev.* 14:599–662.

Ingersoll, A. P.; Münch, G.; Neugebauer, G.; Diner, D.; Orton, G. S.; Schupler, B.; Schroeder, M.; Chase, S. C.; Ruiz, R. D.; and Trafton, L. M. 1975. Pioneer 11 infrared radiometer experiment: the global heat balance of Jupiter. *Science* 188:472–473.

Lewis, J. S. 1969. Observability of spectroscopically active compounds in the atmosphere of Jupiter. *Icarus* 10:393–409.

———. 1972. Low temperature condensation from the solar nebula. *Icarus* 16:241–252.

———. 1974. The temperature gradient in the solar nebula. *Science* 186:440–443.

Null, G. W.; Anderson, J. D.; and Wong, S. K. 1974. Gravitational coefficients of Jupiter and its Galilean satellites determined from Pioneer 10 spacecraft data. Presented at IAU Colloquium No. 28 on Planetary Satellites, Ithaca, N.Y.

———. 1975. The gravity field of Jupiter from Pioneer 11 tracking data. *Science* 188:476–477.

Ostriker, J. P., and Peebles, P. J. E. 1973. A numerical study of the stability of flattened galaxies: or, can cold galaxies survive? *Astrophys. J.* 186:467–480.

Perri, F., and Cameron, A. G. W. 1973. Hydrogen flash in stars. *Nature* 242:395–396.

———. 1974. Hydrodynamic instability of the solar nebula in the presence of a planetary core. *Icarus* 22:416–425.

Podolak, M., and Cameron, A. G. W. 1974a. Models of the giant planets. *Icarus* 22:123–148.

———. 1974b. Possible formation of meteoritic chondrules and inclusions in the precollapse Jovian protoplanetary atmosphere. *Icarus* 23:326–333.

———. 1975. Further investigations of Jupiter models. *Icarus* 25:627–634.

Pollack, J. B., and Reynolds, R. T. 1974. Implications of Jupiter's early contraction history for the composition of the Galilean satellites. *Icarus* 21:248–253.

Salpeter, E. 1973. On convection and gravitational layering in Jupiter and in stars of low mass. *Astrophys. J.* 181:L83–L86.

ter Haar, D., and Cameron, A. G. W. 1963. Historical review of theories of the origin of the solar system. *Origin of the Solar System.* (R. Jastrow and A. G. W. Cameron, eds.) pp. 1–37. New York: Academic Press.

Ward, W. R. 1975. Some remarks on the accretion problem. Submitted to *Proceedings of the International Meeting on Planetary Physics and Geology.* Rome, 1974.

INTERIOR MODELS OF JUPITER

D. J. STEVENSON and E. E. SALPETER
Cornell University

Understanding the interior of Jupiter depends upon our knowledge of the thermo-dynamics and transport properties of hydrogen-helium mixtures at high pressures and temperatures. The current status of this knowledge is reviewed, and attention is given to the metallic-molecular hydrogen transition and the limited solubility of helium in hydro-gen. Models of Jupiter are constructed which are consistent with all the observations to date, but which make various assumptions about the thermodynamics and composi-tion of the interior. These models typically consist of a rocky core surrounded by a nearly-solar fluid mixture. In contrast to the models of Podolak and Cameron, a large enhancement of water or helium is not found to be essential. It is concluded that further progress in constructing interior models requires a better understanding of the thermo-dynamics of dense molecular hydrogen.

Among the planets, Jupiter is of special interest, not merely because of its physical dimensions, but because it appears to represent a well preserved sample of the early solar system. Unlike that for the terrestrial planets, the escape time of any gas from the Jovian atmosphere greatly exceeds the age of the universe. Furthermore, the low average density of Jupiter (1.33 g cm^{-3}) greatly limits the possible choices of interior composition. In Fig. 1, the mass-radius relationship for zero temperature bodies (Zapolsky and Salpeter 1969) is shown. This figure shows that if Jupiter is *cold* then it is predomi-nantly hydrogen. In this context, a cold body is one for which the thermal energy content is much less than the total internal energy. Alternatively, it could be supposed that Jupiter is composed of heavier elements, but is *hot* (i.e., greatly expanded from the zero temperature state). Even aside from cosmogonic considerations, this can be discounted. First, it would require a very large photospheric radiative temperature of at least 10^3 °K, in strong disagreement with the observed effective temperature of 125°K (Hubbard and Smoluchowski 1973). Second, it would lead to an oblateness that is substantially smaller than that observed (Öpik 1962).

The predominance of hydrogen is thus firmly established. Since there is no known cosmogonic process for separating hydrogen and helium before the formation of Jupiter it is reasonable to assume that Jupiter also contains a solar abundance of helium — about 25% by mass (Cameron 1973).[1] This is roughly consistent with the position of Jupiter in Fig. 1. However, the curves in Fig. 1 are only approximate; they are based on Thomas-Fermi-Dirac theory and do not incorporate details of the high pressure chemistry (such as phase transitions) or finite temperature corrections. In Sec. I, we review the current knowledge of the thermodynamics of hydrogen and helium at pressures and temperatures relevant to Jupiter. Particular attention is given to assessing the current uncertainties.

A model of the interior of Jupiter also requires specification of the internal temperature distribution. Jupiter emits about 1.4×10^4 erg cm^{-2} sec^{-1} in the infrared, of which about half is generated internally within the planet (Ingersoll et al. 1975).[2] In Sec. II we show that this internal heat flux is too great to be carried by conduction at a subadiabatic temperature gradient, but too small for convective heat transport to require significant superadiabaticity. To a first approximation, Jupiter is then adiabatic. In Sec. II, we also consider the dynamic effects of first-order phase transitions, and helium separation and precipitation. These effects may invalidate the simple adiabatic model, and amply demonstrate that fluid dynamical, as well as thermodynamical parameters, are relevant for determining compositional and thermal gradients.

Acceptable models of Jupiter must have not only the correct mass and radius, but also the correct rotational response (as characterized by the oblateness, and the gravitational moments J_{2n}) and the correct matching atmosphere. The thermal structure of the Jovian atmosphere is well constrained by a variety of recent infrared observations (Orton;[3] Lacy et al. 1975) and the deep atmosphere can be characterized by an adiabat which passes through $T_s = (175 \pm 20)°$K at $P = 1$ bar. (The estimated error in T_s assumes no unforeseen important opacity sources in the atmosphere.) In Sec. III we construct interior models for T_s in the range 140°K to 210°K. In contrast, the helium abundance in the atmosphere is not usefully constrained by observations. In our models, the helium abundance is treated as an adjustable parameter, but is not allowed to deviate too greatly from current estimates of the solar abundance.

Finally, in Sec. IV, we discuss the implications of our model calculations for future observations and theoretical work.

---

[1] See also p. 266.
[2] See p. 202.
[3] See p. 207.

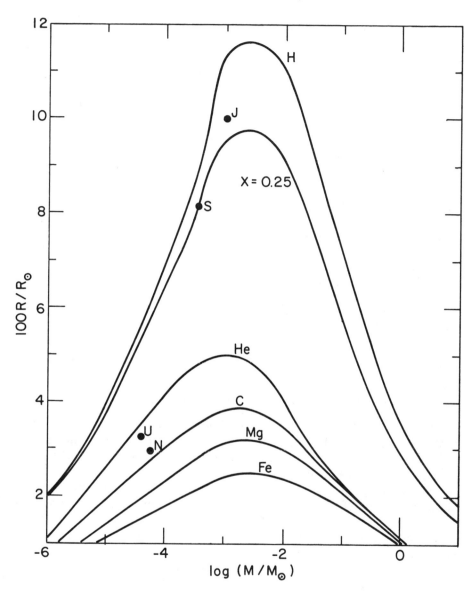

Fig. 1.   Mass-radius relation for zero-temperature sphere of various chemical compositions. $X = 0.25$ refers to a 75% H and 25% He mixture by mass. The symbols J, S, U and N represent the giant planets. $R_\odot$ and $M_\odot$ are the solar radius and mass.

I.   THERMODYNAMICS

According to the virial theorem (Clayton 1968),

$$\Omega = -3\int P\mathrm{d}V \tag{1}$$

where Ω is the total gravitational energy, P is the pressure and V the volume of the planet. An "average" internal pressure in Jupiter is then

$$\bar{P} \sim \frac{GM_J^2}{4\pi R_J^4} \sim 10 \text{ megabars} \qquad (2)$$

where G is the gravitational constant, and M_J, R_J are respectively the mass and radius of Jupiter. Later in this section (see Figs. 2 and 3) we will see that the corresponding "average" internal temperature and density are $\bar{T} \sim 10^4\,°K$ and $\rho \sim 1$ g cm^{-3} respectively. Under these circumstances, the interior is *cold* since $k_B\bar{T}$ is small compared to the internal energy per particle:

$$k_B\bar{T} \simeq \frac{1}{20} \frac{GM_Jm_P}{R_J} \qquad (3)$$

where m_P is the proton mass. Thus, models of Jupiter require a thermo-dynamic description of relevant constituents over a range of conditions that extend from ideal gas behavior (in the atmosphere) to dense, degenerate matter (in the deep interior).

For simplicity, we first discuss the thermodynamics of pure hydrogen in some detail. Later in this section, the thermodynamics of helium and minor constituents are briefly discussed.

We first consider the well understood high-density metallic (atomic) phase of hydrogen and then consider the less understood molecular phase.

(a) Metallic Hydrogen

At very high densities ($\rho \gg 1$ g cm^{-3}), the Pauli exclusion principle pre-cludes the existence of molecules or localized electronic states, and dense hydrogen is a Coulomb plasma: protons immersed in an almost uniform sea of degenerate electrons. (A discussion of the low pressure transition of this metallic phase to the molecular allotrope is postponed until the next section.)

In most respects, metallic hydrogen is the simplest of the alkali metals, and the equation of state can be calculated accurately from first principles. In particular, the most recent calculations for a static metallic hydrogen lattice by the augmented-plane-wave method (Ross 1974) and by high-order perturbation theory (Hammerberg and Ashcroft 1974) are in excellent agreement. The finite temperature and zero point motion corrections are not so well understood, since they depend on details of the screening of the proton-proton interaction. As we explain below, the liquid state is relevant to the present interior of Jupiter, and calculations by Hubbard and Slattery (1971) and by Stevenson (1975) have been made for this state. The calcula-tion by Stevenson includes exchange and correlation in the electron gas, the non-linear response of electrons to the ionic distribution, quantum correc-tions for the protons, the small nondegeneracy of the electron gas, and a perturbation theory of fluids for approximating the ionic distribution. The resulting equation of state is expected to be accurate to 1% at the pressures and temperature relevant to Jupiter:

$$P = \frac{51.6}{r_s^5} (1 - 0.654 r_s + 0.085 r_s^2 - 0.008 r_s^3) \tag{4}$$

where P is the pressure in megabars at $6000°K$ (an arbitrary reference temperature) and r_s is the usual measure of electron spacing:

$$n = \left(\frac{4}{3} \pi r_s^3 a_0^3\right)^{-1} \tag{5}$$

where n is the electron density and $a_0 = 0.529 \times 10^{-8}$ cm. In Jupiter, $r_s \sim 1$. The density ρ in g cm^{-3} is related to r_s by

$$\rho = \frac{2.685}{r_s^3}. \tag{6}$$

As $r_s \to 0$, the pressure in equation (4) is dominated by the Pauli exclusion principle pressure of an ideal non-relativistic Fermi gas. The dominant corrections at finite r_s are the exchange energy and the electrostatic energy (Madelung energy) for ions immersed in a uniform electron gas. The last two terms in parentheses in equation (4) result primarily from thermal corrections and the non-uniformity of the electron gas.

At other temperatures the pressure can be evaluated using the parameter C, where

$$C = \frac{1}{nk_B} \left(\frac{dP}{dT}\right)_v \tag{7}$$

is tabulated by Stevenson (1975). Two other thermodynamic parameters of importance are

$$\gamma = \left(\frac{d\ell n\, T}{d\ell n\, \rho}\right)_s \tag{8}$$

and

$$C_v = \frac{k_B C}{\gamma} \tag{9}$$

where C_v is the heat capacity per proton, and the thermodynamic derivatives in equations (7) and (8) are at constant volume and constant entropy, respectively. For a high-temperature Debye solid, $\gamma \simeq 0.5$ (neglecting screening), $C = 3\gamma$ and $C_v = 3k_B$. In liquid metallic hydrogen, the calculations of Hubbard and Slattery (1971) and Stevenson (1975) agree that for the pressures and temperatures of interest, $0.61 \lesssim \gamma \lesssim 0.64$, $C \sim 2\gamma$ and $C_v \sim 2k_B$. Unlike the equation of state, which is rather insensitive to structure, it is clear that some of the thermodynamic derivatives are sensitive to the assumption of fluidity. Consequently, it is important to estimate the melting temperature, T_M. One common method is Lindemann's rule, but this method is unreliable for a substance such as metallic hydrogen, where T_M is *less* than the Debye temperature (Stevenson and Ashcroft 1974). The only rigorous way to evaluate T_M is to compare the free energies of the solid and liquid phases. The Monte Carlo calculations of Pollock and Hansen (1973) give, where ρ is in g cm^{-3}

$$T_M \simeq 1500\, \rho^{\frac{1}{3}}\ {}^{\circ}\text{K} \tag{10}$$

when this procedure is used. This is an *upper bound* since it does not include the effects of screening, which weaken the ion-ion interaction and lower T_M. The screening may be sufficient to reduce the melting point to zero. Although this problem is unresolved, the upper bound above is sufficient to ensure that for any reasonable model of the present Jupiter (see Sec. III), the metallic core of Jupiter is fluid. In asserting this we have neglected the effect of alloying the metallic hydrogen with helium, but this effect appears to be small (Stevenson and Ashcroft 1974) and may even reduce the melting point (Smoluchowski 1971).

Apart from properties that depend on subtle free energy differences (e.g., melting point, metastability at low pressures, superconductivity) the thermodynamics of metallic hydrogen seem to be well understood. In particular, the 1% accuracy in the equation of state is very helpful in constraining the models discussed in Sec. III.

(b) Molecular Hydrogen

At a pressure of less than a few megabars, the molecular phase of hydrogen is thermodynamically favored over the metallic phase. The estimated transition pressure is based primarily on theoretical arguments (Ross 1974), since the claimed experimental verification (Grigoryev *et al.* 1972) is unconvincing.

The molecular phase extends from ideal gas densities to the densities of strongly interacting matter ($\rho \sim 1$ g cm^{-3}) where the distance between molecules is comparable to the size of the molecule. There are no first order phase transitions in this extended density range, at the high temperatures relevant to Jupiter. For example, at $\rho = 0.09$ g cm^{-3} (the density of zero pressure and zero temperature H_2 solid) the temperature within Jupiter is several thousand degrees, which exceeds the melting point of the solid by over two orders of magnitude. Furthermore, the gas-liquid critical point for H_2 occurs at $T = 22^{\circ}$K and $P \sim 10$ bars, whereas the actual temperature at 10 bar pressure is about 500°K.

The traditional approach used to investigate the equation of state for H_2 is to approximate the interaction energy by a sum of pairwise-additive intermolecular potential energies. It has recently been demonstrated (Ree and Bender 1974) that this approach is invalid because triplet and higher order interactions become increasingly important near the transition pressure. It now seems probable that the only rigorous way to calculate the properties of H_2 at megabar pressures is to use methods that have previously been applied to metallic hydrogen—in particular, band structure calculations. Some calculations of this kind have been made (Ramaker *et al.* 1975; Friedli and Ash-croft 1975) but for the present, we are forced to resort to semi-empirical pair potentials that are compatible with the experimental shock data (Ross 1974), yet are also plausible modifications of first principles calculations (McMahan *et al.* 1974). Ross considers several pair potentials that differ substantially in their hardness and yet are compatible with the shock data. We have considered three of these (in atomic units):

$$\varphi_I(r) = 7.0e^{-1.65r} + \varphi_{vdw}(r) \tag{11}$$

$$\varphi_{II}(r) = 7.5e^{-1.69r} + \varphi_{vdw}(r) \tag{12}$$

$$\varphi_{III}(r) = 8.2e^{-1.74r} + \varphi_{vdw}(r) \tag{13}$$

$$\varphi_{vdw}(r) = -\left(\frac{13}{r^6} + \frac{116}{r^8}\right) \exp\left(\frac{-400}{r^6}\right). \tag{14}$$

These potentials are given in order of increasing softness. The constant contribution $\varphi_{vdw}(r)$ represents a truncated van der Waals attraction. Since the real interaction depends on the relative orientations of the molecules, we have already assumed an orientational average.

To evaluate the thermodynamic properties of H_2, the free energy is calculated. We assume that the fluid state is appropriate for Jupiter and follow the method used by Ross (1974) to evaluate the energy. The assumption of fluidity can be checked later by comparing the value of the packing fraction η with the critical value of about 0.45 that is required for solidification in a classical fluid (Wainwright and Alder 1958). In all the models considered in Sec. III, it was found that $\eta < 0.45$ at all densities.

The total free energy is written as

$$F = F_{hs} + F_{int} + F_{mol} + \Delta F \tag{15}$$

where F_{hs} is the free energy of an equivalent hard sphere gas, F_{int} is the molecular interaction energy, F_{mol} accounts for the rotation and vibration of individual molecules, and ΔF contains all other corrections (dissociation, electronic excitation, quantum effects). F_{hs} and F_{int} are evaluated as by Ross (1974). F_{mol} is expressed as

$$F_{mol} = - Nk_BT \, \ell n \, Z_r Z_v \tag{16}$$

where N is the number of molecules and Z_r, Z_v are the partition functions for rotation and vibration respectively. At low temperatures ($T \lesssim 2\theta_R$) the quantum partition function for rotation is used (Landau and Lifshitz 1969). Otherwise,

$$Z_r = \frac{T}{2\theta_R}$$

$$Z_v = [1 - \exp(-\theta_v/T)]^{-1} \tag{17}$$

where θ_R, θ_v are the characteristic rotation and vibrational temperatures. At low densities, $\theta_R = 85.4°$K and $\theta_v = 6100°$K. At high densities, calculations by Neece *et al.* (1971) indicate that $\theta_R \sim 2000°$K at $\rho = 0.7$ g cm^{-3}. Monte Carlo calculations for the fluid phase (Slattery and Hubbard 1973) indicate a lower θ_R. In contrast, θ_v is not a strong function of density and may even decrease at high density because of the "softening" of the intramolecular bonds (Silver and Stevens 1973). In the thermodynamic models described below, we assume $\theta_v \equiv 6100°$K, but try different values of θ_R.

At low densities ($\rho \lesssim 0.1$ g cm^{-3}), the modified Saha equation (DeWitt and Rogers 1972) predicts negligible dissociation and electronic excitation of molecules in the present Jupiter. At higher densities (and hence temperatures), the Saha equation is inapplicable and dissociation or ionization is neither readily calculable nor negligible. For example, preliminary band structure calculations indicate that the electronic band gap vanishes at or near the molecular-metallic transition.

Quantum corrections for the translational molecular motion (Wigner 1932) have been estimated and are found to be negligible unless both the temperature and pressure are low (Krumhansl and Wu 1968).

Rather than attempt to model realistically the free energy contributions F_{mol} and ΔF, we consider three different models that are expected to encompass the widest range of conceivable circumstances. (Together with the three choices of interaction potential, this gives us a matrix of nine different thermodynamic models.) The three models are labelled A, B, and C in order of increasing entropy:

A.　Low Entropy Model: $\theta_v = 6100°$K, $\theta_R = (85.4 + 3000\rho^{\frac{2}{3}})°$K where ρ is in g cm^{-3} (so that θ_R is comparable to the Debye temperature); and $\Delta F \equiv 0$.

B.　Medium Entropy Model: $\theta_v = 6100°$K, $\theta_R = 85.4°$K and $\Delta F \equiv 0$.

C.　High Entropy Model: $\theta_v = 6100°$K, $\theta_R = 85.4°$K and $\Delta F = -2.6Nk_BT$ $\exp(-T_0/T)$ where $T_0 = 6400\rho^{\frac{1}{2}}$. We shall not attempt to justify this choice (although it is appropriate for electronic excitation in an intrinsic semiconductor with band gap $2k_BT_0$).

These models differ mainly in the high density region and are essentially indistinguishable for $\rho \lesssim 0.05$ g cm^{-3} at the temperatures relevant for Jupiter. At these low densities, *all* the models agree (to within experimental error) with the fluid-phase measurements of Presnall (1969).

In Table I, the properties of molecular hydrogen at the molecular-metallic transition are listed for each of the nine thermodynamic models. In each case, the transition is found by equating the Gibbs energies of the two phases. The temperature is determined by requiring that the entropy in the molecular phase be equal to that of hydrogen gas at $T = 180°$K and $P = 1$ bar (an appropriate choice of adiabat for typical Jupiter models). Notice that the entropy change at the transition ranges from -1 to $+1$ in units of k_B per proton. Notice also that the transition pressure and temperature are uncertain by a

TABLE I

Thermodynamic Properties of H_2 at the Phase Transition for a Molecular Adiabat that passes through $T = 180°K$ at $P = 1$ bar

Thermodynamic Model[a]	Pressure (megabars)	Temperature (°K)	Molecular Volume (a_0^3/proton)	Δv[b] (a_0^3/proton)	Δs[c] (k_B/proton)
IA	2.0	13900	17.0	3.2	−1.0
IB	2.3	9300	15.5	3.2	−0.1
IC	2.5	5400	14.1	3.1	+1.0
IIA	2.5	15000	15.6	2.9	−1.0
IIB	3.0	9600	13.7	2.8	−0.1
IIC	3.4	5800	12.5	2.7	+0.9
IIIA	3.8	15900	12.7	2.4	−0.9
IIIB	4.4	9900	12.0	2.25	+0.1
IIIC	4.8	6200	11.3	2.2	+1.1

[a]The models are labeled by a Roman numeral for the potential [Eq. (11)−(14)] and by a letter (A, B, or C) for the entropy (see text).
[b]Δv = molecular volume − metallic volume
[c]Δs = molecular entropy − metallic entropy

factor of two. In Figs. 2 and 3, some of the corresponding adiabats $P(\rho)$ and $T(\rho)$ are shown for pure hydrogen. The effects of these uncertainties on interior models are discussed in Secs. II and III.

(c)   Helium and Hydrogen-Helium Mixtures

At high pressures ($r_s \gtrsim 1$), the theory of helium is a simple generalization of the theory of metallic hydrogen. At lower pressures, an approach similar to that described for molecular hydrogen is appropriate, with a pair potential such as that given by Trubitsyn (1967). The merging of the two limiting theoretical approaches is uncertain, but the low abundance of helium relative to hydrogen in Jupiter makes the choice of interpolation rather unimportant. For Jupiter, it is more relevant to determine the properties of hydrogen-helium mixtures.

It has been traditional to calculate the equation of state for hydrogen-helium mixtures by invoking volume additivity (Hubbard and Smoluchowski 1973). Since this makes unwarranted assumptions about the nature of the interactions between different chemical species, we have chosen to avoid this. Instead, we explicitly calculate the thermodynamic properties of mixtures.

In the metallic region, Stevenson (1975) has treated the H-He system as a non-ideal Coulomb plasma while incorporating many higher order effects. This calculation does not explicitly consider bound (or resonant) electronic states associated with the alpha particles, but nevertheless goes to high

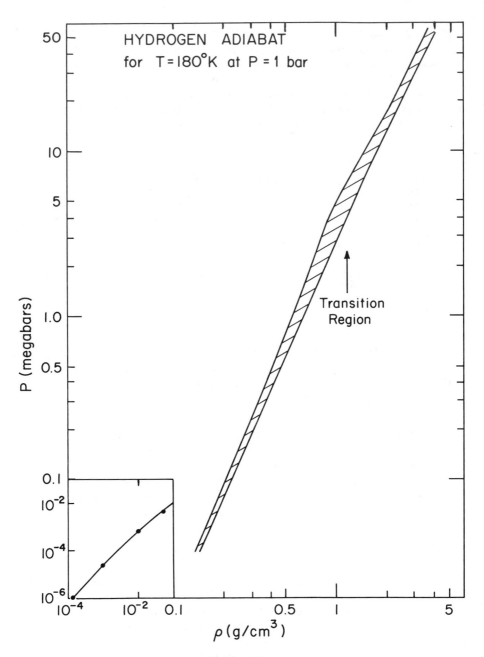

Fig. 2.   Hydrogen adiabat $P(\rho)$: shaded region indicates the uncertainty in the adiabat given by the matrix of thermodynamic models from Table I. This is not a true adiabat because of the entropy discontinuity at the transition region.

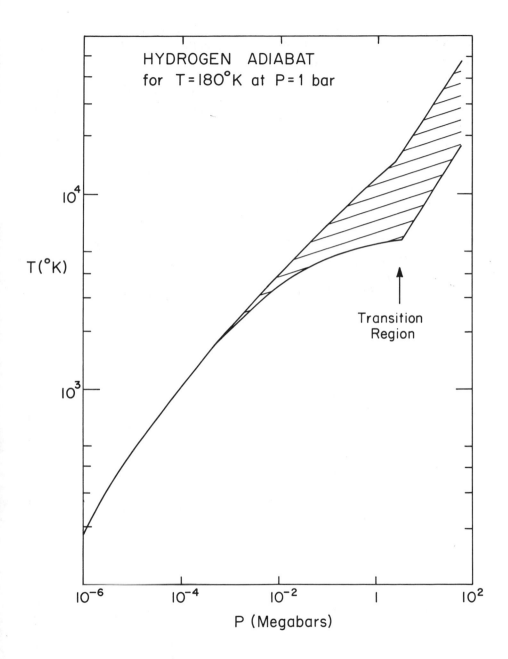

Fig. 3.   Hydrogen adiabat $T(\rho)$: shaded region indicates the uncertainty in the adiabat given by the matrix of thermodynamic models from Table I. This is not a true adiabat because of the entropy discontinuity at the transition region.

enough order in the electron-ion interaction to include such effects approximately. These detailed calculations are used in the metallic hydrogen region of the interior models discussed in Sec. III.

We comment here on two important consequences of those calculations. One is the failure of the volume additivity approximation. Let $\Omega(x, P)$ be the volume per ion, where x is the number fraction of helium ions and P is the pressure. Thus,

$$[1 + \delta(x, P)] \, \Omega \, (x, P) = x\Omega(1, P) + (1 - x) \, \Omega \, (0, P) \qquad (18)$$

where $\delta(x, P)$ is a measure of the deviation from volume additivity in the mixture. Figure 4 shows that δ is significantly non-zero, in contrast to the Thomas-Fermi theory where it is natural to assume $\delta \equiv 0$ (Salpeter and Zapolsky 1967). Although the calculation is for finite temperature, this non-additivity is an electronic effect. Since $\delta > 0$, the alloy occupies less volume than the separated phases. This means that the density of a given hydrogen-

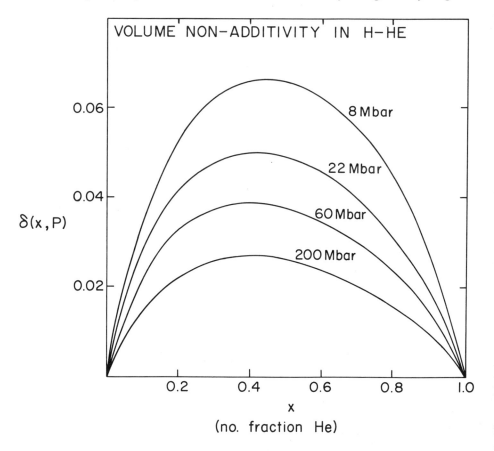

Fig. 4.   Volume non-additivity in a fully-ionized H-He mixture at $T = 6000°K$.

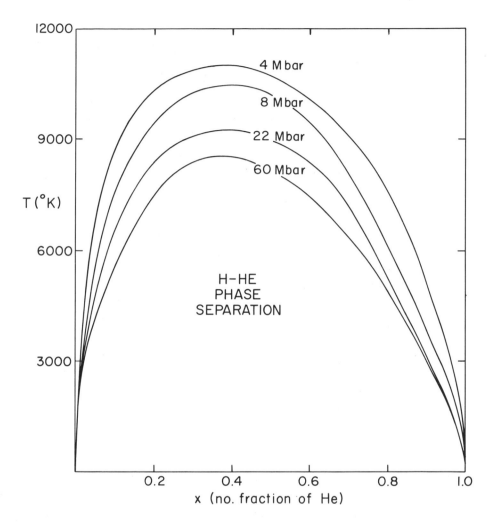

Fig. 5.   Phase separation curves for various pressures. The miscibility gap is below the curve
in each case.

helium alloy is greater than volume additivity would predict. Consequently,
interior models of Jupiter require *less* helium than volume additivity would
predict. This effect is small but not negligible (typically a 10% reduction in
the required total mass of helium).

A more dramatic effect is the prediction of a phase separation in a hydrogen-helium mixture at sufficiently low temperatures. In Fig. 5, the calculated
phase separation curves are shown. For each pressure, the phase-excluded
region (miscibility gap) is below the curve. These results indicate that a H-
He mixture of solar composition ($x \sim 0.1$) separates into hydrogen-rich and
helium-rich phases at temperatures less than about 8000°K and pressures in

the range 2-200 megabars. However, it must be emphasized that, since the calculation of these phase boundaries depends on small energy differences (comparable to the thermal energy), errors on the order of 20% could be expected (especially at the lower pressures). The implications of this phase transition are described in Sec. II.

In the molecular region, the properties of hydrogen-helium mixtures were evaluated by an extension of the calculations for pure molecular hydrogen, using hard sphere pair distribution functions for a mixture (as by Stevenson 1975). The H_2-He and He-He interaction potentials were taken from Shafer and Gordon (1973) and Trubitsyn (1967) respectively. These potentials are no better known than the H_2-H_2 potential, but are less important for Jupiter models. For each of the H_2-H_2 potentials previously discussed, volume additivity was satisfied to within 1%. No attempt was made to calculate possible phase separation in the H_2-He fluid because of the uncertainties in the interaction potentials. Smoluchowski (1973) has scaled the low pressure phase separation observations of Streett (1974) to infer that helium is only partially soluble in molecular hydrogen at the pressures and temperatures appropriate for the molecular-metallic transition in Jupiter. This scaling could be misleading because of the failure of a simple pairwise-additive potential at high pressures. The possibility of significant immiscibility in the molecular region of Jupiter will remain an open question until the thermodynamics are better understood or more relevant experiments are carried out.

(d)   Rocks and Ices

In the models discussed in Sec. III, the "ices" (H_2O, NH_3, CH_4) constitute about 1% of the planetary mass, and an approximate Thomas-Fermi equation of state (Salpeter and Zapolsky 1967) is adequate. Even in the models of Podolak and Cameron (1974, 1975), where there is a large water enhancement, they conclude that their results are not significantly affected by a more careful calculation of the H_2O equation of state.

The rocky materials (presumed to be SiO_2, MgO, Ni and Fe in their relative solar abundances) constitute as much as 5% of the planetary mass in our models [as much as 10% in models by Podolak and Cameron (1974)] so it is relevant to ask whether a Thomas-Fermi equation of state is adequate. The Salpeter-Zapolsky (1967) calculations (including correlation) appear to be inaccurate by roughly 10% at the lower pressures ($P \lesssim 10$ megabars). The percentage error is expected to decrease as P increases, since it is caused by band structure effects. Nevertheless, it is evident that underestimates of the order of $5 - 10\%$ in the "rock" pressure could lead to small but significant miscalculations of the required helium or some other adjustable parameter. In our calculations, the Salpeter-Zapolsky equation of state was used since higher order corrections are difficult to evaluate and no better alternative exists.

II.   HEAT TRANSPORT AND DYNAMIC PROCESSES

In addition to the equation of state, the temperature gradient and distribution of constituents must also be specified in constructing an interior model. We consider these two constraints separately.

(a)   Convection and the Adiabatic Approximation

Infrared observations from Pioneers 10 and 11 indicate that Jupiter has an internal heat source of about 5×10^{24} erg sec^{-1}. For the corresponding heat flux in the deep interior, the models of Bishop and DeMarcus (1970) show that there is no significant subadiabatic conductive zone in the metallic region, unless the thermal conductivity exceeds 10^{10} erg deg^{-1} cm^{-1} sec^{-1}. The calculations of Stevenson and Ashcroft (1974) predict a conductivity of less than 10^9 erg deg^{-1} cm^{-1} sec^{-1}. Accordingly, the metallic region is convectively unstable and nearly all the heat flux is transported by convection.

In the molecular region, electronic or molecular conduction is negligible. Furthermore, the bound-free radiative opacity can be shown to be prohibitive, unless the molecular band gap exceeds $20\,k_B T$. Since the band gap is *less* than 10 eV (the low density value), it follows that radiative conduction is negligible for $T \gtrsim 6000°$K. For $400°$K $\lesssim T \lesssim 2000°$K, the calculations of Trafton (1967) show that pressure-induced opacity is sufficient to reduce the radiative heat flux to less than a few percent of the total heat flux. We conclude that a subadiabatic radiative zone in the molecular region at temperatures $2000°$K $\lesssim T \lesssim 6000°$K is highly unlikely but not rigorously excluded.

These considerations led Hubbard (1973) to propose that Jupiter is convective almost everywhere, with the consequence that the specific entropies of the deep atmosphere and metallic interior should be almost equal. There are at least two possible objections to this. One possibility is that the convective heat transport requires substantial superadiabaticity. A convenient measure of this is ϵ, where

$$\epsilon = \left[\left(\frac{dT}{dr}\right)_{actual} - \left(\frac{dT}{dr}\right)_{ad} \right] \bigg/ \left(\frac{dT}{dr}\right)_{ad}. \qquad (19)$$

All the models we consider are fluid throughout so that large-scale fluid flows are not inhibited by viscosity. Simple mixing length theory (Hubbard and Smoluchowski 1973) would then predict $\epsilon \sim 10^{-8}$ for a mixing length of 10^9 cm.

However, the simple mixing length theory does not allow for the influence of planetary rotation on the fluid flow. A measure of this is the ratio of inertial to Coriolis forces, $v/\Omega\ell$, where v is the convective velocity, Ω the planetary angular velocity, and ℓ a length scale characterizing the convection. Simple mixing length theory predicts $v \sim 10$ cm sec^{-1} so that this ratio (known as the Rossby number) is $R_o \sim 10^{-4}$. It is clear that large scale convection is strongly affected by rotation, since the Coriolis force is large.

There is no direct generalization of simple mixing length theory to a rotating fluid. One possibility is to suppose that ℓ becomes small enough so that $R_o \sim 1$. Equating the convective heat flux to the actual heat flux then gives $\epsilon \sim 10^{-1}$ or 10^{-2} and $v \sim 1$ cm sec^{-1}. In this case, the superadiabaticity can be large enough to significantly modify interior models. However, there are more efficient convective modes in which the dominant force balance is between the Coriolis force and hydrodynamic pressure gradients. For these modes, the convective pattern is elongated along the rotation axis, and $R_a \simeq T_a^{\frac{2}{3}}$ (Chandrasekhar 1961), where R_a is the Rayleigh number and T_a the Taylor number. This result is for laminar flow, but it is plausible that a similar result applies to turbulent flow, provided the kinematic viscosity is reinterpreted as a self-consistently determined "eddy viscosity." This procedure minimizes ϵ and the results are $\epsilon \sim 10^{-4}$, $v \sim 1$ to 10 cm sec^{-1} and a mixing length perpendicular to the rotation axis of roughly 10^5 cm. There is some evidence to suggest that turbulent convection does not always choose the most efficient mode (Spiegel 1971) but this last example does give some justification for the assumption of $\epsilon \ll 1$. Notice also, that the convective velocity is rather weakly dependent on the geometry of the convective flow.

Since the deep interior of Jupiter is likely to have a magnetohydrodynamic dynamo, it should be mentioned that Lorentz forces are likely to be comparable to Coriolis forces. This may change the above conclusions.

A second possible objection to Hubbard's "fully adiabatic" hypothesis is the influence on the convection of a first order phase transition in the fluid. In a recent paper, Salpeter and Stevenson (1975) consider a self-gravitating fluid, stratified into two phases of appreciably different densities and heated from within. It is assumed that, away from the interface between the phases, the heat flux is mainly carried by turbulent convection with a very small superadiabaticity. Different modes are investigated for transporting the heat flux across the interface, and both possible signs for the phase transition latent heat L are considered. Under a wide range of conditions, it is found that the transition region near the interface is thin, with a small change in temperature across it. The entropy difference between the two phases is then L/T, where T is the temperature of the transition. Two other assumptions are needed to reach this conclusion:

1. The heat flux is determined by conditions in the surface layers of the planet, and its *average* is not affected by the dynamics of the phase transition region.
2. The phase transition is first order with a finite positive surface energy between the phases.

Salpeter and Stevenson (1975) conclude that these assumptions and considerations are probably applicable to the molecular-metallic hydrogen transition in Jupiter. The main uncertainty is the nature of the fluid molecular to fluid metallic hydrogen transition. It has often been stated that this fluid-

fluid transition is not first order (see, for example, Trubitsyn 1972). Others do not share this view (Landau and Zeldovich 1943). It is clear that there must be a critical temperature T_c beyond which no first order transition can occur, since the free energy of mixing must dominate the interaction energies if the temperature is high enough. In the case of hydrogen, it is not known whether T_c exceeds the melting point of either phase at the transition pressure. Apparently, no first-order fluid-fluid transition, analogous to the proposed hydrogen transition, has been observed yet in nature. Nevertheless, there is no statistical mechanical justification for excluding the possibility of such a transition. Furthermore, the existence of a first-order phase transition in the *solid* phases is strongly implied by the predicted large density discontinuity at the transition (Neece *et al.* 1971). Since the electronic structures of the solid and fluid in either phase are almost identical, presence of fluidity does not necessarily prevent a first-order phase transition. The resolution of this controversial point requires a very careful energy calculation of an arbitrary mixture of the two phases.

If the considerations of Salpeter and Stevenson (1975) are applicable to Jupiter then the temperature drop in the transition region, ΔT, is found to be much less than the temperature T. This has an important consequence for interior models. In contrast to Hubbard's hypothesis, the temperature at the center of the planet is *not* independent of the poorly known thermodynamics of the dense molecular region. Instead, we have

$$S_c + \Delta S = S_{atm} \tag{20}$$

where S_c, S_{atm} are the specific entropies of central and atmospheric regions of the planet respectively, and $\Delta S = L/T$ is the entropy change at the transition. As Table I shows, ΔS is not well known. Since the specific heat in the deep interior is of order 2 k_B per nucleus, it follows that if the central temperature of the planet is evaluated by assuming $S_c = S_{atm}$, then it is incorrect by a *multiplicative* factor of about $\exp(-\Delta S/2)$. This factor could be as small as 0.6 or as large as 1.7, according to the crude thermodynamic models considered in Sec. I(b). In Sec. III, we construct some models indicating the possible effects of this.

(b)   Helium Separation in a Convecting Fluid

For interior models in which the central temperature is of order $2 \times 10^4 °K$, the thermal energy content of Jupiter is about 10^{41} ergs. An even larger energy is available, however, if Jupiter changed from a chemically homogeneous structure to one where the denser helium resides in a central core (Kiefer 1967; Flasar 1973). It is not yet clear whether primordial heat alone is sufficient to maintain Jupiter's heat flux over 4.5 billion years (Graboske *et al.* 1975) and this motivated consideration of helium separation as an alternative energy source (Smoluchowski 1967, 1970, 1973; Salpeter 1973).

It might be supposed that chemical separation and gravitational layering are impossible in the presence of fully developed turbulent convection because diffusion times are enormously long compared with convective times. Salpeter (1973) pointed out that layering may nevertheless take place in the presence of convection, if helium becomes insoluble in hydrogen. Droplets of helium-rich fluid can then nucleate and grow, and fall towards the center of the planet.

This insolubility may take place in the molecular region (Streett 1974) as well as the metallic region. We confine our discussion to the metallic region, for which an estimated theoretical phase diagram exists (Stevenson 1975). Figure 5 shows these theoretical predictions, indicating that the insolubility is greatest at low pressures and that the critical helium concentration x_c (the value of x at the peak of each curve) substantially exceeds the helium cosmic abundance $x_0 \sim 0.1$. As Jupiter cools down from an initially hotter, homogeneous state, a time is reached when the actual temperature becomes less than the phase transition temperature at $x \simeq x_0$ and $P = P_{tr}$ the molecular-metallic transition pressure. At this level, droplets of helium-rich fluid then begin to nucleate and grow at a rate determined by the surface tension and the degree of supercooling. These droplets eventually grow to have a radius of the order of 1 cm, at which stage their terminal downward Stokes velocity is comparable to typical convective velocities of 10 cm sec^{-1}. For a helium diffusivity of 10^{-2} cm^2 sec^{-1} (Hubbard and Smoluchowski 1973), the growth time for these droplets is of order 10^2 sec, which is much less than convective time scales of 10^8 sec. Thus, helium separation is not highly inefficient and there is a net downward stream of helium droplets, even in the presence of convection. These droplets eventually redissolve at a deeper level in the planet, where the temperature is higher, but $x < x_c$ still. At a much later stage in the evolution, a predominantly helium core begins to form at the center. The net effect of helium separation is to release gravitational energy in the form of heat, and set up helium gradients within the planet. The temperature estimates of Fig. 5 indicate that helium differentiation is already in progress, or will begin within a few billion years. Work on the details of the helium distribution is in progress and will be included in a subsequent publication. Once the thermodynamics in both molecular and metallic hydrogen regions are fully understood, it should be possible to *predict* the helium abundance in the Jovian atmosphere. Alternatively, an observation of the helium abundance could be used as a very important diagnostic for the interior dynamics.

III. INTERIOR MODELS

An excellent review of interior models (pre-1973) is to be found in Hubbard and Smoluchowski (1973). We confine ourselves to a brief summary of these models and of more recent models.

The most complete early models were those of DeMarcus (1959). His zero temperature models were predominantly hydrogen, but had a total helium abundance of about 22% by mass, and a very small high density (rocky) core.

The next significant advance was by Peebles (1964) who found that external observations such as the gravitational moments, are better fitted by a planet with an *adiabatic* outer envelope. This deduction, which implies a thermally active planet, was made before the discovery of an excess infrared flux. Peebles also confirmed DeMarcus's conclusions about the abundances of major constituents. However, Peeble's models did not include a significant thermal perturbation to the deep interior.

Smoluchowski (1967, 1971) emphasized the expected insolubility of helium in *solid* metallic hydrogen and the consequent non-uniform distribution of helium. Our discussion in Sec. I indicates that Jupiter is probably entirely fluid, so these calculations are most likely not applicable to the present Jupiter.

Subsequent to the discovery of the infrared excess of Jupiter (Low 1966), Hubbard (1968, 1969, 1970) constructed several models that were analogous to extremely low mass M-dwarf stars. These models were fully adiabatic, with a uniform chemical composition and sufficient internal heat content to explain the infrared emission. In contrast to previous models, the temperature in the deep interior ($\sim 10^4$ °K) was high enough to significantly modify the interior thermodynamics and thermally expand the planet. As a consequence of the thermal expansion, a higher helium mass fraction of $30-35\%$ was required, which is in excess of the estimated solar abundance (Cameron 1973).

More recent models by Podolak and Cameron (1974, 1975) have a similar thermal structure to those of Hubbard. However, they constrain the helium abundance to the solar value and estimate the enhancement of heavier elements that is then required. These models have a rocky core of ten or twenty Earth masses, and a comparable amount of icy material (mainly H_2O) uniformly mixed with the hydrogen and helium. As Podolak and Cameron discuss, the initial formation of Jupiter probably required a core of refractory material so as to initiate the collapse of the hydrogen-helium envelope. Thus, an enhancement of refractory material is expected from cosmogonic considerations.

Recent models by Zharkov et al. (1975)[4] and Hubbard and Slattery[5] agree that models with a solar abundance of helium require a substantial enhancement of refractory materials over the solar abundance. (These models require several tens of Earth masses of heavier elements.) Like the models we describe below, these most recent models also calculate the

[4]See also p. 133.
[5]See p. 185.

rotational response (i.e., J_2, J_4, J_6) to a higher accuracy than previous models.

The main aim of our model calculations is to assess the effect of uncertainties in the thermodynamics. These models incorporate the best available equations of state for the metallic region and correctly allow for volume non-additivity. The following considerations are common to all the models.

Rocky Core. An isothermal rocky core with constituents in their relative solar abundances (cf. Podolak and Cameron 1974).

Hydrogen-Helium-Ice Mantle. The rocky core is surrounded by a hydrogen-helium mixture augmented with a solar abundance (1.24% by mass) of "ices." We do not consider enhancement of ice content above the solar value. The solar abundance of helium is not known to sufficient accuracy so the helium concentration is an adjustable parameter. According to Cameron (1973), the solar number fraction of helium is in the range 0.07 < x < 0.09.

Molecular-Metallic Hydrogen Phase Transition. The effect of helium on the phase transition pressure was neglected, but the calculation is otherwise self-consistent (i.e., the Gibbs energies of the two phases are equated for each thermodynamic model; see Table I).

Surface Boundary Condition. To construct internal adiabats it is necessary to specify the entropy in the deep atmosphere. This is parameterized by T_s, the temperature at $P = 1$ bar. Since external observations and atmospheric models are rather uncertain on the value of T_s, we treat T_s as an adjustable parameter in the range $140°K \lesssim T_s \lesssim 210°K$.

Hydrostatic Equilibrium and the Shape of the Planet. The equation of hydrostatic equilibrium is numerically integrated:

$$\frac{dP}{ds} = -\frac{GM(s)\rho(s)}{s^2} + \frac{2}{3}\Omega^2 s \rho(s) \qquad (21)$$

$$\frac{dM(s)}{ds} = 4\pi s^2 \rho(s)$$

where the parameter s labels equipotential surfaces and is a mean radius (the radius of a sphere containing an equal volume). The value of s at the "external surface" ($P = 1$ bar) is taken to be $s_0 = 6.979 \times 10^9$ cm. This value is derived from the observed equatorial radius (7.140×10^9 cm) and the observed gravitational moments J_2, J_4, using the theory of Zharkov and Trubitsyn (1974). It should be emphasized that s_0 is a function of the planetary shape, so the calculation of the gravitational moments cannot be treated independently of the integration of the equation of hydrostatic equilibrium. The mass of Jupiter is taken to be 1.900×10^{30} g.

The gravitational moments J_2, J_4 (and J_6, very approximately) are evaluated using the theory of Zharkov *et al.* (1973), which is correct to third order in the rotation parameter $q = \Omega^2 R_{eq}^3 / GM$ where R_{eq} is the equatorial radius. For Jupiter, $q = 0.0888$. The procedure adopted is to first solve for the rotational response, using the much simpler second-order theory of Peebles (1964). These results are then substituted into the Zharkov equations. With experience, it is possible to empirically modify the Peebles solutions so that the Zharkov integral equations require only one iteration. For example, the Peebles theory systematically underestimates J_2 by approximately $4-5\%$. Our computer algorithm for J_{2n} was tested by evaluating a Jupiter model for an $n = 0.95$ polytrope equation of state. This polytrope is of particular interest because it gives gravitational moments that are compatible with the observations (Hubbard et al. 1975). Table II shows that the Peebles theory is inadequate, but the Zharkov theory is satisfactory and in good agreement with the zonal harmonic theory of Hubbard et al. (1975).

TABLE II
The Rotational Response of an n = 0.95 Polytropic Model of Jupiter

Source	J_2	J_4	J_6
Zonal Harmonics Hubbard et al.	1.474×10^{-2}	-5.80×10^{-4}	3.4×10^{-5}
Present Work to 2nd Order (Peebles)	1.396×10^{-2}	-6.15×10^{-4}	0
Present Work to 3rd Order (Zharkov)	1.469×10^{-2}	-5.75×10^{-4}	3.9×10^{-5}
Observations: Pioneer 10 (Anderson et al. 1974)	$(1.472 \pm 0.004) \times 10^{-2}$	$-(6.5 \pm 1.5) \times 10^{-4}$	Assumed zero
Observations: Pioneer 11 (Null et al. 1975)	$(1.475 \pm 0.005) \times 10^{-2}$	$-(5.8 \pm 0.4) \times 10^{-4}$	$(5 \pm 6) \times 10^{-5}$

Computational Procedure. First, the surface condition (T_s) and helium concentration are chosen. The atmospheric specific entropy is evaluated and the central temperature estimated using Eq. (20). A guess is made for the size of the rocky core and central density, and Eq. (21) is then integrated to give a mass and radius. Subsequent iterations of the core size, central density and central temperature are made to obtain the correct mass and effective radius s_0 for Jupiter, each to within 0.1%. The gravitational moments are then evaluated: J_2 to 0.2%, J_4 to 3% and J_6 to about 20%. The whole process is repeated for a different helium concentration until

TABLE III
Jupiter Models
(H and He Uniformly Mixed)

Thermodynamic Model	x	T_s (°K)	$\dfrac{M_{rock}}{M_{earth}}$	T_c (°K)	$-10^4 J_4$
IIB	0.073	150	14.5	17900	5.9
IIB	0.085	160	11.0	19800	5.8
IIB	0.102	185	8.6	24300	5.5
IIB	0.110	200	6.9	27900	5.2
IIIB	0.068	150	15.4	17900	5.9
IIIB	0.089	170	13.0	21300	5.7
IB	0.078	150	11.9	17800	6.0
IB	0.095	170	8.0	21300	5.7
IIC	0.070	160	10.2	11700	5.8
IIC	0.085	180	5.7	13300	5.7
IIC	0.091	190	5.0	14900	5.3
IIIC	0.066	160	11.4	11800	5.8
IIIC	0.082	180	7.4	13400	5.6
IC	0.075	160	8.9	11700	5.9
IC	0.088	180	5.2	13200	5.6
IIA	0.115	140	14.1	33500	6.3
IIA	0.120	150	11.3	35100	5.8
IIA	0.123	180	6.5	42200	5.6
IIIA	0.110	140	16.2	33600	6.2
IIIA	0.117	160	10.9	37300	5.7
IA	0.118	140	11.8	33300	6.4
IA	0.123	160	9.7	37200	5.7

J_2 is also correct. The range of values for T_s that we chose ensured that J_4 was usually within the acceptable observational limits (see Table II). The observational constraint on J_6 is not yet strong enough to be useful. (In our models, J_6 was typically 3×10^{-5} and never exceeded 4.5×10^{-5}.)

Results. We first consider models in which the hydrogen and helium are uniformly mixed. The results are summarized in Table III. This table demonstrates a very significant point: Given our ignorance of the thermodynamics of dense H_2, there is a wide range of possible values of T_s and x that are compatible with the observations of mass, radius, J_2 and J_4. It is also evident that variations of the H_2 adiabat (i.e., choice of models A, B or C) have more effect than variations of the intermolecular potential (i.e., choice of models I, II, III). Models with a positive latent heat (Type C) have a lower entropy in the center than in the atmosphere. They consequently require a larger T_s for a given central temperature T_c and helium abundance x. In contrast, models with a negative latent heat at the molecular-metallic transition require a *lower* T_s. The effect of changing the H_2 intermolecular potential is smaller and less obvious. Some of the models are described in more detail in Table IV. These models are chosen

TABLE IV
Details for Some Jupiter Models

	Model (1)	Model (2)	Model (3)	Model (4)	Model (5)
Thermodynamic Model	IIB	IIIB	IB	IIC	IIA
T_s (°K)	160.0	170.0	170.0	180.0	150.0
Helium Content X	0.085	0.089	0.093	0.085	0.119
PRESSURE (megabars)					
Center	104.0	116.4	87.4	75.8	106.1
Rock Core/H Boundary	44.8	46.3	43.25	42.9	45.1
H/H$_2$ Boundary	3.09	4.07	2.03	3.57	2.49
DENSITY (g cm^{-3})					
Center	26.0	27.3	24.1	22.7	26.25
Rock Core/H Boundary	18.1/4.325	18.3/4.405	17.8/4.283	17.7/4.344	18.1/4.343
H/H$_2$ Boundary	1.33/1.09	1.48/1.23	1.15/0.92	1.46/1.16	1.20/1.04
TEMPERATURE (°K)					
Center	19760	21380	21352	13277	35371
Rock Core/H Boundary	19760	21380	21352	13277	35371
H/H$_2$ Boundary	9258	10621	9133	6597	15423
RADIUS (10^9 cm)					
Rock Core/H Boundary	0.904	0.947	0.825	0.742	0.911
H/H$_2$ Boundary	5.307	5.065	5.583	5.175	5.482
MASS (10^{30} gm)					
Rock Core	0.0655	0.0776	0.0477	0.0337	0.0674
H/H$_2$ Boundary	1.460	1.352	1.590	1.416	1.538
J_4	-5.8×10^{-4}	-5.7×10^{-4}	-5.7×10^{-4}	-5.7×10^{-4}	-5.65×10^{-4}

because they have a value of J_4 that is in excellent agreement with the Pioneer 11 observations and because they are a representative selection of the thermodynamic models.

IV. DISCUSSION AND CONCLUSION

The main conclusion of these model calculations is that the thermodynamics of dense molecular hydrogen are not yet sufficiently well known for an accurate determination of the abundance of helium or other minor constituents. Firmer conclusions could be made if we had accurate values of x and T_s from external observations. Infrared observations (Orton[6]; Lacy *et al.* 1975) indicate that $T_s = (175 \pm 20)°K$. However, the helium abundance x is not usefully constrained by observations.

Since there is no known cosmogonic process for efficient differentiation of hydrogen and helium before the formation of Jupiter, the average abundance should equal the cosmic value. The latest estimate for this is (Cameron 1973) $0.070 < x < 0.090$, with the lower value favored. Even with $x = 0.070$, Table III shows that models with $T_s \sim 170°K$ may be possible. In general, however, a larger value of T_s requires a larger value of x. If T_s is actually $180°K$ or more, then our models are not compatible with $x = 0.070$. In this respect, our models are not in strong disagreement with those of Podolak and Cameron (1974, 1975). Their thermodynamics most nearly approaches our model II B, for which a value of $x = 0.07$ gives an unacceptably low T_s. It must be emphasized, however, that there are important differences between the calculations: We have used more accurate thermodynamics for the metallic region and we have evaluated J_2 to higher order. Given the present uncertainties in the H_2 thermodynamics, helium abundance and atmosphere, we feel that there is no clear necessity for the water enhancement proposed by Podolak and Cameron.

Some other general conclusions can be made from our model calculations:

1. Models with a large helium abundance generally require a large T_s.
2. For any reasonable value of x, the observed value of J_4 constrains the atmospheric temperature to the range $140°K \lesssim T_s \lesssim 200°K$. (This conclusion is qualitatively related to the conclusions of gravitational inversion theory: see Hubbard and Slattery.)[7]
3. Models are not very sensitive to the value of the molecular-metallic hydrogen transition pressure, at least in the range $2-5$ megabars.
4. Models are definitely sensitive to the form of the molecular hydrogen adiabat.
5. Models which incorporate helium separation generally require a larger helium content (for given T_s) than models without helium separation.

We conclude by listing, in order of importance, the observations and

---

[6]See p. 207.   [7]See p. 182.

theoretical calculations needed for further progress in understanding the interior of Jupiter.

Observations
1. The temperature profile and composition in the deep atmosphere; especially the helium abundance and T_s.
2. More accurate values for J_4, J_6.
3. The internal heat flux, including possible latitude dependence.
4. More details on the intrinsic magnetic field.

Theory
1. Thermodynamic properties of molecular hydrogen and its transition to the metallic phase.
2. Properties of hydrogen-helium mixtures, especially in the molecular region, and especially the solubility of He in H_2.
3. Radiative conductivity of dense H_2. (An upper bound is probably sufficient.)
4. High temperature and high pressure chemistry of minor constituents. (For example, is the rocky core soluble in metallic hydrogen?)
5. Better theories of turbulent convection in the presence of rotation and magnetic fields. (This probably requires a model of the Jovian dynamo.)

Acknowledgments. Discussions with A. G. W. Cameron, W. C. DeMarcus, and W. B. Hubbard are gratefully acknowledged. This work is supported by the NSF Grants MPS72-05056-A02, MPS74-17838 and NASA Grant NGR-33-010-188.

DISCUSSION

R. Smoluchowski: In processes of nucleation the value of the surface energy plays an important role especially for very small embryos. What value did you use?

D. J. Stevenson: We used the bulk value as a first approximation.

W. H. McCrea: How do you infer the existence of a rocky core and why do you assign a mass of ten or so Earth masses to it? Why would one Earth mass not suffice?

D. J. Stevenson: Our models *assume* a refractory core surrounded by a solar composition mixture. We then find that we cannot fit the observed mass, radius, and values of J_2 and J_4, unless this core contains ten or so Earth masses. This refractory material could alternatively be uniformly distributed within the planet. In either case, roughly an order of magnitude enhancement above the solar abundance of refractory materials seems to be required.

W. Kaula: Can you place any meaningful constraints on the amount of H_2O or the H_2O/silicates ratio? This would be of considerable interest relevant to the hypothesis that the Jupiter embryo reached significant size before the terrestrial planets.

D. J. Stevenson: The thermodynamic uncertainties prevent us from separately estimating the H_2O and silicate contents. However, our models do indicate that, in contrast to the refractory material, an enhancement of H_2O above that expected for solar abundance material is not essential.

W. Kaula: The olivine-spinel phase transition in the Earth's mantle apparently does not prevent convection (Schubert, G. *et al.* 1970. *Science* 169:1075–1077). What is different about the hydrogen molecular-metallic transition?

D. J. Stevenson: The main difference is that in Jupiter, the Prandtl number is about twenty orders of magnitude smaller. This, together with the low internal heat flux in Jupiter, ensures highly subsonic convection and a temperature gradient that is only slightly superadiabatic, except very near the phase transition. Our criteria for convective inhibition near a phase transition are described in detail in Salpeter and Stevenson (1975).

J. B. Pollack: Jupiter's evolutionary history may help resolve the question of the value of the central temperature: The predicted luminosity after 4.5 billion years depends on this value. From our calculations, I suspect that in order to match the observed excess luminosity, a current central temperature will be required that is close to your lower bound of 13,000°K.

D. J. Stevenson: In the degenerate cooling phase of Jupiter, the cooling time is proportional to the heat content. Thus, a *large* central temperature — or, more exactly, a *large* average value of $(dT/dP)_{ad}$ — is needed to prolong the evolution and match the observed excess luminosity. Evolution is an important constraint on models for the present Jupiter, and will be the subject of a future study. The possibility of a non-uniform helium distribution may complicate such evolutionary considerations.

REFERENCES

Anderson, J. D.; Null, G. W.; and Wong, S. K. 1974. Gravity results from Pioneer 10 Doppler data. *J. Geophys. Res.* 79:3661–3664.

Bishop, E. V., and DeMarcus, W. C. 1970. Thermal histories of Jupiter models. *Icarus* 12: 317–330.

Cameron, A. G. W. 1973. Abundances of elements in the solar system. *Space Sci. Rev.* 15: 121–146.

Chandrasekhar, S. 1961. *Hydrodynamic and hydromagnetic stability.* Ch. III. Oxford: Clarendon Press.

Clayton, D. D. 1968. *Principles of stellar evolution and nucleosynthesis.* p. 138. New York: McGraw-Hill.

DeMarcus, W. C. 1959. The constitution of Jupiter and Saturn. *Astron. J.* 63:2–28.

DeWitt, H. E., and Rogers, F. J. 1972. Quantum statistical mechanics of dense partially ionized hydrogen. *Phys. Earth Planet. Interiors* 6:51–59.

Flasar, F. M. 1973. Gravitational energy sources in Jupiter. *Astrophys. J.* 186:1097–1106.

Friedli, C., and Ashcroft, N. W. 1975. Band structure of molecular hydrogen. In draft.

Graboske, H. C.; Pollack, J. B.; Grossman, A. S.; and Olness, R. J. 1975. The structure and evolution of Jupiter. *Astrophys. J.* 199:265–281.

Grigoryev, F. V.; Kormer, S. B.; Mikhailova, O. L.; Tolochko, A. P.; and Urlin, V. D. 1972. Experimental determination of the compressibility of hydrogen at densities $0.5-2$ g/cm^3. *J. Exptl. Theoret. Phys. Lett.* 16:201–204.

Hammerberg, J.; and Ashcroft, N. W. 1974. Ground state energies of simple metals. *Phys. Rev.* 9B:409–424.

Hubbard, W. B. 1968. Thermal structure of Jupiter. *Astrophys. J.* 152:745–754.

——. 1969. Thermal Models of Jupiter and Saturn. *Astrophys. J.* 155:333–344.

——. 1970. Structure of Jupiter. *Astrophys. J.* 162:687–697.

——. 1973. Observation constraint on the structure of hydrogen planets. *Astrophys. J.* 182: L35–L38.

Hubbard, W. B., and Slattery, W. L. 1971. Statistical mechanics of light elements at high pressure. I. *Astrophys. J.* 168:131–139.

Hubbard, W. B.; Slattery, W. L.; and DeVito, C. 1975. High zonal harmonics of rapidly rotating planets. *Astrophys. J.* 199:504–516.

Hubbard, W. B., and Smoluchowski, R. 1973. Structure of Jupiter and Saturn. *Space Sci. Rev.* 14:599–662.

Ingersoll, A. D.; Münch, G.; Neugebauer, A.; Diner, D. J.; Orton, G. S.; Schupler, B.; Schroeder, M.; Chase, S. C.; Ruiz, R. D.; and Trafton, L. M. 1975. Pioneer 11 radiometer experiment: the global heat balance of Jupiter. *Science* 188:472–473.

Kiefer, H. H. 1967. Calculated physical properties of planets in relation to composition and gravitational layering. *J. Geophys. Res.* 72:3179–3197.

Krumhansl, J.; and Wu, S-W. 1968. Solid molecular hydrogen as a quantum crystal. *Phys. Lett.* 28A:263–264.

Lacy, J. H.; Larrabee, A. I.; Wollman, E. R.; Geballe, T. R.; Townes, C. H.; Bregman, J. D.; and Rank, D. M. 1975. Observations and analysis of the Jovian spectrum in the 10 micron band of NH_3. *Astrophys. J.* 198:L145–L148.

Landau, L., and Lifshitz, E. 1969. *Statistical mechanics* p. 133. Reading, Mass.: Addison-Wesley.

Landau, L.; and Zeldovich, G. 1943. On the relation between the liquid and the gaseous states of metals. *Acta. Phys. Chim. USSR* 18:194–196.

Low, F. J. 1966. Observations of Venus, Jupiter and Saturn at 20 microns. *Astron. J.* 71:391.

McMahan, A.; Beck, H.; and Krumhansl, J. 1974. Short range interaction between hydrogen molecules. *Phys. Rev.* 9A:1852–1864.

Neece, G. A.; Rogers, F. J.; and Hoover, W. G. 1971. Thermodynamic properties of compressed solid hydrogen *J. Comput. Phys.* 7:621–636.

Null, G. W.; Anderson, J. D.; and Wong, S. K. 1975. Gravity field of Jupiter from Pioneer 11 tracking data. *Science* 188:476–477.

Öpik, E. J. 1962. Jupiter: Chemical composition, structure and origin of a giant planet. *Icarus* 1:200–257.

Peebles, P. J. 1964. Structure and composition of Jupiter and Saturn. *Astrophys. J.* 140:328–347.

Podolak, M., and Cameron, A. G. W. 1974. Models of the giant planets. *Icarus* 22:123–148.

——. 1975. Further investigations of Jupiter models. *Icarus* 25:627–634.

Pollock, E. L., and Hansen, J. P. 1973. Statistical mechanics of dense ionized matter. II. *Phys. Rev.* 8A:3110–3122.

Presnall, D. C. 1969. Pressure-volume-temperature measurements on hydrogen from 200°C to 600°C and up to 1800 bar. *J. Geophys. Res.* 74:6026–6033.

Ramaker, D. E.; Kumar, L.; and Harris, F. E. 1975. Exact-exchange crystal Hartree-Fock calculations of molecular and metallic hydrogen *Phys. Rev. Lett.* 34.812–815.

Ree, F. H., and Bender, C. F. 1974. Nonadditive interaction in molecular hydrogen at high pressure. *Phys. Rev. Lett.* 32:85–88.

Ross, M. 1974. A theoretical analysis of the shock compression experiments of the liquid hydrogen isotopes and a prediction of the metallic transition. *J. Chem. Phys.* 60:3634–3644.

Salpeter, E. E. 1973. On convection and gravitational layering in Jupiter and in stars of low mass. *Astrophys. J.* 181:L83–L86.

Salpeter, E. E., and Stevenson, D. J. 1975. Heat transport in a stratified two-phase fluid. Submitted to *Physics of Fluids.*

Salpeter, E. E., and Zapolsky, H. S. 1967. Theoretical high pressure equations of state, including correlation energy. *Phys. Rev.* 158:876–886.

Shafer, R., and Gordon, R. G. 1973. Quantum scattering theory of rotational relaxation in H_2-He mixtures. *J. Chem. Phys.* 58:5422–5443.

Silver, D. M., and Stevens, R. M. 1973. Reaction paths on the H_4 potential energy surface. *J. Chem. Phys.* 59:3378–3394.

Slattery, W. L., and Hubbard, W. B. 1973. Statistical mechanics of light elements at high pressure. III. *Astrophys. J.* 181:1031–1038.

Smoluchowski, R. 1967. Internal structure and energy emission of Jupiter. *Nature* 215:691–695.

――――. 1970. Solid state convection on Jupiter. *Phys. Rev. Lett.* 25:693–695.

――――. 1971. Metallic interiors and magnetic fields of Jupiter and Saturn. *Astrophys. J.* 166: 435–439.

――――. 1973. Dynamics of the Jovian interior. *Astrophys. J.* 185:L95–L99.

Spiegel, E. A. 1971. Convection in stars. I. Basic Boussinesq convection. *Ann. Rev. Astron. Astrophys.* 9:323–352.

Stevenson, D. J. 1975. Thermodynamics and phase separation of dense fully-ionized hydrogen-helium fluid mixtures. *Phys. Rev. B* (Nov. 15).

Stevenson, D. J., and Ashcroft, N. W. 1974. Conduction in fully-ionized liquid metals *Phys. Rev.* 9A:782–789.

Streett, W. B. 1974. Phase equilibria in molecular H_2-He mixtures at high pressures. *Astrophys. J.* 186:1107–1125.

Trafton, L. M. 1967. Model atmospheres of the major planets. *Astrophys. J.* 147:765–781.

Trubitsyn, V. P. 1967. Equation of state of solid helium at high pressures. *Sov. Phys. Solid State* 8:2593–2598.

――――. 1972. Adiabatic model for Jupiter. *Soviet Astron.-A.J.* 16:342–347.

Wainwright, T., and Alder, B. 1958. Molecular dynamics computations for the hard sphere system. *Il Nuovo Cimento Suppl.* 116–132.

Wigner, E. 1932. On the quantum correction for thermodynamic equilibrium. *Phys. Rev.* 40: 749–759.

Zapolsky, H. S., and Salpeter, E. E. 1969. The mass-radius relationship for cold bodies of low mass. *Astrophys. J.* 158:809–813.

Zharkov, V. N.; Makalkin, A. B.; and Trubitsyn, V. P. 1973. Integration of equations of the theory of planetary figures. *Soviet Astron.-A.J.* 17:97–104.

――――. 1975. Models of Jupiter and Saturn. *Soviet Astron.-A.J.* 18:768–774.

Zharkov, V. N., and Trubitsyn, V. P. 1974. Determination of the equation of state of the molecular envelopes of Jupiter and Saturn from their gravitational moments. *Icarus* 21:152–156.

THE GRAVITY FIELD OF JUPITER

J. D. ANDERSON
Jet Propulsion Laboratory

Preliminary analysis of two-way Doppler data from Pioneers 10 and 11 has provided the first detailed model of the Jovian gravity field. A review of the determination of the zonal harmonic coefficients through the sixth degree is presented, and the results are used to derive a number of geodetic parameters in the atmospheric region of the planet. On a level surface at a pressure of one bar, the net acceleration of gravity g is found to vary from a maximum of 2707 cm sec^{-2} at the poles to a minimum of 2322 cm sec^{-2} at the equator. The large dynamical flattening $(a - b)/a$ of 0.064841 at the one-bar level produces a significant deviation of the local vertical from the Jovicentric radius vector. The angular difference is as much as 3.83 degrees of arc in the high temperate zones of the planet. These considerations are important for an accurate modeling of the atmosphere of Jupiter and for an interpretation of occultation data.

The first objective of this chapter is to provide an intuitive notion for the determination of the gravity fields of the giant planets from the precise Doppler tracking of flyby spacecraft. This is accomplished in Sec. I by appealing to the law of conservation of energy. The second objective is to review previously published results on the gravity field of Jupiter as determined by Pioneers 10 and 11, and to use those results to derive geodetic information at the surface of the planet.

Of course, the planetary gravity information available to us at any time depends on the observations of the orbital motions of bodies that are significantly affected by the gravity field of the planet. In this sense it is not totally unexpected that coherent Doppler data from a flyby of Pioneer 11 at 1.62 Jupiter radii (R_J), yields more information than many years of astrometric data on the natural satellites at greater distances. Similarly, we might expect even higher resolution in the gravity field by means of the long-term Doppler tracking of planetary orbiters with periapsis distances comparable to those of Pioneers 10 and 11. However, this is a subject for future outer planet missions. For now we must concentrate on the Doppler data obtained from the two Pioneer spacecraft as our best source of information on the Jovian gravity field.

[113]

J. D. ANDERSON

I. DATA AND TRAJECTORY CHARACTERISTICS

Data were obtained from both Pioneer 10 and 11 by the three 64-m antennas of the NASA/JPL Deep Space Network at Goldstone, California, Canberra, Australia, and Madrid, Spain. This allowed almost continuous two-way tracking of the spacecraft during the encounter period. Typically, the rms Doppler noise, as determined by residuals from least-squares fits, was 0.005 Hz (0.3 mm sec^{-1}) for an integration time of 60 sec in the cycle count. The only interruption of the two-way Doppler occurred during a period of about one hour for each spacecraft during their respective occultations by Jupiter.

For Pioneer 10 the immersion into occultation occurred about one hour after closest approach at 2.8 R_J. For Pioneer 11, the closest approach of 1.62 R_J occurred near the center of the occultation period. However, two-way data were obtained up to 1.75 R_J on the incoming leg of the Pioneer 11 trajectory and were picked up again at 2.0 R_J on the out-going leg. Therefore, Pioneer 11 was tracked at closer distances to Jupiter than Pioneer 10, and hence the gravity field determined with the Pioneer 11 data is more accurate. Also, the trajectory of Pioneer 11 was retrograde and much more inclined to the equator of Jupiter. Not only was the gravity field probed at a closer distance, but also at higher latitudes in both hemispheres and over a wider range of longitudes.

An idea of the importance of the trajectory to a determination of the gravity field can be obtained as follows. Consider a two-body system (Jupiter plus spacecraft), and write down the law of conservation of energy with the kinetic energy $\frac{1}{2} v^2$ on the left hand side and with the total energy ϵ and potential function Φ on the right:

$$\frac{1}{2} v^2 = \epsilon + \Phi (r, \phi') . \tag{1}$$

In general, Φ will be expressed as an infinite series of spherical harmonics of all degrees and orders. However, for purposes of illustration, assume that Jupiter is in hydrostatic equilibrium so that all harmonics except the even zonals are zero. Then

$$\Phi (r, \phi') = \frac{GM}{r} \left[1 - \sum_{i=1}^{\infty} J_{2i} \left(\frac{a}{r} \right)^{2i} P_{2i} (\mu) \right] \tag{2}$$

where M is the mass of the planet, a is its equatorial radius, and the coefficients J_{2i} describe the deviations of the gravitational field from spherical symmetry. The radial coordinate of a point in space is given by r, and μ is equal to the sine of the Jovicentric latitude ϕ' of the same point.

Now, at a very great distance from Jupiter, the velocity of the spacecraft v is determined almost exclusively by the constant ϵ. However, as the spacecraft approaches the planet, the function Φ becomes more and more important in the determination of v. In fact, small errors in the gravity coefficients J_{2i} will propagate into the spacecraft velocity in the following way:

$$v \, \Delta v = - \, GM \sum_{i = 1}^{\infty} \frac{a^{2i} \, \Delta J_{2i}}{r^{2i + 1}} P_{2i} \, (\mu) \, . \qquad (3)$$

The projection of this velocity perturbation onto the line of sight is measured by the Doppler system, and hence the harmonic coefficients J_{2i} can be determined. Equation (3) is a useful equation for purposes of studying the effects of the trajectory on the determination of the coefficients. In effect, we want to maximize the sensitivity of the velocity to variations ΔJ_{2i}, and at the same time we want the sensitivity coefficients for each harmonic to be as different in character as possible, so as to minimize the correlations between the coefficients. It is easy to see that both these ends are met by making the coordinate r as small as possible; the sensitivity coefficient is maximized and the differing power laws in r for each harmonic are emphasized. Detailed covariance analyses have shown that it is this differing power law that is most effective in reducing the correlations. The effect of a large inclination to the equator, which results in a wider range in $P_{2i} \, (\mu)$ near closest approach is important, but not nearly as important as having a close flyby to emphasize the different response of the velocity to various harmonics as a function of distance. Of course, this argument does not hold for the terrestrial planets and the satellites, because they support large deviations from hydrostatic equilibrium in their surface layers, and hence it is important to probe the entire gravitational field of the planet in order to locate local gravity variations. For flybys of Jupiter and Saturn, however, the most important consideration is to approach the surface of the planet as closely as possible.

Pioneer 10 has flown close enough to Jupiter to provide the first definitive measurement of J_4 and to improve the accuracy of J_2. Pioneer 11, by flying even closer, has provided an even more precise value of J_4 and has given us some precursory information on J_6.

II.   RESULTS

The zonal harmonic coefficients J_2, J_3, J_4, and J_6 in Jupiter's gravity field are given in Table I for Pioneers 10 and 11 for an assumed equatorial radius of 71,398 km. These results and the details of the data analysis which produced them, have been published previously (Anderson et al. 1974b; Null et al. 1975). The small values obtained for J_3 and J_6 from Pioneer 11 are significant in that they are consistent with the assumption that Jupiter is in hydrostatic equilibrium at all levels. Also, no gravity anomalies have been detected on Jupiter, although at a flyby distance on the order of 2 R_J only large scale anomalies would be detectable above the Doppler noise levels of 0.005 Hz. Current fits to Pioneer 10 and Pioneer 11 data are bounded

TABLE I

Jupiter Gravity Harmonics from an analysis
of Doppler Data from Pioneers 10 and 11.
Values are based on an assumed equatorial
radius of 71,398 km (Null et al. 1975)

Coefficient $\times\ 10^6$	Pioneer 10	Pioneer 11
J_2	$14{,}720 \pm 40$	$14{,}750 \pm 50$
J_3	< 150	10 ± 40
J_4	-650 ± 150	-580 ± 40
J_6	Assumed Zero	50 ± 60

within the limits of ± 0.01 Hz over the several days that the data are fit. No anomalies, gravity or otherwise, are evident. Nevertheless, because the analysis of spacecraft gas leaks and other possible systematic errors is incomplete, conservative standard errors have been assigned to the Pioneer 11 results. Values given in Table I are considered realistic errors at this time. In the future the accuracy of the Pioneer 11 results will be improved, but the Pioneer 10 results probably will not change very much.

Values for the masses of the four Galilean satellites have been obtained from Pioneer 10 (Anderson *et al.* 1974).[1] It is doubtful that Pioneer 11 will add any fundamentally new information on satellite masses, with the possible exception of Callisto. The mass of Jupiter itself can be determined from both Pioneers 10 and 11, but the realistic error in the result is comparable to the realistic error from analyses of satellite and asteroid data. The current value for the ratio of the mass of the sun to the mass of Jupiter from Pioneer is 1047.342 ± 0.02 (Anderson *et al.* 1974), which is consistent with the value from natural bodies (Greenberg[2]).

The results for the harmonic coefficients J_2 and J_4 from Pioneer can be compared to advantage with results from the natural satellites, even though independent values of J_2 and J_4 are not available from the satellites to an accuracy comparable with Pioneer. However, the precession of the node of Jupiter's innermost satellite J V places a very tight constraint on J_2 and J_4 (Brouwer and Clemence 1961). This constraint is plotted in Fig. 1 along with the values of J_2 and J_4 from Pioneers 10 and 11. Recently, the effect of the Galilean satellites on the motion of the node of J V has been taken into account (Chao 1975; Greenberg[3]). This will change the constraint computed from Brouwer and Clemence's analysis slightly, but we plot their older curve here for purposes of illustration. The point of the figure is that the determi-

[1]See p. 994.
[2]See p. 123.
[3]See p. 127.

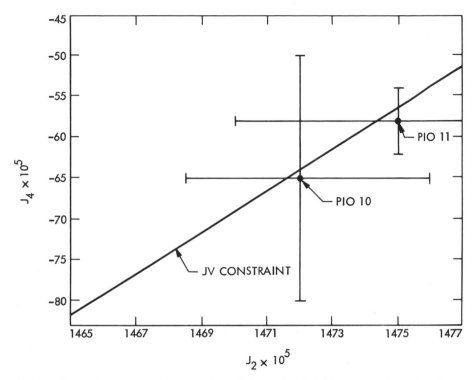

Fig. 1.   Constraint on the zonal harmonic coefficients J_2 and J_4 imposed by the motion of the node of J V, the innermost satellite of Jupiter. The solutions for J_2 and J_4 from Pioneer 10 and Pioneer 11 are shown with their respective realistic one sigma error bars.

nation of the Jovian gravity field is completely consistent with the orbital motion of J V, even though no attempt has been made to assure this consistency.

III.   GEODETIC PARAMETERS

There are a number of geodetic parameters which can be derived from the Jovian gravity field. These parameters are important to atmospheric modeling and to the analysis and interpretation of occultation data. Accurate model atmospheres at all latitudes are of theoretical interest, and also are important to the design of entry probes for possible future missions to Jupiter.

For purposes of planetary geodesy on Jupiter, it is satisfactory at present to assume hydrostatic equilibrium and to write the gravitational potential at the surface of the planet as follows (Heiskanen and Moritz 1967):

$$V(r, \phi') = \frac{GM}{r} \left\{ 1 - \sum_{i=1}^{\infty} J_{2i} \left(\frac{a}{r}\right)^{2i} P_{2i}(\mu) + \frac{q}{3} \left(\frac{r}{a}\right)^3 [1 - P_2(\mu)] \right\}. \quad (4)$$

This is the same function as given by Eq. (2), but now a rotation term has been added which is characterized by the small parameter q. In terms of the angular velocity of rotation ω of the planet, the parameter q is defined by

$$q = \frac{\omega^2 \, a^3}{GM} . \tag{5}$$

It is equal to 0.08885 for a rotation rate given by System III ($9^h \, 55^m \, 29\overset{s}{.}75$) and for an equatorial radius a of 71,400 km. Now in Eq. (4), the parameter r refers to the surface of the planet along which the sum of the gravitational and rotational potential is constant. At the equator ($\mu = 0$), the radius $r = a$, while at the poles ($\mu = 1$), $r = b$, the polar radius. Even though a/r is nearly unity, the series for the gravitational potential converges fairly rapidly because the harmonic coefficients J_{2i} are of order q^i.

The radius at any latitude can be expressed as a series in Legendre polynomials:

$$r = b \left[1 + \sum_{i=0}^{\infty} D_{2i} \, P_{2i} \, (\mu) \right] \tag{6}$$

where D_0 and D_2 are both of order q and the remaining D_{2i} are of order q^i. Explicit expressions for the coefficients D_0, D_2, D_4, and D_6 have been developed as functions of q, J_2, J_4, and J_6 to the third order in q (Zharkov and Trubitsyn 1970).[4] The resulting expression for r given by Eq. (6) will then yield an equipotential surface along which $V(r, \phi')$ is equal to a constant $V(b, 90°)$ to third order in q. Recently, a new fourth order theory for planetary figures has been developed (Hubbard et $al.$ 1975).[5] This permits a calculation of $r(\mu)$ to order q^4, which for Jupiter results in a potential which is constant to one part in 10.

An important parameter for atmospheric modeling is the acceleration of gravity g. On the level surface defined by $r(\mu)$, this acceleration will be normal to the surface and will have two components g_r and $g_{\phi'}$. These components are in the plane of the meridian along the radial and transverse directions, respectively. They are given by

$$g_r = \frac{\delta V}{\delta r} = -\frac{GM}{r^2} \left\{ 1 - \sum_{i=1}^{\infty} (1 + 2i) \, J_{2i} \left(\frac{a}{r} \right)^{2i} P_{2i} \, (\mu) \right.$$

$$\left. - \frac{2}{3} q \left(\frac{r}{a} \right)^3 [1 - P_2 \, (\mu)] \right\} \tag{7}$$

and

---

[4]See p. 170.
[5]See also p. 181.

$$g_{\phi'} = \frac{1}{r}\frac{\delta V}{\delta\phi'} = -\frac{GM}{r^2}\cos\phi'\left\{\sum_{i=1}^{\infty} J_{2i}\left(\frac{a}{r}\right)^{2i} P'_{2i}(\mu) + \frac{1}{3}q\left(\frac{r}{a}\right)^3 P'_2(\mu)\right\}. \quad (8)$$

The angle α between the local vertical to the level surface and the geocentric radius is important to an interpretation of occultation data (Kliore and Woiceshyn[6]). In geodetic terms it is the difference between the Jovicentric latitude ϕ' and the geodetic latitude ϕ, and for $g_{\phi'} \ll g_r$ it is given simply by the ratio of $g_{\phi'}$ and g_r;

$$\tan(\phi - \phi') = \frac{g_{\phi'}}{g_r}. \quad (9)$$

Also,

$$g = (g_r^2 + g_{\phi'}^2)^{\frac{1}{2}}. \quad (10)$$

The new fourth-order theory has been applied to a calculation of D_0, D_2, D_4, D_6, and D_8 from the gravity field determined by Pioneer 11. The calculation has been carried out for values of the equatorial radius ranging from 71,400 km to 72,000 km, corresponding to various heights in the atmosphere of Jupiter. The best current estimate for the equatorial radius corresponding to a pressure in the atmosphere of 1 bar is 71,400 from the occultation of β Scorpii (Hubbard and Van Flandern 1972; Anderson et al. 1974a); this number probably will be refined in the future by the final analysis of the Pioneer radio occultation data.

TABLE II
Radius r, Acceleration of Gravity g, and
Angle of the Local Vertical $\alpha = \phi - \phi'$
as a Function of μ on an Equipotential
Surface in the Atmosphere of Jupiter

$\mu = \sin\phi'$	r (km)[a]	g (cm sec^{-2})	α (deg)
0.0	71,400.0	2322.21	0.000
0.1	71,347.3	2326.70	0.840
0.2	71,190.3	2340.07	1.644
0.3	70,931.5	2362.05	2.375
0.4	70,575.3	2392.21	2.996
0.5	70,127.5	2430.00	3.472
0.6	69,595.0	2474.75	3.764
0.7	68,985.4	2525.72	3.823
0.8	68,306.7	2582.12	3.573
0.9	67,566.2	2643.00	2.840
1.0	66,770.3	2707.13	0.000

[a]An equatorial radius of 71,400 km is used.

[6]See p. 217.

The calculation of the geodetic parameters is displayed in Table II as a function of μ for an equatorial radius of 71,400 km. Similar calculations at other radii have been made, and the numerical results have been fit in the least-squares sense by simple polynomials. These expressions are accurate to five or six figures in the interpolation region of 71,400 km to 72,000 km, and are probably fairly accurate for a few hundred kilometers outside of this region. Current uncertainties in the gravity field of Jupiter limit the accuracy of the two components of the acceleration of gravity (g_r, $g_{\phi'}$) to about 0.5 cm sec^{-2}, so that a five-place calculation is sufficient.

The polynomials which fit the numerical data are given as follows, with the altitude parameter H, in km, defined by $H = a - 71,400$:

$$f = \frac{a - b}{a} = [64840.8 + 1.0831\ H$$
$$+ 3.52 \times 10^{-5}\ H^2 - 1.0 \times 10^{-9}\ H^3] \tag{11}$$

$$r = b \left[1 + \sum_{i=0}^{4} D_{2i}\ P_{2i}\ (\mu) \right] \tag{12}$$

with

$$D_0 = (45041.8 + 0.7833\ H) \times 10^{-6}$$

$$D_2 = -\ (46967.4 + 0.8781\ H) \times 10^{-6}$$

$$D_4 = (2053.7 + 0.1013\ H) \times 10^{-6}$$

$$D_6 = -\ (127.3 + 0.0078\ H) \times 10^{-6}$$

$$D_8 = -\ (1.2 - 0.0013\ H) \times 10^{-6}$$

and the polynomials for g_r and $g_{\phi'}$ are

$$g_r = \sum_{i=0}^{6} \alpha_{ri}\ \mu^i + H \sum_{i=0}^{6} \beta_{ri}\ \mu^i \tag{13}$$

$$g_{\phi'} = \cos\phi' \left[\sum_{i=0}^{6} \alpha_{\phi i}\ \mu^i + H \sum_{i=0}^{6} \beta_{\phi i}\ \mu^i \right] \tag{14}$$

where in units of cm sec^{-2},

$$\alpha_{ri} = -\ 2322.15, -\ 0.36, -\ 418.41, -\ 31.82, 126.64, -\ 98.00, 37.03$$

$$\beta_{ri} \times 10^3 = 75.10, 0.03, -\ 12.84, 2.55, -\ 2.73, 7.19, -\ 3.88$$

$$\alpha_{\phi i} = 0.00, -\ 343.02, -\ 2.83, 29.09, -\ 42.02, 55.66, -\ 30.05$$

$$\beta_{\phi i} \times 10^3 = 0.00, 3.98, 0.25, -\ 2.46, 3.75, -\ 4.96, 2.85$$

and $i = 0, 1, 2, \cdots 6$.

Thus, from a given value of a or H in units of km, the dynamical flattening f and hence the polar radius can be computed from Eq. (11), and the radius for any value of $\mu = \sin \phi'$ can be computed from Eq. (12). The components g_r and $g_{\phi'}$ for the acceleration of gravity can be computed from Eqs. (13) and (14), and then $\alpha = \phi - \phi'$ and g can be computed from Eqs. (9) and (10), respectively.

Acknowledgements. This paper presents the results of one phase of research carried out at the Jet Propulsion Laboratory, California Institute of Technology, under NASA Contract NAS 7-100. I would like to thank W. B. Hubbard for several helpful discussions on the calculations of geodetic parameters and for the use of his computer subroutine *SHAPE*. I would also like to acknowledge the two Co-Investigators G. W. Null and S. K. Wong on the Pioneer Celestial Mechanics Experiment, who played a major role in obtaining results, and E. K. Lau who provided valuable mathematical analysis and computer support for the experiment.

DISCUSSION

R. Smoluchowski: Hydrostatic equilibrium requires that the directly observed oblateness be equal to the oblateness calculated from the gravitational coefficients. Is this the case here?

J. D. Anderson: Yes, the agreement is very good.

REFERENCES

Anderson, J. D.; Hubbard, W. B.; and Slattery, W. L. 1974a. Structure of the Jovian envelope from Pioneer 10 gravity data. *Astrophys. J.* 193:L149.

Anderson, J. D.; Null, G. W.; and Wong, S. K. 1974b. Gravity results from Pioneer 10 Doppler data. *J. Geophys. Res.* 79:3661–3664.

Brouwer, D., and Clemence, G. M. 1961. Orbits and masses of planets and satellites. *Planets and satellites.* (G. P. Kuiper and B. M. Middlehurst, eds.) Ch. 3, pp. 70–73. Chicago: Univ. of Chicago Press.

Chao, C. C. 1975. A general perturbation method and its application to the motions of the four massive satellites of Jupiter. p. 75. Ph.D. Dissertation. University of California at Los Angeles.

Heiskanen, W. A., and Moritz, H. 1967. *Physical geodesy.* San Francisco: W. H. Freeman Publ.

Hubbard, W. B., and Van Flandern, T. C. 1972. The occultation of Beta Scorpii by Jupiter and Io. III. Astrometry. *Astron. J.* 77:65–74.

Hubbard, W. B.; Slattery, W. L.; and De Vito, C. L. 1975. High zonal harmonics of rapidly rotating planets. *Astrophys. J.* 199:504–516.

Null, G. W.; Anderson, J. D.; and Wong, S. K. 1975. Gravity field of Jupiter from Pioneer 11 tracking data. *Science* 188:476–477.

Zharkov, V. N., and Trubitsyn, V. P. 1970. Theory of the figure of rotating planets in hydrostatic equilibrium — a third approximation. *Sov. Astron. — A.J.* 13:981–988.

THE MOTIONS OF SATELLITES AND ASTEROIDS: NATURAL PROBES OF JOVIAN GRAVITY

R. J. GREENBERG
University of Arizona

Before the recent Pioneer probes, our knowledge of Jupiter's gravitational field was obtained from the motions of satellites and asteroids. The study of orbital perturbations of asteroids near the 2:1 commensurability yielded a value of the mass of the Jupiter system at least as precise as that obtained by the artificial probes. Precession of the inner satellites' orbits placed constraints on the harmonic coefficients J_2 and J_4. A new correction to the satellite determination of J_4 lowers its mean value closer to the Pioneer result. The orbital grouping among the outer satellites and the resonance among the Galilean satellites are described in detail but the origins of these phenomena are not understood. However, recent research suggests that the explanation will be intimately associated with models of the origin and evolution of the planet itself.

In the midst of the deluge of data returned by Pioneers 10 and 11, we should recall the importance of natural bodies as probes of the Jovian environment. Such a review is not merely a sentimental exercise. Several examples show that information from the satellites and asteroids is by no means obsolete. The value of Jupiter's mass as determined from the behavior of certain asteroids is at least as precise as the Pioneer determination. A number of properties of the satellites' orbits offer clues to the long-term environment in which they have evolved. And as the technology of Earth-based observation continues to develop, the natural satellites may again play a role in the study of Jupiter's gravitational field.

In this paper, I shall describe the dynamical properties of satellites and asteroids under the influence of Jupiter, and review the state of theoretical interpretation of this behavior. Areas in which future work may yield significant results shall also be discussed. The emphasis shall be on applications to the study of Jupiter, rather than on detailed analysis of the dynamics of satellites and asteroids. For such details, the reader is referred to Gehrels (1971) and Burns (1976).

[122]

I. ASTEROIDS

The asteroid belt spans distances from the sun at which orbital periods form ratios of small whole numbers with Jupiter's period. If an asteroid is near one of these commensurabilities, the points in its orbit at which conjunctions with Jupiter occur will vary slowly. Thus the perturbing effects of Jupiter will be enhanced. Such resonances are presumed to be responsible for the Kirkwood gaps and for certain concentrations in the asteroid belt which correspond to various commensurabilities (Brouwer 1963). The mechanism responsible for these gaps and concentrations probably involves collisional properties of the asteroids (Wiesel 1974). It is thus outside the subject of this volume.

However, the existence of resonances in the asteroid belt does provide important information about Jupiter: By studying the orbits of individual asteroids which experience the enhanced perturbations, Jupiter's mass can be derived. For example, Klepczynski (1969) used the motions of four minor planets near the 2:1 commensurability to obtain the value 1047.360 ± 0.004 for M_S/M_J, the ratio of the mass of the sun to that of the Jupiter system. Duncombe et al. (1974) have combined all asteroid determinations to obtain a value of 1047.364 ± 0.005. They note that the scatter among the individual determinations is so great that more realistic error brackets are five times as large as this formal value. These results are compared with the Pioneer determination, 1047.342 ± 0.02, by Anderson.[1]

II. OUTER SATELLITES

Proceeding closer to Jupiter, we next consider the outer satellites. These bodies are in highly irregular orbits, with substantial eccentricities and inclinations (see Table I). They each experience substantial solar perturbations. However, they divide into two distinct groups. Group A has direct orbits and semi-major axes of 1.1 to 1.2×10^7km and group B has retrograde orbits and semi-major axes of 2.1 to 2.4×10^7km. For comparison, the radius of Jupiter's sphere of influence is $\sim 4.5 \times 10^7$km. Since the discovery of J XIII (Kowal et al. 1975), there are now four known satellites in each group.

The origin of these two groups is still not understood. It is quite likely that the explanation will be intimately interwoven with the history of the planet. Kuiper (1956) suggested that each group consists of the fragments of a comet-like body which broke up when it encountered a gaseous envelope around the early planet. Morrison and Burns[2] discuss the observed physical properties of the satellites which bear on the plausibility of this scenario.

Colombo and Franklin (1971) also attributed the formation of the groups to the break-up of larger bodies. They considered a model which involves one event, a collison of two bodies within Jupiter's sphere of influence. They

---

[1]See p. 116.
[2]See p. 1026.

TABLE I
Dynamical Properties of Jupiter's Satellites[a]

	Satellite	Semi-major Axis (km)	Sidereal Period (days)	Eccentricity	Inclination To Equator (degrees)	Year Discovered	Discoverer
V	Amalthea	1.81×10^5	0.4982	0.003	0.4	1892	Barnard
I	Io	4.22×10^5	1.769	0.000	0.0	1610	Galileo[b]
II	Europa	6.71×10^5	3.551	0.000	0.5		
III	Ganymede	1.07×10^6	7.155	0.001	0.2		
IV	Callisto	1.88×10^6	16.69	0.01	0.2		
XIII	Leda	1.11×10^7	239	0.147	26.7	1974	Kowal
VI	Himalia	1.15×10^7	250.6	0.158	27.6	1904	Perrine
VII	Elara	1.17×10^7	259.7	0.207	24.8	1905	
X	Lysithea	1.19×10^7	263.6	0.130	29.0	1938	Nicholson
XII	Ananke	2.12×10^7	631.1	0.169	147	1951	
XI	Carme	2.26×10^7	692.5	0.207	164	1938	
VIII	Pasiphae	2.35×10^7	738.9	0.378	145	1908	Melotte
IX	Sinope	2.37×10^7	758	0.275	153	1914	Nicholson

[a]After Wilkins and Sinclair (1974) with additions and corrections. *Note:* The orbits of the outer satellites vary considerably. The eccentricities and inclinations of the Galilean satellites are also variable; proper elements from Marsden (1966) are given here. Elements for J XIII from Kowal *et al.* (1975).

[b]Simon Marius (or Mayer) may have observed the major satellites contemporaneously with Galileo.

could find no obvious objections to such a hypothesis. Long-term orbital integration showed that, going back in time, it is possible for all members of group B to penetrate into the region of group A. (The orbital variation is due to solar perturbations combined with variation of Jupiter's orbit due to the influence of other planets.) Energy and angular momentum considerations are not violated by reasonable relative velocities and masses of the colliding bodies. Moreover, the velocity dispersion within each present group is comparable (~ 100m/sec) to that within asteroid families which are believed to result from collisions. Colombo and Franklin found their hypothesis to be most reasonable for a collision between a satellite and an asteroid, although a collision between two satellites or two asteroids could not be ruled out. They also concluded that, given the proximity of group B to the boundary of the sphere of influence and the long-term variation of orbits, group B may represent the low energy particles of an original group which largely escaped the planet.

The hypotheses by Kuiper and by Colombo and Franklin both involve break-up of larger bodies. Thus both models led to predictions by these authors that many smaller undetected fragments exist in each group. Pioneer observations, by the imaging system and the small particle detectors, have been inconclusive on this point. However, the discovery of J XIII has borne out the predictions.

Bailey (1971a, 1971b, 1972) suggested an alternative explanation for the origin of the outer satellites: Objects originally in solar orbit moved with nearly zero relative velocity through the Sun-Jupiter inner Lagrangian (L_2) point into Jupiter's sphere of influence. The L_2 point is analogous to a potential barrier between solar orbits and orbits bound to Jupiter. If a particle barely surmounted this obstacle, Bailey apparently reasoned, it would not return. The L_2 point itself has zero velocity relative to the sun and planet only when Jupiter is at aphelion or perihelion. Thus, Bailey supposed, capture could only have occurred at those times. By a confusing treatment of the elliptic three-body problem, he claimed to have shown that a particle captured while Jupiter is at aphelion would have the same Jacobi integral (one of the integrals of motion) as a planetary satellite in a direct orbit of semi-major axis 1.15×10^7km. Similarly, a particle captured when Jupiter is at perihelion would have the same Jacobi integral as a retrograde satellite with semi-major axis 2.17×10^7km. The numerical agreement with the observed orbits in groups A and B lent striking support to the notion that the satellites were divided into groups by capture at either aphelion or perihelion.

There appear to be a number of conceptual flaws in Bailey's argument. Davis (1974) noted that Bailey's theory accurately predicts the orbits of groups A and B only if Jupiter had its present orbital eccentricity (0.048) at the time of capture. Since the eccentricity varies between about 0.03 to 0.06 on a time scale of about 5×10^4yr due to the effects of other planets, capture of the present outer satellites appears to be confined to the past 10^4yr. But the numerical integration by Colombo and Franklin showed that the satellites have been bound to Jupiter over this period. Thus Bailey's theory breaks down. Another flaw noted by Davis involves the notion of capture through the L_2 point. Unless some energy is dissipated, a particle once crossing this potential barrier into Jupiter orbit might very well escape again into heliocentric orbit. Bailey gave no consideration to how long a satellite would remain captured. Another objection to the Bailey theory is that no convincing explanation was given for restricting the possibility of capture to those instants when Jupiter was at aphelion or perihelion. This choice seems to be a matter of calculational convenience rather than a physical necessity.

Besides these conceptual problems with the Bailey hypothesis, Heppenheimer (1975) has pointed out several analytical errors. His reworked analysis overcomes some of Davis' objections. Heppenheimer found that the predicted size of satellite orbits does not depend on Jupiter's eccentricity after all. The size does depend on M_S/M_J. Capture of group A satellites would require $M_S/M_J = 1730$. Heppenheimer suggested that a decrease of M_S/M_J to its present value, through growth of the planet or mass loss by the sun, might have prevented subsequent escape of the group A satellites. He found that the corrected analysis does not yield capture of the retrograde satellites. Since the only evidence in favor of Bailey's entire approach had been the agreement with observed orbits, this type of capture does not seem viable as an explanation of the grouping of the outer satellites.

A basic difficulty with Bailey's capture model, even after correction by Heppenheimer, is the use of only one integral of the motion to identify the present orbits with hypothesized original orbits. In fact, integration of motion from originally heliocentric orbit, through capture and into Jovicentric orbit, has been performed by Everhart (1973). He found that, with no dissipation mechanism, capture is rare and temporary. Moreover, capture orbits do not pass near the L_2 point.

We are left with two plausible hypotheses for the origin of the outer satellite groups: (1) capture and break-up in a gaseous envelope (Kuiper 1956) and (2) a collision within Jupiter's sphere of influence (Colombo and Franklin 1971). Resolution of the problem will require further studies of the dynamics of these models combined with improved understanding of the planet's past.

The satellite JIX has been used to obtain a value of Jupiter's mass. De Polavieja and Edelman (1972) employed statistical filtering to compare observed positions with numerical integration of the orbit. They obtained $M_S/M_J = 1047.352 \pm 0.002$. However, they strongly emphasized that their formal error brackets represent internal consistency of their solution but are not a good measure of the precision of the mass determination. Nevertheless, Duncombe et al. (1974) weighted this determination very heavily when they combined all determinations based on satellite and planet motions with the asteroid results (Sec. I). Their combined result is thus 1047.357 ± 0.005, where again we must note their caveat that the true value probably lies within error brackets five times as large as these.

III.   REGULAR SATELLITES

The five inner satellites have nearly circular, equatorial orbits with a fairly orderly progression of semi-major axes (Table 1). The regularity strongly suggests that these satellites formed in a nebula around the planet, analogous to the solar nebula which yielded the planets (e.g., Kuiper 1956; Cameron and Pollack[3]). The motions of the inner satellites also provide information on the harmonics of Jupiter's gravitational field.

The oblateness of a planet causes the pericenter and node of a satellite's orbit to precess at the following respective rates, in addition to the precession caused by other planets and the sun:

$$\frac{d\tilde{\omega}}{dt} = + n\left(\frac{JR^2}{a^2} - \frac{1}{2}\frac{J^2R^4}{a^4} + \frac{KR^4}{a^4}\right) \tag{1}$$

$$\frac{d\Omega}{dt} = - n\left(\frac{JR^2}{a^2} - \frac{3}{2}\frac{J^2R^4}{a^4} + \frac{KR^4}{a^4}\right) \tag{2}$$

where $\tilde{\omega}$ is longitude of pericenter, Ω is longitude of the node, R is the planet's equatorial radius, a is the orbital semi-major axis, n is the mean motion, and terms of second order in eccentricity have been neglected. The coefficients J and K are $(3/2)J_2$ and $-(15/4)J_4$, respectively, where J_2 and J_4 are the Legendre coefficients in the spherical harmonic expansion of the planet's gravitational field.

Theoretically, measurement of $d\tilde{\omega}/dt$ and $d\Omega/dt$ for one satellite would allow solution of (1) and (2) for J and K. In practice, however, the J^2-term is so small that for no single satellite are the precession rates measurable to sufficient precision to obtain solutions. For Amalthea (J V), the satellite most sensitive to Jovian oblateness and least perturbed by other influences, $d\Omega/dt = n(-0.0034653 \pm 0.0000014)$ (van Woerkom 1950). Because of the small eccentricity of J V, $d\tilde{\omega}/dt$ has not been determined. It is necessary to use, in addition, observations of other satellites to separate the values of J and K. De Sitter (1931) determined J from the Galilean satellites, which are far enough from the planet that K-terms are negligible. He found $J = 0.02206 \pm 0.00022$ or $J_2 = (14.71 \pm 0.14) \times 10^{-3}$. Brouwer and Clemence (1961) substituted this value into Eq. (2) and, using the known values of $d\Omega/dt$ and R/a for J V, obtained $K = 0.00253 \pm 0.00141$ or $J_4 = (-6.7 \pm 3.8) \times 10^{-4}$. For comparison of these values of J_2 and J_4 with Pioneer results, see the chapter by Anderson.[4]

In using Eq. (2) to evaluate K, Brouwer and Clemence ignored the effect of the Galilean satellites on the precession of J V. Incorporation of this effect into Eq. (2) yields $J_4 = (-6.2 \pm 3.8) \times 10^{-4}$. This new mean value is in strik-

---

[3]See p. 79.
[4]See p. 116.

ing agreement with the Pioneer results, although the error brackets are large.

Sudbury (1969) obtained a more recent determination of $d\Omega/dt$ for J V using data obtained in 1967. He found that, during the periods 1892–1918 and 1949–1967, $d\Omega/dt = -0.003459n$. Use of this value rather than van Woerkom's would yield $J_4 = -5.5 \times 10^{-4}$. However, it is apparent that the precession did not continue at this slow rate during the interval 1918–1949 when no usable observations were made. Sudbury offered two interpretations of the irregular precession: (1) The rate undergoes periodic variation about a value close to van Woerkom's or (2) the orbit suffered a single sudden perturbation. He noted that if sufficient observations are made during the present decade it will be possible to distinguish between (1) and (2). If either of these interpretations is confirmed, theorists may have to find more plausible explanations than Sudbury's. He suggested that (1) might be due to variation in Jupiter's rotation rate or (2) might have resulted from a close encounter with a body of mass comparable to J V. I must note that I have found a sign error in Sudbury's work which indicates that the precession for the periods 1892–1918 and 1949–1967 is faster, not slower than van Woerkom's value. However, the irregularity of the precession rate still needs to be explained.

Besides providing reasonable pre-Pioneer values of the harmonic coefficients, the motions of the inner satellites also permitted determination of the orientation of the planet's gravitational field. In the reduction of Pioneer data, the nominal orientation of the pole was taken from Sampson's (1921) theory of the Galilean satellites. It appears to be accurate to within $0°.1$ (Anderson *et al.* 1974).

The process of extracting information from the motions of the Galilean satellites is complicated by their strong mutual perturbations. These mutual effects are enhanced by the commensurabilities among orbital periods. The ratios of orbital periods are nearly 1:2 for the pairs J I–J II and J II–J III and approximately 3:7 for the pair J III–J IV. Moreover, the mean motions obey the Laplace relation, $n_\mathrm{I} - 3n_\mathrm{II} + 2n_\mathrm{III} = 0$, so that the combination of mean longitudes $\Theta \equiv \lambda_\mathrm{I} - 3\lambda_\mathrm{II} + 2\lambda_\mathrm{III}$ is constant. The observed value of Θ is 180°. This result implies that whenever J II and J III are in conjunction (i.e., when $\lambda_\mathrm{II} = \lambda_\mathrm{III}$), J I is 180° away. These three satellites never line up all on one side of the planet. Moreover, conjunction of J I and J II occurs 180° from the longitude of conjunction of J II and J III.

Laplace showed that the value of Θ is stable at 180°. The three-satellite interaction is quite complicated, so this stability is usually demonstrated mathematically. However, the mechanism can also be described qualitatively in the following manner based on the analysis of Souillart (Tisserand 1896). This description is necessarily simplified to include only those effects which contribute significantly to the behavior of Θ.

First consider the effect of J III on J II. If the ratio of their orbital periods were exactly 2:1, the longitude of conjunction of this pair would be fixed. In fact, because the ratio is only approximately 2:1, conjunction regresses at a rate of about $0°74/$day. As a result of the planet's oblateness and the secular effects of the other satellites, J II's pericenter advances. Thus conjunction circulates in the retrograde direction relative to the apsides of J II's orbit.

We can approximate the effect of J III on J II as an impulsive force exerted radially outward from Jupiter whenever conjunction occurs. It is well known in celestial mechanics that such a radial force tends to cause a regression of the apsides if exerted near pericenter and an advance of apsides if exerted near apocenter. As conjunction passes J II's pericenter the advance of the apsides slows down; as conjunction passes J II's apocenter, the advance of apsides speeds up. Thus J II's pericenter spends more time near the longitude of conjunction than apocenter does.

Next consider the effect of J II on J I. Suppose conjunction of these two satellites occurs after J II's pericenter, but before apocenter. Since the velocities are diverging at such a conjunction, the two satellites are closest to one another shortly before conjunction. Thus when the satellites are closest (and the perturbing force is greatest), J II is slightly ahead of J I. Therefore J I gains orbital energy and its period increases. As a result, conjunction is accelerated forward towards J II's apocenter. Similarly, if conjunction occurs after apocenter and before pericenter, conjunction is accelerated back towards J II's apocenter. This effect is rather weak due to the small eccentricity. Therefore it operates over such a long time scale that J II's apocenter may be considered to be, on the average, 180° from conjunction of J II and J III. Thus conjunction of J I and J II tends to be restored to a longitude 180° from conjunction of J II and J III.

Similar consideration of the effects of each of the three satellites on each of the others supports the same conclusion: Θ is stable at 180°. Observation of the libration of Θ about 180° should, in principle, yield important information on the satellites' masses, but the amplitude of libration is too small to detect (De Sitter 1931). The difficulty is analogous to trying to tell time with a clock whose pendulum has not been started. On the other hand, the very fact that the amplitude is nearly zero, if interpreted properly, may provide information about the evolution of the satellite orbits. This problem is related to the question of why commensurabilities exist at all.

It is conceivable that the commensurabilities are simply a result of chance. However, the large number of commensurabilities among satellites of all the major planets indicates that some mechanism has tended to favor such relations. For example, the Laplace relation is nearly satisfied by three satellites of Uranus as well as by the Galilean satellites (Greenberg 1975). Goldreich (1965) suggested that satellites which were originally non-resonant might have evolved independently due to tidal dissipation until stable commensurabilities were reached and maintained. For tidal evolution to have significantly altered the orbits of the Galilean satellites without driving them beyond their present distances from Jupiter, the planet's tidal dissipation parameter, Q, must be $\sim 10^5$ (Goldreich and Soter 1966). This value, however, is not consistent with interior models of Jupiter which suggest that the extent of possible tidal evolution was considerably less (Hubbard 1974). Moreover, Sinclair (1975) has investigated the process of capture of Θ into libration and subsequent damping of the libration amplitude. He finds that for his simplified model tidal evolution does not yield the Laplace relation. Perhaps the origin of the commensurabilities among the Galilean satellites involved drag effects in the primeval gaseous envelope; perhaps changes in precession rates accompanied past changes in Jupiter's effective oblateness and promoted development of the Laplace relation; or perhaps satellites have a tendency to form in such commensurabilities. These are topics for future research.

Whether satellite studies will play a role in post-Pioneer investigation of Jupiter's gravitational field remains to be seen. The theory of their motion is quite complicated and, because they have no radio transmitters aboard, they cannot be tracked as precisely as an artificial probe. On the other hand, the satellites provide the advantage of a long time available for observations as compared to a fly-by mission. And future observations will provide much more precise position determination than was possible in the past. In a few years, radar should be capable of yielding range and radial velocity accuracies of ~ 20 km and ~ 5 m/sec, respectively, (G. H. Pettengill, personal communication) compared with past position accuracy of $\sim 0\overset{''}{.}03$ (or 100km) at best (Ferraz-Mello 1975; Peters 1973). Thus, continued work will be needed in the area of the details of satellite motions as well as the origin and evolution of general orbital characteristics.

Acknowledgements. I am grateful to D. R. Davis, L. Andersson and P. K. Seidelmann for helpful discussions. This research was supported by a grant from NASA.

DISCUSSION

T. Gehrels: The imaging system on Pioneers 10 and 11 cannot observe small fragments near Jupiter. If the aperture and gain were set for faint light (Zodiacal Light mode, see p. 560) and, by mistake, Jupiter would enter the field of view, the detectors would be overexposed to destruction.

The work of C. Kowal for the discovery of *J* XIII indicates that the outer satellites are not fragments of collision(s) that occurred in a medium without resistance. For such collisional fragments the magnitude-frequency law probably is the one we know for the Asteroid Belt; it gives twice as many objects per fainter magnitude. There were seven outer satellites; in 1974 Kowal improved the detection limit by at least one magnitude, but, instead of finding 14 or more satellites, he found only one.

REFERENCES

Anderson, J.; Null, G. W.; and Wong, S. K. 1974. Gravity results from Pioneer 10 Doppler data. *J. Geophys. Res.* 79:3661–3664.

Bailey, J. M. 1971*a*. Jupiter: its captured satellites. *Science* 173:812–813.

———. 1971*b*. Origin of the outer satellites of Jupiter. *J. Geophys. Res.* 76:7827–7832.

———. 1972. Studies on planetary satellites. *Astron. J.* 77:177–182.

Brouwer, D. 1963. The problem of the Kirkwood gaps in the asteroid belt. *Astron. J.* 68:152–159.

Brouwer, D., and Clemence, G. M. 1961. Orbits and masses of planets and satellites. *The Solar system.* (G. P. Kuiper and B. M. Middlehurst, eds.) Vol. III, pp. 31–94. Chicago: Univ. of Chicago Press.

Burns, J. ed. 1976. *Planetary satellites.* Tucson: University of Arizona Press. In press.

Colombo, G., and Franklin, F. 1971. On the formation of the outer satellite groups of Jupiter. *Icarus* 15:186–189.

Davis, D. R. 1974. Secular changes in Jovian eccentricity: effect on the size of capture orbits. *J. Geophys. Res.* 79:4442–4443.

De Polavieja, M. G., and Edelman, C. 1972. Determination of the mass of Jupiter using the motion of its ninth satellite and a Kalman-Bucy filter. *Astron. and Astrophys.* 16:66–71.

De Sitter, W. 1931. Jupiter's Galilean satellites. *Mon. Not. Roy. Astron. Soc.* 91:706–738.

Duncombe, R.; Klepczynski, W. J.; and Seidelmann, P. K. 1974. The masses of the planets, satellites and asteroids. *Fund. Cosmic Phys.* 1:119–165.

Everhart, E. 1973. Horseshoe and Trojan orbits associated with Jupiter and Saturn. *Astron. J.* 78:316–328.

Ferraz-Mello, S. 1975. Problems of the Galilean satellites of Jupiter. *Celestial Mechanics.* 12:27–38.

Gehrels, T. ed. 1971. *Physical studies of minor planets.* NASA SP-267. Washington, D. C.: Government Printing Office.

Goldreich, P. 1965. An explanation of the frequent occurrence of commensurable mean motions in the solar system. *Mon. Not. Roy. Astron. Soc.* 130:159–181.

Goldreich, P., and Soter, S. 1966. Q in the solar system. *Icarus* 5:375–389.

Greenberg, R. 1975. The dynamics of Uranus' satellites. *Icarus* 24:325–332.

Heppenheimer, T. A. 1975. On the presumed capture origin of Jupiter's outer satellites. *Icarus* 24:172–180.

Hubbard, W. B. 1974. Tides in the giant planets. *Icarus* 23:42–50.

Klepczynski, W. J. 1969. The mass of Jupiter and the motion of four minor planets. *Astron. J.* 74:774–775.

Kowal, C. T.; Aksnes, K.; Marsden, B. G.; and Roemer, E. 1975. The thirteenth satellite of Jupiter. *Astron. J.* 80:460–464.

Kuiper, G. P. 1956. On the origin of the satellites and the Trojans. *Vistas in astronomy.* (A. Beer, ed.) Vol. 2, pp. 1631–1666. New York: Pergamon Press.

Marsden, B. G. 1966. The motions of the Galilean satellites of Jupiter. Ph.D. dissertation, Yale Univ. New Haven, Conn.

Peters, C. F. 1973. Accuracy analysis of the ephemerides of the Galilean satellites. *Astron. J.* 78:951–956.

Sampson, R. A. 1921. Theory of the four great satellites of Jupiter. *Mem. Roy. Astron. Soc.* 63:1–270.

Sinclair, A. T. 1975. The orbital resonance amongst the Galilean satellites of Jupiter. *Mon. Not. Roy. Astron. Soc.* 171:59–72.

Sudbury, P. V. 1969. The motion of Jupiter's fifth satellite. *Icarus* 10:116–143.

Tisserand, F. 1896. *Traité de mécanique céleste,* Vol. 4. Paris: Gauthier-Villars.

van Woerkom, A. J. J. 1950. The motion of Jupiter's fifth satellite. *Astron. Papers Amer. Ephem.* 13:1–77.

Wiesel, W. E. 1974. A statistical theory of the Kirkwood gaps. Center for Astrophysics. Preprint no. 204.

Wilkins, G. A.; and Sinclair, A. T. 1974. The dynamics of the planets and their satellites. *Proc. R. Soc. Lond.* A. 336:85–104.

STRUCTURE, COMPOSITION,
AND GRAVITATIONAL FIELD
OF JUPITER

V. N. ZHARKOV

and

V. P. TRUBITSYN
Institute of Physics of the Earth, Moscow

We review the physics of the Jovian interior, making a number of comparisons with Saturn. We discuss techniques for constructing equations of state for the principal chemical constituents of Jupiter, and we use the results to set limits on the enrichment of the rock-ice component relative to the hydrogen-helium component in Jupiter and Saturn. We show that the present thermal state of these planets requires a molten interior. Finally, we show how measurements of the external gravity field can be used to constrain models.

Jupiter, like Saturn, is a hydrogen-helium planet, and this circumstance is not accidental. Hydrogen is the most abundant element in the solar system, in stars and in the interstellar medium, and the gravitational field of the giant planets is such that they can retain a hydrogen atmosphere through the lifetime of the planet. The second most abundant element in the universe is helium. The abundance of helium by particle number is such that the ratio H/He is on the order of 10. Because of this, it is natural to expect in both Jupiter and Saturn the existence of a significant helium admixture. The determination from models of the helium concentration in both planets is one of the most important tasks of planetary physics, and it has great significance for cosmogony. The abundances of remaining elements in the solar system— for example, oxygen, carbon, nitrogen, silicon, iron—in comparison with the abundance of hydrogen and helium, are small quantities and their content in Jupiter and Saturn at present is not as well determined as is the content of hydrogen and helium.

The first work on construction of models of Jupiter was carried out by Jeffreys (1923, 1924) and Fesenkov (1924). The hydrogen theory as we now understand it had its beginning with the work of Wigner and Huntington (1935) on the metallization of hydrogen; in essence that paper also started work on the problem of phase transitions from dielectrics to metals. In 1937 Goldschmidt published the first modern table of cosmic distributions of elements, where it followed that hydrogen is the most abundant element in the solar system and in the universe. In 1938 Wildt and Kothari presented the first hydrogen models of Jupiter and in 1951 Fesenkov and Masevich (1951), Ramsey (1951) and DeMarcus (1951), all three independently, introduced the concept of a hydrogen-rich Jupiter.

Modern investigations have their origin in the work published in 1958 by Wildt's student, DeMarcus (1958). DeMarcus used the experimental data of Stewart (1956) and determined interpolated equations of state of hydrogen and helium so well that they have changed very little since.

Earlier, for determination of I from J_2, the Radau-Darwin formula was used,

$$\frac{C_1}{Ma_1^2} \approx \frac{I}{MS_1^2} = \frac{2}{3}\left(1 - \frac{2}{5}\sqrt{1 + \eta_1}\right) \qquad (1)$$

$$\eta_1 = \frac{5}{2}\frac{m}{e_1} - 2 , \quad e_1 = \frac{3}{2}J_2 + \frac{m}{2} \qquad (2)$$

$$m = \frac{\omega^2 S_1^3}{GM} = \frac{3\pi}{G\bar{\rho}\tau^2} \qquad (3)$$

which is a poor approximation for a planet with a strong concentration of mass towards the center. In Eqs. (1), (2) and (3), C_1 is the polar moment of inertia, a_1 is the equatorial radius, ω and τ are the angular velocity and period of rotation of the planet, and G is the gravitational constant. The work of DeMarcus was expanded by Peebles (1964) who used an electronic calculator and considered a larger number of models. Subsequent deeper studies of the equations of state of hydrogen and helium and their phase diagram were carried out by Trubitsyn (1965a,b, 1966a,b, 1971) and Hubbard (1968, 1969, 1970, 1972). At the end of the 1960s in the United States there were systematic publications of papers on Jovian physics by Hubbard and Smoluchowski; the results are described in a review paper (Hubbard and Smoluchowski 1973).[1] In the Soviet Union, at the O. Yu. Schmidt Institute of Physics of the Earth of the Academy of Sciences of the USSR, work on the question of the giant planets was begun by Zharkov and Trubitsyn about ten years ago. Approaching the problem from a geophysical point of view, the authors concluded that all of the giant planets should be hot and adiabatic (Zharkov and Trubitsyn 1969; Zharkov et al. 1971). Hubbard (1968, 1969)

---

[1]See pp. 8 and 176.

independently deduced an adiabatic model of Jupiter, although Peebles (1964) already formally considered adiabatic models of the molecular envelope of Jupiter. The adiabatic concept was developed further in joint work carried out in 1973 during Hubbard's stay in Moscow at the Institute of Physics of the Earth (Hubbard *et al.* 1974). Later progress in the study of Jupiter is connected with the development of the theory of figures of rotating liquid planets to terms of third order, and with application of this basic theory to the interior structure of the giant planets and their gravitational field (Zharkov and Trubitsyn 1969; Zharkov *et al.* 1971; Zharkov *et al.* 1973). These papers led to the development of methods of gravitational sounding of Jupiter (Zharkov and Trubitsyn 1974*b*; Hubbard 1974). As a result of these latter papers, the problems of the structure, the gravitational field and the gravitational sounding of Jupiter appear to be closely connected. The theory of figures was constructed with accuracy to terms of the fifth order by Zharkov and Trubitsyn (1975). At the present time we have good equations of state available for the basic cosmochemical elements and their compounds which allow one to investigate not only the outer envelopes of the giant planets, but also their deep interiors (Zharkov *et al.* 1974*d*; Zharkov *et al.* 1974*a,b*).

I. EQUATIONS OF STATE OF COSMOCHEMICAL COMPOUNDS

Groups of Cosmochemical Substances

The abundance of elements in simple compounds in the proto-planetary cloud is discussed in a review by Makalkin (1974); Table I shows the abundance of the most abundant compounds. In columns *SS1* and *SS2* are given the abundances of elements according to spectroscopic observations of the solar atmosphere and in column *SCH* the abundance obtained from the sun and from chondritic meteorites. For iron we give the various versions of its possible compounds and correspondingly the changed abundances.

All substances are collected into three groups according to volatility. Substances of Group II are customarily called icy components and are designated by the term "ice" since, under the pressure and temperature conditions of the proto-planetary cloud in the zone of the giant planets, they apparently were in the solid state. In Table I these substances are given in order of decreasing volatility. The ones in Group III are called rock (or heavy) components; the substances in this group are given in order of increasing density. The rocky materials together with the ice-component form the condensate in the region of the cloud of interest here. Substances of the first group were in the gaseous state in the proto-planetary cloud. We refer to these with the designation *G* (gas component).

In the case where the temperature in the part of the cloud under consideration is higher, it is necessary to also consider a different division of substances into groups where part of the material of the ice component converts to a gaseous component. Then the following ice variants are possible: ice-I (CH_4 + NH_3 + H_2O), ice-II (NH_3 + H_2O), ice-III (H_2O). The corresponding G component variants are: GI (H_2 + He + Ne), GII (GI + CH_4), $GIII$ (GII + NH_3). There is also a version of the G component that corresponds to the case when only the rock component is in the condensate: GIV ($GIII$ + H_2O).

We will consider planetary models with various types of chemical compositions of core and envelope. For a chemically uniform core, we adopt the composition rock + ice with relative abundances according to Table I. For the planetary envelope we assume a composition either only of gaseous component abundances from Table I, or of gases with the addition of rock +

TABLE I

Percentage Mass Fractions of the Most Abundant Substances[a]

				% Mass Fraction		
Group	No.	Substance	Molecular wt.	*SS1*	*SS2*	*SCH*
(Version 1: All sulfur combined with iron [FeS]; remaining iron in oxide [FeO]).						
I	1	H_2	2.016	78.3	78.1	76.8
	2	He	4.003	19.9	20.0	21.2
	3	Ne	20.18	0.118	0.04	0.167
				Sum ~ 98.3	~ 98.1	~ 98.2
II	4	CH_4	16.04	0.446	0.446	0.453
	5	NH_3	17.03	0.114	0.152	0.153
	6	H_2S	34.08	0	0	0
	7	H_2O	18.02	0.663	0.776	0.774
				Sum ~ 1.22	~ 1.37	~ 1.38
III	8	SiO_2	60.09	0.167	0.167	0.144
	9	MgO	40.32	0.096	0.110	0.102
	10	(a) FeS	87.92	0.113	0.112	0.106
		(b) FeO	71.85	0.072	0.050	0.057
				Sum ~ 0.45	~ 0.44	~ 0.41
(Version 2:[b] All iron oxidized [FeO])						
	6	H_2S	34.08	0.044	0.044	0.041
	7	H_2O	18.02	0.644	0.753	0.751
	10	FeO	71.85	0.160	0.142	0.143
(Version 3:[b] All iron in metallic state)						
	6	H_2S	34.08	0.044	0.044	0.041
	7	H_2O	18.02	0.683	0.789	0.787
	10	Fe	55.85	0.125	0.110	0.111

[a]The total mass fraction of substances not included in the table is ~ 0.07–0.08%.
[b]No change for substances other than numbers 6, 7, and 10.

ice. The relationship of the rock + ice abundances and G is obtained from the calculation of models.

The relative helium to hydrogen abundance (Y/X) is obtained from spectroscopic solar data (Hirayama 1971; Lambert 1967): $Y/X = 0.26 \pm 0.06$. This is very close in value to average cosmic proportions (Burbidge 1971).

The equation of state of a mixture of substances can be calculated for both the envelope and the core using the approximation of additive partial volumes:

$$\frac{1}{\rho(p)} = \sum_i \frac{X_i}{\rho_i(p)} \tag{4}$$

where X_i is the abundance by mass of each substance, $\sum_i X_i = 1$. Formula (4) is exact in the region of ideal gases ($p \lesssim 10^2$ bar) and for high pressures ($p \gtrsim 100$ Mbar), i.e., in the atmospheres and cores of Jupiter and Saturn. In the envelopes ($10^3 \lesssim p \lesssim 10^6$ bar) non-additivity for the fundamental component of the mixture does not exceed a few percent (Hubbard 1972).

In accordance with the assumption of convection in the envelope (see Sec. II) we assume its composition to be chemically uniform. We shall assume that the core is also chemically uniform. This assumption is obviously a good first approximation for an adiabatic core with possible convection.

The equation of state of the materials is constructed in two steps (Zharkov and Kalinin 1971): at first the null isotherm is calculated (equation of state at $T = 0$), and then the thermal part, i.e., the thermal pressure.

Equations of State at $T = 0$, $p \lesssim 1$ Mbar. In the geophysical pressure range (up to ~ 3 Mbar) for description of the zero isotherm (potential pressure) one uses both experimental data directly, and also a number of analytic dependences which are smoothly fitted to the data. In the method of potentials the potential pressure has the form (Zharkov and Kalinin 1971)

$$p_p(x) = Ax^{-\frac{2}{3}}e^{b(1 - x^{1/3})} - Kx^{-\frac{4}{3}}, \quad x = \rho_0/\rho \tag{5}$$

where ρ_0 is the density under normal conditions, and the constants A, b, K are determined with the help of experimental data. Equation (5) describes well the compressibility of silicates, oxides and metals with exponential forces of repulsion up to ~ 1 Mbar. In the case of the outer planets, the equations of state of the substances H_2, He, H_2O, CH_4, NH_3 are of the greatest interest. These substances belong to the molecular crystals, where the forces of repulsion have an exponential character, but the attraction is caused by van der Waals forces (Zharkov and Kalinin 1971; Trubitsyn 1965a,b):

$$p_p(x) = Ax^{-\frac{2}{3}}e^{b(1 - x^{1/3})} - Kx^{-3} \tag{6}$$

where the constants A, b, and K are determined from experimental data.

In contrast to the method of potentials, one may use either the Birch (1952) or the Murnaghan (1951) equations. However, these equations [as well as Eq. (6)] are fundamentally deficient because they have an incorrect asymptotic form at high pressures; at pressures greater than ~ 1 Mbar and often at lower pressures they begin to systematically deviate from the real compression curve. This important circumstance has not so far attracted sufficient attention either in geophysics or in the physics of high pressures.

Equation of State at $T = 0$ and $p > 100$ Mbar. For determination of the potential pressure of strongly compressed material one uses the statistical model of the atom. All statistical models have their beginning with the well-known Thomas-Fermi model (*TFM*) of the atom (Landau and Lifshitz 1958).

The quasi-classical approximation, on which the *TFM* is based, is not valid in the neighborhood of the turning points, close to the nucleus and at the boundary of the atom. The behavior of the electron density close to the boundaries was clarified by Dirac by means of an exchange correction, and close to the nucleus by Weizsäcker using a quantum correction.

Kirzhnits (1957, 1963) solved the Hartree-Fock equation in powers of $\hbar$. In the zero approximation he obtained the quasi-classical *TFM* and in the second approximation ($\propto \hbar^2$) Dirac's exchange correction is added to the *TFM* as well as a quantum correction which coincides in form with Weizsäcker's correction, but which has a numerical coefficient nine times smaller. This latter model was studied in detail by Kalitkin (1974). This model is used in two versions: In the case where the sum of the exchange correction of Dirac and the Weizsäcker-Kirzhnits quantum correction are calculated, using the electron distribution in the zero approximation, one speaks of the *TFM* with quantum and exchange corrections, or the *TFCM*. If, on the other hand, these corrections are included directly in the equation for determination of the electron density, this model is called the quantum-statistical model, or *QSM*. As a result of somewhat fortuitous circumstances the *QSM* describes well the behavior of compressed materials at moderate pressures. Thus, for determination of the potential pressure of strongly compressed material by Zharkov *et al.* (1974c, 1975) the calculations of Kalitkin and Kuz'mina (1971) on the *QSM* were used, which are strictly valid only at pressures greater than 1 atomic unit,[2] but they actually give good accuracy also at pressures 1–2 orders smaller. The *QSM* lowers the pressure for a given density and serves to give a correct matching with data extrapolated from below, at substantially smaller pressures than is possible by using the *TFM* and the *TFCM*.

In the *QSM* the electron density close to the nucleus is finite and at infinity it falls exponentially in correspondence with quantum mechanics, while

---

[2]Atomic unit of pressure = 2.942×10^8 bar
   Atomic unit of energy   = 27.2 eV = 4.36×10^{-11} erg
   Atomic unit of distance = 0.529×10^{-8} cm

in the TFM $\rho(r) \to \infty$ as $r \to 0$; far from the nucleus $\rho(r)$ decreases according to a power law, which is qualitatively incorrect.

The fact that at super-high pressures the electron density within the atom is approximately constant, $\rho(r) \approx Z/V$, while for the free atom the value of $\rho(r)$ exponentially decreases away from the nucleus, allowed Kalitkin and Kuz'mina to derive a convenient interpolation formula

$$\rho(R) \approx Z/V \exp[-\alpha(Z)R - \beta(Z)R^2]$$

$$\alpha(Z) = 0.1935Z^{0.495} - 0.039\ell nZ \tag{7}$$

$$\beta(Z) = 0.068 + 0.078\ell nZ - 0.086(\ell nZ)^2$$

where $V = \dfrac{4\pi R^3}{3}$ is the atomic volume, R is the radius of the Wigner-Seitz sphere, Z is the atomic number of the nucleus, and all quantities are expressed in atomic units. The potential pressure is calculated with the help of an approximate formula

$$p \approx \left[\frac{1}{5}(3\pi^2)^{\frac{2}{3}}\rho^{\frac{5}{3}} - \frac{13}{36}\left(\frac{3}{\pi}\right)^{\frac{1}{3}}\rho^{\frac{4}{3}} \right]_{Z\,=\,R}. \tag{8}$$

The inaccuracy of Eqs. (7) and (8) in pressure is less than 3% for $p \gtrsim 0.1$ $Z^{\frac{1}{2}}$. These formulas also allow one to calculate the potential pressure of mixtures and compounds. For this one uses the conditions

$$p(Z_1, V_1) = p(Z_k, V_k), \quad k > 1, \quad V = \sum_{k=1}^{n} V_k. \tag{9}$$

The first relations of Eqs. (9) state the equality of the pressure in the cells of the various components, and the second relation requires the additivity of partial volumes.

Interpolated Potential Curves. At pressures up to ~ 1 Mbar, the relationship which gives the compression curve for any substance can be established with the help of experimental data and equations such as (5) and (6).

At high pressures, $p > 100$ Mbar, this relationship can be established theoretically by using the QSM. In order to determine the zero isotherm of a specific cosmochemical substance at all pressure intervals, it is necessary to interpolate or blend the experimental data at low pressures with calculated values at high pressures (Zharkov et al. 1974c; Zharkov et al. 1974d).

The validity of the blending can be verified for the example of metallic hydrogen. A large number of papers are devoted to calculations of the equation of state of metallic hydrogen. Metallic hydrogen has a number of possible phases which have very similar energies and equations of state. The energy of the most stable phase of metallic hydrogen at $T = 0$ and $p > 1$ Mbar coincides closely with the energy calculated in analytic form by Trubitsyn (1966a). Data for metallic hydrogen are shown in Fig. 1.

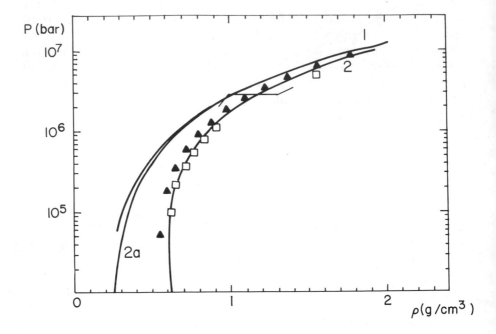

Fig. 1.   Curve 1: Quantum-statistical model calculation for hydrogen; Curve 2: Trubitsyn's
(1966*a*) metallic hydrogen equation of state; Curve 2*a*: Trubitsyn's (1965*a*) molecular hydrogen equation of state; Triangles: DeMarcus' (1958) metallic hydrogen equation of state;
Squares: metallic hydrogen from Neece *et al.* (1971). The discontinuity shows the transition
from molecular to metallic hydrogen estimated from the data of Grigor'ev *et al.* (1972).

Comparison of the equation of state of metallic hydrogen calculated purely from theory with the equation of state of hydrogen from the QSM shows
that both equations coincide at pressures greater than 150 Mbar and deviate
only slightly in the pressure region 5–150 Mbar.

For still higher pressures, of the order of 10^9 bar and higher, the properties
of all materials unify, and they can be calculated using the Thomas-Fermi-
Dirac model (TFD). In this pressure region, the $p(\rho)$ relationship calculated
using the QSM blends with the $p(\rho)$ curves from the TFD model, although
the blend again tends to fall on the low side, i.e., for a given pressure the
QSM equation gives a higher density than the TFD model.

The second principle, used for constructing zero isotherms of high-pressure phases, was the assumption of the monotonicity of the respective behavior of the $p = p(\rho)$ curve for related classes of materials.

The equation of state for the QSM was calculated using the assumption
of additive volumes of elements. For determining the $p(\rho)$ relationship for
hydrogen compounds H_2O, CH_4, NH_3 and H_2S, in the high-pressure region

we calculated the partial volume $V_H(p)$ for H using the equation of state of metallic hydrogen which has somewhat higher density than the QSM.

It was further assumed that the $p(\rho)$ relationship calculated from the QSM corresponds to the most dense phase at high pressure for a given substance.

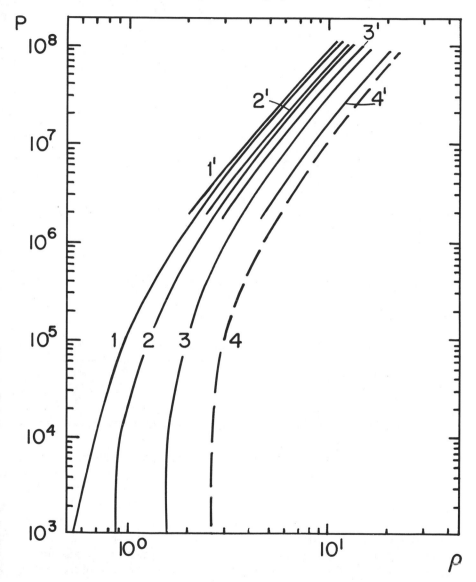

Fig. 2. Null isotherms of hydrogen compounds. Curves $1'$, $2'$, $3'$ and $4'$ are for CH_4, NH_3, H_2O, and H_2S respectively, using the quantum-statistical model. Curves 1, 2, and 3 are interpolated null isotherms for CH_4, NH_3, and H_2O respectively, while Curve 4 is a hypothetical null isotherm for H_2S. Pressure and density are in same units as in Fig. 1.

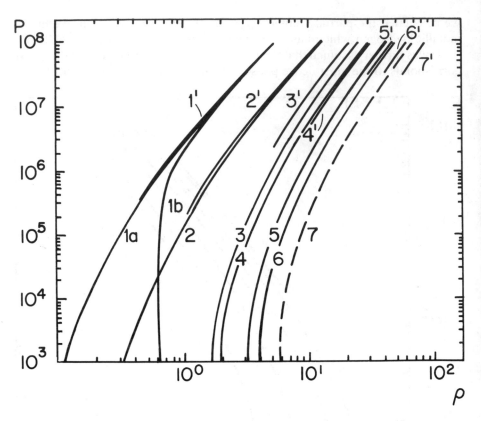

Fig. 3.   Null isotherm of hydrogen and the solid phase of inert gases. Curves 1, 2, 3, 4, 5, 6, and 7 are for H, He, Ne, Ar, Kr, Xe, and Rn respectively, using the quantum-statistical model. Curves $1a$ and $1b$ are for H_2 and H respectively; Curves 2, 3, 4, 5, and 6 are interpolated null isotherms for He, Ne, Ar, Kr, and Xe. Curve 7 is a hypothetical null isotherm for Rn.

At lower pressures, experimental data were analyzed where possible to also find the densest phase. For a number of substances this phase obviously will be unstable or metastable. The potential pressures of high-pressure phases of cosmochemical substances are shown in Figs. 2, 3 and 4 (Zharkov *et al.* 1974*c*; Zharkov *et al.* 1974*d*).

Melting Curves

By definition the melting curve $T_M = T_M(p)$ divides the region of the crystalline state of solid matter from the region of the liquid state. The melting curves are obtained from Lindemann's assumption: "The ratio of the average quadratic amplitude of thermal oscillation to the lattice constant squared is constant along the melting curve." This assumption leads to the result (Zharkov and Kalinin 1971):

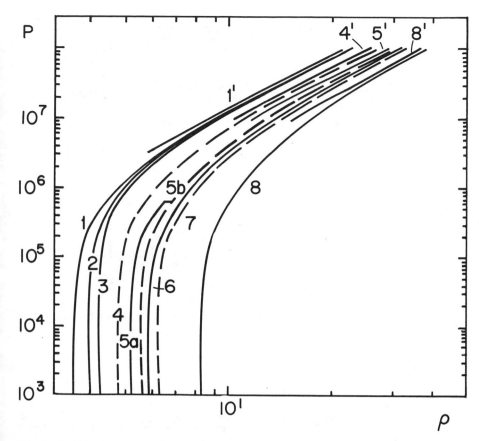

Fig. 4. Null isotherms of oxides and iron. Curve 1': MgO, Al_2O_3, SiO_2 (quantum-statistical model); Curve 4': CaO (quantum-statistical model); Curve 5': Fe_2O_3 (quantum-statistical model); Curve 8': Fe (quantum-statistical model). Interpolated null isotherms are shown for: MgO (Curve 1), Al_2O_3 (Curve 2), SiO_2 (Curve 3), Fe_2O_3 low pressure phase (Curve 5a), FeO (Curve 6), Fe (Curve 8). For hypothetical high pressure phases, we have CaO (Curve 4), Fe_2O_3 (Curve 5b), and FeS (Curve 7).

$$\left(\frac{<U^2>}{a^2}\right)^{\frac{1}{2}} \approx 0.075 \tag{10}$$

where $<U^2>$ is the average quadratic amplitude of thermal oscillation at melting, and a is the distance between nearest neighbors. The Lindemann constant changes with increasing pressure. Thus, in order to construct good melting curves, additional experimental data are necessary. Figures 5 and 6 show the phase diagrams of hydrogen and helium (Trubitsyn 1971; Zharkov et al. 1971).

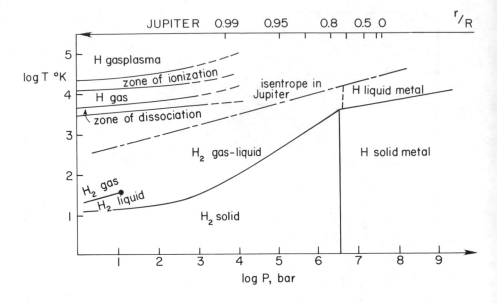

Fig. 5. Hydrogen phase diagram. The quantity r/R is the relative distance from the center of Jupiter.

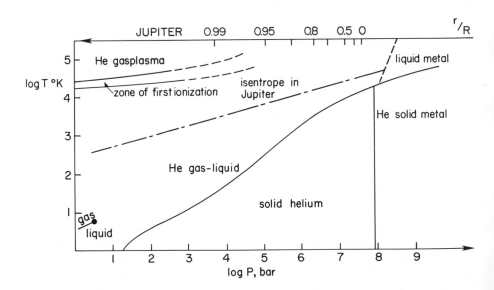

Fig. 6. Helium phase diagram. The Jovian adiabat is for hydrogen.

II. MODELS OF JUPITER AND SATURN

The construction of Saturn models is so closely related to construction of Jupiter models that it is natural to consider both planets together. This will not increase the volume significantly, and on the other hand a separate discussion of the structure of Saturn would require a duplication of all of the material which is presented here. Comparison of Jupiter and Saturn models is most instructive, and the planned flyby of Pioneer 11 past Saturn in 1979 will make possible a comparative analysis of both planets.

Gas-Liquid Adiabatic Model

In this section, we discuss experimental and theoretical work from which it follows that both Jupiter and Saturn are gas-liquid, with the exclusion of a possible central region (Hubbard et al. 1974). It is obvious that any process which could have produced such large bodies as Jupiter and Saturn would lead to initially hot planets. Thus we arrive at the problem of the manner by which such hot planets could lose their initial heat during the subsequent $\sim 5 \times 10^9$ yr. The rate of transport of heat to the surface depends essentially on whether the heat transport is by convection or by molecular conduction. The critical condition for the existence of convection is the existence of a super-adiabatic temperature gradient. If the present temperature distribution inside Jupiter and Saturn is adiabatic, then the temperatures in the center of these planets will be of the order of 20–30 thousand degrees (Trubitsyn 1972a; Hubbard 1973; Zharkov et al. 1974b). Such temperatures are about ten times higher than the melting temperature of metallic hydrogen at pressures of 10–100 Mbar. The temperatures cannot exceed these adiabatic values since they would be quickly reduced by convection.

On the other hand, the thermal conductivity of molecular hydrogen of the envelope is so low and the dimensions of the planet are so large, that the process of molecular heat conduction is in no way capable of significantly reducing the initial heat content. The thermal skin depth λ can be estimated using the dimensional relationship

$$\lambda \sim (\chi\tau)^{\frac{1}{2}} \tag{11}$$

where χ is the thermal diffusivity and τ is the cooling time. Setting $\tau \sim 5 \times 10^9$ yr and χ at 10^{-2}–10^{-3} cm^2 sec^{-1} (Zharkov et al. 1971), we find that $\lambda \sim 10^2$ km. Only such a thin layer can lose its initial heat in the absence of convection. Thus we conclude that the interiors of Jupiter and Saturn must have been in a hot liquid convective state since the time of their formation. We can also conclude from independent data that at present these planets are in a convective state. The value of the heat flow q due to the molecular mechanism of heat transport can be easily estimated using the formula

$$q = K\nabla T \leqslant K(\nabla T)_{ad} \tag{12}$$

where the adiabatic gradient for an ideal gas is equal to

$$\left(\frac{dT}{d\ell}\right)_{ad} = \frac{(\gamma - 1)}{\gamma} \left(\frac{T_0}{T}\right) \left(\frac{p}{p_0}\right)^{(\gamma - 1)/\gamma} \frac{\mu}{R} g \, . \tag{13}$$

Here ℓ is the depth, T_0 and p_0 are the temperature and pressure at the surface (start) of the adiabat, μ is the molecular weight, $\mu = 2$ for hydrogen, $R = 8.3 \times 10^7$ erg/degree-mole is the gas constant, g is the force of gravity, $g_J = 2.7 \times 10^3$ cm sec^{-2}, $g_S = 1.15 \times 10^3$ cm sec^{-2}, γ is the adiabatic index, and $\gamma = 7/5$ for hydrogen.

On calculating the adiabatic gradient $(\nabla T)_{ad}$ in the outer molecular regions using Eq. (13), and setting $K \sim 10^4$ erg cm^{-1} sec^{-1} deg^{-1} (Hubbard 1968; Zharkov et al. 1971), we find

$$q_J \lesssim 0.2 \text{ erg cm}^{-1} \text{ sec}^{-1} \tag{14}$$

for Jupiter, and

$$q_S \lesssim 0.1 \text{ erg cm}^{-2} \text{ sec}^{-1} \tag{15}$$

for Saturn.

Broadband infrared measurement (Aumann et al. 1969) allows one to determine the heat flux from the interior of both planets. This flux is $\sim 10^4$ erg cm^{-2} sec^{-1} for Jupiter and $\sim 3 \times 10^3$ erg cm^{-2} sec^{-1} for Saturn. For Jupiter these data have been confirmed by Pioneers 10 and 11.[3] Consequently, regardless of the actual source of the heat which produces this flux, the result itself shows that both planets are in a convective state at present. The convection would be localized in the outer layers of the molecular hydrogen envelopes of these planets, if the sources of the heat were concentrated in the outer layers. However, we know that the interior regions must also be in the convective state. This is because we know that Jupiter has a strong intrinsic magnetic field which shows that its metallic core must be in a convective state. The metallic envelope of Saturn is situated in the depth interval 0.3 to 0.5 R_S, where R_S is the radius of the planet. Evidence for radiation belts at Saturn has been reported (Brown 1975) and thus the existence of an intrinsic magnetic field would indicate the convective state of the metallic envelope of Saturn. It follows that the thermal sources which keep both planets in the convective state must be situated at great depth.

The intensity of Jupiter's magnetic field is about an order of magnitude greater than the intensity of the earth's field. In order to excite and support such a large field over cosmic time, the interior of the planet must possess a convective region of significant radial extent and be composed of a substance which has a sizable electrical conductivity and low viscosity. It is natural to suppose that Jupiter's magnetic field is produced in the liquid convective zone composed of metallic hydrogen.

---

[3]See p. 202.

Observations of the microwave radiation of Jupiter and Saturn (Gulkis and Poynter 1972)[4] give additional support to our model. Because of the increased transparency of the atmospheres of Jupiter and Saturn for wavelengths ~ 10 cm and greater, one can "see" quite well to depths below the cloud layer in this region of the spectrum. After making a correction to obtain the thermal radiation in the case of Jupiter, Gulkis and Poynter concluded that Jupiter and Saturn both may be characterized by an adiabatic temperature distribution below the cloud layer with the temperatures reaching values of at least $\sim 500°K$ at pressures ~ 100 bar. The values of the specific dissipation function Q for Jupiter and Saturn (Goldreich and Soter 1966) give additional indications about the state of the interior of both planets. Goldreich and Soter found $Q \gtrsim 1 \times 10^5$ for Jupiter and $\gtrsim 6 \times 10^4$ for Saturn. In their calculation they adopted the value for the Love number $k_2 = 1.5$, corresponding to the model of a liquid uniform planet. Calculation of the Love number for models which are closer to reality give a value about three times smaller, $k_2 \sim 0.5$ (Gavrilov *et al.* 1975). Since the torque which affects the evolution of the satellite orbits is directly proportional to k_2 and inversely proportional to Q of the planet, the values obtained by Goldreich and Soter must be reduced[5] by a factor of 3:

$$Q_J \gtrsim 4 \times 10^4, \quad Q_S \gtrsim 2 \times 10^4. \tag{16}$$

These Q-values are about two orders of magnitude greater than typical for substances in the solid state, in particular for the earth's mantle (Zharkov *et al.* 1971) and for terrestrial planets (Goldreich and Soter 1966).

Finally, when Pioneers 10 and 11 measured the gravitational field of Jupiter, they did not detect the imprint of the first odd moment J_3, to an accuracy of 10^{-5} in the gravitational field of the planet. The absence of odd moments in the gravitational potential expansion shows that the planet is close to hydrostatic equilibrium, i.e., it is essentially in the liquid state.[6]

We now expand further on the concept of the giant planets as gas-liquid bodies. The critical pressure and critical temperature of hydrogen are equal to 13 atm and 33°K, respectively. At pressures and temperatures which exceed the critical values, there exists no boundary between the gas and liquid phases of molecular hydrogen. Jupiter and Saturn consist largely of hydrogen. Consequently, due to the increase of pressure with depth, the gaseous atmosphere becomes denser under the pressure of overlying layers and continuously transforms to a comparably dense liquid state; there is no boundary between the gaseous atmosphere and the underlying liquid region. A graphical illustration of this point is seen in Fig. 5, where on the hydrogen phase

---

[4]See p. 650.
[5]For the same reason the value $Q \gtrsim 7.2 \times 10^4$ for Uranus should also be reduced by about a factor of 3; $Q_U \gtrsim 3 \times 10^4$.
[6]See p. 115.

diagram we show an adiabat which is close to the Jovian adiabat and lies entirely in the super-critical region.

Observational Data and Boundary Conditions

A discussion of observational data for giant planets (as of May, 1972) is given by Newburn and Gulkis (1973). However, some of these parameters require further discussion. For example, it is quite difficult to choose the correct value of the rotation period of the planet, τ. The atmospheres of Jupiter and Saturn are in a state of differential rotation. The depth of the region involved in differential rotation is not established. It is plausible that the true rotation period of Jupiter as a whole is the same as the period of rotation of its magnetosphere (Carr 1971), since the sources of the large intrinsic magnetic field should be in the metallic envelope of the planet which is 0.2 R_J below the cloud layer. This period is almost identical with the rotation period of temperate and polar regions of the cloudy layer of the planet.

New boundary conditions resulting from the Pioneer 10 and 11 missions are $M = 1.901 \times 10^{33}$ g, $J_2 = 0.01475 \pm 0.00005$, $J_4 = -(5.8 \pm 0.4) \times 10^{-4}$, $J_6 = (5 \pm 6) \times 10^{-5}$, $a_1 = 71,398$ km; $|J_3| < 10^{-5}$, tesseral moments $|C_{22}|$, and $|S_{22}|$ are less than 10^{-6}. (Null et al. 1975; Anderson et al. 1974; Anderson p. 116; Hubbard and Slattery p. 183.)

Along with the mass, radius, and moments J_2 and J_4, the most important parameter for construction of model planets is the small parameter of the theory of figures m, inversely proportional to the square of the period τ^2,

$$m = \frac{3\pi}{G\bar{\rho}\tau^2} \qquad (17)$$

where $\bar{\rho}$ is the average density of the planet ($\bar{\rho} = 3M/4\pi R^3$) and G is the gravitational constant. An uncertainty in m (due to uncertainties in $\bar{\rho}$ and τ) propagates into calculations of the theory of figures for J_{2n} ($n = 1, 2, \ldots$) which are used to compare model planets with observations. The dynamical oblateness e_d is calculated using de Sitter's formula

$$e_d = \frac{1}{2}(3J_2 + m)\left(1 + \frac{3}{2}J_2\right) + \frac{5}{8}J_4 . \qquad (18)$$

Using values of the parameters given above we find $m = 0.0830$, $e_d = 0.0651$, and $\bar{\rho} = 1.334$.

For construction of models of the interior structure of planets, it is necessary to prescribe p_1, T_1, M_1 and S_1 on some level surface which is adopted as a boundary. Here M_1 is the mass of material enclosed within the surface; S_1 is its average radius. The value of the density on the boundary surface ρ_1 is related to p_1 and T_1 by the equation of an ideal gas. The relationship between pressure, temperature, mass and radius is obtained from theoretical models of the atmosphere, in which one also calculates the composition, cloud structure and other properties of the atmosphere.

The customary choice of the boundary is the surface on which $p_1 = 1$ bar. From model atmospheres, one finds that the quantity S_1 for this surface should be within a scale height of the average radius of the planet R at the outer boundary of the visible cloud layer: $|S_1 - R| \lesssim 40$ km for Jupiter and $\lesssim 70$ km for Saturn. The quantity $|S_1 - R|$ is less than the uncertainty in R. The contribution of mass above the arbitrarily chosen boundary is a negligible fraction of the total mass of the planet. At the same time, the surface adopted is sufficiently deep in the atmosphere so that at 1 bar one is already in the region of an adiabatic temperature distribution.

For $p_1 = 1$ bar, T_1 for Jupiter is estimated to lie between 130 and 230°K, and for Saturn between 90 and 150°K (Trafton 1967; Newburn and Gulkis 1973). As a fundamental model for both planets we choose $T_1 = 140$°K. In order to find the sensitivity of models to boundary conditions, calculations were also carried out for other values of T_1. For $T_1 = 250$°K for Jupiter, and $T_1 = 90$°K for Saturn, limiting models were obtained. Calculations show that the uncertainty in T_1, which is about 1.5 order of magnitude greater than the uncertainty in S_1, has a significant effect on the uncertainty of model planets.

Equation of State Used for Construction of Models

Molecular Hydrogen. The null isotherm $p(\rho, T = 0)$ has been calculated (Trubitsyn 1965a, 1966a) and tabulated (Zharkov et al. 1971; Trubitsyn 1971). It was slightly shifted to higher densities in Zharkov et al. (1974c). It agrees well both with theoretical calculations of the potential of molecular interaction, and also with experiments on adiabatic pressure of hydrogen (Hoover et al. 1972; Hawke et al. 1972). Experimental data in the megabar region are still insufficiently precise. In particular, we cannot use these data to estimate the pressure of transition to the metallic phase, but we can only say that the transition pressure exceeds 2.2 Mbar.

For solid and liquid hydrogen at low temperatures one may use the equation of state in the Debye approximation, with the Grüneisen lattice parameter γ_{lat}, which does not depend on temperature. The adiabatic index of the material coincides with the thermodynamic Grüneisen parameter γ_T.

From the virial equation of state of a weakly non-ideal gas, it follows that γ_T should increase with increasing density (Trubitsyn 1972a) and decrease with increasing temperature. Trubitsyn gave estimates for the upper limit of the value of γ_T and the adiabatic temperatures for a gas-liquid molecular hydrogen. In the paper of Slattery and Hubbard (1973) the first direct calculations of the thermodynamic functions of gas-liquid hydrogen were given for a range of values of pressure and temperature. The Monte Carlo method was used to calculate statistical sums on a computer. Because of insufficient calculating time, these calculations may still contain some inaccuracies, but, if the effective potential of interaction between molecules is known, in principle this method may give a substantially greater accuracy than the Debye approximation for liquids. In the paper by Kerley (1972) the energy and density of deuterium were calculated over a large range of pressure and temperature using liquid perturbation theory.

Taking into account the results of the papers by Trubitsyn (1972a), by Slattery and Hubbard (1973) and by Kerley (1972) in the region of pressure and temperature corresponding to the assumed adiabat, one may write the equation of the adiabat of molecular hydrogen in the form

$$\log T = \log T_{id} = \log T_1 + 0.4(x+2), \quad x < -2;$$

$$\log T = \log T_1 + 0.76(x+2) + 0.09(x^2-4), \quad x > -2$$

(19)

where T_1 is the boundary temperature, and $x = \log \rho$.

For pressures $p > 10^5$ bar in the Debye approximation for rotating molecular hydrogen at $T > \theta$ we have

$$p(\rho, T) = p(\rho, 0) + \frac{3RT}{\mu} \rho \gamma_{lat}$$

(20)

where γ_{lat} is the lattice Grüneisen parameter for a Debye spectrum, μ is the molecular weight and R is the gas constant. For $\gamma_{lat} = 5/3\gamma_T$ (Zharkov et al. 1971; Trubitsyn 1972a), the calculated temperature distribution and pressure along an adiabat of hydrogen with starting conditions $\rho_1 = 1.7 \times 10^{-4}$ g cm^{-3}, $p_1 = 1$ bar and $T_1 = 140°$K are close to the results of the paper by Slattery and Hubbard (1973).[7]

Metallic Hydrogen. In the Debye approximation, the equation of state of metallic hydrogen is of a form analogous to (20) where $\gamma_{lat} = \gamma_T = \gamma$ (Zharkov et al. 1971; Trubitsyn 1971).

The most detailed calculations for various types of crystals of metallic hydrogen are given by Brovman et al. (1972). It turns out that for most lattices the difference in energies may be quite small (Yordanskii et al. 1973). The calculations of Brovman et al. for $p(\rho, 0)$ for stable lattices essentially coincide with the equations of state of Trubitsyn (1971) tabulated by Zharkov et al. (1971, 1974c).

The biggest uncertainty may be in the value of the Grüneisen parameter. For a lattice composed of Coulomb nuclei $\gamma = 0.5$. Calculations from the Thomas-Fermi model give $\gamma \simeq 2/3$ for $p > 10^9$ bar. Brovman et al. obtained values of $\gamma = 0.5$ to 0.9 for pressures of the order of one megabar.

In the paper by Hubbard and Slattery (1971) numerical calculations were carried out statistically for the thermodynamic functions of metallic hydrogen using a pseudo-potential. It turns out that in the pressure region of several Mbar, γ is almost independent of temperature and $\gamma \simeq 0.6$.

Using the value of Hubbard and Slattery for γ, we obtain the equation of an adiabat of metallic hydrogen in the form

$$T_a(\rho) = T_2 (\rho/\rho_2)^\gamma$$

(21)

---

[7]See Fig. 2 of Smoluchowski, p. 16.

where ρ_2 is the density at the point of the metallic transition, and T_2 is the final temperature on the molecular hydrogen adiabat.

The equation of an adiabat of metallic hydrogen in Jupiter may also be calculated independently of data on molecular hydrogen, which are greatly uncertain in the pressure region 10^3 bar $< p < 10^6$ bar. Since an adiabat is by definition an equation of constant entropy, it is sufficient to equate the entropy of metallic hydrogen to the entropy of the molecular hydrogen at the boundary point on the surface of Jupiter (Hubbard 1969). In this case, expressions for the entropy of both phases can be written directly using the corresponding chemical constants.

In liquid hydrogen the metallic transition should not occur suddenly, but should be spread out over some range of pressure or temperature with an anomalous increase of density.[8] But in order to simplify calculations, it is convenient to assume that the transition is associated with some average pressure p^* and an effective density jump $\Delta\rho$. So far, the pressure p^* is not well known. According to Zharkov *et al.* (1971) and Trubitsyn (1971) it may be on the order of 2–10 Mbar. The dependence of p^* on temperature is weak (Trubitsyn 1971; Kerley 1972) and as yet cannot be accurately determined. The dependence of the effective density jump on temperature in the region of the transition can be estimated if one knows the coefficient of thermal expansion $\alpha(p, T)$ of molecular and metallic hydrogen. It turns out that the effect of temperature on $\Delta\rho$ is weak and depends sensitively on the value of γ of both phases which is still not sufficiently known.

For computational purposes we have calculated adiabats of hydrogen with various boundary temperatures for $p_1 = 1$ bar, $T_1 = 140°K$ (Model 1), $T_1 = 250°K$ (Model 2) and $T_1 = 90°K$ (Model 3), as well as the null isotherm of hydrogen. Part of adiabat 1 on both sides of the phase transition is shown in Fig. 7. This adiabat assumes that $\gamma = 0.6$. For $\gamma = 0.9$, the density jump at the phase transition is equal to zero. The corresponding model is designated as $1'$. For investigation of the effect of uncertainty in the equation of state of hydrogen, a modified or limiting adiabat of molecular hydrogen is considered, Curve 1a, which continuously transforms to an adiabat of metallic hydrogen with $\gamma = 0.6$. Adiabat 1a can also be obtained from adiabat 1 by adding $\sim 17\%$ helium to the hydrogen. A slight enhancement of the helium concentration in the molecular hydrogen layer with respect to the metallic hydrogen could correspond to a greater solubility of helium in the molecular hydrogen.

Hydrogen-Helium Adiabat. The adiabat of a mixture of hydrogen and helium for a relatively weak concentration of helium is close to the hydrogen adiabat calculated from (20) and (21). The equation of state of helium along an adiabat for $p < 10^3$ bar corresponds to a weakly non-ideal gas, and for $p > 10^5$ bar can be taken in the Debye approximation analogous to (21).

---

[8]See also p. 88.

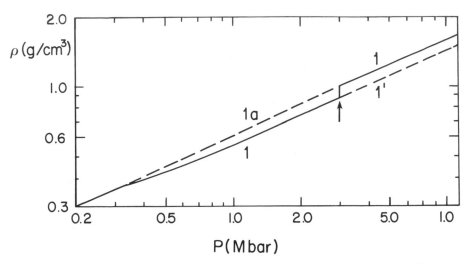

Fig. 7.   Details of the $T_1 = 140°K$ hydrogen adiabat in the vicinity of the phase transition near 3 Mbar. Curve 1 is calculated with $\gamma = 0.6$. The dashed curves are modified adiabats with no density discontinuity; Curve 1′ sets $\gamma = 0.9$ in the metallic phase; Curve 1a is a modified molecular hydrogen adiabat.

As a result it turns out that along an adiabat $\rho_{He} (p,T_a) \simeq k\rho_{H_2} (p, T_a)$ where $k \simeq 2$ for $p \leqslant 3 \times 10^5$ bar, and then increases proportional to the logarithm of the pressures to $k = 2.5$ at $p = 3$ Mbar. For $p > 3$ Mbar $\rho_{He} (p, T_a) \simeq 2.5\rho_H$ (p, T_a).

Rock-Ice Adiabats. Null isotherms for water, methane, ammonia, oxides and other cosmochemical compounds are given in Figs. 2, 3 and 4. The equation of state of a mixture is calculated by using formula (4). The dependence of density on temperature for cosmochemical compounds is still only approximately known. Thus, for the mixture at temperatures greater than the Debye temperature, we use the Mie-Grüneisen equation with effective parameters

$$p(\rho, T) = p(\rho, T = 0) + \frac{3RT}{A} \rho\gamma \qquad (22)$$

where A is the average atomic weight.

We adopt various values of A for different versions of rock-ice, using the abundances of Table I. For rock-ice I, $A = 5.6$; for rock-ice II, $A = 7.8$; for rock-ice III, $A = 8.4$; and for rock, $A = 25$. In the models, we assume two versions: $\gamma = 0.5$ and $\gamma = 1.0$.

Envelope Models

As a first approximation, one may assume that the mass concentration of the envelope is entirely hydrogen-helium, i.e., $X + Y = 1$ where X and Y are the mass concentrations of hydrogen and helium respectively. The density distribution in this case differs little from the distribution which is obtained by including a Z component (other substances besides hydrogen and helium). In hydrogen-helium models, the ratio Y/X in the envelope of the planet is taken to satisfy the parameters of the gravitational field. It turns out that the chemical composition of the core has no effect on estimates of the composition of the envelope of Jupiter and has almost no effect for Saturn. This is due to the fact that the distribution density in the envelope is basically constrained by the values of the gravitational moments J_2 and J_4, while the dimensions and mass of the core determine the value of the mean density of the planet. The density distribution ρ (β) in a pure hydrogen-helium model of the envelope can be regarded as a phenomenological relation. It is used below for construction of models and corresponds to various versions of the chemical composition.

Hydrogen-Helium (Phenomenological) Models of the Envelope. Figure 8 shows a selection of Y/X relations in the envelopes of planets as a function of J_2. The number of the model in Fig. 8 corresponds to the number of the adiabat. The observed values of the moments J_2 and J_4 are shown with horizontal lines, and the errors are denoted with vertical error bars. The derived value of X is equal to 0.68 for Model 1 for both Jupiter and Saturn (boundary temperature $T_1 = 140°K$ at $p_1 = 1$ bar). In Model 2 of Jupiter ($T_1 = 250°K$) $X = 0.5$, and in Model 3 of Saturn ($T_1 = 90°K$) $X = 1.0$. Envelope models for both Jupiter and Saturn consist of two layers: an outer one in which the hydrogen is in the molecular phase, and an interior layer of metallic hydrogen.

The values of the dimensionless radius β^* of the phase transition surface (for $p = 3$ Mbar) and the dimensionless mass q^* enclosed within the surface depend little on the boundary temperature and they vary in Models 1 and 2 of Jupiter by about 1.5% and in Models 1 and 3 of Saturn $\sim 2\%$. The values of β^* and q^* for Model 1 are equal respectively to 0.76 and 0.76 for Jupiter and 0.47 and 0.43 for Saturn.

Model 1' of Jupiter, in which the thermal part of the equation of state of metallic hydrogen is changed (the Grüneisen parameter γ_H is taken equal to 0.9 instead of $\gamma_H = 0.6$ in Model 1) has an abundance X in the envelope about 0.04 smaller than in Model 1 ($\Delta X = -0.04$). The comparatively large value of ΔX for Jupiter is related to the fact that the metallic hydrogen region comprises the major part of the planet. In contrast, for Saturn the metallic hydrogen region is so small that it has little effect on the parameters of the model. On the other hand, Model 1a for Saturn is important, since in this model we modify the molecular hydrogen adiabat in the region of the phase transition (Fig. 7) with no jump in density at the transition. In Saturn, the molecular hydrogen envelope comprises the majority of the planet (from 1 to 0.5 R). For Jupiter in contrast, the region of the molecular hydrogen en-

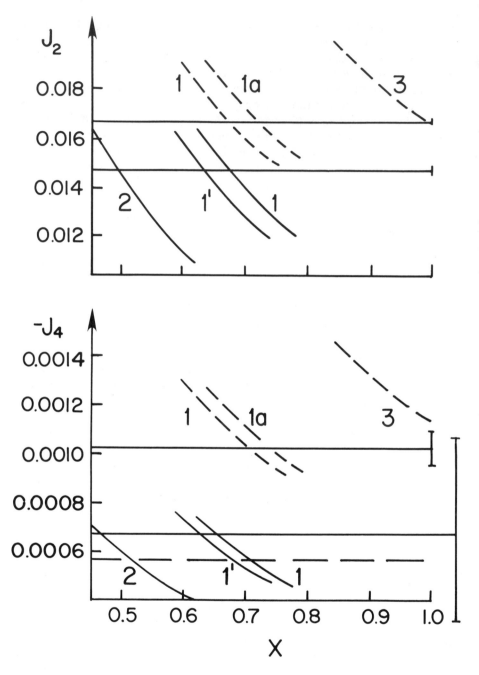

Fig. 8. Gravitational moments J_2 and J_4 as a function of free hydrogen abundance X in the envelope of Jupiter (solid curves) and Saturn (dashed curves), for the various types of models. Horizontal lines with error bars are observed values for J_2 and J_4; the horizontal dashed line is the new Pioneer 11 result for Jupiter's J_4.

velope is a relatively thin shell above the layer $\sim 0.8\ R_J$. Model 1a differs from Model 1 for Saturn by $\Delta X = 0.05$.

We test the validity of models by calculating the gravitational moment J_4. In Fig. 8 it is evident that all models, except for Model 3 of Saturn, have a satisfactory agreement with the observed value of J_4 within the limits of error of the observations.

From inspecting hydrogen-helium models it appears that Model 3 for Saturn is a limiting model for small temperatures T_1, for two reasons. First, the concentration of helium in the envelope is zero. In models with still lower boundary temperatures, to satisfy J_2 one would have to have a substance with a density lower than that of hydrogen. Secondly, the calculated and observed values of J_4 begin to differ significantly as T_1 decreases. For Jupiter, the limiting model is Model 2, in which the rock-ice core disappears. For models with still higher values of T_1, one finds a mass deficit in the central region of the planet. Thus the permissible models for Jupiter are bounded by T_1 from above and for Saturn from below. We have not attempted to bound this parameter from the other side (on the low side for Jupiter, and on the high side for Saturn) since the corresponding values of T_1 lie beyond bounds on temperature that are permissible from model atmospheres.

Types I and II Models for Envelopes. For convenience in constructing models, we have divided substances into classes according to volatility. The first group is the gas-components. For our purposes, we are interested in four possible gas compositions: GI (H_2 + He + Ne), GII (GI + CH_4), $GIII$ (GII + NH_3) and GIV ($GIII$ + H_2O). Substances which are in principle condensed at Jupiter and Saturn's orbits are divided into two groups: the ice-component and the rock-component. We will consider three versions of the ice-component corresponding to three gas versions: ice I (CH_4 + NH_3 + H_2O), ice II (NH_3 + H_2O) and ice III (H_2O). The fourth version of the gases (GIV) corresponds to the case when the condensate consists only of rock. In the latter we have SiO_2, MgO, Al_2O_3, FeS, FeO, and Fe. A uniform mixture of rock with ice is termed rock-ice-I, rock-ice-II, or rock-ice-III, corresponding to the ice-component version. In constructing models for giant planets, the ratio between gases and condensates is varied, but within each of these components the abundance of elements is taken in the solar proportion. The temperature in the cloud corresponding to each version is approximately the following (at pressures 10^{-5} to 10^{-7} bar): for GI-rock-ice-I, $T < 60°K$; for GII-rock-ice-II, $60 < T < 100°K$; for $GIII$-rock-ice-III, $120 < T < 160°K$; and for GIV-rock-ice, $T > 160°K$.

We construct two types of envelope models with Z-components. In models of Type I the envelope of the planet consists only of the gaseous component. The abundance by mass of free hydrogen, helium, and other gases is equal to respectively, X, Y, and Z_1; $X + Y + Z_1 = 1$. This may be regarded as a limiting model. It corresponds to a process of accumulation of the planet, where the material which at the time of the formation of the planet was in a condensed state falls to the core, and in the envelope there remains only the more volatile material corresponding to the G-component.

In models of Type II, we have in the envelope not only gases but also core material — one of the rock-ice components with abundance Z_2; $X + Y + Z_1 + Z_2 = 1$. At present we have no information which would allow us to prefer a given model of the envelope. An experimental determination of the ratio Y/X in the envelope of Jupiter or Saturn would allow one to determine $Z = Z_1 + Z_2$. Since $Z_1/(X + Y)$ should be in solar proportion, one could also determine Z_2. Since there are insufficiently precise data on the helium abundance, one must consider both types of models.

We proceed to the determination of various components and models of the envelopes of both types. Prescribing Z_1 in solar proportion ($Z_1 \sim 0.001$ for $G\mathrm{I}$, ~ 0.006 for $G\mathrm{II}$ and $G\mathrm{III}$, 0.014 for $G\mathrm{IV}$) permits one to find the abundance of the other component. In models of Type I, Y/X can be determined from a knowledge of J_2. Here Y/X can be different from the solar proportion (Table I) (Hirayama 1971; Lambert 1967). In models of Type II there is another parameter (Z_2) and thus it is necessary to introduce an additional relationship between components. We fix the ratio Y/X in the solar proportion; $Y/X = 0.26$. In order to obtain the correct value of J_2, in models of Type II we adopt the value $Z_2/(X + Y + Z_1)$.

For calculation of X, Y, Z_1, and Z_2 we use the equation of state (4) of mixtures, where the density is taken from a pure hydrogen-helium model of the envelope with a prescribed value Y/X (from the phenomenological model). The density of the various components is conveniently expressed in terms of the hydrogen density ρ_H:

$$\rho_\mathrm{He} = \kappa_1 \rho_\mathrm{H}, \quad \rho_{G'} = \kappa_2 \rho_\mathrm{H}, \quad \rho_{rock\text{-}ice} = \kappa_3 \rho_\mathrm{H}. \tag{23}$$

Here $\rho_{G'}$ is the density of the mixture of all the elements of the G-component (with the exception of hydrogen and helium) with abundance Z_1. The coefficients κ_1, κ_2, and κ_3 are functions of pressure, which may approximately be set constant in the entire envelope. For this, the possible effective coefficients are chosen from extremes in the range of possible values for $p = 10^4$–10^7 bar. This procedure is adequate to obtain the abundance to an accuracy of a few percent. For κ_1 we adopt a range from 2–2.5. The coefficients κ_2 and κ_3 change depending on the version of the G- and rock-ice-components. For each version, the value of κ_2 was varied for the purpose of estimating the uncertainty in the equation of state of the rock-ice-component and hydrogen.

The abundance in the envelope models of Type I is determined from the equation

$$\frac{X}{\rho_\mathrm{H}} + \frac{(1 - Z_1) - X}{\kappa_1 \rho_\mathrm{H}} + \frac{Z_1}{\kappa_2 \rho_\mathrm{H}} = \frac{X_0}{\rho_\mathrm{H}} + \frac{1 - X_0}{\kappa_1 \rho_\mathrm{H}}. \tag{24}$$

Here X_0 is the abundance of hydrogen in the phenomenological hydrogen-helium model; κ_2 varies from 2 to 4 for various versions of the G-component.

On solving (24), we find to an accuracy of 1% that $X \approx X_0, Y \approx Y_0 - Z_1$ for the versions GI–$GIII$. For GIV, we find $X \approx X_0 + Z_1, Y \approx Y_0 - 2Z_1$.

In envelopes of Type II, the concentration of hydrogen is found from the equation

$$\frac{X}{\rho_H} + \frac{0.26X}{\kappa_1\rho_H} + \frac{Z_1}{\kappa_2\rho_H} + \frac{1 - Z_1 - 1.26X}{\kappa_3\rho_H} = \frac{X_0}{\rho_H} + \frac{1 - X_0}{\kappa_1\rho_H}. \tag{25}$$

Z_1 for various versions of the G-component is chosen as in models of Type I.

The results of calculations for models of Type II are given in Table II. Varying κ_1, κ_2 and κ_3 in this case has a much stronger effect on the abundance than in the calculation of Type I models. The values of the abundances X, Y, and Z_2 in each line of Table II correspond to the values $\kappa_1 = 2$ and 2.5 respectively. For κ_3 we chose the most extreme value for each version of G-rock-ice. The uncertainty of κ_2, as for the Type I model, is unimportant. In the last line, for each model (G-rock-ice) we show the largest and smallest values of the abundances X, Y, and Z_2.

TABLE II
Abundance Parameters in Type II Envelopes

Boundary temperature	Envelope composition	κ_2	κ_3	X	Y	Z_1	Z
				Model 1			
$T_1 = 140°K$	GI–	4	3.5	0.70–0.72	0.18–0.19	0.001	0.12–0.09
$X_0 = 0.68$	rock-ice-I		4.5	0.71–0.73	0.18–0.19		0.11–0.08
(Jupiter and	GII–	2	4	0.71–0.72	0.18–0.19	0.006	0.10–0.09
Saturn)	rock-ice-II		5	0.71–0.73	0.18–0.19		0.10–0.07
	GIV – rock	3	8	0.72–0.73	0.19	0.014	0.08–0.07
			10	0.72–0.73	0.19		0.08–0.07
	G – rock			0.70–0.73	0.18–0.19		0.12–0.07
				Model 2			
$T_1 = 250°K$	GI–	4	3.5	0.56–0.60	0.15–0.16	0.001	0.29–0.24
$X_0 = 0.5$	rock-ice-I		4.5	0.58–0.62	0.15–0.16		0.27–0.22
(Jupiter)	GII–	2	4	0.57–0.64	0.15–0.16	0.006	0.28–0.23
	rock-ice-II		5	0.58–0.62	0.15–0.16		0.26–0.21
	GIV – rock	3	8	0.60–0.64	0.16–0.17	0.014	0.23–0.18
			10	0.61–0.64	0.16–0.17		0.22–0.18
	G – rock			0.56–0.64	0.15–0.17		0.29–0.18

Core Models from Rock-Ice Material

Chemically homogeneous core models were calculated for rock-ice I, rock-ice II, rock-ice III, and rock. It is quite possible that the core material is differentiated: separate layers of rock component and ice, iron compounds (or free iron) and silicates, and silicates and ice. Core parameters and abundances of the elements comprising the core are obtained on the basis of uniform core models which should be close to the corresponding values for non-

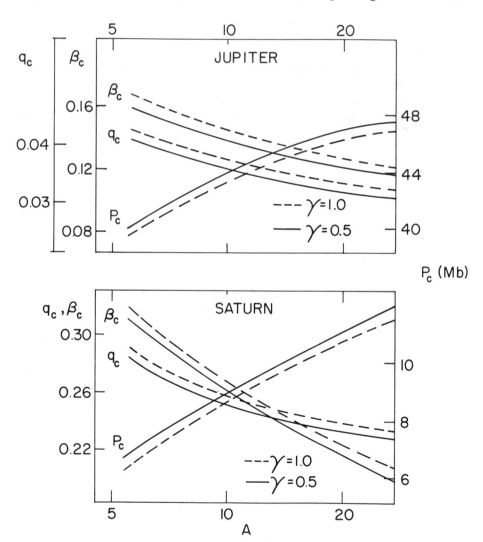

Fig. 9. Dimensionless core mass q_c and radius β_c, and pressure at core surface P_c, for rock-ice cores as a function of average atomic weight A for the core material. The value of γ varies between 0.5 and 1.0; adiabat 1 is assumed.

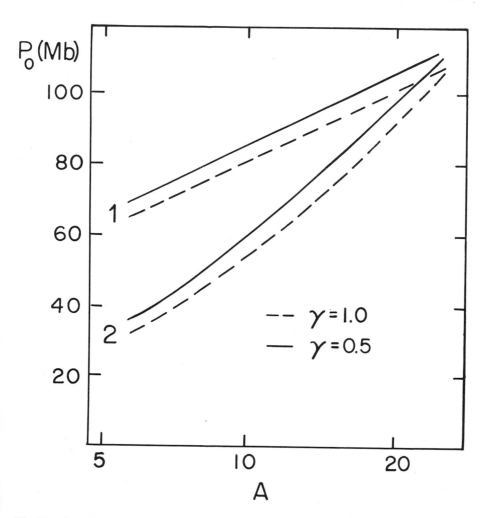

Fig. 10.   Pressure at the center of Jupiter (1) and Saturn (2) as a function of mean core atomic weight A, for various values of γ (adiabat 1).

uniform models. In differentiated core models, p and T in the center of the planet (p_0 and T_0) are higher than for the uniform core because of the steep behavior of $p(\beta)$ in rock material. Various versions of iron compounds in the cores of Jupiter and Saturn can be realized, depending upon the temperature of the protoplanetary cloud in the regions of these planets and on the character of processes of condensation in the cloud (Makalkin 1974). However, core models depend little on the chosen version for iron. (1) FeS + FeO; (2) FeO; (3) Fe.

Results for the rock-ice cores in Model 1 ($T_1 = 140°K$) are shown in Figs. 9 and 10. The core parameters are the dimensionless radius β_c and mass q_c, and also the pressure at the core boundary and at the center of the

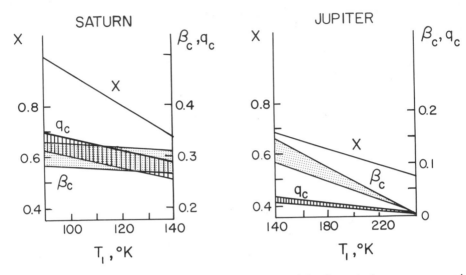

Fig. 11. Abundance of free hydrogen in envelope X, and the dimensionless core mass and radius q_c and β_c, as functions of boundary temperature T_1. The shaded regions correspond to changing the core composition from rock-ice-I (upper boundary) to rock-ice-III for Saturn, and from rock-ice-II to rock-ice for Jupiter.

planet as a function of the average atomic weight A for rock-ice mixtures. For each mixture the adiabat was calculated using a Grüneisen parameter $\gamma = 0.5$ and $\gamma = 1$, where $\gamma = 1$ is the more likely. Figure 10 shows that in the case of a rock core, the pressure at the center of Saturn is greater than 100 Mbar, as it is for Jupiter.

Figure 11 shows the mass and core dimensions as a function of the boundary temperature T_1. The temperature extremes correspond to Models III and I for Saturn and Models I and II for Jupiter. The shaded areas correspond to various core compositions from rock-ice I to rock-ice III for Saturn and from rock-ice II to rock for Jupiter (for $\gamma = 1$). In order to show the relationship between the mass of the core and the chemical composition of the envelope, the same diagram shows the hydrogen abundance in the envelope X for pure hydrogen-helium models.

Structure of Jupiter and Saturn and p-T Conditions in Their Interior

Table III gives the pressure p, density ρ, temperature T and dimensionless mass q at four important layers: 1) at the cloud level ($p_1 = 1$ bar), 2) at the level of hydrogen metallization, 3) at the boundary of the rock-ice core and the envelope, and 4) at the center of the planet. Results are given for various possible envelope models and core compositions. At levels 2 and 3 we give two densities separated by a semi-colon. These refer to the densities above and below the given layer since at this layer there is a discontinuity in the density. Two values for the temperature at the center are given corre-

TABLE III
Interior Characteristics at Selected Levels

Model No.	Level No.	β	p(Mbar)	ρ(g cm^{-3})	$T(10^3\,°K)$	q
JUPITER						
1($T_1 = 140°$K)	1	1	10^{-6}	2.02×10^{-4}	0.14	1
	2	0.765	3	1.12; 1.22	9.5	0.766
rock-ice II core	3	0.15	42	4.4; 13	18.5	0.037
	4	0	76	16	21-23	0
1; rock core	3	0.12	47	4.7; 20	19	0.030
	4	0	110	28	23-27	0
1'($\gamma = 0.9$);	3	0.19	38	3.8; 13	26	0.06
rock-ice II core	4	0	110	20	30-34	0
2($T_1 = 250°$K),	1	1	10^{-6}	1.27×10^{-4}	0.250	1
no core	2	0.76	3	1.16; 1.28	14	0.76
SATURN						
1($T_1 = 140°$K)	1	1	10^{-6}	2.02×10^{-4}	0.14	1
	2	0.465	3	1.12; 1.22	9.5	0.427
rock-ice II core	3	0.27	8	1.9; 6.6	12	0.26
	4	0	46	13	17-25	0
1; rock-ice I	3	0.3	6.6	1.7; 5.1	11.7	0.28
core	4	0	34	10	16-22	0
1; rock core	3	0.21	12	2.3; 11	13.5	0.23
	4	0	110	28	21-36	0
3($T_1 = 90°$K)	1	1	10^{-6}	2.64×10^{-4}	0.09	1
	2	0.448	3	1.0; 1.1	6	0.44
rock-ice II core	3	0.30	7	1.9; 5.2	7.5	0.33
	4	0	50	12	11.2-16	0
3; rock-ice I	3	0.33	5.5	1.3; 4.9	7	0.35
core	4	0	38	11	10.5-15	0

sponding to Grüneisen parameters $\gamma = 0.5$ and $\gamma = 1$. The central temperature T_0 for various Saturn models differs by a larger amount (from 10,000–36,000°) than for Jupiter models (20,000–30,000°). Note that in Model 1 with a rock core where $\gamma = 1$, the temperature T_0 at the center of Saturn is greater than it is for Jupiter by 8000 or 9000 degrees (for equal values of p_0) because of the large dimensions of the region with a steep adiabat. For rock cores with $\gamma = 0.5$, T_0 is similar for both planets.

Model 1 ($T_1 = 140°$K) for Jupiter and Saturn is given in Table IV. For this model, the core consists of rock-ice II for both planets. The gravitational fields coefficients J_2, J_4, and J_6 have been calculated using the methods of

TABLE IV

Interior Models for Jupiter and Saturn

β	p(Mbar)	ρ(g cm^{-3})	g(m sec^{-1})	q
		JUPITER		
0	78	16.5	0	0
0.055	72	16.4	17.9	0.0021
0.095	62	15.2	29.6	0.0104
0.146	43.6	13.2	41.8	0.0347
0.150	42.3	13.0	42.5	0.0369
0.150	42.3	4.45	42.5	0.0369
0.193	37.4	4.18	34.2	0.0495
0.241	33.0	3.90	31.5	0.071
0.291	29.1	3.64	30.7	0.101
0.337	25.6	3.40	31.1	0.137
0.385	22.1	3.16	32.0	0.184
0.443	18.2	2.86	33.1	0.252
0.490	15.3	2.61	33.9	0.316
0.541	12.3	2.33	34.5	0.392
0.588	9.89	2.08	34.9	0.467
0.641	7.43	1.82	34.9	0.556
0.690	5.46	1.59	34.7	0.639
0.732	3.98	1.38	34.3	0.712
0.765	3.00	1.22	33.8	0.766
0.765	3.00	1.121	33.8	0.766
0.800	2.18	0.953	32.9	0.816
0.850	1.26	0.738	31.5	0.881
0.874	0.910	0.644	30.7	0.910
0.899	0.602	0.559	29.9	0.937
0.919	0.397	0.479	29.2	0.957
0.938	0.238	0.390	28.5	0.973
0.959	0.108	0.291	27.7	0.987
0.968	0.0610	0.233	27.3	0.992
0.976	0.0331	0.176	27.0	0.996
0.988	0.00611	0.0751	26.4	0.9992
0.9917	0.00223	0.0430	26.2	0.9997
0.9953	0.00028	0.0114	26.0	0.99996
1.0	0.000001	0.00020	25.8	1.0

Zharkov *et al.* (1973). The response functions $\Lambda(n)$ which were introduced in this paper allow one to estimate the values of the higher gravitational moments J_8 and J_{10}. The values of the gravitational moments and the moments of inertia are given in Table V and they correspond to observed values for Jupiter (Klepczinski *et al.* 1971; Hubbard and Van Flandern 1972; Carr 1971) and for Saturn (Klepczinski *et al.* 1971; Dollfus 1970; Brouwer and Clemence 1961). In Table VI we give the changed values of the abundances

TABLE IV (Continued)
Interior Models for Jupiter and Saturn

β	p(Mbar)	ρ(g cm^{-3})	g(m sec^{-1})	q
		SATURN		
0	48	13.5	0	0
0.055	45	13.4	12	0.0032
0.091	41	12.9	19.5	0.0143
0.147	31.5	11.5	29.3	0.0556
0.198	21.1	9.89	36.2	0.125
0.247	11.6	7.76	40.2	0.216
0.267	8.12	6.78	40.9	0.257
0.267	8.12	1.89	40.9	0.257
0.298	6.91	1.76	35.4	0.276
0.338	5.63	1.61	30.3	0.306
0.388	4.41	1.45	26.3	0.348
0.445	3.32	1.28	23.2	0.406
0.465	3.00	1.22	22.4	0.427
0.465	3.00	1.121	22.4	0.427
0.500	2.54	1.030	21.1	0.465
0.549	1.99	0.910	19.6	0.521
0.599	1.54	0.804	18.4	0.581
0.650	1.16	0.710	17.3	0.644
0.699	0.849	0.626	16.5	0.708
0.749	0.588	0.554	15.7	0.774
0.799	0.374	0.468	14.9	0.839
0.848	0.210	0.371	14.2	0.899
0.873	0.145	0.326	13.8	0.926
0.899	0.0890	0.270	13.4	0.952
0.917	0.0564	0.225	13.1	0.968
0.934	0.0331	0.176	12.8	0.981
0.960	0.00964	0.0946	12.2	0.994
0.969	0.00470	0.0644	12.0	0.997
0.979	0.00171	0.0368	11.8	0.9989
0.9864	0.000471	0.0163	11.7	0.9997
0.9919	0.000111	0.00604	11.5	0.99992
0.9958	0.000023	0.00199	11.4	0.99998
1.0	0.000001	0.00020	11.3	1.0

in the envelope and the corresponding parameters of the gravitational field for the average density of Jupiter using the results of Dollfus (1970) and the rotation period of Saturn of Dollfus (1963). From Table VI we see that the new value of the average density of Jupiter leads to a reduction of X in the envelope by -0.03. Increasing the period of rotation of Saturn leads to a reduction of X in the envelope of -0.03 and to a reduction of J_4 which improves the agreement of the calculated value with that observed.

TABLE V

Gravitational Moments, Moments of Inertia, and Constant of Precession

Planet	Model	$J_2 \, 10^3$	$J_4 \, 10^4$	$J_6 \, 10^5$	$J_8 \, 10^6$	$J_{10} \, 10^7$	$\dfrac{A}{Ma_1^2}$	$\dfrac{C}{Ma_1^2}$	H
JUPITER	1 ($T_1 = 140°K$)	14.7	−6.1	4.1	−3.0	2.2	0.247	0.262	0.0562
	2 ($T_1 = 250°K$)	14.5	−5.6	3.3	−2.2	1.5			
SATURN	1 ($T_1 = 140°K$)	16.6	−10.8	11.5	−14	18			
	2 ($T_1 = 90°K$)	16.7	−11.4	12.6	−16	21	0.199	0.216	0.0770

TABLE VI

*Corrected value of X in the envelope and gravitational moments
upon changing observational data,
for phenomenological Model 1 of Jupiter and Saturn*

Planet	$\bar{\rho}(g\ cm^{-3})$	$\tau(hr)$	X	$J_2\ 10^3$	$J_4\ 10^4$	$J_6\ 10^5$
JUPITER	1.357	9.92	0.65	14.7	-6.2	4.0
SATURN	0.703	10.63	0.65	16.6	-10.6	11

Chemical Composition and Cosmogonical Conclusions

In Table VII we give the abundance of free hydrogen, helium, and rock-ice component for the planet as a whole: X_p, Y_p and $Z_{rock\text{-}ice}$. They are determined by using the formulas $X_p = X(1 - q_c)$, $Y_p = Y(1 - q_c)$, $Z_{rock\text{-}ice} = Z_2 (1 - q_c) + q_c$. The abundance of Z_p for the gas component is not included in the table, because it is negligible. The three entries at each position in Table VII refer to the different chemical composition version (left to right): GI-rock-ice I, GII-rock-ice II, GIV-rock. Composition III differs negligibly from that of version II.

In Table VII note that the ratio of rock-ice mass in Jupiter to that in Saturn is about 1.0–1.2 for models of Type II with $T_1 = 140°K$. When $T_1 = 190°K$ for Jupiter and $140°K$ for Saturn, this ratio increases to 1.2–1.5. We assume that all or almost all of the condensate which was present in the zone of formation of a given planet was accumulated into the planet. Then one may use the estimates of the rock-ice mass in Jupiter and Saturn to derive the corresponding mass in the zones of the protoplanetary clouds.

In models of Type I, the rock-icc mass in Jupiter relative to Saturn is about 0.5–0.15. This could be interpreted as being caused by different temperature conditions in the zone of accumulation during the formation of these planets.

From a knowledge of the rock-ice abundance one may estimate the minimum mass of the protoplanetary cloud in the zone of formation. The mass of all of the material in the zone relative to the mass of the planet is given by $M_Z/M = (Z_{rock\text{-}ice}/Z_{rock\text{-}ice}°)$, where $Z_{rock\text{-}ice}°$ is the assumed rock-ice abundance, in the protoplanetary cloud, assumed identical with solar abundance in percentage, taken from Table I. The mass lost from the zone will be equal to M_Z/M-1. The mass of gas which is dissipated from Jupiter's zone differs considerably for various models. For Saturn, this quantity is between 15 and 24 planetary masses for models with rock-ice I to rock-ice III type cores.

Gravitational Field of Jupiter

Referring to Fig. 8 and Table V, one may conclude that within the limits of error the model for Jupiter should lie between Curves 1 ($T_1 = 140°K$) and 2 ($T_1 = 250°K$), and that a good estimate would be $T_1 \sim 200°K$. Param-

TABLE VII
Chemical Composition of Jupiter and Saturn

T_1 (°K)	Model type	X_p			Y_p			$Z_{rock-ice}$			$M_{rock-ice}$ ($\oplus = 1$)			M_z/M		
								JUPITER								
140	I	0.65	0.65	0.67	0.31	0.30	0.29	0.04	0.04	0.03	13	13	10	2	3	7
	II	0.69	0.69	0.71	0.18	0.18	0.18	0.13	0.12	0.10	41	38	32	7	9	22
250	I	0.51	0.51	0.51	0.47	0.47	0.47	0.018	0.013	0.0045	5.7	4.1	1.4	1	1	1
	II	0.60	0.61	0.64	0.16	0.16	0.17	0.24	0.23	0.18	76	73	57	13	18	40
								SATURN								
140	I	0.49	0.50	0.54	0.23	0.23	0.23	0.28	0.26	0.22	27	25	21	16	20	49
	II	0.51	0.53	0.56	0.14	0.14	0.14	0.35	0.33	0.29	33	31	28	19	25	64
90	I	0.66	0.68	0.72	0.	0.	0.	0.34	0.32	0.28	32	30	27	19	25	63

eters for this model can easily be obtained by interpolating the numbers given for Models 1 and 2 in the tables. Clearly, a measurement of J_6 for Jupiter to two significant figures would be a decisive gravitational experiment. This would allow one to precisely define the temperature T_1 and the corresponding adiabatic model.

III. FIGURE AND GRAVITATIONAL FIELD

The theory of figures of self-gravitating bodies in a state of hydrostatic equilibrium is applicable to the liquid bodies of the giant planets. The problem reduces to finding the equations of the equipotential (level) surfaces of the body, $r = r(\theta)$, as determined from the condition

$$U(r, \theta) = V(r, \theta) + Q(r, \theta) = \text{const} \tag{26}$$

where U, V and Q are the total, gravitational and centrifugal potentials respectively

$$V = G \, \frac{\rho(r')}{|\mathbf{r} - \mathbf{r}'|} \, d^3r'$$
$$Q = \frac{\omega^2}{2} \, r^2 \sin^2 \theta \,. \tag{27}$$

In this chapter we limit the discussion to equilibrium figures which differ only slightly from spheres; we shall call such figures *spheroids*. In this approach we develop the theory in powers of a small parameter which depends on the departure from spherical symmetry. For this purpose one may use q, the ratio of the equatorial centrifugal acceleration to the gravitational acceleration, at least in first approximation (see the chapter by Hubbard and Slattery, p. 178). In our theory, it is more convenient to use the parameter m. These quantities are defined by

$$q = \omega^2 a_1^3 / GM \tag{28}$$
$$m = (S_1/a_1)^3 q = 3\omega^2/(4\pi G\bar{\rho})$$

where M is the mass, $\bar{\rho}$ is the average density of the body, a_1 is the equatorial radius of the body and S_1 is its average radius. In first approximation, q and m are equivalent. It is also useful to define the flattening or oblateness of the body, $e_1 = (a_1 - b_1)/a_1$, where b_1 is the polar radius. The oblateness e_1 and the parameters q and m are all of first order in smallness.

The first-order theory of equilibrium figures of nonuniform planets was originated by Clairaut in 1743. Clairaut derived an integro-differential equation for the oblateness of level surfaces as a function of depth. In this approximation, where terms of order e^2 are discarded, the level surfaces are all ellipsoids of revolution.

Laplace addressed the problem of finding the equation for an equilibrium spheroid in terms of powers of q. For this, Legendre and Laplace separated the potential V into a sum of potentials produced by material exterior and interior, respectively, to a level surface passing through a given point. They then expanded the quantity $|\mathbf{r} - \mathbf{r}'|^{-1}$, which appears in Eq. (27), in a series of Legendre polynomials. This procedure has been repeatedly criticized since the expansion actually diverges in a small region between the sphere and the level surface passing through the point (Appell 1911).

Darwin (1899) derived a system of two integro-differential equations for the functions e and k, where k characterizes the second-order departure of the spheroid from an ellipsoid. Darwin used an artificial technique in his derivation which is still essentially equivalent to the Legendre-Laplace technique. The second-approximation theory of figures is discussed in detail in a paper by DeMarcus (1958) and in Kopal's (1960) book.

In 1903, Lyapunov derived a rigorous method for solving the Clairaut-Laplace problem without resorting to the Legendre expansion. In place of the variables r and θ, Lyapunov introduced the variables s and θ and expanded the potential in a convergent Taylor series in terms of powers of an auxiliary parameter. In this way, he proved the existence of solutions and produced a technique which in principle allows one to solve for figures with arbitrary accuracy. Although Lyapunov's approach is less convenient for practical calculations, one may show (Trubitsyn 1972*b*) that the respective expansions of Laplace and Lyapunov turn out to be equivalent. This shows that the Legendre polynomial expansion, which diverges in a small region, becomes convergent after integration, and corresponds precisely to an expansion in powers of the small parameter m.

For current study of the giant planets it is necessary to carry the theory to still higher order. Lanzano (1962) obtained a third-order expression for the potential but did not explicitly calculate the coefficients. Zharkov and Trubitsyn (1969, 1970) and Zharkov *et al.* (1972, 1973) derived a third-order theory and a corresponding numerical computation scheme, tested it with simple analytic models, and used it for analyses of the gravitational fields of the giant planets. As discussed elsewhere in this volume (p. 000), Hubbard *et al.* (1975) have developed a fourth-order technique which does not use level surfaces. Kopal (1973), and Kopal and Mahanta (1974) have independently derived third and fourth-order schemes, but without generating explicit solutions. Ostriker and Mark (1968) derived yet another scheme for astrophysical applications which does not use perturbation theory at all.

For a planet in hydrostatic equilibrium, the interior density distribution can be expressed as a one-dimensional array of the values of the density on level surfaces, as characterized by some index, e.g., their average radius s. At the same time, the external gravity field is expressed in terms of a one-dimensional array of coefficients in a Legendre expansion. Thus one may pose the inverse problem of determining the density distribution for a planet in hydrostatic equilibrium from a knowledge of its external gravity field (Zharkov and Trubitsyn 1974*b*; Hubbard 1974). Alternatively (and more straightforwardly), one may test a given interior model by calculating its external gravity field and comparing it with observations.

Expansion of the Gravitational Potential and Equations for Level Surfaces

We express the gravitational potential $V(r, \theta)$ within the planet as

$$V(r, \theta) = \frac{G}{r} \sum_{n=0}^{\infty} [r^{-n}D_n + r^{n+1}D_n']p_n(t)$$

$$t = \cos\theta$$

$$D_n = \int_1 \rho(r', t')r'^n p_n(t')d^3r'$$

$$D_n' = \int_2 \rho(r', t')r'^{-(n+1)}p_n(t')d^3r'$$

(29)

where we have used the expansion of $|\mathbf{r} - \mathbf{r}'|^{-1}$ in Legendre polynomials, and volumes 1 and 2 are interior and exterior respectively to a level surface passing through (r, θ). Questions of the convergence of this expansion were discussed in the previous section. The quantities D_n and D_n' are multipole moments of the interior and exterior mass distributions, respectively.

We now transform from the variables r, t to the new variables ℓ, t, where ℓ is a generalized radius and is constant for a given level surface, e.g., its polar, equatorial, or average radius. Then (Trubitsyn 1972b; Kopal 1973)

$$D_n(\ell) = \frac{2\pi}{n+3} \int_0^s d\ell'\rho(\ell') \int_{-1}^1 p_n(t)\frac{dr^{n+3}}{d\ell'}\,dt$$

$$D_n'(\ell) = \frac{2\pi}{2-n} \int_s^{s_1} d\ell'\rho(\ell') \int_{-1}^1 p_n(t)\frac{dr^{2-n}}{d\ell'}\,dt, \quad (n \neq 2)$$

(30)

$$D_2'(\ell) = 2\pi \int_s^{s_1} d\ell'\rho(\ell') \int_{-1}^1 p_n(t)\ell\,n\,r\,dt.$$

Next, we expand

$$r(\theta) = \ell\left[1 + \sum_{n=0}^{\infty} \ell_{2n}(\ell)p_{2n}(t)\right]$$

(31)

for a given level surface. In general, $\ell_0 \sim m$ and $\ell_2 \sim m^n(n \neq 0)$. In our work, we use $\ell = s$, the average radius, defined by

$$\frac{4\pi}{3}s^3 = \frac{2\pi}{3}\int_{-1}^1 r^3(\theta)dt$$

(32)

i.e., s is the radius of a sphere with volume equal to that enclosed within the level surface.

Writing

$$r(\theta) = s\left[1 + \sum_{n=0}^{\infty} s_{2n}(s)p_{2n}(t)\right] \tag{33}$$

and substituting in Eq. (32), we find

$$2 = \int_{-1}^{1}\left[1 + \sum_{n=0}^{\infty} s_{2n}p_{2n}(t)\right]^3 dt \tag{34}$$

i.e., the s_{2n} are not all independent.

Next, we substitute expansion (34) in Eq. (29) and (26), expand in powers of m, and retain all terms up to the given level of approximation n. We group coefficients of a given Legendre polynomial in the form

$$U(s, \theta) = \sum_{k=0}^{n} A_{2k}(s)p_{2k}(t). \tag{35}$$

Now on a level surface, $U(s, \theta) = U(s)$ by definition. So

$$A_{2k}(s) = 0 \quad (k = 1, 2, \ldots, n) \tag{36}$$

and

$$A_0(s) = U(s). \tag{37}$$

Equations (36) and (37) are a system of equations which determine the figure of the planet. The system is solved self-consistently in that the $\ell_{2n}(\ell)$ are determined by the $D_n(\ell)$ and $D_n{'}(\ell)$. Once the solution has been found, the gravitational moments are obtained from

$$J_n = \frac{-1}{Ma_1{}^n} D_n(\ell_1) \tag{38}$$

where ℓ_1 refers to the outside level surface of the planet. From the structure of the governing equations, it turns out that $J_{2n} \sim m^n$, so that a third order theory ($\sim m^3$) is sufficient to calculate J_2 to third order, J_4 to second order, and J_6 to first order. At present, a general theory has been developed by Zharkov and Trubitsyn (1969, 1970, 1975) to order m^5, and numerous planetary models have been calculated.

Table VIII gives results of a fifth-order calculation for the linear density model.

Estimate of Higher Moments by Extrapolation

The equations of figure for a variety of mathematical and realistic giant-planet models have been solved to third and fifth approximation. As we have

TABLE VIII
Gravitational Moments of Jupiter and Saturn[a]

	Approx.	$J_2 \times 10^{-2}$	$-J_4 \times 10^{-4}$	$J_6 \times 10^{-5}$	$-J_8 \times 10^{-6}$	$J_{10} \times 10^{-7}$
Jupiter,	1	1.45				
$m = 0.0830$	2	1.476	5.64			
$\rho = \rho_c(1 - s/s_1)$	3	1.4806	5.884	3.53		
	4	1.48055	5.935	3.464	2.78	
	5	1.48055	5.935	3.504	2.52	2.4
Model of Zharkov,	1	1.45				
Makalkin and	2	1.473	5.62			
Trubitsyn (1974)	3	1.4782	5.857	3.51		
	4	1.47820	5.908	3.443	2.76	
	5	1.47819	5.908	3.483	2.41	2.4
Model of Zharkov,	1	1.62				
Makalkin and	2	1.654	10.0			
Trubitsyn (1974)	3	1.6653	10.66	10.4		
Saturn,	4	1.66466	10.881	10.25	13.8	
$m = 0.143$	5	1.66454	10.881	10.58	12.07	2.0

[a]Calculated by Yefimov, Zharkov, Makalkin, and Trubitsyn using theory
of Zharkov and Trubitsyn (1975).

noted, the leading contribution to the moment J_{2n} is of order m^n. Thus one may define a function of index n:

$$\Lambda_{2n} = |J_{2n}|/m^n . \tag{39}$$

(Note that our definition of the quantities Λ_{2n} differs slightly from that in the chapter by Hubbard and Slattery, p. 183.) The Λ_{2n} are smooth functions of n and thus may be easily extrapolated to higher n. Comparison of subsequent fifth-order calculations with an earlier estimate of Λ_{2n} by extrapolation of third-order results indicates that estimating higher moments through this technique gives good results (Zharkov et al. 1972).

IV. CONCLUSION

We have summarized the basic physical considerations which are used to construct models of Jupiter. Presently available equations of state, together with a high-order theory of figures, allow us to place useful limits on the relative abundances of elements in Jupiter's deep interior. We have also shown that the models are consistent with the present thermal state of Jupiter. Further progress is likely to result from improved measurements of the fundamental boundary conditions on interior models, in particular from accurate measurement of the high-order gravity field components.

Acknowledgements. The authors wish to thank A. B. Makalkin and I. A. Tsarevskii for assistance with this work, and W. B. Hubbard for discussions of all of these problems during his stay in Moscow in 1973. This paper was translated and edited by W. B. Hubbard.

REFERENCES*

Anderson, J. D.; Null, G. W.; and Wong, S. K. 1974. Gravity results from *Pioneer 10* Doppler data. *J. Geophys. Res.* 79:3661–3664.

Appell, P. 1911. *Traité de mécanique rationelle.* Paris: Gauthier-Villars.

Aumann, H. H.; Gillespie, Jr., C. M.; and Low, F. J. 1969. The internal powers and effective temperatures of Jupiter and Saturn. *Astrophys. J.* 157:L69–L72.

Birch, F. 1952. Elasticity and constitution of the earth's interior. *J. Geophys. Res.* 57:227–286.

Brouwer, D., and Clemence, G. M. 1961. *Planets and satellites.* (G. P. Kuiper and B. Middlehurst, eds.) p. 31. Chicago: University of Chicago Press.

Brovman, E. G.; Kagan, Yu.; and Kholas, A. 1972. Properties of metallic hydrogen under hurst, eds.) p. 31. Chicago: University of Chicago Press.

Brovman, E. G.; Kagan, Yu.; and Kholas, A. 1972. Properties of metallic hydrogen under pressure. *Sov. Phys. JETP* 35:783–787 [*Zh. Eksp. Teor. Fiz.* 62:1492–1501].

Brown, L. W. 1975. Saturn emission near 1 Mhz. *Astrophys. J.* 198:L89–L92.

Burbidge, G. 1971. *Highlights in astronomy* Vol. 2, (C. DeJager, ed.), p. 328. Dordrecht-Holland: D. Reidel Publ. Co.

Carr, T. D. 1971. Jupiter's magnetospheric rotation period. *Astrophys. Lett.* 7:157–162.

Darwin, G. H. 1899. The theory of the figure of the earth carried to the second order of small quantities. *Mon. Not. Roy. Astron. Soc.* 60:82–124.

DeMarcus, W. C. 1951. Ph.D. dissertation. Yale University, New Haven, Connecticut.

———. 1958. The constitution of Jupiter and Saturn. *Astron. J.* 63:2–28.

Dollfus, A. 1963. Mouvements dans l'atmosphère de Saturne en 1960. *Icarus* 2:109–114.

———. 1970. New optical measurements of the diameters of Jupiter, Saturn, Uranus, and Neptune. *Icarus* 12:101–117.

Fesenkov, V. G. 1924. [*Astron. Zh.* 1:102].

Fesenkov, V. G., and Masevich, A. G. 1951. On the question of the structure and chemical composition of the major planets. [*Astron. Zh.* 28:317–337].

Gavrilov, S. V.; Zharkov, V. N.; and Leont'ev, V. V. 1975. [*Astron. Zh.* In press.]

Goldreich, P., and Soter, S. 1966. Q in the solar system. *Icarus* 5:375–389.

Grigor'ev, F. V.; Kormer, S. B.; Mikhailova, O. L.; Tolochko, A. P.; and Urlin, V. D. 1972. Experimental determination of the compressibility of hydrogen at densities 0.5–2 g/cm³. Metallization of hydrogen. *Sov. Phys. JETP (Letters)* 16:201–204 [*Zh. Eksp. Teor. Fiz. (Pis. Red.)* 16:286–290].

Gulkis, S., and Poynter, R. 1972. Thermal radio emission from Jupiter and Saturn. *Phys. Earth Planet. Interiors* 6:36–43.

Hawke, R. S.; Duerre, D. E.; Huebel, J. G.; Keeler, R. N.; and Klapper, H. 1972. Isentropic compression of fused quartz and liquid hydrogen to several Mbar. *Phys. Earth Planet. Interiors* 6:44–47.

Hirayama, T. 1971. The abundance of helium in prominences and in the chromosphere. *Solar Phys.* 19:384–400.

Hoover, W. G.; Ross, M.; Bender, C. F.; Rogers, F. J.; and Olness, R. J. 1972. Correlation of theory and experiment for high-pressure hydrogen. *Phys. Earth Planet. Interiors* 6:60–64.

Hubbard, W. B. 1968. Thermal structure of Jupiter. *Astrophys. J.* 152:745–754.

———. 1969. Thermal models of Jupiter and Saturn. *Astrophys. J.* 155:333–344.

*Dates are given for the original publication of the paper. References in brackets are in Russian.

――――. 1970. Structure of Jupiter: chemical composition, contraction, and rotation. *Astrophys. J.* 162:687–697.

――――. 1972. Statistical mechanics of light elements at high pressure. II. *Astrophys. J.* 176: 525–531.

――――. 1973. Observational constraint on the structure of hydrogen planets. *Astrophys. J.* 182:L35–L38.

――――. 1974. Inversion of gravity data for giant planets. *Icarus* 21:157–165.

Hubbard, W. B., and Slattery, W. L. 1971. Statistical mechanics of light elements at high pressure. I. *Astrophys. J.* 168:131–139.

Hubbard, W. B.; Slattery, W. L.; and De Vito, C. 1975. High zonal harmonics of rapidly rotating planets. *Astrophys. J.* 199:504–516.

Hubbard, W. B., and Smoluchowski, R. 1973. Structure of Jupiter and Saturn. *Space Sci. Rev.* 14:599–662.

Hubbard, W. B.; Trubitsyn, V. P.; and Zharkov, V. N. 1974. Significance of gravitational moments for interior structure of Jupiter and Saturn. *Icarus* 21:147–151.

Hubbard, W. B., and Van Flandern, T. C. 1972. The occultation of Beta Scorpii by Jupiter and Io. III. *Astron. J.* 77:65–74.

Jeffreys, H. 1923. The constitution of the four outer planets. *Mon. Not. Roy. Astron. Soc.* 83:350–354.

――――. 1924. On the internal constitution of Jupiter and Saturn. *Mon. Not. Roy. Astron. Soc.* 84:534–538.

Kalitkin, N. N. 1974. Materials at high energies. [Doctoral dissertation: State University, Moscow].

Kalitkin, N. N., and Kuz'mina, L. V. 1971. Curves of cold compression at high pressures. *Sov. Phys.-Solid State* 13:1938–1942 [*Fiz. Tver. Tel.* 13:2314–2318].

Kerley, G. J. 1972. Equation of state and phase diagram of dense hydrogen. *Phys. Earth Planet. Interiors* 6:78–82.

Kirzhnits, D. A. 1957. Quantum corrections to the Thomas-Fermi equation. *Sov. Phys. JETP* 5:64–71 [*Zh. Eksp. Teor. Fiz.* 32:115–123].

――――. 1963. *Field methods in many-body theory.* Moscow: Nauka Press.

Klepczynski, W. J.; Seidelmann, P. K.; and Duncombe, R. L. 1971. The masses of the principal planets. *Cel. Mech.* 4:253–272.

Kopal, Z. 1960. *Figures of equilibrium of celestial bodies.* Madison, Wisconsin: University of Wisconsin Press.

――――. 1973. On secular stability of rapidly rotating stars of arbitrary structure. *Astrophys. Space Sci.* 24:145–174.

Kopal, Z., and Mahanta, M. K. 1974. Rotational distortion of the stars of arbitrary structure: fourth approximation. *Astrophys. Space Sci.* 30:347–360.

Lambert, D. L. 1967. Estimates of the solar helium abundance. *Observatory* 87:199–200.

Landau, L. D., and Lifshitz, E. M. 1958. *Quantum mechanics.* Reading, Massachusetts: Addison-Wesley Publ. Co.

Lanzano, P. 1962. A third-order theory for the equilibrium configuration of a rotating planet. *Icarus* 1:121–136.

Makalkin, A. B. 1974. Abundances of common substances in the protoplanetary cloud. *Sov. Astron.* 18:243–247 [*Astron. Zh.* 51:417–424].

Murnaghan, F. D. 1951. *Finite deformation of an elastic solid.* New York: Wiley.

Neece, G. A.; Rogers, F. J.; and Hoover, W. G. 1971. Thermodynamic properties of compressed solid hydrogen. *J. Computational Phys.* 7:621–636.

Newburn, Jr., R. L., and Gulkis, S. 1973. A survey of the outer planets. *Space Sci. Rev.* 3:179–271.

Null, G. W.; Anderson, J. D.; and Wong, S. K. 1975. Gravity field of Jupiter from *Pioneer 11* tracking data. *Science* 188:476–477.

Ostriker, J. P., and Mark, W.-K. 1968. Rapidly rotating stars. I. *Astrophys. J.* 151:1075–1088.

Peebles, P. J. E. 1964. The structure and composition of Jupiter and Saturn. *Astrophys. J.* 140:328–347.

Ramsey, W. H. 1951. On the constitutions of the major planets. *Mon. Not. Roy. Astron. Soc.* 111:427–447.

Slattery, W. L., and Hubbard, W. B. 1973. Statistical mechanics of light elements at high pressure. III. *Astrophys. J.* 181:1031–1038.

Stewart, J. W. 1956. Compression of solidified gases to 20,000 kg/cm^2 at low temperature. *J. Phys. Chem. Solids* 1:146–158.

Trafton, L. M. 1967. Model atmospheres of the major planets. *Astrophys. J.* 147:765–781.

Trubitsyn, V. P. 1965a. Equation of state of solid hydrogen. *Sov. Phys.-Solid State* 7:2708–2714 [*Fiz. Tver. Tel.* 7:3363–3371].

——. 1965b. Van der Waals forces at high pressures. *Sov. Phys.-Solid State* 7:2779–2780 [*Fiz. Tver. Tel.* 7:3443–3445].

——. 1966a. Phase transition in a hydrogen crystal. *Sov. Phys.-Solid State* 8:688–690 [*Fiz. Tver. Tel.* 8:862–865].

——. 1966b. Equation of state of solid helium at high pressures. *Sov. Phys.-Solid State* 8:2593–2598 [*Fiz. Tver. Tel.* 8:3241–3247].

——. 1971. Phase diagrams of hydrogen and helium. *Sov. Astron.* 15:303–309 [*Astron. Zh.* 48:390–398].

——. 1972a. Adiabatic model for Jupiter. *Sov. Astron.* 16:342–347 [*Astron. Zh.* 49:420–426].

——. 1972b. The equation of the figure of planets in Clairaut's problem. [*Izv. Akad. Nauk SSSR – Fiz. Zem.* No. 4:10–15].

Wigner, E., and Huntington, H. B. 1935. On the possibility of a metallic modification of hydrogen. *J. Chem. Phys.* 3:764–770.

Yordanskii, S. V.; Lokutsievskii, O. V.; Vul, E. B.; Sidorovich, L. A.; and Finkelstein, A. M. 1973. Instability of metallic hydrogen structure to small changes in the allowance for the electron-electron interaction. *Sov. Phys. JETP (Letters)* 17:383–386 [*Zh. Eksp. Teor. Fiz. (Pis. Red.)* 17:530–534].

Zharkov, V. N., and Kalinin, V. A. 1971. *Equations of state for solids at high pressures and temperatures.* New York: Consultants Bureau.

Zharkov, V. N.; Makalkin, A. B.; and Trubitsyn, V. P. 1973. Integration of equations of the theory of planetary figures. *Sov. Astron.* 17:97–104 [*Astron. Zh.* 50:150–162].

——. 1974a. Models of Jupiter and Saturn. I. *Sov. Astron.* 18:492–498 [*Astron. Zh.* 51:829–840].

——. 1974b. Models of Jupiter and Saturn. II. *Sov. Astron.* 18:768–773 [*Astron. Zh.* 51:1288–1297].

Zharkov, V. N.; Pan'kov, V. L.; Kolachnikov, A. A.; and Osnach, A. I. 1969. *Introduction to lunar physics.* Moscow: Nauka Press.

Zharkov, V. N., and Trubitsyn, V. P. 1969. Theory of the figure of rotating planets in hydrostatic equilibrium – a third approximation. *Sov. Astron.* 13:981–988 [*Astron. Zh.* 46:1252–1263].

——. 1970. The figure of planets with a uniform or two-component density distribution. *Sov. Astron.* 14:1012–1018 [*Astron. Zh.* 47:1268–1276].

——. 1974a. The gravitational field of giant planets. *Sov. Astron.* 15:465–472 [*Astron. Zh.* 48:590–601].

——. 1974b. Determination of the equation of state of the molecular envelopes of Jupiter and Saturn from their gravitational moments. *Icarus* 21:152–156.

——. 1975. The system of equations of the theory of a figure of the fifth approximation. [*Astron. Zh.* 52:599–613].

Zharkov, V. N.; Trubitsyn, V. P.; and Makalkin, A. B. 1972. The high gravitational moments of Jupiter and Saturn. *Astrophys. Letters* 10:159–167.

Zharkov, V. N.; Trubitsyn, V. P.; and Samsonenko, L. V. 1971. *Physics of the earth and planets*. Moscow: Nauka Press.

Zharkov, V. N.; Trubitsyn, V. P.; and Tsarevskii, I. A. 1974c. [*DAN — SSSR* 214:557–560].

——. 1975. The equation of state of hydrogen, hydrogen compounds, inert gas crystals, oxides, iron, and FeS. *Geodynamic Res.* 3:5–45 Moscow: Nauka Press.

Zharkov, V. N.; Trubitsyn, V. P.; Tsarevskii, I. A.; and Makalkin, A. B. 1974d. Equations of state of cosmochemical substances and the structure of the major planets. [*Izv. Akad. Nauk SSSR — Fiz. Zem.* No. 10:7–18].

INTERIOR STRUCTURE OF JUPITER:
THEORY OF GRAVITY SOUNDING

W. B. HUBBARD and W. L. SLATTERY
University of Arizona

Using relatively simple interior models and a fourth-order theory of figures, we find that there are basically two extremes of interior structure which agree with current gravity data. One extreme is a "solar" composition envelope with 10 – 15 Earth masses of heavy material in a core; the other extreme has nearly uniform "solar" composition but with an additional ~ 30 Earth masses of heavy material distributed essentially uniformly; thus Jupiter is not of "solar" composition. We show how additional gravity data and improvement in knowledge of the molecular hydrogen equation of state will permit a significant reduction in the number of possible models.

As a result of the close encounter of the *Pioneer 11* spacecraft with Jupiter in 1974, the harmonic components of the Jovian gravity field have now been determined to a level of accuracy of about 10^{-5} (Null *et al.* 1975).[1] It is hoped that the same spacecraft will survive until the time of a grazing encounter with Saturn in 1979, when even more detailed gravity data for Saturn should result.

The theory of the interpretation of such gravitational data has barely kept pace with the new data, and indeed, the scientific community now finds itself in the position of planning future planetary missions without an adequate understanding of the information contained in gravitational data. The purpose of this chapter is to provide a current, but possibly not yet exhaustive, assessment of the constraints which are placed upon Jovian models by the data of Null *et al.*, and to provide a basis for evaluating the utility of still more detailed gravity data.

The first point which we must understand is that the gravitational potential of a giant planet such as Jupiter differs quite fundamentally from the potential of a terrestrial planet such as the earth. To consider this point,

---

[1]See p. 116.

we will now define a number of dimensionless parameters which will be used repeatedly in subsequent discussion.

The external gravitational potential of a planet can be fully represented by the harmonic expansion

$$
V_g = \frac{GM}{r} \left\{ 1 - \sum_{\ell=2}^{\infty} (a/r)^{\ell} J_{\ell} P_{\ell}(\mu) \right.
$$

$$
\left. + \sum_{\ell=2}^{\infty} \sum_{m=1}^{\ell} (a/r)^{\ell} P_{\ell}^{m}(\mu) \, (C_{\ell m} \cos m\phi + S_{\ell m} \sin m\phi) \right\} \qquad (1)
$$

where G is the gravitational constant, M is the total mass of the planet, a is the equatorial radius (or some other standard radius), r is the distance from the center of mass of the planet, P_{ℓ} are the Legendre polynomials and P_{ℓ}^{m} are the associated Legendre polynomials, μ is the cosine of the colatitude measured from the rotation axis, and ϕ is the longitude measured from some standard meridian. The gravitational "shape" of the planet is characterized by the dimensionless numbers J_{ℓ} (the *zonal harmonics*) and by $C_{\ell m}$ and $S_{\ell m}$ (the *tesseral harmonics*). It is easy to show that the gravitational harmonics are related to the mass-density distribution $\rho(r, \mu, \phi)$ in the planet:

$$
M a^{\ell} J_{\ell} = - \int d^3 r \, r^{\ell} \, \rho(r, \mu, \phi) \, P_{\ell}(\mu) \qquad (2)
$$

$$
M a^{\ell} C_{\ell m} = \frac{2(\ell - m)!}{(\ell + m)!} \int d^3 r \, r^{\ell} \, \rho(r, \mu, \phi) \, P_{\ell}^{m}(\mu) \cos m\phi \qquad (3)
$$

and

$$
M a^{\ell} S_{\ell m} = \frac{2(\ell - m)!}{(\ell + m)!} \int d^3 r \, r^{\ell} \, \rho(r, \mu, \phi) \, P_{\ell}^{m}(\mu) \sin m\phi \qquad (4)
$$

where $d^3 r$ is an increment of volume in spherical polar coordinates, and the integration is carried out over $0 \leqslant r \leqslant a$, $-1 \leqslant \mu \leqslant 1$, and $0 \leqslant \phi \leqslant 2\pi$. Equations (2), (3), and (4) show that a given gravitational harmonic of index ℓ, m is produced only by the corresponding density component projected on the spherical harmonic Y_{ℓ}^{m}. As the density component (ℓ, m) becomes larger in absolute value, i.e., as the planet deviates more from spherical symmetry, the corresponding gravitational harmonic increases in absolute value. Note also that the larger the value of ℓ, the more the contribution to the harmonic is weighted toward the surface of the planet. Because Eqs. (2), (3), and (4) measure the angular dependence of the density distribution as well as the radial dependence, they alone cannot give any information about the possible radial density variations.

Each of the gravity harmonic coefficients in Expansion (1) represents a contribution to the planet's gravity potential from: (a) response to a perturbing potential, and (b) intrinsic distortion due to departure from hydrostatic equilibrium. It is clear that even if we are dealing with a perfectly fluid planet which will precisely assume a figure of equilibrium in some external potential, some of the gravity harmonics will be finite if the perturbing potential is not spherically symmetric.

In the case of the major planets and the earth and Mars, the most important perturbation to the gravity potential is the rotation potential of the planet. For uniform rotation, this perturbation can be expressed as

$$Q = \frac{1}{3}\omega^2 r^2 \left[1 - P_2(\mu)\right] \tag{5}$$

where ω is the angular rotation velocity, and the z-axis is assumed to coincide with the rotation axis. At the equator ($\mu = 0$ and $r = a$), we have

$$Q = \frac{1}{2}\omega^2 a^2 \equiv \frac{1}{2}q \, GM/a \tag{6}$$

where the dimensionless parameter

$$q \equiv \omega^2 a^3 / GM \tag{7}$$

is a measure of the importance of the rotation potential with respect to the unperturbed gravitational potential of the planet.

The perturbing potential Q has certain symmetry properties which are important for understanding the gravity field of a rotating planet in hydrostatic equilibrium. Note that Q is (a) axially symmetric (no ϕ-dependence), and (b) invariant with respect to change of sign of μ. In hydrostatic equilibrium, surfaces of constant potential and constant density coincide, and as long as $q \ll 1$, the planet's gravity field should have the same symmetry properties as Q, i.e., it should have tesseral harmonics and odd zonal harmonics equal to zero. Moreover, in hydrostatic equilibrium, it is possible to express the even zonal harmonics as a power series in q, in the form

$$J_{2\ell} = \sum_{n=0}^{\infty} \beta_{n\ell} q^{\ell+n} \tag{8}$$

where the $\beta_{n\ell}$ are dimensionless response coefficients which depend on the pressure-density equation of the planet. The basic problem of gravity sounding is to invert the $\beta_{n\ell}$ to obtain a pressure-density relation.

We are now in a position to qualitatively understand the difference between the gravity field of the earth and Jupiter. For the earth, q is about 10^{-3} and we would expect, since $\beta_{01} \sim 1$ for plausible models, $J_2 \sim q$. This is in fact true; the quadrupole component of the geopotential is of order q, dominates the other gravity harmonics, and is very close to the value expected in hydrostatic equilibrium. Using Eq. (8) with $\ell = 2$, we would then expect that the terrestrial J_4 would be $\sim 10^{-6}$. The earth's J_4 is indeed of this order, but it is even smaller than J_3, which we know to be a non-hydrostatic term. Many of the earth's tesseral harmonics are also of order 10^{-6} (Gaposchkin 1974), and so we conclude that β_{01} is the only equilibrium response coefficient which we can reliably estimate for the earth and relate to an equation of state. This conclusion holds even though the terms in the earth's potential have now been determined to a very high degree, and to a precision better than 10^{-10} (Gaposchkin 1974).

In contrast, Jupiter's potential has only been measured to a precision of about 10^{-5} (Anderson et al. 1974b; Null et al. 1975),[2] but the measured terms clearly show the dominance of the equilibrium response to Q. For Jupiter, q is almost 10^{-1}, which makes it possible to measure the response coefficients in Eq. (8) to a much higher order than for the earth. As expected, one observes $J_2 \sim q$, $J_4 \sim q^2$, $J_6 \sim q^3$, and the remaining gravity harmonics are zero to within the precision of the measurement. As we shall see below, the information contained in the observed gravity harmonics supports the hypothesis that Jupiter is a liquid body, and that the limiting level of nonhydrostatic terms in the potential may not be reached until a very high order in q.

The foregoing discussion makes it clear why gravity sounding has not been carried to high order until very recently. As long as one's perspective is limited to the terrestrial planets, there is no point in calculating beyond first or at most second order in q.

I.   THE DIRECT PROBLEM

Computational Techniques

It is clear that one cannot address the inverse problem until one can solve the direct problem, i.e., calculate the $\beta_{n\ell}$ of Eq. (8) for a given pressure-density relation. We proceed as follows. Let us suppose that we have available a pressure-density relation

$$P = P(\rho) \tag{9}$$

where P is the total pressure. The equation of hydrostatic equilibrium relates the density and pressure to the total (gravitational and rotational) potential:

$$\rho^{-1}dP = dV . \tag{10}$$

Integrating Eq. (10) once, we obtain

$$\zeta(\rho) = V - V_s = \int_{\rho_s}^{\rho} \frac{dP}{d\rho'} \frac{d\rho'}{\rho'} \tag{11}$$

---

[2]See p. 120.

where ρ_s and V_s are evaluated at the planet's surface. We assume that $\zeta(\rho)$ is a known function. Next, we calculate the total potential by direct integration:

$$V = V_s + \zeta(\rho) = Q + G \int d^3r' \frac{\rho\,(r',\,\mu')}{|\underline{r} - \underline{r}'|}. \tag{12}$$

The solution to Eq. (12) can then be used to find the angular and radial dependence of the density, and thus, via Eq. (2), the zonal harmonics. Despite its simplicity, Eq. (12) is very difficult to solve in generality. The obvious approach is to use perturbation analysis; express the potential in units of GM/a. Then

$$aV/GM \equiv \psi = \frac{1}{3} qx^2 \left[1 - P_2\,(\mu)\right] +$$

$$\int\limits_{-1}^{1} d\mu' \int\limits_{0}^{1} 4\pi x'^2 dx' \frac{\rho\,(x',\,\mu')}{<\rho>} |\underline{x} - \underline{x}'|^{-1} \tag{13}$$

where $x = r/a$ and $<\rho> = 3M/4\pi a^3$. The dimensionless function ψ, which is of order unity, is related to $\rho/<\rho>$ through the equation of state, Eq. (9), and the equation of hydrostatic equilibrium, Eq. (11). Since q is assumed to be a small quantity, we may solve Eq. (13) by successive approximations, first with $q = 0$ to generate a spherical non-rotating model, and then using the previous density distribution to include a higher power of q in the next approximation. In this way, a perturbation expansion such as Eq. (8) arises naturally.

We have two alternative procedures for implementing the perturbation technique. The older theory involves the introduction of *level surfaces*. A level surface is a surface of constant potential—and, by Eqs. (9) and (10), of constant density and pressure—whose equation can be expressed in the form

$$r = s \left[1 + \sum_{\ell = 0}^{\infty} s_{2\ell}\,(s)\,P_{2\ell}\,(\mu)\right] \tag{14}$$

where s is a parameter which corresponds to the mean radius of the level surface. For a suitable choice of definition of s, the coefficients $s_{2\ell}\,(s)$ can be shown to fall in an ascending order in q (Zharkov *et al.* 1971).[3] We assume that $\rho\,(s)$ is a known function. Thus ψ is a known function of s. The variable r is eliminated through Eq. (14), and Eq. (13) is then decomposed into a system of integral equations for the parameters $s_{2\ell}\,(s)$, which are calculated in ascending orders in q. This procedure has been carried to order q^3 by Zharkov *et al.* (1972),[4] and some steps to extend the theory to q^4 have been taken by Kopal and Mahanta (1974).

---

[3] See p. 169.   [4] See p. 168.

A newer theory for solving Eq. (13) has been introduced by Hubbard *et al.* (1975). Instead of introducing level surfaces, we retain the spherical polar coordinates in Eq. (13). The density is expanded in multipole components:

$$\rho = \sum_{\ell = 0}^{\infty} \rho_{2\ell} \, (r) \, \rho_{2\ell} \, (\mu) .$$ (15)

After $|\underline{r} - \underline{r}'|^{-1}$ and $\zeta(\rho)$ have been similarly decomposed into multipole components, Eq. (13) can be solved in ascending powers of q by noting that $\rho_{2\ell} \sim q^{\ell}$. The method of Hubbard *et al.* (1975) has been used to generate solutions to order q^4, which we will discuss below.

Scaled Maclaurin Spheroids

Before we discuss solutions for the gravity field of a rotating body with an arbitrary pressure-density relation, it is instructive to consider a special model which exhibits some of the properties of more general models. Consider a planet with a small, high-density core of mass M_c, overlain with a uniform envelope of constant density ρ and equatorial radius a. Above the envelope there is a massless atmosphere of mean thickness d. The core is assumed to be sufficiently small and dense that it remains undistorted by the rotation. Under these assumptions, only the $\ell = 0$ term of Eq. (15) enters the problem, and the multipoles of the gravity field are produced entirely by the shell integrals (defined by Hubbard 1974) of the envelope.

First, assuming $d = 0$, and evaluating Eq. (2) directly, we obtain

$$Ma^2 J_2 = - \frac{4\pi}{5} \, a^5 s_2 \rho$$ (16)

to first order in q, where s_2 is a coefficient of expansion [Eq. (14)], evaluated at the surface of the envelope. We can solve $V_g + Q = const$ evaluated on the surface, to find

$$s_2 = - (J_2 + \frac{1}{3} q)$$ (17)

to first order in q. Now, since the mass of the envelope is given by

$$M_e = \frac{4\pi}{3} \, a^3 \rho$$ (18)

to lowest order in q, we obtain

$$J_2 = \frac{\frac{1}{5} \delta}{(1 - \frac{3}{5} \delta)} \, q$$ (19)

where $\delta = M_e / M$. For $\delta = 1$, we obtain $J_2 = \frac{1}{2} q$, the usual first-order result for a Maclaurin spheroid.

Equation (19) illustrates a characteristic feature of the theory of figures. We are considering a model which is a superposition of a Maclaurin spheroid ($\rho = const$) and a Roche model (a point-mass core and a massless envelope). The linear response coefficient which we have calculated,

$$\beta_{01} = \frac{1}{5} \delta \left/ \left(1 - \frac{3}{5} \delta\right)\right. \tag{20}$$

reduces to the proper limit for $\delta = 0$ and $\delta = 1$, but it is in no way a linear combination of the response coefficients for the two models. The reason for this is that the perturbation which excites the response includes a contribution from the response; i.e., the problem must be solved self-consistently.

We can scale the solution (20) for the case of a massless atmosphere by noting that solution (19) still holds, but for a different q. The q which applies to the surface of the envelope is

$$q_e = q \left(\frac{a}{a+d}\right)^3 \tag{21}$$

where q is the rotation parameter defined for the top of the atmosphere. Thus, in this case we obtain

$$\beta_{01} = \frac{\frac{1}{5} \delta}{1 - \frac{3}{5} \delta} \left(\frac{a}{a+d}\right)^5 . \tag{22}$$

The lowest-order hexadecapole response coefficient can be calculated in an entirely analogous manner. For $d = 0$, we obtain

$$Ma^4 J_4 = -\frac{4\pi}{9} a^7 \left(s_4 + \frac{54}{35} s_2^2\right)\rho \tag{23}$$

and after carrying out a self-consistent solution, we get

$$\beta_{02} = \frac{-\delta(4 - 2\delta)}{35(1 - \frac{1}{3}\delta)(1 - \frac{3}{5}\delta)^2} \left(\frac{a}{a+d}\right)^{10} \tag{24}$$

valid for any d. For $d = 0$, $\delta = 1$, we get $\beta_{02} = -15/28$, the result for a Maclaurin spheroid.

It is possible to derive similar expressions for higher-order response coefficients, which we shall not exhibit here. Equations (22), (24), and the analogous higher-order expression permit us to conveniently map out the range of variation of the multipole response coefficients in terms of the dimensionless ratios δ and d/a.

II. THE INVERSE PROBLEM

Observational Data

This section of the chapter concerns itself with the interpretation of currently available gravity data for Jupiter. We wish to explore the range of interior models which satisfy these data, and to find the properties which they have in common.

The best available values for the zonal harmonics of Jupiter are obtained from the Pioneer 11 encounter (Null *et al.* 1975).[5] For interpretation of the gravity data, we also need the rotation parameter q, the mass, and the equatorial radius a. The equatorial radius is taken to be the equatorial radius of the level surface at one bar pressure, and the rotation period is the radio rotation period of Jupiter, $9^h 55^m 5$. Finally, we define the general response coefficients:

$$\Lambda_{2\ell} = |J_{2\ell}|/q^\ell = \left| \sum_{n=0}^{\infty} \beta_{n\ell} q^n \right| \approx |\beta_{0\ell}| . \tag{25}$$

The general response coefficients, which were first introduced by Zharkov *et al.* (1972),[6] are convenient for relating observed multipole gravity terms to interior models, since they are independent of the size of the perturbation to lowest order.

The data for Jupiter are summarized in Table I.

TABLE I
Jovian Parameters for Gravity Modeling

a (km)	71,398
M (g)	1.901×10^{33}
q	0.0888
J_2	$.01475 \pm .00005$
Λ_2	$.1661 \pm .0006$
J_4	$-5.8 \times 10^{-4} \pm .4 \times 10^{-4}$
Λ_4	$.0735 \pm .0050$
J_6	$5 \times 10^{-5} \pm 6 \times 10^{-5}$
Λ_6	$.072 \pm .086$

The Pioneer 10 values for the gravity harmonics (Anderson *et al.* 1974*b*)[7] are consistent with the results of Pioneer 11, but with somewhat larger error bars for J_4, and J_6 was not detected. Pioneer 10 found $|C_{22}|$ and $|S_{22}|$ to be less than 10^{-6}, and Pioneer 11 placed a limit of 10^{-5} on $|J_3|$, all consistent with hydrostatic equilibrium.

Comparison with Models

As is evident from Table I, J_2 for Jupiter is now known to a few parts in a thousand, and J_4 is known to within 10%. First we wish to know the possible

---

[5]See p. 116.
[6]See also p. 171; note that their definition differs slightly from Eq. (25).
[7]See p. 116.

range in J_4 for plausible planetary models, and compare this with the present uncertainty in J_4. This will provide some estimate of the present utility of J_4 for constraining models. For this purpose, we adopt two limiting scaled Maclaurin spheroids. Model (1) has $\delta = 0.555$ and an envelope of constant density $\rho = 0.7$ g cm^{-3} which extends out to the surface ($d = 0$). Model (4) has no core ($\delta = 1$), an envelope of constant density $\rho = 4$ g cm^{-3} which has a radius of 48,000 km, and an "atmosphere" of zero density with $d = 22,000$ km. These models probably bracket the range of J_4 consistent with the observed J_2, in hydrostatic equilibrium. The response coefficients Λ_n are then estimated from Eqs. (22), (24) and from analogous expressions for β_{03} and β_{04}. The results are plotted in Fig. 1, together with observed values for Jupiter. We note that the Pioneer 11 value for Λ_6 is still far too uncertain to constrain models, although the value is clearly consistent with hydrostatic equilibrium. Before the Pioneer measurements, Brouwer and Clemence (1961) estimated Λ_4 to be in the range 0.133 to 0.037, a result which excludes

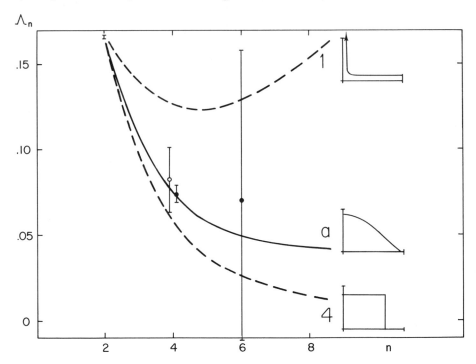

Fig. 1.   Response coefficients for Jovian models. *Open circle,* Pioneer 10 measurement; *closed circles,* Pioneer 11 measurements; *error bar without circle,* Pioneer 10 and 11 measurements of Λ_2. The *dashed curves* are scaled Maclaurin Models (1) and (4), with the response coefficients connected by a smooth line. The *solid curve* connects the response coefficients of a polytrope of index 0.95, Model (a). The density distribution as a function of radius is shown for each model, to scale. Models designated by letters have more realistic equations of state which are shown in Figs. 2 and 3.

no plausible model in hydrostatic equilibrium, as noted by De Marcus (1965). The Pioneer 10 result for Λ_4 reduced the range considerably but still left a large ensemble of possible models. With the Pioneer 11 value, we are now able to discriminate usefully between models; for example, De Marcus' (1958) model of Jupiter with a cold, solid hydrogen envelope and $\Lambda_4 = 0.085$ is finally ruled out.

Since the Pioneer 11 results are the most accurate available for Jupiter, we shall now concentrate our attention on the range of models that are consistent with these data. Because the scaled Maclaurin models depend on two parameters, we can simultaneously satisfy a constraint on Λ_2 and Λ_4. Maclaurin Models (2) and (3) bracket the uncertainty in Λ_4; Model (2) has $\delta = 0.8$, an envelope of $\rho = 1.95$ g cm^{-3}, $d = 11,000$km, and Model (3) has $\delta = 0.9$, an envelope of $\rho = 2.80$ g cm^{-3}, $d = 15,400$km. Here we illustrate an important point: Models (2) and (3) are clearly inconsistent with any plausible equation of state, yet they are consistent with all available gravity data. However, Models (2) and (3) do bear a qualitative resemblance to more plausible models, as we shall see.

After the scaled Maclaurin spheroids, the simplest interior model is a polytrope of index n_P. It is clear that, for a given q, the Λ_n are functions only of n_P. If we match Λ_2 to the observed value for Jupiter, we obtain $n_P = 0.95$, and the value of the equatorial radius then fixes the pressure-density relation

$$P = 1.9545\rho^{2.05263} \tag{26}$$

for ρ in g cm^{-3} and P in Mbar. Equation (26) corresponds to Model (a), which is also plotted in Fig. 1. We note that Model (a) fits the Pioneer 11 data quite precisely with $\Lambda_2 = 0.1656$ and $\Lambda_4 = 0.0733$, although only one parameter has been adjusted. Figure 2 shows that Model (a) is a plausible equation of state; if we take a solar composition adiabat in the pressure range $1-40$ Mbar, and enhance the density by $\sim 10\%$ due to addition of a relatively incompressible, dense component, we recover Model (a). This would imply roughly a ten-fold increase in the relative mass abundance of heavy materials. The equation of state of a mixture can be written, approximately, as

$$\frac{1}{\rho} = \frac{X}{\rho_H(P)} + \frac{Y}{\rho_{He}(P)} + \frac{Z}{\rho_z(P)} \tag{27}$$

where X, Y, Z are respectively the fractional mass abundances of hydrogen, helium, and heavier elements, and $\rho_\alpha(P)$ is the density of pure substance α at pressure P. Now since ρ_H is normally much smaller than the density of any other substance at the same pressure, we can write

$$\frac{1}{\rho} \approx \frac{X}{\rho_H(P)} \tag{28}$$

as long as $X \gtrsim 0.5$. We have just seen that Model (a) requires that ρ be about ten per cent larger than solar composition density, which means that X must be reduced from 0.73 to, roughly, 0.6. The corresponding increase in mass fraction could be either in Y or Z; the equation of state is not very sensitive to which we choose. Podolak and Cameron (1974)[8] argue that it would be difficult to envision a bulk fractionation of hydrogen from helium in a proto-Jupiter, therefore the enrichment is probably in Z rather than in Y as proposed by Hubbard (1969). In this case, we note that Model (a) would require $Z \sim 0.1$, to be compared with a "solar" value of $Z \sim 0.015$.

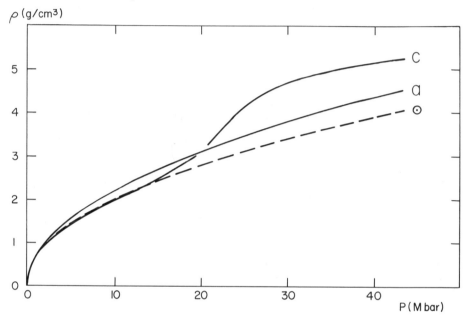

$\rho \ (g/cm^3)$

$P \ (M \, bar)$

Fig. 2.   Model equations of state: $\odot$ is a solar composition adiabat with 73% hydrogen, 27% helium by mass, and a starting temperature of 230°K at one bar pressure. The possible phase transition in the vicinity of 3-5 Mb has been smoothed out. Models (a) [Eq. (26)] and (c) [Eq. (29)] are also plotted.

We have seen that Model (a) is plausible in the range 1 Mbar $\leqslant P \leqslant 40$ Mbar. However, the value of Λ_4 tends to be quite sensitive to the equation of state in the pressure range $P \sim 100$ kbar (Hubbard 1974), for polytrope-like models. Fig. 3 shows a comparison of Eq. (26) with calculated adiabats in this pressure range. The adiabats are calculated for $X = 0.73$, $Y = 0.27$, using the physics described by Slattery and Hubbard (1975). We see that Model (a) is in reasonable agreement with a 230°K adiabat, which shows that Model (a) is physically reasonable over the entire pressure range. To test this hypothesis further, we constrained the $P(\rho)$ relation to follow the 230°K adiabat exactly up to a pressure of 150 kbar, and then to follow the 0.95 polytropic

---

[8]See p. 66.

ρ (g/cm³)

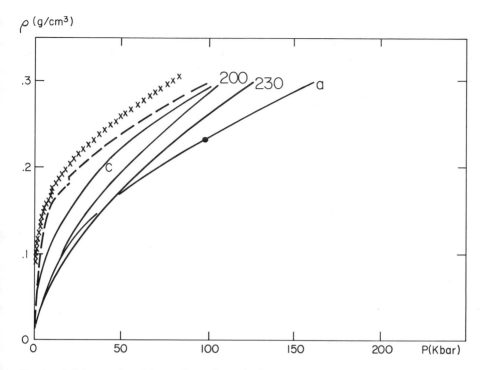

Fig. 3.   Adiabats and model equations of state in the outer envelope of Jupiter. Curves (*a*) and (*c*) correspond to Eqs. (26) and (29) respectively. Two solar composition adiabats with start-ing temperatures of 200°K and 230°K at one bar are shown: *dashed curve,* De Marcus' (1958) Jupiter model; *crossed curve,* equation of state for solid hydrogen (Zharkov *et al.* 1971). *Dot* is obtained from the inversion theory of Hubbard (1974), as applied to Pioneer 11 gravity data.

$P(\rho)$ from 1000 kbar to 40 Mbar. A satisfactory fit to the gravity data results from this procedure, which yields Model (b). Model (b) differs from Model (a) only in that a "real" adiabat is substituted in the outer envelope; the Λ_n profiles are virtually identical.

Next, we apply a perturbation to Model (a) in order to test the uniqueness of this model. Radau-Darwin theory shows that models which are con-strained to the same value of Λ_2 will have essentially the same dimensionless moment of inertia about the rotation axis, C/Ma^2. We can keep C/Ma^2 con-stant by shifting mass in the middle envelope to both the outer envelope and the core. This is accomplished by constructing equation-of-state Model (c) (Figs. 2, 3), which resembles a planet with a solar composition envelope and an adiabat with a starting temperature somewhat lower than 200°K. This model has a "core" with a radius of about 30,000km. The equation of state for Model (c) is

$$P = 0.4656\rho^2 + 2.6958\rho^3 - 1.0406\rho^4 + 0.1081\rho^5 \qquad (29)$$

where P is in Mbar and ρ is in g cm⁻³. Model (c), which is clearly rather different from Models (a) or (b), fits available gravity data equally well, with $\Lambda_2 = 0.1656$ and $\Lambda_4 = 0.0705$. If we identify the extra density of $\sim 1 \text{ g cm}^{-3}$ in the core with a Z-component, we obtain a mass of material heavier than hydrogen or helium of $\sim 1 \times 10^{29}\text{g}$, corresponding to $Z \sim 0.06$ for the planet as a whole. Thus we see that Models (a), (b), and (c) all require a considerable enhancement of the Z-component over solar composition, although the location of this component in the planet varies considerably.

High-Order Components of the Gravity Field; Inversion Techniques

Although Models (a), (b), and (c) all agree with data which are presently available, we may ultimately hope to discriminate between such models by sufficiently accurate measurement of higher-order gravitational multipoles. Figure 4 shows the 2^6- and 2^8-pole terms in the model gravity fields. We see that there are clear differences between the models in the higher-order terms, although rather accurate measurements will be necessary to discriminate between models. Model (c) exhibits a "turn-up" in the response profile at Λ_8. Such behavior appears to be characteristic of models which have a sufficiently rapid and non-linear increase of density with depth near the surface. It would not be possible to predict a "turn-up" on the basis of a simple extrapolation of the trend in Λ_2, Λ_4, and Λ_6. Maclaurin Model (1) shows another example of this type of behavior (Fig. 1).

We now consider the question of direct inversion of the Λ's to reconstruct the interior density profile. There are at present roughly three distinct techniques for inverting gravity data. The first technique is the *shell integral* method proposed by Hubbard (1974). Hubbard notes that for models which are continuous near the surface, the 2^4-pole and higher harmonics tend to be produced mostly by the shell integrals, for which the lowest-order expressions are

$$J_4 \cong \frac{4\pi}{35} s_2{}^2 \frac{b^3}{M} b\rho' \tag{30}$$

and

$$J_6 \cong \frac{4\pi}{1001} \frac{b^3}{M} \left[(15qs_2{}^2 + 33s_2{}^3 + 35s_2 s_4)b\rho' - 3s_2{}^3 b^2 \rho'' \right] \tag{31}$$

where the s_2 and s_4 shape coefficients [Eq. (14)] are known at the surface of the planet from the external gravity field, and b is the polar radius. The first and second derivatives of density with radius ρ' and ρ'' are evaluated for the equivalent non-rotating planet at a "sounding" depth below the surface equal to $-bs_2$, or about 3000 km in the case of Jupiter. Now since we know J_2, J_4, and q for Jupiter, we can solve Eq. (30) for ρ' at this depth, or, using the equation of hydrostatic equilibrium, we can determine the quantity

$$K = \frac{1}{2} \frac{1}{\rho} \frac{dP}{d\rho} = -\frac{2\pi}{35} \left(J_2 + \frac{1}{3} q \right)^2 \frac{b^2 G}{J_4} \tag{32}$$

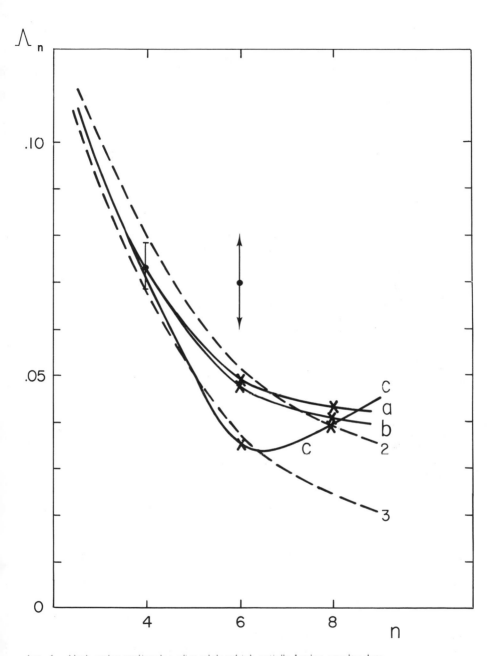

Fig. 4.   High-order multipoles of models which satisfy Jovian gravity data.

at the sounding level (Anderson *et al.* 1974a). Using the Pioneer 11 results, we obtain $K \simeq 1.8$ Mbar/(g cm^{-3})^2 at a depth of 3000 km. To infer an equation of state from Eq. (32), we need a boundary condition ($\rho \simeq 0$ at the surface), and we need to know $K(\rho)$. If we make the additional assumption $K = const$ throughout the envelope, the equation of state is determined trivially. However, in this case we have already considerably restricted the class of models, as is evident from Fig. 3: Model (c) has $K = 1.6$ at the sounding level, a significantly different equation of state, and a correct value of J_4. We conclude then, that setting $K = const$ will generate an acceptable model but by no means a unique one.

The situation can be improved if we also have available J_6. As we see from Eq. (31), J_6 is also sensitive to ρ'' at the sounding level. Therefore, an accurate measurement of J_6 offers the possibility of checking the assumption $K = const$. Comparison of the calculations for Models (a) and (c) illustrates this point (Fig. 4). At the sounding level for Model (a), we have $\rho = 0.22$, $b\rho' = -4.9$, $b^2\rho'' = 0.8$ (g cm^{-3}), while for Model (c) we have $\rho = 0.34$, $b\rho' = -5.9$, $b^2\rho'' = -77.6$. As we see from Fig. 4, J_6 for Model (c) falls substantially below J_6 for Model (a); this is due to the large negative value of ρ'' which goes into Eq. (31). It also illustrates the point that Λ_4 and Λ_6 are not trivially re-lated to the mean density of the outer envelope; Model (c) is denser than Model (a) in the outer envelope. Proceeding in the inverse direction, we might try to calculate ρ'' for a given value of J_6 by eliminating ρ' through Eq. (30). This procedure does not work well because the coefficient of $b\rho'$ in Eq. (31) is about ten times larger than the coefficient of $b^2\rho''$. Since Eq. (30) is only valid to about 15%, it is clear that we can only estimate the order of magnitude of ρ'' in this way. In summary: Hubbard's technique will generate a physically plausible model, such as Model (b). The model is nonunique and must be checked through comparison with Λ_6 and Λ_8 when these quantities become available.

The second technique for inverting gravity data, the *Taylor-series tech-nique,* was proposed by Zharkov and Trubitsyn (1974). In this method, we express the density on level surfaces in the form

$$\rho(s) = \rho(\bar{s}) + (s - \bar{s})\,\rho'(\bar{s}) + \frac{1}{2}(s - \bar{s})^2\,\rho''(\bar{s}) + \ldots, \qquad (33)$$

where $\bar{s}$ is some reference level surface. It is assumed that any discontinuities are small. The number of terms included in the expansion (33) is equal to the number of constraints; the mean density of the planet and J_2 allow us to determine $\rho(\bar{s})$ and $\rho'(\bar{s})$. We can then determine $\rho''(\bar{s})$ as well, once we know J_4, etc. The Taylor-series method has an advantage over the shell-integral method in that there is no need for an assumption about the smallness of interior contributions to external multipoles. However, many physically real-istic density distributions could not be well represented by a few terms in a Taylor series over more than a very thin shell. The Taylor-series method of inversion has not yet been tested extensively. Like Hubbard's technique, with limited information about the gravity field, it gives a possible model but not a unique one.

The third inversion technique is suggested by a generalization of the scaled Maclaurin spheroids. At present, we are considering a family with two adjustable parameters: relative core mass and thickness of the massless atmosphere, allowing us to fit Λ_2 and Λ_4. To fit Λ_6 we would add another parameter, presumably another constant density shell with a fixed density and adjustable thickness, or else adjustable density and fixed thickness. With Λ_8, another such shell is added, etc. As the number of constraints and shells increases, we eventually approach a continuous model. Whether such a model is unique is an unsolved problem of theoretical physics. It is also clearly an academic one since the number of constraints practically available is less than six.

III. CONCLUSIONS

Figure 5 summarizes a family of interior models which agree with current Jovian gravity data. Models (2) and (3) correspond to no real equation of state but have one feature worth noting: about 10–20% of their mass is concentrated in a high-density core. Models (a), (b), and (c) roughly bracket the principal types of interior models which have been proposed. Models (a) and (b) have no core, and have a heavy element mass fraction of $Z \cong 0.1$, distributed more or less uniformly throughout the metallic hydrogen portion of the planet. The molecular envelope is hot and has a relatively low density, with a temperature at 1 bar of $\sim 230°K$. Model (c) is essentially solar composition, but with a heavy-element core which contributes about 0.06 of the total mass. Model (c) has a cooler molecular envelope, with a temperature at 1 bar of $\lesssim 200°K$.

Can we rule out Models (a) and (b) on the basis of an independent measurement of the temperature at one bar? This would be possible if our confidence in the $P(\rho)$ adiabats in the 100 kbar range were high. However, as we shall discuss elsewhere (Slattery and Hubbard 1975), the starting temperature which we assign to a particular $P(\rho)$ adiabat is uncertain by about 40% due to residual uncertainties in the experimentally determined intermolecular potentials (Hoover et al. 1972). Thus the 230°K adiabat which was calculated for a given intermolecular potential might have the same $P(\rho)$ as a 180°K adiabat with a somewhat stiffer potential, while in the metallic phase the effect of the difference in initial temperatures is almost negligible. It follows that improved information of molecular hydrogen adiabats from experimental data should permit a better discrimination between models.

Can we rule out the possibility of a solid hydrogen planet on the basis of the gravity field? Actually, the strongest arguments for a liquid Jupiter are

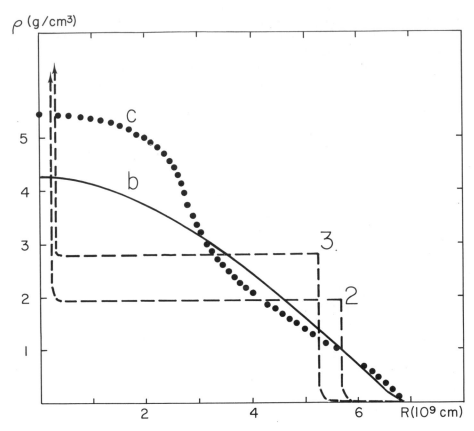

Fig. 5. Density distributions for Jovian models which satisfy Pioneer 11 constraints. The density distributions shown are calculated with zero rotation from the same equation of state as for the rotating model. Models (a) and (b) have virtually identical density profiles.

obtained from a consideration of the planet's heat flux and energy budget, tidal Q, thermal microwave emission spectrum, and reasonably secure theoretical calculation of the thermodynamics of hydrogen at high pressure (Hubbard *et al.* 1974). The observed gravity field, with its apparent absence of tesseral harmonics and odd zonal harmonics, and its agreement with reasonable interior structure, certainly supports the concept of a liquid planet. Although De Marcus' model is now ruled out, it may still be possible to contrive a model with a solid hydrogen envelope and a reasonable interior equation of state which fits the gravity data; we will leave this task to other investigators.

We have seen that, although Models (a) and (c) both fit present data, one could discriminate between them with accurate measurements of J_6 and J_8. Can such accuracy be practically achieved, for example with a geodetic Jupiter orbiter? Ideally, one would wish to know Λ_2, Λ_4, Λ_6, and Λ_8 to within an uncertainty of 10^{-3}, which corresponds to an uncertainty in J_2, J_4, J_6, and

J_8 of 10^{-4}, 8×10^{-6}, 7×10^{-7}, 6×10^{-8} respectively. These requirements are at least two orders of magnitude less precise than current terrestrial geodesy (see Table II).

What are the reasonably secure conclusions that we can reach upon the basis of present Jupiter models? The principal characteristic of both types of models (with and without core) is that they contain 5 to 10% heavy material by mass. Hubbard (1969) showed that Jupiter would require some 40% helium by mass if the planet were composed of pure hydrogen and helium. The 1969 Hubbard models were analogous to Models (a) and (b). Podolak and Cameron (1974)[9] instead adopted a model with solar hydrogen-to-helium ratio and increased Z in the form of a core, corresponding roughly to Model (c). Models ranging qualitatively from (a) to (c) have also been calculated by Stevenson and Salpeter[10] and Zharkov et al. (1974).[11] We can safely conclude:

1. A "solar" composition, $X \approx 0.73$, $Y \approx 0.255$, $Z \approx 0.015$, is ruled out for any adiabatic or nearly adiabatic model of Jupiter. Because of the accuracy with which we know the equation of state of hydrogen, X for Jupiter must be lower than the "solar" value.
2. If we require a "solar" ratio of hydrogen to helium in Jupiter, then the Z-component is enhanced over solar by a factor of ~ 3 (for c-type models) to ~ 10 (for a- and b-type models).

TABLE II
Comparison of Zonal Harmonics of Earth and Jupiter

	Jupiter Model (b)	Earth (Gaposchkin 1974)
J_2	1.476×10^{-2}	1.083×10^{-3}
J_4	-5.74×10^{-4}	-1.618×10^{-6}
J_6	3.28×10^{-5}	5.52×10^{-7}
J_8	-2.5×10^{-6}	-2.05×10^{-7}

There is clearly some point in trying to acquire both more gravity data and more experimental data on the molecular hydrogen equation of state in order to discriminate between a-type and c-type models. Model (a) suggests that the Z-component is soluble in metallic hydrogen and remains mixed in

---

[9]See p. 67.
[10]See p. 106.
[11]See p. 166.

the metallic core over the lifetime of Jupiter, as well as implying a total Z-component in Jupiter of ~ 30 Earth masses. Model (c) suggests that Jupiter accreted onto a Z-composition core of some 10–15 Earth masses, and that the core is insoluble in the metallic hydrogen at present.

Finally, although we have concluded that Jupiter's bulk composition is non-"solar", it should perhaps be noted that the composition and distribution of Z in Model (c) is quite similar to the chemical makeup of an *ad hoc* solar model proposed by Hoyle (1975) to explain the low solar neutrino emission.

Acknowledgment. This research was supported by NASA Grant NSG-7045.

REFERENCES

Anderson, J. D.; Hubbard, W. B.; and Slattery, W. L. 1974*a*. Structure of the Jovian envelope from Pioneer 10 gravity data. *Astrophys. J.* 193:L149–L150.

Anderson, J. D.; Null, G. W.; and Wong, S. K. 1974*b*. Gravity results from Pioneer 10 Doppler data. *J. Geophys. Res.* 79:3661–3664.

Brouwer, D., and Clemence, G. M. 1961. *Planets and satellites* (G. P. Kuiper, B. Middlehurst, eds.) p. 31. Chicago: Chicago Univ. Press.

De Marcus, W. C. 1958. The constitution of Jupiter and Saturn. *Astron. J.* 63:2–28.

———. 1965. Models of Jupiter and Saturn. *Magnetism and the cosmos* (W. R. Hindmarsh *et al.* eds.), pp. 352–364. New York: Elsevier.

Gaposchkin, E. M. 1974. Earth's gravity field to the eighteenth degree and geocentric coordinates for 104 stations from satellite and terrestrial data. *J. Geophys. Res.* 79:5377–5411.

Hoover, W. G.; Ross, M.; Bender, C. F.; Rogers, F.; and Olness, R. J. 1972. Correlation of theory and experiment for high-pressure hydrogen. *Phys. Earth Planet. Interiors* 6:60–64.

Hoyle, F. 1975. A solar model with low neutrino emission. *Astrophys. J.* 197:L127–L131.

Hubbard, W. B. 1969. Thermal models of Jupiter and Saturn. *Astrophys. J.* 155:333–344.

———. 1974. Inversion of gravity data for giant planets. *Icarus* 21:157–161.

Hubbard, W. B.; Slattery, W. L.; and De Vito, C. 1975. High zonal harmonics of rapidly rotating planets. *Astrophys. J.* 199:504–516.

Hubbard, W. B.; Trubitsyn, V. P.; and Zharkov, V. N. 1974. Significance of gravitational moments for interior structure of Jupiter and Saturn. *Icarus* 21:147–151.

Kopal, Z., and Mahanta, M. K. 1974. Rotational distortion of the stars of arbitrary structure: fourth approximation. *Astrophys. Space Sci.* 30:347–360.

Null, G. W.; Anderson, J. D.; and Wong, S. K. 1975. Gravity field of Jupiter from Pioneer 11 tracking data. *Science* 188:476–477.

Podolak, M., and Cameron, A. G. W. 1974. Models of the giant planets. *Icarus* 22:123–148.

Slattery, W. L., and Hubbard, W. B. 1975. Thermodynamics of a solar mixture of molecular hydrogen and helium at high pressure. In draft.

Zharkov, V. N.; Makalkin, A. B.; and Trubitsyn, V. P. 1974. Models of Jupiter and Saturn. II. Structure and composition. *Astron. Zh.* 51:1288–1297.

Zharkov, V. N., and Trubitsyn, V. P. 1974. Determination of the equation of state of the molecular envelopes of Jupiter and Saturn from their gravitational moments. *Icarus* 21:152–156.

Zharkov, V. N.; Trubitsyn, V. P.; and Makalkin, A. B. 1972. The high gravitational moments of Jupiter and Saturn. *Astrophys. Lett.* 10:159–161.

Zharkov, V. N.; Trubitsyn, V. P.; and Samsonenko, L. V. 1971. *Physics of the earth and planets.* Moscow: Nauka.

PART III

Atmosphere and Ionosphere

RESULTS OF THE INFRARED RADIOMETER EXPERIMENT ON PIONEERS 10 AND 11

A. P. INGERSOLL, G. MÜNCH
G. NEUGEBAUER, and G. S. ORTON
California Institute of Technology

The infrared radiometers on the Pioneer 10 and 11 spacecraft have mapped Jupiter in two broad spectral channels centered at wavelengths of 20 and 45 μm. Comparison of Pioneer 10 and 11 data and the results of in-flight and laboratory calibrations indicate an absolute accuracy of ± 8% in the measured intensities. Relative accuracy, for comparing different regions on the planet, is ± 2%. Within the precision of observations, the derived value of effective temperature is independent of assumptions about the Jovian emission spectrum. The global effective temperature is 125 ± 3°K, implying a value of 1.9 ± 0.2 for the ratio of planetary thermal emission to solar energy absorbed. The effective temperatures of belts are greater than those of zones by as much as 3°.5K. On a global scale, the poles and equator have the same effective temperature.

Data from the infrared radiometers on the Pioneer 10 and 11 spacecraft have been used to obtain estimates of the global heat balance, the heat balance of different regions on the planet (Chase *et al.* 1974; Ingersoll *et al.* 1975), atmospheric temperature profiles in the region 0.1 to 1.0 bar, and an estimate of the hydrogen–helium ratio (Orton 1975).[1] We summarize these results below, emphasizing the basic data themselves and those results which follow more or less directly from the data, independent of the atmospheric model.

There are two channels on the radiometer, centered at wavelengths $\lambda = 20.0$ and $\lambda = 45.4$ μm, with equivalent widths of 11.6 and 22.7 μm, respectively. The filter transmission curves are shown in Fig. 1, along with equivalent whole disk spectra computed from an atmospheric model (Orton 1975). The field of view is $1°0 \times 0°3$, with the long axis aligned approximately north-south on Jupiter and covering about $\frac{1}{30}$ of the planetary diameter. Pioneer 10 (Chase *et al.* 1974) gave one image of Jupiter in each channel, each image centered at a latitude of $-11°$. Pioneer 11 gave two images in each channel, an inbound image centered at $-41°$ and a partial outbound image at $52°$ (see Fig. 2). Data were lost in the outbound viewing period as a result of false commands during passage through the radiation belts.

_____
[1]See p. 206.

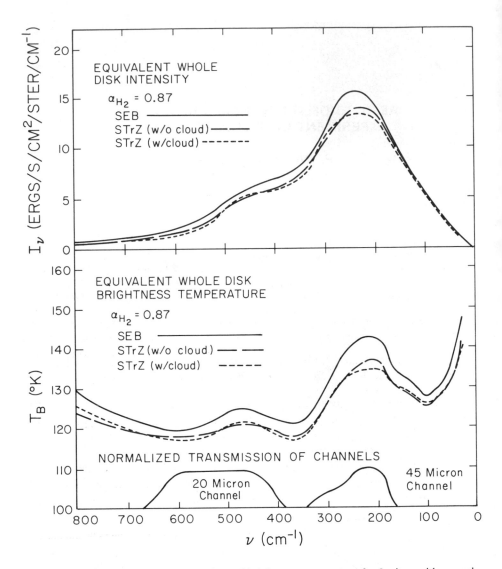

Fig. 1. Equivalent whole-disk intensity and brightness temperature for Jupiter, with normalized transmission functions for the 20 and 45 μm channels. The Jupiter spectra are from the atmospheric model of Orton (1975), based on Pioneer 10 infrared radiometer observations of the South Equatorial Belt (SEB: latitudes − 7° to − 5°) and South Tropical Zone (STrZ: latitudes − 16° to − 14°).

Calibration of the radiometer was done in the laboratory before the flight and monitored for changes during the flight. In the laboratory, two measurements were made. First, the response of the instrument to a blackbody emitter was measured. Second, the response to a calibrator plate, which is a part of the instrument and can be moved into the field of view during the flight, was also measured. The first measurement provides the absolute calibration;

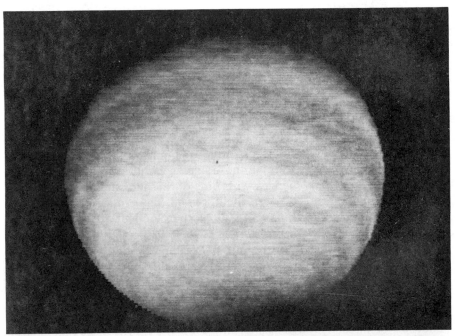

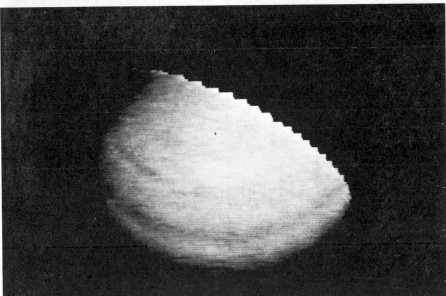

Fig. 2.   Two images of Jupiter at 45 μm wavelength from Pioneer 11 data. North is at the top. Light and dark bands in visible light (zones and belts) are regions of low and high emission temperature, respectively. The upper right portion of the outbound image was lost because of false commands during passage of Pioneer 11 through Jupiter's radiation belts. The apparently greater limb darkening at the poles is an artifact of the changing aspect of the planet during the 2.5-hour interval during which the images were made.

the second provides a means of detecting changes in the instrument during the flight. The calibrator plate is moved into the field of view every 12 seconds; its temperature is measured on the spacecraft and transmitted to earth.

The results of the in-flight calibration were encouraging. In no case did the response to the calibrator plate differ from the laboratory value by more than 3%. In addition, Pioneer 10 and Pioneer 11 gave good agreement when looking at the same latitude on Jupiter at the same value of the emission angle. Figure 3 shows curves of intensity versus latitude for 3 values of the emission angle cosine, namely $\mu = 0.95, 0.75, 0.55$. Data from all longitudes and from all values of μ differing from the nominal values by as much as ± 0.05 have been used in forming these averaged curves. In spite of variations with longitude and changes with time during the one-year interval between Pioneer 10 and 11, the agreement is good. The Pioneer 11 curves generally show shallower maxima and minima than Pioneer 10, but this can be attributed to the greater foreshortening of the belts and zones as viewed by Pioneer 11. Our estimates of the errors of calibration fall in the range ± 4 to $\pm 8\%$, and we have adopted the latter value throughout our analysis.

In estimating the effective temperature of a region on Jupiter, additional uncertainties arise because of the integrations over emission angle and over wavelength. Since each region was viewed only once in each image, the dependence of intensity on emission angle must be inferred by comparing

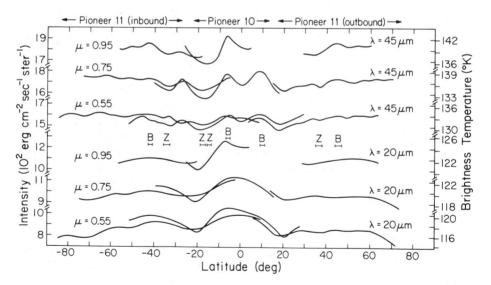

Fig. 3.   Intensity (left-hand ordinate) and equivalent brightness temperature (right-hand ordinate) as a function of latitude for three values of the emission angle cosine $\mu = 0.95, 0.75, 0.55$. The ordinate scale has been divided, and each curve shifted upward by two divisions. Significant features of these curves include: (1) the agreement between Pioneer 10 and 11 data, (2) the magnitude of belt-zone contrasts, and (3) the lack of pronounced equator-to-pole contrasts.

TABLE I
*Intensity Versus the Cosine, μ,
of the Emission Angle Observed by Pioneers 10 and 11.*

Latitudes	Values of μ					
	0.45	0.55	0.65	0.75	0.85	0.95
20 μm Intensity (10^2 erg cm^{-2} sec^{-1} $ster^{-1}$)						
$-43°$ to $-41°$ (B)	8.52	9.36	9.99	10.08	11.02	11.33
$-36°$ to $-33°$ (Z)	8.68	9.02	9.52	9.74	10.55	10.92
$-16°$ to $-14°$ (Z)	8.88	9.27	9.43	10.00	10.29	10.51
$-7°$ to $-5°$ (B)	10.22	10.38	11.43	11.40	11.88	12.67
35° to 38° (Z)	8.08	8.86	8.96	9.89	10.24	10.77
44° to 47° (B)	7.86	8.58	9.33	9.80	10.46	10.99
45 μm Intensity (10^2 erg cm^{-2} sec^{-1} $ster^{-1}$)						
$-43°$ to $-41°$ (B)	15.16	16.10	16.70	17.46	17.99	18.60
$-36°$ to $-33°$ (Z)	14.87	15.88	16.40	16.49	17.59	17.89
$-16°$ to $-14°$ (Z)	14.01	14.49	15.22	15.47	16.16	16.33
$-7°$ to $-5°$ (B)	15.75	16.24	16.83	17.68	18.45	19.15
35° to 38° (Z)	14.52	15.50	15.88	16.44	17.16	17.63
44° to 47° (B)	14.74	15.95	16.54	17.26	17.88	18.58

different regions in the same latitude band observed at different emission angles. Latitudes where there was obvious longitudinal structure, inferred either from infrared or visible data, were excluded or treated specially. This approach also underlies Orton's (1975) inferences concerning the thermal structure and He/H_2 ratio in the atmosphere. Table I lists the measured intensities as functions of μ for six distinct latitude bands, all of which are identified as belt (B) or zone (Z) in Fig. 3. Data for $\mu < 0.4$ are not included because of resolution problems close to the limb.

Several models were used to estimate the ratio $F/(F_{20} + F_{45})$ of the total emitted flux $F = \sigma T_e^4$ to the sum of the fluxes $F_{20} + F_{45}$ observed in the two channels of the radiometer. Here σ is the Stefan-Boltzmann constant and T_e is the effective temperature. These methods are summarized by Ingersoll *et al.* (1975). First, the emission spectrum is assumed to be a blackbody at temperature T_e such that the sum $F_{20} + F_{45}$ computed for the blackbody spectrum is equal to that observed. Second, an atmospheric model is assumed in which the opacity is due to H_2 and NH_3, mixed with He, and the temperature profile satisfies constraints of radiative equilibrium. Vapor equilibrium controls the amount of NH_3 at the levels of interest, so that the model is characterized by just two constants, T_e and the molar fraction of hydrogen α_{H_2} (Trafton 1973; Trafton and Stone 1974). The values of these constants are determined from the values of $F_{20} + F_{45}$ and F_{45}/F_{20} (Hunten and Münch 1973). The third approach is to use the μ-dependence as given in Table I to

infer the temperature profile without constraints of radiative equilibrium
(Orton 1975). The opacity model is the same as in the second approach.
The flux at wavelengths shorter than 12.5 μm was assumed by Ingersoll *et
al.* and by Orton to be 500 erg cm^{-2} sec^{-1} for all models. This avoids having
to calculate opacities in the window regions at shorter wavelengths and is in
agreement with groundbased measurements at 4.5–5.5 μm and at 8–14 μm
wavelength.

As shown by Ingersoll *et al.* (1975), all of the above methods are consistent with the value

$$F/(F_{20} + F_{45}) = 1.72 \pm 0.03 . \tag{1}$$

Thus the effective temperatures can be computed from the data of Table I,
by first integrating with respect to μ to get $F_{20} + F_{45}$, and then multiplying
by 1.72 to get F and hence T_e. The values computed in this way are not significantly different from those found by Orton and Ingersoll, which range
from 124°1K for the South Tropical Zone (lat. $-16°$ to $-14°$) to 127°7K for
the South Equatorial Belt (lat. $-7°$ to $-5°$). For the whole planet, the global
mean effective temperature is

$$T_e = 125 \pm 3°K . \tag{2}$$

The $\pm 3°$K uncertainty is mainly due to calibration, although all other known
sources have been included in the estimate. In computing the global mean
effective temperature, data at all latitudes and emission angles from both
Pioneer 10 and 11 have been used in the analysis, as summarized by Ingersoll
et al. (1975).

The above value of effective temperature is lower than the 134°K estimate obtained from aircraft observations (Aumann *et al.* 1969; Armstrong
et al. 1972), and it is also lower than an estimate based on a preliminary analysis of the Pioneer 10 infrared radiometer data (Chase *et al.* 1974). The former
discrepancy is of unknown origin, and barely significant. The latter is smaller
yet and is due to the preliminary nature of the Pioneer 10 analysis.

For $T_e = 125 \pm 3°$K, the ratio of emitted radiation to absorbed sunlight
for the planet as a whole is 1.9 ± 0.2. This is based on an assumed Bond
albedo $A = 0.425$, which is the product of the geometric albedo $p = 0.28 \pm$
0.05 (Taylor 1965) and the phase integral $q = 1.52 \pm 0.05$ (Tomasko *et al.*
1974). The average internal heat flux is then about 6.5×10^3 erg cm^{-2} sec^{-1}.
In contrast, for $T_e = 134°$K the ratio of emitted to absorbed power is 2.5 and
the average internal heat flux is 1.1×10^4 erg cm^{-2} sec^{-1}. It may be significant that interior models of Jupiter can barely account for the energy implied
by the larger value of the internal heat flux integrated over 4.5×10^9 yrs, but
can more easily account for the lower value (Hubbard 1970; Graboske *et al.*
1975; Pollack and Reynolds 1974).[2]

---

[2]See p. 76; also see p. 508.

Acknowledgements. We thank the members of the Pioneer Project for their cooperation and support. We also thank J. Bennett, S. C. Chase, and B. Schupler for their advice and assistance.

DISCUSSION

F. J. Low: Figure 4 summarizes the available data on the thermal emission of Jupiter and its wavelength dependence. Short of 13 microns the solid curve is based on groundbased work, primarily that of Gillett *et al.* (1969). There are also groundbased points at 21 and 34 μm. The longer wavelength data are from the NASA Lear Jet (Armstrong *et al.* 1972), except for the 1-mm observations of Low and Davidson (1965). The two pairs of circles represent Pioneer 10 and 11 data (Ingersoll *et al.* 1975). It is noteworthy that such excellent agreement exists between the results obtained by quite different instruments operated over a span of several years.

The horizontal bar at the top indicates the wavelength range over which a 125°K blackbody emits half of its energy, with equal amounts at shorter and longer wavelengths.

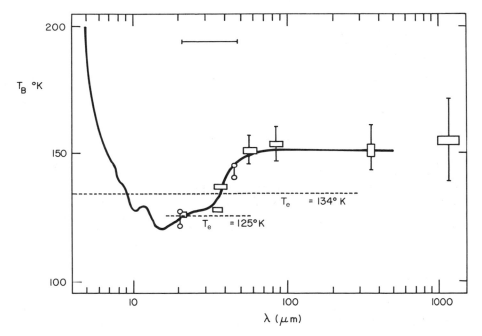

Fig. 4. Brightness temperature T_B as a function of wavelength. The rectangular points and solid line are from groundbased and airborne observations. The circles are from Pioneer 10 and 11 data. Note the excellent agreement. Effective temperature T_e is also indicated for Lear Jet and Pioneer results. Horizontal bar at top shows wavelength range containing half of 125°K blackbody radiation.

The effective temperature T_e is shown for two cases: (1) the Lear Jet experiment of Aumann *et al.* (1969) in which all wavelengths out to ~ 300 μm were included and, (2) the Pioneer 10 and 11 results (reported by Ingersoll *et al.* above) including only about half the thermal emission within the relatively narrow band of their two filters. From the data now available it seems that the value of 134°K is somewhat too high since the spacecraft results yield 125°K and there is excellent agreement between the ground-based, airborne and spacecraft data where there is overlap. Because T_B increases to 150°K at 100 mm and beyond, I prefer a value of 129 ± 4°K. Techniques are now readily available for improving the accuracy of far-infrared radiometry. This is needed for other areas of research at these wavelengths.

Finally, it should be noted that there is at least some experimental support for significant variations in the infrared emission of Jupiter. In particular, at those wavelengths where the emission originates deep in the cloud layers, such as at 5 μm, existing results (see Keay *et al.* 1973) show variations as large as a factor of 2 integrated over the disk.

J. B. Pollack: Even at Lear Jet altitudes, there is a significant amount of water vapor absorption at wavelengths longward of 35 μm. The stars used to calibrate the broadband observations of Jupiter have a very different spectral distribution from that of Jupiter. There could be a systematic error introduced into the deduced effective temperature of Jupiter from the broadband aircraft observations.

F. L. Low: This is quite true. However, the effect is to make T_e higher, not lower.

REFERENCES

Armstrong, K. R.; Harper, Jr., D. A.; and Low, F. J. 1972. Far-infrared brightness temperatures of the planets. *Astrophys. J.* 178:L89–L92.

Aumann, H. H.; Gillespie, Jr., C. M.; and Low, F. J. 1969. The internal powers and effective temperatures of Jupiter and Saturn. *Astrophys. J.* 157:L69–L72.

Chase, S. C.; Ruiz, R. D.; Münch, G.; Neugebauer, G.; Schroeder, M.; and Trafton, L. M. 1974. Pioneer 10 infrared radiometer experiment: preliminary results. *Science* 183:315–317.

Gillett, F. C.; Low, F. J.; and Stein, W. A. 1969. 2.8 – 14 micron spectrum of Jupiter. *Astrophys. J.* 157:925–934.

Graboske, H. C., Jr.; Pollack, J. B.; Grossman, A. S.; and Olness, R. J. 1975. The structure and evolution of Jupiter: the fluid contraction state. *Astrophys. J.* In press.

Hubbard, W. B. 1970. Structure of Jupiter: chemical composition, contraction, and rotation. *Astrophys. J.* 162:687–697.

Hunten, D. M.; and Münch, G. 1973. The helium abundance on Jupiter. *Space Sci. Rev.* 14: 433–443.

Ingersoll, A. P.; Münch, G.; Neugebauer, G.; Diner, D. J.; Orton, G. S.; Schupler, B.; Schroeder, M.; Chase, S. C.; Ruiz, R. D.; and Trafton, L. M. 1975. Pioneer 11 infrared radiometer experiment: the global heat balance of Jupiter. *Science* 188:472–473.

Keay, C.; Low, F. J.; Rieke, G. H.; and Minton, R. B. 1973. High-resolution maps of Jupiter at five microns. *Astrophys. J.* 183:1063–1073.

Low, F. J., and Davidson, A. W. 1965. Lunar observations at a wavelength of 1 mm. *Astrophys. J.* 142:1278–1282.

Orton, G. S. 1975. The thermal structure of Jupiter. I. Implications of Pioneer 10 infrared radiometer data. *Icarus* 26:125–141.

Pollack, J. B., and Reynolds, R. T. 1974. Implications of Jupiter's early contraction history for the composition of the Galilean satellites. *Icarus* 21:248–253.

Taylor, D. J. 1965. Spectrophotometry of Jupiter's 3400–10000 Å spectrum and a bolometric albedo for Jupiter. *Icarus* 4:362–373.

Tomasko, M. G.; Clements, A. E.; and Castillo, N. D. 1974. Limb darkening of two latitudes of Jupiter at phase angles of 34° and 109°. *J. Geophys. Res.* 79:3653–3660.

Trafton, L. M. 1973. On the He − H$_2$ thermal opacity in planetary atmospheres. *Astrophys. J.* 179:971–976.

Trafton, L. M., and Stone, P. H. 1974. Radiative-dynamical equilibrium states for Jupiter. *Astrophys. J.* 188:649–655.

PIONEER 10 AND 11 AND GROUND-BASED INFRARED DATA ON JUPITER: THE THERMAL STRUCTURE AND He−H₂ RATIO

G. S. ORTON

and

A. P. INGERSOLL
California Institute of Technology

Temperature profiles for low- and mid-latitude regions on Jupiter are obtained from Pioneer 10 and 11 and ground based infrared data. Temperatures at 1.0 bar are near 165°K and drop to 105−110°K near 0.10 bar, for an overlying thermal inversion which reaches 133−143°K near 0.03 bar. The molar fraction of H_2 most consistent with the Pioneer data is 0.88 ± 0.06 (He/H_2 = 0.14 ± 0.08), conditional on the perfect validity of the model and lack of systematic errors in the data. The local effective temperatures of various regions range from 124°.2 to 127°.6K, consistent with the global average of 125°K in the model-independent results in the previous chapter. It is probable that the temperature profiles of various latitudes on the planet are quite similar and that differences in observed intensity reflect the presence or absence of an optically thick cloud near 0.70 bar.

Temperature profiles for the 1.0−0.1 bar pressure regime and an estimate of the molar fraction of H_2 have been derived previously from Pioneer 10 infrared radiometer data for low-latitude regions by Orton (1975a). These results are not constrained by considerations of radiative-convective equilibrium, but they are direct inversions of the limb structure data. Further information about the cloud structure and about the temperature profile overlying the 0.1 bar pressure level have been determined from spectrally and spatially resolved data in the 8 − 14 μm wavelength region (Orton 1975b). This report presents a recovery of temperature profiles for mid-latitude regions and an improvement in the determination of the molar mixing ratio of H_2, α_{H_2}, from Pioneer 11 data.

Orton (1975a) determined a value of $\alpha_{H_2} = 0.91 \pm 0.07$ as a best fit to data from both infrared channels (near 20 and 45 μm). The thermal structures recovered from the data are summarized in Fig. 1, taken from that report, and assume $\alpha_{H_2} = 0.87$, expected from the assumption of solar composition

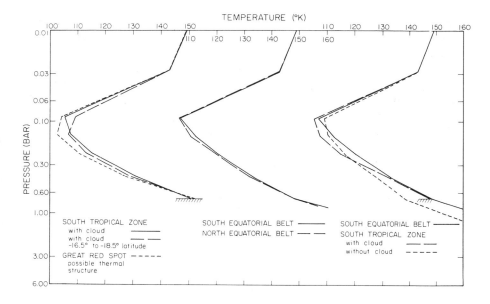

Fig. 1.   Temperature profiles derived from Pioneer 10 infrared radiometer data, from Orton (1975*a*). The profiles displayed assume a thermal inversion overlying the 0.10 bar level, as determined by Orton (1975*b*). A value of 0.87 is assumed for the H_2 mixing ratio. Where not otherwise noted, the latitude bin for the South Tropical Zone is −16°.0 to −14°.0.

(Weidenschilling and Lewis 1973). The temperature lapse rate in the deep atmosphere is assumed to be adiabatic. Two types of models are shown for zone regions. The first assumes pure gaseous absorption, as in the models for the belt regions. Near 0.70 bar, however, such models are some 10°K cooler than for the adjacent belt. The second model includes the presence of an optically opaque cloud top with unit emissivity. The temperature of the zones in the models with clouds is assumed to be nearly the same as for the belts at the level of the cloud top.

The deviation from constant flux is demonstrated in Fig. 2 which shows the emitted net flux, flux divergence, and cooling rate for a temperature profile which is an average of those derived for the South Equatorial Belt and the "cloudy" model of the South Tropical Zone. The effects of the cloud are not included in these calculations. The departure from constant flux near 0.45 bar is due to the fact that the temperature lapse rate is much less than that required for radiative equilibrium where the mean infrared optical depth is near unity. The flux divergence near 0.03 bar is associated with the thermal inversion, and is probably due to deposition of solar radiative energy.

The Pioneer infrared radiometer data cannot reliably recover temperature profile information above the 0.10 bar level. Information about the thermal structure above the 0.10 bar level is obtained from spectral and limb structure data in the 7.7 μm CH_4 fundamental band (Orton 1975*b*). As an *a priori* assumption therefore, a thermal inversion has been adopted for the mid-latitude regions which is similar to those shown in Figs. 1 and 2 and is consistent with the 8 μm analysis for equatorial regions. However, the temperatures in the inversion have been scaled so that, for a region at latitude λ, the flux divergence is a factor of cos λ smaller than that shown in Fig. 2. This assumes that the flux divergence can be accounted for solely by deposition of solar energy and it neglects any changes in the distribution of solar energy with altitude which may take place at increasing latitudes. For completeness, the alternative assumption has also been made that the temperature profile above the 0.10 bar level is isothermal.

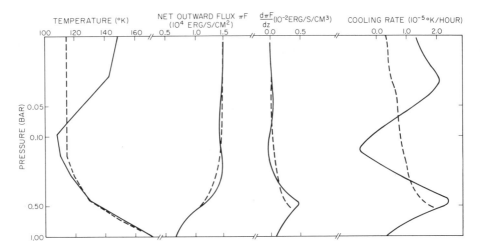

Fig. 2. Thermal structure, net emitted flux, flux divergence, and cooling rate of model similar to South Equatorial Belt model of Fig. 1, from Orton (1975*b*). Dashed lines refer to temperature profile (left) which is similar to South Equatorial Belt Model, assuming an isotherm above the 0.10 bar level, as derived by Orton (1975*a*).

Two mid-latitude regions in the southern hemisphere and two in the northern hemisphere were chosen for analysis as they respectively represented the widest latitude features available in the Pioneer 11 inbound and outbound thermal maps (Ingersoll *et al.*).[1] In each hemisphere the two regions consist of an adjacent belt and zone.

As shown by Orton (1975*a*), the assumed α_{H_2} has been varied in order to determine the one value whose fit to the data provides the minimum rms residual. The variation of the fit with α_{H_2} is shown in Fig. 3 for the northern hemisphere belt region (lat. $+ 44.0°$ to $+ 47.0°$). No data for $\mu < 0.40$ have been considered, where μ is the cosine of the angle between the direction of emission and the local vertical. The inversion of the data follows Smith's

[1]See p. 199.

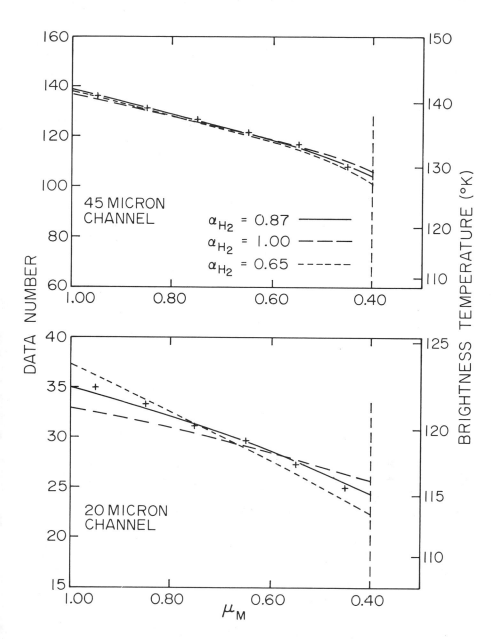

Fig. 3.   Fit of northern hemisphere belt (+ 44°0 to + 47°0 latitude) models to averaged data for the rising limb. The data were acquired during Pioneer 11 outbound passage; no data are available for the setting limb.

(1970) iterative technique as used in the inversion of the Pioneer 10 data. A plot of the rms residual versus α_{H_2} for this region and two models for a north-

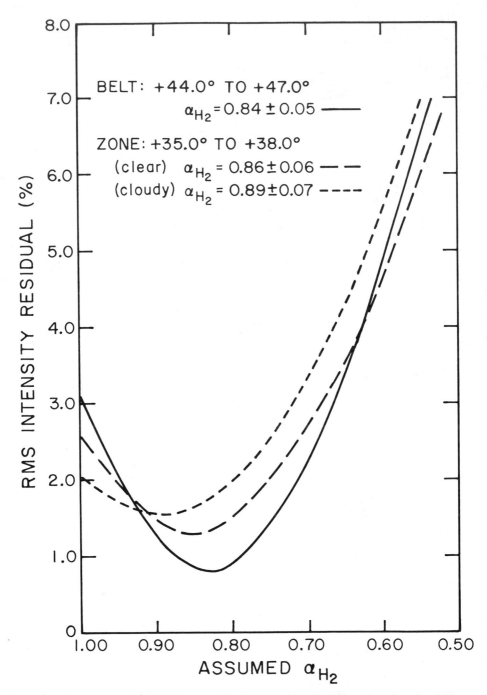

Fig. 4. Rms residuals for the fit of models to the data for the indicated northern hemisphere regions. Formal errors shown are computed according to Eq. 1 in the text.

ern hemisphere zone are shown in Fig. 4. The uncertainty associated with each determination of α_{H_2} is calculated using (Bevington 1969)

$$\sigma = \left(\frac{m-n}{r} \frac{\partial^2 r}{\partial \alpha_{H_2}^2} \right)^{-\frac{1}{2}} \tag{1}$$

A conservative estimate of the degrees of freedom, $m - n = 3$, is assumed by setting $m = 4$ (independent data from $\mu = 1.00$ and $\mu = 0.40$ in each channel, corresponding to a temperature and lapse rate determination for each channel in overlapping regions of the atmosphere) and $n = 1$ (for the single parameter α_{H_2}). The rms residual is represented by r.

The values for α_{H_2} determined for each region investigated are summarized in Table I. Averaging the results obtained with clear and cloudy atmospheric models for zones, the weighted mean of all results shown in Table I is $\alpha_{H_2} = 0.88$ ($He/H_2 = 0.14$). The average of individual uncertainties is 0.07, the standard deviation from the mean is 0.06, and the error of the mean is 0.04. It is emphasized that this result and the associated estimates of the formal uncertainty depend on: (1) the perfect validity of the opacity model (which assumes gaseous absorption by H_2 and NH_3); (2) the accuracy of the relative calibration of both radiometer channels; (3) the accuracy of the assumed geometry (assignment of values of μ to each datum); and (4) the absence of systematic effects from unresolved details of the planetary structure and from adjacent regions. On the other hand, the good quality of the fit is an indication of: (a) the good internal consistency of the data which is averaged as in Fig. 3, and (b) at least the approximate validity of the opacity model assumed (or the presence of self-cancelling systematic errors).

TABLE I
Derived Values for H_2 Mixing Ratio

	α_{H_2}	
Region	Clear Model	"Cloudy" Model
$-43.0°$ to $-41.0°$ (belt)	0.87 ± 0.07	——
$-36.0°$ to $-33.0°$ (zone)	0.92 ± 0.08	1.00 ± 0.09
$-16.0°$ to $-14.0°$ (zone)	0.89 ± 0.07	0.89 ± 0.07
$-7.0°$ to $-5.0°$ (belt)	0.93 ± 0.09	——
$+35.0°$ to $+38.0°$ (zone)	0.85 ± 0.06	0.89 ± 0.07
$+45.0°$ to $+48.0°$ (belt)	0.84 ± 0.05	——
Weighted average	0.87	0.89
Average individual uncertainty	0.07	0.07
Standard deviation from the mean	0.06	0.06
Error of the mean	0.04	0.03

Figure 5 shows the recovered thermal structure of each region, for $\alpha_{H_2} = 0.87$, assuming solar composition. The models shown in Fig. 5 assume the thermal inversion which has been devised as described above for mid-latitude regions. The temperature difference between the belt and the clear atmosphere model of the zone near 0.70 bar is not so great as for the South Equatorial Belt and the South Tropical Zone (Fig. 1). Nevertheless, the location of the optically opaque cloud top in the "cloudy" zone model is less than 4 km deeper than the corresponding cloud in the case of the South Tropical Zone (Fig. 1). The correlation between such a cloud and the level where NH_3 saturation is expected to begin is maintained for mid-latitude as well as low-latitude regions.

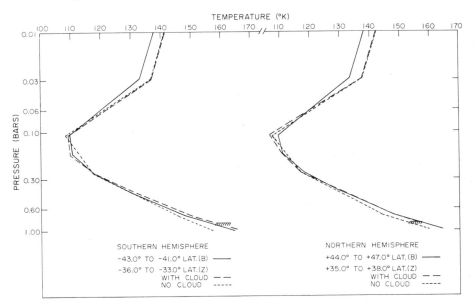

Fig. 5.   Derived temperature profiles for mid-latitude regions. The profiles displayed are derived assuming a thermal inversion overlying the 0.10 bar pressure level, as explained in the text. "Cloudy" models for zone regions have optically opaque cloud layer with unit emissivity and are nearly the same temperature as adjacent belt regions at the cloud top level.

Table II displays the local effective temperatures for each region determined from the spectra of the models shown in Figs. 1 and 5. These values do not change significantly for the assumption of an overlying isotherm instead of a thermal inversion. Furthermore, all the values are close to the global value of 125°K determined by Ingersoll et al. (1975) independently of specific models.

Finally, Fig. 6 shows a plot of temperatures along three isobars for the regions from which temperature profile recovery was attempted. In order to remove any restrictions imposed by a priori assumptions regarding the over-

TABLE II
Derived Values for Local Effective Temperatures

Region	$T_{eff}(K)$ Clear Model	"Cloudy" Model
− 43.0° to − 41.0° (belt)	125°9	——
− 36.0° to − 33.0° (zone)	125.2	125°2
− 16.0° to − 14.0° (zone)	124.2	124.2
− 7.0° to − 5.0° (belt)	127.6	——
+ 35.0° to + 38.0° (zone)	124.5	124.6
+ 45.0° to + 48.0° (belt)	125.2	——

lying thermal structure, the temperature above the 0.10 bar level was as-sumed constant and equal to the temperature determined at the 0.10 bar level. Furthermore, the clear atmosphere model was assumed for zone re-gions. The local effective temperature of each region, as given in Table II, is also plotted in Fig. 6. Adjacent regions are connected with a dashed line.

Several comments may be made with respect to Fig. 6.

1. If one averages together the temperatures of adjacent belts and zones, the temperature variation between equator and pole is two degrees or less. This is in agreement with similar observations made by Ingersoll et al. (1975)[2] with respect to the behavior of the observed intensity variation with latitude.

2. Temperature differences between adjacent belts and zones near 0.70 bar are greater at the equator than at mid-latitudes. This may be a conse-quence of inadequate resolution of the mid-latitude features which are somewhat narrower than those near the equator, such as the South Equatorial Belt and the South Tropical Zone. However, the fairly con-stant belt temperatures at 0.70 bar tend to argue against this.

3. The apparent temperature difference between adjacent belts and zones which appear at 0.30 and 0.70 bar has vanished near 0.10 bar. This raises the possibility that the real temperature structure in the 0.70 to 0.10 bar pressure regime may be quite similar for a wide range of lati-tudes and that all variations in outgoing intensity are simply reflections of the presence or absence of an optically thick cloud near the 0.70 bar level.

As discussed in Orton (1975a), such a cloud correlates well with the visible and 5 μm appearance of the planet,[3] as well as with the expected depth at which NH_3 begins to saturate. The slightly different altitude of the cloud top as a function of latitude may be attributed to dynamical processes which vary in latitude.

---

[2]See p. 201.   [3]See p. 407 and p. 351.

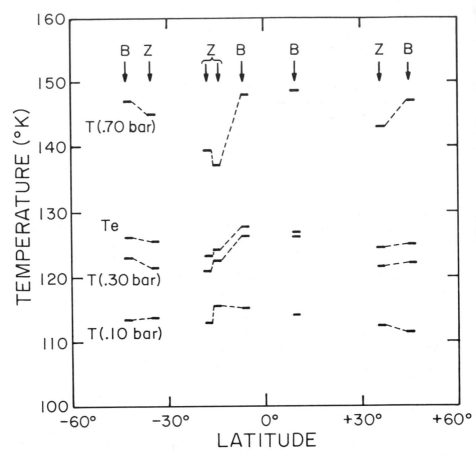

Fig. 6. Variation of thermal structure with latitude along indicated isobars for regions where temperature profiles have been recovered. These models assume no clouds and an isotherm overlying the 0.10 bar level. The local effective temperatures listed in Table II are also displayed.

Acknowledgments. The authors gratefully acknowledge the advice and support of G. Münch. This research is supported in part by Pioneer Project funds and by NASA grant NGL 05-002-003. One of the authors (GSO) is currently a NASA-NRC Resident Research Associate at the Jet Propulsion Laboratory.

DISCUSSION

J. B. Pollack: The opacity of ammonia is not well known in the 45 μm spectral region. Does ammonia make an important contribution to the opacity in your 45 μm band and if so, how does this affect the error bars of your helium-to-hydrogen ratio determination?

G. S. Orton: The opacity contribution of the NH_3 pure rotational band significantly affects the intensity detected by the 45 μm channel, as the model spectra in Orton (1975*a*) show. Until the ammonia opacity can be checked in the laboratory, the uncertainty in the 45 μm channel intensity originating from any given atmospheric model is nearly impossible to estimate. There are also other differences between your spectral observations and the spectra of the models presented here which probably contribute to the uncertainty of the helium-to-hydrogen ratio determination, because of the unknown limb structure of such features.

L. Wallace: Are the clouds of your cloudy model opaque in the thermal infrared or only in visible light?

G. S. Orton: The tops of the clouds in these models are opaque at thermal infrared wavelengths, and they are modeled crudely as blackbody surfaces radiating at the ambient temperature. The correlation of cloud presence or absence with the visible albedo structure is discussed by Orton (1975*a*).

REFERENCES

Bevington, P. R. 1969. *Data reduction and error analysis for the physical sciences.* pp. 242–245. New York: McGraw-Hill Book Co.

Ingersoll, A. P.; Münch, G.; Neugebauer, G.; Diner, D. J.; Orton, G. S.; Schupler, B.; Schroeder, M.; Chase, S. C.; Ruiz, R. D.; and Trafton, L. M. 1975. Pioneer 11 infrared radiometer experiment: the global heat balance of Jupiter. *Science* 188:272–273.

Orton, G. S. 1975*a*. The thermal structure of Jupiter: I. Implications of Pioneer 10 infrared radiometer data. *Icarus* 26:125–141.

Orton, G. S. 1975*b*. The thermal structure of Jupiter: II. Observations and analysis of 8–14 micron radiation. *Icarus* 26:142–158.

Smith, W. L. 1970. Iterative solution of the radiative transfer equation for the temperature and absorbing gas profile of an atmosphere. *Appl. Opt.* 9:1993–1999.

Weidenschilling, S. J., and Lewis, J. S. 1973. Atmospheric and cloud structure of the Jovian planets. *Icarus* 20:465–476.

STRUCTURE OF THE ATMOSPHERE OF JUPITER FROM PIONEER 10 AND 11 RADIO OCCULTATION MEASUREMENTS

A. J. KLIORE and P. M. WOICESHYN
Jet Propulsion Laboratory

Radio occultation data recorded from Pioneers 10 and 11 during their encounters with Jupiter have been analyzed using a technique that takes into account the oblateness of Jupiter's atmosphere, which had been neglected in previous analyses. The new center of refraction, which differs from the center of figure, is located by the radius of curvature and the normal direction at the closest approach of point of the ray. The large variation of oblateness with altitude in Jupiter's atmosphere causes the location of the instantaneous center of refraction to vary. However, it must be held constant at some point between its extremes of variation in order to invert the data using the Abel integral transform. The temperature-pressure profiles so derived have been shown to be sensitive to the refractivity bias computation for all data, as well as to the modeling of the oscillator drift function for the case of Pioneer 11 exit data, in which the oscillator drift rate was changing abruptly. The three measurements described in this chapter (Pioneer 10 entry and exit, and Pioneer 11 exit) have produced results that are mutually consistent, all showing a temperature inversion between 10 and 100 millibars, with temperatures between 130° and 170°K at 10 mbar and 80°–120°K at 100 mbar. The radio occultation results are also in close agreement with models derived from infrared radiometer data of Pioneer 10 and those calculated on the basis of radiative-convective balance.

The data for the Pioneer 10 and 11 radio occultation measurements were obtained during the flybys of Pioneer 10 on December 4, 1973 and of Pioneer 11 on December 3, 1974. The previously reported analyses of these data (Kliore *et al.* 1974*a*; Kliore *et al.* 1974*b*; Kliore *et al.* 1975) produced results indicating temperatures in the Jovian atmosphere far in excess of those expected on the basis of theoretical models and other observations. The unusual nature of these results was explained by the realization (Hubbard *et al.* 1975*a*) that the data inversion technique was very sensitive to the oblateness of the atmosphere of Jupiter, which had been neglected in previous analyses. This paper describes a technique for incorporating the effects

[216]

of the oblateness of the atmosphere of Jupiter into the data analysis procedure making use of a spherical harmonic representation of the gravity field
of Jupiter to compute the shape of the planet, and presents intermediate results derived from Pioneer 10 and 11 data obtained by applying the Abel
integral inversion method with consideration of the oblateness of the planet.

I. PROCEDURE FOR THE INVERSION OF DATA CONSIDERING THE OBLATE SHAPE OF JUPITER

The procedure used for inversion of radio occultation data obtained from
Mariner spacecraft to Mars and Venus was based on a technique which relied on the assumption of sphericity, namely, that the center of refraction is
also the center of figure (Phinney and Anderson 1968; Fjeldbo and Eshleman
1968; Kliore 1972; Kliore *et al.* 1972).

The technique that is used in this chapter makes use of our knowledge
of the shape of the Jovian atmosphere to compute the location of a new center of refraction based on the radius of curvature and the normal direction
at the closest approach point. The new center of refraction, together with
the directions to the earth and the spacecraft, are then used to define a new
plane of refraction, and the analysis then is carried on in the same manner
as in the spherical case. This is an approximation, inasmuch as the path of
a ray refracted through a non-spherical atmosphere does not lie in a plane,
and the plane defined by the incoming and outgoing asymptotes will not, in
general, contain precisely the normal at the closest approach point. However,
it has been shown (Hubbard 1976) that the angle between the normal and the
plane of the asymptotes does not exceed 10^{-5} radians for the Pioneer occultations.

The geometry pertinent to oblate inversion is shown in an exaggerated
view in Fig. 1. C is the center of symmetry of the figure and C' is the new
center of refraction determined by the normal vector $\hat{n}$ at the point of closest
approach 0 and the radius of curvature ρ, which is the radius of curvature of
the atmosphere at the point of closest approach of the ray, in the plane defined by the incoming and outgoing asymptotes. The plane of refraction is
determined by the direction to the earth and the vector M_n describing the
location of the spacecraft relative to the center of refraction C', and r_0 is the
distance of the closest approach point from the center of figure C.

The computation for each point proceeds approximately as follows:

1. A new value of the vector from the center of the planet to the point of
    closest approach r_0 is computed from the previously determined values
    of the bending angle ϵ and the ray asymptote parameter p.
2. The radius of curvature ρ and the normal direction $\hat{n}$ are computed at
    r_0 using a spherical harmonic representation of the shape of the planet
    (using a modification of the subroutine *SHAPE* kindly provided by W. B.
    Hubbard).

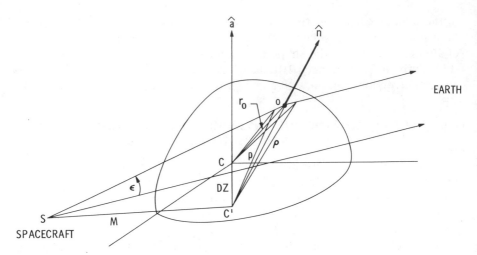

Fig. 1.   Geometry of refraction for an oblate planet. It is assumed that rays are refracted in a plane determined by the location of the spacecraft S, the center of refraction C' and the earth. The center of refraction C' is determined by the normal $\widehat{\mathbf{n}}$ and the radius of curvature ρ at the point of closest approach of the ray 0.

3.   The center of refraction C' is then defined by the quantity $\mathbf{DZ}$, and the bending angle ϵ and the ray asymptote parameter p are then computed as in the spherical case using the new center of refraction.

This procedure is repeated until the difference between the previous value of $\mathbf{r}_0$ and the one computed in the current iteration is arbitrarily small, in which case computation then proceeds to the next data point.

In more detail, the closest approach point is defined by the vector

$$\mathbf{r}_0 = \mathbf{R} - \mathbf{DZ} \tag{1}$$

where

$$\mathbf{R} = \left(p \, \cos\!\left(\frac{\epsilon}{2}\right) \right) \widehat{\mathbf{PM}} + \left(p \, \sin\!\left(\frac{\epsilon}{2}\right) \right) \widehat{\mathbf{E}} \tag{2}$$

and $\mathbf{DZ}$ is the vector representing the location of the new center of refraction C' with respect to the center of the body C. In Eq. (2) p is the ray asymptote parameter, ϵ is the refractive bending angle, $\widehat{\mathbf{PM}}$ is the unit vector from C' in the direction of $\mathbf{p}$, and $\widehat{\mathbf{E}}$ is the unit vector in the direction of the earth.

Having determined $\mathbf{r}_0$, the magnitude, latitude and longitude corresponding to this vector are inputs to a program defining the shape of the planet as follows (Hubbard *et al.* 1975*b*):

$$r_0 = b\left[1 + \sum_{i=0}^{4} D_{2i} P_{2i}(\mu) \right], \qquad \mu = \cos\theta . \tag{3}$$

In Eq. (3) b is the polar radius of a level surface, θ is the colatitude, $P_{2i}(\mu)$ are Legendre polynomials, and D_{2i} are coefficients derived from the second-through sixth-order zonal harmonics of the gravity field of Jupiter (Null *et al.* 1975). Then, if r_0' is the first derivative and r_0'' is the second derivative of Eq. (3), the radius of curvature ρ at the point r_0 is given by:

$$\rho(r_0) = (r_0^2 - r_0'^2)^{\frac{1}{2}} \left| \frac{r_0'^2 + r_0^2 + (A/B)^2 r_0^2 \sin^2 \theta}{r_0 r_0'' + 2r_0'^2 - r_0^2 + (A/B)^2 r_0 \sin \theta (r_0' \cos \theta - r_0 \sin \theta)} \right| \qquad (4)$$

where

$$A = (r_0'^2 + r_0^2) \left[\sin \phi (x_E + x_T) - \cos \phi (y_E - y_T) \right]$$

and

$$B = r_0 \sin \theta \{ (r_0' \sin \theta + r_0 \cos \theta) \left[(x_E - x_T) \cos \phi + (y_E - y_T) \sin \phi \right] +$$
$$(r_0' \cos \theta - r_0 \sin \theta) (z_E - z_T) \} \, .$$

In Eq. (3) ϕ is the longitude of the $\mathbf{r}_0$ point, θ is the colatitude, x, y and z values subscripted with E are the coordinates of the earth relative to the center of Jupiter, and the variables subscripted with T are the coordinates of the spacecraft with respect to the center of Jupiter.

The direction of the unit vector $\widehat{\mathbf{n}}$, which is a normal to the equipotential surface at the point $\mathbf{r}_0$, is then given as follows:

$$\widehat{\mathbf{n}}(r_0) = \frac{1}{(r_0^2 + r_0'^2)^{\frac{1}{2}}} \begin{bmatrix} \cos \phi (r_0 \sin \theta - r_0' \cos \theta) \\ \sin \phi (r_0 \sin \theta - r_0' \cos \theta) \\ r_0 \cos \theta + r_0' \sin \theta \end{bmatrix} . \qquad (5)$$

The quantities ρ and $\widehat{\mathbf{n}}$ allow one to determine the new center of refraction as follows:

$$\mathbf{DZ} = \mathbf{r}_0 - \rho \widehat{\mathbf{n}} . \qquad (6)$$

Finally, the new vector $\mathbf{M}_n$ from the center of refraction C' to the spacecraft, which is used in the computation of the plane of refraction, is defined as follows:

$$\mathbf{M}_n = \mathbf{M} - \mathbf{DZ} \qquad (7)$$

where $\mathbf{M}$ is the vector to the spacecraft from the center of the planet.

The refractive bending angle ϵ and the ray asymptote parameter p are then computed as in the spherical case (Kliore 1972). With reference to Fig. 2, the computations are carried out in a plane defined by the new center of refraction C', the vector $\mathbf{M}_n$, and the direction to the earth. In Fig. 2, $\tilde{V}$ is the velocity vector of the spacecraft relative to Jupiter projected on the plane of refraction and V_E is the component of that velocity in the direction of the earth. The angles ψ and ψ_E of Fig. 2 are defined below:

$$\psi_E = \cos^{-1}(V_E / \tilde{V})$$

and

$$\psi = \cos^{-1}\left[\frac{c}{f\tilde{V}}(\Delta f_E + \Delta f_i)\right] \tag{8}$$

In Eq. (8) the quantity Δf_E represents the Doppler frequency that would be observed without refraction. It is given by

$$\Delta f_E = \frac{f}{c} V_E . \tag{9}$$

The quantity Δf_i in Eq. (8) is the observable, namely the frequency residual due only to the refraction of the ray in the planetary atmosphere.

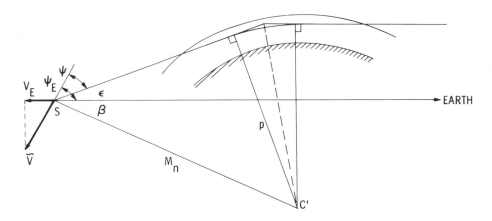

Fig. 2. Quantities used in obtaining the refractive bending angle ϵ and the ray asymptote parameter p. $\tilde{V}$ is the projection of the spacecraft velocity in the plane of refraction and the angles ψ and ψ_E are the angles between the velocity vector and the refracted and unrefracted rays, respectively.

Thus, having the observable frequency residuals and the geometry relative to the planet and the earth, all quantities in Eqs. (8) and (9) can be computed, and the bending angle ϵ as well as the ray asymptote parameter p can be expressed as shown below:

$$\epsilon = \psi_E - \psi$$
$$p = M_n\sin(\epsilon + \beta) \tag{10}$$

where the angle β is the angle subtended at the spacecraft by the earth and the center of refraction C'.

Having computed the quantities ϵ and p, one can in general obtain the index of refraction as a function of the radial distance from the center of refraction by means of the Abel integral transform (Fjeldbo and Eshleman 1968; Phinney and Anderson 1968; Kliore 1972). However, when one per-

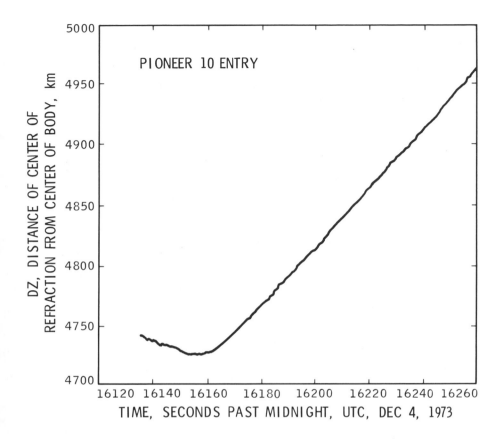

Fig. 3.   Variation of DZ, the distance between the center of refraction and the center of the body, with time for the Pioneer-10 entry data. The segment that is sloping up represents the data in the lower atmosphere, where refraction is significant.

forms the computations outlined by Eqs. (1) through (10) for the Jupiter data of Pioneers 10 and 11, one finds that the value of the ray asymptote parameter p increases as the ray enters the lower atmosphere of Jupiter where the refraction occurs, and this non-monotonic nature of the ray asymptote parameter makes ordinary Abel integral transform inversion impossible.

The reason for this behavior of the ray parameter is shown in Fig. 3. This figure displays the magnitude of the distance DZ from the center of the planet to the center of refraction. This quantity is seen to increase from a value of $\sim$ 4725 just after entry into the neutral atmosphere to a value of $\sim$ 4960 at the deepest penetration point, for a total excursion of $\sim$ 235 km. For this reason, the value of the ray asymptote parameter p increases because, while the closest approach distance with respect to the center of the

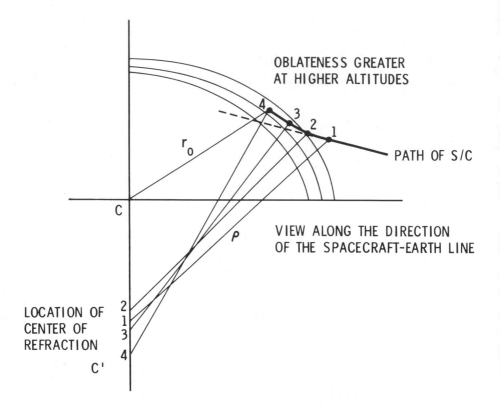

Fig. 4. Varying oblateness of the atmosphere with altitude as the cause for the migration of the center of refraction.

planet r_0 monotonically decreases as the ray enters the atmosphere, the center of refraction C' recedes at a faster rate causing the net change in the ray asymptote parameter to be positive.

The reason for this behavior can be explained with reference to Fig. 4, showing the view of the limb of Jupiter and the spacecraft along the spacecraft-Earth line. The basic mechanism at work here is related to the fact that the oblateness of Jupiter is greater at higher altitudes, which is a consequence of the higher centrifugal force at higher altitudes for a rapidly rotating planet. While the spacecraft is outside of the atmosphere and there is no refraction, as in points 1 and 2 in Fig. 4, the path of the spacecraft takes the ray to levels of lesser oblateness, causing the center of curvature C' to move closer to the center of the planet. However, when refraction occurs, the ray traverses higher altitudes than it would along the straight line-path, and the higher values of oblateness move the center of curvature away from the center of the planet.

For these reasons, it is evident that the Abel integral transform inversion method cannot be used when the center of refraction is allowed to move with the instantaneous center of curvature computed for each data point. However, as an approximation, the center of refraction can be fixed at some position intermediate between those corresponding to, respectively, the top and the bottom of the neutral atmosphere. For the data from Pioneers 10 and 11, the consequence of the relative geometry between the earth and the Jupiter-spin-axis geometry is such that the center of curvature moves along the spin axis, which is also the z-axis for all computations. (The x and y components are on the order of 0.1 km.) Thus, the approximation was made by fixing the center of refraction on the z-axis at some distance from the center of the planet, within limits of the excursion of the quantity DZ. Figure 5 shows

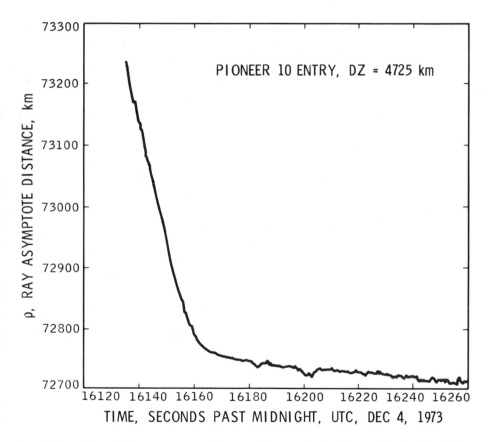

Fig. 5.   Variation of the ray asymptote distance with time for the Pioneer-10 entry data. Note the uneven nature of the flatter portion of the curve corresponding to the lower atmosphere of Jupiter. These irregularities are probably caused by turbulence effects in the neutral atmosphere.

the variation of the ray asymptote distance p as a function of observation time, for the Pioneer 10 entry data with $DZ = 4725$ km. It is immediately apparent that the ray parameter is now generally monotonic with time, which allows application of the Abel integral transform. However, it should be noted that because of fluctuations in the observable frequency, thought to be caused primarily by turbulence in the lower atmosphere of Jupiter, as well as possibly by inhomogeneities along the path due to departure from spherical symmetry, the fine structure of the curve is not monotonic. This can cause certain problems in the inversion, as discussed in the following section.

II. DISCUSSION OF DATA ANALYSIS AND RESULTS

The techniques of the acquisition of data as well as the processing of the analog recordings to obtain frequency versus time histories have been discussed in previous publications (Kliore 1972; Kliore et al. 1974a; Kliore et al. 1974b; Kliore 1975) and therefore will not be discussed in this chapter. The discussion below applies only to the analysis and inversion of the frequency residual data from the Pioneer 10 entry and exit and the Pioneer 11 exit. The Pioneer 11 entry data, obtained while the spacecraft was in two-way communication with the earth, have characteristics different from those of the three sets of data analyzed here and lead to results that are not consistent with the other three. The inconsistency may have been caused by cycle-slipping in the transponder as the frequency and amplitude of the uplink signal changed rapidly when the radio ray entered the atmosphere. It may also have been caused by the fact that the uplink and downlink rays took paths separated by some 24 km in the atmosphere because of the round-trip propagation time between the limb of the planet and the spacecraft. The treatment of the Pioneer 11 entry data must await results from scheduled testing of the Pioneer transponder to determine the possibility of the existence of cycle-slipping, and an inversion technique specifically for two-way data.

When tables of the bending angle ϵ and the ray asymptote parameter p are compiled for all data points as shown in the previous section, the refractive index is computed for each point by applying the Abel integral transform, using a numerical Gaussian quadrature integration method,

$$n_i = \exp\left[\frac{1}{\pi}\int_{p_i}^{p_{max}} \frac{\epsilon(p)\mathrm{d}p}{(p^2 - p_i^2)^{\frac{1}{2}}}\right].$$ (11)

$$r_i = p_i/n_i$$

In Eq. (11), p_{max} is the ray asymptote distance corresponding to the data point farthest from the planet and r_i is the radial distance from the center of refraction. It should be pointed out that because of the local non-monotonicity of the ray asymptote parameter, complications may arise in the denominator of Eq. (11) when $p < p_i$. If such a case is encountered, the point in question is discarded.

The atmospheric pressure is obtained from a table of index of refraction and radius by application of the hydrostatic equation, as shown below:

$$P_i = -\frac{m}{Rq} \int_{r_{i-1}}^{r_i} N_i(z)g(z)dz + P_{i-1}.$$ (12)

The temperature at each point is computed from the perfect gas law:

$$T_i = \frac{P_i q}{N_i},$$ (13)

$$N = (n-1) \times 10^6$$

where m is the mean molecular weight, q is the mean refractivity of the gas mixture and R is the universal gas constant. The acceleration of gravity g is computed for each point; however, in this version of the data-analysis program, it is approximated by the acceleration of gravity at the closest approach point of the straight-line path between the spacecraft and the earth, which would make it higher by a maximum of $\sim$ 4 to 5% than the true acceleration of gravity at the point of closest approach. All refractivity points non-monotonic in radius are also removed before the integration for pressure is performed. All computations are performed assuming a composition of 85% H_2 and 15% He.

A representative profile of refractivity as a function of the distance from the center of refraction derived from the Pioneer 10 entry data with $DZ = 4842.5$ (the mid-range of the total variation in DZ in the lower atmosphere) is shown in Fig. 6. Features that are readily apparent are the effects of the upper ionosphere, extending for some 3000–4000 km, a disturbed region in the lower ionosphere where multipath and ray-crossing effects are known to occur (Fjeldbo et al. 1975) and another uneven region below the ionosphere and above the very abrupt beginning of the neutral atmospheric effects. This region, between the bottom of the sensed ionosphere and the top of the sensed neutral atmosphere, has in the past been used for the computation of bias in refractivity arising from such possible causes as deviation from spherical symmetry (which is implicit in the Abel inversion technique) and from the effects of lateral inhomogeneities. Such a method of refractivity bias removal was used in previous applications of the radio occultation method (Kliore 1972) and it makes use of the fact that the refractivity should be zero between the bottom of the ionosphere and the top of the sensible lower atmosphere. The procedure that is followed is to compute the average refractivity in the chosen region, and to refer all refractivity values to a straight line drawn from zero refractivity at the largest-radius data point

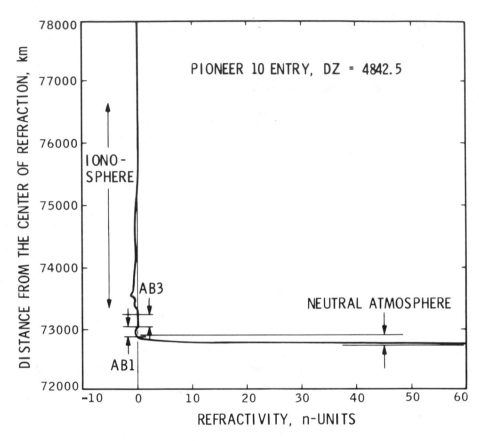

Fig. 6. Profile of refractivity as a function of the distance from the center of refraction derived from the Pioneer-10 entry data, showing the possible ambiguities resulting from the removal of refractivity bias prior to analyzing the lower atmosphere data.

through the computed bias refractivity at the midpoint of the bias interval. For the Pioneer 10 entry data, this procedure is complicated by the roughness of the interval in question, possibly caused by perturbations from the sharp layers in the lower ionosphere. The rather significant effect of the choice of the bias-removal interval on the temperature profiles is demonstrated in Fig. 7. By moving the bias-removal computation interval from the region *AB3* to the region *AB1*, shown in Fig. 6, one produces a change of approximately 50° in the temperature at the 10 mbar level, although the change at the 100 mbar level is only about 10°. This large effect is due mostly to the presence of a peculiar bulge in the refractivity profile immediately above the onset of the lower atmosphere effect. Because it utilizes a flatter region of the profile for bias computation, bias interval *AB3* is thought to be more reliable. The effect of the selection of the refractivity bias-removal for the other two sets of data is not as large, but it is still a significant source of uncertainty.

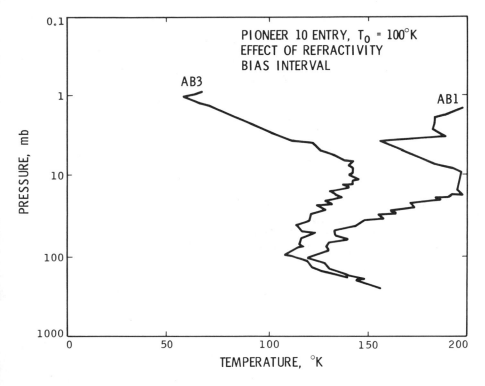

Fig. 7. Temperature-pressure profiles derived from the Pioneer-10 entry data, illustrating the large effect that the selection of the refractivity bias computation interval has upon the final result. This effect is due to unevenness of the refractivity profile above the neutral atmosphere, which most probably can be attributed to effects of the sharp layers in Jupiter's lower ionosphere.

Another source of uncertainty in the derived temperatures is the modeling of the drift characteristics of the spacecraft oscillator during occultation. It has been shown that unmodeled drifts in the spacecraft oscillator can have large effects upon the derived results (Eshleman 1975). In the processing of the Pioneer 10 and 11 data, oscillator drift, as well as other sources of biases and drifts in the data, are removed by fitting a least-squares straight line to an interval of frequency residual data chosen to lie outside the regions affected by the atmosphere and ionosphere. The uncertainty introduced into the results by any remaining unmodeled or nonlinear drifts is assessed by varying the bias interval over the baseline data region and observing the effect in the final result. The interval of frequency residual data used for removal of the bias in the case of the Pioneer 10 entry data is shown in Fig. 8. The data to the right of the time of 15,890 are perturbed by the effects of the ionosphere and are not used. The data to the left also show a slight amount of nonlinearity, possibly due to the continuing warmup of the spacecraft oscillator, which had been switched on some 25 minutes earlier. Two intervals were

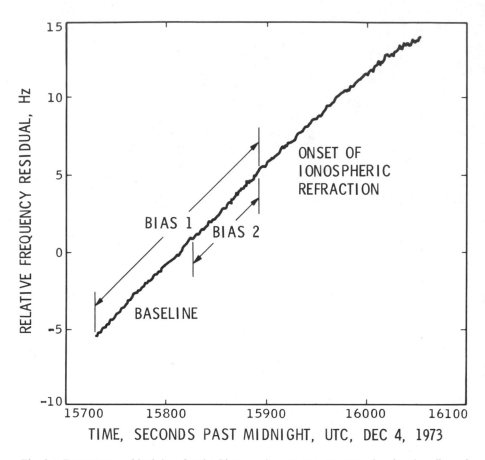

Fig. 8.   Frequency residual data for the Pioneer-10 entry measurement showing the effect of oscillator drift. The two bias-computation intervals that were used for the assessment of drift modeling accuracy are indicated as *BIAS 1* and *BIAS 2*.

chosen, labeled *BIAS 1* and *BIAS 2* in Fig. 8. The root-mean-square deviation from the straight-line fit over both intervals is approximately 0.09 Hz corresponding to an oscillator stability, defined in this sense, of about 4 parts in 10^{11}.

The effects of frequency-bias removal uncertainties are shown in Fig. 9. The profile resulting from the *BIAS 1* interval is represented by a solid line, while the one derived with *BIAS 2* is shown dotted. There is obviously very little difference, indicating that for data taken with well-behaved oscillators, as was the case during the Pioneer 10 entry measurements, the uncertainty, after modeling, in the oscillator drift rate produces a very small uncertainty in the final temperature profiles, much lower than that caused by the uncertainty connected with the choice of the refractivity bias-computation interval.

The situation in the case of the Pioneer 10 exit data is similar. Because the oscillator had by that time completed its warmup transient, the uncer-

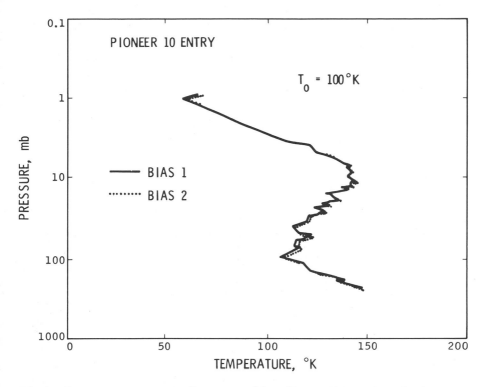

Fig. 9.   Temperature-pressure profiles computed from Pioneer-10 entry data using frequency bias-computation intervals *BIAS 1* and *BIAS 2*. The small change in the profiles indicate that the results are relatively independent of the location of the bias-computation interval used for modeling of the oscillator drift.

tainties connected with modeling of the oscillator drift rate are even lower than they are in the case of the Pioneer 10 entry data. This is not, however, the case for the Pioneer 11 exit. An examination of Fig. 10 shows that the frequency residual baseline, namely the frequency data immediately after exit of the radio beam from the atmosphere and ionosphere, shows several discontinuous changes in the oscillator drift rate. These are thought to be due to the effects of intense radiation upon the crystal of the spacecraft oscillator while Pioneer 11 was traversing the most intense part of Jupiter's radiation belts near its closest approach to the planet. Again, two bias intervals were chosen, lying on that portion of the curve nearest to the ionospheric and atmospheric data and prior to the large discontinuity in drift rate occurring at $\sim 23{,}430$.

The resulting effects on the final temperature-pressure profiles are displayed in Fig. 11. In this case, because of the multiple discontinuities in the drift rate within the bias interval, the effect of using different intervals is not negligible. There is a change of $\sim 30°$ at the level of 10 mbar, but again only about $10°$ at 100 mbar. Thus, it is obvious that modeling of oscillator drift is greatly affected by abrupt changes in the oscillator drift rate, and that radiation shielding of spacecraft oscillators will be necessary in order to obtain precise radio occultation data with spacecraft approaching Jupiter closer than Pioneer 10 did.

In this method of analysis, another source of uncertainty is the choice for the location of the center of refraction. The excursion in the distance DZ in the lower atmosphere amounts to about 5.0% in the case of Pioneer 10 entry, 1.3% for Pioneer 10 exit, and 3.2% for the Pioneer 11 exit. This excursion is dependent on the latitude of the occultation and on the trajectory of the spacecraft, and the effect on the results is proportional to the magnitude of the total excursion in DZ. The pressure-temperature profiles derived from

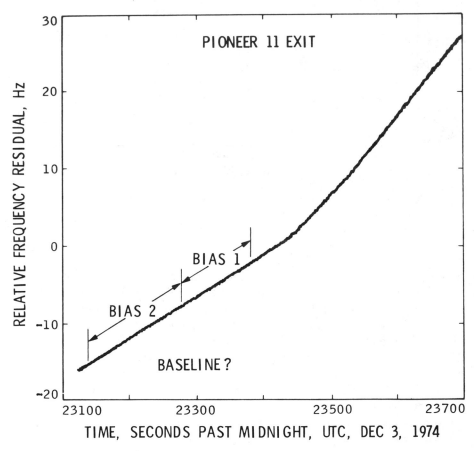

Fig. 10.   Portion of the exit frequency residual data from Pioneer 11 outside of the atmospheric effect, showing some abrupt changes in the oscillator drift rate apparently caused by the irradiation of the spacecraft oscillator crystal by Jupiter's trapped radiation belts. The two intervals used to test the accuracy of oscillator drift removal in this case are labelled *BIAS 1* and *BIAS 2*. *BIAS 2* is preferred because of its proximity to the atmospheric portion of the data.

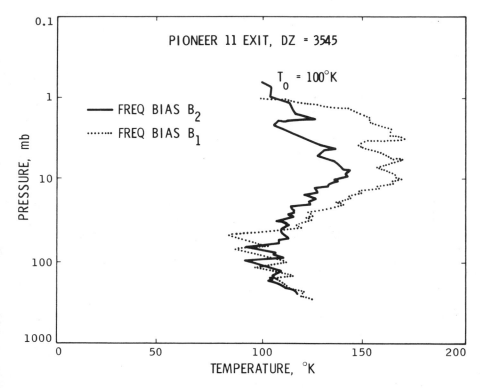

Fig. 11.   Temperature-pressure profiles derived from Pioneer-11 exit data using the intervals *BIAS 1* and *BIAS 2* to compute the drift function. Unlike the case of the Pioneer-10 entry data, the changing oscillator drift rate causes an appreciable variation in temperature between the two profiles.

the Pioneer 10 entry data are shown in Fig. 12 for three different values of *DZ*, representing the lowest, highest, and mean values. No temperatures are given for pressures lower than 1 mbar because of the obviously unreliable nature of the results in that region. In order to produce a temperature from the refractivity profile, an initial temperature must be assumed at the start of the integration of the hydrostatic equation. For the Pioneer 10 entry data, the nature of the refractivity profile near zero refractivity is such that it makes very little difference whether one assumes initial temperatures of 100° K or 200°K. For that reason, only the profiles for an initial temperature of 100°K are shown in Fig. 12. It is interesting to note that the profiles for the cases for a *DZ* of 4725 km and 4842.5 km are coincident in the region between 1 and 10 mbar while the profile for a *DZ* of 4960 km diverges. Since in all cases the refractivity bias was removed in the same interval, this implies that the choice of location for the center of refraction can influence the refractivity profile at the top of the neutral atmosphere. The profiles are nearly identical in the region between 10 and 100 mbar, and diverge by about

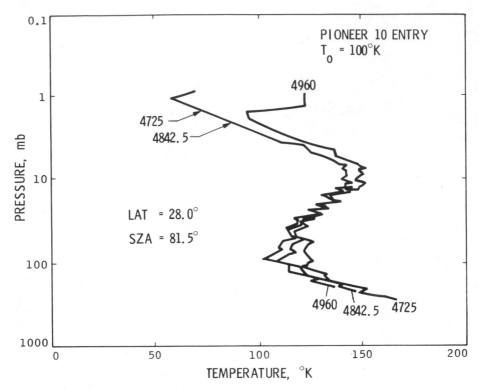

Fig. 12. Temperature-pressure profiles derived for Pioneer-10 entry data using three different values for the location of the center of refraction, corresponding to the minimum, maximum, and mean values. There is little difference in the temperatures between ~ 10 and 100 millibars of pressure. Below 100 mbar the most reliable profile is thought to be the one for $DZ = 4960$ km.

40° at the lowest level of penetration. Clearly, the choice of the location of the center of refraction is quite important in the lower region of the atmosphere. The value of $DZ = 4960$ km is most appropriate for this region, because it is derived for the lowest measurement, leading to a final temperature of about 140°K at a pressure of about 210 mbar.

The corresponding profiles for the Pioneer 10 exit data are shown in Fig. 13. Here, because the percent variation in DZ is only 1.3, its effect upon the profiles is very slight. The profiles for $DZ = 8390$ km are shown in solid lines while those for $DZ = 8280$ km are shown dotted. Between 0.1 and 1 mbar the curves diverge because of the different initial conditions, but from 10 to 100 mbar they are almost identical, differing slightly in the region of 100 mbar. Because the Pioneer 10 spacecraft was farther from the limb of the planet when the exit data were taken, the defocusing attenuation was greater, and the signal could not be followed deeper than a pressure level of about 100 mbar, where the temperature was found to be ~ 105°K.

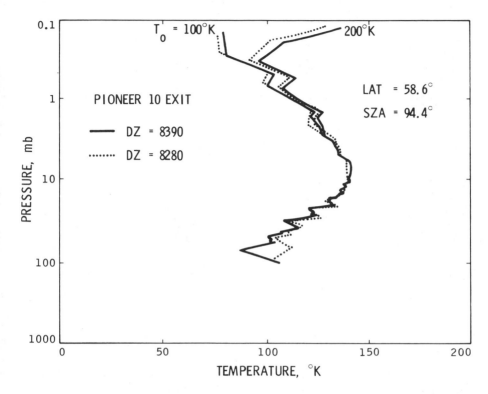

Fig. 13.   Temperature-pressure profiles derived from the Pioneer-exit data showing the effect of the extreme values of *DZ* and initial temperatures of 100°K and 200°K. For pressures higher than ~ 7 mbar there is very little difference between the profiles.

The pressure-temperature profiles derived from the Pioneer 11 exit data are shown in Fig. 14. In this case the variation in *DZ* does make a considerable difference, as does the choice of the initial temperature. Again, all of the profiles are within 10° to 15° of one another between the pressure levels of 10 and 100 mbar. The Pioneer 11 spacecraft was closer to the limb of the planet than was Pioneer 10, and hence the signal could be followed to a pressure level of about 250 mbar, where the temperature was about 120°K. It should be pointed out that for the Pioneer 11 exit data the uncertainties due to the choice of the initial temperature, as well as of the location of the center of refraction, are quite large between the 1- and 10-mbar levels.

Finally, Fig. 15 shows temperature-pressure profiles from the Pioneer 10 entry, Pioneer 10 exit and Pioneer 11 exit data plotted for comparison together with temperature-pressure profiles derived from other observations and models. The model represented by the solid angular line is one derived from the Pioneer 10 infrared radiometer data (Orton 1975).[1] The radiative-convective equilibrium model of Trafton (1973) for an effective planetary

[1]See p. 207.

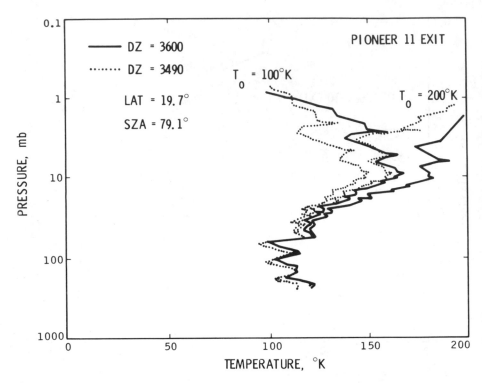

Fig. 14. Temperature-pressure profiles derived from the Pioneer-11 exit data for extreme values of *DZ* and initial temperatures of 100°K and 200°K. In this case the effects of initial temperature and a different center of refraction are about equal and the profiles are substantially different above the 10-mbar level.

temperature of 130°K is represented by the dotted curve. The dashed curve represents the radiative-convective equilibrium model *c* of Wallace *et al.* (1974),[2] which takes into account the effect of solar heating of the atmosphere and is computed for an effective planetary temperature of 134°K. A model computed by Cess and Chen (1975), based on radiative-convective equilibrium, but including the cooling effects of ethane and acetylene is shown as a dot-dashed curve. In addition, a microwave spectral measurement by Gulkis *et al.* (1974)[3] at 20 to 24 GHz is shown by a point corresponding to a pressure of 0.48 atm at 130°K, with 2-σ and 3-σ error bars.

The profiles presented in Fig. 15 suggest several conclusions. First, it is evident that the three temperature-pressure profiles, derived from Pioneer 10 and 11 data, which have been selected for their diversity between the 1- and 10-mbar levels, show a high degree of consistency among the three sets of results. The three profiles are almost identical between ∼ 10 and 50 mbar,

---

[2]See p. 291.
[3]See p. 670.

and differ by only ~ 30° between 50 and 100 mbar. The uncertainty is relatively large for pressures lower than about 10 mbar, primarily because of the roughness of the refractivity profiles, probably caused by the effects of multipath propagation in the lower ionosphere. For pressures higher than 100 mbar the differences between profiles are also greater, possibly due to increasing effects of turbulence in the lower atmosphere.

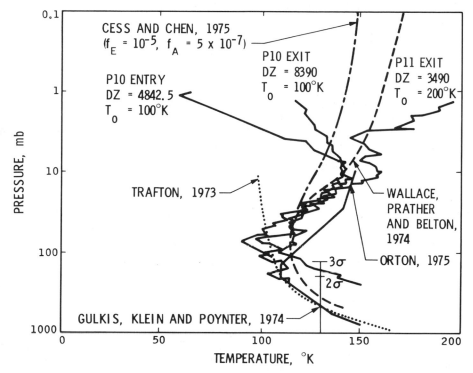

Fig. 15. Temperature-pressure profiles derived from the Pioneer-10 entry, Pioneer-10 exit, and Pioneer-11 exit data compared to temperature-pressure profiles of models derived from radiative-convective balance calculations and from the analysis of Pioneer-10 infrared radiometer data.

A feature that is common to all three profiles is the temperature inversion between the 10- and 100-mbar levels. At the 10-mbar level the temperatures lie between 130° and 170°K, and at 100 mbar the temperatures are between about 80° and 120°K. Because the inversion is observed consistently in all three independent measurements, this feature is inferred to be a property of the atmosphere of Jupiter, which also confirms the conclusions of Gillett *et al.* (1969).

In addition, comparison of the radio occultation profiles with the radiative-convective models, as well as with the profile based on the Pioneer 10 infrared radiometer data, indicates good agreement between the models and the occultation results. This rather close agreement with the theoretical models, which have a high degree of mutual consistency, should increase confidence in the validity of this analysis.

The method for the analysis of radio occultation data from oblate planets described in this chapter, consisting of the use of the radius of curvature and normal vector direction at the point of closest approach of the ray to define the plane of refraction, and a subsequent approximation involving the use of a constant center of refraction, followed by the Abel transform inversion method derived for spherical planets, is only an intermediate step in the analysis of the Pioneer 10 and 11 data. Although it produces internally consistent results which also agree well with theoretical and observational models for the atmosphere, it is felt that a completely general treatment of the inversion of data from highly oblate planets must make use of techniques such as three-dimensional ray tracing instead of the Abel integral transform.

Acknowledgement. The authors wish to thank I. Patel and D. Sweetnam for their work with the digital computer programs, and W. Hubbard for helpful and stimulating discussions as well as the kind permission to use his subroutine *SHAPE* which was adapted for our use by J. Anderson and E. Lau. Special thanks are due to H. Lass for his help with the mathematical formulation which forms the basis of this chapter, and to D. M. Hunten for his interest and guidance. This work represents one phase of research carried out at the Jet Propulsion Laboratory, California Institute of Technology, under NASA Contract NAS 7-100.

REFERENCES

Cess, R. D., and Chen, S. C. 1975. The influence of ethane and acetylene upon the thermal structure of the Jovian atmosphere. *Icarus.* In press.

Eshleman, V. R. 1975. Jupiter's atmosphere: problems and potential of radio occultation. *Science* 189:876–878.

Fjeldbo, G., and Eshleman, V. R. 1968. The atmosphere of Mars analyzed by integral inversion of the Mariner 4 occultation data. *Planet. Space Sci.* 16:1035–1059.

Fjeldbo, G.; Kliore, A.; Seidel, B.; Sweetnam, D.; Cain, D. 1975. The Pioneer 10 radio occultation measurements of the ionosphere of Jupiter. *Astron. Astrophys.* 39:91–96.

Gillett, F. C.; Low, F. J.; and Stein, W. A. 1969. The 2.8-14 micron spectrum of Jupiter. *Astrophys. J.* 157:925–934.

Gulkis, S.; Klein, M. J.; and Poynter, R. L. 1974. Jupiter's microwave spectrum: implications for the upper atmosphere. *Exploration of the planetary system.* (A. Woszczyk and C. Iwaniszewska, eds.) pp. 367–374. Dordrecht, Holland: D. Reidel Publishing Company.

Hubbard, W. B. 1976. Ray propagation in oblate atmospheres. *Icarus* (special Jupiter issue). In press.

Hubbard, W. B.; Hunten, D. M.; and Kliore, A. 1975a. Effect of the Jovian oblateness on Pioneer 10/11 radio occultations. *Geophys. Res. Lett.* 12:265–268.

Hubbard, W. B.; Slattery, W. L.; and DeVito, C. L. 1975b. High zonal harmonics of rapidly rotating planets. *Astrophys. J.* 199:504–516.

Kliore, A. 1972. Current methods of radio occultation data inversion. *Mathematics of profile inversion.* (L. Colin ed.) *NASA TM X-62, 150, 3-2/3-16.*

Kliore, A.; Cain, D. L.; Fjeldbo, G.; Seidel, B. L.; and Rasool, S. I. 1974a. Preliminary results on the atmospheres of Io and Jupiter from the Pioneer 10 S-band occultation experiment. *Science* 183:323–324.

———. 1974b. The atmospheres of Io and Jupiter measured by the Pioneer 10 radio occultation experiment. *Preprint of paper no. II-VII-1-4*, presented at the *17th Plenary Meeting of COSPAR*, São Paulo, Brazil, June 24–July 1, 1974.

Kliore, A. J.; Cain, D. L.; Fjeldbo, G.; Seidel, B. L.; Sykes, M. J.; and Rasool, S. I. 1972. The atmosphere of Mars from the Mariner 9 radio occultation measurements. *Icarus* 17: 484–516.

Kliore, A.; Fjeldbo, G.; Seidel, B. L.; Sesplaukis, T. T.; Sweetnam, D. N.; and Woiceshyn, P. M. 1975. Preliminary results on the atmosphere of Jupiter from the Pioneer 11 S-band occultation experiment. *Science* 188:474–476.

Null, G. W.; Anderson, J. D.; and Wong, S. K. 1975. Gravity field of Jupiter from Pioneer 11 tracking data. *Science* 188:476–477.

Orton, G. 1975. The Jovian thermal structure from Pioneer 10 infrared radiometer data. *Part 2 of Ph.D. Thesis*. California Inst. of Technology, Pasadena, California.

Phinney, R. A., and Anderson, D. L. 1968. On the radio occultation method for studying planetary atmospheres. *J. Geophys. Res.* 73:1819–1827.

Trafton, L. M. 1973. A comment on Jovian greenhouse models. *Icarus* 19:244–246.

Wallace, L.; Prather, M.; and Belton, M. J. S. 1974. The temperature structure of the atmosphere of Jupiter. *Astrophys. J.* 193:481–493.

THE PIONEER 11 RADIO OCCULTATION
MEASUREMENTS OF THE JOVIAN IONOSPHERE

G. FJELDBO, A. KLIORE, B. SEIDEL,
D. SWEETNAM and P. WOICESHYN
Jet Propulsion Laboratory

Radio occultation data obtained with the Pioneer 11 spacecraft are utilized to study Jupiter's ionosphere. The ingress measurements, which were conducted by using a stable Earth-based frequency reference for the tracking link, yielded ionospheric data near the morning terminator at about 79° south latitude. Data were also taken during egress on the evening side near 20° north latitude. The latter measurements were conducted in the one-way mode; i.e., an on-board crystal oscillator was employed as a frequency reference for the downlink (spacecraft-to-Earth) signal. The new data confirm previous results obtained with Pioneer 10 and show that Jupiter has a multilayered ionosphere extending over an altitude range of more than 3000 km. The topside scale-height near 79° south latitude was 540 ± 60 km. Assuming a topside electron, H^+ distribution controlled by diffusion yields a plasma temperature of 850 ± 100 °K in this region. The radio data indicate that the upper atmosphere is either warmer or more dissociated into atomic hydrogen than previously anticipated.

The first radio occultation measurements of the Jovian ionosphere were conducted with the Pioneer 10 spacecraft on December 4, 1973 (Fjeldbo *et al.* 1975). A year later Jupiter was revisited with Pioneer 11 and the new flyby has yielded more occultation data. The purpose of this report is to present the initial results of the analysis of these latest ionospheric measurements. A companion chapter by Kliore and Woiceshyn[1] outlines the results for the lower neutral atmosphere.

In describing the new occultation measurements, we have organized the material into two parts. The first part contained in Sec. I, briefly outlines how the experiment was conducted. The results inferred from the observations, including the vertical profiles for the electron density, the temperature, and the atomic and molecular hydrogen densities, are summarized in Sec. II.

---

[1]See p. 216.

I. DESCRIPTION OF EXPERIMENT

Pioneer 11 passed by Jupiter on December 3, 1974 at a distance of 1.1×10^5 km from the center of mass (Hall 1975). Approximately 20 minutes before periapsis passage, the spacecraft was occulted by the planet. During ingress, the S-band radio tracking link was utilized to probe the ionosphere of Jupiter near the morning terminator at about 79° south latitude. Ionospheric data were also obtained during egress which occurred near 20° north latitude. The spacecraft to planetary limb distance was about 10^5 and 1.2×10^5 km at the time of ingress and egress, respectively. The transverse velocity component of the radio beam perpendicular to the limb was 36 to 39 km sec^{-1}.

The tracking signal from Pioneer 11 is transmitted from a paraboloidal reflector with a diameter of about 2.7 meter and a feed system for right-handed circular polarization. The antenna beam is pointed along the spin axis of the spacecraft.

During ingress, the 2.3 GHz radio frequency transmitted from Pioneer 11 was coherently derived from the 2.1 GHz uplink frequency received from NASA's Deep Space Station 14 located at Goldstone, California. The uplink signal was in turn derived from the station rubidium vapor oscillator, stable to about one part in 10^{11}, and swept in frequency so that the spacecraft's radio transponder received a frequency approximately equal to its "best lock frequency". The frequency sweep rate employed during the ingress measurements was about -50 Hz sec^{-1}.

The uplink signal became progressively weaker as Pioneer 11 moved in behind Jupiter—causing the spacecraft's phase locked loop receiver to go out of lock. An on-board crystal oscillator is automatically switched in to provide the reference frequency for the downlink when loss of lock occurs. This operating mode is denoted one-way tracking and it was used during emersion from occultation.

The prime tracking station involved in the reception of the radio signal from Pioneer 11 was Deep Space Station 43 located near Canberra, Australia. This station is equipped with a 64-meter diameter paraboloidal antenna and a triple-conversion superheterodyne receiver with a liquid helium cooled travelling wave maser amplifier at the front end. Following the last mixer, the receiver was connected to a magnetic tape recorder. The recording bandwidth was approximately 16 KHz.

II. INTERPRETATION OF MEASUREMENTS

The analysis of the signals received from Pioneer 11 was carried out in the same manner as for the previous measurements with Pioneer 10 (Fjeldbo et al. 1975). The first step in this procedure consisted of determining the amplitude and frequency perturbations imposed on the radio tracking carrier by the Jovian ionosphere. Subsequently, the ionospheric Doppler frequency perturbations were inverted in order to obtain vertical electron density profiles. In computing these profiles, it was assumed that the electron density distribution was spherically symmetric in the regions probed by the radio link.[2] Figure 1 shows the results obtained from the ingress data.[3] Measurements were also made during egress; however, these data are of lower quality due to oscillator instabilities and severe ionospheric scintillation and multipath effects.

As illustrated by the computed electron density profile, the Jovian ionosphere consists of a number of layers and ledges distributed over an altitude range of more than 3000 km. Based on previous usage, we have labeled these features L_1 through L_8. The top layer L_1 has a peak density of approximately 4×10^4 electrons per cm^3 near 1500 km altitude.[4] The topside plasma scale height is 540 ± 60 km. Following earlier studies (Gross and Rasool 1964; Hunten 1969; McElroy 1973), we will assume that the topside plasma consists of electrons and H$^+$ ions in ambipolar diffusive equilibrium. This assumption requires a plasma temperature of about 850 ± 100 °K in order to explain the observations.

Approximately 200 km below the L_1 layer, Fig. 1 shows a ledge which we have denoted L_2. At the 1000, 800, and 700 km altitude levels, we see three distinct layers marked L_3, L_4, and L_5, respectively. In addition, the data indicate that a sharp ionospheric layer, denoted L_6 and located near 400 km altitude, caused multipath propagation on the downlink.

Ionospheric multipath propagation causes ambiguities in the computed electron density profiles because we have no way of knowing which signal modes the spacecraft receiver was tracking. Thus, one can not determine precisely what the electron density distribution was below the L_5 layer in the region probed during ingress. The L_6 layer is, therefore, only indicated by an arrow in Fig. 1.

In addition, there might be two more layers, denoted L_7 and L_8, in the

---

[2] The center of spherical symmetry was put at the local center of curvature. The center of curvature was assumed to lie in a plane containing the ray-path and the local vertical at the lowest point on a ray grazing the top of the ionosphere. Detailed discussions of inversion algorithms are available elsewhere (see, for example, Fjeldbo *et al.* 1971). The computed ionization profiles are insensitive to errors in the planetary oblateness. For example, a 10% uncertainty in the oblateness causes an uncertainty of about 0.1% in the topside plasma temperature near 79° south latitude.

[3] The complication introduced by having two different ray-paths during the ingress measurements, one uplink path and one downlink path separated in altitude by about 1 km, was handled by dividing the ionospheric Doppler residuals by a factor of two and treating the problem as one-way propagation at approximately 2.1 GHz along a mean path located between the uplink and the downlink. Ray-tracing calculations for the up and downlink paths have verified that this approach is applicable in the ionospheric regions where the scale height is large compared with the altitude separation of the two radio links.

[4] Zero altitude was selected at the level where the atmosphere has a refractivity of 1. At this atmospheric level, the uplink signal is believed to have been sufficiently strong to keep the spacecraft's phase-locked loop receiver in good lock. Pure H$_2$ gas with a refractivity of 1 contains 2×10^{17} molecules per cm^3.

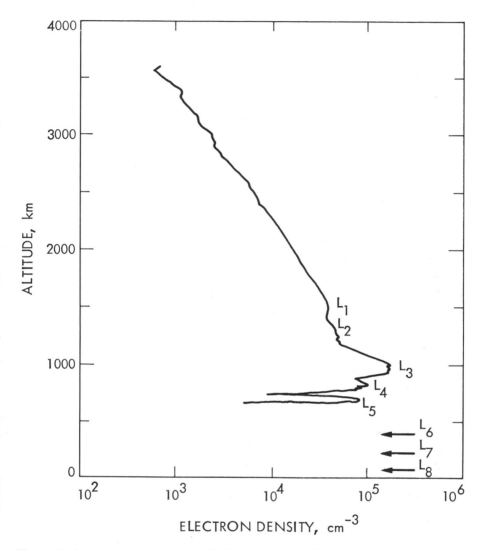

Fig. 1. Early morning electron density distribution computed from the ingress data. The measurements were made near 79° south latitude. The solar zenith angle was approximately 93°. Zero altitude was selected at the level where the lower neutral atmosphere has a refractivity of 1. (A pure H_2 gas with a refractivity of 1 contains 2×10^{14} molecules cm^{-3}). The peak density in the L_1 layer was detected at a distance of 68.650 km from the planetary center of mass.

lower ionosphere. However, these layers do not have very distinct signatures in the radio recordings. It is, therefore, possible that these lower features in the data may have been produced by scintillation noise or by non-spherical symmetry, i.e., local increases in the electron density occurring at a higher altitude.

The egress measurements near 20° north latitude show a high altitude layer (L_1) near 1800 km. In addition, there appears to be a sharp layer at approximately 750 km altitude. This is probably the L_3 layer. At still lower levels, the scintillation and multipath effects are so severe that we have so far been unable to resolve the various layers. From these egress measurements it might be tempting to suppose that the low-latitude ionosphere differs substantially from the high-latitude results shown in Fig. 1. However, this supposition does not seem to be generally applicable because the Pioneer 10 ingress measurements at 26° north latitude yielded results similar to those shown in Fig. 1. Thus, all we can conclude from these data is that the lower Jovian ionosphere changes considerably with time and location.

The ionization profile obtained from the ingress data may be utilized to put plausible limits on the hydrogen density in the upper atmosphere. Figures 2 and 3 illustrate two alternative interpretations. Below 300 km altitude,

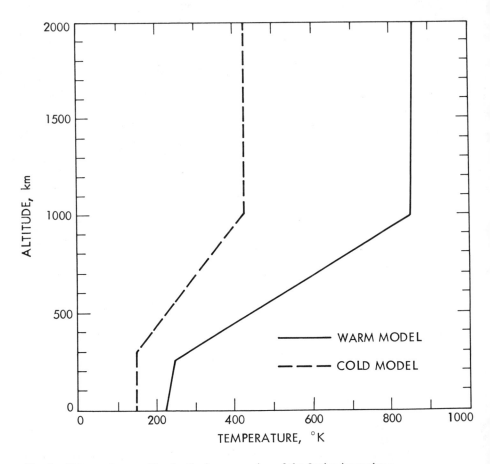

Fig. 2.   Temperature profiles for the ingress region of the Jovian ionosphere.

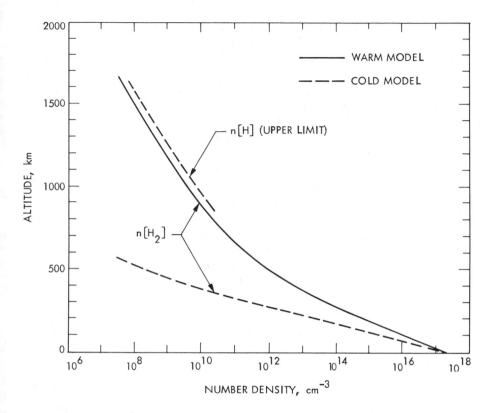

Fig. 3. Number density profiles for the ingress region of the Jovian ionosphere.

we have adopted atmospheric temperatures consistent with data obtained from star occultation measurements (Hubbard *et al.* 1972; Sagan *et al.* 1974) and infrared measurements (Wallace *et al.* 1974). On the topside of the ionosphere we assume a gas temperature equal to the plasma temperature for the warm atmospheric model. Of course, the plasma temperature might be considerably higher than the temperature of the neutral gas. This possibility is considered in what we call the cold atmospheric model which has a temperature equal to half the plasma temperature above 1000 km altitude. Between 300 and 1000 km altitude, we assume constant temperature gradients.

Strobel and Smith (1973) have shown that heating by solar extreme ultraviolet cannot alone explain the exospheric temperatures proposed in the warm model. However, other heat sources may also be important. French and Gierasch (1974), for example, find evidence of inertia gravity waves in the Beta Scorpii occultation data. They show that the upward energy flux received by the Jovian thermosphere in the form of gravity waves may be some 260 times greater than the solar extreme ultraviolet energy flux absorbed above the mesopause level. Similar conclusions have been reached by Shimizu (personal communication) who also includes the effect of electron precipitation from the magnetosphere in his thermal calculations. Shimizu's analysis appears to yield plasma temperatures in agreement with the results reported here.

Additional clues regarding the Jovian thermosphere have been obtained from the Beta Scorpii occultation data. A number of the temperature profiles derived from these data indicate that there may be a temperature rise of several hundred degrees Kelvin above the Jovian mesopause (Hubbard *et al.* 1972; Sagan *et al.* 1974; Vapillon *et al.* 1973).

The H_2 density profiles for the warm and the cold atmospheric models are shown in Fig. 3. These profiles were obtained by assuming that H_2 is the only atmospheric constituent contributing significantly to the observed refractivity at the zero altitude level.[5] The altitude distribution was computed by assuming hydrostatic equilibrium. The inclusion of some He in these models tends to reduce the H_2 density at higher altitudes since it increases the effective molecular mass below the turbopause. The exact level of the turbopause is not known. However, results obtained from the Pioneer 10 ultraviolet photometer experiment indicate that it might be located at a density level as low as 10^{10} cm^{-3} (Carlson and Judge 1974).

If the warm H_2-profile in Fig. 3 is applicable, one may interpret the L_3 ionization peak as being produced by solar extreme ultraviolet. The principal ion production process may be dissociative photoionization of H_2; i.e., $H_2 + h\nu \rightarrow H^+ + H + e$ (McElroy 1973). Radiative recombination of H^+ ions with electrons and vertical transport of plasma appear to proceed sufficiently slowly at the 1000 km altitude level so that the ionization peak can persist through the Jovian night. The electron density peak occurs near the altitude level where the diurnally averaged ion production rate has its maximum. In the vicinity of the ionization peak, the plasma scale height is approximately equal to twice the scale height of H_2. At greater altitudes, ambipolar diffusion controls the plasma distribution and the scale height is given by $k(T_e + T_i)/mg$ where k is Boltzmann's constant, T_e is the electron temperature, T_i is the temperature of the H^+ ions, m is the proton mass, and g is the acceleration of gravity.

In the absence of any significant latitudinal temperature gradients, the warm model outlined above predicts that the altitude of the L_3 layer should increase with increasing latitude. Such a latitude effect is also seen in the data. However, the model does not explain the detailed structure of the observed electron density profile in the vicinity of the L_2 ledge. Another deficiency of the warm model is that it predicts too large a plasma scale height in the vicinity of the L_3 peak.

If the cold H_2-profile shown in Fig. 3 is applicable, one can not interpret the L_3 layer as being produced by dissociative photoionization of H_2 by solar

---

[5]Assuming 100% H_2 at the zero altitude level (where the refractivity is 1) yields a H_2 density of 2×10^{17} cm^{-3}. Adding 20% He by volume reduces the H_2 density by about 6% at the reference level.

extreme ultraviolet radiation because the H_2 density is too low at 1000 km altitude. Instead one might assume that photoionization of atomic hydrogen by solar extreme ultraviolet is the process forming the L_3 peak. This assumption yields what we regard as a probable upper limit for the topside atomic hydrogen density[6] (see Fig. 3). As in the warm ionospheric model, one might interpret the abrupt change in the plasma scale height near 1150 km altitude as marking the level above which ambipolar diffusion controls the plasma distribution. The topside temperatures of the plasma and the neutral gas are 850 and 425 °K, respectively, in the cold atmospheric model.

A major deficiency of the cold model is that it requires far more atomic hydrogen in the Jovian upper atmosphere than theoretical model studies and Lyman-α measurements require. For a review of the theoretical and experimental work on the atomic hydrogen distribution, the reader is referred to Hunten (1969).

The lower electron density layers are presumably produced by ionizing agents such as UV, X-rays, and bombarding protons and electrons at various energy levels. The chemistry of the lower ionosphere is probably controlled by minor atmospheric constituents. The development of detailed models for the lower ionization layers may, therefore, require additional data on ion composition. Possible ion candidates include Na^+, H_3^+ (Atreya *et al.* 1974) and CH_3^+ (Prasad and Tan 1974).

III. CONCLUDING REMARKS

The major sources of error affecting the electron density profile shown in Fig. 1 are:

1. Spectral broadening of the radio signal due to ionospheric scintillations.
2. Multipath propagation producing ambiguities in the radio data.
3. Non-spherical plasma distribution near the terminator.

In spite of the uncertainties involved in interpreting the data, we believe the four computed density peaks, denoted L_1, L_3, L_4, and L_5, represent real ionospheric layers. Based on earlier theoretical studies of the Jovian ionosphere, we interpret the L_3 layer as being a Chapman-type layer produced by solar extreme ultraviolet radiation, and we obtain what we regard as plausible empirical limits for the hydrogen density distribution in the upper atmosphere. The resulting profiles indicate that Jupiter's ionosphere is either warmer or more dissociated into atomic hydrogen than was previously anticipated.

Acknowledgments. The radio occultation measurements discussed in this paper were accomplished through the combined efforts of the Pioneer Project

---

[6]Even more atomic hydrogen would be needed in this model if the same photoionization process were to produce the L_1 peak.

staff, the Deep Space Net personnel, the TRW Company, and the Radio Science Team. The Radio Team consisted of D. L. Cain, G. Fjeldbo, A. J. Kliore (team leader), and B. L. Seidel from the Jet Propulsion Laboratory and S. I. Rasool from NASA headquarters. We thank G. W. Null, T. T. Sesplaukis and S. K. Wong for invaluable assistance in the analysis of the data. The work reported here was supported by the National Aeronautics and Space Administration.

REFERENCES

Atreya, S. K.; Donahue, T. M.; and McElroy, M. B. 1974. Jupiter's ionosphere: prospects for Pioneer 10. *Science* 184:154–156.

Carlson, R. W., and Judge, D. L. 1974. Pioneer 10 ultraviolet photometer observations at Jupiter encounter. *J. Geophys. Res.* 79:3623–3633.

Fjeldbo, G.; Kliore, A. J.; and Eshleman, V. R. 1971. The neutral atmosphere of Venus as studied with the Mariner 5 radio occultation experiments. *Astron. J.* 76:123–140.

Fjeldbo, G.; Kliore, A.; Seidel, B.; Sweetnam, D.; and Cain, D. 1975. The Pioneer 10 radio occultation measurements of the ionosphere of Jupiter. *Astron. Astrophys.* 39:91–96.

French, R. G., and Gierasch, P. J. 1974. Waves in the Jovian upper atmosphere. *J. Atmos. Sci.* 31:1707–1712.

Gross, S. H., and Rasool, S. I. 1964. The upper atmosphere of Jupiter. *Icarus* 3:311–322.

Hall, C. F. 1975. Pioneer 10 and Pioneer 11. *Science* 188:445–446.

Hubbard, W. G.; Nather, R. E.; Evans, D. S.; Tull, R. G.; Wells, D. C.; van Citters, G. W.; Warner, B.; and Vanden Bout, P. 1972. The occultation of Beta Scorpii by Jupiter and Io. I. Jupiter. *Astron. J.* 77:41–59.

Hunten, D. M. 1969. The upper atmosphere of Jupiter. *J. Atmos. Sci.* 26:826–834.

McElroy, M. B. 1973. The ionospheres of the major planets. *Space Sci. Rev.* 14:460–473.

Prasad, S. S., and Tan, A. 1974. The Jovian ionosphere. *Geophys. Res. Lett.* 1:337–340.

Sagan, C.; Veverka, J.; Wasserman, L.; Elliot, J.; and Liller, W. 1974. Jovian atmosphere: structure and composition between the turbopause and mesopause. *Science* 184:901–903.

Strobel, D. F., and Smith, G. R. 1973. On the temperature of the Jovian thermosphere. *J. Atmos. Sci.* 30:718–725.

Vapillon, L.; Combes, M.; and Lecacheux, J. 1973. The Beta Scorpii occultation by Jupiter. II. The temperature and density profiles of the Jovian upper atmosphere. *Astron. Astrophys.* 29:135–149.

Wallace, L.; Prather, M.; and Belton, M. J. S. 1974. The thermal structure of the atmosphere of Jupiter. *Astrophys. J.* 193:481–493.

STELLAR AND SPACECRAFT OCCULTATIONS BY JUPITER: A CRITICAL REVIEW OF DERIVED TEMPERATURE PROFILES

D. M. HUNTEN
Kitt Peak National Observatory

and

J. VEVERKA
Cornell University

The basic ideas involved in the processing of occultation data are reviewed, with emphasis on sources of uncertainty. The results from the β Scorpii event are discussed in detail, and those from Pioneers 10 and 11 in as much detail as possible at present. It is suggested that stratospheric and mesospheric temperatures are probably between 170° and 200°K. The implications of an exospheric temperature as hot as 800°K are briefly discussed.

Although occultations of stars by objects in the planetary system occur frequently (O'Leary 1972), it is only recently that planetary astronomers have come to a widespread realization of their utility. Almost simultaneously, beginning with Mariner 4 at Mars in 1965, a related and even more powerful method, occultation of the spacecraft, came into use. Since the two methods explore the atmosphere at different heights, their results are complementary. This chapter reviews both methods, pointing out their strengths and weaknesses, and makes a first attempt at combining the results that have been obtained for Jupiter. The radio method also gives the electron density in the ionosphere, which is briefly discussed.

In both optical and radio occultations, the intensity of the source is reduced by differential refraction in the density gradient of the planetary atmosphere. Derivations of this effect have been given by Pannekoek (1903), Fabry (1929), and Baum and Code (1953), whose observations of Jupiter are briefly discussed below. Data from a Venus occultation were given by de Vaucouleurs and Menzel (1960) and critically discussed by Hunten and

McElroy (1968), who pointed out the large, and previously neglected, errors that can arise from the difficulty of determining the initial and final levels. The same ideas were briefly applied to the Baum and Code results by Mc-Elroy (1969). Several more recent and more detailed discussions of such errors are discussed below. We do not discuss the 1968 Neptune event.

It is readily found (see Sec. I) that half intensity occurs when the bending angle θ is equal to the angle subtended by the atmospheric scale height H at the distance D of the observer. For the spacecraft occultation, D is replaced by m, the distance from spacecraft to limb; strictly one should use $(m^{-1} + D^{-1})^{-1}$. Typically the best results on the atmosphere are obtained somewhat below this height, usually called the *occultation level*. With great care, it is possible to extend the results considerably deeper, especially in the radio case; much of this chapter is devoted to the question "how much deeper?" There is at present no fully agreed-on answer, and a lively debate is in progress. For an isothermal atmosphere of scale height H, and approximately for any atmosphere, the bending angle (if small) is

$$\theta = \nu_1 \left(2\pi a / H\right)^{\frac{1}{2}} \tag{1}$$

where a is the planetary radius and ν_1 is the refractivity (refractive index minus unity) at the minimum height of the ray. Radio workers usually multiply ν by 10^6 and call it by the same name, with the symbol N. At standard conditions, a solar mixture of hydrogen and helium has a refractivity of $\sim 1.2 \times 10^{-4}$; the number density n is therefore approximately $2.2 \times 10^{23} \nu$. For Jupiter's stratosphere the temperature is found to be close to 200°K, and H is about 30 km. The factor in Eq. (1) is therefore 121, and typical conditions at the occultation level are shown in Table I. The density ratio for the two methods is 3700, and the corresponding height difference 8 scale heights or 240 km.

Until recently, the only method used to analyze optical data has been the fitting of curves computed for various scale heights (sometimes even for a variety of scale-height gradients). The method of integral inversion, expound-

TABLE I

Typical Bending Angle θ, Refractivity ν,
Number Density n, and Pressure P
at the Occultation Level for an Assumed Scale Height of 30 km

Object	D, m[a] (km)	θ	ν	n (cm^{-3})	P (μ bar)
Star	7.5×10^8	4×10^{-8}	3.3×10^{-10}	7×10^{13}	2
Spacecraft	2×10^5	1.5×10^{-4}	1.2×10^{-6}	3×10^{17}	7400

[a]The characteristic distance is D for the stellar case and m for the spacecraft case.

ed in the radio context by Phinney and Anderson (1968), was first applied to optical data by Kovalevsky and Link (1969). As discussed in more detail below, this method allows much better use of the data, though it does have one pitfall: the seeming objectivity of a computer can give a false sense of certainty to the derived results. The only remedy is to vary the uncertain parameters in a systematic way, as has been done in several recent papers (e.g., Vapillon *et al.* 1973; Wasserman and Veverka 1973). In the past this procedure has been less applied to radio than to optical data.

Processing of radio data is discussed in many papers (e.g., Fjeldbo and Eshleman 1965; Phinney and Anderson 1968; Fjeldbo *et al.* 1971; Eshleman 1973). A helpful conference record, with a particularly useful paper by Kliore, is available as a NASA report (Colin 1972). The primary output of a radio experiment is a Doppler frequency residual f (Eq. 21 below) due to the change of the ray path by the atmosphere. This frequency may also be regarded as the time derivative of the extra path, or phase lag, due to transit through the atmosphere and diversion of the ray. Intensity data can also be used, just as for the optical case, but for Jupiter these seem to be too greatly perturbed by the ionosphere to be usable. Processing is usually done by integral inversion of f, although a layer-by-layer matrix inversion has also been used. The direct result is a refractivity or density profile. Further processing to obtain temperatures is identical in optical and radio cases. The uncertainties that occur at the top of the atmosphere are among the topics discussed below.

The results initially obtained from Pioneers 10 and 11 were so strange that many people were reluctant to accept them, and indeed an error in the processing has since been found (Hubbard *et al.* 1975). The problem is basically an unexpected sensitivity to the exact geometry of the trajectory relative to the limb. A general discussion is given in Sec. III, and in the chapter by Kliore and Woiceshyn elsewhere in this volume.[1] The data from Pioneers 10 and 11 give temperatures between 10 and 100 mb that agree very well with other lines of evidence. The great sensitivity warns us that significantly deeper penetration may be difficult, even with the higher signal-to-noise ratio to be expected on future missions.

Finally, we shall give a brief description of the ionospheric results.

I. BASIC CONCEPTS

To illustrate the ideas involved, we consider an occultation of a star by a planet with an isothermal atmosphere of constant composition. Such an atmosphere is characterized by a single parameter, its scale height $H = RT/\mu g$, where R is the universal gas constant, T is the temperature, μ is the mean molecular weight and g is the acceleration due to gravity. From the perfect

[1]See p. 216.

gas law ($P = \rho RT/\mu$) and the equation of hydrostatic equilibrium ($dP = -\rho g dr$) the pressure profile of such an atmosphere is seen to be exponential with height

$$P = P_0 \, e^{-(r - r_0)/H} \tag{2}$$

as is the density profile:

$$\rho = \rho_0 \, e^{-(r - r_0)/H} . \tag{3}$$

Here r_0 represents some reference level in the planet's atmosphere.

The optical properties of the atmosphere are specified by its index of refraction as a function of wavelength: $n = n(\lambda)$. For all gases n is close to unity and it is convenient to define the *refractivity*, $\nu \equiv \nu(\lambda) = n(\lambda) - 1$. As noted above, a solar mixture of hydrogen and helium at standard conditions has a refractivity of about 1.2×10^{-4}.

The refractivity at a given wavelength is proportional to the density for non-polar molecules,

$$\frac{\nu(\lambda)}{\nu_s(\lambda)} = \frac{\rho}{\rho_s} \tag{4}$$

where the subscript s refers to conditions at standard temperature and pres-

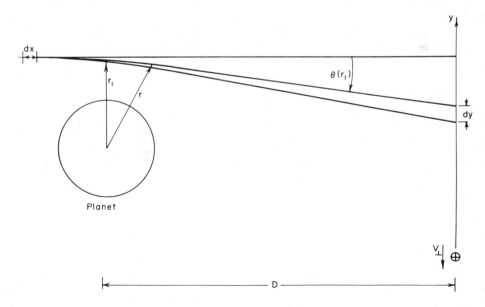

Fig. 1. Occultation geometry for immersion. A light ray from the star passes at a distance r_1 from the center of the planet and is refracted through an angle $\theta(r_1)$. A ray passing a little closer to the planet's center will be bent a little more. At Earth the two rays will be spread by a distance dy. The star and planet are considered at rest relative to each other and the earth is moving with velocity $v_\perp$ in the $-y$ direction.

sure (STP). Therefore the refractivity profile of an isothermal atmosphere of constant composition is also exponential with height.

Following Baum and Code (1953) it is easy to see how the observed light-curve will depend on the atmospheric scale height H. The best way to think of the occultation is in terms of the sketch in Fig. 1 where we consider the planet-star system to be at rest and the earth to be moving with a velocity $-v_\perp$ in the y-direction.

In this discussion we consider explicitly the star's immersion: the star disappears behind the planet's limb and a decrease in stellar flux is observed. (On emersion the process is reversed.) As the occultation proceeds (Fig. 1), a point is reached at which the rays from the star begin to be bent significantly by the atmosphere. In this context, "significantly" means that the bending produces a detectable effect on the observed star flux due to differential refraction (see below); the total bending of the ray during a stellar occultation is always small. The important quantities are θ and r_1, the total bending of the ray and its distance of closest approach to the planet, respectively. As r_1 decreases, the ray encounters greater atmospheric densities and suffers more bending since the ray's curvature is given by

$$K \equiv \frac{1}{n}\frac{dn}{dr} \simeq \frac{dv}{dr}. \tag{5}$$

The total bending of a ray passing within a distance of r_1 from the center of the planet is given by

$$\theta = \int_{-\infty}^{+\infty} K\,dx \tag{6}$$

where dx is an increment of path length along the direction of propagation (θ is always small). For an isothermal atmosphere, $K = -\dfrac{r}{H}$; substituting into Eq. (6) and using the geometry in Fig. 1, we have for the total bending of a ray in an isothermal atmosphere

$$\theta_0 = v_1(2\pi a/H)^{\frac{1}{2}} \tag{7}$$

where a is the radius of the planet and v_1 is the refractivity at the level of closest approach (r_1). The sign convention used is such that if the ray is bent down (as shown in Fig. 1), θ is assigned a negative sign. This is the common case in stellar occultations since the refractivity and hence the curvature increases toward the planet's center.

A.   The Flux Equation

The diminution of starlight during an occultation is due to differential refraction and not to extinction (absorption + scattering). The flux equation describing the observed lightcurve is derived easily by reference to Fig. 1.

The light in a small increment dr at the planet is spread out by differential refraction into an increment dy at the earth. Since $dy > dr$, a decrease in stellar flux is observed. The general result, derived by Baum and Code (1953), is

$$\frac{\phi_*}{\phi} - 1 = D \frac{d\theta}{dr_1} \tag{8}$$

where ϕ_* is the unocculted star flux, $\phi = \phi(t) =$ flux at time t and D is the Earth-planet distance.

For an isothermal atmosphere, the occultation lightcurve is given by

$$\left(\frac{\phi_*}{\phi} - 2\right) + \ln\left(\frac{\phi_*}{\phi} - 1\right) = \frac{v_\perp t}{H} \tag{9}$$

where $t = 0$ at $\phi_*/\phi = 0.5$ and $v_\perp$ is the apparent velocity of the star perpendicular to the planet's limb.

An interesting result was first derived by Goldsmith (1963) for an atmosphere with a linear scale-height gradient β:

$$H(r) = H_0 + \beta (r - r_0). \tag{10}$$

Provided that β is not too large, the occultation lightcurve can be approximated by

$$\left(\frac{\phi_*}{\phi} - 2\right) + \ln\left(\frac{\phi_*}{\phi} - 1\right) = \frac{v_\perp t}{H_0}\left(1 + \frac{3}{2}\beta\right). \tag{11}$$

Thus an atmosphere with a linear gradient can be mistaken for an isothermal atmosphere with $H = H_0 (1 + \frac{3}{2}\beta)^{-1}$, and vice versa.

The lightcurves represented by Eqs. (9) and (10) are smooth; unfortunately, observed lightcurves are not (see Fig. 2, for example), indicating that $v(r)$ is usually not a simple function of height. It is therefore impractical in analyzing real occultation lightcurves to proceed as we did above, that is, by assuming a refractivity profile $v(r)$, calculating the corresponding lightcurve, and comparing the result with the observations. It is preferable to try to determine the behavior of $v(r)$ directly from the observed lightcurve.

B. Inversion of Lightcurves

Several methods of analyzing observed lightcurves have been reviewed by Wasserman and Veverka (1973). The most reliable procedure seems to be that of formally inverting Eq. (6):

$$\theta(r_1) = \int_{-\infty}^{+\infty} \left(\frac{1}{n}\frac{dn}{dr}\right) dx \simeq \int_{-\infty}^{+\infty} \frac{dv}{dr} \cdot dx \tag{12}$$

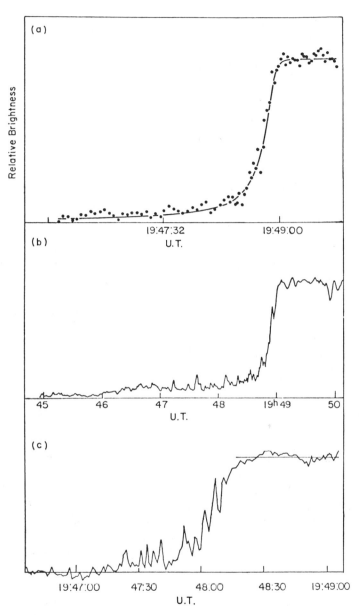

Fig. 2. Lightcurves of the emersion of β Sco A. (a) is from Hubbard *et al.* (1972), their *J* 3 curve at full time-resolution. A fitted model lightcurve (solid line) is included. The base of the graph corresponds to the zero level. The upper straight segment of the fitted lightcurve indicates the unit level. Jovian latitude is −57°3. (b) is from Vapillon *et al.* (1973), at full time resolution. The base of the graph indicates the zero level; the unit level is not specified. Jovian latitude is −57°3. (c) is from Veverka *et al.* (1974a), 0.01 sec data averaged over 1 sec intervals. The base of the graph indicates the zero level; the unit level is indicated by a thin straight line. Jovian latitude is −57°9.

which is an Abelian integral equation for $\theta(r_1)$ in terms of $\nu(r)$, and can be inverted to give $\nu(r_1)$ as a function of $\theta(r)$:

$$\nu(r_1) = \frac{1}{\pi(2a)^{\frac{1}{2}}} \int_{\infty}^{r_1} \frac{\theta(r')dr'}{(r'-r)^{\frac{1}{2}}}. \tag{13}$$

At a given time t in the lightcurve, a certain flux $\phi(t)$ is observed. $\phi(t)/\phi_*$ is directly related to the total bending $\theta(r_1)$ at this point [see Eq. (8)]. Thus values of $\theta(r)$ can be obtained from the occultation record, and substituted into the above integral equation to give the refractivity $\nu(r)$ as a function of height in the atmosphere (i.e., the refractivity profile of the atmosphere). Wasserman and Veverka (1973) have shown that this scheme works well even in the presence of spikes (see also Wallace 1975).

In practice the integration in (13) is not started at ∞ but at some finite level r_* above which the refractivity is assumed to be negligible. This boundary condition affects the derived refractivity profile near the beginning of calculation, but its effects are negligible after about 3 scale heights (Wasserman and Veverka 1973). The next step is to deduce the temperature structure from the refractivity profile. The error just mentioned becomes greatly magnified, as will be discussed next.

C. Effect of Starting Conditions

It is widely — but not widely enough — realized that there are large, unavoidable uncertainties in the upper parts of a derived temperature profile in both optical and radio cases. We shall give a simplified exposition of the question; more details are given by Wasserman and Veverka (1973). Vapillon et al. (1973) have also given many illustrations, obtained by parametrically varying the starting conditions, and a brief report of a similar treatment is given by Wallace (1975).

The actual result of an occultation experiment is a density (actually refractivity) profile as a function of relative height. The temperature profile must then be obtained by further operations. The principal difficulty is that the density is known only to within an additive constant. (There are some exceptions in the radio case, but not so far for Jupiter.) For illustration, we assume that the atmosphere is isothermal with a true scale height H; a family of density curves, differing only by a constant, is shown in Fig. 3. The corresponding temperature profiles, obtained by logarithmic differentiation, are shown at the right. We use unprimed quantities for the (unknown) actual atmosphere, and primed for the observed quantities. Thus

$$\rho' = \rho + a = \rho_0 e^{-z/H} + a \tag{14}$$

where a is the unknown constant and ρ_0 is the density at $z = 0$. The observed density scale height is therefore

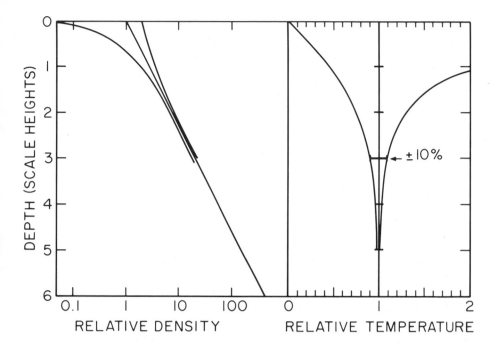

Fig. 3.   Illustration of the effect of uncertain starting conditions in conversion of a density profile to a temperature profile. The model atmosphere is isothermal, and the true density profile is the straight line. The other density profiles indicate the range likely to be consistent with an observation, and the corresponding deduced temperature profiles are shown at the right.

$$-\rho'\left(\frac{d\rho'}{dz}\right)^{-1} = H'\rho = H + \frac{aH}{\rho_0}\, e^{z/H} = H(1 + y)\,. \tag{15}$$

To find the temperature, we need the pressure scale height $H' = kT'/mg = H'_\rho\,(1 + \beta')$, where $\beta' = dH'/dz$. The observed quantity is dH'_ρ/dz, which may be called β'_ρ; it turns out that, to an adequate approximation for this illustration,

$$H' \doteq \frac{H'_\rho}{1 - \beta'_\rho}\,. \tag{16}$$

From Eq. (15) $\beta'_\rho = (a/\rho_0)e^{z/H} = y$, and thus

$$\frac{H'}{H} = \frac{T'}{T} \doteq \frac{1 + y}{1 - y}\,. \tag{17}$$

The reasonable limits on y are ± 1, but without further information its value at the upper boundary usually cannot be determined within those limits. A

10% uncertainty in T', which requires that $y = 1/20$, requires a depth below the boundary of $H \ln 20 = 3.0 \, H$, as illustrated in Fig. 3. If the uncertainty is less, one merely cuts a bit off the top of the curves.

Frequently the temperature is determined not by differentiation but by use of hydrostatic equilibrium, which is more suitable for computer reduction. The pressure can be obtained by integration of the density profile

$$P(z) = g \int_z^\infty \rho \, dz' . \qquad (18)$$

The gas law then gives

$$T = \frac{P}{nk} . \qquad (19)$$

This case has been studied by Wasserman and Veverka (1973) for atmospheres with temperature gradients as well as the isothermal case. The conclusion is similar, although different in minor details.

Use of (18) and (19) brings in a second uncertainty, since the integral in (18) actually does not reach infinity. Thus, the pressure remains uncertain for 1–2 scale heights below the boundary; the effect is illustrated in the first report from the Pioneer 10 radio experiment (Kliore et al. 1974a). The corresponding uncertainty with the differentiation method used here is that the slope is very poorly determined near the end of a noisy density curve. In any case, this class of uncertainty, though perfectly real, is usually much less than the one due to the initial density, and dies away considerably faster. It can therefore be neglected, except in the occasional favorable case for the radio experiment where the ionospheric part of the signal can be subtracted exactly.

D. The Doppler Frequency

Analysis of the radio experiment is greatly facilitated by the simple expressions derived here. These are limited to certain model atmospheres with constant (or zero) temperature gradients, and are therefore not particularly useful for analyzing observational data. In the form given here, they are limited to small bending angles, an excellent approximation for Pioneer 10 where the angle did not exceed 8 milliradians. Figure 4 shows a projection in the plane containing the local vertical and the earth. With some arbitrary choice of origin and a corresponding initial time t_0 the geometry gives

$$z = m \, \theta - v_s \, (t - t_0) . \qquad (20)$$

By definition, the Doppler frequency f is zero in the absence of atmospheric bending. An immediate consequence of the bending of the beam through the

Fig. 4.   Occultation geometry for a radio experiment at immersion. The velocity v_s is the component in the plane of the figure and at right angles to path P_2 of the actual spacecraft velocity.

angle θ is to project a fraction $v_s\theta$ back into the line of sight. The corresponding Doppler shift is therefore (Phinney and Anderson 1968)

$$f = \frac{f_0}{c}\, v_s \theta = \frac{v_s}{\lambda}\, \theta \tag{21}$$

where f_0 is the radiated frequency and λ the corresponding wavelength, 13.05 cm. The barometric equation for number-density n with a scale-height gradient $dH/dz = \beta$ is

$$\frac{n_0}{n} = \left(\frac{H}{H_0}\right)^{1+\frac{1}{\beta}} = \left(\frac{H_0 + \beta z}{H_0}\right)^{1+\frac{1}{\beta}}. \tag{22}$$

The bending angle for a ray grazing a level with scale-height H is proportional to $nH^{-\frac{1}{2}}$ (Goldsmith 1963). Therefore

$$\frac{f}{f_0} = \frac{\theta}{\theta_0} = \frac{n}{n_0}\left(\frac{H_0}{H}\right)^{\frac{1}{2}} = \left(\frac{H_0}{H}\right)^{\frac{3}{2}+\frac{1}{\beta}}. \tag{23}$$

With $H = H_0 + \beta z$ as in (22), Expression (23) gives f as a function of z. Substitution of (21) and (23) into (20) gives

$$t - t_0 = \frac{\lambda m}{v_s^2}\, f - \frac{H_0}{\beta v_s}\left(\frac{f_0}{f}\right)^{\frac{2\beta}{2+3\beta}}. \tag{24}$$

If the atmosphere is isothermal, with $\beta = 0$, (24) becomes

$$t - t_0 = \frac{\lambda m}{v_s^2}\, f - \frac{H}{v_s}\ln f \tag{25}$$

where a term involving $\ln f_0$ has been absorbed into t_0; the same thing can be done with (24) if desired. Differentiation of either (25) or, after a substitution, (24) gives

$$\frac{dt}{df} = \frac{\lambda m}{v_s^2} + \frac{H}{v_s f\left(1 + \frac{3}{2}\beta\right)}. \tag{26}$$

A more direct derivation of (26) can be made by differentiation of (20) followed by substitution of a hydrostatic equation in its differential form. These equations, for the isothermal case, were first given by Hubbard *et al.* (1975) and Eshleman (1975).

II. STELLAR OCCULTATIONS

Jupiter occultations of four stars have been observed photoelectrically: σ Arietis, β Scorpii A_1, β Scorpii A_2 and β Scorpii C. The occultation of σ Arietis on 20 November 1952 was observed by Baum and Code (1953). The β Scorpii system which consists of at least four stars (Hubbard *et al.* 1972; Elliot and Veverka 1976), was occulted on 13 May 1971. Lightcurves of the occultation of the spectroscopic binary β Sco A_1 and β Sco A_2 ($V = 2.63$) were obtained by a number of observers including Hubbard *et al.* (1972), Veverka *et al.* (1974a) and Vapillon *et al.* (1973). The first two groups also observed the occultation of β Sco C ($V = 4.94$). No detection of β Sco B, a 10th magnitude companion of β Sco A, has been reported.

In this chapter β Sco A_1 and β Sco A_2 will be dealt with as one star, although the individual contributions to the lightcurve of the two components can be identified in the high time-resolution observations of Veverka *et al.* (1974a). At the time of the occultation, the projected separation of the two components was 1.496 ± 0.018 10^{-3} arcsecs, or about 3 km at the distance of Jupiter (Elliot *et al.* 1975a).

Since β Sco C is over 2 magnitudes fainter than β Sco A, the lightcurves for the faint star are of considerably lower quality than those for β Sco A. A conservative view would hold that β Sco C lightcurves are inadequate to determine the atmospheric scale height. For example, while Hubbard *et al.* (1972) quote errors of ± 2 km for scale heights derived from the β Sco A lightcurves, the errors quoted for those derived from the β Sco C observations are typically ± 10 km!

The problems with the σ Arietis lightcurve are even more serious. Like β Sco C, this star is faint ($V \sim 5$). In addition, the observations seem to be affected by systematic errors (see below), and it is imprudent to draw any conclusions about Jupiter's atmosphere from the σ Arietis lightcurve.

This chapter will therefore concentrate on observations of the bright spectroscopic binary β Sco A. Lightcurves for the β Sco C occultation are discussed by Hubbard *et al.* (1972) and by Elliot *et al.* (1975b).

A. *The Occultation of Beta Scorpii*

The occultation of 13 May 1971 was a widely observed event; observations by at least six independent groups have been published. Of the various efforts, that at the University of Texas (Hubbard *et al.* 1972) was the most

TABLE II

Observations of the 13 May 1971 Occultation of Beta Scorpii

Reference	Type of Observations	Reduction Procedure	Scale Height H (km)	Comments
Bhattacharyya (1972)	Photoelectric	Curve Fitting	3 km	Low signal-to-noise ratio
Freeman and Stokes (1972)	Photoelectric	Curve Fitting	8 km	Low signal-to-noise ratio
Larson (1972)	Photographic	Curve Fitting	$\geqslant 8$ km	Low photometric accuracy
"Meudon Group" Combes *et al.* (1971)	Photoelectric	Inversion	30 ± 5 km	Upper portion of refractivity profile
"Texas Group" Hubbard *et al.* (1972)	Photoelectric	Inversion	31 ± 2 km 23 ± 2 km 24 ± 2 km	Three best lightcurves
"Cornell/Harvard Group" Veverka *et al.* (1974*a*)	Photoelectric	Inversion	25 ± 2 km	Emersion β Sco A

extensive with three expeditions to various parts of the world. The discrepancy among some of the published results should dispel the attitude that stellar occultations by planets are easy to observe and easy to analyze. For example, reported scale heights (Table II) differ by a factor of 10! Most of the erroneous results can be blamed on very low-quality observations combined with inadequate analysis.

The major observational pitfall is scattered light from Jupiter's limb. To keep this contribution to the total signal constant, guiding must be very accurate, especially if large apertures are used. In this respect the β Sco occultation presented much more of a challenge than the occultation of BD–17° 4388 by Neptune in 1968 (Kovalevsky and Link 1969; Freeman and Lyngå 1970). In the Neptune occultation the star and the planet were of comparable magnitude and both could be included in the same aperture. Thus it was easy to assure a constant contribution from the planet to the total signal. It is note-worthy that observations of the Neptune occultation at 3 different observa-tories give concordant results (Rages *et al.* 1974).

In the case of the β Sco occultation, it was impractical to include both β Sco and the whole of Jupiter in one aperture, since the star's contribution to the total signal would then be less than 10% (Freeman and Stokes 1972). By use of a small aperture (13 arcsec) the star's contribution to the total signal could be increased to 96% (Veverka *et al.* 1974*a*), but this procedure places a premium on good guiding, or on some method of calibrating the instanta-

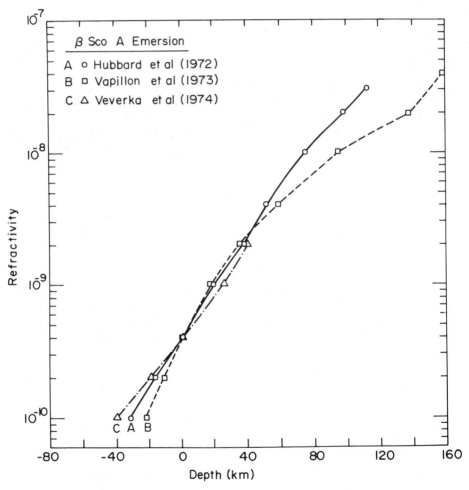

Fig. 5.   Refractivity profiles derived from the emersion lightcurves shown in Fig. 2. The indivi-dual symbols indicate points read from published graphs. The zero of the depth scale is arbitrary and has been chosen so that $\nu = 4 \times 10^{-10}$ at the zero level for each curve.

neous contribution of light from Jupiter's limb to the total signal (Sec. II.B).

The atmospheric scale heights determined from the three best sets of observations (Hubbard *et al.* 1972; Combes *et al.* 1971; and Veverka *et al.* 1974a) are concordant near the $10^{14}-10^{15}$ cm^{-3} number-density level (Table II and Figs. 2 and 5). Since the atmospheric structure is not isothermal, scale heights quoted in this chapter usually refer to effective values obtained by fitting a straight line to a segment of the refractivity profile (a plot of log ν versus z, where ν is the refractivity and z is a linear depth scale).

The remaining scale-height determinations listed in Table II cannot be given any weight. They are based on lightcurves of very low photometric quality reduced by curve-fitting. The observations of Freeman and Stokes (1972) were obtained under poor atmospheric conditions: "The seeing was unusually bad and we were forced to include Jupiter and the star in a large aperture . . . ; β Sco gave only 0.08 of the total signal." The lightcurve published by Bhattacharyya (1972) is of very low quality, and no significance can be attached to the deduced scale height of 3 km. The observations of Larson (1972) were made photographically; although high-speed photography provides an interesting qualitative record of the occultation (Fairall 1972; Larson 1972), the photometric accuracy of film is insufficient to define accurate lightcurves.

Not included in Table II is the 8-km scale height determined by Baum and Code (1953) from the occultation of σ Arietis. Continuous guiding was not possible during that event, and the lightcurve showed periodic fluctuations caused by telescope tracking errors. Although an attempt to remove these errors was made, large uncertainties in the corrected lightcurve remain. These guiding problems, combined with the faintness of the occulted star, cast doubt on the quality of the published lightcurve. It is probable that the 8-km scale height inferred is in error, and should not be construed as evidence for changes in the temperature structure of Jupiter's upper atmosphere between 1952 and 1971.

B. Methods of Correcting for Jupiter's Contribution to the Total Signal

Jupiter's contribution to the observed signal must be determined accurately if a precise "zero level" for the occultation curve is to be defined. In turn, a precise zero level is required if an accurate scale height is to be determined from the lightcurve (Hunten and McElroy 1968; Hubbard *et al.* 1972; Wasserman and Veverka 1973; Vapillon *et al.* 1973). A 10% error in the zero level can lead to a 100% error in the scale height (see, for example, Hubbard *et al.* 1972).

It is interesting that the three groups who obtained extensive observations of the β Sco occultation (Fig. 2) used different techniques for dealing with the troublesome problem of accurate guiding during the event. The Texas group (Hubbard *et al.* 1972) performed rapid area scans through the limb and the star. The Cornell-Harvard group (Veverka *et al.* 1974*a*) used a beam splitter and guided visually throughout the occultation. The Meudon group (Berezne *et al.* 1975) used a double aperture technique (described below) to monitor tracking errors.

The instrumentation and observing techniques used by the Texas group are described in detail by Hubbard *et al.* (1972). The observations were made at a single wavelength centered around the calcium K-line (λ 3934 Å). A narrow slit, 1 by 10 arcsec, was scanned tangentially to the limb (Fig. 6) at a rate of ~ 8 complete cycles (or 16 traverses of the star image) per second. Clearly, this technique is very powerful for monitoring accurately the limb contribution to the total signal. Its main disadvantage is that the star is observed only intermittently.

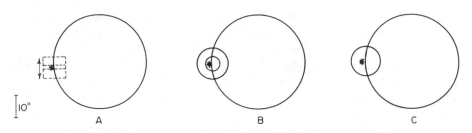

Fig. 6. Schematic representation of different techniques used to observe the β Sco occultation. *A* shows the area-scanning method of Hubbard *et al.* (1972). A rectangular aperture 1 by 10 arcsec is scanned rapidly as indicated by the double arrow. *B* shows the double-aperture method of Vapillon *et al.* (1973). The inner aperture is 5.5 arcsec in diameter; the outer aperture is annular with inner and outer diameters of 5.5 and 16.5 arcsec, respectively. *C* shows the single-aperture method of Veverka *et al.* (1974*a*). A beam-splitter allows continuous guiding during the occultation.

During the β Sco A events Hubbard *et al.* recorded data in "frames" of either 4 or 8 scans, corresponding to time intervals of 0.5 or 1 sec. In the reduction procedure, groups of 3 frames were averaged yielding one data point every 1.5 to 3 sec. The typical time resolution of the published data can be judged from the Johannesburg emersion curve: there are 37 data points in the 88-sec interval between 19^h 47^m 32^s and 19^h 49^m 00^s, or an average of one point every 2.4 sec. Due to the intermittent sampling and also perhaps due to some of the smoothing procedures applied, some of the lightcurve details are lost. For example, if the high time-resolution data of Veverka *et al.* (1974*a*) are degraded by averaging to a time resolution of 2.5 sec, spikes are not completely averaged out, whereas they do not appear convincingly in any of the data of Hubbard *et al.* (1972).

Although area scanning does not yield a good record of the occultation at the highest possible time resolution it does provide a mechanical method of accurately monitoring the limb contribution to the signal.

The method used by the Cornell-Harvard group (Veverka *et al.* 1974*a*) relies on the skill of the observer to perform any necessary guiding corrections during the event (Fig. 6). The instrumentation and observing procedures are described in detail by Veverka *et al.* (1974*a*). Since the star was observed at all times, continuous measurements of the star at a time resolution of 0.01 sec were obtained (Liller *et al.* 1974).

The observations were made simultaneously at three wavelengths: 0.35 μm (Channel 1), 3934 Å (Channel 2) and 0.62 μm (Channel 3). Except for 600 Å between 0.59 and 0.65 μm, all light beyond 0.46 μm was transmitted (with 50% efficiency) to a guiding eyepiece. The defining aperture of 13 arcsec was clearly illuminated by light from Jupiter, and was positioned so that during the occultation about half of it was filled by Jupiter's limb (Fig. 6). On emersion, typical contrasts (β Sco/Jupiter limb) were about 25, 7 and 2 to 1 in Channels 1, 2 and 3 respectively. Since a large fraction of the signal in Channel 3 came from the limb, the output of this channel provides a sensitive test to the accuracy of guiding during the occultation. Fluctuations in the Channel 3 base-level average less than $\pm 10\%$, which means that the fluctuations in the base level of Channel 2 are probably less than $\pm 2\%$, and those in the base level of Channel 1 less than $\pm 1\%$.

The method used by the Cornell-Harvard group has the great advantage that continuous observations of the star at very high time resolution are possible. Since observations are made simultaneously in three channels, one of which (Channel 3) is very sensitive to the limb contribution, a powerful color-correction method was applied to the β Sco C data (Elliot *et al.* 1975*b*), but not in the original reduction of those of β Sco A (Veverka *et al.* 1974*a*) since it was deemed unnecessary. However, it appears that the color-correction method can be as powerful a technique for determining the limb contribution as the area-scanning technique.

The instrumentation and observing techniques used by the Meudon group have been described by Berezne *et al.* (1975), although some important details remain unclear. Vapillon *et al.* (1973) state: "Visual guiding was continuously performed . . . When the star was too faint to allow visual guiding, offset guiding was performed based on predictions for a Jovian atmospheric scale height of 25 km." According to Vapillon (personal communication), only those portions of the lightcurves based on visual guiding were inverted. A 3-channel photometer was used (Berezne *et al.* 1975); one of the channels monitored transparency by measuring the signal from the whole disk of Jupiter. The other two measured light (near 3934 Å), from two concentric areas centered about β Sco (Fig. 6). Channel A measured the signal from a circular area 5.5 arcsec across, centered on β Sco. Channel B measured the signal from an annular area (inner diameter 5.5, outer diameter 16.5 arcsec, respectively) centered on the inner aperture. This procedure was intended to provide an accurate method of monitoring the limb contribution. However, it does have an important disadvantage. It would appear that under normal seeing conditions a significant portion of the star's flux must fall outside of the 5.5 arcsec aperture. Thus the output from this channel is probably quite erratic, and one would wish to see the uncorrected occultation curves. It is not clear how such a signal can be corrected accurately by using the output from Channel B. For example, how does one distinguish between an increase in Channel B output caused by a drifting in of Jupiter's limb, and one caused by the spilling in of more starlight from the inner aperture?

Vapillon informed us that the signal from Channel A was averaged over 0.5 sec, digitized and printed on paper-tape. As a back-up, the signals were also recorded on a chart recorder, with an effective time resolution of 1-2 sec. Unfortunately, the raw occultation lightcurves as seen by Channel A have not been published; therefore, an independent evaluation of the accuracy of the correction method applied cannot be made. Vapillon et al. (1973) believe that their zero level is determined within confidence limits of ± 1%.

All detailed analyses to date have assumed that the refractivity gradient in Jupiter's atmosphere is parallel to the gravity gradient. Elliot and Veverka (1976) have labeled this convenient simplifying assumption the "gravity-gradient model." If for simplicity we refer to the direction defined by the local gravity gradient as the local vertical, the model assumes that only vertical refractivity gradients exist, and that horizontal gradients on scales appropriate to single occultation events are absent. Elliot and Veverka (1976) estimate that for a typical occultation by Jupiter this assumption demands that horizontal refractivity gradients be absent on lateral scales of less than ∼ 3000 km over an altitude of about 100 km around the occultation level. Young (1976) has discussed how strong turbulence might invalidate this assumption.

From a detailed review of available data, Elliot and Veverka (1976) conclude that no observational evidence has been produced to show that the simple and convenient gravity gradient model cannot be applied to Jupiter's upper atmosphere. However, they also stress that no evidence exists to prove that the assumptions of the model are valid over lateral scales of more than a few kilometers.

The inherent inaccuracy in the upper parts of the refractivity and temperature profiles has been discussed in Sec. I [see also Hubbard et al. (1972); Wasserman and Veverka (1973); Vapillon et al. (1973); Veverka et al. (1974a); Veverka et al. (1974b); Rages et al. (1974); Wallace (1975)]. At least two fundamental problems are involved: (1) The initial dimming of the occulted star involves very little differential refraction, hence small errors in the upper portions of the lightcurve translate into large errors in the low refractivities at these upper levels (Veverka et al. 1974b); (2) The upper part of the refractivity profile is very sensitive to the boundary conditions used to start the inversion (Hubbard et al. 1972; Wasserman and Veverka 1973; Vapillon et al. 1973). Veverka et al. (1974a) judged that the refractivity profiles become tolerably accurate only at levels of about $\nu \geqslant 10^{-10}$. This estimate is generally consistent with the results of Hubbard $et\ al.$ (1972) (cf. their Fig. 19), and Vapillon $et\ al.$ (1973). The best accuracy is found in the neighborhood of $\nu = 10^{-9}$, where the pressure is about 5 microbar.

C. Refractivity Profiles

In Fig. 5, typical refractivity profiles obtained by the three groups are compared. These profiles are all based on observations of the emersion of β Sco A near latitude $-60°$. The depth scale, z, is arbitrary and the origin has been chosen so that $z = 0$ occurs at $\nu = 4 \times 10^{-10}$. For reasons discussed in Sec. IIB, for $\nu \leqslant 10^{-10}$ the refractivity profiles are very uncertain and are not shown. We note that the profiles in Fig. 5 were derived assuming that the gravity-gradient model (Elliot and Veverka 1976) is valid for Jupiter's atmosphere at the levels in question.

Note that Veverka $et\ al.$ (1974a) stopped their inversion procedure near $\nu \sim 2 \times 10^{-9}$, since at this point the residual lightcurve level was comparable to the estimated uncertainties in the baseline. The original reduction of the β Sco A data by the Cornell-Harvard group did not involve any corrections for the small baseline fluctuations that might exist in their data. If the color-correction method used by Elliot $et\ al.$ (1975b) to reduce the β Sco C observations were applied to the β Sco A data, the refractivity profile could probably be extended to deeper levels.

Hubbard $et\ al.$ (1972) and Vapillon $et\ al.$ (1973) extended their inversions to about $\nu \simeq 3 \times 10^{-8}$ and $\nu = 10^{-7}$ respectively. The area scanning technique of Hubbard $et\ al.$ provides a powerful method for determining the accurate baselines essential for reaching these levels of the atmosphere. It is difficult to evaluate fairly the method of Vapillon $et\ al.$ in view of its limited description in the literature. Around $\nu = 10^{-9}$ there is reasonable agreement among the three data sets (Fig. 5). Approximate local scale heights at this level are about 25 km (Hubbard $et\ al.$ 1972), 25 km (Veverka $et\ al.$ 1974a) and 22 km (derived from Vapillon $et\ al.$ 1973). For refractivities greater than 4×10^{-9}, the profiles of Vapillon $et\ al.$ begin to diverge significantly from those of Hubbard $et\ al.$ For example, the local scale height between $\nu = 10^{-8}$ and 2×10^{-8} is about 33 km in the Texas data, but about 63 km in the Meudon data. These large differences are discussed in the following section.

D. Temperature Profiles

In principle the refractivity profiles discussed in the previous section can be converted easily into temperature versus number-density profiles (Wasserman and Veverka 1973) provided that the atmosphere is well mixed at these levels and that the mean molecular weight is known. The following three subsections summarize current evidence about the composition of Jupiter's atmosphere and the location of the homopause. The conclusion is that at the levels probed by the occultations, the atmosphere is well mixed, and undissociated, and that the $[He]/[H_2]$ ratio is probably about 0.1 (by number).

Location of the Homopause. The eddy-diffusion coefficient in Jupiter's upper atmosphere is now known to be high. Carlson and Judge (1974)[2] have found a value of $K_v = 3 \times 10^{8 \pm 1}$ cm^2 sec^{-1} from Pioneer 10 Lyman-α observations. This determination places the homopause near the level of $n = 10^{11 \pm 1}$ cm^{-3}. Detectable refraction of the starlight begins at about 10^{12} cm^{-3}, and the solutions cease to be dependent on boundary conditions below about 2–5×10^{13} cm^{-3}. Thus the atmosphere at the levels probed by the occultation is almost certainly well mixed. The Pioneer result contradicts the previous opinion (discussed by Combes *et al.* 1971) that in order to reach well-mixed levels of the atmosphere one must get down to and below the 0.02 level in the lightcurves.

Chemical Composition. For our purposes the only important atmospheric constituents are helium and molecular hydrogen (Elliot *et al.* 1974). There is no reason to assume that the helium-hydrogen ratio on Jupiter differs from the solar or cosmic value. Within the last two years a number of definite determinations have been published by Elliot *et al.* (1974), Carlson and Judge (1974), Orton (1975) and by Houck *et al.* (1975) (Table III). The most restrictive determination is that of Houck *et al.* (1975) based on infrared spectra between 16 and 40 μm. In what follows, we adopt a nominal value of $[He]/[H_2] = 0.1$ by number, consistent with the determinations listed in Table III.

TABLE III
Jovian Helium-Hydrogen Ratio by Number

Method	Reference	$[He]/[H_2]$
β Sco occultation (spike time delays)	Elliot *et al.* (1974)	$0.16^{+0.19}_{-0.16}$
Pioneer 10 ultraviolet photometer	Carlson and Judge (1974)	$0.18^{+0.46}_{-0.12}$
Pioneer 10 infrared radiometer	Orton (1975)	0.09 ± 0.08
Infrared spectrum	Houck *et al.* (1975)	0.11 ± 0.11
Solar value	Pagel (1973)	~ 0.13

---

[2]See also p. 426.

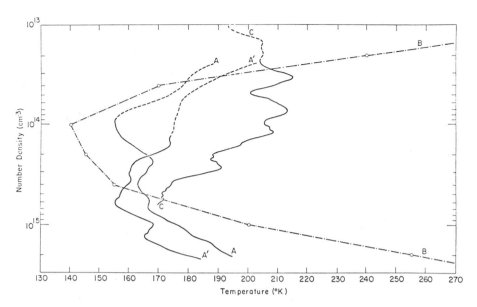

Fig. 7. Temperature versus number density profiles derived from the emersion lightcurves of β Sco A, assuming a well-mixed atmosphere with [He]/[H₂] = 0.1, by number. Curve *A* is the emersion profile calculated by Wasserman (1974) from the data of Hubbard *et al.* (1972). Curve *B* is the emersion profile from Vapillon *et al.* (1973). The symbols indicate individual points read from the published graph. Curve *C* is the emersion profile from Ve-verka *et al.* (1974a). Also shown for comparison is curve *A'*, the immersion profile calculated by Wasserman (1974) from the data of Hubbard *et al.* (1972). The dashed portions of profiles *A*, *A'* and *B* are affected by assumed boundary conditions and are probably unreliable [cf. Wasserman (1974), and Fig. 3].

Degree of Dissociation at the Occultation Level. Judging from Fig. 2 in the paper by Wallace and Hunten (1973), the atomic hydrogen concentration is less than 1% of the total number density even at 10^{10} cm³; these calculations were made for $K_v = 10^6$ cm² sec⁻¹. Hence the actual values are less, and at the occultation level the degree of dissociation is insignificant.

Figure 7 shows temperature versus number-density profiles corresponding to the refractivity profiles of Fig. 5. All curves were calculated under the assumption of a well-mixed atmosphere with [He]/[H₂] = 0.1. Curves *A*, *A'* and *C* are taken from Wasserman (1974); *A* and *A'* were calculated by Wasserman using the data of Hubbard *et al.* (1972). Curve *B* is drawn from data published by Vapillon *et al.* (1973). Unfortunately the latter authors published graphs of *T* versus *z*, and *n* versus *z*, but not of *n* versus *T*. Curve *B* was obtained by reading off a dozen points from the published graphs, and is adequate for our purposes.

A comparison of Curves *A*, *B* and *C* (all obtained from the emersion light-curves of β Sco A) shows tolerable agreement near $n \sim 5 \times 10^{14}$ cm³ ($\nu \sim 2 \times 10^{-9}$). Since the three observations probed slightly different points on

Jupiter's periphery it is impossible to be sure (although it is probable, as indicated below) that the $\pm 15°K$ divergence among the curves near this level is a measure of the systematic errors involved. As can be seen from Table I, the region of best agreement is about 2 scale heights below the occultation level. The discrepancies at higher and lower levels are now discussed separately.

The unavoidable error in the refractivity profile at the top of the atmosphere has been discussed above. It is virtually impossible to assign meaningful error bars to calculated temperature profiles, a situation which unfortunately has sometimes led to the conclusion that such errors are negligibly small. Such errors do exist in both the vertical *and* horizontal directions on an *n* versus *T* graph. An estimate of these errors is given below.

By inverting model curves, Wasserman and Veverka (1973) concluded that temperature profiles have little chance of converging to the correct temperature for at least 3 or 4 scale heights from the beginning of the calculation. Below this level, the model calculations suggest that calculated temperatures tend to approach the correct values, but that large errors in the temperature gradients can persist. This was noted especially by Veverka *et al.* (1974b) in their analysis of the Mount Stromlo observations of the occultation of BD $-$ 17°4388 by Neptune. Two different facsimiles of the *same* chart recorder trace were digitized: the resulting temperature profiles showed significantly different temperature gradients, but converged to similar final temperatures. Another implication of the modeling done by Wasserman and Veverka (1973) is that it is impossible to make general statements such as "inversion always overestimates temperatures and temperature gradients." The situation can go either way [cf. Figs. 16 and 17 in the paper by Wasserman and Veverka (1973), and the discussion in the present Sec. I].

In regions deeper than 10^{15} cm^{-3} the major error is due to the uncertainty of the zero level. Vapillon *et al.* (1973) and Combes *et al.* (1975) have shown an envelope obtained by changing their zero by $\pm 1\%$. This envelope contains all the curves shown here in Fig. 7, but still becomes rather warm, around 250–280°K, just below the bottom of the figure. If the error is increased only slightly, the data would probably be consistent with an atmosphere that is not warmer than 200°K, and would fit in much better with the radio data from Pioneers 10 and 11.

A reasonable conclusion would seem to be that temperatures between 10^{14} and 10^{13} cm^{-3} (Fig. 7) can be trusted much better than the temperature gradients. For levels higher than 10^{13} cm^{-3}, there certainly are significant uncertainties in refractivity profiles and large errors in the temperature profiles. For regions deeper than $n \sim 10^{15}$ cm^{-3} one is putting a great premium on the accuracy of the zero level of the lightcurves, not to mention possible difficulties due to ray crossing discussed by Young (1976).

We suspect that the temperature gradients shown in Fig. 7 have little or no significance. However, a reasonable estimate of the likely errors in the temperatures near 5×10^{14} cm^{-3} can be made. Curves A, B and C

differ by $\sim \pm 15°$K. On the other hand, if we compare the temperature profiles calculated from the Johannesburg emersion and immersion lightcurves of β Sco A published by Hubbard *et al.* (1972)—that is, Curves A and A' — the difference in temperatures is typically only about $\pm 10°$K, even though these two lightcurves correspond to points on Jupiter which are much farther apart than those for Curves A, B, and C. This suggests that much of the divergence among the curves is due to systematic errors in the three different methods used. By examination of Fig. 7 one concludes that near $n = 5 \times 10^{14}$ cm^{-3}, $T \simeq 170°$K, with estimated uncertainty of about ± 10–$20°$K. From an independent review of these results Combes *et al.* (1975) arrive at $T = 170 \pm 30°$K. The validity of the overall temperature gradients in this region is questionable. There is no doubt, however, that small-scale temperature fluctuations are present.

The high temperatures inferred by Vapillon *et al.* (1973) for $n > 10^{15}$ cm^{-3} can be tested in two ways. First, although the temperatures near $n \sim 10^{16}$ cm^{-3} from the Pioneer 10 and 11 S-band data unfortunately suffer from the same difficulty with starting conditions, at 5×10^{18} cm^{-3} the temperature is almost certainly not over $200°$K. Secondly, the three-color high time-resolution observations of β Sco A obtained by Veverka *et al.* (1974*a*) can be corrected for small baseline fluctuations using the color-correction method of Elliot *et al.* (1975*b*). It should be possible to obtain a reliable baseline using this technique and extend the refractivity profile to $\nu \sim 10^{-8}$ or deeper.

E. Irregularities

Prominent features of lightcurves that are taken with adequate time resolution are the spikes seen in Fig. 2*c*. The obvious interpretation is refocusing of the light by small density irregularities (e.g., Baum and Code 1953; Wasserman and Veverka 1973; Wallace 1975). It has usually been assumed that the irregularities are layers of large horizontal extent (Elliot and Veverka 1976). They could be static, as inferred for the earth's lower stratosphere (cf., Dütsch 1971) or dynamic (French and Gierasch 1974). Recently Young (1976) has urged that the proper interpretation requires scintillation or refraction by a large number of small irregularities. These too could be static or dynamic, though Young's preference is for turbulence. Further work with the data will be required before a consensus is possible. It may be that layers dominate at the higher light levels, with a transition to scintillation deeper in the atmosphere. Perhaps this work will also elucidate some of the questions about radio propagation, such as the possibility of ray-crossing.

F. Diurnal Temperature Variations

The emersion and immersion events of the β Sco occultation took place at the sunrise and sunset limb, respectively. Wasserman (1974) found no evidence of diurnal temperature changes at the occultation level ($n \sim 10^{14}$ cm³) in the best data of Hubbard *et al.* (1972) and Veverka *et al.* (1974*a*); furthermore he argued that there is no reason to expect them to be measurable at this level. A comparison of the Johannesburg curves of Hubbard *et al.* (1972) (Curves *A* and *A'* in Fig. 2) shows that the diurnal temperature range at the occultation level may be less than ± 10°K, an amount that is within the uncertainty of the temperature determination. Combes *et al.* (1975) also concluded that there is little noticeable difference between immersion and emersion results.

G.   *Temperature Variations with Latitude*

All bright star events occurred between − 49° and − 60° latitude, while the faint star events clustered near − 10°. The suggestion that the scale height decreases with increasing latitude (Hubbard *et al.* 1972) is one possible way of accounting for the data (Fig. 8). Unfortunately, all of the low-latitude

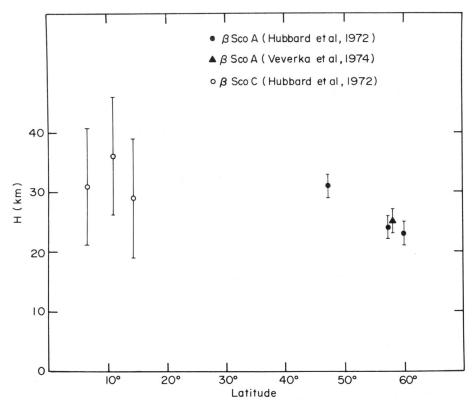

Fig. 8.   Distribution of scale-height determinations with latitude. Not included is a 25 km scale height at −12° obtained by Elliot *et al.* (1975*b*) from their analysis of the β Sco C emersion lightcurve (see Fig. 9*b*). This point is not plotted because the authors felt that they could not assign a meaningful error bar to this result.

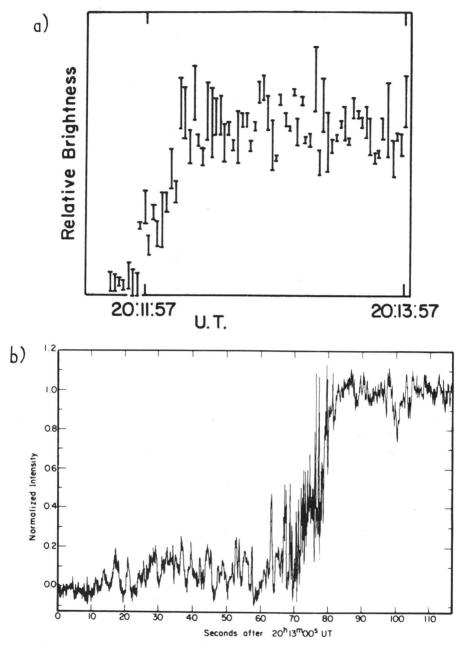

Fig. 9.   Representative lightcurves for the occultation of β Sco C. (a) Emersion lightcurve observed from Naini Tal using the area-scanning technique of Hubbard *et al.* (1972). Jovian latitude is −6°.5. (b) Emersion lightcurve observed from Boyden after removal of the Jovian limb component by the color-correction method of Elliot *et al.* (1975*b*). Time-resolution is 0.1 sec; Jovian latitude is −12°.

points are based on observations of β Sco C; the quality of the lightcurves for this faint star is fairly low (Fig. 9) and it is impossible to calculate accurate temperature profiles from them. The three scale heights derived by Hubbard *et al.* from the β Sco C observations have a mean value of 32 km; this value is higher than most of the β Sco A scale heights, but the uncertainties are large (± 10 km). One must also bear in mind that Elliot *et al.* (1975) obtained a lower scale height (~ 24 km) from their observations of the emersion of β Sco C (latitude $-12°$).

However, the scale height observed at $-47°.2$ (NT3 curve of Hubbard *et al.* 1972) seems to be significantly greater than the scale height observed near $-57°$ (Fig. 8). Whether or not this is evidence for the type of latitude variation proposed by Hubbard *et al.* remains unclear, in view of the large error bars associated with the β Sco C points near $-10°$. The effective acceleration due to gravity, g, changes from about 2300 cm sec^{-2} near latitude 10°, to about 2600 near latitude 60°. Other things being equal, this effect increases scale heights ($H = kT/\mu g$) at 10° latitude to about 110% of their value at 60° latitude, and cannot explain the difference between the scale height measured at 47° latitude and those obtained near latitude 57°.

H. Conclusions

The best observations of the occultation of β Sco A by Jupiter indicate atmospheric scale heights of about 25 km near the 5×10^{14} cm^{-3} level at latitudes of 60°. A single high-quality observation at latitude 47° gives 31 km (Hubbard *et al.* 1972). It is clear that high-quality lightcurves with a high signal-to-noise ratio and good photometric stability are required to determine accurate refractivity profiles. The crucial task is to determine the zero levels of the lightcurves accurately, which means that the contribution of Jupiter's limb to the total signal must be kept constant by careful guiding or be monitored continuously. Since high photometric accuracy is necessary the data should be recorded on a medium such as magnetic tape, not on a chart recorder.

The agreement between the results of Hubbard *et al.* (1972) and those of Veverka *et al.* (1974a) is reasonable. The results of Vapillon *et al.* (1973) appear to give excessive scale heights and temperatures at levels near 10^{16} cm^3, and they are in disagreement with other data as well as with theoretical models. The lower limit of their error band is more nearly acceptable, and could be pushed lower still if the baseline is slightly more uncertain than they believe.

The area scanning method used by Hubbard *et al.* (1972) is very powerful for determining precise baselines, but is wasteful of light. The color-correction method introduced by Elliot *et al.* (1975b) is probably just as powerful and has the advantage that the star is being observed continuously throughout the event.

The advantages of simultaneous observations in several colors and at high time resolution deserve to be stressed. From observations at a time resolution of 0.01 sec in three colors, Elliot *et al.* (1974) were able to determine a value of the $[He]/[H_2]$ ratio ($0.16^{+0.19}_{-0.16}$) which is in agreement with other more recent determinations (Table III). In addition, Elliot and Veverka (1976) were able to show that near the 10^{14} cm^{-3} density level the refractivity gradient is parallel to the local gravity gradient over scales of at least several kilometers.

Refractivity profiles obtained from high-quality lightcurves are reliable at refractivity levels $\geq 10^{-10}$, unless the assumptions of the gravity gradient model turn out to be invalid as suggested by Young (1976). Since the mean molecular weight of the atmosphere at these levels is now known with tolerable accuracy (Table III), and since it is also now known that the atmosphere is well mixed at these levels, temperatures can be inferred from the refractivity profiles. Typical temperatures near $n \sim 5 \times 10^{14}$ cm^{-3} indicated by the data are about 170°K with an uncertainty of probably ±10–20°K. The observations also indicate that small-scale temperature fluctuations occur at these levels (Veverka *et al.* 1974a).

The signal-to-noise ratio of the β Sco C lightcurves is generally low, and it is unlikely that these curves can be used to infer accurate scale heights or temperature profiles. The refractivity profile of Veverka *et al.* (1974a) derived from the occultation of β Sco A should be extended to deeper levels by using the color-correction method of Elliot *et al.* (1975b).

III. RADIO OCCULTATIONS

Since the first experiment at Mars in 1965 by Mariner 4, the radio-occultation method has become a workhorse of planetary exploration. It gives information on the electron density in the ionosphere, and the total density in the neutral atmosphere at a much deeper level where pressures range from one to a thousand millibars. At Jupiter, both parts of the experiment ran into difficulties. In the lower ionosphere, the data were confused by multipath propagation, until they were laboriously sorted out by hand (Fjeldbo *et al.* 1975). Temperatures for the lower atmosphere were unacceptably high, and only recently has the explanation been found (Hubbard *et al.* 1975) in a small error of the geometry used for the data reduction. Corrected results are given in this volume by Kliore and Woiceshyn.[3] Here we shall discuss the general principles, focusing on an explanation of the cause of the difficulty, and on other possible problems.

---

[3]See p. 232.

A. The Neutral Atmosphere

The original temperature profiles are illustrated by Kliore *et al.* (1974*a*,*b*, 1975). Typically they reach temperatures of 600–700°K at pressures of 1–3 bars; these were greeted with general disbelief by nearly everyone familiar with our prior knowledge of Jupiter. Attempts were nevertheless made to reconcile the older data with high temperatures (e.g., McElroy 1975; Hunten 1975).

Eventually, the most telling contradiction of all came to the fore, involving the microwave emission of Jupiter. At 13 cm, which is the wavelength of the occultation experiment, the observed brightness temperature is 280°K. Gulkis and Poynter (1972) have shown that the spectrum from 1 to 30 cm is naturally explained by the opacity of ammonia, in the solar abundance, in the temperature profile of a conventional model. But whether or not this be strictly true, the lower atmosphere must be opaque to 13-cm radiation at the 280°K level. For an occultation experiment, which probes the atmosphere tangentially, the opaque level must be higher and cooler; but with allowance for measuring error, one can say that the occultation cannot probe deeper than the 280°K level. Any higher temperatures would make themselves highly visible to a microwave radiometer. (Optically thin regions, such as a stratosphere or ionosphere, are exempt from this prohibition but they are not in question here. Moreover, such a region normally enhances the microwave emission whereas a suppression would be required for Jupiter.) Because of this argument, suggested by Gulkis, it was known that an error must exist in the occultation results, long before that error was found.[4]

We can be almost as sure that some other error persists in the Pioneer-11 entry, which still shows high temperatures. (It is barely conceivable that the polar troposphere *is* very hot, since it is essentially invisible to a radio telescope on the earth.)

The basis of the error pointed out by Hubbard *et al.* is an extreme sensitivity to the exact geometry of the occultation. The principle is conveniently illustrated by Eq. (26), which for constant temperature may be written

$$H = \frac{kT}{\mu g} = v_s f \left(\frac{dt}{df} - \frac{\lambda m}{v_s^2} \right). \tag{27}$$

The scale height H, or temperature T, for an isothermal atmosphere is obtained from the Doppler frequency f and its time derivative. The spacecraft velocity, projected into the plane of the sky and then normal to the limb, is v_s; λ is the radio wavelength, and m the distance from limb to spacecraft. The difficulty is the cancellation in the parentheses of (27). Deep in the atmosphere, the difference of the two terms is only 5–10% of their value, and is very sensitive to the exact value of v_s. The same effect is discussed

---

[4]See p. 681.

from a slightly different viewpoint by Eshleman (1975). As he points out, $v_s^2/\lambda m$ is the rate of change of frequency that would be produced by bending of the ray around a fixed knife-edge (or an atmosphere with a scale height of zero). The information about atmospheric structure is all contained in the small difference.

It is helpful to look at some actual data, which are reproduced for Pioneer 10 entry in Fig. 10. The mean frequency for the previous 40 sec was taken as baseline, and the effect of the changing path through the ionosphere removed by use of a model derived from the earlier data. Three computed curves are also shown, displaced by a small, arbitrary amount, for different values of the velocity component v_s. The model atmosphere has a linear temperature gradient varying from 110°K at 100 mb to 244°K at 1 mb, adapted from the work of Wallace et al. (1974); the curves were obtained from Eq. (24). The steepest line is for the velocity v_s derived without allowance for Jupiter's oblateness. The value 18.4 km sec^{-1}, a 6% reduction, is that derived in the preliminary analysis of Hubbard et al. (1975). A further, totally arbitrary, reduction of 2% gives a curve nearly indistinguishable from the observations. This must not be taken as a serious suggestion; it is intended only as a further illustration of sensitivity. The mean geometry of the atmosphere can be calculated accurately, but there may still be problems with local deviations from spheroidal stratification.

Deep in the atmosphere, df/dt becomes very nearly equal to the constant $v_s^2/\lambda m$, either in (27) or its generalization (26). In Fig. 10, this behavior is seen in the tendency of the lines to become straight. The observed curve

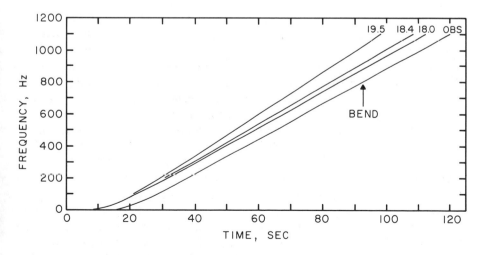

Fig. 10. Curves of frequency as a function of time for Pioneer 10 entry. The upper three were calculated from Eq. (26) for the velocities v_s shown. The fourth is the observed curve. (Data courtesy of A. Kliore.)

has a very slight downward bend at the point indicated, presumably corresponding to some feature of the temperature profile. No doubt this tiny deviation from straightness carries important information. The demands on the quality of the data, as well as the accuracy of the encounter geometry, are extreme.

A number of other difficulties are known. Fjeldbo *et al.* (1975), Eshleman (1975), and Kliore and Woiceshyn[5] discuss the changes and jumps in the drift rate of the spacecraft oscillator on Pioneers 10 and 11. (Steps are being taken to improve this situation on future missions.) The effect was particularly serious for the measurements of the ionosphere because of its great extent, but not negligible for the neutral atmosphere. Closed-loop operation on Pioneer 11 entry solved the problem for the ionosphere but ruined the lower-atmosphere experiment. Signal-to-noise ratio was a problem, particularly for the lowest part of the measurements, and is further discussed below. This too is being improved, but the problem must inevitably recur at lower levels of the atmosphere. A major difficulty is the need to allow for the ionosphere, which is traversed in two regions separated from the one measured by the experiment. The only satisfactory solution is the use of two frequencies, which allow unambiguous separation of the neutral and ionized components.

Other somewhat speculative perturbations have recently been suggested as offshoots from the search for the cause of the primary problem. They can come from atmospheric turbulence, from the characteristics of the data link (especially at low signal-to-noise ratio), and interactions between the two. Effects of turbulence, or multipath propagation, have been mentioned as a possibility by Kliore *et al.* (1974*b*), and have been studied by Hunten (unpublished), Hubbard and Jokipii (1975), and Young (1976). Woo *et al.* (1974) have already shown that some characteristics of turbulent density fluctuations can be obtained from occultation data. It is convenient to think in terms of ray theory, which is a useful approximation for scales larger than the Fresnel zone. In free space the radius of this zone is $\sqrt{\lambda m}$ or 5 km for Pioneer 10 entry. During occultation, the Fresnel zone is flattened, as has recently been emphasized by Young (1976). (The axial ratio of the flattened zone is the square-root of the intensity reduction due to the atmospheric defocusing; the height is therefore on the order of one kilometer. As Young points out, the ray is thus more and more sensitive to small scales of disturbance as it penetrates deeper.) The suggestion of Hubbard and Jokipii may be summarized as follows. As in the Fresnel construction, the theoretical ray path is one of minimum phase. Any refractive disturbance that causes the ray to take a different path must therefore cause an increased phase shift. Such disturbances are likely to be a function of height and can therefore in principle cause a perturbation of the Doppler frequency. One

---

[5]See p. 227.

possible consequence of this mechanism is a considerable spread in frequency of the observed signal, a question that is taken up below.

One can also imagine the ray being forced out of its "natural" path by some sort of obstacle. Gulkis originally made this suggestion, with the idea that the obstacle might be formed by ammonia absorption, but he abandoned it because the effect was too small, and of the wrong polarity, to cause the large discrepancy. Jokipii and Hubbard (1976) have shown that their effect can be regarded as caused by a diversion of the centroid of the ray bundle, which tends to avoid regions of stronger scattering. Thus, a gradient of scattering increasing downwards will force the ray upwards and increase the frequency in accord with Eq. (21). The important role of this relation has been stressed by Eshleman in unpublished criticism of Hubbard and Jokipii's (1975) original interpretation. In any case, as these authors point out, the effect is small for reasonable amounts of turbulence.

It should be noted that the effects commonly ascribed to turbulence may instead be due to stratified, static perturbations of refractive index, which are entirely possible in regions of high static stability. The duration of an occultation event is so short that it can only give a "snapshot" of a situation. Whether it is static or dynamic must be inferred from other evidence.

An understanding of possible problems in the radio link requires some knowledge of the nature of a phase-locked loop, or ϕLL. This device is the heart of the transponder, which was used for the Pioneer 11 entry, and is simulated in a computer for the reduction of all the data. A brief description of the latter process is given by Kliore *et al.* (1972) (see their Fig. 8, where the block is called *LLR*, or locked-loop receiver). As discussed in Appendix 1, a ϕLL can be regarded as a single-component Fourier analyzer in which information derived from the signal is used as feedback to control the frequency of a "local oscillator." This frequency is then the output of the device; it is also multiplied into the signal to give the amplitude of the latter. The derived information is therefore very limited; in particular, the rate of change of frequency cannot exceed some value, determined by a low-pass filter. The great virtue of this system is its ability to track a signal that is almost buried in noise. The limitation on slewing rate is largely removed by "steering" the local oscillator according to a prior estimate, or the equivalent process of heterodyning the signal against such an estimate to a near-constant beat frequency. This process can be iterated. The hard-wired ϕLL on Pioneer 11 could not use the "steering" procedure. It had to try to follow a signal like that shown in Fig. 10, but changing even more rapidly because of the closer encounter. It is strongly suspected that the ϕLL simply could not keep up, given the low signal strength from Earth due to attenuation by Jupiter's atmosphere. We understand that the Pioneer Project is undertaking some tests of this idea.

If there is a spread of frequencies due to multipath propagation, the ϕLL will normally lock to the strongest component, or a moderately strong one that is closest to its previous frequency (cf., Fjeldbo and Eshleman 1969; Fjeldbo 1973). If there is a momentary dropout, it will hold its frequency. If there is a ray crossing due to some kind of layering in the atmosphere, three valid frequencies will be simultaneously present for a moment, and the above comments still apply; only one will be tracked. Ray crossings are a common feature in stellar occultations (Wallace 1975; Young 1976) and there is no obvious reason why they should be absent at the deeper levels of the radio experiment.

A valuable guide to such problems is point-by-point Fourier analysis of the signal (Fjeldbo 1973; Fjeldbo et al. 1975). The ϕLL gives a better rejection of noise, but it gives too little information about what could be going wrong. When the signal-to-noise ratio is very low, there is no other recourse and the limitations must simply be accepted. When spectral analysis can be used the three frequencies of a ray crossing can be sorted out, but the required signal-to-noise ratio is considerably higher.

The data for pressures $\lesssim 20$ mb suffer from the starting-point errors discussed in Sec. I; the problem is well illustrated by Fig. 15 in the chapter by Kliore and Woiceshyn.[6] The difficulty of recovering absolute density from a noisy amplitude signal has been analyzed by Wallace (1975), and his results apply equally well to phase data. Under the best conditions the radio experiment actually measures absolute density, but this was not achieved by the Pioneers because absolute phase was lost as the ray traversed the ionosphere. Even at Mars, a much more tractable case, it has been found necessary to subtract a so-called "bias" (Kliore et al. 1972), a process equivalent to setting the atmospheric density equal to zero at some height. Probable causes are deviations from spherical (or spheroidal) symmetry due to meteorological effects, and similar asymmetries in the ionosphere.

The temperature profiles derived by Kliore and Woiceshyn[7] are still not the final ones, but major changes are not expected. They suggest a minimum (or tropopause) temperature of 105°K at 80–100 mb and a rise to 150°K at 10 mb. This is entirely consistent with Earth-based and infrared data and with computations of radiative balance.

The various results from the β Sco occultation can be used as a further source of information. At the pressure where they agree best (~ 0.01 mb, the half-intensity point) they give $\sim 170°$K, and the discussion in Sec. II suggests that the temperature rises at greater depths. A slow rise to the vicinity of 200°K at 0.1–1 mb is reasonable and a maximum as high as 250°K is conceivable; but most likely the entire region is nearly isothermal between 160 and 200°K. More details appear in this volume in the chapters by Wallace[8] and Hunten.[9]

---

[6]See p. 235.   [8]See p. 284.
[7]See p. 233.   [9]See p. 25.

B. The Ionosphere

This section is only a brief supplement to the chapter by Atreya and Donahue in this volume.[10] The data processing for the ionosphere was done separately from the neutral atmosphere and did not omit the oblateness of the limb (Fjeldbo *et al.* 1975). In any case it appears that the cancellation in the analog of (27) would not have been serious. Strong layering in the lower part gave multipath propagation, which was sorted out as well as possible with the aid of frequency spectra. Some ambiguities remain; in particular, the authors suggest that horizontal non-uniformities may be responsible for some of the effects observed. About half a dozen layers seem to be present, spaced 100–200 km.

The layering immediately calls to mind the terrestrial phenomenon of sporadic E, a major form of which is due to metallic ions, concentrated by wind shears with the aid of the geomagnetic field (Whitehead 1961; Axford and Cunnold 1966). Moreover, the presence of sodium at and around Io has already led to the suggestions that Na^+ forms the ionosphere of Io and could be present on Jupiter as well (Yung and McElroy 1975; Atreya *et al.* 1974). Although sporadic E usually shows only a single layer, the basic mechanism seems capable in principle of producing multiple layers. On the whole, this idea is more attractive than anything involving different ions in different layers. The segregation into layers is much more naturally explained by a mechanism involving atmospheric motions than by any kind of static equilibrium between ion production and loss.

The ionosphere as a whole is enormously extended. Its detectable thickness is more than 3000 km, and the topside scale height is around 600 km. The maximum electron density (over 10^5 cm^{-3}) is large for a planet so far from the sun. There was considerable difficulty with the Pioneer 10 data, because of possible jumps in drift rate of the oscillator in the radiation belt. Pioneer 11, at entry, eliminated this problem by operating in closed-loop mode, and gave an unambiguous scale height. The corresponding temperature for a plasma of protons and electrons is 750°K (Kliore *et al.* 1975).

This temperature is a big surprise indeed and will certainly require a lot of new thinking. The exospheric temperature predicted for the neutral atmosphere is only 10–20°K above the mesopause temperature, or not much more than 200°K (Strobel and Smith 1973). The ion and electron temperatures are expected to be the same, according to Henry and McElroy (1969). Even if this is not so, another heat source must be present to supplement the solar ultraviolet, and comparison of the observed and predicted temperature rises suggests that it may be some 40 times larger, or 0.5 erg cm^{-2} sec^{-1} global average. The radiation belt is one possibility, but certainly not the high-energy particles measured by the Pioneers, which would penetrate far

---

[10] See p. 304.

too deeply. Perhaps there is a low-energy population as well. Another energy source is waves and turbulence propagated from the lower atmosphere; this thought is encouraged by the irregularities that seem to be present in the stratospheric temperature profile (French and Gierasch 1974) and by the large eddy coefficient inferred from the Lyman-α experiment (Carlson and Judge 1974).[11] The Jovian thermosphere would thus be an analog of the solar corona. Prasad (1975) has published a model of the thermosphere heated by dissipation of postulated turbulent energy. Still other possibilities are ion drag due to the different rotational speeds of the neutral atmosphere and magnetic field, and meteoroid impact (if the particles lose their energy at a high enough altitude).

Acknowledgements. We are grateful to J. Elliot, V. Eshleman, P. Gierasch, A. Kliore, L. Vapillon, and L. Wallace for helpful comments, and to A. Kliore for access to unpublished data. This work was supported in part by NASA Grant NGR 33-010-082. Kitt Peak National Observatory is operated by the Association of Universities for Research in Astronomy, Inc., under contract with the National Science Foundation.

APPENDIX

Principles of a Phase-Locked Loop

The components of a phase-locked loop are a multiplier, a filter, and an oscillator whose frequency is controlled by the output of the filter. Applied to the multiplier are an input signal $A \sin \omega t$ and the oscillator output $2 \cos (\omega t + \phi)$, where ϕ is a phase difference, normally small. After filtration, or averaging, the output of the multiplier is

$$< 2A \sin \omega t \cos (\omega t + \phi) > \; = \; < A \sin (2\omega t + \phi) + A \sin (-\phi) > \quad (28)$$
$$= -A \sin \phi .$$

This quantity is an error signal that is used to control the frequency of the "local" oscillator. Normally ϕ remains small, fluctuating about zero, and the phase of the oscillator is locked to that of the input signal. The output frequency of the oscillator therefore represents accurately the signal frequency. If a measure of the amplitude is also required, the oscillator output is shifted 90° in phase, and applied, along with the signal, to a second multiplier and filter. The result is

$$< 2A \sin \omega t \sin (\omega t + \phi) > \; = \; < A \cos (2\omega t + \phi) - A \cos \phi >$$
$$= -A \cos \phi \quad (29)$$
$$\approx -A \text{ when } \phi \text{ is small.}$$

---

[11]See also p. 426.

The resemblance to a Fourier analysis is obvious; the operations (28) and (29) are identical, and the novelty is the use of (28) as a feedback control to track the incoming frequency.

REFERENCES

Atreya, S. K.; Donahue, T. M.; and McElroy, M. B. 1974. Jupiter's ionosphere: prospects for Pioneer 10. *Science* 184:154–156.

Axford, W. I., and Cunnold, D. M. 1966. The wind-shear theory of temperate zone sporadic E. *Radio Sci.* 1:191–198.

Baum, W. A., and Code, A. D. 1953. A photometric observation of the occultation of σ Arietis by Jupiter. *Astron. J.* 58:108–112.

Berezne, J.; Combes, M.; Laporte, R.; Lecacheux, J.; and Vapillon, L. 1975. The occultation of Beta Scorpii by Jupiter. III. Discussion of the photometric results. *Astron. Astrophys.* 40: 85–90.

Bhattacharyya, J. C. 1972. Occultation of Beta Scorpii by Jupiter on May 13, 1971. *Nature Phys. Sci.* 228:55–56.

Carlson, R. W., and Judge, D. L. 1974. Pioneer 10 ultraviolet photometer observations at Jupiter encounter. *J. Geophys. Res.* 79:3623–3633.

Colin, L. (ed.) 1972. *Mathematics of profile inversion.* NASA Tech. Mem. TM X-62, 150, Washington, D.C.

Combes, M.; Lecacheux, J.; and Vapillon, L. 1971. First results of the occultation of β Sco by Jupiter. *Astron. Astrophys.* 15:235–238.

Combes, M.; Vapillon, L.; and Lecacheux, J. 1975. The occultation of β Scorpii by Jupiter. IV. Divergences with other observers in the derived temperature profiles. *Astron. Astrophys.* 49:399–403.

de Vaucouleurs, G., and Menzel, D. H. 1960. Results of the occultation of Regulus by Venus, July 7, 1959. *Nature* 188:28–33.

Dütsch, H. U. 1971. Photochemistry of atmospheric ozone. *Advances in Geophysics, 15.* (H. E. Landsberg and J. Van Mieghem, eds.), pp. 219–322. New York: Academic Press.

Elliot, J.; Rages, K.; and Veverka, J. 1975a. The occultation of Beta Scorpii by Jupiter. VI. The masses of Beta Scorpii A_1 and A_2. *Astrophys. J.* 197:L123–L126.

Elliot, J., and Veverka, J. 1976. Jupiter: Occultation lightcurve spikes as atmospheric probes. *Icarus* (special Jupiter issue). In press.

Elliot, J.; Wasserman, L. H.; Veverka, J.; Sagan, C.; and Liller, W. 1974. The occultation of Beta Scorpii by Jupiter. II. The hydrogen-helium abundance in the Jovian atmosphere. *Astrophys. J.* 190:719–729.

––––––. 1975b. Occultation of β Scorpii by Jupiter. V. The emersion of β Scorpii C. *Astron. J.* 80:323–332.

Eshleman, V. R. 1973. The radio occultation method for the study of planetary atmospheres. *Planet. Space Sci.* 21:1521–1531.

––––––. 1975. Jupiter's atmosphere: problems and potential of radio occultation. *Science* 189: 876–878.

Fabry, Ch. 1929. Le rôle des atmosphères dans les occultations par les planètes. *J. des Observateurs* 12:1–10.

Fairall, A. P. 1972. Symmetry of flashes during the Jovian occultation of β Scorpii. *Nature* 236:342.

Fjeldbo, G. 1973. Radio occultation experiments planned for Pioneer and Mariner missions to the outer planets. *Planet. Space Sci.* 21:1533–1547.

Fjeldbo, G., and Eshleman, V. R. 1965. The bistatic radar occultation method for the study of planetary atmospheres. *J. Geophys. Res.* 70:3213–3226.

———. 1969. The atmosphere of Venus as studied with the Mariner 5 dual radio frequency occultation experiment. *Radio Sci.* 4:879–897.

Fjeldbo, G.; Kliore, A. J.; and Eshleman, V. R. 1971. The neutral atmosphere of Venus as studied with the Mariner V radio occultation experiments. *Astron. J.* 76:123–140.

Fjeldbo, G.; Kliore, A.; Seidel, B.; Sweetnam, D.; and Cain, D. 1975. The Pioneer 10 radio occultation measurements of the ionosphere of Jupiter. *Astron. Astrophys.* 39:91–96.

Freeman, K. C., and Lyngå, G. 1970. Data for Neptune from occultation observations. *Astrophys. J.* 160:767–780.

Freeman, K. C., and Stokes, N. R. 1972. The occultation of β Sco by Jupiter. *Icarus* 17: 198–201.

French, R. G., and Gierasch, P. J. 1974. Waves in the upper Jovian atmosphere. *J. Atmos. Sci.* 31:1707–1712.

Goldsmith, D. W. 1963. Differential refraction in planetary atmospheres with linear scale height gradients. *Icarus* 2:341–349.

Gulkis, S., and Poynter, R. 1972. Thermal radio emission from Jupiter and Saturn. *Phys. Earth Planet. Interiors* 6:36–43.

Henry, R. J., and McElroy, M. B. 1969. The absorption of extreme ultraviolet solar radiation by Jupiter's upper atmosphere. *J. Atmos. Sci.* 26:912–917.

Houck, J. R.; Schaack, D.; Reed, R. A.; Pollack, J.; and Summers, A. 1975. Jupiter: its infrared spectrum from 16 to 40 microns. *Science* 189:720–722.

Hubbard, W. B.; Hunten, D. M.; and Kliore, A. 1975. Effect of the Jovian oblateness on Pioneer 10/11 radio occultations. *Geophys. Res. Lett.* 2:265–268.

Hubbard, W. B., and Jokipii, J. R. 1975. Effects of turbulence on radio-occultation scale heights. *Astrophys. J.* 199:L193–L196.

Hubbard, W. B.; Nather, R. E.; Evans, D. S.; Tull, R. G.; Wells, D. C.; van Citters, G. W.; Warner, B.; and Vanden Bout, P. 1972. The occultation of Beta Scorpii by Jupiter and Io. I. Jupiter. *Astron. J.* 77:41–59.

Hunten, D. M. 1975. Planetary atmospheres. *Atmospheres of earth and the planets.* (B. M. McCormac, ed.) Dordrecht-Holland: D. Reidel Publishing Co.

Hunten, D. M., and McElroy, M. B. 1968. The upper atmosphere of Venus: the Regulus occultation reconsidered. *J. Geophys. Res.* 73:4446–4448.

Jokipii, J. R., and Hubbard, W. B. 1976. Effects of turbulence on radio occultations. In draft.

Kliore, A.; Cain, D. L.; Fjeldbo, G.; Seidel, B. L.; and Rasool, S. I. 1974a. Preliminary results on the atmospheres of Io and Jupiter from the Pioneer 10 S-band occultation experiment. *Science* 183:323–324.

———. 1974b. The atmospheres of Io and Jupiter measured by the Pioneer 10 radio occultation experiment. Paper no. II-VII · 1 · 4, 17th Plenary Meeting of COSPAR, São Paulo, Brazil, June 1974.

Kliore, A. J.; Cain, D. L.; Fjeldbo, G.; Seidel, B. L.; Sykes, M. J.; and Rasool, S. I. 1972. The atmosphere of Mars from Mariner 9 radio occultation measurements. *Icarus* 17:484–516.

Kliore, A.; Fjeldbo, G.; Seidel, B.; Sesplaukis, T.; Sweetnam, D.; and Woiceshyn, P. 1975. Atmosphere of Jupiter from the Pioneer 11 S-band occultation experiment: preliminary results. *Science* 188:474–476.

Kovalevsky, J., and Link, F. 1969. Occultation par Neptune de BD-17° 4388. *Astron. Astrophys.* 2:398–412.

Larson, S. M. 1972. Photographic observations of the occultation of Beta Scorpii by Jupiter. *Contr. Bosscha Obser.* No. 45.

Liller, W.; Elliot, J. L.; Veverka, J.; Wasserman, L. H.; and Sagan, C. 1974. The occultation of Beta Scorpii by Jupiter. III. Simultaneous high time-resolution records at three wavelengths. *Icarus* 22:82–104.

McElroy, M. B. 1969. Atmospheric composition of the Jovian planets. *J. Atmos. Sci.* 26: 798–812.

——. 1975. Jupiter. *Atmospheres of earth and the planets.* (B. M. McCormac, ed.) Dordrecht-Holland: D. Reidel Publishing Co.

O'Leary, B. 1962. Frequencies of occultations of stars by planets, satellites, and asteroids. *Science* 175:1108–1112.

Orton, G. S. 1975. The thermal structure of Jupiter. I. *Icarus* 26:125–141.

Pagel, B. E. J. 1973. Stellar and solar abundances. *Cosmochemistry.* (A. G. W. Cameron, ed.), pp. 1–21. Dordrecht-Holland: D. Reidel Publ. Co.

Pannekoek, A. 1903. Über die Erscheinungen, welche bei einer Sternbedeckung durch einen Planeten auftreten. *Astron. Nachr.* 164:5–10.

Phinney, R. A., and Anderson, D. L. 1968. On the radio occultation method for studying planetary atmospheres. *J. Geophys. Res.* 73:1819–1827.

Prasad, S. S. 1975. Possible new Jovian thermospheric models. *Astrophys. J.* 200:L171–L174.

Rages, K.; Veverka, J.; Wasserman, L. H.; and Freeman, K. C. 1974. The upper atmosphere of Neptune: an analysis of occultation observations. *Icarus* 23:59–65.

Strobel, D. F., and Smith, G. R. 1973. On the temperature of the Jovian thermosphere. *J. Atmos. Sci.* 30:718–725.

Vapillon, L.; Combes, M.; and Lecacheux, J. 1973. The β Scorpii occultation by Jupiter. II. The temperature and density profiles of the Jovian upper atmosphere. *Astron. Astrophys.* 29:135–149.

Veverka, J.; Wasserman, L. H.; Elliot, J.; Sagan, C.; and Liller, W. 1974a. The occultation of β Scorpii by Jupiter. I. The structure of the Jovian upper atmosphere. *Astron. J.* 79:73–84.

Veverka, J.; Wasserman, L. H.; and Sagan, C. 1974b. On the upper atmosphere of Neptune. *Astrophys. J.* 189:569–575.

Wallace, L. 1975. On the 1968 occultation of BD-17° 4388 by Neptune. *Astrophys. J.* 197: 257–261.

Wallace, L., and Hunten, D. M. 1973. The Lyman-Alpha albedo of Jupiter. *Astrophys. J.* 182:1013–1031.

Wallace, L.; Prather, M.; and Belton, M. J. S. 1974. The thermal structure of the atmosphere of Jupiter. *Astrophys. J.* 193:481–493.

Wasserman, L. H. 1974. The occultation of β Scorpii by Jupiter. IV. Diurnal temperature variations and the methane mixing ratio in the Jovian upper atmosphere. *Icarus* 22:105–110.

Wasserman, L. H., and Veverka, J. 1973. On the reduction of occultation light curves. *Icarus* 20:322–345.

Whitehead, J. D. 1961. The formation of the sporadic-E layer in the temperate zones. *J. Atmos. Terr. Phys.* 20:49–58.

Woo, R.; Ishimaru, A.; and Kendall, W. B. 1974. Observations of small-scale turbulence in the atmosphere of Venus by Mariner 5. *J. Atmos. Sci.* 31:1698–1706.

Young, A. T. 1976. Scintillation during occultations by planets. I. An approximate theory. *Icarus* (special Jupiter issue). In press.

Yung, Y. L., and McElroy, M. B. 1975. The atmosphere and ionosphere of Io. *Astrophys. J.* 196:227–250.

THE THERMAL STRUCTURE OF JUPITER IN THE STRATOSPHERE AND UPPER TROPOSPHERE

L. WALLACE
Kitt Peak National Observatory

The upper tropospheric part of the thermal structure seems to be quite well established because studies of the thermal infrared spectrum, studies of the thermal microwave spectrum, and radiative-convective models all yield very similar results. Currently useful information on the stratosphere, limited to observations of stellar occultations and the 7.8 μm band of CH₄, is not sufficiently potent to establish the stratospheric structure in any detail. Radiative-convective models exist which are compatible with these observations but the models may not contain enough of the relevant physics to be realistic. The Pioneer radio occultation experiment may turn out to be of importance in this region.

In this review of the average thermal structure of Jupiter I will attempt to collate information obtained with a variety of techniques and involving a number of physically separable modes of heat transport at heights above pressures of approximately 3 atmospheres. The most recent review of this topic seems to be contained in a review of the upper atmosphere of Jupiter by Hunten (1969); since then the thermal structure has become much better defined. There is also very clear evidence that the thermal structure varies from point to point across the disk (see, e.g., Ingersoll[1] *et al.* 1975; or Orton[2] 1975*a,b*). This review, however, will be restricted to a consideration of the mean global structure.

I will more or less follow Hunten's (1969) layering nomenclature. The lowest part, at pressures $\gtrsim 10^5$ dynes cm^{-2} (0.1 bar), is the troposphere where the temperatures increase rapidly inward. Radiation by He and H$_2$ seems to be the dominant heat transport mechanism in the upper troposphere down to 5×10^5 or 10^6 dynes cm^{-2}; convection takes over below that point. The tropopause occurs at a temperature minimum of between 100

---

[1]See also p. 197.
[2]See p. 206.

and 120°K. Above that, in the stratosphere, radiative transport is still dominant and the temperature increases with height because of the rapidly decreasing ability of He and H_2 to radiate the solar flux deposited in the atmosphere. The stratopause is taken to occur at $\sim$ 10 dynes cm^{-2} and $T \sim$ 180°K at which point insufficient collisions occur to maintain local thermodynamic equilibrium. Above the stratopause, in the thermosphere, heat is conducted downward to a point where radiation can occur. The term "mesosphere" is not used here because a temperature maximum between the tropopause and the base of the thermosphere has not been definitely established.

Consideration of the thermal structure requires knowledge of the atmospheric composition. Since the composition is not known with precision I assume that the needed Jovian atmospheric abundances are consistent with solar abundances, following current cosmological thinking (e.g., Cameron 1973). I have adopted[3] $[He]/[H_2] = 0.11$, as suggested by Hunten and Münch (1973). Other values of this ratio determined from Hirshberg (1973) and Allen (1975) range from 0.10 to 0.20. The best published measurements of this quantity on Jupiter appear to be $0.19 \, ^{+0.35}_{-0.19}$, obtained by Elliot et al. (1974) from the β Sco occultation and $0.18 \, ^{+0.46}_{-0.12}$, obtained by Carlson and Judge (1974) from observations of the hydrogen Lyman-α line and the He I 584 Å line. More recently, Judge[4] et al. have revised the ultraviolet result down to the range $0.05 - 0.10$, and studies in the thermal infrared have yielded 0.14 ± 0.08 (Orton[5]) and ~ 0.1 (Houck et al. 1975). The stellar occultation and ultraviolet results pertain to the high stratosphere while the infrared results pertain to pressures $\sim 10^6$ dynes cm^{-2} and there is no reason to expect departures from a constant mixing ratio up to the turbopause. For $[CH_4]/[H_2]$ a value of 7×10^{-4} is used, consistent both with solar abundance (Allen 1975) and studies of the $3\nu_3$ methane band and H_2 quadrupole lines on Jupiter (Owen and Mason 1968; Belton 1969; Fink and Belton 1969) at about 10^6 dynes cm^{-2}. I also take this ratio constant through the atmosphere. Deep in the troposphere I have used $[NH_3]/[H_2] = 1.8 \times 10^{-4}$, obtained from solar work (Allen 1975) but consistent with microwave studies of Jupiter (see Sec. VI). NH_3 is considered mixed up to the point at which it would be supersaturated ($5 \times 10^5 - 6 \times 10^5$ dynes cm^{-2}); above that point it is assumed to follow the saturation vapor pressure curve. At some point above the tropopause, photochemical destruction of NH_3 would occur (Strobel 1975) but the present discussion does not require consideration of that problem.

The location of the turbopause is not well established but with Wallace and Hunten's (1973) definition of it as the point at which the molecular and eddy diffusion coefficients are equal and the current estimate that the eddy

---

[3]Square brackets are used to indicate number densities.
[4]See p. 419.
[5]See p. 211.

diffusion coefficient is $\sim 10^8$ cm^2 sec^{-1} (Carlson and Judge 1974), it occurs at pressures much less than 1 dynes cm^{-2}. This is high enough to warrant the assumption that He, H_2 and CH_4 are completely mixed throughout the whole pressure range of interest here. It is conceivable that photochemical destruction could cause the CH_4 mixing ratio to be less than completely mixed below the turbopause (e.g., Hunten, personal communication; Prinn and Owen[6]) but currently there is no evidence for such an effect.

I. THERMAL INFRARED SPECTRUM, 13–200 μm

This region contains the bulk of the thermal radiation which arises almost entirely from the troposphere. The observation of the broad-band flux from 1.5 to 350 μm by Aumann et al. (1969) gave $T_e = 134 \pm 4°$K, where T_e is the effective temperature, a basic atmospheric parameter. Narrower-band observations and a model of the sort calculated by Trafton (1965) (see Sec. III) are compared in Fig. 1. The calculated spectrum is derived from a tropospheric temperature structure for a He-H_2 atmosphere in radiative-convective equilibrium with $T_e = 134°$K. The general temperature level of the model is set by T_e but the variation with frequency is defined by the model temperature variation and the frequency dependence of the pressure-induced absorption coefficient of the He-H_2 mixture of the model. Minimum temperatures occur at 350 and 600 cm^{-1}, positions of maxima in the absorption coefficient corresponding to the S(0) and S(1) rotational H_2 lines. The broadened rotational lines dominate the spectrum from 200 to 900 cm^{-1}; the plateau from 100 to 200 cm^{-1} and the abrupt temperature increase below 100 cm^{-1} are due to the translational part of the absorption coefficient. The discrepancy between the model and the datum labelled "b" at 35 μm is not a problem at this point because NH_3 absorption, omitted from the model, would diminish or eliminate the temperature rise (Taylor 1972) and the plotted brightness temperature, although entirely reasonable, was an assumed calibration for a study of other planets. Without recourse to a model, it is apparent from the figure that the temperature drops to at least 120°K.

The higher resolution observations in this region given by Houck et al. (1975) actually turn down at the short frequency end rather than rising. In addition, their spectrum shows clear temperature minima at the location of the S(0) and S(1) lines and, since the ratio of the rotational to the translational part of the absorption coefficient depends on [He]/[H_2], has yielded a measure of that mixing ratio. An unexplained feature of the spectrum by Houck et al. is a third temperature minimum between the S(0) and S(1) lines at 23 μm.

One of the important results of the Pioneer infrared experiment has been a determination of the global average $T_e = 125 \pm 3°$K (Ingersoll[7] et al. 1975),

---

[6]See p. 356.
[7]See p. 202.

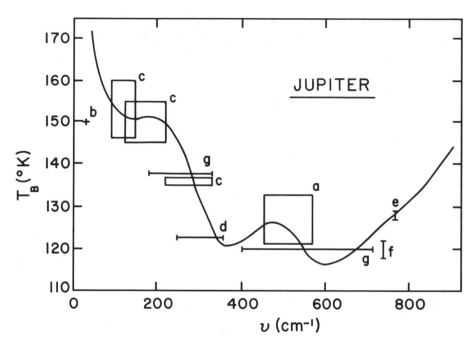

Fig. 1.   Observations of the thermal infrared spectrum compared with a model spectrum. The observations are from *a)* Low (1965), *c)* Armstrong *et al.* (1972), *d)* Low *et al.* (1973), *e)* Aitken and Jones (1972), *f)* Gillett *et al.* (1969), and *g)* Chase *et al.* (1974). The point *b)* was assumed by Harper *et al.* (1972). The observations and model are discussed in Secs. I and III. The figure is from Wallace (1975*b*).

distinctly lower than the groundbased result of Aumann *et al.* (1969) of $134 \pm 4°K$, but Low (verbal communication, and also see p. 204) has suggested that the groundbased result might be revised down to 129°K. Since the Pioneer experiment used a relatively more direct reference to a black body for calibration than the groundbased work it may turn out that 125°K is the better measurement. If it should turn out that way the model spectrum of Fig. 1 would be adjusted downward by 9°K, to a first approximation. The figure then suggests that the narrower-band observations may be a little too hot also. A resolution of this problem will have to wait for further observations. The Pioneer infrared results are further discussed in Sec. IV.

II.   THERMAL INFRARED SPECTRUM, 5–13 μm

The spectrum is complex in this region and has important implications regarding the stratosphere and troposphere. The spectrum obtained by Gillett *et al.* (1969) in this region is given in the lower part of Fig. 2. At longest wavelengths the spectrum is dominantly due to the short wavelength tail of the rotational pressure-induced absorption coefficient of the He-H$_2$

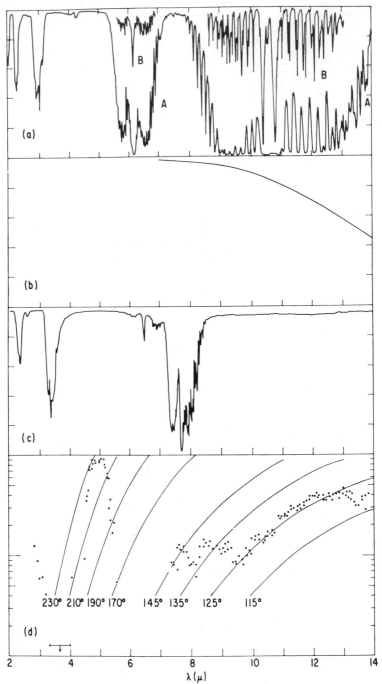

Fig. 2. The spectrum of Gillett *et al.* (1969), given by filled circles in panel *d*), compared with transmission spectra of NH₃ (panel *a*), H₂ (panel *b*), and CH₄ (panel *c*).

atmosphere (cf. Fig. 1). The 10 μm region is strongly influenced by the 10 μm absorption band of NH_3 and the 7.8 μm region is undoubtedly dominated by the 7.8 μm band of CH_4. Other absorption bands of CH_4 and NH_3 are present at short wavelengths. The region at 5 μm appears to be relatively free of band absorption but the brightness variation across the disc (e.g., Westphal et al., 1974) indicates that the emission arises from greater depths in the belts than in the zones.

From this spectrum Gillett et al. (1969) accurately identified the minimum brightness temperature (115°K at 13.5 μm) with an upper limit to the minimum temperature. Also, because of the large opacity in the 7.8 μm band of CH_4, they concluded that the 7.8 μm emission arose from above the temperature minimum and consequently identified the brightness temperature in this band of 145°K as a lower limit to the temperatures reached in the stratosphere. The observation by Gillett and Westphal (1973) of limb brightening at 7.94 μm confirmed the stratospheric origin of this emission.

Although the higher resolution work of Aitken and Jones (1972) clearly shows 10 μm NH_3 absorption lines, the first studies of this band are by Wildey and Trafton (1971) and Trafton and Wildey (1974) from limb darkening curves integrated over the 8 to 14 μm region. The observations were compared with Trafton-type models (see Sec. III) containing He-H_2 pressure-induced opacity and the 10 μm NH_3 opacity in which the NH_3 followed the saturated vapor pressure curve. The end result was a determination of the effective temperature as 135 ±4°K, in excellent agreement with the direct earth-based measurement of 134 ±4°K cited above.

Orton (1975b) has obtained synthetic spectra in the 7−14 μm region which are in good agreement with an improved unpublished version of the spectrum of Gillett et al. (1969) circulated by Gillett, and the spectrum of Aitken and Jones (1972). The synthetic spectra were obtained by first deducing the required thermal structure from the spectrum and the 7.94 μm limb darkening data of Gillett and Westphal (1973) and then using the deduced thermal structure to calculate a more detailed spectrum including gaseous CH_4 and NH_3 and solid NH_3 opacities. An opaque cloud was included in the zones at a pressure of 6×10^5 dynes cm^{-2} to account for the behavior of the 5 μm brightness. The solid NH_3 provided the opacity in the 8.2−9.5 μm region where neither gaseous CH_4 nor NH_3 are sufficient. The agreement with the spectra is quite good but the fit to the center-to-limb scans at 8.15 μm shows a residual problem. Orton's (1975b) deduced thermal structure is discussed in Sec. IV.

The high-resolution data of Lacy et al. (1975) in the 10 μm band of NH_3 are very puzzling. I will demonstrate in later sections of this paper that the thermal structure of the troposphere where this band is formed is quite well defined by infrared observations beyond $\sim$ 13 μm and by the microwave spectrum and that, in the region of saturation, NH_3 follows the saturated vapor pressure curve to probably better than a factor of two. As I understand it, the spectrum at 4 cm^{-1} resolution of Lacy *et al.* is in agreement with this picture, but at 0.5 cm^{-1} resolution discrepancies appear which require either a much slower temperature gradient if NH$_3$ follows the saturation curve or very substantial sub-saturation ($\sim$ 10 times) of NH$_3$ if the temperature structure follows that deduced from the infrared and microwave spectra. As Trafton suggested in discussion of the paper by Lacy *et al.*, the difficulty may be due to errors in the line shape used by Rank in the calculation of his synthetic spectra.

Higher resolution observations in the 11-13 μm region by Ridgway (1974*a,b*) have revealed the 12.2 μm band of C$_2$H$_6$ and the 13.7 μm band of C$_2$H$_2$ in emission giving strong support to the idea that the temperature in the stratosphere must be considerably higher than the minimum temperature. The bands are weak enough that the general run of the continuum level indicated in Fig. 2 is not grossly affected.

III. RADIATIVE–CONVECTIVE MODELS

Trafton's (1965, 1966) calculations of the H$_2$-H$_2$ and He-H$_2$ pressure-induced absorption coefficients and the Local Thermodynamic Equilibrium (LTE) models he obtained with them (Trafton 1965, 1967; Trafton and Stone 1974) have become generally accepted as being good representations of the tropospheric structure. Radiative equilibrium models were calculated for constant flux, defined by the effective temperature, and since it is expected that dynamical effects would wipe out a super-adiabatic gradient, an adiabatic lapse rate is assumed below the point at which the radiative gradient exceeds the adiabatic. Such a model for Jupiter given by Trafton and Stone (1974) is shown as the curve of Fig. 3 which terminates at a pressure of 3×10^3 dynes cm^{-2}. The curve should actually be extended through the stratosphere at a temperature of 106°K. Absorption by NH$_3$ and pressure-induced absorption by CH$_4$-H$_2$ is not significant for these models (Trafton 1967, 1972; Fox and Ozier 1971).

This type of model fails to give the required high stratospheric temperature because it includes no stratospheric heating. It is, however, very good in the troposphere as is indicated by the agreement between the observations and the model spectrum in the 30 to 800 cm^{-1} region (see Fig. 1) which arises almost entirely from the troposphere. Specifically, the comparison shows that the He-H$_2$ opacity assumed in the model is the dominant thermal opacity.

Gillett *et al.* (1969) suggested that the high stratospheric temperatures were maintained by the balance of solar flux deposition in the 3.3 μm band of CH$_4$ and reradiation in the 7.8 μm band. Hogan *et al.* (1969) attempted to include this mechanism in radiative-convective models of He-H$_2$-CH$_4$ atmospheres in the stratosphere but Hogan (personal communication) indicates that errors in their calculations have been found. The good agreement be-

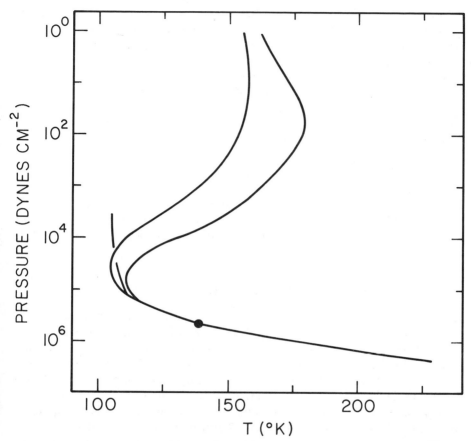

Fig. 3. Radiative-convective models with and without methane heating. The model with no stratospheric temperature inversion is from Trafton and Stone (1974) for $[He]/[H_2] = 0.1$, $[CH_4]/[H_2] = 0$ and $T_e = 135°K$. The convective region for this model falls below 4.7×10^5 dynes cm^{-2}, indicated by the filled circle. The two hotter models are from Wallace *et al.* (1974) for $[He]/[H_2] = 0.11$, $[CH_4]/[H_2] = 7 \times 10^{-4}$ and $T_e = 134°K$. The model exhibiting intermediate stratospheric temperatures includes heating and cooling only in the 3.3 and 7.8 μm bands of CH_4; the hottest model also includes heating in the weaker CH_4 bands.

tween the results of Cess and Khetan (1973) and Wallace *et al.* (1974) for this problem in which only the 3.3 and 7.8 μm bands are included gives confidence that their results are essentially correct, particularly considering their very different approaches to the problem. This model is indicated by the curve at intermediate stratospheric temperatures in Fig. 3. It gives only about $\frac{1}{8}$ the intensity observed in the 7.8 μm band. In order to obtain higher temperatures, Wallace *et al.* (1974) added the heating due to deposition of solar flux in the weaker infrared and visible CH_4 absorption bands. This model, which yields the highest temperature stratosphere of Fig. 3, gives an intensity in the 7.8 μm band only $\sim 25\%$ low.

In the troposphere the He-H_2 thermal opacity dominates over that of CH_4 so that the only difference between the models is that in those with solar heat deposition in the methane bands the lower troposphere does not have to transport quite as much flux and is therefore slightly cooler. This effect, which is not noticeable in Fig. 3, amounts to about $10°K$ at $\sim 10^6$ dynes cm^{-2}. However, since the modelling of the cloud structure by Wallace *et al.* was very crude, the true magnitude of the effect is uncertain. In the stratosphere the model including the 3.3 μm, 7.8 μm, and weaker CH_4 bands is preferred over that which includes only the 3.3 and 7.8 μm bands of CH_4 because it is in better agreement with the observed 7.8 μm band. However, the stratospheric part of the model, as opposed to the tropospheric, remains very uncertain because not all of the heating and cooling mechanisms have been included. Both cooling in the 12.2 μm ethane band and 13.7 μm acetylene band, which have been observed in emission by Ridgway (1974*a,b*), and heating by means of absorption of solar flux by aerosol or dust particles, could be important.

IV. INVERSION OF THERMAL INFRARED SPECTRA

An alternative to comparing observed and model thermal infrared spectra as a test of possible temperature profiles is to perform a numerical inversion of the observed thermal spectrum to obtain the temperature profiles. This method has the advantage that it does not use the assumption of radiative-convective equilibrium. Detailed relaxation methods have been developed for this inversion by Chahine (1968, 1970, 1974) and Smith (1970).

Ohring (1973) has applied Smith's method to observations of Jupiter in the 7.8 μm band of CH_4 and in the 12 μm region assuming that the latter region was dominated by He-H_2 pressure-induced opacity. The CH_4 band emission, arising mostly from the stratosphere, largely constrains the stratospheric part of the profile; observations at $\gtrsim 12$ μm in the He-H_2 continuum constrain the profile below the temperature minimum. Since Ohring used essentially the same opacities and observational constraints as Wallace *et al.* (1974) in their modelling one might expect the two approaches to yield essentially identical profiles but Fig. 4 shows that this is not the case above the temperature minimum. [Wallace and Smith (1975) have noted that the discrepancy below the temperature minimum claimed by Wallace *et al.* is removed if Ohring's profile is plotted on a pressure scale which is appropriate to the details of Ohring's calculation.] The results of additional inversions by Wallace and Smith (1975) and Orton (1975*b*) are also indicated.

The important aspect of Fig. 4 is that all of the profiles shown predict essentially the same fluxes in the 7.8 μm CH_4 band and in the He-H_2 continuum from ~ 14 to 100 μm in good agreement with the published observations (Figs. 1 and 2). The details of the different observed fluxes and inversion methods used are of secondary importance. The question to be addressed here is which, if any, of these stratospheric temperature structures is to be believed.

The cause of the ambiguity is apparently well known to those involved in inverting terrestrial infrared spectra (see, e.g., Chahine 1974). The intensity I emerging at direction cosine μ and frequency ν is

$$I(\mu, \nu) = \int_0^\infty B(\nu, p) \exp[-t(\nu, p)/\mu] \, dt(\nu, p)/\mu$$

where B and t are the Planck function and vertical optical depth at pressure p. Very approximately, if B varies slowly with p in the region $t/\mu \sim 1$, a mean value of B at $t/\mu = 1$ can be removed from the integral to yield the result that the temperature at $t/\mu = 1$ is approximately the observed brightness temperature. The numerical inversion takes care of the temperature gradient. This straightforward behavior is responsible for the good agreement between the different results in Fig. 4 in the troposphere which rely on spectral data at wavelengths $\gtrsim 12 \ \mu m$. But while the ratio of B at 180°K to that at 110°K (the approximate range of Fig. 4) is $\lesssim 70$ at $\lambda \gtrsim 12 \ \mu m$, that ratio

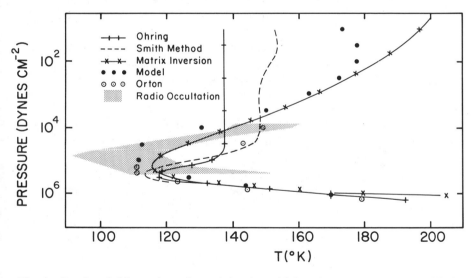

Fig. 4. Results of different inversions of the thermal infrared spectrum compared with the hottest model of Fig. 3 and the current results of the Pioneer radio occultation experiments. The curve labelled "Ohring" gives the result of Ohring's (1973) inversion, those labelled "Smith Method" and "Matrix Inversion" give the results of different inversions by Wallace and Smith (1975) and that labelled "Orton" gives Orton's (1975b) result. The radio occultation results are from Kliore and Woiceshyn.[8] The figure is from Wallace and Smith (1975) with Orton's and Kliore and Woiceshyn's results added.

---

[8]See p. 235.

is 700 at 7.8 μm. Thus at 7.8 μm the bulk of the intensity can come from near $t/\mu = 1$ if the temperature gradients are small *or* from a location of high temperature well away from $t/\mu = 1$. In comparing the results labelled "Ohring" and "Matrix Inversion" note that considerably more 7.8 μm emission comes from "Matrix Inversion" above 10^4 dynes cm^{-2}; and in order to obtain the same total intensity "Matrix Inversion" must be colder than "Ohring" between 10^4 dynes cm^{-2} and the temperature minimum.

Orton (1975*b*) assumed a constant lapse rate at pressures less than 3 $\times$ 10^4 dynes cm^{-2} which extrapolates to 177°K at 10^2 dynes cm^{-2}. If he included the contribution to the intensity from the upper (hot) stratosphere, a definite inconsistency exists between Orton's result and those of Ohring and Wallace and Smith. If, however, Orton did not actually include this contribution there is no conflict. This problem has not been resolved.

Thus while the 7.8 μm band establishes a thermal inversion, the current thermal infrared observations have not been adequate to define the details. However, Wallace and Smith (1975) have argued that the more acceptable of these results are the ones which achieve $\sim$ 180°K at $\sim$ 10 dynes cm^{-2} in essential agreement with the optical occultation results (see Sec. VI). While that consideration favors the "Matrix Inversion" and "Model" profiles, a very irregular run of temperature between 10 and 10^5 dynes cm^{-2} could also satisfy both the infrared and the optical occultation results. It seems unlikely that the ethane and acetylene emissions could be used as a further constraint on the stratospheric structure because they are photochemical products whose variation with height are not known (Strobel, 1973, 1975). High spectral and spatial observations in the 7.8 μm band of CH_4 might, however, be very useful.

The results of inverting longer wavelength data are much less sensitive to the assumed overlying structure and should give a good definition of the profile in the $10^5 - 10^6$ dynes cm^{-2} region. Data much superior to that used in Fig. 4 have been subsequently analyzed; by Houck *et al.* (1975) using a much higher resolution earth-based spectrum of the 16−40 μm region than indicated in Fig. 1, and by Orton[9] (1975*b*) using the Pioneer 10 and 11 limb darkening data in bands at 20 and 45 μm. These results will supersede any others in this pressure range. An intercomparison of these two would be very useful but I have not been able to do that here.

V.   THE THERMAL MICROWAVE SPECTRUM

Gulkis[10] gives a good review of this part of the spectrum but I want to briefly emphasize its importance to the thermal structure. The figures used in this context do not contain the newer data that Gulkis' do, but they will suffice.

---

[9]See p. 209.
[10]See chapter by Berge and Gulkis, p. 662.

At wavelengths up to about 7 cm the Jovian emission is confined to the disk and assumed to be thermal; beyond 7 cm synchrotron and other radiations dominate the spectrum. After measurements of brightness temperatures $\sim 150°$K at $\sim 100\,\mu$m (see Fig. 1), there appears to be a gap in the measurements up to 0.1 cm. Between 0.1 and 0.3 Gulkis and Poynter (1972) indicate mean disk temperatures $\sim 150°$K, and the spectrum beyond 0.3 cm is indicated in Fig. 5. The data at 10 and 20 cm have been corrected for the contribution due to synchrotron radiation.

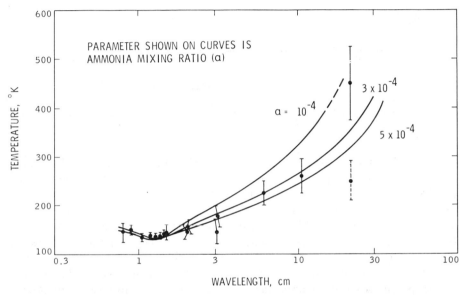

Fig. 5.   Observed and model thermal microwave brightness temperature spectrum from Gulkis and Poynter (1972). The mixing ratio $\alpha = [NH_3]/[H_2]$ applies below the region of NH_3 saturation.

For abundances corresponding to solar, the only important opacity in the $0.3-20$ cm region are the NH_3 inversion lines centered at ~ 1.25 cm. Models by Gulkis and Poynter computed with this opacity for a temperature variation essentially identical to that of Fig. 3 (passing through 210°K at 1.6 $\times 10^6$ dynes cm^{-2}) at pressures $> 3 \times 10^5$ dynes cm^{-2} are compared with the observations in Fig. 5. The differences in the profiles at pressures $< 3 \times 10^5$ dynes cm^{-2} are not important since insignificant emission comes from this region. NH_3 is assumed mixed up to the point where its partial pressure equals the saturated vapor pressure and above that it follows the saturated vapor pressure curve. The result that $[NH_3]/[H_2] \sim 3 \times 10^{-4}$ below the region where NH_3 is saturated, is in good agreement with the value consistent with solar abundances (1.8×10^{-4}) and indicates that the radiative-convective models appear to be valid down to depths where the temperature is at least $\sim 280°$K.

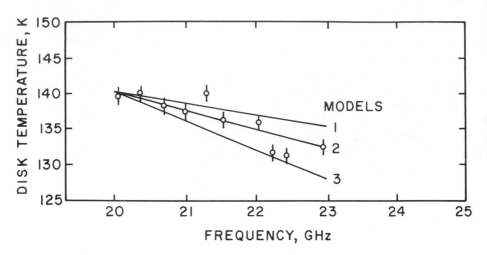

Fig. 6.   Observed and model thermal microwave brightness temperatures from 1.3 to 1.5 cm from Gulkis *et al.* (1974). See text.

The subsequent study by Gulkis *et al.* (1974) limited to near the center of the 1.25 cm absorption is illustrated in Fig. 6. In this study the authors assumed $[NH_3]/[H_2] = 2 \times 10^{-4}$ in the unsaturated region and temperature profiles following an adiabatic lapse rate through 130°K at the three pressures of 1.0, 0.49 and 0.24 atm (Models 1, 2 and 3 of Fig. 6). Model 2, the best fit, is in very good agreement with the results from infrared studies given in Fig. 4.

The good agreement between the infrared and microwave results for the troposphere is very important because they are almost independent. The independence arises from the different opacity sources and different absorbers in the two regions. They are, however, not quite independent since the microwave data require an assumed lapse rate (adiabatic) whereas the infrared data yield a lapse rate which is adiabatic to within the errors of observations.

VI.   STELLAR OCCULTATIONS

This subject is part of the review by Hunten and Veverka.[11] I have no arguments with their conclusions, but in the context that these events offer one of the lines of evidence bearing on the thermal structure, I would like to present my own comments.

There is a considerable body of evidence that the observed change of stellar intensity during immersion and emission is due to refraction of starlight in the planetary atmosphere, and consequently that information about

[11]See p. 258.

the atmospheric scale height can be obtained from the event (see, e.g., Wasserman and Veverka 1973). This type of information is unique in sampling the high stratosphere where the other techniques noted in this review are ineffective. The early attempt by Baum and Code (1953) to interpret the occultation of σ Arietis by Jupiter was not very successful. They compared smooth synthetic light curves for an isothermal atmosphere with noisy observations to obtain upper and lower limits to the average scale height of 12.5 and 6.3 km as opposed to the modern result of $\sim$ 30 km. (To my eye, their observations seem compatible with 30 km.)

Interpretations of various occultation curves of β Sco, which are of much higher quality than that obtained by Baum and Code but show spikiness of varying amounts, have been published by Hubbard et al. (1972), Vapillon et al. (1973), and Veverka, et al. (1974).

The three-color observations considered by Elliot et al. (1974) and Elliot and Veverka[12] demonstrate that in the main body of the light curve, roughly between the points where the stellar signal is between 0.9 and 0.1 of the unocculted stellar signal level, the spikes are due to layering in the Jovian atmosphere. At stellar signal levels below $\sim$ 0.1 the case for layering is not so clear. Wallace (1975a) has indicated that the features called "plateaus" by Vapillon et al. (1973) are of a shape that would naturally result from layering but Young (verbal communication at Jupiter conference) has argued with great persuasiveness that this part of the light curve, at least, may be dominated by scintillation in the Jovian atmosphere.

Thus the interpretation in terms of refraction in a layered atmosphere appears to be a good first approximation. Because of the spikes a direct comparison with the smooth synthetic light curves for an isothermal atmosphere, as attempted by Baum and Code (1953), is very difficult. Instead, Hubbard et al. (1972), Vapillon et al. (1973), and Veverka et al. (1974) have all used essentially the same inversion technique, many of the pitfalls of which have been reviewed by Wasserman and Veverka (1973). The inversion technique hinges on the assumption of a symmetric atmosphere, assumed spherical, in these studies. The real atmosphere is surely not symmetric, but over the path lengths through the atmosphere that are involved it should be symmetric enough in the gross sense to define a mean scale height. The layering, however, may not be so extensive in the horizontal dimension and consequently the structural details which are derived from the spikes may not be accurate.

The inversion method involves two separate sets of point-by-point integrations over the lightcurve from time t to the time when the star was not yet occulted; they yield relative height (miss distance of the ray) and refractivity (the index of refraction minus one) as a function of time. Noise and uncertainty in the signal level of the planet plus the unocculted star produce

---

[12]See Elliot and Veverka 1976, *Icarus* (special Jupiter issue) 27:359–386.

errors which are propagated downward in the integration and spoil the upper part of the refractivity curves. At greater depths, corresponding to times where the star is almost completely occulted, the results are spoiled by noise and uncertainty in the intensity of the planetary signal. The central parts of the plots of log refractivity vs. height given by Hubbard *et al.* (1972) and Veverka *et al.* (1974) yield average scale heights of 27 ±3 km and 30 ±3 km. The corresponding plots given by Vapillon *et al.* (1973) show too much curvature to define a single scale height. With the assumption of hydrostatic equilibrium and the reference atmospheric composition of $[He]/[H_2] = 0.1$ these yield the temperatures 169 ±19°K and 186 ±19°K at a pressure $\sim$ 3 dynes cm^{-2}. The highest temperature model of Fig. 3, which indicates 170°K at this pressure is in good agreement with those results.

It is my opinion, based on the review of Wasserman and Veverka (1973) and my own attempts to interpret the occultation by Neptune (Wallace 1975a) that it is unlikely that a detailed temperature structure or even a temperature gradient can be obtained from the data. However, the data can be further manipulated by assuming a composition and converting the refractivities to number densities, integrating the densities with respect to height to yield pressures. Then temperatures can be obtained with the perfect gas law. It is not obvious that the data can survive the further downward propagation, through the density integration, of the errors already propagated downward in the integrations to obtain refractivity and height. Concern on that point is justified by inspection of Fig. 10 of Veverka *et al.* (1974) which compares three temperature profiles from three colors at the same telescope. The agreement between the three profiles is not bad overall but in the central, best-defined parts of the curves essentially the same temperature structure appears in all three colors but is displaced by as much as a factor of two in pressure.

Both Veverka *et al.* (1974) and Vapillon *et al.* (1973) went through this further step to obtain temperature profiles but Hubbard *et al.* (1972) did not. Consequently Wallace *et al.* (1974) applied this step to the data of Hubbard *et al.* to obtain the comparison in Fig. 7. The high pressure ends of the curves come from the low intensity tails of the lightcurves and are very sensitive to the inaccuracies of the subtraction of the background (planet plus dark level). Consequently the termination of the curves at varying high pressures is a reflection of the various worker's confidence in their ability to track the signal out to very low levels. Additionally, my truncation of the occultation results at 10^2 dynes cm^{-2} is an obvious measure of bias; the curves of Vapillon *et al.* (1973) actually continue to higher pressures, reaching $\sim$ 350°K between $(2-5) \times 10^2$ dynes cm^{-2}.

Fig. 7 also shows the hottest model of Fig. 3, which gives a good match with the observed intensities of the 7.8 μm band of CH_4. It is easy enough to dismiss a radiative-convective model since it is only as good as the assumptions invoked concerning the heating and cooling mechanisms and the radi-

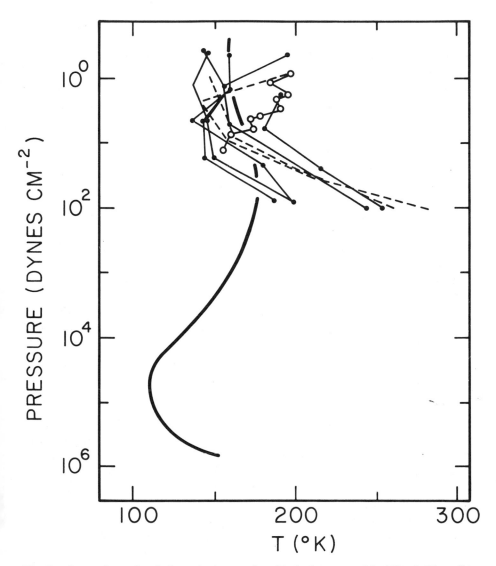

Fig. 7.   Comparison of optical occultation results with the hottest model of Fig. 3. The solid straight line segments connecting filled circles are based on the results of Hubbard *et al.* (1972); the dashed straight-line segments are from Vapillon *et al.* (1974), and the open circles connected by solid lines represent maxima, minima and inflection points of the results of Veverka *et al.* (1974). Adapted from Wallace *et al.* (1974).

ation equilibrium. However, Wallace *et al.* (1974) pointed out that stratospheric temperatures substantially hotter than the model of Fig. 7 would yield a 7.8 μm band of CH_4 more intense than observed. This is because the model gives a close match to the observed intensity, because significant contributions to the emission arise from almost the whole stratosphere, and because the Planck function is a very strong function of temperature at Jovian temperatures and 7.8 μm. To test the high-temperature and high-pressure parts of the occultation curves, Wallace et al. (1974) characterized them by a temperature peak of 240°K in the region $100-200$ dynes cm^{-2} followed by a sharp decrease at higher pressures to 110°K at 10^4 dynes cm^{-2} and then followed by a Trafton-type model (Fig. 3). Wallace et al. (1974) found that the predicted intensity of the band was only a factor of 2 greater than observed and therefore only marginally discrepant. If, however, [He]/[H$_2$] is, for example, 0.5 instead of 0.1 the 240°K temperature must be scaled up according to the mean molecular mass to 320°K and a similar calculation leads to an intensity in the 7.8 μm band 30 times greater than observed. In that case they found that [CH$_4$]/[H$_2$] would have to be reduced to 10^{-7} to remove the discrepancy, i.e., a factor of $\sim 7 \times 10^{-3}$ less than observed in the troposphere. Wallace et al. (1974) did not test the higher temperature results of Vapillon et al. (1973) but since these are almost the same as the 320°K situation just mentioned, essentially the same problem with the intensity and CH$_4$ abundance would result.

No clear resolution of the problem of such high temperatures at these pressures vis-à-vis a stratospheric reduction of CH$_4$ has been reached. However, since there is no reason to think that [He]/[H$_2$] is as large as 0.5, and the high temperatures indicated by Vapillon et al. (1973) are very uncertain, it is likely that no problem exists.

VII. RADIO OCCULTATION

The radio occultations, which have been reviewed by Hunten and Veverka,[13] are similar in many ways to the optical ones but because of the different geometry and the use of frequency rather than intensity data, they probe deeper into the atmosphere and their data are much more sensitive to the detailed geometry of the occultation.

The first reductions of the Pioneer 10 and 11 S-band occultation data by Kliore et al. (1974) and Kliore et al. (1975), assuming a spherical atmosphere, indicated temperatures in the range $450-600$°K at 1 atm pressure, completely at odds with the other lines of evidence presented here. The subsequent reduction by Kliore and Woiceshyn[14] takes into account, in an approximate manner, the oblateness of Jupiter's atmosphere and gives results which are compared with the others given in Fig. 4. The radio results are not given in the figure for pressures $\lesssim 10^4$ dynes cm^{-2} because of their extreme sensitivity in this pressure regime to the details of the analysis. The highest pressures probed are approximately 4×10^5 dynes cm^{-2}.

These results are in contradiction to some of the inversions of the infrared data but, as I have noted above, the infrared data yield ambiguous results

---

[13]See p. 273.   [14]See p. 217.

in this pressure range. The agreement with the infrared inversion labelled "matrix inversion" is acceptable, indicating at least approximate compatibility with the infrared data, and the agreement with the model is not bad.

Unfortunately, the reduction by Kliore and Woiceshyn may not be the final one. These authors have indicated approximations in their method which, although reasonable, should probably be removed. Consequently it might be prudent to take the current version of the radio occultation results as an indication of the type of result to be expected from a more detailed analysis.

Acknowledgment. Kitt Peak National Observatory is operated by the Association of Universities for Research in Astronomy, Inc., under contract with the National Science Foundation.

REFERENCES

Aitken, D. K., and Jones, B. 1972. The 8 to 13 μm spectrum of Jupiter. *Nature* 240:230–232.

Allen, C. W. 1975. *Astrophysical Quantities*, 4th ed. London: The Athlone Press.

Armstrong, K. R.; Harper, D. A. Jr.; and Low, F. J. 1972. Far infrared brightness temperatures of the planets. *Astrophys. J.* 178:L89–L92.

Aumann, H. H.; Gillespie, C. M. Jr.; and Low, F. J. 1969. The internal powers and effective temperatures of Jupiter and Saturn. *Astrophys. J.* 157:L69–L72.

Baum, W. A., and Code, A. D. 1953. A photometric observation of the occultation of σ Arietis by Jupiter. *Astron. J.* 58:108–112.

Belton, M. J. S. 1969. An estimate of the abundance and the rotational temperature of CH_4 on Jupiter. *Astrophys. J.* 157:469–472.

Cameron, A. G. W. 1973. Elemental and isotopic abundances of the volatile elements in the outer planets. *Space Sci. Rev.* 14:392–400.

Carlson, R. W., and Judge, D. L. 1974. Pioneer 11 ultraviolet photometer observations at Jupiter encounter. *J. Geophys. Res.* 79:3623–3633.

Cess, R. D., and Khetan, S. 1973. Radiative transfer within the atmospheres of the major planets. *J. Quant. Spectrosc. Radiat. Transfer* 13:995–1009.

Chahine, M. T. 1968. Determination of the temperature profile in an atmosphere from its outgoing radiance. *J. Opt. Soc. Amer.* 58:1634–1637.

——. 1970. Inverse problems in the radiative transfer: Determination of atmospheric parameters. *J. Atmos. Sci.* 27:960–967.

——. 1974. Remote sounding of cloudy atmospheres. I. The single cloud layer. *J. Atmos. Sci.* 31:233–243.

Chase, S. C.; Ruiz, R. D.; Münch, G.; Neugebauer, G.; Schroeder, M.; and Trafton, L. M. 1974. Pioneer 10 infrared radiometer experiment: preliminary results. *Science* 183:315–317.

Elliot, J. L.; Wasserman, L. H.; Veverka, J.; Sagan, C.; and Liller, W. 1974. The occultation of Beta Scorpii by Jupiter. II. The hydrogen-helium abundance in the Jovian atmosphere. *Astrophys. J.* 190:719–729.

Fink, U., and Belton, M. J. S. 1969. Collision narrowed curves of growth for H_2 applied to new photoelectric observations of Jupiter. *J. Atmos. Sci.* 26:952–962.

Fox, K., and Ozier, I. 1971. The importance of methane to pressure-induced absorption in the atmospheres of the outer planets. *Astrophys. J.* 166:L95–L100.

Gillett, F. C.; Low, F. J.; and Stein, W. A. 1969. The 2.8–14-micron spectrum of Jupiter. *Astrophys. J.* 157:925–934.

Gillett, F. C., and Westphal, J. A. 1973. Observations of 7.9-micron limb brightening on Jupiter. *Astrophys. J.* 179:L153–L154.

Gulkis, S.; Klein, M. J.; and Poynter, R. L. 1974. Jupiter's microwave spectrum – Implications for the upper atmosphere. IAU Symp. 65, *Exploration of the Planetary System* (A. Woszczyk and C. Iwaniszewski, eds.) Dordrecht, Holland: D. Reidel Publ. Co. In press.

Gulkis, S., and Poynter, R. 1972. Thermal radio emission from Jupiter and Saturn. *Phys. Earth Planet. Interiors* 6:36–43.

Harper, D. A. Jr.; Low, F. J.; Rieke, G. H.; and Armstrong, K. R. 1972. Observations of planets, nebulae, and galaxies at 350 microns. *Astrophys. J.* 177:L21–L25.

Hirshberg, J. 1973. Helium abundance of the Sun. *Rev. Geophys. and Space Phys.* 11:115–131.

Hogan, J. S.; Rasool, S. I.; and Encrenaz, T. E. 1969. The thermal structure of the Jovian atmosphere. *J. Atmos. Sci.* 26:898–905.

Houck, J. R.; Pollack, J. B.; Schaack, D.; Reed, R. A.; and Summers, A. 1975. Jupiter: its infrared spectrum from 16 to 40 micrometers. *Science* 189:720–721.

Hubbard, W. B.; Hunten, D. M.; and Kliore, A. 1975. Effect of the Jovian oblateness on Pioneer 10/11 radio occultations. *Geophys. Res. Lett.* In press.

Hubbard, W. B.; Nather, R. E.; Evans, D. S.; Tull, R. G.; Wells, D. C.; Van Citters, G. W.; Warner, B.; and Vanden Bout, P. 1972. The occultation of Beta Scorpii by Jupiter and Io. I. Jupiter. *Astron. J.* 77:41–59.

Hunten, D. M. 1969. The upper atmosphere of Jupiter. *J. Atmos. Sci.* 26:826–834.

Hunten, D. M., and Münch, G. 1973. The helium abundance on Jupiter. *Space Sci. Rev.* 14: 433–443.

Ingersoll, A. P.; Münch, G.; Neugebauer, G.; Diner, D. J.; Orton, G. S.; Schupler, B.; Schroeder, M.; Chase, S. C.; Ruiz, R. D.; and Trafton, L. M. 1975. Pioneer 11 infrared radiometer experiment: The global heat balance of Jupiter. *Science* 188:472–473.

Kliore, A.; Cain, D. L.; Fjeldbo, G.; Seidel, B. L.; and Rasool, S. I. 1974. Preliminary results on the atmospheres of Io and Jupiter from the Pioneer 10 S-band occultation experiment. *Science* 183:323–324.

Kliore, A.; Fjeldbo, G.; Seidel, B. L.; Sesplaukis, T. T.; Sweetnam, D. W.; and Woiceshyn, P. M. 1975. Atmosphere of Jupiter from the Pioneer 11 S-band occultation experiment: Preliminary results. *Science* 188:474–476.

Lacy, J. H.; Larrabee, A. I.; Wollman, E. R.; Geballa, T. R.; Townes, C. H.; Bregman, J. D.; and Rank, D. M. 1975. Observations and analysis of the Jovian spectrum in the 10-micron ν_2 band of NH_3. *Astrophys. J.* 198:L145–L147.

Low, F. J. 1965. Planetary radiation at infrared and millimeter wavelengths. *Bull. Lowell Obs.* 6:184–187.

Low, F. J., Rieke, G. H.; and Armstrong, K. R. 1973. Ground-based observations at 34 microns. *Astrophys. J.* 183:L105–L109.

Ohring, G. 1973. The temperature and ammonia profiles in the Jovian atmosphere from inversion of the Jovian emission spectrum. *Astrophys. J.* 184:1027–1040.

Orton, G. S. 1975*a*. The thermal structure of Jupiter: I. Implication of Pioneer 10 infrared radiometer data. Submitted to *Icarus*.

———. 1975*b*. The thermal structure of Jupiter: II. Observations and analysis of 8 – 14 micron radiation. Submitted to *Icarus*.

Owen, T., and Mason, H. P. 1968. The abundance of hydrogen in the atmosphere of Jupiter. *Astrophys. J.* 154:317–326.

Ridgway, S. T. 1974*a*. Jupiter: Identification of ethane and acetylene. *Astrophys. J.* 187: L41–L43.

———. 1974*b*. Jupiter: Identification of ethane and acetylene. *Astrophys. J.* 192:L51.

Smith, W. L. 1970. Iterative solution of the radiative transfer equation for the temperature and absorbing gas profile of an atmosphere. *Appl. Optics* 9:1993–1999.

Strobel, D. F. 1973. The photochemistry of hydrocarbons in the Jovian atmosphere. *J. Atmos. Sci.* 30:489–498.

———. 1975. The photochemistry of NH_3 in the Jovian atmosphere. *J. Atmos. Sci.* 30:1205–1209.

Taylor, F. W. 1972. Temperature sounding experiments for the Jovian planets. *J. Atmos. Sci.* 29:950–958.

Trafton, L. M. 1965. A study of the energy balance in the atmospheres of the major planets. Ph.D. dissertation, California Institute of Technology.

———. 1966. The pressure-induced monochromatic translational absorption coefficients for homopolar and non-polar gases and gas mixtures with particular application to H_2. *Astrophys. J.* 146:558–571.

———. 1967. Model atmospheres of the major planets. *Astrophys. J.* 147:765–781.

———. 1972. On the methane opacity for Uranus and Neptune. *Astrophys. J.* 172:L117–L120.

Trafton, L. M., and Stone, P. H. 1974. Radiative-dynamical equilibrium states for Jupiter. *Astrophys. J.* 188:649–655.

Trafton, L., and Wildey, R. 1974. Jupiter: a comment on the 8- to 14-micron limb darkening. *Astrophys. J.* 194:499–502.

Vapillon, L.; Combes, M.; and Lecacheux, J. 1973. The β Scorpii occultation by Jupiter. *Astron. Astrophys.* 29:135–149.

Veverka, J.; Wasserman, L. H.; Elliot, J.; Sagan, C.; and Liller, W. 1974. The occultation of β Scorpii by Jupiter I. The structure of the Jovian upper atmosphere. *Astron. J.* 79:73–84.

Wallace, L. 1975a. On the 1968 occultation of BD 17°4388 by Neptune. *Astrophys. J.* 197: 257–261.

———. 1975b. On the thermal structure of Uranus. *Icarus* 25:538–544.

Wallace, L., and Hunten, D. M. 1973. The Lyman-alpha albedo of Jupiter. *Astrophys. J.* 182: 1013–1031.

Wallace, L.; Prather, M.; and Belton, M. J. S. 1974. The thermal structure of the atmosphere of Jupiter. *Astrophys. J.* 193:481–493.

Wallace, L., and Smith, G. R. 1975. On Jovian temperature profiles obtained by inverting thermal spectra. Submitted to *Astrophys. J.*

Wasserman, L. H., and Veverka, J. 1973. On the reduction of occultation light curves. *Icarus* 20:322–345.

Westphal, J. A.; Matthews, K.; and Terrile, R. J. 1974. Five-micron pictures of Jupiter. *Astrophys. J.* 188:L111–L112.

Wildey, R. L., and Trafton, L. M. 1971. Studies of Jupiter's equatorial thermal limb darkening during the 1965 apparition. *Astrophys. J. Suppl.* 23:1–34.

MODEL IONOSPHERES OF JUPITER

S. K. ATREYA

and

T. M. DONAHUE
The University of Michigan

The principal concepts presently involved in modeling the Jovian ionosphere are reviewed. A model ionosphere is developed on the basis of our present knowledge of atmospheric composition, relevant chemical and ion-molecule reactions, with their associated rate constants. The shortcomings of this model are discussed when it is compared with the electron density profile obtained from the Pioneer 10 radio occultation data. It is demonstrated that the apparent great extent of the observed topside ionosphere may imply a hot thermosphere, as if Jupiter sustained a corona. Some of the layers observed in the electron density profile may be due to sporadic-E like clustering of protons and other ions.

Many models of the ionosphere of Jupiter have been developed in the past (Rishbeth 1959; Zabriskie 1960; Gross and Rasool 1964; Hunten 1969; Shimizu 1971; Prasad and Capone 1971; Tanaka and Hirao 1971; McElroy 1973; Capone and Prasad 1973; Atreya *et al.* 1974; Prasad and Tan 1974; and Atreya and Donahue 1975*b*). The recent flights of Pioneer 10 and 11 past Jupiter have contributed immensely to our knowledge of the environment of Jupiter. In view of the nature of the results obtained from the first direct detection of the Jovian ionosphere by the Pioneer 10 radio occultation experiment and their divergence from the models, it would be useless for us to devote much space to discussing each of the above mentioned models individually. We refer readers to articles by Hunten (1969), who has reviewed the progress in Jovian ionospheric modeling until 1969, by McElroy (1973) for progress until 1973, and by Atreya *et al.* (1974). The paper by McElroy is significant in pointing out the importance of dissociative ionization of H_2 among processes neglected in previous work, including the Hunten review. The paper by Atreya *et al.* (1974) draws attention to the three body association reaction

$$H^+ + H_2 + H_2 \rightarrow H_3^+ + H_2 \tag{1}$$

[304]

as an important sink for ionization below 220 km. In this chapter we shall first outline a model of the Jovian ionosphere based on the latest available information on the relevant chemical reactions and rate constants, and a model atmosphere appropriate to a conventionally low value for the exospheric temperature, and then discuss how this model must be modified to conform to the measurements.

NEUTRAL ATMOSPHERE

Jupiter's upper atmosphere consists mostly of H_2 (67 km atm), He ($<$ 34 km atm), CH_4 (45 m atm), C_2H_2 (2×10^{-6} m atm), C_2H_6 (10^{-4} m atm). The abundance of C_2H_2 and C_2H_6 is uncertain. Helium has been directly identified only recently by Judge and Carlson (1974) with the ultraviolet photometer on board Pioneer 10. Carlson and Judge (1974)[1] deduce a mixing ratio for He in the homosphere to be $0.18 \begin{smallmatrix} + 0.46 \\ - 0.12 \end{smallmatrix}$; Elliott et al. (1974)[2] find the ratio to be $0.19 \begin{smallmatrix} + 0.35 \\ - 0.19 \end{smallmatrix}$ from observation of the occultation of β-Scorpii by Jupiter. These results are not grossly different from the solar composition ratio of 0.11 (Hunten and Münch 1973), although the precision of both measurements is insufficient to determine this important quantity. Interpretation of Moos and Rottman's (1972) Lyman-α albedo data of Jupiter indicates an eddy diffusion coefficient K of 10^6 cm^2 sec^{-1} at the turbopause (Wallace and Hunten 1973).[3] Strobel (1973) derives a value of 2×10^4 cm^2 sec^{-1} in the lower atmosphere from consideration of the ultraviolet albedo. These values of K are, however, questionable (Atreya et al. 1974) since the rather high Lyman-α emission rate of 4.4 kR observed by Moos and Rottman (1972) may have had its origin partly in the hydrogen torus around Jupiter at the orbit of Io (McElroy et al. 1974). As a matter of fact Orbiting Astronomical Observatory-*Copernicus* measurements by Jenkins et al. (1973) indicated only 660 ± 350 R of Lyman-α attributable to Jupiter. Carlson and Judge's (1974) Pioneer 10 measurements show only 440 R of Lyman-α coming from the disk of Jupiter itself. From their observations of Lyman-α emission from Jupiter, Carlson and Judge (1974) derive a value for $K = 3 \times 10^{8 \pm 1}$ cm^2 sec^{-1}. Veverka et al. (1974) argue for $K \geqslant 7 \times 10^5$ cm^2 sec^{-1} from their interpretation of the spikes in the temperature profile deduced from the β-Scorpii occultation data. The weight of the evidence, therefore, argues for an eddy mixing coefficient greater than 10^5 cm^2 sec^{-1}, and perhaps as high as 10^9 cm^2 sec^{-1}. The upper atmospheric temperature profile now is derived from the Pioneer radio occultation data in the chapter by Kliore and Woiceshyn.[4] In view of the wide range of possible values for the He/H_2 mixing ratio, for K and for T we take Strobel's (1975) neutral model atmosphere to

[1]See p. 437.   [3]See p. 426.
[2]See p. 266.   [4]See p. 235.

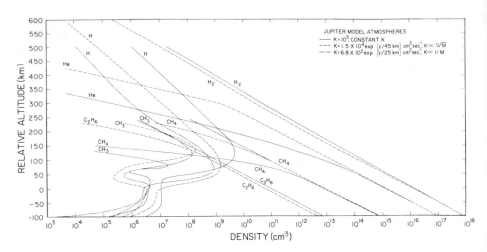

Fig. 1. Model atmospheres of Jupiter (hydrocarbon densities from Strobel 1975), for $K = 10^5$ cm^2 sec^{-1}, $K \propto M^{-\frac{1}{2}}$ and $K \propto M^{-1}$, where M is the atmospheric number density. Thermospheric temperature is assumed constant at 150°K and He/H$_2$ mixing ratio is taken as 0.1. Height scale refers to altitude above (or below) the level at which atmospheric density is 10^{16} cm^{-3}.

be as acceptable as any other for use in the illustrative calculations we shall present. Figure 1, adapted from Strobel (1975), assumes the He/H$_2$ ratio to be 0.1, and T_∞ to be 150°Kelvin and constant. Various values of K are assumed. Only those hydrocarbons which are suspected of playing a role in the ionosphere are included in Fig. 1. Further on in this chapter we shall show how this model atmosphere might be modified to explain the measured electron density profiles.

CHEMICAL MODEL AND IONOSPHERE

Reactions relevant to Jupiter's ionosphere are taken from Atreya and Donahue (1975b) and listed in Table I. Solar extreme-ultraviolet radiation provides the principal source of ionization in this model. (We are aware of a potentially important source from energetic magnetospheric particles, but do not yet know how to model that source.) Continuous absorption of radiation by atomic hydrogen occurs below 911 Å, by H$_2$ below 912 Å and by He below 504 Å. The appropriate photoabsorption and photoionization cross-sections are taken from Cook and Metzger (1964), Stewart and Webb (1963), Samson and Cairns (1965), and Samson (1966). Strobel (personal communication 1974) estimates an upper limit of 2×10^{-17} cm^2 for the photoionization cross-section of the methyl radical CH$_3$ whose ionization potential is 9.82 ± 0.04 eV (Elder *et al.* 1962). Photoionization of the methyl radical as an important process has been discussed recently by Atreya and Donahue (1975b) and also by Prasad and Tan (1974). The latter authors assume unacceptably large values for the ionization cross-section of CH$_3$ and the solar

Lyman-α flux at Jupiter. Photoionization of CH_3 is expected to be caused principally by solar Lyman-α. At the suggestion of Strobel (1975), Atreya and Donahue (1975b) investigated EUV photolysis of methane as an additional source of stratospheric ionization. They found that ionization of CH_4 contributes less than 3% of the total electron density. [At least 90% of the initial positive ions resulting from photoionization of CH_4 by radiation below 945 Å are CH_4^+ and CH_3^+ (Rebbert *et al.* 1973)]. Strong absorption by H_2 in the relevant wavelength region is responsible for the small contribution from this source. As far as ion production rates are concerned, we emphasize the importance of dissociative photoionization of H_2 as a source of H^+ (McElroy 1973). In this chapter we incorporate the recent data of Monahan *et al.* 1974), Browning and Fryar (1973), and Samson (1972) for the branching ratio for production of protons by dissociative photoionization of H_2. Ion production rates for a model in which K varies inversely as the square root of the atmospheric density ($K = 3 \times 10^7$ cm^2 sec^{-1} at the turbopause) are shown in Fig. 2. In this calculation the solar EUV fluxes of Hinteregger (1970), scaled to Jupiter, were employed. In the upper ionosphere where time constants for removal of the major ion (H^+) are much greater than a Jovian day, fluxes were halved in constructing diurnal average models. A solar zenith angle of 60° was assumed. The wavelength interval between 0 and 960 Å was divided into a mesh of 5 Å size, and appropriate cross-sections were averaged in each 5 Å interval. In Fig. 2 it will be seen that the

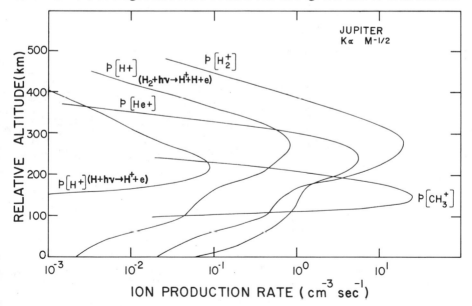

Fig. 2.   Photo-ion production rates, $p(X^+)$ for the model with $K \propto M^{-\frac{1}{2}}$ ($K \simeq 3 \times 10^7$ cm^2 sec^{-1} at the turbopause). Note that the photodissociative ionization of H_2 is the major source of H^+ production. Height scale is the same as in Fig. 1.

TABLE I

Important Reactions in the Ionosphere of Jupiter

Reaction Number	Reaction	Rate Constant[a]	Reference
	Ion Production:		
p1	$H_2 + h\nu \rightarrow H_2^+ + e$		
p2	$\rightarrow H^+ + H + e$		
p3	$H_2 + e \rightarrow H_2^+ + 2e$		
p4	$\rightarrow H^+ + H + 2e$		McElroy (1973)
p5	$H + h\nu \rightarrow H^+ + e$		
p6	$H + e \rightarrow H^+ + 2e$		
p7	$He + h\nu \rightarrow He^+ + e$		
p8	$He + e \rightarrow He^+ + 2e$		
	Ion Exchange:		
e1	$H_2^+ + H_2 \rightarrow H_3^+ + H$	2.0×10^{-9}	Theard and Huntress (1974)
e2	$H_2^+ + H \rightarrow H^+ + H_2$	$\sim 1.0 \times 10^{-10}$	Hunten (1969)
e3	$He^+ + H_2 \rightarrow H_2^+ + He$	$\lesssim 20\%$	
e4	$\rightarrow HeH^+ + H$	1.0×10^{-13}	Johnsen and Biondi (1974)
e5	$\rightarrow H^+ + H + He$	$\gtrsim 80\%$	

(e3, e4, e5 braced together: sum)

	Reaction	Rate constant	Reference
e6	$He^+ + CH_4 \rightarrow CH^+ + H_2 + H + He$	2.4×10^{-10}	Huntress (1974)
e7	$\rightarrow CH_2^+ + H_2 + He$	9.3×10^{-10}	Huntress (1974)
e8	$\rightarrow CH_3^+ + H_2 + He$	6.0×10^{-11}	Huntress (1974)
e9	$\rightarrow CH_4^+ + He$	4.0×10^{-11}	Huntress (1974)
e10	$H^+ + H_2 + H_2 \rightarrow H_3^+ + H_2$	3.2×10^{-29}	Miller et al. (1968)
e11	$H^+ + CH_4 \rightarrow CH_3^+ + H_2$	2.3×10^{-9}	Huntress (1974)
e12	$\rightarrow CH_4^+ + H$	1.5×10^{-9}	Huntress (1974)
e13	$HeH^+ + H_2 \rightarrow H_3^+ + He$	1.85×10^{-9}	Theard and Huntress (1974)
e14	$H_3^+ + CH_4 \rightarrow CH_5^+ + H_2$	2.4×10^{-9}	Huntress (1974)
e15	$CH^+ + H_2 \rightarrow CH_2^+ + H$	1.0×10^{-9}	Huntress (1974)
e16	$CH_2^+ + H_2 \rightarrow CH_3^+ + H$	7.2×10^{-10}	Huntress (1974)
e17	$CH_3^+ + CH_4 \rightarrow C_2H_5^+ + H_2$	8.9×10^{-10}	Huntress (1974)
e18	$CH_4^+ + CH_4 \rightarrow CH_5^+ + CH_3$	1.11×10^{-9}	Huntress (1974)
e19	$CH_4^+ + H_2 \rightarrow CH_5^+ + H$	4.1×10^{-11}	Huntress (1974)

Ion Removal/Electron-Ion Recombination:

	Reaction	Rate constant	Reference
r1	$H_3^+ + e \rightarrow H_2 + H$	3.8×10^{-7}	Leu et al. (1973)
r2	$H_2^+ + e \rightarrow H + H$	$< 1.0 \times 10^{-8}$	Hunten (1969)
r3	$HeH^+ + e \rightarrow He + H$	$\sim 1.0 \times 10^{-8}$	Hunten (1969)
r4	$H^+ + e \rightarrow H + h\nu$	6.6×10^{-12}	Bates and Dalgarno (1962)
r5	$He^+ + e \rightarrow He + h\nu$	6.6×10^{-12}	Bates and Dalgarno (1962)
r6	$CH_5^+ + e \rightarrow$ neutral	1.9×10^{-6}	Rebbert et al. (1973)
r7	$C_2H_5^+ + e \rightarrow$ products	1.9×10^{-6}	Rebbert et al. (1973)

[a]The rate constants are in units of cm^3 sec^{-1} for two-body reactions, and cm^6 sec^{-1} for three-body reactions (after Atreya and Donahue 1975b).

dissociative photoionization of H_2 provides the dominant source of Jovian protons, while production of H^+ due to direct photoionization of H can be ignored. Ion production in the very low altitude regions is due to penetration of X-radiation below 100 Å. We should, however, point out that both the solar EUV flux and the values of appropriate cross-sections are quite uncertain in the hard X-ray region. Therefore, the present picture of very low altitude ion-production may change as better data become available. H^+ ion production due to direct photoionization of H is negligible below 150 km because of absorption by H_2 in the relevant wavelength band. CH_3^+ ion production occurs in a very narrow altitude range since the radiation responsible for CH_3 ionization (principally solar Lyman-α) is strongly absorbed by methane below the turbopause. Neutralization of the ions formed proceeds via various reactions with other atmospheric constituents and eventual electron-ion recombination mechanisms listed in Table I. H_2^+ ions are converted to H_3^+ (reaction $e1$) in the lower atmosphere where the H_2 density is high, while in the upper atmosphere they form H^+ (reaction $e2$) or dissociatively recombine with electrons (reaction $r2$). Above the methane turbopause, He^+ is mainly removed by a series of reactions ($e3-e5$) with H_2, while the bulk of He^+ ends up as H^+. Below the methane turbopause He^+ preferentially combines with CH_4 to give various hydrocarbon ions (reactions $e6-e9$). Protons at and above the altitude of the maximum electron density $[N_e]_{max}$ are removed by radiative recombination with electrons (reaction $r4$); by three body association (reaction $e10$) between the altitude of maximum electron density and the turbopause; and by methane (reactions $e11$ and $e12$) below the turbopause. The importance of the three body reaction ($e10$) as a fast sink of Jovian protons was first recognized independently by Dalgarno (1971) and by Donahue (see McElroy 1973). H_3^+ ions formed in reactions $e1$, $e10$ and $e13$ combine rapidly with CH_4 to give CH_5^+ ions in the vicinity of the turbopause above which they are neutralized in dissociative recombination with electrons (reaction $r1$). Various hydrocarbon ions resulting from chemical reactions eventually form higher-order hydrocarbon ions CH_5^+ and $C_2H_5^+$ which are rapidly neutralized in dissociative recombination with electrons (reactions $r6$ and $r7$). The importance of methane as a potential sink for H^+ ions at low elevation was first recognized by McElroy (1973). Later Atreya and Donahue (1975a) extended the sink to include He^+ and H_3^+ ions. As more hydrocarbon reaction rate constants have become available (Huntress 1974), the role of hydrocarbons in the lower ionosphere of Jupiter has been discussed by Prasad and Tan (1974) and Atreya and Donahue (1975b). Under assumptions of photo-chemical equilibrium, the ion distribution was obtained by numerically solving one-dimensional coupled continuity equations. Results of these calculations are shown in Fig. 3. Electron density (N_e) profiles are shown for a constant eddy coefficient of 10^5 cm^2 sec^{-1} and for a K that varies with altitude. The maximum electron density in the high K model is greater by almost a factor of 3 than in the model with K set at 10^5 cm^2 sec^{-1}. This is

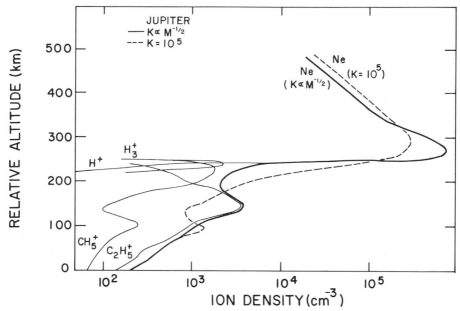

Fig. 3.   Calculated electron density profiles in Jupiter's atmosphere for $K = 10^5$ cm^2 sec^{-1} and $K \propto M^{\frac{1}{2}}$ cases. The secondary peak in Ne profile at lower elevation is due to CH$_3^+$ ions. Ion distribution is shown for $K \propto M^{\frac{1}{2}}$ model only. Height scale is the same as in Fig. 1.

caused by the fact that the He density remains relatively great to a much higher altitude when K is large and becomes an important source of Jovian protons because of reaction $e5$. Precise determination of the rate constant of $e5$ compared to the combined rate of $e3$, $e4$ and $e5$ is essential to an understanding of the relationship between K and the He source. The ion distribution profiles in Fig. 3 correspond to the model in which K varies as previously described. H$^+$ is the major ion down to the altitude of maximum electron density; H$_3^+$ dominates over a small range below the density maximum and then CH$_5^+$ and C$_2$H$_5^+$ prevail. A slight increase in the CH$_5^+$ density between 140 and 220 km is due to eventual conversion of H$_3^+$, H$^+$ and He$^+$ to CH$_5^+$. A secondary peak in electron density near $\sim$ 150 km (in the variable K model) is due to a maximum in CH$_3^+$ ion production at that altitude.

CONCLUSIONS

The electron density profiles shown in Fig. 3 are, in essence, similar to the pre-Pioneer 10 model of Atreya *et al.* (1974); only various refinements discussed above have been included. Inversion of the Pioneer 10 radio occultation data is complicated due to effects of multipath propagation of the S-band signal (Kliore *et al.* 1974). Preliminary analyses of the closed loop Doppler data (Fjeldbo *et al.* 1975) indicate that the electron density profile shows considerably more structure than that given by the model just developed. In Fig. 4, we reproduce, by solid lines, electron density profiles de-

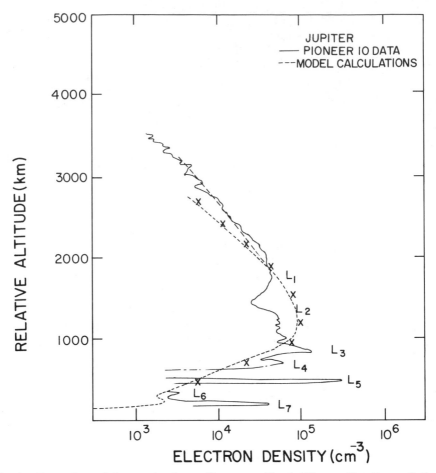

Fig. 4.   Comparison of electron density profile measured by the Pioneer 10 radio occultation
experiment with the model calculations based on a hot thermospheric model. Measured profile corresponds to late afternoon immersion at latitude 26°N and solar zenith angle, $\xi = 81°$.
The short-dashed curve refers to calculations in which $\xi = 81°$, $T_e \simeq T_i \simeq T_n$ and the radiative recombination coefficient, α_r (of H^+ and He^+) varies as $T_e^{-0.5}$. If $\alpha_r \propto T_e^{-0.75}$, a curve
may be drawn through the crosses (x's). Diffusive equilibrium distribution is shown by the
long-dashed curve for the case where $\alpha_r \propto T_e^{-0.5}$. The height scale refers to altitude above
the level where the refractivity is 10 (see Fjeldbo et al. 1975).

duced during immersion by the radio occultation experiment in late afternoon
of December 4, 1973 when the solar zenith angle ξ was 81°. The height scale
refers to the level at which the refractivity is 10. The model ionosphere of
Fig. 3 for variable K has some resemblance to the measured one up to the
altitude of the L5 peak. Beyond this level, however, the electron density
drops to about 10^3 cm^{-3} within 300 km in the model, whereas the measured
electron density remains fairly large for another 3000 km or so. We shall
attempt to explain this gross discrepancy in the topside ionosphere on the

basis of a possible "hot thermosphere" implying that a given density would occur at a much higher altitude in the model atmosphere. The source of this thermospheric heating probably lies in the lower atmosphere in the form of inertia gravity waves. French and Gierasch (1974) first broached this possibility in order to explain the peculiar temperature profile obtained from the inversion of β-Scorpii occultation data by Veverka et al. (1974). French and Gierasch (1974) argue that the observed oscillations in the temperature profile are due to propagating inertia gravity waves and the absence of flashes (lightcurve spikes) at densities lower than 10^{13} cm^{-3} supports the interpretation of upward energy propagation with damping of the waves above the mesopause level. They calculate the associated energy flux to be about 3.4 ergs cm^{-2} sec^{-1}, some two orders of magnitude larger than the solar EUV flux absorption above the mesopause in Jupiter (Strobel and Smith 1973). McElroy (personal communication 1975) has extended Strobel and Smith's (1973) calculations to include dissipation and subsequent absorption of the inertia gravity wave energy in the thermosphere. Energy absorbed at a level z in the thermosphere is conducted downward and radiated in the neighborhood of the mesopause by species like CH_4, C_2H_2, C_2H_4, C_2H_6, etc. (Strobel and Smith 1973). The temperature difference ΔT between z_0 and z is given by the following relationship (after Strobel and Smith 1973):

$$\Delta T = T(z) - T_0 = \frac{\Delta z}{\kappa} \mathscr{F}(z) \tag{2}$$

where $\mathscr{F}(z)$ is the energy flux associated with the inertia gravity waves and κ is the conductivity of the background gas H_2 — about 68.6 $T(z)$ ergs cm^{-1} sec^{-1} °K^{-2} according to Strobel and Smith (1973). In view of the electron density profile observed by Pioneer 10 (shown in Fig. 4) and the model ionosphere of Fig. 3, it appears that the bulk of the energy carried by the inertia gravity waves is dissipated nearly 250 km above the mesopause (i.e., $\Delta z = 250$ km). With the above parameters we have calculated Jupiter's upper atmosphere temperature profile. The results of our calculations are depicted in Fig. 5; we estimate the exospheric temperature T_∞ to be nearly 900°K higher than the mesopause temperature. With the lower boundary at the mesopause ($T_0 = 150$°K, $n(H_2) = 10^{13}$ cm^{-3}) and T_∞ as calculated above, we obtain an atmospheric density profile, essentially H_2 insofar as the topside ionosphere is concerned, from the mesopause through the thermosphere. The electron density profile for this hot thermospheric model we have then calculated in the manner outlined earlier in this chapter. We assume after Henry and McElroy (1969) that thermal equilibrium is maintained between electrons, ions, and neutral molecules (i.e., $T_e \simeq T_i \simeq T_n$). Protons, which are the major ions at thermospheric heights, are lost by radiative recombination with electrons. We assume a $T_e^{-0.5}$ dependence on electron temperature for the radiative recombination rate α_r of the protons (Bates and Dalgarno 1962). The small-dashed curve in Fig. 4 shows the electron density profile calculated for this hot thermospheric model, when we assume $\xi = 81°$, $T_e \simeq T_i \simeq$

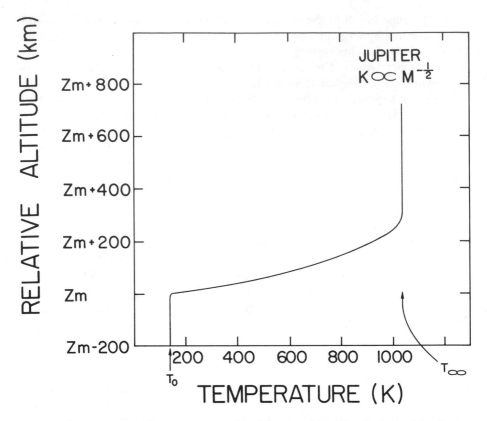

Fig. 5. Upper atmospheric temperature profile of Jupiter calculated on the basis of the dissipation of the upward propagating inertia gravity waves. The height scale refers to altitude above or below a mesopause level, Z_m assumed at density level 10^{13} cm^{-3}. The mesopause temperature, T_0 is assumed to be 150°K and the exospheric temperature T_∞ is estimated at $\sim$ 1050°K.

T_n and $\alpha_r \propto T_e^{-0.5}$. If we assume α_r to vary as $T_e^{-0.75}$ (Bauer 1973) we obtain a profile drawn between the crosses (x's) in Fig. 4. As should be expected, the model in Fig. 4 does not differ appreciably from the one shown in Fig. 3 below the thermosphere. We also find that the transport effects begin to be important above $\sim$ 1500 km (Fig. 4) where the time constant for diffusion is smaller than the photochemical time constant. We show the diffusive equilibrium distribution of protons above 1500 km by the long-dashed curve in Fig. 4. If we ignore the various layers, $L1$–$L7$ which the radio occultation data indicate to be present in the Ne profile shown in Fig. 4, we see reasonably good agreement (certainly within a factor of two at all altitudes) between the measurements and the model depicted by the broken line in Fig. 4. We do not attempt to construct a model that will match identically the measured profile since the measurements are not entirely unambiguous. The purpose of these calculations is to illustrate that our ideas of Jupiter's thermosphere probably must be modified in view of the apparent topside ionospheric profile measured by Pioneer 10. One can easily see, though, that only a slight modification of the thermospheric temperature structure would yield a better fit to the data.

The time constant for removal of protons at the altitude of $L5$ is on the order of 10^4 sec, and it increases to 10^7 at the altitude of $L1$. At these altitudes, long-lived protons will, under the combined influence of Jupiter's magnetic field and probable horizontal winds (or propagating vertical waves), cluster into layers at positions of null points or nodes. This phenomenon may be likened to the sporadic-E layering in the earth's ionosphere. Layer $L7$ appears to be at the level of maximum CH_3^+ ion production. Identification of layer $L7$ with CH_3^+ ion production would, however, be premature. This is due to the fact that a reduction in the ratio, $\sigma_I(CH_3)/r_7$ by as little as a factor of 4 would wipe out this feature in the N_e profile. Preliminary data of Maier and Fessenden (1975) suggest a value of r_7, twice as large as that given by Rebbert *et al.* (1973); and the value of $\sigma_I(CH_3)$ used here is the upper limit. Furthermore, a possibility exists that $C_2H_5^+$ may not be the terminal ion in the reaction e_{17}. New measurements indicate that $C_2H_5^+$ combines rapidly with C_2H_4 and C_2H_2 *at room temperature* to yield higher-order hydrocarbon ions, $C_3H_5^+$ and $C_4H_5^+$:

$$C_2H_5^+ + C_2H_4 \rightarrow (C_4H_9^+)^* \rightarrow C_3H_5^+ + CH_4 \tag{3}$$

and

$$C_2H_5^+ + C_2H_2 \rightarrow (C_4H_7^+)^* \rightarrow C_4H_5^+ + H_2 . \tag{4}$$

Thus it might be the recombination rate of $C_3H_5^+$ and $C_4H_5^+$ that should enter in the ratio $\sigma_I(CH_3)/r$. No information is yet available on the needed rates. In view of this, we regard identification of the peak $L7$ as due to CH_3^+ to be dubious at best. The height of the $L7$ layer coincides with the lower elevation peak (due to CH_3^+) in the variable K model; this layer is nearly 50 km lower in the low $K(K = 10^5 \ cm^2 \ sec^{-1})$ model (Fig. 3). Layers $L6$ and $L7$ could possibly be due to Io-related sodium ions which are expected to end up in Jupiter's lower ionosphere as surmised by Atreya *et al.* (1974) in their pre-Pioneer 10 model. Removal of Na^+ is extremely slow ($\tau \simeq 10^6$ sec) by radiative recombination with a rate constant on the order of $10^{-12} \ cm^3 \ sec^{-1}$. This is the only likely loss mechanism of these ions in Jupiter's atmosphere. Clustering with ammonia in the presence of a third body and subsequent dissociative recombination of the cluster may be an alternate avenue as in the following reactions:

$$Na^+ + NH_3 + M \rightarrow Na^+ \cdot NH_3 + M \quad (k \simeq 10^{-28} \ cm^6 \ sec^{-1}) \tag{5}$$

followed by

$$Na^+ \cdot NH_3 + e \rightarrow Na + NH_3 \quad (k \simeq 10^{-6} \ cm^3 \ sec^{-1}) . \tag{6}$$

However, removal of the sodium ions by ammonia in Jupiter's atmosphere appears unlikely due to the extremely low NH_3 densities at ionospheric heights. Presence of metallic ions other than sodium, either intrinsic to Jupiter or those swept from the torus, is likely. Therefore, more than one stratospheric layer in the electron density profile may be attributed to these potential long-lived metallic ions. However, it has by no means been demonstrated that sodium ions from Io can penetrate into the Jovian atmosphere enough to provide a significant source.

Acknowledgements. We have benefited from enlightening discussion with D. M. Hunten on the Pioneer 10 radio occultation data, with M. B. McElroy on Jovian thermospheric heating and with D. F. Strobel on the EUV photolysis of hydrocarbons. This work was sponsored by the Kitt Peak National Observatory operating under contract with the National Science Foundation, and it was supported by the Jet Propulsion Laboratory, California Institute of Technology under NASA Contract NAS 7-100.

DISCUSSION

D. M. Hunten: The suggestion that Ionian sodium ions would fall down the fieldlines into Jupiter is interesting, but has one big problem. There has been a lot of work on Jupiter as a centrifugal accelerator, based on the idea that the magnetic field carries ions around much faster than the Kepler velocity. The net force on them is strongly outward. Perhaps a way can be found around this difficulty, but the situation is hardly as simple as Atreya *et al.* (1974) originally suggested.

S. K. Atreya: I have two comments on this remark: (1) as we have stated in the paper the source of potential sodium ions in Jupiter's stratosphere is not unequivocally established; and (2) we may be attempting to explain an unreal feature in the electron density profile as inferred from the Pioneer 10 data. Regarding the source of the sodium ions, I think Ionian sodium is only one of the several possibilities. Sodium intrinsic to Jupiter or that deposited by the meteorites, for example, is another likelihood. The validity of these hypotheses remains to be tested. My second comment refers to G. Fjeldbo's analysis of his Pioneers 10 and 11 data. Fjeldbo *et al.* (1975) state that the ionization peaks $L6$ and $L7$ do not have distinct signatures in the radio recordings and instead of being real layers, these computed ionization peaks may have been caused by scintillation noise or by deviations from spherical symmetry. It is for explaining precisely these "layers," $L6$ and $L7$, that we invoked the metallic ion hypothesis. At and beyond the altitude of layer $L5$, the lifetime of the protons is large enough for them to behave in a way similar to the metallic ions in the sporadic E-type layering.

REFERENCES

Atreya, S. K., and Donahue, T. M. 1975a. Ionospheric models of Saturn, Uranus and Neptune. *Icarus* 24:358–362.

——. 1975b. The role of hydrocarbon in the ionospheres of the outer planets. *Icarus* 25: 335–338.

Atreya, S. K.; Donahue, T. M.; and McElroy, M. B. 1974. Jupiter's ionosphere: prospects for Pioneer 10. *Science* 184:154–156.

Bates, D. R., and Dalgarno, A. 1962. *Electronic recombination, atomic and molecular processes*. (D. R. Bates, ed.) p. 245. New York: Academic Press.

Bauer, S. J. 1973. *Physics of planetary ionospheres*. p. 84. New York: Springer-Verlag.

Browning, R., and Fryar, J. 1973. Dissociative photoionization of H_2 and D_2 through the $1s\sigma_g$ ionic state. *Proc. Phys. Soc. London (At. Mol. Phys.)* 6:364–371.

Capone, L. A., and Prasad, S. S. 1973. Jovian ionospheric models. *Icarus* 20:200–212.

Carlson, R. W., and Judge, D. L. 1974. Pioneer 10 ultraviolet photometer observations at Jupiter encounter. *J. Geophys. Res.* 79:3623–3633.

Chase, S. C.; Ruiz, R. D.; Münch, G.; Neugebauer, G.; Schroeder, M.; and Trafton, L. M. 1974. Pioneer 10 infrared radiometer experiment: preliminary results. *Science* 183:315–317.

Cook, G. R., and Metzger, P. H. 1964. Photoionization and absorption cross sections of H_2 and D_2 in the vacuum ultraviolet region. *J. Opt. Soc. Amer.* 54 (8):968–972.

Dalgarno, A. 1971. Applications in aeronomy. *Physics of electronic and atomic collisions, VII ICPEAC.* (T. R. Grover and F. J. deHeer, eds.) pp. 381–398. Amsterdam: North Holland Publishing Co.

Elder, F. A.; Geise, C.; Steiner, B.; and Inghram, M. 1962. Photoionization of alkyl-free radicals. *J. Chem. Phys.* 36:3292–3296.

Elliott, J. L.; Wasserman, L. H.; Veverka, J.; Sagan, C.; and Liller, W. 1974. The occultation of Beta Scorpii by Jupiter. 2. The hydrogen-helium abundance in the Jovian atmosphere. *Astrophys. J.* 190:719–729.

Fink, U., and Belton, M. J. S. 1969. Collision-narrowed curves of growth for H_2 applied to new photoelectric observations of Jupiter. *J. Atmos. Sci.* 26:952–962.

Fjeldbo, G.; Kliore, A. J.; Seidel, B.; Sweetnam, D.; and Cain, D. 1975. The Pioneer 10 radio occultation measurement of the ionosphere of Jupiter. *Astron. Astrophys.* 39:91–96.

French, R. G., and Gierasch, P. J. 1974. Waves in the Jovian upper atmosphere. *J. Atmos. Sci.* 31:1707–1712.

Gross, S. H., and Rasool, S. I. 1964. The upper atmosphere of Jupiter. *Icarus* 3:311–322.

Henry, R. J. W., and McElroy, M. B. 1969. The absorption of extreme ultraviolet solar radiation by Jupiter's upper atmosphere. *J. Atmos. Sci.* 26:912–917.

Hinteregger, H. E. 1970. The extreme ultraviolet solar spectrum and its variation during the solar cycle. *Ann. Geophys.* 26:547–554.

Hunten, D. M. 1969. The upper atmosphere of Jupiter. *J. Atmos. Sci.* 26:826–834.

Hunten, D. M., and Münch, G. 1973. The helium abundance on Jupiter. *Space Sci. Rev.* 14: 433–443.

Huntress, Jr., W. T. 1974. A review of Jovian ionospheric chemistry. *Advances in atomic and molecular physics*. (D. R. Bates and B. Beaderson, eds.) pp. 295–340. New York: Academic Press.

Jenkins, E. B.; Wallace, L.; and Drake, J. F. 1973. Unpublished results from OAO-*Copernicus;* see discussion in Carlson and Judge 1974.

Johnsen, R., and Biondi, M. A. 1974. Measurements of positive ion conversion and removal reactions relating to the Jovian ionosphere. *Icarus* 23:139–142.

Judge, D. L., and Carlson, R. W. 1974. Pioneer 10 observations of the ultraviolet glow in the vicinity of Jupiter. *Science* 183:317–320.

Kliore, A.; Cain, D. L.; Fjeldbo, G.; and Seidel, B. L. 1974. Preliminary results on the atmospheres of Io and Jupiter from the Pioneer 10 S-band occultation experiment. *Science* 183: 323–324.

Leu, M. T.; Biondi, M. A.; and Johnsen, R. 1973. Dissociative recombination of electrons with H_3^+ and H_5^+ ions. *Phys. Rev.* 8:413–419.

Maier, H. N., and Fessenden, R. W. 1975. Electron ion recombination rate constants for some compounds of moderate complexity. *J. Chem. Phys.* 62:4790–4795.

McElroy, M. B. 1973. The ionosphere of the major planets. *Space Sci. Rev.* 14:460–473.

McElroy, M. B.; Yung, Y. L.; and Brown, R. A. 1974. Sodium emission from Io: implications. *Astrophys. J.* 187:L127–L130.

Miller, T. M.; Moseley, J. T.; Martin, D. W.; and McDaniel, E. W. 1968. Reactions of H^+ in H_2 and D^+ in D_2; mobilities of hydrogen and alkali ions in H_2 and D_2 gases. *Phys. Rev.* 173: 115–123.

Monahan, K. M.; Huntress, Jr., W. T.; Lane, A. L.; Ajello, J.; Burke, T. G.; LeBreton, P.; and Williamson, A. 1974. Cross sections for the dissociative photoionization of hydrogen by 584 Å radiation: the formation of protons in the Jovian ionosphere. *Planet. Space Sci.* 22: 143–149.

Moos, H. W., and Rottman, G. J. 1972. The far ultraviolet emission spectrum of Jupiter. *Bull. Amer. Astron. Soc.* 4:360.

Prasad, S. S., and Capone, L. A. 1971. The Jovian ionosphere: composition and temperatures. *Icarus* 15:45–55.

Prasad, S. S., and Tan, A. 1974. The Jovian ionosphere. *Geophys. Res. Lett.* 1:337–340.

Rebbert, R. E.; Lias, S. G.; and Ausloos, P. 1973. Pulse radialysis of methane. *J. Res. Nat. Bur. Stand.* (U.S.), 77A:249–257.

Rishbeth, H. 1959. The ionosphere of Jupiter. *Aust. J. Phys.* 12:466–468.

Samson, J. A. R. 1966. The measurement of the photoionization cross sections of the atomic gases. *Adv. At. Mol. Phys.* 2:177–261.

———. 1972. Observation of double electron excitation in H_2 by photoelectron spectroscopy. *Chem. Phys. Lett.* 12:625–627.

Samson, J. A. R., and Cairns, R. B. 1965. Total absorption cross sections of H_2, N_2 and O_2 in the region 550–200 Å. *J. Opt. Soc. Amer.* 55:1035.

Shimizu, M. 1971. The upper atmosphere of Jupiter. *Icarus* 14:273–281.

Stewart, A. L., and Webb, T. G. 1963. Photoionization of helium and ionized lithium. *Proc. Phys. Soc. London* 82:532–536.

Strobel, D. F. 1973. The photochemistry of hydrocarbons in the Jovian atmosphere. *J. Atmos. Sci.* 30:489–498.

———. 1975. Aeronomy of the major planets: photochemistry of ammonia and hydrocarbons. *Rev. Geophys. Space Phys.* 13:372–382.

Strobel, D. F., and Smith, G. R. 1973. On the temperature of the Jovian thermosphere. *J. Atmos. Sci.* 30:718–725.

Tanaka, T., and Hirao, K. 1973. Structure and time variations of the ionosphere. *Planet. Space Sci.* 21:751–762.

Theard, L. P., and Huntress, Jr., W. T. 1974. Ion-molecule reactions and vibrational deactivation of H_2^+ ions in mixtures of hydrogen and helium. *J. Chem. Phys.* 60:2840–2848.

Veverka, J.; Elliott, J.; Wasserman, L.; and Sagan, C. 1974. The upper atmosphere of Jupiter. *Astron. J.* 179:73–84.

Wallace, L., and Hunten, D. M. 1973. The Lyman alpha albedo of Jupiter. *Astrophys. J.* 182: 1013–1031.

Zabriskie, F. R. 1960. Studies on the atmosphere of Jupiter. Ph.D. dissertation. Princeton Univ., Princeton, New Jersey.

CHEMISTRY AND SPECTROSCOPY
OF THE JOVIAN ATMOSPHERE

R. G. PRINN
Massachusetts Institute of Technology

and

T. OWEN
State University of New York

A comprehensive review of the chemistry and spectroscopy of the Jovian atmosphere is presented. We begin by considering current ideas on atmospheric structure to provide a framework for the subsequent discussion. Thermochemical equilibrium models for determining atmospheric composition can apparently provide a reasonable first-order picture of the visible atmosphere. However, a number of the observed characteristics require consideration of disequilibrating processes. Theoretical methods for modeling disequilibrium phenomena including considerations of atmospheric transport in the troposphere, stratosphere, and thermosphere are reviewed. Studies of the photochemistry of H_2, CH_4, NH_3, H_2S, and PH_3 using these modeling methods are then discussed. The considerable problems associated with attempting to simulate the Jovian atmosphere in a laboratory flask are elucidated. The roles of lightning, thunder, and energetic particle bombardment are secondary to ultraviolet light in inducing disequilibration in the atmosphere. It is shown that inorganic colored species such as sulfur and red phosphorus are produced much more rapidly than colored organic species in Jupiter's atmosphere. Because the Jovian troposphere is rapidly mixed and thermal dissociation is rapid in the hot deep atmosphere, the existence of living organisms on Jupiter appears highly improbable. We present an historical review of the discovery of various gases in the Jovian atmosphere using spectroscopic methods. The many problems involved in interpreting the equivalent widths of observed spectral lines in terms of the abundance of the absorbing gas are emphasized. The various estimates for the abundances of H_2, HD, CH_4, CH_3D, NH_3, C_2H_6, C_2H_2, and PH_3 are critically discussed. Upper limits for the abundances of a number of unobserved gases in the visible atmosphere are also given. The identifications of PH_3, C_2H_4, CO, GeH_4, CO_2, and HCN need further verification. The abundance estimates for C_2H_6 and H_2O also require additional attention. A good deal of knowledge of the Jovian atmosphere has been gained by spectroscopic observations combined with studies of the chemistry of the atmosphere and these disciplines will no doubt continue to make important contributions in the future.

[319]

The recent successful flights of Pioneers 10 and 11 past Jupiter have provided the planetary science community with a plethora of new and significant information. These flights have also served to catalyze renewed effort toward answering the basic questions of structure, composition, circulation, origin, and evolution for this, the largest planet in our solar system.

For those of us interested in the chemistry of the Jovian atmosphere these two planetary flybys did not in fact provide us with any new compositional data other than the first direct detection of helium. We are, therefore, still entirely reliant on Earth-based spectroscopic observations for most of our information on atmospheric composition. It is indeed a tribute to a small and dedicated group of planetary astronomers operating from groundbased telescopes, aircraft, balloons, and small rockets, that we have nevertheless been able to answer at least some of the more fundamental chemical and compositional problems concerning Jupiter.

We have written this chapter with two goals in mind; firstly to present current reviews of the chemistry and spectroscopy of Jupiter's atmosphere and secondly to emphasize a number of unsolved problems which will require attention by future researchers in these disciplines. We begin the chapter with a brief section on atmospheric structure. No comprehensive review of Jovian chemistry has previously appeared in the literature; Secs. II and III by Prinn present an attempt to rectify this situation. This discussion is then followed by Secs. IV and V by Owen, which, used in conjunction with earlier reviews of this subject, provide an up-to-date picture of Jovian spectroscopy. In the final section we shall attempt to paraphrase the major unanswered questions with particular emphasis on current discrepancies between theory and observation.

I. ATMOSPHERIC STRUCTURE

The recent measurements from Pioneer 10 (Chase *et al.* 1974) confirm the earlier Earth-based observations (Low 1966; Aumann *et al.* 1969) suggesting that Jupiter emits at thermal wavelengths about twice the energy it receives from the sun at visible wavelengths.[1] Smoluchowski (1967) suggested that the required internal heat source was provided by gravitational contraction of the planet at a rate of about one millimeter per year. Hubbard (1968) showed that the existence of such an internal heat source implies that Jupiter has a convective fluid throughout most of its interior; a solid interior could not conduct heat outwards efficiently enough. We will see the significance of this conclusion for chemical modeling in Sec. II.

Structural models of Jupiter as a homogeneous completely convective fluid with an essentially adiabatic temperature lapse rate have now been ex-

---

[1]See pp. 202 and 204.

tensively investigated (Hubbard 1968, 1970).[2] These models are not, however, unique and Smoluchowski (1967, 1970a,b) emphasizes the possibility of inhomogeneous models with well-defined phase boundaries. These inhomogeneous models do, however, require the visible atmosphere to be considerably more enriched in helium than present observations suggest. In both types of model there is a transition from the familiar molecular hydrogen envelope to a metallic hydrogen core at about 0.8 of the planetary radius.

In the visible atmosphere we must consider both the upward flux of heat from the interior and the direct heating by the sun. Transport of energy by radiation (in addition to atmospheric motions) now becomes significant. Trafton (1967) computed radiative-convective equilibrium temperature profiles in which atmospheric motions were taken into account by simply altering the superadiabatic lapse rates in a radiative equilibrium calculation to the adiabatic lapse rate. Trafton and Stone (1974) argue that this radiative-convective temperature profile is an accurate representation of the true radiative-dynamical profile since dynamical fluxes (other than convective) are not strong enough to produce significantly subadiabatic lapse rates in Jupiter's troposphere.

The Trafton (1967) computations were designed to fit only the total thermal radiation from the planet. More refined models can be derived by attempting to fit portions of the actual thermal emission spectrum and its variation over the Jovian disk. This refinement has been attempted by Hogan et al. (1969), Cess and Khetan (1973), Wallace et al. (1974) and Orton (1975) with the most recent paper using Pioneer spacecraft data.[3] These various temperature profiles are somewhat model-dependent but all show a pronounced temperature inversion in the stratosphere. The predicted positions and temperatures of the temperature minimum in each model are roughly: 250 mbar, 115°K (Hogan et al. 1969); 100 mbar, 105°K (Cess and Khetan 1973); 80 mbar, 110°K (Wallace et al. 1974); 100 mbar, 105°K (Orton 1975). The latter three models are in reasonable agreement and their results are considered the most reliable. For purposes of illustration we present a typical model temperature profile in Fig. 1. The assumed temperature profile in the visible atmosphere impacts significantly of course on our discussions of upper atmospheric chemistry and spectroscopic abundances which follow.

II. THERMOCHEMICAL EQUILIBRIUM MODELS

If we accept the postulate reviewed above, namely that the Jovian atmosphere is convective to great depths with an essentially adiabatic lapse rate, it is then possible to model the chemical composition of the atmosphere as a

---

[2]See also p. 176.
[3]See also pp. 206 and 23.

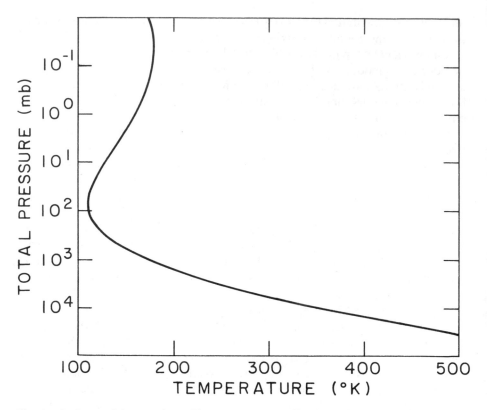

Fig. 1. Jovian model atmosphere. The temperature profile above the 800 mbar level is obtained from Wallace *et al.* (1974) and below the 800 mbar level an adiabatic lapse rate in a solar composition atmosphere is adopted.

function of altitude. The simplest notion to take is that the atmospheric constituents below the tropopause are in thermochemical equilibrium. There are no irreversible reactions occurring and the Gibbs free energy and the entropy of the air parcel are time-invariant. As we shall discuss later, this postulate should increase in validity with increasing atmospheric temperature; that is, with increasing depth into the planet. This approach has been taken by Lewis (1969*a,b*) and has produced a model which serves to explain a significant number of the gross compositional characteristics of the planet.

Of course, chemical modeling of Jupiter requires some hypothesis concerning elemental abundances. From an evolutionary viewpoint we can argue that Jupiter formed in a primitive solar nebula containing the elements in their solar or cosmic abundances, and that the planet is sufficiently massive to prevent any escape of these elements after formation. In addition, spectroscopic observations which we shall discuss later lend considerable credence to the idea that Jupiter is a solar composition planet. Further support is gained from the physical models of Jupiter constructed by Hubbard (1968,

1970) which we mentioned earlier and which imply that a solar composition mixture of the elements provides a satisfactory fit to the observed density and oblateness of the planet.

Lewis (1969a,b) considered a list of about 500 volatile compounds of the 40 cosmically most abundant elements. Recent estimates (Cameron 1974) of the cosmic abundance of the most abundant 50 elements is given in Table I and except for Fe the abundances used by Lewis do not differ in any important manner from these. Beginning at the base of his atmospheric model (2000°K, 2×10^5 bar) Lewis computed the abundance of each compound in thermochemical equilibrium. Those compounds, if any, whose vapor pressures exceeded their saturation vapor pressures over any solid or liquid phase, were removed from the gas phase until their vapor pressures equaled

TABLE I
The Fifty Most Abundant Elements in the Solar System[a]

Element	Abundance $(S_i = 10^6)$	Element	Abundance $(S_i = 10^6)$
H	3.18×10^{10}	B	3.50×10^2
He	2.21×10^9	V	2.62×10^2
O	2.17×10^7	Ge	1.15×10^2
C	1.18×10^7	Se	6.72×10^1
N	3.74×10^6	Li	4.95×10^1
Ne	3.44×10^6	Ga	4.8×10^1
Mg	1.06×10^6	Kr	4.68×10^1
Si	1.00×10^6	Sc	3.5×10^1
Fe	8.3×10^5	Zr	2.8×10^1
S	5.0×10^5	Sr[b]	2.69×10^1
Ar	1.17×10^5	Br	1.35×10^1
Al	8.5×10^4	As[b]	6.6
Ca	7.21×10^4	Te	6.42
Na	6.0×10^4	Rb[b]	5.88
Ni	4.80×10^4	Xe	5.38
Cr	1.27×10^4	Y[b]	4.8
P	9.60×10^3	Ba[b]	4.8
Mn	9.30×10^3	Mo[b]	4.0
Cl	5.70×10^3	Pb[b]	4.0
K	4.20×10^3	Sn[b]	3.6
Ti	2.77×10^3	Ru[b]	1.9
F	2.45×10^3	Cd[b]	1.48
Co	2.21×10^3	Nb[b]	1.4
Zn	1.24×10^3	Pt[b]	1.4
Cu	5.40×10^2	Pd[b]	1.3

[a]Table after Cameron (1974).
[b]Not included in the Lewis (1969a) model (this model did, however, include I, Be, and Hg whose presently accepted abundances are too low to appear in this table).

the saturation values. The temperature profile was then extrapolated upwards with the wet adiabatic lapse rate computed using the abundance and latent heat of condensation of these condensing compounds. The parcel of gas at the lowest level was then moved up an appropriate increment in altitude leaving behind all condensed phases.[4] These computations were then repeated at the new level until the tropopause altitude was reached.

In Fig. 2 we give a schematic representation of the principal cloud layers predicted for Jupiter. The diagram represents roughly the outer 6% of the planetary radius. Below the 2000°K level only refractory oxides such as CaO, MgO, Al_2O_3, Ti_2O_3 and V_2O_3 and iron group metals such as Fe, Ni, Cr, Co, and Cu could possibly precipitate out to form a solid or liquid planetary core. Lewis concludes that any core formed from these materials would constitute considerably less than 0.3% of the planetary mass and clearly such a core even if it existed would have little influence on the gross chemical and physical properties of the planet. Above the 1500°K level quartz (SiO_2) clouds precipitate out together with Na_2SiO_3, K_2SiO_3, and B_2O_3. Near the 460°K level ammonium chloride, bromide, and iodide condense out. The thickest clouds on the planet are predicted to occur near the 300°K level where aqueous ammonia and ice crystals precipitate. Also phosphorus and fluorine are condensed out in this water-bearing cloud layer as $PO_3^=$ and F^- ions respectively. Near the 225°K level H_2S reacts with NH_3 to form solid NH_4SH clouds, and finally above the 165°K level NH_3 condenses out as colorless crystals which are predicted to constitute the bulk of the particulates in the visible atmosphere.

In Fig. 3 we illustrate the predicted cloud structure in the visible atmosphere in more detail. These results are dependent on the assumed atmospheric temperature profile which is apparent upon comparing them with recent computations by Weidenschilling and Lewis (1973) where different temperature profiles were utilized. In the latter paper the ice clouds were much more dominant relative to the aqueous ammonia clouds.

Finally, in Table II we show the predicted mixing ratios of all the important gases at the 300°K (water-cloud) level and the 150°K (ammonia-cloud) level. We will shortly see that the observed abundances of H_2, He, CH_4 and NH_3 in the visible atmosphere are in remarkable agreement with these model predictions. In addition, the predominantly bright white clouds on the planet can be satisfactorily interpreted as the ammonia crystals predicted in this model.

Despite its successes this model does not serve to explain all the observed details of the visible atmosphere. For example, H_2O has been observed but with an abundance considerably less than that predicted here. In addition,

---

[4]With this assumption the *absolute* cloud mass densities become dependent on the vertical altitude increment chosen and therefore should not be interpreted too literally. The *relative* densities in different cloud layers are less affected by this assumption and should be reasonably accurate. Complete modeling of the cloud densities requires knowledge of particle nucleation and growth rates, particle sedimentation velocities, and atmospheric turbulence and mean winds. None of these are known with the desired degree of accuracy.

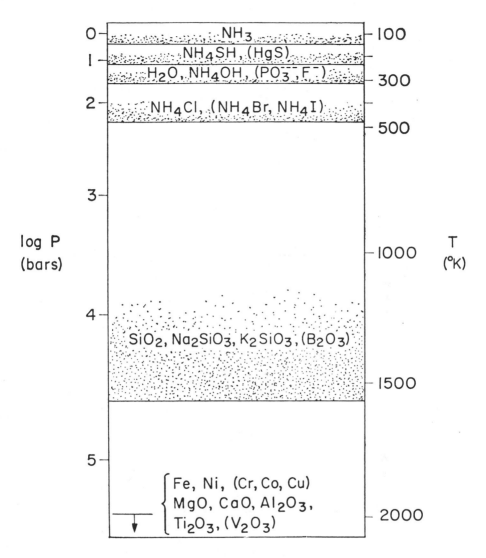

Fig. 2. Predicted cloud layers in an adiabatic solar-composition thermochemical-equilibrium model of the Jovian atmosphere (after Lewis 1969*b*). Minor constituents are parenthesized. The species indicated at the bottom of the figure are those which have precipitated out of the atmosphere below the 2000°K level and which may be present in Jupiter's core, should one exist. The altitude scale is strongly nonlinear; the 100°K level is ~ 500 km below the NH_3 clouds, and the 2000°K level is ~ 3800 km deeper.

H_2S has not been detected while PH_3 (phosphine), CO, C_2H_6, and C_2H_2 have.[5] Finally, these thermochemical calculations provide no candidate for the observed extensive colorations in the Jovian atmosphere. It is apparent

---

[5]See p. 405.

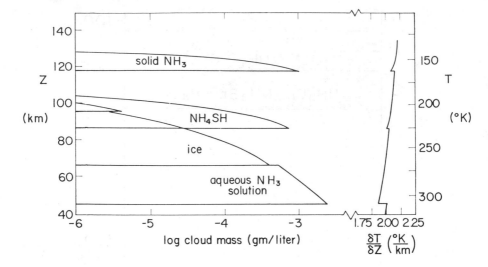

Fig. 3. Cloud masses and wet adiabatic lapse rate as functions of altitude and temperature for NH_3–H_2O and NH_3–H_2S clouds in a solar-composition model (after Lewis 1969a).

TABLE II
Predicted Mixing Ratios of Gases in the Ammonia Clouds ($\sim 150°K$ level) and in the Water Clouds ($\sim 300°K$ level)[a]

Gas	150°K level	300°K level
H_2	0.89	0.89
He	0.11	0.11
H_2O	8.8×10^{-13}	10^{-3}
CH_4	6×10^{-4}	6×10^{-4}
NH_3	1.2×10^{-5}	1.5×10^{-4}
Ne	1.3×10^{-4}	1.3×10^{-4}
H_2S	8.8×10^{-13}	2.9×10^{-5}
Ar	9.7×10^{-6}	9.7×10^{-6}
HF	~ 0	$\lesssim 2.1 \times 10^{-6}$
P_4O_6	~ 0	$\lesssim 2.1 \times 10^{-7}$
GeH_4	~ 0	$\ll 6.3 \times 10^{-9}$
Kr	3.2×10^{-9}	3.2×10^{-9}
Xe	2.5×10^{-10}	2.5×10^{-10}

[a]Table from Lewis (1969a).

that we must next look at possible processes which can disrupt thermochemical equilibrium in the atmosphere and thereby seek possible explanations for these phenomena.

III. DISEQUILIBRATING PROCESSES

When the Gibbs free energy or the entropy of an air parcel changes significantly with time, the assumption of thermodynamic equilibrium becomes invalid. In the Jovian atmosphere chemical disequilibration may be due to the occurrence of irreversible chemical reactions; for example, dissociation caused by photons, by shock waves from thunder, or by lightning discharges, incident on the air parcel. Disequilibration may also be due to a net exchange of material in the air parcel with its surroundings. Some examples of species which are either depleted or produced in the visible atmosphere by such disequilibrium processes are given in Table III.

TABLE III
Predicted Mixing Ratios of Some Gases Resulting from Disequilibration by UV Radiation and Advection

Gas	$[M] = 10^{17}$ cm^{-3} level[e]	150°K level[f]	$\geq$ 800°K level[g]
C_2H_6[a,d]	$\sim 10^{-5}$	$\sim 10^{-5}$	0
C_2H_4[a,d]	10^{-9} to 10^{-8}	~ 0 to 10^{-9}	0
C_2H_2[a,d]	10^{-7} to 10^{-6}	$\sim 10^{-8}$ to 10^{-6}	0
PH_3[b,d]	~ 0 to 10^{-9}	4×10^{-7}	4×10^{-7}
NH_3[c,d]	~ 0	1.2×10^{-5}	1.5×10^{-4}

[a]Strobel (1974).
[b]Prinn and Lewis (1975a).
[c]Strobel (1973a,b).
[d]Lewis (1969a,b).
[e]In the upper atmosphere.
[f]In the ammonia clouds.
[g]In the deep atmosphere where thermochemical equilibrium can be assumed.

The extent and cause of chemical disequilibration for each species i can be readily ascertained from a scale analysis of its continuity equation. This equation is inherently three-dimensional and its solution requires accurate knowledge of atmospheric motions. On Jupiter we unfortunately lack sufficient knowledge of these motions for us to consider anything more complex than a one-dimensional continuity equation. If species i is a minor constituent this is

$$\frac{\partial [i]}{\partial t} \simeq \left(\frac{d[i]}{dt}\right)_{\text{chem}} + \frac{\partial}{\partial z}\left(K[M]\frac{\partial f_i}{\partial z} + D_i[i]_e \frac{\partial f_i^*}{\partial z}\right) \qquad (1)$$

where z is altitude, t is time, $f_i = [i]/[M]$, $f_i^* = [i]/[i]_e$, $[i]$ is the observed molecular number density of i, $[i]_e$ is the molecular number density of i in diffusive equilibrium, and $[M]$ is the molecular number density of all molecules in hydrostatic equilibrium.

Hydrostatic equilibrium here implies zero net vertical mass flux and is defined by

$$\frac{1}{[M]}\frac{\partial [M]}{\partial z} = -\frac{\bar{m}g}{kT} - \frac{1}{T}\frac{\partial T}{\partial z} = -\frac{1}{H_M} \tag{2}$$

where $\bar{m}$ is the mean molecular mass, g is the acceleration due to gravity, k is the Boltzmann constant and T is temperature. Diffusive equilibrium here implies zero net vertical flux of i by molecular diffusion and is defined by

$$\frac{1}{[i]_e}\frac{\partial [i]_e}{\partial z} = -\frac{m_i g}{kT} - \frac{(1 + \alpha_i)}{T}\frac{\partial T}{\partial z} = -\frac{1}{H_i^e} \tag{3}$$

where m_i is the molecular mass of i and α_i is the thermal diffusion parameter of i. We refer to H_M and H_i^e as the density scale heights of M and i under the appropriate equilibrium conditions.

The first term of the right-hand side of Eq. (1) describes the time variation of $[i]$ in the air parcel due to chemical reactions and the second term describes the variation due to advection. The parameter K is the eddy diffusion coefficient intended to describe vertical mixing by both large-scale winds and small-scale eddies and D_i is the molecular diffusion coefficient. The turbopause is defined where $D_i = K$ and we can usually neglect eddy diffusion above the turbopause and molecular diffusion below the turbopause. It is also usual to consider long-term averages in planetary atmospheres and we therefore seek only the steady-state solution for (1) where $\partial [i]/\partial t = 0$.

If we define the time constant due to a particular process as the time during which the process must operate to change $[i]$ by a factor of e, we can use these time constants to determine the dominant terms in (1). Thermochemical equilibrium of i requires the time constant, for thermochemical reactions producing *and* removing i, to be much shorter than those for other reactions and for advection. The advection time constant is simply

$$\left| \frac{K}{H_i}\left(\frac{1}{H_M} - \frac{1}{H_i}\right) \right|^{-1} \tag{4}$$

below the turbopause and

$$\left| \frac{D_i}{H_i}\left(\frac{1}{H_i^e} - \frac{1}{H_i}\right) \right|^{-1} \tag{5}$$

above the turbopause. Here H_i is the actual density scale height of i. For molecular reactants the thermochemical reaction rates are generally very rapid only at temperatures well above room temperature, while for atomic, free radical, or ionic reactants they can be rapid even at temperatures well below room temperature.

A photochemical steady state for i occurs when i is predominantly produced by photodissociation and predominantly removed by a thermochemical reaction or vice versa. A photochemical-dynamical steady state occurs when i is added to the system by advection and removed by photodissociation or vice versa. A thermochemical-dynamical steady state exists when i is produced by thermochemical reactions and removed by advection or vice versa. With these simple concepts in hand we can now proceed to consider disequilibrating processes in the Jovian atmosphere.

A. Advection

In order to assess the disequilibrating role played by advection on Jupiter we require values for D_i and K as a function of altitude. Values for D_i can be readily assessed from the kinetic theory of molecules (e.g., Hirschfelder *et al.* 1954) and in most cases we can write

$$D_i(z) = D_i(0) \left(\frac{T(z)}{T(0)}\right)^{\frac{1}{2}} \frac{[M](0)}{[M](z)}. \tag{6}$$

Thus D_i increases upwards roughly inversely as the atmospheric density.

Stone (p. 591) has given an extensive review of the dynamics of the Jovian atmosphere below the turbopause. Unfortunately, the relationships between these actual atmospheric motions and the hypothetical diffusion coefficient K which we use here to parameterize them is complex if indeed such a relationship exists at all. In some cases we can identify the principal type of motion causing mixing in a particular region and therefore utilize a non-empirical or at least a semi-empirical estimate for K. For example, in the subadiabatic part of the Jovian atmosphere above the ~ 100 mbar level, transient internal gravity waves that are vertically propagating may play an important role in vertical mixing. In this case we can write approximately (Lindzen 1971)

$$K(z) \approx K(0) \left(\frac{[M](0)}{[M](z)}\right)^{\frac{1}{2}}. \tag{7}$$

In the deep Jovian atmosphere below the ~ 10 bar level, free convection is probably the dominant motion involved in vertical mixing and we can write (Stone p. 598)

$$K(z) \approx \left(\frac{R\phi}{C_p \bar{m}[M](z)}\right)^{\frac{1}{3}} H_M \approx 10^9 \text{ cm}^2 \text{ sec}^{-1}. \tag{8}$$

Here ϕ is the upward heat flux carried by the free convection which is approximately equal to the solar constant at Jupiter, R is the gas molar constant, and C_p is the molar gas heat capacity at constant pressure. In the region between the 10 bar and 100 mbar levels where most of the visible clouds exist, $K(z)$ may be similar to that for free convection [Eq. (8)] at low latitudes and similar to that for vertical mixing by baroclinic eddies (Stone 1973) at high latitudes

$$K(z) \approx 0.23 \frac{fH_M^2}{1 + R_i} \gtrsim 10^7 \text{ cm}^2 \text{ sec}^{-1}. \tag{9}$$

Here R_i is the Richardson number which is $\lesssim 100$ (Stone p. 593) and f is the Coriolis parameter ($\simeq 4 \times 10^{-4}$ sec^{-1} at high latitudes on Jupiter).

The most reliable method of obtaining K is, not unexpectedly, empirical. If we can observe the vertical distribution of a diffusing species (or some related quantity) we can essentially invert Eq. (1) to obtain the appropriate K value. As we shall discuss shortly this procedure has been carried out on Jupiter using NH_3 and H_2 as the diffusing species.

B. Photochemical Reactions

1. *Hydrogen*. The photochemistry of H_2 on Jupiter was first discussed by Gross and Rasool (1964) and subsequently by Hunten (1969) and Wallace and Hunten (1973). Molecular hydrogen in the Jovian upper atmosphere is ionized to H_2^+ by ultraviolet light of wavelengths < 804 Å and dissociated into H atoms directly below 845 Å and by predissociation below 1109 Å. The thermochemical charge exchange reaction of He^+ with H_2 also produces H_2^+. Essentially all the H_2^+ is converted to hydrogen atoms resulting in a net dissociation of H_2. The chemical cycle is illustrated in Fig. 4.

The column production rate of H atoms by these various processes is $\sim 10^9$ molecules cm^{-2} sec^{-1} and this equals twice the column destruction

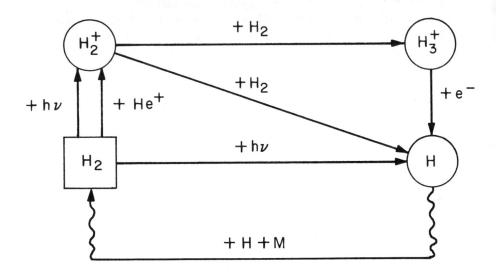

Fig. 4.   A summary of hydrogen photochemistry in Jupiter's thermosphere. In Figs. 4, 5, 6, 8, and 9 stable molecular species are enclosed in squares while reactive atoms, radicals, and ions are enclosed in circles. Vertical transport is represented by a vertical wavy line.

rate of H_2. At the highest altitudes the time constant associated with the thermochemical three-body reaction

$$H + H + M \rightarrow H_2 + M \qquad (10)$$

is much longer than that for vertical advection. Thus H atoms after photo-chemical production are transported downwards (by molecular diffusion above the turbopause and eddy diffusion below the turbopause) until they reach the level where the time constant for Reaction (10), which is inversely proportional to $[M]$, equals the advection time constant. Above this level H and H_2 are in a photochemical-dynamical steady state while below it they are in a thermochemical-dynamical steady state.

The column abundance of H atoms above the level where CH_4 begins to absorb solar H Lyman-α radiation can be obtained from measurements of the Lyman-α emission by the planet. By solving Eq. (1) for the H atom distribution we may thus obtain an estimate of K near the Jovian turbopause (Hunten 1969; Wallace and Hunten 1973). The K values derived in these latter papers depended upon Earth-based measurements of the Lyman-α emission which included emission from the hydrogen torus associated with Io. The recent measurements from Pioneer 10 (Judge and Carlson 1974)[6] suggest that the emission from the Jovian disk alone is about 1/4 that previously assumed. Utilizing Wallace and Hunten's (1973) Fig. 6 for a 150°K thermosphere, the old measurements imply $3 \times 10^4 < K < 10^6$ cm^2 sec^{-1} with a preferred value of 3×10^5 cm^2 sec^{-1}; the new measurements imply $4 \times 10^5 < K < 10^9$ cm^2 sec^{-1} with a preferred value of 6×10^7 cm^2 sec^{-1}.

2.   *Methane.* Photodissociation of CH_4 has been extensively studied by Strobel (1969, 1973a, 1974). It is dissociated by wavelengths < 1600 Å with about 70% of the dissociation attributable to the solar H Lyman-α line at 1216 Å. At the latter wavelength $\sim 8\%$ of the dissociated methane yields CH radicals and 92% yields $^1CH_2$. The CH predominantly reacts with CH_4 to produce C_2H_4 which may then lead to higher hydrocarbons. The $^1CH_2$ predominantly reacts with H_2 to produce CH_3. The CH_3 in turn combines principally with H atoms to recycle CH_4. Roughly 20% of the dissociated methane is converted irreversibly to C_2H_2, C_2H_4, and C_2H_6 (Strobel 1973a). These higher hydrocarbons are transported down to the hot lower atmosphere where thermochemical reactions convert them back to the thermo-dynamically stable CH_4. This chemical cycle for methane is illustrated in Fig. 5.

Ethane and acetylene which both have negligible concentrations in the thermochemical equilibrium model of Lewis (1969a,b) are predicted to have volume mixing ratios $\sim 10^{-5}$ and $\sim 5 \times 10^{-7}$ respectively, after disequilibration by ultraviolet light (Strobel 1974). Both these predicted species have

---

[6]See also p. 418.

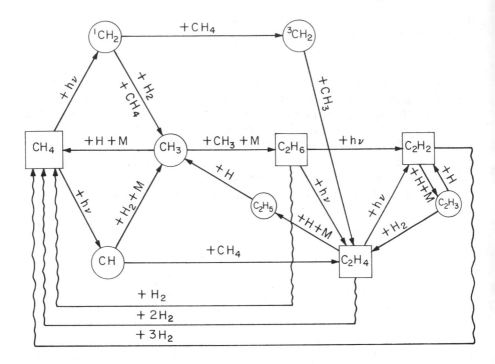

Fig. 5. A summary of methane photochemistry in Jupiter's mesosphere and stratosphere. The recycling of CH_4 from C_2H_6, C_2H_4, and C_2H_2 is accomplished in the hot lower atmosphere by thermochemical reactions. (See Fig. 4 legend for further information.)

recently been observed in the Jovian atmosphere (Ridgway 1974a)[7] although, as we shall discuss later, the acetylene observation needs further verification. The predicted abundances are strongly dependent on the assumed values for K. The mixing ratios quoted above were computed using a K value varying from $\sim 2 \times 10^4$ cm^2 sec^{-1} in the Jovian stratosphere to $\sim 10^8$ cm^2 sec^{-1} at the turbopause.

3. *Ammonia.* Ammonia photochemistry in Jupiter's stratosphere was first discussed qualitatively by Cadle (1962) and McNesby (1969). A more quantitative treatment has recently been presented by Strobel (1973b).

Ammonia absorbs significantly at all wavelengths less than 2300 Å. As we pointed out earlier, condensation depletes this gas in the upper levels of the troposphere. The NH_3 mixing ratio in the upper atmosphere therefore cannot significantly exceed its mixing ratio in a saturated air parcel at the Jovian tropopause or temperature minimum. Some time ago, Greenspan and Owen (1967) showed that if the tropopause temperature is $\sim 88°$K then condensation is sufficient to explain the rapid decrease in the NH_3 mixing

[7]See also p. 405.

ratio with altitude suggested by comparison of ultraviolet and infrared measurements. Multiple scattering computations by Prinn (1970)[8] also showed that the altitude decrease in the NH_3 mixing ratio caused by condensation at an 88°K tropopause necessitated that essentially all radiation of wavelengths less than 1600 Å is absorbed by CH_4, H_2, and He, or scattered out of the atmosphere before reaching the levels where significant NH_3 concentrations existed. One could thus treat CH_4 and NH_3 photochemistry separately and for NH_3 consider only photons in the 1600–2300 Å range.

However, the more recent models of the Jovian atmosphere which we reviewed earlier suggest tropopause temperatures $\gtrsim 110°K$. Tomasko (1974) pointed out that an atmosphere saturated with NH_3 at these higher temperatures could no longer explain the very low NH_3 abundances inferred from ultraviolet observations. Either the gas passing through the tropopause has a very low NH_3 relative humidity or NH_3 gas is being depleted in the stratosphere by photochemistry.

The conclusion mentioned above, that only wavelengths in the 1600–2300 Å range penetrate to the regions where appreciable NH_3 absorption occurs, is independent of the method of NH_3 depletion in the stratosphere, and it therefore still holds. Ammonia is dissociated by radiation in this latter wavelength interval to produce only NH_2 radicals. Strobel (1973b) estimates that about 5/8 of these NH_2 radicals are recycled back to NH_3 by combination with H. The other 3/8 react with NH_2 to produce N_2H_4 which, depending on the local temperature and other factors, will condense as hydrazine crystals or react with H atoms to ultimately produce N_2. There is thus an irreversible destruction of NH_3 in the upper atmosphere which must be compensated for by upward transport of ammonia gas. The N_2H_4 and N_2 produced are then transported down to the hot lower troposphere where thermochemical reactions can convert them back to NH_3. This chemical cycle for NH_3 is illustrated in Fig. 6.

In his numerical computations Strobel (1973b) varied the vertical diffusion coefficient K until, working in conjunction with Tomasko (1974), his computed NH_3 distributions provided a satisfactory fit to the ultraviolet albedos. The best fit was obtained for $K = 2 \times 10^4$ cm² sec⁻¹; the stratosphere in the 20 to 500 mbar region on Jupiter therefore appears to be only weakly mixed and this weak mixing is responsible for the observed depletion of NH_3 in the upper atmosphere. The latter point is most easily illustrated by using an analytical solution to Eq. (1) (see Prinn 1975) which gives for the scale height h of the NH_3 volume mixing ratio, f_{NH_3}, in an isothermal region

$$\frac{1}{f_{NH_3}} \frac{\partial f_{NH_3}}{\partial z} = -\frac{1}{h} - \frac{1}{2H_M} - \sqrt{\frac{1}{4H_M^2} + \frac{1}{Kt_0}}. \tag{11}$$

---

[8]There is an inadvertent error in the vertical axis in the figure 7 of this paper. The actual volume absorption rates should be 1/50 those indicated.

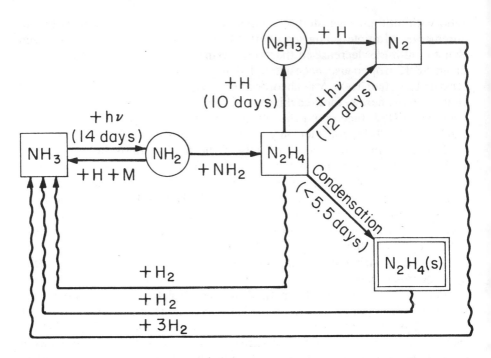

Fig. 6.   A summary of ammonia photochemistry in the Jovian stratosphere. In Figs. 6, 8, and 9 condensed particulate material is enclosed in two squares. The recycling of NH_3 from N_2 or hydrazine occurs through thermochemical reactions in the hot lower atmosphere. Condensation as hydrazine dominates the fate of N_2H_4 gas in the stratosphere providing the N_2H_4 condensation time is <5.5 days. (See Fig. 4 legend for further information.)

Here t_0 is the lifetime of NH_3 for irreversible conversion to N_2 or N_2H_4 which is about 38 days (Strobel 1973b). Using $H_M = 17$ km and $K = 2 \times 10^4$ cm^2 sec^{-1}, we have $h = 2.8$ km; the NH_3 mixing ratio therefore decreases by a factor of e every 2.8 km in altitude in the upper atmosphere.

Strobel (1973b) emphasized that his computed N_2H_4 concentrations exceeded by several orders of magnitude the maximum concentrations of N_2H_4 allowed by the hydrazine vapor pressure equation. Under these conditions homogeneous nucleation of hydrazine crystals should be very rapid and in addition micrometeorite dust and ultraviolet and cosmic ray produced ions may provide the means for heterogeneous nucleation.

In order to assess the thickness of this proposed hydrazine layer Prinn (1974) constructed a model including irreversible dissociation of NH_3 to form N_2H_4 crystals above the 250 mbar level, vertical eddy transport of NH_3, vertical transport of particles by both eddies and sedimentation, and finally irreversible evaporation of hydrazine particles below the 1 bar level. He assumed $K = 2 \times 10^4$ cm^2 sec^{-1} above the 250 mbar level and 10^5 cm^2 sec^{-1} below it. The computational procedure used was the same as that used for the Venus sulfuric acid clouds (Prinn 1973, 1975) and for the Jovian phosphorus clouds (Prinn and Lewis 1975a,b).

Combining these computed hydrazine particle densities with Mie scattering calculations for various particle radii, Prinn (1974) obtained the results in Fig. 7 for the opacity of the hydrazine haze layer at 3000 Å. This opacity depends critically on both the NH_3 concentrations in the lower stratosphere (i.e., on the tropopause temperature and pressure) and on the particle sedimentation velocity and extinction efficiency (i.e., on the particle radius). In order to fit the observed geometric albedo at 3000 Å on Jupiter, Axel (1972) required a high-altitude haze layer (of then unknown composition) with an absorbing optical depth ~ 0.2 to 0.25. From Fig. 7 this opacity is precisely what would be expected from a layer of small hydrazine particles (radii $\ll 1$ μm) with a tropopause temperature between 110 and 120°K.

4. *Hydrogen Sulfide.* Photodissociation of H_2S on Jupiter has been discussed by Lewis and Prinn (1970, 1971) and Prinn (1970). Hydrogen sulfide is dissociated by all wavelengths < 2700 Å. The condensation of H_2S with NH_3 to form NH_4HS causes this gas to be strongly depleted above the 230°K level on Jupiter and the multiple scattering computations by Prinn (1970) imply that only radiation in the 2200–2700 Å radiation penetrates to levels where H_2S concentrations are significant. Absorption in this wavelength region produces only HS and H.

The details of the subsequent chemistry are currently under investigation. Destruction of H_2S is caused by both photodissociation and hydrogen abstraction

$$\left. \begin{aligned} H_2S &\to H + HS \quad (2200 \text{ Å} < \text{wavelength} < 2700 \text{ Å}), \\ H + H_2S &\to H_2 + HS. \end{aligned} \right\} \quad (12)$$

The primary fate of the HS radicals appears to be

$$HS + HS \to H_2S + S \qquad (13)$$

$$HS + H \to H_2 + S \qquad (14)$$

$$HS + HS \to H_2 + S_2 \qquad (15)$$

$$HS + H + M \to H_2S + M \qquad (16)$$

The S atoms can react with HS or combine with other S atoms to give S_2 and the S_2 and S are both ultimately converted to elemental sulfur (S_8), ammonium polysulfides $[(NH_4)_xS_y]$, and hydrogen polysulfides (H_xS_y). These latter compounds are yellow, orange and brown. A tentative summary of the sulfur cycle on Jupiter is given in Fig. 8.

Photodissociation of H_2S should be most rapid in regions where the upper NH_3 clouds are thin or absent as this allows more ultraviolet radiation to penetrate down to the region of the NH_4HS clouds (Prinn 1970). With this in mind, Lewis and Prinn (1970) hypothesized that the NH_3 clouds over the North Equatorial Belt were essentially absent. The yellow and brown colorations in this belt (and other similarly colored regions of the

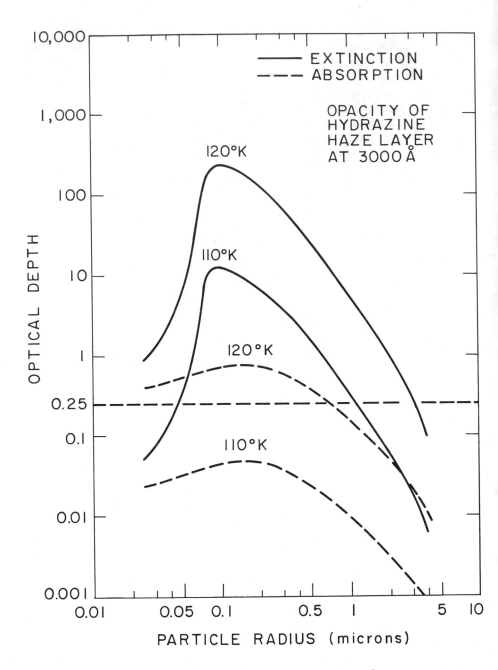

Fig. 7.   Predicted absorption and extinction optical depths at 3000 Å of a hydrazine haze layer in the Jovian stratosphere for tropopause temperatures of 110 and 120°K. The absorption optical depth required by Axel (1972) at 3000 Å to fit the observed ultraviolet spectrum is shown as a horizontal dashed line.

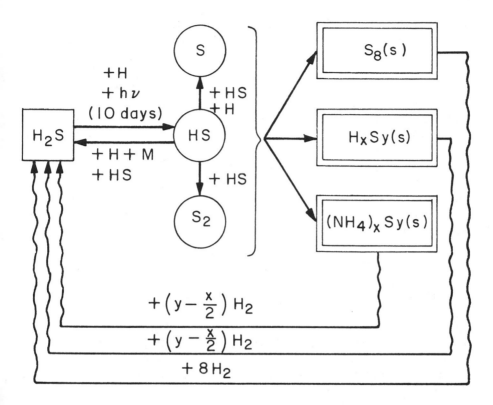

Fig. 8.   A tentative summary of hydrogen sulfide photochemistry in the region of the Jovian ammonium hydrosulfide clouds. The recycling of H_2S from elemental sulfur, and ammonium and hydrogen polysulfides takes place in the hot lower atmosphere. (See legends of Figs. 4 and 6 for further information.)

planet) were then a natural consequence of the ultraviolet dissociation of H_2S. This cloud-clearing hypothesis also provides an explanation for the variations in emission of 5 μm radiation between this belt and adjacent zones (Terrile and Westphal 1975). We should add that yellow and brown colorations have also been produced by direct irradiation of solid NH_4HS in the laboratory (Lebofsky 1974). This suggests that photochemical reactions within the NH_4HS particles themselves may also contribute to the colorations on Jupiter.

As a first approximation we can assume that the rate of irreversible dissociation of H_2S equals its photodissociation rate. This implies that the lifetime of H_2S for irreversible dissociation above the 230°K level, $t_{H_2S} \gtrsim 10^6$ sec (Prinn 1970). Depletion of H_2S in the atmosphere above the 230°K level over and above that expected from H_2S condensation is possible if vertical transport rates are slow. As mentioned earlier, Eq. (11) for NH_3 (or H_2S) is valid only in isothermal layers. We can divide the atmosphere into arbitrarily thin layers to satisfy this requirement and it is then apparent from Eq. (11) that depletion of H_2S by photodissociation in each layer is significant if

$$t_{H_2S} < \frac{4H_M^2}{K}. \tag{17}$$

This condition is satisfied for $K \lesssim 3 \times 10^7$ cm² sec⁻¹ and from our earlier discussion we suggested $10^7 \lesssim K \lesssim 10^9$ cm² sec⁻¹ in this atmospheric region. Depletion of H_2S above the 230°K level by photodissociation may therefore be significant.

If we assume thermochemical equilibrium, depletion of H_2S by condensation as NH_4HS particles is also very significant. In this case the mixing ratio scale height, h^*, of H_2S above the 230°K level is simply obtained by differentiating the thermochemical equilibrium equation between H_2S, NH_3, and solid NH_4HS (Prinn 1970)

$$-\frac{1}{h^*} = \frac{1}{f_{H_2S}} \frac{df_{H_2S}}{dz} = \frac{2.303 \times 4705}{T^2} \frac{dT}{dz} + \frac{2\bar{m}g}{kT}. \tag{18}$$

For $\dfrac{dT}{dz} = -2°K$ km⁻¹ (roughly the adiabatic lapse rate) we have $h^* \simeq 3$ km at the 230°K level; the equilibrium mixing ratio of H_2S decreases by a factor of e every 3 km above the 230°K level. This depletion of H_2S in the upper atmosphere suggests we should search for this gas in regions such as the North Equatorial Belt, where observation of the atmosphere at and below the 230°K level may be possible.

5. *Phosphine.* As we mentioned earlier the detection of PH_3 on Jupiter (if correct) is not in agreement with the thermochemical equilibrium model. This model indicated that phosphorus should be present as PH_3 only below the 800°K level in the deep unobservable portion of the atmosphere. Between the 800 and 300°K levels phosphorus is oxidized by H_2O to form P_4O_6 while above the 300°K level P_4O_6 dissolves in the water clouds and phosphorus compounds are therefore predicted to be absent in the visible atmosphere. A reasonable explanation for the observation of PH_3 is possible if atmospheric vertical mixing is much faster than the thermochemical reactions oxidizing PH_3 to P_4O_6. Similar arguments would apply to the thermochemical reduction reactions for CO and oxidation reactions for SiH_4 and GeH_4 if these gases were observed in the low-temperature visible atmosphere. Carbon monoxide, silane and germane have significant concentrations in thermochemical equilibrium only at high temperatures and pressures. Referring to Eq. (11) we see that this situation is possible (i.e., $\partial f_{PH_3}/\partial z \simeq 0$) if

$$t_{PH_3} \gg \frac{4H_M^2}{K} \sim 6 \text{ days} \tag{19}$$

where t_{PH_3} is the lifetime of PH_3 for irreversible oxidation to P_4O_6 in the vicinity of the 800°K level. In Eq. (19) we use $H_M = 114$ km and $K = 10^9$ cm^2 sec^{-1}. Simple arguments from reaction rate theory imply that $t_{PH_3} \gg 6$ days is not implausible. Alternatively if $f_{H_2O} < 10^{-6}$ at the 800°K level then there may simply be insufficient H$_2$O to oxidize all the PH$_3$ (we expect $f_{H_2O} \sim 10^{-3}$ in the solar composition model).

The photochemistry of PH$_3$ in the visible atmosphere has been recently considered by Prinn and Lewis (1975a,b). Phosphine is dissociated by wavelengths less than 2350 Å. For $K < 4 \times 10^6$ cm^2 sec^{-1} PH$_3$, like NH$_3$ [see Eq. (11)], is severely depleted in the upper atmosphere. Only wavelengths > 1600 Å penetrate to regions where PH$_3$ concentrations are appreciable and radiation in the 1600–2350 Å interval produces only PH$_2$ and H. Prinn and Lewis (1975a,b) estimated that about half of the PH$_2$ produced is recycled back to PH$_3$ and the other half reacts with PH$_2$ to ultimately produce P$_4$ which condenses to form triclinic red phosphorus crystals. The PH$_3$ chemical cycle is illustrated in Fig. 9.

In order to obtain quantitative estimates of red phosphorus particulate densities Prinn and Lewis (1975a,b) constructed a model including: (1) irreversible destruction of PH$_3$ with a lifetime $t_{PH_3} \simeq 38$ days above the 80 mbar, 112°K level, (2) vertical transport of PH$_3$ and red phosphorus particles, and (3) irreversible evaporation of the red phosphorus followed by thermochemical conversion of the resultant P$_4$ to PH$_3$ below the 20 bar, 430°K level. One run was intended to simulate average conditions on the planet and had $K = 10^4$ cm^2 sec^{-1} above the 80 mbar level, $K = 10^5$ cm^2 sec^{-1} in the 80 mbar to 20 bar region, and $K = 10^6$ cm^2 sec^{-1} below 20 bar. This run produced a particle optical depth of red phosphorus above the ammonia cloud tops (situated roughly at the tropopause or 80 mbar level) of only 0.02. Optical depths ≥ 1 are required to produce a visible coloration and, as observed, there is no distinct red coloration over the whole planet. A second run was intended to simulate Jupiter's Great Red Spot which has been pictured by Ingersoll (1973) as a long-lived storm system. Using $K = 10^6$ cm^2 sec^{-1} throughout in order to simulate this condition, the particle optical depth obtained above the ammonia cloud tops was 0.7 implying significant red coloration. If the Great Red Spot is a region of considerable dynamical activity, the Prinn and Lewis (1975a,b) model provides a self-consistent explanation for the redness of this region in comparison to the rest of the planet.

6. *Laboratory Simulations.* The accurate simulation of atmospheric photochemistry in the laboratory is an exceedingly difficult task. Studies in our own stratosphere have taught us that the best approach is to identify and measure the rates of each elementary reaction involved and then to apply these laboratory data to the construction of a computer model. It is this approach which we have just described for H$_2$, CH$_4$, NH$_3$, H$_2$S, and PH$_3$ photochemistry on Jupiter. However, it is almost inevitable that our

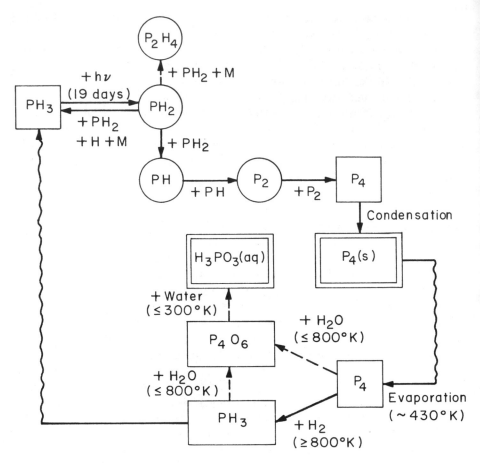

Fig. 9. A summary of phosphorus chemistry and photochemistry in the Jovian atmosphere. When vertical transport is very rapid, or H_2O concentrations are very low, the formation of P_4O_6 and H_3PO_3 (aq) from PH_3 is inhibited. In this case PH_3 photodissociates in the stratosphere yielding principally red phosphorus particles. (See legends of Figs. 4 and 6 for further information.)

selection of elementary reactions in these computer models is incomplete and laboratory experiments if properly conceived and interpreted may lead to identification of important omissions from these chemical schemes.

Because the primary products of photodissociation are strongly dependent on the incident ultraviolet radiation spectrum, the spectrum used in the laboratory should closely approximate that encountered in the region of the atmosphere under consideration. There is similarly no obvious validity in equating the results obtained from electrical-discharge experiments (Woeller and Ponnamperuma 1969) or high-temperature experiments (Sagan *et al.* 1967) with the chemistry induced by ultraviolet radiation in Jupiter's upper atmosphere. It is also mandatory that the temperature, pressure, and gas composition used in the experiment should closely mimic the real atmosphere. Temperature is important both for controlling reaction rates and also for determining whether condensation of a particular product (which essentially removes it from further reaction) will occur. The pressure is important for it controls three-body reaction rates and also the quenching rates for excited atoms and radicals. Care should also be taken to ensure that reactions on the walls of the apparatus are not influencing the results. All experimental simulations of Jovian photochemistry to date have neglected the enormous excess of H_2 and He over CH_4 and NH_3 and must be criticized on this basis.

The most difficult aspect of the atmosphere to simulate in the laboratory is the effect of advection and indeed the experimental studies to date have not even attempted to include it. It is therefore not surprising that these experiments have yielded complex molecules; successively more complicated species can build up without being removed, or without replacement of their precursors, by transport processes. Imaginative use of a flow system in which the photodissociation and transport times are comparable to those in the real atmosphere seems a feasible way to overcome this problem.

The most serious attempts at simulating Jovian photochemistry have been those of Sagan and Khare (1971) and Khare and Sagan (1973, 1975). These experiments are subject to certain of the shortcomings mentioned above but nevertheless have produced results worthy of discussion. They irradiated mixtures of CH_4, C_2H_6, NH_3, H_2S, and in some cases H_2O with 2537 Å radiation. This particular wavelength should penetrate down to levels where these gases have reasonable abundances (Prinn 1970). In the gas phase they produced $(C_2H_5S)_2$, $(C_2H_5)_2S$, CH_3CN, and possibly $(CH_3S)_2$ and H_2S_2. In runs intended to simulate the presence of the aqueous ammonia clouds (Lewis 1969a) by exposing the reactants to this solution, polymeric sulfur and α-amino acids were subsequently identified in the solution phase. Also a dark reddish-brown polymeric solid was produced which they suggested may be a dominant cause for the red colorations in the atmospheres of Jupiter, Saturn, and Titan.

This polymer was initially presumed to be organic but subsequent analysis revealed that it was primarily composed of elemental sulfur with only a few percent by weight of H, C, O, and N (Khare and Sagan 1975). Sulfur and sulfur compounds had of course been suggested earlier by Lewis and Prinn (1970) as the dominant yellow and brown coloring agents in the Jovian clouds. We should also stress here that the C, O, and N compounds identified in these particular experiments result from dissociation of CH_4, C_2H_6, H_2O, and NH_3 by translationally excited H atoms. These atoms are produced from H_2S photodissociation at 2537Å. However, the experiments did not contain the factor of 10^3 to 10^4 excess of H_2 and He over the other reactants which would be required to simulate Jupiter's atmosphere. Since H_2 and He are very efficient quenchers of translationally excited H atoms one should therefore scale down the yields of C, O, and N compounds in these experiments by a factor of 10^3 to 10^4.

This finally leads us to the question of the relative importance of inorganic (e.g., sulfur, polysulfides, red phosphorus) and organic (e.g., the colored fraction free of elemental sulfur in the above experiments) compounds in producing the observed colorations in the Jovian atmosphere. The quenching arguments given above certainly suggest that organic material produced by using the hot H atoms from H_2S photodissociation is entirely negligible compared to sulfur production from H_2S. Organic material produced from direct photodissociation of methane is limited by the methane column dissociation rate. The column photodissociation rates of NH_3, H_2S, and PH_3 are 10^3 to 10^4 times greater than that of CH_4 (Prinn 1970; Prinn and Lewis 1975a,b) and it is therefore expected that colored inorganic compounds such as sulfur, polysulfides, and red phosphorus will have production rates far in excess of those for organic material from CH_4 most of which will be colorless. Therefore, in order for this organic material to be an important coloring agent its average visible extinction coefficient would have to be $> 10^3$ times that for, say, red phosphorus.

C. *Other Processes*

Based on laboratory experiments it has been suggested that dissociation of molecules like CH_4, NH_3, H_2S, and H_2O on Jupiter may be accomplished by the high temperatures and electron impacts associated with lightning discharges (Woeller and Ponnamperuma 1969; Chadha *et al.* 1971), by the extreme heating induced by thunder shock waves (Bar-Nun 1975), and finally by bombardment of the upper atmosphere by energetic ions derived from the Jovian radiation belts (Scattergood *et al.* 1975).

These experimental simulations are unfortunately subject to many of the same criticisms which were voiced above concerning the laboratory simulations of Jovian photochemistry; they do not include the correct proportions of H_2 and He and more importantly they neglect atmospheric advection. Nevertheless, until precise chemical-dynamical models of these phenomena have been constructed, these laboratory simulations should still be regarded as indicating *possible* tendencies in the overall chemistry no matter how weakly these tendencies may manifest themselves in the actual Jovian atmosphere.

The question which is central to the theme of this chapter is the effect of these processes on the overall chemical composition of the planet. We must therefore examine the energy available for dissociating chemical bonds in each case. This problem has been approached by Lewis and Prinn (1971) and more recently by Lewis (1976). On Earth about 10^{-5} of the incident solar energy is converted to energy in lightning as electric currents and acoustic waves. A similar conversion factor on Jupiter implies that roughly 2×10^{-5} of the incident solar flux at Jupiter is converted to electrical and acoustic energy if we include the planet's internal heat source. Lewis (1976) estimates that the peak energy flux (principally proton kinetic energy) in Jupiter's radiation belts is $\sim 10^{-3}$ of the incident solar flux at Jupiter and that $\sim 10^{-2}$ of this energy flux may conceivably be dumped into the Jovian upper atmosphere. Therefore, charged particle bombardment of the upper atmosphere corresponds to an energy flux no larger than 10^{-5} of the incident solar flux at Jupiter.

The above figures can be compared with the fraction of the incident solar flux in the form of ultraviolet radiation which is $\sim 10^{-2}$. For breaking chemical bonds there is therefore $\sim 10^3$ times more energy available as ultraviolet radiation than as electrical currents and acoustic waves in lightning, or as ion fluxes from the radiation belts. In addition, in contrast to the ultraviolet energy, a significant portion of the energy associated with the lightning strokes and particle bombardment is simply involved in heating the atmosphere without causing chemical decomposition.

It is apparent then that photodissociation and advection constitute the major disequilibrating processes in Jupiter's atmosphere and that lightning and charged particle bombardment are relatively minor factors in the *bulk* chemistry of the planet. They do, however, have possibilities for influencing the chemistry in special localities and for production of interesting chemical compounds. For example, a number of biologically interesting compounds were produced in the electrical discharge experiments including alanine, glycine, sarcosine, aspartic acid, iminodiacetic acid, and iminoacetic α- and β-propionic acid (Chadha *et al.* 1971). In addition, the proton irradiation experiments (Scattergood *et al.* 1975) yielded long chain alkyl hydrocarbons, alkyl amines, alkyl sulfides, and amino acids.

The production of complex organic compounds in these experiments, and also in the photochemical experiments discussed earlier, forces us to address at least briefly here the controversial question of the existence or nonexistence of life on Jupiter. This subject has received considerable airing among biologists and biochemists presumably because the chemically reducing atmosphere on Jupiter is at least qualitatively similar to that supposed to have existed on the primitive earth. For an extensive review of "exobiology" in the outer solar system the reader is referred to a paper by one of the strongest protagonists of the "right to life" on Jupiter (Sagan 1971).

It is, however, important to emphasize that there is no surface on Jupiter at a low enough temperature so that complex organic material might accumulate on it. All nonequilibrium species produced in the visible atmosphere are ultimately destroyed in the deeper warmer region well below the visible clouds. In contrast, the primitive earth apparently possessed vast lakes and oceans in which organic material produced by photochemical and other processes could build up, and in which, after some 10^4 to 10^9 yr of complex organic chemistry, the first life forms presumably could evolve. We must therefore enquire whether the lifetime of any organic compound in the biologically interesting portion of Jupiter's atmosphere (say above the 500°K level) is long enough for significant evolution to occur.

If ℓ is the altitude of a particular organic species above the level where its thermochemical destruction begins, the lifetime of that species in the atmosphere is roughly ℓ^2/K. Typically $\ell \sim 100$ km, and from our earlier discussion $10^4 \lesssim K \lesssim 10^9$ cm^2 sec^{-1} with the larger values of K applying to the biologically interesting part of the troposphere. We therefore have 1 day $\lesssim \ell^2/K \lesssim 300$ yr. Evolution of life on the time scale of one day to a few hundred years would appear highly unlikely and the biologists' case for evolving life on Jupiter must therefore be viewed with considerable skepticism. One is left only with the remote possibility of an extra-Jovian source for living organisms and even these alien life-forms would be required to reproduce themselves from nonliving material on the time scale of ~ 1 day if the troposphere is as strongly mixed as $K \sim 10^9$ cm^2 sec^{-1} implies.

The arguments presented above for Jupiter should apply qualitatively to Saturn, Uranus, and Neptune if their atmospheric structures are reasonably similar. These very pessimistic conclusions concerning the existence of life do not, however, apply to the satellites of the outer planets where surfaces do exist and where the environment at or below this surface may be conducive to the evolution of primitive organisms. Titan would appear to be a much more hopeful target for future searches for extra-terrestrial life than Jupiter or indeed any of the major planets. Even on Titan, however, it is difficult to be very optimistic about these possibilities.

IV. COMPOSITION FROM REMOTE SPECTROSCOPY

A review of spectroscopic investigations of Jupiter was published five years ago (Owen 1970). Since that time, the most significant advances have come about as a result of the development of the Fourier Transform Spectrometer (FTS) and the growing accessibility of increasingly larger portions of the infrared and ultraviolet regions of the spectrum. The ultimate weapon in remote investigations is the use of spacecraft, and Pioneer 10 and 11 made significant contributions, as reported elsewhere in this volume. The development of the FTS as an immensely powerful tool for astronomical observations is primarily the result of pioneering work by Connes and Connes (1966). The general methodology of planetary spectroscopy has been covered by a number of authors (e.g., Hunten 1972; Owen 1972) and will not be repeated here. In this chapter we shall concentrate on the new information obtained about the composition of the Jovian atmosphere (Sec. V) after giving a historical introduction of the subject.

The first studies of the spectrum of Jupiter were carried out visually during the latter part of the 19th century. Rutherfurd (1863) reported the detection of features at 6190 and 6450 Å which were not present in the solar spectrum. He attributed them to gases in the Jovian atmosphere, but the identity of these gases was not established for more than sixty years. Meanwhile, the development of the photographic plate with emulsions that were sensitive to wavelengths beyond the red limit of the human eye led to the discovery of additional bands near 7250, 7900, 8650, and 8990 Å (Wildt and Meyer 1931; Slipher 1933). Using published laboratory spectra of methane and ammonia and deriving molecular vibrational frequencies to predict absorptions at shorter wavelengths, Wildt (1932) was able to establish the identifications shown in Table IV. These identifications were subsequently confirmed by laboratory studies of the two gases (Dunham 1933a,b; Adel and Slipher 1934).

TABLE IV
Identifications
(Wildt 1932)

Band (Å)	Absorbing Gas
8990	Methane
8650	"
7250	"
6190	"
7900	Ammonia
6450	"

Additional photographic studies led to the discovery of an increasing number of absorption bands, all of which turned out to be caused by either methane or ammonia (Dunham 1952; Kuiper 1952; Owen 1963; Owen et al. 1964; Owen 1965a,b). Most of these bands were discovered in the photographic infrared region of the spectrum (8000–12,000 Å), but new bands have also been found in the visible region despite the fact that this part of the Jovian spectrum has been studied for more than a century (Owen 1971; Owen and Cess 1975). A sample of the green to red region of the planet's spectrum is reproduced in Fig. 10.

Since these absorptions are all vibration-rotation bands, extension of the study of the planetary spectrum to wavelengths beyond the capability of the photographic plate would clearly permit one to study the stronger bands arising from the fundamental vibrational frequencies and their lowest overtones and combinations. Such studies became practical after the development of the lead-sulfide detector during World War II. The application of PbS detectors to low resolution scanning spectrometers led to the discovery of many additional absorptions in the planet's spectrum all of which again turned out to be caused by methane or ammonia (Kuiper 1952, 1964; Moroz 1966; Cruikshank and Binder 1968, 1969) (Fig. 11).

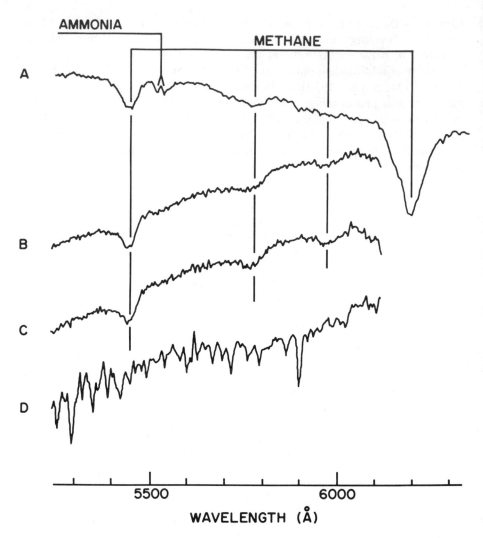

Fig. 10.   Part of the visible region of the Jovian spectrum (A) showing weak bands of methane and ammonia. Spectra of Saturn (B) and Titan (C) are included for comparison. All have been divided by the lunar spectrum (D). The zero intensity baselines for these spectra are not shown in order to fit them all on one page (Lutz *et al.* 1976).

Extension of the observations to shorter wavelengths led to essentially the same result. The presence of a substantial amount of hydrogen and/or helium could be inferred from the brightness of Jupiter in the ultraviolet, as deduced from rocket observations of the spectrum between 2000 and 3000 Å (Stecher 1965). This conclusion required the use of a model atmosphere and the assumption that Rayleigh scattering made the largest contribution to the ultraviolet albedo. The planet's spectrum at these wavelengths

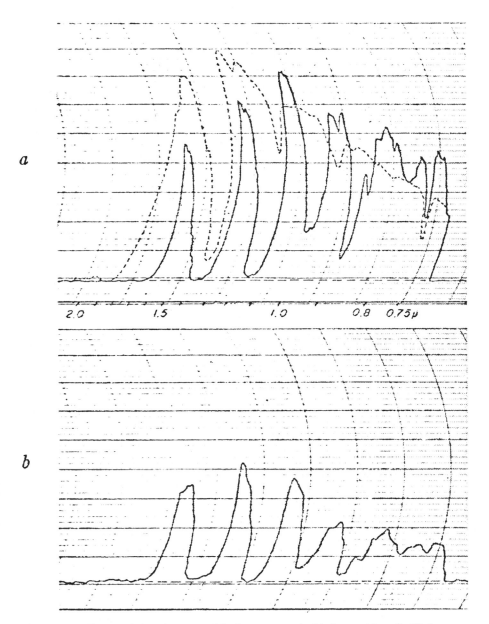

Fig. 11. (*a*) The near infrared spectra of Jupiter compared with the sun (dotted); (*b*) the same spectral region for Saturn (Kuiper 1952).

shows strong absorption increasing toward 2000 Å that can be interpreted as caused by a small amount of ammonia (Greenspan and Owen 1967). The identification is still not conclusive, however, since high-resolution spectra are lacking. Hydrogen sulfide could contribute to this absorption, as could the particles responsible for the generally low ultraviolet albedo (Axel 1972) which we discussed earlier.

The dominance of CH_4 and NH_3 in the spectrum of Jupiter led occasionally to the mistaken conclusion that they must be the most abundant atmospheric constituents. The reason that hydrogen and helium were not discovered more quickly is readily apparent from an examination of the absorption bands of these two gases. Molecular hydrogen, being homonuclear, has no permanent dipole moment so only quadrupole absorptions are permitted. These are exceedingly weak. Although predicted theoretically in 1938, these absorptions were not observed in the laboratory until much later (Herzberg 1938, 1952). The first detection of H_2 on Jupiter was achieved by Kiess *et al.* (1960) through the observation of several lines in the 3–0 band. The S(1) line of the 4–0 band of H_2 was subsequently discovered by Spinrad and Trafton (1963). The collision-induced dipole spectrum of H_2 was observed by Danielson (1965) in spectra obtained from a balloon-borne telescope and by Moroz (1966) from groundbased observations.

The helium resonance line occurs at 584 Å, a region of the spectrum that was totally inaccessible until very recently. In fact, the first direct detection of helium on Jupiter was made by Judge and Carlson (1974)[9] with an ultraviolet photometer on the Pioneer 10 spacecraft. Until that time, the existence of helium in the Jovian atmosphere was inferred from cosmic abundance arguments and attempts were made to measure it indirectly by photometry of stellar occultations and from line broadening studies. Reviews of these efforts have been given by Hunten and Münch (1973) and by Gautier (1974).

Throughout the decades since the spectrum of Jupiter first came under careful scrutiny, efforts have been made to discover traces of other gases whose existence in the atmosphere was postulated for various reasons. Cosmic abundance arguments indicated that H_2O, H_2S, and SiH_4 might all be expected, but as we have seen, only the former two would exist in thermodynamic equilibrium. Higher hydrocarbons such as C_2H_2, C_2H_6 and other simple organic compounds like HCN and CH_3NH_2 were suggested by various investigators as possible photochemical products. Repeated attempts failed to discover any trace of these substances until Ridgway (1974a) reported the detection of C_2H_2 and C_2H_6 in spectra of Jupiter obtained with a Fourier Transform instrument near 12 μm. These discoveries were soon followed by detection of PH_3 (Ridgway 1974b) and H_2O (Larson *et al.* 1975). Still newer discoveries have been announced and are described elsewhere in this book.[10] One may expect the list to grow as these new techniques are applied with ever greater finesse over an increasing range of the spectrum.

A separate but similar problem is posed by the search for uncommon isotopes. The cosmological significance of deuterium made it an early topic of

---

[9]See p. 419.
[10]See p. 405.

interest. In fact, the first discovery of this isotope in an astronomical source was the detection of CH_3D in the spectrum of Jupiter by Beer *et al.* (1972) (Fig. 12). The subsequent detection of HD by Trauger *et al.* (1973) (Fig. 13) permitted a more straightforward determination of D/H in the Jovian atmosphere. A search for $^{13}CH_4$ was carried out by Fox *et al.* (1972) who reported tentative detection with a large uncertainty in the abundance. A new determination has been presented by de Bergh *et al.* (1976).

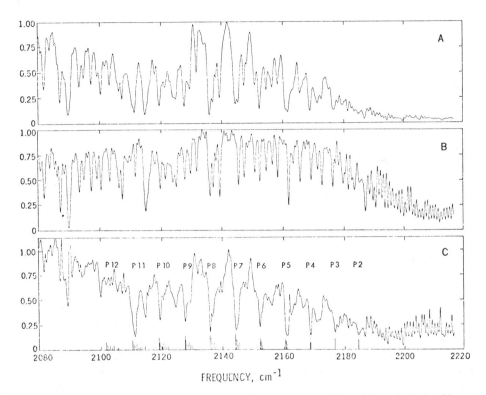

Fig. 12. The detection of CH_3D. Spectra of Jupiter (A), the sun (B) and the ratio Jupiter/Sun (C) with ν_2 P-branch lines of CH_3D indicated (Beer *et al.* 1972).

V.   SPECTROSCOPIC ABUNDANCES

The determination of abundances of gases in the atmosphere of Jupiter is complicated by many factors. Among these are the thermal structure of the atmosphere which may cause some of the constituents to condense and others to manifest themselves only through emission lines, the presence of clouds or hazes, and the lack of a well-defined lower boundary for reflection of incident sunlight. There is no doubt that all of these complications are present; what is lacking is a comprehensive model atmosphere that permits

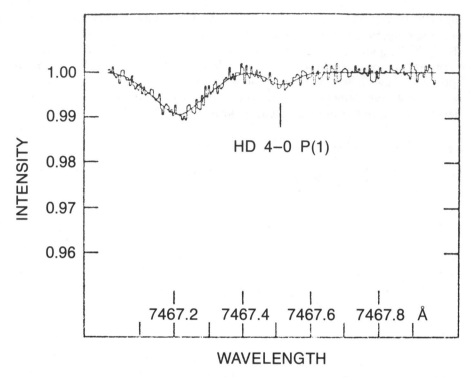

Fig. 13.   The detection of HD. The 4–0 P(1) line of HD at 7467.52 Å. Note that the depth of
this feature is approximately 0.3% of the continuum level (Trauger *et al.* 1973).

each irregularity to be accounted for in a quantitative fashion. The conventional approach to this problem is to ignore it, and to analyze the observations in terms of a simple reflecting layer model that assumes direct optical paths for incident and reflected light. In practice, this convenient fiction works remarkably well as long as its limitations are recognized (Regas and Sagan 1970; Margolis 1971), but it is by no means adequate to explain the details of the observations. In fact, its use should be restricted to observations made near the center of the disk, at the same time, at the same location, and at approximately the same wavelength using bands of similar strength and shape (Chamberlain 1970; Owen and Westphal 1972; Lutz *et al.* 1976). Proximity in wavelength should provide a similar continuum albedo for the species being compared. Deviations from this protocol lead to increasing inconsistencies in the results. Even the apparently safe procedure of studying global averages at low spatial and spectral resolution begins to reveal departures from the predictions of the model.

The fact that we are only looking into a cloudy atmosphere and not seeing down to a solid surface is strikingly evident from visual observations of the planet (Peek 1958) as well as from the long record of photographs (e.g.,

Humason 1961; Slipher 1964). The variability of this cloud cover is manifested spectroscopically in a variety of ways. Changes in the appearance of the spectrum with time were first reported by Bobrovnikoff (1934) who studied the ammonia band at 6450 Å. He also noted variations in the strength of this band across the disk of the planet. Such variations in both the temporal and spatial domains were illustrated in high-dispersion spectrograms published by Spinrad and Trafton (1963). The variations appear as changes in the relative strengths of individual lines of the band on spectra recorded at different times or different places on the planet. Hess (1953) pointed out that neither the 6190 Å methane band nor the 6450 Å ammonia band exhibits the increase in absorption toward the limb of the planet that is expected in a clear atmosphere above a reflecting layer. This absence of limb darkening has been variously interpreted as resulting from (1) a systematic increase in the height of the cloud layers toward the limb (Hess 1953), (2) the presence of a haze of ammonia cirrus above the main clouds that is more deeply penetrated near the center of the disk (Kuiper 1952), and (3) the change in viewing angle on a reflecting surface consisting of cumulus cloud towers (Squires 1957).

At the present time, there seems to be good observational evidence[11] for three distinct cloud layers: a high-altitude haze (above or at the tropopause) that absorbs sunlight in the near UV (Greenspan and Owen 1967), a layer of ammonia cirrus in the upper troposphere revealed by studies of the strong methane bands (Münch and Younkin 1964; Owen 1969) and a lower layer of thick clouds that forms the optical boundary in the near infrared (Owen 1969). It is possible that the high-altitude haze is simply the upper part of the ammonia cirrus, contaminated by photochemical reaction products (e.g., hydrazine, red phosphorus) to produce the observed low UV albedo. In this case there would only be two distinct layers (Danielson and Tomasko 1969; Axel 1972; Tomasko 1974). One must also remember that none of these layers is likely to be globally uniform; there is clear evidence for the inhomogeneity of the cirrus clouds in photographs taken with interference filters centered on strong methane bands (Owen 1969, 1970) and similar conclusions can be reached from studies of stronger bands at longer wavelengths (Moroz and Cruikshank 1969). The lower cloud layer seems to be more uniform at the scale accessible to these spectroscopic techniques, but the fact that it contains numerous holes is attested to by observations at 5 μm where thermal flux from deeper levels of the atmosphere has been detected (Gillett *et al.* 1969). The distribution of these holes is very non-uniform, but they appear to occur preferentially in the regions of the dark belts (Keay *et al.* 1973; Westphal *et al.* 1974).[12]

In an atmosphere that is this complex, the determination of abundances will obviously not be a straightforward procedure. One may expect *a priori*

---

[11]See also p. 495.
[12]See also p. 407.

that the stronger absorption bands will give systematically lower values for abundances, since light scattered back by the aerosols will "hide" the lower part of the atmosphere. This effect was pointed out by van de Hulst (1952) and is supported by those observations of strong bands from which an attempt has been made to derive abundances (Cruikshank and Binder 1969). One also expects that abundances derived from ultraviolet observations will be systematically lower than those obtained in the near infrared because of the absorbing haze, and this also appears to be borne out by the available data (Greenspan and Owen 1967; Anderson *et al.* 1969; Tomasko 1974). The longest pathlengths into the atmosphere will occur at 5 μm where the holes in the lower cloud deck can be utilized (Beer and Taylor 1973*a,b*). Keeping these results in mind, it is possible to derive mixing ratios (or relative abundances) that have some meaning if one accepts all the constraints imposed by the reflecting-layer model as previously described.

As if this picture were not dark enough, there is the additional problem frequently posed by the absence of appropriate laboratory data. To convert an observed absorption line into an abundance, it is necessary to know the intrinsic strength of the molecular transition responsible for the line. Expressed mathematically, we can write (for weak lines)

$$W = \int (1 - e^{-\alpha_\nu u}) d\nu \simeq u \int \alpha_\nu d\nu = uS = \eta \omega S \qquad (20)$$

where W is the equivalent width of the line (a measure of the amount of absorption), ν is the frequency, α_ν is the absorption coefficient in (cm amagat)$^{-1}$ [13], η is the air mass factor (dimensionless), S is the strength of the line, and ω is the column abundance (in cm amagat) of the gas. In this expression, W is the quantity we measure in the planet's spectrum. The factor η allows us to take into account the difference in the effective pathlength between the center of the disk and the limb. As we have seen, this difference is not very large. For weak lines in a clear atmosphere over a reflecting layer, $\eta = \pi$ if the slit of the spectrograph extends across the entire disk and there is no limb darkening (Owen and Kuiper 1964); in practice a value near 2 or 2.5 is usually adopted for Jupiter. If one is simply interested in mixing ratios, it is not necessary to specify η, provided the conditions described above are met. This leaves the factor S, and here some outside help is needed. In principle, S can be measured in the laboratory for a given transition. In practice, this often proves difficult to accomplish, either because the line occurs in an absorption band that has not been well enough analyzed to permit identification of the responsible transition, or because the necessary laboratory conditions cannot be met. In the case of Jupiter, we only have line strengths for one band of methane, $3\nu_3$ at 11,057 Å (Margolis and Fox 1968) and one band of hydrogen, 3–0 at 8200 Å (Rank *et al.* 1966). There are strengths available

---

[13]A column abundance of one cm amagat is equivalent to 2.68×10^{19} molecules cm^{-2}.

for bands further in the infrared, but these bands have not yet been used for abundance determinations because they are too strong and badly overlapped by other features (cf. Cruikshank and Binder 1968). The detailed study of the ammonia bands is still in its early stages (McBride and Nicholls 1972a). The 4–0 hydrogen quadrupole intensities have been calculated (Birnbaum and Poll 1969; Dalgarno et al. 1969) but these lines have not yet been observed in the laboratory, even at a pathlength of 44 km amagat (D. H. Rank personal communication). However, with improved photometric precision and spectroscopic resolution, experimental verification of the theoretical line strengths should be possible, since they predict an equivalent width of 1 mÅ for the $S_4(1)$ line at a pathlength of 15 km amagat. That kind of sensitivity is currently achievable (see, e.g., the detection of $P_4(1)$ of HD by Trauger et al. 1973) as well as the pathlength.

As the absorption lines become stronger, we need to worry about the effects of pressure. Even for weak lines formed in a dense atmosphere, the line shape is determined by collisions and a knowledge of the line width becomes important. Direct measurements of line widths and determinations of curves of growth (the change in equivalent width as a function of abundance and pressure) can again be obtained in the laboratory.

Lacking data on lines, we can use an entire absorption band, provided that it is observed under the same constraints. Various mathematical models for absorption bands may be found in the literature (Goody 1964). One which was recently applied to methane bands in spectra of the outer planets has the following form (Owen and Cess 1975):

$$W_b = 2 A_0 \ln \left\{ 1 + \frac{x}{2 + \left[x \left(1 + \frac{1}{\beta} \right) \right]^{\frac{1}{2}}} \right\} \qquad (21)$$

where W_b is the equivalent width of the band, A_0 is the bandwidth parameter and

$$x = \frac{\eta S \omega_{\text{gas}}}{A_0} \qquad (22)$$

$$\beta = \frac{4 \gamma_0 \tilde{P}}{d}. \qquad (23)$$

In the expression for β, γ_0 is the line half width at STP, d is the mean line spacing and $\tilde{P}$ is the effective pressure. A_0, S (now the band strength), γ_0, and d are all to be determined from laboratory observations.

This approach is especially helpful for methane and ammonia where we now have laboratory-determined band strengths for several of the features identified in the Jovian spectrum (Giver et al. 1975; Lutz et al. 1976). It has also been used to interpret ethane and acetylene emission features observed near 12 μm (Ridgway 1974a,c; Combes et al. 1974; Varanasi et al. 1974).

With this background in mind, we can discuss the abundance determinations of individual gases.

A. Hydrogen

Early attempts to determine the abundance of hydrogen after the discovery of the quadrupole absorptions by Kiess *et al.* (1960) were severely hampered by the absence of reliable values for the line strengths (Zabriskie 1962). Improved planetary observations and laboratory measurements of line strengths for the S(1) lines in the 1–0, 2–0, and 3–0 bands led to steadily increasing values for the column abundance of H_2 (Spinrad and Trafton 1963; Rank *et al.* 1966). Using the laboratory curve of growth developed by Rank *et al.* (1966) and a theoretical value for the line strength of $S_4(1)$, Owen and Mason (1968) derived a line of sight abundance of 190 ± 30 km amagat from photographic observations of $S_3(1)$ and $S_4(1)$ in the Jovian spectrum. With an airmass of 2, this becomes 85 ± 15 km amagat for the column abundance. Fink and Belton (1969) pointed out that the pressure narrowing of the quadrupole lines predicted by Dicke (1953) must be taken into account for the $S_3(1)$ line, although for $S_4(1)$ the effect is negligible. They derived a line-of-sight abundance of 140 ± 30 km amagat, from photoelectric scans of $S_4(1)$, $S_3(1)$ and $S_3(0)$. Using an airmass of 2.1, a column abundance of 67 ± 17 km amagat was obtained.

The abundance of HD rests on the observation of the $P_4(1)$ line by Trauger *et al.* (1973). The equivalent width of this line was measured as 0.29 ± 0.06 mÅ (Fig. 13), a remarkable observation. The abundance was evaluated by using McKellar's (1973) laboratory value for the $R_4(0)$ line strength to correct theoretical predictions for $P_4(1)$. Instead of calculating abundances directly, Trauger *et al.* (1973) derived a value of $D/H = (2.1 \pm 0.4) \times 10^{-5}$ for the Jovian atmosphere, using their own observation of the equivalent width of $S_4(1)$ of H_2 (8.1 ± 0.2 mÅ) to obtain the abundance of H_2. For consistency with our other numbers, this ratio can be converted into line-of-sight abundances for the two gases. The results are 5.27×10^{-3} km amagat for HD and 125 km amagat for H_2. The latter value employs a saturation correction of 1.10 for the $S_4(1)$ line, somewhat less than that used by previous workers. This correction was based on the preliminary measurement (and first detection) of the $Q_4(1)$ line, which was found to have an equivalent width of 1.7 ± 0.2 mÅ, close to the value predicted by theory if no saturation occurred. However, the theoretical line strengths are themselves uncertain, so this argument should not be given undue weight.

In evaluating these results we find that the precision with which the $S_4(1)$ line has now been measured is much greater than that for the 3–0 lines, hence we would be inclined to give abundances derived from this line more weight [Trafton (1972) summarized observations of the 3–0 lines with precision comparable to that attained by Trauger *et al.* (1973) for the $S_4(1)$ line but the details have not yet been published]. Unfortunately the $S_4(1)$ has never been observed in the laboratory so there are uncertainties regarding both the value of the line strength and the curve of growth. Under the circumstances, we feel that the prudent recourse is to adopt an average value of 70 ± 15 km amagat for the column abundance of H_2, recognizing that the quoted error may still be somewhat conservative.

An alternative approach to the problem is to use the collision-induced 1–0 absorption of H_2 at 2.4 μm. The first attempt at such an analysis was made by Danielson (1965) from spectra obtained from a balloon-borne telescope (Stratoscope II). The derived abundance was ~ 45 km amagat, assuming a temperature near the reflecting level of 200 to 225°K. Moroz (1966) concentrated his attention on groundbased observations of the 1.97 μm region, which would correspond to the S(3) line of this transition, and he derived a value of 30 ± 15 km amagat. Martin (1975) has recently obtained spectra of this region with an interferometer at the Mauna Kea Observatory, taking advantage of the reduction in telluric water vapor afforded by the high altitude of the telescope. Using the most up-to-date values for the H_2 absorption coefficient, he computed synthetic profiles for the S(1) line at 2.1 μm and derived an abundance on the order of 25 km amagat. The corresponding temperature was 248°K.

One is obviously confronted with a substantial paradox here. In order to match the planetary observations, it is necessary to assume that the reflecting layer is at a reasonably high temperature, implying long optical paths into the atmosphere. Yet the derived abundances are less by a factor of two or three than those determined from the quadrupole lines. Other things being equal, one would assume *a priori* that the low abundance results from line formation higher in the atmosphere, as a result of scattering by cloud particles. The temperature that is deduced from the band profile must then be an artifact of the spectrum fitting procedure, perhaps resulting from the presence of an as yet unidentified continuum absorber in this spectral region (Martin 1975).

B. Methane

It was clear when methane was detected in the spectrum of Jupiter that the amount in the planet's atmosphere was greater than the pathlengths used in the laboratory up to that time, i.e., on the order of tens of meter amagats (Wildt 1932). Subsequent laboratory studies gradually refined this number. Using a multiple path absorption cell to obtain long optical paths at low pressures, Kuiper (1952) deduced an abundance of 150 m amagat by comparing medium-resolution photographic spectra of Jupiter with laboratory spectra of the red bands of methane. In effect, he employed the band intensity method referred to above, but in a qualitative way, relying on a visual comparison of the two sets of data to obtain the estimate.

There was no further progress on this problem until Owen (1965a) identified the $3\nu_3$ band of CH_4 in the Jovian spectrum at 1.1057 μm. This band had been carefully studied in the laboratory by Childs and Jahn (1939) so R, Q and P branches had been identified and the major lines in each had been assigned rotational quantum numbers. The bands thus offered the opportunity for deducing a rotational temperature, an effective pressure, and a methane abundance, once appropriate intensity measurements had been carried out in the laboratory. Indeed, a rotational temperature could be deduced on the basis of statistical weights derived by Childs and Jahn (1939), and this led to a value of $200 \pm 25°K$ (Owen 1965a). Subsequent observations of Jupiter at higher resolution indicated problems of blending with telluric water-vapor lines and led to a more reliable estimate of the rotational temperature of $180 \pm 20°K$ (Walker and Hayes 1967; Owen and Woodman 1968).

Meanwhile a new theoretical and laboratory study of the band was carried out by Margolis and Fox (1968) which led to the values for line strengths required for an abundance analysis. Surprisingly, the initial applications of these data led to much lower values (~ 50 m amagat) for the methane abundance than those derived by Kuiper (1952) (Belton 1969; Margolis and Fox 1969a,b,c). A large amount of observational data on the $3\nu_3$ band is now available and this result seems quite well established (Bergstralh 1970; Margolis 1971). Nevertheless, it must be acknowledged that there are strong lines occurring in the R branch whose identity is not yet understood (the so-called interlopers) and the behavior of this band at large optical depths has not yet been studied in the laboratory.

These cautionary remarks are required by a new study of the visible methane bands which essentially confirms Kuiper's original abundance estimates (Lutz et al. 1976). The discovery of the 4860, 5430 and 5760 Å methane bands in the spectrum of Jupiter offered the possibility of using these very weak bands to determine the abundance directly, once they had been calibrated in the laboratory (Owen and Cess 1975). Laboratory curves of growth demonstrated that these bands are all in the linear region in the Jovian atmosphere, and the resulting column abundance for CH_4 is 150 m amagat (Lutz et al. 1976). This disagreement with the abundance of 50 m amagat derived from the $3\nu_3$ band seems likely to be caused by a combination of the greater effect of particle scattering at the shorter wavelengths and some uncertainty in the continuum definition in the region of $3\nu_3$. It is not clear at this writing which effect dominates, but Chamberlain (1970) has shown that scattering can lead to overestimates of abundances by as much as an order of magnitude. Pilcher et al. (1973) have computed the continuum single scattering albedos as a function of wavelength on Jupiter and thus an interpretation of the methane bands using a multiple scattering model is feasible.

Meanwhile, we have a certain dilemma in trying to decide which of these two methane abundances is the appropriate one to use in deriving a mixing ratio for H_2/CH_4. Since the H_2 abundance is derived from lines rather than an unresolved band, and since the H_2 lines are slightly saturated, the abundance of CH_4 derived from $3\nu_3$ would seem more appropriate. This ratio then becomes $H_2/CH_4 = 70 \times 10^3/50 = 1400$. This in turn corresponds to a value of $H/C = 2800$, in good agreement with the solar value of 2700 ± 300 (Cameron 1974). This agreement has frequently been cited as support for the idea that Jupiter's atmosphere represents a solar mixture of the elements (Owen 1970), but the uncertainties in both the H_2 and CH_4 abundances indicate that this conclusion should be viewed with caution. We should add that there is another aspect of the observations of the $3\nu_3$ methane band which lends some weight to its use as a "true" abundance indicator. The rotational temperature of $180 \pm 20°K$ and the mean effective pressure of one atmosphere derived from the relative intensities of the R manifolds and the widths of $R(0)$ and $R(1)$ are in good agreement with the revised temperature profile developed from the Pioneer 10 and 11 observations (Owen and Mason 1968, 1969; de Bergh et al. 1976; Kliore and Woiceshyn p. 235). This certainly supports, although it does not prove, the contention that a simple reflecting model gives an adequate interpretation of this band (cf. Margolis 1971).

Mono-deuterated methane, CH_3D, was shown to be present in the Jovian atmosphere by Beer et al. (1972) who published spectra showing evidence for the ν_2 band appearing in absorption at 4.7 μm. The abundance of this isotopic form of methane was subsequently determined by Beer and Taylor (1973a) who had to obtain a band strength by their own laboratory observations. They found a value for $\omega_{CH_3D} = 1.3$ cm amagat. To proceed from this number to an evaluation of the D/H ratio in the Jovian atmosphere, it was necessary to determine the equivalent CH_4 column abundance at the level of the atmosphere being probed and to derive the fractionation factor F that accounts for the slight preference for deuterium to be bound in methane rather than in hydrogen. Carrying out these calculations with an appropriate model atmosphere, Beer and Taylor (1973a) found

$$D/H = 5.1 \pm 2.2 \times 10^{-5}. \tag{24}$$

This is more than a factor 2 higher than the value derived by the direct measurement of Trauger et al. (1973) described above. Beer and Taylor (1973b) showed that for the smaller value to be reconciled with the CH_3D observations, the exchange reaction

$$HD + CH_4 \rightleftarrows CH_3D + H_2 \tag{25}$$

must be catalyzed, since the extreme fractionation implied would require a time greater than the age of the universe by a factor 10^3. They suggested that the cloud particles in the NH_4SH cloud might serve as the necessary catalyzing agents.

The other isotopic form of methane that has been sought in the Jovian spectrum is $^{13}CH_4$ (Fox et al. 1972). In this case it was the $3\nu_3$ band at 1.1

μm that was used with the expectation that the substitution of ^{13}C for ^{12}C would not affect the band structure or the line strengths. Isotopically enriched methane was observed in the laboratory to verify these predictions and to establish the wavelengths of the lines (Pugh *et al.* 1974). The result for Jupiter is less satisfactory than in the case of CH_3D because there is so much confusion with weak lines of ordinary methane in this spectral region. It is at least clear that there is no great enrichment of $^{13}CH_4$ in the Jovian atmosphere compared with values of $^{13}C/^{12}C$ in the sun, the meteorites, or the earth.

C. Ammonia

The identification of ammonia absorption in the Jovian spectrum led immediately to the realization that the abundance of this gas would be controlled by its vapor pressure in the cold atmosphere of the planet (Wildt 1932). Assuming a temperature of 144°K, an abundance of 0.03 gram cm^{-2} was derived, corresponding to a column abundance of 40 cm amagat. Wildt recognized that this result also indicated that the upper clouds on Jupiter must be ammonia cirrus.

Dunham (1933*a*) attempted to match the Jovian spectrum with a laboratory spectrum of ammonia instead of assuming a temperature to deduce the abundance. He noted that laboratory pathlengths of about 40 m amagat provided a good fit to high-resolution observations of the 6450 Å band, suggesting abundances of 5 to 10 m amagat in the planet's atmosphere. This amount of gas implied a temperature on the order of 150 to 170°K on Jupiter. Since the saturation vapor pressure curve is such a steep function of temperature, small changes in the latter parameter will have a great effect on the derived ammonia abundance. The first measurement of the Jovian atmospheric temperature gave a value of 130°K, but it was assumed that this corresponded to a rather high level in the atmosphere compared to the level being probed by the spectroscopists (Menzel *et al.* 1926; for a more recent determination, see Gillett *et al.* 1969).

Kuiper's (1952) discussion of the ammonia absorption bands was similar to his work on methane. He obtained a value of 7 m amagat for the single column abundance and pursued the cloud problem in more detail, deriving the first semi-empirical model for the Jovian atmosphere. Kuiper pointed out that roughly 1 cm amagat of ammonia would be sufficient to provide unit optical depth in the strong bands at 10 μm that were responsible for the emission from which the 130°K temperature was deduced. Assuming that the atmosphere was close to saturation from the tropopause down to the ammonia clouds, Kuiper calculated a cloud-top temperature of 150°K corresponding to his abundance of 7 m amagat of NH_3.

This same approach was used by Owen (1969) who pointed out that the temperature corresponding to the saturation value for the column abundance of NH_3 actually represented the bottom of the ammonia cirrus layer, rather than the top. The limit to the penetration of near infrared radiation on Jupiter is not provided by the ammonia cirrus, but by a lower cloud level at a temperature of $\sim 225°K$ in this simple heuristic model. These empirically derived values are in good accord with the theoretical model proposed by Lewis (1969a) which we have already discussed. The important point for our purposes here is that the near-infrared observations are apparently sampling a region of the atmosphere where the ammonia abundance is no longer controlled by its saturation vapor pressure, so the mixing ratio can provide useful information about the true abundance of this gas in the Jovian envelope. This statement is not strictly true, however, since the lower cloud deck is probably composed of NH_4SH, $(NH_4)_2S$, and other sulfur-ammonia compounds, so that one cannot predict a priori exactly what the vapor pressure of ammonia will be over this mixture.

Unfortunately, we are still not in an ideal position to determine the ammonia abundance directly. Laboratory curves of growth for individual lines in the 6450 Å band were obtained by Mason (1970) at room temperature. Taking an average value for the abundances derived from the various lines he had studied, Mason found a value of 12 m amagat for the ammonia abundance. To make further progress, some means of applying a temperature correction is required. A preliminary analysis of the 6450 Å band has been attempted (McBride and Nicholls 1972a), but it is not yet completely satisfactory. This means that one is still constrained to the use of the integrated band intensity for quantitative work. Fortunately the 6450 Å band has been well-studied in the laboratory (Rank et al. 1966; Giver et al. 1975) so that the required parameters are available. Using both approaches (line-by-line and integrated band), Encrenaz et al. (1974) derived an abundance of 13 ± 3 m amagat, essentially identical with previous results.

At the present time, the best hope for improving this determination seems to lie in the application of the integrated band strengths to observations of the 6450 Å and the recently discovered 5520 Å band (Owen 1971), both of which can be observed with very high spectrophotometric accuracy. It is important to recognize, however, that these bands vary in intensity with time and with position on the planet, in ways that are not yet fully understood. For example, it seems possible that the extreme variations in the equivalent width of the 6190 Å *methane* band (Teifel 1969; Pilcher et al. 1973), may be caused by fluctuations in the underlying *ammonia* absorption that occurs in this spectral region (McBride and Nicholls 1972b).

If we follow our previous guidelines for determining mixing ratios, it would seem to be most appropriate to compare the ammonia abundance with the methane abundances deduced from the weak visible bands. This gives a value of $C/N = 11.5$, rather different from the solar value of 4 (Cameron 1974). If we were to use the $3\nu_3$ methane abundance instead, however, we would find $C/N = 3.8$. The value of $H/N = 1.08 \times 10^4$ may be compared with the solar value of 1.17×10^4.

TABLE V

Near Infrared Abundances in the Jovian Atmosphere

Molecule	Band	Wavelength	Abundance (m amagat)	Ratio{H_2/molecule}
H_2	3–0, 4–0 Quadrupole	8200 Å, 6400 Å	$70 \pm 15 \times 10^3$	1
CH_4	$3\nu_3$	11,057 Å	50 ± 10	1400
	Unassigned	4860 Å, 5430 Å	150 ± 30	470
NH_3	$5\nu_1$	6450 Å	13 ± 3	5380

D. Summary–H_2, CH_4, NH_3

The near infrared abundances for these gases are summarized in Table V, in accord with the discussion given above. It should be stressed again that these values were obtained on the assumption of a simple reflecting layer model, so the absolute values have little meaning. Mixing ratios should be more significant, but even here the differences in line formation processes for the different types of molecular transitions could be misleading. There is clearly more work to be done on the analysis of additional methane and ammonia bands as well as more laboratory studies of the hydrogen quadrupole absorptions.

E. Trace Constituents

The discoveries of H_2O, HCN, GeH_4, and CO are described elsewhere in this volume (see Ridgway *et al.* p. 405). That chapter should be consulted for a discussion of the abundance derivations. The detection of these gases is made possible by the very strong absorption bands they possess near 5 μm, where methane and ammonia are transparent. This window and a similar one near 2.7 μm hold the promise of additional discoveries of this type as investigations proceed.

At still longer wavelengths, we encounter the 8–14 μm window which has recently begun to yield its share of information about the composition of the Jovian atmosphere. As already mentioned, this region is dominated by the ν_2 band of NH_3 centered at 10.5 μm. However, as soon as the spectrum was recorded (as opposed to simply integrating all the flux in this wavelength interval) it became evident that methane emission (ν_4 at 7.8 μm) and hydrogen absorption (pressure induced rotational) were also present{Gillett *et al.* 1969) (Fig. 14). The first *FTS* spectrogram yielded still more surprising results, *viz.* the detection of ethane (C_2H_6) and acetylene (C_2H_2) at the long wavelength end of the window (Ridgway 1974*a,c*) (Fig. 15). The mixing ratios were stated as 4×10^{-4} and 8×10^{-5}, respectively. The presence of

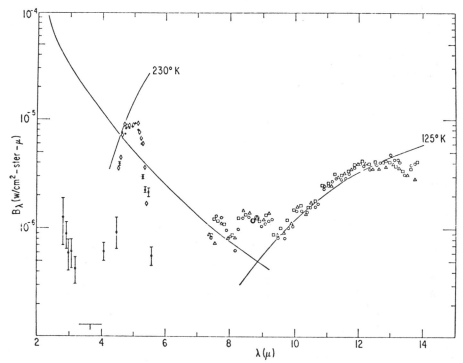

Fig. 14. The middle infrared. Symbols represent observations at different times. Blackbody curves for 125°K and 230°K are shown, as well as the surface brightness of a perfectly diffuse reflecting sheet at Jupiter's distance. Note the 5 μm window and the evidence for a temperature inversion near 8.5 μm (Gillett *et al.* 1969).

these two gases had been expected as a result of photolysis of methane. However, the theoretical predictions which we discussed earlier implied much lower mixing ratios than observed (Strobel 1973*a*, 1974).

The evaluation of abundances of gases in emission is complicated by the need to know the temperature in the upper atmosphere with considerable precision. Other evaluations of the same data and independent observations of the spectrum have led to the suggestion that the abundances might be lower by more than an order of magnitude (Combes *et al.* 1974; Varanasi *et al.* 1974). Most recently, the acetylene observations have been called into question and it now appears that the abundance of this gas in the Jovian atmosphere may be close to the theoretical prediction (Tokunaga *et al.* 1975).

This downward revision of the acetylene abundance carries with it a question about the precision of another intriguing observation which seems to indicate the presence of phosphine (Ridgway 1974*b*). The absence of phosphine absorptions in the 5 μm spectrum of Jupiter, despite the fact that these bands should be as strong as the 10 μm features, raises some question

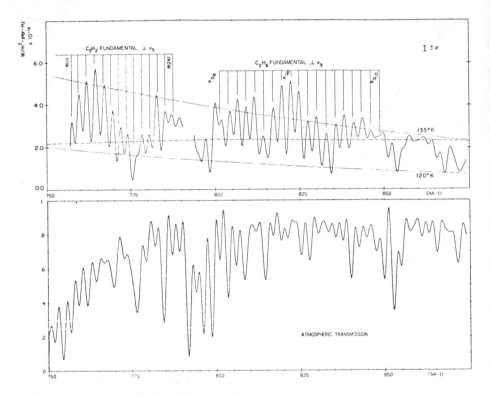

Fig. 15.   The detection of C_2H_2 and C_2H_6. (*a*) The thermal emission spectrum of Jupiter corrected for absorption in the earth's atmosphere. The dashed line is the predicted form of the H_2 continuum. (*b*) The atmospheric transmission spectrum at the time of the Jupiter observations (Ridgway 1974*a*).

about the reality of the latter (Robertson and Fox 1928; Fink, personal communication, 1975). One hopes that this very important observation will be repeated in the near future.

Another type of ambiguity is posed by the existence of several absorption features that are still unidentified. Perhaps the most prominent of these is a broad absorption centered near 4.7 μm, originally reported by Münch and Neugebauer (1971). Part of this absorption is caused by lines in the P branch of the ν_2 band of CH_3D (see discussion above) but some absorption persists even when this contribution is taken into account (Beer and Taylor 1973*a*). The high-resolution study of this spectral region also revealed a number of lines that have not yet been identified (Beer and Taylor 1973*a*), and the broadening of the spectral window by means of high-altitude observations has added to the roster (Larson *et al.* 1975). The broad absorption is of particular interest because of its possible association with the chromophores responsible for the colors in the cloud deck. Neither H_2S nor NH_4HS exhibit absorptions here in the solid state, but this is a well known frequency for absorption by $-C \equiv N$. Sulfur-containing compounds produced in laboratory simulations of Jovian atmospheric conditions also absorb in this region (Woeller and Ponnamperuma 1969; Kieffer and Smythe 1974; Scattergood *et al.* 1975).[14]

F. Upper Limits

The revival of interest in high-resolution planetary spectroscopy in the last decade included a number of deliberate searches for gases considered to be possible constituents of the atmosphere of Jupiter. Much of this work has been superceded by the actual detection of these constituents (see above), but there are still some likely candidates that have not been found. Special reference is made to compilations by Cruikshank and Binder (1969), Gillett *et al.* (1969), and Owen and Sagan (1972). A congeries of current limits on the abundances is given as Table VI, which is based on the cited references. Note that these limits are wavelength dependent in the sense that ultraviolet limits correspond to a high, cold region of the atmosphere where the vapor pressures for many of these substances will be too low to permit detection, and the longest pathlengths cited here are those corresponding to the 0.8–1.0 μm range, which is not as favorable as the 2.7 μm and 5.0 μm windows.

It is perhaps worthwhile to call special attention to the case of H_2S, since this gas is probably the most prominent of those that are still undetected. The vapor pressure of H_2S approaches that of NH_3 at the temperatures in the accessible region of the Jovian atmosphere, but its rotation-vibration absorptions are neither as intense nor as numerous as those of ammonia. They also have an unfortunate tendency to overlap the (stronger) NH_3 and CH_4 absorptions, making them especially difficult to discover. The 1.53 μm region initially appeared to show some promise (Cruikshank and Binder 1969), but subsequent study has indicated that this region is filled with weak ammonia lines, and no trace of H_2S absorptions has been found (Encrenaz *et al.* 1974). At the present time, the best hope for a detection seems to lie in high-resolution observations of the long-wavelength wing of the fundamental ν_3 in the 4–5 μm window, or the combination band $\nu_2 + \nu_3$ in the 2.7 μm window — a region that has not yet been probed owing to strong telluric CO_2 and H_2O absorptions, or possibly the longward wing of ν_2 near 9 μm.

For the electronic absorptions in the ultraviolet, conditions are hardly better. Here the H_2S again is overlapped by NH_3. Now, however, the structure of the two absorptions is radically different, so that spectra obtained with moderate resolution and high photometric precision might be able to detect the presence of both absorbers. At wavelengths below 2000 Å, further discrimination becomes possible but the observations become more

---

[14]See also p. 377.

TABLE VI

Upper Limits on Mixing Ratios for Minor
Constituents in the Jovian Atmosphere

Molecule	0.2–0.3 μm (5 km amagat H_2)	0.6–1 μm (70 km amagat H_2)	1–2.5 μm (20 km amagat H_2)	8–14 μm (12 km amagat H_2)
SiH_4		$< 3 \times 10^{-4}$		$< 8 \times 10^{-7}$
CH_3NH_2			$< 1 \times 10^{-6}$	
H_2S	$< 6 \times 10^{-9}$		$< 1.3 \times 10^{-5}$	$< 2.5 \times 10^{-4}$
HCl	$< 6 \times 10^{-8}$			
Benzene and its derivatives	$< 5 \times 10^{-10}$			
Purines, pyrimidines and their derivatives	$< 2 \times 10^{-10}$			

difficult owing to the low light levels. There is the additional handicap posed by the very low expected concentrations of H_2S in the upper part of the atmosphere that is probed in the ultraviolet. As discussed earlier, this high-altitude depletion of H_2S results from both condensation and photodissociation.

VI. CONCLUDING REMARKS

It is appropriate to end this chapter with a brief summary of the major unsolved problems and a discussion of important areas of controversy. In reviewing computer models of atmospheric chemistry we see the need for ensuring inclusion of the full complement of elementary reactions and for accurate laboratory rate constants for these reactions. This is particularly true for PH_3 and H_2S. Atmospheric dynamics need further study as inputs to these models. The laboratory simulations of atmospheric chemistry need to be tailored more closely to the known atmospheric structure, composition, and transport times before they become believable at least to aeronomers familiar with our own atmosphere. We are now at a sufficiently sophisticated level in atmospheric chemistry that qualitative discussions of possible constituents without quantitative computations of their expected mixing ratios are unsatisfactory.

We have pointed out two important problems in interpretation of spectroscopic measurements: firstly, there is an unfortunate lack of laboratory data on line strengths for many important lines; secondly there is a need for multiple-scattering interpretations of the observed line equivalent widths (at least in terms of idealized atmospheric models) for comparison with existing reflecting-layer interpretations. The identifications of PH_3, C_2H_2, CO, CO_2, and HCN need careful verification by independent astronomers. The abundance estimates for C_2H_6 and H_2O need to be examined. The search for H_2S attains very high priority, although the problems involved are far from trivial.

In closing, it is interesting to contemplate a hint of symbiosis in the relationship between the atmospheric modeler and the planetary observer; the former relies on the latter for a good deal of his basic premises while the latter may be stimulated toward those observations which may prove (or disprove) the former's contemplations. We hope this chapter will serve as a catalyst for this surely beneficial process.

Acknowledgements. This work was supported by the Atmospheric Sciences Section of the National Science Foundation under a grant to Massachusetts Institute of Technology and by the National Aeronautics and Space Administration under a grant to State University of New York.

REFERENCES

Adel, A., and Slipher, V. M. 1934. The constitution of the atmospheres of the giant planets. *Phys. Rev.* 46:902–906.

Anderson, R. C.; Pipes, J. G.; Broadfoot, A.; and Wallace, L. 1969. Spectra of Venus and Jupiter from 1800 to 3200 Å. *J. Atmos. Sci.* 26:874–888.

Aumann, H.; Gillespie, C.; and Low, F. 1969. The internal powers and effective temperatures of Jupiter and Saturn. *Astrophys. J.* 157:L69–L72.

Axel, L. 1972. Inhomogeneous models of the atmosphere of Jupiter. *Astrophys. J.* 173:451–468.

Bar-Nun, A. 1975. Thunderstorms on Jupiter. *Icarus* 24:86–94.

Beer, R.; Farmer, C. B.; Norton, R. H.; Martonchik, J. V.; and Barnes, T. G. 1972. Jupiter: observation of deuterated methane in the atmosphere. *Science* 175:1360–1361.

Beer, R., and Taylor, F. W. 1973a. The abundance of CH_3D and the D/H ratio in Jupiter. *Astrophys. J.* 179:309–327.

———. 1973b. The equilibration of deuterium in the Jovian atmosphere. *Astrophys. J.* 182: L131–L132.

Belton, M. J. S. 1969. An estimate of the abundance and rotational temperature of CH_4 on Jupiter. *Astrophys. J.* 157:469–472.

Bergstralh, J. T. 1970. An observational test of absorption line formation models in the Jovian atmosphere. Doctoral dissertation, University of Texas, Austin, Texas.

Birnbaum, A., and Poll, J. D. 1969. Quadrupole transitions in the H_2, HD and D_2 molecules. *J. Atmos. Sci.* 26:943–945.

Bobrovnikoff, N. T. 1934. Note on the spectrum of Jupiter. *Publ. Astron. Soc. Pacific* 45: 171–174.

Cadle, R. D. 1962. The photochemistry of the upper atmosphere of Jupiter. *J. Atmos. Sci.* 19:281–285.

Cameron, A. G. W. 1974. Abundances of the elements in the solar system. *Space Sci. Rev.* 15:121–146.

Cess, R. D., and Khetan, S. 1973. Radiative transfer within the atmospheres of the major planets. *J. Quant. Spec. Rad. Trans.* 13:995–1009.

Chadha, M.; Flores, J.; Lawless, J.; and Ponnamperuma, C. 1971. Organic synthesis in a simulated Jovian atmosphere. *Icarus* 15:39–44.

Chamberlain, J. W. 1970. Behavior of absorption lines in a hazy planetary atmosphere. *Astrophys. J.* 159:137–158.

Chase, S.; Ruiz, R.; Münch, G.; Neugebauer, G.; Schroeder, M.; and Trafton, L. 1974. Pioneer 10 infrared radiometer experiment: preliminary results. *Science* 183:315–317.

Childs, W. H. J., and Jahn, H. A. 1939. A new coriolis perturbation in the methane spectrum. III. Intensities and optical spectrum. *Proc. Roy. Soc. London* A 169:451–463.

Combes, M.; Encrenaz, Th.; Vapillon, L.; Zeau, Y.; and Lesqueren, C. 1974. Confirmation of the identification of C_2H_2 and C_2H_6 in the Jovian atmosphere. *Astron. and Astrophys.* 34:33–35.

Connes, P., and Connes, J. 1966. Near-infrared planetary spectra by Fourier spectroscopy. I. Instruments and results. *J. Opt. Soc. Amer.* 56:896–910.

Cruikshank, D. P., and Binder, A. B. 1968. The infrared spectrum of Jupiter, 0.95–1.60 microns, with laboratory calibrations. *Comm. Lunar and Planet. Lab.* 6:275–288.

———. 1969. Minor constituents in the atmosphere of Jupiter. *Astrophys. Space Sci.* 3: 347–356.

Dalgarno, A.; Allison, A. C.; and Browne, J. C. 1969. Rotation-vibration quadrupole matrix elements and quadrupole absorption coefficients of the ground electronic states of H_2, HD and D_2. *J. Atmos. Sci.* 26:946–951.

Danielson, R. E. 1965. The infrared spectrum of Jupiter. *Astrophys. J.* 143:949–960.

Danielson, R. E., and Tomasko, M. G. 1969. A two-layer model of the Jovian clouds. *J. Atmos. Sci.* 26:889–897.

de Bergh, C.; Maillard, J. P.; Lecacheux, J.; and Combes, M. 1976. A study of the $3v_3$-CH_4 region in a high-resolution spectrum of Jupiter recorded by Fourier transform spectroscopy. *Icarus* (special Jupiter issue).

Dicke, R. H. 1953. The effects of collisions on the Doppler width of spectral lines. *Phys. Rev.* 89:472–473.

Dunham, T. 1933a. Note on the spectra of Jupiter and Saturn. *Publ. Astron. Soc. Pacific* 45:42–44.

———. 1933b. The spectra of Venus, Mars, Jupiter, and Saturn under high dispersion. *Publ. Astron. Soc. Pacific* 45:202–204.

———. 1952. Spectroscopic observations of the planets at Mount Wilson. *The Atmospheres of the Earth and Planets* (G. P. Kuiper, ed.) pp. 288–305. Chicago, Illinois: University of Chicago Press.

Encrenaz, Th.; Owne, T.; and Woodman; J. H. 1974. The abundance of ammonia on Jupiter, Saturn and Titan. *Astron. and Astrophys.* 37:49–55.

Fink, U., and Belton, M. J. S. 1969. Collision narrowed curves of growth for H_2 applied to new photoelectric observations of Jupiter. *J. Atmos. Sci.* 26:952–962.

Fox, K.; Owen, T.; Mantz, A.; and Rao, K. 1972. A tentative identification of $^{13}CH_4$ and an estimate of $^{12}C/^{13}C$ in the atmosphere of Jupiter. *Astrophys. J.* 176:L81–L84.

Gautier, D. 1974. Hydrogen mixing ratio in the giant planets. *Exploration of the planetary system* (A. Woszczyk and C. Iwaniszewska, eds.), pp. 321–328. Dordrecht, Holland: D. Reidel Publ. Co.

Gillett, F. C.; Low, F.; and Stein, W. A. 1969. The 2.8–14-micron spectrum of Jupiter. *Astrophys. J.* 157:925–934.

Giver, L. P.; Boese, R. W.; and Miller, J. H. 1975. A laboratory atlas of the $5v_1$ NH_3 absorption band at 6475 Å with applications to Jupiter and Saturn. *Icarus* 25:34–48.

Goody, R. M. 1964. *Atmospheric radiation. I. Theoretical basis.* London: Oxford University Press.

Greenspan, J. A., and Owen, T. 1967. Jupiter's atmosphere: its structure and composition. *Science* 156:1489–1493.

Gross, S. H., and Rasool, S. I. 1964. The upper atmosphere of Jupiter. *Icarus* 3:311–322.

Herzberg, G. 1938. On the possibility of detecting molecular hydrogen and nitrogen in planetary and stellar atmospheres by their rotation-vibration spectra. *Astrophys. J.* 87:428–437.

———. 1952. Laboratory absorption spectra obtained with long paths. *The atmospheres of the earth and planets* (G. P. Kuiper, ed.) pp. 406–416. Chicago, Illinois: University of Chicago Press.

Hess, S. L. 1953. Variations in atmospheric absorption over the disks of Jupiter and Saturn. *Astrophys. J.* 118:151–160.

Hirschfelder, J. O.; Curtis, C. F.; and Bird, R. B. 1954. *Molecular theory of gases and liquids.* New York: J. Wiley and Sons.

Hogan, J.; Rasool, S. I.; and Encrenaz, Th. 1969. The thermal structure of the Jovian atmosphere. *J. Atmos. Sci.* 26:898–905.

Hubbard, W. B. 1968. Thermal structure of Jupiter. *Astrophys. J.* 152:745–754.

———. 1970. Structure of Jupiter: chemical composition, contraction, and rotation. *Astrophys. J.* 162:687–697.

Humason, M. L. 1961. Photographs of planets with the 200-inch telescope. *Planets and satellites* (G. P. Kuiper and B. M. Middlehurst, eds.), p. 572. Chicago, Illinois: University of Chicago Press.

Hunten, D. M. 1969. The upper atmosphere of Jupiter. *J. Atmos. Sci.* 26:826–834.

———. 1972. Lower atmospheres of the planets. *Physics of the solar system* (S. I. Rasool, ed.), pp. 197–241. NASA SP-300, Washington, D.C.

Hunten, D. M. and Münch, G. 1973. The helium abundance on Jupiter. *Space Sci. Rev.* 14:433–443.

Ingersoll, A. P. 1973. Jupiter's Great Red Spot: a free atmospheric vortex? *Science* 182:1346–1348.

Judge, D. L. and Carlson, R. W. 1974. Pioneer 10 observations of the ultraviolet glow in the vicinity of Jupiter. *Science* 183:317–318.

Keay, C. S. L.; Low, F. J.; Rieke, G. H.; and Minton, R. B. 1973. High-resolution maps of Jupiter at five microns. *Astrophys. J.* 183:1063–1073.

Khare, B. N., and Sagan, C. 1973. Red clouds in reducing atmospheres. *Icarus* 20:311–321.

———. 1975. S_8: a possible infrared and visible chromophore in the clouds of Jupiter. *Science* 189:722–723.

Kieffer, H. H., and Smythe, W. D. 1974. Frost spectra: comparison with Jupiter's satellites. *Icarus* 21:506–512.

Kiess, C. C.; Corliss, C. H.; and Kiess, H. K. 1960. High dispersion spectra of Jupiter. *Astrophys. J.* 132:221–231.

Kuiper, G. P. 1952. Planetary atmospheres and their origin. *The atmospheres of the earth and planets* (G. P. Kuiper, ed.), pp. 306–405. Chicago, Illinois: University of Chicago Press.

———. 1964. Infrared spectra of planets and cool stars; introductory report. *Mém. Soc. Roy. Sci. Liège* 9:365–391.

Larson, H. P.; Fink, U.; Treffers, R.; and Gautier, T. N. 1975. Detection of water vapor on Jupiter. *Astrophys. J.* 197:L137–L140.

Lebofsky, L. A. 1974. Chemical composition of Saturn's rings and icy satellites. Doctoral dissertation, Massachusetts Institute of Technology, Cambridge, Mass.

Lewis, J. S. 1969a. The clouds of Jupiter and the NH_3-H_2O and NH_3-H_2S systems. *Icarus* 10:365–378.

———. 1969b. Observability of spectroscopically active compounds in the atmosphere of Jupiter. *Icarus* 10:393–409.

Lewis, J. S. 1976. Equilibrium and disequilibrium chemistry of adiabatic, solar-composition planetary atmospheres. First College Park Colloquia on chemical evolution—giant planets. In press.

Lewis, J. S., and Prinn, R. G. 1970. Jupiter's clouds: structure and composition. *Science* 169: 472–473.

———. 1971. Chemistry and photochemistry of the atmosphere of Jupiter. *Theory and experiment in exobiology* (A. Schwartz, ed.), pp. 123–142. Groningen, Netherlands: Wolters-Noordhoff.

Lindzen, R. S. 1971. Tides and gravity waves in the upper atmosphere. *Mesospheric models and related experiments* (G. Fiocco, ed.), pp. 122–130. Boston: D. Reidel.

Low, F. J. 1966. Observations of Venus, Jupiter, and Saturn at 20 microns. *Astron. J.* 71: 391–398.

Lutz, B. L.; Owen, T.; and Cess, R. D. 1976. Laboratory band strengths of methane and their application to the atmospheres of Jupiter, Saturn, Uranus, Neptune, and Titan. *Astrophys. J.* In press.

Margolis, J. S. 1971. Studies of methane absorption in the Jovian atmosphere. III. The reflecting-layer model. *Astrophys. J.* 167:553–558.

Margolis, J. S., and Fox, K. 1968. Infrared absorption spectrum of CH_4 at 9050 cm^{-1}. *J. Chem. Phys.* 49:2451–2452.

———. 1969a. Studies of methane absorption in the Jovian atmosphere. I. Rotational temperature from the $3\nu_3$ band. *Astrophys. J.* 157:935–943.

———. 1969b. Studies of methane absorption in the Jovian atmosphere. II. Abundance from the $3\nu_3$ band. *Astrophys. J.* 158:1183–1188.

———. 1969c. Extension of calculations of rotational temperature and abundance of methane in the Jovian atmosphere. *J. Atmos. Sci.* 26:862–864.

Martin, T. Z. 1975. Saturn and Jupiter: a study of atmospheric constituents. Doctoral dissertation, University of Hawaii, Honolulu, Hawaii.

Mason, H. P. 1970. The abundance of ammonia in the atmosphere of Jupiter. *Astrophys. Space Sci.* 1:424–436.

McBride, J. O. P., and Nicholls, R. W. 1972a. The vibration-rotation spectrum of ammonia gas. II. A rotational analysis of the 6450 Å band. *Can. J. Phys.* 50:93–102.

———. 1972b. The vibration-rotation spectrum of ammonia gas. I. *J. Phys. B: Atomic Molec. Phys.* 5:408–417.

McKellar, A. R. W. 1973. Intensities of the 3–0 and 4–0 rotation-vibration bands of HD. *Astrophys. J.* 185:L53–L55.

McNesby, J. R. 1969. The photochemistry of Jupiter above 1000 Å. *J. Atmos. Sci.* 26:594–599.

Menzel, D. H.; Coblentz, W. W.; and Lampland, C. O. 1926. Planetary temperatures derived from water-cell transmissions. *Astrophys. J.* 63:177–187.

Moroz, V. I. 1966. The spectra of Jupiter and Saturn in the 1.0–2.5 μ region. *Soviet Astron. A.J.* 10:457–468.

Moroz, V. I., and Cruikshank, D. P. 1969. Distribution of ammonia on Jupiter. *J. Atmos. Sci.* 26:865–869.

Münch, G., and Neugebauer, G. 1971. Jupiter: an unidentified feature in the 5-micron spectrum of the North Equatorial Belt. *Science* 174:940–941.

Münch, G., and Younkin, R. L. 1964. Molecular absorptions and color distributions over Jupiter's disk. *Astron. J.* 69:553.

Orton, G. S. 1975. The structure of the Jovian atmosphere from ground-based and spacecraft observations in the thermal infrared. *Bull. Am. Astron. Soc.* 7:380.

Owen, T. 1963. Comparisons of laboratory and planetary spectra. I. The spectrum of Jupiter from 9000 to 10,000 Å. *Publ. Astron. Soc. Pacific* 75:314–322.

———. 1965a. Comparisons of laboratory and planetary spectra. II. The spectrum of Jupiter from 9700 to 11,200 Å. *Astrophys. J.* 141:444–456.

————. 1965b. Comparisons of laboratory and planetary spectra. III. The spectrum of Jupiter from 7750 to 8800 Å. *Astrophys. J.* 142:782–786.

————. 1969. The spectra of Jupiter and Saturn in the photographic infrared. *Icarus* 10:355–364.

————. 1970. The atmosphere of Jupiter. *Science* 167:1675–1681.

————. 1971. The 5520 Å ammonia band in the spectrum of Jupiter. *Astrophys. J.* 164:211–212.

————. 1972. The composition of planetary atmospheres. *Physics of the solar system* (S. I. Rasool, ed.), pp. 243–267. NASA SP-300. Washington, D.C.

Owen, T., and Cess, R. D. 1975. Methane absorption in the visible spectra of the outer planets and Titan. *Astrophys. J.* 197:L37–L40.

Owen, T., and Kuiper, G. P. 1964. A determination of the composition and surface pressure of the Martian atmosphere. *Comm. Lunar Planet. Lab.* 2:113–132.

Owen, T., and Mason, H. P. 1968. The abundance of hydrogen in the atmosphere of Jupiter. *Astrophys. J.* 154:317–326.

————. 1969. New studies of Jupiter's atmosphere. *J. Atmos. Sci.* 26:870–873.

Owen, T.; Richardson, E. H.; and Spinrad, H. 1964. A new ammonia band in the Jovian spectrum. *Astrophys. J.* 139:1374–1377.

Owen, T., and Sagan, C. 1972. Minor constituents in planetary atmospheres: ultraviolet spectroscopy from the Orbiting Astronomical Observatory. *Icarus* 16:557–568.

Owen, T., and Westphal, J. A. 1972. Jupiter's clouds: observational characteristics. *Icarus* 16:392–396.

Owen, T., and Woodman, J. H. 1968. On the atmospheric temperature of Jupiter derived from the $3\nu_3$ methane band. *Astrophys. J.* 154:L21–L23.

Peek, B. M. 1958. *The planet Jupiter*. London: Faber and Faber.

Pilcher, C. B.; Prinn, R. G.; and McCord, T. 1973. Spectroscopy of Jupiter: 3200 to 11,200 Å. *J. Atmos. Sci.* 30:302–307.

Prinn, R. G. 1970. UV radiative transfer and photolysis in Jupiter's atmosphere. *Icarus* 13:424–436.

————. 1973. Venus: composition and structure of the visible clouds. *Science* 182:1132–1135.

————. 1974. Stratospheric haze layers on Jupiter. *Bull. Am. Astron. Soc.* 6:375–376.

————. 1975. Venus: chemical and dynamical processes in the stratosphere and mesosphere. *J. Atmos. Sci.* 32:1237–1247.

Prinn, R. G., and Lewis, J. S. 1975a. Photochemistry of phosphine in the atmospheres of Jupiter and Saturn. *Bull. Am. Astron. Soc.* 7:381.

————. 1975b. Phosphine on Jupiter and implications for the Great Red Spot. *Science* 190:274–276.

Pugh, L. A.; Owen, T.; and Rao, K. N. 1974. $3\nu_3$ band of $^{13}CH_4$. *J. Chem. Phys.* 60:708–709.

Rank, D. H.; Fink, U.; and Wiggins, T. A. 1966. Measurements on spectra of gases of planetary interest. II. H_2, CO_2, NH_3, and CH_4. *Astrophys. J.* 143:980–988.

Regas, J., and Sagan, C. 1970. Line formation in planetary atmospheres. I. The simple reflecting model, (No. 3). Line formation in planetary atmospheres. II. The scattering model, (No. 4). Line formation in planetary atmospheres. III. Refinements due to inhomogeneity and to anisotropic scattering, (No. 5). *Comments Astrophys. Space Phys.* 2:116–120; 2:138–143; 2:161–166.

Ridgway, S. T. 1974a. Jupiter: Identification of ethane and acetylene. *Astrophys. J.* 187:L41–L43.

————. 1974b. The infrared spectrum of Jupiter, 750–1200 cm($^{-1}$). *Bull. Am. Astron. Soc.* 6:376.

————. 1974c. Erratum. *Astrophys. J.* 192:L51.

Robertson, R., and Fox, J. J. 1928. Studies in the infrared region of the spectrum. *Proc. Roy. Soc. London* 120:128–176.

Rutherfurd, L. M. 1863. Astronomical observations with the spectroscope. *Amer. J. Sci.* 85: 71–77.

Sagan, C. 1971. The solar system beyond Mars: an exobiological survey. *Space Sci. Rev.* 11: 827–866.

Sagan, C., and Khare, B. N. 1971. Experimental Jovian photochemistry: initial results. *Astrophys. J.* 168:563–569.

Sagan, C. E.; Lippincott, E. R.; Dayhoff, M. O.; and Eck, R. V. 1967. Organic molecules and the coloration of Jupiter. *Nature* 213:273–274.

Scattergood, T.; Lesser, P.; and Owen, T. 1975. Production of organic molecules in the outer solar system by proton irradiation: laboratory simulations. *Icarus* 24:465–471.

Slipher, E. C. 1964. *A photographic study of the brighter planets.* (Lowell Observatory Report, Flagstaff, Arizona): 75–107.

Slipher, V. M. 1933. Spectrographic studies of the planets. *Mon. Not. Roy. Astron. Soc.* 93, No. 9:657–668.

Smoluchowski, R. 1967. Internal structure and energy emission of Jupiter. *Nature* 215:691–695.

———. 1970a. Solid-state convection on Jupiter. *Phys. Rev. Letters* 25:693–695.

———. 1970b. Jupiter's convection and its red spot. *Science* 168:1340–1342.

Spinrad, H., and Trafton, L. 1963. High dispersion spectra of the outer planets. I. Jupiter in the visual and red. *Icarus* 2:19–28.

Squires, P. 1957. The equatorial clouds of Jupiter. *Astrophys. J.* 126:185–194.

Stecher, T. P. 1965. The reflectivity of Jupiter in the ultraviolet. *Astrophys. J.* 142:1186–1190.

Stone, P. H. 1973. The effect of large scale eddies on climatic change. *J. Atmos. Sci.* 30: 521–529.

Strobel, D. F. 1969. The photochemistry of methane in the Jovian atmosphere. *J. Atmos. Sci.* 26:906–911.

———. 1973a. The photochemistry of hydrocarbons in the Jovian atmosphere. *J. Atmos. Sci.* 30:489–498.

———. 1973b. The photochemistry of NH_3 in the Jovian atmosphere. *J. Atmos. Sci.* 30: 1205–1209.

———. 1974. Hydrocarbon abundances in the Jovian atmosphere. *Astrophys. J.* 192:L47–L49.

Teifel, V. G. 1969. Molecular absorption and the possible structure of the cloud layers of Jupiter and Saturn. *J. Atmos. Sci.* 26:854–859.

Terrile, R. J., and Westphal, J. A. 1975. The vertical cloud structure of Jupiter from 5 micron measurements. *Bull. Am. Astron. Soc.* 7:380.

Tokunaga, A.; Knacke, R. F.; and Owen, T. 1975. Ethane and acetylene abundances in the Jovian atmosphere. *Astrophys. J.* In press.

Tomasko, M. G. 1974. Ammonia absorption relevant to the albedo of Jupiter. II. Interpretation. *Astrophys. J.* 187:641–650.

Trafton, L. M. 1967. Model atmospheres of the major planets. *Astrophys. J.* 147:765–781.

———. 1972. Quadrupole H_2 absorption in the spectra of Jupiter and Saturn. *Bull. Am. Astron. Soc.* 4:359.

Trafton, L. M., and Stone, P. 1974. Radiative-dynamical equilibrium state for Jupiter. *Astrophys. J.* 188:649–655.

Trauger, J. T.; Roesler, F. L.; Carleton, N. P.; and Traub, W. A. 1973. Observation of HD on Jupiter and the D/H ratio. *Astrophys. J.* 184:L137–L141.

van de Hulst, H. C. 1952. Scattering in the atmospheres of the earth and planets. *The atmospheres of the earth and planets* (G. P. Kuiper, ed.), pp. 49–111. Chicago, Illinois: University of Chicago Press.

Varanasi, P.; Cess, R. D.; and Bangaru, B. R. P. 1974. The ν_3 fundamental of ethane: integrated intensity and band absorption measurements with application to the atmospheres of the major planets. *J. Quant. Spec. and Rad. Trans.* 14:1107–1114.

Walker, M. F., and Hayes, S. 1967. Image tube observations of the $3\nu_3$ band of CH_4 in the spectrum of Jupiter. *Publ. Astron. Soc. Pacific* 79:464–472.

Wallace, L., and Hunten, D. M. 1973. The Lyman-alpha albedo of Jupiter. *Astrophys. J.* 182: 1013–1031.

Wallace, L.; Prather, M.; and Belton, M. 1974. The thermal structure of the atmosphere of Jupiter. *Astrophys. J.* 193:481–493.

Weidenschilling, S. J., and Lewis, J. S. 1973. Atmospheric and cloud structures of the Jovian planets. *Icarus* 20:465–476.

Westphal, J. A.; Matthews, K.; and Terrile, R. J. 1974. Five-micron pictures of Jupiter. *Astrophys. J.* 188:L111–L112.

Wildt, R. 1932. Absorptions Spektren und Atmosphären der Grossen Planeten. *Veröff. Univ. Sternwarte Göttingen* 2, No. 22:171–180.

Wildt, R., and Meyer, E. J. 1931. Das Spektrum des Planeten Jupiter. *Veröff. Univ. Sternwarte Göttingen* 2, No. 19:141–156.

Woeller, F., and Ponnamperuma, C. 1969. Organic synthesis in a simulated Jovian atmosphere. *Icarus* 10:386–392.

Zabriskie, F. R. 1962. Hydrogen content of Jupiter's atmosphere. *Astron. J.* 67:168–170.

THE CHEMISTRY OF THE JOVIAN CLOUD COLORS

G. T. SILL
University of Arizona

The various hypotheses which have been advanced to explain the cloud colorations of Jupiter are reviewed. These hypotheses may be divided into two classes corresponding to the classical divisions of chemistry, namely inorganic and organic. Emphasis is placed on the earliest paper for each separate kind of hypothesis. Shortcomings of these various hypotheses are noted. In the inorganic class, Wildt proposed blue and red-brown solutions of sodium-ammonia to explain some colors on Jupiter; Rice interpreted a greater variety of colors through frozen trapped free-radicals; Papazian proposed color centers produced through high-intensity bombardment from the radiation belt; Lewis and Prinn called on elemental sulfur and polysulfides to explain yellow and orange-brown colors; and Prinn and Lewis stated that the color of the Great Red Spot may be due to red phosphorus. In the organic class, Urey reported that acetylene could polymerize into red cuprene, Miller and Sagan proposed various types of polymers of C and N, while Moeller and Pannamperuma, and Sagan and Khare produced red, brown and yellow polymers through arc discharge or ultraviolet radiation of CH_4, NH_3 and in some cases H_2S.

There is general agreement that the main constituent of the clouds of Jupiter is frozen ammonia. As Kuiper (1952) pointed out, the amount of NH_3 gas must be limited by its vapor pressure, since the quantity observed in the Jovian atmosphere is much less than its solar abundance. Where the ammonia pressure exceeds the vapor pressure, condensation occurs to produce ammonia cirrus-type clouds. Kuiper showed that the upper level where this condensation occurs is about 165°K (Fig. 1). The existence of ammonia ice clouds as the topmost layer has not been challenged by any investigator after Kuiper.

Ammonia ice, however, is colorless, and it has been known for centuries that Jupiter has various shades of color: red (or pink, red-orange); brown (or red-brown and tan); blues (or blue-gray, purple-gray); grays; yellows (or yellow-brown, ocher, cream, greenish yellow); and perhaps even green. Numerous high-quality color photographs, such as those taken at the Lunar and Planetary Laboratory in Tucson, at New Mexico State in Las Cruces, and at the Lowell Observatory in Flagstaff, exhibit enough contrast in color to convince us that these colors really exist.

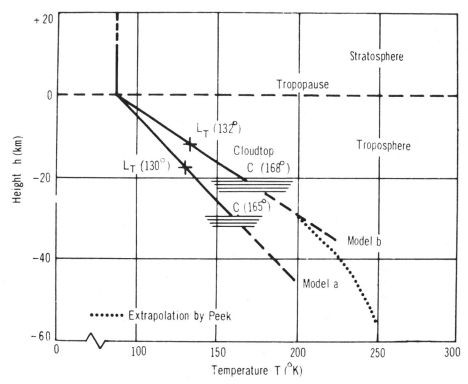

Fig. 1.   Condensation of NH_3 to produce cirrus-type clouds in Jovian atmosphere (after Kuiper 1952).

What are the causes of the colors? Since a simple scattering process would have difficulty in explaining such a wealth of color, most investigators place the cause in the chemical composition of the pigmented areas. The light zones of the planet often look pure white; these are presumably the ammonia cirrus clouds. As for the dark colored belts and the Great Red Spot, there are numerous possible explanations.

For purposes of classification the various hypotheses can be divided into organic and inorganic, depending upon the constitution of the chromophore. If the chromophores contain carbon or are produced in processes utilizing carbon compounds, they are referred to as organic. If the colored materials are produced by processes utilizing other elements besides carbon, they are classed as inorganic.

The problem is not simple because almost any color can be produced by some chemical compound or mixture of compounds, and virtually identical colors can be found in both inorganic and organic substances, if one searches hard enough. The only constraint is to devise some mechanism to produce the desired compound from constituents presumably present in the atmosphere of Jupiter. It would be desirable if firmer constraints could be placed on the proposed coloring materials, such as those of photometry, spectroscopy or polarimetry. Data from Pioneer 10 and 11 appear to offer no additional constraints to those from previous telescopic observations. Photometry of the whole planet indicates an overall ultraviolet (UV) darkening, but high spatial resolution comparing bright zones with dark belts reveals little more than a slight change in slope in the visible spectrum, which is to be expected from subtle color differences. At the present time narrowband photography seems to reveal the contrasts in color better than any other method but, so far, the spatial resolution is low.

I. INORGANIC HYPOTHESES

Wildt (1939) proposed the first inorganic hypothesis. When he became acquainted with the fascinating properties of sodium-liquid ammonia solutions with their brilliant blue and bronze colors, it reminded him of Jupiter's colors. Whereas a conventional chemist might hesitate to propose such a reactive substance for a planet's atmosphere, Wildt did not think this was an insuperable problem, inasmuch as elemental sodium is present in the tails of comets. Perhaps elemental sodium is similarly produced in Jupiter's atmosphere and would color the clouds.

A deep, royal blue color is characteristic of dilute solutions of Na in NH_3. The color is due to the solvated electron (Jolly 1964):

$$Na + NH_3(l) = Na^+ + e^- \text{ (ammonia)} . \tag{1}$$

The electron is apparently captured in a cavity in the ammonia solvent. More concentrated solutions have a bronze color and are due to a pair of solvated electrons:

$$2Na + NH_3(l) = 2Na^+ + 2e^- \text{ (ammonia)} . \tag{2}$$

The phase diagram of Fig. 2 shows that a solution of intermediate composition (5%), when cooled, separates into two components, the bronze floating on the denser blue solution. Upon further cooling, the blue solution freezes at 195°K into a gray-white solid. The bronze solution becomes more concentrated as solid Na freezes out until the eutectic is reached at 162°K. Wildt proposed the following scheme: the gray areas of Jupiter exist at any temperature up to 195°K; the blue dilute solution from 195 to 270°K, depending on the NH_3 vapor pressure; and the more concentrated bronze solutions from 162 to 270°K. While Kuiper (1952) originally thought this scheme feasible, he revised this opinion later (1972) when the spectral properties of these solutions became better known. The blue color is due to the visible absorption wing of an extremely deep, broad absorption band centered on 1.5 μm (Fig. 3). Kuiper thought the blue festoons showed more contrast (were darker) at 0.6 μm than at 0.8 μm. This would be contrary to the spectral properties of the blue liquid.

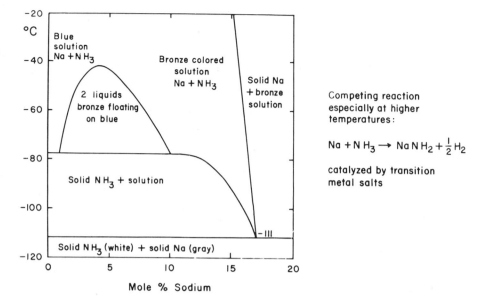

Competing reaction
especially at higher
temperatures:

$$Na + NH_3 \rightarrow NaNH_2 + \frac{1}{2}H_2$$

catalyzed by transition
metal salts

Fig. 2.   Phase diagram of Na-NH₃ solutions.

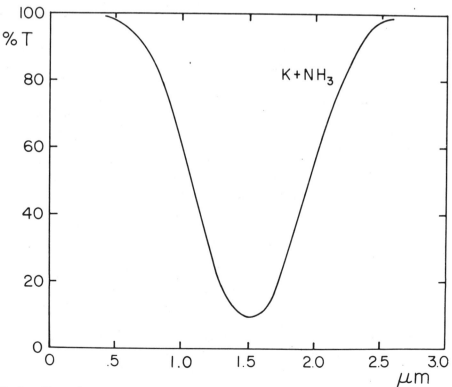

Fig. 3.   Absorption spectrum of K-NH₃ blue solution.

   The reactivity of alkali metals in a dense atmosphere cannot be lightly dismissed. Self decomposition of sodium-ammonia solutions occur, even at 240°K:

$$2Na + 2NH_3 = 2NaNH_2 + H_2 \, . \tag{3}$$

   The product, sodium amide, is white. Sodium would also react in the upper atmosphere with hydrogen and amide radicals, the latter having been produced by solar UV:

$$Na + H^{\cdot} = NaH$$
$$Na + {}^{\cdot}NH_2 = NaNH_2 \, . \tag{4}$$

Neither elemental sodium, nor the hydride or amide would survive contact with water vapor in the lower atmosphere:

$$2Na + 2H_2O = 2NaOH + H_2$$
$$NaH + H_2O = NaOH + H_2 \tag{5}$$
$$NaNH_2 + H_2O = NaOH + NH_3 \, .$$

Once turned into the hydroxide, the sodium is removed from further reactions with ammonia. The elemental sodium would have to be provided by a source outside Jupiter, and the intensity of colors observed would be a steady-state result of this influx.

   Some colors on Jupiter do not correspond to the alkali metal-ammonia solutions, namely, cream, yellows and oranges in particular. At times these dominate on the planet. Furthermore, since small temperature changes can give rise to a separation of blue and bronze solutions, as well as a freezing to a gray solid, these relatively rapid changes should be observed. Such dramatic changes are, however, not observed. Another difficulty is the fact that the colors are distributed in belts at various latitudes. If the sodium or other alkali metals are captured by the planet from outside the atmosphere, it is hard to see why the colors are not uniformly distributed over the planet's surface, or perhaps enhanced in the equatorial regions, if the source region were similar to Io's cloud, on the equatorial plane.

   The next step in inorganic chemistry was made by a physical chemist, Rice (1956a,b). While trapping free radicals at cryogenic temperatures, he noticed that brilliant colors often appeared. The trapped free-radicals are stable at low temperatures and they remain stable until the transition temperature is reached, at which point they revert to more stable molecules and lose their colors. A list of some species is seen in Table I. The energy source for dissociation is solar UV and electrical discharge in the Jovian atmosphere. The chief criticism that should be leveled at this hypothesis is that it neglects the possibility of recombination, especially with abundant H atoms while the free-radicals are in the gaseous state, and that it uses an unrealistic cold trap to stabilize the free-radicals. The transition temperatures are much lower than most models of the Jovian atmosphere would permit, namely 165°K according to Kuiper's (1952) model, 135°K on Weidenschilling and Lewis' (1973) model, and 125°K measured by the Pioneer 10 and 11 radiometer (Ingersoll *et al.* 1975).[1] If the free radicals are produced by electrical discharges, these would occur in the lower portion of the atmosphere near 300°K, thus precluding the trapping of any free radicals.

TABLE I
Colors of Trapped Free Radicals[a]

Species	Color	Transition temp. °K	Remarks
$(NH)_n$	blue	148	changes to ammonium azide, NH_4N_3
$(NH_2NH)_n$	yellow	95	may be tetrazane, $NH_2NHNHNH_2$; gives off N_2 at trans. temp.
$(CH_3N)_n$	colorless	78	thermal decomp. of methylazide, CH_3N_3
$(C_2H_5)_2N$	light yellow	113	elec. discharge on $(C_2H_5)_2NH$
$(CH_3)_2N$	deep green	103	from $(CH_3)_2N - N = N - N(CH_3)_2$ in elec. discharge
CH_3S	yellow	$103-113$	from $CH_3S - SCH_3$
$(CH_3)_3CS$	red	$103-133$	from $(CH_3)_3CS - SC(CH_3)_3$
S_2	purple		changes from purple to yellow in a few seconds at room temp. and in a few hours at 193°K.

[a]Table after Rice (1956*b*).

A variation on Rice's theme was proposed by Papazian (1959). Charged particles falling into the Jovian atmosphere from its radiation belt could produce colors by creating voids in crystalline solids. This effect of radiation is well known. He calculated that a 100 MeV proton could penetrate 10 g cm^{-2} of CH_4 and NH_3, but if hydrogen and helium are abundant, he noted that they would effectively shield the clouds from the particles.

Another inorganic hypothesis was proposed by Lewis (1969*a*). Considering the cosmic abundance of sulfur and the reducing state of the atmosphere, he concluded that hydrogen sulfide should be present and should react with ammonia to produce ammonium hydrogen sulfide, NH_4SH, a volatile solid which should condense into a cloud around the 200°K level (Fig. 4). Lewis

[1]See p. 202.

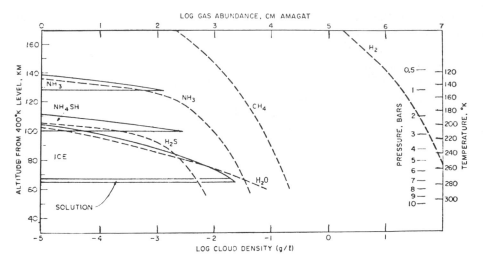

Fig. 4.   Cloud layers on Jupiter (after Weidenschilling and Lewis 1973).

and Prinn (1970) pointed out that H_2S would decompose under solar UV radiation to produce H_2S, sulfur and ammonium polysulfide, $(NH_4)_2S_x$; the polysulfide can take on yellow, orange, or brown color depending on the length of the chain. Under the influence of cumulus-type convection the ammonium sulfide would rise to the top of the atmosphere where photolytic decomposition could occur, or alternatively according to Prinn (1970), the solar UV would penetrate through holes in the ammonia cirrus clouds to photolyze ammonium sulfide, or its gaseous component, H_2S.

Following this suggestion, W. Wiśniewski and Sill (unpublished data) attempted to photolyze NH_4SH, condensed at 100°K, with UV from both a Hg emission lamp and a broadband UV deuterium lamp. In both cases a pale green polymer was produced, which turned out to be elemental sulfur. The photolysis produced complete decomposition to sulfur without producing any notable quantity of $(NH_4)_2S_x$. Apparently it is difficult to produce orange or brown ammonium polysulfide from irradiation of cold, white $NH_4SH(s)$. In another experiment, hydrogen sulfide gas was irradiated at 300°K with a Hg emission lamp. Again the main product of decomposition was elemental sulfur, with approximately 1% yellow liquid (H_2S_x) after 24 hours of irradiation. Ammonium polysulfide, however, can be produced by dissolving elemental S_8 in ammonium hydroxide solution, substance of a cloud layer proposed by Lewis (1969a). This solution can have colors ranging from yellow to brown depending on the quantity of sulfur dissolved. Upon evaporation of the solution, S_8 reappears. A model for the yellow-orange-brown clouds would involve convection of the ammonium polysulfide solution from the level of the ammonium hydroxide clouds, 275°K, to the top of the atmosphere, with the polysulfide frozen into the solution particles. As ammonia and water evaporated, the sulfur would reform, bleaching the clouds to pale yellow or green. However, one might expect more water vapor to be detected in the atmosphere if this were the case (Larson *et al.* 1975).[2]

Following the discovery by Ridgway (1974) of phosphine on Jupiter, Prinn and Lewis (1975)[3] proposed the following reaction scheme:

$$PH_3 = PH_2 + H \ (UV \text{ photolysis})$$

$$PH_2 + PH_2 = PH + PH_3$$

$$PH + PH = P_2 + H_2 \tag{6}$$

$$P_2 + P_2 = P_4$$

$$P_4(g) = P_4(s) \text{ red}.$$

Phosphine would be regenerated at depth by reacting with hydrogen. The P_4 solid has a dark red color in one of its allotropic forms. A complication arises from the fact that the mass densities of the ammonia crystal at the cloud base are about 10^4 greater than the predicted phosphorus mass densities. An increase of phosphorus versus ammonia could occur if there is a strong vertical mixing, such as occurs in the Great Red Spot. In this case they compute a maximum optical depth of the red phosphorus at ~ 0.7. Phosphorus particles might be visible where ammonia clouds are largely absent, such as in the North Equatorial Belt. They reject the importance of yellow P_2H_4, because the reaction rate for

$$PH_2 + PH_2 + M = P_2H_4 + M \tag{7}$$

is too slow.

II. ORGANIC CHEMICAL HYPOTHESES

Urey (1952) was the first to propose that photolytic reaction in the Jovian atmosphere could produce colored species from carbon compounds. Earlier Wildt (1939) had predicted that acetylene could be produced in small amounts in a methane atmosphere and UV photolysis could polymerize this and produce a solid product, but at the same time surplus H atoms would attack complex hydrocarbons and reduce them to methane. However, Urey believed acetylene would survive and produce a red complex polymer called cuprene, and that this could be the source of yellow or red colors on Jupiter.

Laboratory simulations began with Sagan and Miller (1969) when they subjected mixtures of H_2, CH_4 and NH_3 to a semi-corona discharge and produced acetylene, ethene, and cyanides. They again noted that C_2H_2 could polymerize and produce a colored solid. Small amounts of heterocyclic compounds, like purines, could be formed as well. Sagan (1960) also went on to postulate that using their experimental results as a guide, approximately 1 kg cm^{-2} of organic matter would be produced in the four-aeon history of Jupiter.

---

[2]See p. 406.   [3]See p. 340.

Sagan *et al.* (1967), expanding on theoretical calculations advanced by Eck *et al.* (1966), employed chemical thermodynamic "equilibria" (actually quenched high-temperature reactions) to predict the formation of complex hydrocarbons and nitrogen compounds on Jupiter. The calculations were made for 1500°K and 1 atm pressure, corresponding to an electric discharge followed by immediate cooling of the reaction products. Among the substances produced would be gases like C_2H_2, C_2H_4, C_2H_6, HCN, CH_3CN, and the solids napthalene, asphalt (yellow) and azulene (blue). Using a radiofrequency generated plasma, Eck *et al.* (1966) produced some polynuclear aromatics like pyrene, coronene (yellow), and chrysene (red-violet fluorescence).

Further experimental simulations on CH_4 and NH_3 were carried out by Woeller and Ponnamperuma (1969), as well as by Sagan and Khare (1971) who added H_2S to the above gases. Woeller and Ponnamperuma utilized a semi-corona or arc discharge for the high-energy source. With 20 W of electric power on exposures of a few hours they collected, on a cold finger, a solid which turned red-brown when warmed; it apparently was a polymer of cyanogen $(CN)_x$. Sagan and Khare used UV for the required energy source, namely a Hg-vapor lamp emitting at 2537 and 1849 Å. In these initial experiments only the short-wavelength radiation was able to produce colored material on the walls of the container, after 14 days of exposure. The longer wavelength radiation (2537 Å) produced sulfur and various alkyl sulfides, but nothing with a red or brown color. Further experiments (Khare and Sagan 1973) produced a brown polymer, with a methanol-extracted yellow component. The yellow polymer appeared to be a long-chain aliphatic substance, with some polyene character, containing OH and NH_2 substituents. The brown polymer was not analyzed, but a transmission spectrum was obtained in visible light. Recent analysis of this brown material indicates that it is 96% sulfur.

Ponnamperuma and Molton (1973) analyzed the red tar produced in an arc-discharge experiment and they proposed that it was a linear, unsaturated red nitrile polymer; the reaction steps are shown in Fig. 5. They also simulated high-pressure effects of hydrogen and helium on the rate of production of material by sparking a mixture of these gases (plus CH_4 and NH_3) under a total pressure of 50 atm. The spark generated 504 Å radiation from the ionization of helium. This apparently caused hydrogenation of the products, resulting in a much reduced yield that was too small to analyze. They believed, however, that there could be regions in Jupiter where organic synthesis would occur.

The hypotheses of organic synthesis are not without shortcomings, and various authors have criticized the proposed mechanisms. Strobel (1969) estimated that about 10^{-4} of CH_4 could be condensed into heavier hydrocarbons by solar UV, but Prinn (1970) stated this is not likely because H atoms produced (mostly from NH_3) by ultraviolet radiation would attack C-C bonds readily and destroy complex hydrocarbons. Lewis (1969b) pointed

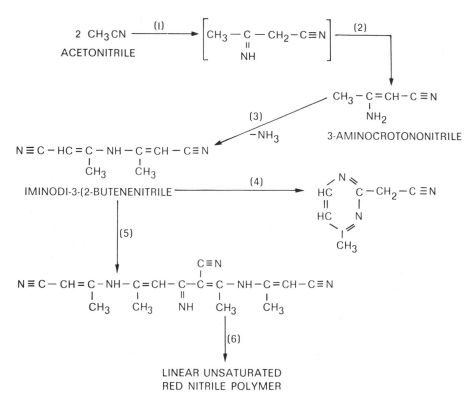

Fig. 5. Reaction mechanism for producing colored nitrite polymer (after Ponnamperuma and Molton 1973).

out that equilibration exists in the Jovian atmosphere, otherwise there would be no gaseous ammonia present since it would be decomposed into N_2 and H_2. But gaseous ammonia is present and thermodynamic recombination therefore exists at depth. The same process would occur with complex organic compounds, reconstituting methane.

The rates of reaction in the simulation experiments are undoubtedly enhanced by concentration effects as well as by high-energy sources operating in close proximity to the reactants. For example, in Sagan and Khare's (1971) experiment, if the UV produced by the Hg lamp had a power of 0.1 W, running for 14 days, this would be the equivalent of 40 years of exposure in the Jovian atmosphere. Likewise Woeller and Ponnamperuma's (1969) arc discharge utilized about 10 W in a restricted apparatus.

Lewis (1974) examined various rates of energy production on Jupiter, including solar UV and lightning discharges. C_2 hydrocarbons are produced at the rate of 10^{-14} g cm^{-2} sec^{-1} by solar UV. If the internal heat source of Jupiter is converted into electric discharges, vast amounts of energy would be produced, enough to produce 10^{-10} g cm^{-2} sec^{-1}, presuming 100% efficiency in the process. More realistically, however, he estimated the production rate as 10^{-17} g cm^{-2} sec^{-1}, less than that of the UV. But this is much less than the production of inorganic chromophores, which he estimated as 5×10^{-11} g cm^{-2} sec^{-1}.

III. CONCLUSIONS

There is no conclusive proof for any hypothesis on the production of colored material in Jupiter's clouds. Each hypothesis can be criticized for leaving unanswered questions. The calculations used to predict inorganic chromophores, particularly those of sulfur, require corroborative laboratory simulations. Polysulfides are difficult to produce in simulated Jovian exposures to UV. The organic hypotheses have abundant laboratory simulations to show that the colors can be produced, but are unable to satisfactorily explain the energy sources which would be required on Jupiter to produce the quantity of coloring agents required. However, little is known of the deep atmosphere of Jupiter, and perhaps the needed energy sources are available. The discovery of hydrogen cyanide deep in the planet's atmosphere would enhance the organic synthesis hypothesis, due to the ease of polymerizing cyanide into the red linear polymer. More experimental observations of Jupiter at high spatial resolution, at long infrared wavelengths, will be necessary to identify the colored material.

REFERENCES

Eck, R. V.; Lippincott, E. R.; Dayhoff, M. O.; and Pratt, Y. T. 1966. Thermodynamic equilibrium and the inorganic origin of organic compounds. *Science* 153:628–633.

Ingersoll, A. P.; Münch, G.; Neugebauer, G.; Diner, D. J.; Orton, G. S.; Schupler, B.; Schroeder, M.; Chase, S. C.; Ruiz, R. D.; and Trafton, L. M. 1975. Pioneer 11 infrared radiometer experiment: the global heat balance of Jupiter. *Science* 188:472–473.

Jolly, W. L. 1964. *The inorganic chemistry of nitrogen.* New York: W. A. Benjamin.

Khare, B. N., and Sagan, C. 1973. Red clouds in reducing atmospheres. *Icarus* 20:311–321.

Kuiper, G. P. 1952. Planetary atmospheres and their origin. *The atmospheres of the earth and planets.* (G. P. Kuiper, ed.) pp. 306–405. Chicago, Illinois: University of Chicago Press.

———. 1972. Lunar and Planetary Laboratory studies of Jupiter. II. *Sky and Telescope* 43: 75–81.

Larson, H. P.; Fink, U.; Treffers, R.; and Gautier, T. N. 1975. Detection of water vapor on Jupiter. *Astrophys. J.* 197:L137–L140.

Lewis, J. S. 1969a. The clouds of Jupiter and the NH_3-H_2O and NH_3-H_2S systems. *Icarus* 10: 365–378.

———. 1969b. Observability of spectroscopically active compounds in the atmosphere of Jupiter. *Icarus* 10:393–409.

———. 1974. Equilibrium and disequilibrium: chemistry of adiabatic solar-composition planetary atmospheres. *MIT Planet. Astron. Lab.* Contrib. No. 119.

Lewis, J. S., and Prinn, R. G. 1970. Jupiter's clouds: structure and composition. *Science* 169: 472–473.

Papazian, H. A. 1959. The colors of Jupiter. *Publ. Astron. Soc. Pacific* 71:237–239.

Ponnamperuma, C., and Molton, P. 1973. The prospect of life on Jupiter. *Space Life Sci.* 4: 32–44.

Prinn, R. G. 1970. UV radiative transfer and photolysis in Jupiter's atmosphere. *Icarus* 13: 424–436.

Prinn, R. G., and Lewis, J. S. 1975. Phosphine on Jupiter and implications for the Great Red Spot. *MIT Planet. Astron. Lab.* Contrib. No. 132.

Rice, F. O. 1956a. The chemistry of Jupiter. *Sci. Amer.* 194, No. 6:119–128.

———. 1956b. Colors on Jupiter. *J. Chem. Phys.* 24:1259.

Ridgway, S. T. 1974. The infrared spectrum of Jupiter, 750–1200 cm^{-1}. *Bull. Amer. Astron. Soc.* 6:376.

Sagan, C. 1960. Production of organic molecules in planetary atmospheres. *Astron. J.* 65:499.

Sagan, C., and Khare, B. N. 1971. Experimental Jovian photochemistry: initial results. *Astrophys. J.* 168:563–569.

Sagan, C.; Lippincott, E. R.; Dayhoff, M. O.; and Eck, R. V. 1967. Organic molecules and the coloration of Jupiter. *Nature* 213:273–274.

Sagan, C., and Miller, S. L. 1960. Molecular synthesis in simulated reducing atmospheres. *Astron. J.* 65:499.

Strobel, D. F. 1969. The photochemistry of methane in the Jovian atmosphere. *J. Atmos. Sci.* 26:906–911.

Urey, H. C. 1952. *The planets.* New Haven, Connecticut: Yale University Press.

Weidenschilling, S. J., and Lewis, J. S. 1973. Atmospheric and cloud structure of the Jovian planets. *Icarus* 20:465–476.

Wildt, R. 1939. On the chemical nature of the colouration in Jupiter's cloud forms. *Mon. Not. Roy. Astron. Soc.* 99:616–623.

Woeller, F., and Ponnamperuma, C. 1969. Organic synthesis in a simulated Jovian atmosphere. *Icarus* 10:386–392.

THE INFRARED SPECTRUM OF JUPITER

S. T. RIDGWAY
Kitt Peak National Observatory

H. P. LARSON AND U. FINK
University of Arizona

The principal characteristics of Jupiter's infrared spectrum are reviewed with empha-sis on their significance to our understanding of the composition and temperature struc-ture of the Jovian upper atmosphere. The spectral region from 1 μm to 40 μm divides naturally into three regimes: the reflecting region, the thermal emission from below the cloud deck (5-μm hot spots), and the thermal emission from above the clouds. Opaque parts of the Jovian atmosphere further subdivide these regions into windows, and each is discussed in the context of its past or potential contributions to our knowledge of the planet. Recent results are incorporated into a table of atmospheric composition and abundance which includes positively identified constituents as well as several which require verification. The limited available information about spatial variations of the infrared spectrum is presented. Almost every topic deserves further investigation, but we have tried to indicate the most promising areas.

Historically, the study of infrared spectra of astronomical sources has proceeded less rapidly than the study of visible spectra. The reasons include low flux levels, low quantum efficiency of detectors, and obscuration of part of the infrared spectral region by the terrestrial atmosphere. Yet the rewards for planetary observations in the infrared are considerable. Planetary atmo-spheres are cool so that they consist almost exclusively of molecules, whose fundamental absorption frequencies lie in the infrared. Since the fundamental bands have much larger band strengths than their overtone or combination bands, the infrared region of the spectrum offers the best chance for detecting constituents in planetary atmospheres. During the last decade a number of technological developments have become available and helped to overcome many of the experimental difficulties. Most influential in this breakthrough were improved detector systems, Fourier transform multiplexing spectrome-ters, and in some cases the availability of airplane, balloon and spacecraft observing stations.

The convergence of these technologies in recent years has had great impact on our knowledge of the planet Jupiter. Until recently several regions of the infrared had no spectra at all, while others suffered from low resolution and excessive noise. Now we have spectra of very high resolution to about 1.6 μm (Connes and Michel 1974; de Bergh et al. 1976), good quality spectra of the 5 μm region from the ground (Beer et al. 1972), and from airplanes (Larson et al. 1975), medium resolution spectra in the 10 μm region from the ground (Ridgway 1974a, Lacy et al. 1975), and from airplanes (Encrenaz et al. 1976), and low resolution spectra from 20 to 40 μm from aircraft (Houck et al. 1975).

In his review of the material presented on Jupiter at the Third Arizona Conference on Planetary Atmospheres, Goody (1969) noted that no new chemical species had been identified. Since 1932 when methane and ammonia were identified (Wildt 1932) and nearly three decades later when proof of the existence of molecular hydrogen was provided by Kiess et al. (1960), progress had indeed been slow. In 1969 the composition of the atmosphere of Jupiter was therefore limited to H_2, CH_4, NH_3 with helium presumed but not subject to observational verification. By contrast, at the 1975 Jupiter Colloquium the list of proposed atmospheric constituents for Jupiter had dramatically increased, now including H_2O, C_2H_6, GeH_4, C_2H_2, PH_3, CO, HCN, and CO_2, as well as isotopic species $C^{13}H_4$, HD and CH_3D. The evidence for some of these new molecules requires additional confirmation but the length and variety of this list, which except for HD is drawn entirely from infrared observations, is indicative of the current activity in the field.

While the identification of new molecular species is an important end in itself, distribution of line intensities, line profiles, etc. can serve as sensitive probes into the physical conditions of the atmosphere. Thermal radiation of the planet, which can only be detected in the infrared, provides a natural technique for determining the temperature structure of the atmosphere at wavelengths beyond 4 μm. In addition, the low opacity of Jupiter's atmospheric constituents and clearings in its cloud cover, in the spectral region near 5 μm, permits deep penetration into Jupiter's atmosphere and probing below its visible cloud tops. Very little analysis has been done on the above aspects of the observations so that opportunities there are only beginning.

INTRODUCTION TO THE INFRARED SPECTRUM

For the purposes of this review we take the infrared region to begin at about 1 μm, the limit of photographic and photomultiplier detection techniques. The region below 1 μm is discussed in an accompanying review by Prinn and Owen.[1] Since spectroscopic observations of Jupiter do not presently exist beyond about 40 μm, we take this as the upper limit of our review.

---

[1]See p. 319.

We exclude from consideration the far infrared which extends out to 1 mm.

Spectroscopy and photometry from nine different instruments have been combined to form a nearly complete overview of Jupiter's infrared spectrum. This is shown in Fig. 1a. In order to cover the large range of intensity it was necessary to use a logarithmic plot. The region from 1 to 3 μm was put on an approximate photometric scale by using the balloon data of Danielson (1966) and groundbased spatial spectrophotometry by Binder and McCarthy (1973). No JHKL photometry of Jupiter exists, mostly because its diameter exceeds present photometer apertures. We hope that suitable photometry or spectrophotometry of Jupiter can be obtained sometime in the future. The absolute intensity scale for that region is therefore only approximate. The spectral shape of the windows in the 1–3 μm region were taken from data by Danielson (1966), Fink and Larson (1971), and Maillard et al. (1973). The photometry and infrared maps of Westphal et al. (1974) were used to determine intensities in the 5 μm window, while the photometry of Gillett et al. (1969) was used for the 10 μm window and the data of Houck et al. (1975) for the region 20–40 μm. The detailed appearance of the window at 5 μm came from Larson et al. (1975, 1976), and at 10 μm from Ridgway (1974a).

The infrared spectrum can be divided into three distinctly different regions, each with important consequences for physical observations of Jupiter's atmosphere. For frequencies higher than 3000 cm^{-1}, the observed flux from Jupiter is due entirely to reflected sunlight. The curve labeled *ALBEDO = 1.0* indicates the expected intensity for a diffusely reflecting planet with no absorption. The appearance of the spectrum in this region is dominated by strong absorption bands of methane and ammonia. The penetration of the reflected solar radiation is limited by the top of the visible cloud deck where the pressure, while uncertain, is probably $\sim$ 1 atm and the temperature is $\sim$ 180°K. For frequencies less than 2500 cm^{-1}, the spectrum is dominated by thermal emission. The curve labeled *BB 125K* shows the expected intensity for a blackbody Jupiter with a temperature of 125°K. The continuum absorption of the pressure-induced translational and rotational dipole spectrum of molecular hydrogen is responsible for the bulk of the infrared opacity in that region from about 10 μm to 40 μm and prevents us from seeing deeper into the atmosphere. The deviations of the observed spectrum from the 125°K curve are due to a combination of molecular absorption and emission. These spectral features originate in the upper atmosphere ($P < 0.5$ atm, $T < 150$°K) and they can provide valuable information on the thermal structure of this region of the atmosphere. The integrated flux in the thermal emission spectrum exceeds the total flux absorbed in the reflecting region by about a factor of 2, hence the deduction that Jupiter has an internal energy source (cf. Ingersoll et al. 1975).[2]

---

[2]See pp. 202 and 508.

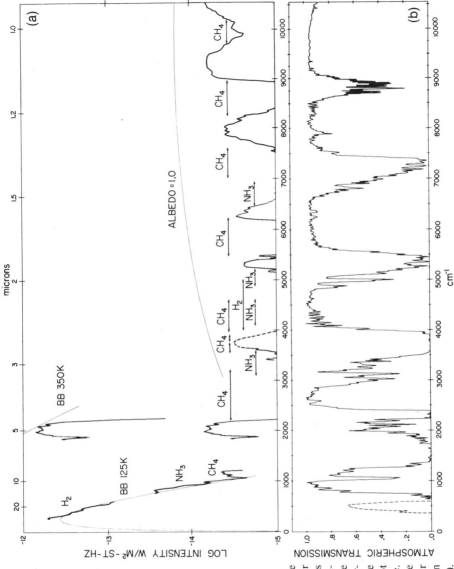

Fig. 1a and b. *Above:* composite infrared spectrum of Jupiter assembled from many sources (see text). In the 2000 cm⁻¹ region two curves are given: the lower curve is for the disk average brightness temperature; the upper curve is for a possible hot spot brightness temperature. *Below:* transmission of the terrestrial atmosphere under typical good conditions from Kitt Peak (elevation 2064 m).

The region near 2000 cm^{-1}, usually called the 5 μm window, is rather special.[3] Here there are no strong methane or ammonia bands. The high disk-average intensity (lower curve in Fig. 1a) indicates that we are seeing especially deep into the atmosphere. The disk-average brightness temperature of about 230°K would suggest that we are seeing to a depth where the pressure is about 2 atm. Actually, spatially resolved studies of Jupiter's 5 μm radiation show that the intensity is strongly peaked in small regions comprising, at most, a few percent of the surface area. The upper curve in Fig. 1a indicates the intensity of such a hot spot having a brightness temperature of 350°K. This temperature is approximately consistent with observed molecular rotational temperatures from spectra in this region (Larson *et al.* 1975), and indicates that the flux may come from a pressure level as great as 5 atm.

It is quite evident in Fig. 1 that infrared observations of Jupiter can be made only in selected windows. These regions of transparency are defined by the combined spectral characteristics of the planetary atmosphere and, except for space observations, the earth's intervening atmosphere (Fig. 1b). For a number of experimental reasons, observations are usually optimized for specific spectral regions. We describe below the observations and results obtained for different windows in Jupiter's atmosphere.

RECENT OBSERVATIONS AND RESULTS

1.0–4.0 μm: The Reflecting Region

In this region of the infrared spectrum the observed flux is reflected sunlight. Because this spectral interval is covered by the sensitivity of the PbS detector, long the mainstay of near-infrared detection, it is also sometimes referred to as the lead-sulfide region. It includes a number of telluric transmission windows (Fig. 1b) separated by the water-vapor bands at 1.14, 1.4, 1.9, and 2.7 μm and the carbon dioxide band at 4.2 μm. Only the first three windows show reasonable flux on Jupiter because of strong absorption bands in its spectrum.

The first spectrum of Jupiter in that region (0.75–2.0 μm) was obtained by Kuiper (1952) using a prism spectrometer and the newly developed Cashman PbS cells. The resolution was low ($\lambda/\Delta\lambda \sim 80$) and the spectra showed only the expected methane and ammonia bands. Further spectra with improved instrumentation and detectors were taken by Kuiper (1964) (0.95–1.8 μm, $\lambda/\Delta\lambda \sim 450$), Moroz (1964, 1966) (1–1.8 μm, $\lambda/\Delta\lambda \sim 250$), and Cruikshank and Binder (1968) (0.95–1.6 μm, res ~ 700). Additional low resolution spectra by means of Fourier spectroscopy were published by Delbouille *et al.* (1964), and at somewhat higher resolution (~ 1200) by Johnson (1970) and Fink and Larson (1971). A major advance was made

---

[3]See p. 28.

by Connes, who produced the first high resolution spectrum of Jupiter (1.2–1.7 μm, res $\sim$ 20,000) (Connes et al. 1969) using a new stepping technique for driving a Fourier interferometer. This resolution was improved slightly by Maillard et al. (1973) using a similar instrument, and considerably surpassed (res $\sim$ 140,000) by a new Connes interferometer used at the Coudé focus of the 5-meter reflector on Palomar Mountain (Connes and Michel 1974; de Bergh et al. 1976) (see Fig. 2).

1.0–1.14 μm Window: The $3\nu_3$ Methane Band. The wavelength interval just beyond 1 μm up to the telluric water band at 1.14 μm has been one of the most actively studied regions of the spectrum of Jupiter. This is mostly due to the presence of the $3\nu_3$ methane band at 9050 cm^{-1} ($\sim$ 11,050 Å). The assignment as the $3\nu_3$ band of methane and an analysis of its rotational structure was already accomplished in 1936 by Childs (1936) who recognized that the band had a relatively simple structure similar to that of the ν_4 fundamental. Even today this band is the highest energy transition in methane for which reliable rotational quantum number assignments exist. Being the second overtone, this band is sufficiently weak so that its absorptions on spectra of Jupiter are not heavily saturated as is the case for the lower overtone or combination bands. It is therefore the only methane band from which a rotational temperature can be derived, and one of the few bands for which identification of the $C^{13}H_4$ isotopic molecule could be made (Fox et al. 1972).

Although the band is accessible with a variety of instrumental techniques, observations up to now were hampered by the low sensitivity of the detectors available in this wavelength region. It lies at the long wavelength limit of photographic plate sensitivity (Kodak Z plates) and photo-multiplier response (S–1 photocathode), and at the short wavelength end of the detectivity of typical infrared detectors such as PbS or bolometers. Recently, however, the development of germanium and InSb photovoltaic detectors, together with the application of Fourier transform spectroscopy have produced much higher quality spectra in that region (Connes and Michel, 1974; de Bergh et al. 1976). The observations of the $3\nu_3$ band and its analysis are summarized in Table I and are briefly reviewed below.

Owen (1965) first utilized the $3\nu_3$ band for planetary research by analyzing plates of Jupiter taken by G. P. Kuiper in 1955 at the McDonald Observatory. Using the quantum number assignments by Childs (1936) and the intensity weighting factors by Childs and Jahn (1939), he derived a temperature of around 200°K. Sometime afterwards, Walker and Hayes (1967) obtained improved spectra of the $3\nu_3$ region on Jupiter using a modified Lallemand image intensifier having an S–1 photocathode and recording the spectra on electronographic plates. They concluded that no definite temperature could be derived from their measurements and estimated a range between 100° and 300°K. Owen and Woodman (1968) thereupon re-analyzed the data of Walker and Hayes using their own measured laboratory strengths

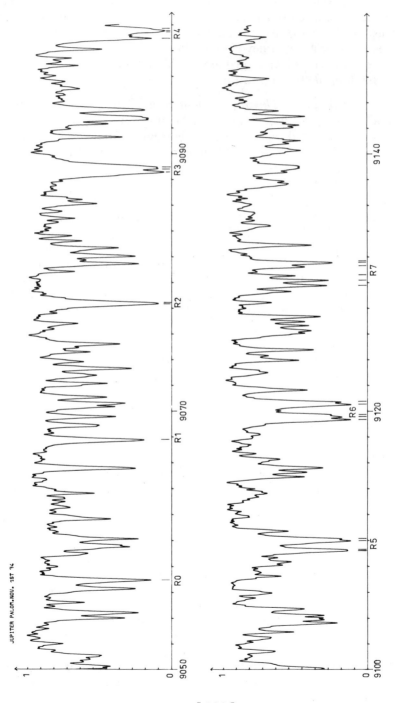

Fig. 2. Section of the Jupiter spectrum (res 0.06 cm^{-1}) covering the region of the R branch of the $3\nu_3$ methane band, taken by Connes and Maillard at the Coudé focus of the 5-meter telescope on Palomar Mountain in October 1974.

TABLE I

Studies of the $3v_3$ Band of Methane

Analysis	Spectra	Resolution (cm^{-1})	Abundance (m amagat)	Temperature (°K)	Line width [$\gamma_0 p$ (cm^{-1})]	Pressure (atm)
Owen (1965).	Kuiper 1955	~ 2		200 ± 25		
Walker & Hayes (1967).	Walker & Hayes 1967	~ 0.8		100 – 300		
Owen & Woodman (1968).	Walker & Hayes 1967	~ 0.8		180 ± 20		
Belton (1969).	Walker & Hayes 1967	~ 0.8	30	163 ± 20		~ 0.6
Margolis & Fox (1969a,b,c).	Walker & Hayes 1967	~ 0.8	40–80	185 ± 15		
Farmer (1969),	Farmer 1969	(0.1)			0.18	2.3
Bergstralh (1973a,b).	Bergstralh 1973	0.09			0.07	0.7
Maillard et al. (1973).	Maillard 1973	0.2–0.4	38 ± 8	150 ± 15	0.10	1.0
de Bergh et al. (1976).	Connes & Maillard 1974	0.05	52 ± 6	175 ± 10	0.09	0.9

instead of the theoretical weighting factors of Childs and Jahn, and concluded that a temperature of $180° \pm 20°K$ could be derived. Belton (1969), though using the same measured relative laboratory strengths of Owen and Woodman, disagreed with this analysis since saturation and possibly scattering effects had been neglected. He obtained a slightly lower temperature of $163° \pm 20°K$.

Further progress was made by Margolis and Fox in a series of laboratory studies (Margolis and Fox 1968; Margolis 1970, 1971; Bergstralh and Margolis 1971), and a series of papers applying these to Jupiter (Margolis and Fox 1969*a*, 1969*b*, 1969*c*). Following the treatment of Hecht (1960) and Fox (1962) they gave quantum assignments to the *J*-manifolds in the $3\nu_3 R$ branch and measured their splitting and relative intensities. They then applied their measurements to the data of Walker and Hayes using the correct synthesis of line profiles for the *J*-manifolds, the theoretical statistical weights for the rotational lines, and taking saturation into account. They determined an abundance of 40–80 m amagat and a temperature of $185° \pm 15°K$.

Since the resolution of the Jupiter observations was not high enough to measure the line profile, a pressure could not be determined and Margolis and Fox left the actual line width of the Jovian methane lines as a parameter. Farmer (1969) tried to measure the line half-widths by taking new plates with an S–1 image tube at the 2.5-meter Mount Wilson reflector. However, his value of 0.18 cm^{-1} for the line half-width is considerably higher than subsequent measurements. Better measurements were obtained by Bergstralh (1973*a,b*) using the photoelectric spectrum scanner at the McDonald Observatory 2.7-meter telescope. He measured a half-width for the $R(1)$ line of 0.08 cm^{-1}, which can be translated into an effective pressure of 0.7 atm using the pressure broadening coefficient of 0.075 cm^{-1} atm^{-1} obtained by Rank *et al.* (1966) for the $2\nu_3$ band of methane.

The development of Fourier spectroscopy provided the next advance for the observations of the $3\nu_3$ band of methane. Maillard *et al.* (1973) extended the spectrum of Jupiter in the *Connes Atlas* to the $3\nu_3$ region using a new Connes interferometer, with which he obtained an instrumental resolution of 0.22 cm^{-1} broadened to about 0.40 cm^{-1} by Jupiter's rapid rotation. Analysis of their data (Maillard *et al.* 1973) following the treatment of Margolis and Fox, gave an abundance of 38 m amagat, at a temperature of 150°K and a pressure of ~ 1 atm. More recently, even better spectra of Jupiter were obtained by Connes and Maillard (e.g., see Connes and Michel 1974) at a resolution of 0.06 cm^{-1}. A section of their beautiful spectrum showing the lines in the R-branch of the $3\nu_3$ methane band is given in Fig. 2. These data were analyzed by de Bergh *et al.* (1976) giving an improved temperature, abundance, and pressure determination.

From Table I it can be seen that even the higher quality spectral data obtained over the last few years show considerable scatter and uncertainty for the derived temperature, pressure, and abundance. All of the analyses, except Belton's, used a reflecting layer model so that they should be internally comparable. A methane abundance of 50 m atm at a temperature of 160°K and a pressure near 1 atm seems to be representative of the results obtained for this band. It appears to us that interference from the telluric absorptions and the complexity of the methane absorption in that region explains the scatter in the results obtained. Further progress must come from a better understanding of the methane spectrum through laboratory data, so that the effects of interlopers (unidentified lines), pressure broadening, temperature intensity anomalies, etc. can be taken into account. With this information, analyses taking cloud structure and scattering into account may then be applied and will hopefully lead to a better definition of the real atmospheric structure on Jupiter.

In addition to the $3\nu_3$ methane band, there is a weak ammonia band at 1.08 μm which has been studied by Cruikshank and Binder (1968, 1969) and Mason (1970). The first authors used a laboratory match to the whole band and obtained an abundance of 5 m atm while Mason used a three point curve of growth for seven lines in the band, obtaining 15 ± 8 m amagat for the ammonia abundance. Since no quantum number assignments are available for this band and, compared to the $3\nu_3$ methane band, very few laboratory measurements have been made, the derived ammonia abundance is quite uncertain. In addition Mason (1970) points out that spectra of the 6450 Å NH_3 band gave definite variations both with time and position on the disk of Jupiter. Obviously much more work is needed in that area.

1.14–3.0 μm. The region below 1.7 μm has been the subject of a number of studies already mentioned above. Although this interval of the Jupiter spectrum is potentially very interesting, the strength and saturation of the observed bands have precluded an analysis that might provide data for atmospheric structure, and have made the search for possible minor constituents very difficult so that no new molecules have been identified there. The main analyses have concerned themselves with the broad absorptions due to the pressure-induced molecular hydrogen spectrum. A careful comparison with laboratory spectra of methane and ammonia, particularly by Kuiper (1964) and Cruikshank and Binder (1968) confirmed the original conclusion of Kuiper (1952) that all the absorptions in that region could be accounted for by methane and ammonia. Matching the room-temperature laboratory spectra of methane and ammonia to the Jupiter absorptions gave considerably different abundances for different bands (Cruikshank and Binder 1968) with a methane abundance of about 50 m amagat (at 1 atm) and an ammonia abundance of about 1.5 m amagat being representative. In addition, Cruikshank and Binder obtained upper limits for C_2H_2, H_2S, and HCN of 8, 50, and 10 cm amagat respectively.

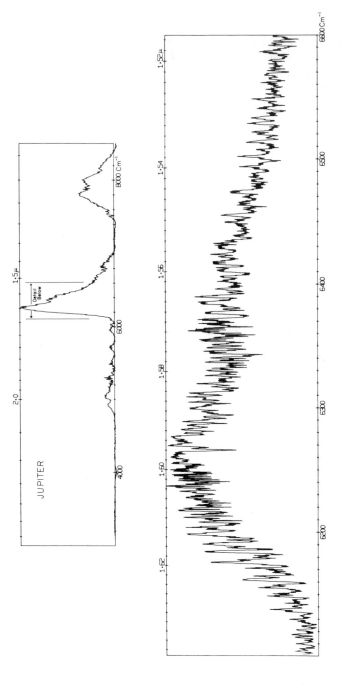

Fig. 3. *Above:* a low resolution (8 cm⁻¹) spectrum of Jupiter showing three of the windows contained in the reflecting region of its infrared spectrum. The transmission of the terrestrial atmosphere has not been removed. *Below:* a high resolution detail of the Jovian spectrum illustrating the complexity of the absorption features. Both spectra are unpublished data from the Lunar and Planetary Laboratory.

Special mention should be made of a spectrum of Jupiter obtained with a balloon-borne spectrometer at an altitude of 25.6 km (Danielson 1966). Although the resolution of the spectrum was low ($\lambda/\Delta\lambda \sim 40$) it was almost completely free of telluric absorptions, and shows two windows in the Jupiter atmosphere centered at 1.9 μm (~ 5300 cm^{-1}) and at 2.7 μm (3700 cm^{-1}). Both of these windows are unfortunately obscured by strong telluric water vapor and carbon dioxide absorptions (see Fig. 1b). As can be seen from the observational data in Fig. 3 a portion of the 1.9 μm window is visible, but the 2.7 μm window is completely blanked out. The spectrum of Jupiter recorded by Martin (1975) from a high-altitude mountain-top site (Mauna Kea) under conditions of low atmospheric water content represents about the best that can be achieved with groundbased facilities. His spectrum permitted an analysis of the S (1) line of the broad pressure-induced 1-0 band of molecular hydrogen. Spectroscopic observations of both these windows from aircraft are quite feasible at medium resolution and will greatly reduce the intensities of the telluric H_2O lines although the CO_2 bands may remain moderately strong. Since both windows may permit a search for further trace constituents, an effort should be made to acquire better data in these two regions.

In trying to fit all of the observed Jovian absorption features in his balloon spectra, Danielson (1966) pointed out the importance of the pressure-induced hydrogen dipole spectrum, which was first identified for higher overtones in spectra of Uranus and Neptune by Herzberg (1952). Danielson attributed most of the absorption in the region 3900–5000 cm^{-1} (around 2.25 μm) to the fundamental (1-0) band of the pressure-induced hydrogen spectrum and a good fraction of the absorption near 8700 cm^{-1} (~ 1.15 μm) to its first overtone (2-0). He felt that a hydrogen abundance of 45 km amagat at an abundance-density product of 50 km amagat^2 was consistent with his data. Further work on the 2-0 band was undertaken by de Bergh et al. (1974) by analyzing the high resolution spectra taken with a Connes interferometer (Maillard et al. 1973). Since this band is almost coincident with strong methane absorptions at 1.15 μm (see Cruikshank and Binder 1968), the authors came to the conclusion that it is difficult to disentangle the different absorptions. They derived a hydrogen abundance of approximately 34 km amagat.

Very strong methane absorptions from 4100–4600 cm^{-1} also overlie the pressure-induced fundamental (~ 3900–5000 cm^{-1}). However, the (1-0) line complex centered at 4750 cm^{-1} is reasonably free of strong methane absorptions. Martin (1975) has attempted an analysis of that line using both a reflecting layer and a scattering model. He obtained no clear differentiation between the models and obtained conflicting temperature results, with an abundance-density product of 18.2 km amagat^2 (roughly equivalent to a hydrogen abundance of 24 km amagat) giving the best fit. Although the pressure-induced hydrogen spectrum is capable of yielding important information on the Jupiter atmospheric structure, it appears that the methane absorptions, particularly at temperatures appropriate for Jupiter, must be much better understood for a precise quantitative analysis.

5 μm Region: Thermal Emission from Beneath the Upper Cloud Deck

The success of spectroscopic observations in Jupiter's 5 μm window is due to a special combination of factors. We have already commented on the origin of the Jovian 5 μm radiation in localized hot spots. The observed flux represents radiation from deep atmospheric levels beneath the visible cloud tops as a consequence of the high transmission of H_2, CH_4, and NH_3 at these wavelengths. This flux may contain the spectroscopic signatures of chemical species that for various reasons (line strength, condensation, nonequilibrium reactions) do not exist in spectroscopically detectable amounts in the colder regions of Jupiter's atmosphere above the visible clouds and which would therefore remain undetected in observations in the other Jovian windows. Furthermore, the longer absorption paths in Jupiter's atmosphere at 5 μm permit a much more sensitive search for trace atmospheric constituents or upper limits to the presence of any molecule with absorptions at these wavelengths.

The 5 μm Jovian window is contained within a somewhat broader terrestrial window. Carbon dioxide at 4.3 μm and H_2O at 6.3 μm define the telluric window while CH_4 at 4.5 μm and NH_3 at 5.2 μm limit the extent of the Jovian window. Unfortunately, many strong water lines in the wing of the 6.3 μm H_2O band occur throughout the Jovian 5 μm window so that spectra are best recorded from a high, dry mountain-top site or, even better, from an aircraft. Figure 4 provides a striking contrast of the quality of the 5 μm window in groundbased and airborne observations. Both of the lunar spectra in Fig. 4 were recorded with the same spectrometer at the same resolution (0.6 cm^{-1}). The lower spectrum represents the typical atmospheric transmission at 5 μm at a mountain-top observatory. It is dominated by many strong telluric water-vapor lines that would greatly interfere with the interpretation of astronomical data. The upper spectrum, on the other hand, demonstrates the nearly complete elimination of telluric water-vapor absorption at a sufficiently high altitude, in this case 12.4 km, to clear the tropopause.

At this altitude the precipitable water-vapor abundance above the aircraft is typically less than 10 μm compared to abundances of the order of millimeters at mountain-top observatories, a reduction by a factor between 100 and 1000. At an altitude of $\sim$ 12 km the abundances of other terrestrial atmospheric constituents that are uniformly mixed are reduced by $\sim$ 80%. Thus, in Fig. 4 absorptions by telluric CO_2, CO, and N_2O remain but they are clearly no problem. The abundance of O_3 at these altitudes is, of course, unchanged.

The first spectroscopic observations of the Jovian 5 μm window were those of Gillett et al. (1969) which revealed the surprisingly high brightness temperature of this region. They were able to account for all of the features seen in their spectrum through absorptions by the known atmospheric con-

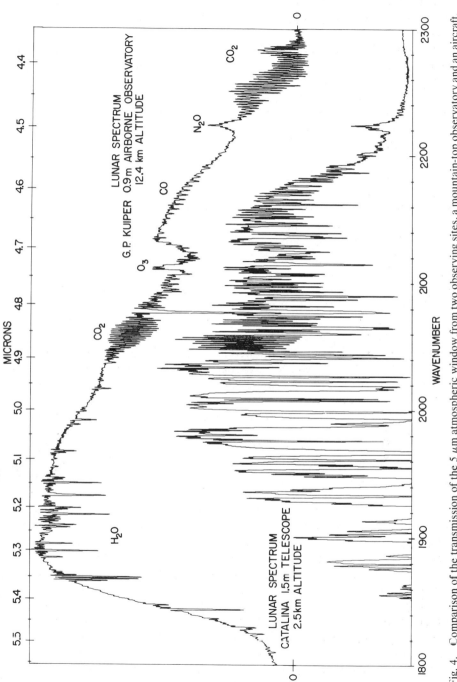

Fig. 4. Comparison of the transmission of the 5 μm atmospheric window from two observing sites, a mountain-top observatory and an aircraft. Both spectra were recorded with the same spectrometer at the same resolution (0.6 cm^{-1}).

stituents H_2, CH_4, and NH_3. Their spectral resolution ($\sim$ 40 cm^{-1} at 2000 cm^{-1}) was too low to permit a sensitive search for trace atmospheric constituents.

Münch and Neugebauer (1971) recorded a 5 μm spectrum of Jupiter at higher spectral resolution ($\sim$ 11 cm^{-1} at 2000 cm^{-1}) in which they found an unknown Jovian absorber at 4.73 μm. They listed a number of gases (HCN, C_2N_2, CH_3D, SiH_4, GeH_4) as possible candidates for these absorptions but they could not distinguish between these various possibilities. Alternatively, their spectrum could have been explained by absorptions in solids such as the ammonium salts predicted in chemical modeling of Jupiter's atmosphere (Lewis and Prinn 1970) or in organic molecules synthesized in Jupiter's atmosphere (Woeller and Ponnamperuma 1969).

The first high-resolution (0.55 cm^{-1}) spectrum of Jupiter at 5 μm was obtained by Beer et al. (1972) using a Fourier spectrometer. The ν_2 band of CH_3D was identified providing the first detection of deuterium in a planetary atmosphere. Subsequent analysis of this spectrum (Beer and Taylor 1973a, b) established a D/H ratio of about 5×10^{-5} on Jupiter, significantly less than the terrestrial value of 1.57×10^{-4} but close to recent estimates of the primordial value of about 2×10^{-5}. As the authors themselves point out, the calculation of the D/H ratio from the CH_3D abundance is model-dependent and a more reliable value of 2.1×10^{-5} was found by Trauger et al. (1973) from measurements of HD at 7467 Å.

Beer and Taylor also saw evidence of the same Jovian absorption at 4.73 μm reported by Münch and Neugebauer but the very much higher spectral resolution revealed so many more telluric and Jovian absorptions that identification of the unknown Jovian absorber(s) was still impossible.

Observations of Jupiter during 1974 produced the detection of water vapor by Larson et al. (1975) in a moderate resolution spectrum (2.4 cm^{-1}) recorded from the G. P. Kuiper Airborne Observatory (NASA C 141 airplane). Figure 5 summarizes the observational data of that study. The middle curve is the spectrum of Jupiter that is dominated by strong lines of CH_3D and H_2O. The synthetic spectrum of H_2O at the bottom is included not only to support the identification of water vapor on Jupiter, but also to emphasize that the observed intensities of the water lines require temperatures of 300°K or higher. This is consistent with the expectation that at 5 μm one is looking deep into the Jovian atmosphere in the regions of the hot spots. The surprising result of this analysis is the low mixing ratio of 10^{-6} for the water vapor, suggesting a depletion of oxygen at this depth on Jupiter by about a factor of 10^3 compared to its solar abundance. Several explanations are possible but further observations and model atmosphere calculations will be required to determine the significance of this result. As a consequence of the lack of telluric absorptions at aircraft altitudes, the high-altitude spectrum of Jupiter in Fig. 5 very clearly displays the broad 4.73 μm depression noted by previous observers. An explanation of this feature in terms of absorptions by

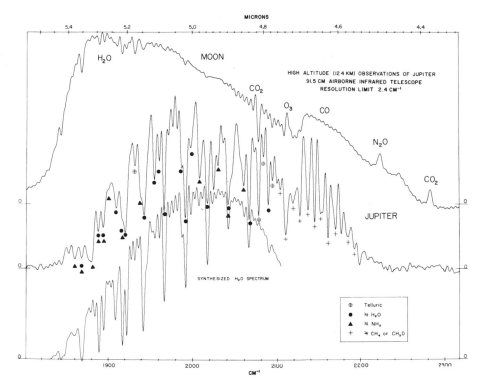

Fig. 5. The 5 μm spectrum of Jupiter recorded with the G. P. Kuiper Airborne Observatory. The lunar comparison spectrum at the top is the same data as in Figure 4 *(top)*, but transformed at the reduced resolution of the Jovian spectrum. A synthetic spectrum of water vapor is included at the bottom to illustrate the identification of water vapor on Jupiter and to emphasize that high temperatures ($T \geq 300°K$) are required to explain the observed rotational line intensities.

HCN and GeH$_4$ was presented by Larson *et al.* (1976) who combined new high resolution groundbased observations and laboratory comparison spectra with the high-altitude spectrum in Fig. 5 to support their identifications. In the same study the possible presence of CO$_2$ on Jupiter was suggested.

An improved groundbased spectrum of Jupiter in the 5 μm region has recently been recorded by Beer (1975). With this spectrum Beer reported evidence of CO on Jupiter. From chemical equilibrium arguments both CO and CO$_2$ are not expected in spectroscopically detectable amounts in the upper atmosphere of Jupiter. If confirmed, their presence may represent a nonequilibrium process whose identification can have important consequences for chemical models of Jupiter's atmosphere.

Improved spectra of Jupiter in the 5 μm region, especially observations from the C 141 airplane, should be able to confirm or disprove the presence of the above molecules. Higher-resolution spectra combined with the deep penetration into Jupiter's atmosphere may allow the identification of further constituents in that region.

10–40 μm: The 125°K Thermal Emission Spectrum

The appearance of the spectrum in the region 400–1100 cm^{-1} (Fig. 1) resembles a 125°K blackbody. The apparently good match is partly an illusion due to the steep slope and the logarithmic scale of the figure. The flux levels in this region are in fact determined by highly non-gray molecular line opacities. A broad band, medium resolution spectrum of the 10 μm window is shown in Fig. 6 (Ridgway 1974a; Ridgway et al. 1975). Reference to this figure will clarify some of the detailed discussion below.

The first Jovian spectrum in the 10 μm window was reported by Low (1965) who employed a filter-wheel photometer. A greatly improved spectrum was obtained by Gillett et al. (1969) with a similar instrument. They obtained a resolution $\lambda/\Delta\lambda \sim 50$. Their analysis yielded several important results, and their data has remained a unique asset to this time (see Fig. 7).

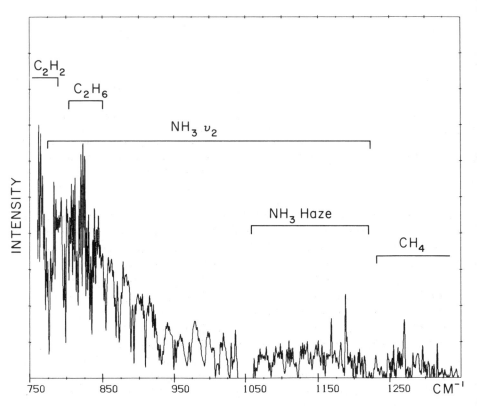

Fig. 6. The spectrum of Jupiter in the terrestrial 10 μm window. This spectrum has been corrected for terrestrial absorption.

Figure 7 shows the original Gillett *et al.* data incorporating a recalibration of the photometry. The same figure also includes more recent data from Gillett (personal communication) and spectra by Aitken and Jones (1972) plotted at reduced resolution. Although very similar to the 1969 publication, Fig. 7 is preferable for critical work.

From 850–1250 cm^{-1} the Jovian spectrum is depressed substantially below the computed H$_2$ continuum. By detecting the two Q branches of the ν_2 ammonia band Gillett *et al.* (1969) confirmed Kuiper's (1952) suggestion that NH$_3$ is an important opacity source near 10 microns. Gillett *et al.* deduced a column abundance of about 3 cm atm (with considerable uncertainty) and they suggested that ammonia is probably saturated near the 125°K level. Aitken and Jones (1972) studied this region with a grating spectrometer. Their resolution ($\lambda/\Delta\lambda \sim 140$) revealed a strong rotational structure in the ν_2 band consistent with the analysis of Gillett *et al.* The Aitken and Jones data is plotted at reduced resolution in Fig. 7. Lacy *et al.* (1975) and Ridgway *et al.* (1975) obtained spectra of the ν_2 band with resolutions $\lambda/\Delta\lambda \sim 225$ to 6000. They used spectral synthesis to confirm Orton's (1975 *a,b*) band model result that NH$_3$ opacity in fact suffices to account for the flux levels in the 850 to 1050 cm^{-1} region. Analysis of the spectrum of the relatively simple ν_2 band of NH$_3$ can be used as a diagnostic to evaluate models of the atmospheric temperature structure if the ammonia abundance profile is known, or to study the ammonia abundance if the temperature structure is known. At present, neither abundance nor temperature profiles are known with sufficient confidence, but Lacy *et al.* and Ridgway *et al.* describe some of the constraints imposed on possible models by their observations.

In the 1050 to 1200 cm^{-1} region it appears that the ammonia opacity is insufficient to produce the observed depression of the spectrum. Ammonia clouds are certain to be present near the 125° level (see Prinn and Owen p. 000), and ammonia frost is known to absorb in the 1100 cm^{-1} region (Taylor 1973). Orton (1975*a,b*) found that an absorbing haze with the absorption coefficient of ammonia frost could match the low resolution spectrum.

Ridgway *et al.* (1975) have studied this portion of the spectrum with a resolution $\lambda/\Delta\lambda \sim 800$. They found evidence for phosphine (PH$_3$) approximately consistent with a solar P/H ratio, but noted that PH$_3$ opacity did not suffice to produce the observed depression of the spectrum. They have concluded that the appearance of the spectrum is qualitatively consistent with Orton's mixed absorbing haze hypothesis. The possible detection of PH$_3$ urgently requires further investigation since its presence in detectable abundance is unexpected (Prinn and Owen p. 338).

An intensity peak occurs near 1280 cm^{-1}. Gillett *et al.* who first observed this feature, proposed that an atmospheric temperature inversion, maintained by absorption of solar flux in near-infrared methane bands, supports emission from the 7.8 μm methane band. Strong confirming evidence was found by

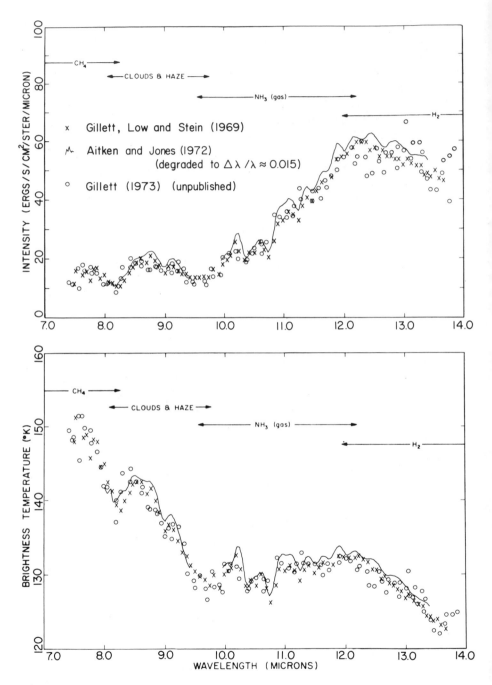

Fig. 7. Spectrophotometry in the 10 μm window. Data are from Gillett *et al.* (1969) with recent recalibration; unpublished data from Gillett; and data from Aitken and Jones (1972) at reduced resolution. (Figure from Orton 1975*b*).

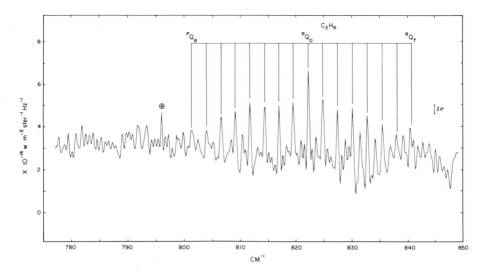

Fig. 8. The emission spectrum of ethane at high resolution (Tokunaga *et al.* 1975). Terrestrial absorption has been removed by ratioing to a lunar spectrum.

Gillett and Westphal (1973) in the observation of limb brightening in the 7.8 μm band. Interestingly, however, as of this writing six years later, no one has yet resolved this band sufficiently to identify any line structure; in a large part this is due to severe contamination of the 7.8 μm region by terrestrial H_2O.

The reality of the temperature inversion was verified with the discovery of ethane—the ν_9 band at 825 cm^{-1} appears in emission (Ridgway 1974a,b; Combes *et al.* 1974). Figure 8 illustrates a recent spectrum by Tokunaga *et al.* (1975) clearly showing the ethane band structure. Acetylene emission has also been reported near 775 cm^{-1} (Ridgway 1974a; Combes *et al.* 1974) but was not found by Tokunaga *et al.* This is a difficult observation, due partly to interference from terrestrial CO_2. Observational difficulties may be responsible for the discrepant results. A high-resolution spectrum from the C 141 aircraft is entirely feasible and would be very valuable. The existing medium-resolution aircraft spectrum of Encrenaz *et al.* (1976) is, unfortunately, inconclusive.

The mixing ratios proposed by Ridgway (1974a,b) for ethane and acetylene of 4×10^{-4} and 8×10^{-5}, respectively, are probably too high due to choice of model parameters. Tokunaga *et al.* (1975) have initiated an abundance analysis for C_2H_6 and C_2H_2 based on improved laboratory and Jovian spectra. The preliminary results indicate that the abundances may be consistent within an order of magnitude with the photochemical computations of Strobel (1974). Additional laboratory and theoretical study of the ν_9 band of C_2H_6 is still needed since the band has not yet been adequately analyzed.

The first spectrum of Jupiter near the 125°K thermal emission peak was obtained just recently. In an airborne observation with a liquid helium cooled spectrometer, Houck *et al.* (1975) studied the spectral range 250 to 600 cm⁻¹. They found the expected pressure-induced molecular hydrogen spectrum with lines at 357 and 555 cm⁻¹, but also detected an unidentified absorption at 425 cm⁻¹. Pollack used the spectrum of Houck *et al.* (1975) to determine the upper atmospheric temperature structure near 0.5 atm, and the He/H_2 ratio. Radiometric data in this spectral region obtained by Pioneers 10 and 11 are discussed below.

Thermal emission spectra of Earth and Mars have been successfully analyzed to determine temperature profiles. The rich molecular spectra of hydrogen, ammonia, methane, and minor constituents in the range 500–2000 cm⁻¹ raise the possibility of inverting the observed Jovian spectra (Taylor 1972), and this has in fact been attempted (Ohring 1973; Wallace and Smith 1975). The technique takes advantage of the fact that spectral regions of different opacity may reach "well defined" optical depth of unity at different depths in the atmosphere, and hence emit at different brightness temperatures. Wallace and Smith (1975) have noted an ambiguity in the problem associated with the temperature inversion. They employed additional constraints, including the β Scorpii occultation (e.g., Hubbard *et al.* 1972) to derive a more reliable result. Their analysis draws attention to the limitations of the technique and emphasizes the importance of incorporating additional available information into the analysis. Both the computations by Ohring and by Wallace and Smith are based on the original data of Gillett *et al.* (1969) and do not incorporate the recalibration mentioned by Orton (1975*b*).

These attempts to invert the spectrum used low-resolution data. Naturally a high-resolution spectrum should exhibit both higher and lower brightness temperature extremes, and in principle permit inversion of the spectrum over a greater pressure range. Although suitable spectra can now be obtained, there is some question as to whether the molecular spectra are sufficiently well understood for such high-resolution inversion (the methane 7.8 μm band is especially complex). Precise knowledge of line shapes and widths is also a crucial problem. Another difficulty, as mentioned by Wallace,[4] is knowledge of atmospheric abundances. Complex photochemistry at high altitudes, saturation of ammonia at intermediate altitudes, possible nonequilibrium chemistry in the convective region are serious, possibly insurmountable problems for the minor constituents. Of course, a given temperature profile, computed or derived from other observations, can be combined with infrared spectra to determine variation of abundance with altitude and to shed light on the difficult problems just enumerated.

---

[4]See p. 285.

The pressure-induced H_2 transition is probably the most reliable probe for temperature sounding. The abundance is adequately known, and the strong pressure dependence of the opacity results in a sharply peaked contribution function for each wavelength, which unfortunately also restricts the pressure range of applicability.

ATMOSPHERIC COMPOSITION: A SUMMARY

A knowledge of the composition of the atmosphere of Jupiter is an important end in itself in understanding the nature of this giant planet. Extra significance may be attached to these studies if Jupiter's present atmosphere is close to its primordial composition. A comparison of Jupiter's relative chemical abundances with solar values thus provides important information not only for Jupiter itself but also for the solar system in general.

Table II summarizes the known atmospheric constituents on Jupiter at the time of the Tucson Jupiter Colloquium in May 1975. The first group represents those molecules for which the spectroscopic evidence is firm. The case for helium depends partly on indirect measurements but its presence as a major atmospheric constituent seems secure. The second group comprises six gases for which spectroscopic evidence has very recently been reported. Some of these will require independent confirmation before they can be fully accepted.

[Note added in proof: In December 1975 Larson, Fink, and Treffers obtained a high resolution (0.6 cm^{-1}), 5-micron spectrum of Jupiter from the Kuiper Airborne Observatory. These observations test the uniqueness of several molecular assignments (GeH_4, CO_2, and HCN) tentatively made to single, unknown Jovian absorptions in lower resolution, high altitude data. For GeH_4 the detailed agreement between the observational data and laboratory comparison spectra remains excellent, strengthening the evidence for this molecule in Jupiter's atmosphere. The strong Jovian absorption assigned to CO_2, however, split at higher resolution into two unknown lines, neither of which can any longer be associated with this molecule. The strong Jovian absorption tentatively assigned to HCN persists in the new observations, but its position differs from that of a laboratory spectrum of HCN by about 1 cm^{-1}, sufficient to reject its identification. There is at this time no alternative explanation for the Jovian absorptions previously associated with CO_2 and HCN.]

There is good reason to believe that the gross composition of Jupiter's atmosphere is based on chemical thermodynamic equilibrium. This condition can be used to predict the abundance of any molecule for comparison with observation. A gross disparity between observation and prediction points to some nonequilibrium process which may be responsible for additional properties of the atmosphere. Much of the significance of many of the recently identified molecules on Jupiter depends on the fact that they should not exist

TABLE II

Composition of the Jovian Atmosphere

Gas	Abundance relative to H_2	Region (μm) where detected	Reference to first detection
1. Identification firm			
H_2	1	0.8	Kiess *et al.* 1960.
HD	2×10^{-5}	0.7	Trauger *et al.* 1973.
He	$.05 - .15$	–	Orton & Ingersoll p. 206; Houck *et al.* 1975; Carlson and Judge p. 418.
CH_4	7×10^{-4}	0.7	Wildt 1932.
CH_3D	3×10^{-7}	5.0	Beer and Taylor 1973*a,b*.
$C^{13}H_4$	$C^{12}/C^{13} \sim 110 \pm 35$	1.1	Fox *et al.* 1972.
NH_3	2×10^{-4}	0.7	Wildt 1932.
H_2O	1×10^{-6}	5.0	Larson *et al.* 1975.
C_2H_6	[a]4×10^{-4}	10.0	Ridgway 1974*a*.
2. Recent reports			
CO	2×10^{-9}	5.0	Beer 1975.
GeH_4	7×10^{-10}	5.0	Larson *et al.* 1976.
HCN	1×10^{-7}	5.0	Larson *et al.* 1976.
C_2H_2	[a]8×10^{-5}	10.0	Ridgway 1974*a*.
PH_3	4×10^{-7}	10.0	Ridgway *et al.* 1975.

[a]Recent work by Tokunaga *et al.* (1975) indicates lower abundances.

in spectroscopically detectable amounts under conditions of thermodynamic equilibrium applied to the observable regions of the Jovian atmosphere. Ethane (C_2H_6) and acetylene (C_2H_2) are readily explained by photolysis, a non-equilibrium mechanism active in the upper atmosphere of Jupiter. The possible presence of HCN may indicate additional non-equilibrium reactions leading to the production of complex hydrocarbons that might explain Jupiter's coloration.[5] Carbon monoxide and CO_2 should be thoroughly reduced on Jupiter. Evidence for these two gases is a complete surprise that has no immediate explanation.

The measured abundances of each atmospheric constituent can provide additional information on the structure and evolution of a planetary atmo-

---

[5]See p. 372.

sphere. While water vapor is expected in Jupiter's atmosphere on the basis of chemical equilibrium, the measured abundance (10^{-6} mixing ratio with respect to H_2) is about a factor of 10^3 lower than the solar value (10^{-3}). It is not known at this time whether to interpret this as a true oxygen deficiency on Jupiter, to other as yet unidentified sinks for oxygen, or to some atmospheric phenomena in the region of the hot spots where the water molecule is observed.

In addition to the known and suspected atmospheric constituents in Table II, a very much larger list of upper limits to other molecules can be made. These negative results are often valuable in checking the consequences of various chemical theories of Jupiter's atmosphere. The recent spectroscopic studies of Jupiter have rendered all previous lists obsolete and new compilations can be expected in the near future.

SPATIAL RESOLUTION

High spatial resolution may very well be crucial for satisfactory interpretation of high-resolution spectra. Little of this has yet been done in the infrared region. However, mapping and disk scans with medium and narrow band filters have been obtained in some cases. The most extensive activity has centered on the 5 μm window.

Shortly after the Jovian 5 μm emission was detected, Westphal (1969) found that this thermal radiation originated primarily in a dark band, the North Equatorial Belt. Subsequent spatially resolved studies of Jupiter's thermal radiation showed that the 5 μm radiation is strongly peaked in local regions comprising at most a few percent of the surface area (Keay et al. 1973; Westphal et al. 1974). These regions, often called hot spots, are strongly correlated with the dark visual markings on Jupiter usually described as blue or purple; lighter areas described as red, yellow and orange are not sources of the 5 μm radiation. Figure 9 contains examples of the high resolution thermal maps of Jupiter produced by Westphal et al. (1974) showing the discrete nature of the emitting regions and their general alignment along the equatorial bands of Jupiter.[6]

The current interpretation is that this 5 μm radiation escapes through holes in Jupiter's nearly universal cloud cover. Since infrared thermal maps show that these holes constitute only a few percent of the observed area of the planetary disk, they must be quite hot ($T > 300°K$) to produce the observed disk-average brightness temperature of 230°K since the brightness temperature measured at the top of the visible cloud deck is only about 150°K. In fact, the observed brightness temperatures of the hot spots are as

[6]See also pp. 502 and 555.

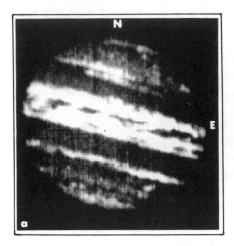
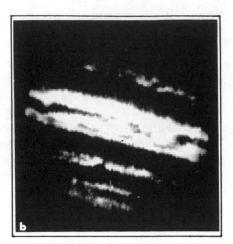
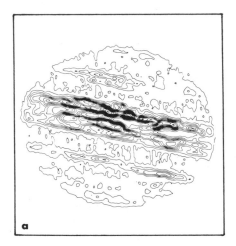
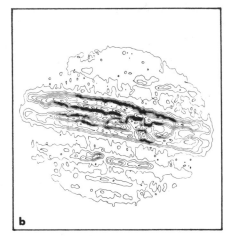

Fig. 9.   5 μm disk map assembled from scans with a single detector. *Above:* reconstructed 5 μm picture, and *below:* intensity contours from Westphal *et al.* (1974).

high as 310°K, and may be even higher owing to limitations on spatial resolution in the mapping experiments. The upper curve in Fig. 1*a* shows the thermal emission spectrum of these hot spots assuming a brightness temperature of 350°K. At this temperature they need constitute only one percent of the area of the planetary disk to produce a disk-average brightness temperature of 230°K. These high temperatures for the hot spots imply observations below the (presumed) lower cloud decks to regions where the pressure may be 5 atm or higher.

Further support of this picture of Jupiter at 5 μm is provided by recent high-resolution spectroscopic observations made in this window. The detection of water vapor on Jupiter in itself requires temperatures of 300°K or higher and in addition the observed H_2O rotational line intensities require temperatures of at least 300°K (Larson *et al.* 1975).

These hot spots are somewhat unique to the 5 μm window. At longer wavelengths the increased opacity of molecular hydrogen becomes a limiting factor. At shorter wavelengths where reflected sunlight dominates the flux received from Jupiter, high spatial resolution will have to be combined with spectroscopic observations in order to isolate the very weak contribution from Jovian thermal emission in the hot spots. At present, the 5 μm window is the only convenient region of Jupiter's infrared spectrum where one can seek detailed information about its atmosphere below the visible cloud layers by remote analysis. The only other measurements probing to comparable depths are microwave observations.[7]

Armstrong *et al.* (1976) have recently presented elaborate observational studies which permit a detailed and statistical comparison between 5 μm and photographic maps. They find significant long-term variations in the statistical properties of the hot spots, but they also detect a tendency for the distribution of brightness temperatures to peak near 190°K or 205°K. The authors suggest that these temperatures may represent persistent, successively deeper cloud decks. The result also indicates that their spatial resolution (1.5 arcsec) is resolving at least some of the spots. Armstrong *et al.* also cite observational evidence that the hot spots are unique to Jupiter – they find no evidence for a similar phenomenon on Saturn, Uranus or Neptune.

Both disk mapping and spatial scans have been obtained at other wavelengths in the thermal infrared, but far less systematically. Wildey (1968) obtained full disk maps with the 8–13 μm filter bandpass. As may be seen from Fig. 6, the flux in this region is strongly peaked toward the red, so the observation will be dominated by the flux in the range 800–900 cm^{-1}. H_2 continuum and NH_3 absorption are significant in this region, and should show limb darkening, while C_2H_6 emission will show limb brightening. An example of Wildey's maps is shown in Fig. 10. Generally, these maps show only slight contrast and little correlation with visible features. The lack of strong contrast is attributable to the strong temperature dependence of the vapor pressure of NH_3. The limited correlation with visible features is probably due, in part, to the greater depth of the visible clouds ($\sim$ 1 atm) relative to the level of H_2-NH_3 "continuum" formation ($\sim$ 0.5 atm).

Wildey and Trafton (1971) (also Trafton and Wildey 1974) obtained a

---

[7]See p. 650.

Fig. 10.   10 μm disk map assembled from scans with a single detector. Filter bandpass is 8–13 μm. Intensity contours are plotted over retouched simultaneous photograph (from Wildey 1968).

large number of disk scans with the 8–13 μm filter used by Wildey for mapping. Many scans were averaged in an attempt to improve signal/noise and to minimize the effect of transient features. They show that, subject to several assumptions, observed limb darkening in the ν_2 band of NH_3 can be used to determine the Jovian effective temperature T_e which enters as a parameter in their model computation of limb darkening. Their 1974 result gave $T_e = 135° \pm 4°K$, consistent with other determinations available at that time, but somewhat higher than the recent result $T_e = 125° \pm 3°K$ from Pioneer data (Ingersoll *et al.* 1975). It is interesting to note that Wildey and Trafton (1971) found a systematic asymmetry in the observed limb darkening in the sense that an intensity shoulder appeared at the east (sunrise) limb and they gave low weight to the east limb in their analysis. The origin of this shoulder is unknown.

The Wildey and Trafton data, which represented an impressive observational accomplishment when obtained in 1965, is now somewhat obsolete since they used an 8–13 μm filter which covered an undesirably large spectral interval. Trafton and Wildey argue that C_2H_2, C_2H_6, and CH_4 emission will introduce only small errors in their analysis. Still it would be desirable to repeat the investigation with narrow-band filters, perhaps in the core and wing of an isolated ammonia line, in order to reduce uncertainties in the opacities.

Several studies have employed narrow-band filters and disk scanning for specific, restricted purposes. Westphal (1971) recorded pole-to-pole scans with an 8.2–9.2 μm filter. He found strong correlation of these scans to broadband 8.2–13.5 μm scans, and also to cloud patterns. These scans appear to indicate latitude variations of cloud structure and/or cloud composition. In order to test the temperature inversion hypothesis of Gillett *et al.* (1969), Gillett and Westphal (1973) made equatorial scans at 7.04 and 8.44 μm. They found limb brightening at 7.94 μm in the methane band, and limb darkening at 8.44 μm outside the methane band (in the ammonia "haze" region discussed above). The interpretation is that the strong CH_4 band center is emitting from above the temperature minimum in a region where the temperature increases with altitude, and the weaker "unknown" opacity at 8.44 μm emits below the temperature minimum in a region where the temperature still decreases with altitude. The 7.94 μm scan interestingly showed an asymmetry with larger intensity at the east limb. This suggests that the asymmetry observed by Wildey and Trafton (1971) might be associated with the CH_4 emission which they did not explicitly include in their modeling. Orton (1975b) obtained additional scans at 8.15 μm in the wing of the 7.9 μm methane band. The result, in Fig. 11, shows asymmetrical brightening at the limbs with a suggestion of darkening nearer disk center. The interpretation would be that in scanning from center to limb, optical depth "unity" at 8.15 μm rises to the temperature minimum (limb darkening) and then passes above it (limb brightening). This interesting interpretation is plausible, but cannot be considered certain without additional data. The asymmetry of Orton's 8.15 μm scan is in the opposite sense from the asymmetries noted above, indicating that the effect may be wavelength or time dependent. More data are needed to understand even qualitatively what is happening, but it appears that data recorded at different times and/or of different spatial regions should be compared with caution.

The Pioneer radiometers produced disk maps at 20 and 45 μm. Analysis of these data has determined the ratio $He/H_2 = 0.14 \pm 0.08$ (Orton 1975a) and the effective temperature (Ingersoll *et al.* 1975)[8] $T_e = 125° \pm 3°$K. Orton (1975a,b)[9] has made a careful study of the temperature structure implied by

_____

[8]See pp. 202 and 204.   [9]See p. 266.

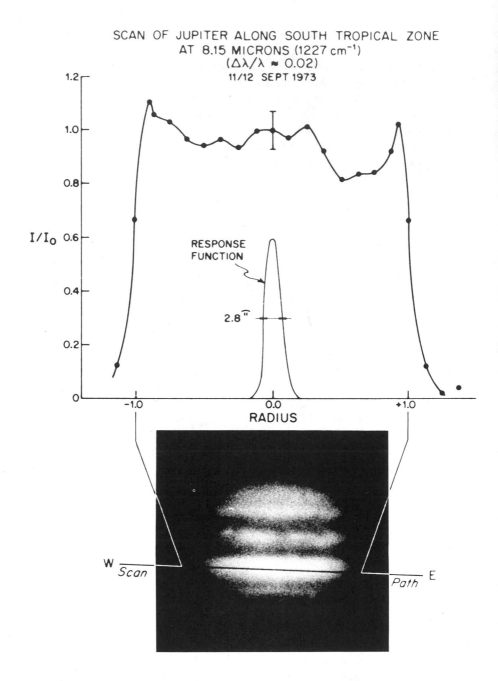

Fig. 11. Disk scan with a narrow filter centered at 8.14 μm in the wing of the 7.8 μm methane band (from Orton 1975*b*).

the Pioneer data. He has also incorporated the groundbased disk scans to obtain temperature profiles, both for belts and zones, in the pressure range 0.01–1.0 bar. This subject is treated in more detail by Wallace.[10]

The possible applications of mapping and spatial scans in the 10 μm window have not been exhausted. Improved capability for spectral and spatial resolution promises further developments in the use of these techniques for the study of the temperature structure and cloud distribution.

Very little has been done in the way of spatially resolved infrared spectroscopy. Bergstralh (1973a,b) studied the equatorial center-to-limb behavior of the $3\nu_3$ band of methane. He found that the observations were best understood in terms of an inhomogeneous scattering layer model such as employed by Danielson and Tomasko (1969).[11] Pilcher (1973) compared spectra of several belts and zones for the range 6000–6500 cm^{-1} at a resolution of 1 cm^{-1}. Several differences were found, but could not be interpreted. Larson and Fink obtained similar spectra at a resolution of 8 cm^{-1} (unpublished) and likewise found no useful differences.

Spatially resolved spectroscopy in the 5 and 10 μm windows has only recently been attempted. Of course the 5 μm spectra automatically yield spatial resolution in the sense that the hot spots contribute essentially all of the flux. But to obtain, for example, meaningful brightness temperatures for the 5 μm spectra it is necessary to resolve the hot spots ($\lesssim$ 1 arcsec). This might be possible from Earth orbit, or it might be possible only from Jupiter orbit or near approach.

In the 10 μm window, Lacy et al. (1975) reported that high-resolution spectra of a ν_2 Q branch of NH_3 were the same in the north polar region as at disk center. In view of the Trafton and Wildey limb scans it is perhaps not surprising that no difference was detected. Spectra nearer the limb, however, would be expected to show at least a reduced brightness temperature. Ridgway has obtained low resolution limb spectra (unpublished) which show that ethane emission is enhanced relative to the H_2-NH_3 "continuum" at the limb, as would be expected.

FUTURE PROSPECTS FOR JOVIAN INFRARED SPECTROSCOPY

Much remains to be done in observational work and in supporting laboratory studies and interpretation. The continuing rapid development of infrared detector technology and the increasing availability of large aperture telescopes will permit important observational contributions, and particularly improved spatial resolution. High-altitude and spacecraft spectrometers will continue to explore in increasing detail the portions of the spectrum obscured by telluric absorption. We are confident that infrared spectroscopy

---

[10]See p. 285.   [11]See p. 493.

will continue to provide important and, at times, exciting contributions to our knowledge of the solar system.

Acknowledgements. We are grateful to the following people for discussions and, in some cases, communicating results or figures prior to publication: R. Beer, M. Combes, C. de Bergh, T. Encrenaz, F. C. Gillett, J. R. Houck, J. H. Lacy, J. P. Maillard, T. Z. Martin, G. S. Orton, J. B. Pollack, R. C. Prinn, D. F. Strobel, A. T. Tokunaga, and L. Wallace.

Kitt Peak is operated by the Association of Universities for Research in Astronomy, Inc., under contract with the National Science Foundation. This research was supported in part by NASA Grants 03-002-332 and NSG 7070.

REFERENCES

Aitken, D. K., and Jones, B. 1972. The 8 to 13 μm spectrum of Jupiter. *Nature* 240:230–232.

Armstrong, K. R.; Minton, R. B.; Rieke, G. H.; and Low, F. J. 1976. Jupiter at five microns. *Icarus* (special Jupiter issue).

Beer, R. 1975. Detection of carbon monoxide in Jupiter. *Astrophys. J.* 200:L167–L169.

Beer, R.; Farmer, C. B.; Norton, R. H.; Martonchik, J. V.; and Barnes, T. G. 1972. Jupiter: observation of deuterated methane in the atmosphere. *Science* 175:1360–1361.

Beer, R., and Taylor, F. W. 1973a. The equilibration of deuterium in the Jovian atmosphere. *Astrophys. J.* 182:L131–L132.

———. 1973b. The abundance of CH_3D and the D/H ratio in Jupiter. *Astrophys. J.* 179: 309–327.

Belton, M. J. S. 1969. An estimate of the abundance and the rotational temperature of methane in the Jovian atmosphere. *Astrophys. J.* 157:469–472.

Bergstralh, J. T. 1973a. Methane absorption in the Jovian atmosphere. I. The Lorentz half width in the $3\nu_3$ band at 1.1 μ. *Icarus* 19:499–506.

———. 1973b. Methane absorption in the Jovian atmosphere. II. Absorption line formation. *Icarus* 19:390–418.

Bergstralh, J. T., and Margolis, J. S. 1971. Recomputation of the absorption strengths of the methane $3\nu_3$ J-manifolds at 9050 cm^{-1}. *J. Quant. Spectr. Rad. Transfer* 11:1285.

Binder, A. B., and McCarthy, D. W. 1973. IR spectrophotometry of Jupiter and Saturn. *Astron. J.* 78:939–980.

Childs, W. 1936. The structure of the near infrared bands of methane. I. General survey and a new band at 1,1050 Å. *Proc. Roy. Soc. London* A 153:555–567.

Childs, W., and Jahn, H. A. 1939. A new Coriolis perturbation in the methane spectrum. III. Intensities and optical spectrum. *Proc. Roy. Soc. London* A 169:451–463.

Combes, M.; Encrenaz, Th.; Vapillon, L.; Zéau, Y.; and Lesqueren, C. 1974. Confirmation of the identification of C_2H_2 and C_2H_6 in the Jovian atmosphere. *Astron. Astrophys.* 34: 33–35.

Connes, J.; Connes, P.; and Maillard, J. P. 1969. Near infrared spectra of Venus, Mars, Jupiter, and Saturn. *Edition du Centre National de la Recherche Scientifique, Paris.*

Connes, P., and Michel, G. 1974. High resolution Fourier spectra of stars and planets. *Astrophys. J.* 190:L29–L32.

Cruikshank, D. P., and Binder, A. B. 1968. The infrared spectrum of Jupiter, 0.95–1.60 microns with laboratory calibrations. *Comm. Lunar Planet. Lab.* 6:275–288.

———. 1969. Minor constituents in the atmosphere of Jupiter. *Astrophys. Space Sci.* 3:347–356.

Danielson, R. E. 1966. The infrared spectrum of Jupiter. *Astrophys. J.* 143:949–960.

Danielson, R. E., and Tomasko, M. G. 1969. A two-layer model of the Jovian clouds. *J. Atmos. Sci.* 26:889–897.

de Bergh, C.; Lecacheux, J.; and Combes, M. 1974. The first overtone pressure-induced H_2 absorption in the atmospheres of Jupiter and Saturn. *Astron. Astrophys.* 35:333–337.

de Bergh, C.; Maillard, J. P.; Lecacheux, J.; and Combes, M. 1976. A study of the $3\nu_3$ CH_4 region in the high-resolution spectrum of Jupiter recorded by Fourier transform spectroscopy. *Icarus* (special Jupiter issue). In press.

Delbouille, L.; Roland, G.; and Gebbie, H. A. 1964. Practical possibilities of planetary spectroscopic observations by interferometry. *Mém. Soc. Roy. Soc. Liège* 9:125–129.

Encrenaz, Th.; Gautier, D.; Michel, G.; Zéau, Y.; Lecacheux, J.; Vapillon, L.; and Combes, M. 1976. Airborne interferometric measurement of the infrared spectrum of Jupiter. *Icarus* (special Jupiter issue). In press.

Farmer, C. B. 1969. An estimate of line width and pressure from the high resolution spectrum of Jupiter at 1,1000 Å. *J. Atmos. Sci.* 26:860–861.

Fink, U., and Larson, H. 1971. Fourier spectroscopy at the Lunar and Planetary Laboratory of the University of Arizona. Aspen International Conference on Fourier Spectroscopy, 1970. *AFCRL Special Report No. 114.*

Fox, K. 1962. Vibration-rotation interactions in infrared active overture levels of spherical top molecules; $2\nu_3$ and $2\nu_a$ of CH_4, $2\nu_3$ of CD_4. *J. Mol. Spectros.* 9:381–420.

Fox, K.; Owen, T.; Mantz, A. W.; and Rao, K. R. 1972. A tentative identification of $^{13}CH_4$ and an estimate of C^{12}/C^{13} in the atmosphere of Jupiter. *Astrophys. J.* 176:L81–L84.

Gillett, F. C.; Low, F. J.; and Stein, W. A. 1969. The 2.8–14 micron spectrum of Jupiter. *Astrophys. J.* 157:925–934.

Gillett, F. C., and Westphal, J. A. 1973. Observations of 7.9 micron limb brightening on Jupiter. *Astrophys. J.* 179:L153–L154.

Goody, R. 1969. The atmospheres of the major planets. *J. Atmos. Sci.* 26:997–1001.

Hecht, K. T. 1960. The vibration-rotation energies of tetrahedral XY_4 molecules. *J. Mol. Spectros.* 5:355–390.

Herzberg, G. 1952. Laboratory absorption spectra obtained with long paths. *The atmospheres of the earth and the planets.* (G. P. Kuiper, ed.) Chicago, Illinois: University of Chicago Press.

Houck, J. R.; Pollack, J.; Schaack, D.; Reed, R. A.; and Summers, A. 1975. Jupiter: its infrared spectrum from 16 to 40 microns. *Science* 189:720–722.

Hubbard, W. B.; Nather, R. E.; Evans, D. S.; Tull, R. G.; Wells, D. C.; van Citters, G. W.; Warner, B.; and van den Bout, P. 1972. The occultation of Beta Scorpii by Jupiter and Io. I. Jupiter. *Astron. J.* 77:41–59.

Ingersoll, A. P.; Münch, G.; Neugebauer, G.; Diner, D. J.; Orton, G. S.; Schupler, B.; Schroeder, M.; Chase, S. C.; Ruiz, R. D.; and Trafton, L. M. 1975. Pioneer 11 infrared radiometer experiment: the global heat balance of Jupiter. *Science* 188:472–473.

Johnson, H. L. 1970. The infrared spectrum of Jupiter and Saturn at 1.2 to 4.2 microns. *Astrophys. J.* 159:L1–L5.

Keay, C. S.; Low, F. J.; Rieke, G. H.; and Minton, R. B. 1973. High resolution maps of Jupiter at five microns. *Astrophys. J.* 183:1063–1073.

Kiess, C. C.; Corliss, C. H.; and Kiess, H. K. 1960. High dispersion spectra of Jupiter. *Astrophys. J.* 132:221–231.

Kuiper, G. P. 1952. Planetary atmospheres and their origin. *The atmospheres of the earth and the planets* (G. P. Kuiper, ed.) Chicago, Illinois: University of Chicago Press.

————. 1964. Infrared spectra of planets and cool stars; introductory report. *Mém. Soc. Roy. Soc. Liège* 9:365–391.

Lacy, J. H.; Larrabee, A. I.; Wollman, E. R.; Geballe, T. R.; Townes, C. H.; Bregman, J. D.; and Rank, D. M. 1975. Observations and analysis of the Jovian spectrum in the 10-micron ν_2 band of NH_3. *Astrophys. J.* 198:L145–L148.

Larson, H. P.; Fink, U.; and Treffers, R. R. 1976. High-altitude spectroscopic observations of Jupiter at 5 μm. In draft.

Larson, H. P.; Fink, U.; Treffers, R. R.; and Gautier, T. N. 1975. Detection of water vapor on Jupiter. *Astrophys. J.* 197:L137–L140.

Lewis, J. S., and Prinn, R. C. 1970. Jupiter clouds: structure and composition. *Science* 169: 472–473.

Low, F. J. 1965. Planetary radiation at infrared and millimeter wavelengths. *Lowell Obs. Bull.* 6:184–187.

Maillard, J. P.; Combes, M.; Encrenaz, Th.; and Lecacheux, J. 1973. New infrared spectra of the Jovian planets from 1,2000 to 4000 cm⁻¹ by Fourier transform spectroscopy. I. Study of Jupiter in the $3\nu_3$ CH₄ band. *Astron. Astrophys.* 25:219–232.

Margolis, J. S. 1970. Measurement of the absorption strength of the methane $3\nu_3$ J manifold at 9050 cm⁻¹. *J. Quant. Spectr. Rad. Transfer* 10:165–174.

———. 1971. Self broadened half widths and pressure shifts for the R-branch J-manifolds of the $3\nu_3$ methane band. *J. Quant. Spectr. Rad. Transfer* 11:69–73.

Margolis, J., and Fox, K. 1968. Infrared absorption spectrum of CH₄ at 9050 cm⁻¹. *J. Chem. Phys.* 49:2451.

———. 1969a. Studies of methane absorptions in the Jovian atmosphere. I. Rotational temperature from $3\nu_3$ band. *Astrophys. J.* 157:935–943.

———. 1969b. Studies of methane absorptions in the Jovian atmosphere. II. Abundance from the $3\nu_3$ band. *Astrophys. J.* 158:1183–1188.

———. 1969c. Extension of calculations of rotational temperature and abundance of methane in the Jovian atmosphere. *J. Atmos. Sci.* 26:862–864.

Martin, T. Z. 1975. Saturn and Jupiter: a study of atmospheric constituents. Ph.D. dissertation, Univ. of Hawaii, Hawaii.

Mason, H. P. 1970. The abundance of ammonia in the atmosphere of Jupiter. *Astrophys. Space Sci.* 7:424–436.

Moroz, V. I. 1964. Recent observations of infrared spectra of planets (Venus 1–4μ, Mars 1–4μ, Jupiter 1–1.8μ). *Mém. Soc. Roy. Soc. Liège* 9:406–419.

———. 1966. Spectra of Jupiter and Saturn in the 1.0 to 2.5μ region. *Sov. Astron. J.* 10: 457–468.

Münch, G., and Neugebauer, G. 1971. Jupiter: an unidentified feature in the 5-micron spectrum of the north equatorial belt. *Science* 174:940–941.

Ohring, G. 1973. The temperature and ammonia profiles in the Jovian atmosphere from inversion of the Jovian emission spectrum. *Astrophys. J.* 184:1027–1040.

Orton, G. S. 1975a. The thermal structure of Jupiter: I. Implications of Pioneer 10 infrared radiometer data. *Icarus* 26:125–141.

———. 1975b. The thermal structure of Jupiter: II. Observations and analysis of 8–14 micron radiation. *Icarus* 26:142–158.

Owen, T. 1965. Comparison of laboratory and planetary spectra. II. The spectrum of Jupiter from 9700 to 11,200 Å. *Astrophys. J.* 141:444–456.

Owen, T., and Woodman, J. H. 1968. On the atmosphere temperature of Jupiter derived from the $3\nu_3$ methane band. *Astrophys. J.* 154:L21–L23.

Pilcher, C. B. 1973. Interferometric observations of Jupiter and the Galilean satellites. Ph.D. dissertation, Massachusetts Inst. of Technology, Cambridge, Mass.

Rank, D. H.; Fink, U.; and Wiggins, T. A. 1966. Measurements on spectra of gases of planetary interest. II. H₂, CO₂, NH₃, and CH₄. *Astrophys. J.* 143:980–988.

Ridgway, S. T. 1974a. Jupiter: identification of ethane and acetylene. *Astrophys. J.* 187: L41–L43.

———. 1974b. Erratum—Jupiter: identification of ethane and acetylene. *Astrophys. J.* 192:L51.

Ridgway, S. T.; Wallace, L.; and Smith, G. 1975. The 800–1200 cm⁻¹ absorption spectrum of Jupiter. In draft.

Strobel, D. F. 1974. Hydrocarbon abundances in the Jovian atmosphere. *Astrophys. J.* 192: L47–L49.

Taylor, F. W. 1972. Temperature sounding experiments for the Jovian planets. *J. Atmos. Sci.* 29:950–958.

———. 1973. Preliminary data on the optical properties of solid ammonia and scattering parameters for ammonia cloud particles. *J. Atmos. Sci.* 30:677–683.

Tokunaga, A. T.; Knacke, R. F.; and Owen, T. 1975. Ethane and acetylene abundances in the Jovian atmosphere. *Bull. Amer. Astron. Soc.* 7:468.

Trafton, L., and Wildey, R. 1974. Jupiter: a comment on the 8- to 14-micron limb darkening. *Astrophys. J.* 194:499–502.

Trauger, J. T.; Roesler, F. L.; Carleton, N. P.; and Traub, W. A. 1973. Observation of HD on Jupiter and the D/H ratio. *Astrophys. J.* 184:L137–L141.

Walker, M. F., and Hayes, S. 1967. Image tube observations of the $3\nu_3$ band of CH_4 in the spectrum of Jupiter. *Publ. Astron. Soc. Pacific* 79:464–472.

Wallace, L., and Smith, G. R. 1975. On Jovian temperature profiles obtained by inverting thermal spectra. Submitted to *Astrophys. J.*

Westphal, J. A. 1969. Observations of localized 5-micron radiation from Jupiter. *Astrophys. J.* 157:L63–L64.

———. 1971. Observations of Jupiter's cloud structure near 8.5 μ. *Planetary atmospheres.* (C. Sagan, T. C. Owen, H. J. Smith, eds.), pp. 359–362. New York: D. Reidel, Springer Verlag.

Westphal, J. A.; Matthews, K.; and Terrile, R. J. 1974. Five-micron pictures of Jupiter. *Astrophys. J.* 188:L111–L112.

Wildey, R. L. 1968. Structure of the Jovian disk in the ν_2-band of ammonia at 100,000 Å. *Astrophys. J.* 154:761–770.

Wildey, R. L., and Trafton, L. M. 1971. Studies of Jupiter's equatorial thermal limb darkening during the 1965 apparition. *Astrophys. J. Suppl.* 23:1–34.

Wildt, R. 1932. Absorptionsspektren und Atmosphären der grossen Planeten. *Nachr. Ges. Akad. Wiss. Göttingen* 1:87–96.

Woeller, F., and Ponnamperuma, C. 1969. Organic synthesis in a simulated Jovian atmosphere. *Icarus* 10:386–392.

PIONEER 10 ULTRAVIOLET
PHOTOMETER OBSERVATIONS OF JUPITER:
THE HELIUM TO HYDROGEN RATIO

R. W. CARLSON

and

D. L. JUDGE
University of Southern California

Pioneer 10 ultraviolet photometer observation of 5.1 Rayleighs (R) of HeI 584 Å emissions from the disc of Jupiter verifies the presence of Jovian atmospheric helium. The excitation process is resonance scattering of incident solar helium radiation. Quantitative estimates of the He-to-H_2 mixing ratio in the lower atmosphere of Jupiter can be obtained by using Jovian atmospheric models, radiative transfer calculations, and the strength of the solar line; an initial determination has been obtained by Carlson and Judge (1974). We have recently measured the profile and intensity of the solar line and found, as have other experimenters, that it is more intense than was originally supposed, implying a lower albedo and less helium. The various measurements of the solar line are reviewed along with the two parameters – the eddy mixing rate and thermospheric temperature – which appear in the model atmosphere. In addition, the radiative transfer calculations are refined to include the anisotropic phase function for helium resonance scattering, and the accuracy of the coherent scattering assumption is verified. Using the Pioneer 10 helium emission rate and eddy diffusion coefficient, the range of helium-to-hydrogen ratios by volume is found in the present analysis to be $0 < [He]/[H_2] \lesssim 0.28$ for an atmospheric temperature of 150°K. If the high electron temperatures observed by the Pioneer radio occultation experiment reflect equally high neutral temperatures, then more helium is indicated but in this case the Lyman-α albedo and eddy diffusion coefficient must be re-evaluated. For the present, however, we consider the cooler thermosphere to be most realistic since ample terrestrial evidence exists for electron temperatures much greater than the corresponding neutral temperature.

Comparison of the present and other helium-to-hydrogen mixing ratio measurements suggests that the best current estimate is the solar abundance, ($[He]/[H_2] = 0.11$), in which case the eddy diffusion coefficient for the Jovian atmosphere is $K \simeq 3 \times 10^7$ cm^2 sec^{-1}.

The Jovian helium abundance and its possible variations in the atmosphere, envelope, and deep interior are of considerable cosmogonic interest. Unfortunately, the high excitation potential of helium precludes its direct observation using groundbased instrumentation. One method to discern the presence of helium is through spacecraft observation of the He resonance line, which occurs at 584 Å in the vacuum ultraviolet region. This line will appear in emission from Jupiter, being formed by resonance scattering of the solar HeI 584 Å radiation incident on the upper atmosphere. The detection of this line and characterization of the Jovian He-H_2 abundance ratio was one of the major objectives of the ultraviolet photometer experiment on Pioneer 10.

A description of the experiment and planetary results have been presented by Carlson and Judge (1974, 1975) and Judge and Carlson (1974). The instrument is a two channel photometer with one channel employing an Al thin film filter and LiF photocathode to isolate a spectral band containing the helium resonance line. The second channel is for observation of the atomic hydrogen Lyman-α resonance line (1216 Å). Mechanical collimation defines the field of view and the entrance aperture. The axis of the collimator is oriented 20° to the spacecraft spin axis, consequently an annular field 40° in diameter is swept out during each spacecraft spin period. Each 12-sec spin produces ~ 32 samples, each with a tangential and sagittal angular width of $\sim 1°$ by 10°.

Jupiter was in the field of view during two periods of the Pioneer 10 encounter, although the influence of high-energy magnetospheric particles allowed useful data to be obtained only during the first viewing opportunity. Positive identification of signals in the short wavelength channel, identified as HeI 548 Å radiation, verify the existence of helium in the atmosphere of Jupiter. The average emission rate over the disc observed at 30° phase angle was found to be 5.1 R. Using this emission rate a quantitative estimate of the atmospheric He-H_2 abundance ratio can be determined by comparison with theoretical resonance scattering models. These models require (1) knowledge of the source of radiation being scattered by the helium atoms – that is, the intensity and profile of the solar HeI line; (2) a model of the atmosphere which describes the mixed and gravitationally separated regions and the atmospheric temperature; and (3) radiative transfer calculations which include multiple scattering by He and absorption by H_2.

A previous estimate of the number-density ratio in the mixed region of the atmosphere $[He]/[H_2] \sim 1/5$ was obtained by Carlson and Judge (1974) using the Pioneer 10 observations, the solar flux values available at the time, an assumed 150°K atmosphere, and multiple scattering calculations which assumed coherent isotropic scattering and neglected polarization effects.

We have recently remeasured the intensity and profile of the solar He line (Maloy et al. 1976) and found, along with other experimenters, that it is more intense than originally had been supposed, implying a lower value for the diffuse 584 Å reflectivity of Jupiter and therefore less He than previously estimated. The improved solar flux value and the substantial reduction in helium abundance inferred from these measurements suggest that the accuracy of the assumptions employed in the model calculations should be verified. In addition, the Pioneer 10 occultation experiment indicates high plasma temperatures for the outer ionosphere which may imply a higher neutral temperature. In this chapter we discuss measurements of the solar line and improve the radiative transfer calculations by including the anisotropic scattering phase function for helium and the frequency redistribution (i.e., non-coherence) which occurs in resonance scattering. Higher temperatures and their impact on the resulting He-H_2 abundance are also discussed.

I. RESONANCE SCATTERING BY HELIUM

The resonance transition of He($1s2p\ {}^1P_1^0 - 1s^2\ {}^1S_0$) occurs at $\lambda_0 = 584.334$ Å (21.21 eV). For a gas kinetic temperature T, the Doppler width of the gas is

$$\frac{\Delta\lambda_D}{\lambda_0} = \frac{\Delta\nu_D}{\nu_0} = \frac{U}{c} \tag{1}$$

where U is the most probable speed,

$$U = \sqrt{\frac{2kT}{M}}. \tag{2}$$

At frequencies ν displaced $\xi = (\nu - \nu_0)/\Delta\nu_D$ Doppler units from line center, the atomic absorption cross section is

$$\sigma(\xi) = \sigma_0 \phi(\alpha, \xi) \tag{3}$$

with α the ratio of the natural to Doppler width, $\phi(\alpha, \xi)$ the Voigt profile normalized to unity at line center for $\alpha = 0$, and

$$\sigma_0 = \frac{\pi e^2}{mc} \frac{f}{\sqrt{\pi}\Delta\nu_D}. \tag{4}$$

The oscillator strength for the helium transition is $f = 0.2762$ Wiese et al. 1966) and at 150°K, $\Delta\lambda_D = 1.55$ mÅ, $\alpha = 1.05 \times 10^{-2}$, and $\sigma_0 = 3.0 \times 10^{-13}$ cm^2. For moderate optical thicknesses, or if there is sufficient absorption by another species (as in the case here with H_2 photoionization) then one can, to good approximation, neglect the natural radiation damping wings of the line and use the Doppler profile,

$$\phi(\alpha, \xi) \simeq e^{-\xi^2}. \tag{5}$$

A small fraction of the photons absorbed produce the infrared $1s2p-1s2s$ line while the remainder re-emit the resonance line with probability

$$\widetilde{\omega}_0 = \frac{A(1s2p-1s^2)}{A(1s2p-1s2s) + A(1s2p-1s^2)}. \tag{6}$$

The transition rates A are given by Wiese *et al.* (1966) from which $\tilde{\omega}_0 = 0.9989$.

The scattered photons are not distributed isotropically; rather, since $J = 0$ and $\Delta J = 1$, they are distributed according to the Rayleigh phase function (Hamilton 1947)

$$p(\Theta) = \frac{3}{4}(1 + \cos^2 \Theta) \qquad (7)$$

where Θ is the angle between the incident and scattered photon. In resonance scattering this represents the maximum possible anisotropy.

After scattering, the photons do not in general possess the same frequency incident owing to the Doppler effect in scattering (Henyey 1940). This effect is simply demonstrated by the following. Consider an atom moving with velocity v. It will absorb a photon at frequency ν propagating in the same direction if the Doppler shifted frequency is at resonance $\nu(1 - v/c) = \nu_0$ (approximating the natural line shape with a delta function). Since the atom radiates coherently in its rest frame (neglecting recoil), the observed emission frequency will depend upon the angle of emission. In forward scattering, the frequency is unchanged whereas it will be oppositely situated from line center in backward scattering. For intermediate scattering angles the frequencies will be shifted closer to line center.

Hummer (1962) has evaluated the angle-integrated probability that a photon of relative frequency ξ' is absorbed and re-emits at ξ, $R(\xi' \to \xi)$. In the case of the phase function for Rayleigh scattering and neglecting the natural wings, the probability distribution function in frequency is

$$P(\xi' \to \xi) = \frac{R(\xi' \to \xi)}{e^{-\xi'^2}} \qquad (8)$$

$$= \frac{3}{8}e^{\xi'^2} \left\{ \frac{\sqrt{\pi}}{2} \, erfc(|\bar{\xi}|)[3 + 2(\xi^2 + \xi'^2) + 4\xi^2\xi'^2] - e^{-|\bar{\xi}|^2} |\bar{\xi}|(2|\underline{\xi}|^2 + 1) \right\}$$

where $|\bar{\xi}|(|\underline{\xi}|)$ is the greater (lesser) of $|\xi|$ and $|\xi'|$. This distribution is illustrated in Fig. 1. Later considerations [Eq. (28)] make use of the relations (Hummer 1962)

$$R(\xi' \to \xi) = R(\xi \to \xi') \qquad (9)$$

and

$$\int_{-\infty}^{\infty} R(\xi' \to \xi)d\xi' = e^{-\xi^2} \qquad (10)$$

where $R(\xi' \to \xi)$ here differs from Hummer's value by a normalization constant.

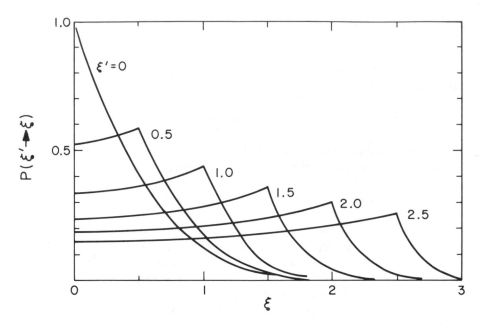

Fig. 1. The probability density function for frequency redistribution in resonance scattering by He. It is assumed that the damping wings of the line are important. The incident and scattered frequencies are measured in Doppler widths displaced from the center of the line. The distributions are symmetric about line center.

II. THE SOLAR HELIUM LINE

The strength of an airglow feature produced by resonance scattering depends upon the spectral intensity of the solar line over the interval of the planetary absorption profile. The width of the planetary absorption feature is generally small compared to the width of the solar line (a few mÅ compared to a solar He line width of roughly one hundred mÅ), and therefore direct measurement of the solar line in the appropriate wavelength interval requires very high resolution measurements. However, the required high resolving powers are difficult to attain in rocket and satellite spectrometers, particularly in the extreme ultraviolet where losses in the optical elements are considerable, so at present one must resort to a less direct determination of the spectral intensity.

Two methods are available. The first is determination of the total solar line intensity and use of a measured or inferred line profile, while the second method uses a known quantity of helium gas (e.g., terrestrial thermospheric He) to absorb and re-radiate the solar line. Measurement of the intensity (i.e., of the He airglow) allows one to infer the solar spectral intensity at the center of the line.

Measurements of the total intensity of the solar line have been performed by several experimenters, perhaps the most widely used being those of the Air Force Cambridge Research Laboratory (AFCRL) group (e.g., Hinteregger 1970). These data, which show an intensity variation with solar activity and the solar 10.7 cm radio flux, were used in our previous analysis of the Jovian He abundance with extrapolation to the low level of solar activity at the time of Pioneer 10 encounter. Recently Hinteregger's values have been questioned since those fluxes may not be sufficient to produce the observed exospheric and ionospheric properties (Roble and Dickinson 1973; Swartz and Nisbet 1973). While the question is not yet resolved, the AFCRL group has remeasured the solar spectrum (Heroux et al. 1974) and found that certain lines, including the He line, are more intense than indicated previously. Our own measurements are in substantial agreement with both the latter and with observations of other experimenters. A summary of these and other 584 Å flux measurements, including our own, is presented in Fig. 2.

The profile and width of the solar helium line has received less experimental effort than the intensity. A curve-of-growth analysis of the solar line transmitted through varying amounts of terrestrial helium gave only an approximate lower limit to the width: ~ 140 mÅ field width at half maximum ($FWHM$) when a Gaussian line profile was assumed (Carlson 1973). The first direct measurements of the profile (Doschek et al. 1974) were obtained photographically at a resolution of 60 mÅ and indicated an approximate Gaussian line shape (but with a slightly flatter central region) with a width $FWHM = 140 \pm 15$ mÅ. Photographic spectra of a small region of the quiet solar disc by Cushman et al. (1975) show an $FWHM$ of 80 mÅ and 100 mÅ for an active region.

Line shape measurements by our laboratory (Maloy et al. 1976) employed a helium filled spectrometer and the curve-of-growth technique. It was found that the line shape could be accurately represented by a Gaussian with $FWHM = 122 \pm 10$ mÅ. A similar experiment (Delaboudinière and Crifo 1975) provides a lower limit of 110 mÅ $FWHM$ in their preliminary analysis. Theoretical calculations by Milkey et al. (1973) predict a flatter profile with $FWHM \simeq 150$ mÅ. In this chapter we have used our measured line width, an assumed Gaussian profile, and various total intensity measurements to generate the line center spectral intensities shown in Fig. 2.

The second method of inferring the central intensity of the solar line, comparison of airglow intensities with that predicted by known He distributions and radiative transfer calculations, has been summarized by Meier and Weller (1972) for several rocket measurements of the terrestrial airglow and for their own satellite measurements (Meier and Weller 1974). The latter measurements are preferred, not only because they are free of atmospheric absorption effects, but they also allow an intercomparison with the present measurements of Jupiter and the sun through mutual observations of the interplanetary helium glow, as discussed in Sec. IV.

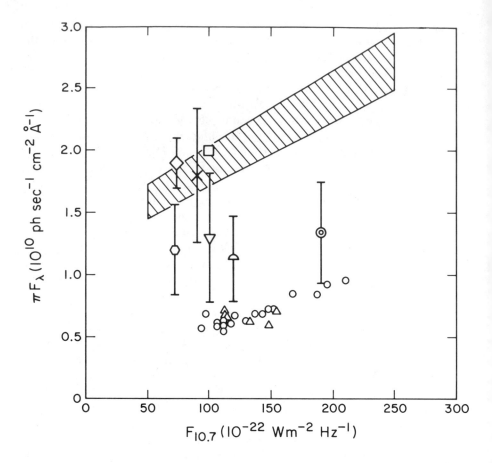

Fig. 2. The solar HeI 584 Å spectral flux at line center as a function of the solar 10.7 cm (2800 MHz) radio flux. The values displayed are based upon direct measurements, integrated intensity measurements normalized with our observed line shape (Maloy *et al.* 1976), and values inferred from airglow data. The normalized integral values are represented by the circles (Hinteregger 1970; Hall and Hinteregger 1970), the double circle (Timothy *et al.* 1972), hexagon (Hinteregger *et al.* 1965), half circle (Heroux *et al.* 1974), cross (Maloy *et al.* 1976), and triangles (apex up) (Reeves and Parkinson 1970; Dupree *et al.* 1973; Dupree and Reeves 1971). In addition, the shaded area is from the regression analyses of total intensity data from Orbiting Solar Observatory, OSO-6, by Woodgate *et al.* (1973). The direct photographic measurements of Cushman *et al.* (1975) are indicated with the triangle (apex down) while the terrestrial (Meier and Weller 1974) and Venus Mariner 10 (Kumar and Broadfoot 1975) values obtained from airglow measurements are represented by the square and diamond respectively.

Mariner 10 data on the Cytheran helium glow (Kumar and Broadfoot 1975) can also be used for determination of the solar intensity, even though the helium content of the Venus atmosphere is unknown, because the reflected intensity of a resonantly-scattering atmosphere that is moderately optically thick is almost independent of the total column abundance in the absence of dissipative absorption (*not* the case for Jupiter where H_2 absorption is present at all altitudes). These two inferred values are also indicated in Fig. 2.

In our original analysis of the Jovian He glow (Carlson and Judge 1974), a line center flux of 0.5×10^{10} photons $\sec^{-1}$ cm^{-2} $\overset{\circ}{A}^{-1}$ (at 1 A.U.) was used, which was based on Hinteregger's total flux and a provisional value for the effective line width. It can be seen from Fig. 2 that even at the low level of solar activity at Pioneer 10 encounter ($F_{10.7} = 72 \times 10^{-22}$ Wm^{-2} Hz^{-1}) the line center flux is probably greater than that which was used. It is difficult to obtain an accurate value from the rather scattered experimental values displayed in Fig. 2, but we suggest (biased somewhat toward our own measurement, the airglow values and the recent AFCRL results) that choosing $\pi F_\lambda = 1.45 \pm 0.4 \times 10^{10}$ photons $\sec^{-1}$ cm^{-2} $\overset{\circ}{A}^{-1}$ (at 1 A.U.) offers a reasonable representation of the HeI 584 $\overset{\circ}{A}$ solar flux at the time of the Pioneer 10 encounter.

Since the solar hydrogen ultraviolet lines exhibit self-reversal, it might be imagined that the same may be true for the helium lines. Three lines of evidence argue against significant self-reversal, however. First are the line-center flux values inferred from airglow measurements. If there were strong reversal, these values would be lower than the normalized integral flux values; in fact, they are among the highest values shown in Fig. 2. Secondly, any reversal effects from the outer chromosphere would be expected to display absorption widths that are characteristic of relatively high temperatures and velocities and could have been seen at the resolving power employed by Cushman *et al.* (1975). No such absorption is evident in their spectra. Finally, the theoretical model of Milkey *et al.* (1973) shows that the line can be formed with little self-reversal although other models (Zirin 1975) do produce strong self-reversal (Milkey 1975).

III. JOVIAN ATMOSPHERIC MODELS

In addition to helium atoms which scatter the resonance line, the atmosphere of Jupiter contains molecular hydrogen which will absorb this radiation through photoionization. Other components such as methane will also absorb, but the effect of these minor constituents is small relative to H_2 absorption. It is the simultaneous absorption by H_2 which determines the strength of the airglow signal, and allows a determination of the relative He-H_2 composition through a measurement of the intensity of the glow.

The scattering and absorptive properties are not constant throughout the atmosphere, but vary with altitude due to the gravitational separation of the molecular constituents. In contrast to atmospheres of the terrestrial planets, where the absorbing component is generally heavier and produces an absorbing lower boundary because of its smaller scale height, the atmospheres of the major planets become relatively more absorbing with increasing altitude since the scale height of the absorbing H_2 is larger than that for He. In the lower regions, mechanical mixing by large scale turbulent motion produces uniform mixing and height-independent scattering and absorbing properties. Calculations of the diffusely reflected helium glow must include the altitude-dependent absorption and the transition from the mixed and gravitationally separated regions using a suitable model of the atmosphere.

The transition level between the mixed region, where eddy diffusion dominates, and the region where molecular diffusion produces gravitational (diffusive) separation is termed the turbopause. At that level, the eddy and molecular diffusion coefficients are equal. The latter varies inversely with number density, so if the eddy coefficient is K, in $cm^2 sec^{-1}$, the number density of H_2 at the turbopause is

$$n_0 = b/K \qquad (11)$$

where $b = 1.81 \times 10^{19}$ $cm^{-1} sec^{-1}$ at 150°K (Wallace and Hunten 1973) and varies as the square root of the temperature. Referring the altitude z to the turbopause and with H denoting the scale height for H_2 ($H = kT/m_{H_2}g = 26.3$ km at 150°K), then treating He as a minor constituent, the H_2 and He number densities are (Wallace and Hunten 1973)

$$n(H_2|z) = n_0 e^{-z/H} \qquad (12a)$$

$$n(He|z) = rn_0 e^{-z/H}/(1 + e^{z/H}). \qquad (12b)$$

The vertical column densities are

$$N(H_2|z) = n_0 H e^{-z/H} \qquad (13a)$$

$$N(He|z) = rn_0 H[e^{-z/H} - \ln(1 + e^{-z/H})]. \qquad (13b)$$

Here, r denotes the volume ratio of He to H_2 in the mixing region far below the turbopause. Estimation of the mixing ratio r from airglow measurements requires knowledge of the remaining two parameters of the model, the eddy diffusion coefficient and temperature, which are discussed below.

Few methods are presently available to determine the eddy diffusion coefficient. One means is measurement of the Lyman-α airglow, since the abundance of atomic hydrogen depends upon the rate of downward flow by eddy diffusion to regions of higher density where recombination occurs (Hunten 1969; Wallace and Hunten 1973). Ultraviolet photometer measurements of this glow from Pioneer 10 gave values of $K \simeq 3 \times 10^{8 \pm 1}$ $cm^2 sec^{-1}$ for temperatures of 150°K (Carlson and Judge 1974), and are reasonably well supported by the Copernicus Lyman-α measurements of Wallace, Jenkins and Drake (personal communication) which imply a value of $K \simeq 10^{8^{+2}_{-1}}$ cm^2 sec^{-1}. Rocket measurements of the Lyman-α glow from Jupiter (Rottman et al. 1973) have indicated a higher albedo and lower eddy diffusion coefficient, $K \simeq 10^6$ $cm^2 sec^{-1}$.

The abundance of other aeronomic species can be used to infer the magnitude of the eddy coefficient. Photochemical calculations by Strobel (1973) based on the lower values for K ($= 5 \times 10^5$ cm^2 sec^{-1}) predicted C_2H_2 and C_2H_6 abundances some two orders of magnitude less than the experimentally observed values of Ridgway (1974). Subsequent calculations by Strobel (1974) employed an altitude dependent K, consistent with the Pioneer 10 and Copernicus values, with results compatible with the observational estimates so long as Ridgway's observations probe down to the 10^{-1} atm level.

A lower limit to K is provided by a stellar occultation measurement of density inhomogeneities in the atmosphere, which are presumably due to atmospheric turbulence (Veverka *et al.* 1974).[1] These fluctuations were seen to persist up to at least the 3×10^{13} cm^{-3} density level which corresponds to $K > 7 \times 10^5$ cm^2 sec^{-1}.

Theoretical discussions of the sources of atmospheric turbulence and their diffusivities are presented by French and Gierasch (1974) and Shimizu (1974). Inertia gravity waves are considered in both analyses, and the diffusion coefficient is calculated as a function of the horizontal wavelength. French and Gierasch (1974) have assumed a vertical wavelength of 13 km, based on prominent occultation features, and find values of K up to 7×10^6 cm^2 sec^{-1} for horizontal wavelengths of 10 km. They suggest that $K \simeq 3 \times 10^6$ cm^2 sec^{-1} and that the horizontal wavelength is ~ 50 km. Shimizu (1974) has performed essentially the same calculation but chooses different characteristic wavelengths based on terrestrial analogy, suggesting that $K \simeq 10^7 - 10^8$ cm^2 sec^{-1}. French and Gierasch also considered acoustic waves and found for the corresponding diffusivity $K \simeq 7 \times 10^7$, but they point out that the energy carried in such waves seems prohibitive and that continuous wave motion of this type seems improbable.

The above estimates of the eddy diffusion coefficient are summarized in Table I. For the present we leave K as a free parameter but note that the majority of the evidence favors relatively high values, $K \gtrsim 10^7$ cm^2 sec^{-1}.

The choice of the thermospheric temperature and temperature profile is presently rather uncertain. The earliest works on this subject were theoretical calculations by Gross and Rasool (1964), McGovern (1968), Shimizu (1971), McGovern and Burke (1972), Gross (1972), and Strobel and Smith (1973). All of these models assumed solar ultraviolet heating only, but they treated the cooling mechanisms and heat conduction processes somewhat differently. The Pioneer 10 encounter occurred near solar minimum for which an isothermal atmosphere at 150°K was considered representative (Carlson and Judge 1974) based on the computations of Shimizu (1971).

Attempts have been made to determine the temperature of the upper atmosphere based on inversion of occultation light curves (Hubbard *et al.*

---

[1]See also pp. 269 and 273.

TABLE I
*Estimates of the Eddy Diffusion Coefficient K for
the Upper Atmosphere of Jupiter*

Method	K ($cm^2\ sec^{-1}$)	Reference
Pioneer 10 Lyman-α	$3 \times 10^{8\ \pm\ 1}$	a
Copernicus Lyman-α	$10^{8^{+2}_{-1}}$	a, b
Sounding Rocket Lyman-α	10^6	c
Stellar Occultation	$> 7 \times 10^5$	d
Theory: Inertia Gravity Waves	$10^7 - 10^8$	e
Theory: Inertia Gravity Waves	$3-7 \times 10^6$	f
Theory: Acoustic Waves	7×10^7	f

$a.$ Carlson and Judge (1974).
$b.$ Wallace, Jenkins, and Drake (personal communication).
$c.$ Rottman *et al.* (1973).
$d.$ Veverka *et al.* (1974).
$e.$ Shimizu (1974).
$f.$ French and Gierasch (1974).

1972; Vapillon *et al.* 1973; Veverka *et al.* 1974), but the results for the same event are not mutually consistent and it appears that only a temperature characteristic of the upper stratosphere can be derived. The combination of the various β-Sco occultation data suggests that the temperature is $\sim 170°K$ at the 0.01 mb level (p. 23). It seems quite impossible to make any inferences about the thermosphere directly from stellar optical occultation data.

The only other experimental temperature datum pertains to the opposite extreme of the atmosphere (relative to the He and H line formation region). The ionospheric profiles observed by the Pioneer occultation experiments (Fjeldbo *et al.* 1975)[2] found topside-ionosphere electron temperatures of $900 \pm 500°K$. In the lower regions, substantial layering was found which does not lend itself to ionospheric modeling based on terrestrial analogy. The question of concern here is, what implications does this observed electron temperature hold for the neutral temperature in the lower thermosphere? Certainly there is no compelling reason to assume equal neutral, ion, and electron temperatures since ample terrestrial evidence exists for electron temperatures larger by a factor of two or more than the corresponding neutral temperature. In fact, Goertz (1973) has calculated electron, ion, and neutral temperatures for the upper atmosphere of Jupiter and found electron temperatures of a factor of three greater than the neutral gas temperature.

The high plasma temperatures observed by Pioneer radio occultations refer to a region far above the altitudes where resonance scattering by H and

---

[2]See p. 240.

He occur. The altitudes where the Pioneer plasma temperature determinations were obtained were some 2000 km above the cloud tops, whereas the region of airglow line formation is roughly 500–800 km above the clouds depending upon the specific atmospheric model. Temperatures between these two regions can be quite different. In calculations by Goertz, peak plasma temperatures are attained at altitudes above 1000 km and the largest temperature gradients occur in the 700–1000 km interval. Jovian ionospheric electron temperatures have also been calculated by Henry and McElroy (1969), who find that the onset for temperature gradients occurs in the region where H_2 densities are $\sim 10^7$ cm^{-3}, an altitude significantly above the density levels where line formation takes place. At lower altitudes and higher densities, where significant resonance scattering occurs, Henry and McElroy's model is isothermal with a temperature of 150°K, identical to that used in the previous analysis of the Lyman-α glow. Goertz' model exhibits neutral temperatures of ~ 275°K in the line formation region. If this temperature were appropriate at the time of Pioneer 10 encounter, then the observed Lyman-α glow would mean $K \gtrsim 3 \times 10^8$.

If one does assume that the neutral thermospheric temperature is elevated up to the plasma temperature observed by the Pioneer radio occultation, then one is faced with the problem of identifying the heating mechanism, since the solar ultraviolet would seem insufficient. Particle precipitation is a possible mechanism, but such a process must also meet the low flux requirement for any Lyman α glow produced by particle excitation. A viscous interaction between the ionosphere (which tends to rotate with the magnetic field at System III velocity) and the neutral atmosphere rotating at a different velocity could be a possible heat source. Terrestrial analogies of ionospheric winds are known and are thought to be responsible for the sporadic-E layer. While it is possible that some of the layering observed in the Jovian ionosphere could be a manifestation of this effect, we have made crude estimates of the magnitude of the heat source and found it insufficient. Hunten (p. 26) has independently suggested this process as a candidate energy source. The dissipation of wave energy associated with eddy diffusion has been suggested by Prasad (1975) and Atreya and Donahue (p. 313). The calculations performed so far indicate that this mechanical heating could be a significant process, but the numerical values are still very uncertain.

A problem introduced by neutral temperatures elevated to the observed plasma temperature is in the interpretation of the low Lyman-α albedo observed by Pioneer 10 and Copernicus. At higher temperatures the albedo will increase (for a fixed-eddy diffusion coefficient) due to increased H-atom column densities and to variation of the Doppler width. The low albedo, if reinterpreted for higher temperatures, implies much greater (and perhaps unrealistic) values for K. The previous analysis may become inapplicable, however, due to other processes becoming important at high temperatures (e.g., increased H-atom recombination rates and diurnal variations in the H-atom abundance owing to the diffusion time constant becoming comparable to the rotation period). In the present work we have assumed two temperatures, 150°K and 450°K, but prefer the lower value as being more appropriate for the neutral temperature in the region of line formation.

IV. RADIATIVE TRANSFER
OF THE JOVIAN HELIUM AIRGLOW LINE

We have discussed the physics of scattering by He, the incident solar line, and a model of the atmosphere of Jupiter. Combining these with multiple-scattering calculations will give the predicted airglow intensity which is to be compared to observation. The previous analysis (Carlson and Judge 1974) assumed coherent scattering as a reasonable approximation since numerical calculations by Wallace (1971) of the diffusely reflected airglow of terrestrial atomic oxygen showed little difference between coherent scattering and scattering with frequency redistribution. This assumption may be unwarranted for the present case due to the different altitude distribution of absorbing species and radiation imprisonment effects. To examine this possibility, we derive the radiative transfer equation for isotropic scattering which includes frequency redistribution, and then develop approximate solutions. A first-order correction for the phase function of anisotropic scattering is then applied and the disc-averaged albedo is calculated for coherent and incoherent scattering.

In the description of atmospheric densities, Eqs. (12) and (13), we define x to be proportional to the H_2 density,

$$x = e^{-z/H} \tag{14}$$

and denote the H_2 absorption cross sections at 584 Å as σ_d (6.2×10^{-18} cm²). Then at a frequency that is ξ Doppler widths from line center, the vertical optical depth is

$$\tau(x, \xi) = \tau_0\{x + \eta(\xi) \ [x - \ln(1 + x)]\} \tag{15}$$

where

$$\tau_0 = n_0 H \sigma_d \tag{16a}$$

and

$$\eta(\xi) = r \frac{\sigma_0}{\sigma_d} \phi(\alpha, \xi) . \tag{16b}$$

The fraction of all photons absorbed in unit volume at x, that are re-emitted as a resonance line photon, is

$$\widetilde{\omega}(x, \xi) = \frac{\widetilde{\omega}_0 \sigma(\xi) n(\mathrm{He}|z)}{\sigma(\xi) n(\mathrm{He}|z) + \sigma_d n(\mathrm{H_2}|z)} \tag{17}$$

or

$$\tilde{\omega}(x, \xi) = \frac{\tilde{\omega}_0 \eta(\xi) x}{1 + x + x\eta(\xi)}. \tag{18}$$

Assume a plane parallel atmosphere and let $S(x', \xi')$ be the emission rate per unit x per cm^2 per Doppler width. $S(x)$ is related to the volume emission rate $Q(z)$ [cm^{-3} sec^{-1} Doppler width^{-1}] through $xS(x) = HQ(z)$. The number of photons passing through a plane at x is

$$\frac{1}{4\pi} \int_0^{2\pi} d\phi \int_0^1 d\mu \, e^{-\frac{|\tau(x, \xi') - \tau(x', \xi')|}{\mu}} S(x', \xi'). \tag{19}$$

The number absorbed per unit x, at x is

$$-\frac{1}{2} \frac{\partial}{\partial x} \int_0^1 d\mu \, e^{-\frac{|\tau(x, \xi') - \tau(x', \xi')|}{\mu}} S(x', \xi') \tag{20}$$

or

$$\frac{1}{2} \frac{\partial \tau(x, \xi')}{\partial x} \int_0^1 \frac{d\mu}{\mu} e^{-\frac{|\tau(x, \xi') - \tau(x', \xi')|}{\mu}} S(x', \xi'). \tag{21}$$

Of these absorbed photons, the fraction $\tilde{\omega}(x, \xi')$ reappear as photons at frequency ξ with probability distribution $P(\xi' \rightarrow \xi)$. Integrating over μ, we find the contribution at ξ is

$$\frac{1}{2}\tilde{\omega}(x, \xi') \, P(\xi' \rightarrow \xi) \frac{\partial \tau(x, \xi')}{\partial x} E_1[|\tau(x, \xi') - \tau(x', \xi')|] \, S(x', \xi') \tag{22}$$

where $E_1(|\tau - \tau'|)$ is the first exponential integral. Integration of the above over altitude x' and frequency ξ' gives the net contribution to the source function at x due to photons previously scattered. The total source function includes those and also the incident solar photons initially scattered at x, $S_0(x, \xi)$. The resulting source function is

$$S(x, \xi) = S_0(x, \xi) + \frac{1}{2} \int_{-\infty}^{\infty} d\xi' \, \tilde{\omega}(x, \xi') \, P(\xi' \rightarrow \xi) \frac{\partial \tau(x, \xi')}{\partial x} \int_0^{\infty} dx' \, S(x', \xi')$$

$$E_1[|\tau(x', \xi') - \tau(x, \xi')|] \tag{23}$$

with

$$S_0(x, \xi) = \Delta\lambda_D \ \pi F_\lambda \int_{-\infty}^{\infty} d\xi' \ \widetilde{\omega}(x, \xi') \ P(\xi' \rightarrow \xi) \ e^{-\tau(x, \xi')/\mu_0} \left(\frac{\partial\tau(x, \xi')}{\partial x}\right) \tag{24}$$

where solar radiation is incident at $\cos^{-1}\mu_0$ and πF_λ is the spectral flux at line center in photons $\sec^{-1}$ cm^{-2} $Å^{-1}$. While machine calculations can be used to solve Eq. (23) to any required degree of accuracy, the added frequency dimension necessitates lengthy computations and it is desirable to develop an approximate solution. We first note that in the coherent case, $P(\xi' \rightarrow \xi) = \delta(\xi - \xi')$ and so

$$S(x, \xi) = S_0(x, \xi) + \frac{1}{2}\widetilde{\omega}(x, \xi)\frac{\partial\tau}{\partial x} \int_0^{\infty} dx' \ E_1[|\tau(x, \xi)$$

$$-\tau(x', \xi)|] \ S(x', \xi) \tag{25a}$$

where,

$$S_0(x, \xi) = \Delta\lambda_D \ \pi F_\lambda \ \widetilde{\omega}(x, \xi) \ \frac{\partial\tau(x, \xi)}{\partial x} \ e^{-\tau(x, \xi)/\mu_0} \tag{25b}$$

can be accurately described by the following approximation due to Biberman [see Van Blerkom and Hummer (1969)]. Since $s(\tau, \xi) \ d\tau = S(x, \xi) \ dx$ where $s(\tau, \xi)$ is the number of emissions per unit optical depth at ξ (i.e., photons $\sec^{-1}$ cm^{-2} per optical depth per Doppler width), then if $s(\tau, \xi)$ varies over τ much more slowly than the first exponential integral, it can be taken outside the integral [i.e., $s(\tau, \xi) \simeq s(\tau', \xi)$] resulting in

$$S(x, \xi) = \frac{S_0(x, \xi)}{\epsilon(x, \xi) + L(x, \xi)} \tag{26}$$

where

$$\epsilon(x, \xi) = 1 - \widetilde{\omega}(x, \xi) \tag{27a}$$

$$L(x, \xi) = \frac{1}{2}\widetilde{\omega}(x, \xi) \ E_2[\tau(x, \xi)] \tag{27b}$$

and E_2 is the second exponential integral.

The same reasoning can be applied to the incoherent case and simplified further by assuming that the frequency dependence of the source function is proportional to the absorption cross section. From Eqs. (9), (10), and (23) we then find

$$S(x, \xi) = \frac{S_0(x, \xi)}{<\epsilon(x, \xi)> + <L(x, \xi)>} \tag{28}$$

in which

$$<\epsilon(x, \xi)> = \int_{-\infty}^{\infty} P(\xi \to \xi') \left[1 - \tilde{\omega}(x, \xi')\right] d\xi' \qquad (29a)$$

and

$$<L(x, \xi)> = \frac{1}{2} \int_{-\infty}^{\infty} P(\xi \to \xi') \tilde{\omega}(x, \xi') E_2[\tau(x, \xi')] d\xi' . \quad (29b)$$

These functions physically represent the loss rate of photons of frequency ξ, scattered to ξ'. The first term represents subsequent local reabsorption and dissipative loss while the second gives the rate of loss by escape from the medium without scattering.

The correction for Rayleigh scattering can be made by noting that repeated scattering of a photon produces near-isotropy, i.e., the photon loses memory of the initial direction. In fact, most of the anisotropy in the source function is due to the initial scattering of the solar photon, i.e., $S_0(x, \xi)$ (Chandrasekhar 1960). If observed at an angle Θ to the incident radiation, the first scattering contributes $p(\Theta)S_0$ to the emerging radiation. Higher order scattering, which can be assumed isotropic, contributes $S(x, \xi) - S_0(x, \xi)$. Therefore the emerging flux at $\mu = \cos \Theta$ is approximately

$$4\pi I(\mu) = \int_{-\infty}^{\infty} d\xi \int_0^{\infty} \frac{dx}{\mu} e^{-\tau(x, \xi)/\mu} \{S(x, \xi) + [p(\Theta) - 1] S_0(x, \xi)\} . \quad (30)$$

This method of approximation was checked by comparing it and Chandrasekhar's exact H function solutions for the Rayleigh case and conservative, coherent scattering. The agreement between the two was found to be excellent, with errors of only a few percent.

The disc averaged emission rate is given by

$$<4\pi I> = 2 \int_0^1 \mu \, d\mu \, 4\pi I(\mu, \mu_0) . \qquad (31)$$

We have used $\mu = \mu_0$ (but $\Theta = 30°$) and a single term Gaussian quadrature to evaluate the average brightness, i.e.,

$$<4\pi I> \approx 4\pi I\left(\mu = \mu_0 = \frac{2}{3}\right). \qquad (32)$$

These seem justifiable assumptions since the variation of $4\pi I$ with μ was found to be quite slow.

With the above formulae one can evaluate the importance of frequency redistribution in formation of the Jovian He airglow line and the adequacy

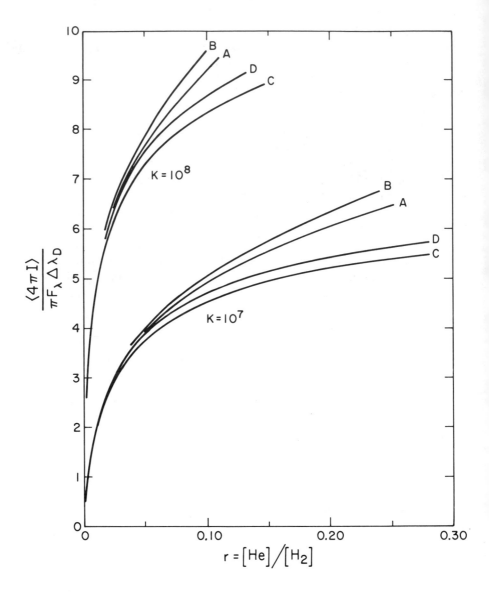

Fig. 3. The average emission rate from Jupiter as a function of the $[He]/[H_2]$ ratio for a 150°K atmosphere computed with various assumptions. Curve A is an exact iterative solution to Eq. (25) assuming coherent scattering and the Voigt profile. Curve B is the Biberman approximation [Eq. (26), (27)] with the above assumptions. Curve C includes frequency redistribution with the Doppler profile [Eq. (5)] while Curve D uses the same formulation as C but for coherent scattering. It is concluded that the influence of frequency redistribution is negligible for interpretation of the present observations, and the solutions corresponding to Curve A provide sufficient accuracy.

of the coherent approximation. We have calculated solutions for an isothermal $150°K$ atmosphere with $K = 10^7$ and 10^8 cm^2 sec^{-1} and the following cases: A — an exact numerical solution (by iteration) of Eq. (25) for coherent scattering using the Voigt profile; B — the Biberman approximation [Eqs. (26) and (27)] for coherent scattering with the Voigt profile; C — the approximate solution [Eqs. (28) and (29)] to non-coherent scattering using the Doppler profile and the redistribution function given by Eq. (8); D — the same as the preceding case except substituting $P(\xi' \rightarrow \xi) = \delta(\xi - \xi')$ in the numerical code, i.e. coherent scattering. Results for these cases are shown in Fig. 3. The scant difference between solutions A and B shown in the figure demonstrates that Biberman's approximation is excellent, while the dissimilarities of B and D indicate the minor (but not insignificant) importance of including the Lorentzian wings of the line. Note the small difference between the solutions for cases C and D, indicating that the coherent approximation overestimates the albedo by only $\sim 3\%$. Even though we have neglected the wings in this comparison, the conclusion that the coherent approximation is quite adequate still holds since scattering in the wings is itself very nearly coherent (Carlson and Judge 1971). For comparison with the observations then, we have assumed coherent scattering with the Voigt absorption profile, and solved Eq. (25) by iteration. The results are presented in the following section.

V. RESULTS AND DISCUSSION

The disc-averaged 584 Å apparent emission rate was calculated for various values of K and two different temperatures, $150°K$ and $450°K$. For a given He-H$_2$ abundance and K the albedo is generally less at higher temperatures since the increased scale height places more H$_2$ in and above the region of He resonance scattering. The albedos are placed on an absolute scale, expressed in Rayleighs (R), using the central flux (at 1 A.U.) of $\pi F_\lambda = 1.45 \times 10^{10}$ photons sec^{-1} cm^{-2} Å^{-1} as previously discussed. These results are shown in Fig. 4 along with the observed datum of 5.1 R. The combination of the absolute calibration errors, measurement uncertainties, and uncertainties in the solar flux, combine to give an estimated error limit of $\pm 50\%$.

A literal interpretation of the Pioneer values ($<4\pi I> = 5.1$ R, $K = 3 \times 10^8$, assuming $150°K$) implies a low helium-to-hydrogen ratio, on the order of one percent, which might indicate depletion of helium in the Jovian atmosphere (Stevenson and Salpeter p. 97). But the case for such depletion is weak since the range of uncertainties in both the albedo and K admit $0 <$ [He] / [H$_2$] $\leqslant 28\%$. This range of values is consistent with other determinations of the [He] / [H$_2$] ratio obtained by a variety of methods: the β-Sco stellar occultation, and infrared measurements from aircraft and Pioneer. Values obtained by those methods are summarized in Fig. 5 along with the solar value (Hunten and Münch 1973) of [He] / [H$_2$] $= 0.11$, although even

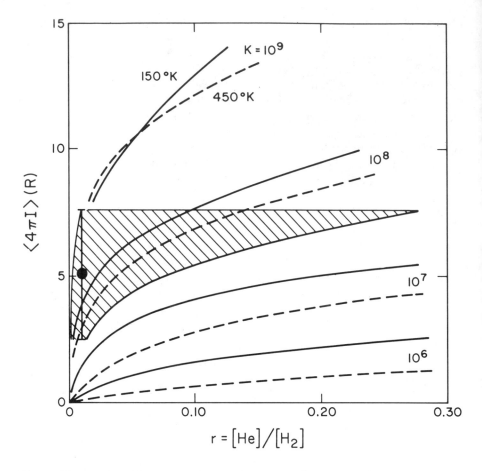

Fig. 4.   The experimental and theoretical average HeI 584 Å emission rates from Jupiter. The model calculations were performed for coherent scattering and the Voigt profile by exact iterative solution of Eq. (25), and normalized using $\pi F_\lambda = 1.45 \times 10^{10}$ photons sec^{-1} cm^{-2} Å^{-1} (at 1 A.U.). The Pioneer 10 datum (5.1 R) is shown and the combined range of uncertainties is indicated. Using the eddy diffusion coefficient determined from the Pioneer-10 Lyman-α albedo ($K = 3 \times 10^{8 \, + \, 1}$ cm^2 sec^{-1}), indicated by the shaded area, the [He]/[H$_2$] volume abundance ratio is found to be less than 28% but greater than zero.

the solar value is quite uncertain (Hirschberg 1973). Consideration of these available measurements suggests that presently the most reasonable number to adopt corresponds to the solar value. If one does indeed assume solar composition, the inferred value of K is quite reasonable: $K \simeq 3 \times 10^7$ cm^2 sec^{-1}, and consistent with both the Pioneer and Copernicus values. The relatively low $K \simeq 10^6$ cm^2 sec^{-1} from sounding rocket Lyman-α observations seems difficult to reconcile with the various estimates of the He abundance and the Pioneer helium albedo.

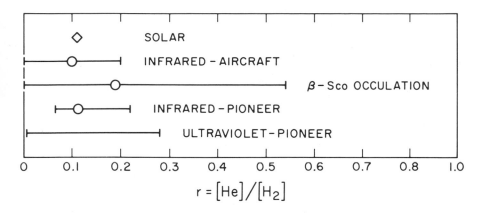

$$r = [\text{He}] / [\text{H}_2]$$

Fig. 5. Summary of various determinations of the Jovian atmospheric $[\text{He}]/[\text{H}_2]$ abundance ratio. Shown is the solar value from Hunten and Münch (1973), the aircraft data from Pollack (1976), a stellar occultation determination (Elliot *et al.* 1974), the Pioneer infrared data (Orton p. 206) and the present range for the ratio. The combination of these data suggest solar composition.

Since the ultraviolet determination of the helium abundance depends upon relative measurements of two quite different radiation sources — the sun and Jupiter — with two quite different instruments, it is important to be able to assess their relative consistency. In fact, this can be done by intercomparison with a third set of measurements for which mutual observations of the interplanetary helium glow have been made. Weller and Meier (1974) observed the interplanetary He 584 Å glow as have the Pioneer photometers. Their measurements, obtained from Earth orbit at $\alpha = 18^h$ show emission rates of 1.5 ± 0.5 R while our own Pioneer 10 values for the same geometry give ~ 1.8 R. Meier and Weller (1974) also measured terrestrial He airglow with the same instrument and were able to deduce the solar flux value $\pi F \lambda = 2 \times 10^{10}$ photons sec^{-1} cm^{-2} Å^{-1} since the thermospheric helium content is rather well known. When variations with solar activity are taken into account, their solar flux and that used in this work agree to within 25%. Not only do these values of brightness and solar flux agree well in the relative sense, which is the only necessary requirement, but they also agree well in absolute magnitude. This adds confidence to the Lyman-α observations on Pioneer 10 since the relative calibration between the two channels is more accurate than the absolute calibration.

The He-H$_2$ ratio found above holds only if the Jovian thermospheric temperature is not significantly higher than the assumed 150°K. At 450°K, the calculations indicate increased He abundances (for the same K). However, if these higher neutral temperatures are found, the Lyman-α albedo and eddy diffusion coefficient will have to be reconsidered.

VI. SUMMARY

Observations of 5.1 R diffusely reflected HeI 584 Å emissions from the atmosphere of Jupiter by the ultraviolet photometer on Pioneer 10 establish the expected presence of He in the giant planet. In order to make a quantitative estimate of the relative abundance, we have measured the width and intensity of the solar line and compared these measurements with other data in order to estimate a line-center flux (at the time of Jupiter encounter) of $1.45 \pm 0.4 \times 10^{10}$ photons sec^{-1} cm^{-2} $Å^{-1}$ at 1 A.U. The assumption of coherent scattering in formation of the Jovian airglow line was investigated and found to be an accurate approximation. The anisotropic angular scattering function for He was included in model calculations of the average disc brightness. The present status of knowledge of the atmospheric temperature and eddy diffusion coefficient, which are required in the resonance scattering model calculations, have been briefly reviewed. Assuming an isothermal 150°K atmosphere and the range of eddy diffusion values from the Lyman-α observations of Pioneer 10, the atmospheric helium-to-hydrogen abundance ratio is found to be greater than zero but less than approximately 28%. Calculations performed for a 450°K atmosphere indicate higher He abundances if such a high neutral temperature is truly characteristic. However, this case would require a reassessment of the eddy diffusion coefficient and Lyman-α albedo.

Acknowledgement. This work was supported by Contract NAS2-6558 with the NASA Ames Research Center.

REFERENCES

Carlson, R. W. 1973. Resonance absorption of the solar HeI 584 Å line by telluric helium. Unpublished analysis of OSO-6 observations.

Carlson, R. W., and Judge, D. L. 1971. The extreme ultraviolet dayglow of Jupiter. *Planet. Space Sci.* 19:327–343.

———. 1974. Pioneer 10 ultraviolet photometer observations at Jupiter encounter. *J. Geophys. Res.* 79:3623–3633.

———. 1975. Pioneer 10 ultraviolet photometer observations of the Jovian hydrogen torus: the angular distribution. *Icarus* 24:395–399.

Chandrasekhar, S. 1960. *Radiative transfer.* pp. 145–149. New York: Dover.

Cushman, G. W.; Farwell, L.; Godden, G.; and Rense, W. A. 1975. Solar line profiles of HeI 584 Å and HeII 304 Å. *J. Geophys. Res.* 80:482–486.

Delaboudinière, J. D., and Crifo, J. F. 1975. The profile of the Helium I 584 Å line from the sun. XVIII COSPAR meeting, Varna, Bulgaria.

Doschek, G. A.; Behring, W. E.; and Feldman, V. 1974. The widths of the solar HeI and HeII lines at 584, 537, and 304 Å. *Astrophys. J.* 190:L141–L142.

Dupree, A. K.; Huber, M. C. E.; Noyes, R. W.; Parkinson, W. H.; Reeves, E. M.; and Withbroe, G. L. 1973. The extreme ultraviolet spectrum of a solar active region. *Astrophys. J.* 182:321–333.

Dupree, A. K., and Reeves, E. M. 1971. The extreme ultraviolet spectrum of the quiet sun. *Astrophys. J.* 165:599–613.

Elliot, J. L.; Wasserman, L. H.; Veverka, J.; Sagan, C.; and Liller, W. 1974. The occultation of Beta-Scorpii by Jupiter. II. The hydrogen helium abundance in the Jovian atmosphere. *Astrophys. J.* 190:719–729.

Fjeldbo, G.; Kliore, A.; Seidel, B.; Sweetnam, D.; and Cain, D. 1975. The Pioneer 10 radio occultation measurements of the ionosphere of Jupiter. *Astron. Astrophys.* 39:91–96.

French, R. G., and Gierasch, P. J. 1974. Waves in the Jovian upper atmosphere. *J. Atmos. Sci.* 31:1707–1712.

Goertz, C. K. 1973. Jupiter's ionosphere and magnetosphere. *Planet. Space Sci.* 21:1389–1398.

Gross, S. H. 1972. On the exospheric temperature of hydrogen-dominated planetary atmospheres. *J. Atmos. Sci.* 29:214–218.

Gross, S. H., and Rasool, S. I. 1964. The upper atmosphere of Jupiter. *Icarus* 3:311–322.

Hall, L. A., and Hinteregger, H. E. 1970. Solar radiation in the extreme ultraviolet and its variation with solar rotation. *J. Geophys. Res.* 75:6959–6965.

Hamilton, D. R. 1947. The resonance radiation induced by elliptically polarized light. *Astrophys. J.* 106:457–465.

Henry, R. J. S., and McElroy, M. B. 1969. The absorption of extreme ultraviolet radiation by Jupiter's upper atmosphere. *J. Atmos. Sci.* 26:912–917.

Henyey, L. G. 1940. The Doppler effect in resonance lines. *Proc. Natl. Acad. Sci. U.S.* 26:50–54.

Heroux, L.; Cohen, M.; and Higgins, J. E. 1974. Electron densities between 110 and 300 km derived from solar EUV fluxes of August 23, 1972. *J. Geophys. Res.* 79:5237–5244.

Hinteregger, H. E. 1970. The extreme ultraviolet solar spectrum and its variation during a solar cycle. *Ann. Geophys.* 26:547–554.

Hinteregger, H. E.; Hall, L. A.; and Schmidtke, G. 1965. Solar XUV radiation and neutral particle distribution in the July 1963 thermosphere. *Space Research V* (D. G. King-Hele, P. Muller, and G. Righini, eds.), pp. 1175–1190. Amsterdam: North Holland.

Hirschberg, J. 1973. Helium abundance of the sun. *Rev. Geophys. Space Phys.* 11:115–131.

Hubbard, W. B.; Nather, R. E.; Evans, D. S.; Tull, R. G.; Wells, D. C.; van Citters, G. W.; Warner, B.; and Vanden Bout, P. 1972. The occultation of Beta Scorpii by Jupiter and Io. I. Jupiter. *Astron. J.* 77:41–59.

Hummer, D. G. 1962. Non-coherent scattering. I. The redistribution functions with Doppler broadening. *Mon. Not. Roy. Astron. Soc.* 125:21–37.

Hunten, D. M. 1969. The upper atmosphere of Jupiter. *J. Atmos. Sci.* 26:826–834.

Hunten, D. M., and Münch, G. 1973. The helium abundance on Jupiter. *Space Sci. Rev.* 14:433–443.

Judge, D. L., and Carlson, R. W. 1974. Pioneer 10 ultraviolet photometer observations in the vicinity of Jupiter. *Science* 183:317–318.

Kumar, S., and Broadfoot, A. L. 1975. He 584 Å airglow emission from Venus. Mariner 10 observations. *Geophys. Res. Lett.* 2:357–360.

Maloy, J. O.; Carlson, R. W.; Hartmann, U.; and Judge, D. L. 1976. Observations of the profile and intensity of the solar helium λ 584 Å resonance line. In draft.

McGovern, W. E. 1968. Exospheric temperatures of Jupiter and Saturn. *J. Geophys. Res.* 73:6361–6363.

McGovern, W. E., and Burke, S. D. 1972. Upper atmosphere thermal structure of Jupiter with convective heat transfer. *J. Atmos. Sci.* 29:179–189.

Meier, R. R., and Weller, C. S. 1972. EUV resonance radiation from helium atoms and ions in the geocorona. *J. Geophys. Res.* 77:1190–1204.

——— 1974. EUV observations of the latitudinal variation of He. *J. Geophys. Res.* 79:1575–1578.

Milkey, R. W. 1975. Comments concerning the photoionization model for excitation of resonance lines of HeI and HeII in the solar chromosphere. *Astrophys. J.* 199:L131–132.

Milkey, R. W.; Heasley, J. N.; and Beebe, H. A. 1973. Helium excitation in the solar chromosphere: HeI in a homogeneous chromosphere. *Astrophys. J.* 186:1043–1052.

Pollack, J. B. 1976. Estimates of Jupiter's helium to hydrogen ratio and vertical temperature structure from aircraft observations in the 16–40 micron region. In draft.

Prasad, S. S. 1975. Possible new Jovian thermospheric models. *Astrophys. J.* 200:L171–L174.

Reeves, E. M., and Parkinson, W. H. 1970. An atlas of extreme ultraviolet spectro heliograms from OSO-IV. *Astrophys. J. Suppl.* vol. 21.

Ridgway, S. T. 1974. Jupiter: identification of ethane and acetylene. *Astrophys. J.* 187:L41–L43. (Erratum 192:L51).

Roble, R. G., and Dickinson, R. E. 1973. Is there enough solar extreme ultraviolet radiation to maintain the global mean thermospheric temperature? *J. Geophys. Res.* 78:249–257.

Rottman, G. J.; Moos, H. W.; and Freer, C. S. 1973. The far ultraviolet spectrum of Jupiter. *Astrophys. J.* 184:L89–L92.

Shimizu, M. 1971. The upper atmosphere of Jupiter. *Icarus* 14:273–281.

——. 1974. Atmospheric mixing in the upper atmospheres of Jupiter and Venus. *J. Geophys. Res.* 79:5311–5313.

Strobel, D. S. 1973. The photochemistry of hydrocarbons in the Jovian atmosphere. *J. Atmos. Sci.* 30:489–498.

——. 1974. Hydrocarbon abundances in the Jovian atmosphere. *Astrophys. J.* 192:L47–L49.

Strobel, D. F., and Smith, G. F. 1973. On the temperature of the Jovian thermosphere. *J. Atmos. Sci.* 30:718–725.

Swartz, W. E., and Nisbet, J. S. 1973. Incompatibility of solar EUV fluxes and incoherent scatter measurements at Arecibo. *J. Geophys. Res.* 78:5640–5657.

Timothy, A. F.; Timothy, J. G.; and Willmore, A. P. 1972. The ion chemistry and thermal balance of the E- and lower F-regions of the daytime ionosphere: an experimental study. *J. Atm. Terr. Phys.* 34:969–1035.

Van Blerkom, D., and Hummer, D. G. 1969. The normalized on-the-spot approximation for line transfer problems. *J. Quant. Spectrosc. Radiat. Transfer* 9:1567–1571.

Vapillon, L.; Combes, M.; and Lecacheux, J. 1973. The β Scorpii occultation by Jupiter. II. The temperature and density profiles of the Jovian upper atmosphere. *Astron. Astrophys.* 24:135–149.

Veverka, J.; Wasserman, L. H.; Elliot, J.; and Sagan, C. 1974. The occultation of β Scorpii by Jupiter. I. The structure of the Jovian upper atmosphere. *Astron. J.* 79:73–84.

Wallace, L. 1971. The effect of non-coherency on the intensity of resonantly scattered sunlight in a planetary atmosphere. *Planet. Space Sci.* 19:377–389.

Wallace, L., and Hunten, D. M. 1973. The Lyman-alpha albedo of Jupiter. *Astrophys. J.* 182:1013–1031.

Weller, C. S., and Meier, R. R. 1974. Observation of helium in the interplanetary/interstellar wind—the solar wake effect. *Astrophys. J.* 193:471–476.

Wiese, W. L.; Smith, M. W.; and McGlennon, B. M. 1966. *Atomic transition probabilities vol. I. Hydrogen through neon.* NSRDS-NBS 4. Washington, D.C.: U.S. Government Printing Office.

Woodgate, B. E.; Knight, D. E.; Uribe, R.; Sheather, P.; Bowles, J.; and Nettleship, R. 1973. Extreme ultraviolet line intensities from the sun. *Proc. Roy. Soc. A.* 332:291–309.

Zirin, H. 1975. The helium chromosphere, coronal holes, and stellar X-rays. *Astrophys. J.* 199:L63–L66.

MORPHOLOGY OF MOLECULAR ABSORPTION
ON THE DISK OF JUPITER

V. G. TEIFEL
Astrophysical Institute, Alma-Ata

The general characteristics of the molecular absorption distribution across the Jovian disk are derived from analysis of current observational data. Most of the observed peculiarities of the longitudinal and latitudinal variations of this absorption can be explained by a two-layer model, which supposes that molecular bands and lines are formed within a multiply-scattering, optically-thick cloud or haze layer, with a modest amount of absorption also occurring in a clear atmospheric layer overlying the cloud. The more complicated multilayer models give nearly identical results: the accuracy of available band and line intensity measurements is not high enough to discriminate between two-layer and multilayer models.

The reflectivity of clouds and the distribution of brightness across the Jovian disk are reproduced satisfactorily by theoretical calculations for the Henyey-Greenstein scattering phase function with asymmetry parameter $g \approx 0.5 \pm 0.1$ in red light. However, at shorter wavelengths it is necessary to take into account Rayleigh scattering in the atmosphere above the clouds. The upper limit to the H_2 abundance in the upper atmosphere is 9 ± 3 km amagat, and the CH_4 abundance upper limit derived from measurements of the 6190 Å and 7250 Å bands is ~ 35 m amagat. The mean aerosol scattering coefficient is estimated to be $\sim (2-4) \times 10^{-6}$ cm^{-1}.

The effective upper boundary of the aerosol haze is highest for observations in the ultraviolet and in the centers of strong absorption bands, and it varies with latitude by about 8 km; the haze boundary is ~ 19 km higher in the polar caps than in the equatorial belt. The effective reflecting layer lies deeper, and has smaller altitude fluctuations in the visual continuum, in weak absorption bands and in the wings of strong bands.

In recent years considerable attention has been given to studies of variations of the intensities of molecular absorption lines and bands across the disk of Jupiter. In conjunction with the results of photometry and polarimetry of Jupiter's visible surface, data on the characteristics of the changes of absorption from the center of the disk to the limb can provide information on the probable structure of the cloud cover of this planet, and on the most probable gaseous composition of the atmosphere.

[441]

Here we review briefly recent morphological studies of molecular absorptions, and some related questions concerning the optical properties of the clouds and the upper atmosphere of Jupiter. Unfortunately limited space permits reproduction of only a selected sample of pertinent figures from the cited literature.

I.   OBSERVATIONAL DATA ON THE DISTRIBUTION OF MOLECULAR ABSORPTION ACROSS THE JOVIAN DISK

Following the initial attempts to establish how the intensity of absorption in the CH_4 6190 Å band varies over the disk of Jupiter (Elvey and Fairley 1932; Bobrovnikoff 1933; Eropkin 1933), interest in this question waned for 20 years. It was revived again thanks to the work of Hess (1953), who measured the equivalent widths of the 6190 Å CH_4 and 6450 Å NH_3 bands on spectrograms of several regions of the planet's disk. Subsequently, from low-dispersion spectra Teifel (1959, 1963, 1964, 1967a, 1968), and Teifel and Priboeva (1963) obtained results in general agreement with those of Hess. They confirmed the extremely small dependence of the equivalent width, W, and the central depth, R, of the 6190 Å band upon position on Jupiter's disk. From observations made in 1962 and 1963, the variations of W and R with latitude appeared to vary no more than 8–10% and were on the order of the statistical errors of measurement. A statistical analysis of Hess' data also reduces the estimated precision of his measurements to the point where it appears that he saw no significant variations in the intensity of the 6190 Å band.

During the period 1963–1970, a large number of spectrograms of the equatorial region, the central meridian, and various other regions of Jupiter were obtained, from which the variations of the CH_4 6190 Å and NH_3 6450 Å bands have been studied (Avramchuk 1968, 1970, 1972; Teifel 1967b, 1969a,b,c; Aksenov et al. 1972; Aksenov 1973). These studies show that the intensity of the 6190 Å band decreases from the center of the disk to the limb at the equator and along the central meridian. The characteristics of the variations of the bands are not uniform; the quantities W and R at the center of the disk vary by almost 40% during the period covered by the observations. It is tempting to seek the reason for these differences in actual variations of the structure of Jupiter's cloud cover. However, photographic measurements are subject to significant errors due to photometric calibration and other factors related to the photographic process and therefore time-dependent variations determined from photographic observations must be regarded with caution. This also significantly affects measurements of changes in the intensity of individual molecular lines, which we shall discuss below.

Aksenov (1973) found an interesting peculiarity, namely, variations along the equator of the relative depth of the 6190 Å CH_4 band in the band center, R_0, and at wavelength displacements of 25, 50, and 75 Å from the band center

(R_1, R_2, and R_3, respectively) occur unequally. Out to a distance r_x of 0.4 of the radius from the center of the disk, the steepness increases with transition from R_0 to R_3. On approaching the limb, where $r_x > 0.4$, the opposite effect is observed; that is, R_0 has the greatest steepness and R_3 the least. This peculiarity is seen in both short and long wavelength wings of the bands. Confirmation of the reality of this effect would provide a basis for its interpretation in terms of vertical inhomogeneity of Jupiter's cloud cover.

Studies of the more intense CH_4 absorption bands at 7250 Å carried out by photographic spectroscopy (Teifel 1966; Avramchuk *et al.* 1972; Ibragimov 1974) show that in the equatorial belt a decrease of W and R is observed close to the limb, although absorption in this band is almost unchanged out to a distance $r_x \simeq 0.8$ from the center. On the other hand, measurements along the central meridian in the same band (Teifel 1966) show an increase in absorption from the equator toward middle latitudes, and then a sharp weakening of absorption near the poles. The same is observed in the strong bands of CH_4 at 8860 Å and 9900 Å.

Although still only qualitative, the results of photographs of Jupiter made through narrowband interference filters centered on the 8860 Å CH_4 band (Owen 1969; Owen and Mason 1969; Kuiper 1972; Minton 1972) are very interesting. In agreement with the data cited above, photographs in this band show a significant decrease in brightness toward the temperate latitudes, indicating increased absorption compared with that in the equatorial band. The most interesting features of these pictures are the bright arcs which appear near the poles of the planet. These formations are not visible in the continuum adjacent to the absorption bands, but are well displayed in both the CH_4 bands at 8860 Å and 9900 Å on spectrograms obtained in 1974 by Vdovichenko and Kuratov using an image intensifier tube at the Kazakh Academy of Sciences-Astrophysical Institute Observatory. The appearance of bright polar arcs must therefore be connected with a sharp decrease in methane absorption near the poles.

Owen and Westphal (1972) have published photometric profiles of Jupiter's central meridian in the region of the CH_4 bands at 8920–8969 Å and in the neighboring continuum (8200–8400 Å) from observations made on 24 August 1970. Using these profiles, we have attempted to evaluate the latitude variation of the depth of the band. On the basis of spectrophotometric observations (Taylor 1965; Pilcher *et al.* 1973) of the depth of the band at the center of the disk, we found that R_1 is 0.86–0.90. Figure 1 illustrates the latitude dependence of depth, $R(\phi)$, that we calculated. The absorption increases slowly with latitude up to $\simeq 50°-55°$ to the north and south, but then decreases sharply at higher latitudes.

We turn now to the results of photoelectric measurements of the absorption bands on the Jovian disk. The advantage of these measurements lies in their linearity, which facilitates their reduction and statistical analysis. This allows better judgement as to the reality of temporal variations of the

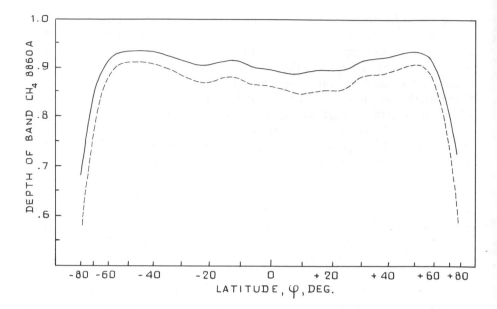

Fig. 1.   The latitudinal variations of the central depth of the CH$_4$ 8860 Å absorption band on Jupiter derived from the observations of Owen and Westphal (1972). The value of R_1 on the equator is assumed as 0.86 (dashed line) and 0.90 (solid line).

absorptions, although in this connection photoelectric observations near the edge of the disk are somewhat susceptible to errors due to poor seeing. Estimates of the continuum level, especially in the neighborhood of infrared absorption bands, contain a major source of uncertainty for photoelectric as well as photographic spectrophotometry.

The first observations of this kind were made by Münch and Younkin (1964) who scanned the disk of Jupiter in a series of fixed spectral regions. These observations indicated that the intensity of the weak 6190 Å CH$_4$ and 6450 Å NH$_3$ bands diminish, at a fixed latitude, towards the limbs. Absorption in the strong 8860 Å CH$_4$ band increased with latitude, except in the polar regions where a "bright cusp" of radiation reflected by the planet was detected. Narrowband photographs later revealed the cusp as a bright polar arc.

From several measurements of the equivalent width of the 6190 Å CH$_4$ band at the center of the disk and at distances $r_x \simeq 0.5$–0.7 from the center, Boyce (1968) concluded that the absorption *increases* towards the limbs. However, his results stand apart from all the others, which show the opposite effect. Bugaenko *et al.* (1972) found, from observations made in 1970, that absorption in the 6190 Å and 7250 Å CH$_4$ bands decreased near the western and eastern limbs, compared with the center of the disk. At middle latitudes ($\sim 50°$), the 6190 Å band shows a slight decrease, but the 7250 Å band displays a small increase in strength. Differences in absorption over the bright zones and dark belts of Jupiter are lost in the observational errors. Pilcher *et al.* (1973) also found no significant variation of absorption in spectral scans of four belts on the Jovian disk, although they found differences in W amounting to 4 Å in the 6190 Å band for the Equatorial Zone and South Equatorial Belt (there were only two measurements in each).

The equivalent widths and depths of the 6190 Å CH_4 band measured at several points along the Jovian central meridian in April 1968 (Teifel 1968) show only a small latitude variation, lying almost within the estimated error limits, although a slight increase of absorption is evident at high latitudes (Fig. 2). Observations of the 7250 Å band made in 1970 (Teifel 1971) reveal a more obvious increase in absorption at latitudes between 40° and 60° (Fig. 3). The latitude variation of the absorption in this band along the central meridian is similar to that of the 8860 Å CH_4 band (Owen and Westphal 1972), as illustrated in Fig. 1. The increase of absorption in the 7250 Å band with latitude was clearly seen in the observations made by Avery *et al.* (1974), which also showed decreases in the strength of absorption in both the 6190 Å and 7250 Å bands towards the limb in the equatorial belt. Figure 4 illustrates our estimate of the most probable range of variations in the depths of the 6190 Å and 7250 Å CH_4 bands in Jupiter's equatorial belt, from all references cited.

According to the observations made by Avery *et al.* (1974), absorption in the 6450 Å NH_3 band decreases more sharply from the center to the limbs than in the CH_4 bands, as Avramchuk (1970) also noted. However, this band which consists of widely-spaced discrete lines, is difficult to observe and to interpret. First, the conditions under which the NH_3 absorptions are formed are different from those for CH_4 since NH_3 condenses in the visible part of Jupiter's atmosphere. The NH_3 partial pressure in the upper atmosphere is therefore limited by the NH_3 vapor-pressure curve, which drops sharply with decreasing temperature. Secondly, by contrast to the closely-packed lines in the central regions of the CH_4 bands which can be treated as a single absorption, the discrete line structure of the NH_3 band requires computation of a statistical band model. It is desirable to study the variations across the Jovian disk of each line in the NH_3 band separately, but the modest spectroscopic resolution of the existing data does not permit this.

Binder and McCarthy (1973) observed large variations in the intensities of the CH_4 and NH_3 bands at 1.3–1.6 μm, similar to the results described above for the 7250 Å CH_4 band. Although some differences in specific regions along the central meridian are observed in these bands, it would be risky to attach much significance to them without an analysis of possible errors. Absorption in both bands shows a general tendency to increase with latitude. Observations of strengthening of the 1.53 μm NH_3 band towards the poles and weakening towards the equatorial limbs were described earlier by Moroz and Cruikshank (1969) and Binder (1972).

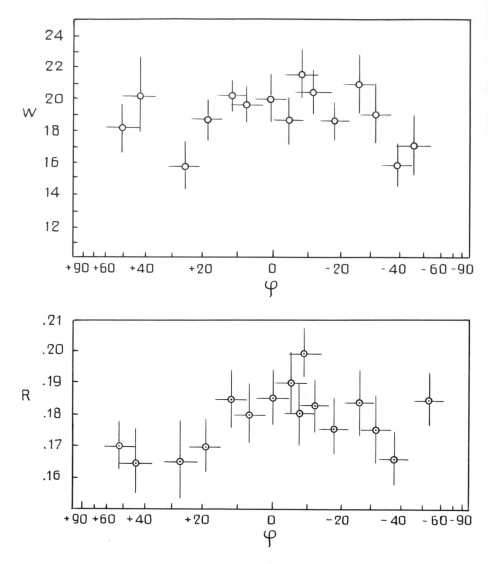

Fig. 2. The latitudinal variations of the CH_4 6190 Å absorption band equivalent width W and central depth R on Jupiter from the photoelectric measurements in 1968.

As we have already noted, the measured intensities of the absorption bands along the limb are subject to large errors since the brightness gradient (limb darkening) is maximum in red light and hence small oscillations of the image due to seeing cause significant noise in photoelectric tracings of band contours. It is therefore preferable in most cases to make measurements close to the center of the disk and encompassing the entire equatorial region, or other regions near the planet's equator. When the spectrometer slit is placed

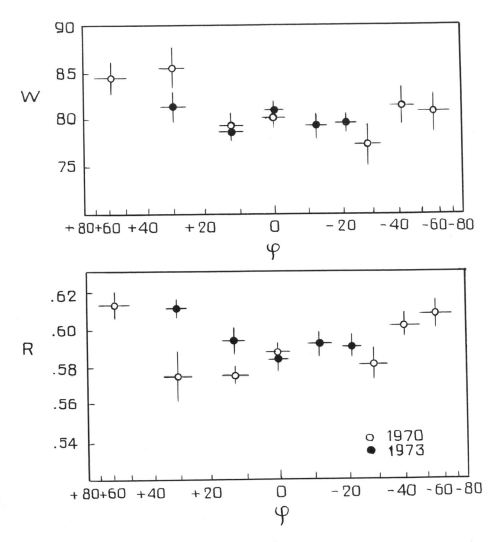

Fig. 3. The latitudinal variations of the CH_4 7250 Å absorption band equivalent width W and central depth R on Jupiter from photoelectric measurements in 1970 (open circles) and 1973 (filled circles).

along the equator the depths of the band profiles (or the residual intensities) are far more accurate than those determined for separate regions near the limb. We shall demonstrate below that a comparison of the depths of the CH_4 absorptions at the center of the disk, R_1, and for the entire equator, R_e, establishes the inapplicability of the simple reflecting-layer model, and permits as precise an evaluation of the optical thickness, τ, of the atmosphere above the visible cloud tops as do measurements made near the limb. We

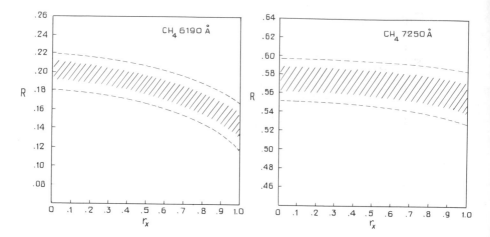

Fig. 4. The areas of the most probable center-to-limb variations of the central depths of CH$_4$ 6190 and 7250 Å absorption bands for the equatorial belt of Jupiter.

refer here to results of photoelectric spectrophotometric observations made by the author in 1968–1974 (Fig. 5).

We now examine measurements of individual absorption lines. The most intense and best resolved lines are observed in the $3\nu_3$ band of CH$_4$ at 1.1 μm. Studies of this band have comprised several papers, although there have been few studies of the center-to-limb variations of the line intensities. We shall therefore discuss only two papers here. Walker and Hayes (1967) measured equivalent widths of lines in the $3\nu_3$ band at the center of the disk and at a distance $r_x \simeq 0.8$ from the center. They concluded that there was no detectable center-to-limb variation in line intensity. Detailed studies of the center-to-limb variations of five lines of the same band were reported by Bergstralh (1973) from spectrograms made with the entrance slit placed along several chords across the disk, oriented parallel to the equator and to the central meridian. Measurements of W show no systematic variation in the longitudinal direction larger than the estimated errors of measurement. Lines $R(3)$ and $R(5)$ have a slight tendency to increase towards the limbs, but this result is not certain. No large variation of W from pole to pole is apparent either, but the data on the latitude variation of line intensity are inconclusive.

Variations in the very narrow H$_2$ quadrupole lines, in the (4, 0) band near 6735 Å and in the (3, 0) band near 8150 Å, are difficult to study by conventional spectroscopic techniques. However, by using an interferometric technique, Carleton and Traub (1974) have obtained very precise photoelectric measurements of the profiles of the S(1) lines in both bands over several regions on the Jovian disk. They noted that, along with a temporal fluctuation in the equivalent widths of these lines, there is a tendency for W to decrease by roughly 10% from the center to the east and west limbs, and to increase slightly ($< 10\%$) in the mid-latitude regions. In other words, the H$_2$ quad-

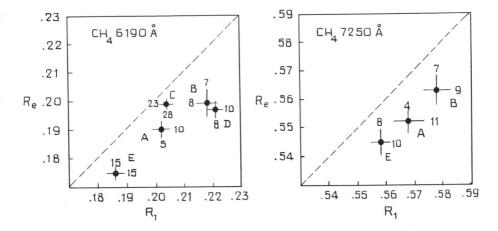

Fig. 5. A comparison of the central depths of the CH_4 6190 and 7250 Å absorption bands from measurements for the center of the Jovian disk (R_1) and for the entire equator (R_e). The numbers of the measured band profiles are written near each point. The observations were made in 1968 (A,B), 1971 (C), 1973 (D) and 1974 (E).

rupole lines appear to display qualitatively similar behavior to that noted for the CH_4 absorptions.

Hunt and Bergstralh (1974) made many spectroscopic observations of the H_2 quadrupole lines over the central regions of the Jovian disk during 1972. They found considerable fluctuation in the equivalent widths: 4 to 15 mÅ for the (4, 0) S(1) and 34 to 57 mÅ for (3, 0) S(1). It is difficult to say just how significant these fluctuations were, since the authors presented no analysis of possible sources of error in their data. They simply attributed their results to variations in the effective level of line formation at various latitudes.

With these, we exhaust the observational data available at the present time on the distribution of molecular absorption across the disk of Jupiter. In spite of some quantitative differences in the values of W and R associated with different observational techniques and precision of measurement and also some possible real changes in the atmosphere of Jupiter, the general character of spatial variation of absorption is reasonably well determined:

1. In the equatorial band of Jupiter, molecular absorption decreases toward the limb by 10–20%.
2. To the north and south of the equator absorption in weak bands decreases with latitude; in medium and strong bands absorption increases in the latitude range 40–60°, but not more than 5–7%, sharply decreasing close to the poles.
3. Differences in absorption between the dark and light cloud bands of Jupiter, according to measurements by various authors, are contra-

dictory in character but they are most likely absent or occur at the level of the errors of measurement.

4. Time variations of absorption in the central region of the disk of Jupiter are noted by the majority of investigators; however, in the absence of systematic observations and control, it is impossible to determine exactly how much of the variation is real and how much can be attributed to methodological error.

In June and July 1971 we made photoelectric observations in the CH_4 6190 Å band along the equatorial band of Jupiter (Egorov *et al.* 1971) consisting of 878 measurements on 12 nights. A statistical analysis showed that the influence of systematic effects, namely possible time-variable absorption, was not more than ± 3%. An analysis of other observations further indicates that time variations are probably less than 10%.

Studies of the CH_4 6190 Å band and NH_3 6450 Å band in spectra of the Great Red Spot (GRS) do not disclose any significant differences in absorption compared with the South Tropical Zone (STrZ) (Teifel 1969*b*, 1970). Photoelectric measurements by Bugaenko *et al.* (1972) agree with this result. On the other hand, the GRS appears brighter than other planetary details in narrowband photographs centered on the strong 8860 Å CH_4 band, which indicates a decrease in CH_4 absorption above the Spot. We shall return to a detailed discussion of this point in Sec. IV.

II. SPECTRAL REFLECTIVITY OF
THE CLOUD DECK OF JUPITER

Theoretical calculations show that the intensity of molecular absorption bands and their change from the center to the limb are not very sensitive to such characteristics of the aerosol medium as particle albedo and the scattering phase function. Therefore it is pointless to use observations of absorption bands to determine these parameters. However, in order to interpret the absorption data and to achieve some conformity with the theoretical calculations, it is necessary to assign approximate values to the asymmetric phase function and the coefficient of diffuse reflection in the continuous spectrum.

Unfortunately if we disregard the rather large number of determinations of the stellar magnitudes and geometric albedo of Jupiter, there are very few measurements of the absolute spectral reflectivity of separate regions of the cloud deck of the planet. In the papers by Pilcher and McCord (1971) and Younkin and Münch (1963) we find only the relative spectral reflectivity for a few regions of Jupiter's disk. Absolute photometric data (Litkevich and Khodyachikh 1973) and spectrophotometry (Teifel 1971; Bugaenko 1972; Pilcher *et al.* 1973) have been published recently. While these data are in satisfactory agreement in the relative sense of $r_\lambda / r_{\lambda_0}$, the absolute values show some systematic differences. Thus the coefficient of diffuse reflection,

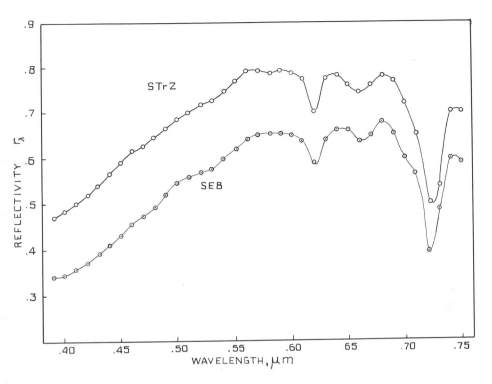

Fig. 6. The absolute spectral reflectivity $r(\lambda)$ for STrZ and SEB obtained from photoelectric measurements in 1973.

r_λ, measured for the center of the Jovian disk in April 1971 by Pilcher et al. (1973) is, on the average, 15% higher than that measured in April 1969 by Bugaenko (1972). These differences can be attributed only partially to local fluctuations of albedo, while the majority of the disagreement is most likely connected with insufficient knowledge of the absolute energy distribution of the stellar standards.

Figure 6 illustrates r_λ for the STrZ and the South Equatorial Belt (SEB) of Jupiter, as derived from spectrophotometric measurements which we made in September 1973. The values of r_λ were obtained by correcting the spectral measurements for atmospheric dispersion from photometric profiles of Jupiter's central meridian, and fitting the maximum and minimum reflectivities of the planet's disk during this period.

From all absolute and relative measurements of the reflectivity of the Jovian clouds, there appears to be a maximum in the region 0.57–0.68 μm and a significant decrease towards shorter wavelengths, with a minimum in the region 0.32–0.36 μm. Further in the ultraviolet ($\lambda < 0.3 \mu$m), according to observations made above the earth's atmosphere (Wallace et al. 1972), there is an increase in the reflection coefficient and the geometric albedo owing to Rayleigh scattering in the atmosphere above the visible cloud top. At $\lambda > 0.7$ μm the albedo of the cloud material cannot be determined reliably because of the very strong CH_4 absorption bands, the wings of which overlap and affect the spectrum even between the bands.

The decrease of r_λ at shorter wavelengths is more marked in the dark clouds, a fact which is manifested in greater contrast on Jupiter's disk in violet light than in red. This does not, however, depend on the cloud structure; relatively bluish details are observed on Jupiter, most often in the form of small projections on the edges of the dark equatorial belt. These details are seen in red light, but they vanish in the blue and violet. We may have information here on the stratification of the cloud cover: above the highly turbulent and relatively bluish cloud band there is a yellowish haze or cloud veil, which absorbs strongly at short wavelengths and has sharp boundaries in latitude.

Figure 7a illustrates the spectral variation of the relative intensity of dark and bright materials (D/B) from photographic and photoelectric photometric data. In Fig. 7b, the variation with wavelength of cloud reflectivity is shown. It is interesting to note that as a rule we never find a case when the maximum value of r_λ in red light corresponds to the minimum value in violet light within the limits of our spatial resolution on the planet. The only exception is the GRS, which in red light blends in with surrounding background while in violet light there is great contrast against the background (Teifel 1969; Reese 1971). Therefore, variations of the integrated brightness and geometric albedo of Jupiter are connected not only with changes in the relative brightness and relative areas of bright and dark material on the disk but also with changes of the planet-wide level of reflectivity. According to Harris' (1961) measurements of the integrated brightness of Jupiter in visible light, the greatest amplitude of the variations is $\Delta V(1, 0) = 0.45$ mag, which corresponds to a change in visual albedo, A_g, by a factor of 1.5. For a planetary surface with homogeneous optical properties

$$A_g = \frac{1 + 2\beta}{3} r_0 \tag{1}$$

where

$$\beta = \frac{r(\mu = 0.5)}{r(\mu = 1.0)} \tag{2}$$

is the limb darkening factor and r_0 is the coefficient of diffuse reflection at the center of the disk. It can be demonstrated that for a semi-infinite, diffusely reflecting atmosphere, a decrease of r_0 due to an increase in real absorption in the aerosol is accompanied by an increase of β. According to Eq. (1), a change in A_g by a factor of 1.5 must therefore correspond to a larger change in r_0, and in the same direction. In the case of the Henyey-Greenstein phase function:

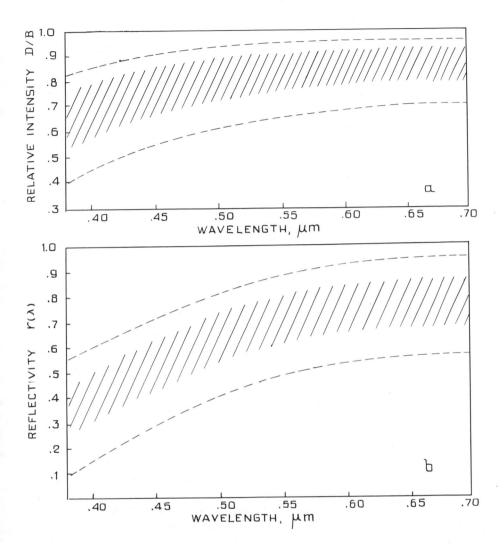

Fig. 7.   The areas of the most probable variations of contrast (or relative intensity) of dark and
bright clouds and of the spectral absolute reflectivity on Jupiter.

$$\chi(\gamma) = \frac{1 - g^2}{(1 + g^2 - 2g \cos \gamma)^{\frac{3}{2}}} \tag{3}$$

the dependence between A_g and r_0, according to the calculations of Dlugach
and Yanovitskii (1974), takes the form shown in Fig. 8. A change in A_g by a
factor of 1.5 corresponds to a change in r_0 by a factor of 1.65. Suppose that
the visible Jovian surface consisted entirely of dark "belt material" at the

time of one albedo measurement, and entirely of bright "zone material" at the time of another measurement. The contrast of dark to bright material would then be $D/B = 1.65^{-1} = 0.6$. However, we know from observation that Jupiter's visible surface is never uniformly bright or dark; i.e., there is always sensible contrast between belts and zones. In other words, the fraction of the visible surface represented by, say, bright material is never 0 or 1. Suppose then that the albedo of the bright material was always approximately 1, and that the only variables were the albedo of the dark cloud material and the fraction of the visible surface represented by dark material. In that case, higher contrast $(D/B < 0.5)$ would be required to explain the observed changes of Jupiter's visual geometric albedo. By inspection of Fig. 7a, however, such values of D/B are not found on Jupiter at visible wavelengths (0.55 μm). It follows therefore that there was global variation in the reflectivity of the cloud cover in addition to changes in the relative area and contrast of the dark belts and bright zones.

III. LIMB DARKENING, THE PHASE FUNCTION, AND SINGLE-SCATTERING ALBEDO

The determination of the scattering phase function of the cloudy atmosphere is the most complex problem of planetary photometry and polarimetry.[1] It is particularly difficult from Earth-based observations which, for the outer planets, cover only a very narrow range of solar phase angles. Since the components of these atmospheres are not singly-scattering, the problem becomes still more complex. One must determine the parameters of the aerosol particle-size distribution, the choice of which (Junge distribution, normal exponential law, gamma dispersion, etc.) is rather subjective. An estimate of the refractive index of the particles is also necessary. The particles are usually assumed spherical which, strictly speaking, applies only to liquid droplets. The observed photometric properties of a planet (albedo, brightness distribution on the disk, degree and angle of polarization, absorption band profiles) represent some unknown combinations of these parameters. So far it has been impossible in practice to derive the parameters of the scattering aerosol particles by inversion of the observed photometric properties. The only practical approach to the problem has been to carry out rather vast theoretical computations of observed characteristics on the basis of solutions of the equation of transfer for different combinations of atmospheric parameters. The fact that results obtained from independent methods are non-contradictory serves as a good control. It is obvious from comparison of given observations with results of the solutions, however, that we have only a general idea of the probable values of the parameters within the framework of models of the aerosol and gaseous atmospheric components. Unfortunately we are restricted so far to comparatively mixed data obtained

[1]See also pp. 491 and 518.

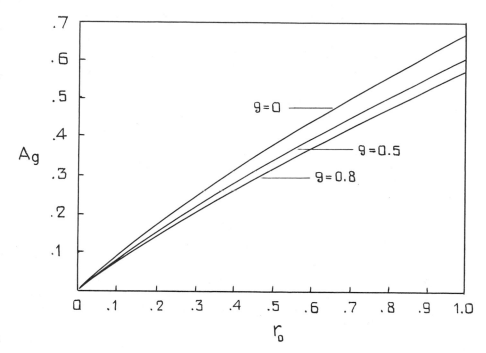

Fig. 8.   The dependence between the geometric albedo A_g and the reflectivity r_0 for the planet covered by the optically infinite aerosol layer with the Henyey-Greenstein phase function characterized by asymmetry parameter g.

at different times, which therefore correspond to different states of the Jovian atmosphere.

From photometric observations we obtain a simultaneous determination of the diffuse reflection coefficient, r_0, and the relative distribution of brightness across the planetary disk, $B(\mu)$, where μ is the cosine of the angle of reflection. The scattering phase function and single-scattering albedo, ω, can then be evaluated from among a selection of possible candidate phase functions [e.g., isotropic; Euler's function, $\chi(\gamma) = 1 + x_1 \cos \gamma$; the more general expansion in terms of Legendre polynomials, $\chi(\gamma) = 1 + x_1 P_1(\cos \gamma) + x_2 P_2(\cos \gamma)$; the Henyey-Greenstein function, and others].

Binder (1972) noted that the Minnaert function

$$B(\mu) = B_0 \mu_0^k \mu^{k-1} \qquad (4)$$

adequately describes the observed brightness variation across Jupiter's disk. As is conventional, μ_0 is the cosine of the incidence angle and μ is the cosine of the emergence angle of diffusely reflected radiation, while k is an empirically determined limb-darkening parameter. Equation (4) can also be written in linear form as

$$\log (B\mu) = \log B_0 + k \log (\mu\mu_0), \qquad (5)$$

from which k is readily determined. Clearly the Minnaert function provides a convenient single-parameter description of the brightness variation.

Loskutov (1971) and Dlugach and Yanovitskii (1974) have computed theoretical brightness distributions at 0° phase angle for a planet with a homogeneous, semi-infinite atmosphere for the selection of scattering phase functions listed above. We have compared these theoretical variations with the empirical Minnaert function [Eq. (4)]. It appears that for all tabulated values of the asymmetry parameter in the phase functions and for any value of the single-scattering albedo, ω, Eq. (4) provides a good approximation to the brightness variation for $\mu > 0.3$, i.e., for all well-measured regions of the planet's disk (Teifel 1975a). The observed brightness variation in both the visual and near-infrared spectral regions satisfies Eq. (4) at all solar phase angles.

The limb-darkening coefficient, k, constrains one class of phase functions through a well-defined empirical relationship with the reflection coefficient, r_0, for given values of the asymmetry parameter, g, and single-scattering albedo, ω. Figure 9 illustrates the dependence of k on r_0 for different values of g and ω for the Henyey-Greenstein phase function. Inspection of this figure indicates the precision to which k and r_0 must be determined to derive g and ω with sufficient accuracy.

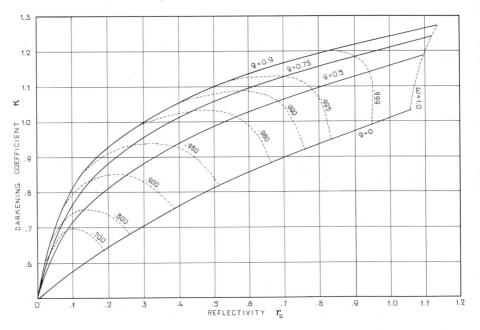

Fig. 9.   The dependence between the darkening coefficient k and reflectivity r_0 for a planet with an optically infinite cloudy atmosphere characterized by single scattering albedo ω and Henyey-Greenstein phase function at values of the asymmetry parameter g.

We evaluated the limb-darkening coefficient, k, in a series of spectral regions from photographs of Jupiter made in 1962 at phase angles between $2°1$ and $9°4$, and from spectra obtained in 1964. The 1962 photographs were especially interesting because in that period the equatorial band of Jupiter appeared considerably filled with dark material. For comparison with the theoretical relationship for $k(r_0)$, we used approximate values of r_0 determined from isophotometric tracings of the photographs and data on the absolute reflectivity of the entire disk of Jupiter in 1962 taken from Taylor's (1965) spectrophotometric studies. The results are illustrated in Fig. 10a.

Spectrograms of the Jovian equatorial belt were obtained in good seeing with exposures of 6 to 20 sec. Photometric profiles were made crosswise to the dispersion in relatively narrow spectral intervals, $\Delta\lambda$, from 32 Å at 0.388 μm to 200 Å at 0.596 μm. The reflection coefficients were calculated from absolute spectrophotometric measurements made by Bugaenko (1972) in 1969 and by the author in 1970 and 1973, with allowances for the differences in the relative brightness of the equatorial belt in those years. Figure 10b shows that for $\lambda > 0.5$ μm the observed and theoretical values of $k(r_0)$ agree for $g \simeq 0.45$. For wavelengths of 0.53 μm and 0.625 μm the 1962 observations yielded values of k and r_0 close to the theoretical curves for $g \simeq 0.55 - 0.65$, i.e., for moderate asymmetry of the scattering phase function. From photometric observations made in 1971 Binder and McCarthy (1973) found a value of 1 (± 0.02 to 0.05) for the limb-darkening coefficient, k, at wavelengths 0.59, 0.64 and 0.72 μm, from which we derive $g \simeq 0.35 \pm 0.05$ in red and orange light.

According to Tomasko et al. (1974), the distribution of relative intensity along the Jovian equator in red light, from Pioneer 10 measurements at phase angles from 34° to 109°, satisfactorily fits calculations for $g \simeq 0.5$.[2] Although the best agreement comes from a combination of the Henyey-Greenstein function for the cloud layer with some small degree of continuous absorption in the atmosphere above the cloud, one cannot consider this an argument in favor of the reality of absorption in the continuum. Most likely the real phase function of the aerosol differs slightly from the Henyey-Greenstein function. Nonetheless it is interesting to note that the value of the asymmetry parameter is remarkably different from values which correspond to strong elongation of the phase function in the forward direction. Therefore, both the groundbased and spacecraft photometric data indicate that the particles in the Jovian cloud deck are not very large compared to the wavelength of yellow and red light.

The best agreement of theoretical calculations of the degree of polarization of the Jovian clouds with the polarimetric measurements made by Morozhenko and Yanovitskii (1973) is obtained with a mean geometric particle

---

[2]See, however, p. 504.

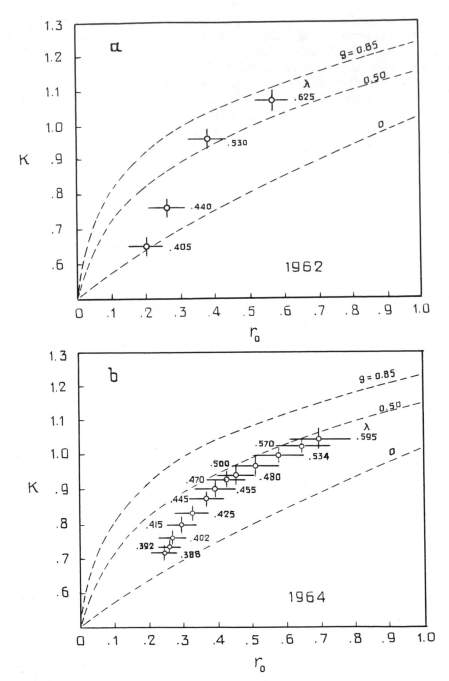

Fig. 10. The observed dependence of the darkening coefficient k on reflectivity r_0 which were measured at various wavelengths for the equatorial belt of Jupiter in 1962 (*a*) and 1964 (*b*). Dashed lines are the theoretical curves at values of the asymmetry parameter g for the Henyey-Greenstein phase function.

radius $a_0 \simeq 0.2$ μm, refractive index $n = 1.36 \pm 0.01$, and the parameter of the normal-logarithmic law of particle-size distribution $\sigma^2 \simeq 0.3$. At 0.59 μm wavelength the polydispersing phase function calculated with these parameters has the asymmetry coefficient

$$g = <\cos\gamma> = \frac{1}{2} \int_{-1}^{+1} \chi(\gamma) \cos\gamma \, d(\cos\gamma) \simeq 0.76 \,. \tag{6}$$

Loskutov (1971a,b) found similar results. Polarization measurements therefore indicate a more forward-elongated phase function but not too large mean radius of the Jovian cloud particles. The discrepancy should not discourage us, however, since the interpretation of the polarimetric measurements depends on several assumptions. Furthermore, as we have already noted, the possibility of real temporal variability in the optical properties of the cloud particles cannot be excluded owing to the lack of simultaneous polarimetric and photometric observations.

Along with the uncertainty in the value of the asymmetry parameter, g, an uncertainty in the single-scattering albedo, ω_c, also remains. It is well established that scattering by the aerosol component in Jupiter's atmosphere is significantly non-isotropic. For the various observed reflection coefficients, r_0, ω_c must be larger than the corresponding value for isotropic scattering. It is most likely that ω_c in red light for Jupiter lies between 0.993 and 0.997. However, these limits, which are related to variations in the visual albedo of the clouds, might be significantly wider with the result that greater certainty in the value of ω_c can be obtained only when both r_0 and k can be determined simultaneously. The value of ω_c is still less reliably known at shorter wavelengths, especially at $\lambda < 0.5$ μm, where Rayleigh scattering in the atmosphere above the cloud tops becomes significant.

IV. RAYLEIGH SCATTERING AND THE OPTICAL THICKNESS OF THE ATMOSPHERE ABOVE THE CLOUDS

As Fig. 10 illustrates the observed behavior of $k(r_0)$ at short wavelengths does not follow the theoretical calculations for any single value of the asymmetry parameter, g. Instead $k(r_0)$ deviates towards decreasing degrees of forward extension of the scattering phase function with decreasing wavelength. This effect cannot be associated with the properties of the aerosol particles since, for a given dispersion of particle sizes, the parameter g would either increase quickly with decreasing wavelength or would remain almost constant. The reason for the deviation must be sought in Rayleigh scattering, which can decrease the asymmetry of the phase function if the gas and aerosol form a homogeneous mixture, or decrease the limb darkening by superposing the brightness of a pure gaseous atmosphere above the cloud surface.

In the first case, we can evaluate the relative contribution of Rayleigh scattering

$$\delta = \frac{\sigma_R}{\sigma_R + \sigma_a} \tag{7}$$

where σ_R and σ_a are the Rayleigh and aerosol volume scattering coefficients. Yanovitskii has kindly made available his computations based on the phase function obtained from polarimetric observations of Jupiter. From his values we find that the observed limb darkening and reflectivity may be obtained at $\delta \leqslant 0.05$ between 0.40 μm and 0.45 μm, and that we can reproduce the observed limb darkening. This gives us a rough estimate for the lower limit of the aerosol volume scattering coefficient if we assume a pressure $P \simeq 0.5$ atm at the effective level at which the short wavelength radiation is reflected. At this pressure, $\sigma_R \simeq 4.6 \times 10^{-8}$ cm^{-1} and for $\delta = 0.05$ we have

$$\sigma_a = (\frac{1}{\delta} - 1)\, \sigma_R \simeq 20\sigma_R \simeq 9 \times 10^{-7}\ \text{cm}^{-1}. \tag{8}$$

In the second case, we suppose that Rayleigh scattering is significant only in the atmosphere above the cloud top. The visual coefficient of reflection, $r(\mu)$, is then related to the characteristics of the underlying cloud surface, r_{a1} and k_1, and to the optical thickness of the Rayleigh-scattering atmosphere, τ_R, by the following approximate expression (assuming $\mu = \mu_0$):

$$r(\mu) = r_{a1}\mu^{2k_1 - 1}\, e^{-2\tau_R/\mu} + r_R(\mu, \tau_R) + \Delta r \tag{9}$$

where

$$\Delta r = \left[\frac{A_s}{1 - A_s C(\tau_R)}\, M^2(\mu) - A_s\, e^{-2\tau_R/\mu} \right] \mu. \tag{10}$$

Here A_s is the spherical albedo of the underlying surface, $C(\tau_R)$ is the spherical albedo of the planet with the surrounding Rayleigh atmosphere (supposing the underlying surface has $A = 0$), $M(\mu)$ is an auxiliary function, and r_R is the coefficient of reflection of the Rayleigh atmosphere for $A = 0$ (taken from the tables of Coulson *et al.* 1960). The quantity Δr is calculated from the tables of Morozhenko and Yanovitskii (1964). From Eqs. (9) and (10), we have calculated a series of graphs which show the dependence between the limb darkening factor [see Eq. (2)] and the brightness coefficient, r_1 ($\mu = 1$), for different values of r_{a1}, k_1, and τ_R. Figure 11 illustrates such a series of curves for a cloud deck phase function with $g = 0.5$. Using this graph with an observed value of r_1 ($\mu = 1$) and β, as arguments, it is easy to find the quantities τ_R and r_{a1}. At 0.40 μm we find from our 1964 measurements that $\tau_R \simeq 0.08 \pm 0.03$ which corresponds to an equivalent H$_2$ abundance $U(\text{H}_2) = 9 \pm 3$ km amagat in the atmosphere above the clouds. Krugov (1972) found a similar result from measurements of photographs at wavelengths of 0.36 μm and 0.39 μm. It must be noted that analogous computa-

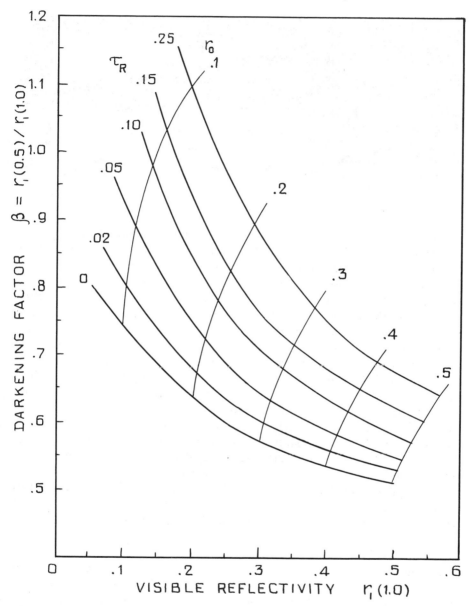

Fig. 11. The dependence between the darkening factor β and the visible reflectivity r_1 for a planet with outer Rayleigh atmosphere and cloud surface which has reflectivity r_0 and asymmetry parameter for phase function $g = 0.5$. τ_R is the optical thickness of the outer atmosphere.

tions with phase functions obtained by Morozhenko and Yanovitskii (1973) lead to still smaller values of τ_R and $U(H_2)$, although the same authors derived from polarization data models with a Rayleigh atmosphere above the

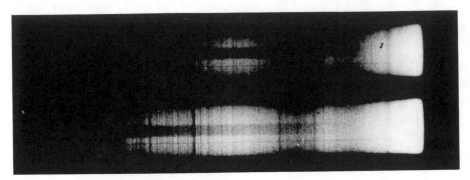

Fig. 12. Spectrograms of the South Tropical Zone with the Great Red Spot obtained in 1964. The thin line under the GRS spectrum is caused by a slit defect.

clouds of thickness $\tau_R \simeq 0.1$ at 0.375 μm, which corresponds to $\tau_R \simeq 0.08$ at 0.40 μm, and hence $U(H_2) \simeq 8.5$ km amagat. A larger value of $U(H_2)$ can be obtained only if the correction for Δr is not taken into account in Eq. (9). Then $\tau_R \simeq 0.12 \pm 0.03$ at 0.40 μm and $U(H_2) \simeq 13 \pm 3$ km amagat. Thus the limb darkening measurements imply that the abundance of molecular hydrogen in the atmosphere above the clouds cannot exceed 10 km amagat, in contrast to the results obtained from measurements of the H_2 quadrupole absorption lines, as interpreted with the simple reflecting-layer model.

Polarimetric measurements from Pioneer 10 (Coffeen 1974) give $\tau_R \simeq$ 0.05–0.12 at 0.44 μm wavelength in a model which supposes a Lambert reflection law for the underlying cloud surface. The large amplitude of the fluctuations in τ_R (which are directly correlated with fluctuations in the albedo of the underlying cloud surface) raises some doubts; Coffeen emphasizes that these preliminary results are strongly model-dependent, however. Nonetheless it is interesting that the range of heights thus derived for the upper boundary of the clouds is close to the range of variation with latitude as derived from the CH$_4$ 8860 Å band. On the other hand, Owen and Sagan (1972) derived $U(H_2) \simeq 5$ km amagat from data obtained with the ultraviolet spectrometer on the Orbiting Astronomical Observatory's OAO-2.

The dependence of the limb darkening coefficient k on wavelength was also studied for the GRS from measurements of its relative intensity at different distances from the central meridian. For this study the mean of 40 spectrograms obtained by Teifel (1967c) on 13–14 December 1964 were reduced according to the Minnaert function [Eq. (4)]. The quality of the spectrograms can be evaluated from the reproduction of one of them in Fig. 12. In 5 regions of the spectra and in the limits of the spectral intervals $\Delta\lambda$, the limb darkening coefficient of the STrZ (k_z), and the quantity $k_s - k_z$ were found, from which it was possible to determine k_s for the GRS, its relative intensity at $\mu = \mu_0 = 1$, and the reflection coefficient r_{s_0} by comparison with r_{z_0} for the STrZ. All of these data are presented in Fig. 13 which shows that $k_s - k_z$ is always less than zero; i.e., the limb darkening of the GRS is less

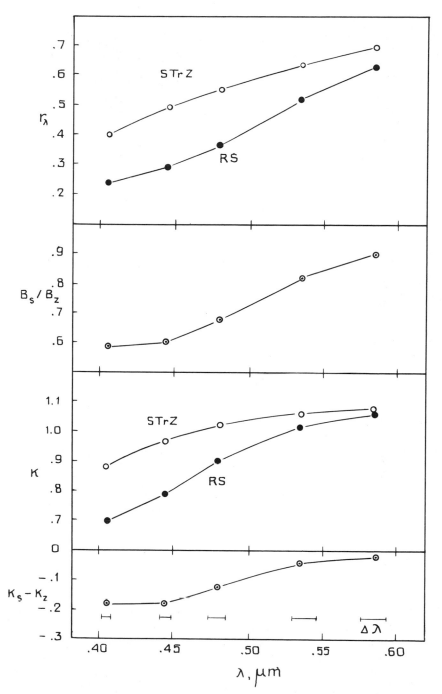

Fig. 13. Some characteristics of STrZ and the GRS derived from spectral measurements in 1964 (see text).

than that in neighboring regions, as it must be since true absorption in the Spot increases toward short wavelengths faster than it does for other features on Jupiter. The dependence of $k_s(r_{s_0})$ given by computations gives an analogous result; τ_R is greater over the GRS than over its surroundings. A similar result is also obtained from photometric measurements of the relative intensity of the GRS when it is near the limb, as has been determined from a series of 62 photographs of Jupiter in blue light (21–22 September 1963). Of course the absolute value of τ_R is not known with sufficient reliability for many cases, but qualitatively these results may be considered as evidence that the surface of the GRS as seen in visible light does not rise above the surrounding cloud layer.[3]

As an alternative, we might speculate that the scattering phase function in the GRS differs from that of the surrounding cloud surface, but this does not solve the problem. We need measurements of the contrast of the GRS in ultraviolet radiation in different positions on the Jovian disk. So far, measurements similar to those described above have only been obtained in the visible part of the spectrum (Kartashov 1972).

V. MOLECULAR ABSORPTION IN THE METHANE BANDS AND THE VOLUME DENSITY OF THE AEROSOL COMPONENT

The discussion and interpretation of existing data regarding molecular absorption on Jupiter's disk has been the subject of several papers in recent years. For the most part, they begin from the theoretical problem of line formation in homogeneous cloud layers and proceed to analytical expressions for computing line equivalent widths at various points on the disk (e.g., Chamberlain 1965, 1970; Lenoble 1968; McElroy 1969; Pilcher *et al.* 1973), or give computational results obtained from homogeneous or multilayer models with isotropic or simple nonspherical phase functions for the scattering aerosol (Danielson and Tomasko 1969; Hunt 1973; Avery *et al.* 1974; Bergstralh 1973; Teifel 1969*a,b,c*; Teifel and Usoltzeva 1973; Michalsky *et al.* 1974). For weak lines and absorption bands in two-layer models, analytical expressions were derived by Sobolev (1972, 1973) and Rozenberg (1962), and used by Loskutov (1974), Teifel (1971), Anikonov (1974), and others.

As with other photometric and spectral studies, the primary step in interpreting the data consists of computing a series of variants of the basic model, following a selection of more or less reasonable estimates of the structure and optical properties of the medium in which the absorption occurs, and then comparing the observations with the computations. Without taking time for a complete exposition, we present here a rather simple and straightforward method for calculating absorption for a model in which lines and bands

---

[3]See, however, p. 534.

are formed within a homogeneous, semi-infinite cloud layer as well as in a pure gaseous layer overlying the clouds. This method incorporates existing computations of the brightness distribution across the planetary disk with the Henyey-Greenstein phase function, but it could also be applied to other phase functions for which diffuse reflection coefficients and coefficients of limb darkening have been computed.

Molecular absorption in a cloud with multiple scattering results in a lowering of the visual albedo (i.e., the coefficient of diffuse reflection) of the cloud layer at the wavelengths of the absorption bands because the single scattering albedo at these wavelengths, ω_ν, decreases as compared with the single scattering albedo in the continuum, ω_c:

$$\omega_\nu = \frac{\sigma_a}{\sigma_a + \kappa_a + \kappa_\nu} = (\omega_c^{-1} + \beta_\nu)^{-1} \tag{11}$$

where

$$\beta_\nu = \frac{\kappa_\nu}{\sigma_a} \tag{12}$$

and κ_ν is the volume coefficient of true molecular absorption.

Calculation of the central depths of absorption lines or bands formed in a scattering medium at various points on the disk

$$R'(\mu, \mu_0) - 1 - \frac{r_\nu(\omega_\nu, g, \mu, \mu_0)}{r_c(\omega_c, g, \mu, \mu_0)} \tag{13}$$

can be greatly simplified if we recall that the change in brightness across a planetary disk in a semi-infinite atmosphere is well represented by the Minnaert function [Eq. (4)]. From this formula it follows that

$$1 - R'(\mu, \mu_0) = \frac{r_{\nu 0}}{r_{c0}} (\mu\mu_0)^{k_1 - k_c} = (1 - R'_1)(\mu\mu_0)^{k_1 - k_c} \tag{14}$$

or when $\mu = \mu_0$,

$$1 - R'_\mu = (1 - R'_1)\mu^{2(k_1 - k_c)} \tag{15}$$

where k_1 and k_c are the coefficients of limb darkening in the center of the absorption band and in the continuous spectrum, respectively, and R'_1 is the central depth of absorption at $\mu = \mu_0 = 1$. The quantities $1 - R'_1$ and $2(k_1 - k_c)$ can be derived empirically as functions of $\log \beta_\nu$ (Fig. 14) for given g and r_c from the theoretical limb darkening computations for various phase functions (e.g., Loskutov 1971a,b; Dlugach and Yanovitskii 1974). These empirical relationships can be used to compute profiles of Lorentz lines formed entirely within the cloud layer. At any point in the line

$$\beta_\nu = \frac{\beta_{\nu 0}}{1 + \left(\dfrac{\nu - \nu_0}{\alpha_0 P_e}\right)^2} \tag{16}$$

or

$$\log \beta_\nu = \log \beta_{\nu_0} - \log (1 + a^2) \,, \qquad (17)$$

where

$$a = \frac{\nu - \nu_0}{\alpha_0 P_e} \qquad (18)$$

and where α_0 is the halfwidth of the line at standard temperature and pressure (NTP) and P_e is the effective pressure. Figure 14 is entered with $\log \beta_\nu$ as the argument to obtain $(1 - R'_1)_\nu$ and $2(k_1 - k_c)_\nu$, from which $(1 - R'_{\mu,\nu})$ is then computed [cf. Eq. (15)]. Figure 15 compares profiles computed this way for $R_1 = 0.5$, at $\mu = 1.0$ and 0.5, with the corresponding Lorentz profiles.

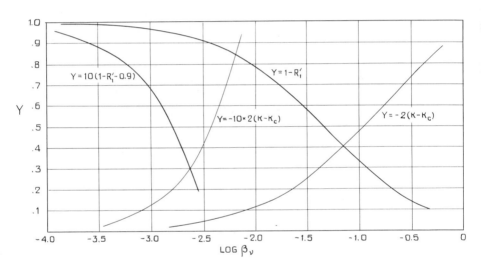

Fig. 14.   The dependence between $1-R'_1$, $2(k - k_c)$ and $\log \beta_\nu$ calculated for an optically infinite and homogeneous aerosol layer with $g = 0.5$ and $r_c = 0.785$.

The majority of lines in Jupiter's spectrum are in the red and infrared region, where r_c is maximum. It is evident from inspection of Fig. 9 that the variation of absorption from center to limb must depend only weakly on the form of the phase function, since the curves of $k(r_0)$ are almost parallel where $r_0 > 0.5$.

In a real planetary atmosphere, some part of the molecular absorption must occur in the clear (i.e., non-scattering) atmosphere above the clouds. We therefore generalize Eq. (14) to account for this absorption as follows:

$$1 - R (\mu, \mu_0) = (1 - R'_1) (\mu\mu_0)^{(k_1 - k_c)} e^{-\tau_\nu (1/\mu + 1/\mu_0)} \qquad (19)$$

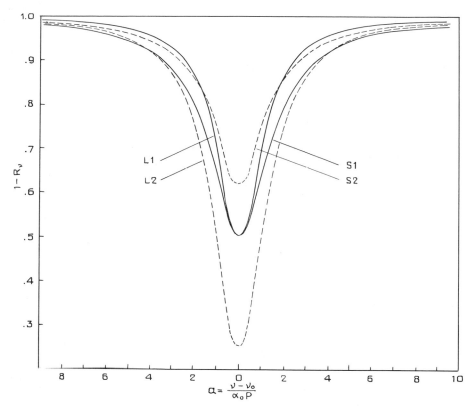

Fig. 15.   The absorption line profiles calculated for the absorption formation in a clear gas L, and in the scattering aerosol layer S, with $g = 0.5$ and $r_c = 0.785$, at the center of disk (solid lines) and at $\mu = 0.5$ (dashed lines).

or, assuming for simplicity that $\mu = \mu_0$,

$$1 - R_\mu = (1 - R_1)\, \mu^{2(k_1 - k_c)}\, e^{-2\tau_\nu(1/\mu - 1)}. \qquad (20)$$

R_1 denotes the observed band depth at $\mu = \mu_0 = 1$, which is a function of molecular absorption within the clouds as well as in the clear atmosphere above them, in contradistinction to R_1' which depends only on absorption within the clouds. From Eqs. (19) and (20), $(1 - R_1) = (1 - R_1')e^{-2\tau_\nu}$. A few results of computations for $g = 0.5$ and $r_c = 0.785$ are illustrated in Fig. 16.

It is not difficult to demonstrate that the ratio

$$D = \frac{1 - R_2'}{(1 - R_1')^2} = \frac{1 - R_2}{(1 - R_1)^2} \qquad (21)$$

is independent of the optical depth τ_ν of the atmosphere above the clouds, and that for a given scattering function, D can also be calculated from

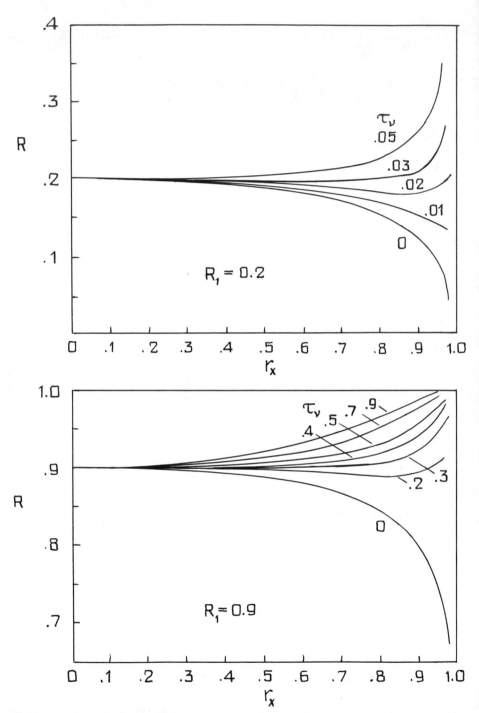

Fig. 16.   The variations of the central depth of the absorption band R with the distance from the center of disk calculated for the two-layer model. The cloud layer is infinite and homogeneous. Optical thickness of the outer atmosphere is varied and the value of R_1' is changed correspondingly for the constancy of the value of R_1.

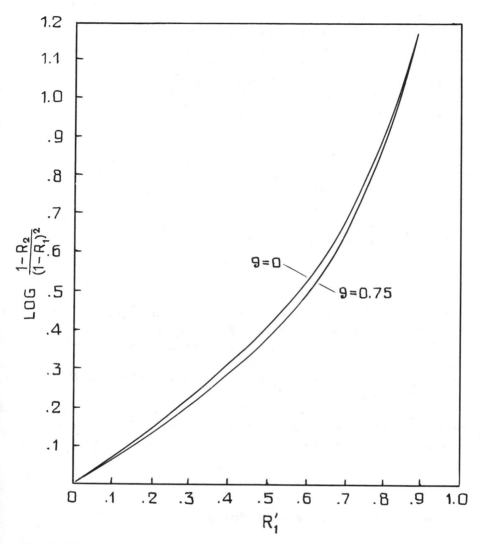

Fig. 17. The value of log D versus R_1' for two asymmetry parameters of phase function (see text).

$$D = \frac{0.5^{2(k_1 - k_c)}}{1 - R_1'}.$$ (22)

$D(R_1')$ is not strongly dependent on the asymmetry parameter, g, of the phase function (see Fig. 17). Measurements of R_1 and R_2 (i.e., absorption band depth at $\mu = 1.0$ and 0.5) give us D through Eq. (21) and then from Eq. (22), we can derive $(1 - R_1')$. In turn, we find τ_ν from the formula

$$\tau_\nu = -\frac{1}{2} \ln \frac{1 - R_1'}{1 - R_1}.$$ (23)

As noted earlier, the most reliable measurements are generally made near the center of the disk and along the entire equator. Clearly, the depth of an absorption band profile can be determined much more precisely if the spectrometer slit covers the entire equator than if separate regions of the disk close to the limb ($\mu \simeq 0.5$) are measured, especially if seeing conditions are poor. Let us measure band depths R_1 and R_e, where

$$R_e = 1 - \frac{B_{ev}}{B_{ec}}. \tag{24}$$

B_{ev} and B_{ec} represent the brightness of the entire equatorial belt included by the spectrometer slit, in the band center and in the continuum interpolated to the band center, respectively. Using our earlier theoretical calculations, we can determine R_e as a function of τ_ν and the apparent depth of the band at the center of the disk, R_1. Assuming $\mu = \mu_0$ for simplicity, we find

$$1 - R_e = (1 - R_1) \frac{\int_0^1 \mu^{2k_1 - 1} \, e^{-2\tau_\nu(k/\mu - 1)} \, dr_x}{\int_0^1 \mu^{2k_c - 1} \, dr_x} \tag{25}$$

where r_x is the distance from the center of the disk in units of the radius ($\mu = \sqrt{1 - r_x^2}$ along the equator). Figure 18 shows the results of calculating R_e (τ_ν) for the 6190 and 7250 Å CH$_4$ bands, assuming a constant value of R_1, and $g = 0.5$. This figure also illustrates the relationship between R_1' and τ_ν. Obviously we can determine τ_ν, given measurements of R_1 and R_e and an assumed value for g.

Let us now consider some observational results. A simple reflecting-layer model which supposes that all methane absorption occurs in the clear atmosphere above the cloud tops, predicts an equatorial band depth equal to

$$R_e = 1 - \frac{\int_{-1}^{+1} \mu^{2k_c - 1} \, e^{-2\tau_\nu/\mu} \, dr_x}{\int_{-1}^{+1} \mu^{2k_c - 1} \, dr_x} \tag{26}$$

while the depth for the center of the disk is

$$R_1 = 1 - e^{-2\tau_\nu}. \tag{27}$$

Our observed value of $R_1 = 0.200$ for the 6190 Å band gives $\tau_\nu^* \simeq 0.12$ and hence $R_e = 0.240$. However, our photoelectric measurements made in 1968–1974 indicate that the depths of both methane absorption bands are always *less* for the whole equator than for the center of the disk (see Fig. 5), contrary to the prediction of the reflecting-layer model. We conclude that the simple reflecting model is not applicable.

We have calculated and plotted graphically the quantity R_1 for each data point in Fig. 5, in a manner analogous to Fig. 18, and then determined the corresponding values of τ_ν. The optical depth of the atmosphere above the cloud tops was found to be $0.0 < \tau_\nu < 0.02$ for CH$_4$ 6190 Å, and $0.04 < \tau_\nu < 0.07$ for CH$_4$ 7250 Å.

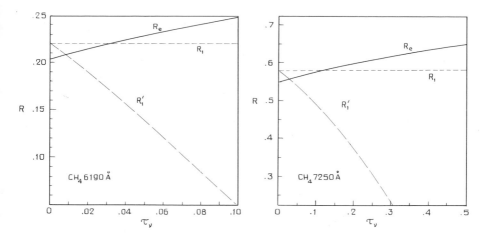

Fig. 18. The change of R_e and R_1' versus τ_ν at two values of the visible absorption band depth R_1 at the center of the disk (see text).

The majority of absorption observed in both methane bands therefore occurs within the cloud layer and not in the overlying clear atmosphere. We have reached the same conclusion as a result of our measurements of the depths of the same bands at the center and at the limb of the Jovian disk (Teifel 1969, 1971; Bugaenko *et al.* 1972).

We can also compare the variation of CH_4 absorption along the Jovian equator, computed theoretically from our two-layer model, with central band depths calculated from photoelectric measurements by Avery *et al.* (1974). The probable errors quoted for the band depths correspond to the estimated relative errors of the band equivalent widths. Figure 19 demonstrates that the observed variation can be matched within the estimated error limits by our "semi-infinite cloud layer plus pure gaseous upper atmosphere" model [cf., Eqs. (19) and (20)]; the match is no worse than that given by the inhomogeneous (multilayer) model used by Michalsky *et al.* (1974) to interpret the same observations. The modest spatial resolution and the uncertainties in W and R of the existing observational material do not permit us to discriminate between homogeneous and inhomogeneous models, since both classes of models fit the observations equally well. Moreover, the available data set does not permit a choice among different variants of the multilayer model (i.e., variations of the optical depth of the upper cloud layer and/or the optical thickness of the clear gas layer between the cloud decks), as calculated by Michalsky *et al.* (1974).

Taking all observations into consideration, we can calculate upper limits of 0.03 and 0.10 for τ_ν, the optical depth in the 6190 Å and 7250 Å bands respectively, of the clear atmosphere above the cloud tops. We can then estimate approximate absorption coefficients in these bands by the following line of reasoning. The methane path-length in these bands (assuming a re-

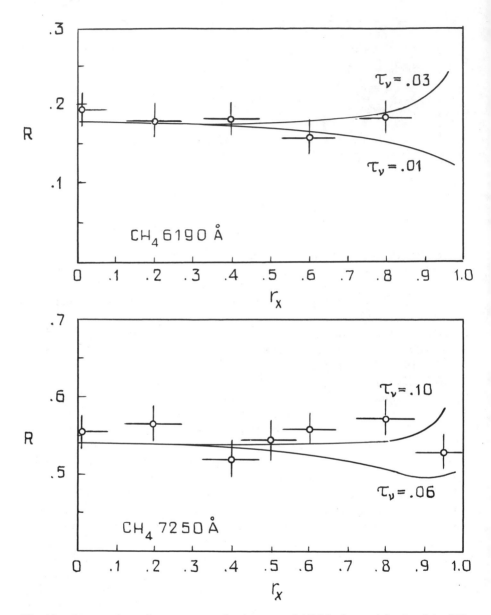

Fig. 19. A comparison of measurements by Avery *et al.* (1974) of central depths of the CH_4 absorption bands on Jupiter's equator (circles) with the calculated variations of R for the two-layer model of band formation (lines).

flecting-layer model) is $L_{CH_4} \simeq 133$ m amagat (Owen 1969). For $R_{6190} = 0.20$ and $R_{7250} = 0.58$, we obtain $\tau^*_{6190} \simeq 0.112$ and $\tau^*_{7250} \simeq 0.434$ in the reflecting-layer model. Since

$$\tau_\nu^* = \kappa_\nu^* L \tag{28}$$

where L is the path length, it follows that

$$\kappa_{6190}^* \simeq 8.42 \times 10^{-4} \text{ (m amagat)}^{-1}$$

$$\kappa_{7250}^* \simeq 3.26 \times 10^{-3} \text{ (m amagat)}^{-1}. \tag{29}$$

The upper limit to the reduced methane abundance in the Jovian upper atmosphere, derived from the upper limits to τ_ν in the two bands, is then

$$U_{CH_4} \lesssim 36 \pm 2 \text{ m amagat from 6190 Å}$$

$$U_{CH_4} \lesssim 31 \pm 2 \text{ m amagat from 7250 Å}. \tag{30}$$

If the upper limit to the H_2 concentration in the upper atmosphere is taken to be 10 km amagat, and if $n(H_2)/n(He) = 5$, the relative methane concentration is

$$A(CH_4) = \frac{n(CH_4)}{n(H_2) + n(He)} \simeq 3 \times 10^{-3}. \tag{31}$$

The latitude variation of absorption in the 8860 Å band of methane (Owen and Westphal 1972) can be interpreted in terms of our simple two-layer model as the combined result of absorption within the cloud layer and in the clear atmosphere above the clouds. We can estimate the probable optical depth of the upper atmosphere in this band through Bugaenko's interpretation of Cruikshank's laboratory data on the relative absorption coefficients in the centers of the 8860 Å and 7250 Å methane bands, namely

$$\frac{\kappa_{8860}^*}{\kappa_{7250}^*} \simeq 7.8. \tag{32}$$

Thus, if $\tau_\nu \lesssim 0.10$ in the 7250 Å band, we obtain the upper limit $\tau_{8860} \lesssim 0.78$. The most probable value of τ_{7250} is approximately 0.05, from which it follows that $\tau_{8860} \simeq 0.4$.

Let us suppose momentarily that the observed variations of R_{8860} along the central meridian are due solely to variations in the optical depth of the atmosphere above the cloud tops. Absorption within the cloud layer is characterized by the quantity R_1'. If we choose the values $g = 0.5$ and $r_c = 0.785$ for the optical parameters of the cloud particles, we can compute [Eq. (20)] the quantity τ_ν at various points along the central meridian (Fig. 20). The upper limit $\tau_\nu \lesssim 0.78$ is probably unrealistic because very strong limb darkening would be observed along the Jovian equatorial belt for such an optical depth (Fig. 21). No one has published quantitative photometric measurements along the equator in the 8860 Å band, but narrowband photographs (Owen 1969; Owen and Mason 1969; Kuiper 1972; Minton 1972) appear to show, qualitatively, weaker limb darkening in the 8860 Å band than in the

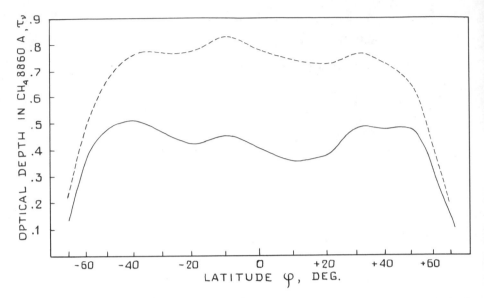

Fig. 20.   The latitudinal variations of the optical thickness τ_ν of the outer atmosphere derived from the observations of Owen and Westphal (1972) in the CH_4 8860 Å absorption band for two cases: $R_1 = 0.86$, $\tau_{\nu_1} = 0.78$ (dashed line) and $R_1 = 0.90$, $\tau_{\nu_1} = 0.40$ (solid line).

neighboring continuum. Binder and McCarthy (1973) found a limb darkening coefficient of approximately 1 in the wings of the strong CH_4 band at 0.9 μm. We shall therefore adopt the "most probable" value derived above, $\tau_{8860} \simeq 0.4$, for further analysis. In this case, the variations of τ_ν are on the order of 45% near latitudes $\pm 60°$ and in general, τ_ν is maximum near $|\phi| \simeq 45°$. The maximum height difference in the effective cloud top near $|\phi| \simeq 45°$ would be $\Delta z \simeq 8.6$ km for an adiabatic, or $\Delta z \simeq 7.5$ km for an isothermal atmosphere (assuming scale height $H = 20$ km). The cloud boundary would be lower at middle latitudes than in the equatorial region. By this interpretation the cloud boundary must be 19 km higher in the atmosphere in the polar regions than at the equator. The latter supposition is consistent with ultraviolet photographs of Jupiter (Fountain 1972), in which the polar regions appear dark even at 3000 Å, while slight limb brightening due to Rayleigh scattering is evident at low and middle latitudes. On the other hand, Gehrels *et al.* (1969)[4] concluded from polarimetry that the atmosphere above the cloud tops is nearly 10 times *deeper* in the polar regions than at the equator. Evidently these measurements are actually related to a latitude near 50° rather than the poles. In that case, the polarimetric results agree qualitatively with the absorption band data, although large quantitative discrepancies remain.

---

[4]See p. 534.

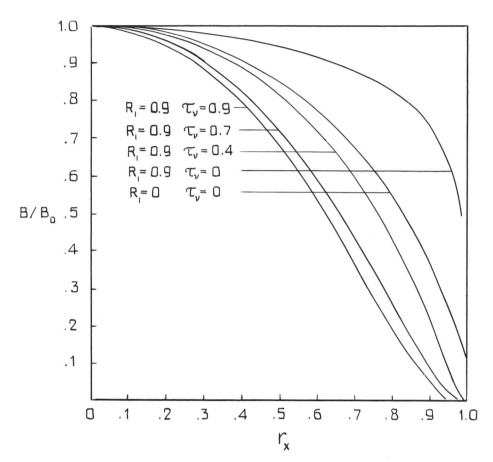

Fig. 21. The relative brightness B/B_0 distribution over the planetary disk at some values of the central absorption band depth R_1 and optical thickness of outer atmosphere τ_ν for the cloud layer with $r_c = 0.785$ and $g = 0.5$.

Saturation of the stronger lines near the center of the 8860 Å band would correspond to values of Δz approximately 1.5 times larger than those estimated above. These estimates are upper limits to the amplitude of the height variations, since we have attributed *all* differences in the absorption to differences in the thickness of the upper atmosphere, and have so far neglected possible variations of the density of the underlying cloud layer which might manifest themselves (through variations in R_1') in the absorption measured near the centers of strong bands (Teifel 1975b).

Woodman (1974) recently measured residual intensities in the near-infrared methane bands over the GRS, $(1 - R_s)$, and over the STrZ, $(1 - R_z)$. He found that the ratio $(1 - R_z)/(1 - R_s)$ was approximately equal to 0.65 when $(1 - R_z) = 0.1$. By inspection of Figs. 1 and 20, we find $(1 -$

$R_z) \simeq 0.1$ and $\tau_z \simeq 0.44$ for the STrZ in the 8860 Å band. If we suppose that absorption within the cloud layer is the same for the GRS and the STrZ, i.e., $(1 - R_s') = (1 - R_z')$, we find $\tau_s \simeq 0.22$ and thus conclude that the effective upper cloud boundary must be higher by $\Delta z \simeq 14$ km in the GRS than in the STrZ. However, Woodman found that when $(1 - R_z) = 0.5$, the residual intensities over the GRS and STrZ are equal $[(1 - R_z)/(1 - R_s) = 1]$; that is, medium and weak absorption bands, and the wings of strong bands, are not significantly different over the GRS and STrZ. As noted earlier, our own observations of the 6190 Å band showed no difference in absorption over the GRS and over its surroundings. Since the effective depth of absorption must be *less* in the core of a strong band than in its wings or in weak bands, it appears that the higher residual intensity observed in the core of the 8860 Å band over the GRS corresponds to decreased R_s' (due to, say, vertical distribution of aerosol density in the GRS near the effective level of formation of the band core) rather than to an elevated effective cloud boundary (decreased τ_s).

Let us therefore consider the supposition that it is absorption within the cloud layer (characterized by R_1'), and not τ_v, which varies with latitude. By assuming constant τ_v we can estimate limits to the fluctuations in R_1' and hence in the density of the cloud material. Our information on τ_v in the 6190 Å and 7250 Å bands permits an approximate evaluation of the mean volume scattering coefficient of the cloud layer at the effective depth of formation of bands that are not too strong. We calculate

$$(1 - R_1') = (1 - R_1) \, e^{2\tau_v} \tag{33}$$

and thence determine β_v by inspection of Fig. 14. We then compute the volume absorption coefficient κ_v, assuming $P_e \simeq 1$ atm and $A_{CH_4} \simeq 3 \times 10^{-3}$. Finally,

$$\sigma_a = \frac{\kappa_v}{\beta_v}. \tag{34}$$

The quantity σ_a is insensitive to variations in τ_v with amplitude on the order of the precision to which τ_v is determined. On the average

$$\sigma_a \simeq (2 \text{ to } 4) \times 10^{-6} \text{ cm}^{-1}. \tag{35}$$

From the particle parameters determined by polarimetry, the scattering cross section of a single particle is $S_a \simeq 5 \times 10^{-9}$ cm^2 and the particle concentration is therefore $N_a \simeq 6 \times 10^2$ cm^{-3}. The corresponding volume density of the aerosol is approximately

$$Q_a = \frac{4}{3}\pi \bar{a}^3 \, \rho_a N_a \tag{36}$$

where $\bar{a}$ is the mean particle radius and $\rho_a = 0.82$ g cm^{-3} is the density of solid NH$_3$. If $\bar{a} \simeq 0.2$ μm, we obtain

$$Q_a \simeq (1 \text{ to } 2) \times 10^{-11} \text{ g cm}^{-3} . \tag{37}$$

Cloud densities at the effective depth of formation of the cores of strong bands can vary with latitude by a factor of two if $\tau_\nu(\phi)$ is constant. However, τ_ν must decrease at the polar caps to avoid an absurd result.

Although mechanisms of cloud formation from NH_3 crystals have been considered extensively (Lasker 1963; Lewis 1969; Sorokina *et al.* 1972; Khodyachikh 1972; Weidenschilling and Lewis 1973), existing models of the upper cloud layers do not account for a number of other possible factors. Specifically, the role of meteoritic dust in the formation of an aerosol haze at high altitudes has never been discussed. Spectrophotometry of Jupiter at short wavelengths and observations of satellite eclipses (Price and Hall 1971; Price *et al.* 1972; Greene and Smith 1974) indicate the presence of such a haze. These observations are consistent with a considerable quantity of aerosols at the atmospheric level where the molecular number density is $n_M \simeq 10^{18}$ cm^{-3}, which corresponds to $U(H_2) \simeq 1$ km amagat. The relationship governing the variation of the aerosol number density with altitude may therefore differ from earlier estimates by a factor related to the sublimation conditions (Teifel 1973). A new candidate for the high-altitude haze has been suggested by Prinn (1974),[5] who considered a hydrazine aerosol formed by photochemical action of solar ultraviolet on NH_3 in the Jovian stratosphere.

A multi-layered structure of Jupiter's upper cloud cover is not excluded. Along this line, let us consider one result of the photometric studies of Jovian atmospheric activity at different wavelengths. Using photometric coefficients of atmospheric activity determined from observations made in 1962–63 (Aksenov *et al.* 1967; Sorokina and Priboeva 1972; Sorokina 1973; Petrova and Sorokina 1973), we analyzed the asymmetry of activity in the northern and southern hemispheres from two characteristics, namely, a coefficient of activity asymmetry

$$L_a = \frac{R_N - R_S}{R_N + R_S} \tag{38}$$

where R_N and R_S are the photometric coefficients of activity for each hemisphere (Focas and Baños 1964), and a coefficient of the asymmetry of instability

$$L_s = \frac{S_N - S_S}{S_N + S_S} \tag{39}$$

where

$$S_N = \frac{\sigma_N}{\bar{R}_N}, \qquad S_S = \frac{\sigma_S}{\bar{R}_S} \tag{40}$$

[5]See p. 333.

and where $\bar{R}_N$ and $\bar{R}_S$ are the mean values of the activity coefficients for the entire observing season, and σ_N and σ_S are the dispersions of R_N and R_S.

The asymmetry of the activity coefficient, expressed in terms of L_a, keeps the same sign at all wavelengths (Fig. 22). However, the trend of increased L_a towards shorter wavelengths is accompanied by a decrease in the asymmetry of instability, L_s. In fact, the shapes of the two curves are approximately opposite to one another for the period 1962–1968: L_a is negative, but L_s changes sign from negative to positive in the transition from long to short wavelengths. A kind of transition is apparent in 1966–1967, with both L_s and L_a roughly independent of wavelength. It is interesting to note that no correlation is evident between the values of the activity coefficients at short and long wavelengths, while a correlation is observed in neighboring parts of the spectrum. Evidently, fluctuations of the relative quantity of dark and light material observed at different wavelengths have a different character and are related to different levels in the atmosphere; i.e., processes visible at short wavelengths originate at higher altitudes than those seen at longer wavelengths. We may speculate that ultraviolet observations ($\lambda < 0.35$ μm) and narrowband observations in the centers of strong absorption bands sample only the upper reaches of the extended aerosol haze. They may therefore reflect greater (relative) spatial and/or temporal fluctuations of aerosol density than do observations related to deeper layers of the aerosol haze, or clouds, where weak bands and the spectral continuum are formed (Fig. 23).

VI. REMARKS FOR FUTURE STUDIES

Existing data on the character of the variations of molecular absorptions across the Jovian disk are clearly inconsistent with the simple reflecting-layer model, although the latter is still frequently used for quantitative estimates of the molecular composition of the planet's atmosphere. It is not really our aim here to criticize the reflecting-layer model and its variants since this has already been done adequately (McElroy 1969). However, we do wish to propose a simple two-layer model with a homogeneous, semi-infinite aerosol layer, one which accounts for molecular absorption in the clear atmosphere above the visible cloud tops as well as within the cloud layer. We have developed this model to the point of obtaining numerical results for anisotropic scattering. More complicated inhomogeneous models have not yet reached the stage of development where they can be considered better approximations to reality. In fact, the trend is not towards true inhomogeneous models but rather towards 3 and 4-layer models, each layer of which is homogeneous; the parameters of each layer can be varied within established limits. It would be equally important to consider an inhomogeneous model with an arbitrarily assigned expression for the variation of aerosol density with depth. With such a model, it would be desirable to examine the sensitivity of the relative intensities and profiles of weak and strong absorption bands to the parameters

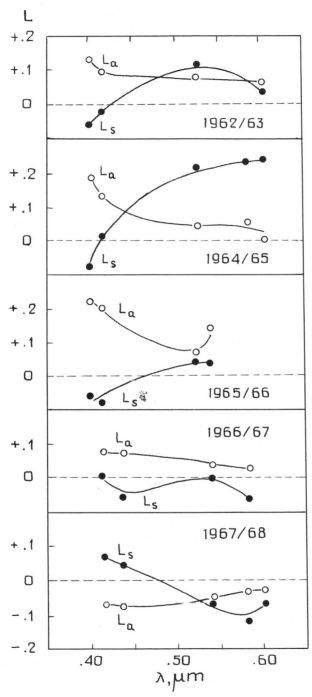

Fig. 22. The spectral and temporal changes of the asymmetry factors for activity L_a and for instability L_s from photometric observations of Jupiter during 1962–1968.

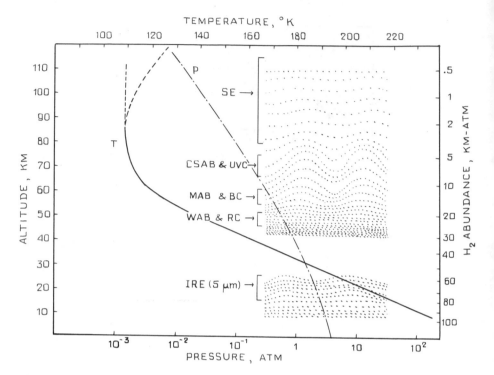

Fig. 23. The probable vertical distribution and altitude variations of aerosols in the Jovian atmosphere, and suggested levels for the effective reflection of radiation noted as *SE* (the satellite eclipses), *CSAB* and *UVC* (centers of strong absorption bands and ultraviolet continuum), *MAB* and *BC* (moderate absorption bands and blue continuum), *WAB* and *RC* (weak absorption bands and red continuum), *IRE* (upward 5-μm infrared emission). The temperature and pressure profiles are calculated for a "grey" atmospheric model with $T_e =$ 134°K and $H_2/He = 5$.

of the vertical distribution of the aerosol, and to specify what experiments might give the most information on the distribution. However, analytical solutions to this problem are very complex. The most expedient computational method is through application of numerical (Monte Carlo) techniques, in which model parameters can be adjusted at will. Unfortunately, application of these techniques presents some special problems.

In the majority of cases, existing measurements are not sufficiently precise to permit discrimination between two-layer homogeneous and multilayer inhomogeneous models, and do not afford answers to questions about the optical thickness of the upper cloud cover of Jupiter. Evidently the most urgent problem of experimental (observational) work is to improve the precision and spatial resolution of the data. The greatest shortcoming of the existing data lies in their incompleteness and in the absence of a systematic, homogeneous series of observations. Ideally, one would like to see regular and parallel determinations of a variety of planetary characteristics (e.g., intensities of molecular bands and lines at a large number of points on the disk, coefficients of diffuse reflection from absolute measurements, coefficients of limb darkening over a wide range of spectral regions including the ultraviolet, spectrophotometric data on the contrasts on the planet's disk, polarization measurements, and so on).

Such a complex investigation would best be undertaken with international cooperation, especially within the framework of the International Decade of Solar System Studies which has been proposed (Pickering 1975) for the period 1976–1986. Comparisons of groundbased data obtained in a systematic observational program with data obtained at Jupiter by space probes can provide us with a quantity of information the importance of which cannot be overestimated.

Acknowledgements. I am grateful to D. P. Cruikshank for the translation of this chapter from Russian and to J. T. Bergstralh for his help in editing the text.

REFERENCES

Aksenov, A. N. 1973. The change of the methane absorption band CH_4 6190 Å profile from the center to limb on the Jovian disk. *Astron. Tsirkular USSR Acad. Sci.* No 782:3–5 (in Russian).

Aksenov, A. N.; Grigorjeva, Z. N.; Pribocva, N. V.; Romanenko, Z. G., and Teifel, V. G. 1967. The experience of the photometric study of the Jovian atmospheric activity in 1962–1965. *Trudy Astrophys. Inst. Acad. Sci. KazSSR.* 9:7–21 (in Russian).

Aksenov, A. N.; Grigorjeva, Z. N.; Teifel, V. G.; and Kharitonova, G. A. 1972. The study of the peculiarities of molecular absorption in the Jovian spectrum. *Physics of the moon and planets.* pp. 433–438. (D. Ya. Martynov and V. A. Bronshten, eds.) Moscow: Nauka Press (in Russian).

Anikonov, A. S. 1974. The interpretation of weak absorption bands in Jupiter's spectrum. *Astron. Vestnik* 8:223–228 (in Russian).

Avery, R. W.; Michalsky, J. J.; and Stokes, R. A. 1974. Variation of Jupiter's CH_4 and NH_3 bands with position on the planetary disk. *Icarus* 21:47–54.

Avramchuk, V. V. 1968. On the distribution of methane absorption in the band 6190 Å on the Jovian disk. *Astrometry and Astrophys.* 1:126–160 (in Russian).

———. 1970. On the results of the spectrophotometry of the methane (6190 Å) and ammonia (6441 and 6478 Å) absorption bands on the disk of Jupiter. *Astron. Zh.* 47:577–585 (in Russian).

———. 1972. The results of three-years observations of the methane (6190 Å) and ammonia (6441 and 6478 Å) absorption bands on Jovian disk. *Physics of the moon and planets,* pp. 439–443. (D. Ya. Martynov and V. A. Bronshten, eds.) Moscow: Nauka Press (in Russian).

Avramchuk, V. V.; Ibragimov, N. B.; and Slivenko, E. F. 1972. On some results of the spectrophotometry of the methane absorption band (7250 Å) on Jupiter's disk. *Astron. Vestnik* 6:218–222 (in Russian).

Bergstralh, J. T. 1973. Methane absorption in the Jovian atmosphere. II. Absorption line formation. *Icarus* 19:390–418.

Binder, A. B. 1972. Spectrophotometry of 1.5 μm window of Jupiter. *Astron. J.* 77:93–99.

Binder, A. B., and McCarthy, D. W. 1973. IR spectrophotometry of Jupiter and Saturn. *Astron. J.* 78:939–950.

Bobrovnikoff, N. T. 1933. Note on the spectrum of Jupiter. *Publ. Astron. Soc. Pacific* 45: 171–174.

Boyce, P. B. 1968. Preliminary observations of the Jovian 6190 Å methane band. *Lowell Obs. Bull.* 146:93–97.

Bugaenko, L. A. 1972. Monochromatic coefficients of the brightness of giant planets. *Astron. Vestnik* 6:19–21 (in Russian).

Bugaenko, L. A.; Galkin, L. S.; and Morozhenko, A. V. 1972. The investigation of molecular absorption in the atmospheres of major planets. *Astron. Vestnik* 6:223–227 (in Russian).

Carleton, N. P., and Traub, W. A. 1974. Observations of spatial and temporal variation of the Jovian H_2 quadrupole lines. *Exploration of the planetary system.* (A. Woszczyk and C. Iwaniszewska, eds.) pp. 345–349. Dordrecht-Holland: D. Reidel Publ. Co.

Chamberlain, J. W. 1965. The atmosphere of Venus near her cloud tops. *Astrophys. J.* 141: 1184–1205.

———. 1970. Behavior of absorption lines in a hazy planetary atmosphere. *Astrophys. J.* 159:137–158.

Coffeen, D. L. 1974. Optical polarization measurements of the Jupiter atmosphere at 103° phase angle. *J. Geophys. Res.* 79:3645–3652.

Coulson, K.; Dave, J. V.; and Sekera, Z. 1960. *Tables related to radiation emerging from a planetary atmosphere with Rayleigh scattering.* Berkeley, California: University of California Press.

Danielson, R. E., and Tomasko, M. G. 1969. A two-layer model of Jovian clouds. *J. Atmos. Sci.* 26:889–897.

Dlugach, J. M., and Yanovitskii, E. G. 1974. The optical properties of Venus and the Jovian planets. II. Methods and results of calculations of the intensity of radiation diffusely reflected from semi-infinite homogeneous atmosphere. *Icarus* 22:66–81.

Egorov, Yu. A.; Teifel, V. G.; and Kharitonova, G. A. 1971. On the methane absorption changes in the equatorial belt of Jupiter. *Astron. Tsirkular USSR Acad. Sci.* No. 656, 3–5 (in Russian).

Elvey, C. T., and Fairley, A. S. 1932. On the absorption band at 6191 Å in the spectrum of Jupiter. *Astrophys. J.* 75:373–378.

Eropkin, D. I. 1933. On the spectra of polar and equatorial zones of Jupiter. *Doklady USSR Acad. Sci.* 4:143–145.

Focas, J. H., and Baños, C. J. 1964. Photometric study of the atmospheric activity on the planet Jupiter and peculiar activity in its equatorial area. *Ann. Astrophys.* 27:36–45.

Fountain, J. W. 1972. Narrow-band photography of Jupiter and Saturn. *Comm. Lunar Planet. Lab.* 9:327–337.

Gehrels, T.; Herman, B. M.; and Owen, T. 1969. Wavelength dependence of polarization. XIV. Atmosphere of Jupiter. *Astron. J.* 74:190–199.

Greene, T. F., and Smith, D. W. 1974. The aerosol extinction wavelength dependence of the Jovian cloud tops. *Bull. Amer. Astron. Soc.* 6:376.

Harris, D. L. 1961. Photometry and colorimetry of planets and satellites. *Planets and satellites* (G. P. Kuiper and B. M. Middlehurst, eds.), pp. 272–342. Chicago, Illinois: University of Chicago Press.

Hess, S. L. 1953. Variation in atmospheric absorption over the disks of Jupiter and Saturn. *Astrophys. J.* 118:151–160.

Hunt, G. E. 1973. Interpretation of hydrogen quadrupole and methane observations of Jupiter and the radiative properties of the visible clouds. *Mon. Not. Roy. Astron. Soc.* 161:347–363.

Hunt, G. E., and Bergstralh, J. T. 1974. Structure of visible Jovian clouds. *Nature* 249: 635–636.

Ibragimov, N. B. 1974. Investigation of Jupiter and Saturn with the 2-m reflector of the Shemakha astrophysical observatory. *Astron. Zh.* 51:178–186 (in Russian).

Kartashov, V. F. 1972. Photometry of the Red Spot on Jupiter. *Astron. Tsirkular USSR Acad. Sci.* No. 742, 1–3 (in Russian).

Khodyachikh, M. F. 1972. On the structure of Jovian clouds. *Vestn. Kharkov Univ. Astronomy* 7:69–75 (in Ukrainian).

Krugov, V. D. 1972. Spectrophotometry of Jupiter and Saturn in the shortwave region of spectrum. Ph.D. dissertation. Kiev (in Russian).

Kuiper, G. P. 1972. Interpretation of the Jupiter Red Spot. I. *Comm. Lunar Planet. Lab.* 9:249–310.

Lasker, B. M. 1963. Wet adiabatic model atmospheres for Jupiter. *Astrophys. J.* 138:709–719.

Lenoble, J. 1968. Absorption lines in the radiation scattered by a planetary atmosphere. *J. Quant. Spectr. Transfer.* 8:641–654.

Lewis, J. S. 1969. The clouds of Jupiter and the NH_3-H_2O and NH_3-H_2S systems. *Icarus* 10: 365–378.

Litkevich, N. G., and Khodyachikh, M. F. 1973. Photometry of Jupiter with interference filters. *Vestn. Kharkov Univ. Astronomy.* 8:18–26 (in Russian).

Loskutov, V. M. 1971a. On the interpretation of the photometric observations of Jupiter. *Astron. Vestnik* 5:153–158 (in Russian).

———. 1971b. On the interpretation of polarimetric observations of Jupiter. *Astron. Zh.* 48: 1046–1050 (in Russian).

———. 1974. The variations of the absorption bands intensity on the Jovian disk. *Astron. Zh.* 51:1060–1063 (in Russian).

McElroy, M. B. 1969. Atmospheric composition of the Jovian planets. *J. Atmos. Sci.* 26: 798–812.

Michalsky, J. J.; Stokes, R. A.; Avery, R. W.; and DeMarcus, W. C. 1974. Molecular band variations as a probe of the vertical structure of a Jovian atmosphere. *Icarus* 21:55–65.

Minton, R. B. 1972. Latitude measures of Jupiter in the 0.89 μ methane band. *Comm. Lunar Planet. Lab.* 9:339–351.

Moroz, V. I., and Cruikshank, D. P. 1969. Distribution of ammonia on Jupiter. *J. Atmos. Sci.* 26:865–869.

Morozhenko, A. V., and Yanovitskii, E. G. 1964. *Tables for calculations of the radiation intensity of planetary atmospheres.* Kiev: Naukova Dumka Press (in Russian and Ukrainian).

———. 1973. The optical properties of Venus and Jovian planets. I. The atmosphere of Jupiter according to polarimetric observations. *Icarus* 18:583–592.

Münch, G., and Younkin, R. L. 1964. Molecular absorption and color distributions over Jupiter's disk. *Astron. J.* 69:553.

Owen, T. 1969. The spectra of Jupiter and Saturn in the photographic infrared. *Icarus* 10: 355–364.

Owen, T., and Mason, H. P. 1969. New studies of Jupiter's atmosphere. *J. Atmos. Sci.* 26: 870–873.

Owen, T., and Sagan, C. 1972. Minor constituents in planetary atmospheres: ultraviolet spectroscopy from the orbiting astronomical observatory. *Icarus* 16:557–568.

Owen, T., and Westphal, J. A. 1972. The clouds of Jupiter: observational characteristics. *Icarus* 16:392–396.

Petrova, N. N., and Sorokina, L. P. 1973. Photometric study of the Jovian atmospheric activity in 1962–1969. *Astron. Vestnik* 7:9–15 (in Russian).

Pickering, W. H. 1975. International solar system decade. XVIII Cospar Meeting, Varna, Bulgaria.

Pilcher, C. B., and McCord, T. B. 1971. Narrow-band photometry of the bands of Jupiter. *Astrophys J* 165:195–201

Pilcher, C. B.; Prinn, R. G.; and McCord, T. B. 1973. Spectroscopy of Jupiter: 3200 to 11,200 Å. *J. Atmos. Sci.* 30:302–307.

Price, M. J., and Hall, J. S. 1971. The physical properties of the Jovian atmosphere inferred from eclipses of the Galilean satellites. I. Preliminary results. *Icarus* 14:3–12.

Price, M. J.; Hall, J. S.; Boyce, P. B.; and Albrecht, R. 1972. The physical properties of the Jovian atmosphere inferred from eclipses of the Galilean satellites. II. 1971 apparition. *Icarus* 17:49–56.

Prinn, R. G. 1974. Stratospheric haze layers on Jupiter. *Bull. Amer. Astron. Soc.* 6:375.

Reese, E. J. 1971. Jupiter: its Red Spot and other features in 1969–1970. *Icarus* 14:343–354.

Rozenberg, G. V. 1962. The light characteristics of thick layers of scattering medium with a small specific absorption. *Dokl. Acad. Sci. USSR.* 145:775–779 (in Russian).

Sobolev, V. V. 1972. On the theory of planetary spectra. *Astron. Zh.* 49:397–404 (in Russian).

——. 1973. The spectrum of the planet with two-layer atmosphere. *Dokl. Acad. Sci. USSR.* 211:63–66 (in Russian).

Sorokina, L. P. 1973. On the long-periodic variations of Jovian atmospheric activity. *Astron. Tsirkular USSR Acad. Sci.* No. 747, 3–4 (in Russian).

Sorokina, L. P., and Priboeva, N. V. 1972. Photometric study of the atmospheric activity of Jupiter. *Physics of the moon and planets.* (D. Ya. Martynov and V. A. Bronshten, eds.) pp. 443–444. Moscow: Nauka Press (in Russian).

Sorokina, L. P.; Teifel, V. G.; and Usoltzeva, L. A. 1972. Optical properties and the structure of Jupiter's atmosphere. V. A possible structure of the ammonia aerosol layer. *Astron. Vestnik* 6:77–84 (in Russian).

Taylor, D. J. 1965. Spectrophotometry of Jupiter's 3400–10000 Å spectrum and a bolometric albedo for Jupiter. *Icarus* 4:362–373.

Teifel, V. G. 1959. On the intensity distribution over the Jovian disk in the methane absorption bands. *Izv. Komissii po fizike planet.* 1:93–104 (in Russian).

——. 1963. Some data on the intensity of methane absorption in the atmosphere of Jupiter. *Mém. Soc. Roy. Sci. Liège.* 7:589–592.

——. 1964. On the latitudinal distribution of the methane absorption on Jovian disk. *Astron. Tsirkular USSR Acad. Sci.* No. 296, 1–4 (in Russian).

——. 1966. Spectrophotometry of the methane absorption bands on Jovian disk in the near infrared (0.7–1.0 μm). *Astron. Zh.* 43:154–156 (in Russian).

——. 1967a. Molecular absorption of light in the atmosphere of Jupiter and a structure of the cloud layer surface of planet. *The study of the planet Jupiter.* pp. 3–20 (in Russian). Moscow.

——. 1967b. A comparison of the absorption distribution in the band CH_4 6190 Å on the disks of Jupiter and Saturn. *Trudy Astrophys. Inst. Acad. Sci. KazSSR.* 9:59–62 (in Russian).

——. 1967c. Spectrophotometry of Red Spot on Jupiter. *Trudy Astrophys. Inst. Acad. Sci. KazSSR.* 9:52–58 (in Russian).

——. 1968. Optical properties and the structure of Jupiter's atmosphere. I. Photometric contrasts, molecular absorption and a probable structure of the dark equatorial belt in 1962–1963. *Astron. Vestnik* 2:229–238 (in Russian).

——. 1969a. Molecular absorption and the possible structure of the cloud layers of Jupiter and Saturn. *J. Atmos. Sci.* 26:854–859.

——. 1969b. The continuous spectrum and molecular absorption in the Red Spot on Jupiter. *Trudy Astrophys. Inst. Acad. Sci. KazSSR.* 13:3–12 (in Russian).

——. 1969c. Optical properties and the structure of Jupiter's atmosphere. II. The influence of the multiple scattering in a cloud layer on planetary absorption line profiles. *Astron. Vestnik* 3:85–95 (in Russian).

——. 1970. Optical properties and the structure of Jupiter's atmosphere. III. Spectrophotometric peculiarities of the Red Spot. *Astron. Vestnik* 4:34–42 (in Russian).

——. 1971. Optical properties and the structure of Jupiter's atmosphere. IV. The results of photoelectric spectrophotometry at 6300–8100 Å. *Astron. Vestnik* 5:222–231 (in Russian).

——. 1973. Optical properties and the structure of Jupiter's atmosphere. VI. The thickness and density of the upper cloud layer. *Astron. Vestnik* 7:143–149 (in Russian).

——. 1975a. On the wavelength dependence of the Jovian limb darkening. *Astron. Zh.* 52:615–622 (in Russian).

————. 1975*b*. On the latitudinal change of the Jovian cloud cover top height. *Astron. Zh. (Lett.)* 1, No. 10:34–38 (in Russian).

Teifel, V. G., and Priboeva, N. V. 1963. On the intensity of the methane absorption in the band CH_4 6190 Å on Jupiter's disk. *Izv. Acad. Sci. KazSSR.* 16:61–73 (in Russian).

Teifel, V. G., and Usoltzeva, L. A. 1973. The profiles and curves of growth for the absorption lines formed in the scattering medium. *Astron. Zh.* 50:568–575 (in Russian).

Tomasko, M. G.; Clements, A. E.; and Castillo, N. D. 1974. Limb darkening of two latitudes of Jupiter at phase angles of 34° and 109°. *J. Geophys. Res.* 79:3653–3660.

Walker, M. F., and Hayes, S. 1967. Image tube observations of the $3\nu_3$ band of CH_4 in the spectrum of Jupiter. *Publ. Astron. Soc. Pacific* 79:464–472.

Wallace, L.; Caldwell, J. J.; and Savage, B. D. 1972. Ultraviolet photometry from the orbiting astronomical observatory. III. Observations of Venus, Mars, Jupiter, and Saturn longward of 2000 Å. *Astrophys. J.* 172:755–769.

Weidenschilling, S. J., and Lewis, J. S. 1973. Atmospheric and cloud structures of the Jovian planets. *Icarus* 20:465–476.

Woodman, J. H. 1974. Methane absorption over Jupiter's Great Red Spot and South Tropical Zone. *Bull. Amer. Astron. Soc.* 6:376.

Younkin, R. L., and Münch, G. 1963. Spectral energy distribution of the major planets. *Mém. Soc. Roy. Sci. Liège.* 7:125–136.

PHOTOMETRY AND POLARIMETRY OF JUPITER

M. G. TOMASKO
University of Arizona

The limb darkening studies in molecular absorption features and nearby continuum portions of the spectrum at small phase angles indicate the presence of a relatively clear region above an extended portion of Jupiter's atmosphere in which multiple scattering is required. It seems possible to explain the data of this type, which are presently available, by treating the scattering portion of the atmosphere by a variety of models ranging from two cloud layers having short mean-free scattering paths separated by an extensive clear space, to a single diffuse cloud in which the mean-free scattering path is very long (kilometers). Similar studies in absorptions of different strengths should constrain the mean-free scattering path and limit the range of acceptable models. Of these possible models, only the "two-cloud" idealization that has a short mean-free scattering path in the clouds below the upper clear space has been used to date to analyze spectroscopic data. It has been generally successful in reconciling the widths of strong and weak absorptions and their center-to-limb variation. When analyzed using this model, the limb-darkening data in molecular absorption features indicate a thinner upper cloud at high latitudes; the polarimetric data also suggest an increased penetration of radiation at high latitudes.

The presence of small absorbing aerosols in the gas above the scattering clouds has been suggested to explain the variation of Jupiter's geometric albedo with wavelength, and some indication of their presence may exist in the large phase-angle Pioneer photometry. Nevertheless, the shapes of the limb darkening curves at large phase angle suggest that Jupiter's belts are darker than its zones, principally because of darker cloud particles in the belts rather than because of increased aerosol concentrations above belts. The analysis of Pioneer data, while incomplete, also suggests that particles that are strongly forward scattering will be required to explain Jupiter's limb darkening at large phase angles. The groundbased polarimetry of Jupiter suggests that the cloud particles are not spherical—a fact that will greatly complicate the analysis of Jovian polarimetry.

Because the first entry probe is yet to be sent into the atmosphere of Jupiter, essentially all that we know of the Jovian atmosphere results from the study of the intensity and polarization of the radiation coming from the planet (i.e., photometry and polarimetry) as a function of wavelength, position on the planet, time, and scattering geometry. Needless to say, our first concern will not be whether or not to limit the scope of this review but where to place these limits.

With regard to wavelength, we are fortunate to have separate reviews for the long wavelength end of the spectrum in this volume — decameter and hectometer wavelengths by Smith[1] and by Carr and Desch,[2] millimeter to meter by Berge and Gulkis,[3] thermal by Wallace,[4] and 1 to 25 μm by Ridgway et al.[5] We shall not concern ourselves, therefore, with the radio observations and shall mention only briefly some of the infrared results. At the short wavelength end of the spectrum we shall set a limit of $\sim$ 0.2 μm, thus omitting studies at shorter wavelengths that generally are concerned with processes occurring in the upper atmosphere such as scattering and emission in resonance lines.

Generally we shall mention only in passing the large number of studies of relative photometry at high spectral resolution such as studies of absorption line profiles or equivalent widths. We shall rely primarily on the reviews of spectroscopy by Prinn and Owen[6] and Ridgway et al.[7] in this volume to cover this work thoroughly. Also, we will rely on the review of occultation data by Hunten and Veverka[8] to cover that important aspect of Jovian photometry.

Within the above restrictions on wavelength and resolution, we shall attempt to summarize some of the more recent photometric and polarimetric observations as functions of wavelength, position on the planet, and scattering geometry. We shall begin by reviewing some basic definitions and theoretical considerations of photometry in the next section. This will be followed by a section reviewing photometric data for the integrated disk at wavelengths from 0.2 to 1.0 μm. Next we shall consider the photometric data regarding the variation of the brightness of individual belts and zones with wavelength, and then the spatial distribution of intensity across belts and zones. Finally, we shall turn our attention to the much smaller set of polarimetric data for both the integrated disk and for selected regions.

We should mention that space missions to Jupiter are now beginning to extend the range of phase angles of photometric and polarimetric observations of the planet to the entire range from 0° to 180° compared to 0° to 12° available from the earth. This extension will have a great impact on the information that can be extracted from photometric and polarimetric data. Although some of the work reviewed here may be superseded in the near future, it may prove useful to summarize briefly our current knowledge before the flood of new data from space missions descends upon us.

---

[1]See p. 1146.
[2]See p. 693.
[3]See p. 621.
[4]See p. 284.

[5]See p. 384.
[6]See p. 319.
[7]See p. 384.
[8]See p. 247.

I. PHOTOMETRY

Basic Definitions

Much of the notation relating to radiation reflected from a planetary atmosphere can be found in Chandrasekhar (1950). The amount of radiant energy, dE_ν in a specified frequency interval (ν to $\nu + d\nu$), that crosses an element of area $d\sigma$ traveling in directions within a solid angle $d\omega$ in a time dt is given in terms of the intensity I_ν by

$$dE_\nu = I_\nu \cos\theta \; d\nu d\sigma d\omega dt \qquad (1)$$

where θ is the angle between the outward normal to $d\sigma$ and the direction in which the energy is traveling. Note that $\cos\theta d\sigma$ is the area perpendicular to the beam direction. The units of intensity I_ν are, for example, watts cm^{-2} μm^{-1} sr^{-1}.

The net flow of energy crossing the surface $d\sigma$ in all directions is given by the integral of Eq. (1) over solid angle, or

$$d\nu d\sigma dt \int I_\nu \cos\theta d\omega . \qquad (2)$$

The quantity

$$\mathscr{F}_\nu = \pi F_\nu = \int I_\nu \cos\theta d\omega \qquad (3)$$

is called the net flux and is the net flow of energy across $d\sigma$ per unit area per unit time per unit frequency interval.

Next consider the radiation field at the top of a planetary atmosphere at a location where the cosine of the zenith angle of the sun is μ_0 and the azimuth of the sun is defined as ϕ_0. The solar energy striking the atmosphere per unit area of atmosphere is $\mu_0 \pi F$. The intensity reflected at a zenith angle whose cosine is μ and at an azimuth angle ϕ is given by Chandrasekhar in terms of a scattering function S as

$$I(\mu, \phi; \mu_0, \phi_0) = \frac{F}{4\mu} S(\mu, \phi; \mu_0, \phi_0) . \qquad (4)$$

Many techniques are available for computing scattering functions of planetary atmospheres composed of various types of scatterers as reviewed by Hansen and Travis (1974). Scattering particles are classified according to their single scattering albedo, ω, defined as the ratio of the cross section of the particle for scattering to the sum of its cross sections for scattering and absorption and by the angular distribution $P(\cos\theta)$ of radiation scattered through an angle θ in a single scattering event. Extensive tables of functions exist for evaluating S for several specific phase functions; for isotropic scattering $[P(\cos\theta) = \omega]$ by Carlstedt and Mullikan (1966), scattering according to the Rayleigh phase function $[P(\cos\theta) = (3/4)\omega(1 + \cos^2\theta)]$ by Sweigart (1970) and for the phase functions $\omega(1 + a \cos\theta)$ where a is a parameter taking values from -1 to $+1$ (Chandrasekhar 1950). In addition, numerical methods (Hansen and Travis 1974) exist for computing S for arbitrary phase functions.

In cases where observations of $I(\mu, \phi; \mu_0, \phi_0)$ exist over a wide range of scattering geometry, one may hope to derive information on the single scattering albedo and phase function of the particles. This information is needed for the interpretation of spectroscopic observations. In addition, given a reasonable model scattering function, the spherical or Bond albedo of the planet can be calculated. The spherical albedo is defined as the total light reflected from the planet divided by the total light incident on the planet in some frequency interval. In terms of the Chandrasekhar scattering function it is simply

$$A = \int_0^1 \int_0^1 S^{(0)}(\mu, \mu_0) d\mu d\mu_0 \tag{5}$$

where $S^{(0)}(\mu, \mu_0)$ is the zero order term in the Fourier series expansion for $S(\mu, \phi; \mu_0, \phi_0)$; that is, where

$$S(\mu, \phi; \mu_0, \phi_0) = \sum_{m=0}^{\infty} S^{(m)} \cos[m(\phi - \phi_0)] . \tag{6}$$

The spherical albedo at each wavelength together with the thermal emission from the planet indicates the overall energy balance of a planet.

In practice, observations are generally lacking over a wide range of scattering geometries with high spatial resolution. In these cases, observational quantities are often presented relative to scattering from a Lambert surface. A Lambert surface of reflectivity A is one that reflects A times the flux incident on it in such a way that the reflected intensity is independent of the direction to the observer. For an incident flux πF normal to itself from a direction specified by μ_0, we have the reflected flux given by

$$2\pi \int_0^1 I(\mu, \mu_0) \mu d\mu = A \mu_0 \pi F \tag{7}$$

or

$$I_L(\mu_0) = A \mu_0 F . \tag{8}$$

Note that while the intensity of the Lambert surface I_L is independent of the position of the observer, it does depend on the direction from which the incident flux comes, and so a spherical planet whose surface scatters according to Lambert's law will show a decrease in intensity from the center to the limb. The limb darkening observed for Jupiter in the continuum at small phase angles turns out to be approximated quite well by scattering from a Lambert surface.

Observationally, a useful quantity is the geometric albedo p defined as the flux from a planet at $0°$ phase angle (sun-planet-observer angle) divided by the flux that would be measured from a flat disk of the same diameter as that of the planet when oriented normal to the direction to the sun and located at the position of the planet. In addition, let the flux from a planet at phase angle α relative to the flux at $\alpha = 0$ be denoted by $\phi(\alpha)$. If r is the distance from the sun to the planet, R is the radius of the planet, Δ is the distance from the planet to the earth (all measured in astronomical units), and the flux of the sun at the earth's orbit is πF, then the intensity of the Lambert disk is F/r^2, and the flux received from the Lambert disk at $0°$ phase angle is

$$\mathscr{F}_{LD,\, \alpha = 0} = \frac{F}{r^2}\, \pi \left(\frac{R}{\Delta}\right)^2. \tag{9}$$

The flux from the planet at phase angle α, $\mathscr{F}_p(\alpha)$, relative to the flux from the Lambert disk at zero phase angle is

$$\frac{\mathscr{F}_p(\alpha)}{\mathscr{F}_{LD,\, \alpha = 0}} = p\phi(\alpha) \tag{10}$$

so

$$\mathscr{F}_p(\alpha) = p\phi(\alpha)\pi F \left(\frac{R}{r\Delta}\right)^2. \tag{11}$$

In terms of the observed magnitude of the planet at phase angle α, $m_p(\alpha)$, and the magnitude of the sun $m_\odot$, the flux of the planet is given by

$$\log \mathscr{F}_p = 0.4[m_\odot - m_p(\alpha)] + \log \pi F \tag{12}$$

so

$$\log p + \log \phi(\alpha) = 0.4[m_\odot - m_p(\alpha)] + 2\log\left(\frac{r\Delta}{R}\right). \tag{13}$$

Observations at $\alpha = 0$ thus give p, and observations at other phase angles give $\phi(\alpha)$.

The spherical albedo of the planet can be obtained from p and $\phi(\alpha)$ if it is assumed that all parts of the surface of the planet obey the same scattering law. If the flux reflected by the planet in a direction making an angle α with the incident direction is given by Eq. (11), then the total energy reflected by the planet relative to the energy incident on it is

$$A = \frac{p\pi F r^{-2} R^2 \Delta^{-2}}{\pi F \pi R^2 r^{-2}}\, 2\pi \int_0^\pi \phi(\alpha)\Delta \, \sin(\alpha)\Delta d\alpha \tag{14}$$

or

$$A = p \left[2 \int_0^\pi \phi(\alpha) \sin \alpha \, d\alpha \right] \qquad (15)$$

The quantity in parentheses is defined as the phase integral q, or

$$q \equiv 2 \int_0^\pi \phi(\alpha) \sin \alpha \, d\alpha . \qquad (16)$$

For a spherical planet scattering according to Lambert's law, $q = 1.50$, whereas for other scattering laws q takes on other values (Harris 1961). In practice, the determination of the spherical albedo, A, of a planet like Jupiter which has a non-uniform cloud cover is a difficult task even with information from a space flight. We instead begin by considering the variation with wavelength of Jupiter's geometric albedo, p, which is observable from the earth.

Photometry of the Integrated Disk

Much of the recent photometry of the integrated disk of Jupiter between 0.2 μm and 1.0 μm has been summarized by Axel (1972), who offered a model for explaining the general trend of brightness of the planet with wavelength. Figure 1 summarizes the variation of geometric albedo with wavelength. The solid points are from the broadband UBVRI photoelectric photometry given by Harris (1961), and the bars (indicating the widths of the filter used) are the narrowband ($\sim$ 100 Å) measurements described by Young and Irvine (1967) and Irvine et al. (1968a,b). The solid curve is the scanner spectrophotometry of Jupiter by Taylor (1965) converted to geometric albedo by Axel using the solar flux given by Labs and Neckel (1968). The albedos given by Harris and by Irvine et al. increase by $\sim$ 10% if the solar magnitude is adjusted to $-$ 26.71 as suggested by Labs and Neckel from the value of $-$ 26.81 used in this plot. Below 3000 Å, the circles in Axel's figure give the rocket data of Kondo (1971), and the dashes give the Orbiting Astronomical Observatory (OAO) data of Wallace et al. (1972).

Earlier observations in ultraviolet light (uv) of Jupiter have been reported by Anderson et al. (1969), Moos et al. (1969), Jenkins et al. (1969), Evans (1967), Stecher (1965), and Boggess and Dunkelman (1959). The observations of Evans and those of Jenkins et al. were obtained photographically whereas the others were photoelectric measurements.

Analyses of the data from 2200 Å to 3000 Å by Stecher (1965) and by Greenspan and Owen (1967) emphasized the low value of Jupiter's albedo in this spectral region by showing that it was consistent with conservative Rayleigh scattering from only 10 to 12 kilometer amagats[9] of H_2 above even a black cloud layer. Moreover, the most likely constituent for the cloud par-

---

[9]An abundance of one kilometer amagat (km am) of a gas above a particular level in an atmosphere corresponds to 2.69×10^{24} molecules of the gas above every square centimeter of atmosphere at that level.

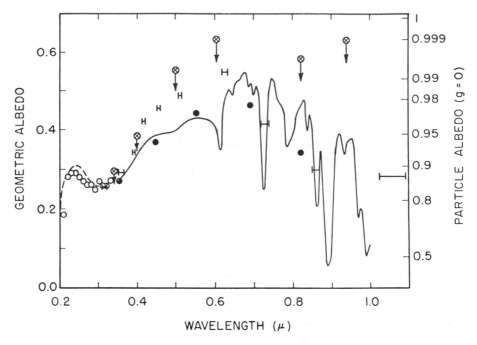

Fig. 1.   Observed reflectivity of Jupiter as a function of wavelength, with corresponding particle single-scattering albedo, ω_0, for isotropic scattering. The sources of the data are: *filled
circles*, Harris (1961); *bars* (indicating half-widths of filters used), Irvine *et al.* (1968b); *open
circles*, Kondo (1971); *broken line*, OAO data (Wallace *et al.* 1972); *solid line*, Taylor (1965);
crosses, Pilcher and McCord (1970). The latter (measured for the equatorial zone) have been
normalized to agree with other observations in the ultraviolet as shown by the arrows (from
Axel 1972).

ticles, NH_3 crystals, would not be expected to absorb longward of 2100 Å
(Dressler and Schnepp 1960) and should be a bright rather than a dark surface under the gas. Rayleigh scattering above a bright cloud would make the
planet increasingly bright at shorter wavelengths rather than increasingly
dark as observed. Further, infrared absorption-line data were consistent with
some 70 or more (rather than 10) km am of H_2 above a cloud deck when analyzed with a model of gas absorption above a discrete cloud deck (Owen and
Mason 1969; Fink and Belton 1969). Homogeneously mixing the cloud particles with the gas did not seem to reduce the discrepancy between ultraviolet
and infrared abundance estimates (Anderson *et al.* 1969).

Recent work has led to models that seem promising for understanding
Jupiter's general variation of albedo with wavelength. Lewis (1969) has suggested that Jupiter's clouds might occur in layers, and Danielson and Tomasko (1969), and others since [e.g., Axel (1972); Bergstralh (1973a,b); Hunt
(1973); and Hunt and Bergstralh (1974)] have found a two-layer cloud model (see Fig. 2) useful in reconciling the equivalent width of strong and weak

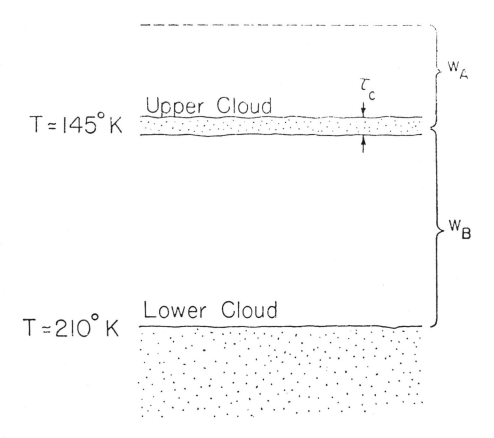

Fig. 2.   Model of the Jovian clouds. The upper cloud has an optical thickness τ_c; the abundance of H_2 above the upper cloud is denoted by w_A, while w_B is the abundance of H_2 in the clear space between the clouds (from Danielson and Tomasko 1969).

absorption lines formed on Jupiter. The upper cloud in these models is semi-transparent in infrared light (ir) and allows photons to sample large abundances of gas as required by the H_2 quadrapole lines. It also occurs relatively high in the atmosphere, so if some process can be found to darken the cloud particles toward the uv it may by itself be consistent with the uv albedo. Otherwise, Danielson and Tomasko (1969) have suggested the presence of absorbing particulates mixed with gas above the upper cloud.

Such a "dust" model for the behavior of Jupiter's albedo from the near ir to the uv has been developed by Axel (1972). Axel's model consists of small particles mixed with the gas above the clouds. He points out that very small particles will be dark and have an optical depth that varies as $1/\lambda$ even if their index of refraction is independent of wavelength. Thus, the "dust" can have a small effect at near-ir wavelengths and becomes increasingly dark toward the uv. At some wavelength the λ^{-4} variation of Rayleigh scattering

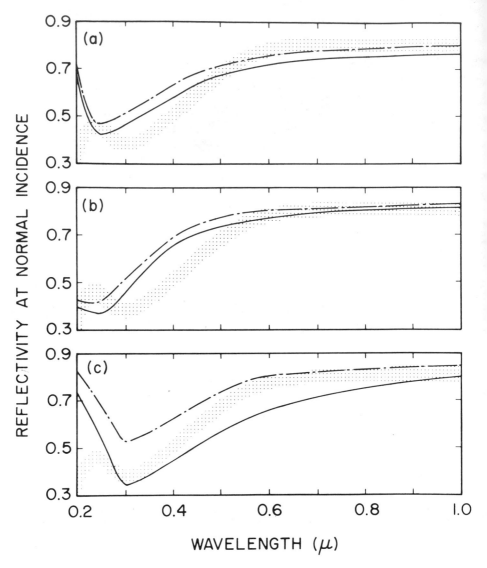

Fig. 3. *(a)* Calculated reflectivity for normal incidence and scattering as a function of wavelength for a layer of 30 km-amagat H_2 containing two amounts of dust over a Lambert surface of reflectivity $A = 0.9$. The dust is small compared to the wavelength and has a constant index of refraction. The values of the optical depth in dust at 0.3 μm are $\tau_D = 0.20$ *(broken line)* and $\tau_D = 0.25$ *(solid line)*. τ_H (0.3 μm) = 0.97. Absorption by NH_3, which has not been included in the calculations, will lower the curves below 2300 Å. For comparison, the stippled curve gives an (approximate) average of the observations. *(b)* Same as Fig. 3a, but for graphite grains. *(c)* Same as Fig. 3a, but for 0.05-μm graphite spheres, with relative number densities $n_D/n_H = 1.55 \times 10^{-17}$ *(dot-dash line)*; $n_D/n_H = 3.0 \times 10^{-17}$ *(solid line)* (from Axel 1972).

begins to dominate, and the albedo of the planet increases with decreasing wavelength beyond this point.

Some of Axel's models are given in Fig. 3 where he plots the reflectivity at normal incidence of his models compared with 1.5 times the observed geometric albedo (the reflectivity at normal incidence if the planet behaves as a Lambert surface). Even without invoking a variation of index of refraction with wavelength, the model seems to have promise for understanding the general variation of reflectivity with wavelength. In addition, several other investigations using polarimetry (Gehrels *et al.* 1969), satellite eclipse observations (Price and Hall 1971), and large phase angle photometry (Tomasko *et al.* 1974) also suggest the need for aerosols mixed with the gas above the clouds.

Axel has suggested absorption by molecular NH_3 as a way to turn the albedo of the planet downward shortward of about 2300 Å as indicated by the more recent observations (see Fig. 4) but did not include the effect in his calculations. NH_3 is also mentioned as a candidate for causing the decrease in albedo shortward of 2300 Å by Wallace *et al.* (1972), Owen and Sagan (1972), and Anderson and Pipes (1971), who emphasize the possible role of solid NH_3 at the shorter wavelengths.

Tomasko (1974) computed the albedo of Jupiter in this region of the spectrum for various distributions of NH_3 with altitude above a bright cloud and including absorption by "dust" suspended in the gas. He found that the decrease in mixing ratio predicted by the NH_3 vapor pressure curve was still too slow to prevent the planet from being much darker than observed at 2100 Å. When the profile calculated by Strobel (1973) including the effects of photodestruction of NH_3 at high altitudes is used, however, the albedo can fall within the error limits of the observations.

There is still some question as to whether some other absorber is required to flatten the albedo between 2200 and 2400 Å where NH_3 absorbs very little, and indeed, whether the effect of this other absorber might not be important throughout the region from 2000 to 2400 Å. Higher spectral resolution in this spectral region could be used to hunt for the approximately 30 Å vibrational structure expected for NH_3 in order to clarify its role.

In any case, Axel's model of absorbing dust above the upper of two cloud layers seems to allow a way of reconciling a large body of uv and ir observations. Observations of the distribution of brightness over the disk, especially at ultraviolet wavelengths, would seem capable of giving much information on the distribution and properties of the absorbers. While these observations do not yet exist at uv wavelengths, they are planned for the International Ultraviolet Explorer satellite in 1977. Meanwhile we turn our attention to available observations of the variation of brightness of Jupiter's belts and zones at visible and near-ir wavelengths.

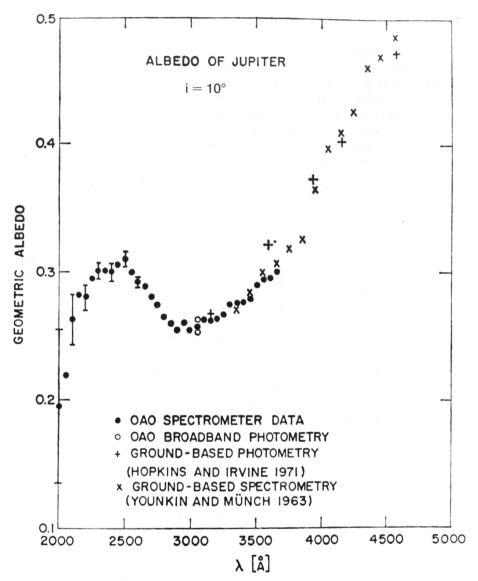

Fig. 4.   Continuous albedo of Jupiter at a phase angle of 10°. Error bars indicate possible error due only to the background correction. The absolute level of the OAO curve has been set by the OAO broadband photometry data at 3062 Å. The groundbased data of Hopkins and Irvine (1971) for a phase angle of 10°, as well as the relative groundbased data of Younkin and Münch (1963) adjusted to the data of Hopkins and Irvine, are also shown (from Wallace *et al.* 1972).

Photometry of Belts and Zones

Spectrophotometry of individual belts and zones of Jupiter has been carried out in a number of investigations. Younkin and Münch (1963) (see also

Münch and Younkin, 1964*a*,*b*) obtained spectra with some 10Å resolution of various belts and zones. Pilcher and McCord (1971) used a set of 53 narrow (~ 200 to 500 Å wide) interference filters to obtain the relative spectral reflectivities of the South Tropical Zone (STrZ), the North Tropical Zone (NTrZ), and the combination of the North Equatorial Belt (NEB) and Equatorial Zone (EZ) from 0.30 to 1.60 μm. Binder and McCarthy (1973) used a set of 10 interference filters from 0.59 μm to 2 μm to explore the distribution of reflectivity on the disk (see next section). Pilcher *et al.* (1973) used a 5-arcsec diaphragm on the central meridian of the NTrZ, NEB, EZ, and STrZ to obtain absolute spectra at better than 10 Å resolution from 0.32 μm to 1.12 μm. Orton (personal communication 1975) carried out a similar study from 3390 Å to 8400 Å at an even greater number of latitudes.

The data of Pilcher, Prinn, and McCord, and of Orton are often described by giving the single scattering albedo for a semi-infinite atmosphere of isotropically scattering particles that would produce the observed reflected absolute intensity. This is an especially useful form because it avoids the necessity to know the precise scattering geometry at which these data were obtained. To use the data to predict the brightness of the region of interest at any new, slightly different geometry, only the new geometry (μ, μ_0) must be known. This can be significant because the details of the scattering geometry under which these photometric data were obtained are not always given. Fortunately, for this purpose, the range of scattering geometry possible for viewing Jupiter from the earth is not large.

Tables I and II and Figs. 5, 6, and 7 summarize the isotropic single scattering albedos obtained by Pilcher *et al.* (1973) at various wavelengths and locations on the planet compared with single-scattering albedos that have been derived from integrated disk data. As these authors point out, the

TABLE I
Isotropic Single-Scattering Albedos for Four Areas on Jupiter[a]

Area	Wavelength[b] (Å)		
	3400	4000	5000
NTrZ	0.91	0.97	0.994
NEB	0.87	0.92	0.981
EZ	0.86	0.91	0.990
STrZ	0.91	0.96	0.994
Whole disk[c]	0.86	0.93	0.981

[a]From Pilcher *et al.* (1973).
[b]The error for all values larger than 0.990 is ± 0.003. For values near 0.90 the error is ± 0.02.
[c]Whole disk values are derived from the data of Irvine *et al.* (1968).

newer values are substantially higher than the older ones and will affect the scattering calculations for absorption line formation. In addition, the improved values for absolute brightness of various belts and zones will provide important constraints on both the limb darkening models discussed in the next section and on models for the chemical cause of the colors of the belts and zones.

Several investigators have measured the variation of equivalent width of the prominent CH_4 bands at 6190, 7250, and 8900 Å and the NH_3 bands at 6450 and 7900 Å (Pilcher *et al.* 1973; Orton 1975; Bugaenko *et al.* 1971; Teifel 1968, 1969; Moroz and Cruikshank 1969; Avery *et al.* 1974). However, as the quantum numbers for lines of these bands have not yet been determined, quantitative analysis is difficult. Also, molecular band models generally do not include scattering, which is important in the formation of these Jovian absorption features. An exception is the attempt by Michalsky *et al.* (1974) to analyze the center-to-limb variation of the 6190 and 7250 Å CH_4 bands by using a two-layer cloud model. In modeling the band they have selected a smooth variation of single scattering albedo and optical depth with frequency to match the profile of the band when observed at low resolution. It is somewhat unclear that the center-to-limb variation of this model will be the same as one in which the variation of single-scattering albedo and optical depth with frequency are determined from the very rapid and strong variations of the actual band. Multiple scattering would be expected to remove rapidly in a non-linear way the light at frequencies where the absorption coefficient is large and to remove relatively less light where the absorption coefficient is small. Thus, it is difficult to choose an "average"

TABLE II

Continuum Isotropic Single-Scattering Albedos at the Center of the Jovian Absorption Bands[a]

	Wavelength[c] (Å)					
Area[b]	CH_4 6190	NH_3 6450	CH_4 7250	NH_3 7900	CH_4 8900	$CH_4 + NH_3$ 10,000
NTrZ	0.997	0.995	0.991	0.987	0.990	0.983
NEB	0.994	0.994	0.990	0.987	0.991	0.990
EZ	0.996	0.996	0.994	0.991	0.993	0.991
STrZ	0.998	0.998	0.997	0.995	0.996	0.996
Whole disk[d]	0.992	0.990	0.97	0.94	0.89	0.88

[a]From Pilcher *et al.* 1973.
[b]The continuum levels at the band centers were interpolated between the levels at either side of the bands.
[c]The error for all values larger than 0.990 is ± 0.003.
[d]Whole disk values calculated from photometric data of Irvine *et al.* (1968) are also shown.

single-scattering albedo that produces the same reflected intensity. When different scattering geometries are involved, such as when center-to-limb variations are investigated, it is not clear that it is possible to define an average single-scattering albedo over frequency in a consistent way for each geometry that gives the same reflected intensity produced by the rapidly varying actual absorption coefficients.

It may be that laboratory studies such as those by Fink and Dick (personal communication 1975) will yield models for the decrease of transmission

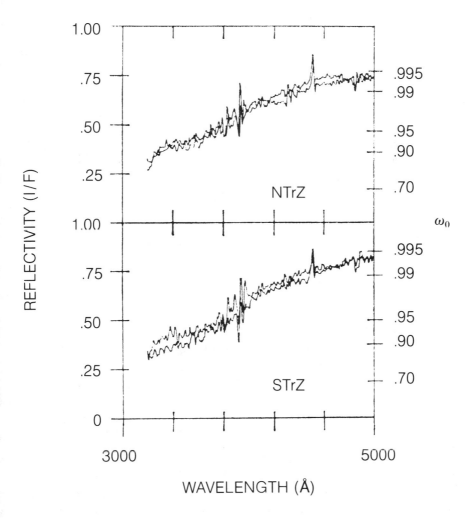

Fig. 5. Reflectivities of two areas on Jupiter, successive measurements shown superimposed. The reflectivities are given in units of I/F, where I is the specific intensity observed from Jupiter, πF the solar flux, and ω_0 the isotropic single scattering albedo. The narrow features in both reflectivities are artifacts of the data reduction process (from Pilcher *et al.* 1973).

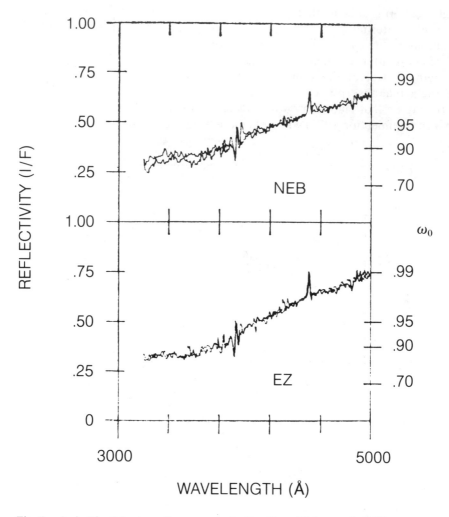

Fig. 6.   As in Fig. 5 for two other areas on Jupiter (from Pilcher *et al.* 1973).

in portions of these bands with increasing path lengths and pressure, which, together with scattering models, can aid in the extraction of some of the information contained in this portion of the spectral data. Indeed, some work is being done along these lines such as that by Clements (1974) discussed in the next section.

For the moment, most investigators simply look for trends in the observed band-equivalent widths. These are somewhat difficult to see in large accumulations of the data, perhaps partly because of time variation. In general, the 7250 and 6190 Å CH_4 bands decrease by $\sim$ 10% from center to limb along the equator, and the 7250 Å band increases by 10% from center

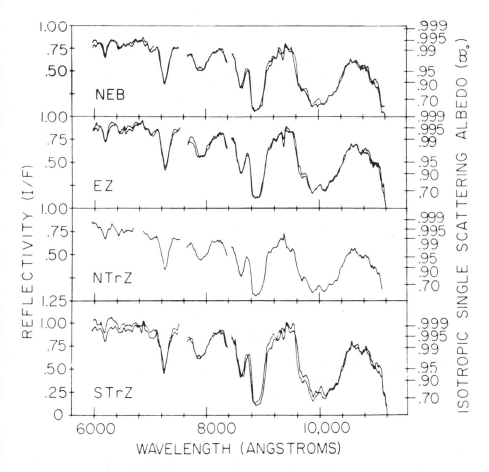

Fig. 7.   Reflectivities of four areas on Jupiter, successive measurements shown superimposed. Units and symbols are the same as for Fig. 5. The gaps in the spectra occur at positions corresponding to changes in the gain of the data system amplifier. The skewing of the trace at either side of the gap near 7600 Å is an artifact of the data reduction process and the gap near 8400 Å occurs in the methane band (from Pilcher *et al.* 1973).

to poles (Avery *et al.* 1974). The 6450 Å NH_3 band also decreases from the center to the limb along the equator and increases (by an amount approaching 40%) along the central meridian from the center to the poles (Avery *et al.* 1974).

These kinds of effects have been taken as evidence for the cloud deck occurring at a lower altitude at higher latitudes (Moroz and Cruikshank 1969). However, narrowband images obtained in strong CH_4 bands (Owen and Mason 1969) show bright areas over the poles, which may indicate a high deck of scatterers over the poles (Owen and Mason 1969; Fountain and

Larson 1971; Fountain 1972; Minton 1972). Primarily, these data demonstrate the need for scattering models to be used in the analysis of Jovian photometric data, although a comprehensive band model including scattering remains to be constructed.

More successful to date have been scattering models of individual H_2 quadrupole lines and of the $3\nu_3$ band of CH_4 at very high resolution when the absorption coefficient can be accurately taken as constant over each of several narrow frequency intervals and where the quantum levels are known (Fink and Belton 1969; Danielson and Tomasko 1969; Axel 1972; Margolis and Hunt 1973; Hunt 1973; Bergstralh 1973a,b). In general, scattering models consisting of absorbing gas above and between two cloud layers have been more successful than single homogeneous cloud models for understanding the relatively flat center-to-limb variations of these spectral features, although it is possible that a model consisting of clear gas above a very diffuse haze could also reproduce the data.

Narrowband photometric images of Jupiter at high resolution are also being made at longer wavelengths. For example, the work of Westphal (1969), Keay et al. (1973), Westphal et al. (1974) at 5 μm,[10] which approaches the spatial resolution of visual images, shows a striking structure with hot spots at temperatures of $> 300°$K. While there is little apparent correlation of these images with images obtained in the 8900 Å CH_4 band, there may be a general correlation of the 5-μm data and visual color (Minton 1973). Generally, the whitest regions on visual images, the STrZ for example, are coolest at 5 μm whereas the bluish features, such as portions of the NEB, are hottest at 5 μm. Scattering models for the data at these wavelengths are being constructed (Orton).[11]

Limb Darkening

To compute the distribution of intensity across a planet observed at some phase angle, the phase function for single scattering and the single-scattering albedo of the scatterers must be known as functions of altitude. In principle, limb darkening observations contain information on these quantities. In turn, the phase function contains information on the size of the particles.

It is desirable to have such a scattering model for an atmosphere for a number of reasons. First, scattering often can play an important part in the formation of spectral lines, and some model for scattering is often required to extract and interpret the temperature, pressure, and abundance information contained in such data. This is very much the case for Jupiter.

Also such a model, if sufficiently sophisticated, allows computation of the spherical albedo of the planet, which is needed together with measures of the planet's thermal emission in order to obtain the overall energy balance of the

---

[10]See p. 407.
[11]See p. 213.

planet. It may seem that direct measures of the variation of planetary brightness with phase as seen from a space probe would give this quantity directly, but this is true only to the extent that the scattering properties of the atmosphere are independent of position on the planet. For Jupiter this is far from true, and the brightness observed by a space probe will depend on the latitude of the sub-spacecraft point as well as on the phase angle. Scattering models of belts and zones, constrained by the space probe measurements at large phase angles, are needed to provide the spherical albedo of the planet.

Finally, the scattering models are of some interest in themselves. For example, the images of the planet at various wavelengths show great detail. We wish to know the extent to which bright and dark bands are caused by variations in single-scattering albedo, phase function, or variations with height of these parameters. Are we seeing bright clouds on a dark background, or dark clouds on a bright background, for example?

A few recent studies of the limb darkening of Jupiter at visible and near-infrared wavelengths have been made from the earth. Smith (1972) studied limb darkening in high-resolution photographs made in several different colors. The distribution of intensity across the apparently homogeneous bands looks remarkably like that predicted by a Lambert surface for the small phase angles observable from the earth. In particular, Smith found that while the data did not fit the values predicted by isotropic scattering, the phase function $\omega(1 + a \cos\theta)$ gave acceptable agreement with values of $a \simeq 0.3$ for both belts and zones in 1971. That is, Smith's data were consistent with the hypothesis that the difference between belts and zones was due only to changes in single-scattering albedo, ω. Smith also analyzed some data from 1969 that suggested a different (negative) value for a for both belts and zones.

Binder (1972) and Binder and McCarthy (1973) have performed a similar analysis on narrowband photoelectric measures of various portions of Jupiter's disk at wavelengths from 0.6 to 2.0 μm. They found values of a in the range from 0.42 to 0.85 depending on latitude and wavelength (see Figs. 8 and 9).

There is, however, a great danger in assuming that simple scattering models, which reproduce the limb darkening at small phase angles, will also hold at large phase angles. For example, Tomasko et al. (1974) reduced for red light the Pioneer 10 limb darkening at phase angles of 34° and 109°. They compared the models to data consisting of a thin absorbing layer over a semi-infinite cloud of particles scattering according to the Henyey-Greenstein phase function and found the best agreement of this limited portion of the Pioneer data with phase functions having asymmetry parameters from 0 to 0.25. The asymmetry parameter is defined as the average value of the cosine of the scattering angle for singly scattered radiation. For the phase function used by Smith and by Binder and McCarthy, the asymmetry parameter is $a/3$, so the asymmetry parameters found by these investigators are in the same range as that obtained in the preliminary Pioneer analysis.

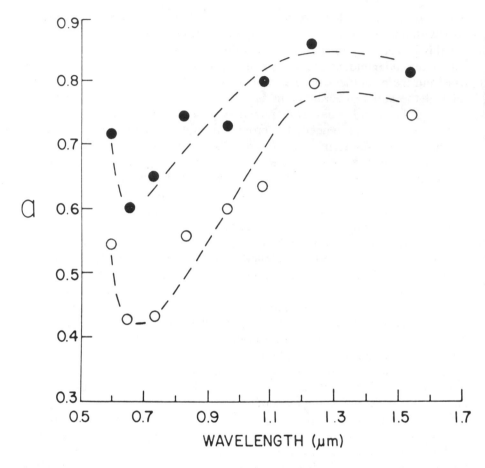

Fig. 8.   Plot of a for the phase function $P(\cos\theta) = \omega(1 + a\cos\theta)$, which fit the limb darkening at small phase angles at various wavelengths. The open circles are for the South Tropical Zone (STrZ) and the filled circles are for the Equatorial Region (from Binder and McCarthy 1973).

Recently, however, additional Pioneer photometry has been reduced at even larger phase angles of 120° and 127° by Tomasko and Castillo (1975). These data are in poor agreement with models having such small asymmetry parameters (see Fig. 10). On the other hand, the data for the STrZ in the red, for example, is best fit overall by a cloud of particles each having a large asymmetry parameter (~ 0.75) above which a small amount of absorption and scattering according to the Rayleigh phase function is permitted to occur (see Fig. 11). A similar model is in fair agreement with the blue data for the STrZ.

In the north component of the South Equatorial Belt (SEBn), however, a slightly different structural model may be required. The data seem to show a drop at the bright limb of the largest phase angle images and a rather flat

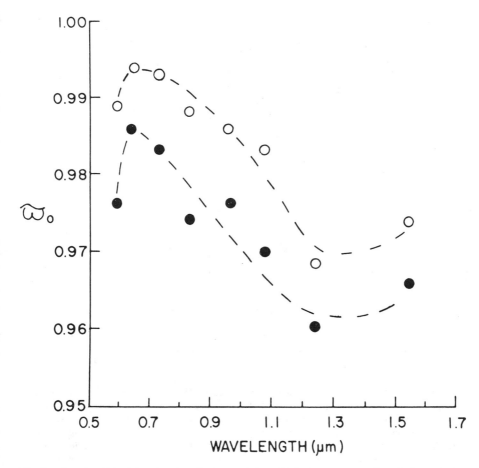

Fig. 9. Same as Fig. 8 but showing single-scattering albedo ω versus wavelength (from Binder and McCarthy 1973).

shape at the small phase angles (see Fig. 12). The modeling of these images has not been completed, but the data suggest a lower single-scattering albedo for the cloud particles as the main source of the greater absorption in the belt (to account for the flat shape at low phase angles), while also requiring the presence of an increased concentration of absorbers at high altitude to produce the decrease at the bright limb of the large phase angle images.

The values of q for the models constructed to date of individual belts and zones show considerable variation—from ~ 1.1 to ~ 1.5. The planet-wide average of q is yet to be determined. The limb darkening data over a large variation of phase angle should provide an important new set of constraints on scattering models of the Jovian clouds.

Another type of data that holds promise for yielding information on the cloud structure of Jupiter's atmosphere is limb darkening in and out of ab-

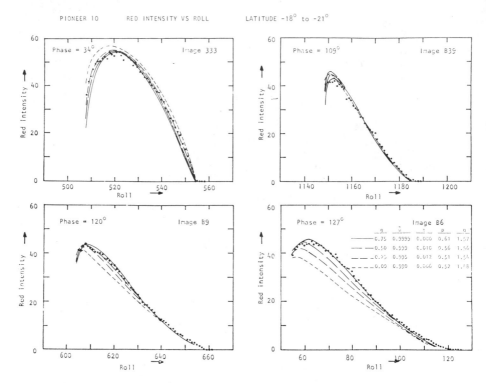

Fig. 10. Red (5950 to 7200 Å) intensity in the STrZ (− 18° to − 21° latitude) versus position (called roll number) across four Pioneer 10 Jupiter images at the phase angles indicated. The dots are the observations and the curves are four models consisting of a layer of gaseous extinction having an optical depth τ above a semi-infinite cloud of particles scattering according to the Henyey-Greenstein phase function with asymmetry parameter g and single scattering albedo ω. The geometric albedo p and phase integral q of a spherical planet everywhere scattering according to each model is also shown.

sorption bands. Clements (1974) has applied layer-adding and doubling techniques to the analysis of limb darkening scans of Jupiter's STrZ and North Polar Region (NPR)—he actually scanned ∼ + 40° latitude using filters ∼ 100 Å wide centered in the CH_4 band at 8800 Å and in the nearby "continuum" at 9215 Å. He compared the data with the limb darkening predicted from families of two-layer cloud models having gas above and between the cloud layers. While the optical depth of the upper cloud and the CH_4 abundances in the clear spaces depended somewhat on cloud phase function and absorption band model, Clements did find good agreement between the data and the types of two-layer models recently used to analyze equivalent widths of H_2 quadrupole lines and the $3\nu_3$ band of CH_4. Furthermore, Clements found that the optical thickness of the upper cloud layer was thinner by a factor of approximately 5 in the NPR than in the STrZ, almost independent of his other model assumptions. It would be most inter-

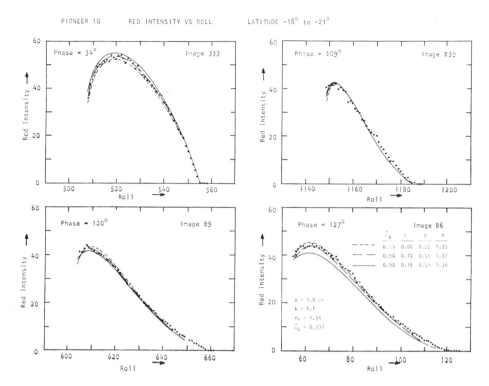

Fig. 11.   Same as Fig. 10 but for a size distribution of spherical particles characterized by parameters a and b as shown (see Hansen and Travis 1974) with index of refraction 1.34 and single scattering albedo $\omega_c = 0.997$. Above this semi-infinite cloud deck are three different optical depths τ of particles with single scattering albedo $\widetilde{\omega}_R = 0.90$ scattering according to the Rayleigh phase function.

esting to repeat his analysis on data for other regions to investigate belt and zone differences in general.

Another interesting feature of Clement's analysis is his exploration of the range of scattering models that can fit the limb darkening in the 8900 Å CH_4 band. He found a family of two-layer models (of the type shown in Fig. 2) that could fit the data ranging from models having a large gas abundance (W_B in Fig. 2) between cloud layers that have short mean-free scattering paths, to models in which the clear space (W_B) between the layers vanished and that had a long mean-free scattering path in one semi-infinite cloud.

The gas abundance above the cloud or clouds (W_A in Fig. 2) was relatively constant throughout the range of acceptable models. It may be hoped that similar analysis using data obtained in bands of different strengths can reduce the range of acceptable scattering models by constraining the mean-free scattering paths in the clouds.

In summary, the analyses that have been performed to date indicate the presence of a relatively clear gas above an extended region in which multiple

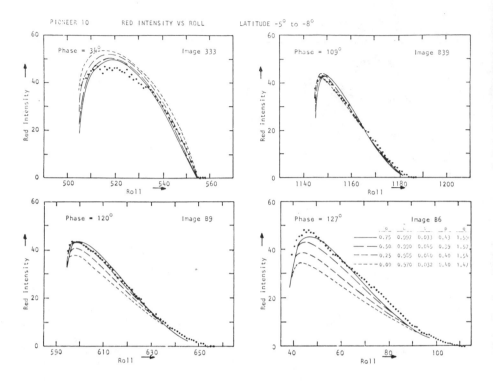

Fig. 12.   Same as Fig. 10 but for latitudes − 5° to − 8°.

scattering is required. Apparently, however, much more effort will be required before the vertical and horizontal distributions of scatterers in this region is known.

Limb darkening has also been studied to advantage at longer wavelengths in order to yield a different type of information. Wildey and Trafton (1971) (also, Trafton and Wildey 1974) have studied the limb darkening at 8 to 14 μm where NH_3 is an important source of thermal opacity. They took advantage of the sensitive dependence of NH_3 vapor pressure on temperature (and hence abundance above a given level) to use the limb darkening to constrain the temperature structure and thus the effective temperature of the planet. They obtained 135 ± 4°K (Trafton and Wildey 1974) as the effective temperature in good agreement with other Earth-based determinations of that quantity. Also, recently Murphey and Fesen (1974) have studied 20 μm limb darkening curves and estimated $T_{eff} = 136 ± 4°K$. However, the infrared radiometers aboard the Pioneer 10 and 11 spacecrafts yielded a somewhat lower value of about 125°K (Ingersoll et al.).[12] In any case, Jupiter emits between about 2 and 2.5 times as much radiation as it receives

---

[12]See pp. 202 and 204.

from the sun. The precision to which the size of Jupiter's internal heat source is known is certain to improve as additional groundbased and space-craft data are analyzed.

II. POLARIMETRY

Recent reviews exist of both the theoretical technique of computing the polarization of light reflected from planetary atmospheres (Hansen and Travis 1974), and of the observations and interpretation of the polarization of the planets, including Jupiter (Coffeen and Hansen 1974). We refer the reader to these excellent reviews, and accordingly we shall briefly summarize only some of the recent work on Jupiter in this area.

Groundbased studies of the polarization of Jupiter are handicapped by the small range of phase angle (0° to 12°) available from the earth. Nevertheless, several studies have been made of the polarization of the integrated disk and of the center of the disk with phase angle (Lyot 1929; Morozhenko and Yanovitskii 1973) and of the variation across the disk (Hall and Riley 1969, 1974; Dollfus 1957) including the wavelength dependence of several regions on the disk (Gehrels et al. 1969). The polarization of the center of the disk is increasingly negative at larger phase angles (negative polarization meaning that the electric vector maximum is in the scattering plane). These data have been used by Morozhenko and Yanovitskii (1973) to deduce a mean particle radius of 0.2 μm and an index of refraction of 1.36. This analysis (using Mie theory) was based on the assumption that the particles are spheres, which is not at all certain for Jupiter. Coffeen and Hansen (1974) point out further that the broad glory that gives rise to negative polarization at small phase angles in the Mie theory persists for values of the index of refraction of from 1.25 to at least 2.0, and they urge caution in attempts to derive particle size and index of refraction solely from the small phase angle data even if the particles are spherical. Recently, Kawabata and Hansen (1975) have compared groundbased polarization data as a function of wavelength, phase angle and position on the disk with numerical "doubling and adding" calculations. They find the cloud particle phase matrix to be approximately independent of wavelength for $0.35 \leq \lambda \leq 8$ μm, and suggestive of nonspherical particles larger than the wavelength. They interpret the variation of polarization over the disk as indicating a diffuse vertical distribution of particles, as opposed to a sharp opaque cloudtop. Coffeen and Hansen (1974) also point out that negative polarization is observed at small phase angles for cirrus crystal clouds and emphasize the need for polarization measurements at large phase angles from space probes.

Polarization measurements of Jupiter at large phase angles have been made by the recent Pioneer 10 and 11 flybys of Jupiter (Gehrels et al. 1974; Coffeen 1974; Baker et al. 1975) and more are planned for the Mariner

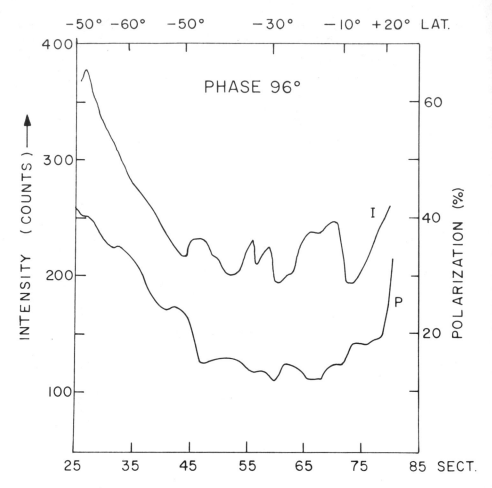

Fig. 13.   Intensity and polarization of blue light as a function of sector (representing one field-of-view) and latitude along a scan in a roughly north-south direction for the southern hemisphere. The data were received 2^h 16^m before pericenter by Pioneer 11 at a phase angle of ~ 96° (from Baker *et al.* 1975).

Jupiter/Saturn mission. The proper analysis of these new data will probably be a lengthy task, particularly if the particles are not spherical.

Nevertheless, some quantities are determined from even relatively simple modeling. For example, in the Pioneer blue data at large phase angles the polarization is large (see Fig. 13) and probably dominated by Rayleigh scattering. A model consisting of Rayleigh scattering over a Lambert surface agrees moderately well with the observed polarization across the STrZ (see Fig. 14) as well as with the limb darkening for this region at a phase angle of 96°. Such a model has been used by Coffeen (1974) and by Baker *et al.* (1975) to estimate the Rayleigh scattering optical depth of the gas

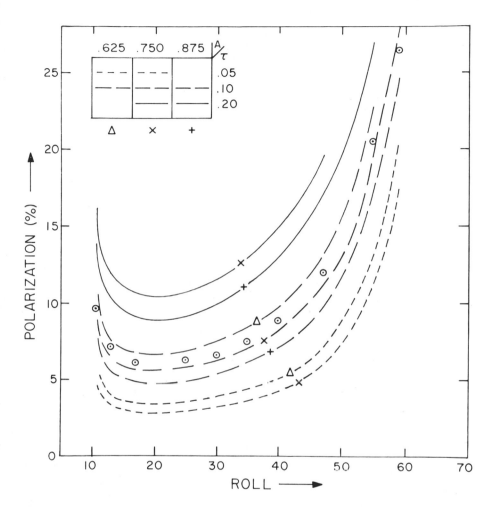

Fig. 14. Observed polarization of blue light scattered by Jupiter's STrZ (circled dots) is plotted against spacecraft roll (numbered from an arbitrary starting roll). Small roll numbers correspond to observations near the limb, and large numbers correspond to observations near the terminator. The observations were made between 2^h 43^m and 2^h 6^m before pericenter by Pioneer 11 at a phase angle of 96°. The data are compared to seven models having different combinations of optical depth τ for conservative Rayleigh scattering above Lambert surfaces of reflectivity A. Close fits can be found for various values of A by interpolation for τ (from Baker *et al.* 1975).

above the "cloudtops." The optical depths derived at 0.44 μm are shown in Table III. The increase from $\tau \simeq 0.1$ at latitudes less than 30° to ~ 0.3 at 60° corresponds to Rayleigh scattering from ~ 15 and 45 km am of H_2. While great caution is needed in interpreting these model optical depths in view of the well known shortcomings of the reflecting layer model for Jupiter, they do seem to argue for increased penetration of the radiation at higher

TABLE III
Latitudinal Dependence of Optical Depth
and Surface Reflectivity at 0.44 μm[a]

Latitude	τ	A	Latitude	τ	A
− 60.4	0.36	0.51	+ 13.2	0.10	0.53
− 49.9	0.22	0.46	+ 20.4	0.10	0.68
− 40.1	0.13	0.49	+ 32.8	0.14	0.51
− 31.0	0.11	0.58	+ 40.0	0.17	0.52
− 19.5	0.10	0.67	+ 49.9	0.23	0.50
− 8.9	0.10	0.64	+ 59.8	0.28	0.50
+ 1.0	0.10	0.57	+ 71.3	0.29	0.50

[a]From Baker et al. (1975).

latitudes — a hypothesis offered by Gehrels et al. (1969) to explain the different wavelength dependence of polarization for equatorial and polar regions. The application of more sophisticated scattering models to the unique large phase angle data is planned.

Circular polarization has been observed on Jupiter by Kemp et al. (1971a,b). Hansen (1971) has shown that the magnitude and symmetry of handedness agree with those expected for multiple scattering. Summaries of the observations have recently appeared (Kemp 1975; Wolstencroft 1976) and Hansen[13] reviews especially the interpretations in this volume.

REFERENCES

Anderson, R. C., and Pipes, J. G. 1971. Jovian ultraviolet reflectivity compared to absorption by solid ammonia. J. Atmos. Sci. 28:1086–1087.

Anderson, R. C.; Pipes, J. G.; Broadfoot, A. L.; and Wallace, L. 1969. Spectra of Venus and Jupiter from 1800 to 3200 Å. J. Atmos. Sci. 26:874–888.

Avery, R. W.; Michalsky, J. J.; and Stokes, R. A. 1974. Variation of Jupiter's CH_4 and NH_3 bands with position on the planetary disk. Icarus 21:47–54.

Axel, L. 1972. Inhomogeneous models of the atmosphere of Jupiter. Astrophys. J. 173:451–468.

Baker, A. L.; Baker, L. R.; Beshore, E.; Blenman, C.; Castillo, N. D.; Chen, Y. P.; Coffeen, D. L.; Doose, L. R.; Elston, J. P.; Fountain, J. W.; Gehrels, T.; Kendall, J. H.; KenKnight, C. E.; Norden, R. A.; Swindell, W.; and Tomasko, M. G. 1975. The imaging photopolarimeter experiment on Pioneer 11. Science 188:468–472.

Bergstralh, J. T. 1973a. Methane absorption in the Jovian atmosphere. I. The Lorentz half-width in the $3\nu_3$ band at 1.1 μm. Icarus 19:499–506.

Bergstralh, J. T. 1973b. Methane absorption in the Jovian atmosphere. II. Absorption line formation. Icarus 19:390–418.

[13]See p. 517.

Binder, A. B. 1972. Spectrophotometry of the 1.5 μm window of Jupiter. *Astron. J.* 77:93–113.
Binder, A. B., and McCarthy, D. W. Jr. 1973. IR spectrophotometry of Jupiter and Saturn. *Astron. J.* 78:939–950.
Boggess, A., and Dunkelman, L. 1959. Ultraviolet reflectivities of Mars and Jupiter. *Astrophys. J.* 129:236–237.
Bugaenko, L. A.; Galkin, L. S.; and Morozhenko, A. V. 1971. Spectrophotometric studies of the giant planets. *Sov. Astron.-A. J.* 15:473–477.
Carlstedt, J. L., and Mullikan, T. M. 1966. Chandrasekhar's X- and Y-functions. *Astrophys. J. Suppl.* 12:449–585.
Chandrasekhar, S. 1950. *Radiative transfer.* Oxford, England: Clarendon Press.
Clements, A. E. 1974. Cloud structure in the South Tropical Zone, Red Spot and North Polar Region of Jupiter. Ph.D. dissertation, Univ. of Arizona.
Coffeen, D. L. 1974. Optical polarization measurements of the Jupiter atmosphere at 103° phase angle. *J. Geophys. Res.* 79:3645–3652.
Coffeen, D. L., and Hansen, J. E. 1974. Polarization studies of planetary atmospheres. *Planets, stars and nebulae studied with photopolarimetry.* (T. Gehrels, ed.) 518–581. Tucson, Arizona: University of Arizona Press.
Danielson, R. E., and Tomasko, M. G. 1969. A two-layer model of the Jovian clouds. *J. Atmos. Sci.* 26:889–897.
Dollfus, A. 1957. Étude des planètes par la polarisation de leur lumière. *Ann. Astrophys. Suppl.* 4. English Transl. NASA-TT F-188, 1964. National Aeronautics and Space Administration, Washington, D.C.
Dressler, K., and Schnepp, O. 1960. Absorption spectra of solid methane, ammonia, and ice in the vacuum ultraviolet. *J. Chem. Phys.* 33:270–274.
Evans, D. C. 1967. Ultraviolet reflectivity of Venus and Jupiter. *The moon and planets.* p. 135. Amsterdam, Netherlands: North-Holland Press.
Fink, U., and Belton, M. J. S. 1969. Collision narrowed curves of growth of H_2. *J. Atmos. Sci* 26:952–962.
Fountain, J. W. 1972. Narrow-band photography of Jupiter and Saturn. *Comm. Lunar and Planetary Lab.* 9:327–338.
Fountain, J. W., and Larson, S. M. 1971. Multicolor photography of Jupiter. *Comm. Lunar and Planetary Lab.* 9:315–326.
Gehrels, T.; Coffeen, D.; Tomasko, M.; Doose, L.; Swindell, W.; Castillo, N.; Kendall, J.; Clements, A.; Hämeen-Anttila, J.; KenKnight, C.; Blenman, C.; Baker, R.; Best, G.; and Baker, L. 1974. The imaging photopolarimeter experiment on Pioneer 10. *Science* 183: 318–320.
Gehrels, T.; Herman, B. M.; and Owen, T. 1969. Wavelength dependence of polarization. XIV. Atmosphere of Jupiter. *Astron. J.* 74:190–199.
Greenspan, J. A., and Owen, T. 1967. Jupiter's atmosphere: its structure and composition. *Science* 156:1489–1494.
Hall, J. S., and Riley, L. A. 1969. Polarization measures of Jupiter and Saturn. *J. Atmos. Sci.* 26:920–923.
———. 1974. Polarization measurements of Jupiter and the Great Red Spot. *Planets, stars and nebulae studied with photopolarimetry.* (T. Gehrels, ed.) pp. 595–598. Tucson, Arizona. University of Arizona Press.
Hansen, J. E. 1971. Circular polarization of sunlight reflected by clouds. *J. Atmos. Sci.* 28: 1515–1516.
Hansen, J. E., and Travis, L. D. 1974. Light scattering in planetary atmospheres. *Space Sci. Rev.* 16:527–610
Harris, D. L. 1961. Photometry and colorimetry of planets and satellites. *Planets and satellites.* (G. P. Kuiper and B. M. Middlehurst, eds.) pp. 272–342. Chicago, Illinois: University of Chicago Press.

Hopkins, N. B., and Irvine, W. B. 1971. *Planetary atmospheres*. (T. Owen and H. Smith, eds.) p. 349. Dordrecht, Holland: D. Reidel Publishing Co.

Hunt, G. E. 1973. Formation of spectral lines in planetary atmospheres. IV. Theoretical evidence for structure of the Jovian clouds from spectroscopic observations of methane and hydrogen quadrupole lines. *Icarus* 18:637–648.

Hunt, G. E., and Bergstralh, J. T. 1974. Structure of the visible Jovian clouds. *Nature* 249: 635–636.

Irvine, W. M.; Simon, T.; Menzel, D. H.; Charon, J.; Lecomte, G.; Griboval, P.; and Young, A. T. 1968a. Multicolor photoelectric photometry of the brighter planets. II. Observations from Le Houga Observatory. *Astron. J.* 73:251–264.

Irvine, W. M.; Simon, T.; Menzel, D. H.; Pikoos, C.; and Young, A. T. 1968b. Multicolor photometry of the brighter planets. III. Observations from the Boyden Observatory. *Astron. J.* 73:807–828.

Jenkins, E. B.; Morton, D. C.; and Sweigart, A. V. 1969. Rocket spectra of Venus and Jupiter from 2000 to 3000 Å. *Astrophys. J.* 157:913–924.

Kawabata, K., and Hansen, J. E. 1975. Interpretation of ground-based observations of the polarization of Jupiter. *Bull. Amer. Astron. Soc.* 7:382–383.

Keay, C. S. L.; Low, F. J.; Rieke, G. H.; and Minton, R. B. 1973. High-resolution maps of Jupiter at five microns. *Astrophys. J.* 183:1063–1073.

Kemp, J. C. 1975. Circular polarization of the planets. *Planets, stars and nebulae studied with photopolarimetry*. (T. Gehrels, ed.) pp. 607–616. Tucson, Arizona: University of Arizona Press.

Kemp, J. C.; Swedlund, J. B.; Murphy, R. E.; and Wolstencroft, R. D. 1971a. Circularly polarized visible light from Jupiter. *Nature* 231:169–170.

Kemp, J. C.; Wolstencroft, R. D.; and Swedlund, J. B. 1971b. Circular polarization: Jupiter and other planets. *Nature* 232:165–168.

Kondo, Y. 1971. Ultraviolet reflectivity of Jupiter observed from a rocket. *Icarus* 14:269–272.

Labs, D., and Neckel, H. 1968. The radiation of the solar photosphere from 2000 Å to 100 μ. *Zs. f. Astroph.* 69:1–73.

Lewis, J. S. 1969. The clouds of Jupiter and the NH_3-H_2O and NH_3-H_2S systems. *Icarus* 10:365–378.

Lyot, B. 1929. Recherches sur le polarisation de la lumière des planètes et de quelques substances terrestres. *Ann. Ob. Paris* (Meudon) VIII. English Transl. NASA-TT F-187, 1964. National Aeronautics and Space Administration, Washington, D.C.

Margolis, J. S., and Hunt, G. E. 1973. On the level of H_2 quadrupole absorption in the Jovian atmosphere. *Icarus* 18:593–598.

Michalsky, J. J.; Stokes, R. A.; Avery, R. W.; and De Marcus, W. C. 1974. Molecular band variations as a probe of the vertical structure of a Jovian atmosphere. *Icarus* 21:55–65.

Minton, R. B. 1972. Latitude measures of Jupiter in the 0.89 μ methane band. *Comm. Lunar and Planetary Lab* 9:339–352.

———. 1973. A correlation between the colors of Jovian clouds and their 5 μm temperatures. *Comm. Lunar and Planetary Lab* 9:389–396.

Moos, H. W.; Fastie, W. G.; and Bottema, M. 1969. Rocket measurement of ultraviolet spectra of Venus and Jupiter between 1200 and 1800 Å. *Astrophys. J.* 155:887–897.

Moroz, V. I., and Cruikshank, D. P. 1969. Distribution of ammonia on Jupiter. *J. Atmos. Sci.* 26:865–869.

Morozhenko, A. V., and Yanovitskii, E. G. 1973. The optical properties of Venus and the Jovian planets. I. *Icarus* 18:583–592.

Münch, G., and Younkin, R. L. 1964a. Wavelength dependence of the band structure of Jupiter and of Saturn. *Astron. J.* 69:565.

———. 1964b. Molecular absorptions and color distributions over Jupiter's disk. *Astron. J.* 69:553.

Murphy, R. E., and Fesen, R. A. 1974. Spatial variations in the Jovian 20-micrometer flux. *Icarus* 21:42–46.

Orton, G. S. 1975. Ph.D. dissertation in preparation, California Inst. of Technology, Pasadena, California; personal communication.

Owen, T., and Mason, H. P. 1969. New studies of Jupiter's atmosphere. *J. Atmos. Sci.* 26: 870–873.

Owen, T., and Sagan, C. 1972. Minor constituents in planetary atmospheric ultraviolet spectroscopy from the Orbiting Astronomical Observatory. *Icarus* 16:557–568.

Pilcher, C. B., and McCord, T. B. 1970. Narrowband photometry of the bands of Jupiter. Paper presented at the American Geophysical Union meeting, Dec. 7-10, 1970 in San Francisco, Calif.

———. 1971. Narrow-band photometry of the bands of Jupiter. *Astrophys. J.* 165:195–201.

Pilcher, C. B.; Prinn, R. G.; and McCord, T. B. 1973. Spectroscopy of Jupiter: 3200 to 11,200 Å. *J. Atmos. Sci.* 30:302–307.

Price, M. J., and Hall, J. S. 1971. The physical properties of the Jovian atmosphere inferred from eclipses of the Galilean satellites. I. Preliminary results. *Icarus* 14:3–12.

Smith, B. A. 1972. Observation of atmospheric limb darkening in the visual continuum and an analysis of multiple scattering in the atmospheres of Jupiter and Saturn. Ph.D. dissertation, New Mexico State University.

Stecher, T. P. 1965. The reflectivity of Jupiter in the ultraviolet. *Astrophys. J.* 142:1186–1190.

Strobel, D. F. 1973. The photochemistry of NH_3 in the Jovian atmosphere. *J. Atmos. Sci.* 30: 1205–1209.

Sweigart, A. V. 1970. Radiative transfer in atmospheres scattering according to the Rayleigh phase function with absorption. *Astrophys. J. Suppl.* 22:1–80.

Taylor, D. J. 1965. Spectrophotometry of Jupiter's 3400–10000 Å spectrum and a bolometric albedo for Jupiter. *Icarus* 4:362–373.

Teifel, V. G. 1968. Optical properties and structure of the Jovian atmosphere. I. Photometric contrasts, molecular absorption, and possible structure of the dark equatorial belt during 1962–1963. *Astron. Vestnik* 2:229–238.

———. 1969. Molecular absorption and the possible structure of the cloud layers of Jupiter and Saturn. *J. Atmos. Sci.* 26:854–859.

Tomasko, M. G. 1974. Ammonia absorption relevant to the albedo of Jupiter. II. Interpretation. *Astrophys. J.* 187:641–650.

Tomasko, M. G., and Castillo, N. D. 1975. Photometry of Jupiter at large phase angles. *Bull. Am. Astron. Soc.* 7:378–379.

Tomasko, M. G.; Clements, A. E.; and Castillo, N. D. 1974. Limb darkening of two latitudes of Jupiter at phase angles of 34° and 109°. *J. Geophys. Res.* 79:3653–3660.

Trafton, L. M., and Wildey, R. 1974. Jupiter: a comment on the 8- to 14-micron limb darkening. *Astrophys. J.* 194:499–502.

Wallace, L.; Caldwell, J. J.; and Savage, B. D. 1972. Ultraviolet photometry from the Orbiting Astronomical Observatory. III. Observations of Venus, Mars, Jupiter, and Saturn longward of 2000 Å. *Astrophys. J.* 172:755–769.

Westphal, J. A. 1969. Observations of localized 5-micron radiation from Jupiter. *Astrophys. J.* 157:L63–L64.

Westphal, J. A.; Matthews, K.; and Terrile, R. J. 1974. Five micron pictures of Jupiter. *Astrophys. J.* 188:L111–L112.

Wildey, R. L., and Trafton, L. M. 1971. Studies of Jupiter's equatorial thermal limb darkening during the 1965 apparition. *Astrophys. J. Suppl.* 23:1–34.

Young, A. T., and Irvine, W. M. 1967. Multicolor photoelectric photometry of the brighter planets. I. Program and procedure. *Astron. J.* 72:945–950.

Younkin, R. L., and Münch, G. 1963. Spectral energy distributions of the major planets. *Mém. Soc. Roy. Sci. Liège* 7:125–136.

Wolstencroft, R. D. 1976. The circular polarization of Jupiter. *Icarus* (special Jupiter issue). In press.

CIRCULAR POLARIZATION OF SUNLIGHT REFLECTED BY JUPITER

Y. KAWATA

and

J. E. HANSEN
Goddard Institute for Space Studies

The detection of circular polarization of the sunlight reflected by planets is a new development which has yet to be exploited for remote investigations of the planets. In this chapter, the circular polarization observations of Jupiter are described, and the circular polarization of other planets is discussed to the extent that it aids interpretation of the Jupiter data. The evidence strongly supports the interpretation that the circular polarization arises from scattering by aerosols in a gaseous atmosphere.

Accurate calculations of the circular polarization are made for multiple scattering by an atmosphere with spherical aerosols, as a function of the particle size and refractive index and the mixing ratio of aerosols and gas. The calculations for spheres and the few available circular polarization observations of Jupiter permit only very limited constraints to be placed on the haze and cloud properties of the atmosphere of Jupiter. However, multispectral circular polarization observations, combined with measurements of the linear polarization and of intensity, would permit detailed analysis of atmospheric aerosol properties.

HISTORICAL DEVELOPMENT

Observations

Kemp *et al.* (1971*a*) discovered circular polarization of sunlight reflected by Jupiter, and Kemp and his colleagues have been responsible for most of the measurements of the circular polarization of the planets. Kemp *et al.* (1971*b*) measured circular polarization on Venus, and Swedlund *et al.* (1972) reported the first such measurements for Saturn. Circular polarization of sunlight scattered in the earth's atmosphere was sought at least as early as 1811 (by Arago), but apparently the first unambiguous measurements were obtained by Eiden (1970); measurements of the circular polarization of twilight in the earth's atmosphere were reported by Angel *et al.* (1972). Kemp *et al.* (1971*b*) also found circular polarization for Mercury and Mars, planets with little atmosphere, and Lipskii and Pospergelis (1967) earlier reported circular polarization for the Moon.

Following the notation of Kemp, let us use q for the circular polarization, which is the ratio of Stokes parameters V/I (cf. Hansen and Travis 1974). The observed circular polarization of Jupiter, as well as that of the other planets, has the following basic characteristics:

1. The northern and southern hemispheres of the planetary disk have approximately equal magnitudes for the circular polarization, but opposite signs (the so-called polar effect):

$$q_N(\rho) \simeq -q_S(\rho) \tag{1}$$

where ρ is the phase (sun-planet-earth) angle, which is defined to be positive after opposition.

2. The magnitude of the circular polarization in a given hemisphere is similar before and after opposition, but the sign changes upon passing through opposition (the so-called opposition effect):

$$q(\rho) \simeq -q(-\rho) . \tag{2}$$

3. The magnitude of the circular polarization is small,

$$|q| \lesssim 10^{-3} - 10^{-5} . \tag{3}$$

Fig. 1 shows the observations of the circular polarization of Jupiter made by Kemp *et al.* (1971*a*) and Michalsky and Stokes (1974). All of the observations were for wavelengths approximately 630–680 nm. The apertures employed varied in size, but in each case a large fraction of a hemisphere was observed. The data of Kemp *et al.* were obtained in 1971 and those of Michalsky and Stokes in 1973.

Interpretation

The basic characteristics of the circular polarization, Eqs. (1)–(3) described above, are sufficient to indicate that the basic mechanism producing the circular polarization is scattering by atmospheric aerosols, as was pointed out by Hansen (1971*a*). This follows from the fact that Jupiter is known to have a substantial gaseous atmosphere and from basic properties of the phase matrices of gases and aerosols.

The intensity and state of polarization of sunlight reflected by a planetary atmosphere can be specified in terms of a reflection matrix, **R**, such that (cf. Hansen and Travis 1974):

$$
\left\{ \begin{array}{c} I \\ Q \\ U \\ V \end{array} \right\} = \mu_0 \left\{ \begin{array}{cccc} R^{11} & R^{12} & R^{13} & R^{14} \\ R^{21} & R^{22} & R^{23} & R^{24} \\ R^{31} & R^{32} & R^{33} & R^{34} \\ R^{41} & R^{42} & R^{43} & R^{44} \end{array} \right\} \left\{ \begin{array}{c} F_0 \\ 0 \\ 0 \\ 0 \end{array} \right\} \tag{4}
$$

where μ_0 is the cosine of the zenith angle of the incident sunlight, πF_0 is the incident flux per unit area perpendicular to the incident beam, and I, Q, U and V are the Stokes parameters of the light reflected in an arbitrary direction. Thus $I \propto R^{11}$ and $V \propto R^{41}$. All 16 elements of $\mathbf{R}$ can be obtained from multiple scattering theory if the single scattering (phase) matrices of the scatterers are known.

The phase matrices for gases (Rayleigh scattering) and for aerosols (spheres or randomly oriented non-spherical particles, each with a plane of symmetry) have the forms (cf. Hansen and Travis 1974):

$$
P(\text{gases}) = \left[\begin{array}{cccc} c & c & s & 0 \\ c & c & s & 0 \\ s & s & c & 0 \\ 0 & 0 & 0 & c \end{array} \right] \tag{5}
$$

$$
P(\text{aerosols}) = \left[\begin{array}{cccc} c & c & s & 0 \\ c & c & s & s \\ s & s & c & c \\ 0 & s & c & c \end{array} \right] \tag{6}
$$

where c represents an even matrix element (containing only cosine terms in a Fourier expansion in the azimuth angle, $\phi - \phi_0$) and s represents an odd matrix element (containing only sine terms).

Multiple scattering consists of taking products of the phase matrices. Thus, second-order scattering involves the product of the two phase matrices involved (with an integration over angle), third-order scattering involves the product of the three participating phase matrices, and so on. It is clear that as long as only Rayleigh scattering is involved (Eq. 5) $R^{41} = 0$, no matter how many scatterings there are. However, if the multiple scattering includes even one scattering by aerosols R^{41} in general becomes non-zero. In addition, for the assumed (quite general) type of phase matrix, R^{41} is always an odd function of $\phi - \phi_0$.

Thus, for an atmosphere containing aerosols, the opposite signs for q in the two planetary hemispheres and the switch in signs as the planet passes through zero phase angle can be easily understood. Two points located symmetrically about a planet's intensity equator have the same zenith angles but azimuth angles $\phi - \phi_0$ and $-(\phi - \phi_0)$. A similar conclusion holds for a given point on the planetary disk at phase angles ρ and $-\rho$. Thus, if the physical conditions are the same in the northern and southern hemispheres, q will have the properties expressed by the polar and opposition effects since $q = R^{41}/R^{11}$ is an odd function of azimuth angle.

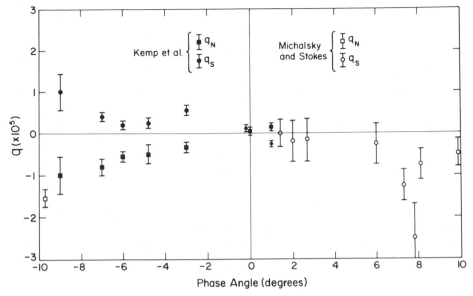

Fig. 1. Circular polarization of Jupiter measured by Kemp *et al.* (1971*b*) and Michalsky and Stokes (1974). The observations of Kemp *et al.* were made with a filter centered at 680 nm with a bandwidth ~ 50 nm. The observations of Michalsky and Stokes were made at ~ 633 nm with a filter width ~ 5 nm, except the point at $\rho = -9.7°$ which was made with a filter centered at 645 nm having a width of ~ 120 nm.

The conclusion that aerosols introduce the circular polarization is further substantiated by quantitative calculations in the following section and in more detail in a paper by Kawata in preparation. We will thus mention only briefly other proposed mechanisms for the production of the circular polarization; more extended discussion has been given by Kemp (1974). A magnetic field origin for the circular polarization is ruled out by the opposition effect. A solid surface could produce the polar and opposition effects, but we know from spectroscopic studies that, even if Jupiter has a solid surface, the atmosphere above it is too substantial to permit the surface to be viewed. Double (or multiple) Rayleigh scattering by absorbing atoms gives zero circular polarization, because the phase matrix is of the type in Eq. (5), even for anisotropic Rayleigh scattering. Absorption does not modify the symmetry of the phase matrices in Eqs. (5) and (6) or the symmetry conditions discussed above (although absorption does change the quantitative circular polarization introduced by aerosols, as illustrated by Hansen and Travis 1974).

COMPUTATIONS FOR SPHERES

In order to illustrate the production of circular polarization by multiple scattering we make computations for spherical particles. We first illustrate the form of the phase matrix as a function of size parameter and scattering

angle. We then perform multiple scattering computations both with and without Rayleigh scattering included. Finally we compare the results to the observations of Jupiter.

Single Scattering

For the special case of isotropic homogeneous spheres, the phase matrix
has the form

$$\mathbf{P}(\alpha) = \left\{ \begin{array}{cccc} P^{11}(\alpha) & P^{21}(\alpha) & 0 & 0 \\ P^{21}(\alpha) & P^{11}(\alpha) & 0 & 0 \\ 0 & 0 & P^{33}(\alpha) & -P^{43}(\alpha) \\ 0 & 0 & P^{43}(\alpha) & P^{33}(\alpha) \end{array} \right] \tag{7}$$

where α is the scattering angle. The components of the phase matrix can be
obtained from the well-known Mie theory. They depend on the size distribution of the spheres, $n(r)$, the size parameter x (equal to $2\pi r/\lambda$, where r is the
particle radius and λ the wavelength), and the optical constants n_r and n_i,
which are the real and imaginary parts of the refractive index.

We use the particle size distribution (cf. Hansen 1971b; Hansen and Travis 1974)

$$n(r) \propto r^{(1 - 3v_{eff})/v_{eff}} \, e^{-r/(r_{eff}v_{eff})} \tag{8}$$

which depends on two parameters, the effective radius (r_{eff}) of the size distribution and the effective variance (v_{eff}). Contour diagrams for the elements
of $\mathbf{P}(\alpha)$ as a function of the effective size parameter $x_{eff} = \dfrac{2\pi r_{eff}}{\lambda}$ ($0 \leqslant x_{eff} \leqslant$
30) and the phase angle $\rho(= \pi - \alpha)$ are shown in Fig. 2 for $n_r = 1.44$ and
$v_{eff} = 0.07$.

In the diagram of linear polarization, $-P_{21}(\alpha)/P_{11}(\alpha)$, the region of positive polarization at small size parameter is due to Rayleigh scattering. The
long peninsula of positive polarization at scattering angle $\sim 15°$ is due to
Fresnel reflection. For scattering angle $\alpha \sim 20°$ and size parameter $x_{eff} \sim$
11, there is a hill of positive polarization, known as anomalous diffraction.
The region of positive polarization at scattering angle $\alpha \sim 160°$ is the primary
rainbow. The features appearing in Fig. 2 are discussed in more detail by
Hansen and Travis (1974) and by Coffeen and Hansen (1974).

Multiple Scattering

We make contour diagrams of the linear and circular polarization as a
function of size parameter x_{eff} and phase angle ρ for the apparent northern
hemisphere of a planetary disk, assuming a spherical but locally plane-parallel
atmosphere. The reflection matrix for the atmosphere is found by using the
doubling method in the manner described by Hansen and Travis (1974), and
the results are integrated over the upper half of the planetary disk using the
integration method of Horak (1950).

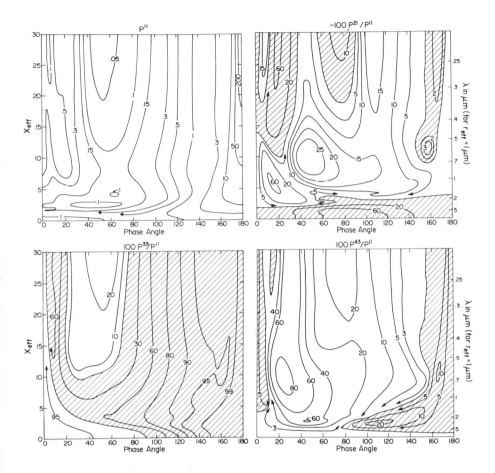

Fig. 2. Phase matrix for a size distribution of spheres with the real refractive index $n_r = 1.44$. The size distribution is given by Eq. (8) with $v_{eff} = 0.07$. P^{11} is positive everywhere; for the other matrix elements positive regions are cross-hatched.

The contour diagrams of the linear and circular polarizations are shown in Fig. 3 for three different refractive indices $n_r = 1.33$, 1.44 and 1.6. For these results integrated over the planetary disk, positive linear polarization corresponds to the electric vector of the radiation perpendicular to the intensity equator. For the sign of the circular polarization we use the same convention as Kemp (1974; cf. also p. 53 of that book), i.e. V is positive if the electric vector rotates counter-clockwise for an observer facing the planet. The contour diagrams of the circular polarization are for the northern hemisphere of the planetary disk at positive phase angles $\rho > 0$ (after opposition). For the northern hemisphere at negative ρ (before opposition) or for the southern hemisphere at positive ρ, the results would be the same as in Fig. 3, but with the opposite sign.

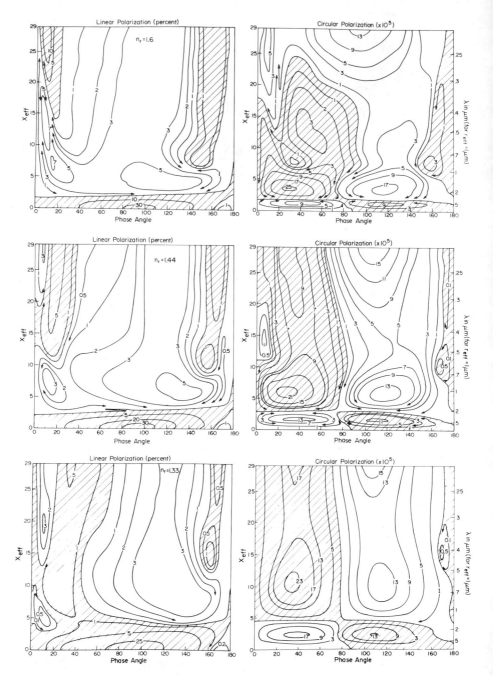

Fig. 3.   Linear and circular polarization $q = V/I$ in units of 10^{-5} for the northern hemisphere of a planet at positive phase angles. The results are for multiple scattering by spherical aerosols having the size distribution given by Eq. (8) with $v_{eff} = 0.07$; the optical thickness is 128 and the single scattering albedo is unity. Results are shown for three values of the refractive index n_r.

The features in the linear polarization contour diagrams are qualitatively similar to those for single scattering. The effect of multiple scattering on the linear polarization is primarily to dilute the magnitude.

We note several significant features in the contour diagrams of the circular polarization:

1. There is a sign inversion as a function of size parameter at small x_{eff}.
2. There is a sign inversion as a function of phase angle at $\rho \sim 80°$.
3. There is a region of positive polarization at $\rho \sim 170°$.
4. As n_r increases, a region of negative polarization appears at large x and small ρ, and it increases in extent.

By comparing the features of the circular polarization with those in the linear polarization, we can see that the sign inversion at small size parameters essentially delineates the region of Rayleigh scattering. The Rayleigh region is compressed for larger n_r, since $n_r x \ll 1$ is one of the required conditions for Rayleigh scattering (van de Hulst 1957). The sign inversion at $\rho \sim 80°$ is apparently of geometrical origin, as shown by a simple interpretive second-order scattering model which will be described in detail in Kawata's paper in preparation. The region of positive circular polarization at large ρ is related to the anomalous diffraction in single scattering.

The circular polarization of second- and higher-order scatterings arises from the linear polarization of single scattering (cf. Eqs. 5 and 6 and the discussion that follows). Thus it could be anticipated that many of the features would be qualitatively similar in the linear and circular polarizations.

Mixture with Rayleigh Scattering

It is important to quantitatively examine the effect of Rayleigh scattering on the circular polarization introduced by aerosols. The product of an aerosol phase matrix and the Rayleigh phase matrix often produces a larger circular polarization than the product of two aerosol phase matrices, since the linear polarization for Rayleigh scattering is large.

Figure 4 illustrates how a contribution from Rayleigh scattering affects the circular polarization. In Fig. 4a, f_R is the fraction of the phase matrix due to Rayleigh scattering,

$$P(\alpha) = f_R P(\text{Rayleigh}) + (1 - f_R) \times P(\text{Mie}) \qquad (9)$$

for a homogeneous mixture of Rayleigh and Mie scatterers. In Fig. 4b, τ_R is the optical thickness of Rayleigh scattering above an optically thick layer of Mie particles. In both cases the Mie particles are the same as in Fig. 2 with $n_r = 1.46$, $x_{eff} = 4$, $v_{eff} = 0.07$. Figure 4a shows that for substantial values of f_R the Rayleigh contribution can change the sign of the circular polarization and even increase its magnitude, as compared to the case of aerosols alone. This is a result of the fact that Rayleigh scattering and the aerosol phase matrix considered have opposite signs for P^{21}, and Rayleigh scattering has a high degree of linear polarization. Rayleigh scattering above

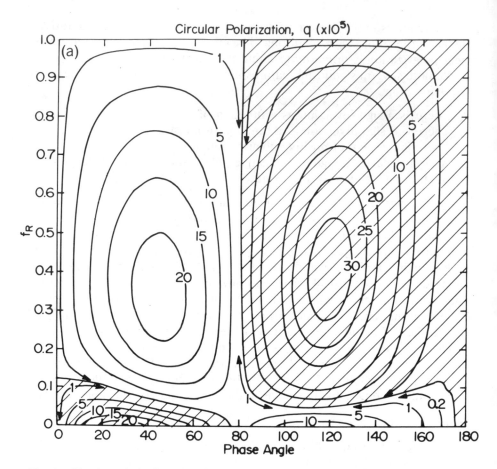

Fig. 4. Circular polarization $q = V/I$ in units of 10^{-5} for the northern hemisphere of a planet at positive phase angles. In (a) the atmosphere is a homogeneous mixture of aerosols and gas with phase matrix $f_R P_R + (1 - f_R)P_A$. P_A, the aerosol phase matrix, is for spheres having $n_r = 1.46$, $x_{eff} = 4$ and $v_{eff} = 0.07$. The total optical thickness is 128 and the single scattering albedo is chosen to produce the observed geometric albedo (~ 0.47) of Jupiter at $\lambda = 0.68$ μm for the case $f_R = 0$. In (b) a Rayleigh layer of optical thickness τ_R is located above the same type of cloud as in (a), but with $f_R = 0$ within the cloud.

the cloud affects the circular polarization quite differently than in the homogeneous case; for example, in Fig. 4b the circular polarization decreases monotonically with increasing τ_R. However, this result depends upon the phase matrix of the cloud particles; we have also found cases in which the circular polarization increases with increasing τ_R until $\tau_R \sim 1$.

From the linear polarization of Jupiter it is known [cf. the chapters by Tomasko[1] in this book and by Coffeen and Hansen (1974), and a paper by

---

[1]See p. 492.

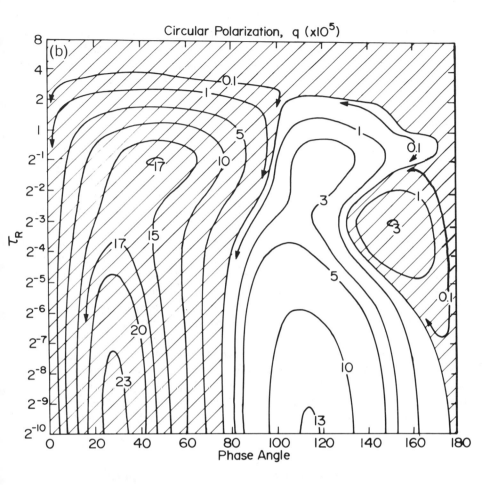

Kawabata and Hansen in preparation] that the aerosols in the visible atmosphere of Jupiter are distributed diffusely, more closely resembling a homogeneous mixture of particles and gas than pure gas above a dense reflecting layer. The presently available linear polarization observations of the equatorial region are adequately reproduced by a homogeneous model with $f_R \sim$ 0.3 at $\lambda = 0.365$ μm. This corresponds to $f_R \sim 0.024$ at $\lambda = 0.68$ μm. Based on Fig. 4 we anticipate that such a value of f_R is sufficient to significantly affect the circular polarization, but not to modify qualitative conclusions based on calculations with $f_R = 0$.

It is interesting to note from Fig. 4 that the opposite signs for the circular polarization on Jupiter and Saturn could arise simply from different values for f_R. Thus, the observed circular polarizations do not necessarily require different cloud compositions on the two planets.

Note also from Fig. 4 that *multispectral* observations of the circular polarization could provide a tool for readily distinguishing between a diffuse mixture of gas and haze particles, or gas above a dense cloud layer. In the former case multispectral observations would provide information on the parameter f_R and in the latter case on τ_R. It is particularly important to have observations at long wavelengths in order to establish the properties of the cloud particles themselves.

Comparison to Jupiter

To indicate the nature of potential applications of circular polarization observations we consider the few available measurements of Jupiter. We compare the data to our computations for multiple scattering by spherical particles. Thus, any quantitative conclusions derived must be based on the assumption that either the cloud particles are spherical or the circular polarization behaves in similar ways for spherical and non-spherical particles.

The observations indicate positive circular polarization for the northern hemisphere at positive phase angles ($0 < \rho < 10°$) in the wavelength range $\lambda \simeq 0.63–0.68$ μm. Judging only from the sign of the circular polarization (since that is not very sensitive to the assumed optical thickness or to the presence of a small amount of absorption) we can determine from Fig. 3 the ranges of particle size parameter that are consistent with the observed circular polarization for each refractive index. These ranges of size parameter are shown in the second column of Table I, and the corresponding ranges of particle radius are shown in the third column. If we add the condition that the linear polarization should be negative for $0.4 < \lambda < 0.8$ μm, as indicated by the observations of Morozhenko and Yanovitskii (1973), the constraints on the particle radius become those shown in the fourth column of Table I.

TABLE I
Constraints on Cloud Particle Size on Jupiter
Imposed by the Observed Signs of Circular and Linear
Polarization, Assuming Spherical Particles.[a]

	Circular Polarization		Circular and Linear Polarization
n_r	x_{eff}	r_{eff}	r_{eff}
1.33	$5 \lesssim x_{eff}$	$0.5\mu\text{m} \lesssim r_{eff}$	none
1.44	$3 \lesssim x_{eff} \lesssim 10$ $20 \lesssim x_{eff}$	$0.3 \lesssim r_{eff} \lesssim 1\mu\text{m}$ $2\mu\text{m} \lesssim r_{eff}$	$0.3 \lesssim r_{eff} \lesssim 0.8\mu\text{m}$ $2.5\mu\text{m} \lesssim r_{eff}$
1.60	$2 \lesssim x_{eff} \lesssim 15$	$0.2 \lesssim r_{eff} \lesssim 1.5\mu\text{m}$	$0.2 \lesssim r_{eff} \lesssim 1.5\mu\text{m}$

[a]The requirements imposed are positive circular polarization at $0.63 < \lambda < 0.68\mu$m for the northern hemisphere at positive phase angles, and negative linear polarization for $0.4 < \lambda < 0.8\mu$m.

The above ranges for the particle size would tend to be increased by the fact that the shape of the particle size distribution (effective variance and higher moments of the size distribution) is unknown, but they could probably be reduced by quantitative model fitting which includes consideration of the magnitudes of the linear and circular polarizations. However, we do not believe that unique properties of the cloud particles can be deduced from the available observations and the existing computations, because of the limited range of wavelengths and phase angles for the observations and the unknown effects of particle shape.

It is possible to quantitatively test the specific particle sizes and refractive indices proposed by Loskutov (1972) and Morozhenko and Yanovitskii (1973; cf. also Bugaenko *et al.* 1974) on the basis of single scattering analyses of the linear polarization, because they specifically proposed spherical particles. Figure 5 gives an indication of the circular polarization that could be anticipated for different n_r and x_{eff} with Rayleigh scattering and absorp-

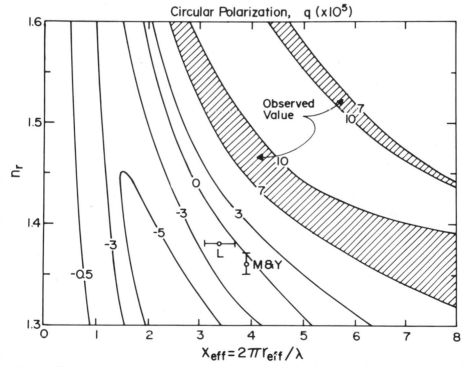

Fig. 5.   Circular polarization $q = V/I$ in units of 10^{-5} for the northern hemisphere of a planet at positive phase angles. The results are for spherical aerosols of refractive index n_r and effective size parameter $x_{eff} = 2\pi r_{eff}/\lambda$; the size distribution is given by Eq. (8) with $v_{eff} = 0.07$. The optical thickness is 128 and the single scattering albedo is unity. The shaded regions indicate the observed value $(7 - 10 \times 10^{-5})$ of circular polarization. Open circles mark the combinations of n_r and r_{eff} (for $\lambda = 0.68\ \mu$m) corresponding to the particle properties proposed by Loskutov (1972) and Morozhenko and Yanovitskii (1973).

tion neglected (i.e., $f_R = 0$, $\widetilde{\omega}_0 = 1$). At $\lambda = 0.68$ μm and $\rho = 8°$ we find for the model of Loskutov ($n_r = 1.38$, $r_{eff} = 0.37$ μm) a circular polarization of approximately -1.2×10^{-5}, and for the model of Morozhenko and Yanovitskii ($n_r = 1.36$, $r_{eff} = 0.43$ μm) a circular polarization of about -0.5×10^{-5}, in both cases for $f_R = 0$. With a more plausible value for f_R (0.3 at $\lambda = 0.365$ μm, or ~ 0.024 at $\lambda = 0.68$ μm) the discrepancy with observations increases, with q of about -3×10^{-5} for the n_r and x_{eff} proposed by Morozhenko and Yanovitskii. Allowing a small amount of absorption ($\widetilde{\omega}_0 < 1$) to match the observed geometric albedo does not significantly alter the results.[2] Thus these models do not even yield the observed sign for the circular polarization. Evidently either a larger refractive index or a larger particle size would be required to provide agreement with the observed circular polarization.

Recently Wolstencroft (1976) suggested that the circular polarization of Jupiter may be produced by secondary scattering by aerosols after first-order Rayleigh scattering, which is a portion of what we have called aerosol scattering above. Figure 4 shows that only a small amount of Rayleigh-aerosol scattering is sufficient to substantially change the circular polarization from that for pure aerosol scattering. Based on the calculations above it seems probable (but not certain) that the observed polarization at ~ 650 nm is due predominantly to the aerosol-aerosol contribution rather than the Rayleigh-aerosol contribution. If that interpretation is correct then the circular polarization of Jupiter may change signs at shorter wavelengths. It would be very interesting and informative to have observations at shorter wavelengths to test that possibility.

Finally we should mention that the observations of circular polarization of Jupiter have been described as containing a sharp spike as a function of phase angle, with a width $\sim 1°$ (Michalsky and Stokes 1974). In principle such a spike could be caused by very large particles ($x_{eff} \gtrsim 100$), but it would also require either a very narrow particle size distribution or preferentially oriented non-spherical particles. The spike could also be caused by variation of the atmospheric structure with time, or it could be the result of observational problems. It would be better to have firm observational confirmation of the feature before making any detailed consideration of it.

CAVEATS AND FUTURE WORK

The limitations of the circular polarization observations to phase angles $\rho \lesssim 10°$ and wavelengths ~ 650 nm prevents their use in any definitive analysis of cloud properties on Jupiter. Extension of the phase angle range

---

[2] The primary effect of small amounts of absorption is to reduce I, with a lesser effect on V and $\sqrt{Q^2 + U^2}$. Thus for comparison with the observations of Jupiter the computed magnitudes of q should be increased by a factor < 2, since the observed geometric albedo is ~ 0.5 (Taylor 1965) while typical geometric albedos for conservative scattering are 0.6–0.9. We have quantitatively verified the effect of small amounts of absorption with actual calculations; some results are shown, for example, by Hansen and Travis (1974).

beyond $\sim 12°$ would require observations from a spacecraft, which is a possibility that will not be realized for at least several years. There were linear polarization observations made from Pioneers 10 and 11, but detailed analyses of those data have not yet been published.[3] Extension of the wavelength coverage of the circular polarization observations could be more readily achieved, and measurements for the range $0.4 \ \mu m \lesssim \lambda \lesssim 2 \ \mu m$ would be practical and useful for studies of the cloud properties.

We emphasize that analysis of circular polarization data is most useful and reliable if it is made together with analysis of linear polarization and intensity (radiance) data. Because there are major changes in the cloud and haze structure on Jupiter with time, it is highly desirable to have near simultaneity of the different observations. And because of the apparent complexity of the atmospheric structure it would be valuable to have polarization measurements coordinated with other types of observations such as visible absorption band measurements and thermal infrared spectra.

We also emphasize that it is important to achieve a better understanding of the circular (and linear) polarization for scattering by non-spherical particles. Some information on cloud heights and broad limits on particle sizes can be deduced without such knowledge, but accurate derivation of particle refractive indices and particle sizes will require a quantitative understanding of scattering by non-spherical particles.

Acknowledgements. We are indebted to L. Travis and K. Kawabata for substantial help during the course of this work and to R. Wolstencroft for helpful suggestions concerning the manuscript. Y. Kawata is supported by a National Research Council Resident Research Associateship.

REFERENCES

Angel, J. R. P.; Illing, R.; and Martin, P. G. 1972. Circular polarization of twilight. *Science* 238:389–390.
Bugaenko, O. I.; Morozhenko, A. V.; and Yanovitskii, E. G. 1974. Polarization investigations of the planets carried out at the Main Astronomical Observatory of the Ukrainian Academy. *Planets, stars and nebulae studied with photopolarimetry.* (T. Gehrels, ed.) pp. 599–606. Tucson, Arizona: University of Arizona Press.
Coffeen, D. L., and Hansen, J. E. 1974. Polarization studies of planetary atmospheres. *Planets, stars and nebulae studied with photopolarimetry.* (T. Gehrels, ed.) pp. 518–581. Tucson, Arizona: University of Arizona Press.
Eiden, R. 1970. Influence of the atmospheric aerosol on the elliptical polarization of skylight. *Radiation including satellite techniques.* World Meteorological Organization Tech. Note 104, pp. 275–278, Geneva.
Hansen, J. E. 1971a. Circular polarization of sunlight reflected by clouds. *J. Atmos. Sci.* 28: 1515–1516.
———. 1971b. Multiple scattering of polarized light in planetary atmospheres. II. Sunlight reflected by terrestrial water clouds. *J. Atmos. Sci.* 28:1400–1426.

---

[3]See pp. 509 and 534.

Hansen, J. E., and Travis, L. D. 1974. Light scattering in planetary atmospheres. *Space Sci. Rev.* 16:527–610.

Horak, H. G. 1950. Diffuse reflection by planetary atmospheres. *Astrophys. J.* 112:455–463.

Kemp, J. C. 1974. Circular polarization of planets. *Planets, stars and nebulae studied with photopolarimetry.* (T. Gehrels, ed.) pp. 607–616. Tucson, Arizona: University of Arizona Press.

Kemp, J. C.; Swedlund, J. B.; Murphy, R. E.; and Wolstencroft, R. D. 1971a. Circular polarized visible light from Jupiter. *Nature* 231:169–170.

Kemp, J. C.; Wolstencroft, R. D.; and Swedlund, J. B. 1971b. Circular polarization: Jupiter and other planets. *Nature* 232:165–168.

Lipskii, Yu. N., and Pospergelis, M. M. 1967. Some results of measurements of the total Stokes vector for details of the lunar surface. *Sov. Astron. A. J.* 11:324–326.

Loskutov, V. M. 1972. Interpretation of polarimetric observations of Jupiter. *Sov. Astron. A. J.* 15:828–831.

Michalsky, J. J., and Stokes, R. A. 1974. A note on Jupiter's circular polarization. *Publ. Astron. Soc. Pacific* 86:1004–1006.

Morozhenko, A. V., and Yanovitskii, E. G. 1973. The optical properties of Venus and the Jovian planets. I. The atmosphere of Jupiter according to polarimetric observations. *Icarus* 18:583–592.

Swedlund, J. B.; Kemp, J. C.; and Wolstencroft, R. D. 1972. Circular polarization of Saturn. *Astrophys. J.* 178:257–265.

Taylor, D. J. 1965. Spectrophotometry of Jupiter's 3400–10,000 Å spectrum and a bolometric albedo for Jupiter. *Icarus* 4:362–373.

van de Hulst, H. C. 1957. *Light scattering by small particles.* New York: Wiley.

Wolstencroft, R. D. 1976. The circular polarization of the light from Jupiter. *Icarus* (special Jupiter issue). In press.

THE RESULTS OF THE IMAGING PHOTOPOLARIMETER ON PIONEERS 10 AND 11

T. GEHRELS
University of Arizona

The imaging photopolarimeter is a small telescope for observations of Jupiter and its satellites, and of interplanetary dust and background starlight. The techniques are those of spin-scan imaging, brightness and color measurements, and measurements of amount and orientation of linear polarization. In the analyses, models are constructed for the surface texture of the Galilean satellites, for the optical properties and spatial distribution of the interplanetary dust, and for the cloud layers at various longitudes, latitudes and heights in the Jovian atmosphere.

The Jupiter observations generally fit a two-layer cloud model with particles that are larger than the wavelength of light. The cloudtops in the zones were found to be higher than those in the belts. The Great Red Spot rises a few kilometers above the surrounding South Tropical Zone. The optical depth at the poles is much larger than that between about $+45°$ and $-45°$ latitudes; this should be taken into account when interpreting the infrared observations of similar temperatures at poles and equator.

Images of Jupiter and of the Galilean satellites, with resolution down to 190 km, show details never seen before especially in the transition regions of Jupiter near $\sim 45°$ latitude. Evidence of vertical and latitudinal motion is found in the pictures. A description is made of Jupiter on the basis of the Pioneer images.

A combination of results on small particles indicates that cometary grains accreted on Io may explain the sodium emission.

The imaging photopolarimeter (IPP) is a small multi-purpose astronomical telescope on Pioneers 10 and 11. It is pointable over 160° (as specified in the Appendix of this chapter), and it has a wheel-sector near the focal plane that allows selection of various apertures and calibration devices. The two instruments have obtained about half of the $\sim 1.6 \times 10^9$ data bits of Pioneers 10 and 11 during the encounters with Jupiter; because of the pointing and the selection possibilities of the aperture sector, the instrument requires about 90% of the commands. For the two encounters 21,000 commands were sent that had to be carefully prepared and tested.

The Pioneer program has been operated under stringent financial limitations which required us to maintain a relatively small team for the IPP experiment. It is a remarkable crew (named at the end of this chapter) with general versatility for technical tasks such as observing and image reconstruction as well as for analysis and scientific discussion of the data. Until 1975, our attention had to be directed towards the encounters with Jupiter and only preliminary results therefore are available at this time. Immediately after the Jupiter encounter of Pioneer 10, the first findings were published in a team report (Gehrels et al. 1974), and the same was done for Pioneer 11 (Baker et al. 1975).

Section I describes some of the complexities and limitations of the instrument. Section II summarizes the results of the photometry and polarimetry of Jupiter; a detailed review may be found in the chapter of Tomasko.[1] Section III reviews our spin-scan technique and its results for imaging of Jupiter; Sec. IV has a description of Jupiter as I have learned to see it from our pictures; the public interest in the pictures is mentioned in Sec. VII. Present and expected results for the Galilean satellites are given in Sec. V. Section VI gives a brief summary of the IPP observations of the zodiacal light by Weinberg et al., while I also refer to the observations of Humes[2] and Brown and Yung[3] in order to make some remarks about the accretion of cometary grains by Io.

I. THE EXPERIMENT

There are not many polarimeters on spacecraft (they are reviewed by Coffeen 1974a), and careful development of new techniques was therefore necessary; we started in 1962 (Fig. 3 of Gehrels and Teska 1963; Pellicori and Gray 1967). Some testing on high-altitude balloons was accomplished and we had a feasible instrument concept when the flight opportunity for Pioneers F and G was announced in 1968. I determined three basic concepts: 1) the telescope was to be *pointable* so that observations could be made over a range of scattering angles; 2) we would pay particular attention to the *calibration* of the photometry and polarimetry; and 3) it was to be a *versatile* astronomical telescope to be used also during the long interplanetary cruise. These three requirements added complexity, but the success seems to have proven the versatility appropriate.

We wrote a proposal (Gehrels et al. 1968) with the Santa Barbara Research Center as a team member, which company contributed flight experience with spin-scan imaging devices particularly on the Applications Technology Satellites ATS I and ATS III, and SBRC subsequently built both Pioneer 10 and 11 IPP instruments. Financial, mass and power restrictions

---

[1]See p. 486.   [3]See p. 1102.
[2]See p. 1052.

set tight limits on the data bit stream, the size of the telescope and the number of color channels. The SBRC engineers did, however, an exquisite job of matching the parameters of the IPP to those available on the spacecraft. A summary of properties of the instrument is given in the Appendix; the optical design and performance are described by Pellicori *et al.* (1973); a picture of the instrument is shown by Pellicori *et al.* and also by Coffeen (1974*a*) and by Fimmel *et al.* (1974, 1976). In addition to the amount of linear polarization, we also obtain the position angle by inserting a half-wave plate occasionally into the beam.

The nonlinearity of the detectors occasionally amounts to 5%, due to trimming resistors, but it is calibrated to within ±1% (KenKnight 1976). The precision of the polarimetry during the encounters is on the order of ±0.1%; the instrumental polarization effects are ~0.6%, known to about ±0.1%. The precision of the zodiacal light polarimetry is on the order of a few percent, except when bright stars are in the field and the large field of view can make the instrumental effects unwieldy.

The instruments have performed well in space and in the Jupiter environment, and both instruments still are in operation. Some problems have occurred as follows. On Pioneer 10 (see Coffeen 1974*b*) we have lost a few pictures and an appreciable fraction of the photopolarimetry because there were uncommanded gain decrements and one of the two red channels is too noisy to be of use. The aperture cycling was irregular so that much hand reduction is needed. The roll position of the spacecraft was uncertain, which requires special procedures in the zodiacal light analysis. On Pioneer 11 our principal problem is that of irregular stepping of the telescope which occurs especially in the forward direction (look angle ~160°, where the best-resolution images of Saturn will have to be observed). The stepping problem has caused little loss of Jupiter data, but it does require much image restoration. On both instruments, for Pioneers 10 and 11, there have been irradiation effects near Jupiter, probably mostly due to spacecraft charging (Scarf),[4] that cause the loss of a few pictures. A slight decrease with time is noted in the sensitivity of the detectors, or in the transmission of the optics.

II. PHOTOMETRY AND POLARIMETRY OF JUPITER

A general review chapter on polarimetry of planetary atmospheres has been written by Coffeen and Hansen (1974). The linear polarization of the planet Venus when plotted as a function of phase angle shows remarkable features including "cloudbow" effects which are explained as scattering by spherical aerosols. If such effects would be found on Jupiter also, they would be diagnostic for the type of aerosol present in the upper atmosphere. For this purpose, extensive observations with the IPP on Pioneers 10 and 11 have

---

[4]See p. 892.

been made of the integrated disk of Jupiter during the interplanetary cruise, and this is continuing in order to complete the phase coverage; no specific evidence for spherical particles has yet been found (Baker and Coffeen 1976).

The presence of the aerosols in the upper atmosphere is reviewed by Tomasko,[5] together with the modeling from IPP photometry of the cloud layers in the Jovian atmosphere. A large number of photometric models were computed for comparison with the Pioneer data at various phase angles for the South Tropical Zone and also for dark belts. His conclusions on the size of the cloud particles supercede those in a previous paper (Tomasko *et al.* 1974); the data seem much better fitted by particles having asymmetry parameters in the neighborhood of 0.75 rather than < 0.25 as previously reported. This value of the asymmetry parameter is consistent with particles larger than the wavelength of light. Secondly, the data seem much more sensitive to small amounts of absorbing material in the region above the clouds than to the size of the cloud particles. The modeling of the cloud layers is still at an early stage considering the large amount of data that the IPP obtained on Pioneers 10 and 11, and further improvements in new iterations are to be expected (Tomasko 1976).

The phase integral, q, has a variety of values for scattering from various regions of the planet (Tomasko).[6] For some scattering models of the northern component of the South Equatorial Belt in blue light q is ~ 1.1, while for the South Tropical Zone in red light, q as large as 1.5 is found. The apparently large variations found in these early models make the determination of accurate photometric models of the planet's various belts and zones even more important. The average of q over wavelengths and over the planet enters into the determination of the Bond albedo of the planet, and therefore affects estimates of Jupiter's internal energy source.[7]

The polarimetric derivation of various heights in the Jovian atmosphere by Coffeen (1974*b*; his Fig. 9 is reproduced here as Fig. 1) is reviewed in the Tomasko chapter, and additional results are forthcoming (Coffeen 1976). The visible cloud height in the dark belts lies lower than that of the cloudtops of the bright zones. This polarimetric derivation of cloud height is model dependent, but the approximate validity is shown in the paper by Baker *et al.* (1975).

The Great Red Spot has a cloudtop that is a few kilometers above the surrounding top of the South Tropical Zone (Coffeen 1974*b*), and these effects are further studied by Doose (1976; also see Swindell and Doose 1974).

In the polar regions we find much greater optical depths than elsewhere on Jupiter (see Table III of Tomasko on p. 512). The onset of greater optical depth is rather sudden; the change usually occurs between 40 and 55° lati-

---

[5]See p. 492.   [6]See p. 505.   [7]See p. 202.

tude (Gehrels *et al.* 1969). To what extent the actual cloud heights in the polar regions can be derived from the polarization measurements of Pioneer 11 remains to be seen; a simple interpretation in terms of clear molecular scattering is insufficient and the scattering by aerosols will have to be taken into account. Lyot (1929) made a remarkable observation regarding the upper level haze phenomena of Jupiter, namely, he observed on occasions the strong polarization of the polar regions and a grayish haze both extending to nearly the equatorial regions. The detailed Pioneer observations seem essential to unravel the various contributions to the light scattering. It will be a long time before such high sub-spacecraft latitudes are observed again; future missions mostly go by equatorially.

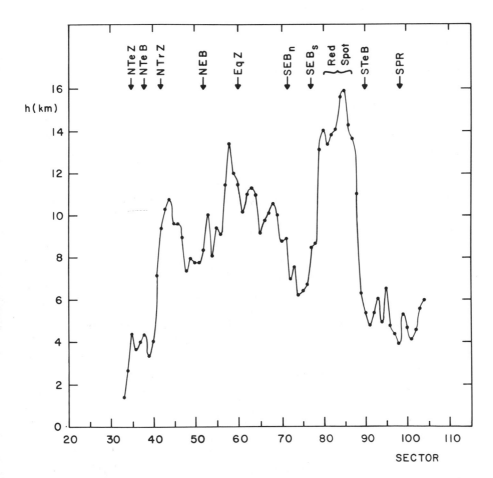

Fig. 1.   Cloud-top altitudes above the 650-mbar level for various belts and zones of the Jovian atmosphere (Coffeen 1974*b*).

III. IMAGING OF JUPITER

NASA decided, at an early stage of Pioneer F/G Mission preparations, in favor of the imaging by spin-scan rather than by television technique. Television on the Mariner Jupiter/Saturn Missions will give much better resolution, but that of spin-scan can also be high, $\sim 10\,km$, with high data rates of large spacecraft (Russell and Tomasko 1976). Several advantages of the spin-scan technique over that with television, such as photopolarimetric linearity, together with the fundamentals of spin-scan imaging and an overview of applications have been published by Gehrels *et al.* (1972).

With the smallest aperture of 0.5 milliradian (0°03), a scan line is made across the planet and the output of the detectors from segments of 0.001 sec duration is sent back to Earth where scans and segments are assembled in a computer to make the spin-scan image. A general description of the IPP imaging is made by Swindell and Doose (1974), who include examples of the *cosmetics:* substitution for missing data, and other reconstruction at an early stage in the image processing. Figure 2 is an example of an image that had a large number of bits missing in the reception at the antennas of the Deep Space Network. Also seen are positional shifts of scan lines, and gain changes that were made in order to obtain the best signal level. Figure 3 shows how the blanks have been filled in, by interpolation from adjacent data, and by other improvements.

Swindell and Doose also show and describe *rectification* of the scan lines, namely the re-arrangement in proper geometrical configuration. The original scan lines are generally not straight but have curvature due to the cone described by the telescope axis during the spin of the spacecraft. Figure 4 is an example of original data receipt while Fig. 5 gives the same data after the proper rectification.

A variety of techniques is available for *enhancement* of selected image characteristics. A rather simple example, using darkroom techniques, is shown — with Fig. 6 for the original data — in Fig. 7 (other enhancement procedures are referred to in Sec. V).

Some 200 images have better resolution than is typical for good ground-based photography. A few pictures have the highest resolution that could be obtained, namely $\sim 190\,km$ (in Figs. 4 and 5, for instance). Regarding colors, all images come in pairs of the red and blue channels (see the Appendix), and various features are seen differently in these two channels; the most striking example is the Great Red Spot which is hardly seen in the red (Fig. 6) but strongly in the blue (Fig. 8). For our usual color compositing a third bandpass is simulated; these colors on issued prints must therefore be considered uncertain. The reliability of our imaging data is discussed in some detail by Frieden and Swindell (1976). There is, however, a large amount of information in the colors, as is strikingly demonstrated in the color-ratio imaging of Fountain (1976) of which the present Fig. 9a is an example.

The Great Red Spot, and some other red spots that will be mentioned in the next section, actually are not a deep red, but rather of a brownish-reddish color; in any case, they are unique in color, differing from the brown and yellow shades seen in the belts. What is the material that condenses out at a height above and a temperature below that of the South Tropical Zone? These problems are discussed by Sill[8] and by Prinn and Owen.[9]

Blue colors I shall also mention in the next section, namely those of the polar regions, the terminator and the North Equatorial Belt. Based on our photopolarimetry we consider these regions to have greater optical depth, i.e., through the molecular atmosphere down to a lower cloud deck. Alternately, with a model of two cloud layers, the upper one may be broken and we observe through gaps down to the thicker bottom layer (Baker *et al.* 1975). The greater Rayleigh scattering, $\propto \lambda^{-4}$, presumably causes the blue and purple intensities and polarizations to be enhanced; physical interpretations of multiple molecular scattering have been made for the terrestrial blue sky (Gehrels 1962) and the Jovian polar regions (Gehrels *et al.* 1969).

Fountain *et al.* (1974) discovered a marked anvil-and-plume structure of the clouds near the equator (Fig. 10); it may have been present many years (Smith and Hunt).[10] Other similar formations in the Equatorial Zone have now been recognized; the plume structure apparently is common in the EZ. The head or anvil usually occurs just a little north of the equator while the plumes are generally inclined with respect to the east-west direction, possibly showing transport towards the equator (Stone).[11] Figures 11 and 12 show new detail in another EZ anvil. This is one of our finest global picture pairs showing a wealth of information also in EZ plumes (especially in the red channel; Fig. 12) and in detailed structures everywhere.

The polar regions that have large optical depth (Sec. II) allow deeper penetration by the red than by the blue channel; this is seen in Figs. 11 and 12, and also in Fig. 13 which is the blue-channel version of Fig. 5. The deeper penetration also occurs in the infrared and this should be taken into account with the results of the infrared radiometer; Ingersoll (1976) finds nearly the same temperature at the poles as at the equator, but that measurement is not at the same height in the atmosphere. That the 40°–60° latitudes are remarkable is mentioned by Teifel[12] and Stone.[13] We knew beforehand also from polarimetry (Gehrels *et al.* 1969) that the polar atmosphere would appear drastically different from that between $\pm 45°$ latitudes, and for this reason we pleaded for the high-latitude passage of Pioneer 11.

It is sometimes believed[14] that cloud motion cannot be gleaned from the Pioneer images because they are snapshots, and that one needs a movie film. This seems incorrect because such a film will show the fastest motion pre-

---

[8]See p. 372.   [11]See p. 606.   [14]See, for example, p. 28.
[9]See pp. 325 and 493.   [12]See p. 474.
[10]See p. 577.   [13]See p. 604.

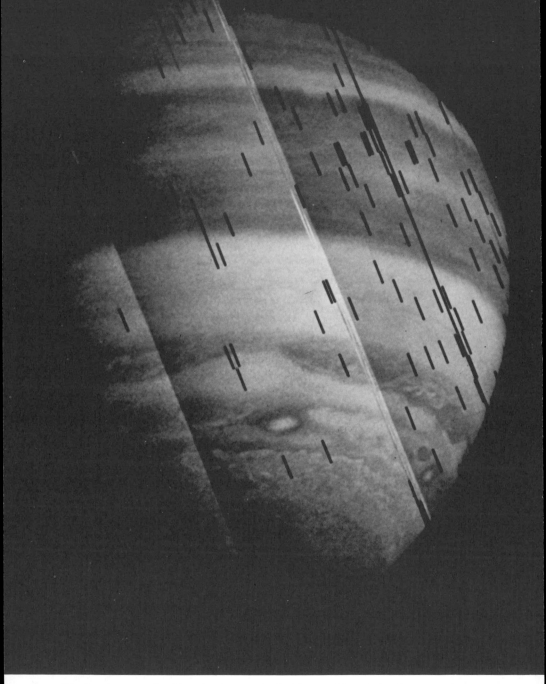

Fig. 2. An example of imaging data received with many bits missing, lost probably during the reception on Earth. Also seen are the discontinuities in sensor sensitivity as the gain had been set deliberately for best imaging of the dark polar regions. Lighter and darker lines show places where the imaging is misplaced.

Fig. 3.   The data of Fig. 2 with missing data interpolated, gain changes allowed for, and image lines put in their proper place. ERT (Earth received times UT) $20^h 54^m 29^s$–$22^h 03^m 40^s$ on Dec. 2, 1974. Range (i.e., the distance of spacecraft from the center of Jupiter) 765,000 km. Phase (i.e., the Sun-Jupiter center-spacecraft angle) 63°. Sub-spacecraft latitude $-25°$. Longitude of the central meridian (LCM; System II) 263°. Blue channel.

Fig. 4.   Before rectification and before allowing for gain change. ERT $9^h\,02^m\,28^s$–$9^h\,51^m\,37^s$ on Dec. 3, 1974. Range 375,000 km. Phase 64°. Lat. +52°. LCM 232°. Red channel.

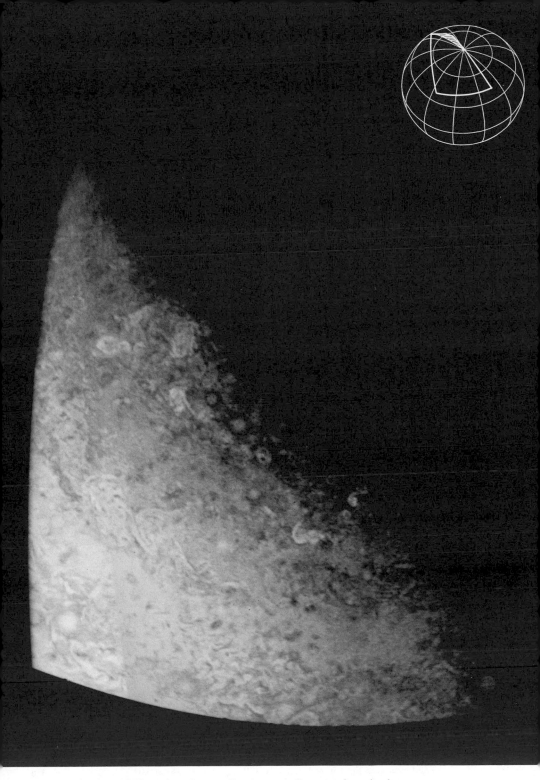

Fig. 5. The data of Fig. 4, but after rectification and allowance for gain change.

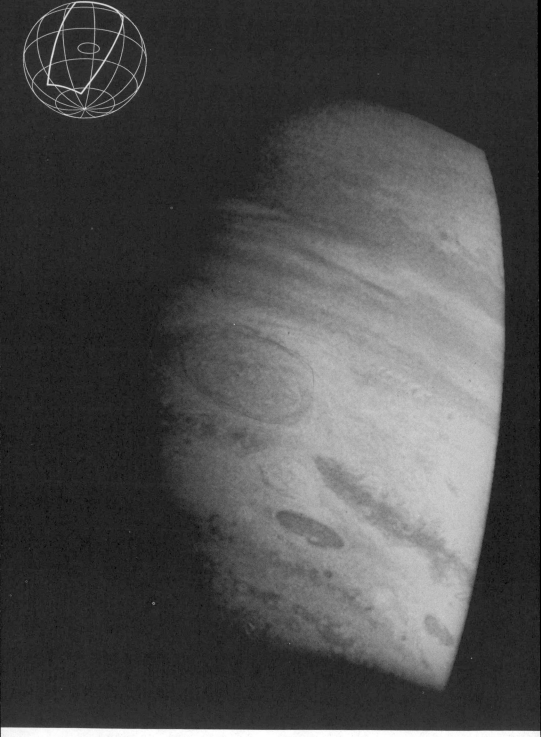

Fig. 6.   The Great Red Spot in red light. ERT 0ʰ 03ᵐ 32ˢ–0ʰ 56ᵐ 04ˢ on Dec. 3, 1974. Range 545,000 km. Phase 70°. Lat. −31°. LCM 21°.

[542]

Fig 7   The data of Fig. 6, after enhancement by unsharp masking. A low contrast, unsharp
negative was placed under the positive transparency. A high contrast contact negative was
made to produce this image. The effect is to increase the contrast of the high frequencies with
respect to the low ones (Fountain, personal communication).

[543]

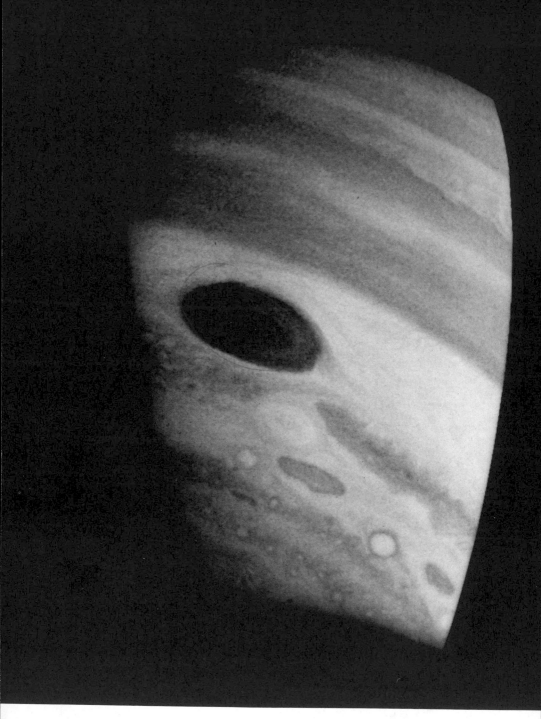

Fig. 8.   The Great Red Spot in blue light (no enhancement). Counterpart of Fig. 6.

dominantly, that is of the rotation, obliterating the effects of cloud streaming and vertical lift. Typical time scales for the changes on Jupiter are months and years, not minutes or hours.

I see four ways to study motions from the Pioneer images: 1) Views of the same aspect (same sub-spacecraft longitude and latitude) but of different times can be compared; we have not done this as yet. 2) Known velocity vectors (shown on p. 566, for instance) can be compared with cloud morphology; this is done to some extent in Sec. IV, and more is planned. 3) Where there is not a sharp velocity gradient, i.e., where there may be a more quiescent condition as far as east-west motion is concerned, Coriolis effects may be seen. A good example is in the North Tropical Zone (see below), and other cases will be mentioned in Sec. IV. 4) Bright white regions may be indicators of updraft. Presumably the white crystals are of NH_3 condensing near the top of the visible atmosphere, and the greater the cumulus effects are the more crystals would be observed per unit area. Several examples will be given in Sec. IV.

In Fig. 14 one sees in the North Tropical Zone what appears to be a streaming motion that is slanted upwards to the right. Coriolis forces are expected to be dominant since the Rossby number is small.[15] It is straightforward to compute the Coriolis effect for the latitude differences of the southern and northern edge of the NTrZ, and one finds that the slant angle would be 20° which agrees precisely with the slant seen in Fig. 14. It is, therefore, a demonstration of motion in latitude within the NTrZ. Examples of less quiescent conditions are found in the spectacular billows and whirls near the north edge of the North Equatorial Belt where there are steep velocity gradients.[16]

All our images are being delivered to, and can be obtained from, the National Space Science Data Center (Goddard Space Flight Center, Greenbelt, Md. 20771). A listing with aspect data for our images is given for Pioneer 10 by Swindell and Doose (1974) and by Fimmel *et al.* (1975) and for Pioneer 11 by Fimmel *et al.* (1976); these publications show many of our best images, and color pictures are given by Fimmel *et al.* (1975, 1976).

IV. A DESCRIPTION OF JUPITER

The following is a personal view of Jupiter based on Pioneer images. The reader is referred to the chapter by Smith and Hunt[17] for other description and time variation, while my descriptions are only for 1973/1974. In all of the following, north is up and the rotation of the planet is from left to right, as is the case for all pictures in this book.

---

[15]See p. 592.
[16]See pp. 566 and 590.
[17]See p. 564.

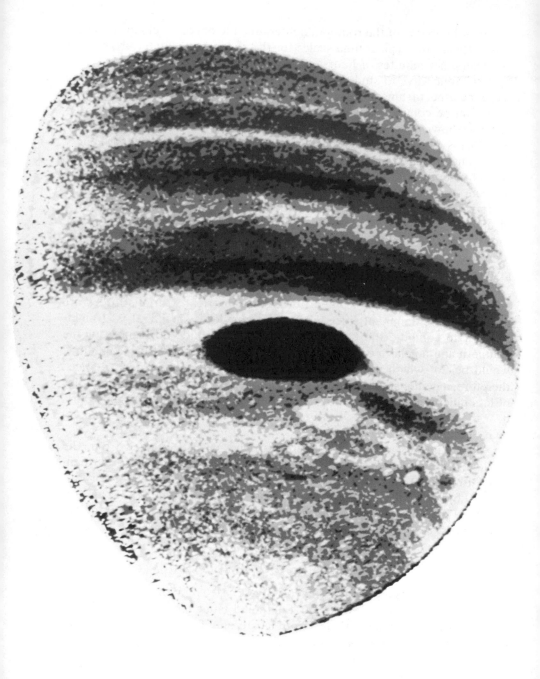

Fig. 9a. Color ratio image. Intensity in the original display was proportional to the logarithm of the ratio of the red to the blue data (Fountain, personal communication). Here red is dark and blue is light. ERT 15^h 07^m 06^s–15^h 49^m 54^s of Dec. 2, 1974. Range 1,144,000 km. Phase 56°. Lat. −19°. LCM 42°.

Fig. 9*b*.   For comparison with Fig. 9*a*, this is the blue-channel image of the original data.

Fig. 9c. As for Fig. 9b, the red channel.

Fig. 10.   This image shows the equatorial plume. ERT 23^h 05^m 41^s–23^h 41^m 33^s on Dec. 3, 1974. Range 1,310,000 km. Phase 40°. Lat. +41°. LCM 59°.

[549]

Fig. 11. ERT 18^h 52^m 13^s–20^h 10^m 42^s of Dec. 3, 1974. Range 1,079,000 km. Phase 42°. Lat. +43°. LCM 275°. Blue channel.

Fig. 12. Counterpart of Fig. 11. Red channel.

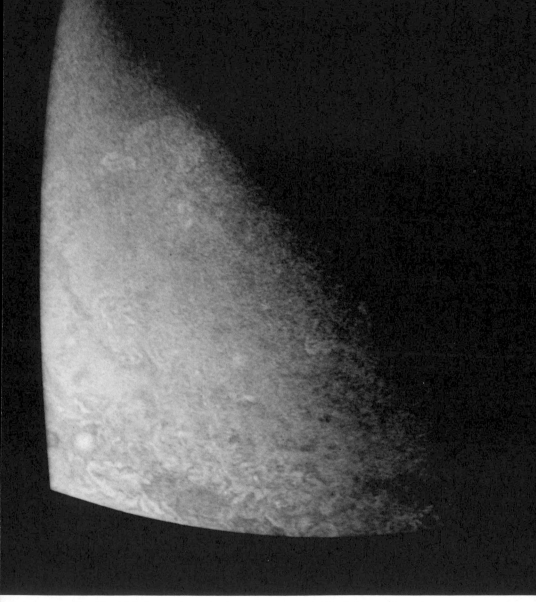

Fig. 13.   This image shows the lesser penetration in blue light near the pole. It is the blue counterpart of Fig. 5.

Fig. 14.   The image shows the Coriolis motion in the North Tropical Zone, the billows south of
it in the North Equatorial Belt, and the remarkable cloud patterns north of the North Tropical
Belt. ERT 11ʰ 54ᵐ 05ˢ–13ʰ 00ᵐ 41ˢ of Dec. 3, 1974. Range 610,000 km. Phase 52°. Lat. +49°.
LCM 3°. Blue channel. Unrectified image; the shape after rectification would be as shown on
inset figure.

In the south polar regions there are some details of individual clouds, best seen in red light (see Sec. III), but they are not organized in zones and belts. A discontinuity of the general brightness level in the blue images is seen near 60° latitude and perhaps another one near 40°. Near about −50° we have not seen spectacular detail as in the north, but that is, most likely, because our images did not have the high resolution as in, for instance, Fig. 14. A mottled appearance seems to indicate vertical motion with ammonia crystallization at the top. The first banded structures of dark belts and bright zones are noticed in the temperate regions; sudden and drastic changes with *longitude* do, however, occur (Fig. 3).

The south temperate regions are unique in showing, especially in blue light, large and bright ovals that are uniformly white and usually have a dark perimeter. They give the impression of a slow convectional rise at the center — until the condensation level of ammonia crystals, that are white, is reached — followed by a subsidence at the perimeter where the ammonia crystals evaporate, allowing us to see deeper down into the atmosphere. Some of the ovals, however, do not have the dark perimeter and are simply large white areas. Occasionally the opposite is seen, namely a bright perimeter with a darker ring inside it, while within that a brighter oval may occur.

The South Temperate Zone is clearly distinguished by generally a soft but rather straight southern boundary; i.e., large-scale irregularities do not occur here. The white of the zone is interspersed at some longitudes with dark reddish spots which may combine into streaks that almost split the STZ in two. Then comes its strongest feature, namely the wavy structure of the northern edge, sharply delineating the South Temperate Belt by a thin brown part of the STB. It is also noticed in the Pioneer 11 pictures that the STZ may be interrupted by dark ellipsoidal features that have a sharp bright rim and bright white spots inside (Fig. 6); in blue light (Fig. 8) there is a thin bright ellipse inside.

The STB is tremendously varied in contrast and brownish coloring. At lower resolution some of the structures appear to show billows, reminding us of the descriptions of solitons, the solitary waves of Maxworthy and Redekopp (1976; reviewed by Stone[18]). The STB shows strenuous shears in its irregular shapes and cloud streaks, while there must also be strong vertical motion seen in occasional red spots and frequently occurring ovals. The bright ovals, sometimes intruding upon and interacting with the South Tropical Zone, are seen even at low resolution because of their large size and exceptionally high reflectivity.

Going north from the STB we come to a wide bright band. It was Pioneer's luck to have that striking feature, with a strongly developed Great Red Spot, at this time. It consists of two parts: the South Tropical Zone and the southern component of the South Equatorial Belt (SEB_s) and they are

---

[18]See p. 612; also see p. 582.

clearly separated by a thin brown streak that has a wavy structure especially in the pictures of Pioneer 11. The STrZ and SEB$_s$ differ in another manner: fine detail in the STrZ occurs in mostly east-west directions with a slight slant, reminiscent of a Coriolis effect (Sec. III). Finely detailed streaks are also seen in the SEB$_s$, with somewhat denser spacing, but they are slanted ($\sim 30°$ with respect to the east-west direction) not in the Coriolis direction but indicating a decrease in the equatorial jet stream (p. 566) velocity between north and south boundaries of the SEB$_s$.

The Great Red Spot appears embedded in the STrZ. The GRS seems to push the demarcation of STrZ and SEB$_s$ towards the north, and the south rim of the STrZ also seems pushed a little southward by the GRS; the STrZ is thin there but clearly distinguishable. To the left of the GRS, the STrZ-SEB$_s$ demarcation comes south again quickly; in this region the colors are bluish (in 1974), as if the high cloud of the SEB$_s$ may be breaking up, allowing our view to penetrate deeper into the molecular atmosphere. Over considerable longitude, to the left of the GRS, the STrZ appears to be broken in two by a thin dark streak.

The GRS itself has an outer perimeter of deeper red-brown, while more towards the inside usually a concentric brighter ring is seen. In some of the best pictures, a spiraling effect appears there with variation of deep and less-deep red (brown) coloring.

The bright white SEB$_s$ has a rather sharp northern edge, going over into a dark brown narrow belt. This narrow strip may have cloud streaks in it, with some slant as described for SEB$_s$. Next to the north, in the remainder of SEB$_n$, there is the appearance of alternating brighter and darker streaks that are oriented in an east-west direction without slant. In the Pioneer 11 pictures the effect of these streaks is so strong that it almost looks like a whole belt-zone structure, with the belts having some bluish hue.

The Equatorial Zone has anvil-and-plume structures (Sec. III) that are bright white, with occasionally bluish areas preceding (to the right of) the anvil; these often are infrared hot spots (Armstrong et al. 1976).[19] The bluish hue is probably due to Rayleigh scattering in a deep molecular atmosphere through which our vision penetrates to the deeper levels where the infrared experiments find water vapor. The anvils have exceptionally high reflectivity and they usually are sharply defined (Figs. 11 and 12), but most often the slanted plumes appear to have no marked anvil head. The plumes, but not the anvils, are best resolved in the red, as are most SEB$_n$-NEB clouds, which probably means that they are at a lower level than the anvils (but they still appear to lie above the EZ generally).

The North Equatorial Belt has marvelous detail in contrast and color. Here we see billows, apparently solitons again, with great variation in their color detail, while there also are some million-km long dark brown streaks

---

[19]See pp. 351, 407 and 502.

and, intruding upon the North Tropical Zone, some great ovals may be seen. The overall direction, seen at low resolution, of the features in the NEB may have some slant, somewhat similar, but in opposite direction, to that noted in the SEB$_s$, perhaps again indicating a decrease, this time from south to north, of the equatorial jet stream velocity. The transition to the NTrZ is very irregular with deep insertions of dark and brown material into the NTrZ.

Where we can see structure in the white material of the NTrZ we notice Coriolis streaks (Sec. III). Small red spots in the NTrZ may be seen in our blue images (for instance in Fig. 14), with little or nothing noted in the red in this area. These spots usually have two long dark peaked streaks reaching into the NTrZ, the one on the right joining the spot more south, the one on the left more north; these peaks also are slanted in the Coriolis direction. The outer perimeter of the spot usually is dark and wide, while diffusely at its center a brighter island may be present.

North of the NTrZ much less detail is seen in our red images than in the blue ones, but in the blue the detail is so indescribable that I have to refer you to the figures. There are multiple cloud layers, and changes in longitudinal direction that are surprisingly sudden while occurring over a large range in latitude (Fig. 14).

The north rim of the NTrZ is sharp and bluish, and then suddenly and sharply there is the transition to the North Temperate Belt and a region where the general direction of motion appears to be east-west without slant.

Next, in the north temperate regions ovals may occur as I described before for the south temperate regions, namely with a bright perimeter and darker interior, these are seen only in the blue image, or with dark perimeter; but there are not as many and they are not as striking as in the southern hemisphere. The features here are more wisp-like billows, reminding us of solitons again, with bright thin clouds down to the resolution limit ($\sim$ 190 km). At low resolution all of this detail seems to combine into a mottled appearance due to variations in reflectivity, indicating vertical motion and subsequent condensation.

I have the impression that the clouds near latitudes $>45°$ lie deeper; in any case, the atmosphere looks darker than at lesser latitudes. Just before the banded structures stop there occurs a strongly marked brown-orange dark belt, not going all around in longitude. In this region again the wavy structure of belt-zone transitions is seen as it was at similar southern latitudes. The bright zone just south of the dark belt, incidentally, appears to show a Coriolis slant.

Finally, the banded structure stops, at latitude $\sim 60°$, but there still is great detail to be seen in the North Polar Region, together with a slightly bluish hue again, as is the case near the South Pole and the terminator. Note in Fig. 12 the large oval, to the left of the central meridian in the NPR, which in blue light (Fig. 11) is nearly absent. In the polar regions we apparently witness cloud formations deep down in the molecular atmosphere.

V. SATELLITES AND ASTROMETRY

Several enhanced images of the Galilean satellites are shown by Fimmel *et al.* (1976), a series for Ganymede is published by Frieden and Swindell (1976), and some of the best are in the Frontispiece of Burns (1976).

The enhancement exercises of Frieden and Swindell and their improvement in the resolution, by a factor 2, teach us to see detail on Ganymede not noticed before. For instance, in Fig. 3 of Morrison and Burns,[20] there is a bright spot, below center and to the right, surrounded by a wide dark ring which is surrounded by another ring that is bright and wide so that its outer diameter is 2/3 the disk. The configuration is reminiscent of the concentric ring structures on the moon where they are interpreted in terms of impacts. On another image of Ganymede, obtained by Pioneer 11, shadowing appears on one side of the bright spot which suggests that it has an appreciable elevation above the surrounding surface. Ganymede probably has a thin mantle (Consolmagno and Lewis)[21] that may, in fact, show in these images the straining effect of supporting the elevated bright spot. This, however, does not explain why the spot has relatively high reflectivity.

Let us now continue inspection of the outer ring: on its interface with the large dark region to the upper left, Frieden and Swindell describe three small dark mare depressions of which one is seen even in the lower resolution picture reproduced by Morrison and Burns.

Little color is seen on Ganymede. One of the advantages of spin-scan imaging is that the detector outputs are linear (Sec. III) so that the colors can be studied from the simple quotient of the *R*ed and *B*lue outputs. For Ganymede, the R/B quotient ranges between 1.2 and 1.7 with an overall average of 1.44 (Gehrels 1976). No correlation of color and reflectivity is found. Fringes have been seen in the imaging of Ganymede that are tentatively interpreted as due to a partial layer of ice crystals that have a characteristic dimension of ~ 4 mm (Frieden 1976).

As for photometry and polarimetry, scattered light has been a problem in the observations of the satellites, especially for Io because of the proximity of Jupiter, and the reductions have had to follow special procedures. There now is, however, a rather complete set of polarimetry and relative photometry with the blue and red filters over the range of phase angle from ~ 40° to 110°.

There are about 200 low resolution images of the Galilean satellites from the Pioneer 10 and 11 encounters that are suitable to improve the ephemerides for use of missions in the future (Duxbury 1975). These observations are of interest particularly because the spacecraft were below, above, or inside the satellite orbits with aspects that are impossible from Earth. Positional accuracies of a few hundred kilometers can be obtained.

---

[20]See p. 1005.
[21]See p. 1042.

It appears possible from a detailed study of the imaging of the satellites (Smith 1976) to obtain the diameters to a precision of a few tens of kilometers so as to improve the density determinations especially for Callisto.

Observations of J V Amalthea have not been possible because of the proximity to Jupiter causing a large amount of scattered light. A careful search was made by the Jet Propulsion Laboratory for an opportunity to observe an asteroid during the interplanetary trips, but none was found bright enough to be observable (also see KenKnight 1971), and the same holds for the periodic comets; the ephemerides of ~ 3000 asteroids and of ~ 200 comets were checked.

Bejczy of the Jet Propulsion Laboratory, assisted by KenKnight, looked for oscillation or drift in the inertial orientation of the spin axis of Pioneer 10. The IPP was locked on Jupiter in the imaging mode for a total of 36 hours in May and June of 1974, but no motion of the spin axis, to a precision of ~ 40 arcsecs, was detected. If desired, the precision could be appreciably improved in future measurements.

VI. INTERPLANETARY PARTICLES

Our original concern that contamination by light-scattering particles around the spacecraft would occur has not materialized. The spacecraft apparently are clean and the hydrazine gas bursts for orientation of the spacecraft have been carefully avoided at the time of observations. Sunlight scattered by parts of the spacecraft has been mostly eliminated, thanks to special baffling and a cone-shaped shield at the end of the telescope (see Leinert and Klüppelberg 1974).

The various components of the light of the night sky are reviewed by Weinberg (1974). IPP observations from Pioneers 10 and 11 have made it possible to separate these components with confidence, by mapping starlight and zodiacal light periodically during the cruise phase of both missions (Weinberg *et al.* 1976).

Soon after launch of Pioneer 10, the counterglow or "Gegenschein" was observed at 1.011 A.U. from the sun when the spacecraft-antisun direction was 3°4 from the Earth-antisun direction (Weinberg *et al.* 1973). Earlier interpretations had the faint counterglow due to a gas or dust tail of the earth, or to material near the L_3 libration point in the system Sun-Earth-particle, or due in some other way to a geocentric phenomenon. Roosen (1971) showed these interpretations to be wrong, but the Pioneer observations yielded the final proof because they were made *in situ*. The counterglow is therefore primarily a phase effect in the zodiacal light; i.e., the interplanetary dust has an increased efficiency in the backscattering region (an "opposition effect"). Appreciable polarizations and their striking time variations have been reported by Roach *et al.* (1974) and by Wolstencroft and Brandt (1974).

Brightnesses in the ecliptic at elongations greater than 90° indicate a drop in the dust density near 3.3 A.U., beyond which the zodiacal light is $\leq 1\%$ of the background starlight; i.e., the zodiacal light is negligible beyond the asteroid belt (Hanner *et al.* 1974).

From beyond the asteroid belt, Pioneers 10 and 11 are observing background starlight (integrated starlight plus diffuse galactic light) in the absence of zodiacal light for the first time (Weinberg *et al.* 1974); star counts further allow the separation of the stellar and galactic components. These observations are being used between 1 and 3.3 A.U. to derive the change in zodiacal-light brightness with heliocentric distance. Based on the assumption that the scattering properties of the dust do not change significantly with heliocentric distance, the observed change in brightness suggests that the spatial distribution can be represented by a power law, $R^{-\nu}$ ($\nu \approx 1$), or by a two-component model ($\nu \approx 1.5$, with increased dust in the asteroid belt). It is not possible from these results alone to distinguish between asteroidal or cometary origins for the particles.

The particles that Humes[22] measured have sizes 10–50 μm (size depending on the assumed density); particles of lesser mass would not penetrate the walls of his impact detectors. Since Humes finds no correlation with the asteroid belt, these grains must be produced by *comets*. After reading Secs. VII B and D of Brown and Yung,[23] three conclusions appear to fall into place: 1) the source of Io's sodium emission may well be accreted grains; 2) their composition is that of cometary particles, rather than of chondritic meteorites; and 3) their composition indicates a tensile strength appreciably greater than the exceedingly low one of carbonaceous chondrites. The meteoritic explanation for Io's sodium cloud[24] seems supported by these conclusions; "meteoritic" at this distance from the sun means "cometary."

VII. PUBLIC RELATIONS

Because of the general interest in the pictures of Jupiter, we made a fairly large effort in showing them, namely in some 40 colloquia and lectures in the United States and Europe, including 2 COSPAR meetings (Blenman *et al.* 1976).

Two NASA Special Publications are devoted to the results from Pioneers 10 and 11 (Fimmel *et al.* 1974, 1976 respectively). Our pictures have appeared in various articles and on front covers of journals (for example, see Wolfe 1975).

During the encounters of Jupiter the Pioneer Image Converter System (PICS; see Baker 1975) was an essential facility to monitor the operation of the instrument while producing color television pictures for the broadcast media and for a closed circuit network in the San Francisco area.

---

[22]See p. 1054.   [23]See p. 1138.   [24]See p. 1140.

Acknowledgements. We thank Pioneer Project Manager C. F. Hall and his team at the Ames Research Center for close support and particularly for flexibility during encounter operations; without their special assistance some of our most reproduced images would not have been made; these were salvaged, through previously unknown conditions, with last-minute changes in detector gain and telescope pointing. The cooperation with R. F. Hummer and his team at Santa Barbara Research has been close, and we similarly thank our partners for the zodiacal-light observations, J. L. Weinberg and M. S. Hanner. I thank the following — presently team members at the University of Arizona — for their assistance with this chapter (their main areas of expertise are indicated): M. G. Tomasko (encounter preparations, photometry), D. L. Coffeen (instrument development, launch tests, encounter operations, polarimetry), W. Swindell (imaging, special publications), C. Blenman (image production, administration), L. R. Doose (encounters, trajectories, instrument performance, image processing), C. E. KenKnight (encounters, instrument calibrations), J. W. Fountain (photographic processing and quality control), N. D. Castillo (encounters, programming for photometry), J. H. Kendall (encounters, image restoration, programming), J. O. Hämeen-Anttila (instrument design and testing), A. L. Baker (polarimetry), L. R. Baker (PICS), Y. P. Chen (image programming), P. H. Smith (image programming), while there are still others who give us essential support.

APPENDIX
Characteristics of the Imaging Photopolarimeter

Telescope: Maksutov type with 2.5 cm aperture and 8.6 cm focal length

Modes of operation:
  1 — Standby
  2 — Zodiacal light
  3 — Polarimetry
  4 — Imaging

Instantaneous field of view:
  Mode 2: $2°3$ square
  Mode 3: $0°5$ square
  Mode 4: $0°03$ square

Spectral bandpass:
  Blue: 390 to 500 nm
  Red: 595 to 720 nm

Detectors: Two dual-channel Bendix Channeltrons, S-11 and Extended-Red S-20 spectral responses

Polarization analyzer:
  Mode 2:   symmetrical Wollaston prism
  Mode 3:   same Wollaston prism; an achromatic half-wave retardation plate in and out of the beam for determination of position angle of polarization

Calibration:
Mode 2: radioisotope-activated phosphor light source
Mode 3: solar diffuser (in the antenna disk of the spacecraft); tungsten filament
lamp and Lyot depolarizer

Look angle range: 10° to 170° from earthward spin axis (the solar diffuser is viewed
at 10°, and the antenna from 10° to 29°)

Size: 18 × 19 × 47 cm

Mass: 4.2 kg

REFERENCES

Armstrong, K. R.; Minton, R. B.; Rieke, G. H.; and Low, F. J. 1976. Jupiter at five microns. *Icarus* (special Jupiter issue). In press.

Baker, A. L.; Baker, L. R.; Beshore, E.; Blenman, C.; Castillo, N. D.; Chen, Y.-P.; Coffeen, D. L.; Doose, L. R.; Elston, J. P.; Fountain, J. W.; Gehrels, T.; Kendall, J. H.; KenKnight, C. E.; Norden, R. A.; Swindell, W.; and Tomasko, M. G. 1975. The imaging photopolarimeter experiment on Pioneer 11. *Science* 188:468-472.

Baker, A. L., and Coffeen, D. L. 1976. The polarization-phase angle curve for Jupiter. In preparation.

Baker, L. R. 1975. The Pioneer/Jupiter real-time display system. *J. Motion Pict. and Television Eng.* 84:481-485.

Blenman, C.; Coffeen, D. L.; Gehrels, T.; KenKnight, C. E.; Swindell, W.; and Tomasko, M. G. 1976. The imaging photopolarimeter experiment on Pioneer 11. *Space research XVI.* (M. J. Rycroft, ed.) Berlin, GDR: Akademie-Verlag.

Burns, J. A. ed. 1976. *Planetary satellites.* Tucson, Arizona: University of Arizona Press.

Coffeen, D. L. 1974a. Optical polarimeters in space. *Planets, stars and nebulae studied with photopolarimetry.* (T. Gehrels, ed.) pp. 189-217. Tucson, Arizona: University of Arizona Press.

———. 1974b. Optical polarization measurements of the Jupiter atmosphere at 103° phase angle. *J. Geophys. Res.* 79:3645-3652.

———. 1976. Pioneer 10 and 11 polarimetry of the atmosphere of Jupiter. In preparation.

Coffeen, D. L., and Hansen, J. E. 1974. Polarization studies of planetary atmospheres. *Planets, stars and nebulae studied with photopolarimetry.* (T. Gehrels, ed.) pp. 518-581. Tucson, Arizona: University of Arizona Press.

Doose, L. R. 1976. Light scattering properties of Jupiter's red spot. Ph.D. Dissertation, Univ. of Arizona, Tucson, Ariz. In preparation.

Duxbury, T. C. 1975. Pioneer imaging of the Galilean satellites. *Bull. Amer. Astron. Soc.* 7:379.

Fimmel, R. O.; Swindell, W.; and Burgess, E. 1974. *Pioneer Odyssey, encounter with a giant.* NASA SP-349 (National Aeronautics and Space Administration). Washington, D.C.: U.S. Government Printing Office.

———. 1976. *Pioneer Odyssey, encounter with a giant.* NASA SP (National Aeronautics and Space Administration). Washington, D.C.: U.S. Government Printing Office. In press.

Fountain, J. W. 1976. Color maps of Jupiter from Pioneers 10 and 11. *Bull. Amer. Astron. Soc.* In preparation.

Fountain, J. W.; Coffeen, D. L.; Doose, L. R.; Gehrels, T.; Swindell, W.; and Tomasko, M. G. 1974. Jupiter's clouds: Equatorial plumes and other cloud forms in the Pioneer 10 images. *Science* 184:1279-1281.

Frieden, B. R. 1976. Apparent interference fringes across an image of Ganymede: can they be due to ice crystals of a characteristic thickness? In preparation.
Frieden, B. R., and Swindell, W. 1976. Restored pictures of Ganymede, moon of Jupiter. *Science.* In press.
Gehrels, T. 1962. The wavelength dependence of polarization of the sunlit blue sky. *J. Opt. Soc. Am.* 52:1164–1173.
——. 1976. Picture of Ganymede. *Planetary satellites.* (J. A. Burns, ed.) Tucson, Arizona: University of Arizona Press.
Gehrels, T.; Coffeen, D. L.; Hartmann, W. K.; Hummer, R. F.; KenKnight, C. E.; and Weinberg, J. L. 1968. Polarimetry, photometry and imaging experiment for the Pioneer F/G asteroid-Jupiter missions. *Proposal.* (National Aeronautics and Space Administration, Washington, D.C.), pp. 525.
Gehrels, T.; Coffeen, D.; Tomasko, M.; Doose, L.; Swindell, W.; Castillo, N.; Kendall, J.; Clements, A.; Hämeen-Anttila, J.; KenKnight, C.; Blenman, C.; Baker, R.; Best, G.; and Baker, L. 1974. The imaging photopolarimeter experiment on Pioneer 10. *Science* 183:318–320.
Gehrels, T.; Herman, B. M.; and Owen, T. 1969. Wavelength dependence of polarization. XIV. Atmosphere of Jupiter. *Astron. J.* 74:190–199.
Gehrels, T.; Suomi, V. E.; and Kraus, R. J. 1972. On the capabilities of the spin-scan imaging technique. *Space research XII.* (A. C. Strickland, ed.) Berlin, GDR: Akademie-Verlag.
Gehrels, T., and Teska, T. M. 1963. The wavelength dependence of polarization. *Appl. Opt.* 2:67–77.
Hanner, M. S.; Weinberg, J. L.; deShields, L. M.; Green, B. A.; and Toller, G. N. 1974. Zodiacal light in the asteroid belt: the view from Pioneer 10. *J. Geophys. Res.* 79:3671–3675.
Ingersoll, A. P. 1976. Pioneer 10 and 11 observations and the dynamics of Jupiter's atmosphere. *Icarus* (special Jupiter issue). In press.
KenKnight, C. E. 1971. Observations in the asteroid belt with the imaging photopolarimeter of Pioneers F and G. *Physical studies of minor planets.* (T. Gehrels, ed.) pp. 633–637. NASA SP-267. Washington, D.C.: U.S. Government Printing Office.
——. 1976. Real polarimeters, Pioneer 11. In preparation.
Leinert, C., and Klüppelberg, D. 1974. Stray light suppression in optical space experiments. *Appl. Opt.* 13:556–564.
Lyot, B. 1929. Recherches sur la polarisation de la lumière des planètes et de quelques substances terrestres. *Ann. Obs. Paris (Meudon)* 8. (in English, NASA TT F-187).
Maxworthy, T., and Redekopp, L. G. 1976. A solitary wave theory of the Great Red Spot and other observed features in the Jovian atmosphere. *Icarus* (special Jupiter issue). In press.
Pellicori, S. F., and Gray, P. R. 1967. An automatic polarimeter for space applications. *Appl. Opt.* 6:1121–1127.
Pellicori, S. F.; Russell, E.; and Watts, L. A. 1973. Pioneer imaging photopolarimeter optical system. *Appl. Opt.* 12:1246–1258.
Roach, F. E.; Carroll, B.; Aller, L. H.; and Roach, J. R. 1974. The linear polarization of the counterglow region. *Planets, stars and nebulae studied with photopolarimetry.* (T. Gehrels, ed.) pp. 794–803. Tucson, Arizona: University of Arizona Press.
Roosen, R. G. 1971. The Gegenschein. *Rev. Geophys. Space Phys.* 9:275–304.
Russell, E. E., and Tomasko, M. G. 1976. Spin-scan imaging—application to planetary missions. *First College Park colloquium on chemical evolution—giant planets.* New York: Academic Press. In press.
Smith, P. H. 1976. The diameters of the Galilean satellites as determined from Pioneer data. *Bull. Amer. Astron. Soc.* In preparation.
Swindell, W., and Doose, L. R. 1974. The imaging experiment on Pioneer 10. *J. Geophys. Res.* 79:3634–3644.

Tomasko, M. G. 1976. Photometry of Jupiter from Pioneer 10. *Bull. Amer. Astron. Soc.* In preparation.

Tomasko, M. G.; Clements, A. E.; and Castillo, N. D. 1974. Limb darkening of two latitudes of Jupiter at phase angles of 34° and 109°. *J. Geophys. Res.* 79:3653–3660.

Weinberg, J. L. 1974. Polarization of the zodiacal light. *Planets, stars and nebulae studied with photopolarimetry.* (T. Gehrels, ed.) pp. 781–793. Tucson, Arizona: University of Arizona Press.

Weinberg, J. L.; Hanner, M. S.; Beeson, D. E.; deShields, L. M.; and Green, B. A. 1974. Background starlight observed from Pioneer 10. *J. Geophys. Res.* 79:3665–3670.

Weinberg, J. L.; Hanner, M. S.; Mann, H. M.; Hutchison, P. B.; and Fimmel, R. 1973. Observations of zodiacal light from the Pioneer 10 Asteroid-Jupiter probe: preliminary results. *Space research XIII.* (M. J. Rycroft and S. K. Runcorn, eds.) pp. 1187–1193. Berlin, G. D. R.: Akademie-Verlag.

Weinberg, J. L.; Hanner, M. S.; Schuerman, D.; Giouane, F.; Clarke, D.; and Sparrow, J. 1976. The Pioneer 10/11 zodiacal light experiment. *Space Sci. Instr.* In preparation.

Wolfe, J. H. 1975. Jupiter. *Scientific American* 233:118–129.

Wolstencroft, R. D., and Brandt, J. C. 1974. Multicolor polarimetry of the night sky. *Planets, stars and nebulae studied with photopolarimetry.* (T. Gehrels, ed.) pp. 768–780. Tucson, Arizona: University of Arizona Press.

MOTIONS AND MORPHOLOGY OF CLOUDS IN THE ATMOSPHERE OF JUPITER

B. A. SMITH
University of Arizona

and

G. E. HUNT
British Meteorological Office

The dominating axisymmetric regime in the global circulation of Jupiter's atmosphere is responsible for the familiar banded appearance of the planet, although it is seen to be here and there disrupted by non-axisymmetric features ranging in size from the Great Red Spot to the smallest observable discrete clouds. Such features serve as tracers for determining the latitude dependence and time variability of Jupiter's zonal wind. It is noted that zonal currents tend to exhibit long-term stability in their latitudinal positions and average wind speeds; however, it is the time-variable aspects of cloud motions and morphology that we deal with in much of this chapter.

We discuss such phenomena as the vorticity and zonal oscillation of the Great Red Spot and its interactions with the surrounding South Tropical Zone; periodic disturbances in the South Equatorial Belt which appear to have their origins in three "hot spots" located within the deeper layers of the Jovian atmosphere; a westerly subtropical zonal jet with velocities of up to 163 m sec^{-1} and the longevity of several atmospheric features with lifetimes in excess of a decade. We then comment upon the meteorological significance of the observed morphology and motions of the clouds and conclude with a look at the potential for spacecraft studies of the dynamics of the Jovian atmosphere over the next decade.

For more than a century Jupiter has been a favored observational object for many astronomers — both amateur and professional. Because of its relatively great distance from the earth, Jupiter varies far less in angular diameter than the terrestrial planets and thus remains continuously observable (except for intervals of a few months centered on solar conjunctions, which occur every 13 months). Although quasi-systematic photography of Jupiter extends back to the closing decades of the nineteenth century, very few quantitative data relating to cloud motion and morphology have emerged from photographic studies until the start of the program at New Mexico State University in the early 1960's. That we know as much as we do about the long-term behavior of the Jovian atmosphere is due primarily to the many conscientious visual observations collected by the British Astronomical Association and the (primarily American) Association of Lunar and Planetary Observers. Of equal importance are the assimilation and interpretation of these visual data which are largely the work of B. M. Peek, T. E. R. Phillips, E. J. Reese and C. R. Chapman. For more than a decade now, Reese has been the nucleus of an analysis group at New Mexico State University where photographic techniques have made possible a quantum step in the detection of many complex properties of the Jovian global circulation. Similar photographic measuring programs are now in progress at the Lowell Observatory Planetary Research Center and the Lunar and Planetary Laboratory of the University of Arizona.

In this chapter we shall review some of the results of the studies mentioned above, discuss their meteorological significance and describe some new and far reaching observations to be made by planetary spacecraft near the end of the decade.

I. CLOUD MOTIONS AND MORPHOLOGY

The dominating axisymmetric regime in the global circulation of Jupiter is, of course, responsible for the familiar banded appearance of this rapidly rotating planet. The variation of this large-scale zonal flow with latitude has been well studied, notably by Peek (1958), Chapman (1969) and Reese (1972c). These studies have shown that, although some temporal variation in velocity is observed in all zonal currents, each latitude of Jupiter is characterized by a particular zonal motion that remains relatively constant over many decades. This long-term stability is also found in the coarse morphology of the atmospheric cloud structure. The bright and dark bands, referred to classically as *zones* and *belts* respectively, have latitudinal boundaries which tend to remain remarkably stationary over time intervals of a decade or longer (Reese 1972c). Therefore, while much of this chapter is devoted to the time-variable aspects of Jupiter's atmospheric circulation, it is important to remember that large-scale properties tend to be rather constant over long periods of time.

In the past it has been a common practice to describe the zonal motions of atmospheric currents on Jupiter either by their sidereal rotation periods or by their velocities relative to one of two quasi-arbitrary rotating coordinate systems known as System I and System II. We have abandoned both of these

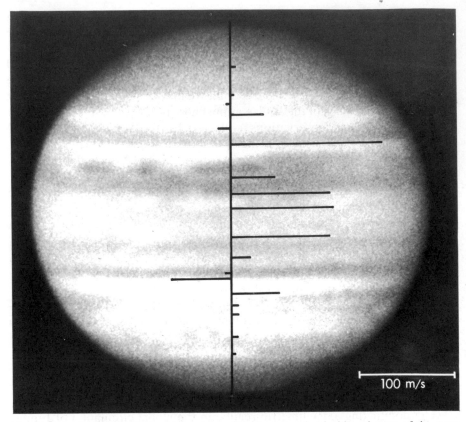

Fig. 1. Zonal velocities in the atmosphere of Jupiter. Zonal velocities of some of the more prominent currents on Jupiter are shown here relative to System III (see text). The scale in meters per second is indicated at the lower right. Note the 100 m sec⁻¹ westerly jet in the equatorial region and the even stronger North Temperate Belt jet in the northern hemisphere. This photograph was taken in blue light on 6 December 1965 with a 30-cm Cassegrain reflector at New Mexico State University. North is at the top.

methods, because they yield numbers which are either physically meaningless or, at best, poorly suited to the study of atmospheric motions. Instead, all of the velocities given in this chapter are referred to System III, a rotating coordinate system defined by the mean rotational period of decametric radio sources, and one which presumably is co-rotating with the deeper, more viscous regions of the planet's fluid interior. System III as defined by the International Astronomical Union in 1962 has a rotation period of $9^h55^m29^s37$. However, we have adopted a "modified System III" with a rotation period of $9^h55^m29^s71$, as it seems better able to represent the observed rotation of the decametric radio sources (Riddle;[1] Riddle and Warwick 1975; Carr and Desch[2]).

[1]See p. 526.
[2]See p. 710.

Zonal velocities as we have now defined them are depicted in Fig. 1 and are listed in Tables I through III for three observational epochs selected by Reese (1972c). Tables IV through VI list the latitudes of the more prominent belts and zones over the same time intervals.

Great Red Spot

Among the non-axisymmetric morphological features in the Jovian atmosphere, the Great Red Spot is the most prominent and the longest lived. Its observed existence can almost certainly be traced back 300 years (Peek 1958) and its lifetime could be on the order of 10^5 years (Golitsyn 1970). The Great Red Spot (GRS) is totally immersed within the usually bright South Tropical Zone (STrZ) which is bounded by the southern component of the South Equatorial Belt (SEBs) to the north and the South Temperate Belt (STB) to the south. The size of the GRS tends to vary with time; at present its length and width are 26,200 and 13,800 km, respectively (personal communication, Reese 1975). Its width, however, exceeds that of the STrZ, causing the SEBs to be deflected toward the equator along the northern edge of the GRS. The deep orange-red hue of the Red Spot also varies with time, occasionally disappearing altogether. However, the deflection of the SEBs (sometimes referred to as the "Red Spot Hollow") remains, even when the GRS is invisible, to remind us that the Red Spot is not merely a superficial feature of the Jovian cloud tops.

The contrast or conspicuousness of the GRS seems to be related to the activity of the SEB (Peek 1958; Solberg 1968b; Solberg 1969a). South Equatorial Belt disturbances, of which more will be said later, are often followed by a slow fading of color in the Red Spot until it finally disappears altogether, blending into the surrounding South Tropical Zone. At those times when the color of the GRS is especially intense, however, SEB disturbances seem to have little or no effect on its chromatic conspicuousness. At present, the Red Spot is continuing in an unprecedented interval of deep orange color which began in 1960 and, throughout a 15-year interval, has survived four outbreaks of SEB disturbances.

The mean zonal velocity of the Great Red Spot is -3 m sec^{-1}, although it has been found to exhibit a superimposed zonal oscillation with an average peak-to-peak amplitude of 1800 km and a period of 89.85 ± 0.10 days (Solberg 1968a; Solberg 1969b; Reese 1970; Reese 1972a; personal communication, Reese 1975). This quasi-sinusoidal motion is apparently a fundamental natural or driven frequency of an atmospheric oscillation associated with the GRS. The oscillatory motion has been in evidence continuously since its discovery in 1962–63 (Reese and Solberg 1966), although the *mean zonal velocity* of the GRS has occasionally changed quite abruptly (see, for example, Reese 1972a). (Although there is every reason to believe that the 90-day oscillation existed prior to 1962, observations of sufficient quality and frequency to establish the oscillatory mode had not been made before that time.) The cause of the GRS 90-day oscillation remains unknown; how-

TABLE I

Summary of Jovian Zonal Velocities from 1898 to 1948
From B. M. Peek and the British Astronomical Association[a]

Atmospheric current	Approximate zenographic latitude range in degrees		Mean zonal velocity m sec⁻¹	Extremes of zonal velocity m sec⁻¹		Number of apparitions
N. Polar Current	+90	+47	−3.1	−1	−6	19
N.N.N. Temp. Current	+43		+2.5	0	+5	9
N.N. Temp. Current *A*	+40	+36	−3.6	−2	−7	35
N.N. Temp. Current *B*	+35		+29.1	+26	+30	6
N. Temp. Current *A*	+33	+29	−11.3	−9	−18	24
N. Temp. Current *B*	+27		+43.0	+34	+53	6
N. Temp. Current *C*	+23		+126.6	+123	+130	6
N. Trop. Current *A*	+22	+14	+0.2	−3	+5	52
Middle of NEB	+13		+28.0	+10	+42	7
N. Equatorial Current	+10	+3	+107.6	+102	+113	47
Central Equat. Current	+3	−3	+108.1	+106	+122	14
S. Equat. Current *A*	−3	−10	+107.0	+95	+115	39
S. Equat. Current *B*	−3	−10	+87.6	+78	+95	11
S. Component SEB	−19		−3.2	−7	+1	13
S. Edge SEBs[b]	−19		−65.0	−49	−75	6
S. Tropical Zone	−21	−26	−2.1	−6	+2	7
Red Spot	−22		−2.7	−5	+1	64
N. Edge STB[c]	−27		+49.5	+41	+56	5
S. Temp. Current	−29		+3.0	+1	+5	47
S.S. Temp. Current	−38	−45	+6.2	+5	+7	42
SSSTB and SPR	−45	−90	0.0	−2	+1	5

[a] After Reese (1972c)
[b] Northern branch of circulating current
[c] Southern branch of circulating current

TABLE II

Summary of Jovian Zonal Velocities from 1946 to 1964 From Association of Lunar and Planetary Observers[a]

Atmospheric current	Approximate zenographic latitude range in degrees		Mean zonal velocity m sec⁻¹	Extremes of zonal velocity m sec⁻¹		Number of apparitions
N.N.N. Temp. Current	+46	+44	+2.4	+2	+4	5
N.N. Temp. Current A	+41	+35	-2.7	-4	0	9
N. Temp. Current A	+32	+28	-11.2	-9	-13	7
N. Trop. Current A	+21	+14	+1.1	-2	+5	16
N. Trop. Current B	+21	+14	+8.5	+8	+9	2
Middle of NEB	+12		+46.2	+35	+66	5
N. Equat. Current A	+8	+4	+107.4	+103	+111	17
N. Equat. Current B	+8		+99.2		+111	1
S. Equat. Current A	-6	-8	+107.2	+106	+111	7
S. Equat. Current B	-6	-8	+94.2	+87	+99	7
S. Edge SEBn	-11		+45.6	+43	+48	3
Middle of SEB	-14		+23.7	+13	+30	5
S. Component SEB	-16	-22	-3.5	-2	-6	6
S. Edge SEBs[b]	-18	-22	-54.6	-51	-61	3
S. Tropical Zone A	-23		+13.8	+8	+22	7
S. Tropical Zone B	-23		-6.6	-4	-10	5
Red Spot	-22	-23	-4.0	-3	-4	17
STrZ Disturbances	-23		+1.8	0	+3	4
N. Edge STB[c]	-26		+35.1			1
Middle of STB	-29	-31	+6.7	+5	+7	11
S. Temp. Current	-32	-35	+5.9	+4	+7	17
S.S. Temp. Current	-39	-45	+6.1	+5	+7	9

[a] After Reese (1972c)
[b] Northern branch of circulating current
[c] Southern branch of circulating current

TABLE III

Summary of Rotation Periods from 1962 to 1970
From Measurements of Photographs Taken at NMSU Observatory[a]

Atmospheric current	Range of measured zenographic latitudes in degrees		Mean zonal velocity m sec^{-1}	Extremes of zonal velocity m sec^{-1}		Number of apparitions
N. Polar Region	+ 55.0		+ 3.9			1
N.N.N. Temp. Current	+ 45.9	+ 42.8	+ 2.7	+ 1.4	+ 4.6	3
N.N. Temp. Current A	+ 40.6	+ 37.3	− 2.3	− 1.5	− 3.2	3
N.N. Temp. Current B	+ 35.8	+ 35.1	+ 30.8	+ 30.3	+ 31.9	3
N. Temp. Current A	+ 31.4	+ 30.2	− 11.5	− 9.9	− 12.9	3
N. Temp. Current C	+ 24.2	+ 23.8	+ 144.3	+ 122.1	+ 163.4	2
N. Trop. Current A	+ 19.6	+ 15.5	+ 2.5	+ 1.2	+ 4.2	4
Middle of NEB	+ 12.4		+ 46.2			1
N. Equat. Current A	+ 8.6	+ 6.6	+ 105.3	+ 104.4	+ 105.8	3
N. Equat. Current B	+ 8.6	+ 6.6	+ 99.9	+ 99.3	+ 100.3	2
S. Equat. Current A	− 5.8	− 7.6	+ 105.2	+ 103.0	+ 107.2	4
S. Equat. Current B	− 5.8	− 7.6	+ 94.5	+ 93.2	+ 97.2	3
Middle of SEB	− 12.6	− 13.7	+ 20.1	+ 15.7	+ 29.4	4
S. Component SEB	− 16.1	− 20.3	− 5.7	− 5.2	− 5.8	2
S. Edge SEBs[b]	− 20.3	− 21.7	− 49.6	− 48.4	− 53.1	4
S. Tropical Zone A	− 22.7	− 24.6	+ 7.3	− 2.3	+ 17.8	4
S. Tropical Zone B	− 21.8		− 8.7			1
Red Spot	− 22.1	− 22.9	− 3.6	− 3.5	− 3.8	8
STrZ Disturbance	− 22.7		+ 1.9			1
N. Edge STB[c]	− 25.2	− 26.2	+ 45.2	+ 36.0	+ 48.7	4
Middle of STB	− 29.5	− 30.4	+ 5.3	+ 4.7	+ 6.6	5
S. Temp. Current	− 32.6	− 33.7	+ 4.6	+ 4.3	+ 5.0	6
S.S. Temp. Current	− 38.8	− 41.3	+ 6.7	+ 6.3	+ 7.1	6
S.S. Temp. Zone	− 47.3	− 50.3	− 2.0	− 0.9	− 2.6	3
S.S.S. Temp. Zone	− 59.6	− 59.9	+ 3.4	+ 3.1	+ 3.8	2

[a] After Reese (1972c)   [b] Northern branch of circulating current   [c] Southern branch of circulating current

TABLE IV
Mean Zenographic Latitudes of Jupiter's Belts 1908–1947
From B. M. Peek
and the British Astronomical Association[a]

Feature	Mean latitude (deg)	Range of apparitional means (deg)		Number of apparitions
NNNTB	—	—	—	—
NNTB	+ 36.4	+ 39.5	+ 33.1	17
NTB	+ 27.8	+ 32.2	+ 24.5	33
N. edge NEB	+ 17.3	+ 22.0	+ 12.7	34
S. edge NEB	+ 7.2	+ 9.0	+ 4.5	34
N. edge SEB	− 7.1	− 4.4	− 9.0	33
S. edge SEB	− 18.9	− 16.4	− 22.6	30
STB	− 29.0	− 26.7	− 30.5	36
SSTB	− 41.6	− 37.0	− 45.4	20

[a]After Reese (1972c)

TABLE V
Mean Zenographic Latitudes of Jupiter's Belts 1949–1962
From Association of Lunar and Planetary Observers[a]

Feature	Mean latitude (deg)	Range of apparitional means (deg)		Number of apparitions
NNNTB	+ 45.3	+ 45.7	+ 44.8	2
NNTB	+ 36.3	+ 41.0	+ 33.7	10
NTB	+ 26.7	+ 30.0	+ 23.6	8
N. edge NEB	+ 17.9	+ 21.3	+ 13.7	12
S. edge NEB	+ 6.8	+ 7.9	+ 4.1	12
N. edge SEB	− 7.3	− 5.5	− 7.9	9
S. edge SEB	− 20.8	− 18.2	− 22.6	11
STB	− 31.0	− 29.8	− 32.7	12
SSTB	− 44.9	− 42.1	− 47.1	7

[a]After Reese (1972c)

ever, we note that the characteristic velocity of the "circulating current" — within which the Red Spot is immersed — is approximately ± 55 m sec^{-1} with respect to the GRS, and that any disturbances moving within the circulating current would re-encounter the GRS at intervals of 90 days.

TABLE VI

Mean Zenographic Latitudes of Jupiter's Belts 1960–1970
From measurements of photographs
taken at NMSU Observatory[a]

Feature	Mean latitude (deg)	Range of apparitional means (deg)		Number of apparitions
NPB	+ 56.0	+ 59.5	+ 54.3	2
NNNTB	+ 44.8	+ 45.9	+ 44.3	7
NNTB	+ 37.1	+ 38.0	+ 35.9	7
N. edge NTB	+ 31.3	+ 31.6	+ 31.0	6
NTB	+ 27.8	+ 28.7	+ 26.9	6
S. edge NTB	+ 24.3	+ 25.8	+ 22.8	6
N. edge NEB	+ 19.9	+ 21.4	+ 16.9	10
S. edge NEB	+ 7.4	+ 8.6	+ 6.6	10
EB	− 0.4	+ 0.2	− 1.2	4
N. edge SEBn	− 6.9	− 5.8	− 7.6	8
S. edge SEBn	− 11.0	− 10.6	− 11.4	3
N. edge SEBs	− 16.0	− 14.5	− 17.9	3
S. edge SEBs	− 20.9	− 19.9	− 21.7	10
N. edge STB	− 26.3	− 24.8	− 27.6	10
STB	− 29.9	− 29.3	− 31.0	10
S. edge STB	− 33.6	− 32.9	− 34.5	10
STZB	− 38.3	− 37.6	− 38.8	7
SSTB	− 44.4	− 43.1	− 46.3	10
SSSTB	− 55.9	− 54.9	− 58.0	5
SPB	− 65.9	− 64.9	− 67.4	3

[a]After Reese (1972c)

The Red Spot is not the only feature in the Jovian atmosphere to exhibit an oscillatory component in its zonal motion. Several oscillating dark spots have been reported by Peek (1958) while Reese and Smith (1966) have called attention to an oscillatory mode of a spot moving in the narrowly confined zonal jet at 24°1 N latitude. Reese and Solberg (1969) also describe a dark spot in the southern component of the North North Temperate Belt (NNTBs) at latitude 35°5 N which showed a well-defined sinusoidal zonal oscillation with an amplitude of 7000 km and a period of 66.4 days.

Another interesting property of the Red Spot is its vorticity. During 1966–67, as reported by Reese and Smith (1968), several dark spots associated with the circulating current interacted with the perimeter of the GRS, some apparently attaching themselves to its outer edge and revolving about the Red Spot in a counterclockwise direction with a period of about 12 days. The attached dark spots (and other observed features of unknown origin imbedded within the Red Spot) did not exhibit the $1/r$ dependence of a simple vortex. Tangential velocities with respect to the center of the GRS were 19 and 72 m sec^{-1}, respectively, at the major and minor axes of Red Spot ellipse. Another interaction between the GRS and an external feature occurred in March 1967 when one of the three long-enduring, bright South Temperate Ovals (STO) passed south of the Red Spot. The northern edge of the STO extended northward almost to the southern edge of the GRS and may actually have impinged upon the vortex. For a few days a bright ring surrounded the GRS as though material from the bright oval was drawn into the vortex.

Whether vorticity is a permanent property of the GRS is not known. Reese (1971) reported a bright, cloud-like structure within the Red Spot in 1971 which apparently remained stationary. The distribution of these bright patches tended to be spread out along an ellipse, concentric with the perimeter of the GRS and approximately halfway out from the center. These observations suggest that the vortical motion of the GRS may be restricted to a zone near its perimeter.

A satisfactory meteorological explanation for the Great Red Spot remains to be found, although its suggested identification as an anticyclonic (high-pressure) system seems reasonable. The GRS has remained a relatively isolated atmospheric system for several centuries at least; yet, as stated above, some interactions with other atmospheric features have been observed. For example, a noteworthy interaction occurred in 1970 when a dusky disturbance in the South Tropical Zone made contact with the GRS. First seen in June 1970 (Reese 1971), the disturbance advanced on the Red Spot with a relative velocity of $+ 5$ m sec^{-1}. As Jupiter approached conjunction with the sun in November, it was predicted that the disturbance would arrive at the western (following) end of the GRS on 24 December. The disturbance did indeed reach the Red Spot on 23 December and apparently imparted an instantaneous increase of 3.2 m sec^{-1} to the zonal velocity of the GRS (Reese 1972a). Since the date of that incremental velocity increase, the Red Spot has undergone a constant deceleration, and presently has a zonal velocity 5 m sec^{-1} less than its motion in early 1971. In another example, Minton (1975) reports that a white spot became temporarily drawn out in longitude as its differential motion brought it into conjunction with the GRS. Perhaps the most remarkable interaction between the GRS and its surroundings occurred in 1969 when a series of dark elongated spots were ejected from the eastern (preceding) tip of the Red Spot and were then observed to drift away with a relative zonal velocity of $+ 17$ m sec^{-1} (Reese 1970). It is clear that any satisfactory explanation for the GRS must be able to account for its many peculiarities.

South Tropical Zone Circulating Current

Along the north and south boundaries of the South Tropical Zone flow the northern and southern components of the circulating current. As can be seen in Fig. 1 and Tables I–III, the two components are easterly and westerly, respectively, and both have nearly the same zonal speed, approximately 50–60 m sec^{-1}. It is between these components that the Red Spot is immersed, but it was an apparent interaction with another large feature in the STrZ, the South Tropical Disturbance (STD), that has given the circulating current its name. Although the STD has not been recorded in recent years, spots have been observed approaching the STD along the northern (easterly) component, only to move suddenly southward and into the westerly component as they came in contact with the eastern (preceding) end of the South Tropical Disturbance. The South Tropical Disturbance was last seen in 1939. This feature, a most interesting object in itself, will not be discussed here. The reader is referred to Peek (1958).

Thus, the spots would first approach, cross the STrZ then recede from the STD. Now moving along the southern component, the spots would approach the Red Spot with a relative velocity of 60 m sec^{-1}, but would then disappear as they came within ten thousand kilometers of the western end of the GRS. Until 1964 no spot in the westerly branch of the circulating current had ever been observed to survive the GRS. The exception (Reese and Smith 1968) was a small dark spot in 1965 which approached the GRS along the southern component of the circulating current, moved into the narrow channel which separates the Red Spot from the South Temperate Belt, attached itself to the perimeter of the GRS and revolved around it, thus revealing the GRS vorticity. Several months later, various other dark spots moving along the northern component approached the GRS and interacted in a similar manner.

South Equatorial Belt Disturbances

Since 1919 the South Equatorial Belt has exhibited a series of semiperiodic fadings followed by cataclysmic returns to prominence. These revivals, which tend to recur at intervals or multiples of three years, always begin with a sudden localized outburst of bright and dark spots which soon spread out in longitude until the entire belt is in a state of turmoil (Reese 1972a). In all, a total of 12 SEB disturbances have been observed, occurring in the following years: 1919, 1928, 1943, 1949, 1952, 1955, 1958, 1962, 1964 and 1971. The years 1943 and 1971 each saw two disturbances, and the appearance of the SEB at the present time would suggest that another is imminent. (Since this writing, no less than three separate SEB disturbances have occurred over a six week interval in July–August 1975. The three, in order, appeared at the intense, medium and weak sites, respectively.)

Reese (1972a) has pointed out that the disturbances can be grouped according to their observed intensity with the most active occurring in 1919, 1928 and 1971A and the weakest in 1943B, 1955, 1962 and 1971B. Although the locations of the various outbursts are scattered randomly in System II, the intensity grouping gave rise to the idea that there might be a coordinate system in which there would exist a small number of fixed sources for the eruptions (Chapman and Reese 1968). Reese (1972a) has found that, when grouped by the magnitude of their activity, the disturbances can all be traced to three sources which remain stationary in a coordinate system that rotates with a period of $9^h55^m30\overset{s}{.}11 \pm 0\overset{s}{.}03$, which is only 0.40 seconds longer than that of the "modified System III." It will be recalled that we have assumed the "modified System III" to be co-rotating with the viscous interior of Jupiter. Thus, there seems to be good reason to believe that SEB disturbances have their origins in three "hot spots" located deep within the Jovian atmosphere. The three sources are widely separated in longitude and their zonal drift with respect to our defined coordinate system is only -14 cm sec^{-1}.

South Temperate Ovals

In 1939–40 three bright clouds, each extending more than 90° in length, appeared in the southern edge of the South Temperate Belt (Peek 1958). Since their formation, each has contracted exponentially, very rapidly at first and more slowly in recent years. At present these South Temperate Ovals have lengths and widths of 1400 km and 8800 km, respectively (personal communication, Reese 1975). If the present trend continues, the mean lengths of the STO's will contract to 5200 km in 1980 and 3100 km by 1985. Reese (1971) points out that at some critical size, the disruptive effects of atmospheric turbulence will dominate and that final disintegration may be quite sudden. With their present zonal velocity of approximately $+ 8$ m sec^{-1} with respect to the GRS, each STO requires 530 days to move from one GRS conjunction to the next. On the average the STO's tend to be spaced approximately 120° apart in longitude and are so morphologically similar that it is usually difficult or impossible to recognize a given oval by appearance alone (Reese 1971). Occasionally two STO's have drifted close to one another (10,000 km in August 1950) due to the slight randomness of their individual zonal velocities; however, coalescence is prevented by an apparent repelling force of magnitude inversely proportional to the square of their separation (Sato 1969). Both Reese (1971) and Beebe (personal communication 1975) have attempted to find a relationship between the zonal velocity of the STO's and their longitude relative to the GRS. At this time it appears that short-term correlations do exist but tend to break down over longer time intervals.

Did the STO's really appear for the first time 35 years ago, and how long will they last? It seems unlikely that we should be witnessing a unique event and, indeed, Reese (1972b) has found evidence that three early long-enduring bright ovals observed between 1914 and 1935 may be precursors of the present South Temperate Ovals. If such is the case, we have evidence for three additional Jovian features with lifetimes in excess of half a century and quite possibly much longer.

North Temperate Belt Jet

Evidence for a rapid jet in the southern component of the North Temperate Belt (NTBs) has been given by Peek (1958). Visual observations of this zonal current, moving at velocities of 120–140 m sec^{-1} with respect to adjacent latitudes, have been reported periodically since 1880. It should be noted that this current has a velocity well in excess of the "equatorial jet," although it is located in a subtropical region at 24°.1 N latitude.

Photographically the NTBs jet was first detected in 1964. Appearing dark on blue-sensitive plates, but invisible to visual observers, a 6000 km spot was recorded at the latitude of the jet moving with a mean zonal velocity of $+ 123$ m sec^{-1} (Reese and Smith 1966). Superimposed on its mean zonal motion was a zonal oscillation having an amplitude of 10,000 km and a period of approximately 300 days. The rapid motion of the NTBs spot carried it completely around Jupiter once very 40 days relative to several round white spots which nearly filled the entire width of the adjacent North Tropical Zone. When passing the white spots, the dark NTBs spot was twice observed to encroach upon them, thereby suggesting that the NTBs feature was located at a higher level in Jupiter's atmosphere.

In 1970 another rapidly moving NTBs feature was photographed, this time a spot bright in blue and ultraviolet light at latitude 23°.8 N. It represented the sixth outbreak of activity in the NTBs jet and established a record for the highest zonal velocity ever recorded on Jupiter, $+ 163$ m sec^{-1} (Reese 1971).

Because the NTBs jet can be detected only when it produces an observable disturbance sufficiently large to disrupt its axial symmetry, we do not know whether it is a permanent feature of the Jovian atmosphere. Systematic high-resolution imaging with the Space Shuttle Sortie Telescope or the Large Space Telescope may provide the answer to this question.

The Equatorial Region

Lying between latitudes 7°N and 7°S is the great equatorial region of Jupiter, more properly known as the Equatorial Zone (EZ). The entire region, approximately one eighth of the surface area of Jupiter, is characterized by a 100 m sec^{-1} westerly zonal wind. The width of the EZ is variable, becoming as small as 11° in 1924 and as great as 18° in 1914 (Peek 1958). At present the width of this region is 14°.5, or 17,500 km (personal communication, Reese 1975).

Changes in cloud structure within the EZ are often quite rapid. On occasion, the morphology of a specific feature has been observed to undergo noticeable change within a single rotation of the planet, i.e., about 10 hours. Large changes have also been observed at planetary dimensions over intervals of less than one year. In 1964–65, the entire equatorial region changed from dark to bright in approximately ten months (as recorded in blue light), suggesting the formation of an obscuring high-level, global haze on a time scale which was very short compared to other changes in Jupiter's banded structure. At this high level the haze itself was probably an ammonia-crystal, cirrus-type layer, and its rapid planetwide distribution suggests that a major change in the thermal structure of the equatorial region took place in a matter of a few months.

Although the EZ seems to be a region of Jupiter's atmosphere associated with rapid and widespread change, there exist other observations which suggest localized long-term stability. Several discrete cloud features have been observed to persist for more than a year and a few for more than four years. The record for longevity appears to be established by two plume-like features discussed by Reese and Beebe (1976). These curious objects, seen prominently in the Pioneer 10 and 11 imaging and located almost diametrically apart at 7°N, have been traced as far back as 1964 and are still in evidence twelve years later (Minton 1975; personal communication, Reese 1975).

Cloud motions in the equatorial region have been poorly studied in recent years. This unfortunate inattention can be attributed to problems encountered in measuring the images; the EZ clouds themselves usually show up best in red light, while the planet's limb (necessary for a measuring reference) is not well defined at longer wavelengths due to strong limb darkening. However, new methods for circumventing this problem (e.g., the use of color emulsions) are currently being employed at the University of Arizona and New Mexico State University and a better understanding of equatorial zonal winds should soon follow.

II. METEOROLOGICAL SIGNIFICANCE OF THE VISIBLE APPEARANCE OF JUPITER

Structure of the Belts and Zones

In the preceding section we have described the visible appearance of Jupiter from the wealth of observations compiled throughout a century of observation. We shall now attempt to translate these observations into a form that enables us to understand the basic meteorology of the Jovian atmosphere.

We note that Jupiter's banded structure seems to persist, with some slight deviation, for centuries. The zones are light colored regions, generally thought to be composed of ammonia clouds, which mark the top of a region of anticyclonic vorticity and therefore rising motion. Correspondingly, the belts form the other link to conserve the mass flow and are regions of cyclonic vorticity and sinking motions. The difference in altitude between the belts and zones is small compared with the pressure scale height of the planet (which is approximately 25 km), and we find that the North and South Tropical Zones are the highest cloud bands on Jupiter (Kuiper 1972). Coffeen (1974)[3] has concluded that, at the time of the Pioneer 10 flyby of Jupiter, these zones differed in altitude by 4 and 10 km, respectively, from the tops of the adjacent belts. The high altitude feature in the equatorial region of Coffeen's scan is identified as one of the two long-lived plumes which we discussed in the previous section. At the highest spatial resolution presently available, the Pioneer observations show less structure in a zonal direction

[3]See p. 535.

than is immediately evident in the meridional scans. This, of course, may be a gross oversimplification; certainly the 5 μm observations of Keay et al. (1972)[4] and Westphal et al. (1974) show evidence of local hot spots which may well indicate localized regions of large altitude differences.

The overall picture of Jupiter is one of alternating light and dark cloud bands parallel to the equator with regions of shear at the interfaces between these cloudy regions. The dominant zonal feature is the high-velocity, westerly equatorial jet. In spite of the visible turbulence in this region, certain features can apparently exist there for more than a decade (Reese and Beebe 1976). Other high-velocity currents are found at 35°N, 24°N, 21°S and 26°S (see Fig. 1 and Tables I–III). The westerly jets at 24°N and 26°S suggest a type of hemispheric symmetry, but it would appear that the GRS, extending from 15°S to 27°S, is responsible for the asymmetric easterly current at 21°S. Thus, it seems possible that a Jovian planet without a Red Spot would be truly symmetric. This indeed appears to be the case for Saturn.

Atmospheric Driving Mechanisms and the Stability of Jovian Features

In order to understand the axisymmetric structure and other features of the planet's meteorology such as the visible spots, we need to understand the mechanisms which drive the Jovian atmosphere and the reasons for the stability of these large scale flow patterns.

The weather systems of the terrestrial planets are driven by the sun in the form of differential solar heating. In the cases of the earth and Mars, the atmospheric circulations are influenced by interactions between the atmosphere and surface. If we look more closely at the earth's system, we see that it falls into a hybrid category. In the extratropical regions it is baroclinic, while in the tropical region, convection is the primary mode of heat transfer.

Jupiter is approximately 5 A.U. from the sun so that it receives about 0.04 of the solar flux incident upon the earth. At first sight this may suggest that the sun has less influence on the Jovian atmospheric motions than it does on those of the earth. Nevertheless, several authors (Stone 1967, 1969; Gierasch and Stone 1968), have attempted to explain the Jovian circulation by baroclinic mechanisms. A major weakness of these analyses is their failure to account for the equatorial jet (Stone 1972).[5] Saturn receives a quarter as much solar radiation as Jupiter, and yet possesses an equatorial jet which is four times stronger.

While these equatorial jets may be created by special atmospheric circumstances in which the source of momentum is found in stratospheric waves as Maxworthy (1975) suggests, there are mechanisms other than the sun which may be dominant in driving the Jovian and Saturnian motions.

We have no precise information on the local thermal structure of the atmosphere beneath the visible cloud layers. Ingersoll and Cuzzi (1969) com-

---

[4]See p. 407.
[5]See p. 603.

bined the assumption of a thermal zonal wind with the observed correlation between the banded structure and vorticity, and deduced that the light zones were systematically warmer than the dark belts at a corresponding level beneath the cloud tops. The amount of the temperature difference depends on the depth to which the thermal wind extends. Dynamically, a difficulty with the model occurs at the equator, since the thermal wind becomes infinitely large there, although this method cannot be relied upon to explain the equatorial jet. Barcilon and Gierasch (1970) have suggested that H_2O condensation below the visible cloud tops may be a possible source for this local heating, and their analysis suggests that this mechanism is a possible source for zonal winds. Stone has commented that condensation at levels of approximately 250°K may have only an indirect effect upon the visible cloud layers.

The observations by Chase et al. (1974)[6] confirm the many groundbased observations that Jupiter emits approximately twice the energy it receives from the sun, thus establishing conclusively that the planet has an internal heat source. This is important to our discussions of atmospheric driving mechanisms. Williams and Robinson (1973) have developed a convectively driven dynamical model which they applied to Jupiter. While they have been able to produce the westerly jet, their model requires an internal heat source which is too large when compared to the measured value. Also, this work does seem to create a paradox, since it has shown that an atmosphere which is statically unstable everywhere can organize itself into large-scale flow patterns. Convective models of this type tend to leave high longitudinal wavenumbers suggesting that the zonal flow may be unstable. Indeed, Fig. 21 of Williams and Robinson (1973) shows that their model develops temperature instabilities, and we do not know whether these instabilities decay or tend toward growth which would dominate and destroy the zonal flow.

Ingersoll and Pollard (1975) have attempted to explain the stability of Jupiter's zonal motion with a baroclinic model and an infinitely deep atmosphere. In this way there is no surface interaction, as in the case of the earth, which would destroy the stability of the system. Certainly the high resolution pictures from Pioneer 10 and 11 (Gehrels et al. 1974; Fountain et al. 1974; Swindell and Doose 1974; Baker et al. 1975)[7] show some evidence of small-scale wave disturbances such as the spiral structures, which are possible examples of baroclinic instabilities. Yet, it is not known whether these are minor perturbations on the basic zonal structure or an essential part of it. Nor was the observation time base long enough to tell us whether they are growing instabilities or decaying waves.

We do not at this time possess observations which would enable us to differentiate between the convective and baroclinic modes. Indeed, such a distinction may be meaningless anyway, since there is every possibility that Jupiter's atmosphere is a complex hybrid in which both mechanisms play a

---

[6]See p. 508.
[7]See p. 545.

role. It seems probable that the modes of heat transfer in the Jovian atmosphere will not be fully understood until observations from orbiting spacecraft and entry probes become available.

Meteorological Properties of Jovian Spots

Although our discussions of the zonal motions of portions of the Jovian
disk have been given with relatively high numerical precision, they have been
compiled from measurements of visible features. It is unlikely that these features actually move with the velocity of the fluid in which they are imbedded,
so that to interpret them as a measure of the local zonal velocity could be
misleading. Macdonald (1968) analyzed the motions of cyclones and anti-
cyclones in the earth's atmosphere and compared their motions to the mean
zonal motions measured directly. In mid-latitudes errors of 100% were not
uncommon although better agreement was found in latitudes nearer to the
equator and the poles. Generally, small cyclones and anticyclones are found
to be better markers of the terrestrial mean zonal flow than the larger systems. On the other hand, friction with the terrestrial surface, a condition
which does not exist on Jupiter, may play a significant role in producing the
differences between mean fluid velocity and the observed motions of the disturbances. In any case, this potential source of error must be borne in mind
when interpreting high-resolution pictures of Jupiter.

Spots of various colors and albedos have been seen at differing Jovian
latitudes for centuries and a comprehensive account of their properties is
given by Peek (1958). Starr (1973) has carried out a statistical analysis of the
distribution with Jovian latitude of a number of transient visible spots. Unfortunately, his study relates to only two months of observation between 1
December 1966 and 20 January 1967, so the results are only an indication
of the spot properties. This study did discuss two rows of dark spots which
were concentrated along ± 9° latitude. Rosen (1972) described these configurations as *convective vortex sheets*, since a shear of some 100 m sec^{-1} is
established over a meridional distance of only about two latitude degrees.
Indeed, during this one-month period the dark spots do seem to appear primarily at the interfaces of the belts and zones which are regions of intense
shear. White spots were apparently found at different latitudes than the
darker variety and during the period of Starr's analysis the maxima occurred
at 5° N and 20° N.

In the high-resolution Pioneer pictures (see, for example, Gehrels *et al.*
1974 and Baker *et al.* 1975), many light spots were seen with dark edges
which suggested descending motion at the periphery of the features.

These transient Jovian features generally have rather short lifetimes of a
few days for the smallest to a few weeks or even months for the larger ones.
This generalization does, of course, exclude the South Temperate Ovals and
the Great Red Spot, where their individual special properties require a separate discussion. In providing an explanation for the visible features we return
to the more familiar environment of the terrestrial tropics.

In the terrestrial atmosphere the low level convergence of the NE and SE trade winds, which occur on the equatorward side of the Hadley cells, gives rise to a cloud band commonly referred to as the Intertropical Convergence Zone (ITCZ). Studies of the ITCZ have revealed that it is made up of groups of convecting cells which generate upper and medium level cumulonimbus clouds that merge together. Many of the systems vary in size between 500 and 1500 km and are called *cloud clusters*.

Some clusters are associated with upper-level vortices. These develop within large oceanic troughs which occur in the upper atmosphere over the Pacific and Atlantic Oceans in summer. They are most in evidence at the 200 mbar level although about 40% of them penetrate down to 700 mbar and 10% to the surface. The amount and organization of the cloud is dependent on the depth of penetration of the vortex and on low-level factors which affect convection; for example, sea-surface temperature, moisture content and stability. Significant amounts of cloudiness and large-scale rainfall do not occur unless the vortex extends down at least to 850 mbar, where low-lying moisture can be drawn into the system. Cloud clusters have been found generally to last for periods ranging from one to eight days.

It is indeed possible for large-scale convective storms of this type to occur in the Jovian atmosphere. Local characteristics of the atmospheric flow pattern could set up the convective instability to create a cumulonimbus type of system. Such a feature would be more stable in Jupiter's atmosphere than in the earth's. Chase *et al.* (1974) have confirmed that there are no systematic diurnal temperature variations in the Jovian atmosphere, but convection may be maintained through the latent heat released from condensing ammonia and water vapor, both of which would be recycled in the atmospheric circulation. The recent discovery of water vapor in the Jovian atmosphere (Larson *et al.* 1975)[8] tends to support this view.

The differences in color of the spots may be the result of the composition of the cloud layers where the spots form, and consequently would represent the interaction between the atmospheric motions and the complex chemistry of the Jovian atmosphere. To characterize the transient spots in a more quantitative fashion will require the high-resolution observations from the imaging and infrared experiments of Mariner Jupiter/Saturn mission, which will become available in 1979.

The most spectacular feature in the Jovian atmosphere is the Great Red Spot, whose properties we outlined in Sec. I. It is the most long-lived non-axisymmetric feature on the visible surface of the planet, and as a result it has received considerable attention. Attempts to explain this feature fall into three main categories: (a) floating objects; (b) Taylor columns; and (c) Jovian storm systems.

There have been numerous reviews of this topic in recent years (see, for example, Hide 1961, Kuiper 1972, Newburn and Gulkis 1972, Ingersoll

[8]See p. 398.

1973, and Maxworthy 1973). The strong vorticity within the spot and the complex interaction between the spot and the clouds moving around it make the floating-object and Taylor-column explanations most unlikely. Furthermore, major evidence against the Taylor-column theory is the longitudinal wandering of the GRS, which has varied by more than 1000° in periods of 50 years.

The observations at New Mexico State University (Reese and Smith 1968), together with the recent Pioneer 10 and 11 images (see, for example, Swindell and Doose 1974 and Baker *et al.* 1975) provide further evidence of cell-like structures within the GRS itself. Although there is a suggestion of local convection by the mottled appearance within the spot, there is no evidence yet of an eye or spiral cloud bands which one normally associates with a terrestrial hurricane. However, there is the strong possibility that the feature may be a region of intense local convective activity driven by the release of latent heat, as suggested by Kuiper (1972). The South Temperate Ovals may also have similar properties, but photography in the 890 nm absorption band of methane shows that the tops of the STO's — unlike the GRS — extend only slightly into the upper levels of the Jovian atmosphere (personal communication, Minton 1975).

Maxworthy and Redekopp (1976) have suggested that solitary waves (solitons) may play an important role in the motions observed in the Jovian atmosphere. A soliton is an isolated permanent wave which propagates without change of shape and feeds on the horizontal shear of the flow. When solitons interact, there is a phase shift that looks like an acceleration, they emerge with identical forms as before and the overtaking soliton reforms in the place of the overtaken one. The measurements of Reese and Smith (1968) show that the Great Red Spot is situated in a region of anticyclonic shear while equatorwards of this region the shear changes to cyclonic to match the equatorial jet velocity. Maxworthy and Redekopp (1976) have suggested a soliton morphology which produces flow pattern similar to that observed in the vicinity of the GRS, and also one that can account for the observed motions associated with the South Tropical Disturbance.

The equatorial plumes observed by Reese and Beebe (1976) throughout a twelve year period, could also be accounted for by a soliton. Throughout the observational period the Equatorial Zone has appeared abnormally dark and has contained many dark projections along its northern edge. When the plume approaches to within 25–30° of these projecting features they are deflected in the direction of the plume's motion and then dissolve or become obscured as the plume passes. After passage of the plume, normal features are again observed. This behavior fits Maxworthy and Redekopp's description of a soliton.

However, in all these observational interpretations, we cannot be confident yet of a unique interpretation. To characterize the GRS and the STO's and to study possible soliton motions more quantitatively, we need to know the flow patterns of fluid particles within and around these features, the growth (decay) rate of small spots to changes in the shear flow on which they feed. These dynamical characteristics of the Jovian atmosphere are likely to be obtained by the Mariner Jupiter / Saturn (*MJS*) imaging experiment when the spacecraft encounter Jupiter in early 1979.

III.   SPACECRAFT STUDIES OF THE JOVIAN CLOUD SYSTEMS

In previous sections we have occasionally made reference to spacecraft studies of Jovian cloud motions and morphology. The first useful data to be obtained from spacecraft were those transmitted back from Jupiter by Pioneer 10 in 1973 and Pioneer 11 in 1974. The next spacecraft mission to Jupiter will be *MJS*, in which two spacecraft will fly by Jupiter in 1979 and Saturn in 1980–81. *MJS* will achieve both higher resolution and a longer observing time base than Pioneer; in each case the improvement is approximately a factor of 40. Furthermore, the potential exists for obtaining a total of approximately 30 thousand photographs of these two remote planets and their satellites.

Although *MJS* is the only approved mission to Jupiter at this time, proposals are being made for Jupiter orbiting spacecraft and atmospheric entry probes to be launched within the next ten years.

In addition to studies to be made by planetary spacecraft, there are many useful investigations which can be accomplished with Earth orbiting telescopes. Those telescopes expected to be flown within the decade are a one-meter class diffraction-limited telescope to be carried aboard the Space Shuttle on sortie missions and the 2.4-meter, free-flying, orbiting Large Space Telescope (*LST*). Even the one-meter Space Shuttle telescope will be able to reach a resolution on Jupiter comparable to that obtained by Pioneer and, because of its use in a sortie mode, will be able to make use of a wide variety of state-of-the-art instrumentation. The *LST* should be able to attain even higher resolution than the sortie telescope and, together with a potentially longer observing time base, will supplement spacecraft observations in establishing the long-term behavior of the Jovian atmosphere.

REFERENCES

Baker, A. L.; Baker, L. R.; Beshore, E.; Blenman, C.; Castillo, N. D.; Chen, Y. P.; Doose, L. R.; Elston, J. P.; Fountain, J. W.; Gehrels, T.; Kendall, J. H.; Kenknight, C. E.; Norden, R. A.; Swindell, W.; and Tomasko, M. G. 1975. The imaging photopolarimeter experiment on Pioneer 11. *Science* 188:468–472.

Barcilon, A., and Gierasch, P. 1970. A moist Hadley cell model for Jupiter's cloud bands. *J. Atmos. Sci.* 27:550–560.

Chapman, C. R. 1969. Jupiter's zonal winds: variation with latitude. *J. Atmos. Sci.* 26:986–990.

Chapman, C. R., and Reese, E. J. 1968. A test of the uniformly rotating source hypothesis for the south equatorial belt disturbances on Jupiter. *Icarus* 9:325–335.

Chase, S. C.; Ruiz, R. D.; Münch, G.; Neugebauer, G.; Schroeder, M.; and Trafton, L. M. 1974. Pioneer 10 infrared radiometer experiment: preliminary results. *Science* 183:315–317.

Coffeen, D. L. 1974. Optical polarization measurements of the Jupiter atmosphere at 103° phase angle. *J. Geophys. Res.* 79:3645–3652.

Fountain, J. W.; Coffeen, D. L.; Doose, L. R.; Gehrels, T.; Swindell, W.; and Tomasko, M. 1974. Jupiter's clouds; equatorial plumes and other cloud forms in the Pioneer 10 images. *Science* 184:1279–1281.

Gehrels, T.; Coffeen, D.; Tomasko, M.; Doose, L.; Swindell, W.; Castillo, N.; Kendall, J.; and Baker, L. 1974. The imaging photopolarimeter experiment on Pioneer 10. *Science* 183: 318–320.

Gierasch, P., and Stone, P. 1968. A mechanism for Jupiter's equatorial acceleration. *J. Atmos. Sci.* 25:1169–1170.

Golitsyn, G. S. 1970. A similarity approach to the general circulation of planetary atmospheres. *Icarus* 13:1–24.

Hide, R. 1961. Origin of Jupiter's Red Spot. *Nature* 190:885–890.

Ingersoll, A. P. 1973. Jupiter's Great Red Spot: a free atmospheric vortex? *Science* 182: 1346–1348.

Ingersoll, A. P., and Cuzzi, J. N. 1969. Dynamics of Jupiter's cloud bands. *J. Atmos. Sci.* 26: 981–985.

Ingersoll, A., and Pollard, D. 1975. Stable baroclinic flows in an atmosphere without solid boundaries: application to Jupiter. In draft.

Keay, C. S. L.; Low, F. J.; and Rieke, G. H. 1972. Infrared maps of Jupiter. *Sky and Telescope* 44:296–297.

Kuiper, G. P. 1972. Interpretation of the Jupiter Red Spot. I. *Comm. Lunar and Planetary Laboratory* 9:249–313.

Larson, H. P.; Fink, U.; Treffers, R.; Gautier, T. N. 1975. Detection of water vapor on Jupiter. *Astrophys. J.* 197:L137–L140.

Macdonald, N. 1968. Estimates of the seasonal variation of the general circulation from easily identifiable features. *Tellus* 20:300–304.

Maxworthy, T. 1973. A review of Jovian atmospheric dynamics. *Planet. Space Sci.* 21:623–641.

Maxworthy, T. 1975. Stratospheric waves as a momentum source for the Jovian equatorial jet. Submitted to *Planet. Space Sci.*

Maxworthy, T., and Redekopp, L. G. 1976. A solitary wave theory of the Great Red Spot and other observed features in the Jovian atmosphere. *Icarus* (special Jupiter issue). In press.

Minton, R. B. 1975. Measures of Jupiter photographs – 1973 apparition. In draft.

Newburn, R. L., and Gulkis, S. 1973. A survey of the outer planets: Jupiter, Saturn, Uranus, Neptune, Pluto and their satellites. *Space Sci. Rev.* 14:179–271.

Peek, B. M. 1958. *The planet Jupiter.* London: Faber and Faber.

Reese, E. J. 1970. Jupiter's Red Spot in 1968–1969. *Icarus* 12:249–257.

———. 1971. Jupiter: its Red Spot and other features in 1969–1970. *Icarus* 14:343–354.

———. 1972a. Jupiter: its Red Spot and disturbances in 1970–1971. *Icarus* 17:57–72.

———. 1972b. An earlier generation of long enduring South Temperate Ovals on Jupiter. *Icarus* 17:704–706.

———. 1972c. Summary of Jovian latitude and rotation period observations from 1898 to 1970. *Contrib. Observ. of New Mexico State University* 1:83–94.

Reese, E., and Beebe, R. 1976. Velocity variations of an equatorial plume throughout a Jovian year. *Icarus* (special Jupiter issue). In press.

Reese, E. J., and Smith, B. A. 1966. A rapidly moving spot on Jupiter's North Temperate Belt. *Icarus* 5:248–257.

———. 1968. Evidence of vorticity in the Great Red Spot of Jupiter. *Icarus* 9:474–486.

Reese, E. J., and Solberg, H. G., Jr. 1966. Recent measures of the latitude and longitude of Jupiter's Red Spot. *Icarus* 5:266–273.

———. 1969. Latitude and longitude measurements of Jovian features. *TN-701-69-28*, The Observatory, *New Mexico State University*.

Riddle, A. C., and Warwick, J. W. 1976. Redefinition of System III Longitude. *Icarus* (special Jupiter issue). In press.

Rosen, R. D. 1972. On the possibility of a convective vortex sheet in fluid heated from below. *Pure and Appl. Geophys.* 101:205–207.

Sato, T. 1969. Statistical detection of a repulsive force between the white ovals in the STeZ on Jupiter. *The Heavens* 50:187–190.

Solberg, H. G., Jr. 1968a. Jupiter's Red Spot in 1965–1966. *Icarus* 8:82–89.

———. 1968b. Jupiter's Red Spot in 1966–1967. *Icarus* 9:212–216.

———. 1969a. Jupiter's Red Spot in 1967–68. *Icarus* 10:412–416.

———. 1969b. A 3-month oscillation in the longitude of Jupiter's Red Spot. *Space Sci.* 17: 1573–1580.

Starr, V. 1973. A preliminary dynamic view of the circulation of Jupiter's atmosphere. *Pure and Appl. Geophys.* 110:2108–2129.

Stone, P. 1967. An application of baroclinic stability theory to the dynamics of the Jovian atmosphere. *J. Atmos. Sci.* 24:642–652.

———. 1969. The meridional structure of baroclinic waves. *J. Atmos. Sci.* 26:376–389.

———. 1972. A simplified radiative dynamical model for static stability of rotating atmospheres. *J. Atmos. Sci.* 29:405–418.

Swindell, W., and Doose, L. R. 1974. The imaging experiment on Pioneer 10. *J. Geophys. Res.* 79:3634–3644.

Westphal, J. A.; Matthews, K.; and Terrile, R. J. 1974. Five-micron pictures of Jupiter. *Astrophys. J.* 188:L111–L112.

Williams, G., and Robinson, J. B. 1973. Dynamics of a convectively unstable atmosphere: Jupiter? *J. Atmos. Sci.* 30:684–717.

THE METEOROLOGY OF THE JOVIAN ATMOSPHERE

P. H. STONE
Massachusetts Institute of Technology

The only direct observational evidence available concerning the motions in the Jovian atmosphere are remote measurements of apparent zonal motions. The rapid rotation of the planet makes it likely that these motions are in geostrophic balance and satisfy the thermal wind relation. These relations enable one to deduce from the observed latitudinal variation of zonal velocities that the zones are regions of higher temperatures, anti-cyclonic vorticity, rising motions and enhanced cloudiness, while the belts are regions of lower temperatures, cyclonic vorticity, sinking motions and relatively cloud-free. The belt and zone temperature differences imply that baroclinic energy sources are present. The condition for barotropic (shear) instability appears to be satisfied at the edges of at least some of the belts, and this mechanism is likely to account for at least some of the larger-scale (~ 5000 km) non-symmetric features present in the atmosphere.

Current suggestions for explaining other non-symmetric features and the banded structure and zonal currents require ad hoc *assumptions about the atmospheric structure. To help evaluate such suggestions, it would be particularly valuable to measure meridional velocities; to measure the lapse rate and static stability of the visible cloud layers and the layers immediately below; and to carry out a high-resolution imaging experiment capable of resolving small-scale convection. Theoretical models of the dynamics and structure of the visible cloud layers have not been very successful so far, but models have not been developed for all possible dynamical regimes—e.g., for forced convection and inertial instability regimes. To put theoretical calculations on a firm foundation it will be necessary to determine the latitudinal variation of the internal heating.*

The meteorology of planetary atmospheres other than the earth's is a subject which barely existed ten years ago. Pioneering studies of the circulations of planetary atmospheres, including Jupiter's, were undertaken by the Lowell Observatory in 1948–1952 (Slipher *et al.* 1952) and by Hess and Panofsky (1951) but these studies had few successors. Extra-terrestrial meteorology did not attract widespread interest until the development of spacecraft technology in the mid 1960's provided the necessary stimulus. The subsequent steady increase in spacecraft observations has been accompanied by a steady increase in the number of papers published on extra-terrestrial meteorology. The review article by Goody (1969) and the Conference on the Motions of Planetary Atmospheres sponsored by the Kitt Peak National Observatory (Gierasch 1970) marked the general acceptance of the subject as one of crucial importance not only to meteorologists, but to all atmospheric scientists. Studies of the meteorology of Mars are already having an impact on our understanding of the earth's atmosphere and climate (Stone 1972b, 1973b; Sagan et al. 1973; Hartmann 1974). Studies of Jovian meteorology can be expected to have a similar impact. The substantial differences between Jupiter's atmosphere and the atmospheres in the inner solar system, e.g., the presence of an internal heat source and the great size of the planet, will provide a stringent test for our understanding of atmospheric motions and how they depend on external parameters.

The study of Jovian meteorology is not yet as advanced as the study of the meteorology of the inner planets. Pioneers 10 and 11 have just begun the collection of the necessary data. As with the inner planets, atmospheric orbiters and probes will be necessary to determine crucial meteorological parameters such as the lapse rate and the latitudinal distribution of the internal heating. Consequently this review will necessarily focus on ideas more than on facts, but these ideas do provide a useful framework for future observations and theoretical studies. Similarly, we will concentrate most of our attention in this chapter on the layers of the atmosphere where we have the most information — namely, the visible cloud layers, where most solar heating occurs, roughly those layers between 300 millibar and 3 bar pressure.

In Sec. I we will review the observations relevant to Jovian meteorology; in Sec. II, the theory of dynamical regimes on rotating planets; in Sec. III, integrated theories of Jupiter's atmospheric structure and dynamics; and in Sec. IV, the dynamics of particular features, such as the zonal currents, the banded structure, and the Great Red Spot. In addition to the review by Goody (1969), earlier reviews of Jovian meteorology have been presented by Hide (1969), Stone (1973a), and Maxworthy (1973).

It is convenient to define here the symbols we will be using. They are:

c_p specific heat at constant pressure.
D characteristic vertical scale of the motions.
F flux through the atmosphere.
$\bar{F}_s$ mean flux carried by small scale convection.
f Coriolis parameter $= 2\Omega\sin\phi$.
g acceleration of gravity.
H scale height $= RT/g$.
K_h horizontal eddy diffusion coefficient.
K_v vertical eddy diffusion coefficient.
L characteristic horizontal scale of the motions.
P pressure.
P_0 reference pressure.
R gas constant.

R_0 radius of the planet.
S static stability $= \partial\theta/\partial z$.
T temperature.
u zonal velocity.
v meridional velocity.
w vertical velocity.
x west to east (zonal) coordinate.
y south to north (meridional) coordinate.
z vertical coordinate.
α thermal expansion coefficient $= 1/T$.
β beta parameter $= df/dy$.
Γ adiabatic lapse rate $= g/c_p$.
$\Delta\theta$ potential temperature difference over a depth D $[\theta(z+D) - \theta(z)]$.
ΔT temperature difference between zones and belts.
θ potential temperature $= T(P_0/P)^{R/c_p} \cong T + \Gamma Z$.
ρ density.
τ radiative relaxation time.
ϕ latitude.
Ω angular rate of rotation.

I. OBSERVATIONS

The wealth of detail in the visual appearance of Jupiter exceeds that for any other planet, and has been the subject of investigation for centuries. The observations have been summarized by Peek (1958). The most prominent visual features are the banded structure and the Great Red Spot. These are illustrated schematically in Fig. 1. The bands are symmetric about the axis of rotation, with the lighter bands generally being referred to as "zones," and the darker bands as "belts." The belts are indicated by cross-hatching in Fig. 1, and are labeled according to Peek's nomenclature; e.g., $SSTB$ refers to the South-South Temperate Belt. The positions indicated for the edges of the belts in Fig. 1 are the mean positions given by Peek for many apparitions. The equatorial zone is about 20,000 km wide, and the width of the other bands generally decreases with increasing latitude. The groundbased observations suggest that the banded structure disappears in the North and South Polar Regions (labeled NPR and SPR in Fig. 1), and this has been essentially confirmed by Pioneer 11. Its imaging experiment showed that there was, at most, only vestigial remnants of a banded structure in high latitudes, far less prominent than in low latitudes (Swindell et al. 1975).[1]

The Great Red Spot, located in the South Tropical Zone, is the most prominent and most long-lived of the non-symmetric features. Although its

---

[1]See p. 554.

appearance and visibility vary, typically it is oval, about 12,000 km wide in latitude, and 25,000 km long in longitude. As far as is known, it has been present ever since telescopic observations started in the 17th century. Many smaller non-symmetric features also occur, ranging in size from that of the Great Red Spot, all the way down to the limit of resolution, currently the few hundreds-of-kilometers achieved by Pioneer 11. The Pioneer 10 and 11 imaging experiments (Gehrels *et al.* 1974, Swindell *et al.* 1975)[2] have been especially valuable for obtaining information about these smaller scale non-symmetric features. Particularly noteworthy is the presence of lines of vortices at the boundaries between zones and belts. These have typical diameters of 5000 to 10,000 km, and often give the edges of the bands a wavy appearance.

The presence of all these visual features makes it possible to deduce information about motions in the atmosphere, simply by measuring the features' displacements over a period of time. Peek (1958) has summarized the apparent motions deduced in this fashion. The displacements are almost invariably zonal, i.e., parallel to the equator. The rotation periods deduced for various latitudes are indicated by the X's in Fig. 1, with the scale given on the vertical axis. These periods are again mean values over many apparitions. In some latitudes the measured rotation periods vary considerably, and for these latitudes the X's in Fig. 1 have a vertical bar superimposed, whose length indicates the range of the periods measured. Chapman (1969) has given a particularly useful summary of the apparent rotation periods, showing the latitudinal and time variations in much greater detail than Fig. 1.

The most notable feature in the measured rotation periods is the more rapid rotation of the equatorial regions. The five-minute shorter rotation periods in these regions imply westerly winds—i.e., winds in the direction of rotation—about 100 m sec^{-1} greater than winds in high latitudes. This difference led originally to the introduction of two different systems for measuring longitudinal positions (Peek 1958). In System I positions are based on a rotation period of 35,430.0 sec, corresponding to equatorial regions, and in System II they are based on a rotation period of 35,740.6 sec, corresponding to mid and high latitudes. Subsequently, observations of radio emissions from Jupiter's magnetosphere found that the mean rotation period of the emitting regions was 35,729.8 sec (Newburn and Gulkis 1973).[3] This led to the introduction of a third system for measuring longitudinal positions, System III, which is based on the period of the radio emissions. The rotation period of System III is often interpreted as the "true" rotation period of the planet (Newburn and Gulkis 1973).[3] If we adopt this convention, then the rotation period of System I corresponds to an apparent westerly wind at the equator of 106 m sec^{-1} and the rotation period of System II corresponds to an ap-

[2]See p. 536.
[3]See p. 709.

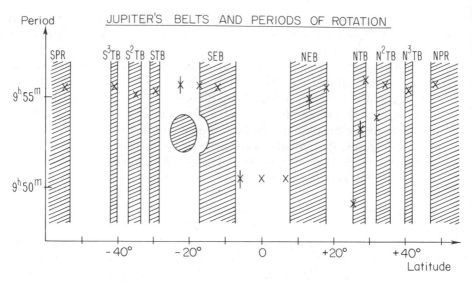

Fig. 1.   Jupiter's belts and periods of rotation.

parent easterly wind at higher latitudes, with a speed at 30° latitude, for example, of 3.3 m sec⁻¹. However, there are substantial variations in the apparent velocities at higher latitudes; most notably, the current at the south edge of the North Temperate Belt has an apparent westerly velocity of 123 m sec⁻¹.

Observations of meridional displacements have been very rare (Peek 1958) and it is not yet possible to say anything about mean north-south motions.[4] One particularly noteworthy set of observations which included meridional displacements has been reported by Reese and Smith (1968).[5] They measured the motions of several spots in the vicinity of the Great Red Spot, and found that they tended to circulate around the Spot in a counter-clockwise direction, corresponding to anti-cyclonic vorticity in the southern hemisphere. The measured periods of the circulating spots imply circulatory velocities around the Great Red Spot in the range of 20 to 60 m sec⁻¹.

Measurements of the thermal balance of the planet are particularly important for the meteorology, since the balance tells us something directly about the drives for atmospheric motions. Groundbased observations show that the planet radiates more heat than it absorbs from the sun (Aumann *et al.* 1969) and Pioneers 10 and 11 have confirmed the presence of an internal heat source (Chase *et al.* 1974; Ingersoll *et al.*).[6] The Pioneer 11 measurements should be the most accurate, since they had the best phase coverage. The results indicate that the total heating from the interior is approximately

---

[4]See, however, p. 554.   [6]See p. 202.
[5]See p. 572.

equal to the total amount of absorbed solar radiation, calculated for a Bond albedo of 0.45. Therefore, the solar heating and internal heating, when averaged over the whole surface of the planet, each supply the atmosphere with about 7×10^3 erg cm^{-2} sec^{-1}. The Pioneer 11 experiment also found that the large scale thermal emissions in high and low latitudes do not differ by more than a few percent (Ingersoll 1976).[7] Since the solar heating is a maximum at the equator, and virtually zero at the poles, this result implies that there is a dynamical transport of heat from equator to pole at some level in the planet.

Knowledge of the atmosphere's temperature structure is also very valuable for analyzing the motions. If the temperature structure were known, one could attempt to solve the equations of motion alone, without having to solve a simultaneous equation for the temperature. Many temperature measurements have been made, but the temperature structure inferred from them is highly model-dependent (Newburn and Gulkis 1973).[8] Brightness temperatures of the belts are generally higher than those of the zones and the Great Red Spot, but there is very little systematic variation of brightness temperatures with latitude or longitude (Westphal 1971; Chase *et al.* 1974; Ingersoll *et al.*).[9] This latter result implies that seasonal and diurnal variations are very small. This is to be expected since the radiative relaxation time of the visible cloud layers of the atmosphere is on the order of 10 Earth years (Gierasch and Goody 1969). The brightness temperatures increase with wavelength, and this implies that the temperature in the atmosphere increases with depth (Newburn and Gulkis 1973). The wavelength dependence of the brightness temperatures is perhaps most useful as a check on the viability of any given model for the atmospheric structure. The models that give the best agreement with the data have near-adiabatic lapse rates in the visible cloud layers, with the clouds being much thicker in the zones than in the belts (Newburn and Gulkis 1973).

II. DYNAMICAL REGIMES ON ROTATING PLANETS

The traditional approach to determining the structure of a planetary atmosphere neglects any differential heating and assumes radiative-convective equilibrium. In this situation, the atmosphere can have two distinct regimes. If the radiative equilibrium temperature profile is statically unstable, small scale convection will arise and prevent the lapse rate from being appreciably superadiabatic. If, on the other hand, the radiative equilibrium temperature profile is statically stable, the atmosphere will be quiescent. This approach is not self-contained, because it is not possible to use the solutions to check the consistency of neglecting the differential heating. Differential heating gives rise to horizontal temperature gradients and large scale motions, and

[7]See p. 213.   [9]See p. 213.
[8]See p. 666.

these motions tend to stabilize the lapse rate (Stone 1972a). The effect of this stabilization is illustrated by the atmospheric structure of the earth and Mars. In mid and high latitudes the lapse rates are substantially subadiabatic, in spite of the fact that the radiative equilibrium profiles are superadiabatic (Stone 1972b, Kliore et al. 1973). In general, the vertical structure and dynamics of an atmosphere cannot be deduced independently of each other. Of course the large-scale motions also transport heat horizontally and are potentially of crucial importance in determining the horizontal structure of the atmosphere (Gierasch et al. 1970).

The importance of Coriolis forces in shaping the motions can be measured by the Rossby number (Ro),

$$Ro = \frac{u}{fL}. \tag{1}$$

In particular, Coriolis forces are dominant if, $Ro \ll 1$. Hide (1963) first pointed out that Ro is generally very small for the large scale motions on Jupiter. For example, at 20° latitude, if we choose $u = 10$ m sec^{-1} as typical, $Ro \ll 1$ provided $L \gg 100$ km. Consequently, we can expect that the large-scale flows will be geostrophic—i.e., horizontal pressure gradients will be balanced by Coriolis forces. The one likely exception to this conclusion is the equatorial current. Taking as typical 4° of latitude, $u = 100$ m sec^{-1}, and $L = 8000$ km, we find $Ro = \frac{1}{2}$ for the equatorial current.

In hydrostatic equilibrium, horizontal pressure gradients imply the presence of horizontal temperature gradients. Consequently, Stone (1967) suggested that the zonal motions on Jupiter arise from meridional temperature gradients and that they are related by the thermal wind relation,

$$f\frac{\delta u}{\delta z} = -\alpha g \frac{\delta T}{\delta y}. \tag{2}$$

Since the derivation of this equation only depends on the assumptions of hydrostatic equilibrium and geostrophy, this relation is a very plausible one for Jupiter. Thermal winds are at right angles to the temperature gradient and, therefore, do not transport heat. However, in geophysical situations these winds are usually unstable, and the resulting instabilities do transport heat and affect the temperature structure. There are a variety of possible instabilities besides small-scale convection that can arise depending on the static stability. Thus, there is a much richer variety of possible atmospheric regimes than is encompassed in the theory of radiative-convective equilibrium.

Figure 2 summarizes these various regimes for a rapidly rotating planet as a function of the Richardson number (Ri), which is a dimensionless measure of the static stability of the atmosphere. Ri is defined as

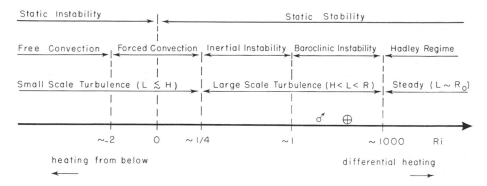

Fig. 2.   Dominant heat transporting modes on rapidly rotating planets and their properties.

$$Ri = \frac{\alpha g S}{\left(\dfrac{\delta u}{\delta z}\right)^2} \sim \frac{\alpha g D^2 S}{u^2} . \tag{3}$$

Figure 2 lists the heat-transporting modes which dominate at different values of Ri, and their properties. If the static instability is sufficiently great, i.e., if Ri is sufficiently negative, then small-scale turbulent convection of the kind invoked in the theory of radiative-convective equilibrium is dominant. In this regime the convection is sufficiently vigorous that its properties are not affected by the horizontal motions arising from differential heating, and by analogy with the theory of turbulence in the terrestrial boundary layer (Priestly 1959), we will refer to this regime as a free convection regime. The characteristic horizontal scales of free convection are generally comparable to the characteristic vertical scales of the atmosphere, although they can be even smaller as, for example, when rotation is very rapid (Chandrasekhar 1961; Roberts 1968).

If the atmosphere is only slightly statically unstable, then the motions driven by differential heating influence the small-scale turbulent convection significantly, although the turbulence is still primarily thermally driven. If the atmosphere is slightly statically stable, then the turbulence is primarily mechanically driven, but still essentially small scale. These two situations, where the static stability is small in magnitude, we will refer to collectively as forced convection, again by analogy with the theory of turbulence in the terrestrial boundary layer (Priestly 1959). The transition point from free to forced convection is not well determined. In the terrestrial boundary layer the most unstable conditions which have been well observed correspond to $Ri \sim -2$, and for this value the convection is still forced rather than free (Businger et al. 1971). The value of Ri for this transition in an atmosphere like Jupiter's is of course completely unknown, and we have entered the value -2 in Fig. 2 somewhat arbitrarily. The value of Ri at which flow with vertical shear ceases to be unstable to small-scale instabilities is predicted by theory to be 0.25 (Drazin and Howard 1966). Observations in the terrestrial boundary layer yield a value for this transition of 0.21 (Businger *et al.* 1971) in good agreement with the theoretical value.

For more stable conditions, $0.25 < Ri < 0.95$, inertial instability is dominant (Stone 1966). This instability draws its energy primarily from the kinetic energy of the thermal wind by means of vertical eddy stresses which tend to destroy the vertical shear of the flow. The unstable perturbations transport heat down the horizontal temperature gradient and upward, and therefore tend to stabilize the atmosphere by increasing the static stability and Ri (Stone 1972a). The motions generated are axially symmetric, i.e., independent of longitude.

Under even more stable conditions, $Ri > 0.95$, the dominant instability mechanism is baroclinic instability (Stone 1966). This is the mechanism which produces the extra-tropical cyclones and anti-cyclones observed in the terrestrial and Martian atmospheres. This instability depends on the existence of the horizontal temperature gradients which accompany the thermal winds. These horizontal gradients imply that potential energy is available for perturbations to feed on, even though the atmosphere is statically stable. In particular, a gas parcel which is displaced horizontally into regions of cooler temperatures at the same time that it is displaced upwards, will be buoyant relative to its surroundings and will have its displacement enhanced. Consequently, these instabilities necessarily transport heat down the horizontal temperature gradient and upward, and they also tend to stabilize the atmosphere by increasing S and Ri (Stone 1972a). The values of Ri characteristic of the baroclinic instability regimes on Earth and Mars have been calculated by Stone (1972b), and are indicated in Fig. 2 by the astronomical symbols for the earth and Mars. In the inertial and baroclinic instability regimes, the characteristic horizontal scales of the motions exceed the characteristic vertical scale of the atmosphere, and systematically increase as Ri increases (Stone 1966).

Finally, for sufficiently large values of Ri, thermal winds are stable, and heat-transporting motions will arise only to the extent that the thermal wind relation is not exact. In particular, dissipative effects will lead to some cross-isobaric flow driven by the differential heating, with poleward flow in the upper layers and equatorward flow in the lower layers, accompanied by rising motions in low latitudes and sinking motions in mid and high latitudes. This kind of flow is referred to as a Hadley regime, and its characteristic horizontal scale is the planetary scale. This stable regime could occur if the characteristic scale of baroclinic instabilities exceeds the planetary scale; if dissipative effects are strong enough; or if the vertical shear of the thermal wind is sufficiently small. Gierasch *et al.* (1970) have estimated that the first two conditions are highly unlikely for Jupiter. The latter condition is given in quantitative form by Phillips (1951) and can be written in terms of the Richardson number as

$$\frac{\beta u}{f^2} Ri > 2 \ . \tag{4}$$

For 30° latitude, if we choose $u \sim 10$ m sec^{-1}, this condition becomes $Ri > 1000$, and this is the value of Ri we have entered in Fig. 2 for the transition from a baroclinic instability regime to a Hadley regime. The highly turbulent appearance of Jupiter's atmosphere strongly suggests that its regime is not a Hadley regime.

It should be noted that Fig. 2 does not make any allowances for clouds and condensation. Any of the motions included in Fig. 2 can be expected to lead to clouds and condensation under appropriate conditions. Correspondingly a greater variety of regimes than those indicated in Fig. 2 can be imagined. For example, the earth's tropics are basically a moist convection regime. Barcilon and Gierasch (1970) investigated the thermodynamic consequences of condensates in the Jovian atmosphere. Using Lewis' (1969) solar composition model, they concluded that water was the most important condensate. Current models indicate that the water vapor condensation level is near the 280°K level, about 60 km below the visible cloud layers (Weidenschilling and Lewis 1973). Barcilon and Gierasch showed that the scale height of the saturation vapor pressure near the condensation level is only about 8 km. Consequently the condensation in these layers is unlikely to play a direct role in the dynamics of the visible cloud layers. In the terrestrial atmosphere there are basically two kinds of cumulus clouds (Ogura and Cho 1973). The common variety is low level cumulus, which typically has a vertical extent comparable to the scale height of the saturation vapor pressure. The less frequent variety is deep cumulonimbus towers, which typically have a vertical extent about four times the scale height of the saturation vapor pressure. Even this latter kind of cloud on Jupiter would only reach up to about the 220°K level, still well below the visible cloud layers. Stratus clouds are thermodynamically less active than cumulus clouds.

In the visible cloud layers the only appreciable condensate is ammonia (Lewis 1969). For a solar composition atmosphere, ammonia condensation has a much smaller effect than water vapor condensation in the earth's atmosphere. For example, ammonia condensation only changes the adiabatic lapse rate by 5% and the ammonia saturation vapor pressure curve at the condensation level has a scale height only about one tenth of the atmospheric scale height (Weidenschilling and Lewis 1973), whereas the same factors for water vapor in the earth's atmosphere are 50% and one third.

Consequently the array of regimes presented in Fig. 2 provides a convenient framework for discussing Jupiter's dynamical regime. This array of regimes is consistent with laboratory experiments with rotating fluids (Stone *et al.* 1969; Hadlock *et al.* 1972). Non-heat-transporting modes such as barotropic instability have been excluded from Fig. 2, because they do not depend explicitly on Ri and can occur in any of the regimes listed.

Heating from below tends to decrease the static stability of an atmosphere and so an internal heat source tends to drive the dynamical regime towards the left in Fig. 2. On the other hand, differential heating drives motions which tend to stabilize the atmosphere, and so differential solar heating tends to drive the dynamical regime towards the right in Fig. 2. Differential solar heating is stronger in high latitudes than in low latitudes so more stable regimes may occur in high latitudes. The regime will also depend on any differential heating present in the interior heat source. All the models, that we shall discuss in the next section, assume that the atmosphere is heated uniformly from below. The recent Pioneer 11 observations of thermal emissions (Ingersoll 1976) cast doubt on this assumption.

Figure 2 suggests two kinds of observations that would be particularly valuable in determining Jupiter's dynamical regime. One would be measurements of the static stability of the atmosphere so that Ri can be estimated. The other would be a high-resolution imaging experiment, capable of resolving scales $L \sim H$, so as to determine whether small-scale convection is present.

It is convenient to summarize here the characteristics of some of the dynamical modes that are included in Fig. 2 and that we shall be referring to frequently in the next two sections. This summary is presented in Table I. The tabulated criteria for instability are actually the necessary conditions for instability, but for geophysical fluids they are usually also sufficient. Note that these conditions do not necessarily correspond to the conditions under which each mode will be dominant, i.e., different modes can occur simultaneously. The characteristic time scales given in Table I are just the inverse growth rates for the most unstable eigenfunction for each kind of instability.

III.   INTEGRATED THEORIES OF JUPITER'S ATMOSPHERIC STRUCTURE AND DYNAMICS

Since dynamical transports influence Ri, the atmospheric regime resulting from specified differential and internal heating is determined by highly nonlinear processes. Consequently, it is usually not possible to obtain a simultaneous solution to the equations of motion and the equations for the temperature structure without restricting the values of Ri *a priori*. In this section we will discuss those models which calculate both temperature structure and properties of the motions after making such a restriction. We shall discuss in separate subsections calculations of radiative-convective equilibrium, calculations of large scale motions in the free convection regime, and calculations for a baroclinic instability regime.

TABLE I
Properties of Dynamical Modes

Dynamical Mode (source)	Criteria for Instability	Structure	Characteristic Horizontal Scale	Characteristic Time Scale
barotropic instability (Kuo 1949)	$\dfrac{d^2u}{dy^2} > \beta$	three dimensional	$\pi \left(\dfrac{u}{\beta}\right)^{\frac{1}{2}}$	$\dfrac{10}{(u\beta)^{\frac{1}{2}}}$
baroclinic instability (Stone 1966, 1970)	$Ri > 0.84$	three dimensional	$2\,\dfrac{u}{f}(1+Ri)^{\frac{1}{2}}$	$\dfrac{3}{f}(1+Ri)^{\frac{1}{2}}$
inertial instability (Stone 1971)	$Ri < 1$	axially symmetric	$2\left(\dfrac{u}{\beta}\right)^{\frac{1}{2}}\left(\dfrac{Ri^2}{1-Ri}\right)^{\frac{1}{4}}$	$\dfrac{1}{f}\left(\dfrac{Ri}{1-Ri}\right)^{\frac{1}{2}}$
radiative instability (Gierasch 1973)	(see text)	axially symmetric	$2\,\dfrac{(\alpha g H^2 S)^{\frac{1}{4}}}{\beta^{\frac{1}{2}}}$	$\dfrac{\tau}{2}\,\alpha\, HS$
free convection (Priestley 1959; Spiegel 1971)	$S < 0$	three dimensional	$\sim H$	$\sim (\alpha\, g\, |S|)^{-\frac{1}{2}}$

(a) Radiative-convective Equilibrium

Calculations of radiative-convective equilibrium for Jupiter have been published by Trafton (1967), Hogan *et al.* (1969), Divine (1971), and Trafton and Stone (1974). In these calculations the radiative equilibrium structure is modified by supressing superadiabatic lapse rates. Since the effect of large scale motions is not included, these calculations implicitly assume that the dynamical regime is a free convection regime. The convection could be either dry or moist; in the calculations the adiabatic lapse rate is taken to be the moist adiabatic lapse rate in regions where a significant condensate is present.

Typically in these calculations the lapse rate becomes adiabatic at $T \sim 140°K$, $P \sim 500$ mb. Consequently, in these models the visible cloud layers of the atmosphere are mainly convecting, with lapse rates $\sim 2°$ per km. The horizontal temperature gradients in radiative-convective equilibrium are the same as in radiative equilibrium and these latter gradients have been calculated by Stone (1972b). If the internal heating is comparable to the solar heating, in the visible layers at 45° latitude the gradient is 3° per 10^4 km. This value represents an upper bound to the mean gradients that would result if dynamical transports were taken into account and is dependent on the assumption that the heating from below is uniform.

If the convection is dry, its properties can be estimated from free convection theory and mixing length theory (Priestly 1959; Spiegel 1971). The characteristic space and time scales are given in Table I, and from these the characteristic velocity can be estimated to be

$$w \sim (\alpha g |S|)^{\frac{1}{2}} H . \tag{5}$$

According to mixing length theory, the flux is given by

$$F \sim - \rho c_p w H S . \tag{6}$$

Solving for S and w in terms of F, we find

$$w \sim \left[\frac{R}{c_p} \frac{F}{\rho} \right]^{\frac{1}{3}} \tag{7}$$

$$S \sim - \frac{w^2}{HR} . \tag{8}$$

Barcilon and Gierasch (1970) used these equations to estimate S in the visible cloud layers. They found $|S| \sim 6 \times 10^{-10}°K$ per cm. They combined this value with the u velocity scale observed in the visible cloud layers and estimated Ri. With values of $u = 30$ m sec^{-1}, $D = 80$ km, they calculated $Ri = -4 \times 10^{-3}$ and concluded that the visible cloud layers could not be in free convection. This calculated value of Ri is so small compared to the minimum critical value of -2, that it appears to be a very strong conclusion.

It is still possible that in lower layers, below the regions of differential solar heating, $\partial u / \partial z$ becomes small enough that free convection can occur.

As Barcilon and Gierasch point out, in the layers immediately below the visible cloud layers, near the 300°K level, condensation of water vapor is important. One would expect convection in these layers to be essentially moist convection, with a significant fraction of the heat flux carried as latent heat rather than as sensible heat. The moist adiabatic lapse rate in these layers is near 1°5K per km, and the dry adiabatic lapse rate is near 1°8K per km. Consequently, these layers could exhibit a static stability ~ 0°3K per km. Clearly the determination of the lapse rates in the visible cloud layers, and in the layers where water condensation can be expected is a prime objective of a probe mission. A static stability of 0°3K per km should be measurable by such an experiment.

At lower layers free of condensates, dry free convection might still occur. Near the 400°K level, Eqs. (7) and (8) yield estimates of $w \sim 2$ m sec^{-1} and $S \sim -2 \times 10^{-10}$ °K per cm. The corresponding characteristic time scale would be 7 hours. Since the characteristic horizontal and vertical scales are the same in free convection, the eddy diffusion coefficients can be estimated from the relation

$$K_h \sim K_v \sim wH .\qquad (9)$$

Their value at these levels would be ~ 10^9 cm^2 sec^{-1}. Since the characteristic time scale at this level is comparable to the period of rotation, free convection would be influenced by Coriolis forces. However, at these levels this influence would not be strong enough to change the order of magnitude of the above estimates.

At deeper levels, as ρ increases, the characteristic time scale increases, and the influence of Coriolis forces would become more and more dominant. Robert's (1968) and Busse's (1970a) analyses of convection in a rapidly rotating deep atmosphere indicate that the amplitude of the convection on a spherical surface is a strong function of latitude. This result suggests the possibility that the flux of heat from the interior into the visible cloud layers is not uniform in latitude, and emphasizes the need for determining its distribution. Such differential heating would imply horizontal temperature gradients, horizontal motions, and horizontal heat transports in the deep atmosphere. These properties could explain the Pioneer 11 observation that thermal emissions in high latitudes are just as strong as those in low latitudes, in spite of differential solar heating. Clearly, an analysis of the finite amplitude, rotating convection problem for a deep atmosphere would make an important contribution to our understanding of Jupiter's atmosphere.

(b) Large-scale Motions in Free Convection

Williams and Robinson (1973) studied the hypothesis that the large-scale motions and structure of Jupiter's atmosphere are convectively driven — i.e., that small-scale convection driven by the internal heat source has a large-scale organization that can explain the observations. To check this hypothesis they integrated the Boussinesq equations of motion and temperature in a statically unstable, rotating spherical shell. The dominant small-scale convection was parameterized with simple eddy diffusion laws, and only the large-scale motions were explicitly calculated.

Williams and Robinson had four parameters in their model which they regarded as free parameters: the depth of the spherical shell, D; the amount by which the potential temperature of the lower surface exceeded that of the upper surface, $\Delta\theta$; and the vertical and horizontal eddy diffusion coefficients, K_v and K_h, respectively. Consequently, they performed a parameter study in which they sought values for these parameters that would give solutions resembling Jupiter's circulations. In this parameter study they neglected differential heating, and therefore were studying essentially a free convection regime. They also assumed that the motions were axially symmetric. They found that there were values of these four parameters which enabled them to match the observed strength and width of the equatorial current, and the observed scale and extent of the banded structure. In particular, in the model which fit the observations best, their model PBJ, the parameter values were $D = 50$ km, $\Delta\theta = 133°$, $K_v = 10^7$ cm^2 sec^{-1}, and $K_h = 10^{13}$ cm^2 sec^{-1}. In addition, the solutions showed features which agreed with the observations even though the parameter values were not specifically picked to fit them. For example, the scale of the banded structure decreased poleward, and the vorticity had the correct correlation with the belts and zones. The considerable agreement between many of the details of their solution and the observations has been cited by Gehrels (1974) in support of the hypothesis that the motions are essentially convectively driven.

However, there is one observable quantity which Williams and Robinson did not use to constrain their parameter study, namely, the energy flux through the atmosphere. The flux in their model is given by

$$F = \rho c_p(w\theta - K_v S).\tag{10}$$

As one would expect for free convection, the main term contributing to this flux in their solutions is the second term, which gives the flux carried by the small-scale parameterized convection. If we average the second term vertically, then we obtain

$$\bar{F}_s = \rho c_p \frac{K_v \Delta\theta}{D}.\tag{11}$$

If we use the parameter values from their model PBJ, and choose $\rho = 1.6 \times 10^{-4}$ g cm^{-3} and $c_p = 1.4 \times 10^8$ erg per °K per gram as typical of the visible cloud layers, we calculate $\bar{F}_s = 6 \times 10^6$ erg cm^{-2} sec^{-1}. This is three orders of magnitude larger than the observed value $\sim 10^4$ erg cm^{-2} sec^{-1}. If one attempts to change the values of K_v, $\Delta\theta$, and D in order to reduce this flux, then as Williams and Robinson showed, in order to maintain the dynamical similarity of their solutions one must keep $D\Delta\theta$ and D/K_v constant, and $D <$

500 km. Consequently, one can reduce the flux in their solutions by only one order of magnitude, without losing the resemblance to Jupiter's circulations.

In the face of this discrepancy, it is difficult to argue that Jupiter's dynamical regime is basically a convective regime driven by the internal heat source. Williams and Robinson's calculations indicate that the known strength of this heat source is too weak to account for the observed circulations. This suggests that differential heating must play a role in driving the motions, and that Jupiter's dynamical regime will be one of the more stable ones listed in Fig. 2.

This conclusion is in accord with Barcilon and Gierasch's (1970) conclusion cited above, that free convection is not consistent with the observed horizontal velocity magnitudes. In fact, the parameter values calculated in the preceding section from free convection theory are quite different from the values that Williams and Robinson found necessary in order to match the observed circulations. In particular, the values estimated from free convection theory were $K_h \sim K_v \sim 10^9$ cm² sec⁻¹ and $\Delta\theta \sim 2 \times 10^{-3}$ °K, in contrast to the necessary values $K_h \sim 10^{13}$ cm² sec⁻¹, $K_v \sim 10^7$ cm² sec⁻¹ and $\Delta\theta \sim 133$°K. According to Williams and Robinson's results, the former values would make the large scales convectively stable—i.e., the Rayleigh number would be subcritical and there would be no large scale motions driven by uniform internal heating.

There are also theoretical difficulties with the convective model. One is the prescribed axial symmetry. Other analyses of convection in a rotating, spherical system show that the dominant convective modes are not axially symmetric (Roberts 1968; Busse 1970a,b). In fact, Williams and Robinson reported that when they perturbed their axially symmetric solutions which matched the observations best, the symmetric convection cells broke up into non-symmetric cells. Another difficulty, pointed out by Williams and Robinson, is the very large values of K_h needed to match the observations. Since the velocities in the solutions were $u \lesssim 100$ m sec⁻¹, the value $K_h \sim 10^{13}$ cm² sec⁻¹ corresponds to mixing lengths $L \geq 10,000$ km. Since these are comparable to the characteristic scales of the large-scale motions, it is not consistent to parameterize the small-scale motions by eddy diffusion laws.

(c) Baroclinic Instability Regime

Stone (1972b) has calculated the atmospheric structure that would occur in the visible cloud layers if baroclinic instabilities were the main source of dynamical heat transport. His model gives qualitatively good results when applied to the earth or Mars (Stone 1972b), but it neglects deep atmosphere effects and, therefore, has to be interpreted cautiously when applied to Jupiter. McIntyre (1972) has shown that baroclinic instability in general still occurs in a deep atmosphere, but the growth rates are reduced. In particular, some of the available potential energy is diverted into driving motions in the lower, non-baroclinic layers, with the division of energy between the baro-clinic and non-baroclinic layers being inversely proportional to the static stability of the respective layers. If the static stability of the non-baroclinic layers is comparable to, or greater than, that of the baroclinic ones, then Stone's model should give qualitatively reasonable results. However, if the static stability of the non-baroclinic layers is smaller, then the instability of the baroclinic ones will be reduced, and perhaps even eliminated in extreme cases. In this case, the stabilizing effect of the instabilities on the temperature structure of the baroclinic layer could be decreased or even removed, and correspondingly, Stone's model can only be interpreted as giving upper bounds to S and Ri in the baroclinic layer.

Stone (1972b) calculated equilibrium states for 45° latitude and found that the value of Ri was so sensitive to external parameters that no conclu-sions could be drawn about the dynamical regime. For example, decreasing the internal heat source by 50% changed Ri from values < 1 to values ~ 10. However, it was also found that the static stability is extremely small, even if one allows for the uncertainty in the external parameters. For example, in the most stable possible baroclinic instability regime Stone calculated S ~ 3×10^{-3} °K per km. Since this very small value represents an effective upper bound on the static stability of the visible cloud layers, one can conclude that the lapse rate in these layers should be very nearly adiabatic, and calcula-tions of radiative-convective equilibrium should give good results for the vertical structure of these layers. In addition, these very small values suggest that deep atmosphere effects may not seriously affect the predictions of Stone's model after all. Judging from the earth's tropics, regions in which moist convection occurs act dynamically as though they were statically sta-ble. In other words, the areas of active moist convection cover only a few percent of the total area (Ogura and Cho 1973), but this is sufficient to pro-duce a mean lapse rate over the whole area which is nearer the moist adia-batic lapse rate than the dry adiabatic lapse rate. Such a layer of moist convection on Jupiter near the 300°K level would exhibit a static stability ~ 0°3 K and this would be sufficient to make Stone's model valid for the visible cloud layers.

Trafton and Stone (1974) pointed out that the sensitivity of Ri to the external parameters implied a strong change in the equilibrium value of Ri with latitude. They used Stone's model to calculate Ri as a function of lati-tude and found that Ri < 1 in low latitudes and Ri > 1 in high latitudes, and that uncertainty in the external parameters primarily introduces uncertainty in the latitude of the critical value, $Ri = 1$. Using the value of Aumann et al. (1969) for the strength of the internal heat source, 1.3×10^4 erg cm^{-2} sec^{-1}, they estimated that this latitude was ~ 70°. If their calculation is re-done using the Pioneer 11 value for the strength of the internal heat source, 7×10^3 erg cm^{-2} sec^{-1}, (Ingersoll)[10] and assuming that the 170°K level is typical

---

[10]See p. 202.

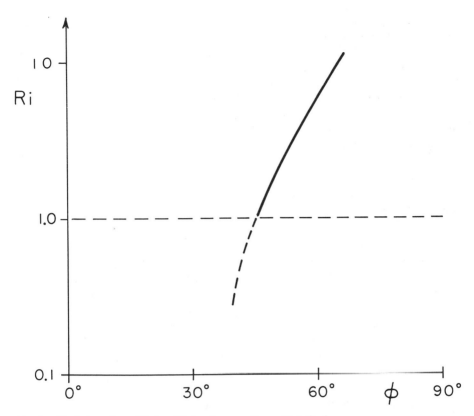

Fig. 3. Variation of equilibrium Richardson number with latitude.

of the visible cloud layers, then the variation of Ri with latitude changes to that given in Fig. 3. Now the critical latitude is $\sim 45°$. The part of the curve in Fig. 3 corresponding to values $Ri < 1$ is dotted to indicate that Stone's model formally breaks down for these values, because baroclinic instabilities cease to be the dominant heat-transporting mechanism (see Fig. 2).

Trafton and Stone's calculation shows that in low latitudes baroclinic instabilities, even if they could occur, could not be sufficiently stabilizing to produce a baroclinic instability regime. This result implies that Jupiter's dynamical regime in low latitudes is one of the less stable ones listed in Fig. 4. This destabilization of Jupiter's atmosphere compared to the earth's and Mars' is a consequence of the much weaker horizontal temperature gradients in Jupiter's atmosphere. These gradients are smaller than on the earth or Mars mainly because the solar differential heating is spread out over a much greater planetary scale, and secondarily, because of the (assumed) lack of differential heating associated with the internal heat source. These gradients are the primary drive for baroclinic instabilities, and thus on Jupiter they would be relatively weaker and less efficient at stabilizing the temperature structure. The breakdown of the baroclinic instability model in low latitudes combined with the difficulties inherent in assuming a free convection regime suggest that Jupiter's dynamical regime in low latitudes is one of the intermediate regimes listed in Fig. 2. Consequently it would be highly desirable to develop integrated models for the dynamics and temperature structure of forced convection and inertial instability regimes. A start has been made toward such a model for an inertial instability regime by Walton (1975), but his analysis is concerned only with the monotonic symmetric instability mode, which is subdominant when $Ri < 1$.

Trafton and Stone's calculation still leaves open the possibility of a baroclinic instability regime in high latitudes. The prediction inherent in Fig. 3, that the regime should change over to a more stable, non-symmetric regime poleward of 45° latitude, is in good agreement with the observed change-over from a predominantly symmetric to a predominantly non-symmetric regime at 50° latitude (Gehrels, p. 537). Figure 3 implies that a typical value of Ri in high latitudes would be ~ 10. The corresponding value for the characteristic scale given in Table I for baroclinic instabilities at 60° latitude, with $u \sim 10$ m sec^{-1}, would be 800 km. This scale refers specifically to the half-wavelength in the downstream direction of the most rapidly growing baroclinic perturbations. This value appears to be typical of many of the features seen in the high-latitude pictures obtained by Pioneer 11.

However, Stone's (1972b) calculations also show that horizontal heat fluxes by baroclinic instabilities in the visible cloud layers have a negligible effect on meridional temperature gradients. This result does not rule out the possibility that there are significant meridional transports in deeper layers, and in fact, the Pioneer 11 observation that the thermal emissions are independent of latitude requires just such a transport. This implies that the visible cloud layers are subjected to greater heating from the interior in high latitudes than in low latitudes. Since all the calculations with Stone's model assumed that the heating from the interior was uniform in latitude, one must consider how an internal heat source that increases with latitude affects the conclusions based on his model. Such an increase with latitude will oppose differential solar heating, and, therefore, will lead to weaker mean meridional temperature gradients than one would expect if the internal heating were uniform. This will weaken the drive for baroclinic instabilities, and in particular, will decrease the stabilizing effect of the instabilities on S and Ri. Therefore, the conclusions that the visible cloud layers will be very nearly adiabatic and that the low latitude regime is less stable than a baroclinic instability regime are merely reinforced if the internal heating does indeed increase with latitude. On the other hand, such an increase with latitude shifts the curve in Fig. 3 toward the right and this decreases the chance that the high latitude regime is a baroclinic instability regime. To assess this possibility more accurately, information about the latitudinal variation of the internal heating in high latitudes is needed.

IV. DYNAMICS OF INDIVIDUAL FEATURES

In this section we review dynamical investigations which attempt to explain particular features of the Jovian circulations without placing them within the context of a complete theory of the general circulation and structure of the atmosphere. We shall consider these investigations under four general headings: zonal motions, the banded structure, the Great Red Spot, and nonsymmetric features. It will be convenient again to refer to Table I for the important characteristics of the dynamical modes to which we will be referring frequently.

(a) Zonal Motions

As we discussed in Sec. II, the zonal winds are likely to be thermal winds, obeying the thermal wind relation, Eq. (2). Stone (1967) pointed out that this relation implies that vertical shear is present in the zonal flow, and this complicates the interpretation of measured zonal wind profiles such as that shown in Fig. 1. The apparent latitudinal variations may be caused by vertical variations as well as by horizontal ones. Another difficulty in interpreting the measured motions has been emphasized by Maxworthy (1973) and Stone (1974). We cannot be sure whether the measured velocities refer to phase velocities or to true mass motions. For example, barotropic Rossby waves with characteristic horizontal scale, L, propagate in the zonal direction with a phase speed given by (Holton 1972)

$$c = u - \frac{\beta L^2}{2\pi^2} .$$ (12)

Thus, a wave at 30° latitude with $L = 5000$ km would propagate 5 m sec^{-1} more slowly than the actual mean flow. Stone (1974) has suggested that this lag could account for the mean lag of System II motions behind System III.

These uncertainties in interpreting the measured velocities indicate that one should not put too much credence in the details of measured velocity profiles. Only the major features, such as the equatorial current and the current at the edge of the North Temperate Belt are likely to be reliable. The equatorial current has attracted particular interest because it appears to have greater angular momentum than either the underlying atmosphere or the atmosphere in higher latitudes (Hide 1970). This raises the problem of how the current is maintained against small scale diffusion. In particular, the diffusion of momentum must be balanced either by an equatorward transport or by an upward transport of momentum by atmospheric motions. An equatorward transport would imply a negative correlation between u and v on the north edge of the current and a positive correlation on the south edge. Observations of meridional motions that can yield information about such correlations would be particularly valuable.

Three mechanisms for supplying an equatorward transport have been suggested: inertial instability (Gierasch and Stone 1968); baroclinic instability (Stone 1972a); and convection under the influence of rotation (Busse 1976). The viability of the first mechanism for supplying the required transport has been discussed extensively (McIntyre 1970; Hide 1970; Stone 1972a; Maxworthy 1972; Gierasch and Stone 1972). Here it is sufficient to note that inertial instability can supply the required transport only under the rather limited circumstance that $0 < Ri < \frac{1}{3}$, and it cannot account for any acceleration relative to the underlying atmosphere. Baroclinic instability can supply the required transport so long as $Ri = 0(1)$ and it can account for an acceleration relative to the underlying atmosphere; it therefore is a more plausible mechanism than inertial instability. However, our ignorance of actual values of Ri in the Jovian atmosphere still makes this mechanism a speculative one.

Perhaps the most plausible for an equatorward transport is the convection mechanism. This has also been invoked to explain the solar differential rotation (Busse 1970b), and it has been observed to operate in laboratory experiments (Busse 1976). The dominant convective modes in a rotating, self-gravitating system are non-axisymmetric cells (Roberts 1968; Busse 1970 a,b). Coriolis forces act to deflect the zonal motions in these cells towards the right in the northern hemisphere, and towards the left in the south, thereby creating an eddy flux of zonal momentum towards the equator. Since the visible cloud layers are predominantly axisymmetric, Busse (1976) suggests that the non-axisymmetric cells operate in the deep atmosphere to produce a deep equatorial current, and that the equatorial acceleration observed in the visible cloud layers is forced by this deep current. Again, the need for an analysis of the finite amplitude, rotating convection problem in a deep atmosphere is apparent.

One mechanism for generating the equatorial current by means of an upward transport of momentum at the equator has also been proposed. Maxworthy (1975) has suggested that equatorially-trapped Kelvin waves and mixed Rossby-gravity waves are generated in the lower atmosphere and propagate upwards carrying zonal momentum. Then they are dissipated in the upper atmosphere, deposit their momentum, and generate a mean current. This mechanism has been invoked previously to explain the equatorial currents in the terrestrial stratosphere (Holton and Lindzen 1972). Invoking this mechanism to explain the Jovian equatorial current does require, however, *ad hoc* assumptions about the generation and dissipation of equatorial waves.

Some other recent studies in fluid dynamics have interesting implications for Jupiter's equatorial current. Whitehead (1975) performed experiments with a barotropic rotating fluid on a β-plane. He found that *any* localized stirring of the fluid produced a westerly jet at the latitude of the stirring. Lorenz (personal communication) has found the same result in a numerical study of a barotropic atmosphere on a rotating, spherical surface. In particu-

lar, even an equatorial jet can be produced if the stirring occurs at the equator. The mechanism at work here is not clear, but the results are in accord
with terrestrial experience, where the stirring mechanism for generating the
mid-latitude westerly jet can be regarded as baroclinic instability. These
results suggest that the fundamental problem in explaining Jupiter's equatorial current may not be explaining the current *per se*, but rather determining what stirring mechanism operates near the equator.

One feature of the equatorial current can be explained without relying on
any particular dynamical mechanism for generating it. Hide (1966) pointed
out that the width of the equatorial current (and similarly of the equatorial
zone) can be deduced simply from scaling analysis. It should have a half-
width of order

$$L \sim \left(\frac{u}{\beta}\right)^{\frac{1}{2}} \tag{13}$$

For $u = 100$ m sec^{-1}, this formula yields $L \sim 4000$ km, in reasonable order-
of-magnitude agreement with the observed half-width, ~ 8000 km.

(b)   Banded Structure

Hess and Panofsky (1951) presented the first dynamical discussion of the
bands. They noted that the measurements of apparent zonal motions indicate
that $\frac{1}{f}\frac{\partial u}{\partial y}$ is positive in zones and negative in belts; i.e., the zones tend to be
regions of anticyclonic vorticity, and the belts, regions of cyclonic vorticity.
This correlation is particularly clear between 20°S and 30°N latitudes (see
Fig. 1). Since the zonal motions appear to be geostrophic, i.e., they satisfy
the relation

$$fu = -\frac{1}{\rho}\frac{\delta P}{\delta y}, \tag{14}$$

this correlation implies that the zones are regions of high pressure, and the
belts regions of low pressure. This, in turn, implies that there will be meridional flow from the zones to the belts, and since presumably we are looking at high levels in the atmosphere, continuity requires that the zones be
regions of rising motions, and the belts regions of sinking motions. Since rising motions favor cloud formations, one would expect from these dynamical
arguments that the zones would be relatively cloudy and the belts relatively
clear. This picture of the meridional circulations is illustrated schematically
in Fig. 4. Hess and Panofsky's picture is convincing, since it is supported by
observations of lower albedos in the belts, as one would expect for relatively
cloud free areas; by observations of higher-brightness temperatures in the
belts, as one would expect if the thermal radiations were coming from deeper
layers; and by theoretical calculations of infrared spectra (Newburn and
Gulkis 1973).

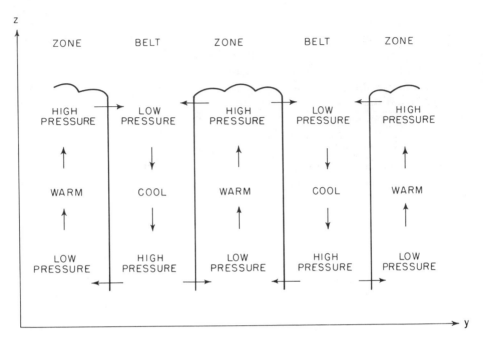

Fig. 4.   Schematic diagram of Jupiter's meridional circulations.

Ingersoll and Cuzzi (1969) showed that the correlation between the vorticity and the banded structure also implies a correlation between temperature and the banded structure because of the thermal wind relation. In particular, they deduced that the zones are warmer than the belts. This property is consistent with the picture of rising motions in zones and sinking motions in belts, and has also been included in Fig. 4. Ingersoll and Cuzzi calculated that the temperature difference between the zones and belts is given by

$$\alpha \Delta T D \sim 1.2 \text{ km}. \tag{15}$$

For example, if $\alpha = (180°\text{K})^{-1}$ and $D = 50$ km, then $\Delta T = 4°\text{K}$. They also noted that these local variations appear to dominate any systematic variation of temperature with latitude.

Barcilon and Gierasch (1970) suggested that the higher temperatures in the zones could be caused by condensation there. They found, using Lewis' (1969) solar composition model for the condensates, that this condensation could produce temperature differences on the order of those deduced by Ingersoll and Cuzzi—specifically, $\Delta T \sim 2°$ at the condensation level. They then constructed a model in which the meridional circulations and banded structure were generated by this condensation. Their model was not closed, so they could not solve for the temperature contrast or the distance between

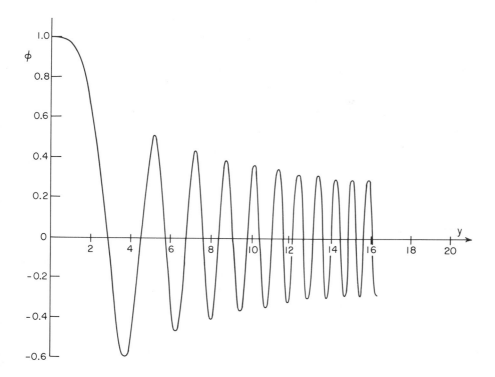

Fig. 5.   Amplitude of meridional stream function vs. latitude for inertial instability (after Stone 1971).

zones and belts. In addition, they had to assume the axial symmetry, and the motions they describe are unstable—for example, the condition for inertial instability is satisfied. Also, as discussed in Sec. II, it is not clear how the thermodynamic effect of the water vapor condensation could be felt in the visible cloud layers, some 60 km above the condensation level.

Two instability mechanisms have been suggested for forming the banded structure. Inertial instability, described in Sec. II, generates motions which are axially symmetric, and consequently Stone (1967) suggested that this instability might be the cause of the banded structure. The meridional circulations generated by this instability tend to parallel the isentropes, and have very rapid oscillations perpendicular to the isentropes. On a spherical surface these oscillations have a large scale modulation (Stone 1971). Fig. 5 illustrates the amplitude of the modulation for the inertial instability mode that dominates in low latitudes. The characteristic scale given in Table I for inertial instability is the distance from the equator of the first zero in this modulation function.

These motions generated by inertial instability can be described as a kind of small-scale turbulence, modulated as a function of latitude. Where the modulation has a large amplitude, turbulence is strong and cloud formation will be favored, and where the modulation has a small amplitude the turbulence and cloud formation will be suppressed. Fig. 5 shows that the resulting cloud structures would closely resemble Jupiter's banded structure — i.e., the bands become systematically smaller and less pronounced as one moves away from the equator. The prime assumption that one has to make in invoking this mechanism is that atmospheric conditions will favor this kind of instability, i.e., $\frac{1}{4} < Ri < 0.95$ (Stone 1966). Given this restriction on Ri, the theory predicts a banded structure that is correct quantitatively as well as qualitatively. For example, choosing $u = 100$ m sec^{-1}, the corresponding range of scales for inertial instability (see Table I) is 9000 to 32,000 km, in good agreement with the observed distance from the equator to the center of the equatorial belts, $\sim 15,000$ km. Comparing the formula for this scale in Table I with Eq. (5), we see that this scale is essentially the one deduced by Hide (1966) from scaling analysis.

The results quoted above for inertial instability strictly apply to the case of no horizontal shear in the mean flow. Some effects of horizontal shear are known. For example, a more accurate condition than that given in Table I for when inertial instability occurs is (Stone 1971)

$$Ri < \frac{1}{1 - \dfrac{1}{f}\dfrac{\partial u}{\partial y}}. \tag{16}$$

Therefore, the instability is favored in regions of anticyclonic vorticity. The observed correlation between the vorticity and the banded structure then suggests that inertial instability might be producing or contributing to the banded structures *locally,* in contrast to the global view suggested above. In particular, given a zonal wind profile like the one observed, regions of anticyclonic vorticity will favor inertial instability, which will tend to generate turbulence and cloudiness consistent with the observed correlation of bright zones with regions of anticyclonic vorticity. In any case, ignorance of actual values of Ri makes this mechanism a speculative one. Again we are reminded of the need to develop models for inertial instability regimes capable of predicting temperature structure as well as motions.

Another instability mechanism for generating the bands has been proposed by Gierasch (1973). This is the radiative instability mechanism first discussed by Gierasch *et al.* (1973). If the clouds contain a condensate which enhances the greenhouse effect, then instability is possible; rising motions lead to condensation, which warms the atmosphere locally by increased thermal blanketing and this warming will enhance the rising motions. Gierasch's (1973) analysis of this instability showed that the dominant modes in low latitudes were axially symmetric, with structures very similar to that illustrated in Fig. 5. In addition, w and θ are necessarily positively correlated so the cloudy regions are warmer, consistent with Ingersoll and Cuzzi's deduction.

The scale given in Table I for radiative instability is again the distance of the first node from the equator. This distance scale is characteristic of low-latitude disturbances in a motionless atmosphere (Lindzen 1967). The scale depends on the static stability, which is unknown, but we can estimate what value of the static stability would give the correct one for the banded structure. In particular, we know that the scale is given correctly, in order of magnitude, by Eq. (3). If we multiply this scale by two, so that it agrees with the observed one, and equate it to the scale given in Table I for the radiative instability, we find that the required static stability is

$$S \sim \frac{u^2}{\alpha g H^2}.$$ (17)

Substituting this value into Eq. (3), and assuming that $H \sim D$, we find that the corresponding value of the Richardson number is $Ri \sim 1$. Thus the radiative instability mechanism can account for the observed scale of the banded structure only under conditions such that inertial instability is also likely to occur. Since the two mechanisms generate very similar structures, they may well act simultaneously and reinforce each other. The relative efficiency of the two mechanisms may be compared by calculating their characteristic time scales, given in Table I. If we choose $Ri = \frac{1}{2}, \tau = 10^8$ sec (Gierasch and Goody 1969), $u = 100$ m sec^{-1}, $H = 20$ km, and evaluate f at 4° latitude, we find that the time scale for inertial instability is 4×10^4 sec and for radiative instability 10^6 sec. Apparently inertial instability would be a much more efficient mechanism for generating the banded structure. In fact, the very long radiative time scales in Jupiter's atmosphere mean that the time scales associated with the radiative instability are invariably much longer than those associated with the other instabilities listed in Table I, and therefore, it is doubtful whether it would be a dominant mechanism.

(c) Great Red Spot

Hide (1961, 1963) presented the first dynamical discussion of the Great Red Spot. He noted that the motions in the vicinity of the Spot resembled those seen in laboratory experiments with Taylor columns. These experiments show that, if a fluid is homogeneous and sufficiently rapidly rotating, even a shallow topographic feature will give rise to a disturbance in the flow pattern that propagates upward and appears at the top of the fluid (Taylor 1923). Such disturbances are sometimes referred to as "Taylor columns" and Hide suggested that the Great Red Spot was a manifestation of such a disturbance.

Hide's suggestion has stimulated a number of studies aimed at elucidating the properties of Taylor columns and the conditions under which they can occur. Recent experimental studies (Hide and Ibbetsen 1968) and theoretical studies (Ingersoll 1969) reinforce the conclusion that the observed flow patterns resemble those around a Taylor column. Calculations of the Rossby number in the vicinity of the Spot show that the atmosphere is sufficiently rapidly rotating for Taylor columns to occur (Hide 1963; Hess 1969). Stone and Baker (1968) criticized the Taylor column hypothesis on the grounds that the atmosphere's stratification and horizontal temperature gradients were not small enough for the atmosphere to be considered homogeneous. Recently Hogg (1973) analyzed the flow of a non-homogeneous fluid past an obstacle and found that Taylor columns can occur if

$$\frac{\alpha g S H^2}{f^2 L^2} \lesssim Ro . \tag{18}$$

This condition is plausible, since even if the atmosphere is isothermal, the Great Red Spot is so large that the left-hand side of Eq. (18) is still only ~ 0.03. In practice, it is likely that the atmosphere is nearly adiabatic, $S \ll \Gamma$, and then the condition is met, even though $Ro \sim 0.01$.

However, even if Eq. (18) is satisfied, Hogg's analysis does not demonstrate that Taylor columns can occur in a real atmosphere. He assumed that the flow was steady, yet his flow is baroclinically unstable. Since these instabilities transport heat, they will modify the flow pattern in his analysis. It is not clear how a Taylor column can be maintained under such conditions.

Other difficulties with the Taylor column hypothesis are the observation that the Great Red Spot does not have a fixed longitude (Peek 1958) and the likelihood that there is no solid surface (Hubbard 1973). To get around these difficulties, Streett (1969) suggested that the Spot was caused by a Taylor column attached to a mass of solid hydrogen floating in helium-rich liquid hydrogen deep in the atmosphere. Streett et al. (1971) attempted to analyze the dynamics of such a floating raft, but their analysis omitted Coriolis forces and, therefore, is not relevant to Jupiter.

Golitsyn (1970) suggested that the Spot was a long-lived eddy, i.e., that there need not be any immediate driving mechanism for the Spot. This idea receives some support from Ingersoll's (1973) analysis, showing that the equations of motion for large-scale flows on Jupiter contain free (i.e., unforced) solutions which resemble the flow patterns in the vicinity of the Spot. However, the idea that no forcing is present currently seems implausible since the radiative and dynamical time scales in Jupiter's atmosphere range from 10 years down to days (Gierasch and Goody 1969; Stone 1973a) and are considerably shorter than the time during which the Spot has been observed. Indeed, many changes in the circulations elsewhere in the atmosphere have been observed (Peek 1958). Ingersoll's (1973) analysis does lessen the attractiveness of the Taylor column hypothesis because it shows that one need not rely on a topographic feature to explain the observed flow patterns.

Another model of the Spot which treats it as a long-lived artifact has been presented by Maxworthy and Redekopp (1976). Their model is based on the concept of "solitons" (Scott *et al.* 1973). These are finite amplitude solitary waves that persist without change in the absence of any dissipation, because dispersive effects are balanced by non-linear effects. Maxworthy and Redekopp found that solitons can exist in a stratified zonal flow that is incompressible, barotropic, and quasi-geostrophic, provided that

$$\frac{\pi^2 f^2 L^2}{\alpha g S D^2} \leqslant 1 . \tag{19}$$

L in their solution corresponds to about one third of the meridional extent of the Spot. Maxworthy and Redekopp found that the flow patterns accompanying these solitons closely resemble the flows observed in the vicinity of the Spot, including the flow observed when the Spot interacts with the South Tropical Disturbance. This latter interaction is not reproduced well by other theories of the Spot.

Eq. (19) in effect places a lower bound on the static stability which must be met if the soliton theory is to be viable. If we choose $\alpha = (130°K)^{-1}$, $H = 20$ km, $3L = 12000$ km, and evaluate f at 17° latitude, then Eq. (19) implies that $S \geqslant 20°/$km; i.e., that the temperature must be increasing with height by at least 18°/km in the vicinity of the Spot. This stability is so much greater than that at the 130°K level in any proposed model of the atmosphere, that the soliton model becomes implausible. It is worth noting that the soliton and Taylor column models require conditions [Eqs. (18) and (19)] which are mutually incompatible so long as $Ro \ll 1$. The fact that many different models can produce flow patterns resembling those around the Great Red Spot [i.e., Maxworthy and Redekopp's; Hide and Ibbetsen's (1968); Hogg's (1973); and Ingersoll's (1973)] suggests that such a resemblance *per se* does not provide strong support for any particular model.

Kuiper (1972) has suggested that the Great Red Spot is a region of organized cumulus convection driven by water vapor condensation. As pointed out in Sec. II, this energy source occurs down near the 300°K level and one would have to assume that the effect of the condensation is being felt in a region 10 or more vapor pressure scale-heights above the condensation level. In any case, the general idea that the Spot represents a disturbed region is plausible. In particular, Ingersoll (1973) has pointed out that the Spot appears to share the characteristics of the zones—i.e., cloudiness, anticyclonic vorticity and rising motions—and Prinn and Lewis (1975) have demonstrated that the red color of the Spot can be explained if the vertical mixing in the Spot is enhanced relative to the rest of the atmosphere. As yet, there seems to be no compelling theory for the nature of this disturbance.

(d) Non-symmetric Features

Two mechanisms have been proposed for explaining the many large-scale non-symmetric cloud features in the Jovian atmosphere. Stone (1967) has suggested that they are analogous to extra-tropical cyclones in the terrestrial atmosphere, i.e., that they are caused by baroclinic instability. The discussion in Sec. III shows that it is very likely that horizontal temperature gradients occur in the visible cloud layers, at least on the scale of the belts and zones, if not on the planetary scale. Consequently, potential energy should be available for baroclinic perturbations to feed on. Thus it is quite possible that baroclinic instability occurs locally, even if it is not the dominant dynamical mode. The main uncertainty in postulating its existence in the Jovian atmosphere is again our ignorance of the static stability of the visible cloud layers and of the layers immediately below them.

The characteristic scale for baroclinic instability (see Table I) depends on the (unknown) value of Ri, but we can, nevertheless, place bounds on it by choosing extreme values. The range of Ri values for which baroclinic instability can occur is roughly $1 < Ri < 1000$ (see Fig. 2). If we choose $u \sim 10$ m sec^{-1}, then the bounds on the characteristic scale, at 30° latitude, are 150 km and 4000 km. This range includes the scales of many observed non-symmetric features. The corresponding range of time scales for baroclinic instability is 6 hours to 150 hours.

Ingersoll and Cuzzi (1969) have suggested that barotropic instability may be an important mechanism in the Jovian atmosphere. This instability draws its energy from the kinetic energy of the mean flow by means of horizontal eddy stresses and thereby tends to limit the horizontal shear in the mean flow. The most unstable modes are three-dimensional, so this mechanism is another possible source of the non-symmetric features seen in the Jovian atmosphere. The scale for this instability given in Table I again refers to the half-wavelength in the down-stream direction of the most unstable mode.

The barotropic instability criterion (see Table I) can be checked directly from observations. For example, if we use the mean distribution of u given by Chapman (1969), we find that d^2u/dy^2 exceeds β on the edges of the equatorial current and the north temperate current by a factor of ten. This excess is sufficient that uncertainties in interpreting the apparent horizontal shear as a real horizontal shear at a constant pressure level are unlikely to reverse the conclusion that barotropic instability should be occurring. In fact, the edges of the bands and currents often are perturbed, as, for example, in the Pioneer 10 photographs (Gehrels et al. 1974). If we choose 10 m sec^{-1} as a typical velocity, then the characteristic scale for barotropic instability is 5000 km. This scale is, in fact, about the scale of the larger perturbations observed. The corresponding characteristic time scale is 400 hours. Also the vortex-like features seen in the Pioneer 11 pictures (Swindell et al. 1975) resemble barotropic (shear) instabilities seen in laboratory experiments. Thus, barotropic instability appears to be a very plausible mechanism for generating at least some of the larger-scale non-symmetric features seen in the Jovian atmosphere.

V. SUMMARY

Theoretical calculations so far shed little light on the nature of Jupiter's circulations. About all one can say is that the regime in the visible cloud layers in low latitudes is likely to be an inertial instability or forced convection regime, with lapse rates very nearly adiabatic. Analytical models of these regimes and of finite amplitude rotating convection in the deep atmosphere are needed to further the deductive approach to Jovian meteorology. Measurements of the latitudinal variation of the internal heating are particularly important for developing realistic models.

On the other hand, it is possible to say something about the kinematics. The plausible assumption, that the zonal motions are in geostrophic balance and satisfy the thermal wind relation, enables one to deduce that the zones are regions of higher pressure, warmer temperatures, rising motions and enhanced cloudiness, while the belts are regions of lower pressure, cooler temperatures, sinking motions and relatively cloud-free. These temperature differences imply that baroclinic energy sources are present. The strong shear at the edges of some belts indicate that barotropic instability should occur, and this could account for at least some of the larger-scale non-symmetric features of the belts. To evaluate different speculations for the causes of the banded structure and other non-symmetric features, it would be valuable to have measurements of meridional velocities, measurements of the lapse rates, and static stability in both the cloud layers and the layers immediately below, and to have high-resolution visual observations capable of resolving small-scale convection.

REFERENCES

Aumann, H. H.; Gillespie, C. M., Jr.; and Low, F. J. 1969. The internal powers and effective temperatures of Jupiter and Saturn. *Astrophys. J.* 157:L69–L72.

Barcilon, A., and Gierasch, P. 1970. A moist, Hadley cell model for Jupiter's cloud bands. *J. Atmos. Sci.* 27:550–560.

Businger, J. A.; Wyngaard, J. C.; Izumi, Y.; and Bradley, E. F. 1971. Flux-profile relationships in the atmospheric surface layer. *J. Atmos. Sci.* 28:181–189.

Busse, F. H. 1970a. Thermal instabilities in rapidly rotating systems. *J. Fluid. Mech.* 44:441–460.

————. 1970b. Differential rotation in stellar convection zones. *Astrophys. J.* 159:629–639.

————. 1976. A simple model of convection in the Jovian atmosphere. *Icarus* (special Jupiter issue). In press.

Chandrasekhar, S. 1961. *Hydrodynamic and hydromagnetic stability.* Oxford: Clarendon Press.

Chapman, C. R. 1969. Jupiter's zonal winds: variation with latitude. *J. Atmos. Sci.* 26:986–990.

Chase, S. C.; Ruiz, R. D.; Münch, G.; Neugebauer, G.; Schroeder, M.; and Trafton, L. M. 1974. Pioneer 10 infrared radiometer experiment: preliminary results. *Science* 183:315–317.

Divine, T. N. 1971. The planet Jupiter. NASA SP-8069. Washington, D.C.: Government Printing Office.

Drazin, P. G., and Howard, L. N. 1966. Hydrodynamic stability of parallel flow of inviscid fluid. *Advanc. Appl. Mech.* 9:1–89.

Gehrels, T. 1974. The convectively unstable atmosphere of Jupiter. *J. Geophys. Res.* 79: 4305–4307.

Gehrels, T.; Coffeen, D.; Tomasko, M.; Doose, L.; Swindell, W.; Castillo, N.; Kendall, J.; Clements, A.; Hämeen-Anttila, J.; KenKnight, C.; Blenman, C.; Baker, R.; Best, G.; and Baker, L. 1974. The imaging photopolarimeter experiment on Pioneer 10. *Science* 183: 318–320.

Gierasch, P. J. 1970. The fourth Arizona Conference on planetary atmospheres: motions of planetary atmospheres. *Earth Extraterr. Sci.* 1:171–184.

———. 1973. Jupiter's cloud bands. *Icarus* 19:482–494.

Gierasch, P., and Goody, R. M. 1969. Radiative time constants in the atmosphere of Jupiter. *J. Atmos. Sci.* 26:979–980.

Gierasch, P.; Goody, R.; and Stone, P. 1970. The energy balance of planetary atmospheres. *Geophys. Fluid Dyn.* 1:1–18.

Gierasch, P. J.; Ingersoll, A. P.; and Williams, R. T. 1973. Radiative instability of a cloudy planetary atmosphere. *Icarus* 19:473–481.

Gierasch, P. J., and Stone, P. H. 1968. A mechanism for Jupiter's equatorial acceleration. *J. Atmos. Sci.* 25:1169–1170.

———. 1972. Reply. *J. Atmos. Sci.* 29:1008.

Golitsyn, G. S. 1970. A similarity approach to the general circulation of planetary atmospheres. *Icarus* 13:1–24.

Goody, R. 1969. Motions of planetary atmospheres. *Annual Rev. Astron. Astrophys.* 7:303–352.

Hadlock, R. K.; Na, J. Y.; and Stone, P. H. 1972. Direct thermal verification of symmetric baroclinic instability. *J. Atmos. Sci.* 29:1391–1393.

Hartmann, W. K. 1974. Martian and terrestrial paleoclimatology: relevance of solar variability. *Icarus* 22:301–311.

Hess, S. L. 1969. Vorticity, Rossby number, and geostrophy in the atmosphere of Jupiter. *Icarus* 11:218–219.

Hess, S. L., and Panofsky, H. A. 1951. The atmospheres of the other planets. *Compendium of Meteorology,* pp. 391–400. Boston: American Meteorological Society.

Hide, R. 1961. Origin of Jupiter's Great Red Spot. *Nature* 190:895.

———. 1963. On the hydrodynamics of Jupiter's atmosphere. *Mém. Soc. Roy. Sci. Liège (Sér. V)* VII:481–505.

———. 1966. On the circulation of the atmospheres of Jupiter and Saturn. *Planet. Space Sci.* 14:669–675.

———. 1969. Dynamics of the atmospheres of the major planets with an appendix on the viscous boundary layer at the rigid bounding surface of an electrically-conducting rotating fluid in the presence of a magnetic field. *J. Atmos. Sci.* 26:841–853.

———. 1970. Equatorial jets in planetary atmospheres. *Nature* 225:254.

Hide, R., and Ibbetsen, A. 1968. On slow transverse flow past obstacles in a rapidly rotating fluid. *J. Fluid Mech.* 32:251–272.

Hogan, J.; Rasool, S. I.; and Encrenaz, T. 1969. The thermal structure of the Jovian atmosphere. *J. Atmos. Sci.* 26:898–905.

Hogg, N. G. 1973. On the stratified Taylor column. *J. Fluid Mech.* 58:517–537.

Holton, J. R. 1972. *An introduction to dynamic meteorology.* New York: Academic Press.

Holton, J. R., and Lindzen, R. S. 1972. An updated theory for the quasi-biennial cycle of the tropical atmosphere. *J. Atmos. Sci.* 29:1076–1080.

Hubbard, W. B. 1973. The significance of atmospheric measurements for interior models of the major planets. *Space Sci. Rev.* 14:424–432.

Ingersoll, A. P. 1969. Inertial Taylor columns and Jupiter's Great Red Spot. *J. Atmos. Sci.* 26:744–752.

———. 1973. Jupiter's Great Red Spot: a free atmospheric vortex? *Science* 182:1346–1348.

Ingersoll, A. P., and Cuzzi, J. N. 1969. Dynamics of Jupiter's cloud bands. *J. Atmos. Sci.* 26: 981–985.

Ingersoll, A. P. 1976. Pioneer 10 and 11 observations and the dynamics of Jupiter's atmosphere. *Icarus* (special Jupiter issue). In press.

Kliore, A. J.; Fjeldbo, G.; Seidel, B. L.; Sykes, M. J.; and Woiceshyn, P. 1973. S-band radio occultation measurements of the atmosphere and topography of Mars with Mariner 9: extended mission coverage of polar and intermediate latitudes. *J. Geophys. Res.* 78:4331–4351.

Kuiper, G. P. 1972. Lunar and Planetary Laboratory studies of Jupiter—II. *Sky and Telescope* 43:75–81.

Kuo, H. L. 1949. Dynamic instability of two-dimensional non-divergent flow in a barotropic atmosphere. *J. Meteor.* 6:105–122.

Lewis, J. S. 1969. The clouds of Jupiter and the NH_3-H_2O and NH_3-H_2S systems. *Icarus* 10: 365–378.

Lindzen, R. D. 1967. Planetary waves on beta-planes. *Mon. Wea. Rev.* 95:441–451.

Maxworthy, T. 1972. Comments on a mechanism for Jupiter's equatorial acceleration. *J. Atmos. Sci.* 29:1007–1008.

————. 1973. A review of Jovian atmospheric dynamics. *Planet. Space Sci.* 21:623–641.

————. 1975. A wave driven model of the Jovian equatorial jet. *Planet. Space Sci.* In press.

Maxworthy, T. and Redekopp, L. G. 1976. A solitary wave theory of the Great Red Spot and other observed features in the Jovian atmosphere. *Icarus* (special Jupiter issue). In press.

McIntyre, M. E. 1970. Diffusive destabilization of the baroclinic circular vortex. *Geophys. Fluid Dyn.* 1:19–57.

————. 1972. Baroclinic instability of an idealized model of the polar night jet. *Quart. J. Roy. Meteor. Soc.* 98:165–174.

Newburn, R. L., Jr., and Gulkis, S. 1973. A survey of the outer planets Jupiter, Saturn, Uranus, Neptune, Pluto, and their satellites. *Space Sci. Rev.* 3:179–271.

Ogura, Y. and Cho, H. R. 1973. Diagnostic determination of cumulus cloud populations from observed large-scale variables. *J. Atmos. Sci.* 30:1276–1286.

Peek, B. M. 1958. *The Planet Jupiter.* London: Faber and Faber.

Phillips, N. A. 1951. A simple three-dimensional model for the study of large-scale extratropical flow patterns. *J. Meteor.* 8:381–394.

Priestly, C. H. B. 1959. *Turbulent transfer in the lower atmosphere.* Chicago: Univ. of Chicago Press.

Prinn, R. G., and Lewis, J. S. 1975. Phosphine on Jupiter and implications for the Great Red Spot. *Science.* In press.

Reese, E. J., and Smith, B. A. 1968. Evidence of vorticity in the Great Red Spot of Jupiter. *Icarus* 9:474–486.

Roberts, P. H. 1968. On the thermal instability of a rotating fluid sphere containing heat sources. *Phil. Trans. Roy. Soc. London* A263:93–117.

Sagan, C.; Toon, O.; and Gierasch, P. 1973. Climatic change on Mars. *Science* 181:1045–1049.

Scott, A. G.; Chu, F. Y. F.; and McLaughlin, D. W. 1973. The soliton: a new concept in applied science. *Proc. IEEE* 61:1443–1483.

Slipher, E. C.; Hess, S. L.; Blackadar, A. K.; Gidas, H. L.; Shapiro, R.; Lorenz, E. N.; Gifford, F. A.; Mintz, Y.; and Johnson, H. L. 1952. The study of planetary atmospheres. *Lowell Obs. Final Report,* Contract No. AF19 122–162.

Spiegel, E. A. 1971. Convection in stars: I. Basic Boussinesq convection. *Ann. Rev. Astron. Astrophys.* 9:323–352.

Stone, P. H. 1966. On non-geostrophic baroclinic stability. *J. Atmos. Sci.* 23:390–400.

————. 1967. An application of baroclinic stability theory to the dynamics of the Jovian atmosphere. *J. Atmos. Sci.* 24:642–652.

Stone, P. H. 1970. On non-geostrophic baroclinic stability: Part II. *J. Atmos. Sci.* 27:721–726.
——. 1971. The symmetric baroclinic instability of an equatorial current. *Geophys. Fluid Dyn.* 2:147–164.
——. 1972a. On non-geostrophic baroclinic stability: Part III. The momentum and heat transports. *J. Atmos. Sci.* 29:419–426.
——. 1972b. A simplified radiative-dynamical model for the static stability of rotating atmospheres. *J. Atmos. Sci.* 29:405–418.
——. 1973a. The dynamics of the atmospheres of the major planets. *Space Sci. Rev.* 14:444–459.
——. 1973b. The effect of large-scale eddies on climatic change. *J. Atmos. Sci.* 30:521–529.
——. 1974. On Jupiter's rate of rotation. *J. Atmos. Sci.* 31:1471–1472.
Stone, P. H., and Baker, D. J., Jr. 1968. Concerning the existence of Taylor columns in atmospheres. *Quart. J. Roy. Meteor. Soc.* 94:576–580.
Stone, P. H.; Hess, S.; Hadlock, R.; and Ray, P. 1969. Preliminary results of experiments with symmetric baroclinic instabilities. *J. Atmos. Sci.* 26:991–996.
Streett, W. B. 1969. Phase equilibria in planetary atmospheres. *J. Atmos. Sci.* 26:924–931.
Streett, W. B.; Ringermacher, H. I.; and Veronis, G. 1971. On the structure and motions of Jupiter's Red Spot. *Icarus* 14:319–342.
Swindell, W.; Doose, L. R.; Tomasko, M.; and Fountain, J. 1975. The Pioneer 11 images of Jupiter. *Bull. Amer. Astron. Soc.* 7:378.
Taylor, G. I. 1923. Experiments on the motion of solid bodies in rotating fluids. *Proc. Roy. Soc. (London)* A104:213–218.
Trafton, L. M. 1967. Model atmospheres of the major planets. *Astrophys. J.* 147:765–781.
Trafton, L. M., and Stone, P. H. 1974. Radiative-dynamical equilibrium states for Jupiter. *Astrophys. J.* 188:649–655.
Walton, I. C. 1975. The viscous non-linear symmetric baroclinic instability of a zonal shear flow. *J. Fluid Mech.* 68:757–768.
Weidenschilling, S. J., and Lewis, J. S. 1973. Atmospheric and cloud structures of the Jovian planets. *Icarus* 20:465–476.
Westphal, J. A. 1971. Observations of Jupiter's cloud structure near 8.5 μ. *Planetary atmospheres.* (C. Sagan, T. C. Owen, H. J. Smith, eds.) pp. 359–362. Dordrecht, Holland: D. Reidel Publ. Co.
Whitehead, J. A., Jr. 1975. Mean flow generated by circulation on a β-plane: an analogy with the moving flame experiment. *Tellus* 26. In press.
Williams, G. P., and Robinson, J. B. 1973. Dynamics of a convectively unstable atmosphere: Jupiter? *J. Atmos. Sci.* 30:684–717.

PART IV

Magnetosphere and Radiation Belts

EARTH-BASED RADIO OBSERVATIONS OF JUPITER: MILLIMETER TO METER WAVELENGTHS

G. L. BERGE
California Institute of Technology

and

S. GULKIS
Jet Propulsion Laboratory

A comprehensive review of earth-based observations of the continuous radio emission from Jupiter in the wavelength range from 1 mm to 5 m is presented. After a brief historical summary of the observations and early theories for the origin of the emission, an up-to-date account of the observations, including total intensity, polarization, brightness distribution, and variability measurements is given. An extensive table of measurements of the total flux density is included.

The principal arguments are presented demonstrating that the observed radiation arises partly from synchrotron emission by relativistic electrons trapped in Jupiter's magnetic field and partly from a thermal process in its atmosphere. Techniques for separating the thermal and nonthermal components are discussed. An elementary review of the characteristics of synchrotron radiation is presented, together with a discussion of the main inferences about the Jovian magnetosphere drawn from the observations.

The interpretation of the thermal emission component in terms of atmospheric emission is illuminated by showing weighting functions and limb darkening curves at a number of frequencies for a specific model. Potential sources of microwave opacities in the Jovian atmosphere are discussed and the radiative transfer equations given. The microwave absorption coefficients of NH_3 (the strongest absorber), H_2O, and H_2 are provided. The main inferences about the Jovian atmosphere drawn from the observations are examined.

Following a brief comparison of the inferences drawn from Earth-based radio observations with the corresponding ones gathered from the Pioneer 10 and 11 measurements, the chapter concludes with a discussion of the future of Earth-based radio observations.

For two decades Jupiter has been one of the most extensively studied celestial radio sources due, in large part, to the fascinating diversity of its emission characteristics. Previous reviews of this topic have been given by

Roberts (1963, 1965), Warwick (1964, 1967), Carr and Gulkis (1969), and others. The wavelength range from one millimeter to five meters, to which this chapter is devoted, is an important window for the investigation of Jupiter. The emission at these wavelengths has been shown to arise partly from a nonthermal process in Jupiter's radiation belt and partly from a thermal process in its atmosphere.

As a radio source, Jupiter is intrinsically feeble, but its close proximity to Earth makes it comparatively easy to observe. The total nonthermal power emitted at radio wavelengths shorter than 5 m is probably no more than a billion watts, less than 10^{-3} of the power generation activity of mankind and less than 10^{-8} of Jupiter's total thermal emission. The behavior of the nonthermal component is very diagnostic of conditions within the radiation belt. The total thermal power emitted at wavelengths longer than 1 mm, while considerably larger, is of little consequence to the energy balance in Jupiter's atmosphere, but it can be used as a probe to study the atmosphere, particularly the deep atmospheric thermal structure beneath the visible clouds.

Although the two components originate in distinctly different regions and from distinctly different mechanisms, they are closely intertwined historically and observationally; hence, the decision to include both in the same review chapter. The organization of topics is as follows: I. Historical Development; II. Observational Data; III. Separation into Thermal and Nonthermal Components; IV. Interpretation of the Nonthermal Component; V. Interpretation of the Thermal Component; VI. Comparison of Earth-Based Radio Measurements with Pioneer 10 and 11 Data; VII. The Future of Earth-Based Radio Observations.

I. HISTORICAL DEVELOPMENT

The discovery of radio emission associated with Jupiter was made quite unexpectedly in 1955 at a frequency of 22.2 MHz (Burke and Franklin 1955). The story of this serendipitous discovery (Franklin 1959) has become a well-known anecdote in the annals of radio astronomy. The emission was very sporadic in character. We now know that it is confined to frequencies less than about 40 MHz and that it exhibits complex structure in time, frequency, and polarization. This complex behavior and the high intensity clearly indicate a nonthermal origin. Shain (1955) was able to identify Jupiter events at 18.3 MHz on records dating back to 1950 and he showed that the probability of reception was highly correlated with the central meridian longitude (System II). By 1962 the data on this correlation had produced a slightly different period, appropriate for radio observations, that was accurate enough to prompt the International Astronomical Union to adopt a provisional longitude system called System III (1957.0) based on a rotation period of $9^h\ 55^m\ 29\overset{s}{.}37$ (I.A.U. 1962). The period has since been further refined, but the I.A.U. definition has, as yet, remained unchanged.[1] Another important discovery was made by Bigg (1964) who showed that the probability of reception was also highly correlated with the geocentric orbital longitude of Io. Further discussion of the decameter emission is beyond the scope of this chapter,[2] but we mention it in passing because it initially focused the attention of radio astronomers to Jupiter and because it has complemented and helped to explain many of the characteristics seen in the Jupiter radio emission at higher frequencies.

The first detection of microwave radiation from Jupiter was accomplished on May 13 and 31, 1956 at a wavelength of 3.15 cm by Mayer *et al.* (1958a,b). The equivalent blackbody disk temperature was 140 ± 56°K (mean error). The same paper also described a more accurate series of measurements made almost a year later. The disk temperature determined from the latter measurements was 145 ± 26°K (mean error). This was numerically consistent with the infrared measurements available at the time. However, the authors were careful to note that "There is considerable uncertainty in the actual level of emission because of possible molecular absorptions in the atmosphere of the planet. As a result, the blackbody temperature for thermal radiation at the 3.15 cm wavelength would not necessarily agree with values derived by other methods . . . " Later measurements of the disk temperature near this wavelength have tended to give somewhat higher results. For example, Giordmaine *et al.* (1959) obtained 171 ± 20°K at 3.03 cm, 173 ± 20°K at 3.17 cm, and 189 ± 20°K at 3.36 cm in 1958 and early 1959. (The quoted uncertainties are probable errors.) The authors state that "The observed radiation is interpreted as thermal emission from ammonia in the region near the top of the cloud layer. The gaseous ammonia radiates through the inversion line at 1.28 cm, pressure broadened in the presence of hydrogen and helium."

From mid-1956 to mid-1958, when the only microwave observations of Jupiter were at wavelengths of 3 to 4 cm, they revealed nothing very surprising about the planet. Indeed, some people may have regarded the demonstration of the sensitive new instruments capable of such measurements to be more interesting than the results of the Jupiter measurements. This quickly changed in the summer of 1958 when measurements at the substantially longer wavelength of 10.3 cm indicated a disk temperature of 640 ± 85°K (standard error) (Sloanaker 1959; McClain and Sloanaker 1959). It was recognized that, if there is a temperature gradient with depth in the Jupiter atmosphere, then one might expect larger observed disk temperatures at longer wavelengths (Mayer 1959). This is because of the reduced absorption at longer wavelengths where one is farther out in the wing of the broad ammonia absorption spectrum. In the same paper Mayer also suggested the

---

[1]Throughout this chapter all longitudes quoted are in System III (1957.0). See p. 826.
[2]See pp. 693 and 1146.

possibility of nonthermal emission to produce the enhanced temperatures at longer wavelengths.

Soon there were observations at still longer wavelengths that showed the trend to be very pronounced. They indicated a disk temperature of 2496 ± 450°K (standard error) at 21 cm (McClain 1959), about 3000°K, also at 21 cm (Epstein 1959), about 3000°K at 22 cm (Drake and Hvatum 1959), about 5500°K at 31 cm (Roberts and Stanley 1959), and 20,000 to 70,000°K at 68 cm (Drake and Hvatum 1959). The flux densities that yield this enhancement are roughly constant with wavelength. As early as May, 1959, Drake is reported to have suggested that synchrotron radiation in a Jovian Van Allen belt may be responsible for the enhanced temperatures (e.g., Roberts and Stanley 1959).

In the following month Field submitted a paper for publication (Field 1959) in which he considered quantitatively four different possible explanations for the enhanced emission at long wavelengths: 1) thermal emission in a deep atmosphere with a temperature gradient and with ammonia as the only absorber, 2) free-free emission in a Jovian ionosphere, 3) cyclotron emission by nonrelativistic electrons in Jupiter's magnetic field, and 4) synchrotron emission by relativistic electrons in Jupiter's magnetic field. He found that atmospheric thermal emission could not reasonably account for the extreme slope of the temperature spectrum at longer wavelengths although it would be important at the shorter wavelengths. It is important to note, as did Giordmaine *et al.* (1959) in the continuation of the quotation given above, that "Thermal radiation from lower levels in the atmosphere probably accounts for an appreciable fraction of the higher temperature radiation reported at longer wavelengths in the centimeter and decimeter region." Field went on to show that the ionosphere explanation was not likely either. The observed spectrum could be obtained only by requiring an unacceptable large electron density with the electron temperature approaching 10^5°K. He found, however, that cyclotron emission and synchrotron emission were both plausible. The former would require a very large magnetic field intensity, about 10^3 Gauss at the surface, while the latter would require a large population of relativistic electrons.

Other calculations regarding synchrotron emission were carried out by Drake and Hvatum (1959) and Roberts and Stanley (1959). The latter paper also considered free-free emission in a Jovian corona. The free-free calculations were similar to those of Field (1959) for the ionosphere except that the larger emitting volume relaxed the requirement for such a high electron density. The authors suggested that sufficient plasma might be trapped by a planetary magnetic field.

Observations of Jupiter having a critical bearing on the question of emission mechanisms were carried out in 1960 by Radhakrishnan and Roberts (1960) with an east-west interferometer at 31.3 cm wavelength. The first of their two important results was that the radiation was linearly polarized by about 30% with the **E**-vector along the plane of the rotational equator to within $\pm 12°$. The large polarization was strong evidence that the mechanism involved a magnetic field. The other result was that the angular distribution was extended beyond Jupiter's disk, at least in the east-west direction, which coincided with the equatorial direction on Jupiter. The equatorial radius was approximately 3 R_J. Only cyclotron emission or synchrotron emission in a Jovian magnetic field would exhibit both an extended size and strong polarization. However, to see 31-cm cyclotron emission as far out as 3 R_J would require a surface field of 10^4 Gauss, an order of magnitude larger than that considered by Field (1959).

Field published two other papers (Field 1960, 1961) in which he developed the cyclotron emission possibility in more detail. He finally rejected this mechanism as being the correct one because he could not explain large time variations in flux density with time scales of months that had been reported. Ironically, these apparent variations later became suspect.

In 1961 there were further interferometric measurements (Morris and Berge 1962) that provided additional insight into the nature of the Jovian decimeter emission. Observations at 31.3 cm with a north-south interferometer, when compared to the results of Radhakrishnan and Roberts (1960), showed that the equatorial to polar size ratio was about 3. Observations at 21.6 cm gave a very similar size and shape: an equivalent Gaussian width to $1/e$ of 2.9 polar diameters in the equatorial direction and 1.2 in the polar direction. Furthermore, it was found that the degree of polarization was not significantly different at the two wavelengths. These observations were in conflict with the thin-shell cyclotron model of Field (1960) which predicted that either the degree of polarization would change greatly between the two wavelengths or else the equatorial to polar size ratio would be less than one. Perhaps this conflict could have been patched up by going to a thick-shell model of exactly the right sort. By this time, however, there were detailed calculations available for synchrotron emission models (Chang 1960; Davis and Chang 1961), and it was clear that this mechanism was much more attractive because it could be made to produce the observed characteristics without the need for any patchwork.

Morris and Berge (1962) also determined that the plane of polarization varied sinusoidally as Jupiter rotates, and they interpreted this as due to a difference in direction between the rotational and magnetic axes that is equal to the amplitude of the sinusoid ($9° \pm 3°$). The crossover points of the sinusoid, corresponding to the times when the rotational and magnetic axes, as seen in projection, appear to have the same direction, occurred at System III central meridian longitudes (CML) of about 20° and 200°. It was pointed out (i.e., Warwick 1963) that since we know the direction of rotation of Jupiter, the fact that the crossover at 200° is a clockwise crossover allows one to deduce that the magnetic pole at 200° is in the northern hemisphere, while the one at 20° is in the southern hemisphere.

The flux density data of Morris and Berge (1962) showed evidence of a doubly periodic variation with rotation such that the minima occurred at 20° and 200°. The authors suggested that the radiation is more intense when observed from Jupiter's magnetic equatorial plane; thus the minima at 20° and 200° when the earth was farthest from the magnetic equator. The accuracy was not adequate to establish definitely the beaming into the plane of the magnetic equator. The proof had to await more accurate measurements.

So far we have described the gross features of the Jovian microwave emission within the historical context of their discovery and interpretation up to about 1962. Since then there have been many further refinements including the discovery of characteristics that may be considered as second-order effects and features that required a longer time base to establish. In the next section we consider the various characteristics or features individually and in detail.

II. OBSERVATIONAL DATA

The microwave radiation from Jupiter, as we have mentioned previously, is dominated by thermal emission in its atmosphere at higher frequencies and by synchrotron emission in a radiation belt at lower frequencies. Both are continuous in time and frequency; that is, there are no bursts such as those characteristic of the decameter emission. Narrow spectral features, which originate in the atmosphere, are predicted but not yet positively confirmed by observation. The crossover point, where the two contributions are equal, occurs at a wavelength near 7 cm. The nonthermal contribution to the total flux density falls off sharply toward shorter wavelengths while the thermal contribution falls off sharply toward longer wavelengths.

Because of its polarization, beaming, and possible dependence on short-term processes, the nonthermal emission is rich in observational detail. In comparison, the data bearing on the thermal radio emission are much more restricted. Almost all of the available data are relevant only to the temperature spectrum. We have almost no information on the angular structure (i.e., brightness or temperature distribution) of the thermal radio emission over the disk of Jupiter except for evidence of limb darkening at 2.04 cm (Bogod *et al.* 1973) and at 3.7 cm (Olsen and Gulkis 1975).

A. Spectrum of the Continuous Microwave Radiation

Probably the most basic measurement that can be made of a discrete celestial radio source, such as Jupiter, is the determination of its integrated flux density S at a given frequency ν and given time t. Flux density is usually expressed in units of Janskys, (Jy) where 1 Jy = 1 flux unit = 10^{-26}W M^{-2} Hz^{-1}. By a long standing convention among radio astronomers, Jovian flux densities are usually reported normalized to an Earth-Jupiter distance of 4.04 A.U.

A convention often used is to convert S to an equivalent disk brightness temperature, T_D. The relationship between S and T_D is derived by making use of the Rayleigh-Jeans approximation to the Planck function:

$$B = \frac{2kT}{\lambda^2}.$$ (1)

In this expression, λ is the wavelength, k is the Boltzmann constant and B is the brightness defined as $dS/d\Omega$ where dS is the flux density from the solid angle $d\Omega$. Throughout the portion of the spectrum that we are considering, this approximation is reasonably accurate except perhaps at the very highest frequencies. Even at one millimeter wavelength the percentage error $\left(\approx 100 \frac{h\nu}{2kT} \right)$ of the approximation is only 5% for Jupiter. Thus we can write

$$S = \int_{source} B d\Omega = \int_{source} \frac{2kT}{\lambda^2} \, d\Omega = \frac{2kT_D}{\lambda^2} \Omega_{disk}$$ (2)

where T_D, the equivalent blackbody disk temperature, is defined as shown. When the source region coincides with the disk solid angle, as with Jupiter's thermal emission, then T_D is a true average over the temperature distribution. T_D is not as meaningful, however, when applied to the extended nonthermal emission. The convention generally adopted for calculating Ω_{disk} is to use the polar (PSD) and equatorial (ESD) semidiameter values quoted in *The American Ephemeris and Nautical Almanac* in the expression

$$\Omega_{disk} = \pi \times PSD \times ESD.$$ (3)

The disk temperature of Jupiter in terms of the flux density is then $T_D = \frac{490 \, S\lambda^2}{PSD \times ESD}$ where S is in Janskys, λ is in cm, and the semidiameters are in seconds of arc. When the flux density is normalized to 4.04 A.U., this expression reduces to $T_D = 0.8846 \, S\lambda^2$.

Since the original, and crude, determination of the radio spectrum of Jupiter described in the previous section, there have been a vast number of measurements of the quantities $S(\nu, t)$. If there is no time dependence then these will define a spectrum. Neglecting possible time variations for the moment, we present Table I and Fig. 1.

Table I shows most of the published values of S and/or T_D with the only change being the normalization of S to the standard distance if not originally done so. Also, in cases where only S or T_D were published we have derived the other quantity. Our purpose in presenting the table is not so much to give a list of the measured values as it is to provide a list of references to the extensive literature on the subject. In fact, the values themselves are very inhomogeneous in terms of quality, technique and assumptions of calibration, etc. Some of the earlier ones did not take account of the effect of polarization or beaming and in some cases the disk size Ω_{disk} has been incorrectly calculated.

TABLE I
Measurements of Jupiter's Radio Emission[1]

λ (cm)	T_D (°K)[2]	S(Jy)[2]	SE[3]	Reference
0.12	155	12200	b	Low and Davidson (1965)
0.14	150	8650	b	Ulich (1975)
0.213	168	4190	b	Ulich (1974)
0.214	178	4400	b	Cogdell *et al.* (1975)
0.225	180	4020	d	Efanov *et al.* (1969)
0.23	140	2990	c	Efanov *et al.* (1970, 1971)
0.309	181	2140	b	Ulich *et al.* (1973)
0.319	111	1230	c	Tolbert (1966)
0.332	153	1390	b	Epstein *et al.* (1970)
0.34	140	1370	b	Epstein (1968)
0.353	174	1570	b	Ulich (1974)
0.387	150	1130	d	Kislyakov and Lebskii (1968)
0.428	182	988	b	Ulich (1975)
0.429	105	645	c	Tolbert (1966)
0.600	184	565	b	Ulich (1975)
0.835	144	233	c	Thornton and Welch (1963)
0.845	157	249	a	Wrixon *et al.* (1971)
0.856	173	267	b	Ulich *et al.* (1973)
0.857	113	174	b	Tolbert (1966)
0.857	149	229	b	Braun and Yen (1968)
0.86	287	439	d	Gibson (1965)
0.86	142	217	b	Kalaghan and Wulfsberg (1968)
0.895	149	210	a	Wrixon *et al.* (1971)
0.95	151	194	a	Ulich (1975)
0.955	152	188	a	Dent (1972)
0.955	157	195	b	Hobbs and Knapp (1971)
0.984	130	152	b	Wrixon *et al.* (1971)
1.051	132	135	b	Wrixon *et al.* (1971)
1.178	140	114	a	Wrixon *et al.* (1971)
1.18	120	97.4	b	Law and Staelin (1968)
1.261	134	95.3	a	Klein (1974)
1.275	136	94.6	a	Wrixon *et al.* (1971)
1.28	120	82.8	b	Law and Staelin (1968)
1.28	137	94.5	b	Klein and Gulkis (1971)
1.30	141	94.3	b	Jones (1972)
1.334	139	88.3	a	Wrixon *et al.* (1971)
1.339	133	83.9	a	Klein (1974)
1.346	136	84.9	a	Janssen and Welch (1973)
1.35	107	66.4	b	Law and Staelin (1968)
1.411	138	78.4	a	Klein (1974)
1.43	112	61.9	b	Law and Staelin (1968)
1.461	145	76.8	b	Wrixon *et al.* (1971)
1.47	130	68.0	b	Jones (1972)
1.473	139	72.4	a	Klein (1974)
1.58	136	61.6	c	Law and Staelin (1968)
1.65	165	68.5	a	Mayer and McCullough (1971)
1.84	153	51.1	d	Haddock and Dickel (1963)
1.9	180	56.4	b	Kellermann and Pauliny-Toth (1966)

TABLE I (continued)

Measurements of Jupiter's Radio Emission[1]

λ (cm)	T_D (°K)[2]	S(Jy)[2]	SE[3]	Reference
1.95	149	44.3	b	Pauliny-Toth and Kellermann (1970)
1.95	182	54.1	b	Dickel *et al.* (1970)
2.0	145	41.0	b	Kellermann (1970)
2.07	173	45.6	b	Gary (1974)
2.07	157	41.4	b	Baars *et al.* (1965)
2.7	201	31.2	b	Mayer and McCullough (1971)
2.7	199	30.9	b	McCullough (1972)
3.02	185	22.9	b	Korol'kov *et al.* (1964) and Gol'nev *et al.* (1965)
3.03	171	21.1	c	Giordmaine *et al.* (1959)
3.12	218	25.3	b	Berge (1968)
3.15	140	16.5	d	Mayer *et al.* (1958a,b)
	145	17.1	c	Mayer *et al.* (1958a,b)
3.17	173	19.5	c	Giordmaine *et al.* (1959)
3.3	193	20.0	c	Bibinova *et al.* (1963)
3.36	189	18.9	c	Giordmaine *et al.* (1959)
3.71	294	20.0	a	Olsen (in preparation)
3.75	213	17.1	b	Haddock and Dickel (1963) and Dickel (1967a)
3.95	267	19.3	b	Soboleva and Timofeeva (1970)
4.52	280	15.5	b	Dickel and Medd (1967)
6.0	290	9.1	b	Hughes (1966)
6.0	344	10.8	b	Dickel (1967b)
6.0	340	10.68	a	Morris *et al.* (1968)
6.0	357	11.22	a	Whiteoak *et al.* (1969)
6.09	289	8.8	b	Wendker and Baars (1965)
6.5	305	8.15	b	Gol'nev *et al.* (1965)
9.4	670	8.57	b	Rose *et al.* (1963)
10.0	672	7.60	a	Bash *et al.* (1974)
10.2	315	3.42	b	Sloanaker and Boland (1961)
10.3	640	6.82	b	Sloanaker and Boland (1961)
10.4	703	7.35	b	Berge (1966)
10.6	825	8.3	b	Morris and Bartlett (1963)
11.1	698	6.4	a	Olsen (in preparation)
11.31	837	7.4	a	Roberts and Komesaroff (1965)
11.31	860	7.60	a	Roberts and Ekers (1968)
12.6	829	5.9	a	Gulkis *et al.* (1973)
13.0	1010	6.76	h	Boischot *et al.* (1963)
13.1	987	6.5	a	Gulkis *et al.* (1973)
13.1	870–1030	5.7–6.8	a	Klein (1976)
18.7	1690	5.45	a	Berge (1974)
20.8	2910	7.60	h	Miller and Gary (1962)
21.0	2860	7.33	b	McClain *et al.* (1962)
	2510	6.43	b	McClain *et al.* (1962)
	2050	5.25	c	McClain *et al.* (1962)
21.0	2570	6.59	b	Boischot *et al.* (1963)
21.0	2780	7.13	a	Roberts and Komesaroff (1965)

TABLE I (continued)
Measurements of Jupiter's Radio Emission[1]

λ (cm)	T_D (°K)[2]	S(Jy)[2]	SE[3]	Reference
	2900	7.43	a	Roberts and Komesaroff (1965)
21.1	2580	6.55	a	Berge (1974)
	2460	6.25	a	Berge (1974)
	1970	5.00	a	Berge (1974)
21.2	2590	6.51	b	Miller and Griffin (1966)
	3250	8.17	b	Miller and Griffin (1966)
21.2	2540	6.4	b	Berge (1966)
21.2	2600	6.5	b	Davies and Williams (1966)
21.2	2550	6.4	b	Conway and Kronberg (1968)
21.3	2650	6.6	b	Branson (1968)
21.4	3080	7.47	a	Gary (1963)
21.6	3380	8.2	b	Morris and Berge (1962)
22	3000	7.0	b	Drake and Hvatum (1959)
31	5500	6.5	b	Roberts and Stanley (1959)
31	5700	6.7	b	Morris and Berge (1962)
31.2	6300	7.3	b	Roberts and Komesaroff (1965)
42.9	12,000	7.4	c	Rzhiga and Trunova (1965)
48.4	13,200	6.39	b	Roberts and Ekers (1968)
49.0	10,200	4.8	c	Barber and Moule (1963)
	14,900	7.0	c	Barber and Moule (1963)
49.0	14,000	6.6	b	Barber and Gower (1965)
49.1	11,100	5.2	b	Dickel (1966)
49.2	15,100	7.05	b	Barber (1966)
68	70,000	17	d	Drake and Hvatum (1959)
	20,000	5	d	Drake and Hvatum (1959)
69.8	31,000	7.2	c	Kazes (1965)
69.8	28,900	6.7	b	Hardebeck (1965a,b)
69.8	31,300	7.26	b	Tiberi (1966)
	32,300	7.50	b	Tiberi (1966)
70.2	20,000	4.6	b	Krotikov et al. (1965)
73.5	< 24,000	< 5	—	Long and Elsmore (1960)
73.5	21,000	4.4	b	McAdam (1966)
74	31,200	6.45	b	Roberts and Komesaroff (1965)
75	30,400	6.1	b	Branson (1968)
154	130,000	6.2	b	Kazes (1965)
168	127,000	5.1	b	Barber and Gower (1965)
169	142,000	5.64	c	Gower (1963)
294	375,000	4.9	c	Artyukh et al. (1972)
368	539,000	4.5	c	Gower (1968)
375	746,000	6.0	b	Slee and Dulk (1972)
719	$< 16 \times 10^6$	< 35	—	Kazes (1965)

[1] This table is intended to be an extensive, but not fully exhaustive, guide to the measurements that define Jupiter's radio spectrum.

[2] Either T_D or S (4.04 A.U.) is given as published and the corresponding quantity calculated from $T_D = 0.8846 S \lambda^2$. The data are very inhomogeneous; no corrections have been made for the beaming, polarization, assumed calibration, assumed disk size, etc. Multiple entries denote independent measurements which in some cases reflect a time variation.

[3] Approximate indication of the quality in terms of the percentage standard error SE. a: $SE \leq 5\%$; b: $5\% \leq SE \leq 15\%$; c: $15\% \leq SE \leq 25\%$; d: $SE \geq 25\%$.

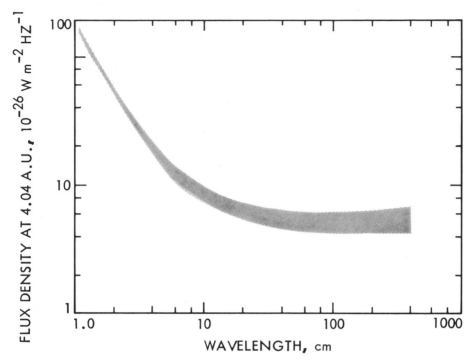

Fig. 1.   Schematic appearance of Jupiter's microwave spectrum from 1 cm to 300 cm. The
shaded area is representative of the observed data over the past decade. In the wavelength
interval from 1 cm to 1 mm the flux density continues to rise following the approximate law
$S \propto \lambda^{-2}$. At wavelengths longer than 750 cm the spectrum takes a sharp upturn due to the
presence of the sporadic component of radiation.

Figure 1 presents a schematic representation of the measurements given
in Table I. The graph clearly shows the general spectral characteristics already discussed. It refers to the total radiation: thermal plus nonthermal.
Attempts to separate the two contributions, in the spectral range where neither is insignificant compared to the other, have met with some success. This
topic will be discussed in Sec. III.

B.   Polarization of the Microwave Radiation

The integrated nonthermal radiation from Jupiter is partially polarized to
a degree that is large compared to that of most celestial radio sources. The
polarization is generally elliptical with a very flattened ellipse. In fact, the
ellipse degenerates into a line twice during each rotation as the sense changes
from right hand to left-hand and vice versa.

First let us clarify the terms that will be used for the polarization discussion. [For an overview of astronomical polarization the reader is referred to
Gehrels (1974).] The complete state of the radiation can be uniquely described by the four Stokes parameters I, Q, U and V where I is the total

intensity of the radiation and is equivalent to the brightness, B, introduced earlier. In this section we wish to consider the integrated polarization properties so that we require the Stokes parameter flux densities obtained by integrating the Stokes parameters over the extent of the source. These are S_I, S_Q, S_U, S_V, the first identical to the flux density S introduced earlier and the last three specifying the polarization ellipse. For fractional polarization we use S_Q/S_I, S_U/S_I, and S_V/S_I. It is convenient to characterize the ellipse by a circular part and a linear part and thus consider V separately from Q and U. The fractional circular polarization is simply $P_C = S_V/S_I$. For the linear part we usually consider the more familiar parameters P_L (fractional linear polarization) and ψ (position angle of the E-vector) related to S_Q and S_U by $P_L = (S_Q^2 + S_U^2)^{\frac{1}{2}}/S_I$ and $\psi = 1/2 \tan^{-1}(S_U/S_Q)$, taking the proper quadrant into account.

P_L, determined for the nonthermal emission alone, appears to be quite constant at about 0.25 (25% referred to the maxima during a rotation) over the wavelength range 6–50 cm. At shorter wavelengths the dilution by the unpolarized thermal emission becomes large and the determination becomes less certain. At longer wavelengths P_L has not been determined accurately, probably because of problems with confusion by background radio sources and the varying Faraday rotation in our ionosphere. On the average over a rotation, ψ is perpendicular to the rotation axis to the accuracy available, perhaps $\pm 1°$ at best.

At 20-cm wavelength, P_C becomes as large as about 0.01 when the earth is at its largest angular distance from the magnetic equator; the sense depends on which side of the magnetic equator the earth is on. There is evidence that P_C increases slowly with increasing wavelength.

These then are the gross features of the polarization. Other interesting features, such as variations with central meridian longitude are discussed later.

C. Brightness Distribution

In contrast with the thermal emission, where there are almost no data on the angular distribution of emission, a great deal of data exist on the brightness distribution of the nonthermal emission. While these data have been extremely useful for constructing crude models of the radiation belts, there is a real need for additional data in order to make substantial improvements in the models. In particular, brightness and polarization distributions at more wavelengths and with higher angular resolution are needed, together with improved coverage on both long and short time scales.

Only three methods have been feasible for obtaining structural information with adequately high angular resolution at centimeter and decimeter wavelengths: interferometry, lunar occultations, and use of a long, essentially one-dimensional aperture (e.g., the Pulkova telescope). Most of our present information comes from the first of these. The primary interferometric mapping projects that have been undertaken are those of Berge (1965a, 1966) at 10.4 and 21.2 cm, Branson (1968) at 21 and 75 cm, and Olsen (in preparation) at 3.7 and 11 cm. Other high-resolution interferometric studies with less completeness have been carried out. These include Radhakrishnan and Roberts (1960) and Morris and Berge (1962) at 31.3 and 21.2 cm, as already mentioned, Barber (1966, 1967) at 49.2 cm, and Berge (1969) at 3.1 cm.

The maps of Berge (1966) at 10.4 cm and Branson (1968) at 21 cm are shown in Fig. 2. They have been used extensively in constraining theoretical models of the Jovian radiation belt. It is important to know just what the maps represent, especially if comparisons between them are to be made, because there are several differences in what is being shown. The main differences are the different beam smoothing, the different polarization accepted, and the inclusion of the thermal emission in the 21-cm contours, but not in the 10-cm contours.

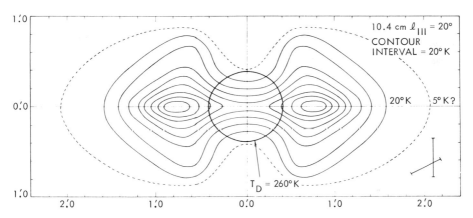

Fig. 2a. Map of the brightness distribution over Jupiter at a wavelength of 10.4 cm (Berge 1966). The contour interval is 20°K; central meridian longitude (System III − 1957.0) is 20°. A disk component of 260°K has been subtracted.

These differences have been discussed in detail by Berge (1972), who converted the maps to the most mutually consistent form possible for purposes of comparison. These are the east-west (virtually equatorial in this particular case) strip scans shown in Fig. 3. In the Cambridge maps of Branson the half-intensity north-south width is almost identical to the half-power north-south width of the synthesized beam. Thus there is apparently little information about the north-south structure of the source, and for purposes of comparison with 10 cm there is little information lost in considering only the equatorial strip scans. The scans of the nonthermal emission representing the accepted E-vector at position angle 90° have a very similar shape in the outer part and may be very similar in the inner part depending on what thermal disk temperature is assumed at 21 cm.

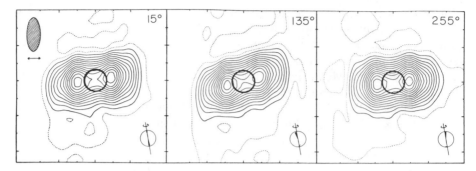

Fig. 2*b*. Maps of the brightness distribution over Jupiter at a wavelength of 21.3 cm (Branson
1968) for central meridian longitudes of 15°, 135°, and 255° respectively. The dark circles
represent the optical limb of the planet. The approximate directions of the magnetic and rotational axes are indicated at the bottom right-hand corner of each map.

The three maps produced by Branson for three different ranges of central
meridian longitude show directly and convincingly that the major axis of
the source, which we would interpret as the magnetic equator, "wobbles"
in position angle, thus supporting the idea that the magnetic and rotational
axes are not parallel. The amplitude and phase are consistent with the variation of E-vector position angle with longitude. Barber (1966) had pointed
out earlier that his interferometer data showed the wobble of the major axis.
He was not able to produce maps to show this directly, but there is no doubt
that the effect was there. In fact, the interferometer data of Berge (1966) also
show the effect of the wobble. In this case, however, the problem was turned
around, and a wobble with a 10° amplitude was assumed as input to the model
fitting procedure.

The 10-cm map, which represents a model fit to interferometer data from
the Owens Valley Radio Observatory, was forced to be symmetric (but
not necessarily centered on the visible disk). The 21-cm maps, which represent a direct Fourier inversion of interferometer data from the one-mile
radio telescope, show a small, but significant, asymmetry in the form of a
hot spot near longitude 200°. For the map centered at CML = 15° the hot
spot is on the far side of the planet and does not distort the contours. It
shows up on the rising limb in the 135° map and on the setting limb in the
255° map.

McAdam (1966) reported measurements of Jupiter at 74 cm with an east-
west beamwidth of 88 arcsec using the east-west arm of the Molonglo Cross.
This instrument provided some resolution of the source, but not a great
deal and only in one dimension. McAdam found that most of the emission
came from a region consistent with the shorter wavelength results, but that
16–20% of it seemed to originate around 6 R_J from the center. However,
the 75-cm results of Branson (1968) obtained with about the same beamwidth
did not reveal this external emission region. In this case the 75-cm distribu-

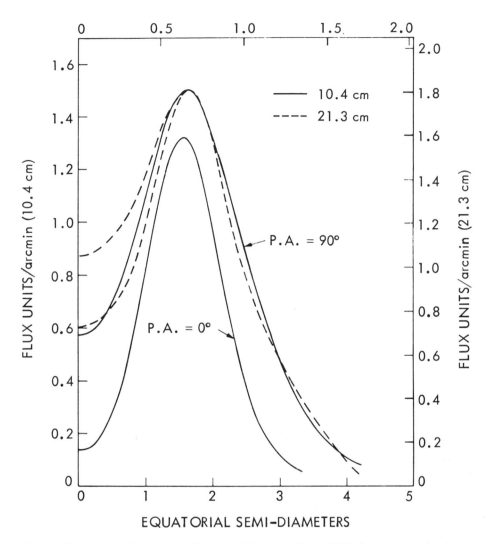

Fig. 3. Equatorial strip scans at 10.4 cm and 21.3 cm (Berge 1972). The emission has been integrated in the polar direction; units are flux density per unit angle. The dashed curves represent the 21.3 cm emission with a 250°K disk removed (upper curve) and a 450°K disk removed (lower curve) as seen by an instrument which is linearly polarized with the E-vector in the east-west direction (*PA* = 90°). The solid curves represent the 10.4 cm emission for two position angles (*PA* = 0° and 90°) with a 260°K disk removed.

tion agreed closely with the 21-cm maps, after taking account of the different beam smoothing, except for an indication of a slightly larger extent at 75 cm, or equivalently, a steeper spectrum in the outer parts.

The interferometer results of Olsen have not yet been published, but the analysis and interpretation are now largely complete. The baseline cover-

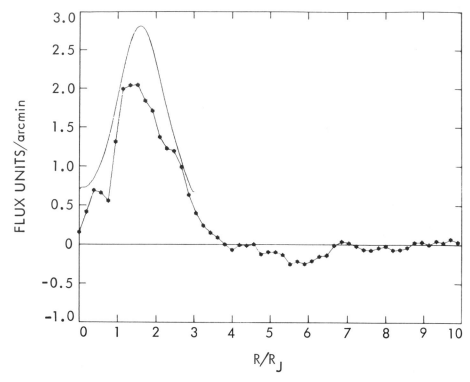

Fig. 4a. Preliminary equatorial strip scan at 3.7 cm (Olsen, in preparation). The scan shows the total brightness I, integrated in the polar direction and averaged over all longitudes. A 212°K limb darkened disk has been removed. Smooth solid curve shows equivalent 10.4 cm scan obtained by adding the two 10.4 cm scans in Fig. 3.

age was rather incomplete in two dimensions, but it was very complete in one dimension parallel to Jupiter's magnetic equator. Therefore Olsen has concentrated on producing strip scans that can be directly compared to those shown in Fig. 3. Two of these are shown in Fig. 4. The first is a strip scan, averaged over all longitudes, at a wavelength of 3.7 cm, and the second is the same thing for 11 cm except that the only data used were from a 90° range of central meridian longitudes centered near 290°. The scans show the total one-dimensional brightness I, and are compared with the equivalent scan at 10.4 cm obtained by adding the E-vector = 0° and E-vector = 90° scans in Fig. 3. There is a very significant difference between the 11-cm and 10.4-cm scans which Olsen interprets as a time variation between 1963.8 (10.4 cm) and 1972.5 (11 cm). As indicated, a separation of the thermal and nonthermal contributions has been made, and the thermal part has already been subtracted from the scans. The interferometer phase data were not sufficiently good at either 3.7 cm or 11 cm to look for asymmetries; thus the scans have been symmetrized by folding at the center and averaging.

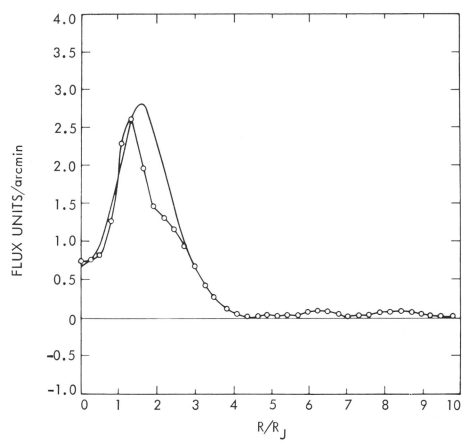

Fig. 4b.   Preliminary equatorial strip scans at 11.1 cm (Olsen, in preparation). The scan shows the total brightness I, integrated in the polar direction and averaged over 90° of longitude centered near 297° (System III − 1957.0). A 280°K limb darkened disk has been removed. Smooth solid curve is equivalent 10.4 cm scan as in Fig. 4a.

Occultations of Jupiter by the moon can, in principle, yield information about the brightness structure of the Jupiter emission. However, in practice, high-quality results are difficult to achieve. The data are generally degraded due to poor signal-to-noise, problems with tracking the limb of the moon, or bad weather. Nevertheless, the technique is useful at longer wavelengths where high angular resolution by other means is more difficult to achieve, and tracking the moon is not so critical. Only two occultation measurements have been reported in the literature: Roberts and Komesaroff (1965) at 74 cm (both disappearance and reappearance) and Gulkis (1970) at 74 cm and 128 cm (disappearance only). The occultations were on April 1, 1962 and October 19, 1968, respectively. Both sets of observations were severely limited by a low signal-to-noise ratio so that detailed interpretation is impossible.

One can say qualitatively, however, that the observations show a double-peaked structure straddling the visible disk with an overall size that appears to be somewhat larger than that found at 10 and 21 cm. There is some evidence for asymmetry, which for the 1968 occultation, at least, was qualitatively similar to that found by Branson.

The 3 m by 122 m east-west transit radio telescope at the Pulkova Observatory has observed Jupiter with moderate angular resolution at 3.02, 3.95, and 6.5 cm (Korol'kov *et al.* 1964; Gol'nev *et al.* 1965, Gol'nev *et al.* 1968; Soboleva and Timofeeva 1970). The best resolution, at 3.02 cm, was 1.1 arcmin and it was correspondingly poorer at the longer wavelengths. The details of the structure were not observable with these resolutions, but the results have been useful in separating the thermal from the nonthermal components. In addition, there was evidence at 3.02 cm that the size of the nonthermal component is less than observed at 10 and 21 cm. At 6.6 cm there was evidence of significant east-west asymmetry of the parameter S_Q.

A topic related to the brightness distribution, and obviously related to the asymmetry in Branson's maps, is the location of the nonthermal emission centroid relative to the center of the visible disk. Berge and Morris (1964) published interferometer phase data that could be interpreted either as a rather large displacement of the polarized emission relative to the thermal emission, or as an effect of circular polarization. They discussed the first possibility in detail, but soon after, Berge (1965*b*) found that there was indeed circular polarization and that the data of Berge and Morris (1964) could be explained by this alone and probably have no bearing on a displacement.

There have been four direct determinations of the radio centroid: two with a pencil beam instrument [Roberts and Ekers (1966) at 11 cm; McCulloch and Komesaroff (1973) at 11 cm] and two with interferometers [Berge (1974) at 21 cm; Stannard and Conway (1976) at 11 cm]. Taken individually, none of them represents a significant detection of a displacement of the radio centroid relative to the visible disk. However, they are all mutually consistent, and taken together they indicate a displacement of about $+0.10\,R_J$ in the polar direction and $\sim 0.08\,R_J$ in the equatorial direction in a longitude plane near 230° (epoch 1975) after removal of the dilution by the thermal emission. Furthermore, these numbers agree with the results of Branson if the centroid displacement is interpreted as due entirely to his observed hot spot (Stannard and Conway 1976).

D. *Rotation-Related Variability*

We have already discussed two apparent variations related to Jupiter's rotation. They are the wobble of the position angle of the major axis and the movement of the hot spot and radio centroid as Jupiter rotates. There are no doubt other apparent variations in the brightness structure related to the rotation, but the available maps are neither accurate enough nor detailed enough to reveal such variations. We can, however, say a good deal about rotation-related variations in the integrated emission.

Figure 5 is an example of the variations that are seen in the integrated emission as Jupiter rotates. The points are unpublished data at 21.1 cm obtained by one of us (GLB) with the interferometer at the Owens Valley Radio Observatory in May of 1974. The interferometer baseline was short so that the instrument did not resolve the source, but rather integrated over the whole angular extent. The top four plots show the complete state of the emission as a function of System III longitude tabulated by Morrison and Meiller (1972). The bottom plot shows the zenomagnetic latitude of the earth, ϕ_M, on the assumption that the magnetic field has axial symmetry and that the magnetic and rotational axes differ by 10° with the magnetic pole in the northern hemisphere at longitude 228°.

The first plot (top), of the flux density, shows the effect of the beaming first suggested by Morris and Berge (1962). This effect was confirmed by Gary (1963) and it has since been observed by many people using a variety of instruments (i.e., Roberts and Komesaroff 1964, 1965; Barber 1966; Roberts and Ekers 1968; Gulkis et al. 1973). The longitude of symmetry of the 1974.4 data is at 228°. It often appears that the peak near 320° is slightly higher than the one near 140°.

The second plot, showing the degree of linear polarization, is very similar in form to the first plot. This means that the integrated net polarized emission is more sharply beamed than the total emission, for if the beaming were proportionally the same then the second plot would show no dependence on longitude. Here again the peak near 320° is higher, but not as much so as observed in 1970 (Berge 1974). Notice that the entire pattern is shifted to earlier longitudes relative to the first plot. In this case the longitude of symmetry is 204°. Many observers have obtained similar data over the years (e.g., Gary 1963; Roberts and Komesaroff 1965; and Berge 1966).

The third plot shows two things: The solid curve is the position angle of the magnetic equator relative to the rotational equator with the same assumptions used for the bottom plot. The data points represent the measured position angle of the E-vector relative to the base level, assumed to be equal to the position angle of the rotational equator. The latter two angles were identical with an uncertainty of $\sim 2°$ due to an uncertainty in the amount of Faraday rotation in our ionosphere and in the position angle calibration. The points have been corrected relative to each other for time variations in the Faraday rotation. Within two years of the discovery of the quasi-sinusoidal variation, Roberts and Komesaroff (1964) presented new data that were accurate enough to display the asymmetric behavior shown in the plot. Other measurements of this variation can be found in Gary (1963), Roberts and Komesaroff (1965), and several others.

The fourth plot shows the variation of P_C with longitude. The data are fitted very well by the curve in the bottom plot if the amplitude is scaled

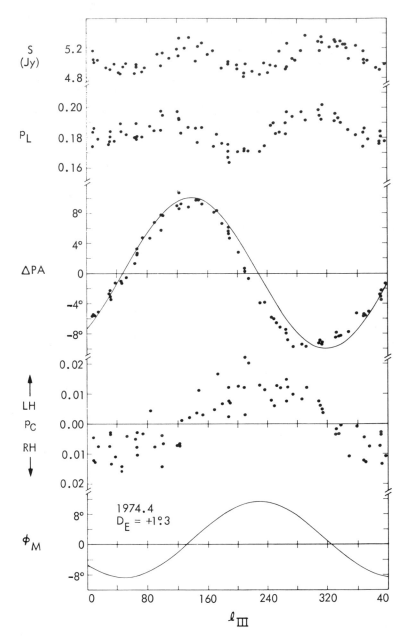

Fig. 5. Variations in the flux density S, degree of linear polarization P_L, change in position angle ΔPA, and degree of circular polarization P_C as a function of System III central meridian longitude. Points are unpublished data at 21.1 cm obtained with the interferometer at the Owens Valley Radio Observatory May 1974 (Berge, unpublished). Bottom plot shows magnetic latitude of the earth, ϕ_M, on the assumption that the Jovian magnetic field is a centered dipole with its axis tilted 10° with respect to the rotation axis and with the magnetic pole in the northern hemisphere at longitude 228°.

to 0.010. Thus P_C is proportional to ϕ_M, to the accuracy available, with the left-hand sense associated with positive magnetic latitude. Observations showing similar results for a variety of wavelengths have been published by Berge (1965*b*), Seaquist (1969), Komesaroff *et al.* (1970), Berge (1974), and Stannard and Conway (1976).

E. Variations Related to the Zenocentric Declination of the Earth (D_E)

There are three variations with D_E that were predicted, and subsequently confirmed, based on the bias that D_E applies to the magnetic latitude ϕ_M. One is a reversal of the deep and shallow minima in the S versus ℓ_{III} relation as D_E goes from $-3°$ to $+3°$ and vice versa (e.g., Gulkis *et al.* 1973). The second is an analogous reversal of minima in the P_L versus ℓ_{III} relation, and the third is a variation in the base level (bias) of P_C versus ℓ_{III}, associated with changes in D_E. These variations, as well as the gross features, at least, of the rotation related variations, are merely apparent variations caused by the change in aspect angle, ϕ_M, rather than intrinsic variations.

Another variation related to D_E emerges in plots of S versus ϕ_M. When D_E is positive, S falls off more slowly at northern latitudes than at southern latitudes (Roberts and Komesaroff 1965; Roberts and Ekers 1968), but when D_E is negative, the reverse is true (Gulkis *et al.* 1973; Berge 1974). When D_E is nearly zero, then S is quite symmetric about $\phi_M = 0$ (e.g., Gulkis *et al.* 1973). This behavior has not yet been explained. Empirically, the variation is such that the magnetic latitude about which S is most symmetric is approximately equal to 0.5 D_E. It is not yet clear whether such an effect also occurs for P_L versus ϕ_M. In this case, the measurement accuracy has usually been poorer, and there are other variations in the shape between different epochs, not yet associated definitely with D_E, that confuse the situation (Berge 1974).

F. Other Time Variations

Variability on other time scales has been reported by several observers, but because of the lack of verification in many cases this has been a rather confusing topic. In many of the early papers that described decimeter observations of Jupiter, the authors considered time variations to be possible or even likely, but not definitely established by the data. Low instrumental sensitivity was a problem, and there were other problems as well. Systematic instrumental effects may have sometimes been unrecognized because of being masked by noise. The potential for confusion effects produced by celestial background emission was sometimes dismissed too lightly. Before the existence and nature of the beaming and polarization were known, they may have produced seemingly anomalous results. Correlations with Jupiter's rotation or with solar activity were sometimes sought. The eventual success in identifying rotation correlations has already been discussed, and correlation with solar activity will be discussed shortly.

When Field (1961) found that the cyclotron mechanism could not accommodate some reported time variations of flux density while satisfying the
other observational constraints, he was using data presented at the General
Assembly of *URSI* (Union Radio Scientifique Internationale) held in London
in 1960. The observations were by Sloanaker and Boland at 10.3 cm in mid-
1958 and October, 1959 (Sloanaker and Boland 1961), by McClain *et al.*
(1962) at 21 cm in May, June, and July of 1959, and by Drake at 68 cm in
May, July, and October of 1959. The first two points at 68 cm were evidently
the same measurements presented by Drake and Hvatum (1959), but revised
considerably. Field also included an unpublished measurement by Drake at
21 cm in April, 1959 and the 31-cm measurement of Roberts and Stanley
(1959) in April–June, 1959.

At all three wavelengths, 10, 21, and 68 cm, the flux density appeared to
decrease over the periods of observation. In retrospect however, the reality
of the variation must be questioned. The 68-cm observations, which showed
the largest decrease, may have been affected by the confusion of background
emission, both extragalactic and galactic. Although Jupiter did not pass
through the galactic plane until 1960, it was close enough to the galactic center during the 1959 observations for the galactic emission to have a brightness temperature of 30–50°K at 68 cm. Since the antenna temperature of
Jupiter (for the 85-foot antenna that was used) was $\lesssim 1°K$, it would have required only small spatial variations in the galactic background to affect the
observations seriously. Confusion effects at 21 and 10 cm would have been
much less severe. The decrease seen at 21-cm wavelength was not very significant in that a constant flux density can be fitted to the data without exceeding the error bars plotted by Field (1961). The decrease at 10 cm between mid-1958 and October 1959 appears to be more significant although
some of it can be explained as an effect of the linear polarization. (The E-
vector of the instrumental response was 67° from Jupiter's equator in 1958
and 79° in 1959.) The flux density measured in 1959 was only slightly larger
than what is expected for the thermal component alone, and thus corresponds
to an almost complete extinction of the nonthermal component. Such an extreme decrease has never been seen subsequently.

In this same context, there are three other sets of observations, made
during this same period or shortly afterwards at a wavelength of 21 cm, that
should be mentioned. Epstein (1959), observing during a 2-week period in
May and June of 1959, reported variations of a factor of 2 in a few hours and
extremes differing by a factor of 5 during the period. Since he was using a
rather small antenna (60-foot diameter), the results can perhaps be attributed
to lack of sensitivity for the short time-scale variations combined with confusion by background emission for the longer time-scale variations. Miller
and Gary (1962) reported on observations made throughout August and September of 1961. The data can be interpreted as showing a slow increase in
flux density at a rate comparable in magnitude to the rate of decrease seen

by McClain *et al.* (1962). However, the significance of the variation was not striking. Roberts (1963) discussed measurements that he made from April through November of 1961. He found a variation in flux density with a maximum-to-minimum ratio, during this period, of 2.5.

Roberts' work leads to a discussion of possible correlations with solar activity. He found what appeared to be a significant correlation, the first that had been detected up to that time. There was a positive correlation, with a time lag of a few days, between the 21-cm Jupiter flux density and the solar indices, sunspot number and 10.7 and 20-cm solar radio emission, that were used. Roberts and Huguenin (1963) extended this work by supplementing the Roberts (1962) observations with all of the other relevant data available at the time and by examining the correlation with the 10.7-cm solar radio emission in more detail. They again found a positive correlation with a time lag of several days for the Roberts (1962) data and for other observations in which the instrumental E-vector had also been approximately parallel to Jupiter's equator. The correlation did not appear, however, in three data sets where the polarization was approximately parallel to Jupiter's axis.

There followed a period of several years in which no variations were observed on time scales other than the rotation period, even though the measurement accuracies had improved (e.g., Gary 1963; Bash *et al.* 1964; Roberts and Komesaroff 1965; Roberts 1965; Berge 1966; Komesaroff and McCulloch 1967; Roberts and Ekers 1968). Interest in possible correlations with solar activity was revived by Gerard (1970). He reported 11.1-cm measurements of Jupiter extending from December 1967 to August 1968 that seemed to show a positive correlation with 10.7-cm solar emission and a phase lag of several days, in qualitative agreement with the results of Roberts and Huguenin (1963). However, the dependence on the instrumental E-vector position angle was opposite that found by Roberts and Huguenin (1963). The maximum variation in flux density throughout the period of observation was 30%.

Klein *et al.* (1972) searched for a similar correlation in Jupiter data that they had obtained at 12.6 cm during the period May through October 1971 (see also Gulkis *et al.* 1973). No correlation was found, and in fact there was no clear detection of any variation, except rotation related variation, over the half year of observation. However, the flux density was 15 to 20% lower than it was during the mid-1960s. They then searched for a long-term correlation with 10.7-cm solar radio emission using Jupiter data from 1963 to 1971. They could not find a convincing correlation of this sort either.

Hence, the existence of a relation between Jupiter's decimeter flux density and solar activity remains in doubt. If a correlation exists, then it appears to occur only at certain times and not at others. Gerard (1974) suggested that a correlation occurs only during periods of solar maximum. This hypothesis explains the Gerard (1970) and Klein *et al.* (1972) data, but is inconsistent with the correlation that Roberts and Huguenin (1963) observed near solar minimum. The existence of variations on time scales of days to a few months, not necessarily associated with solar activity, and perhaps occurring at some times but not at others, cannot be definitely proven or disproven at present.

The long-term decline in the decimeter flux density seen by Klein *et al.* (1972) has been confirmed, and its character examined, by additional measurements (Gulkis *et al.* 1973; Berge 1974; Gerard 1974; Klein 1975; and Klein 1976). The decrease has finally reversed itself, and recently the flux density has been increasing (Klein 1975, 1976). The characteristic "period" of this variation, as well as can be determined from one cycle or less, is at least 14 years, which is longer than both the solar cycle and Jupiter's orbital period. The degree of linear polarization of the nonthermal emission does not exhibit a corresponding variation (Klein 1976).

III. SEPARATION INTO
THERMAL AND NONTHERMAL COMPONENTS

The large overlap in wavelength of the thermal and nonthermal components requires that they must be separated in order to study them independently. Thus far the two spectra, each uncontaminated by the other, have been difficult to determine. There has been notable progress, however, and we examine here the methods that have been used and the results that have been obtained.

One may define nominal wavelength boundaries for the overlap region by specifying the wavelength at which the weaker component is, for example, down to 5% of the total flux density. These boundaries are roughly 1.5–2 cm on the short wavelength side, where the thermal part dominates, and 50–70 cm on the long wavelength side, where the nonthermal part dominates. These numbers are based on a good deal of guesswork because the individual spectra are not known well in the range where the other component is overwhelmingly dominant. They also vary with time because of the long-term variability of the nonthermal flux density. In spite of the large fractional uncertainty in the weaker component near a boundary of the overlap region, the dominant component can be determined without much uncertainty by simply subtracting a rough guess for the weaker component from the total flux density. This is the only separation technique that has yet been possible to determine the thermal spectrum at wavelengths shorter than 2 cm.

The first separation technique we examine, which deals with the problem in the strong overlap region, is based on the polarization properties of the two components. This method makes use of the *a priori* knowledge that the thermal atmospheric emission, S_t, is unpolarized and that all of the observed polarization is nonthermal. A knowledge of the fractional polarization, P_0, of the nonthermal emission, S_{nt}, then allows the separation to be made according to the equations

$$S_t = S(1 - P/P_0),$$
$$S_{nt} = S - S_t$$

(4)

where S and P are the measured flux density and fractional linear polarization, respectively.

Roberts and Komesaroff (1965) showed that for a large range of wavelengths long enough so that the thermal contribution is minor, the emission is essentially constant with wavelength, both in flux density and fractional linear polarization. They found the latter to be 0.22, averaged over all longitudes. If one is willing to assume that the degree of polarization of the nonthermal emission (P_0) remains constant at the same value at shorter wavelengths, then the separation can be made as described above. This procedure has been used by Dickel (1967b) and by Morris et al. (1968) at a wavelength of 6 cm using average values of S and P and assuming $P_0 = 0.22$. Dickel's measured values of $S = 10.8$ Jy and $P = 0.086$ yield $S_{nt} = 4.22$ Jy and $S_t = 6.58$ Jy $\rightarrow T_t = 210 \pm 19°$K where the standard error includes an uncertainty of ± 0.02 in P_0. Morris et al. measured $S = 10.68$ Jy and $P = 0.0763$ to yield $S_{nt} = 3.70$ Jy and $S_t = 6.98$ Jy $\rightarrow T_t = 223 \pm 16°$K.

The same technique can be used on the data of Whiteoak et al. (1969) at 6 cm and Dickel (1967a) at 3.75 cm. Whiteoak et al. obtained $S = 11.22$ Jy and $P = 0.0762$ to give $S_{nt} = 3.89$ Jy and $S_t = 7.33$ Jy, hence $T_t = 233 \pm 15°$K. Dickel measured $S = 17.1$ Jy and $P = 0.056$ to yield $S_{nt} = 4.35$ Jy and $S_t = 12.75$, hence $T_t = 159 \pm 24°$K. The technique could also be applied to data available near 11 cm and 21 cm, but it hardly seems worthwhile because there are measurements with high angular resolution available at these wavelengths.

It is difficult to assess the validity of the assumption that P_0 remains constant at short wavelengths. The values of S_{nt} derived above are all significantly smaller than the longer wavelength values at the same epochs. Thus, S_{nt} and P_0 cannot both remain constant towards shorter wavelengths. These results are consistent with other lines of evidence that indicate a decrease in S_{nt} towards shorter wavelengths. However, if S_{nt} is decreasing because of a decrease in the number of high-energy electrons one might expect that P_0 would change as well. It is clear that the results of this method should be treated with some caution.

We should note that the peak values (during a rotation) of P_0, P, and S are probably better parameters to use than the average values if the observations are sufficiently accurate. For the observations just discussed, however, the accuracy of P and/or S were not sufficient to obtain a good determination of the peak values. In the recent study by Klein (1976), the peak value of P_0 was found to be 0.245 ± 0.010 at 10–13 cm and at 21 cm.

The second method we examine is based on the differences in the angular distributions of the two components. In particular, the thermal component is confined to the disk and the nonthermal component is extended relative to the disk. With sufficiently high resolution the separation is very simple provided a suitable model is used to separate the components where they are superposed (i.e., in the direction of the visible disk).

For example, if the nonthermal emission has azimuthal symmetry about the magnetic axis and is symmetric with respect to the magnetic equator, then the brightness of the nonthermal emission just outside the limb at the equator is twice what it is just inside the limb at the equator. This is a result of the blocking by the planet of the far half of the nonthermal radiation. Just inside the limb the thermal emission contributes to the brightness measured there so that, in terms of brightness temperatures,

$$T_{in} = T_t + \left(\frac{1}{2}\right) T_{out} \tag{5}$$

or

$$T_t = T_{out}\left(\frac{T_{in}}{T_{out}} - \frac{1}{2}\right). \tag{6}$$

If it can be assumed that the thermal emission is uniformly bright across the disk, then T_t is the disk brightness temperature. We are quite certain, however, that there is some limb darkening of the thermal disk; hence the brightness of the thermal component near the limb is not equal to the thermal disk temperature. This can be corrected for, at least crudely, but with a corresponding increase in the value of the uncertainty. We also know that the requirements on symmetry of the nonthermal component are not fully satisfied, but this is not too serious, particularly if measurements are made over the full range of central meridian longitudes.

In practice the diffraction pattern of the radio telescope results in "beam smearing" that degrades the accuracy to which T_{in} and T_{out} can be determined, even with the use of some deconvolution scheme. In fact, for the observations presently available, the resolution is too poor in one, or both, dimensions to allow this determination to be done in any direct fashion. Indirect methods must be used. Eventually we can expect to have high enough angular resolution to utilize the direct method just described; for now, it serves only as a conceptual aid.

One of the indirect methods of separation using data with high angular resolution follows along the lines of the technique developed by Beard and Luthey (1973) for extracting the volume emissivity of the nonthermal component in concentric equatorial shells. These values can then be used to predict, following some simplifying assumptions and convolution with the instrumental response, what the observed strip scan would be if the thermal contribution were zero. Subtracting this from the observed strip scan yields the thermal component. With sufficiently accurate observations the result will contain significant information about the angular structure of the thermal component. This method is essentially the one used by Olsen (in preparation) for separating the two components using his 3.7- and 11-cm data. His results, which may still be subject to minor changes, are at epoch 1972.5:

$$11 \text{ cm:} \quad T_t = 280 \pm 25°\text{K}, \quad S_{nt} = 3.8 \pm 0.4 \text{ Jy} \tag{7}$$
$$\text{(nonthermal is } 60\% \text{ of total)}$$

$$3.7 \text{ cm:} \quad T_t = 212 \pm 10°\text{K}, \quad S_{nt} = 2.6 \pm 0.3 \text{ Jy}. \tag{8}$$
$$\text{(nonthermal is } 13\% \text{ of total)}$$

At 11 cm it was not possible to determine any deviation from a uniform thermal disk. At 3.7 cm, however, there did appear to be significant limb darkening present, although final quantitative results are not yet available.

Another method that has been used is modeling of the nonthermal component. This method has been used by Berge (1966, 1968, 1969) and Branson (1968). In its general form the method consists of constructing a model of the spatial distribution of volume emissivity $J(r, \phi, \ell_{\text{III}})$. The model is specified in such a way that the known geometrical and physical constraints are satisfied. In practice J is assumed to be independent of ℓ_{III} (azimuthal symmetry). Further simplifying assumptions may be made to make J a function of r only. Then the free parameters, if any, are determined by requiring the best fit possible to the observed angular distribution of brightness away from the center where there will not be any contamination from the thermal component. Once the run of emissivities has been found, the method proceeds in the same way as the previous method.

Berge (1966) used a form of this technique to separate his 10.4-cm data (epoch = 1963.8) and found that the thermal disk temperature was considerably higher than at shorter wavelengths. At that time the result was given in the form of a lower limit ($T_t = 260°\text{K}$) without a quantitative upper bound. Berge (1968), after some additional analysis of the same data, found that the thermal disk temperature was $T_t = 290 \pm 30°\text{K}$. This left 4.3 ± 0.5 Jy for the nonthermal component. Further analysis of the 21.2-cm data from 1964.0 led to an upper limit of 450°K for the thermal disk temperature, leaving $> 5.3 \pm 0.5$ Jy for the nonthermal component. A separation analysis was also performed on interferometer data at 3.12-cm wavelength taken in January 1967. The thermal disk temperature was found to be $196 \pm 21°\text{K}$ (Berge 1968) and the nonthermal contribution 2.5 ± 1.0 Jy (Berge 1969). The separation analysis of all three sets of data (3.12, 10.4, 21.2 cm) was aided by having complete observations available for each of two linear polarizations: E-vector of the instrumental response parallel to the axis and parallel to the equator. The angular structure of the nonthermal, but not the thermal, emission is very different for the two responses.

Branson (1968) used modeling of a sort to separate the two components for his 21-cm data (epoch 1967.2). From consideration of a toroidal radiation belt he deduced that, prior to the instrumental beam smoothing, the brightness of the nonthermal emission just beyond the equatorial limb should be twice the nonthermal brightness at the disk center. After allowing for the beam smoothing he determined a thermal disk temperature of $250 \pm 40°K$. The correction for beam smoothing is dependent on details of the model, but this difficulty is presumably reflected in the quoted uncertainty. Branson's result is numerically consistent with the thermal disk temperatures found at 10.4 cm and 11 cm, but it does not continue the upward trend with increasing wavelength that might be expected from a qualitative extrapolation from shorter wavelengths.

Berge (1968) noted two corrections that can be made to Branson's analysis, both involving the deduced brightness ratio of 2. As pointed out earlier the nonthermal brightness at a point just outside the equatorial limb is twice that of a point just inside. However, the brightness ratio between a point just beyond the equatorial limb and the disk center is greater than 2. Just how much greater is model-dependent. In addition one must take account of the effect of emission at high latitudes because the separation was done on measurements in the form of a strip scan with resolution only in the equatorial direction. Even though the high-latitude emission is polarized predominantly with the E-vector perpendicular to the instrumental E-vector response, it is not completely polarized and will make a significant contribution to the strip scan, particularly beyond the equatorial limb, while contributing virtually nothing at the disk center. This too causes the brightness ratio to be larger than 2. Increasing the expected ratio, as both of these corrections do, causes the derived thermal disk temperature to be raised. After including these effects and allowing for beam smoothing, Berge obtained $T_t = 400 \pm 75°K$ from Branson's strip scan leaving 5.6 ± 0.5 Jy for the nonthermal contribution. Although the result for T_t is model dependent, the given uncertainty includes the effect of uncertainties in the model.

Jupiter has been observed with a fan beam instrument at the Pulkova Observatory at 3.02 cm (Korol'kov et al. 1964), 3.95 cm (Soboleva and Timofeeva 1970), and 6.5 cm (Gol'nev et al. 1965; Gol'nev et al. 1968). These authors used model fitting, including some polarization considerations, to separate the thermal and nonthermal components. Their angular resolution was inferior to the other high-resolution observations that have been discussed in this section so accuracy was a problem. In the spectral range covered, the thermal component dominates, and while the fractional uncertainty of the thermal disk temperatures determined is tolerable, it is much worse for the nonthermal component. Here we will give only the thermal results as they appeared in graphical, and presumably final, form in the latest of the papers (Soboleva and Timofeeva 1970): $T_t = 167 \, ^{+30}_{-12}°K$ at 3.02 cm, $198 \, ^{+30}_{-10}°K$ at 3.95 cm, and $205 \pm 15°K$ at 6.5 cm.

A third distinctly different technique was used by Degioanni (1973) and Degioanni and Dickel (1974) to carry out the separation of the thermal and nonthermal components. Model fitting for the nonthermal component was used, but the model was not one of a simple volume emissivity distribution as discussed above. Rather it was a more basic model that used the guiding center approximation for electron motions and synchrotron emission theory to calculate the radiation. Free parameters to be fitted, plus a number of *a priori* assumptions, characterized the electron density, energy, and pitch angle distributions. The model fit relied primarily on observed integrated properties of the emission; high-resolution data played only a minor role.

After the final model was determined, its calculated frequency spectrum was subtracted from the observed one to yield the spectrum of the thermal component. For example, the thermal disk temperature was found to be 180°K at 10-cm wavelength. In general this separation is discordant with most of the other determinations discussed in this section in that the thermal contribution is lower and the nonthermal contribution is higher than for the others.

The model may give some useful insight into, and order-of-magnitude estimates for, the characteristics of the nonthermal emission, but for reasons that follow we must give low weight to some of the detailed results, including the separation of the two components. The model, as initially set up, was clearly over-specified. This was necessary to obtain definite numerical results with the limited number of observational constraints that were applied. Thus the uniqueness of the model is questionable. Some of the available constraints were not applied, particularly those obtained from high-resolution observations. For example, the brightness distribution that would be calculated from the fitted model is probably considerably smaller in extent along the polar direction than was determined observationally by Berge (1966). Finally, there is at least one instance where an observational constraint that was employed (the fractional linear polarization) differs markedly from what more recent analyses of the data show. The model gives a fractional linear polarization of the nonthermal component of 0.33 at 21 cm and 0.32 at 10 cm for the peak values during a rotation. These reflect almost exactly the values used in the model fitting. However, Klein (1976) finds a value of only 0.245 ± 0.010 for this wavelength range for the past 12 years even though he has made a much larger upward adjustment to correct for dilution by the thermal component than Degioanni and Dickel would require.

It is possible to devise other methods for performing this separation using low-resolution data. They have not been considered in the literature, but we mention them here for the sake of completeness. They each require an assumption that may be more suspect than any which have been used to date.

The first of these assumes that the fractional amplitude of the variation with rotation of the nonthermal flux density is independent of wavelength. That is, $A_0 = \Delta S_{nt} / S_{nt} \neq F(\lambda)$. We require that A_0 be known from measurements at longer wavelengths and we assume that the thermal emission does not vary with rotation so that $\Delta S_{nt} = \Delta S$. Because the shape of the beaming curve varies with D_E and because of its asymmetry with respect to $\phi_M = 0$, we must use care in defining ΔS. A good choice might be to use the maximum value of S, which comes at $\phi_M = \phi_0$, and define $\Delta S = S(\phi_0) - S(\phi_{\pm 7})$ where $S(\phi_{\pm 7})$ is the average of $S(\phi_0 \pm 7°)$. We use the measured quantity $A = \Delta S / S$ to find

$$S_t = S(1 - A/A_0).   \tag{9}$$

This method is just like the polarization method except that A and A_0 are used instead of P and P_0.

The next two methods make use of the large long-term variation in flux density, assumed to be due to a variation of the nonthermal component only. In one method we assume that P_0 is independent of time. It need not be independent of wavelength as assumed earlier. We measure the quantities P and S at two epochs, denoted by the subscripts 1 and 2, between which there has been a significant change in S. Then since

$$P_0 = \frac{P_1 S_1}{S_1 - S_t} = \frac{P_2 S_2}{S_2 - S_t}   \tag{10}$$

we can solve

$$S_t = \frac{S_1 S_2 (P_1 - P_2)}{S_1 P_1 - S_2 P_2}.   \tag{11}$$

Since P_0 does not appear in the result, we have the advantage of not having to assume its value in advance. In fact, we can now solve for it and check the agreement with what we think it should be.

The other method is similar except the amplitude of the variation, A_0, is assumed to be independent of time. Since

$$A_0 = \frac{\Delta S_1}{S_1 - S_t} = \frac{\Delta S_2}{S_2 - S_t}   \tag{12}$$

we can solve for

$$S_t = \frac{S_2 \Delta S_1 - S_1 \Delta S_2}{\Delta S_1 - \Delta S_2}.   \tag{13}$$

We can then use the result to determine A_0 rather than having to assume its value in advance.

In principle, there are still other analogous methods of separation which involve assumptions about the constancy of specific parameters of the radiation.

IV. INTERPRETATION OF THE NONTHERMAL COMPONENT

We have seen how the discovery that the decimeter emission is extended in size and is polarized (Radhakrishnan and Roberts 1960) left only two contenders, cyclotron emission and synchrotron emission, as likely emission mechanisms. Detailed predictions for these mechanisms were developed by

Field (1960, 1961) and Chang and Davis (1962). Both mechanisms required the presence of a magnetic field, but this was already rather well established as necessary for the interpretation of the polarized decameter emission. In addition the synchrotron mechanism also required relativistic electrons, and their presence in the Jupiter magnetosphere, implied by the acceptance of the synchrotron process, was a major discovery.

Probably the most compelling reasons which argued for synchrotron emission developed from observations made in 1961 and later. Prior to this time, however, Field (1961) concluded on the basis of the spectrum and the reported time variations in the emission, that the cyclotron model was inadequate to account for the observations. As mentioned previously, the validity of the large intensity variations that had been reported at that time has never been established. Later interferometric measurements by Morris and Berge (1962) at 960 MHz and 1390 MHz yielded the flux density, polarization, and the overall dimensions of the source in both the polar and equatorial direction. They found no significant frequency dependence of any of the measured parameters. They argued that a cyclotron emission source with the observed parameters would have been measurably frequency dependent, in contrast to their observations. Since 1962 there has been no serious challenge to the belief that it is synchrotron emission that we observe.

A. Synchrotron Emission

Synchrotron emission is an important mechanism in astrophysics, and a great deal of theoretical work has been done on the subject (e.g., Ginzburg and Syrovatskii 1965, 1969; Legg and Westfold 1968; and references therein to earlier work). Synchrotron emission in a dipole magnetic field was investigated theoretically by Thorne (1963). Other work on synchrotron emission in a dipole field with particular application to Jupiter has been done by Chang and Davis (1962), Chang (1962), Thorne (1965), Ortwein et al. (1966), Clarke (1970), Gleeson et al. (1969, 1970) and others.

Synchrotron emission, like cyclotron emission, is produced as the result of the acceleration that a moving charge experiences as it moves along a helical path in a magnetic field. It is reasonable to assume that electrons are the charged particles involved because their low mass makes them particularly effective in producing emission. Relativistic effects cause synchrotron emission to be much different in character from cyclotron emission. The emission from a single relativistic electron is sharply beamed, being confined to a cone of apex angle $\epsilon \approx mc^2/E = 0.511/E_{\text{MeV}}$ whose axis at any instant is in the direction of the velocity vector.

An observer in the path of the beam sees a broad frequency spectrum with a maximum at $\nu_{max} \approx 0.3 \, \nu_c$ where ν_c, the critical frequency, is given by

$$\nu_c = 16.08 \, E^2 \, B \sin \alpha \qquad (14)$$

in MHz, where B is in Gauss, E in MeV, and α is the pitch angle of helical orbit (90° in circular orbit). Since E is much larger than the rest mass energy (0.511 MeV), ν_c is much larger than both the relativistic gyrofrequency ν_L ($\nu_c = 11.2\,\nu_L\,E^3\sin\alpha$) and the nonrelativistic gyrofrequency ν_G ($\nu_c = 5.74\,\nu_G E^2\sin\alpha$). If θ is defined as the angle between the direction to the observer and the direction of the magnetic field, then in order for the observer to receive significant emission, the angle θ must lie in the range $\alpha - \epsilon/2 < \theta < \alpha + \epsilon/2$. An approximation often made is that $\theta = \alpha$ so that $B\sin\alpha$ is the magnitude of the component of the magnetic field perpendicular to the line of sight.

The observed emission is elliptically polarized with the major axis of the ellipse normal to the direction of the component of the magnetic field perpendicular to the line of sight. The sense or handedness of the ellipse depends on whether $\theta < \alpha$ or $\theta > \alpha$. The polarization ellipse is often characterized by its decomposition into a linear component, specified by Q and U, and a circular component V.

For a large assembly of relativistic electrons, the observed emission characteristics are determined uniquely by the magnetic field structure (intensity and direction at each point) and by the electron density, energy distribution, and pitch angle distribution at each point. It is not possible in general, to turn this around and determine all source parameters uniquely from the observed characteristics, but certain simplified cases (models) can be considered to see if their predictions agree with the observations. The models for celestial radio sources usually assume simple functional forms for the electron distributions with a magnetic field that is, for example, uniform, ordered (such as a dipole field), or random.

Before proceeding to the particular case of Jupiter, it is instructive to consider synchrotron emission from an assembly of relativistic electrons in a uniform magnetic field as seen by a distant observer. Since the plane of polarization is the same for all of the electrons (perpendicular to the projection of the magnetic field onto the plane of the sky), the linearly polarized components will add in concert to give a large degree of linear polarization for the integrated emission. If all of the electrons have the same pitch angle, then the circularly polarized components also add in concert. However, if there is a smooth pitch angle distribution that is wider than the emission cones, the right and left-hand senses tend to cancel, resulting in a very small degree of circular polarization for the integrated emission, even though the linear polarization is large. As shown by Legg and Westfold (1968) the circular polarization does not go completely to zero even for an isotropic pitch angle distribution. A circular polarization larger than this residual amount can be produced by an anisotropy in the distribution. If we now allow the magnetic field to be nonuniform in direction, then both the linear and circular polarizations will be diminished.

B. *Application to Jupiter*

The various theoretical studies of Jupiter's nonthermal decimeter emission have all assumed that the magnetic field structure is a dipole. To a first approximation, the dipole assumption is supported by the observed shape of the emission region and by the distribution of the plane of polarization over the source as pointed out by Berge (1966). With a dipole magnetic field as a starting point, the model of Jupiter's emission is already constrained in many respects. Even without detailed modeling, several generalizations can be drawn because the magnetic field vector is specified at every point, except for a scale factor (the magnetic moment), and the motion of electrons in the field is well known. The basic motion is essentially helical and guided along a particular fieldline. Since the field intensity varies along the fieldline, increasing monotonically from its value at the magnetic equator, the pitch angle increases from its value at the equator, α_e, until it equals $90°$ (satisfying the adiabatic invariant $B^{-1} \sin^2 \alpha = B_e^{-1} \sin^2 \alpha_e$), at which point the motion along the fieldline reverses direction. This results in a bounce motion between mirror points at equal latitudes north and south of the equator, assuring symmetry of the electron distribution with respect to the magnetic equator. There is also a relatively slow drift in azimuth due to the radial gradient in the magnetic field that assures azimuthal symmetry. There may also be a very slow radial drift, that is, L-shell diffusion. Once the electron distribution, $N(r_e, E, \alpha_e)dE \, d\alpha$, is fixed for any azimuth on the magnetic equator for all radial distances, r_e, it is fixed uniquely everywhere in the dipole field.

One can show by qualitative arguments, involving the symmetry of the field and the electron distribution, that the linear polarization, integrated over the source, will have its E-vector either parallel or perpendicular to the magnetic axis, depending on the electron distribution. The early discovery that the plane of polarization is approximately perpendicular to the rotation axis thus showed that the magnetic axis is either approximately parallel or perpendicular to the rotation axis. The latter case is ruled out, however, because it would result in the degree of polarization varying strongly with rotation, approaching zero when the magnetic axis is approximately along the line of sight. The further discovery that the plane of polarization varies with rotation by $\pm 10°$ in a roughly sinusoidal fashion implies that the projection of the magnetic axis onto the plane of the sky varies similarly. The simplest explanation is that the magnetic axis and rotation axis differ in direction by $10°$. This tilt is not only interesting in itself, but it has proven to be a powerful tool in investigating the nature of the radio source because it allows observations of the source over a limited range of aspect angles.

Before proceeding with this topic, however, we should digress briefly to the subject of the energy distribution of the relativistic electrons. This, of course, is an important part of the physics of the Jovian magnetosphere. In addition, some of the further interpretation to be discussed depends on the energy distribution. A well-known theoretical result is that if the energy distribution of the electrons is a power law of the form $N(E) \, dE = KE^\gamma \, dE$, then the emission spectrum is also a power law of the form $S = Cv^\alpha$ where $\alpha = (\gamma + 1)/2$ and C is a constant. This is an approximation that breaks down if γ is close to zero, but is quite good if γ is not close to zero and if the observed frequency range is well within the range specified by the critical frequencies of the lower and upper cutoff energies of the energy distribution. Since cosmic ray electrons display a characteristic power law energy distribution, and since the spectra of celestial synchrotron emission sources is usually a power law, this result is commonly used to infer that the electron energy distribution in a source is $N(E) \, dE = KE^{2\alpha - 1} \, dE$. When applied to Jupiter, where $\alpha \approx 0$, one finds that $N(E) = KE^{-1} \, dE$.

Various reservations have been expressed over the years regarding the validity of this distribution for Jupiter (e.g., Thorne 1965; Berge 1969; Gleeson et al. 1970; Gulkis 1970). This is because of the complication that arises when there are cutoffs in the energy distribution such that there are no electrons with $E < E_1$ or $E > E_2$, and the power law distribution holds only for $E_1 < E < E_2$. In this case the predicted power law emission spectrum is produced only over the frequency range $3v_{c1} \lesssim v \lesssim 0.1 \, v_{c2}$ where $v_{c1} = 16.08 \, E_1^2 \, B \sin \theta$ MHz and $v_{c2} = 16.08 \, E_2^2 \, B \sin \theta$ MHz. When v_{c1} and v_{c2} vary from place to place throughout the emission region, as would be true for Jupiter, then the relevant frequency range is $3v_{c1} \, (max) \lesssim v \lesssim 0.1 \, v_{c2} \, (min)$. This is a strong condition and there may, in fact, be no frequency range that satisfies it. Then the distribution of $B \sin \theta$ becomes important in determining the resulting emission spectrum as demonstrated by Hoyle and Burbidge (1966). For the lack of something better, the distribution found in the previous paragraph has often been used, but it should be treated cautiously.

The other characteristic of the nonthermal spectrum that can be exploited is the falloff in flux density at short wavelengths beginning in the 10–20 cm range. The interpretation is that there is a deficiency of high-energy electrons, either because of an upper energy cutoff or because $\gamma < -1$, at least at the highest energies. Berge (1969) found that, under the assumptions of a sharp upper cutoff, a uniform magnetic field, and $\gamma = -1$, the best fit to the data occurred for $E_2 = 32/\sqrt{B}$ MeV. Further work on the effect of energy cutoffs was carried out by Gulkis (1970). Gleeson et al. (1970) investigated several combinations of γ, E_1, E_2, and $B_e(3 \, R_J)$ in a model of a thin dipole shell source that could lead to the observed spectrum, including the short wavelength falloff (see also Gleeson et al. 1974).

Returning now to the use of the tilt between the magnetic and rotation axes, we note that the relevant aspect angle is the magnetic latitude ϕ_M of the earth, that is, the angle between the magnetic equator and the direction to Earth. It is given to a close approximation by

$$\phi_M = D_E + \beta \cos (\ell_{\mathrm{III}} - \ell_0) \tag{15}$$

where D_E is the declination of Earth relative to Jupiter's rotational equator $(-3° \lesssim D_E \lesssim 3°)$, $\beta \approx 10°$, $\ell_{\mathrm{III}} = \mathrm{CML}$, and ℓ_0 is the longitude that is at the central meridian when the magnetic axis is tilted towards the earth at the north. For a centered dipole, ℓ_0 would be the longitude of the northern end of the dipole axis. D_E goes through its full cycle in 12 years while ℓ_{III} does so in 10 hours. Thus on any given day we can cover a range in ϕ_M of 20° $(D_E \pm 10°)$ and over a period of many years we can see the range $-13°$ to $+13°$.

It was recognized by Davis and Chang (1961) that to produce the observed linear polarization, the pitch angle distribution of the electrons could not be isotropic; the equatorial pitch angles would have to cluster closer to 90° than for an isotropic distribution (isotropic: $N(\alpha)\, d\alpha = k \sin \alpha\, d\alpha$). Roberts and Komesaroff (1965) carried this a step further by also considering the variation of flux density and linear polarization with ϕ_M. To find a pitch angle distribution that would fit the data they utilized theoretical calculations of Thorne (1963) for synchrotron emission by stars. Thorne (1965), using the same data, modified the Roberts and Komesaroff results with theoretical calculations that were more applicable to Jupiter. The assumed form of the distribution was a sum of terms $X_q \sin^q \alpha$. The model used was an L-shell of negligible thickness with $r_e = 3\ R_J$ and $\gamma = -1$ was used as explained earlier. Emission shadowed by the planet was removed, and equatorial pitch angles resulting in mirror points below the top of the atmosphere were not allowed. Except for the latter two conditions the functional form inferred for the pitch angle distribution is independent of r_e and would be the same for a thick shell.

To explain the 0.22 degree of linear polarization alone required $N(\alpha) \sim \sin^3 \alpha$. However, the observed beaming required a component with much flatter pitch angles. The 21-cm data used were $S(\phi_M = 13°)/S(\phi_M = 0°) = 0.89$ and degrees of polarization of 0.22 and 0.18 at $\phi_M = 0°$ and 13° respectively. The required distribution was found to be $N(\alpha) \sim \sin^2 \alpha + 2.0 \sin^{40} \alpha$. At the time it was thought that there was little or no beaming at 74-cm wavelength so that $N(\alpha) \sim \sin^3 \alpha$ for the electrons emitting at that wavelength. However, it has since been found (e.g., Roberts and Ekers 1968; Tiberi 1966) that the beaming characteristics are, in fact, very similar at the longer wavelengths.

Clarke (1970) extended this work by including circular polarization and by considering thick emission shells as well. He found that the inferred pitch angle distribution is only weakly dependent on the size (r_e) of the L-shell so that going to a thick-shell model has little effect in this respect. Clarke's results for the pitch angle distribution agreed very well with Thorne's.

Since the tilt of the magnetic axis leads to several rotation-related variations, it provides a direct method for determining the rotation period of the magnetic field. With an approximate period and variation observations at two epochs, one can determine the improved rotation period necessary to align the two variation curves. Of the four variations shown in Fig. 5, the first and third (S versus ℓ_{III} and $\Delta\,PA$ versus ℓ_{III}) have been used for determining the period.

The existing data for P_C versus ℓ_{III} have too much scatter for an accurate determination. The same is usually true for the P_L versus ℓ_{III} data, but a more serious problem is that this variation changes shape between different epochs, making the alignment uncertain. The S versus ℓ_{III} variation also changes shape, but in a more predictable fashion, and it is usually assumed that the longitudes of the minima or the longitudes about which there is greatest symmetry are good reference points for alignment. The $\Delta\,PA$ versus ℓ_{III} variation, despite its deviation from a sinusoid, maintains its shape rather well and can be aligned without much difficulty. Table II lists the various published determinations of decimetric rotation period including which variation (S or PA) was used. For the first entry, the variation used was a combination of effects produced by using a linearly polarized response at a fixed position angle; consequently, the significance of the published uncertainty is difficult to assess. It is important to avoid averaging the periods shown or otherwise treating them as independent because many of them were determined with some data in common with others.

The question of the constancy of the rotation rate of Jupiter is an interesting one, and one which can best be answered by continued Earth-based

TABLE II
Decimetric Determination of Jupiter's Rotation Period

λ	Variation	Period	Reference
10 cm	Comb.	$9^h\ 55^m\ 29\overset{s}{.}70 \pm 0\overset{s}{.}05$	Bash, Drake, Gundermann, and Heiles (1964)
21	*PA*	29.37 ± 0.5	Roberts and Komesaroff (1965) and Berge (1966)
21	*S*	29.50 ± 0.29	Davies and Williams (1966)
11	*PA*	29.83 ± 0.26	Komesaroff and McCulloch (1967)
6	*PA*	29.69 ± 0.05	Whiteoak, Gardner, and Morris (1969)
13	*S*	29.72 ± 0.11	Gulkis and Gary (1971)
13	*S*	29.75 ± 0.05	Gulkis, Gary, Klein, and Stelzried (1973)
21	*S*	29.72 ± 0.07	Berge (1974)
11	*PA*	29.76 ± 0.02	Komesaroff and McCulloch (1976)

observations. Changes in the observed radio rotation period can be caused not only by changes in the bulk rotation period, but also by such things as a wandering in longitude of the magnetic pole. It is important to search for variations in the rotation rate since they provide clues on the interior structure of Jupiter. Variations in magnetic field similarly have important implications. There are irregular fluctuations in the length of the terrestrial day, amounting to a few parts in 10^8. These fluctuations are believed to be due to a redistribution of angular momentum within the earth. Whether or not there are similar changes in the rotation of Jupiter is not known. The current estimated uncertainties in the decameter and decimeter rotation periods of Jupiter are approximately one part in 10^6.

We now turn to the subject of circular polarization. That the Jupiter emission might exhibit measurable circular polarization, due to the relatively large magnetic field intensity estimated from the decameter observations and the very large anisotropy in the pitch angle distribution, was anticipated by Roberts and Komesaroff (1965). They derived the approximate result for a uniform magnetic field, namely that

$$P_C = A \frac{N'(\theta)}{N(\theta)} \left(\frac{B \sin \theta}{\nu} \right)^{\frac{1}{2}} \tag{16}$$

where A depends on the energy distribution, and $N(\theta) = N(\alpha)$ evaluated at $\alpha = \theta$. They failed, however, to detect circular polarization. Berge (1965b) used his circular polarization measurements, this theoretical result, and Thorne's pitch angle distribution to solve for the "average" field intensity in the radiating region which was found to be 0.17 Gauss $< \bar{B} < 17$ Gauss. The large uncertainty was due in large part to the fact that the field was not uniform, and model calculations, like those of Clarke for circular polarization in a dipole field, were not available. In a dipole field the magnetic co-latitude $90° - \phi_M$ takes on the role of θ and the constant A becomes some function of $|\phi_M|$. Consider a simplified example in which a uniform magnetic field in the direction of Jupiter's magnetic axis is populated by energetic electrons having the pitch angle distribution described by Thorne (1965). Then one would expect that P_C is approximately proportional to ϕ_M (or $- \phi_M$, depending on the sense of the field) for ϕ_M small such as the $\pm 13°$ range that can be seen from Earth. In fact, to the accuracy currently available this proportionality holds, even though the field is a dipole. The association of left-hand sense with positive magnetic latitude can be shown to indicate that the magnetic field in the northern hemisphere has a north magnetic pole (Berge 1965b).

Legg and Westfold (1968) derived the circular polarization in detail for the case of a uniform magnetic field. They found that for an isotropic pitch angle distribution (in which case $N'(\theta)/N(\theta) = \cot \theta$)

$$P_C \sim \left(\frac{B \sin \theta}{\nu} \right)^{\frac{1}{2}} \cot \theta . \tag{17}$$

If the pitch angle distribution is not isotropic then there is another term, also proportional to $(B \sin \theta / \nu)^{\frac{1}{3}}$, that has the factor $(N'(\theta)/N(\theta) - \cot \theta)$. Legg and Westfold also calculated the relevant coefficients for the case of mono-energetic electrons and for the case of a power-law energy distribution. The second term strongly dominates the first for Thorne's (or Clarke's) pitch angle distribution.

Komesaroff et al. (1970) used new circular polarization measurements at 21 cm, together with model calculations by Clarke (1970) and Gleeson et al. (1969) to infer that 0.4 Gauss $< B <$ 1.9 Gauss in the emitting region. The models were for $r_e = 3\,R_J$, but if the results are valid for $r_e = 2\,R_J$, where the bulk of the emission arises, then the equatorial surface field intensity is between 3 and 15 Gauss.

Circular polarization measurements by Seaquist (1969), Komesaroff et al. (1970), Berge (1974), Stannard and Conway (1976), and in Fig. 5 of this chapter cover the spectrum from 9 cm to 48 cm. The measurements agree where they can be compared, but they do not agree with those of Berge (1965b), which are anomalously high by a factor of 2 or 3. As noted by Komesaroff et al., the frequency dependence of P_C after removing the thermal component is not $\nu^{-0.5}$ as predicted but is more like $\nu^{-0.2}$. In other words, there is a relative deficiency of circular polarization at longer wavelengths.

So far it has been tacitly assumed, for simplicity, that the electron distribution is separable: $N(r_e, E, \alpha_e) = N(r_e)\, P_1(E)\, P_2(\alpha_e)$ where N is number density and P_1 and P_2 are probability distributions. To explain the anomalous frequency dependence of P_C, however, Komesaroff et al. suggested that the pitch angle distribution is correlated with electron energy. If the lower energy electrons, which contribute proportionately more to the longer wavelength emission, have their pitch angles less sharply distributed about $\alpha = 90°$ than do the higher-energy electrons, then there will be less circular polarization than expected at longer wavelengths. This possible correlation has not been investigated quantitatively from a theoretical standpoint, and it is not known whether the resulting decrease of beaming with wavelength would be consistent with the observations.

There is another possible explanation for the frequency dependence of P_C that has not previously been suggested. If the size of the emission region increases slowly with increasing wavelength, a circumstance that has some observational support (Branson 1968; Gulkis 1970) and is theoretically reasonable, then one is sampling a smaller magnetic field intensity, on the average, at longer wavelengths. This means that the frequency dependence of P_C is modified in the right direction to explain the observations. If the emission region were a thin magnetic shell of equatorial radius r_e, then a size dependence of $r_e \sim \nu^{-0.2}$ would result in $P_C \sim \nu^{-0.2}$ for a dipole field.

To complete our consideration of the relativistic electron distribution we now turn to the radial distribution of the number density $N(r_e)$. Although this is difficult to obtain from the decimeter radio observations, many attempts have been made because there are no other types of Earth-based data that are applicable. Several techniques, of various levels of sophistication, utilizing a variety of assumptions have been employed. Divine (1972) describes and summarizes the results of those studies done up to 1971 (Barber and Gower 1965; Branson 1968; Carr and Gulkis 1969; Chang and Davis 1962; Eggen 1967; Haffner 1969; Klopp 1972; Koepp-Baker 1968; Luthey and Beard 1970; Thomas 1967; and Warwick 1970). Divine's plot of the results is shown in Fig. 6. The quantity plotted is the relativistic electron flux $\phi(r_e)$, but since the electrons are all moving at close to the speed of light, the conversion $N(r_e) \approx c^{-1} \phi(r_e)$ can be applied.

In addition, there have been subsequent studies (e.g., Beard and Luthey 1973; Birmingham *et al.* 1974; and Stansberry and White 1974), that are of interest. The latter two both started by constructing a model of the electron radiation belt that included energy gained by radial diffusion inward and loss by synchrotron radiation, and they then fitted to observational data in order to estimate the unknown parameters. Although the assumed models were similar, the fitting techniques were different, and Stansberry and White fitted to a much more complete set of data. Of the various conclusions reached, we mention here only two by Stansberry and White: 1) $\phi(r_e)$ reaches the rather high peak value of 1.4×10^9 cm^{-2} sec^{-1} at $r_e = 2.7\ R_J$ where the characteristic energy is 6.5 MeV, and falls off to 10^8 cm^{-2} sec^{-1} at $r_e = 1.5\ R_J$. 2) To achieve an acceptable fit, an additional but unidentified loss mechanism is required close to the planet.

Related to the second point is the fact that modeling the radial distribution of the volume emissivity to fit the observations, as discussed in the previous section, invariably requires a very low emissivity within about $1.5\ R_J$ in order to match the position and width of the emission peak. Furthermore, Beard and Luthey (1973), using their direct method for determining the radial distribution of the volume emissivity, find that it decreases very fast within about $1.6\ R_J$. If the magnetic field structure becomes more complex than a dipole in the inner regions (as revealed by the presence of higher-order multipole moments), then the electron motion may become complex there compared to the relatively simple motion described earlier for a dipole field. It may be that energetic electrons that have diffused into within, let us say $r_e \approx 1.5\ R_J$, find themselves guided into the atmosphere by this complex field, and thus lost to the radiation belt, on a time scale shorter than the radiative lifetime.

We have already discussed time variations resulting from the variation of magnetic latitude ϕ_M and the value of being able to observe over a modest range of aspect angle. Another topic for discussion would be variations on an intermediate time scale of days to months, perhaps correlated with solar activity. Although such variations have been reported (e.g., Roberts and Huguenin 1963; Gerard 1970), there has been no convincing confirmation, even though they should have been observed in some cases by other

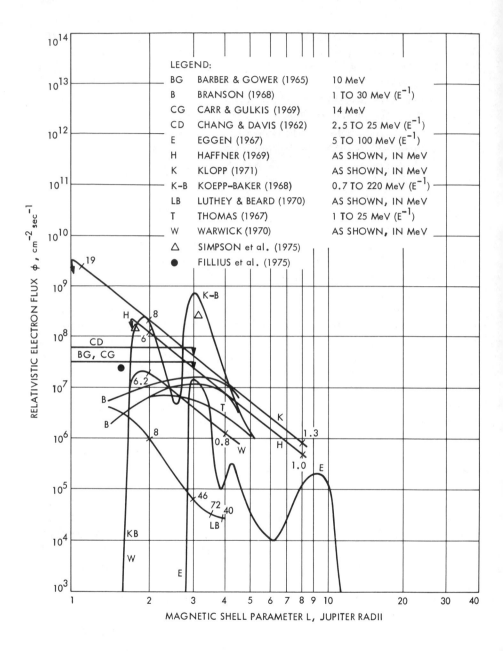

Fig. 6. Solid lines show pre-Pioneer 10 and 11 Jupiter electron models in equatorial plane as compiled by Divine (1972). Heavy circle shows Fillius *et al.* (1975) flux for electron energies greater than 4 MeV near Pioneer 11 periapsis. Open triangles show Simpson *et al.* (1975) data for electron energies greater than 3 MeV.

workers. Gerard (1974) suggested that such variations are significant only near solar maximum. However, since the nature and very existence of the variations are in doubt, it would seem premature to discuss the interpretation at this time.

This still leaves the long-term variations which are now well established (Klein *et al.* 1972; Gulkis *et al.* 1973; Berge 1974; Klein 1976). The "period" of the variation appears too long to be correlated with the solar cycle or with the position of Jupiter in its orbit. Not only is the fundamental cause unknown, but it is also not known whether the variation results from a change in the magnetic moment, electron energy, or electron density, or some combination of the three. Klein (1976) argues that the variation is not caused simply by a change in the magnetic field intensity in the emission region. Although not yet understood, the variation does provide time scales for the decay and regeneration of the underlying physical phenomenon.

Fortunately, there are high-resolution observations available that indicate the change in angular structure associated with the variations. Figure 4*b* shows, in strip-scan form with the thermal component removed, the observations of Berge (1966) near the time of maximum flux density, and those of Olsen (in preparation) near minimum flux density. The observing wavelengths, 10.4 cm and 11.1 cm, respectively, are essentially identical. There is good agreement inward of 2.5 R_J, but from there to 1.3 R_J a large notch appears to have been cut out near the time of minimum flux density. From 1.3 R_J to the center the agreement is good except that, because of superior resolution, Olsen's data more clearly show the discontinuity expected at the limb.

An intriguing point is that, whether by coincidence or not, the onset of the emission deficiency occurs at the orbit of Amalthea. Hess *et al.* (1974) predicted that the effect of Amalthea in sweeping out energetic electrons is negligible compared to that of the Galilean satellites. However, as they point out, the Galilean satellites are so far out that they have little effect in modifying the angular structure of the emission. One might envision a change in the character of the *L*-shell diffusion or the electron distribution that would make Amalthea at times more effective in sweeping out electrons.

There is still one more subject to consider, namely, the asymmetries in structure and in the variations of flux density, degree of linear polarization, and plane of polarization as functions of central meridian longitude or magnetic latitude. These are all usually explained in terms of an anomaly in the magnetic field structure together with the resulting modification of the electron distribution. The exact nature of the anomaly cannot be specified with existing decimeter data, but certain characteristics can be described. To begin with, the brightness maps (Fig. 2*b*) published by Branson (1968) show that the anomalous extra radio emission, or hot spot, occurs at a longitude approximately equal to the longitude ℓ_0 of the magnetic axis, which also coincides roughly with the longitude towards which the radio centroid is displaced. Stannard and Conway (1976) show that the hot spot must be on the order of 10° wide or more in longitude. They find that the excess emission produced by the hot spot (the excess over and above the underlying symmetric structure) is about 3% of the total. They claim that this can fully explain the measured displacement of $\sim 0.08\ R_J$ away from the rotation axis.

Since the beaming and polarization characteristics of the hot spot are as yet unknown it is not clear whether this localized region and its periodic shadowing by the planet are responsible for the other observed asymmetries, whose features were described in Sec. II. The asymmetry of P_L about $\phi_M = 0$ is the most enigmatic because its change with time has not been clearly linked to the variation of D_E. No attempts have been made to explain it. The asymmetry of S about $\phi_M = 0$ is linked to the variation of D_E. The form of the asymmetry changes sense when D_E changes sign and the asymmetry itself disappears when $D_E = 0$. The cause is not understood, however. The two most direct explanations, that β is 15°–18° instead of 10°, or that D_E is really only 0.6 of its tabulated value, have both been ruled out.

Attempts to interpret the departure from a sinusoid of the variation of position angle, of the plane of polarization, with ℓ_{III} date back to Warwick (1964) who tried to fit it to shadowing effects if the dipole is displaced far from the center of Jupiter. Conway and Stannard (1972) explained it as due to a "ripple" in the magnetic equator, centered at ℓ_0. Their approach offers some insight into the problem, but it is over-simplified in that they imagined all of the emission to originate on the longitude plane of the central meridian. Komesaroff and McCulloch (1975) studied the variation in terms of a Fourier series expansion. They found that the first three harmonic components describe the variation very well, with the second and third components representing the departure from a sinusoid. The second component is always nonzero and always has the same sign. They proved that this cannot be explained simply by a departure from a dipole field, but must be caused by shadowing or non-rigid rotation. Indeed, strong shadowing effects can be caused by the hot spot, since it is quite localized. It is clear that improved high-resolution observations can assist greatly in these matters by providing information of the sort shown in Fig. 5 for many different regions separately, and particularly the hot spot, rather than just for an integration over the entire source.

V. INTERPRETATION OF THE THERMAL COMPONENT

A. *Spectrum*

In Sec. II we listed most of the radio brightness temperature measurements which have been reported for Jupiter. Recently Gary (1975) has reviewed these measurements for the express purpose of defining the thermal disk component. Table III gives a listing of the disk thermal spectrum derived by Gary, and Fig. 7 shows these measurements plotted as a function of wavelength. In compiling Table III, Gary rejected certain measurements shown in Table I for one or more of the following reasons: 1) the measurements have been inadequately calibrated; 2) the measurement or calibration technique was inadequately described; or 3) the measurements have been superseded by more precise and accurate measurements. For the remaining measurements, Gary applied normalization factors: 1) to bring the list to a common flux density scale (based on the radio source DR21 wherever possible); 2) to reference all the measurements to a common solid angle for Jupiter; and 3) to account for resolution effects on calibration radio sources. Since the measurements of the flux density at wavelengths $\gtrsim 1.5$ cm include a component due to the synchrotron emission, this component must be removed in order to obtain the thermal disk temperature. For these measurements, Gary gives the reported disk temperature, the estimated thermal component, and the method used to subtract the nonthermal component.

B. Interpretation

The first calculation of the radio emission from Jupiter's atmosphere was carried out by Field (1959) in order to explain the unexpectedly large emission by Jupiter at decimeter wavelengths. Although Field's calculations suggested that nonthermal emission, rather than atmospheric emission, was the primary source of the decimeter radiation, his calculations also showed that the atmospheric contribution to the total emission was substantial. These calculations laid the foundation for interpreting the atmospheric emission once it became possible to separate the thermal and nonthermal components. Subsequent observations and theoretical calculations showed that the nonthermal source was synchrotron emission.

Because the underlying principles of Field's work have been incorporated in all calculations of the microwave thermal spectrum, it is appropriate to outline the assumptions and calculations performed by him. Field assumed that Jupiter's atmosphere was in convective equilibrium with uniformly mixed ammonia supplying the radio opacity. "Model b" of Kuiper (1952) was used for the description of the relative abundances of molecules in the atmosphere. This model contains by number 37.7% hydrogen molecules, 59.5% helium, and 2.8% of other gases. The abundance (mixing ratio) of ammonia by number is 5.8×10^{-3}.

The equations of radiative transfer which relate the disk brightness temperature, T_D, to the physical properties of the medium were given by Field (1959) as

$$T_D = 2 \int_0^1 \int_0^\infty T(\tau) \, e^{-\tau/\mu} \, d\tau \, d\mu \qquad (18)$$

$$\tau(z) = \int_0^z \alpha(z) \, dz \,. \qquad (19)$$

TABLE III

Selected Jupiter Disk Temperatures (Total and Thermal)

λ (cm)	$T_D \pm SE$[a] (reported)	Normalization Factor[b]	$T_D \pm SE$[c] (adjusted)	Separation Method[d]	Reference
0.12	155 ± 15	0.92	143 ± 20		Low and Davidson (1965)
0.14	150 15	0.99	148 16		Ulich (1975)
0.213	168 2	0.99	166 13		Ulich (1974)
0.214	178 1	1.00	178 13		Cogdell et al. (1975)
0.309	181 8	0.96	174 10		Ulich et al. (1973)
0.332	153 2	1.10	168 7		Epstein et al. (1970)
0.353	174 4	0.96	166 6		Ulich (1974)
0.428	182 12	0.99	180 13		Ulich (1975)
0.600	184 13	0.99	182 14		Ulich (1975)
0.835	144 12	1.00	144 15		Thornton and Welch (1963)
0.845	157 4	0.99	156 7		Wrixon et al. (1971)
0.856	173 20	0.99	171 20		Ulich et al. (1973)
0.86	142 11	1.09	154 16		Kalaghan and Wulfsberg (1968)
0.895	149 4	0.99	148 7		Wrixon et al. (1971)
0.95	151 9	0.97	146 11		Ulich (1975)
0.955	152 5	1.00	152 6		Dent (1972)
0.955	157 3	1.07	167 11		Hobbs and Knapp (1971)
0.984	130 4	1.09	142 8		Wrixon et al. (1971)
1.178	139 4	0.99	138 7		Wrixon et al. (1971)
1.261	134 1	1.00	134 4		Klein (1974)
1.275	136 5	0.99	135 7		Wrixon et al. (1971)
1.334	139 5	0.99	138 7		Wrixon et al. (1971)
1.339	133 1	1.00	133 4		Klein (1974)
1.346	136 3	1.00	136 6		Janssen and Welch (1973)
1.411	138 1	1.00	138 4		Klein (1974)
1.461	144 9	0.99	143 9		Wrixon et al. (1971)

λ	T_D	SE	norm.	T_D (adj.)	SE	method	Reference
1.473	139	1	1.00	139	4	A[e]	Klein (1974)
1.65	165	3	1.00	160	7	A[e]	Mayer and McCullough (1971)
1.95	149	7	1.11	158	11	A[e]	Pauliny-Toth and Kellermann (1970)
1.95	182	17	1.00	175	18	A[e]	Dickel et al. (1970)
2.07	173	5	1.00	166	10	A[e]	Gary (1974)
2.7	201	2	1.02	191	15	A[e]	Mayer and McCullough (1971)
2.7	199	3	1.04	193	15	A[e]	McCullough (1972)
3.02	167^{+30f}_{-12}		—	167^{+30}_{-12}		F[f]	Soboleva and Timofeeva (1970)
3.12	196	21[f]	—	196	21	I[f]	Berge (1968)
3.71	212	10[f]	—	212	10	I[f]	Olsen (in preparation)
3.95	198^{+30f}_{-10}		—	$198^{+3)}_{-1)}$		F[f]	Soboleva and Timofeeva (1970)
4.52	280	15	1.00	218	30	P[e]	Dickel and Medd (1967)
6.0	345	6[i]	1.04	217	19	P[g]	Dickel (1967b)
6.0	340	6	1.04	231	16	P[g]	Morris et al. (1968)
6.0	357	5	1.04	242	15	P[g]	Whiteoak et al. (1969)
6.5	205	15[f]	—	205	15	F[f]	Soboleva and Timofeeva (1970)
10.4	290	30[f]	—	290	30	I[f]	Berge (1968)
11.1	280	25[f]	—	280	25	I[f]	Olsen (in preparation)
21.3	2650 ± 160		1.02	400 ± 75		I[h]	Branson (1968)

[a] Reported T_D is for the total emission (thermal plus nonthermal), except for footnote (f). SE due to signal/noise ratio considerations (i.e., precision, not accuracy), except for footnote (f).

[b] Some of the normalization factors may still be subject to minor change.

[c] For λ > 1.5 cm the adjusted T_D is for the thermal emission alone. SE due to signal/noise ratio, plus adopted calibration SE (always ≥ 3%), plus SE for separation procedure.

[d] A: Assumed nonthermal separation (2.0 ± 1.5 Jy nonthermal).
 F: Fan beam spatial resolution separation method.
 I: Interferometer spatial resolution separation method.
 P: Polarization separation method.

[e] Separation by Gary (1975).

[f] Separation by original authors. The reported T_D is for the thermal component alone. The reported SE is assumed to include all uncertainties.

[g] Separation as shown in Sec. III, plus application of normalization factor.

[h] Separation by Berge (1968).

[i] Result as reported subsequently by Dickel et al. (1970), reflecting a disk size correction.

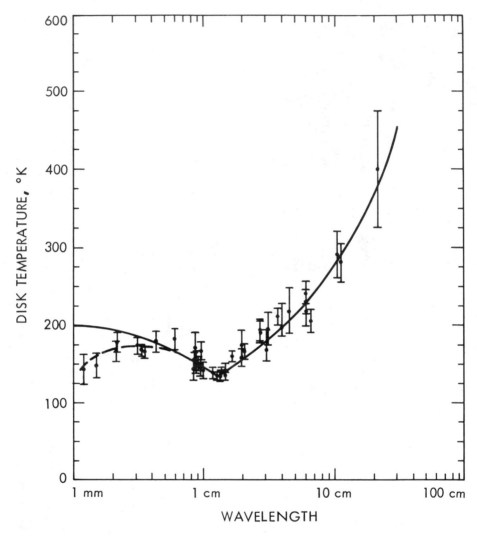

Fig. 7.   Jupiter's microwave thermal spectrum. Solid line shows calculated spectrum for convective model atmosphere (Gulkis *et al.* 1974). Dashed line shows region of apparent discrepancy between model and observations.

In these expressions, $\tau(z)$ is the optical depth measured from the top of the atmosphere ($\tau = 0$) down to the level z, $T(\tau)$ is the physical temperature at τ, and $\alpha(z)$ is the absorption coefficient at z.

The absorption coefficient of ammonia is composed of a series of spectral lines in the microwave region. These spectral lines correspond to the inversion frequencies of the nitrogen atom passing through the plane of the hydrogen atoms. More than 60 of these lines occur between 8 GHz and 40 GHz with those of maximum strength occurring near 24 GHz (0.8 cm^{-1}). Field assumed that the absorption from all the individual lines in the NH_3 spectrum could be represented by a single equivalent line which has a Van Vleck-Weisskopf line shape. The absorption coefficient was given as

$$\alpha = \frac{6.4}{3\,kT}\,\pi^2\,n(\mathrm{NH_3})\,\mu_0^2\,\nu^2\left[\frac{\Delta\nu}{\nu_0^2 + \Delta\nu^2}\right] \qquad (20)$$

where $n(\mathrm{NH_3})$ is the number density of ammonia, μ_0 is dipole moment of ammonia, $\nu_0 \simeq 0.8$ cm^{-1}, and $\Delta\nu$ is the line width of ammonia in a hydrogen-helium mixture.

The two solid lines in Fig. 8 show the results of Field's calculations for two different mixing ratios of ammonia. Also shown are the data available to Field at the time he carried out his calculations. Field concluded from this that there was poor agreement between theory and observations except possibly at the shortest wavelengths. The calculations demonstrated a rather striking transparency of Jupiter's atmosphere at long wavelengths, but not sufficiently transparent to explain the observations. It is of interest to note that if Field had chosen a lower mixing ratio, consistent with a cosmic abundance ratio of nitrogen to hydrogen, his calculations would have shown a much better agreement with the data. This is shown by the dashed curve in Fig. 8, which corresponds to an ammonia mixing ratio of 1.5×10^{-4}.

Field's analysis contains a number of simplifying assumptions appropriate for his estimates. These are pointed out explicitly because many of them form the basis for later papers. His assumptions and simplifications were:

a. All the radio opacity was supplied by ammonia; additional sources of opacity including ionization at great depth were not included.
b. A simplified ammonia absorption coefficient, in which all of the individual lines in the ammonia spectrum could be represented by a single equivalent line, was used.
c. The ammonia concentration remained uniformly mixed throughout the troposphere and stratosphere and did not condense at low temperatures or dissociate at high temperatures.
d. The atmosphere remained adiabatic throughout.
e. High pressure effects were not included.
f. Scattering was not included in the equations of radiative transfer.
g. Horizontal variations in atmospheric parameters were not allowed for.

Winter (1964), Thornton and Welch (1963), and Welch *et al.* (1966) extended Field's work in three areas. First, they computed the spectrum from the middle of the ammonia band at 1.2 cm to millimeter wavelengths. This was an important step because the atmospheric emission dominates the spectrum in this wavelength range, and the theory could be compared directly with the observations. Secondly, they calculated the absorption coefficient by summing the individual lines in the ammonia spectrum rather than assuming a single line profile as Field did. Finally they assumed that the partial pressure of ammonia is limited by its vapor pressure at the corresponding temperature. Figure 9 shows two theoretical curves of the thermal disk temperature of Jupiter as a function of wavelength as calculated by Winter (1964). The upper curve is a calculation based on the "Model b" atmosphere of Kuiper (1952); the lower curve is a calculation based on a model atmosphere in which the cloud top pressure is an order of magnitude larger.

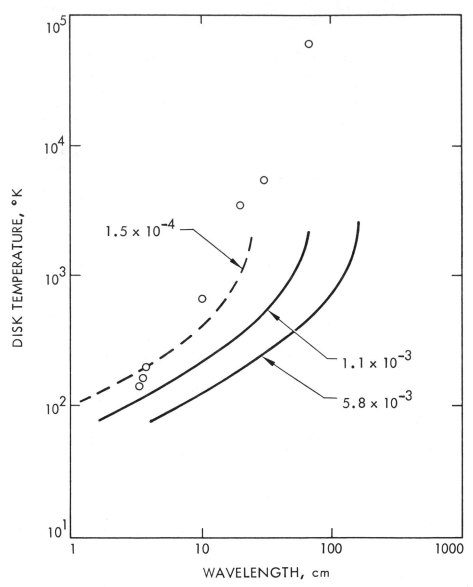

Fig. 8.   Solid and dashed lines show thermal emission from a hot atmosphere for three different mixing ratios of ammonia (Field 1959). Open circles show disk temperatures observed in 1959 or earlier. The disk temperature data are for the combined thermal and nonthermal emission.

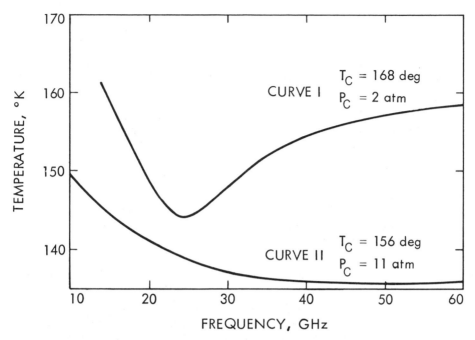

Fig. 9. Thermal emission from a hot atmosphere for two different atmospheric models (Winter 1964). T_C and P_C refer to the temperature and pressure at the ammonia cloud forming region in the two models.

It is evident from these curves that the spectrum varies significantly with pressure and that an absorption feature develops at sufficiently low pressure. Kuiper's "Model b" gave good agreement with the 8-mm to 3-cm data which were available at the time.

Naumov and Khizhnyakov (1965) also calculated the ammonia absorption coefficient by summing over the individual lines in the ammonia spectrum; however, they coupled this with an unrealistic model atmosphere. They assumed that the clouds radiate like a blackbody at 130°K and that overlying, uniformly mixed, ammonia provides the microwave opacity. In regions of small opacity, this model predicts a brightness temperature of 130°K, which is in complete disagreement with the observed spectrum.

Law and Staelin (1968) calculated the thermal emission from 5 model atmospheres and showed that a subadiabatic lapse rate can substantially change the temperature spectrum. Their results also showed that the spectrum changes with pressure. They presented observational data of Jupiter near 1 cm wavelength which suggested that Jupiter was cooler than the convective thermal emission models would allow. Subsequent measurements have not confirmed the measurements of Law and Staelin. The reason for this discrepancy is thought to be related to the calibration technique employed by them. They used the moon for their absolute antenna gain calibrator. Dickel et al. (1970) have pointed out that this technique is rather uncertain since the moon is extended and a difficult correction must be made for the side lobes of the antenna. There is also the problem of determining the spectrum of the moon itself.

Pariiskii (1971) used the results of Field's (1959) calculations, combined with radio observations carried out at Pulkova at wavelengths of 3.02, 3.95, and 6.5 cm, to deduce both the helium to hydrogen ratio and the ammonia to hydrogen ratio. The technique employed by Pariiskii to determine the helium to hydrogen ratio utilized the measured wavelength dependence of the brightness temperature to estimate the specific heat ratio, $\gamma = C_p/C_v$, of a hydrogen-helium mixture. He found that $T \propto \lambda^{0.43}$ which when compared with Field's result $T \propto \lambda^{\gamma-1}$ yielded the result $N_{He}/N_{H_2} = 0.15$. This result is of questionable significance since Field's calculation did not allow for NH_3 saturation and because the line shape is strongly pressure dependent (Winter 1964; Law and Staelin 1968). This same criticism applies to the determination of the ammonia abundance as well.

Wrixon et al. (1971) reported careful measurements of the disk temperature of Jupiter at eight frequencies spanning the range 20.5–35.5 GHz. They found that the observed spectrum agrees well with a model atmosphere by Trafton (1965) in which ammonia is saturated in and above the clouds. In this model the helium to hydrogen ratio is unity and the cloud top level is at a temperature of 158°K and a pressure of 1.8 atm. They showed that the uncertainties in their data allow a factor of 3 uncertainty in the pressure.

Gulkis and Poynter (1972) presented convective models for Jupiter in an attempt to explain both the long-wavelength radiation which originates below the clouds and the short-wavelength radiation which originates within the clouds. They investigated the dependence of the brightness temperature on the mixing ratio of ammonia, the cloud top pressure, and the helium to hydrogen ratio. They found that the brightness temperature is relatively insensitive to the helium-hydrogen ratio. However, the long-wavelength portion of the spectrum is sensitive to both the ammonia mixing ratio and the cloud pressure, while in the center of the ammonia band the spectrum is primarily dependent on the pressure at the clouds. The ammonia mixing ratio has little effect in this spectral region. Because of this independence of mixing ratio, Gulkis et al. (1974) show that the cloud pressure can be determined by measuring the shape of the brightness temperature spectrum in the region of strong ammonia absorption. The reason that the ammonia mixing ratio does not affect the temperature spectrum is that the ammonia clouds are optically thick in the middle of the ammonia absorption band. Hence the ammonia abundance is determined solely by the temperature and not by the mixing ratio. The temperature weighting functions shown in Fig. 10 illustrate that the contribution to the temperature originates in a narrow altitude range within the clouds. Gulkis et al. (1974) presented data from which they estimated that the pressure in the Jovian atmosphere is 0.48 atm at 130°K. The brightness temperature spectrum for the convective model atmosphere shown in Fig. 10 is given superimposed on the data in Fig. 7.

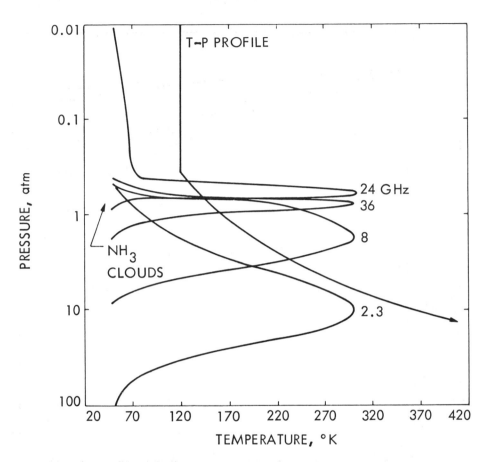

Fig. 10. A typical pressure temperature profile for the Jovian atmosphere and corresponding temperature weighting functions at 24 GHz, 36 GHz, 8 GHz, and 2.3 GHz. The narrow weighting functions lie in the ammonia cloud-forming region.

The ammonia mixing ratio is taken to be 2×10^{-4} for this calculation.

A common feature of the convective model atmospheres used to explain the microwave disk spectrum is the pronounced frequency dependent limb darkening. Limb brightening is also present in some models, but only for narrow spectral lines which originate in the temperature inversion region of the upper atmosphere. The exact shape of the limb darkening curves depends on the distribution of opacity in the atmosphere and on the thermal structure. Figure 11 shows typical limb darkening (brightening) curves expected for Jupiter. The general trend illustrated by these curves is for the limb darkening to be least for wavelengths in the center of the ammonia band and greatest for the long wavelengths where the opacity is the least. Thus far, it has not been possible to measure the limb darkening curves accurately from the earth. In fact, the recent interferometric data on Jupi-

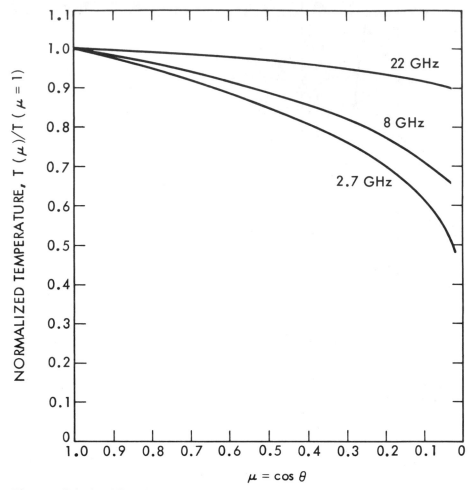

Fig. 11. Calculated limb darkening curves for three frequencies for convective model atmosphere shown in Fig. 10. The parameter θ is the angle a ray makes with the vertical to the atmosphere.

ter obtained by Olsen (see Olsen and Gulkis 1975) and the beam broadening measurements by Bogod *et al.* (1973) are the only data which appear to show the presence of limb darkening. In the future we anticipate that Earth-based and spacecraft instruments will measure the limb darkening, thereby providing additional constraints on Jovian model atmospheres.

Trace amounts of ammonia distributed in and above the NH_3 clouds are expected to produce spectral line features in the microwave spectrum near 1.25 cm. These lines have been searched for by Wrixon (1969), Gulkis *et al.* (1974), and Dickel (1976) but thus far they have not been positively detected. In the region of the strong ammonia lines, Gulkis *et al.* (1974) estimate the upper limit to the lines to be less than 2% of the continuum for line widths between 20 MHz and 100 MHz, while Dickel estimates them to be less than 3% for bandwidths much less than 20 MHz. The absence of narrow ammonia lines is generally interpreted as placing constraints on the high-altitude ammonia abundance rather than on the thermal structure. If the high-altitude abundance is determined by the temperature minimum in the atmosphere, then the observations suggest that the minimum temperature reached is less than 120°K. Photodissociation may also limit the high-altitude ammonia abundance in which case the minimum temperature could be higher and still be consistent with the observations.

C. The Microwave Absorption Coefficient

The potential sources of microwave absorption in the Jovian atmosphere are 1) the polarizable gases such as NH_3, H_2O, CO, and H_2S, 2) the collision-induced dipole gases such as H_2 and CH_4, 3) clouds of NH_3, H_2O, H_2O ice, NH_4SH, and NH_3-H_2O, 4) the ionosphere, and 5) thermal ionization in the deep atmosphere. By far the most important sources of microwave opacity in the upper atmosphere are the gases, with ammonia making the main contribution. Table IV gives the relative absorption coefficients of NH_3, H_2O, H_2, and CH_4 at one point in the Jovian atmosphere. Even though ammonia is a minor constituent in the atmosphere, the fact that its absorption coefficient is more than seven orders of magnitude larger than that of H_2 (at this frequency), explains why it is of primary importance. Clouds are probably important at short wavelengths but their effects have not yet been evaluated. The ionosphere and thermal ionization contribute to the opacity only at very long wavelengths.

TABLE IV
Relative Absorption Coefficients
at $\lambda = 2.5$ cm, $T = 273°K$, $P = 8.55$ atm

Molecule	Absorption Coefficient (cm^{-1} $torr^{-1}$)	Source	Reference
NH_3	2.6×10^{-6} cm^{-1}	Inversion	Goodman (1969)
H_2O	1.6×10^{-8}	Resonant rotation and nonresonant	Goodman (1969)
H_2 + He	$\sim 1 \times 10^{-13}$	Translation and rotation translation	Goodman (1969)
CH_4	10^{-14}	Vibration	Fox (1974)

The Ionosphere. McElroy's (1973) model for the Jovian ionosphere has a peak electron density of $\sim 3 \times 10^5$ cm^{-3} and an emission measure of 2×10^{18} cm^{-5}. This model has a critical or plasma frequency of ~ 5 MHz and an optical depth of 0.02 at 100 MHz scaling as f^{-2} at higher frequencies. Thus the ionosphere does not make a significant contribution to the total attenuation at frequencies greater than ~ 50 MHz.

Thermal Ionization. Thermal ionization will at some deep level in the Jovian atmosphere produce a sufficiently high conductivity to attenuate a propagating radio wave. While this effect has not been included in any models to date, it is important because it will ultimately limit the depth in the atmosphere which can be probed by passive radio sensing techniques. An estimate of the conductivity, σ, for which the effect is important can be determined from the following approximate expression for the optical depth of a weakly absorbing medium

$$\tau = \frac{\pi \sigma}{\lambda \ \omega \epsilon} \ L = 377 \ \sigma \ L \ . \tag{21}$$

In this expression ϵ is the dielectric constant, ω is the angular frequency, λ is the wavelength, L is a typical scale length in meters, and σ is the conductivity in ohm^{-1} m^{-1}. This result assumes that the dielectric constant in the medium is close to the free space value. Taking $L = 5 \times 10^4$ m and $\tau = 1$, we estimate that a conductivity of $\sigma \approx 5 \times 10^{-8}$ ohm^{-1} m^{-1} will cause the atmosphere to be opaque. This ionization level will probably be encountered where the atmospheric temperature is between $1000°$K and $2000°$K.

Clouds. Cloud effects have not been included in existing calculations. While there is some evidence that ammonia clouds are affecting the short millimeter portion of the radio spectrum, there is no evidence of cloud effects at longer wavelengths.

Collision-Induced Dipole Gases. Goodman (1969) has computed the microwave absorption coefficient for a mixture of hydrogen and helium by extrapolating the values computed by Trafton (1965). The absorption coefficient is due to a translational and a rotational-translational component. The absorption coefficient is given as

$$\alpha_{H_2} = 4 \times 10^{-11} \ P_{H_2} \left[P_{H_2} \left(\frac{273}{T}\right)^{2.8} + 1.7 \ P_{He} \left(\frac{273}{T}\right)^{2.61} \right] \frac{1}{\lambda^2} \ \text{cm}^{-1} \ . \tag{22}$$

In this expression P_x is the partial pressure of hydrogen or helium, in atmospheres, and λ is the wavelength in cm.

Fox (1974) has estimated the absorption coefficient of CH_4 to be $\sim 10^{-14}$ cm^{-1} torr^{-1}, which is negligibly small.

Polarizable Gases. Among the gases, ammonia is the most important source of microwave opacity in the Jovian atmosphere. The absorption spectrum of ammonia at microwave frequencies has been extensively studied, both in the laboratory and theoretically. Nevertheless, because of the extremes of pressure and temperature in the Jovian atmosphere, and because ammonia is a minor constituent in the atmosphere, the absorption coefficient is difficult to calculate with certainty. Furthermore, there are no laboratory measurements of the absorption coefficient which duplicate the Jovian environment. The difficulty in the calculation arises from the highly pressure-sensitive characteristics of the absorption coefficient, coupled with uncertainties associated with foreign-gas broadening. This broadening dominates the collisional behavior of the molecule in the Jovian atmosphere. At low pressures (< 1 atm in Jovian atmosphere), the absorption lines are distinct with line shapes resembling a Van Vleck-Weisskopf shape. At higher pressures (≈ 1.5 to 5 atm), the individual lines broaden, overlap, and merge into a single absorption feature. At still higher pressures (> 5 atm) the absorption spectrum takes on a non-resonant Debye shape.

Townes and Schawlow (1955) have given a good discussion of the classical theory of absorption for ammonia. Ben-Reuven (1966) has discussed the general problem of pressure broadening of microwave spectra, while Bleaney and Loubser (1950) have discussed the inversion spectrum of ammonia at moderately high pressures. Recently Morris and Parsons (1970) and Morris (1971) have presented data on the absorption coefficient of ammonia in the presence of hydrogen at very high pressures. Wrixon et al. (1971), Goodman (1969), Kuz'min et al. (1972), and Coombs (1971) have given discussions of the absorption coefficient with particular emphasis on its values in the Jupiter and Saturn atmospheres. Goodman's discussion of the absorption coefficient ignores any high-pressure effects in the spectrum of ammonia and is therefore applicable only in the upper atmosphere. Coombs, Kuz'min et al. and Wrixon all use the Van Vleck-Weisskopf line shape for low pressure and the Ben-Reuven line shape for higher pressures. The analyses of Coombs and Kuz'min differs from that of Wrixon in that the former assumes a single line for the high-pressure case whereas Wrixon et al. sum over all lines.

Gulkis and Poynter (1972) used the line shape of Ben-Reuven throughout the Jupiter atmosphere, summed over the individual ammonia lines, and applied a correction to fit the data of Morris and Parsons. The ammonia absorption coefficient used by Gulkis and Poynter is of the form

$$\alpha(\nu) = C \sum_{J=0} \sum_{K=1} A(J, K)\, F(J, K, \gamma, \delta, \xi, \nu)\ \text{cm}^{-1} \qquad (23)$$

where C is a correction factor (discussed below),

$$A(J, K) = 1.23 \times 10^3 \frac{(2J + 1)K^2}{J(J + 1)} \frac{\nu_0^2(J, K)}{\gamma} S(K) \frac{P_{NH_3}}{T^{\frac{7}{2}}} \qquad (24)$$

$$\times\, e^{-\{[2.98\ J(J\,+\,1)\,-\,1.09\ K^2]\ 4.8/T\}}$$

and $F(J, K, \gamma, \delta, \xi, \nu)$ is the frequency-dependent line shape factor of Ben-Reuven,

$$F(J, K, \gamma, \delta, \xi, \nu) =$$

$$2.0\,\gamma \left(\frac{\nu}{\nu_0}\right)^2 \frac{(\gamma - \xi)\nu^2 + (\gamma + \xi)\ [(\nu_0 + \delta)^2 + \gamma^2 - \xi^2]}{\{[\nu^2 - (\nu_0 + \delta)^2 - \gamma^2 + \xi^2]^2 + 4.0\ \nu^2\ \gamma^2\}}. \qquad (25)$$

The center frequency for the (J, K) transition $\nu_0(J, K)$ is taken from the tabulation of measured frequencies given by Poynter and Kakar (1975). The pressure-broadened line width is given by

$$
\gamma(J, K) = 2.318 \left(\frac{300}{T}\right)^{\frac{2}{3}} P_{H_2} +
$$

$$
0.79 \left(\frac{300}{T}\right)^{\frac{2}{3}} P_{He} + 0.75 \left(\frac{300}{T}\right) \gamma_0(J, K) P_{NH_3} \text{ GHz} . \tag{26}
$$

The coupling element

$$
\xi(J, K) = 1.92 \, P_{H_2} \left(\frac{300}{T}\right)^{\frac{2}{3}} +
$$

$$
0.49 \left(\frac{300}{T}\right) P_{NH_3} \, \gamma_0(J, K) + 0.3 \left(\frac{300}{T}\right)^{\frac{2}{3}} P_{He} \text{ GHz} . \tag{27}
$$

The pressure shift term is given by

$$
\delta = -0.45 \, P_{NH_3} \text{ GHz} . \tag{28}
$$

In these expressions, P_x is the partial pressure of the species x (NH_3, He, H_2) in atmospheres, T is the temperature of the mixture, $S(K) = 3$ for K a multiple of 3, and $S(K) = 1.5$ otherwise, and $\gamma_0(J, K)$ are the self-broadened line widths in MHz torr^{-1} taken from Poynter and Kakar (1975).

The correction factor $C = 1.0075 + (0.0308 + 0.0552 \, P_{H_2}/T)P_{H_2}/T$ was derived by fitting the theoretical absorption coefficient to the data of Morris and Parsons (1970). The computed absorption coefficient, both with and without the correction factor, is compared with the Morris and Parsons data in Fig. 12.

The water vapor absorption coefficient has been discussed by many authors, including Goodman (1969) who calculated its value for Jovian conditions. The absorption coefficient is due to an electric dipole resonance at 22.235 GHz (0.74 cm^{-1}) plus a nonresonant contribution due to a resonance in the infrared. Goodman (1969) gives the absorption coefficient as

$$
\alpha = P_{H_2O}\left(\frac{273}{T}\right)^{\frac{13}{3}} \nu^2 \left\{ 9.07 \times 10^{-9} \left[\frac{\Delta\nu_1}{(\nu/29.97 - 0.74)^2 + \Delta\nu_1^2} + \right. \right.
$$

$$
\left. \left. \frac{\Delta\nu_1}{(\nu/29.97 + 0.74)^2 + \Delta\nu_1^2} \right] + 1.45 \times 10^{-7} \, \Delta\nu_1 \right\} \text{ cm}^{-1} \tag{29}
$$

with

$$
\Delta\nu_1 = 9.88 \times 10^{-2} \left(\frac{273}{T}\right)^{\frac{2}{3}} (0.81 \, P_{H_2} + 0.35 \, P_{He}) . \tag{30}
$$

In this expression P is the pressure in atmospheres, and ν is the frequency in GHz.

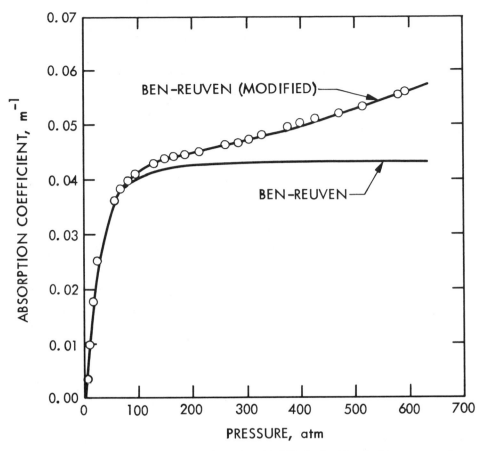

Fig. 12. The calculated absorption coefficient at 9.58 GHz (both with and without a correction factor) of an ammonia-hydrogen mixture at room temperature is compared with experimental data. The mixing ratio of ammonia molecules to hydrogen molecules is 1/228. The open circles are data taken from Morris and Parsons (1970). The curve labeled Ben-Reuven (modified) uses the correction factor $C = 1.0075 + (0.0308 + 0.0552\, P_{H_2}/T\,)P_{H_2}/T$ as discussed in the text.

VI. COMPARISON OF EARTH-BASED RADIO MEASUREMENTS WITH PIONEER 10 AND 11 DATA

As mentioned previously Earth-based radio measurements provided the first evidence that Jupiter has a magnetic field, a population of relativistic electrons trapped in the magnetic field, and that the satellite Io interacted with the magnetosphere (see also the review chapters by Carr and Desch, p. 693, by Hide and Stannard, p. 767 and by Smith, p. 1146). Qualitatively, these aspects of Jupiter were all confirmed by the *in situ* measurements carried out by the Pioneer 10 and 11 spacecraft; in addition the spacecraft measurements gave a substantial new perspective of the conditions and physical processes at work in the Jovian magnetosphere. In this section, we take a closer look at the quantitative agreement between the *in situ* and the Earth-based measurements. Hide and Stannard also discuss this topic in their chapter.

The Jovian magnetosphere, as inferred from the Pioneer 10 and 11 data, is considerably more complicated than it is possible to infer from the Earth-based data. The rapid falloff of the magnetic fieldstrength with radial distance from the planet causes the synchrotron emission to decrease rapidly outside of 4 R_J and little information on the magnetosphere outside this radial distance is contained in the Earth-based data. The Earth-based data are further limited to the high-energy electrons through their synchrotron emission; information on high-energy protons and low-energy plasma is difficult to infer from the Earth-based data. Another difference to keep in mind is that the Earth-based data represent averages in space and time which are very different from the spacecraft data.

The initial magnetospheric modeling work of Chang and Davis (1962) gave energetic electron fluxes, energies, and lifetimes that were consistent with the Earth-based observations. Attempts to explain the origin of the energetic electrons and their distribution outside of their observable range was more speculative. It has generally been assumed that the energetic electrons originate in the solar wind and are diffused inward toward the radiation belt by perturbations which violate the third adiabatic invariant (Hess 1968). The electrons are energized as they move to locations of stronger magnetic fieldstrength while at the same time keeping their magnetic moment constant. This hypothesis leads to the idea that to each L-shell there corresponds a characteristic energy, which increases as L decreases. Figure 13 shows two typical models (Beck 1972) of the characteristic energy variation. In these models an electron reaches 1 MeV near 6 R_J; in more extreme models 1 MeV electrons are found at 20 R_J. In contrast to this simple model, Pioneer 10 and 11 data show that highly energetic electrons and protons exist throughout the magnetosphere. The energization process is therefore clearly different from that anticipated. This difference can result from any one of a number of possibilities. For example, the violation of the first adiabatic invariant may occur when electrons pass through a "neutral sheet" where the magnetic field goes to zero, or when electrons encounter electromagnetic noise near the electron gyrofrequency (Brice and Ioannidis 1970).

Table I of Hide and Stannard (p. 770) shows the quantitative comparison of the magnetic field in the radiation belt as deduced from the Earth-based measurements with that deduced from the *in situ* Pioneer 10 and 11 data. That table summarizes data on field configuration, inclination of dipole axis with respect to Jupiter's rotational axis, offset from planetary center, dipole moment, polarity, and longitude of magnetic pole in the northern hemisphere. It can be seen from that table that there is excellent agreement of the data

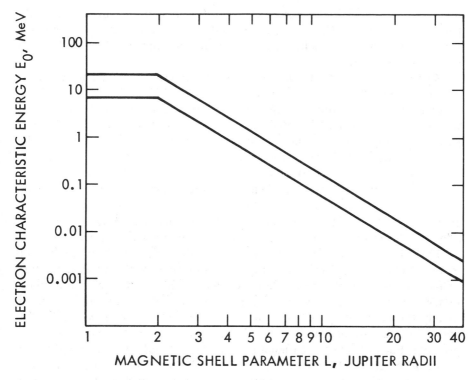

Fig. 13.   Model electron energies as functions of distance from the dipole in Jupiter's magnetic equatorial plane (Beck 1972). Pioneer 10 and 11 data have shown this model to be inadequate.

sets. One point to note is that the *in situ* measurements of the dipole moment remove the large uncertainty which exists in the Earth-based estimates of the magnetic moment. Hide and Stannard point out that both the decameter and decimeter data show clear departures from dipole symmetry as do the Pioneer 10 and 11 data. An outstanding problem at this time is the correlation of asymmetries in the Earth-based data with the spacecraft data. A little work has already been started along this direction but the interpretations are still far from clear. Komesaroff and McCulloch (1976) interpret the measurements of the position angle of the linearly polarized synchrotron emission as implying that the magnetospheric configuration out to a few Jovian radii has a planar (mirror-image) symmetry. The symmetry plane is parallel to a meridian near the System III (1957.0) longitude of 220°.

Pre-Pioneer 10 and 11 attempts to estimate the relativistic electron flux in the radiation belts has been summarized by Divine (1972). Figure 6 shows the results of his study. Also shown are the peak fluxes measured by Fillius *et al.* (1975) near Pioneer 11 periapsis, and Simpson *et al.* (1975) near Pioneer 10 and 11 periapsis. It can be seen from this figure that the model flux levels are in approximate agreement with the *in situ* measurements in the radial range 1.5 to 3 R_J. At larger radial distances, the Earth-based models break down completely since they do not account for all of the high-energy electrons observed.

The Pioneer 10 and 11 data seem to indicate a complex nature for the electron pitch angle distribution. Simpson *et al.* (1975) report, on the basis of Pioneer 10 measurements, that there is some evidence that the electron flux may be described as consisting of two superposed components, one peaked sharply at the magnetic equator and another, observed at latitudes $\geqslant 10°$, that decreased with increasing latitude much less strongly than the first. Van Allen *et al.* (1975), on the basis of Pioneer 11 data, show that a two-component model fits their data. They interpret their result as being due to a resonant interaction between whistler-mode noise and high-energy electrons. In this process, electrons with energies above a critical energy are pitch-angle scattered by whistlers, while lower-energy electrons are not.

Earth-based evidence for an energy dependent pitch angle distribution is inconclusive (see Sec. IV). An early indication that there was little beaming of the 74-cm emission led to the belief that the pitch angle distribution depended on energy. It now appears, however, that the beaming is approximately the same at both long and short decimeter wavelengths. More recently, a qualitatively similar energy dependence has been suggested as a possible, but not unique, explanation for the anomalous frequency dependence of the circular polarization. In both cases the dependence of the pitch angle distribution on energy is opposite in sense from that of the model of Van Allen *et al.* (1975).

Several studies are underway to use synchrotron radiation theory for a comparison of Earth-based data with *in situ* spacecraft data (e.g., Northrop and Birmingham 1974) but the results are not yet complete. The basic idea is to compute the synchrotron emission, including the polarization properties, from the measured relativistic electrons and magnetic fields. By comparing these calculations with the results of the Earth-based observations, it should be possible to intercompare the spacecraft and Earth-based observations, and to infer certain characteristics of the magnetosphere not directly measured by the *in situ* instruments.

Kliore and Woiceshyn (p. 235) and Orton (1975; see also Orton and Ingersoll, p. 206) have compared the Pioneer 10 and 11 results of the radio occultation and infrared experiments with the Earth-based thermal spectrum data. Despite the fact that the Earth-based data refer to global average properties while the spacecraft data reflect localized conditions, there is in general very good agreement of the data sets.

The infrared data provide information on the thermal structure in the 1.0 to 0.1-bar pressure range, a pressure range which directly overlaps the microwave data. Orton and Ingersoll (p. 207) find that temperatures at 1 bar are near 165°K and drop to 105–110°K near 0.1 bar. These profiles pass through the pressure-temperature point ($p = 0.48$ atm, $T = 130°$K) estimated by Gulkis *et al.* (1974), and show the same trend of increasing temperature with depth required to interpret the microwave data. The minimum temperature reached in the atmosphere according to the infrared data is near 105 to 110°K, which is consistent with the upper limit of 120°K set by the microwave data.

The initial interpretation of the Pioneer 10 occultation results for a wavelength of 13 cm suggested a penetration to 2.8 atm for the daytime entry and ~ 2.4 atm for the nighttime exit. The atmospheric temperatures at these pressures were found to exceed 700°K, a value substantially greater than the brightness temperature of the planet at 13-cm wavelength. Since it is unlikely that the horizontally propagated occultation signal, with the measured decibel loss, would have penetrated much deeper than the vertically propagated thermal emission, the occultation result was in conflict with the microwave data. A more recent analysis of the occultation data which takes into account the oblateness of the planet (Hubbard *et al.* 1975; Kliore and Woiceshyn p. 217) indicates a pressure range coverage from 10 to 250 mb rather than the initially reported range. Over this pressure range, there is little overlap with the broadband microwave data. However, the derived temperature at the highest pressure is quite close to that extrapolated from the microwave data.

VII. THE FUTURE OF EARTH-BASED RADIO OBSERVATIONS

While it might be thought that measurements made from flyby spacecraft in the vicinity of Jupiter would obviate the necessity for continuing Earth-based radio observations, further reflection shows this idea to be incorrect. There is, in fact, great potential for increasing our knowledge about Jupiter with Earth-based observations for two principal reasons. One is that there is a strong synergistic relationship between Earth-based and spacecraft measurements because the two approaches measure rather different things and the interpretation of either is strengthened by applying constraints supplied by the other. The interpretation of Earth-based radio observations, for example, is considerably sharpened when constraints supplied by spacecraft data such as magnetic field intensity, energetic particle distributions, and refractivity profiles are brought to bear on the problem. The other reason is that monitoring measurements, aimed at gaining a better understanding of the stability of the radiation belts, the constancy of the Jovian rotation period, long-term Sun-Jupiter relationships, and other effects, will for some time in the future continue to be carried out best from Earth.

Greatly strengthening these arguments is the current development of new Earth-based radio instrumentation that will, in the next few years, provide tremendously improved sensitivity, angular resolution, and spectral coverage, thus permitting much more accuracy and completeness in the Jupiter observations. The Very Large Array (VLA) (e.g., Heeschen 1975) is one example of an instrument, currently under construction, that will greatly increase Earth-based capability. It will consist of 27 antennas, each 25 m in diameter and individually movable along a Y-shaped baseline, of which each leg is 21 km long. Table V gives the operating wavelengths and maximum resolutions of this instrument. The resolution is expressed in three ways: the half-power beam width ($HPBW$), the number of such beam widths along Jupiter's equator, and the number of beam areas [area $= \pi \, (HPBW/2)^2$] on Jupiter's visible disk, the latter two referring to a Jupiter-Earth distance of 4.04 A.U.

TABLE V

Half-Power Beam Widths and Jupiter-Resolutions of the Very Large Array

λ (cm)	HPBW (arcsec)	Disk Diameter/ HPBW	Disk Area/ Beam Area
1.3	0.13	375	1.3×10^5
2.0	0.20	244	5.5×10^4
6.0	0.60	81	6.2×10^3
18.0	1.80	27	680
to			
21.0	2.10	23	503

The disk of Jupiter can be well resolved with this instrument at each wavelength, allowing the thermal emission to be studied in unprecedented detail. Limb darkening, equator-to-pole and morning-to-evening temperature differences, and differences due to belt-zone structure, the Great Red Spot, large white spots, and other features can be studied and related to atmospheric models. High-resolution maps of the radiation belts can be obtained in all polarizations; the importance of this has been discussed above. Tiberi (1966) has suggested a method, using high-resolution circular polarization measurements at two frequencies, to determine, and thus monitor, the magnetic field intensity. Finally, disk temperatures of the satellites will be easily measurable with this instrument, and useful angular resolution will be obtained for the satellites at the three shortest wavelengths.

The VLA is an aperture synthesis instrument that will require 8 to 12 hours of tracking in order to achieve the optimum beam shape. This will result in longitude smearing for a rotating body like Jupiter. However, the smearing is almost inconsequential for studies such as limb darkening, equator-to-pole and morning-to-evening differences, and belt-zone structure. Some studies might be done adequately with the degradated beam obtained in just a few minutes of observing, and thus avoid smearing. For detailed study of features localized in longitude, many days of tracking will be required to get good hour angle coverage in each increment of central meridian longitude. If only one central meridian longitude is desired, the observing time is minimized because only a few minutes per day would be required.

The millimeter spectrum is another area in which instrumentation is rapidly developing. New receivers currently under development for the millimeter region of the spectrum will open up the opportunity of carrying out spectral line observations of molecules in the upper atmosphere of Jupiter. Instrumentation for interferometry at these wavelengths is also developing rapidly and will allow such observations to be made with high angular resolution.

Acknowledgements. We would like to thank authors of chapters in this book and papers in the special Jupiter issues of *Icarus* for providing us with the stimulating environment in which we wrote this chapter. We would also like to thank B. Gary for his assistance in the preparation of Tables I and III. E. Olsen allowed us to use his interferometer data prior to their publication; we are indeed thankful to him both for the data and the useful discussion he provided. We would also like to thank M. Klein, M. Janssen, J. Luthey, E. Olsen, D. Muhleman and G. Orton for many stimulating discussions which ultimately influenced this chapter. Finally, we would like to thank F. DeWindt and M. Owen for their editorial assistance throughout the preparation of this chapter.

Research at the Owens Valley Radio Observatory, California Institute of Technology, is supported by the National Science Foundation, Grant MPS73-04677, and by the Office of Naval Research, Contract N00014-67-A-0094-0019, with additional support for planetary research by the National Aeronautics and Space Administration, Grant NGR 05-002-114. This chapter also presents the results of one phase of research carried out at the Jet Propulsion Laboratory, California Institute of Technology, under Contract NAS7-100, sponsored by the National Aeronautics and Space Administration.

DISCUSSION

L. Trafton: By what factor could NH_3 be supersaturated in Jupiter's atmosphere and still be compatible with the microwave spectra?

S. Gulkis: That is a difficult question to answer. The assumption of saturation leads to predicted brightness temperatures which are in excellent agreement with those observed. Supersaturation therefore implies that either the measured brightness temperatures are too high, the temperature lapse rate used in the models is too shallow, or the ammonia opacity is really less than that computed. I would guess that the data themselves would probably be consistent with supersaturation by a factor of three, but not much more. The temperature lapse rate would have to be considerably super-adiabatic in order to achieve the observed temperatures with large super-saturation. This seems unlikely. As for the ammonia opacity, one always wonders if the temperature dependence and foreign gas broadening is cor-rectly calculated for the Jovian conditions. I think we shall have to await more laboratory data before I can comment further.

D. B. Beard: The plot of planet temperature versus wavelength of obser-vation is made by subtracting out the nonthermal radiation. What was the resolution of the measurement at 21 cm? And how confident can you be that by adjusting the amount of nonthermal radiation within reasonable limits, the disk temperature cannot be reduced by a factor of two to about 250°K, which is what Branson reported for 21 cm?

G. L. Berge: The resolution at 21 cm (Branson 1968), expressed as the width of the beam between half power points (*HPBW*), was 23 arcsec in right ascension, which translates into 1.12 R_J at the time of the observa-tions. In declination it was 2.7 times as large, or 3.0 R_J. Since Jupiter's equatorial plane was aligned approximately in the celestial east-west direc-tion, the observed brightness at the disk center was virtually unaffected by radiation beyond the equatorial limbs.

The underlying assumption used by Branson to derive the 250°K disk temperature was that if there were no thermal emission, then an equatorial strip scan with infinitely high resolution would be twice as great just beyond the equatorial limb as it is at the disk center. However, this ratio is actually greater than 2 for reasons discussed in Sec. III, leading to a higher derived disk temperature. The exact value of the ratio is somewhat model dependent, but this uncertainty is reflected in the generous uncertainty quoted with the result: $400 \pm 75°K$. Thus an adjustment within reasonable limits would seem to preclude a value as small as 250°K.

REFERENCES

Artyukh, V. S.; Kuz'min, A. D.; and Makarov, A. N. 1972. Measurement of Jupiter's radio emission at 2.94 m. *Soviet Astron. —AJ* 16:372–373.

Baars, J. W. M.; Mezger, P. G.; and Wendker, H. 1965. The flux density of the strongest ther-mal radio sources at the frequency 14.5 GHz. *Z. Astrophys.* 61:134–143.

Barber, D. 1966. The polarization, periodicity and angular diameter of the radiation from Jupi-ter at 610 Mc/s. *Mon. Not. Roy. Astron. Soc.* 133:285–308.

———. 1967. Synchrotron radiation from Jupiter at 610 Mc/s. *Magnetism and the Cosmos.* (W. R. Hindmarsh, F. J. Lowes, P. H. Roberts, and S. K. Runcorn, eds.), pp. 310–313. Edinburgh: Oliver and Boyd Ltd.

Barber, D., and Gower, J. F. R. 1965. The spectral index of the radiation from Jupiter between 178 and 610 megacycles/second. *Planet. Space Sci.* 13:889–899.

Barber, D., and Moule, G. L. 1963. Observations at 610 megacycles/second of the radiation from Jupiter. *Nature* 198:947-948.

Bash, F. N.; Drake, F. D.; Gundermann, E.; and Heiles, C. E. 1964. 10-cm observations of Jupiter, 1961-1963. *Astrophys. J.* 139:975-985.

Beard, D. B., and Luthey, J. L. 1973. Analysis of the Jovian electron radiation belts. II. Observations of the decimeter radiation. *Astrophys. J.* 183:679-689.

Beck, A. J. 1972. Conclusion. *Proceedings of the Jupiter radiation belt workshop.* (A. J. Beck, ed.), pp. 473-485. NASA Technical Memorandum 33-543, Washington, D.C.

Ben-Reuven, A. 1966. Impact broadening of microwave spectra. *Phys. Rev.* 145:7-22.

Berge, G. L. 1965a. An interferometric study of Jupiter at 10 and 21 cm. *Radio Sci.* 69D: 1552-1556.

———. 1965b. Circular polarization of Jupiter's decimeter radiation. *Astrophys. J.* 142: 1688-1693.

———. 1966. An interferometric study of Jupiter's decimeter radio emission. *Astrophys. J.* 146:767-798.

———. 1968. Thermal emission from Jupiter at wavelengths of 2 to 21 cm. Symposium on the Atmospheres of Jupiter and the Outer Planets, AAAS meeting, Dallas, Texas, December 1968. Reported by Owen, T. 1970. *Earth and Extraterr. Sci.* 1:89-97.

———. 1969. A new examination of the nonthermal microwave spectrum of Jupiter. *Bull. Amer. Astron. Soc.* 1:233.

———. 1972. Some recent observations and interpretations of the Jupiter decimeter emission. *Proceedings of the Jupiter Radiation Belt Workshop* (A. J. Beck, ed.), pp. 223-242. NASA Technical Memorandum 33-543, Washington, D.C.

———. 1974. The position and Stokes parameters of the integrated 21 cm radio emission of Jupiter and their variation with epoch and central meridian longitude. *Astrophys. J.* 191: 775-784.

Berge, G. L., and Morris, D. 1964. Decimeter measurements relating to the possible displacement of Jupiter's magnetic dipole. *Astrophys. J.* 140:1330-1332.

Bibinova, V. P.; Kuz'min, A. D.; Salomonovich, A. E.; and Shavlovskii, I. V. 1963. Observations of the radio emission from Venus and Jupiter on a wavelength of 3.3 cm. *Soviet Astron.—AJ* 6:840-844.

Bigg, E. K. 1964. Influence of the satellite Io on Jupiter's decametric emission. *Nature* 203: 1008-1010.

Birmingham, T.; Hess, W.; Northrop, T.; Baxter, R.; and Lojko, M. 1974. The electron diffusion coefficient in Jupiter's magnetosphere. *J. Geophys. Res.* 79:87-97.

Bleaney, B., and Loubser, J. H. N. 1950. The inversion spectra of NH_3, CH_3Cl, and CH_3Br at high pressure. *Proc. Phys. Soc. Lond.* 63A:483-493.

Bogod, V. M.; Golubchina, O. A.; Mirovskii, V. G.; Pyatunina, T. B.; Soboleva, N. S.; Strukov, I. A.; and Fridman, P. A. 1973. Preliminary results of observations of discrete sources and of Jupiter with 2-cm wavelength at Pulkova. *Soviet Astron.—AJ* 17:47-53.

Boischot, A.; Ginat, M.; and Kazes, I. 1963. Observations du rayonnement des planètes Vénus et Jupiter sur 13 cm et 21 cm de longueur d'onde. *Ann. Astrophys.* 26:385-390.

Branson, N. J. B. A. 1968. High resolution radio observations of the planet Jupiter. *Mon. Not. Roy. Astron. Soc.* 139:155-162.

Braun, L. D., and Yen, J. L. 1968. Some radio observations at 35 GHz. *Astron. J.* 73:S168.

Brice, N. M., and Ioannidis, G. A. 1970. The magnetospheres of Jupiter and Earth. *Icarus* 13:173-183.

Burke, B. F., and Franklin, K. L. 1955. Observations of a variable radio source associated with the planet Jupiter. *J. Geophys. Res.* 60:213-217.

Carr, T. D., and Gulkis, S. 1969. The magnetosphere of Jupiter. *Ann. Rev. Astron. Astrophys.* 7:577-618.

Chang, D. B. 1960. Synchrotron radiation from the planet Jupiter. *Boeing Sci. Res. Lab. Document* DI-82-0060.

———. 1962. Synchrotron radiation as the source of the polarized decimeter radiation from Jupiter. Ph.D. dissertation, California Institute of Technology, Pasadena, California; *Boeing Sci. Res. Lab. Document* DI-82-0129.

Chang, D. B., and Davis, Jr., L. 1962. Synchrotron radiation as the source of Jupiter's polarized decimeter radiation. *Astrophys. J.* 136:567–581.

Clarke, J. N. 1970. A synchrotron model for the decimetric radiation of Jupiter. *Radio Sci.* 5:529–533.

Cogdell, J. R.; Davis, J. H.; Ulrich, B. T.; and Wills, B. J. 1975. Flux density measurements of radio sources at 2.14 millimeter wavelength. *Astrophys. J.* 196:363–368.

Conway, R. G., and Kronberg, P. P. 1968. A new determination of the position of Jupiter's magnetic axis. *Planet. Space Sci.* 16:445–448.

Conway, R. G., and Stannard, D. 1972. Non-dipole terms in the magnetic fields of Jupiter and the earth. *Nature Phys. Sci.* 239:142–143.

Coombs, W. C. 1971. Estimation of microwave absorption in Jupiter's atmosphere. NASA Technical Memorandum 62–091, Washington, D.C.

Davies, R. D., and Williams, D. 1966. Observations of the continuum emission from Venus, Mars, Jupiter, and Saturn at 21.2 cm wavelength. *Planet. Space Sci.* 14:15–31.

Davis, Jr., L., and Chang, D. B. 1961. Synchrotron radiation as the source of Jupiter's polarized decimeter radiation. *J. Geophys. Res.* 66:2524.

Degioanni, J. J. 1973. The radiation belts of Jupiter. Ph.D. dissertation. University of Illinois, Urbana, Illinois.

Degioanni, J. J., and Dickel, J. R. 1974. Jupiter's radiation belts and upper atmosphere. *Exploration of the planetary system.* (A. Woszczyk and C. Iwaniszewska, eds.), pp. 375–383. Dordrecht-Holland: D. Reidel Publ. Co.

Dent, W. A. 1972. A flux-density scale for microwave frequencies. *Astrophys. J.* 177:93–99.

Dickel, J. R. 1966. Observations of Jupiter at a frequency of 610.5 MHz. *Astron. J.* 71:159–160.

———. 1967*a*. Microwave observations of Jupiter. *Magnetism and the cosmos.* (W. R. Hindmarsh, F. J. Lowes, P. H. Roberts, and S. K. Runcorn, eds.), pp. 296–309. Edinburgh: Oliver and Boyd, Ltd.

———. 1967*b*. 6-cm observations of Jupiter. *Astrophys. J.* 148:535–540.

———. 1976. The microwave spectrum of ammonia in Jupiter's atmosphere. *Icarus* (special Jupiter issue). In press.

Dickel, J. R.; Degioanni, J. J.; and Goodman, G. C. 1970. The microwave spectrum of Jupiter. *Radio Sci.* 5:517–527.

Dickel, J. R., and Medd, W. J. 1967. Unpublished; result presented by Dickel *et al.* (1970).

Divine, N. 1972. Scientific and engineering analyses of Jupiter's energetic electrons and protons. *Proceedings of the Jupiter radiation belt workshop.* (A. J. Beck, ed.), pp. 109–127. NASA Technical Memorandum 33–543, Washington, D.C.

Drake, F. D., and Hvatum, H. 1959. Non-thermal microwave radiation from Jupiter. *Astron. J.* 64:329–330.

Efanov, V. A.; Kislyakov, A. G.; Moiseev, I. G.; and Naumov, A. I. 1969. Radio emission of Venus and Jupiter at 2.25 and 8 mm. *Soviet Astron.—A J.* 13:110–113.

———. 1970. Observations of Jupiter, Venus, and the source 3C273 at wavelengths of 2 and 8 mm. *Radiophys. Quant. Elect.* 13:166–170.

Efanov, V. A.; Moiseev, I. G.; Kislyakov, A. G.; and Naumov, A. I. 1971. Mars and Jupiter: Radio emission at 2.3 mm and 8.15 mm. *Icarus* 14:198–203.

Eggen, J. B. 1967. The trapped radiation zones of Jupiter. *Gen. Dynamics Fort Worth Div. Rep.* FZM-4789.

Epstein, E. 1959. Anomalous continuum radiation from Jupiter. *Nature* 184:52.

———. 1968. Mars, Jupiter, and Saturn: 3.4 mm brightness temperatures. *Astrophys. J.* 151: L149–L152.

Epstein, E. E.; Dworetsky, M. M.; Montgomery, J. W.; and Fogarty, W. G. 1970. Mars, Jupiter, Saturn, and Uranus: 3.3 mm brightness temperatures and a search for variations with time or phase angle. *Icarus* 13:276–281.

Field, G. B. 1959. The source of radiation from Jupiter at decimeter wavelengths. *J. Geophys. Res.* 64:1169–1177.

———. 1960. The source of radiation from Jupiter at decimeter wavelengths. 2. Cyclotron radiation by trapped electrons. *J. Geophys. Res.* 65:1661–1671.

———. 1961. The source of radiation from Jupiter at decimeter wavelengths. 3. Time dependence of cyclotron radiation. *J. Geophys. Res.* 66:1395–1405.

Fillius, R. W.; McIlwain, C. E.; and Mogro-Campero, A. 1975. Radiation belts of Jupiter: a second look. *Science* 188:465–467.

Fox, K. 1974. On the microwave spectrum of methane in the atmospheres of the outer planets. *Exploration of the planetary system.* (A. Woszczyk and C. Iwaniszewska, eds.), pp. 359–365. Dordrecht-Holland: D. Reidel Publ. Co.

Franklin, K. L. 1959. An account of the discovery of Jupiter as a radio source. *Astron. J.* 64:37–39.

Gary, B. 1963. An investigation of Jupiter's 1400 Mc/sec radiation. *Astron. J.* 68:568–572.

———. 1974. Jupiter, Saturn, and Uranus disk temperature measurements at 2.07 and 3.56 cm. *Astron. J.* 79:318–320.

———. 1975. A compilation of planetary disk temperatures. In draft.

Gehrels, T. 1974. *Planets, stars and nebulae studied with photopolarimetry.* (T. Gehrels, ed.) Tucson, Arizona: University of Arizona Press.

Gerard, E. 1970. Long-term variations of the decimeter radiation of Jupiter. *Radio Sci.* 5: 513–516.

———. 1974. Long term variations of the decimetric radio emission of Jupiter (and Saturn?). Neil Brice Memorial symposium on the magnetospheres of the Earth and Jupiter. (Frascati, Italy, May 28–June 1, 1974.)

Gibson, J. E. 1965. Observations of Jupiter at 8.6 mm. *Radio Sci.* 69D:1560.

Ginzburg, V. L., and Syrovatskii, S. I. 1965. Cosmic magnetobremsstrahlung (synchrotron radiation). *Ann. Rev. Astron. Astrophys.* 3:297–350.

———. 1969. Developments in the theory of synchrotron radiation and its reabsorption. *Ann. Rev. Astron. Astrophys.* 7:375–420.

Giordmaine, J. A.; Alsop, L. E.; Townes, C. H.; and Mayer, C. H. 1959. Observations of Jupiter and Mars at 3 cm wavelength. *Astron. J.* 64:332–333.

Gleeson, L. J.; Legg, M. P. C.; and Westfold, K. C. 1969. Circularly polarized synchrotron radiation in dipolar magnetic fields with application to the planet Jupiter. *Proc. Astron. Soc. Australia* 1:274–276.

———. 1970. On the radio frequency spectrum of Jupiter. *Proc. Astron. Soc. Australia* 1: 320–322.

———. 1974. Synchrotron emission from particles having a truncated power-law energy spectrum. *Mon. Not. Roy. Astron. Soc.* 168:379–397.

Gol'nev, V. Ya.; Lipovka, N. M.; and Pariiskii, Yu. N. 1965. Observation of radio emission from Jupiter at a wavelength of 6.5 cm at Pulkova. *Soviet Phys.—Doklady.* 9:512–514.

Gol'nev, V. Ya.; Pariiskii, Yu. N.; and Soboleva, N. S. 1968. Polarized emission at 6.6 cm from Jupiter observed with the large Pulkova telescope. *Radiophys. Quant. Elect.* 11: 821–823.

Goodman, G. C. 1969. Models of Jupiter's atmosphere. Ph.D. dissertation. University of Illinois, Urbana, Illinois.

Gower, J. F. R. 1963. Radiation from Jupiter at 178 Mc/s. *Nature* 199:1273.

———. 1968. The flux density of Jupiter at 81.5 Mc/s. *Observatory* 88:264–267.

Gulkis, S. 1970. Lunar occultation observations of Jupiter at 74 cm and 128 cm. *Radio Sci.* 5:505–511.

Gulkis, S., and Gary, B. 1971. Circular-polarization and total-flux measurements of Jupiter at 13.1 cm wavelength. *Astron. J.* 76:12–16.

Gulkis, S.; Gary, B.; Klein, M.; and Stelzried, C. 1973. Observations of Jupiter at 13 cm wavelength during 1969 and 1971. *Icarus* 18:181–191.

Gulkis, S.; Klein, M. J.; and Poynter, R. L. 1974. Jupiter's microwave spectrum: implications for the upper atmosphere. *Exploration of the planetary system*. (A. Woszczyk, and C. Iwaniszewska, eds.), pp. 367–374. Dordrecht-Holland: D. Reidel Publ. Co.

Gulkis, S., and Poynter, R. 1972. Thermal radio emission from Jupiter and Saturn. *Phys. Earth Planet. Inter.* 6:36–43.

Haddock, F. T., and Dickel, J. R. 1963. The phase variation of Venus and the Jovian polarization rocking effect at λ3.75 cm. *Trans. Am. Geophys. Union.* 44:886.

Haffner, J. W. 1969. Calculated dose rates in Jupiter's Van Allen belts. *Am. Inst. Aeron. Astronaut. J.* 7:2305–2311.

Hardebeck, H. E. 1965a. Radiometric observations of Jupiter at 430 Mc/s. *Astrophys. J.* 141:837.

———. 1965b. Radiometric observations of Venus and Mars at 430 Mc/s. *Astrophys. J.* 142:1696–1698.

Heeschen, D. 1975. The very large array. *Sky and Telescope* 49:344–351.

Hess, W. N. 1968. *The radiation belt and magnetosphere*. Waltham, Mass.: Blaisdell Publ. Co.

Hess, W. N.; Birmingham, T. J.; and Mead, G. D. 1974. Absorption of trapped particles by Jupiter's moons. *J. Geophys. Res.* 79:2877–2880.

Hobbs, R. W., and Knapp, S. L. 1971. Planetary temperatures at 9.55 mm wavelength. *Icarus* 14:204–209.

Hoyle, F., and Burbidge, G. R. 1966. On the nature of the quasi-stellar objects. *Astrophys. J.* 144:534–552.

Hubbard, W. B.; Hunten, D. M.; and Kliore, A. 1975. Effect of the Jovian oblateness on Pioneer 10/11 radio occultations. *Geophys. Res. Lett.* 12:265–268.

Hughes, M. P. 1966. Planetary observations at a wavelength of 6 cm. *Planet. Space Sci.* 14: 1017–1022.

International Astronomical Union. 1962. I.A.U. Information Bull. No. 8, p. 4.

Janssen, M. A., and Welch, W. J. 1973. Mars and Jupiter: radio emission at 1.35 cm. *Icarus* 18:502–504.

Jones, D. E. 1972. A revision of Jupiter brightness temperatures in the frequency interval 18.5–24.0 GHz (1968). *Publ. Astron. Soc. Pacific* 84:434.

Kalaghan, P. M., and Wulfsberg, K. N. 1968. Radiometric observations of the planets Jupiter, Venus, and Mars at a wavelength of 8.6 mm. *Astrophys. J.* 154:771–773.

Kazes, I. 1965. Simultaneous observations of Jupiter on three frequencies. *Radio Sci.* 69D: 1561–1563.

Kellermann, K. I. 1970. Thermal radio emission from the major planets. *Radio Sci.* 5:487–493.

Kellermann, K. I., and Pauliny-Toth, I. I. K. 1966. Observations of the radio emission of Uranus, Neptune, and other planets at 1.9 cm. *Astrophys. J.* 145:954–957.

Kislyakov, A. G., and Lebskii, Yu. V. 1968. Radio emission of the Crab Nebula and Jupiter at 77.5 GHz frequency. *Soviet Astron.—A J.* 11:561–564.

Klein, M. J. 1974. Private communication to Gary (1975).

———. 1975. Recent increase in Jupiter's decimetric radio emission. *Nature* 253:102–103.

———. 1976. The variability of the total flux density and polarization of Jupiter's decimetric radio emission. *J. G. R.* (special Jupiter issue). In press.

Klein, M. J., and Gulkis, S. 1971. Measurements of Jupiter in the spectral range 20–24 GHz. *Bull. Am. Astron. Soc.* 3:276.

Klein, M. J.; Gulkis, S.; and Stelzried, C. T. 1972. Jupiter: new evidence of long-term variations of its decimeter flux density. *Astrophys. J.* 176:L85–L88.

Klopp, D. 1972. Electron and proton flux models for Jupiter's radiation belts. *Proceedings of the Jupiter radiation belt workshop.* (A. J. Beck, ed.), pp. 83–108. NASA Technical Memorandum 33-543, Washington, D.C.

Koepp-Baker, N. B. 1968. A model of Jupiter's trapped radiation belts. *Gen. Electric Missile and Space Div. Rep.* 685D263.

Komesaroff, M. M., and McCulloch, P. M. 1967. The radio rotation period of Jupiter. *Astrophys. Lett.* 1:39–41.

———. 1975. Asymmetries of Jupiter's magnetosphere. *Mon. Not. Roy. Astron. Soc.* 172: 91–95.

———. 1976. Evidence for an unexpected time-stable symmetry of the Jovian magnetosphere. *Icarus* (special Jupiter issue). In press.

Komesaroff, M. M.; Morris, D.; and Roberts, J. A. 1970. Circular polarization of Jupiter's decimetric emission and the Jovian magnetic field strength. *Astrophys. Lett.* 7:31–36.

Korol'kov, D. V.; Pariiskii, Yu. N.; and Timofeeva, G. M. 1964. Radio-astronomical observations of Jupiter with high resolution. *Astron. Circ.* No. 283.

Krotikov, V. D.; Troitskii, V. S.; and Tseitlin, N. M. 1965. Radio emission temperature of the moon and Jupiter at 70.16 cm. *Soviet Astron.—AJ.* 8:761–764.

Kuiper, G. P. 1952. Planetary atmospheres and their origin. *Atmospheres of the earth and planets.* (G. P. Kuiper, ed.), pp. 306–405. Chicago, Illinois: University of Chicago Press.

Kuz'min, A. D.; Naumov, A. P.; and Smirnova, T. V. 1972. Estimate of ammonia concentration in subcloud atmosphere of Saturn from radio-astronomy. *Solar System Res.* 6:10–14.

Law, S. E., and Staelin, D. H. 1968. Measurements of Venus and Jupiter near 1 cm wavelength. *Astrophys. J.* 154:1077–1086.

Legg, M. P. C., and Westfold, K. C. 1968. Elliptic polarization of synchrotron radiation. *Astrophys. J.* 154:499–514.

Long, R. J., and Elsmore, B. 1960. Radio emission from Jupiter at 408 Mc/s. *Observatory* 80: 112–114.

Low, F. J., and Davidson, A. W. 1965. Lunar observations at a wavelength of 1 millimeter. *Astrophys. J.* 142:1278–1282.

Luthey, J. L., and Beard, D. B. 1970. The electron energy and density distributions in the Jovian magnetosphere. University of Kansas, Department of Physics and Astronomy Preprint.

Mayer, C. H. 1959. Planetary radiation at centimeter wavelengths. *Astron. J.* 64:43–45.

Mayer, C. H., and McCullough, T. P. 1971. Microwave radiation of Uranus and Neptune. *Icarus* 14:187–191.

Mayer, C. H.; McCullough, T. P.; and Sloanaker, R. M. 1958a. Measurements of planetary radiation at centimeter wavelengths. *Proc. IRE.* 46:260–266.

———. 1958b. Observations of Mars and Jupiter at a wavelength of 3.15 cm. *Astrophys. J.* 127:11–16.

McAdam, W. B. 1966. The extent of the emission region on Jupiter at 408 Mc/s. *Planet. Space Sci.* 14:1041–1046.

McClain, E. F. 1959. A test for nonthermal radiation from Jupiter at a wavelength of 21 cm. *Astron. J.* 64:339–340.

McClain, E. F.; Nichols, J. H.; and Waak, J. A. 1962. Investigation of variations in the decimeter-wave emission from Jupiter. *Astron. J.* 67:724–727.

McClain, E. F., and Sloanaker, R. M. 1959. Preliminary observations at 10-cm wavelength using the NRL 84-foot radio telescope. *Proceedings IAU Symposium No. 9-URSI Symposium No. 1.* (R. N. Bracewell, ed.) pp. 61–68. Stanford: Stanford University Press.

McCulloch, P. M., and Komesaroff, M. M. 1973. Location of the Jovian magnetic dipole. *Icarus* 19:83–86.

McCullough, T. P. 1972. Phase dependence of the 2.7 cm wavelength radiation of Venus. *Icarus* 16:310–313.

McElroy, M. B. 1973. The ionospheres of the major planets. *Space Sci. Rev.* 14:460–473.

Miller, A. C., and Gary, B. L. 1962. Measurements of the decimeter radiation from Jupiter. *Astron. J.* 67:727–731.

Miller, A. C., and Griffin, J. 1966. 1414 Mc/sec Jupiter observations. *Astron. J.* 71:744–746.

Morris, D., and Bartlett, J. F. 1963. Polarization of the 2840 Mc/s radiation from Jupiter. La physique des planètes (Proceedings of the conference held in Liège, Belgium, July, 1962.) *Mém. Roy. Soc. Sci. Liège, Ser. 5,* 7:564–568.

Morris, D., and Berge, G. L. 1962. Measurements of the polarization and angular extent of the decimeter radiation from Jupiter. *Astrophys. J.* 136:276–282.

Morris, D.; Whiteoak, J. B.; and Tonking, F. 1968. The linear polarization of radiation from Jupiter at 6 cm wavelength. *Australian J. Phys.* 21:337–340.

Morris, E. C. 1971. Microwave absorption by gas mixtures up to several hundred bars. *Australian J. Phys.* 24:157–175.

Morris, E. C., and Parsons, R. W. 1970. Microwave absorption by gas mixtures at pressures up to several hundred bars. *Australian J. Phys.* 23:335–349.

Morrison, B. L., and Meiller, V. 1972. Ephemeris of the radio longitude of the central meridian of Jupiter. *U.S. Naval Obs. Circ.* No. 137.

Naumov, A. P., and Khizhnyakov, I. P. 1965. Thermal emission from Jupiter. *Soviet Astron.— A J.* 9:480–487.

Northrop, T. G., and Birmingham, T. J. 1974. Jovian synchrotron radiation at 10.4 cm as deduced from observed electron fluxes. *J. Geophys. Res.* 79:3583–3587.

Olsen, E. T., and Gulkis, S. 1975. A study of the Jovian atmospheric emission at 3.7 cm wavelength. *Bull. Am. Astron. Soc.* 7:382.

Orton, G. S. 1975. The thermal structure of Jupiter: I. Implications of Pioneer 10 infrared radiometer data. *Icarus* 26:125–141.

Ortwein, N. R.; Chang, D. B.; Davis, Jr., L. 1966. Synchrotron radiation from a dipole field. *Astrophys. J. Supp.* 12:323–389.

Pariiskii, Yu. N. 1971. The atmosphere of Jupiter from radio observations. *Soviet Astron.— A J.* 15:127–128.

Pauliny-Toth, I. I. K., and Kellermann, K. I. 1970. Millimeter-wavelength measurements of Uranus and Neptune. *Astrophys. Lett.* 6:185–187.

Poynter, R. L., and Kakar, R. K. 1975. The microwave frequencies, line parameters, and spectral constants for $^{14}NH_3$. *Astrophys. J. Supp.* 29:87–96.

Radhakrishnan, V., and Roberts, J. A. 1960. Polarization and angular extent of the 960 Mc/sec radiation from Jupiter. *Phys. Rev. Lett.* 4:493–494.

Roberts, J. A. 1963. Radio emission from the planets. *Planet. Space Sci.* 11:221–259.

————. 1965. Jupiter, as observed at short radio wavelengths. *Radio Sci.* 69D:1543–1552.

Roberts, J. A., and Ekers, R. D. 1966. The position of Jupiter's Van Allen belt. *Icarus* 5:149–153.

————. 1968. Observations of the beaming of Jupiter's radio emission at 620 and 2650 Mc/sec. *Icarus* 8:160–165.

Roberts, J. A., and Komesaroff, M. M. 1964. Evidence for asymmetry of Jupiter's Van Allen belt. *Nature* 203:827–830.

————. 1965. Observations of Jupiter's radio spectrum and polarization in the range from 6 cm to 100 cm. *Icarus* 4:127–156.

Roberts, J. A., and Stanley, G. J. 1959. Radio emission from Jupiter at a wavelength of 31 centimeters. *Publ. Astron. Soc. Pacific* 71:485–496.

Roberts, M. S. 1962. Correlation of Jupiter decimeter radiation with solar activity. *Astron. J.* 67:280.

Roberts, M. S., and Huguenin, G. R. 1963. The radiation belt of Jupiter. *La physique des planètes* (Proceedings of the conference held in Liège, Belgium, July, 1962.) *Mém. Roy. Soc. Sci. Liège, Ser. 5,* 7:569–587.

Rose, W. K. 1967. Centimeter radiation from Jupiter and Saturn. *Magnetism and the cosmos.* (W. R. Hindmarsh, F. J. Lowes, P. H. Roberts, and S. K. Runcorn, eds.) pp. 292–295. Edinburgh: Oliver and Boyd Ltd.

Rose, W. K.; Bologna, J. M.; and Sloanaker, R. M. 1963. Linear polarization of the 9.4 cm wavelength radiation from the planets Jupiter and Saturn. Paper presented at 14th General Assembly of URSI, Tokyo. Result is given by Rose (1967).

Rzhiga, O. N., and Trunova, Z. G. 1965. Measurement of Jupiter's intrinsic decimeter-wavelength radiation. *Soviet Astron.—AJ.* 9:93–95.

Seaquist, E. R. 1969. Circular polarization of Jupiter at 9.26 cm. *Nature* 224:1011–1012.

Shain, C. A. 1955. Location on Jupiter of a source of radio noise. *Nature* 176:836–837.

Simpson, J. A.; Hamilton, D. C.; Lentz, G. A.; McKibben, R. B.; Perkins, M.; Pyle, K. R.; Tuzzolino, A. J.; and O'Gallagher, J. J. 1975. Jupiter revisited: first results from the University of Chicago charged particle experiment on Pioneer 11. *Science* 188:455–458.

Slee, O. B., and Dulk, G. A. 1972. 80 MHz measurements of Jupiter's synchrotron emission. *Australian J. Phys.* 25:103–105.

Sloanaker, R. M. 1959. Apparent temperature of Jupiter at a wavelength of 10 cm. *Astron. J.* 64:346.

Sloanaker, R. M., and Boland, J. W. 1961. Observations of Jupiter at a wavelength of 10 cm. *Astrophys. J.* 133:649–656.

Soboleva, N. S., and Timofeeva, G. M. 1970. Results of observations of Jupiter at Pulkova at 3.95 cm. *Soviet Astron.—AJ.* 13:685–688.

Stannard, D., and Conway, R. G. 1976. Recent observations of the decimetric radio emission from Jupiter. *Icarus* (special Jupiter issue). In press.

Stansberry, K. G., and White, R. S. 1974. Jupiter's radiation belts. *J. Geophys. Res.* 79:2331–2342.

Thomas, J. R. 1967. The radiation belts of Jupiter. Unpublished Boeing Company Memorandum.

Thorne, K. S. 1963. The theory of synchrotron radiation from stars with dipole magnetic fields. *Astrophys. J. Supp.* 8:1–30.

———. 1965. Dependence of Jupiter's decimeter radiation on the electron distribution in its Van Allen belts. *Radio Sci.* 69D:1557–1560.

Thornton, D. D., and Welch, W. J. 1963. 8.35 mm radio emission from Jupiter. *Icarus* 2:228–232.

Tiberi, C. F. 1966. Observations of Jupiter at 430 Mc/s. Ph.D. dissertation. University of Florida, Gainesville, Florida.

Tolbert, C. W. 1966. Observed millimeter wavelength brightness temperatures of Mars, Jupiter, and Saturn. *Astron. J.* 71:30–32.

Townes, C. H., and Schawlow, A. L. 1955. *Microwave spectroscopy.* New York: McGraw-Hill.

Trafton, L. M. 1965. A study of the energy balance in the atmospheres of the major planets. Ph.D. dissertation. California Institute of Technology, Pasadena, California.

Ulich, B. L. 1974. Absolute brightness temperature measurements at 21 mm wavelength. *Icarus* 21:254–261.

———. 1975. Private communication to Gary (1975).

Ulich, B. L.; Cogdell, J. R.; and Davis, J. H. 1973. Planetary brightness temperature measurements at 8.6 mm and 3.1 mm wavelengths. *Icarus* 19:59–82.

Van Allen, J. A.; Randall, B. A.; Baker, D. N.; Golitz, C. K.; Sentman, D. D.; Thomsen, M. F., and Flindt, H. R. 1975. Pioneer 11 observations of energetic particles in the Jovian magnetosphere. *Science* 188:459–462.

Warwick, J. W. 1963. The position and sign of Jupiter's magnetic moment. *Astrophys. J.* 137: 1317–1318.

——. 1964. Radio emission from Jupiter. *Ann. Rev. Astron. Astrophys.* 2:1–22.

——. 1967. Radiophysics of Jupiter. *Space Sci. Rev.* 6:841–891.

——. 1970. Particles and fields near Jupiter. NASA Contractor Report. NASA CR-1685, Washington, D.C.

Welch, W. J.; Thornton, D. D.; and Lohman, R. 1966. Observations of Jupiter, Saturn, and Mercury at 1.53 centimeters. *Astrophys. J.* 146:799–809.

Wendker, H., and Baars, J. M. 1965. Private communication to Dickel (1967*a*).

Whiteoak, J. B.; Gardner, F. F.; and Morris, D. 1969. Jovian linear polarization at 6-cm wavelength. *Astrophys. Lett.* 3:81–84.

Winter, S. D. 1964. Expected microwave emission from Jupiter at wavelengths near 1 cm. *Univ. of California Space Sci. Lab. Tech. Note.* Series 5. Issue 23.

Wrixon, G. T. 1969. Atmospheric composition of the Jovian planets. *J. Atmos. Sci.* 26:798–812.

Wrixon, G. T.; Welch, W. J.; and Thornton, D. D. 1971. The spectrum of Jupiter at millimeter wavelengths. *Astrophys. J.* 169:171–183.

RECENT DECAMETRIC AND HECTOMETRIC
OBSERVATIONS OF JUPITER

T. D. CARR
and
M. D. DESCH
University of Florida

The observational results concerning Jupiter's decametric and hectometric radiation which have become available during the past six years are reviewed. Measurements from the Radio Astronomy Explorer RAE-1 and the Interplanetary Monitoring Platform IMP-6 spacecraft have extended the previously known decametric spectrum into the hectometric region, revealing a broad maximum centered at ~8 MHz. Approximately 20 years of synoptic monitoring from ground observatories have more clearly defined the active zones of central meridian longitude and the control by Io. Several characteristics of these source regions are now known to be related to the Jovicentric declination of the earth (D_E), the originally-postulated solar cycle dependence having been disproved. The three principal source regions which are apparent between ~15 and 25 MHz have both Io-controlled and Io-independent components. There is strong evidence that the source regions are due to characteristic beaming patterns, and that the beams for the Io-controlled and Io-independent components of the same source region are of quite different shape. Rotation period determinations from decametric data by several groups yield essentially the same value, $9^h 55^m 29\overset{s}{.}70$, to within a very few hundredths of a second. High-resolution dynamic spectra of the Jovian bursts reveal in some cases an amazing degree of structure. Frequency drift rates range in magnitude from 10ths of an MHz per minute for storms as a whole to tens of MHz per second for S bursts. The drift rates of the so-called modulation lanes, often exhibited by L bursts, are on the order of 100 kHz sec^{-1}, the sign depending to a large extent on central meridian longitude. The drifts of S bursts are consistent with an emission model in which groups of sub-relativistic and low-pitch-angle electrons are emitting at the local gyrofrequency as they ascend the Io-threaded flux tube. Polarization measurements indicate that no appreciable Faraday rotation occurs in the Jovian magnetosphere, and that radiation within the source regions C and D is predominately left elliptically polarized, while that from A and B is of the right-handed sense. Reports of short-term correlation of Jovian decametric activity with solar-related phenomena have continued, but all are based on inadequate data samples and none has been verified. Although the continuation of groundbased observations is essential, a new dimension to the investigation of low-frequency radiation is being provided by spacecraft.

[693]

Jupiter's decametric radiation has been monitored from ground observatories during nearly every apparition since its discovery in 1955. Recent observations from satellites orbiting above the terrestrial ionosphere have extended our knowledge into the hectometric range. The decametric-hectometric radiation consists of noise storms made up of sporadic and often very powerful bursts, the individual storms usually being separated by long periods of inactivity. The radiation is further characterized by a high degree of elliptical polarization, complex dynamic spectra, and an upper cutoff frequency of 39.5 MHz. Its probability of occurrence, intensity, polarization, and dynamic spectral characteristics are related to the central meridian longitude of the planet and to the orbital phase of the satellite Io (and possibly to the phase of Europa at hectometer wavelengths). Pronounced cyclic effects having a period on the order of a decade are also evident. The emission mechanism is still unknown; it will surely prove to be complex. It is generally believed to involve beamed coherent emission at or near the local electron cyclotron frequency, the beam orientations being determined by the co-rotating magnetic field in the lower magnetosphere.

With the exception of the recent hectometric discoveries, these basic facts have long been known and have been discussed to some extent in earlier review papers by Carr and Gulkis (1969), Warwick (1967), Ellis (1965), and Douglas (1964). A considerable number of papers on the subject have appeared in the past six years, however, and highlights of this new information have not yet been assembled and discussed in a single comprehensive review. In this chapter we present such a review. The material covered is that which has become available since the Carr and Gulkis review of 1969. Purely theoretical results have not been included. Theories relating to the mechanism of emission and propagation of the radiation are treated in the chapter by R. A. Smith (p. 1146); decimetric observations are reviewed by Berge and Gulkis (p. 621).

I. SPECTRAL DISTRIBUTION OF ACTIVITY

Until the recent advent of satellite observatories, the portion of the spectrum of Jupiter's low-frequency emission[1] below about 10 MHz was inaccessible to detailed study because of the opacity of the terrestrial ionosphere. It was known from groundbased observations that both the flux densities and the probability of occurrence of the bursts of low-frequency radiation generally increase as the frequency decreases from the upper cutoff of 39.5 MHz, but what happens below that could not be ascertained. However, with the availability of NASA satellites carrying radio telescopes in orbits well above the terrestrial ionosphere, the previously closed window

---

[1]"Low-frequency emission" as used here and elsewhere in this chapter refers to the known Jovian radiation at decameter and hectometer wavelengths.

to the Jovian hectometric spectrum was opened. The first observations of Jovian radiation at frequencies below the ionospheric critical frequency were made by Desch and Carr (1974a) using data from the RAE-1 satellite, and by Brown (1974a) using IMP-6 data. The two sets of observations included measurements down to frequencies of 450 and 425 kHz, respectively.

In Fig. 1 the peak burst flux densities observed by Desch and Carr (1974b) and Brown (1974b) from spacecraft data and those from the ground-station observations of Carr et al. (1964) are displayed as a function of frequency. A new RAE-1 point at 6.55 MHz has also been added to the figure (Desch and Carr, unpublished). The methods used by the three groups of authors in calculating the peak values were sufficiently similar that intercomparison of their results is justified. There is no apparent discontinuity between the spacecraft and ground-station points, and above 2 MHz the agreement between the RAE-1 and IMP-6 points is excellent. The most intense emission occurs at a frequency between 7 and 8 MHz, at about 1/5 the upper cutoff frequency of 39.5 MHz. This ratio indicates (Desch and Carr 1974b) that the most intense radiation is emitted from a region $\sim 1.7\ R_J$ from the magnetic dipole center, assuming that the emission occurs approximately at the electron gyrofrequency and that a pure dipole field configuration exists. It is interesting to note that Jupiter's decimetric (synchrotron) radiation component is most intense at nearly the same radial distance, about $1.6\ R_J$ (Berge 1971). This is probably coincidental, however, since it appears likely that the two regions are widely separated in magnetic latitude and are associated with different magnetic L shells.

At frequencies below 2 MHz, Brown (1974b) has identified two apparently distinct emission components, which he designates as N (normal) and MF (mid-frequency), the latter designation being chosen because of an apparent similarity to the MF component of radiation from the terrestrial magnetosphere (Brown 1973). An MF event is distinguished from an N event by a relatively narrow bandwidth (380–750 kHz), and is usually more intense. Points representing the N and MF peak flux densities are seen to straddle the Desch and Carr curve in Fig. 1. The MF points define a secondary maximum at about 1 MHz, which is almost as prominent as the principal one near 8 MHz. There is only the faintest suggestion of a secondary peak at 1 MHz in the Desch and Carr data.

The fact that one of the two sets of data displays a strong secondary peak and the other does not is perplexing. It might have been due to the relatively wide RAE-1 channel separation (see Fig. 1). The probability was high for the narrow-band MF events to fall largely between RAE-1 frequencies, while this was not so in the case of IMP-6 because of its more closely spaced channels. On the other hand, the difference might be attributable in some way to the lapse of approximately 3 years between the two series of observations. The fact that the Jovicentric declination of the earth, D_E, was essentially the same in the two cases while the sunspot number had decreased

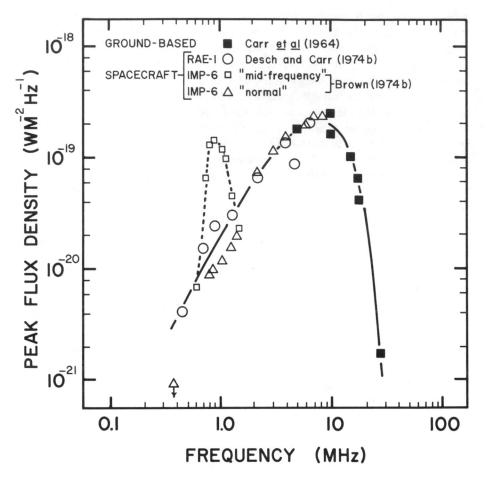

Fig. 1.   Peak flux density spectrum of the radiation, combining groundbased and spacecraft-
derived data.

by a factor of two by the time of the IMP-6 measurements would then suggest that a change in the properties of the solar wind was responsible rather
than a geometrical effect. No reason is known why the spectral characteristics of the radiation should be dependent on the nature of the solar wind,
however, and the evidence that it does is admittedly weak.

   The spectral distribution of activity can also be expressed in other ways,
e.g., by plots of the average flux density as a function of frequency and the
probability of occurrence as a function of frequency. Although the work of
Carr *et al.* (1964) showed that above 10 MHz the peak flux density, the
average flux density and the probability of occurrence vary with frequency
in more or less the same way, this may not be true over the entire spectrum
of the Jovian low-frequency radiation. For example, it is possible that the

frequency at which the probability of occurrence is a maximum may be quite different from that at which the burst peak flux densities are highest. More detailed analyses of the spacecraft data are needed to clarify this point.

II. PERIODIC EFFECTS REVEALED BY SYNOPTIC MONITORING

Synoptic monitoring of the Jovian low-frequency emission consists of the long-term accumulation of occurrence probability data as functions of time and frequency. Intensity and polarization measurements are included when available. Catalogs of synoptic monitoring data have been published by Alexander *et al.* (1975), Bozyan *et al.* (1972) and Warwick *et al.* (1975). Such data can be displayed as functions of the central meridian longitude of the planet, the orbital phases of its satellites, the Jovicentric declination of the earth, sunspot number, and any other periodically varying quantities which might conceivably be correlated in some way with the characteristics of the radiation received on Earth. Results presented in this section are generally consistent with the idea that most of the periodic effects revealed by synoptic monitoring are due to beaming (although a minority of the papers on the subject oppose the beaming hypothesis). These effects presumably result from changing relationships between the earth and complex emission beams which are fixed relative to Jupiter's magnetic field and may in turn be controlled by one or more of the inner satellites. They include periodic variations that are apparent on a time scale of hours (correlations with central meridian longitude, phase of Io, and phase of Europa), of months (correlations with respect to the Sun-Jupiter-Earth-angle), and of years (correlations with quantities depending upon Jupiter's orbital phase). Important developments have occurred in the study of all these effects during the period covered by this chapter. The continuing search for correlations with solar parameters is the subject of a later section.

A. Source Morphology in the λ_{III}-γ_{Io} Plane at the Upper Frequencies

The simultaneous dependence of the Jovian emission probability and intensity on λ_{III} (System III longitude, epoch of 1957.0) and γ_{Io} (departure of Io from superior geocentric conjunction) has been the subject of a number of investigations in recent years. Wilson *et al.* (1968*b*) have presented occurrence probability maps in the λ_{III}-γ_{Io} plane over a wide range of frequencies. By analyzing the data obtained from the University of Colorado radio spectrograph, which had an effective frequency range from 11 to 41 MHz, the authors have delineated the spectral ranges over which the various Jovian "sources" (refer to Fig. 2) can be identified. *Io-B* (or the *early source*) is shown to be well defined at all frequencies between 11 and 39 MHz. *Io-A* (the *main source*) extends from about 14 to 36 MHz. *Io-C* (the *late source*) and *Io-D* (the *fourth source*) cover the ranges from 11 MHz to 26 and 18 MHz, respectively. Thus *Io-B* has the widest frequency spread and *Io-D* the narrowest. Between 11 and 28 MHz a component of Source A emission appears which does not favor any particular Io phase. This is *non-Io-A* (or the *fifth source*), which was recognized soon after the discovery of the Io effect (Bigg 1964) as representing the only part of the emission then detectable not controlled by the satellite. Register and Smith (1969) showed from the University of Florida data that the *Io-A* and *non-Io-A* sources are centered on slightly different central meridian longitudes, and exhibit different long-term drifts (discussed below).

The spectrograph measurements of the University of Colorado have been presented in a somewhat different format by Goertz and Haschick (1972) for comparison with theoretical predictions. The authors examined the highest cutoff frequency of the radiation as a function of the sub-Io longitude (λ_{Io}). They found the peak frequency of the non-Io emission to be independent of λ_{Io} and to be equal to about 30 MHz. The upper cutoff frequency for the Io-related emission, however, was observed to vary considerably as a function of λ_{Io}, a fact for which the authors find that it cannot be explained by the theory of Goldreich and Lynden-Bell (1969) if a pure dipole field is assumed.

By making observations at a single frequency (26.3 MHz) but with a highly directive antenna array, Desch *et al.* (1975) have uncovered an additional non-Io-related source which is generally too weak to be detected by conventional systems. The detection threshold was 2 to 3 orders of magnitude less than that of the radio telescopes usually employed in such monitoring. The new source, *non-Io-B*, is clearly defined in Fig. 2, where it accounts for a surprising 30 to 40% of the total activity time. The flux density of most of this radiation is between 10^2 and 10^4 Jansky.

It may in fact be that every source has an Io-controlled component and a usually weaker Io-independent component (see also Sec. II.G). Since the presence of the appropriate central meridian longitude is a necessary but not sufficient condition for the occurrence of emission from a given non-Io source, the obvious question is what additional parameter is involved in the triggering of the emission. A convincing correlation of the non-Io radiation with some solar-related phenomenon would be most revealing. On the other hand, solar parameters may not be directly involved in the stimulation of emission. For example, Vasyliunas (1975) and Hill and Dessler (1976) have proposed that the *non-Io-A* radiation is stimulated by the same leakage of ionospheric plasma from a certain longitude zone which is believed to be responsible for the observed 10-hour periodicity in the flux of escaped energetic particles. Their evidence is that longitude histograms of the particle flux and the occurrence probability of *non-Io-A* radiation are nearly in phase. However, the significance of this is questionable, since (a) another Io-unrelated source, *non-Io-B*, peaks at a different longitude, and (b) at lower frequencies the Io-unrelated sources peak at still different longitudes.

Miller and Smith (1973) have analyzed data collected at the radio observatories of the University of Florida and of the University of Chile in

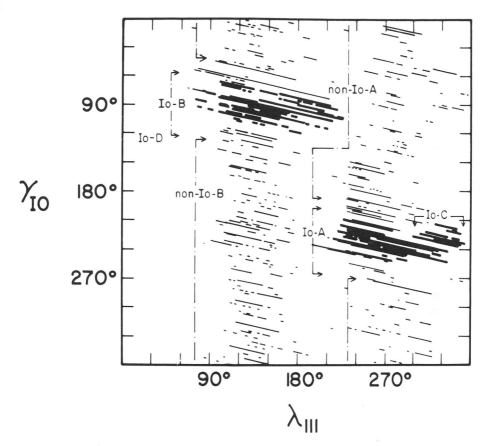

Fig. 2.   Identification of emission sources in the λ_{III}-γ_{Io} plane (Epoch 1957.0). Each line traces the instantaneous λ_{III}-γ_{Io} phase from beginning to end of a single Jupiter storm; recorded with a highly directive array at 26.3 MHz (after Desch *et al.* 1975). The source *Io-D*, which is observable only at lower frequencies, is also indicated. (Heavy lines, $S > 10^5$ Jansky; light lines, $S < 10^5$ Jansky.)

terms of peak flux density as well as occurrence probability over the λ_{III}-γ_{Io} plane. Their map showing the distribution of the mean values of the peak flux densities per 5 min time interval is given in Fig. 3. It is not surprising that the fundamental source structure is maintained at each of the four frequencies plotted. Source A, which generally has the highest probability of occurrence, is shown to be the weakest of the three principal sources in terms of its intensity. Miller and Smith also found some agreement between the locations of sources on their maps and those predicted from models proposed by McCulloch (1971) and Schatten and Ness (1971).

Leacock and Smith (1973) and Thieman *et al.* (1975) have investigated the fine-structure exhibited by the occurrence probability distributions of Sources A, B, and C when plotted as a function of γ_{Io} alone. A large amount

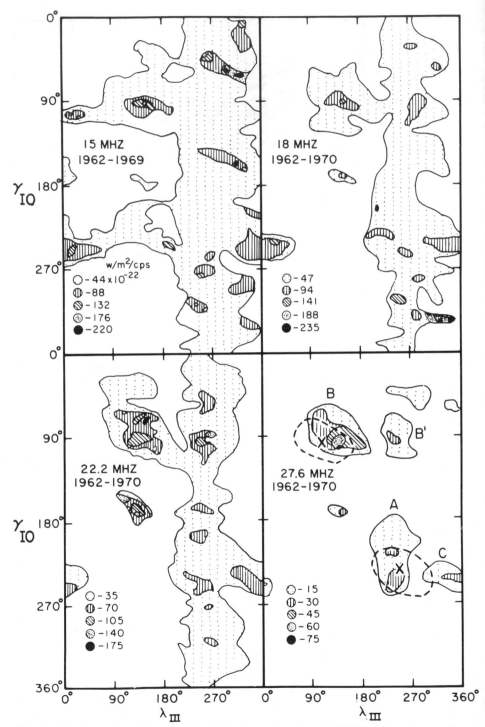

Fig. 3. Contours of peak flux density in the λ_{III}-γ_{Io} plane (epoch 1957.0) at four frequencies (after Miller and Smith 1973).

of data was needed in order to achieve the relatively high resolution in Io phase (2° intervals) required to resolve this structure adequately. They used 22.2 MHz data obtained from several observatories. The most outstanding feature revealed was a bifurcation of Source B. Smaller peaks preceding the principal Source A and C peaks were also apparent. This fine-structure has persisted; it is identifiable in much of the single-apparition data obtained between 1957 and 1972.

Kaiser and Alexander (1973) obtained the power spectrum of the time series representing the presence or absence of Jupiter activity at 22 MHz, using a large body of data. Although their results will be discussed in more detail later, it is interesting to note here that the principal Io-related frequency was found to occur at the γ_{Io}-λ_{III} beat period, which is the period (12^h95) between successive sweeps of a particular System III longitude past Io. Because the spectral power at 12^h95 was significantly greater than the power at Io's orbital period (42^h5), the authors concluded that Io interacts primarily with some feature rotating with Jupiter rather than with a relatively stationary feature such as the planet's shadow or magnetopause, as has been suggested.

Characteristics of the six Jovian source regions recognized at frequencies above $\sim$ 15 MHz are summarized in Table I. Another important characteristic of the Io-related sources which is not included here is their frequency drift pattern (on time scales of tens of minutes) as revealed by dynamic spectra. Such patterns are illustrated by Dulk (1965).

A recent appraisal by Bozyan and Douglas (1976) of some of the topics discussed here may be found in Sec. II.G.

B. The λ_{III}-γ_{Io} Source Regions at the Lower Frequencies

Observations become increasingly difficult as the frequency is reduced below $\sim$ 15 MHz due to the effects of the terrestrial ionosphere. However, some early measurements in the 4 to 10 MHz range during years of minimum solar activity indicated that the morphology of Jupiter's sources is quite different from that at higher frequencies (Lebo 1964; McCulloch and Ellis 1966; Dulk and Clark 1966; Zabriskie 1970). Recent observations made from Earth-orbiting satellites have largely confirmed and considerably extended these earlier findings.

Sources A and B, which are most prominent at high frequencies, are not evident below about 10 MHz. Instead, the two principal source regions which are apparent are near the meridians toward which the north and south magnetic poles are tipped. The idea that this radiation may arise from opposite polar hemispheres is also suggested by the findings of Dowden (1963) and Kennedy (1969) that at 10 MHz the polarizations are of opposite sense (see also Carr and Gulkis 1969). These low-frequency sources are at precisely the longitudes at which little or no activity occurs at the higher decametric frequencies, as can be seen in Fig. 4. In this figure the relative occurrence probabilities versus λ_{III} as observed at 26.3 MHz with a large array (Desch *et al.* 1975) and at 6.55 MHz by the RAE-1 satellite (Desch and Carr

TABLE I

Some Characteristics of the Jovian Source Regions at the Upper Frequencies

Designation	λ_{III} (1965) Span[a] (Deg)	γ_{Io} Span[a] (Deg)	Frequency Range (MHz)	Predominant Polarization Sense	Other Characteristics
Io-A	195-285	220-260	14-36	RH	Often obscured by *non-Io-A* during years of positive D_E. *L* bursts; *S* uncertain.
Non-Io-A	195-285	0-360	11-28	RH	Highest occurrence probability of all sources, for frequencies between 15 and 25 MHz and during years of positive D_E. Greatest change in source position, width, and occurrence probability with change in D_E. *L* bursts only.
Io-B	95-195	65-110	11-39.5	RH	Highest intensity bursts (both *L* and *S*). Principal *S*-burst source. Most predictable source. Bifurcation apparent in high-resolution histograms.
Non-Io-B	95-195	0-360	See Sec. II.G.	Probably RH	Very weak, but relatively high occurrence probability. *L* bursts; *S* uncertain.
Io-C	285-5	220-260	11-26	LH (at least below 23 MHz)	Predictable. Other major *S*-burst source (also *L* bursts).
Non-Io-C	285-5	0-360	See Sec. II.G.	Mixed. See Fig. 8	Moderately distinct from *non-Io-A* (Fig. 7). Abundance of LH polarized emission relative to *non-Io-A* (Fig. 8) may be distinguishing characteristic. Probably *L* bursts only.
Io-D	20-95	95-115	11-18	LH	Very low occurrence probability. Characteristic dynamic spectrum and LH polarization sense important in identification.

[a] Between approximately 15 and 25 MHz.

1975) are compared. While the very sensitive measurements made at 26.3 MHz indicate that the "null" region ($0° \leq \lambda_{III} \leq 70°$) is nearly devoid of activity, there is some evidence that at low frequencies (8.9 and 10 MHz) Jupiter may be continuously active at all central meridian longitudes if observed with a sufficiently sensitive radio telescope (Dulk and Clark 1966).

The Io modulation as ordinarily observed has long been known to be more prominent the higher the frequency, up to the 39.5 MHz cutoff (e.g., Duncan 1966; McCulloch and Ellis 1966; Wilson *et al.* 1968*b*). Non-Io-correlated radiation is not observed at all above 28 MHz. The ratio of that which is correlated to that which is not correlated has been found to decrease toward lower frequencies until, as noted by Dulk and Clark (1966), very little Io control is observed at 8.9 and 10 MHz. Zabriskie (1970) has come to a similar conclusion on the basis of his Earth-based observations made in the

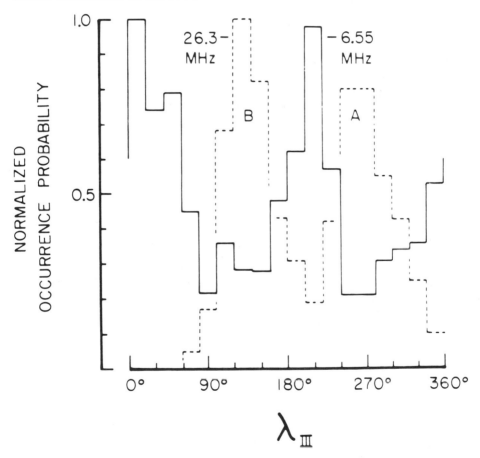

Fig. 4.   Normalized occurrence probability of Jupiter activity as a function of System III longitude (1957.0) showing tendency for low-frequency (6.55 MHz) radiation to occur at longitudes which are generally devoid of activity at higher frequencies. Data taken from Desch and Carr (1975) and from Desch *et al.* (1975).

5 to 7 MHz range. Brown (1974*a*), in analyzing IMP-6 data, found little evidence of Io control in the vicinity of 1 MHz, although he noted that his result might have been affected by selection effects due to insufficient sampling.

Thus, there appears to be inverse relationship between the relative quantity of the non-Io-controlled activity which is observed and the frequency. As mentioned earlier, however, Desch *et al.* (1975) detected significant amounts of radiation, primarily *non-Io-B*, which was *not* coupled to Io at the comparatively high frequency of 26.3 MHz. It was concluded that over most (if not all) of its spectrum the emission consists of an intense Io-controlled component and a much weaker, but relatively abundant non-Io-controlled component. From about 10 MHz to the cutoff near 40 MHz the flux densities of both components fall off steeply. The simple yagi antennas which are generally used detect mostly the strong Io-controlled component above ~ 20 MHz. However, when antennas of comparable directivity are used below 20 MHz, the rise in the spectrum brings the weaker non-Io-controlled component above the detection threshold. For this reason, the apparent Io control becomes increasingly diluted as the frequency is reduced, down to the spectral maximum at ~ 8 MHz. When only the stronger events are selected at the lower frequencies, the Io control should become more apparent.

Additional support in favor of a flux-dependent rather than a frequency-dependent Io effect is provided by the observation of Dulk and Clark that intensity weighting of the 8.9 and 10 MHz activity significantly enhances the Io modulation. Furthermore, at even lower frequencies (2.2 to 6.5 MHz) Desch and Carr (1975) reported a high degree of Io-control over the activity recorded via the RAE-1 spacecraft. In this case the observing frequencies were below that of the spectral maximum (8 MHz), and flux densities were falling off with decreasing frequency. Although a weak non-Io-correlated emission component may still be abundant at such frequencies, the RAE-1 antenna detected mainly the stronger Io-controlled component. It is most interesting to note that at 2.2 to 6.5 MHz the preferred Io positions for emission enhancement were approximately 90° and 240°, essentially the same as at the high frequencies.

C. The Search for Control by Other Satellites

The orbits of five satellites lie within the Jovian magnetosphere. Since Io so profoundly influences Jupiter's emission, it is reasonable to expect that other satellites orbiting within the magnetosphere might also produce effects detectable on Earth. Unfortunately, the orbital period of the innermost satellite, Amalthea, is so close to 12^h that it is practically impossible to separate its effect upon the received radiation from the extreme diurnal modulations inherent in the process of making observations at the surface of the earth. Observations from spacecraft are not subject to this limitation, but as far as we know, no searches for an Amalthea effect have yet been made using spacecraft data.

Several systematic searches for evidence of the modulation of Jovian emission by satellites other than Io have been made from ground observatories (Duncan 1966; Bigg 1966; Tiainen 1967; Wilson *et al.* 1968a,b; Register 1968; Douglas and Bozyan 1970; Kaiser and Alexander 1973). The only significant positive result claimed was that of Tiainen, based on an analysis of the 7 to 15 MHz Boulder dynamic spectra. He reported a joint correlation of Jovian activity with respect to the phases of Io and Europa. The Europa phase at the times of emission was usually near 190° from superior geocentric conjunction.

Desch and Carr (1975), using RAE-1 data, found from a rather limited number of observations recorded in 1969 what appears to be a very strong Europa effect at 1.3, 0.9, and 0.7 MHz. The Europa orbital phase at the times of activity was close to 190°, in apparent agreement with Tiainen. Every 20° interval of Europa phase was sampled between 31 and 38 hours; 20% of the activity time occurred when the Europa phase was between 180° and 200°, and 43% between 160° and 220°. At frequencies down to 2.2 MHz there was a strong and undeniable Io effect. Below 2.2 MHz the Io effect diminished and the apparent Europa effect became dominant. Unfortunately, no evidence of the Europa effect was present in the RAE-1 data of 1970. Although observing conditions and the quality of the radiometer had deteriorated, they were still sufficiently good that a Europa effect should have been seen if its occurrence probability was the same as it had been in the previous year. The Io effect was still apparent at higher frequencies. Although the 1969 evidence of Europa control appears to have been valid, it probably should not be accepted without verification.

D.   The 11.9-Year Periodicities

Jupiter's System III (1957.0) rotation period was determined from measurements obtained during the first few years of observations of the decametric radiation (Douglas and Smith 1963; see also Carr and Gulkis 1969 for background). Later, Gulkis and Carr (1966) showed that this value of the rotation period is ~ 0.3 sec too short. They demonstrated that after the linear drift of the longitude of the Source A maximum (also referred to as the center of Source A) is removed by the use of the corrected rotation period, a sinusoidal residual drift remained. It had an amplitude of about 17° and a period which they believed to be 11.9 years, Jupiter's orbital period. The latter deduction, which was supported by the work of Donivan and Carr (1969), was based on the close correlation of the longitude of the center of Source A with D_E, the Jovicentric declination of the earth. This is the angle of the Jupiter-Earth line with respect to the plane of Jupiter's equator. In one orbital period it ranges between a maximum of $+ 3°3$ and a minimum of $- 3°3$. The alternative possibility that the Source A drift was related in some way to the solar cycle, which has a period of about 10 to 11.5 years, was rejected because the correlation with D_E appeared better than that with sunspot number.

Subsequent work by Carr et al. (1970), Carr (1972), and Mitchell (1974) with more extensive single-frequency data has substantiated the conclusions of Gulkis and Carr. They have shown in addition that although Source A emission most often begins at about the same central meridian longitude regardless of D_E, it tends to continue to higher longitudes during years of more positive D_E (see Fig. 5b). Thus the center of Source A also moves toward higher longitudes with an increase in D_E, and vice versa, giving rise to the sinusoidal drift. Carr et al. (1970) and Carr (1972) suggest that the leading edge of a cross section through the Source A emission beam is nearly parallel to the rotation axis, but that the trailing edge is sloping in such a way that the source is considerably wider when the earth is at a D_E value of $+ 3°3$ than when it is at $- 3°3$. We illustrate such a beam schematically in Fig. 5a. The cross-hatched section indicates the portion of the beam which can sweep past the earth (with Jupiter's rotation) for all possible D_E values. The parts of the beam above and below the cross-hatched area are inaccessible to Earth-bound observers. This model is consistent with the finding of Duncan (1970) that the commencement longitudes of Source A storms do not vary significantly from year to year. Duncan and others have noted that the terminal longitudes of such storms vary considerably more during an apparition than do the starting longitudes. The proposed emission beam in Fig. 5a is perhaps the envelope of narrower beams associated with individual storms.

Dulk (1967) proposed an emission beam model consisting of a single conical sheet, the vertex of the cone lying near the northern foot of the flux tube threading Io. Io-B radiation is supposed to occur only when one side of the conical sheet is aligned with the earth (and certain other conditions are met), and Io-A radiation only when the other side is so aligned. Schatten and Ness (1971) demonstrated that variants of this model can also be adjusted to agree with the relative positions of Io-B and Io-A (and perhaps other Io-related sources as well). Such models are not necessarily inconsistent with the model illustrated in Fig. 5a, since the phenomena upon which the latter is based seem to be associated only with the Io-independent component of Source A. Register and Smith (1969) and Goertz (1971), for example, have shown that it is the non-Io-A component which is mainly responsible for the sinusoidal Source A longitude drift, the Io-A component remaining relatively stationary (when the corrected rotation period is used).

Donivan and Carr (1969) found that the center of Source B also drifts sinusoidally, in phase with the Source A drift (the Io-A and non-Io-A components had not been separated), but with considerably smaller amplitude. They concluded that the fact that the A and B drifts were observed to be in phase rather than antiphased was contrary to the prediction based on Dulk's conical sheet model for Io-B and Io-A. This conclusion now seems to have been premature, however, because the true drift of the Io-A component was probably overshadowed by that of non-Io-A. Thus, while there is no evidence

(a)

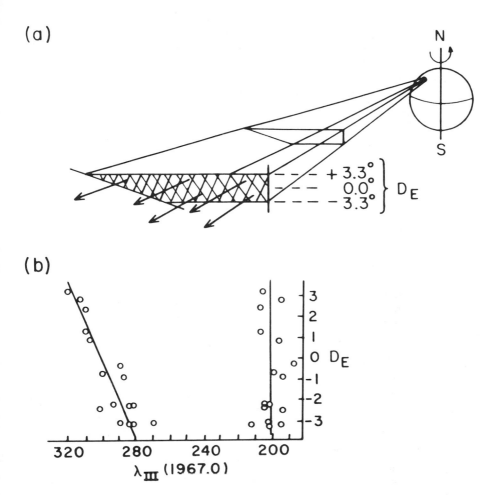

(b)

Fig. 5.   (a) Rotating emission beam model for the source *non-Io-A*. D_E is the Jovicentric declination of the earth at the indicated level of the beam cross section. (b) Data from which model
was derived. Points indicate longitudes of occurrence probability minima preceding and following Source *A*, and the corresponding mean D_E for the apparition. The frequency was 18
MHz. Data are from Carr (1972) and Mitchell (1974).

to date to support this important aspect of the single-cone model, it is still
possible that contamination of *Io-A* by *non-Io-A* activity is biasing the observed D_E dependence. Segregation of the two components based upon
burst type (Olsson and Smith 1966) or dynamic spectrum (Dulk 1965) could
yet lead to the antiphased behavior required by the model.

Apparently the source centroid variations are not restricted to the λ_{III}
coordinate alone. Lecacheux (1974) has uncovered a $\pm 5°$ oscillation of *Io-B*
in the *Io phase* coordinate, which exhibits a slightly improved correlation
when plotted as a function of D_E as opposed to sunspot number. This result

has been verified by Thieman *et al.* (1975) who have, in addition, found that Sources A and C display similar behavior. The motion of Io-C is observed to be in phase with that of Io-B; that is, during years of negative D_E the source centroids of both B and C are displaced toward higher values of γ_{Io}. Lecacheux finds that the strong latitudinal beaming implicit in the observed D_E control is inadequately accounted for by the conical sheet model of Dulk.

Mitchell (1974) found that the width of Source B also increases as D_E increases, although the effect is much smaller than that for Source A. Lecacheux (1974) has arrived at a similar conclusion.

It has long been recognized that there is a marked long-term periodic variation in the integrated probability (i.e., the occurrence probability of activity from all longitudes during an apparition). The period of the variation is on the order of a decade. Since the Jovian activity cycle happened to have been anticorrelated with sunspot number during the first decade after the discovery of the radiation, it was natural to conclude that there was a physical relationship between the two effects. However, Carr *et al.* (1970), Carr (1972), and Mitchell (1974) showed from studies of the *Integrated Source A* probability (the probability of occurrence of activity from all longitudes within Source A during an apparition) that such activity variations are more closely correlated with D_E than with negative sunspot number (or with other measures of solar activity), suggesting that like the longitude drifts this too is a geometrical effect. It may be due to the change in the duration of Source A with D_E. As the source contracts, as seen from the earth, the integrated probability decreases. Although it is still possible that the increased interplanetary scattering during years of high sunspot number may cause some reduction in the amount of Jupiter activity observed, this is certainly not the dominant cause of the Jovian activity cycle.

Kaiser and Alexander (1973) made a power spectrum analysis of the time series representing the presence or absence of Jupiter activity at 22 MHz, using nearly 17 years of data. They found a major peak in the power spectrum at 12.0 ± 0.2 years, in good agreement with the 11.9-yr period of the D_E cycle. There was no peak near 10.5 yr, the period of the sunspot cycle during the 17-yr interval studied. This important result is convincing verification that it is the D_E variation rather than that of solar activity which is responsible for the Jovian activity cycle.

In conclusion, it now appears to be well established that several features of the so-called Jovian sources undergo cyclic changes with a period equal to that of Jupiter's orbital motion of 11.9 years. These phenomena seem to be correlated with the Jovicentric declination of the earth D_E, suggesting that they are all caused by sensitive geometric effects related to the location of the path of the earth through emission beams rotating with the Jovian magnetic field. However, the possibility that the true physical relationship is with respect to the Jovicentric declination of the sun D_S, instead of to D_E cannot be ruled out on the basis of the observational evidence. D_S never differs greatly from D_E. If D_S is truly the controlling factor, the implication is that differences in the direction of arrival of the solar wind at Jupiter as small as 1° can exert an appreciable effect on the emission of radiation. Although this possibility must be considered, it appears much more likely that the true dependence is upon D_E.

E. Decametric Measurements of the Rotation Period

Further background information on decametric measurements of Jupiter's rotation period can be found in Carr (1972). Following the discovery of the sinusoidal 11.9-year drift in Source A position, as described in the preceding section, it became possible to make a precise measurement of the rotation period from data obtained at the same frequency during two apparitions separated by approximately 11.9 yr. The rotation period is obtained by applying whatever correction is needed to the initially assumed period in order to make the center of Source A at the end of the approximately 11.9-year interval appear at the same longitude as it did at the beginning. The D_E effect causes negligible error because the mean values of D_E during two apparitions separated by one orbital period are nearly equal.

The measured rotation period is the average over the 11.9-year interval. Figure 6 shows a series of such measurements obtained by the group at the University of Florida, plotted as a function of the mean epochs of the corresponding intervals. In every case the frequencies used during the initial and final apparitions were essentially the same. This is important, because the Source A position is somewhat dependent on frequency. The error bars are based on a rather subjective appraisal of the possible error in approximating the longitude at which the Source A peak would have occurred if the statistics were not a limitation. The first 6 points are from Carr (1972). These measurements (together with 5 decimetric determinations having lower statistical weights) were used in determining the rotation period upon which a revised System III, epoch of 1967.0, was based. The period used was $9^h\ 55^m\ 29\overset{s}{.}75$. Although this longitude system has been employed extensively by the University of Florida group for several years, it is now superseded by the recently defined System III (1965) which is described in the last paragraph of this section. The points represented by diamonds in Fig. 6 are from measurements made at the Maipu Radioastronomical Observatory in Chile by May (1975, unpublished). The last point is from 26.3 MHz data obtained by Stone et al. (1964) and by Desch et al. (1975).

There is a suggestion in Fig. 6 of a decrease in the measured rotation period from 1962 to 1969. The average of the 7 points between 1962.0 and 1966.0 is $9^h\ 55^m\ 29\overset{s}{.}74$, with the standard deviation of a single measurement σ_i of $0\overset{s}{.}036$, and that of the mean σ_m of $0\overset{s}{.}014$. On the other hand, the average of the 9 points between 1966.0 and 1970.0 is $9^h\ 55^m\ 29\overset{s}{.}72$, with $\sigma_i = 0\overset{s}{.}051$ and $\sigma_m = 0\overset{s}{.}017$. If the unusually high point at 1966.3 (which has a relatively large error bar) is deleted, the average of the 8 remaining points between

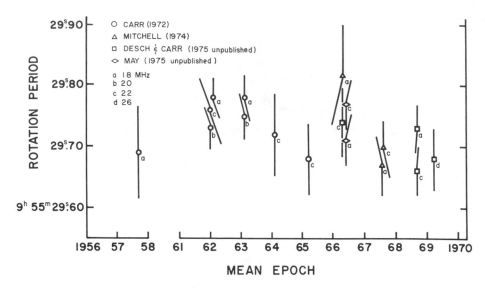

Fig. 6. Mean rotation periods over intervals of approximately 12 years as a function of mean epoch of the interval. Frequencies were from 18 to 26.3 MHz as indicated in the figure.

1966.0 and 1970.0 is $9^h\ 55^m\ 29\overset{s}{.}71$, with $\sigma_i = 0\overset{s}{.}038$ and $\sigma_m = 0\overset{s}{.}013$. It will be possible to add several new points to the plot displayed in Fig. 6 after each Jovian apparition. Their quality should improve for several years because of the improving statistics associated with increasing values of D_E. It will be interesting to determine whether or not changes in the apparent rotation period on the order of the $0\overset{s}{.}03$ drop suggested in Fig. 6 are real and are repeated. If they are, such changes might signify a gradual rearrangement of the magnetic field.

Duncan (1967, 1971) has followed a different approach in measuring the Jovian rotation period from decametric data. Using storm commencement times and not restricting himself to one source region, he arrived at a single value of the rotation period from a superimposed-epoch spectral analysis of all the available data. His value was $9^h\ 55^m\ 29\overset{s}{.}70 \pm 0\overset{s}{.}05$, which is in excellent agreement with the majority of the values displayed in Fig. 6. According to Duncan's conclusions, it is not necessary with the commencement time method to use data spans of 12 years in order to eliminate the D_E effect (although in practice at least this long a data span is needed to attain the statistical precision quoted). Duncan has mixed the data obtained at different frequencies in the same rotation period calculation. Further evidence is needed that this does not bias the result by an appreciable amount, since the earlier data were obtained at a different average frequency than the later data.

Kaiser and Alexander (1972) used their new power spectrum analysis method to determine the rotation period from 22 MHz observations. The period was calculated as a function of the span of the body of data analyzed.

Presumably because of the D_E effect, the measured period was a function of the data span for $<$ 12 years. It converged to the value $9^h\ 55^m\ 29\overset{s}{.}70\ \pm$ $0\overset{s}{.}02$ for 12 years of data, and remained unchanged with a further increase up to the total span of the available data, which was 16.6 years. However, it is not clear why the D_E effect did not cause further variation (although of lesser extent) after the data span exceeded 12 years. The value obtained is exactly the same as that found by Duncan. The precision by this method seems to be significantly higher than that obtained in the earlier determinations.

Alexander (1975) made comparisons of dynamic spectra for pairs of emission events having nearly the same values of λ_{III} and of γ_{Io}, but separated by up to 12 years. Warwick (1963) and Dulk (1965) had previously shown that there is a strong tendency for characteristic dynamic spectral patterns to repeat whenever the same combinations of λ_{III} and γ_{Io} recur. Alexander found that the central meridian longitudes for the characteristic patterns are repeatable only to within $\pm\ 10°$. This inherent variability was attributed to the effect on the emission beam of variations in the field geometry or plasma distribution near the source region. It severely limits the precision to which rotation period values can be determined from individual pairs of such events, even when they are spaced by 12 years.

Lecacheux (1974) prepared histograms of occurrence probability versus central meridian longitude (using an assumed rotation period) for *Io-A*, *Io-B*, *Io-C* and *non-Io* radiation separately for each of 11 apparitions, using University of Colorado data. He then cross-correlated 55 pairs of these histograms for each of the 4 source regions in order to obtain the best longitude shift for each pair, and plotted the shifts (suitably corrected) as a function of opposition date. The result was a sinusoidal variation superimposed upon a small linear drift component. By finding the correction to the initially assumed rotation period which best eliminated the linear drift component, he arrived at the rotation period $9^h\ 55^m\ 29\overset{s}{.}67\ \pm\ 0\overset{s}{.}01$.

This method is a refined version of the method of Gulkis and Carr (1966), in which the visual location of source centers is replaced by cross correlations, and four source regions are employed instead of one, in order to improve the statistics. The sinusoidal variation which was observed is of course due to the differences in D_E over the intervals between the pairs of histograms selected for correlation. The precision claimed for this method is higher than that of any other thus far.

The Lecacheux method is undoubtedly capable of very high precision. The only apparent question about the validity of the present rotation period measurement is that there may be a small error due to frequency bias. The frequency range over which the measurements were made changed somewhat during the 11-year period. The source longitudes are a function of frequency. Histogram peaks are biased toward higher longitudes if a greater proportion of low-frequency data is included, and vice versa. The effect in this case is probably small, but it may be significant.

A resolution has been prepared by Riddle (1976)[2] for submission to the IAU, with the endorsement of a large number of concerned persons, requesting adoption of a revision of the Jovian System III (1957.0) for specifying central meridian longitude. The recommended system is to be designated System III (1965), and is based on a rotation rate of $870°536$ per day, corresponding to a rotation period of approximately $9^h 55^m 29°711$. This value is consistent with the recent decametric and decimetric measurements.

F. Solar Phase Angle Variations

During the 13-month period between oppositions, the Earth-Jupiter-Sun angle α varies between $\pm 12°$. As observed by Gruber (1965) and by Wilson *et al.* (1968*a*), the probability of detecting emission is a rather pronounced function of α, the probability being greatest at or just before opposition ($\alpha \approx 0°$). The authors have shown that this effect is attributable to the increased effectiveness of the earth's ionosphere in absorbing the lower frequency (≤ 20 MHz) radiation during times when the planet is monitored in the day time. Superimposed on this variation, however, is a tendency for the total activity before opposition to exceed that after opposition by a ratio of about 3:2 (Gruber 1965). An effect of this sort was first observed by Six *et al.* (1963) and later by Dulk (personal communication to Gruber). By analyzing the emission probabilities of individual Jovian sources, Gruber and Way-Jones (1972) have found that the post-opposition decline is most pronounced for the non-Io-related sources. They propose an emission beam associated with these sources which is stationary in a Jupiter-Sun frame and directed radially from Jupiter's evening hemisphere.

At the present time there does not seem to be any striking dependence of the various Io-related phenomena on α, although Gruber and Way-Jones have tentatively identified a possible effect based on an inadequate data sample. Also, Alexander (1975) points out that the $\pm 10°$ variation in the beaming of the radiation, as inferred from examination of the Io-related spectra (discussed in Sec. II.E), could possibly be attributed to variations in escape geometry which vary with α.

Gurnett (1974), in characterizing the terrestrial kilometric radiation (*TKR*), noted that it is very similar to some of the gross features of Jovian emission. The major component of the *TKR*, which is beamed primarily from the earth's evening hemisphere, may possibly be compared with the pre-opposition emission enhancement of Gruber and Way-Jones (Kaiser and Stone 1975). Kaiser and Stone have also identified, via the RAE-2 and IMP-6 spacecraft, a late-morning component of the *TKR* which might also be expected to have a Jovian analog. The motivation for investigating such comparisons is, of course, the possibility that the non-Io-related emission is being generated in a manner similar to the aurora-associated *TKR*. Obser-

---

[2]See p. 826.

vations via spacecraft at phase angles greater than $\pm 12°$ would assist in defining any preferred beaming with respect to the Jupiter-Sun line that may exist.

G. Recent Synoptic Monitoring Results of the University of Texas

Bozyan and Douglas (1976) reported on the results obtained from an 18-yr synoptic monitoring program carried out at 22.2, 20.0 and 16.7 MHz. By employing a global network of radiotelescopes it was possible to accumulate between approximately 10,000 and 20,000 hours of effective monitoring, depending upon radiometer frequency. This large volume of data permitted the identification of three distinct Io-independent "sources," one of which (*non-Io-C* or *3n*) had not previously been recognized. They are evident in Fig. 7, where the authors have plotted the occurrence probability of activity as a function of λ_{III} (1965).

As expected, the most prominent Io-independent source is *non-Io-A* (*2n*). The *non-Io-B* source (*1n*) is well-defined at 16.7 MHz, but is barely discernible at 20 and 22 MHz. This is the same source, however, which was found to be a prolific generator of radio bursts at 26.3 MHz by Desch *et al.* (1975), using a highly directive array. Although they are in apparent contradiction with each other, the certainty of the results in both studies strongly implies the operation of a selection effect, permitting some tentative conclusions concerning the nature of the *non-Io-B* radiation. Clearly, the observations are consistent with a source possessing a relatively steep spectral index with a rather low value for its peak flux density at any given frequency. Sufficient data to define a *non-Io-B* profile in the λ_{III} coordinate requires either (a) exceptionally sensitive observations at high decametric frequencies or (b) long-term observations at relatively low frequencies if relatively low-gain antennas are used (as was the case for the data used by Bozyan and Douglas).

Bozyan and Douglas have also identified, for the first time, *non-Io-C* emission, as distinct from that of *non-Io-A*. Data at all three of their frequencies reveal the source plainly, although with decreasing prominence at the higher frequencies. Unlike *non-Io-B,* however, *non-Io-C* appears to be partially merged with *non-Io-A,* perhaps raising questions concerning its classification as a distinct source. It is particularly well camouflaged, for example, when the data are dispersed throughout the λ_{III}-γ_{Io} plane (as in Fig. 1*b* of Bozyan and Douglas 1976). While not appreciated at the time, the polarization studies of Kennedy (1969) offer some clarification in this regard. It is readily apparent from Fig. 8 that when the data are divided according to the occurrence of left-handed (LH) and right-handed (RH) polarized emission, the *non-Io-C* source (as well as the *Io-C* source) becomes much more conspicuous. The high proportion of LH to RH activity for *non-Io-C* relative to that for *non-Io-A* further justifies its classification as a separate source.

In Sec. II. D the D_E variations of the λ_{III} centroids of Sources *A* and *B* were discussed with reference to the restrictions they impose on various

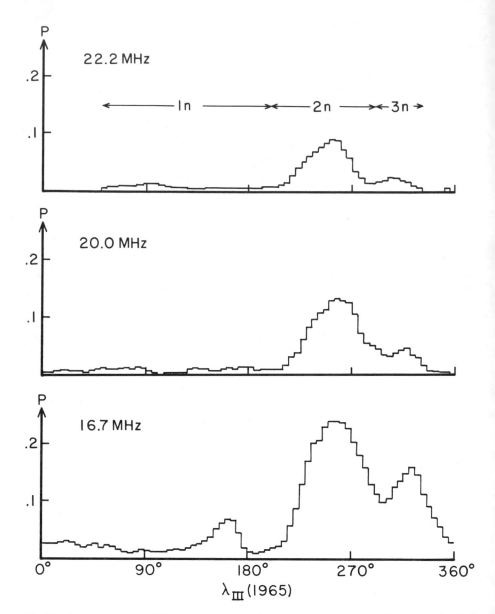

Fig. 7.   Occurrence probability versus λ_{III} for times when $270° < \gamma_{Io} < 60°$ (from Fig. 2 of Bozyan and Douglas 1976). The regions labeled 1*n*, 2*n*, and 3*n* correspond to *non-Io B*, *non-Io A*, and *non-Io C*, respectively.

beaming models. In the light of their extensive catalog, Bozyan and Douglas have generally substantiated the earlier results with respect to Source *A*, and they have extended the investigation to Source *C*. They found that the Io-related component of Source *C* appeared to vary its position with re-

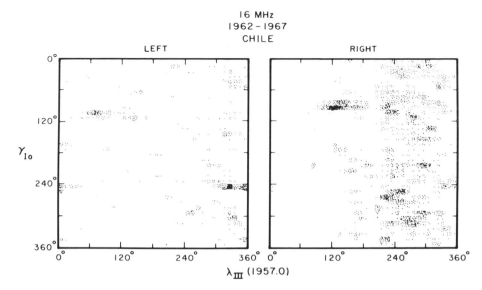

Fig. 8.   The distribution of LH and RH polarized Jupiter bursts in the λ_{III}-γ_{Io} plane at 16 MHz. Each point represents eight bursts (Kennedy 1969).

spect to changes in D_E, in contrast to the generally fixed behavior of *Io-A*. Moreover, the oscillations were in phase with those noted by Donivan and Carr (1969) for *Io-B*. Although the observed D_E dependence of the Io-source centroids is often small and sometimes inconclusive (Bozyan and Douglas found no persistent trend for *Io-B*), such studies are of consequence vis-à-vis the single- and multiple-cone models which attempt to describe the geometry of the Io-controlled emission (see e.g., Dulk 1967). While conclusions concerning the new *non-Io-C* source were somewhat tentative, it evidently varies in amplitude and phase in much the same fashion as *non-Io-A*.

Bozyan and Douglas have further substantiated the oscillation of the Io-dependent sources in the γ_{Io} coordinate, first reported by Lecacheux (1974) and Thieman *et al.* (1975). In addition, the *Io-B* variation was found to be persistent over the frequency range from 16 to 22 MHz, whereas the amplitude of the *Io-C* oscillation decreased significantly toward 16 MHz.

The authors reported further on a number of interesting distinctions concerning the Io versus non-Io phenomenology. The peak occurrence probabilities of the Io-related Sources *A* and *C* were observed to reach a maximum near $0°$ D_E, whereas the non-Io radiation was more abundant during years when D_E was positive. Concerning the latter result, the probabilities seemed to be systematically lower at the maximum D_E values, around $3°$, than they were between $0°$ and $1°$, in apparent disagreement with Carr (1972). This discrepancy is probably related to the fact that Carr used integrated occurrence probabilities, while Bozyan and Douglas used peak probabilities. In addition, Bozyan and Douglas noted a difference, amounting to $\sim 20°$ to $30°$ in λ_{III}, between the centroid of the Io-related component as opposed to the centroid of its non-Io counterpart, both for Sources A and C. A similar disparity in position between $Io\text{-}B$ and $non\text{-}Io\text{-}B$ was not observed by Desch *et al.* (1975). Finally, as illustrated by the gradients in the occurrence probability when plotted as functions of Io phase and sub-Io longitude on Jupiter, the Io-related sources appeared to be defined much more sharply than their non-Io counterparts. On the basis of this and the other observed differences in the Io-related and the non-Io sources, it was suggested by the authors that there may be major differences in their emission mechanisms.

III. BURST WAVEFORMS AND DYNAMIC SPECTRA

When decametric observations of Jupiter are made in the vicinity of 20 MHz using a fixed-frequency receiver having a bandwidth of a few kHz, the type of burst most commonly encountered is composed of random noise with a smoothly rising and falling envelope and a duration of 1 to 10 sec. These are the L bursts. Most noise storms consist of a randomly occurring sequence of L bursts. S bursts are characterized by considerably shorter durations, typically between 1 and 50 millisec each, and are less common. They are often very intense (although the question whether the strongest S bursts are more intense than the strongest L bursts has not yet been answered satisfactorily). They occur only during Io-related storms. Transitions from L to S bursts and back again to L bursts are usual within a storm. As we shall see, the L and S categories are not really definitive in relation to the entire picture, and a more precise classification of burst types can be made only on the basis of characteristic dynamic spectra. However, before the subject of burst types is taken up, we shall discuss briefly the dynamic spectra of Jovian noise storms as a whole.

A.   Dynamic Spectra of Storms

Jupiter's dynamic spectra display an incredible degree of complexity, exhibiting intricate structure on several widely different time scales. On the coarsest scale the characteristic slow drifts of the storm as a whole up or down the spectrum can be observed, periods of minutes or tens of minutes being required for significant changes to take place (Warwick 1963). The rates of such frequency drifts are usually ~ 1 MHz min^{-1} or less, and can be either positive (toward higher frequencies) or negative. Characteristic dynamic spectral patterns are associated with given combinations of central meridian longitude and Io's orbital phase, examples of which have been presented by Dulk (1965). All of the published results on the wide-frequency-range but relatively low-resolution dynamic spectra of storms are from the University of Colorado group.

Warwick (1963) showed that the characteristic dynamic spectral "landmarks" always occurred at the same values of central meridian longitude to within $\pm$ 9°, and he interpreted this as evidence for the narrow beaming of the emitted radiation. It was later discovered that simultaneously with the occurrence of the required central meridian longitude, Io's orbital phase must also lie within a similarly narrow range in order for a given landmark to occur. It is generally believed that the directions and shapes of the emission beams are frequency-dependent, and that they depend also on the orientation of the magnetic field (and perhaps on other factors as well) in the portion of the Io-threaded flux tube within which emission is supposed to take place. This field orientation is continually changing as the planet rotates and as Io progresses in its orbit. Attempts to account for observed features of the dynamic spectra on the basis of plausible postulated beam structures have been unsuccessful.

Apparently the only paper published during the period covered by this review (i.e., since 1969) which deals with the morphology of the dynamic spectra of storms as a whole is that of Alexander (1975). In this work, which has already been referred to in connection with rotation period measurements, the dynamic spectra of intense storms recorded in 1973–74 were paired with events having the same spectral pattern recorded about one Jovian year (11.9 yr) earlier. For the nine such pairs listed by Alexander, the mean difference in λ_{III} (1965.0) was 0°07, with a standard deviation of 6°6, and the mean difference in γ_{Io} was 1°7 with a standard deviation of 2°1. These figures (which were calculated from Alexander's results but do not appear in his paper) suggest that the repeatability of a given dynamic spectral pattern in the λ_{III}-γ_{Io} plane is even better than the $\pm$ 10° quoted in the paper. Such a remarkably close agreement attests to the stability of the physical parameters (presumably the magnetic field and plasma characteristics in the emitting region) which are responsible for the beaming, and is consistent with the degree of constancy of decametric rotation periods averaged over 12-year intervals.

B. L Burst Envelopes and Modulation Lanes

With dynamic spectra of higher resolution (e.g., 0.1 sec in time and 50 kHz in frequency) over a range of at least 2 MHz, the spectral features which are responsible for the fixed-frequency waveforms of L bursts are revealed. These phenomena have been thoroughly investigated by Riihimaa, and the best accounts of them are given in Riihimaa (1971, 1974). (See also Riihimaa et al. 1970, and Riihimaa 1970). The most conspicuous structural features of such dynamic spectra are the burst *envelopes* and the *modulation lanes*, both of which are represented diagrammatically in Fig. 9. The envelope shapes are varied, sometimes being broad and without characteristic form and at other times appearing as vertical or slightly tilted bars. The regions enclosed by the envelopes, representing the frequencies and times of emission, may be continuous or they may show structure. If structure is present, as it most often is, it usually is in the form of tilted, nearly parallel, emission intensity ridges and troughs spanning the envelopes. These are the modulation

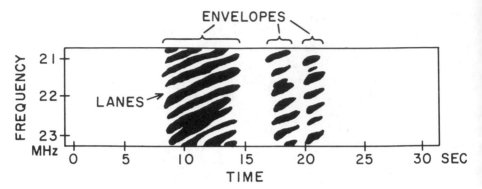

Fig. 9. Dynamic spectrum (idealized) illustrating modulation envelopes and negatively drift-ing modulation lanes (after Riihimaa 1971).

lanes. Within long-duration envelopes the modulation lanes are continuous, but in the case of the more closely spaced vertical-bar envelopes only seg-ments of the lanes are seen, as shown in Fig. 9. The lane slopes may be positive or negative. The lanes are quasi-periodic in time, with a period of about 3 sec at 21 MHz. Adjacent lanes are not necessarily of equal width. Two sets of lanes having slopes of opposite sign may be present simulta-neously, giving rise to a herringbone structure. It is apparent from Fig. 9 that the L burst waveforms observed with a fixed-frequency receiver are due to the combined effect of the envelope regions and of the modulation lanes within them.

The envelopes of the dynamic spectra are largely a consequence of dif-fraction by rapidly drifting inhomogeneities in the solar wind plasma along the propagation path from Jupiter to the earth (Douglas and Smith 1967). The cause of the modulation lanes is unknown. However, the fact that their slopes are strongly correlated with Jupiter's rotation and with the orbital phase of Io indicates that the modulation lanes are either intrinsic to the emission mechanism or are produced by some propagation effect before the radiation escapes from Jupiter's magnetosphere. Most of the negatively drift-ing lanes occur at the central meridian longitudes of Source A, with drift rates between about $- 70$ and $- 140$ kHz sec^{-1}. A considerably smaller number of negative drifts, ranging between about $- 30$ and $- 100$ kHz sec^{-1}, are found at the higher Source B longitudes. Most of the positive drifts are found in the Source B region; these values are largely between 100 and 150 kHz sec^{-1}. Riihimaa states that most of the negative Source B drifts occur simul-taneously with positive drifts, producing the herringbone effect mentioned previously. R. A. Smith (personal communication to Riihimaa) has suggested that these are abnormal events, and that if they are deleted the remainder of the data show positive drifts for Source B and negative drifts for Source A. The crossover is close to λ_{III} (1957) $= 190°$, the approximate longitude toward which the northern-hemisphere magnetic pole was tipped.

Although Riihimaa concludes that only the central meridian longitude controls the variation in drift rates, it is apparent from his plots that Io's orbital phase is at least as important in determining their sign. The drifts are essentially all negative between Io positions of about 90° and 250° from superior geocentric conjunction, and are positive at all other Io positions at which emission was obtained. The reversals in drift sign at the 90° and 250° Io positions are dramatic.

Riihimaa (1974) is unable to decide whether the modulation lanes are primarily lanes of absorption or of enhanced emission. However, there are definite absorption events which are quite distinct from the modulation lanes. They are sharply bounded regions of the dynamic spectra within which radiation is conspicuously but temporarily absent. Riihimaa refers to them as *shadow events*. Some are of irregular shape and are of about the same area as the solar wind diffraction envelopes previously mentioned. They generally appear adjacent to the lower frequency side of emission regions of comparable size. Among other types of shadow events was an isolated negatively drifting dark lane having a drift rate of about -2 MHz sec^{-1}, superimposed upon a background of normal negatively drifting lanes with drifts of about -100 kHz sec^{-1}.

Many variants of the typical L burst emission phenomena have been observed. Riihimaa (1971) described the *narrowband L burst*, which typically is a single long enduring narrowband (about 50 kHz) emission lane with a very low drift rate. Ellis (1974) describes *fast L bursts* which can have either positive or negative drift rates with the constant magnitude of ~ 3 MHz sec^{-1}. They possess the unusual property that the drift rate is independent of frequency between 4 and 17 MHz. Ellis also refers to *shadow drift pairs*, consisting typically of a pair of absorption lanes drifting negatively at about -1.4 MHz sec^{-1}, with a time separation of ~ 0.8 sec.

Torgersen (1972) has made a study of a fairly common special type of L burst characterized by an almost sinusoidal amplitude modulation at a frequency between 20 and 40 Hz. Such bursts have also been recognized by other observing groups by their characteristic sound in the loudspeaker. Torgersen presents an argument that the modulation results from beating of two multipath propagation components, the frequencies of one having been shifted slightly with respect to the other upon arrival at the earth. He offers an explanation for the shift in frequency. Perhaps a more likely explanation, however, is one in which the frequencies in one multipath component are Doppler shifted with respect to the other. Flagg *et al.* (1975) suggest that Torgersen's modulated bursts are fixed-frequency renditions of the complex phenomena shown in Fig. 10c. Torgersen also found many examples of S burst pairs and larger groups which he attributed to multipath propagation in the interplanetary medium.

C. S Bursts

S bursts are characterized by a narrow instantaneous bandwidth (typically 50 kHz, but sometimes as small as 3 kHz), a short single-frequency duration (1 to 10 millisec), and a relatively high negative-frequency drift rate

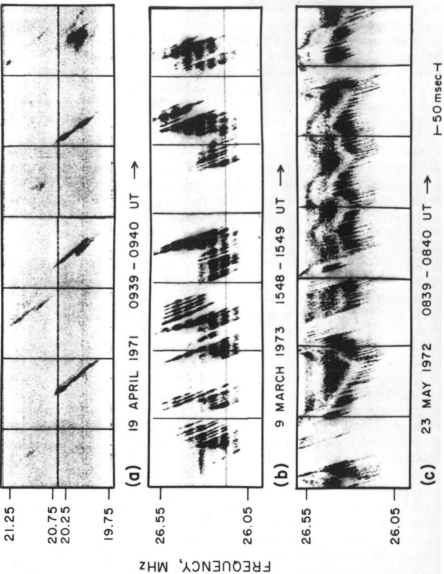

FREQUENCY, MHz

21.25
20.75
20.25
19.75

26.55

26.05

26.55

26.05

19 APRIL 1971   0939 – 0940 UT →

(a)

9 MARCH 1973   1548 – 1549 UT →

(b)

23 MAY 1972   0839 – 0840 UT →

(c)

⊢50 msec⊣

Fig. 10.   High resolution dynamic spectra of S bursts (Krausche *et al.* 1975).

which is approximately proportional to frequency, at least between 5 and 26 MHz (Ellis 1973, 1974, 1975; Krausche *et al.* 1975; Riihimaa 1975). Riihimaa has found them only in *Io-B* and *Io-C* radiation, but Olsson and Smith (1966) report that they may also occur from *Io-A*. One interpretation of the observations is that each *S* burst arises from coherent emission at or near the local gyrofrequency by electrons within the Io-threaded flux tube, the region of coherence ascending the flux tube with a speed which is consistent with the observed frequency drift rate. This speed may or may not represent the actual parallel velocity component of ascending electrons. Ellis (1975) believes that it does. He found a close agreement between his measured *S* burst drift rates plotted versus frequency and the calculated rate of change of gyrofrequency plotted versus gyrofrequency itself for ascending electrons having certain assumed characteristics. The assumed parameters were 4 gauss $R_J{}^3$ for the moment of the Jovian dipole magnetic field, $L = 6R_J$ for the magnetic shell (i.e., approximately the L shell of the Io flux tube during a period of *S* burst activity from Source *Io-B*), $0.1 c$ for the parallel component of velocity of the electrons, and $2°16$ for their equatorial pitch angle. All the available *S* burst drift-rate measurements by Ellis and others are plotted as a function of frequency in Fig. 11. The measurements by Paul and Carr (appearing as unpublished data in Carr and Gulkis 1969) and by Krausche *et al.* (1973, 1975) were for *S* bursts selected to be free of structural complications, as illustrated in Fig. 10a. More complex *S* bursts often have considerably higher drift rates. The curves in Fig. 11, which are similar to those of Ellis but have different parameters, are from Krausche *et al.* (1975). The magnetic moment of the assumed dipole field is 10 gauss $R_J{}^3$. Krausche *et al.* show that a dipole having this moment produces drift rates which differ by less than 10% in the decametric emission region from those calculated from either set of Pioneer 11 field measurements (Acuña and Ness 1975;[3] Smith *et al.* 1975[4]). The assumed parallel component of electron velocity was $0.15 c$. The curves are sensitive to relatively small changes in pitch angle. The pitch-angle values and the parallel component of electron velocity which provide good fits to the measured points do not seem unreasonable. The plotted points due to Paul and Carr, Krausche *et al.*, and Riihimaa all represent observations made during *Io-B* storms. Even so, the scatter in these points may be due largely to the fact that the observations were made at different times, when Io was in slightly different L shells. Future drift rate measurements at a frequency in the vicinity of 35 MHz should prove conclusively whether or not the apparent agreement between observed data of this type and calculated curves is significant, and if so, they may provide a sensitive method for measuring both the velocity and the pitch angle of the emitting electrons if the magnetic field and the L shell are known. Nearly simultaneous measurements at relatively low and high frequencies (e.g., 15 and 35 MHz) would ensure that only one L shell is involved. Similar measurements for *S* bursts

---

[3]See, however, p. 834.   [4]See p. 788.

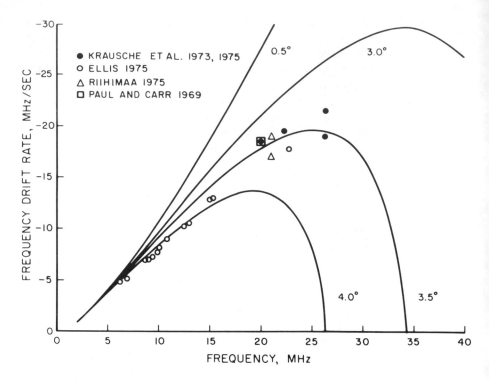

Fig. 11. Measured values of S-burst drift rates. The curves were calculated for a dipole moment of 10 gauss R_J^3, a parallel velocity component of $0.15c$, and the pitch angles indicated
(Krausche *et al.* 1975).

from Source C should be instructive.

All of the thousands of S bursts observed by Krausche *et al.* between 20
and 26.3 MHz had negative frequency drifts. However, Ellis (1975) reports
occasional S bursts at lower frequencies which have positive drifts. Below
10 MHz he has seen U-shaped drift patterns, in which the slope is first negative and later becomes positive. He also mentions inverted U bursts. The
positive S burst drifts might imply that some groups of electrons located
relatively far out are emitting near the local gyrofrequency while descending
the flux tube, rather than while ascending. Ellis' U-shaped S bursts, however,
are difficult to reconcile with the picture of gyro-emission from traveling
groups of electrons. It is possible that the U drifts might indicate emission
while electrons are crossing the magnetic equator (in which case the polarization would be nearly linear), and the inverted U drifts might result from
electrons mirroring not far from the equator. On the other hand, these events
may not be S bursts at all, but instead may be more closely related to one of
the other categories of frequency-drifting bursts in which the drifts certainly
cannot be attributed simply to the motion of the gyroemitting electrons
through regions of changing magnetic field.

Many of the S bursts, perhaps a majority, exhibit one or more types of quasi periodicity. The simple negatively-drifting events often tend to recur regularly at a rate between ~ 10 and 40 sec^{-1}, sometimes for as many as 10 or more recurrences. With higher time resolution, it can often be seen that each drifting burst is striated, consisting of parallel drifting sub-bursts recurring at a rate on the order of 100 sec^{-1}. Examples of such events can be seen in Fig. 10b. Many S-burst events are extremely complex, as illustrated in Fig. 10c. This type of event, which seems to be more or less representative of a class, has been examined by Flagg $et\ al.$ (1975). The event is seen to be divided into three distinct regions. There is an upper-frequency horizontal band of continuous emission, a middle band where no emission is observed, and a diagonally striated lower frequency band consisting of parallel drifting sub-bursts recurring at a rate of several hundred per second. The overall instantaneous emission bandwidth remains essentially constant (400 kHz) as the three bands undulate together in frequency at a mean rate of 26 Hz. Flagg $et\ al.$ believe that the band-like emission is due to an upward-flowing gyroemitting electron stream and that the undulations in the emission bands are caused in part by density variations in the stream. They also suggest that the band-like emission (Fig. 10c) may be responsible for the modulated L bursts described by Torgersen (1972) and discussed in Sec. III.B of this chapter. If a narrowband receiver were tuned to a fixed frequency which intersected the continuous-emission-band undulations, the observed signal would have the appearance of an L-burst, amplitude modulated at the undulation frequency.

Riihimaa (1974) has found that when S and L bursts occur simultaneously, the dark lanes associated with the L bursts are continuous across groups of S bursts. Although there may be other interpretations, this suggests that the modulation lanes are formed by frequency-drifting bands of absorption (or of destructive interference) of the S burst emission, and that the drifts of the lanes are not related to the S burst drifts.

The conclusion is inescapable that the present state of our knowledge of the fine structure of Jovian bursts is chaotic. The situation is probably as confused as it was in the earlier days of the study of solar bursts. This state of confusion results not from any ineptness on the part of the investigators involved but from the inherent complexity of the phenomena. Although some order is beginning to emerge from the study of the simpler S bursts and of the L burst modulation lanes, we are still far from a satisfactory explanation of them. In addition, there is a great variety of other types of burst which defy systematization. It is not yet clear which of the more exotic burst types reported by different observers are the same or are related to one another. One gains the impression that at the moment our view of the forest is being obscured by the trees. However, there can be no doubt that the high-resolution dynamic spectra of the Jovian events are providing us with a wealth of detail related to the intricacies of the emission and propagation process, and that the persistent study of them will eventually be well rewarded.

IV. POLARIZATION

A.   Left-Handed Polarized Sources

It is well known that most of the radiation from Sources A and B is right elliptically or right circularly polarized. However, Kennedy (1969) has shown that Io-C radiation is largely left elliptically polarized at 22.2 and 16 MHz, and left elliptically or circularly polarized at 10 MHz. Io-C was centered at the $(\lambda_{III}, \gamma_{Io})$ coordinates (330°, 250°) at 16 MHz, as is apparent from Fig. 8.[5] The figure represents 16 MHz data obtained at the Maipu Radioastronomical Observatory in Chile during 7 Jovian apparitions. In addition, Kennedy found that Io-D is also strongly left-handed polarized, being centered at about (70°, 110°) at 16 MHz. It can also be seen in Fig. 8. This source was even more conspicuous at 10 MHz (still predominantly left-handed polarized), but did not appear at all at 22.2 MHz. Io-D was first revealed by its dynamic spectral characteristics (Dulk 1965) and has been recognized by several observers on the usual plots of occurrence probability versus λ_{III} and γ_{Io}, but the only determination of its polarization is that by Kennedy.

One interpretation of these results is that the right-handed polarized Io-related sources, Io-B and Io-A, are located in Jupiter's northern hemisphere on either side of the meridian toward which the north magnetic pole is tipped while the left-handed ones, Io-C and Io-D, are located in the southern hemisphere on either side of the meridian of the south magnetic pole. As stated earlier, it is generally believed that Io-B and Io-A represent radiation beamed nearly perpendicularly from opposite sides of the northern-hemisphere foot of the Io-threaded flux tube (Dulk 1965, 1967; Carr et al. 1965; Davis 1966; Piddington 1967; Goldreich and Lynden-Bell 1969; Schatten and Ness 1971). It might therefore be assumed that Io-C and Io-D represent beams on opposite sides of the southern-hemisphere foot of the Io-threaded flux tube. There are geometrical difficulties with this model, however.

Another interpretation is that all four sources are located in the northern hemisphere, near the foot of the Io-threaded flux tube. The radiation is emitted in the extraordinary mode, nearly perpendicular to the flux tube. As that radiation leaving one side on the flux tube propagates outward it enters regions of different base modes and becomes separated into elliptically polarized components of opposite sense following slightly different paths. These two components produce the Io-D (LH polarized) and the Io-B (RH polarized) sources, when the earth is in the proper direction to receive them. Similarly, radiation leaving the other side of the flux tube produces the Io-A (RH polarized) and the Io-C (LH polarized) sources.

Kennedy (1969) also found a variation in the yearly averages (1961–1968) of the axial ratio of the polarization ellipse for the Io-C and Io-D

---

[5]Non-Io C is also apparent in the plot of the LH circular polarization component in Fig. 8 (see Sec. II.G.).

radiation at 16 and 22 MHz, which apparently was correlated with D_E. During years of positive D_E the polarization tended to be more right-handed, and during years of negative D_E more left-handed. If this reported effect can be verified, it should prove very useful in distinguishing between models for emission or propagation of the radiation. Sources A and $Io\text{-}B$ did not show any consistent long-term axial ratio variation.

B. Lack of a Jovian Faraday Effect

Dynamic spectra of Jupiter's noise storms often display horizontal lanes of minimum intensity at widely but regularly spaced frequencies, parallel to the time axis. These are the frequencies at which Faraday rotation in the terrestrial ionosphere has caused the major axis of the polarization ellipse to be perpendicular to the E plane defined by the linearly polarized receiving antenna and the incident ray path. Parker *et al.* (1969) used measured ionospheric data to calculate the amount of Faraday rotation experienced, and deducted this amount from the position angle of the major axis of the polarization ellipse on reception. They found that before reaching the terrestrial ionosphere the polarization ellipse had an orientation independent of frequency over the range 15 to 35 MHz. The position angle of the major axis was $-25° \pm 15°$ from the magnetic dipole axis, for Source B events. Since no appreciable Faraday rotation can occur in the interplanetary medium, this represents the orientation of the polarization ellipse as radiation leaves Jupiter. The measurement indicates that the major axis was nearly perpendicular to the magnetic field in the supposed emission region, corresponding to emission in the extraordinary mode. The implication is that no rotation occurred during the passage through the Jovian magnetosphere. An upper limit of 10 cm^{-3} for Jupiter's magnetospheric electron number density was deduced from these results. However, Goertz (1974) showed that the interpretation leading to this upper limit is not necessarily correct. Frank *et al.* (1975) present Pioneer 10 measurements indicating 100 to 1000 eV proton number densities of ~ 50 cm^{-3} at a distance of $\sim 3R_J$, and it would seem that the number densities might be considerably higher if proton energies below 100 eV were included.

C. Polarization of Burst Fine Structure

No difference has been reported in the average polarization of S and L bursts from the same source. However, Torgersen (1969) has found some interesting effects in an investigation of fixed frequency waveforms of certain classes of bursts as received simultaneously with right and left circularly polarized antennas. He studied two types which he referred to as *random intermittent*, or type Ir, and *quasiperiodic*, or type Iq. His Ir bursts are probably the same as the non-periodic S bursts described previously, while his Iq bursts are the previously discussed sinusoidally-modulated L bursts. The right and left circular components of the fine structure of Ir

bursts were usually well correlated, a given burst having the same shape and structural details on the two polarization channels. For Iq bursts, on the other hand, the correlation of right and left circular components was less pronounced and often absent. Occasionally a rapid and sometimes nearly periodic alternating amplitude variation of the burst fine structure in the two channels was observed, suggesting the *polarization diversity* effect described earlier by Gordon and Warwick (1967). Torgersen concluded that the polarization characteristics of Iq bursts as well as their amplitude modulation are caused by multipath propagation effects in interplanetary space. However, the possibility of the occurrence of the phenomenon within the Jovian magnetosphere or ionosphere certainly cannot be ruled out.

V. HIGH RESOLUTION INTERFEROMETRY

Very long baseline interferometer (VLBI) observations of Jupiter's Source B radiation have been made at 34 MHz using a predominantly east-west baseline of 487,000-wavelengths (Dulk 1970), and at 18 MHz with a predominantly north-south baseline of 462,000 wavelengths (Lynch *et al.* 1972). In all cases the source was unresolved, indicating an angular diameter $< 0\rlap{.}''1$ (assuming incoherent emission), or a linear dimension < 400 km. Although partial resolution of the source was believed to have been achieved in earlier measurements with much shorter baselines, the more recent observations suggest that the source may in fact never have been resolved. Dulk gives an alternative interpretation of his results, based on an assumed coherent source rather than an incoherent one, as indicating an angular diameter < 1 arcsec. Dulk concluded that the observed degree of fringe stability during a 10 sec period of continuous L burst emission implied that the apparent position of the source does not jump about by as much as $0\rlap{.}''2$.

Lynch (1972) and Lynch *et al.* (1972) found that even over the longest baselines nearly all individual S bursts and some L bursts (probably a minority) display high envelope correlation as well as high fringe visibility, implying that their waveforms were not modified to a great extent after leaving the vicinity of Jupiter. These measurements also proved that the sweeping past the earth of narrow beams rotating with Jupiter (or being carried along with Io) could not account for observed burst structure. Lynch *et al.* (1975) showed that successive S bursts in a sequence appear to come from a common source, and that the source does not jump about from burst to burst by as much as $0\rlap{.}''05$.

VI. CORRELATIONS WITH SOLAR ACTIVITY

The long-term (~ 12 yr) modulation of Jupiter's emission probability is now generally regarded as being due mainly to a relatively sharp latitudinal beaming of the non-Io-related Source A emission (see Sec. II.D and references therein) rather than to some effect related to the sunspot number, as had previously been supposed. For example, in the comprehensive analysis by Kaiser and Alexander (1973) of 17 years of 22 MHz data, no significant peaks in Jupiter's power spectrum were found corresponding to either a long-term ($\sim$ 10.5 yr) solar cycle modulation or to a short-term ($\sim$ 25 day) solar rotation modulation. Rather, the long-term spectral power was found to peak at close to the D_E period, as previously mentioned.

It is recognized that the existence of a short-term solar modulation may not reveal itself clearly in a power spectrum presentation. The problem is a complicated one because of the possible masking of any weak solar periodicities by beats between the natural periods of Jupiter, Io, and the earth. Sastry (1968) has demonstrated, for example, that cross correlations between strongly Io-modulated Jupiter activity and either the geomagnetic planetary index Ap or the 10.7 cm solar flux will result in false correlations due to the artificial combination periods involved. This situation may be improved by using non-Io-related activity exclusively. However, at the frequencies where the bulk of such emission occurs ($<$ 20 MHz), there are also greater amounts of terrestrial interference. Further, it is not inconceivable that the solar disturbances required for effective generation of decametric emission may, in fact, be aperiodic in nature.

Many authors have made attempts to overcome the foregoing and other difficulties by selecting the Jupiter data in specific ways and cross correlating the resulting limited data with the more important solar or geomagnetic indices. In a novel approach Kovalenko (1971) and Kovalenko and Malyshkin (1971) have demonstrated what appear to be short-term associations between the occurrence of non-Io activity and both shock waves and high-velocity solar wind streams at Jupiter. Forbush decreases in cosmic ray intensity and fluctuations in the horizontal component of the geomagnetic field are used as indicators of effective interplanetary disturbances. The lag times between the arrival of the streams at the earth and at Jupiter are calculated after assuming a velocity for the shock front which is proportional to the magnitude of the disturbance at the earth (Harang 1968; Neugebauer and Snyder 1967). The total energy in a large shock incident on Jupiter's magnetosphere is shown to be 10^4 times that released by an intense radio burst. The segregation of Io-related from non-Io activity, based upon storms which did or did not exceed 28 MHz, is somewhat uncertain, however. A method by which the selection is made according to the location of storms in the λ_{III}-γ_{Io} plane (Wilson *et al.* 1968*b*; Register and Smith 1969) would be more appropriate.

Following Sastry's (1968) use of the Chree superposed epoch analysis, Barrow (1972) cross correlated the daily geomagnetic character figure C_P with 7 apparitions of Jupiter activity. He minimized the range of delay times between the arrival of particles at the earth and at Jupiter by using only data obtained near Jovian opposition. Peaks in the analysis, which are significant at the 1% level, appear at about $+$ 12 and $-$ 9 days. These are the expected delay times corresponding, respectively, to epochs when the earth is in post-opposition and pre-opposition sectors of its orbit, assuming a solar wind velocity of about 500 km sec^{-1}. The negative delay, according to Barrow, results from an active solar wind stream which has persisted for an entire solar rotation. Jupiter activity may then be recorded *before* the solar wind stream makes its second pass by the earth. The force of the argument is somewhat blunted by the fact that an additional peak at about + 18 days does not appear in the pre-opposition curve, since this would be expected from the *first* passage of the stream across the earth. Io and non-Io-related events were not considered separately in this analysis.

By carefully segregating Io-related from non-Io-related sources and performing individual cross correlations of each with the Zurich sunspot number, Gruber (1975) has found variations in the occurrence of non-Io-emission which are attributed to fluctuations in solar wind strength, but relatively little control of the Io-modulated sources is apparent. When 7 apparitions of Jupiter activity between 1960 and 1967 are merged, *anticorrelations* with lag times of 5 and 27 days (between a change in sunspot number and detection of radio emission) are revealed. The anticorrelation at 5 days delay probably lacks physical credibility because of the unrealistically high solar wind velocity implied (1800 km sec^{-1}). The 27-day lag time is somewhat more significant statistically, and Gruber associates it with a mean solar wind velocity between the sun and Jupiter of about 330 km sec^{-1}. In order to explain the presumed quenching of the Jovian radiation by the solar wind, the author has proposed a model by which the non-Io-emission is driven by Alfvén waves (Goertz 1973) which ultimately derive their energy not from the sun, but from the fact that Jupiter's dipole field is offset in a particular longitude direction (Smith *et al.* 1974). High-latitude fieldlines in this model are emptied of plasma following compression of the magnetosphere by a solar wind shock front, which in turn could suppress the generation of Alfvén waves and explain the anticorrelation observed.

Conseil *et al.* (1971) have found an apparent relationship between the rate of change of the solar wind velocity at Earth and the phase of Io during periods of *Io-A* activity, as shown in Fig. 12. They explain their result on the basis of an assumed plasma bulge in Jupiter's dusk-side hemisphere, in the context of an earlier model (Grebowsky 1970) relating to the terrestrial magnetosphere. The assumed Jovian plasmapause is at about 6 R_J — the orbit of Io — and its location is sensitive to changes in the solar wind velocity gradient. It should be mentioned, however, that there does not appear to be theoretical justification for the existence of a Jovian plasmapause. Unlike the earth, Jupiter's co-rotational electric field is expected to dominate the solar wind-induced convective field completely (Brice and Ioannidis 1970). Nevertheless, Frank *et al.* (1975) believe they have identified a plasmapause-like feature at the orbit of Io based upon *in situ* measurements made by Pioneer 10. The density gradient is less than that for the terrestrial plasmapause by a factor of about 25 (Chappell *et al.* 1971).

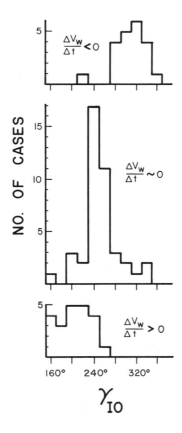

Fig. 12. Histogram showing the number of cases of emission as a function of the geocentric phase of Io for the three different classes of solar wind velocity (V_W) gradients at Jupiter (after Conseil *et al.* 1971).

The data analyzed by Conseil *et al.* extended over a period of about 8 months, during the apparitions of 1964 and 1968. Over this relatively short span of time it is not clear that the Io phase "bins" used in Fig. 12 have been equally monitored. Plotting the occurrence probability of emission instead of the number of cases of emission might have removed this uncertainty.

Although it is quite common to characterize the solar wind velocity at Jupiter through simple extrapolation from earth-orbit measurements, it is clearly not, in general, correct. The interaction that takes place when low- and high-velocity streams collide is not entirely clear (Intriligator 1975, and references therein), but it is surely important for solar correlation studies. Kennedy *et al.* (1974) have investigated the problem in analyzing the occurrence of non-Io-related Source *A* emission during a period of time when there was a stable two-sector solar wind (Sawyer 1974; Wilcox and Ness 1965). A superposed epoch type of analysis indicates the possibility of a correlation. The model is being further investigated in light of interplanetary plasma measurements made by Pioneers 10 and 11.

In summary we see that investigations into possible solar control of the emission have led to results indicating anticorrelations, no influence, and positive correlations. It is obvious that a considerable amount of both theoretical and experimental labor needs to be spent in this area. The situation could be improved through the synthesis of Jovian low-frequency emission data from widely spaced observatories, because the diurnal and seasonal gaps inherent in single-observatory coverage are often a limiting factor. A reliable model of interplanetary plasma interactions is considered essential in characterizing the solar wind velocity, density, and temperature profiles at Jupiter. Extensive studies of the relationship between geomagnetic activity levels and solar wind parameters (see e.g., Garrett *et al.* 1974; Murayama and Hakamada 1975, and references therein) have demonstrated the importance of the southward component of the interplanetary magnetic field (*IMF*). Presumably, northward turnings of the *IMF* might be expected to disturb Jupiter's magnetosphere (Jupiter's magnetic moment is opposite in sign to that of the earth) through the reconnection process first suggested by Dungey (1961). If this process alone were effective in producing emission, only *in situ* measurements of the *IMF* would be appropriate for comparison with the radio data. In this regard, the observations made by Pioneers 10 and 11 and future information to be gained from the Mariner Jupiter/Saturn (MJS-77) mission will be indispensable.

VII. SPACECRAFT EXPERIMENTS
RELATED TO THE JOVIAN LOW-FREQUENCY EMISSION

Groundbased measurements alone might never lead to a satisfactory solution of the perplexing question of the origin of the Jovian low-frequency emission. The final answers will probably come as spacecraft provide the missing pieces of information. This information will be supplied in three stages: first from earth-orbiting and moon-orbiting spacecraft; second from Jupiter flybys; and finally from Jupiter orbiters. As described in Secs. I,II.B, and II.C, we have already benefited from the results of the earth orbiters RAE-1 and IMP-6, and the statistics will probably be improved as the remainder of the RAE-1 data is analyzed. The analysis of Jupiter data from RAE-2, which is orbiting the moon, is in progress. Although the total amount of Jupiter data from RAE-2 may not exceed that from RAE-1, at least a part of it will be of higher quality. The RAE-2 data should improve the statistics of the results from RAE-1 and IMP-6, and among other things may establish whether or not the reported Europa effect (see Sec. II.C.) is real.

The first *in situ* measurements of Jovian magnetospheric parameters were made during the Pioneer 10 and 11 flybys, although low-frequency radio measurements were not included. A number of the results which have been derived from the Pioneer 10 and 11 data have implications with regard to the low-frequency emission. In the first place, the existence of a Jovian magnetic field, the polarity of its dipole component, and the order of magnitude of its strength, all of which were first deduced long ago from decametric observations, have been fully verified. Inner-magnetosphere field measurements have been made by two different methods (Smith et al. 1975;[6] Acuña and Ness[7]); these measurements are of fundamental importance in the effort to account for the observed low-frequency radio phenomena and to establish the mechanism of emission and the mode of propagation. For example, the magnetic field model will substantially influence future discussions of the source locations with respect to the λ_{III} and γ_{Io} coordinates, the frequency ranges of the sources (see e.g., Smith and Wu 1974), and the frequency drift rates on time scales from tenths of seconds to tens of minutes. It is no doubt significant that the central meridian longitude separating Sources B and A, λ_{III} (1965) = 195° (see Table I), agrees within a very few degrees with the longitude of the meridian toward which the north magnetic pole is tipped [as deduced from the Pioneer 10 and 11 results of Smith et al. (1975), and also from the Jovian decimetric results].

Results from other Pioneer 10 and 11 experiments are of comparable importance. The combination of thermal plasma density information (Frank et al. 1975) and the magnetic field data is essential for the identification of plasma resonances and cutoffs in the magnetosphere (Melander and Liemohn 1976). Such plasma and field models are also necessary for ray-tracing computations, which can aid in the determination of source locations, propagation modes, and perhaps beaming patterns (see e.g., Smith 1973; Oya 1974; Goertz 1974). Unfortunately, the accuracy of the low-energy plasma measurements is in question because of the unknown extent of spacecraft charging (Scarf p. 891). Essential to the understanding of Io's role in stimulating the

---

[6]See p. 791.
[7]See p. 830.

low-frequency radiation is the nature of its interaction with the co-rotating magnetosphere. In this regard, the charged particle experiments have identified a number of well-defined effects including a sharp increase in the 0.5 MeV particle flux at the L shell of Io ($L = 6$) (Fillius et al. 1975), and satellite sweeping effects within the $L = 6$ shell (Van Allen et al. 1975).

Two Mariner spacecraft bound for Jupiter (and then Saturn), designated MJS 77, are to be launched about August 1977. They will carry low-frequency radio measuring equipment, the first to be sent to Jupiter. The intensities of the radiation received by right and left circularly polarized antennas are to be recorded over a frequency range extending from 0.02 to 40.5 MHz (in 198 steps). Thus the entire Jovian low-frequency radio spectrum and much of the plasma wave region can be investigated. It is anticipated that the more powerful Jupiter events can be detected reliably from perhaps one year before Jovian encounter until about one year afterward. Nearer encounter, of course, the weaker emission components will also be measurable.

The MJS 77 flights should provide us with a much clearer picture of many of the Jovian low-frequency phenomena. Dynamic spectra covering the entire frequency range will be available, for each polarization sense, with integration times suitable for depicting the gross structure and varying degrees of fine structure of the storms and bursts. Comparisons of dynamic spectra of the same storms as recorded at the spacecraft and on Earth, when the spacecraft is at different Jovicentric phases, should provide new insights into the emission beam structure. The long, almost uninterrupted, data runs during the cruise stages of the flight toward and away from the planet will provide an excellent opportunity for accumulating data to be used in searches for possible control by satellites other than Io, notably Europa and Amalthea. These long runs will also make possible for the first time cross correlations with solar wind parameters and time-series power spectrum analyses which are almost uncontaminated by interruptions. Near Jovian encounter the weakest emission components, perhaps including a low-frequency continuum, can be investigated if there happens to be a lull between major outbursts at that time. Simultaneous observations of bursts from the spacecraft and the earth when the Jovicentric phase of the spacecraft is near zero will provide a direct indication of the modification of burst structure by the intervening interplanetary medium.

Although the MJS 77 results should lead to a considerable increase in our knowledge of the Jovian low-frequency emissions, important if not crucial questions will probably remain unanswered. One of the things we would most like to know is just where the radiation originates. MJS 77 probably cannot provide definite information on the location of the source or sources. The answer to this and perhaps other key questions may have to await the Jupiter orbiter which is under consideration for future space programs.

Acknowledgements. Recognition and thanks are due for the perceptive and helpful suggestions by J. K. Alexander. Support for the preparation of this chapter and for the research contributions by the authors themselves was provided in part by the National Science Foundation and the National Aeronautics and Space Administration.

REFERENCES

Acuña, M. H., and Ness, N. F. 1975. Jupiter's main magnetic field measured by Pioneer 11. *Nature* 253:327–328.

Alexander, J. K. 1975. Note on the beaming of Jupiter's decameter-wave radiation and its effect on radio rotation period determinations. *Astrophys. J.* 195:227–233.

Alexander, J. K.; Kaiser, M. L.; and Vaughan, S. S. 1975. Decameter-wave radio observations of Jupiter: apparitions of 1970–74. NASA-GSFC Report X-693-75-48.

Barrow, C. H. 1972. Decameter-wave radiation from Jupiter and solar activity. *Planet. Space Sci.* 20:2051–2056.

Berge, G. L. 1971. Some recent observations and interpretations of the Jupiter decimetric emission. NASA Report TM33-543:223–242.

Bigg, E. K. 1964. Influence of the satellite Io on Jupiter's decametric emission. *Nature* 203: 1008–1010.

———. 1966. Periodicities in Jupiter's decametric radiation. *Planet. Space Sci.* 14:741–758.

Bozyan, F. A., and Douglas, J. N. 1976. Directivity and stimulation in Jovian decametric radiation. *Icarus* (special Jupiter issue). In press.

Bozyan, F. A.; Douglas, J. N.; and Gopala Rao, U. V. 1972. Catalog of decameter observations of Jupiter. University of Texas Series II, 3, No. 8.

Brice, N. M., and Ioannidis, G. A. 1970. The magnetospheres of Jupiter and earth. *Icarus* 13:173–183.

Brown, L. W. 1973. The galactic radio spectrum between 130 and 2600 kHz. *Astrophys. J.* 180:359–370.

———. 1974a. Jupiter emission observed near 1 MHz. *Astrophys. J.* 192:547–550.

———. 1974b. Spectral behavior of Jupiter near 1 MHz. *Astrophys. J.* 194:L159–L162.

Carr, T. D. 1972. Jupiter's decametric rotation period and the source-A emission beam. *Phys. Earth Planet. Interiors* 6:21–28.

Carr, T. D.; Brown, G. W.; Smith, A. G.; Bollhagen, H.; May, J.; and Levy, J. 1964. Spectral distribution of the decametric radiation from Jupiter in 1961. *Astrophys. J.* 140:778–795.

Carr, T. D., and Gulkis, S. 1969. The magnetosphere of Jupiter. *Ann. Rev. Astron. Astrophys.* 7:577–618.

Carr, T. D.; Gulkis, S.; Smith, A. G.; May, J.; Lebo, G. R.; Kennedy, D. J.; and Bollhagen, H. 1965. Results of recent investigations of Jupiter's decametric radiation. *Radio Sci.* 69D: 1530–1536.

Carr, T. D.; Smith, A. G.; Donivan, F. F.; and Register, H. I. 1970. The twelve-year periodicities of the decametric radiation of Jupiter. *Radio Sci.* 5:495–503.

Chappell, C. R.; Harris, K. K.; and Sharp, G. W. 1971. The dayside plasmasphere. *J. Geophys. Res.* 76:7632–7647.

Conseil, L.; Leblanc, Y.; Antonini, G.; and Quemada, D. 1971. The effect of the solar wind velocity on the Jovian decametric emission. *Astrophys. Letters* 8:133–137.

Davis, Jr., L. 1966. Proceedings of the Cal Tech-JPL Lunar and Planetary Conference. JPL-Cal. Inst. Tech., T.M. 33–266.

Desch, M. D., and Carr, T. D. 1974a. Positive identification of Jupiter bursts at frequencies below the ionospheric cutoff as detected by the RAE-1 satellite. *Bull. Amer. Phys. Soc.* 19: 701.

———. 1974b. Decametric and hectometric observations of Jupiter from the RAE-1 satellite. *Astrophys. J.* 194:L57–L59.

———. 1975. Satellite modulations of Jovian emission below 10 MHz. Submitted to *Astrophys. J.*

Desch, M. D.; Carr, T. D.; and Levy, J. 1975. Observations of Jupiter at 26.3 MHz using a large array. *Icarus* 25:12–17.

Donivan, F. F., and Carr, T. D. 1969. Jupiter's decametric rotation period. *Astrophys. J.* 157: L65–L68.

Douglas, J. N. 1964. Decametric radiation from Jupiter. *IEEE Trans. Mil. Elec.* Mil-8:173–187.

Douglas, J. N., and Bozyan, F. A. 1970. Refutation of Bigg's evidence for new Jovian satellites. *Astrophys. Letters* 4:227–228.

Douglas, J. N., and Smith, H. J. 1963. Decametric radiation from Jupiter. I. Synoptic observations 1957–1961. *Astron. J.* 68:163–180.

————. 1967. Interplanetary scintillation in Jovian decametric radiation. *Astrophys. J.* 148: 885–903.

Dowden, R. L. 1963. Polarization measurements of Jupiter radio bursts at 10.1 Mc/s. *Aust. J. Phys.* 16:398–410.

Dulk, G. A. 1965. Io-related radio emission from Jupiter. Ph.D. Dissertation, University of Colorado, Boulder, Colorado.

————. 1967. Apparent changes in the rotation rate of Jupiter. *Icarus* 7:173–182.

————. 1970. Characteristics of Jupiter's decametric radio source measured with arc-second resolution. *Astrophys. J.* 159:671–684.

Dulk, G. A., and Clark, T. A. 1966. Almost-continuous radio emission from Jupiter at 8.9 and 10 MHz. *Astrophys. J.* 145:945–948.

Duncan, R. A. 1966. Factors controlling Jovian decametric emission. *Planet. Space Sci.* 14:1291–1301.

————. 1967. Jupiter's rotation period. *Planet. Space Sci.* 15:1687–1694.

————. 1970. A theory of Jovian decametric emission. *Planet. Space Sci.* 18:217–228.

————. 1971. Jupiter's rotation. *Planet. Space Sci.* 19:391–398.

Dungey, J. W. 1961. Interplanetary magnetic field and auroral zones. *Phys. Rev. Letters* 6: 47–48.

Ellis, G. R. A. 1965. The decametric radio emission of Jupiter. *Radio Sci.* 69D:1513–1530.

————. 1973. Fine structure of the Jupiter radio bursts. *Nature* 241:387–389.

————. 1974. The Jupiter radio bursts. *Proc. Astron. Soc. Australia* 2:1–8.

————. 1975. Spectra of the Jupiter radio bursts. *Nature* 253:415–417.

Fillius, R. W.; McIlwain, C. E.; and Mogro-Campero, A. 1975. Radiation belts of Jupiter: a second look. *Science* 188:465–467.

Flagg, R. S.; Krausche, D. S.; and Lebo, G. R. 1975. High resolution spectral analysis of the Jovian decametric radiation. II. The band-like emission. *Icarus.* In press.

Frank, L. A.; Ackerson, K. L.; Wolfe, J. H.; and Mihalov, J. D. 1975. Observations of plasmas in the Jovian magnetosphere. *University of Iowa Report* 75-5.

Garrett, H. B.; Dessler, A. J.; and Hill, T. W. 1974. Influence of solar wind variability on geomagnetic activity. *J. Geophys. Res.* 79:4603–4610.

Goertz, C. K. 1971. Variation of source A position of the Jovian decametric radiation. *Nature* 229:151–152.

————. 1973. The Io-controlled decametric radiation. *Planet. Space Sci.* 21:1431–1445.

————. 1974. Polarization of the Jovian decametric radiation. *Planet. Space Sci.* 22:1491–1500.

Goertz, C. K., and Haschick, A. 1972. Variation of highest cutoff frequency of the Io-controlled Jovian decametric radiation. *Nature* 235:91–94.

Goldreich, P., and Lynden-Bell, D. 1969. Io, a Jovian unipolar inductor. *Astrophys. J.* 156: 59–78.

Gorden, M. A., and Warwick, J. W. 1967. High-time resolution studies of Jupiter's radio bursts. *Astrophys. J.* 148:511–540.

Grebowsky, J. M. 1970. Model study of plasmapause motion. *J. Geophys. Res.* 75:4329–4333.

Gruber, G. M. 1965. Possible contribution of Jupiter's magnetospheric tail to the radio emission of the planet in the decametric region. *Nature* 208:1271–1273.

————. 1975. Jupiter's non Io-related decametric radiation. *Astrophys. and Space Sci.* In press.

Gruber, G. M., and Way-Jones, C. 1972. Strong beaming of Jupiter's non-Io-related sources. *Nature* 237:137–139.

Gulkis, S., and Carr, T. D. 1966. Radio rotation period of Jupiter. *Science* 154:257–259.

Gurnett, D. A. 1974. The earth as a radio source: terrestrial kilometric radiation. *J. Geophys. Res.* 79:4227–4238.

Harang, L. 1968. The Forbush-decrease in cosmic-rays and the transit time of the modulating cloud. *Planet. Space Sci.* 16:1059–1101.

Hill, T. W., and Dessler, A. J. 1976. Longitudinal asymmetry of the Jovian magnetosphere and the periodic escape of energetic particles. In draft.

Intriligator, D. S. 1975. In situ observations of the scale size of plasma turbulence in the asteroid belt (1.6-3 astronomical units). *Astrophys. J.* 196:L87–L90.

Kaiser, M. L., and Alexander, J. K. 1972. The Jovian decametric rotation period. *Astrophys. Letters* 12:215–217.

————. 1973. Periodicities in the Jovian decametric emission. *Astrophys. Letters* 14:55–58.

Kaiser, M. L., and Stone, R. 1975. Earth as an intense planetary radio source: similarities to Jupiter and Saturn. *Science* 189:285–287.

Kennedy, D. J. 1969. Polarization of the decametric radiation from Jupiter. Ph.D. Dissertation, University of Florida, Gainesville, Florida.

Kennedy, J. R.; Lebo, G. R.; and Pomphrey, R. B. 1974. A correlation study of solar wind velocity modulation effects and the non-Io-related component of the Jovian decametric radiation. *Bull. Amer. Astron. Soc.* 6:431.

Kovalenko, V. A. 1971. The relationship of Jupiter's decametric radio emission to solar and geomagnetic activity. *Sov. Astron.* 15:478–486.

Kovalenko, V. A., and Malyshkin, V. N. 1971. Decametric radio emission of Jupiter as indicator of high velocity fluxes and shock waves in the solar wind. *Geomagnetism Aeron.* 11: 747–749.

Krausche, D. S.; Flagg, R. S.; and Lebo, G. R. 1975. High resolution spectral analysis of the Jovian decametric radiation. I. Burst morphology and drift rates. *Icarus.* In press.

Krausche, D. S.; Lebo, G. R.; Flagg, R. S.; and Kennedy, J. R. 1973. Recent studies of high-resolution spectral recordings of Jovian decametric emission. *Bull. Amer. Phys. Soc.* 18: 264.

Leacock, R. L., and Smith, A. G. 1973. Fine structure of Jupiter's decametric source B. *Nature* 244:60–61.

Lebo, G. R. 1964. Decameter-wavelength radio observations of the planets in 1962. Ph.D. Dissertation. University of Florida, Gainesville, Florida.

Lecacheux, A. 1974. Periodic variations of the position of Jovian decameter sources in longitude (System III) and phase of Io. *Astron. Astrophys.* 37:301–304.

Lynch, M. A. 1972. Observations of Jupiter's decametric radiation with a very-long-baseline interferometer. Ph.D. Dissertation. University of Florida, Gainesville, Florida.

Lynch, M. A.; Carr, T. D.; and May, J. 1975. In draft.

Lynch, M. A.; Carr, T. D.; May, J.; Block, W. F.; Robinson, V. M.; and Six, N. F. 1972. Long-baseline analysis of a Jovian decametric L burst. *Astrophys. Letters* 10:153–158.

McCulloch, P. M. 1971. Theory of Io's effect on Jupiter's decametric emissions. *Planet. Space Sci.* 19:1297–1312.

McCulloch, P. M., and Ellis, G. R. A. 1966. Observations of Jupiter's decametric radio emissions. *Planet. Space Sci.* 14:347–359.

Melander, B., and Liemohn, H. 1976. CMA propagation diagrams for the Jovian magnetosphere. *Icarus* (special Jupiter issue). In press.

Miller, H. R., and Smith, A. G. 1973. Flux-density maps of Jupiter's decametric radio sources. *Astrophys. J.* 186:687–694.

Mitchell, J. L. 1974. The rotation period of the Jovian magnetosphere. Master's Thesis, University of Florida, Gainesville, Florida.

Murayama, T., and Hakamada, K. 1975. Effects of solar wind parameters on the development of magnetospheric substorms. *Planet. Space Sci.* 23:75–91.

Neugebauer, M., and Snyder, C. W. 1967. Mariner 2 observations of the solar wind. *J. Geophys. Res.* 72:1823–1828.

Olsson, C. N., and Smith, A. G. 1966. Decametric radio pulses from Jupiter: characteristics. *Science* 153:289–290.

Oya, H. 1974. Origin of Jovian decameter wave emissions — conversion from the electron cyclotron plasma wave to the ordinary mode electromagnetic wave. *Planet. Space Sci.* 22:687–708.

Parker, G. D.; Dulk, G. A.; and Warwick, J. W. 1969. Faraday effect on Jupiter's radio bursts. *Astrophys. J.* 157:439–448.

Piddington, J. H. 1967. Jupiter's magnetosphere. *University of Iowa Report* 67–63.

Register, H. I. 1968. Decameter-wavelength radio observations of the planet Jupiter 1957–1968. Ph.D. Dissertation, University of Florida, Gainesville, Florida.

Register, H. I., and Smith, A. G. 1969. A two component model of changes in Jupiter's radio-frequency rotation period. *Astrophys. Letters* 3:209–213.

Riddle, A. C., and Warwick, J. W. 1976. Redefinition of system III longitude. *Icarus* (special Jupiter issue). In press.

Riihimaa, J. J. 1970. Modulation lanes in the dynamic spectra of Jovian L bursts. *Astron. Astrophys.* 4:180–188.

———. 1971. Radio spectra of Jupiter. Report S22, Dept. of Electrical Engineering. University of Oulu, Oulu, Finland.

———. 1974. Modulation lanes in the dynamic spectra of Jupiter's decametric radio emission. *Ann. Acad. Sci. Fenn.* AVI:1–38.

———. 1975. Observations of Jovian S-bursts with an electro-optical radio spectrograph. *Astron. Astrophys.* 39:69–70.

Riihimaa, J. J.; Dulk, G. A.; and Warwick, J. W. 1970. Morphology of the fine structure in the dynamic spectra of Jupiter's decametric radiation. *Astrophys. J. Suppl.* 172 19:175–192.

Sastry, Ch. V. 1968. Decameter radio emission from Jupiter and solar activity. *Planet. Space Sci.* 16:1147–1153.

Sawyer, C. 1974. Semiannual and solar cycle variation of sector structure. *Geophys. Res. Letters* 1:295–297.

Schatten, K. H., and Ness, N. F. 1971. The magnetic field geometry of Jupiter and its relation to Io-modulated Jovian decametric emission. *Astrophys. J.* 165:621–631.

Six, N. F.; Smith, A. G.; and Carr, T. D. 1963. Solar particle stream deviation by the earth's magnetosphere. *Bull. Amer. Phys. Soc.* 8:550.

Smith, E. J.; Davis, Jr., L.; Jones, D. E.; Coleman, Jr., P. J.; Colburn, D. S.; Dyal, P.; and Sonett, C. P. 1975. Jupiter's magnetic field, magnetosphere, and interaction with the solar wind: Pioneer 11. *Science* 188:451–455.

Smith, E. J.; Davis, Jr., L.; Jones, D. E.; Coleman, Jr., P. J.; Colburn, D. S.; Dyal, P.; Sonnett, C. P.; and Frandsen, A. M. A. 1974. The planetary magnetic field and magnetosphere of Jupiter: Pioneer 10. *J. Geophys. Res.* 79:3501–3513.

Smith, R. A. 1973. On the Io-modulated Jovian decametric radiation. Ph.D. Dissertation, University of Maryland, College Park, Maryland.

Smith, R. A., and Wu, C. S. 1974. Implication of the Pioneer 10 measurements of the Jovian magnetic field for theories of Io-modulated decametric radiation. *Astrophys. J.* 193:L101–L102.

Stone, R. G.; Alexander, J. K.; and Erickson, W. C. 1964. Low-level decameter emissions from Jupiter. *Astrophys. J.* 140:374–377.

Thieman, J. R.; Smith, A. G.; and May, J. 1975. Motion of Jupiter's decametric sources in Io phase. *Astrophys. Lett.* In press.

Tiainen, P. O. 1967. Dependence of decametric radio emission from Jupiter on the position of the first two Galilean satellites. *Nature* 215:1467–1468.

Torgersen, H. 1969. Structural differences in the Jupiter millisecond pulses. *Physica Norvegica* 3:195–202.

Torgersen, H. 1972. The quasi-periodic modulation of L-bursts in Jupiter's decametric radiation. Preprint, Phys. Dept. Norway Inst. Tech., University of Trondheim, Trondheim, Norway.

Van Allen, J. A.; Randall, B. A.; Baker, D. N.; Goertz, C. K.; Sentman, D. D.; Thomsen, M. F.; and Flinot, H. R. 1975. Pioneer 11 observations of energetic particles in the Jovian magnetosphere. *Science* 188:459–462.

Vasyliunas, V. M. 1975. Modulation of Jovian interplanetary electrons and the longitude variation of decametric emissions. *Geophys. Res. Lett.* 2:87–89.

Warwick, J. W. 1963. Dynamic spectra of Jupiter's decametric emission, 1961. *Astrophys. J.* 137:41–60.

———. 1967. Radiophysics of Jupiter. *Space Sci. Rev.* 6:841–891.

Warwick, J. W.; Dulk, G. A.; and Riddle, A. C. 1975. Jupiter radio emission, January 1960–March 1975. Report PRA No. 3, Radio Astronomy Observatory of the University of Colorado.

Wilcox, J. M., and Ness, N. F. 1965. Quasi-stationary corotating structure in the interplanetary medium. *J. Geophys. Res.* 70:5793–5805.

Wilson, R. G.; Warwick, J. W.; Dulk, G. A.; and Libby, W. F. 1968*a*. Europa and the decametric radiation from Jupiter. *Nature* 220:1218–1222.

Wilson, R. G.; Warwick, J. W.; and Libby, W. F. 1968*b*. Fifth source of Jupiter decametric radiation. *Nature* 220:1215–1218.

Zabriskie, F. R. 1970. Low-frequency radio emission from Jupiter. *Astron. J.* 75:1045–1051.

DYNAMICS OF THE JOVIAN MAGNETOSPHERE AND ENERGETIC PARTICLE RADIATION

J. A. SIMPSON and R. B. McKIBBEN
University of Chicago

We summarize our current ideas and conclusions concerning the nature of the Jovian magnetosphere as revealed by studies of energetic charged particles during the encounters of Pioneers 10 and 11 with Jupiter. We discuss primarily observations from the University of Chicago experiment covering electron energies from about 3−6 and 6−30 MeV and the integral electron flux $\gtrsim 3$ MeV; and proton energies 0.5−1.8 MeV and the integral flux of protons and heavier nuclei $\gtrsim 35$ MeV/nucleon. The energetic particle observations have confirmed the maintenance of the Jovian trapped radiation by inward diffusion of particles from the outer magnetosphere, and have shown that the electron flux level is in reasonable agreement with that necessary to account for the observed decimetric radio emission. In the outer magnetosphere, a highly compressible region inflated by plasma spun off from Jupiter, we find that observations at high magnetic latitude cast doubt on the confinement of energetic particles to the magnetic equatorial plane, as had been concluded from Pioneer 10 data. We find the observations to be described better by a model in which the population of energetic particles in the outer magnetosphere is renewed every 10 hours. In interplanetary space we find Jovian electrons continually present, occasionally in company with ~ 1 MeV protons of Jovian origin. Collaborative studies with magnetometer investigators have defined the interplanetary conditions necessary for observation of a Jovian electron event and have shown the generation of hydromagnetic waves in the interplanetary medium produced by particles streaming away from Jupiter.

Experiments on the Pioneers 10 and 11 spacecraft have revealed Jupiter's magnetosphere to be a giant laboratory in which we may study some of the most important problems in high-energy astrophysics today; namely the physics of natural acceleration of charged particles and of the interactions between magnetic fields, plasmas, and energetic particles in rapid motion on an astrophysical scale. Jupiter's rapidly rotating magnetosphere in many ways approximates that surrounding a rapidly rotating star, and it appears likely that from investigations of this magnetosphere, so far unique in the solar system, and of its effects on the surrounding interplanetary medium we shall learn of phenomena which can be scaled to other astrophysical systems both in our solar system and in the galaxy.

[738]

In this chapter, we give a summary of our major discoveries and conclusions to date concerning the nature of Jupiter's magnetosphere and the physical processes which are important there. We base our report primarily on conclusions from the analysis performed with data from our own experiments, the University of Chicago charged particle instruments on Pioneers 10 and 11 [described by Simpson et al. (1974a), McKibben et al. (1973), and Chenette et al. (1974)] which measured the fluxes and energy spectra of energetic protons and electrons throughout the spacecrafts' traversals of the magnetosphere. We shall not, however, neglect to mention the work of other experimenters whose data are in many cases similar or complementary to our own. Since this chapter is intended as an introduction to and summary of current ideas about Jupiter's magnetosphere, we shall present only a few new results and instead report what it is that we understand about Jupiter's magnetosphere today, what puzzles remain before us, and how studies of Jupiter's magnetosphere can increase our understanding of basic physical processes in space.

I. OVERVIEW

Careful observations of radio emissions from Jupiter yielded much information about Jupiter's magnetic field even before the launch of Pioneer 10. These radio emissions, which result from cyclotron and synchrotron radiation from electrons spiraling in Jupiter's magnetic field, provided the first evidence for the existence of a magnetic field for a planet other than Earth. Study of the polarization, intensity profile and time variations yielded estimates of the magnetic dipole strength and orientation which were in reasonable agreement with those found by Pioneers 10 and 11 from *in situ* measurements. Estimates of the size and form of the Jovian magnetosphere came primarily from scaling the earth's magnetosphere using the solar wind pressure and dipole strength appropriate to Jupiter, and although most investigators recognized the probable importance of centrifugal stresses from plasma imbedded in the rapidly rotating magnetic field of Jupiter, quantitative models for the magnetosphere including this effect were primitive at best. Several articles summarize the fundamental early work and state of knowledge concerning Jupiter's magnetosphere before Pioneer 10 (e.g., Carr and Gulkis 1969; Kennel 1973; Newburn and Gulkis 1973; Warwick 1967; Ioannidis and Brice 1971; Brice and McDonough 1973).

Since the encounters of Pioneers 10 and 11, our picture of Jupiter's magnetosphere has changed fundamentally. It is now believed that in the outer magnetosphere centrifugal forces lead to concentration of plasma in a sheet near the magnetic equator, which carries a ring current responsible for inflation of the magnetosphere to a radius of about 100 Jovian radii (R_J) (Smith et al. 1974), whereas pressure balance between the measured dipole field and the solar wind alone would yield a radius of only $\sim 40R_J$. Thus, in the outer magnetosphere, balance between the pressures from plasma being spun off from Jupiter and the variable solar wind pressure determines the form and extent of the magnetosphere. Figure 1 contains a highly approximate and idealized sketch of the structure and balancing forces as we presently understand them. The outer plasma-dominated region has no precise analogue in the earth's magnetosphere, but most nearly resembles the situation found in the earth's magnetotail where the magnetic field is also highly distorted and stretched out from a dipole configuration by currents in a central plasma sheet. [See, for example, Hess (1968) for a comprehensive description of Earth's magnetosphere.] On the other hand the innermost part of Jupiter's magnetosphere, at distances less than about 10 R_J from Jupiter, bears a strong similarity to the inner region of Earth's magnetosphere. Here the magnetic field is strong enough to resist deformation by solar wind pressure in the case of Earth, and by centrifugal stresses in the case of Jupiter. In this region we find intense fluxes of energetic protons and electrons stably

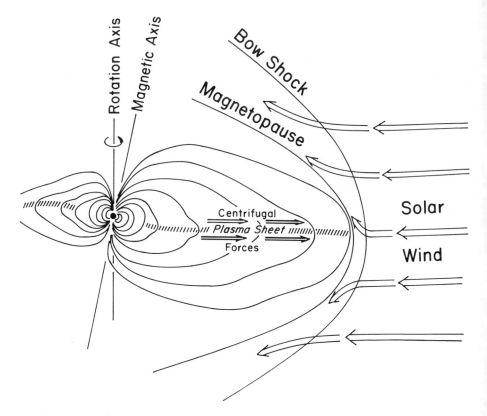

Fig. 1.   Idealized sketch of probable conformation of Jovian magnetosphere. The drawing is approximately to scale for the magnetosphere in a non-compressed state, with a "quiet" solar wind.

trapped in the dipole-like magnetic field, and it is from the inner portion of this region that the radio emissions observed at Earth originate. A significant difference from the earth is the presence of the satellites of Jupiter with orbits deeply embedded in the stably trapped radiation zone. Study of the absorption of the trapped radiation by these moons has provided evidence which virtually proves the nature of the acceleration mechanism which maintains the high-energy radiation belts (Simpson *et al.* 1974*b*; McIlwain and Fillius 1975).

The encounter trajectories of Pioneers 10 and 11 are shown in Fig. 2, with locations of bow shock and magnetopause crossings based on the data of Wolfe *et al.* (1974) for Pioneer 10 and Smith (personal communication) for Pioneer 11. Multiple crossings of the bow shock and magnetopause indicate that the outer, plasma-inflated magnetosphere is highly compressible in response to changes in solar wind pressure. The inbound trajectories for Pioneers 10 and 11 in the outer magnetosphere were nearly identical, whereas outbound, Pioneer 10 stayed near the equatorial plane and exited at the dawn meridian while Pioneer 11 explored the high-latitude region near the nose of the magnetosphere. In the inner magnetosphere, Pioneer 10's trajectory was confined primarily to the equatorial plane, with a closest approach distance of 2.8 R_J from the center of Jupiter while Pioneer 11 remained mainly at high magnetic latitudes, plunging rapidly through the equatorial plane near its closest approach distance of 1.6 R_J. Mead (1974) has published a description of the Pioneer 10 trajectory in Jovi-magnetic coordinates, and representations of the Pioneer 11 trajectory in magnetic coordinates are given by Van Allen *et al.* (1975) and Trainor *et al.* (1975).

In discussing energetic particles associated with Jupiter's magnetosphere, we shall treat first the stably trapped particles in the inner core region, where the physics is reasonably similar to that found in the earth's radiation belts, and then we shall discuss the particles found in the inflated outer region, where the physical situation is quite different from that encountered in the earth's magnetosphere, and is far from fully understood. Finally we shall discuss particles observed in interplanetary space which have escaped from Jupiter's magnetosphere.

Before entering into the detailed discussion of observations, however, it is worthwhile to discuss briefly the significance of energetic particle observations for understanding the magnetosphere. Although interesting in their own right for investigating mechanisms of particle acceleration and energy loss, energetic particle observations can also be used to help define the magnetic field geometry. Magnetometer observations provide the magnitude and direction of a magnetic field at only the point of observation. Large numbers of these single point observations are collected and are fitted into a model which then predicts the magnetic field at points other than those sampled. Charged particle observations can be used to test the validity of such models, for in a magnetic field energetic particles move in a predictable manner, con-

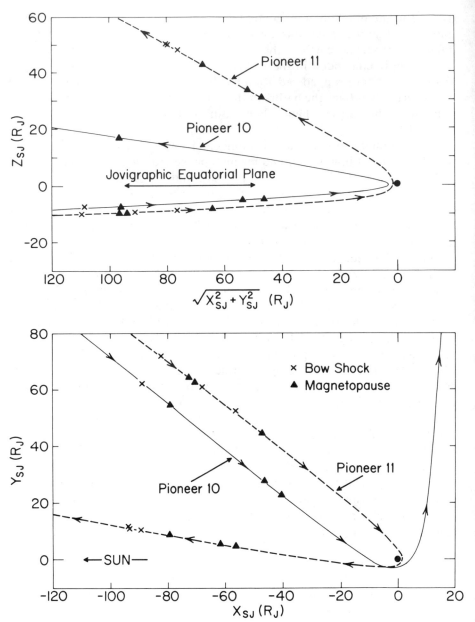

Fig. 2. The trajectories of Pioneer 10 and 11 in the sun-Jupiter system. X_{SJ} is in the ecliptic plane and is positive radially away from the sun along the sun-Jupiter line, Z_{SJ} is positive toward the north pole and is perpendicular to the ecliptic, and Y_{SJ} completes a right-handed coordinate system. Magnetopause and bow shock crossings have been indicated based on the data of Wolfe *et al.* (1974) for Pioneer 10 and Smith (personal communication) for Pioneer 11. The unit of distance is the Jovian equatorial radius ($1R_J = 71,372$ km), and distances are measured from the center of the planet (figure from Simpson *et al.* 1975).

serving their adiabatic invariants if the field varies slowly with time (e.g., Chapters I – III of Roederer 1970). Since, in the course of their trajectories, combining gyration about a magnetic field line, bounce between mirror points along the field, and drift across field lines, the energetic particles sample a large volume of the magnetic field, observations of these particles at different points in the inner magnetosphere can provide a sensitive test of magnetic field models. In the less well ordered magnetic field of the outer region, we may use the fact that motion along a field line is much more rapid than motion across field lines to test the degree of connection of magnetic fields between two regions such as, for example, the regions inside and outside the magnetopause.

II. INNER CORE

The intensity profiles of $\gtrsim 3$ MeV electrons, $\gtrsim 35$ MeV/nucleon protons and heavier nuclei, and $0.5 - 1.8$ MeV protons observed by Pioneer 11 inside $R \approx 10\ R_J$ are shown in Fig. 3, together with similar profiles from Pioneer 10 data for the electrons and high-energy protons. The data are plotted as a function of the McIlwain parameter L defined to be $L = R/\cos^2\lambda$ in a dipole field, where R is radial distance in R_J and λ is the magnetic latitude. L is constant along a field line in a dipole field, and is conserved in the drift motion of a particle across field lines (see ch. IV. 4 of Roederer 1970). The D2 magnetic field model, based on the Pioneer 10 data of Smith et al. (1974),[1] has been used to define L in this figure.

Several conclusions can be drawn immediately from inspection of these data. First, since the fluxes observed on the equatorial trajectory of Pioneer 10 are higher than those observed at high latitudes by Pioneer 11, the stably trapped radiation is confined preferentially to the magnetic equatorial plane. Good agreement between fluxes observed at the same L and magnetic latitude by Pioneer 10 and 11 suggests both that calibration error is not the source of this difference, and that the stably trapped flux is approximately constant over periods as long as one year. Comparing the $\gtrsim 3$ MeV electron flux for Pioneers 10 and 11 in the range $2.9 \leq L \leq 3.5$, we find that if we wish to describe the dependence of the flux on magnetic latitude as a power law in (B/B_{eq}), as would be appropriate for an equatorial directional flux proportional to some power of the sine of the pitch angle (Van Allen et al. 1974), then the flux falls towards high latitudes approximately as $(B/B_{eq})^{-0.8}$ along a given field line for magnetic latitudes $\lesssim 45°$ although there is evidence that near the magnetic equator ($B = B_{eq}$) the dependence upon magnetic latitude is stronger, and that at high latitudes it is weaker than predicted by this power law in B. Such behavior would suggest two components in the trapped flux, one primarily with large pitch angles mirroring near the magnetic equator,

---

[1]See p. 788.

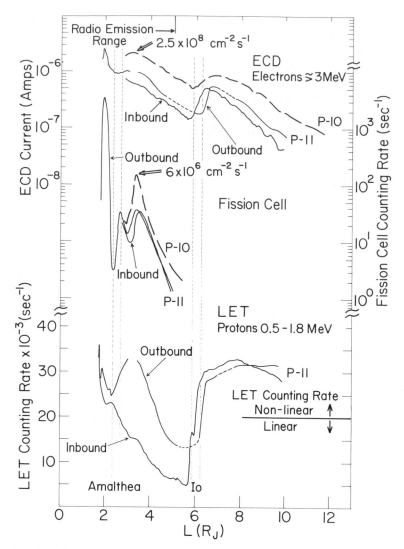

Fig. 3. The intensity profiles of $\gtrsim 3$ MeV electrons, high-energy nuclei from the fission cell, and ~ 1 MeV protons measured by Pioneer 11 as a function of magnetic shell parameter L in the dipole region of Jupiter's magnetosphere. For comparison, inbound data from Pioneer 10 for $\gtrsim 3$ MeV electrons and high-energy nuclei have also been plotted as the dashed curves. The D2 magnetic field model of Smith *et al.* (1974) has been used to compute L. The differential flux of $0.5 - 1.8$ MeV protons (cm^{-2} sec^{-1} MeV^{-1}) may be obtained by multiplying the counting rate of the low-energy telescope (LET) by 20 for rates $< 2 \times 10^4$ counts per second. For the $\gtrsim 3$ MeV electrons, the flux (cm^{-2} sec^{-1}) may be obtained by multiplying the Electron Current Detector (ECD) current by 10^{14}. Large errors (on the order of a factor $2-3$) should be assigned to this flux as a result of spectral dependence of the ECD response and calibration uncertainty. For the fission cell, if the entire response were to protons ≥ 35 MeV, the flux (cm^{-2} sec^{-1}) could be obtained by multiplying the counting rate by 4×10^4. However, for reasons discussed by Simpson *et al.* (1975), we believe heavier nuclei made a significant contribution to the counting rate, so that the flux so derived should be considered an upper limit (figure from Simpson *et al.* 1975).

and the other with a more isotropic equatorial pitch-angle distribution. Van Allen *et al.* (1975)[2] reach a similar conclusion. As discussed below, such a situation could plausibly arise as a result of absorption of trapped particles by the Jovian satellites.

A second conclusion immediately apparent from Fig. 3 is that the profiles of intensity do not show a simple monotonic rise towards low *L* values but indicate a complex, shell-like structure in the trapped radiation. The structure varies depending upon the particle species and energy, but one feature which is common to all species and energies displayed in Fig. 3 is a decrease in the vicinity of the *L* shells occupied by Io and Amalthea. Prior to Pioneer 10 encounter, Mead and Hess (1973) had considered the effect of the satellites embedded in Jupiter's radiation belts and had concluded that, if particles in the belts are maintained and accelerated primarily by inward diffusion with violation of the third adiabatic invariant [see reviews by Shabansky (1971) and Walt (1971)],[3] the satellite Io in particular should remove a large fraction of the particles diffusing inwards across its orbit with a consequent reduction in the intensity of the trapped radiation inside the orbit of Io. The basic correctness of this conclusion is apparent in Fig. 3. Further evidence comes from examination of pitch-angle distributions of the low-energy (0.5 – 1.8 MeV) protons which are strongly absorbed by Io. In Fig. 4 we show the variations of the proton intensity in the course of the spacecraft rotation for various *L* shells from just outside Io inbound (panel *A*) through Pioneer 10's closest approach to Jupiter (panel *E*) to a point just outside Io outbound (panel *I*). Arrows indicate the positions expected for particles traveling along the magnetic field. Clearly, Io preferentially removes particles traveling along the field with small pitch angles, as expected from the nature of Io's motion in magnetic coordinates (Mead and Hess 1973; Simpson *et al.* 1974*b*; Mogro-Campero and Fillius 1975). The persistence of the depleted cone of small pitch-angle particles inwards to closest approach, together with the diminished intensity everywhere inside Io's orbit, seems to offer conclusive proof that the trapped radiation is maintained by inward diffusion of particles from the outer magnetosphere, a process which also results in acceleration of the particles. Further support for this conclusion is found by McIlwain and Fillius (1975),[4] who find the phase-space density of the trapped radiation to be monotonically decreasing towards low *L* shells, thus providing the gradient necessary to drive diffusion. Thus there remains little doubt concerning the primary acceleration and transport mechanism for the stably trapped radiation.

Loss mechanisms have yet to be clearly understood. The need for such mechanisms in addition to absorption by the Jovian satellites is most clearly apparent in the profile of the fission cell counting rate, corresponding to ≳ 35 MeV protons and heavier nuclei. In addition to apparent absorption at the orbit of Amalthea, there are sharp decreases in the intensity inside of

---

[2]See p. 928.
[3]See pp. 36 and 1150.   [4]See p. 1193.

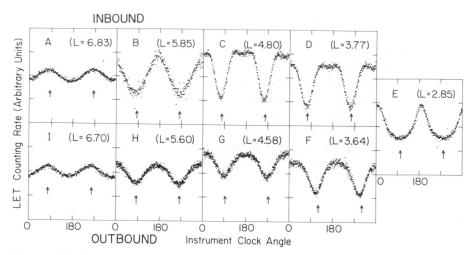

Fig. 4. Distribution of arrival directions of ~ 1 MeV protons observed by Pioneer 10 in a plane approximately perpendicular to the ecliptic. Particles arriving from angles 0° and 180° have trajectories parallel to the ecliptic plane. Twice each spacecraft rotation, at the points indicated by arrows, the lines of magnetic force were in the view cone of the low-energy telescope and reached a minimum angle with respect to the telescope axis, so that particles traveling along field lines appear over the arrows. The D2 magnetic field model of Smith *et al.* (1974) was used to derive the L values (figure from Simpson *et al.* 1974*b*).

$L \approx 3.4$, observed by both Pioneers 10 and 11, and inside of $L \approx 1.9$ for Pioneer 11. The maximum at $L \approx 3.4$ was also found for $0.5 - 1.8$ MeV protons on Pioneer 10 but was much less marked, if present at all, at the low energies for Pioneer 11. The maximum at $L = 1.9$ was also found at the lower energies on Pioneer 11, as is apparent from Fig. 3. The mechanism producing these intensity decreases is one of the largest remaining puzzles confronting us in our attempt to understand the inner radiation zone of Jupiter.

The exact nature of the Jovian magnetic field is also a subject of debate at this time. Pioneer 11 carried two magnetometers, and the best fit models derived from the two experiments, the D4 (Smith *et al.* 1975) and O3 models, differ significantly close to the planet. As explained in Section I above, charged particle observations can be useful in testing the validity of magnetic field models. Since L is conserved in the course of the drift motion of a trapped particle around Jupiter, identifiable features in the particle flux, such as the maximum at $L \approx 3.4$, or the decrease at the orbit of Io should appear at the same value of L on the inbound and outbound portions of the trajectory if the magnetic field model accurately describes the actual field geometry. From Fig. 3, it is clear that the D2 model is not fully successful since, for example, the decrease associated with Io appears at different L values inbound and outbound. In Fig. 5, we have compared the profiles of data from the fission cell on Pioneer 11 when analyzed using the D4 and the O3 models. The D4 model seems superior in organizing these data, and a similar conclusion can be drawn from $\gtrsim 3$ MeV electron and ~ 1 MeV proton profiles, although for $L \lesssim 2.5$ neither model is really successful in organizing the data for ~ 1 MeV protons. (In the period since this chapter was written, Acuña and Ness (p. 830) have refined the calibration of their instrument and, upon reanalyzing their data, obtain results consistent with those of Smith *et al.* This has removed the discrepancies between the models.)

III.   OUTER MAGNETOSPHERE

The characteristics of the outer magnetosphere are quite different from those of the inner regions. As discussed in Sec. I, a ring current in a sheet of plasma flung out from Jupiter as a result of the rapid rotation inflates the mag-

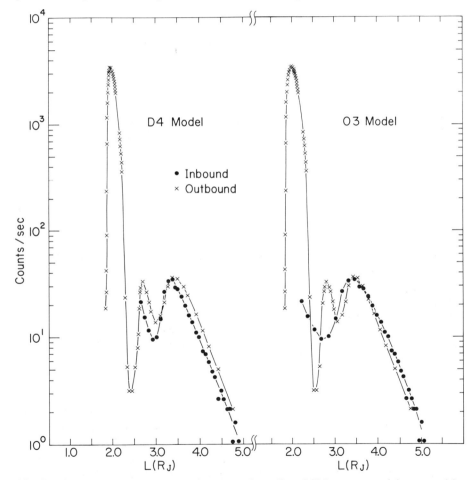

Fig. 5.   A comparison of the intensity versus L profiles of high-energy nuclei measured by the University of Chicago fission cell for L derived from the D4 magnetic field model of Smith *et al.* (1975*a*) and from the O3 magnetic field model of Acuña and Ness (1975). The L values for the O3 model are interpolations based on L values at 10 minute intervals supplied by Acuña and Ness (personal communication).

netosphere to a size far larger than that for which the dipole field alone could account. Like the skirt of a twirling dancer, the outer region is far from rigid, responding sensitively to changes in the solar wind pressure, possibly flapping up and down as a result of the tilt of the dipole axis with respect to the rotation axis, and almost certainly showing significant asymmetries as a function of Jovian local time as it is compressed by the solar wind in front, and drawn out into a Jovimagnetic tail behind (cf. Fig. 1).

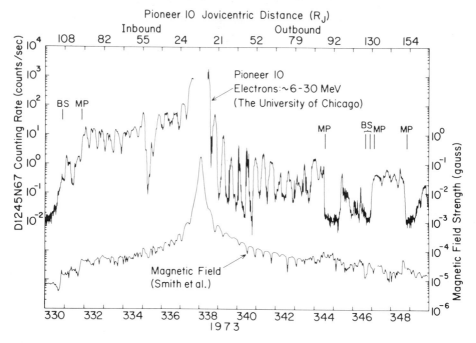

Fig. 6.   A comparison of the intensity profiles of ~ 6–30 MeV electrons (range identification logic code ID5) with the magnetic field intensity measured by Smith (personal communication) for the period when Pioneer 10 was within Jupiter's magnetosphere. To a rough approximation, the differential flux of 6–30 MeV electrons (cm^{-2} sec^{-1} MeV^{-1}) may be considered numerically equal to the plotted counting rate. The data are plotted as a function of spacecraft event time (the time of receipt of data on Earth minus ~ 46 minutes for one-way signal propagation) (figure from McKibben and Simpson 1974).

In Fig. 6 we have compared the electron intensity profile and the magnetic field magnitude for the inbound and outbound passes of Pioneer 10. Several important features are immediately apparent in this figure. First, throughout most of the outer magnetosphere, the electron flux underwent regular variations with a period of about 10 hours. The electron spectrum also varied with a 10-hour period and was steepest (i.e., for a spectral form flux $\propto E^{-\gamma}$, the spectral index γ was greatest) at the times of intensity minima. However, the flux of low-energy (~ 1 MeV) protons did not vary periodically for $R \gtrsim 40\ R_J$ on the inbound pass (Simpson *et al.* 1974*a*). On the outbound pass on the dawn meridian (cf. Fig. 2) the flux variations were much more pronounced than on the inbound pass and were observed for both protons and electrons. The inbound pass differed only by about 60° from the outbound trajectory in its angle with the solar wind. Thus, there are indeed strong asymmetries in the particle distribution in the outer magnetosphere as a function of Jovian local time. On the outbound pass, the association between magnetic field minima and particle intensity maxima is particularly clear. Though not immediately apparent from the figure, we found a similar correlation between magnetic field minima and particle intensity maxima on the inbound pass (McKibben and Simpson 1974). Smith *et al.* (1974) identified the magnetic field minima with entry of the spacecraft into the plasma sheet which supports the outer magnetosphere, and the consensus of most experimenters based on Pioneer 10 data was that in the outer magnetosphere energetic particles were confined primarily to the equatorial plasma sheet, and that the 10-hour variations resulted from the alternate approach and withdrawal of the plasma sheet from the spacecraft due to the inclination of the central dipole. Chenette *et al.* (1974), however, argued for an alternate view of the intensity variations as temporal rather than spatial variations. In this view, every 10 hours the entire outer magnetosphere fills with energetic particles, which then decay away. We shall discuss these alternative models further in Sec. V below.

A second feature of the magnetosphere that is clear from inspection of Fig. 6 is that the magnetopause is a sharp boundary for confinement of the energetic electrons. For every crossing of the magnetopause Pioneer 10 found the flux immediately outside the magnetopause to be at least an order of magnitude less than that just inside. This fact argues for little reconnection of the magnetospheric magnetic field with the interplanetary field in or near the region sampled by Pioneer 10.

Close inspection of Fig. 6 shows that the particle intensity maxima on the outbound pass, in addition to being more pronounced than those on the inbound pass, contain much more fine time structure than is found in the inbound data. In Fig. 7 we have expanded one of the outbound maxima, plotting one minute averages of the energetic electron and ~ 1 MeV proton flux. The rise and fall of the fluxes is extremely rapid, requiring only a few minutes for changes of more than three orders of magnitude for the proton flux. Furthermore, the correlation between proton and electron flux variations on a minute-to-minute basis is very good, especially in the precursor maximum starting at $\sim 01^{h}00^{m}$ on day 340. Strong anisotropies along the field were observed in the proton flux from time to time during this maximum, appearing and disappearing, or changing direction by up to 180° in as little as one minute. The electron flux, however, seldom showed significant anisotropies. The simultaneity of the electron and proton time variations argues that the particles cannot long have been resident on the field lines on which

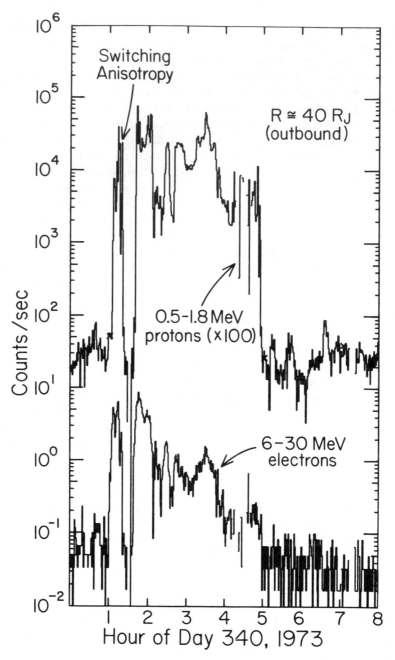

Fig. 7. A comparison of one-minute averages of counting rates of ~ 1 MeV protons and 6–30 MeV electrons during one of the large intensity maxima observed on the outbound pass of Pioneer 10. The double structure (precursor and main maximum) was a recurrent feature in several of these post-periapsis maxima. The data are plotted as a function of spacecraft event time. In the period labeled "switching anisotropy" large anisotropies were observed which appeared and disappeared or changed direction in periods as short as one minute (figure from Simpson *et al.* 1974a).

they were observed, for drift motions would rapidly separate protons and electrons so that intensity features would not remain together long for the two species. Rough calculations of the relative drift velocity based on magnetic field gradients observed in the vicinity of these intensity features (Smith, personal communication) suggest that injection of particles can have taken place at most a few minutes before the particles were observed. Strong anisotropies are also consistent with the recent injection of the particles onto the field lines. It is possible that we are observing particles very recently accelerated, and perhaps even that Pioneer 10 was embedded in the region of acceleration itself, for if there is significant merging of magnetic field lines across the magnetic neutral sheet embedded in the plasma sheet, particle acceleration would be expected. Particle acceleration in such situations has been widely investigated in the search for the solar flare acceleration mechanism (e.g., Petschek 1964; Sonnerup 1973). Retzler and Simpson (1969) have reported observations of electrons in the neutral sheet of Earth's magnetotail and conclude likewise that acceleration is probably occurring in the neutral sheet. Thus we have probably observed a second similar acceleration process for particles in Jupiter's magnetosphere. This process is in addition to that of the violation of the third adiabatic invariant which provides acceleration to maintain the stably trapped inner core as described above. Neither of these mechanisms would appear to explain adequately the large populations of energetic particles undergoing 10-hour variations in the outer magnetosphere.

Pioneer 11 offered a second look at the outer magnetosphere of Jupiter, and a chance to test some of the ideas formed on the basis of data from Pioneer 10. The time-intensity profiles of energetic electrons and ~ 1 MeV protons observed during Pioneer 11's pass through the Jovian magnetosphere are shown in Fig. 8. In some respects, the observations are similar to those made with Pioneer 10. Again the spacecraft crossed the magnetopause several times as a result of expansion and contraction of the magnetosphere in response to changes in solar wind pressure. The confinement of electrons behind the magnetopause was in most cases good, as was that for low-energy protons, leading again to the conclusion that reconnection of magnetic field lines across the front side of the magnetopause is limited, giving magnetospheric particles little access to interplanetary field lines by this route.

Examination of the anisotropy of low-energy protons near the magnetopause shows that an anisotropy produced by co-rotation of the particles persists all the way to the magnetopause, although with a magnitude near the magnetopause much less than that expected for rigid co-rotation of the particles with the planet. Indeed the co-rotation anisotropy was observed at essentially all times that Pioneer 11 was within the magnetosphere, consistent with the observations by Van Allen et al. (1975) for Pioneer 11 and Trainor et al. (1974)[5] for Pioneer 10. Figure 9 shows the behavior of the low-energy proton anisotropy during the period on day 331, 1974, when Pioneer

[5]See p. 967.

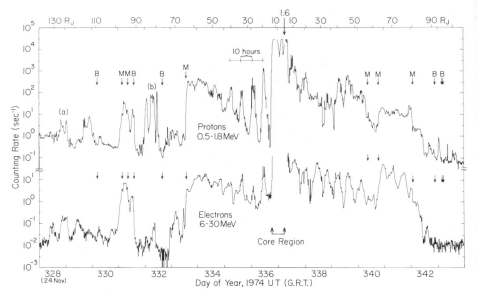

Fig. 8.   Time intensity profiles of ~ 1 MeV protons and ~ 6–30 MeV electrons during Pioneer 11's traversal of the Jovian magnetosphere. Magnetopause crossings (M) and bow shock
crossings (B) are based on the data of Smith (personal communication). Data are plotted as a
function of the time when data were received on Earth (ground received time), which differs
from spacecraft event time by ~ 41 minutes for Pioneer 11 (figure from Simpson *et al.* 1975).

11 was within the magnetosphere for a brief period. The co-rotation anisotropy appeared soon after crossing the magnetopause and persisted nearly
until the second crossing when Pioneer 11 left the magnetosphere.

Once within the magnetosphere we again observed 10-hour variations in
the particle intensity and in the electron spectrum. The variations, however, had a quite different appearance from those found by Pioneer 10. Even
though the inbound trajectories of Pioneer 10 and 11 were nearly identical
in the outer magnetosphere (see Fig. 2), the variations near the magnetopause were much less regular for Pioneer 11 than for Pioneer 10, and inside
$R \equiv 50\ R_J$ the variations took on the character of those observed during the
outbound pass on Pioneer 10. Examination of counting rates on a fine time
scale again shows the presence of rapid time variations that are simultaneous
for both protons and electrons, suggesting particle acceleration (Simpson *et
al.* unpublished). The magnitudes of the intensity changes were much less
than those observed by Pioneer 10, however, and the anisotropies were less
well marked. Comparison with magnetic field data (Smith, personal communication) shows that the large particle maxima were again associated with
well developed magnetic field minima, so that the well developed current
sheet observed on the dawn meridian by Pioneer 10 extended into the morning quadrant at the time of Pioneer 11's flyby. Therefore, the character of the
outer magnetosphere of Jupiter apparently varies significantly with time, on

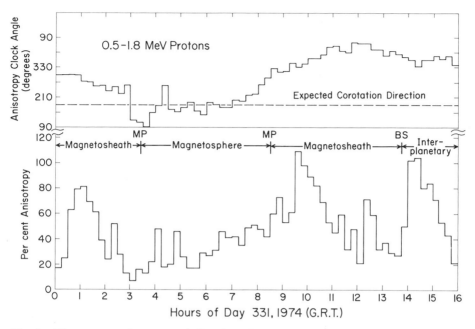

Fig. 9. The percent anisotropy and direction of maximum flux for 0.5 – 1.8 MeV protons measured by the Pioneer 11 low-energy telescope of the University of Chicago. The percent anisotropy is defined to be 100 B/A where the observed distribution of arrival directions has been fitted by the least squares method to a function of the form $A + B \cos (\theta - C)$. The "clock angle" of maximum intensity is equal to C. For $\theta = 0°$ and $180°$ the view cone of the telescope lies approximately in the ecliptic plane, and for $\theta = 90°$ and $270°$, the view is normal to the ecliptic. Magnetopause and bow shock crossings are based on the data of Smith (personal communication).

a scale probably on the order of days, although this remains undetermined. This observation should give us caution in any attempt to build over-elaborate models on the basis of the two penetrations of the magnetosphere now available.

One conclusion based on Pioneer 10 observations which appears to have been refuted by those of Pioneer 11 is that the energetic particles are confined largely to the plasma sheet in the outer magnetosphere. Reference to Fig. 2 shows that the outbound trajectory of Pioneer 11 lay far from the plasma sheet in the equatorial plane which contains the plasma sheet. Nevertheless, the particle fluxes were at least as great on the outbound pass as on the near-equatorial inbound pass, and, for the electrons, showed intensity variations as large as or larger than those observed near the equator. Therefore, the conclusion that the 10-hour variations are due entirely to alternate approach to and recession from the equatorial plasma sheet is open to severe questioning. An alternate hypothesis consistent with this observation is that the intensity variations are a global effect, taking place simultaneously throughout the outer magnetosphere. We will discuss this question further in Sec. V.

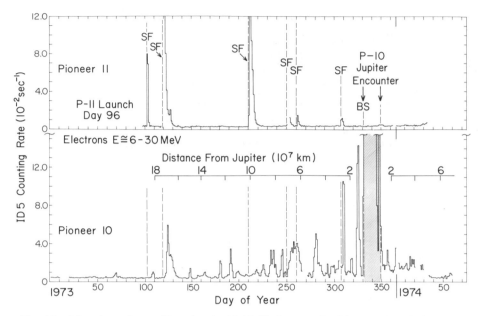

Fig. 10. Time intensity profiles of ~ 6−30 MeV electrons for Pioneer 10 and Pioneer 11 during Pioneer 10's approach to Jupiter. Solar flare (SF) events have been identified. During this period Pioneer 11 was at radial distances from the sun from 1 to ~ 3 A.U. (figure from Chenette *et al.* 1974).

IV. JUPITER AS A SOURCE OF INTERPLANETARY PARTICLES

Not all energetic particles associated with Jupiter are found within the magnetosphere. Bursts of low-energy electrons in the interplanetary medium outside of Earth's bow shock have long been observed (Fan *et al.* 1964; Frank and Van Allen 1964; Anderson 1968) and we made a search for similar events as Pioneer 10 approached Jupiter. In Fig. 10, taken from Chenette *et al.* (1974), we show counting rate time-intensity profiles of energetic electrons from both Pioneer 10 and 11 during Pioneer 10's approach to Jupiter. The Pioneer 11 data, taken far from Jupiter, show the almost complete absence of solar accelerated particle events whereas the counting rate at Pioneer 10 became increasingly active, and the average magnitude of particle events increased as Pioneer 10 approached Jupiter. Proof of the Jovian origin of these particles came with the discovery of a 10-hour period in variations in the intensity and spectrum of electrons within the events. A clear example of such variations is shown in Fig. 11. In common with the variations observed inside the magnetosphere, the steepest spectra are associated with the intensity minima. Furthermore, the phase of the variations is the same as that observed inside the magnetosphere. In Fig. 12 the variations in the electron spectral index observed beginning with the event in Fig. 10 and

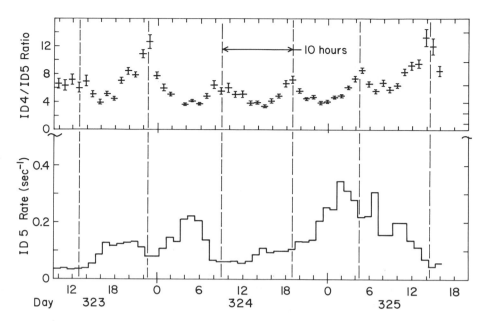

Fig. 11.   The time intensity profile of ~ 6–30 MeV electrons (ID5 electrons) and the ratio of
the flux of ~ 3–6 MeV electrons to that of ~ 6–30 MeV electrons (ID4/ID5 ratio) as a
function of time during an interplanetary electron event observed by Pioneer 10. This ratio
is a measure of the steepness of the slope of the electron differential energy spectrum, or the
spectral index γ. The dashed lines represent times of spectral index maximum for a variation
with a period of 9^h 55^m 30^s and a phase defined by the spectral index maximum observed at
~ 05^h 00^m U.T. (spacecraft event time) on day 332, 1973, when Pioneer 10 was inside the
magnetosphere. Periods when large amplitude hydromagnetic waves were observed are
indicated by brackets. Data are plotted as a function of spacecraft event time (figure from
Smith *et al.* 1976).

continuing until Pioneer 10 was well inside the magnetosphere are displayed,
together with a series of tic marks extending to interplanetary space the
phase of variations observed inside the magnetosphere. The continuity across
the magnetopause is striking and argues strongly that the particles inside and
outside of the magnetosphere have the same origin. Applying the sensitive
technique of power spectral analysis to the intensity profiles, Chenette *et al.*
(1974) found evidence for 10-hour variations as far as 1 A.U. from the planet.
Furthermore, wherever spectral variations were found, they were in phase
with those observed within and near the magnetosphere. Teegarden *et al.*
(1974), from analysis of Earth-based data, have reported observation of Jovian electrons even at the orbit of Earth, although no 10-hour variations have
yet been found in the events near Earth.
    Since the discovery of these electron events, collaborative studies with
the magnetometer investigators on Pioneer 10 have shown significant correlations between electron events and characteristics of the interplanetary

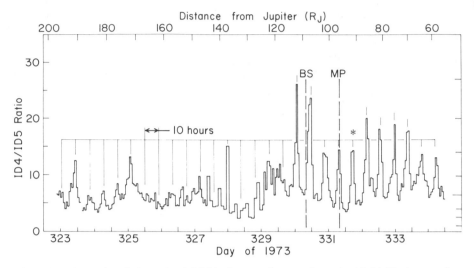

Fig. 12. The ratio of the ~ 3−6 MeV electron flux to the ~ 6−30 MeV electron flux (1D4/1D5 ratio) as a function of spacecraft event time during Pioneer 10's approach to Jupiter. The tic marks correspond to the dashed lines in Fig. 11 (figure from Chenette *et al.* 1974).

magnetic field (Smith *et al.* 1976). In particular, these studies showed that, for events close to Jupiter ($R < 400\ R_J$) a necessary but not sufficient condition for the observation of Jovian electrons was that the interplanetary magnetic field should lie in the direction necessary to provide good connection between the spacecraft and Jupiter. Abrupt changes in the field away from this direction were accompanied by abrupt decreases in the electron flux. Thus, some of the time-intensity structure apparent in the electron counting rate is probably a result of variability in the interplanetary magnetic field.

A second striking discovery to come from this collaborative study is of the association of large-amplitude hydromagnetic waves in the interplanetary magnetic field with the presence of Jovian electrons. Periods during which such waves were observed are marked in Fig. 11 and an expanded view of behavior of the magnetic field during a 3-hour period in which waves were observed is shown in Fig. 13. The simultaneity of electron intensity maxima and the observation of waves argues against a Jovian origin for the waves, with subsequent upstream propagation to the spacecraft in the solar wind. Instead, the most plausible hypothesis is that the waves are generated locally in the solar wind by the electrons themselves, as shown schematically in Figure 14. If the electrons are producing the waves in the magnetic field, this observation can be used to investigate an effect which must be important in the confinement and isotropization of cosmic rays in the galaxy (Wentzel 1974 and references therein).

In addition to electrons from Jupiter we have recently reported the discovery of low-energy (~ 1 MeV) protons in interplanetary space (Simpson

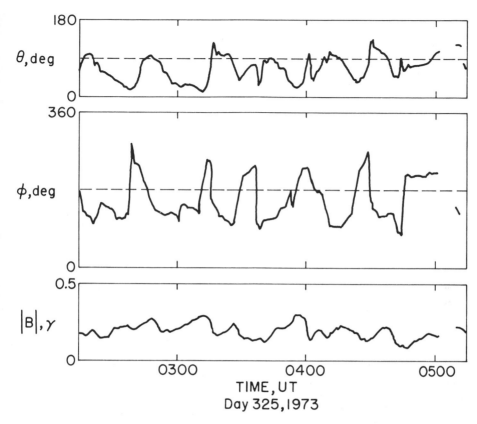

Fig. 13. Example of magnetic waves associated with electron bursts and Pioneer-Jupiter field alignment. One-minute averages of the field magnitude and angles for a 3-hour period on day 325 are shown. These large amplitude waves occurred during a maximum in the $3-6$ MeV electron flux, as can be seen by reference to Fig. 11 (figure from Smith *et al.* 1976).

et al. 1975). One such event is shown in Fig. 8 on day 332 during the period Pioneer 11 was outside of the bow shock after compression of the magnetosphere. Proof of the Jovian origin of interplanetary proton events is difficult since even inside the magnetosphere close to the magnetopause no marked 10-hour variations in the proton intensity or spectrum have been observed. We have noted, however, that high intensities of low-energy protons very near Jupiter have shown steeper energy spectra than normally encountered in quiet or solar active periods in interplanetary space. A search for events with such steep energy spectra has shown the existence of a number of such events, many but not all of which occurred in association with Jovian electron events, often with markedly similar time-intensity profiles for the protons and electrons. Figure 15 contains an example of such an event. In addition, when the anisotropy in such events was significant, the direction of arrival was more consistent with a Jovian than a solar origin for these parti-

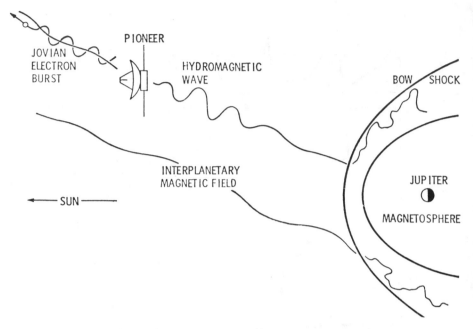

Fig. 14.   Schematic diagram showing interplanetary conditions during a Jovian interplanetary electron event. When the interplanetary magnetic field lies near the Pioneer-Jupiter direction, energetic (MeV) electrons are able to flow up the lines of force from Jupiter to the spacecraft. Since their streaming velocity exceeds the phase velocity of hydromagnetic waves, a cyclotron resonant overstability occurs leading to the generation of interplanetary waves. The hydromagnetic waves propagate upstream in the solar wind frame, but are convected downstream in the inertial frame (figure from Smith *et al.* 1976).

cles, and when hydromagnetic waves were observed with significant circular polarization during proton events, the polarization was in most cases correct for their having been generated by protons rather than by electrons (Chenette *et al.* 1975).

V.   10-HOUR VARIATIONS—SPATIAL OR TEMPORAL?

The observation of phase continuity across the magnetopause for the intensity and spectral variations in the electron flux raises questions which are difficult to answer in terms of the consensus model based on Pioneer 10 data, in which most particles were confined to the plasma sheet associated with the magnetic equatorial plane. In this model, phase continuity between variations for interplanetary and magnetospheric particles requires that the interplanetary particles escape the magnetosphere from one location only, namely that through which Pioneers 10 and 11 entered the magnetosphere. This possibility cannot be excluded, although it seems highly unlikely. Also not easily reconcilable with the plasma sheet confinement model is the fact

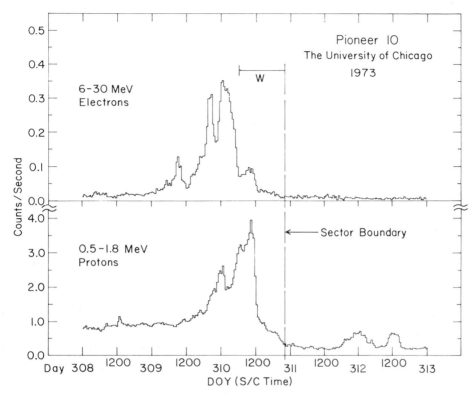

Fig. 15. An example of a Jovian interplanetary particle event containing energetic electrons, low energy protons, and large amplitude hydromagnetic waves. Data are plotted as a function of spacecraft event time.

that near the magnetopause, on both inbound and outbound passes of Pioneer 10, the variations observed had the same phase despite the difference in Jovian local time of the inbound and outbound trajectories (Chenette *et al.* 1974; McKibben and Simpson 1974). Large distortions in the geometry of the outer magnetosphere would be necessary to accommodate these observations under the plasma sheet confinement model, as discussed by McKibben and Simpson (1974) and Northrop *et al.* (1974).

Chenette *et al.* (1974) suggested an alternate model in which the variations were a result of temporal variations in the entire outer magnetospheric particle population. In this case, a space probe anywhere within the magnetosphere, or for that matter outside as well, would see variations with the same phase. With this model, however, it was difficult to reconcile the trend of the phase of the intensity variations observed during Pioneer 10's traversal of the magnetosphere. In Fig. 16 we show the relation between the times expected for intensity minima based on the temporal variation model and the times at which minima were actually observed. In this view, near the mag-

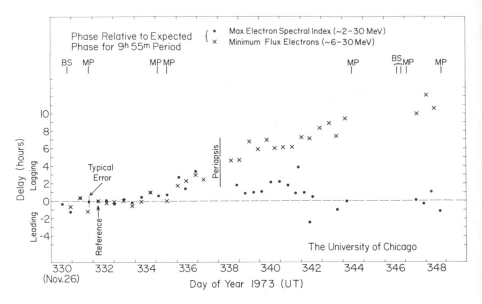

Fig. 16. The times when electron intensity minima (×) and spectral index maxima (•) were observed by Pioneer 10 in relation to the times expected for a model in which the spectral and intensity variations have a period of 9ʰ 55ᵐ. The phase is independent of spacecraft position within the magnetosphere, so that all particles in the magnetosphere vary synchronously. The expected times are indicated by the dashed line. Data are plotted as a function of spacecraft time.

netopause the inbound and outbound variations differed by a phase of ~ 10 hours, or 2π radians. However, the phase of the spectral variations, shown here for the first time, was more nearly consistent with the temporal variation model. A further objection to the temporal model was the correlation observed between magnetic field intensity minima and particle intensity maxima, which McKibben and Simpson (1974) found even near the magnetopause on days 332–334. Thus, based on Pioneer 10 data, neither model appeared fully satisfactory.

Pioneer 11 offered a clear cut test of certain predictions of the two models. The first, which we have already described (cf. Sec. III) showed that, contrary to the expectation based on the plasma sheet model that particle fluxes at high magnetic latitudes should be far lower than those observed near the equatorial plane, the fluxes observed at high magnetic latitudes on Pioneer 11's outbound pass were at least as high as those observed in the equatorial plane inbound. The second test involved the prediction that under the plasma sheet confinement model the phase of the intensity variations observed outbound should differ from that observed inbound by about 4 hours, since Pioneer 11 went from the southern to the northern hemispheres between inbound and outbound passes, inducing a change of ~ 5 hours in the phase, which was partially offset by a change of ~ 1 hour in the Jovian local

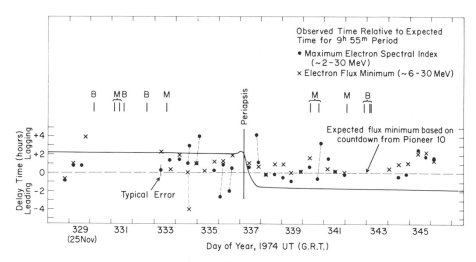

Fig. 17.   Same as Fig. 16 but for Pioneer 11. The dashed line is the locus of intensity minima and spectral index maxima based on extrapolation of the phase defined by the reference spectral maximum in Fig. 16 using the $9^h 55^m 33\overset{s}{.}1$ synodic System III rotation period of Jupiter. Where more than one significant intensity minimum or spectral index maximum occurred in a 10-hour period, both have been plotted and have been joined by a light dashed line. The heavy line is the locus of points of maximum magnetic latitude on Pioneer 11's trajectory. Data are plotted as a function of ground received time (figure from Simpson *et al.* 1975).

time of the trajectory. The temporal model, on the other hand, would predict no change in phase between inbound and outbound variations. The Pioneer 11 data for both intensity and spectral variations are shown in Fig. 17 in a format similar to that of Fig. 16. The heavy line shows the predicted locus of intensity minima (and spectral index maxima) if intensity minima were associated with the points of maximum magnetic latitude on the spacecraft trajectory. The dashed line shows the prediction of the temporal variation model, the result of extrapolating forward the phase of variations observed during the Pioneer 10 encounter using the $9^h 55^m 33\overset{s}{.}1$ rotation period of Jupiter with respect to the sun-Jupiter line (synodic System III rotation period). Clearly the data are more consistent with the temporal variation model. Thus at this point, although the difficulties raised by the Pioneer 10 observations still stand, the weight of the evidence appears to favor the temporal variation model. It is possible still that asymmetries in the magnetosphere between the dawn and noon meridians may explain the apparent deviation of the Pioneer 10 data from the model predictions. It is not clear, however, that there is not a third alternative, of which we have not yet thought, which will be consistent with all observations from both Pioneers 10 and 11.

If we accept for the moment the hypothesis that the variations are temporal in nature, we must find some mechanism that will cause the acceleration or release of a large number of energetic particles every 10 hours. If the magnetosphere were subject to no external stresses, no such periodicity could be established, for one orientation of the dipole would be equivalent to any other. It is clear, therefore, that the variations in particle intensity and spectra owe their existence to the presence of a symmetry breaking force. That the symmetry breaking force is associated with the sun is demonstrated by the fact that in order to extrapolate successfully the phase of the variations established by Pioneer 10 observations in 1973 to the time of Pioneer 11 encounter in 1974 we must use the synodic rather than the sidereal rotation period of Jupiter. (The sidereal period would predict intensity minima about 1 hour earlier than synodic period.) The obvious candidate for such a force is the solar wind pressure.

Even with the symmetry breaking force identified, the mechanism for inducing the variations is not obvious. The small offset in the dipole of $\sim$ 0.1 R_J reported by Smith et al. (1974, 1975) is insignificant on the scale of the magnetosphere, and is well shielded from the solar wind pressure by the "spongy" outer magnetosphere. Without some means of amplifying its effect, perhaps by some such means as the differential plasma loading of field lines suggested by Hill and Dessler (1976), the offset is unlikely to be the source of the variations. A source might be found in work done against the solar wind by flopping motions of the magnetosphere as a result of the dipole tilt but, again, either some interior asymmetry or possibly a north-south asymmetry in the interplanetary medium is required, or a 5-hour period would result. Whatever the mechanism, it must produce copious numbers of energetic electrons every 10 hours. Rough estimates based on counting rates observed in the outer magnetosphere show that $\sim 10^{28}$ electrons with energies $\gtrsim 6$ MeV must be produced and released to populate simultaneously both the northern and southern hemispheres every 10 hours to account for magnetospheric fluxes alone. Furthermore, the mechanism must produce strongest variations at high energies in order to account for the observed spectral variations.

Hill et al. (1974), Hill and Dessler (1976), and Nishida (1975) have recently proposed models to account for the intensity and spectral variations observed in the outer magnetosphere and in interplanetary space. Hill et al. suggest that longitudinal asymmetries in the Jovian magnetic field lead to the development of a "diurnal trapping region" in which charged particles can be stably trapped during part of a Jovian day, but which during the remaining part is open to the Jovimagnetic tail, through which particles escape to the interplanetary medium. In this manner a 10-hour period is imposed on the escaping particles. A primary difficulty with this model is in explaining phase continuity of the variations across the day-side magnetopause if particles are injected into the planetary medium through the night-side tail. Nishida's model avoids this difficulty by having the trapped particles escape to outer magnetospheric and open interplanetary field lines at low altitudes near the magnetic poles, as a result of diffusion violating both the second and third adiabatic invariants. It is not clear in this model, however, how the 10-hour period is induced. In both models, it may also be difficult to achieve the necessary escape rate to populate every 10 hours both the $\sim 200\, R_J$ diameter magnetosphere with the observed $\sim 300\, e\ \text{sec}^{-1}\ \text{cm}^{-2}$ flux and an undetermined volume of interplanetary space with a flux on the order of a factor 10 less. Both models require a trigger for release of particles every 10 hours whose nature is at present undetermined although Hill and Dessler (1976) suggest a plausible mechanism involving asymmetrical loading of magnetic field lines with plasma. The recent recognition of the good correspondence between the System III longitude of the sun-Jupiter line at the time of interplanetary electron intensity minima and the central meridian longitude of the non-Io associated decametric radio bursts (Vasyliunas 1975) provides an interesting addition to the observations, and may help in unravelling the mechanism.

VI. SUMMARY AND UNANSWERED QUESTIONS

As is the case for any good experimental program, the Pioneer 10 and 11 missions through the magnetosphere of Jupiter have yielded some striking discoveries, confirmed some theoretical ideas of general importance, and raised more questions than were answered. Among the striking discoveries are the finding of large numbers of high-energy particles, both protons and electrons, escaping from Jupiter into interplanetary space. If observed at the orbit of Earth, these particles may account for the puzzling "quiet-time electron increases" which have been observed for several years on Earth-orbiting spacecraft (McDonald et al. 1972). Furthermore, by providing a source of particles of known origin propagating inward through the solar wind, they provide a new probe of the interplanetary propagation process which is of great importance in many fundamental questions of cosmic ray physics. The discovery of hydromagnetic waves in the interplanetary medium associated with the electrons provides a means of testing theoretical ideas about the generation of such waves in plasmas by streaming particles, a process thought to be of great importance for confinement of cosmic rays to the galaxy.

A discovery of equal importance, though perhaps less anticipated and thus less able to find its place in existing theories, is of the immense size and complexity of the Jovian magnetosphere. It will require many years, and probably several more penetrations of the magnetosphere before the interplay of forces which govern the behavior of the plasmas, fields, and energetic particles in this large region of space is understood.

In the list of ideas or theories confirmed by observations of Pioneers 10 and 11 we find, perhaps most fundamentally, proof of the maintenance and acceleration of charged particles trapped in Jupiter's magnetic field by inward diffusion in violation of the third adiabatic invariant. In the course of obtaining this proof, the importance of absorption of radiation by the satellite Io in limiting the flux of the trapped radiation in Jupiter's inner radiation zone was confirmed. A probable example of a second acceleration mechanism by merging of magnetic field lines through a neutral sheet was also found although conclusive experimental proof that such acceleration was indeed taking place is not presently available.

Secondly, although we have not discussed this question in detail here, the population of trapped electrons was found to be in rough agreement with that necessary to produce the radio emissions observed at Earth (Luthey *et al.* personal communication) although discrepancies on the order of a factor 3 or so remain in the particle flux data at different electron energies taken by the various experimental groups. A thorough review of calibrations and measurement techniques is under way at the present time, and such discrepancies may shortly be resolved to allow a more definitive test of observation versus theory in relating electron fluxes to the radio emission. Such a test has fundamental importance for astrophysics in general, for much of our knowledge about magnetic fields and charged particle densities outside the solar system is based on deductions from synchrotron emission.

In the category of open questions and remaining puzzles must be placed the nature of the loss mechanism responsible for the proton maxima observed in the inner trapped radiation zone, the manner of escape of energetic particles from Jupiter's magnetosphere, the nature and source of the 10-hour variations observed in the particle flux in interplanetary space and in the outer magnetosphere, and indeed the source of the outer magnetospheric particles themselves, for a large population of high energy electrons was not expected outside the stable trapped radiation zone. These questions are without doubt bound up in a still larger question, that of the nature of the outer magnetosphere itself.

Acknowledgements. We are grateful for the assistance of D. C. Hamilton and K. R. Pyle in preparing data used in this chapter. This work was supported in part by the National Aeronautics and Space Administration Grant NGL 14-001-006, NASA Contract NAS 2-6551, and National Science Foundation Grant DES 75-20407.

REFERENCES

Acuña, M. H., and Ness, N. F. 1975. Jupiter's main magnetic field measured by Pioneer 11. *Nature* 253:327–328.
Anderson, K. A. 1968. Energetic electrons of terrestrial origin upstream in the solar wind. *J. Geophys. Res.* 73:2387–2398.
Brice, N. M., and McDonough, T. R. 1973. Jupiter's radiation belts. *Icarus* 18:206–219.
Carr, T. D., and Gulkis, S. 1969. The magnetosphere of Jupiter. *Ann. Rev. Astron. Astrophys.* (L. Goldberg, ed.) Vol. 7, pp. 477–618. Palo Alto, Calif.: Annual Reviews, Inc.

Chenette, D. L.; Conlon, T. F.; and Simpson, J. A. 1974. Bursts of relativistic electrons from Jupiter observed in interplanetary space with the time variation of the planetary rotation period. *J. Geophys. Res.* 79:3551–3558.

Chenette, D.; Conlon, T.; Simpson, J. A.; Smith, E. J.; and Tsurutani, B. T. 1975. Correlation of Jovian electron bursts with the interplanetary magnetic field (abstract). *Trans. Am. Geophys. Un.* 56:427.

Fan, C. Y.; Gloeckler, G.; and Simpson, J. A. 1964. Evidence for > 30 KeV electrons accelerated in the shock transition region beyond the earth's magnetospheric boundary. *Phys. Rev. Lett.* 13:149–153.

Frank, L. A., and Van Allen, J. A. 1964. Measurements of energetic electrons in the vicinity of the sunward magnetosphere boundary with Explorer 14. *J. Geophys. Res.* 69:4923–4932.

Hess, W. N. 1968. *The radiation belt and magnetosphere.* Waltham, Mass: Blaisdell Publishing Co.

Hill, T. W.; Carbary, J. F.; and Dessler, A. J. 1974. Periodic escape of relativistic electrons from the Jovian magnetosphere. *Geophys. Res. Lett.* 1:333–336.

Hill, T. W., and Dessler, A. J. 1976. Longitudinal asymmetry of the Jovian magnetosphere and the periodic escape of energetic particles. *Icarus* (special Jupiter issue). In press.

Ioannidis, G. A., and Brice, N. M. 1971. Plasma densities in Jupiter's magnetosphere: plasma slingshot or Maxwell demon? *Icarus* 14:360–373.

Kennel, C. F. 1973. Magnetospheres of the planets. *Space Sci. Rev.* 14:511–533.

McDonald, F. B.; Cline, T. L.; and Simnett, G. M. 1972. Multifarious temporal variations of low energy cosmic ray electrons. *J. Geophys. Res.* 77:2213–2231.

McIlwain, C. E., and Fillius, R. W. 1975. Differential spectra and phase space densities of trapped electrons at Jupiter. *J. Geophys. Res.* 80:1341–1345.

McKibben, R. B.; O'Gallagher, J. J.; Simpson, J. A.; and Tuzzolino, A. J. 1973. Preliminary Pioneer 10 intensity gradients of galactic cosmic rays. *Astrophys. J.* 181:L9–L13.

McKibben, R. B., and Simpson, J. A. 1974. Evidence from charged particle studies for the distortion of the Jovian magnetosphere. *J. Geophys. Res.* 79:3545–3549.

Mead, G. D. 1974. Magnetic coordinates for the Pioneer 10 Jupiter encounter. *J. Geophys. Res.* 79:3514–3521.

Mead, G. D., and Hess, W. N. 1973. Jupiter's radiation belts and the sweeping effect of its satellites. *J. Geophys. Res.* 78:2793–2811.

Mogro-Campero, A., and Fillius, R. W. 1975. The absorption of trapped particles by the inner satellites of Jupiter and the radial diffusion coefficient of particle transport. Submitted to *J. Geophys. Res.*

Newburn, R. L., and Gulkis, S. 1973. A survey of the outer planets Jupiter, Saturn, Uranus, Neptune, Pluto, and their satellites. *Space Sci. Rev.* 14:179–271.

Nishida, A. 1975. Outward diffusion of energetic particles from the Jovian radiation belt. In draft.

Northrop, T. G.; Goertz, C. K.; and Thomsen, M. F. 1974. The magnetosphere of Jupiter as observed with Pioneer 10. 2. Nonrigid rotation of the magnetodisc. *J. Geophys. Res.* 79: 3579–3582.

Petschek, H. E. 1964. Magnetic field annihilation. *Physics of Solar Flares.* (W. N. Hess, ed.) pp. 425–439, NASA SP-50. Washington, D.C.: U.S. Government Printing Office.

Retzler, J., and Simpson, J. A. 1969. Relativistic electrons confined within the neutral sheet of the geomagnetic tail. *J. Geophys. Res.* 74:2149–2160.

Roederer, J. G. 1970. *Dynamics of Geomagnetically Trapped Radiation.* New York: Springer-Verlag.

Shabansky, V. P. 1971. Some processes in the magnetosphere. *Space Sci. Rev.* 12:299–418.

Simpson, J. A.; Hamilton, D.; Lentz, G.; McKibben, R. B.; Mogro-Campero, A., Perkins, M.; Pyle, K. R.; Tuzzolino, A. J.; and O'Gallagher, J. J. 1974a. Protons and electrons in Jupiter's magnetic field: results from the University of Chicago experiment on Pioneer 10. *Science* 183:306–309.

Simpson, J. A.; Hamilton, D. C.; Lentz, G. A.; McKibben, R. B.; Perkins, M.; Pyle, K. R.; Tuzzolino, A. J.; and O'Gallagher, J. J. 1975. Jupiter revisited: first results from the University of Chicago charged particle experiment on Pioneer 11. *Science* 188:455–459.

Simpson, J. A.; Hamilton, D. C.; McKibben, R. B.; Mogro-Campero, A.; Pyle, K. R.; and Tuzzolino, A. J. 1974*b*. The protons and electrons trapped in the Jovian dipole magnetic field region and their interaction with Io. *J. Geophys. Res.* 79:3522–3544.

Smith, E. J.; Davis, Jr., L.; Jones, D. E.; Coleman, Jr., P. J.; Colburn, D. S.; Dyal, P.; and Sonett, C. P. 1975. Jupiter's magnetic field, magnetosphere, and interaction with the solar wind. *Science* 188:451–455.

Smith, E. J.; Davis, Jr., L.; Jones, D. E.; Coleman, Jr., P. J.; Colburn, D. S.; Dyal, P.; Sonett, C. P.; and Frandsen, A. M. A. 1974. The planetary magnetic field and magnetosphere of Jupiter: Pioneer 10. *J. Geophys. Res.* 79:3501–3513.

Smith, E. J.; Tsurutani, B. T.; Chenette, D. L.; Conlon, T. F.; and Simpson, J. A. 1976. Jovian electron bursts: correlation with the interplanetary magnetic field direction and hydromagnetic waves. *J. Geophys. Res.* 81:65.

Sonnerup, B. U. O. 1973. Magnetic field reconnection and particle acceleration. *High energy phenomena on the sun.* (R. Ramaty and R. G. Stone, eds.) NASA SP-342, Washington, D.C.: U.S. Government Printing Office.

Teegarden, B. J.; McDonald, F. B.; Trainor, J. H.; Webber, W. R.; and Roelof, E. C. 1974. Interplanetary MeV electrons of Jovian origin. *J. Geophys. Res.* 79:3615–3622.

Trainor, J. H.; McDonald, F. B.; Stilwell, D. E.; Teegarden, B. J.; and Webber, W. R. 1975. Jovian protons and electrons: Pioneer 11. *Science* 188:462–464.

Trainor, J. H.; McDonald, F. B.; Teegarden, B. J.; Webber, W. R.; and Roelof, E. C. 1974. Energetic particles in the Jovian magnetosphere. *J. Geophys. Res.* 79:3600–3614.

Van Allen, J. A.; Baker, D. N.; Randall, B. A.; and Sentman, D. D. 1974. The magnetosphere of Jupiter observed with Pioneer 10. 1. Instrument and principal findings. *J. Geophys. Res.* 79:3559–3578.

Van Allen, J. A.; Randall, B. A.; Baker, D. N.; Goertz, C. K.; Sentman, D. D.; Thomsen, M. F.; and Flindt, H. R. 1975. Pioneer 11 observations of energetic particles in the Jovian magnetosphere. *Science* 188:459–462.

Vasyliunas, V. M. 1975. Modulation of Jovian interplanetary electrons and the longitude variation of decimetric emissions. *Geophys. Res. Letters* 2:87–88.

Walt, M. 1971. The radial diffusion of trapped particles induced by fluctuating magnetospheric fields. *Space Sci. Rev.* 12:446–485.

Warwick, J. W. 1967. Radiophysics of Jupiter. *Space Sci. Rev.* 6:841–891.

Wentzel, D. G. 1974. Cosmic ray propagation in the galaxy: collective effects. *Ann. Rev. Astron. and Astrophys.* (G. R. Burbidge, ed.) Vol. 12, pp. 71–97. Palo Alto, Calif: Annual Reviews, Inc.

Wolfe, J. H.; Mihalov, J. D.; Collard, H. R.; McKibbin, D. D.; Frank, L. A.; and Intrilligator, D. S. 1974. Pioneer 10 observations of the solar wind interaction with Jupiter. *J. Geophys. Res.* 79:3489–3500.

JUPITER'S MAGNETISM: OBSERVATIONS AND THEORY

R. HIDE
British Meteorological Office

and

D. STANNARD
University of Manchester

Observations of non-thermal radio noise from Jupiter on decimeter and decameter wavelengths provided the first evidence of a strong and nearly dipolar Jovian magnetic field and an associated system of Van Allen type radiation belts of electrically charged particles, extending beyond and interacting with the first Galilean satellite Io. Data from Pioneers 10 and 11 corroborate this evidence and confirm and refine previous estimates of the strength, position and orientation of the equivalent Jovian magnetic dipole, and they also provide further information concerning non-dipole components of the field.

Considerations of possible sources of Jupiter's magnetic field indicate that the most likely mechanism is a hydromagnetic (magnetohydrodynamic) dynamo associated with thermally driven fluid motions in the electrically conducting parts of Jupiter's interior. Dynamo theory cannot yet provide precise expressions relating the strength and configuration of the observed magnetic field to other parameters (density, planetary radius, electrical conductivity, rotation rate, etc.) but, as with the earth, it nevertheless is instructive to consider the observations in the light of certain rudimentary ideas based on general considerations of the equations of magnetohydrodynamics of rotating fluids.

I. THE STRUCTURE OF JUPITER'S MAGNETIC FIELD

The Magnetosphere

Experiments on board Pioneers 10 and 11 have provided the first direct measurements of the field structure and particle density in the Jovian magnetosphere (Acuña and Ness 1975; Fillius and McIlwain 1974; Fillius *et al.* 1975; Simpson *et al.* 1974, 1975; Smith *et al.* 1974, 1975; Trainor *et al.* 1974, 1975; Van Allen *et al.* 1974, 1975). The evidence favors a closed magnetosphere, with a well-defined magnetopause, and an overall blunt-shaped configuration. The outer regions of the magnetosphere are strongly influenced by the pressure of the solar wind arriving at the planet, with variations in the ambient conditions causing fluctuations of up to a factor two in the size of the dayside magnetosphere. A prominent feature of the mid-magnetosphere is a thin annular current sheet which co-rotates with the planet like a rigid body. The sheet lies almost parallel to the Jovian equator, but is distorted, so that it lies above the equatorial plane on one side of the planet, and below it on the other. Associated with the current sheet is an equatorial concentration of field lines which confine energetic particles trapped in this region to a relatively thin disk. The structure and dynamics of the Jovian magnetosphere are discussed in more detail elsewhere in this volume (see the chapters by Acuña and Ness,[1] Simpson and McKibben,[2] Smith et al.[3] and Van Allen[4]).

At distances within about 20 R_J of the planet, the magnetic field configuration measured by the Pioneer magnetometers rapidly assumes the characteristics of the main planetary field. There are irregularities in this field close to the orbits of the innermost satellites, but the basic structure seems to be well represented by that of an inclined magnetic dipole which is slightly offset from the center of the planet (Smith et al. 1974, 1975).

The Planetary Field in the Radiation Belts

Information about the structure of the planetary field at distances of about 2–3 R_J is provided by Earth-based observations of the decimetric radio emission. This is synchrotron emission from relativistic electrons which are trapped by the fieldlines to form radiation belts similar to the Van Allen belts which surround the earth. The radiating electrons have an anisotropic distribution of pitch angles, with the majority moving in relatively flat helical orbits and mirroring well within one planetary radius of the magnetic equator. As a result the decimetric emission is strongly beamed into the plane of the magnetic equator, and is linearly polarized with the E-vector perpendicular to the projected direction of the magnetic fieldlines. The variation of the position angle of polarization as Jupiter rotates thus provides a useful trace for the geometry of the magnetic field in the vicinity of the magnetic equator, while the beaming in latitude of the total intensity and of the degree of polarization enable the distributions of electron energy and pitch angle to be deduced. A small amount of circular polarization has also been detected in the emission, and this gives the strength and polarity of the magnetic field.

The properties of the decimetric emission have been extensively reviewed elsewhere (Carr and Gulkis 1969; Stannard 1975; Berge and Gulkis[5]) so that we restrict ourselves here to a comparison of the field structure in the radiation belts deduced from the Pioneer and decimetric observations. As can be seen in Table I, there is good general agreement between the two methods of measurement.

---

[1]See p. 837.   [3]See p. 789.   [5]See pp. 621, 693 and 1146.
[2]See p. 739.   [4]See p. 928; also see p. 32.

For many purposes the field in the radiation belts can be adequately represented by that of an inclined offset dipole, but there is evidence in both the radio and Pioneer data that the real field geometry is more complex. In particular, there seems to be a localized distortion of the main dipole field close to the longitude of the magnetic pole in the northern hemisphere. This anomaly produces asymmetries in the brightness distribution and polarization of the decimetric emission (Branson 1968; Conway and Stannard 1972), and is also apparent in the dual-dipole model fitted to the Pioneer 10 data by Smith *et al.* (1974). Closer to the surface of Jupiter the non-dipole components of the planetary magnetic field become more pronounced.

The Field Near the Surface of the Planet

The structure of the magnetic field close to the surface of Jupiter can be investigated by using observations of the decametric radio emission. The mechanism responsible for these intense bursts of low frequency radio noise is poorly understood, but is generally assumed to be the cyclotron process occurring at or near the local electron gyrofrequency in the Jovian ionosphere. If this is the case, then the sharp cut off in the emission above 39.6 MHz implies a maximum field strength of some 14 Gauss, presumably from the polar regions close to the surface of the planet. Part of the decametric emission is known to be triggered by Jupiter's satellite Io, but there is considerable debate as to the nature of this interaction. For a discussion of this, and of other features of the decametric emission, see the review articles by Carr and Gulkis (1969) and Carr and Desch.[6]

The emission at the lowest decametric frequencies possesses a symmetry which is consistent with a basically dipolar field structure. However, at higher frequencies anomalies become more prevalent. There is a predominance of a single sense of circular polarization and a pronounced asymmetry in the location of the main sources of emission. Warwick (1970) has proposed an explanation for these anomalies in terms of a field structure which is considerably displaced from the center of the planet, but this appears to conflict with both the Pioneer measurements and recent determinations of the centroid of the decimetric emission (Smith *et al.* 1974, 1975; Berge 1974). More tenable seems the explanation of Ellis and McCulloch (1963), which attributes the asymmetries to local deviations in the angle of magnetic dip. There is a tie up here with the decimetric studies, since the strongest of the high-frequency decametric sources (Source A) lies close to the longitude of the field anomaly inferred from the decimetric observations.

The close approach of Pioneer 11 to within 0.6 R_J of the Jovian surface has enabled direct *in situ* measurements to be made of the field structure in the vicinity of the planet. Pioneer 11 carried two experiments designed to investigate the structure of the magnetic field. One involved a vector helium

---

[6]See pp. 621, 693 and 1146.

TABLE I

The Magnetic Field in the Jovian Radiation Belts

	Determined from observations of the decimetric radio emission at a distance of 2–3 R_J	Measurements from the Pioneer 10 Spacecraft. The D2 Model fitted in the region 2.8–6 R_J	Measurements from the Pioneer 11 Spacecraft. The D4 Model fitted in the region 1.7–6 R_J
Field configuration	Inclined dipole, possibly slightly offset	Inclined dipole, slightly offset	Inclined dipole, slightly offset
Inclination to rotation axis	$9°.5 \pm 0°.5$	$10°.6$	$10°.8$
Offset from planetary center	polar: $(+0.1 \pm 0.2)R_J$ equatorial: $\leq 0.05R_J$	$+0.03R_J$ $0.11R_J$ towards longitude 176°	$+0.01R_J$ $0.10R_J$ towards longitude 183°
Dipole moment (Gauss R_J^3)	3–15	4.0	4.2
Polarity of magnetic pole in northern hemisphere	North pole	North pole	North pole
Longitude of magnetic pole in northern hemisphere	230°	222°	228°

magnetometer similar to that on board Pioneer 10; the other used a triaxial fluxgate magnetometer sensitive to stronger fields. There seems to be some disagreement between the preliminary results of the two experiments (Smith *et al.* 1975; Acuña and Ness 1975).[7] Both sets of measurements confirm that the field structure close to the surface of the planet is indeed complex, with marked non-dipole components. There is an asymmetry in the strength of the field at the two poles, with the magnetic pole in the northern hemisphere possessing an appreciably stronger field (14 Gauss versus 11). When the magnetometer data are fully analyzed, they may reveal vital clues as to the nature of the mechanism responsible for the decametric emission.

II. SECULAR VARIATION OF JUPITER'S MAGNETIC FIELD

Although the Earth-based radio astronomy observations provide only limited information about the structure of the Jovian magnetic field, these measurements have now been obtained for over two decades and can thus be examined for evidence of temporal variations in the field structure. Care must be taken, however, to discriminate against a number of features of the radio emission that exhibit apparent variations which are not connected with any long-term secular effects, but are caused by geometrical changes in the aspect of Jupiter during its orbital period of 11.9 years about the sun.

The Rate of Rotation of the Magnetic Field

There is excellent agreement between the rotation periods derived from observations of the motions of the decametric and decimetric radio sources. Early observations of the decametric emission led to the IAU adopting a formal System III period for the radio rotation which was 11.3 seconds shorter than the period of rotation of irregular features of the visible surface of dense cloud in the non-equatorial regions of the planet. The cloud belts nearer the equator spin at a much faster rate.[8] Subsequent observations of stable features in both the decametric and decimetric radio emissions have shown that the System III period is too short by some 0.4 seconds (Table II), and this gives rise to a longitudinal drift of $3°2$ per year, relative to System III, of the radio emission features.

The determination of the decametric rotation period is complicated by a cyclic drift in the apparent source positions which appears to be correlated with Jupiter's orbital position about the sun (Carr *et al.* 1970). If this periodic effect is removed, and a correction applied for the "error" in the System III period, then the longitude of Source *A* shows no further systematic variation with the epoch of observation (Fig. 1*a*). Recent studies of the decametric rotation period by Alexander (1975) suggest that only statistical averages of decametric events should be used to determine an accurate rate

---

[7]Editorial Note: The disagreement no longer exists; see pp. 38 and 747.
[8]See pp. 566 and 590; also see on these topics pp. 709, 656, 826 and 20.

TABLE II
Rotation Periods of Jupiter

ADOPTED IAU ROTATION PERIODS

System I (equatorial belt)	9^h	50^m	30^s00
System II (polar regions)	9^h	55^m	40^s63
System III (radio)	9^h	55^m	29^s37

ACCURATE DETERMINATIONS OF THE TRUE RATE OF RADIO ROTATION

decametric	9^h	55^m	29^s73 ± 0.04
decimetric	9^h	55^m	29^s73 ± 0.02

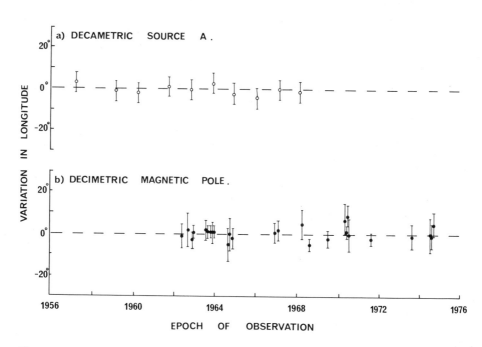

Fig. 1. Departures from the mean position over a complete orbital cycle of (*a*) the longitude of decametric Source *A* (Carr *et al.* 1970), and (*b*) the longitude of the magnetic pole in the northern hemisphere (as deduced from published observations of the decimetric emission). The mean position of both features has been fixed in longitude by the adoption of a radio rotation period of 9^h 55^m 29^s73.

of rotation, rather than measurements of the individual noise bursts themselves.

A useful reference feature in the decimetric data is the longitude of the magnetic pole in the northern hemisphere. This may be determined from the position of the minima in the beaming of the total intensity, from the longitude of the maximum degree of left-handed circular polarization, or from the reproducible shape of the polarization position angle data. [Note that because of the field anomaly near the position of the pole, the true pole position is obtained by adding 15° to the longitude of the half peak-to-peak variation of position angle. This follows from the type of analysis performed by Conway and Stannard (1972).] The three independent methods are in good agreement, and give the mean longitude of the pole as $230 \pm 1°$ in System III coordinates at epoch 1974.0. There is no evidence of any twelve-year periodicity in these determinations of the pole longitude (Fig. 1b, where again a correction has been applied for the known "error" in the System III period).

Any short-term fluctuations in the rate of radio rotation would produce discrepancies between the individual longitude determinations and the mean values shown in Fig. 1. The absence of any marked variation suggests that the rotation in the time scale of about a year is constant to better than ± 0.30 sec, or 1 part in 10^6 of the rotation period.

The theoretical discussion of Jupiter's non-uniform rotation (Hide 1961, 1962, 1965), as revealed by determinations of various rotation periods, involves considerations of the dynamics and magnetohydrodynamics of Jupiter's atmosphere and interior, just as the theoretical interpretation of the average eastward motion of the terrestrial atmosphere relative to the solid earth, of the general westward relative motion of the geomagnetic field, and of fluctuations in the magnitude (and direction) of the rotation of the solid earth, involves the discussion of fluid motions in the earth's oceans, atmosphere and liquid core.

Jovian decametric and decimetric radio sources, which are nearly fixed in System III, are presumably tied to Jupiter's magnetic field. This, in turn, is intimately linked to the upper parts of the electrically conducting region of the interior of the planet, the depth of which is uncertain (Hide 1965; see also Sec. III below). The widely held view that the rotation period of the upper reaches of the main body of the planet is essentially the same as the radio period is but one of several possible interpretations of the data (Hide 1965; Runcorn 1976) and it may not be the best one.

The Inclination of the Magnetic Dipole

If the planetary field in the radiation belts were that of a perfect magnetic dipole, then the position angle of linear polarization of the decimetric emission would be expected to fluctuate in a sinusoidal manner with time, with the amplitude of the sinusoid providing a direct measure of the inclination between the dipole and rotation axes. As was discussed earlier, the observed

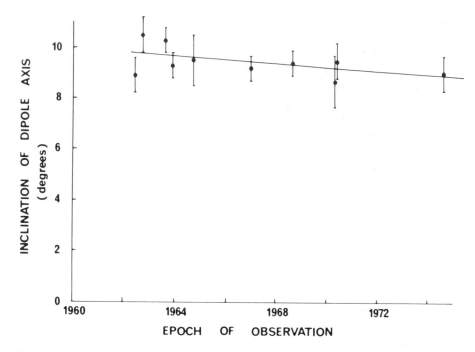

Fig. 2.   Evidence of a possible secular decrease (0.07 ± 0.05 degrees per year) in the inclination of the Jovian magnetic axis (after Stannard and Conway 1976).

variation of position angle is distorted because of the presence of non-dipole
components in the field structure (see also Conway and Stannard 1972). The
resulting asymmetric shape of the position angle curve appears to be extremely stable, and does not change with either the frequency of radiation
(Roberts and Komesaroff 1965) or Jupiter's orbital position about the sun
(Komesaroff and McCulloch 1967; Berge 1974). Because of this constant
shape, we can use estimates of the half peak-to-peak variation in position
angle to search for evidence of any temporal variations in the inclination of
the Jovian dipole (Stannard and Conway 1976). The data of Fig. 2 show that
within the error of measurement there is no marked change in the inclination, although a least squares fit to the data does suggest a secular decrease
of 0.07 ± 0.05 degrees per year (standard error). Clearly observations over
a longer time span are required to confirm the reality of this effect.

The Strength of the Magnetic Field

The magnetic moment of Jupiter measured by Pioneer 11 is some 6%
greater than that determined by Pioneer 10 (Smith *et al.* 1975). It is, however, uncertain whether this discrepancy on the time scale of one year reflects a real change in the field strength, or whether it can be accounted for

by the process of fitting a simple model to different regions of what is in reality a complex field structure.

Variations in the field strength can be investigated using the radio-astronomy measurements of the cut-off frequency of the decametric emission, and of the degree of circular polarization of the decimetric emission. We are not aware of any systematic monitoring of the decametric cut off, but measurements of the circular polarization of the decimetric emission suggest an upper limit of about 15% to any variations in the field strength between epochs 1967 and 1970 (Stannard and Conway 1976).

There have been marked changes in the intensity of the decimetric emission over the past decade (Berge 1974; Klein 1975). The cause of these variations is unknown, but one can speculate that they may be linked to changes in the Jovian magnetic moment (Berge 1974). The variations have occurred over a wide range of wavelengths (Fig. 3), and show no evidence of any twelve-year periodicity which might be connected with either the sunspot cycle or Jupiter's orbital position.

To summarize, there have been unexplained variations in the intensity of the decimetric emission on the time scale of a few years, but observations

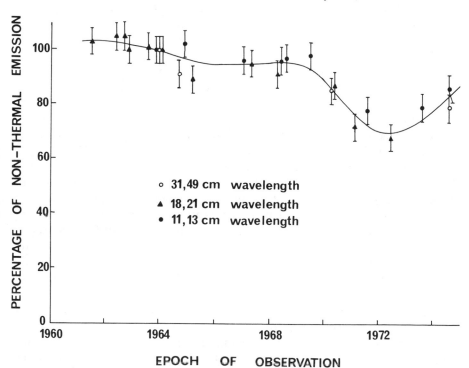

Fig. 3. Variations in the peak intensity of the non-thermal decimetric emission. Published measurements of the flux density have been corrected for an assumed thermal disc temperature of 250°K, and normalized by the spectrum of the non-thermal emission at Epoch 1964.0.

of other properties of the Jovian radio emission suggest that any marked variations in the structure of the magnetic field occur on a time scale which is at least a few tens of years (see below).

III. THE ORIGIN OF JUPITER'S MAGNETIC FIELD

The only serious suggestion concerning the origin of Jupiter's magnetic field closely parallels the ideas developed by geophysicists over the past thirty years towards an explanation of the earth's magnetism on the basis of a hydromagnetic (magnetohydrodynamic) dynamo operating in the electrically conducting fluid part of the interior of the planet (Hide 1965; 1966a; Runcorn 1968; Smoluchowski 1971; Warwick 1970; Stevenson 1974). Accordingly, we suppose that the main magnetic field implied by the foregoing observations is due to ordinary electric currents flowing somewhere within the main body of Jupiter and that these currents are maintained by motional induction, involving the interaction of hydrodynamical motions with the magnetic field itself.

It is now well established (see, e.g., Ingersoll *et al.*, p. 202) that no more than about one half of Jupiter's estimated thermal (infrared) emission can be accounted for in terms of the absorption of incident solar radiation, and that in order to explain the balance it seems necessary to invoke gravitational contraction (and possibly differentiation) of the planet at the very slow rate of ~ 0.1 cm per year, a process which now features in theoretical models of Jupiter's internal structure. Convective motions should contribute substantially to the upward heat transfer within the planet and the Jovian dynamo is presumably engendered by such motions in regions where the planet is a sufficiently good electrical conductor. Thus, the energy of the magnetic field derives ultimately from the thermal and gravitational energy of the planet through the action of buoyancy forces and motional induction.

The main constituent of Jupiter is hydrogen, much of which is compressed to a metallic form, but it is generally accepted that the second lightest element, helium, is also present in substantial amounts, up to $\sim 20\%$ by mass on the average, but not necessarily distributed uniformly. Theoretical models of Jupiter's internal structure (for references see Stevenson and Salpeter, p. 85) give the radial variation of pressure and density within the planet quite precisely, but they fail to give temperatures accurately and cannot therefore predict with confidence which parts of the interior of the planet are fluid. Smoluchowski (1971) considers that the deep metallic part of Jupiter might be fluid — and therefore able to support a dynamo — but only where the pressure is sufficiently high for helium to dissolve in hydrogen and form an alloy. The transition pressure is $\sim 10^{12}$ Pa (and therefore several times the transition pressure at which molecular hydrogen changes into metallic hydrogen and probably higher than the central pressure of Saturn), but its precise value is not known so that the radial extent of Jupiter's liquid He-H alloy region is very uncertain.

However, it is by no means obvious that the field should originate at depths where Jupiter is metallic. An efficient dynamo requires a moderate but not too high value of electrical conductivity [see Eq. (5) below], so it is necessary, as Hide (1965) has pointed out, to consider whether or not the deep metallic interior of Jupiter is the only part of the planet where such a dynamo could reasonably be supposed to operate, and to establish in particular whether or not the outer regions of the planet, though non-metallic, may at some levels be sufficiently conducting to support dynamo action. Smoluchowski (1972, 1975) has examined this suggestion, concluding at first that the conductivity of the molecular hydrogen in the outer fluid layers of Jupiter is probably too low everywhere unless some very unusual impurities are present, but later modifying his conclusions on the basis of further analysis. It is unfortunate that dynamo theory cannot yet provide a basis for determining the location and other properties of the Jovian dynamo from measurements of the strength and configuration of the magnetic field near the surface of the planet; however, it is very likely on geometrical grounds that the deeper the dynamo the more preponderant the dipole contribution to the spherical harmonic components of the field at the surface of the planet (see Appendix and discussion in Sec. I).

Hydromagnetic Dynamos in the Earth and Jupiter

As we have already mentioned, geophysicists now accept that the main geomagnetic field is due to ordinary electric currents within the earth and that these currents flow mainly in the liquid metallic (mainly molten iron) core, where they experience least electrical resistance. The comparatively small (~ 0.2) ratio of the magnitude of the non-dipole field at the earth's surface to that of the dipole field is associated with the great depth of the core, the outer radius of which is 0.55 times that of the whole earth. Owing to its more rapid increase with depth (see Appendix, and especially Table A), the non-dipole field is comparable in magnitude with the dipole at the surface of the core.

Two lines of evidence rule out the possibility that the geomagnetic field is a remnant of a primordial field acquired when the earth was formed, namely the occurrence over geological time of irregular alternations in the sign of the geomagnetic dipole and the comparatively short ($\sim 10^5$ years) time constant τ of free decay of electric currents in the core. This time constant is given by

$$\tau \sim L^2 \mu \sigma \tag{1}$$

where L is a typical length scale, μ the magnetic permeability and σ the electrical conductivity, which are respectively $\lesssim 3 \times 10^6$ m, $4\pi \times 10^{-7} H\,\mathrm{m}^{-1}$ and $\sim 3 \times 10^5\ \Omega^{-1}\mathrm{m}^{-1}$ for the earth. It is, therefore, necessary to suppose that there exists in the core some mechanism capable of: (a) generating the electric currents there and maintaining them against the effects of ohmic dissipation, and (b) altering the configuration of the currents from time to time so as to produce changes in the polarity of the external field. Of the various mechanisms that have been suggested only the "hydromagnetic dynamo" proposed by Bullard and Elsasser and developed by many others is both qualitatively and quantitatively attractive (for references see Roberts 1971; Weiss 1971; Moffatt 1973; Gubbins 1974), and modern research on the origin of the earth's magnetism is concerned with the complex problem of understanding how dynamos work in detail.

The dynamo theory supposes that fluid motions have, by motional induction, amplified and distorted some weak adventitious magnetic field so as to produce the field that is observed. General considerations of the equations of electrodynamics show that one necessary condition for dynamo action is that the "magnetic Reynolds number"

$$S \equiv UL\mu\sigma \tag{2}$$

($\sim \tau U/L$, where U is a characteristic speed of fluid flow) should satisfy

$$S > S^* \tag{3}$$

where, according to theoretical work on specific dynamos, the critical magnetic Reynolds number S^* is typically ~ 10.

Though necessary, Eq. (3) is by no means a sufficient condition for dynamo action. Theorems due to Cowling and others imply that such action cannot occur unless the fluid flow and magnetic field are sufficiently complicated in configuration, and herein lies one of the main obstacles to progress with the theory. This requires the simultaneous solution of the equations of hydrodynamics and the pre-Maxwell equations of electrodynamics, a task with which little headway has been made so far (for references see reviews by Chandrasekhar 1961; Hide and Stewartson 1972; Roberts and Soward 1972; Acheson and Hide 1973). In an attempt to set limits to the strength of the magnetic field produced by dynamo action, Hide (1974) has argued on the basis of general considerations of the equations of motion that the field is unlikely to build up beyond that value for which the Lorentz torque acting on a typical individual fluid element balances the inertial torque; whence the average strength B_i or the magnetic field within the fluid satisfies

$$B_i \lesssim [\mu\rho U(\Omega L + U)]^{\frac{1}{2}} \tag{4}$$

where Ω is the basic rate of rotation and ρ is the density of the fluid. Observe that the ratio of magnetic to kinetic energy implied by this equation is $\lesssim (1 + L\Omega/U)$ and may, therefore, greatly exceed unity when the basic rotation is so rapid that the Rossby number $U/L\Omega$ is much less than unity. Equipartition between magnetic and kinetic energy is only to be expected when $U/L\Omega$ is very large; for the earth and Jupiter the precise value of $U/L\Omega$ is not known but it is certainly very much less than unity.

Now, although the rate of generation of total magnetic energy by the dynamo mechanism, and hence B_i, can be expected to increase with increasing electrical conductivity σ, the strength of the external magnetic field, B_e, produced by the dynamo (and this is the only field we are able to observe) should *decrease* with increasing σ when σ is large, with B_e vanishing altogether when σ goes to infinity; then it is impossible to change the number of magnetic lines of force intersecting the surface of the fluid [see, e.g., Bondi and Gold (1950)]. Hide (1974) therefore has assumed that

$$B_e/B_i \sim S^{-q} \tag{5}$$

where the index q is essentially positive and possibly $\lesssim 1$, and introduces a "scale magnetic field strength"

$$B_s \equiv [\rho(\Omega + UL^{-1})/\sigma]^{\frac{1}{2}} \tag{6}$$

so that equations (5) and (4) can be written as follows:

$$B_i \lesssim B_s S^{\frac{1}{2}} \tag{7}$$

and

$$B_e \lesssim B_s S^{\frac{1}{2} - q} \equiv B_e^* . \tag{8}$$

B_e^* is thus an upper limit to the strength of the magnetic field just above the fluid region where dynamo action is taking place and satisfies

$$B_s S^{-\frac{1}{2}} \lesssim B_e^* \ll B_s S^{\frac{1}{2}} \tag{9}$$

(if $q \lesssim 1$). Corresponding expressions for the equivalent magnetic moment can be obtained from the last two equations by multiplying by the cube of an appropriate length.

Taking as typical values for the core of the earth $\Omega \sim 10^{-4}$ sec^{-1}, $\rho \sim 10^4$ kg m^{-3}, $\sigma \sim 3 \times 10^5 \Omega^{-1}$m^{-1}, $U \sim 10^{-4}$m sec^{-1}, $L \sim 10^6$m, we find that $U/L\,\Omega \sim 10^{-6}$, $B_s \sim (\rho\Omega/\sigma)^{\frac{1}{2}} \sim 2 \times 10^{-3}\ T$ (20 Gauss). Accordingly, by equations (7) and (9) we have $B_i \lesssim 10^{-2}\ T$ (100 Gauss) and $4 \times 10^{-2}\ T$ (4 Gauss) $\lesssim B_e^* \ll 10^{-2}\ T$ (100 Gauss) if for the magnetic Reynolds number S we take a value suggested by various studies of kinematic dynamos of the Bullard-Cellman type, namely ~ 25. The mainly dipolar field of $5 \times 10^{-5}\ T$ (0.5 Gauss) at the surface of the earth implies that the average field strength outside the core, B_e, is $\lesssim 10^{-3}\ T$ (10 Gauss), which falls within the preferred part of the range of B_e^* implied by the foregoing discussion. The value of B_i for the core cannot be inferred directly from geomagnetic observations, but various lines of evidence indicate that the magnetic field throughout the main body of the core might be largely toroidal in configuration, with lines of force lying approximately on horizontal surfaces, and $\sim 10^{-2}\ T$ (100 Gauss) in strength, which is also concordant with the above calculation (Hide 1974).

The values of many of the various quantities appearing in the above equations are even more uncertain in the case of Jupiter. However, on taking $\Omega \sim 10^{-4}$ sec^{-1} supposing that $\rho \sim 10^3$ kg m^{-2}, $L \sim 10^7$m and $U \sim 10^{-2}$m sec^{-1} we find

$$B_i \lesssim (\mu \rho U \Omega L)^{\frac{1}{4}} \sim 10^{-1} \ T \ (1000 \ \text{Gauss}) \tag{10}$$

which indicates that Jupiter's internal magnetic field may greatly exceed the external field (see Secs. I and II above). As in the case of the earth, B_i for Jupiter cannot be determined directly from magnetic observations, but 10^{-1} T is comparable with an estimate given by Hide (1965) from considerations of Jupiter's differential rotation and would be compatible with Equation (5) under plausible assumptions concerning S (see, e.g., Stevenson and Ashcroft 1974), q and the relationship between B_e and the measured field of the planet.

An alternative approach to the problem of finding a useful expression for the magnetic moment produced by a dynamo in a rotating fluid has been given by Levy (1972) and Roberts (1975), who exploited a specific dynamo model due to Parker (1955, 1970). A key parameter in this model is a "turbulence correlation length" λ, which is much less than L, and because the dependence of λ on other quantities is unknown, a useful comparison between the two approaches is hard to provide. Superficially, Hide's expression for B_e indicates a weaker Ω-dependence than the expressions given by Levy and Roberts, but because the dependence of λ on Ω is not known, this difference between the results of the two approaches may turn out to be more apparent than real. In a recent study Busse (1976; see also Busse 1975) analyzes a particular convection-driven model dynamo. Under the *assumption* [*cf.* Eq. (7)] that $B_e \sim B_i$ he finds that $B_e \sim 10^{-2.5} \ L\Omega(\kappa\sigma\mu)^{\frac{1}{2}}$, where κ denotes the thermometric conductivity of the fluid.

Secular Variations

By Faraday's law, the magnetic flux linkage of a perfect conductor cannot change, so that effects due to Ohmic dissipation are central to dynamo theories (see subsection on hydromagnetic dynamos, above) of the generation of the external magnetic field. On the other hand, such effects may not be of primary importance when dealing with secular changes, which occur on much shorter time scales than the time scale of free decay τ [see Eq. (1)], and their neglect leads to a considerable simplification of the governing equations. The mathematical problems involved are still very complex, especially when realistic boundary conditions are taken into account (for references see above-mentioned reviews) and their discussion lies beyond the scope of this chapter.

Fortunately, some of the main processes are exemplified by the properties of plane waves of angular frequency ω and vector wavenumber **k** propagating relative to a fluid which rotates uniformly with steady angular velocity $\mathbf{\Omega}$ and is pervaded by a uniform magnetic field $\mathbf{V}(\mu\rho)^{\frac{1}{2}}$, where **V** is the Alfvén velocity. The dispersion relationship for such waves is the biquadratic equation

$$\omega^4 - \omega^2(2\omega_v^2 + \omega_\Omega^2) + \omega_v^4 = 0 \tag{11}$$

where

$$\omega_v^2 \equiv (\mathbf{V} \cdot \mathbf{k})^2 , \quad \omega_\Omega^2 \equiv (2\mathbf{\Omega} \cdot \mathbf{k})^2/k^2$$

with solutions

$$\omega^2 = \omega_v^2 + \frac{1}{2}\,\omega_\Omega^2 \pm [\omega_\Omega^4/4 + \omega_v^2\omega_\Omega^2]^{\frac{1}{2}} = \omega_+^2 , \omega_-^2 \tag{12}$$

according as the upper or lower sign is taken. For wavelengths so short that $2\omega_v^2 \gg \omega_\Omega^2$, which for the earth's core, where $V \sim 10^{-1}$m sec^{-1}, corresponds to motions on scales much less than ~ 0.1 times the core radius, Eq. (12) reduces to the dispersion relationship for ordinary Alfvén waves, $\omega^2 = \omega_v^2$, which are non-dispersive, linearly polarized and characterized by equipartition between magnetic and kinetic energy. At the other extreme when $2\omega_v^2 \ll \omega_\Omega^2$, and this is the case of interest when dealing with waves in the core on scales of $\sim 10^6$m, Coriolis forces are so strong that the two roots of Eq. (12) have quite different values

$$\omega_+^2 = \omega_\Omega^2 \quad \text{and} \quad \omega_-^2 = \omega_v^4/\omega_\Omega^2 . \tag{13}$$

This extreme "frequency splitting" due to rotation is accompanied by other effects, notably wave dispersion, circular or elliptical polarization of the trajectories of individual fluid elements, and imbalance of kinetic energy, the whole of which is associated with the fast "inertial mode," and magnetic energy, now entirely in the slow "magnetic mode."

When Eqs. (13) are satisfied, the period of the inertial mode $2\pi/\omega_+$ is then typically $\lesssim 2\pi/\Omega$ (i.e., a few days) while that of the magnetic mode, $2\pi/\omega_-$, is $2\pi\Omega L^2/V^2$, which for the earth's core when $L \sim 10^6$m is 10^{10} sec (300 years) and, therefore, comparable with the time scale of the geomagnetic secular variation (see Hide and Roberts 1961). This is the quantitative basis of the theory of the geomagnetic secular variation which interprets the general time scale and westward drift (for references see Hide 1966b; Hide and Stewartson 1972) in terms of free hydromagnetic oscillations of the liquid core. The electrical conductivity of the overlying "solid" mantle, though weak, would be sufficient to prevent magnetic changes in the core on the short time scale of the inertial modes from penetrating to the earth's surface.

If for Jupiter we can take $L \sim 10^7$m and $B_i \sim 10^{-1}T$ (1000 Gauss) (see above) then $V \sim 3$m sec^{-1} and $V/L\Omega$, the "magnetic Rossby number," is $\sim 3 \times 10^{-3}$, about 3 times the corresponding value for the earth and, what is more important, very much less than unity. The corresponding modes of free oscillation of Jupiter's interior range from a few days at one extreme to several decades or even centuries [depending on the value taken for B_i, Eq. (12)] at the other, and it would not be implausible to suppose that the longer-period modes could easily be excited by thermal convection in Jupiter's electrically conducting parts. Whether or not such oscillations manifest themselves in secular changes in Jupiter's magnetism (see Sec. II above) is clearly a subject for future observational and theoretical work.

APPENDIX: THE EARTH'S MAGNETIC FIELD

The magnetic field at the surface of the earth undergoes complicated changes with time, with the most rapid changes occurring on time scales ranging from fractions of a second to several days. These rapid changes, in amplitude usually less than 1% of the total field, are due largely to varying electric currents flowing well above the earth's surface, in the ionosphere and beyond. When, by taking annual mean values, these rapid fluctuations are removed from the magnetic record and effects associated with the 11-year solar cycle and local anomalies are allowed for, the "main geomagnetic field" remains. The main field, which undergoes changes on time scales of decades and longer, originates in the liquid core of the earth.

In configuration and strength the present main field at the surface of the earth can be represented approximately by the field of a hypothetical centered and nearly axial dipole with a moment of 8×10^{25} emu, ($\sim 10^{-4}$ times that of Jupiter's dipole moment) inclined at an angle of about $10°$ to the rotation axis (the corresponding angle for Jupiter being $\sim 170°$; see Table I). According to palaeomagnetic studies, the polarity of the centered dipole has changed in a complicated and erratic way over geological time, the age of the oldest rocks studied to date being $\sim 3 \times 10^{9}$ years. Polarity reversals have occurred comparatively frequently (roughly 5 per million years) since the start some 200×10^{6} years ago of the present era of continental drift and also during certain earlier geological periods; but at other times, such as the Permian which lasted for 50×10^{6} years and ended 230×10^{6} years ago, reversals were rare events (for references see Bullard 1968; Hide and Stewartson 1972; Roberts and Soward 1972). Comparable observations are not, of course, available for Jupiter, whose magnetic polarity has remained the same in the twenty years since Jupiter's magnetism was first discovered.

The "non-dipole" field at the surface of the earth (i.e., the difference between the actual main field and its centered dipole component, as obtained by spherical harmonic analysis) is a good deal weaker ($\sim 20\%$) but undergoes more rapid secular changes than the dipole field, and the respective rms contributions to these two components of the geomagnetic secular variation are roughly comparable, about 5×10^{-4} Gauss per year. On typical magnetic maps of the non-dipole field, lines of equal value of a given magnetic element form a series of sets of oval curves of continental size. The pattern of the non-dipole field shows no striking correlation with topographic features of the earth's surface but there is a general bias away from the Pacific hemisphere and evidence that certain global-scale features of the non-dipole magnetic field are well correlated with corresponding ones of the earth's gravitational field (Hide and Malin 1970, 1971) has recently been adduced (for references see Akasofu and Chapman 1972). A striking feature of the geomagnetic secular variation is the tendency for the magnetic-field pattern at the surface of the earth to migrate westward at a fraction of a degree of longitude per year (10^{-4}m sec^{-1}).

The potential χ in terms of which the main geomagnetic field can be expressed outside its region of origin may be expanded in a spherical harmonic series:

$$\chi(r, \Theta, \phi, t) =$$

$$b \sum_{n=1}^{\infty} \sum_{m=0}^{n} \left\{ [g_n{}^m(t)\cos m\phi + h_n{}^m(t)\sin m\phi] \left(\frac{b}{r}\right)^{n+1} P_n{}^m(\phi) \right\} \quad (14)$$

where t is time; (r, Θ, ϕ) are spherical polar coordinates with the north geographic pole at $\Theta = 0$; $P_n{}^m(\Theta)$ are the Schmidt-normalized associated Legendre functions; while b is the mean radius of the earth. The presence of the term $(b/r)^{n+1}$ ensures that the dipole field (corresponding to $n = 1$) increases more slowly with depth than the non-dipole field ($n > 1$) and is much less dominant just outside the core at $r = a \simeq 0.55b$ than it is at the earth's surface $r = b$. This can be seen in Table A, which lists the magnitude of the maximum coefficient for each degree, up to and including $n = 8$, at the earth's surface and at the upper surface of the core. The non-dipole field is evidently comparable in strength with the dipole where the poloidal field emerges from its region of origin.

TABLE A
Maximum Coefficients of the Geomagnetic Field

n	max ($|g_n^m|$, $|h_n^m|$) (arbitrary units)	$(b/a)^{n+1}$ max ($|g_n^m|$, $|h_n^m|$) (arbitrary units)
1	100	340
2	10	63
3	7	77
4	3	70
5	1.4	47
6	0.7	53
7	0.3	30
8	0.1	17
Total	122.5	697

DISCUSSION

J. K. Alexander: The values of radio rotation period given in Table II suggest that the period derived from the decimeter observations is more accurately determined than that from the decameter observations. It is my impression that the accuracy with which the rotation period may be determined is in fact comparable for the two types of observation.

D. Stannard: The rotation period given for the decimeter observations in Table II is a new determination obtained by fitting a linear drift in longitude to the position of the magnetic pole in the north Jovian hemisphere (*cf.* Fig. 1). Determination of the decameter period is more complicated, both because of the statistical uncertainties inherent in the emission, and the necessity to correct for geometrical beaming effects associated with the orbital parameter D_E (see the chapter by Carr and Desch, p. 711.). The value and error we quote is that given by Carr *et al.* (*Radio Sci.* 5:494–503, 1970), which is typical of most of the decameter determinations. Recently Lecacheux (*Astron. Astrophys.* 37:301–304, 1974) has derived a more accurate period of $9^h\ 55^m\ 29.67^s \pm 0.01^s$. If this value is substantiated, then there may well be a real difference between the rotation periods of the decimeter and decameter emissions.

D. J. Stevenson: Here are some numbers that are important for understanding the dynamics in the Jovian interior:

Characteristic Time Scales

Gravity (free fall, gravity waves, etc.)	$1-5$ hours
Rotation	10 hours
Convection $\ell\ (\rho/F)^{\frac{1}{3}}$	10 years
Magnetic Diffusion ℓ^2/η	10^9 years
Viscous Diffusion ℓ^2/ν	10^{12} years
Thermal Diffusion ℓ^2/k	10^{12} years

where $\rho =$ fluid density, $\ell = 10^9$ cm, $F =$ internal heat flux; ν, η and k are magnetic, viscous and thermal diffusivities respectively. The ratio of magnetic diffusion time to convective time (the magnetic Reynold's number) is 10^8. This highly supercritical value suggests a rich harmonic content for the internal field.

One can also construct characteristic field amplitudes. For example, equating magnetic field energy with fluid kinetic energy gives $B \sim 10^{1.5}$ Gauss. Equating Coriolis and Lorentz forces, gives $B \sim 10^{3.5}$ Gauss. Equating the internal heat flux with the Ohmic dissipation gives $B \sim 10^{5.5}$ Gauss. The latter value is of interest, since it represents an upper bound to the internal field.

R. Hide: Not appearing in your list are the transit times of hydromagnetic waves in a rotating system [see Eq. (13)] and of sound waves.

I agree that an upper limit to the internal magnetic field can be obtained by equating the internal heat flux to Ohmic dissipation, but the more modest upper limit, obtained by supposing that the dynamo cannot build up beyond that magnetic field strength for which gyroscopic and Lorentz torques balance one another, would seem to be more realistic. There is, of course, no good physical basis for invoking equipartition between magnetic energy and kinetic energy when Coriolis forces dominate the inertial terms in the equation of motion.

REFERENCES

Acheson, D. J., and Hide, R. 1973. Hydromagnetics of rotating fluids. *Rep. Prog. Phys.* 36: 159–221.

Acuña, M. H., and Ness, N. F. 1975. The main magnetic field of Jupiter: Pioneer 11. *Nature* 253:327–328.

Akasofu, S. I., and Chapman, S. 1972. *Solar-terrestrial physics.* Oxford: Clarendon Press.

Alexander, J. K. 1975. Note on the beaming of Jupiter's decameter-wave radiation and its effect on radio rotation period determinations. *Astrophys. J.* 195:227–233.

Berge, G. L. 1974. The position and Stokes parameters of the integrated 21-cm radio emission from Jupiter and their variation with epoch and central meridian longitude. *Astrophys. J.* 191:775–784.

Bondi, H., and Gold, T. 1950. On the generation of magnetism by fluid motion. *Mon. Not. R. Astr. Soc.* 110:607–611.

Branson, N. J. B. A. 1968. High resolution radio observations of the planet Jupiter. *Mon. Not. R. Astr. Soc.* 139:155–162.

Bullard, E. C. 1968. Reversals of the earth's magnetic field. *Phil. Trans. Roy. Soc. London* A263:481–524.

Busse, F. H. 1975. A necessary condition for the geodynamo. *J. Geophys. Res.* 80:278–280.
———. 1976. Generation of Jupiter's magnetic field by convection. In preparation.

Carr, T. D., and Gulkis, S. 1969. The magnetosphere of Jupiter. *Ann. Rev. Astron. Astrophys.* 7:577–618.

Carr, T. D.; Smith, A. G.; Donivan, F. F.; and Register, H. I. 1970. The twelve-year periodicities of the decametric radiation of Jupiter. *Radio Sci.* 5:495–503.

Chandrasekhar, S. 1961. *Hydrodynamic and hydromagnetic stability.* Oxford: Clarendon Press.

Conway, R. G., and Stannard, D. 1972. Non-dipole terms in the magnetic fields of Jupiter and the earth. *Nature* 239:142–143.

Ellis, G. R. A., and McCulloch, P. M. 1963. The decametric radio emissions of Jupiter. *Aust. J. Phys.* 16:380–397.

Fillius, R. W., and McIlwain, C. E. 1974. Measurements of the Jovian radiation belts. *J. Geophys. Res.* 79:3589–3599.

Fillius, R. W.; McIlwain, C. E.; and Mogro-Campero, A. 1975. Radiation belts of Jupiter: a second look. *Science* 188:465–467.

Gubbins, D. 1974. Theories of the geomagnetic and solar dynamos. *Rev. of Geophys. Space Phys.* 12:137–154.

Hide, R. 1961. Origin of Jupiter's Great Red Spot. *Nature* 190:895–896.
———. 1962. On the hydrodynamics of Jupiter's atmosphere. *Mém. Soc. Roy. Sci. Liège* 7: 481–505.
———. 1965. On the dynamics of Jupiter's interior and the origin of his magnetic field. *Magnetism and the cosmos.* (W. R. Hindmarsh, F. J. Lowes, P. H. Roberts and S. K. Runcorn, eds.) pp. 378–395, Edinburgh: Oliver and Boyd.
———. 1966a. Planetary magnetic fields, *Planet. Space Sci.* 14:579–586.
———. 1966b. Free hydromagnetic oscillations of the earth's core and the theory of the geomagnetic secular variation. *Phil. Trans. Roy. Soc. London* A259:615–647.
———. 1974. Jupiter and Saturn. *Proc. Roy. Soc. London* A366:63–84.

Hide, R., and Malin, S. R. C. 1970. Novel correlations between global features of the earth's gravitational and magnetic fields. *Nature* 225:605–609.

———. 1971. Bumps on the core-mantle boundary. *Comments on Earth Sciences: Geophysics* 2: No. 1, 1–13..

Hide, R., and Roberts, P. H. 1961. The origin of the main geomagnetic field. *Physics and chemistry of the earth IV*, pp. 27–98. New York: Pergamon Press.

Hide, R., and Stewartson, K. 1972. Hydromagnetic oscillations of the earth's core. *Rev. Geophys. Space Phys.* 10; No. 2, 579–599.

Klein, M. J. 1975. Observations of a recent increase in Jupiter's decimetric radio emission. *Nature* 253:102–103.

Komesaroff, M. M., and McCulloch, P. M. 1967. The radio rotation period of Jupiter. *Astrophys. Letts.* 1:39–41.

Levy, E. H. 1972. Magnetic dynamo in the moon: a comparison with the earth. *Science* 178: 52–53.

Moffatt, H. K., ed. 1973. Report on the NATO Advanced Study Institute on magnetohydrodynamic phenomena in rotating fluids. *J. Fluid Mech.* 57:625–649.

Parker, E. N. 1955. Hydromagnetic dynamo models. *Astrophys. J.* 122:293–314.

———. 1970. The generation of magnetic fields in astrophysical bodies. 1. The dynamo equations. *Astrophys. J.* 162:665–673.

Roberts, J. A., and Komesaroff, M. M. 1965. Observations of Jupiter's radio spectrum and polarization in the range from 6 cm to 100 cm. *Icarus* 4:127–156.

Roberts, P. H. 1971. Dynamo theory. *Lectures in Appl. Math.* 13:129–206.

———. 1975. Stellar winds. *Solar Wind Three.* (C. T. Russell, ed.) Los Angeles, California: Inst. of Geophysics and Planetary Physics, Univ. of California.

Roberts, P. H., and Soward, A. 1972. Magnetohydrodynamics of the earth's core. *Ann. Rev. Fluid Mech.* 4:117–153.

Runcorn, S. K. 1968. Planetary magnetic fields as a test of the dynamo theory. *Geophys. J. Roy. Astron. Soc.* 15:183–189.

———. 1976. The rotation and magnetic field of Jupiter. *Icarus* (special Jupiter issue). In press.

Simpson, J. A.; Hamilton, D. C.; McKibben, R. B.; Mogro-Campero, A.; Pyle, K. R.; and Tuzzolino, A. J. 1974. The protons and electrons trapped in the Jovian dipole magnetic field region and their interaction with Io. *J. Geophys. Res.* 79:3522–3544.

Simpson, J. A.; Hamilton, D. C.; Lentz, G. A.; McKibben, R. B.; Perkins, M.; Pyle, K. R.; Tuzzolino, A. J.; and O'Gallagher, J. J. 1975. Jupiter revisited: first results from the University of Chicago charged particle experiment on Pioneer 11. *Science* 188:455–458.

Smith, E. J.; Davis, Jr., L.; Jones, D. E.; Coleman, Jr., P. J.; Colburn, D. S.; Dyal, P.; and Sonett, C. P. 1975. Jupiter's magnetic field, magnetosphere and interaction with the solar wind: Pioneer 11. *Science* 188:451–455.

Smith, E. J.; Davis, Jr., L.; Jones, D. E.; Coleman, Jr., P. J.; Colburn, D. S.; Dyal, P.; Sonett, C. P.; and Frandsen, A. M. A. 1974. The planetary magnetic field and magnetosphere of Jupiter: Pioneer 10. *J. Geophys. Res.* 79:3501–3513.

Smoluchowski, R. 1971. Metallic interiors and magnetic fields of Jupiter and Saturn. *Astrophys. J.* 166:435–439.

———. 1972. Electrical conductivity of condensed molecular hydrogen in the giant planets. *Phys. Earth Planet. Interiors* 6:48–50; *Astrophys. J.* 185:L95.

———. 1975. Jupiter's molecular hydrogen layer and the magnetic field. *Astrophys. J.* 200: L119–121.

Stannard, D. 1975. The radioastronomy of Jupiter. *The magnetospheres of the earth and Jupiter.* (V. Formisano and C. Kennel, eds.). Dordrecht-Holland: D. Reidel Publ. Co. In press.

Stannard, D., and Conway, R. G. 1976. Recent observations of the decimetric radio emission from Jupiter. *Icarus* (special Jupiter issue). In press.

Stevenson, D. J. 1974. Planetary magnetism. *Icarus* 22:403–415.

Stevenson, D. J., and Ashcroft, N. W. 1974. Conduction in fully-ionized liquid metals. *Phys. Rev.* 9A:782–789.

Trainor, J. H.; McDonald, F. B.; Teegarden, B. J.; Webber, W. R.; and Roelof, E. C. 1974. Energetic particles in the Jovian magnetosphere. *J. Geophys. Res.* 79:3600–3613.

Trainor, J. H.; McDonald, F. B.; Stilwell, D. E.; Teegarden, B. J.; and Webber, W. R. 1975. Jovian protons and electrons: Pioneer 11. *Science* 188:462–465.

Van Allen, J. A.; Baker, D. N.; Randall, B. A.; and Sentman, D. D. 1974. The magnetosphere of Jupiter as observed with Pioneer 10. *J. Geophys. Res.* 79:3559–3577.

Van Allen, J. A.; Randall, B. A.; Baker, D. N.; Goertz, C. K.; Sentman, D. D.; Thomsen, M. F. and Flindt, H. R. 1975. Pioneer 11 observations of energetic particles in the Jovian magnetosphere. *Science* 188:459–462.

Warwick, J. W. 1970. Particles and fields near Jupiter. JPL report NASA-CR 1685. Washington, D.C.: Government Printing Office.

Weiss, N. C. 1971. The dynamo problem. *Quart. J. Roy. Astron. Soc.* 12:432–446.

JUPITER'S MAGNETIC FIELD AND MAGNETOSPHERE

E. J. SMITH
Jet Propulsion Laboratory

L. DAVIS, JR.
California Institute of Technology

and

D. E. JONES
Brigham Young University

The Pioneer 10 and 11 Vector Helium Magnetometers have provided the first direct measurements of the magnetic field of Jupiter and have significantly increased our knowledge of the Jovian field. The two spacecraft have surveyed the magnetosphere over a range of radial distances between 100 and 1.6 R_J between local times from 5^h 30^m to 11^h 30^m and latitudes between -10 and $30°$. The magnetic field measurements reveal the existence of three distinct regions within the magnetosphere. Currents within the planet are dominant in the inner magnetosphere inside 20 R_J. The magnetic dipole moment of Jupiter is 4.2 Gauss $R_J{}^3$, i.e., 10^4 times stronger than the dipole moment of Earth, is tilted $\simeq 10°$ with respect to Jupiter's rotation axis, and is offset from the center of the planet $\simeq 0.1$ R_J. An alternative, more accurate description of the field is possible in terms of a centered dipole, quadrupole and octupole. The quadrupole moment and octupole moment are 20% and 15% of the dipole moment, respectively, or ~ 1.6 times larger than the corresponding ratios for Earth. Use of this representation to compute the magnetic field at the surface leads to maximum field strengths of 14.4 and 10.8 Gauss in the northern and southern hemispheres. The Pioneer model of the magnetic field is consistent with previous radio astronomy observations and eliminates some of the large uncertainties previously associated with them. In the middle magnetosphere, from 30 R_J to within 15 R_J of the outer boundary of the magnetosphere, the planetary field is strongly perturbed by currents associated with low-energy plasma. The field in the equatorial region is dominated by an intense thin current sheet that flows eastward around Jupiter in the form of a ring. The fields adjacent to the current sheet extend nearly radially in direction and are anti-parallel above and below the current sheet. As

[788]

Jupiter rotates, the current layer apparently wobbles up and down causing occasional passages of the current across the spacecraft. Inside the current sheet, the field magnitude decreases to a low value near 1 γ corresponding to a weak southward-directed normal field component. The fields adjacent to the current layer tend to deviate gradually from the local magnetic meridian plane to form a spiral with the outermost field lines lagging behind those nearer Jupiter. The spiral is associated with a radial component of the current flow and is less evident at increasing latitudes above and below the current sheet. At higher latitudes near local noon, the field is much more dipole-like and periodicities associated with the rotating planetary dipole are evident. In the outer magnetosphere, the field varies irregularly in magnitude and direction but exhibits a persistent southward component that is attributable to the planetary field, to the current sheet in the middle magnetosphere and to magnetopause currents flowing along the outer boundary of the magnetosphere. The magnetopause appears to be a relatively thin, well-defined boundary whose distance from Jupiter varies between 50 and 100 R_J. There is no evidence of significant spiraling of the field in the outer magnetosphere or of periodicities associated with the rotation of Jupiter, although the field lines are thought to be closed and to co-rotate with the planet. The observations appear to be consistent with magnetospheric models based on a closed magnetosphere at low latitudes with the possibility of open field lines at higher latitudes. The current sheet is apparently caused by large centrifugal stresses, associated with Jupiter's rapid rotation and the large scale of the magnetosphere, which is exerted on trapped low-energy plasma. The observations do not favor models based on the continual outflow or convection of plasma from Jupiter's magnetosphere in the vicinity of the equator.

The following description of Jupiter's magnetosphere is based on observations made by Vector Helium Magnetometers on board Pioneers 10 and 11. Less comprehensive but more detailed reports of these investigations have already been published elsewhere (Smith *et al.* 1974*a,b*; Smith *et al.* 1975*b*). The magnetometer and the overall scientific rationale of the investigation, including both planetary and interplanetary objectives, have also been described (Smith *et al.* 1975*a*). An attempt is made in this chapter to combine the results obtained from the two encounters without emphasizing details peculiar to either one of them. Emphasis is placed primarily on the observational results, but the latter are presented in such a manner as to be illustrative rather than demonstrative. The detailed arguments and analyses leading to various conclusions are not necessarily reproduced below. Readers interested in such details should refer to the original articles.

Any attempt to describe and understand Jupiter's magnetosphere inevitably begins by comparing it with the earth's magnetosphere. The observations made with the Vector Helium Magnetometer show that, near the planet, Jupiter's field is that of an eccentric, tilted dipole with some admixture of higher-order terms. In this respect it is similar to the earth's field although it is 10 times as strong and has the opposite polarity.

Jupiter has a bow shock, a magnetosheath, and a magnetopause qualitatively similar to Earth's, at least in the regions thus far explored. The scale of those features, on the order of 10^2 R_J, is essentially the same as would be expected from a simple scaling of the earth's field and the solar wind momentum flux if one considers only the innermost crossings. However, what appear to be multiple bow shock and magnetopause crossings are observed on all inbound and outbound traversals spreading over roughly a factor of two in radial position, suggesting that fluctuations are much larger and more frequent than for the earth.

In the region between the innermost 15 to 20 R_J and the outermost 10 to 15 R_J, at least in the part of the magnetosphere roughly within 30° of the equator and between the dawn and noon meridians which the Pioneers have explored, the magnetic field is much more radial and much less dipolar than that of Earth. A model that explains many of the observed features has the topology of the earth's field in the same latitude range and rotates with the planet as does the earth's field but is drastically stretched outward away from the axis of rotation by centrifugal force acting on a thin layer of plasma near the equatorial plane. However, some observations are difficult to reconcile with this model and suggest that fieldline reconnection or an outward flowing plasma might stretch the ends of these fieldlines out into the solar wind. This might give rise to open fieldline configuration having some features of the earth's polar field but at much lower latitudes and much more sunward longitudes. This model might bear little resemblance to any features observed near Earth.

No Pioneer observations were made in or near the Jovian regions that correspond to the earth's tail, plasma sheet, polar regions, or dayside cleft. The extent to which there are similar features near Jupiter is completely unknown. To study these problems and to determine what, if any, secular changes there are in the Jovian magnetosphere will require a Jupiter orbiter that explores the appropriate regions.

From our observations, the magnetosphere at low latitude between the dawn and noon meridians may be divided into three regions which we discuss separately below. In the region we call the *inner magnetosphere,* the field pattern is much like that within 3 radii of the earth and is dominated by the offset dipole term. This region extends to about 20 R_J from the center of Jupiter. Beyond this point there are strong perturbations in both field direction and magnitude due to plasma currents.

In the region within about 15 R_J of the magnetopause, which we call the *outer magnetosphere,* the field direction is mainly southward with substantial fluctuations and is an order of magnitude larger than the vacuum field due to the internal dipole source. The region between the inner and outer parts of the magnetosphere, which we call the *middle magnetosphere,* presumably varies in thickness as the magnetopause moves in and out. In this region the field direction tends to be more nearly radial than dipolar and to show some tendency to spiral westward, particularly at low latitude. The field is predominantly radially outward for northern latitudes and inward for southern latitudes.

In the next three sections, we consider each of these regions in more detail, attempting to explain the observations in terms of plausible physical causes. In the last section we briefly compare the observations with several large-scale models of the entire magnetosphere.

THE INNER MAGNETOSPHERE

The magnetic field within about 10 R_J of the center of Jupiter, in the region which we call the inner magnetosphere, is most easily presented and understood in terms of models that fit the observations reasonably well. The aim is to present all the observations in terms of a small set of relatively easily understood parameters. The accuracy and validity of each model is specified by giving the differences between the observations and the results predicted by the model at each point where an observation was made. A concise characterization of this goodness of fit is given by the square root of the suitably weighted average of the squares of these vector differences.

In the inner magnetosphere we expect that currents in the spherical shell between 1.2 and 10 R_J will produce only a negligible contribution to the total field. We expect that the major contribution will come from sources interior to Jupiter and that there may be a smaller contribution from currents in the middle and outer magnetosphere or in the magnetopause, currents associated with distortions of a vacuum field. Thus the field in the inner magnetosphere should be derivable from a scalar potential. The validity of these assumptions is tested by the degree to which the observations fit the models.

The simplest fairly realistic model, called the offset dipole model, is a point magnetic dipole source located at a fixed point in the interior of Jupiter. It need not be at the center and it need not be parallel to the axis of rotation. The parameters that describe this model are the 3 coordinates of the dipole position and the 3 components of the vector magnetic moment. No sources exterior to the inner magnetosphere are included in this model.

Any magnetic field in a current-free spherical shell can be described in terms of the coefficients used in the expansion of its scalar potential in spherical harmonics. In practice, only a limited number of coefficients can be determined reliably. In our case, we consider up to 15 coefficients (through the octupole terms) associated with interior sources and up to 8 (through the quadrupole terms) associated with currents exterior to the shell. We call this the spherical harmonic model; the parameters that describe it are this set of coefficients.

We evaluate the parameters in any model of interest by the classical method of least squares. The model is essentially a mathematical formula from which the three vector components of the predicted field can be computed at each point where an observation has been made for any specified set of values of the parameters. The parameters are given the values that make the suitably weighted sum of the squares of the differences between the observed and predicted vectors a minimum. The square root of this sum we call the residual. The square root of the quotient of this sum and the sum of the weights we call the root-mean-square (rms) residual. It is desirable to use weights since otherwise the parameters will be so chosen that an error of 10γ at a place where the field is $100,000\gamma$ is taken to be as serious as an error of 10γ at a place where the field is 1000γ.

If the parameters enter linearly into the formula for the predicted field, formulas are available (Bevington 1969) that give the parameters as a solution of linear algebraic equations with constant coefficients. This is the case for the spherical harmonic model. In the offset dipole model, three of the parameters do not enter linearly. Here we compute the residuals for a series of trial values of these parameters, using a computer program that lets us enter new values of the parameter as soon as we see the residual for the previous set. In this way the parameters that give the minimum residual can be determined with reasonable accuracy.

Coordinate Systems

We assume, and the results tend to confirm, that the interior sources of field rotate at a uniform rate that can be deduced from the radio astronomical observations of Jupiter's decimetric radiation. These observations give a rotation rate for a tilted magnetic field that appears to be uniform and is known to within about one part in 10^6. There is no universally adopted precise value for this rate. We use a rotation period, as seen from an inertial frame, of $9^h\ 55^m\ 29\overset{s}{.}7$ recommended by Mead (1974). This is in good agreement with the period of $9^h\ 55^m\ 29\overset{s}{.}71$ recommended by Warwick (personal communication) but is distinctly different from the period of $9^h\ 55^m\ 29\overset{s}{.}37$ adopted in 1962 for System III by the IAU (Information Bulletin 8, Int. Astron. Union, March 1962).[1]

We do not expect the exterior sources, the currents in the magnetopause and outer magnetosphere, to rotate with the core of Jupiter. The only presently reasonable assumption is that on the average they rotate about Jupiter at the same rate that the sun appears to move, i.e., with an angular velocity of $0\overset{\circ}{.}0890$ per day for the period from Dec. 1973 to Dec. 1974.

We define two coordinate systems for use in our analyses. They are spherical polar coordinate systems with their origins at the center of Jupiter and their polar axes parallel to Jupiter's angular momentum. The (r, θ, ϕ_i) System, used for fields due to internal sources, rotates with the core of Jupiter as described above and coincides with System III at $00^h\ 00^m\ 00^s$ UT on Dec. 3, 1974. The connection between this system and the usual System III, with polar coordinates (r, θ, ϕ_{III}) is

$$\phi_i = \phi_{III} + 0\overset{\circ}{.}00804\ \Delta D \tag{1}$$

where ΔD is the number of days from 1974, Dec. 3.0. We call this the Pioneer 11 Jupiter Magnetic Coordinate System, or the $P11J$ System. The

[1]See p. 826.

(r, θ, ϕ_e) System, used for fields due to exterior sources, rotates with the sun as described above; thus $\phi_e = 0$ is always along the Jupiter-Sun line. We call this the *JS* System.

The connection between ϕ_e and ϕ_i during the two Pioneer flybys is given by

$$\phi_e = \phi_i + \omega_{ei}T + \Delta\phi_0 \tag{2}$$

where T is hours after the midnight, UT, immediately preceding periapsis in each case and the parameters are:

For Pioneer 10 and 11, $\omega_{ei} = 36°26863$ per hour

For Pioneer 10, $\Delta\phi_0 = -85°492$ $\qquad\qquad$ (3)

For Pioneer 11, $\Delta\phi_0 = -42°202$

Equations (1) and (2), together with these parameters, may be regarded as a complete definition of the coordinate systems we use in terms of the IAU defined System III. The parameters used are given with more precision than is justified by observation in order that the systems be defined precisely, even though rather arbitrarily. Longitudes, λ, are measured from the same zero as ϕ_i but λ increases in a clockwise direction as seen from the north while the azimuthal angle ϕ_i increases in a counterclockwise direction.

The trajectories of Pioneers 10 and 11 in the (r, θ, ϕ_e) System are shown in Fig. 1. The portions of both trajectories nearest Jupiter are shown in Fig. 2 in which (r, ϕ_i) is plotted for both spacecraft. The variation with latitude of the two trajectories is shown in Fig. 3, which is a plot of Jovigraphic distance and the corresponding magnetic latitude. These three figures provide most of the essential information about the Pioneer trajectories.

The Offset Dipole Model

The magnetic field **B** at the position $\mathbf{r} = \mathbf{r}_0 + \mathbf{r}_d$ produced by an offset dipole is

$$\mathbf{B} = r_d^{-3} (3\,\widehat{\mathbf{U}}\,\widehat{\mathbf{U}} - \widehat{\mathbf{I}})\mathbf{M} \tag{4}$$

where $\mathbf{r}_0$ is the vector from the center of Jupiter to the position of the offset dipole, $\mathbf{r}_d$ is the vector from the dipole to the field point, $\widehat{\mathbf{U}}$ is a unit vector along $\mathbf{r}_d$, $\widehat{\mathbf{U}}\,\widehat{\mathbf{U}}$ is a dyadic, $\widehat{\mathbf{I}}$ is a unit dyadic and $\mathbf{M}$ is the vector dipole moment. We solve (4) for $\mathbf{M}$ to get

$$\mathbf{M} = r_d^3 (3\,\widehat{\mathbf{U}}\,\widehat{\mathbf{U}} - 2\widehat{\mathbf{I}})\,\mathbf{B}/2 . \tag{5}$$

If $\mathbf{B}_i$, $i = 1, 2, \ldots n$ is a series of observations of **B** at different points $\mathbf{r}_i$, then for any selected value of $\mathbf{r}_0$ each observation is completely specified by the

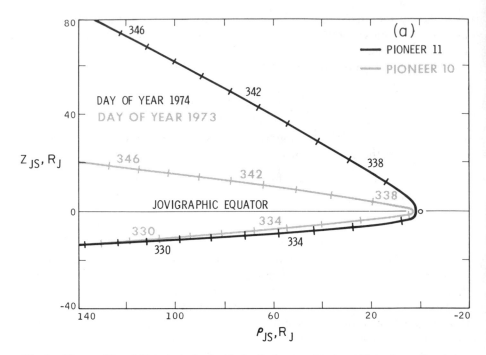

Fig. 1. Pioneer 10 and 11 trajectories inside Jupiter's magnetosphere. Two views of each trajectory are shown in *JS* coordinates (the primary vector **J**, Jupiter's spin axis, defines the *Z* coordinate and **S**, the direction from the planet's center to the sun, defines the prime meridian or *XZ* plane). The Pioneer 10 trajectory is grey and the Pioneer 11 trajectory is black. The panel (a) essentially shows the latitudes of the two spacecraft inbound and outbound. The abscissa ρ_{JS} is the distance perpendicular to *J* and is equal to the square root of the sum of the squares of the *X* and *Y* components. The panel (b) contains the projections onto the equatorial plane to show the local time dependences.

corresponding $\mathbf{M}_i$. By the usual least-squares procedure, we determine the vector $\mathbf{M}_0$ that minimizes

$$R_0^2 = \sum_{i=1}^{n} (\mathbf{M}_i - \mathbf{M}_0) \cdot (\mathbf{M}_i - \mathbf{M}_0). \qquad (6)$$

This is done by an interactive computer program which displays and records R_0^2 and $\mathbf{M}_0$ as soon as $\mathbf{r}_0$ is entered from a keyboard. Guided by this output, successive values of $\mathbf{r}_0$ are entered until the value that minimizes R_0^2 is found to within an uncertainty of about 0.002 R_J. The results of such analyses of the Pioneer 10 and 11 data are compared with those of the spherical harmonic analyses in the section on results, below.

The Spherical Harmonic Model

In any thick spherical shell within which there are no currents, the magnetic field is given by

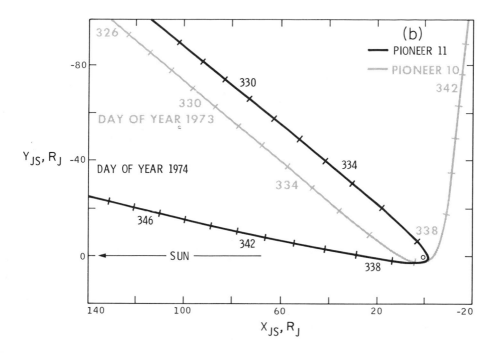

$$\mathbf{B} = -grad\ \phi\,. \tag{7}$$

The scalar potential ϕ can be expanded in a series of spherical harmonics

$$\phi = a\sum_{\ell=1}^{\infty}\sum_{m=0}^{\ell}\left[\left(\frac{r}{a}\right)^{-\ell-1}(g_\ell{}^m\cos m\phi_i + h_\ell{}^m\sin m\phi_i)\right. \tag{8}$$
$$\left.+ \left(\frac{r}{a}\right)^{\ell}(\bar{g}_\ell{}^m\cos m\phi_e + \bar{h}_\ell{}^m\sin m\phi_e)\right]P_\ell{}^m(\cos\theta)\,.$$

Here a is a convenient length, usually the radius of the planet, and $P_\ell{}^m(x)$ is the Schmidt-normalized Legendre function defined, for example, by Chapman and Bartels (1940):

$$P_\ell{}^m(x) = N_{\ell m}(1-x^2)^{m/2}\ \mathrm{d}^m P_\ell(x)/\mathrm{d}x^m \tag{9}$$

where $P_\ell(x)$ is the Legendre polynomial and

$$N_{\ell m} = 1,\ \text{if}\ m = 0,\ \text{or}\ [2(\ell-m)!/(\ell+m)!]^{\frac{1}{2}},\ \text{if}\ m\neq 0\,. \tag{10}$$

If $N_\ell{}^m$ is taken to be 1, $P_\ell{}^m$ has the Ferrer normalization. If $N_\ell{}^m = (-1)^m$, $P_\ell{}^m$ has the Hobsen normalization and, if $N_\ell{}^m = [(2\ell+1)(\ell-m)!/2(\ell+m)!]^{\frac{1}{2}}$, the squares of the $P_\ell{}^m$, averaged over a sphere, are unity. With the Schmidt normalization, the rms of the field over a sphere due to one of the interior coefficients is nearly independent of ℓ and is independent of m.

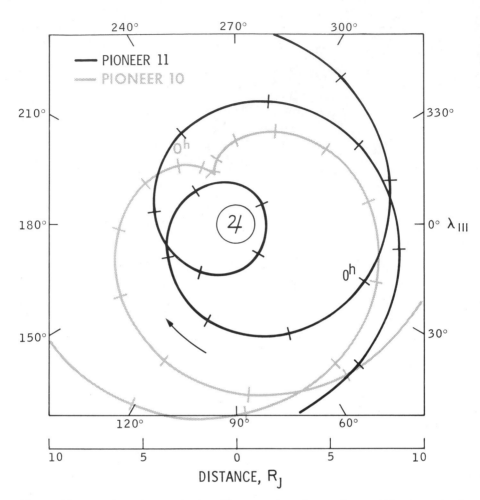

Fig. 2. Pioneer trajectories near Jupiter. The two trajectories are shown within 10 R_J of peri-apsis in a Jupiter-centered, non-rotating system. The radial distance (not the equatorial pro-jection) and the corresponding System III longitude are shown. The tick marks on each trajectory denote the spacecraft locations for each hour of the day. The label 0^h defines the start of the day during which periapsis occurred.

The field is completely defined by the values of the coefficients g_ℓ^m, h_ℓ^m, $\bar{g}_\ell^m$, $\bar{h}_\ell^m$. This analysis is usually derived for the case in which all the sources are constant and static. It will also apply in the case of Jupiter where interior and exterior sources rotate at different rates as long as each system of sources is constant in its own frame.

The terms with $\ell = 1$ are normally called dipole terms, those with $\ell = 2$ are called quadrupole terms, terms with $\ell = 3$ are called octupole terms, etc. Those terms involving negative powers of r are called interior while those corresponding to positive powers are exterior terms. The interior dipole terms produce the familiar dipole field. The exterior dipole terms produce a uniform field over the region inside the external sources.

If one knows two orthogonal components of the magnetic field over any spherical surface, all the coefficients in Eq. (8) are determined by integrals over this surface and in principle the field is completely determined throughout the shell. If one wishes to determine only a finite number of low-order coefficients to provide a convenient approximate description of the field, the coefficients are determined by the same integrals and have the same value whatever the number of coefficients used.

The coefficients are also determined by observations along any curved trajectory for which there are reasonable variations in both radial distance and longitude. In this case, each coefficient is no longer determined by an independent integral. One must decide which coefficients are to be used to provide an approximate description of the field. Using just these terms in (7) and (8), the computed components of **B** at each point at which an observation is made, $B_k{}^c(r_i)$, are expressed in terms of the, as yet undetermined, coefficients. The corresponding observed components are denoted by $B_k{}^0$ (r_i), and the residual R is defined as

$$R^2 = \sum_i w_i \sum_{k=1}^{3} [B_k{}^0(r_i) - B_k{}^c(r_i)]^2 \tag{11}$$

where w_i is an assigned weight of the ith observation. The coefficients are then determined by giving them the values that minimize R^2; this is the usual least-squares procedure. In this situation, the values obtained for the coefficients will depend on just how many coefficients are used in the approximation because the various spherical harmonics are not orthogonal over the trajectory; they are orthogonal only over the surface of a sphere. To determine n coefficients requires the solution of n linear equations in n unknowns, i.e., the inversion of an $n \times n$ matrix.

The values of the coefficients obtained will also vary some with the particular segment of the trajectory used. In part, this is due to the lack of orthogonality. In practice this is also partially due to inaccuracies in the field and position data, to possible changes with time in the sources, and to possible currents in the spherical shell containing the segment of the trajectory used in the analysis. Thus, the least-squares analysis does not provide a unique set of coefficients. It provides as many different sets as one cares to run off, and the decision as to which set gives the most useful representation of field is based partly on the value of the residual, or of the rms residual, and partly on the judgment of the investigator. The results of several such analyses will be given in the section on results below.

Offset Dipoles Deduced From Spherical Harmonic Coefficients

The coefficients obtained in a spherical harmonic analysis depend on the center chosen for the spherical shell within which the data are fitted. There

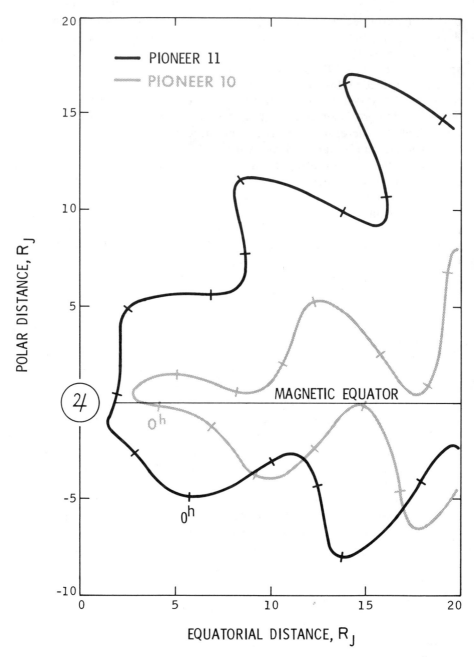

Fig. 3.   Pioneer trajectories in magnetic dipole coordinates. The Pioneer 10 (grey) and Pioneer 11 (black) trajectories are shown within ≃ 20 R_J of Jupiter in coordinates based on the offset dipole D_4. The X axis is the component parallel to the magnetic equator of the radius vector to either spacecraft from the dipole. The Y axis is the instantaneous distance above the magnetic equator. Tick marks are shown every 3 hours with the one corresponding to the beginning of the day of periapsis being labeled 0^h.

is no reason to expect that the magnetic center of the planet precisely coincides with its geometric center. Having carried out an analysis based on one center, it is often of interest to see if use of another center would yield more convenient or more meaningful results. Chapman and Bartels (1940) give formulas determining, from the interior dipole and quadrupole coefficients computed for any particular origin, the coordinates x_0, y_0, z_0 of a new origin, called the magnetic center. With respect to the magnetic center the first three quadrupole coefficients are zero and the square of the part of the magnetic field due to quadrupole terms is a minimum.

Let the interior spherical harmonic coefficients of the original analysis be g_ℓ^m, h_ℓ^m and those for the new center be $\tilde{g}_\ell^m$, $\tilde{h}_\ell^m$. Then the formulas used for the transformation are

$$z_0 = a(L_0 - g_1^0 E)/3H_0^2$$
$$x_0 = a(L_1 - g_1^1 E)/3H_0^2 \tag{12}$$
$$y_0 = a(L_2 - h_1^1 E)/3H_0^2$$

where

$$H_0^2 = (g_1^0)^2 + (g_1^1)^2 + (h_1^1)^2$$
$$L_0 = 2\, g_1^0\, g_2^0 + (g_1^1 g_2^1 + h_1^1 h_2^1)3^{\frac{1}{2}}$$
$$L_1 = -g_1^1 g_2^0 + (g_1^0 g_2^1 + g_1^1 g_2^2 + h_1^1 h_2^2)3^{\frac{1}{2}} \tag{13}$$
$$L_2 = -h_1^1 g_2^0 + (g_1^0 h_2^1 - h_1^1 g_2^2 \mid g_1^1 h_2^2)3^{\frac{1}{2}}$$
$$E = (L_0 g_1^0 + L_1 g_1^1 + L_2 h_1^1)/4H_0^2 .$$

Note that these formulas have been corrected for minor, relatively obvious, typographical errors, as may be verified by comparison with Bartels (1936). The coefficients with respect to the new origin are

$$\tilde{g}_1^0 = g_1^0,\ \tilde{g}_1^1 = g_1^1,\ \tilde{h}_1^1 = h_1^1;\ \tilde{g}_2^0 = \tilde{g}_2^1 = \tilde{h}_2^1 = 0;\ \tilde{g}_2^2 = g_2^2,\ \tilde{h}_2^2 = h_2^2 . \tag{14}$$

Expressing the results of a spherical harmonic analysis in terms of such an offset dipole may provide a concise and physically intuitive description of the major features of the field described by the harmonic coefficients.

Results

The first analyses of the Pioneer 10 and 11 data taken in Jupiter's magnetosphere were based on quick-look data listings provided to the experimenters in nearly real time. These analyses used preliminary trajectory and spacecraft orientation information; later, some gaps in the quick-look data due to operational difficulties at various steps of the data handling chain were filled in. Still later the normal and more reliable data handling chain began to yield data on tapes. A rather time consuming reduction process is then required to get data of the highest possible quality; however, some analyses have been made of data of very good quality before this process was com-

TABLE I

Parameters of Offset Dipoles Computed Either Directly or From Spherical Harmonic Analysis Coefficients[a]

S/C	Orders	Dipole Moment			Position of Dipole					
					Cartesian			Polar		
		M	TILT	LONG	Z_0	X_0	Y_0	R	LAT	LONG
11	3I, 2E	4.156	10.1	227.1	0.002	−0.132	−0.013	0.133	0.9	174.2
11	3I, 1E	4.145	10.2	227.3	0.001	−0.136	−0.013	0.136	0.5	174.6
11	3I, 0E	4.214	11.1	230.4	0.002	−0.120	−0.010	0.121	0.9	175.1
11	2I, 2E	4.230	11.1	228.6	0.011	−0.095	−0.008	0.096	6.8	175.3
11	D_4	4.225	10.8	230.9	0.009	−0.100	−0.001	0.101	5.1	185.7
10	3I, 2E	4.219	9.9	232.7	0.021	−0.065	−0.023	0.072	16.7	160.7
10	3I, 1E	4.155	10.2	234.0	0.025	−0.124	−0.010	0.126	11.3	175.6
10	3I, 0E	3.748	9.9	240.1	0.007	−0.196	+0.166	0.257	1.6	220.3
10	2I, 2E	4.213	10.5	229.5	0.026	−0.095	−0.048	0.110	13.6	153.0
10	D_2	4.000	10.6	225.0	0.030	−0.105	−0.003	0.110	15.9	178.5

[a] In this table the spacecraft is either Pioneer 11 or Pioneer 10. 3I, 2E means that 3 interior orders (dipole, quadrupole, octupole) and 2 exterior orders (dipole, quadrupole) were used in the spherical harmonic analysis from which the equivalent offset dipole representation was derived. D_4 and D_2 are the directly derived offset dipole models. The rms residual for D_4 is 0.196 Gauss R_J^3, that for D_2 is 0.085 Gauss R_J^3. M is the magnitude of the vector dipole moment in Gauss R_J^3; $TILT$ is the polar angle in degrees, and $LONG$ is the longitude in degrees in the $P11J$ System. X_0, Y_0, Z_0 are the Cartesian coordinates in units of R_J of the position of the offset dipole; R, LAT, $LONG$ are its polar coordinates.

pleted. Also, better versions of the trajectory information have become available. The accurate determination of the spacecraft roll orientation is critical in transforming measured field components from the spacecraft system to an inertial frame. This is difficult when the spacecraft is very near Jupiter because of the loss of the direct signal from the spacecraft and the low data rate for the stored data taken during occultation. This is compounded by the fact that the spin rate varies because of drag due to spacecraft currents produced by rotation in Jupiter's strong magnetic field and changes in the spacecraft moment of inertia due to the cooling during solar eclipse. We believe that ultimately the roll orientation can be determined with adequate accuracy but in the preliminary analyses some data had to be omitted because this was not yet done. The results here are based on analyses that we do not regard as final. It is quite possible that there will be minor changes in the various parameters in later, more complex, analyses but we do not expect that they will be large or that they will affect the general character of the models.

The analyses whose results are presented below may be used either as a basis for a study of their accuracy or to supply a model to be used for extrapolating Jupiter's field to regions where it was not measured and for calculating derived properties such as the fieldstrength at the poles or the L-shell passing through a satellite. For those who want only a single recommended model, and not a discussion, we would, at this time, make the following recommendations. If a reasonably accurate model with axial symmetry (albeit an inclined, offset axis) whose characteristics are easily visualized is desired, we recommend use of the D_4 Model of Table I. If more accuracy is desired but one wishes to use a time independent model in a frame that rotates with the interior sources, the (3I, 1E) Model for Pioneer 11 from Table II, discarding the $\bar{g}_1{}^1$ and $\bar{h}_1{}^1$ coefficients, is recommended. If these were to be retained, both the ϕ_i and ϕ_e angles, defined as in Eqs. (1) and (2), would have to be used. When these two exterior source terms are omitted, one may regard the resulting field as including the time average of the external field in the frame of the interior sources.

If one wishes to calculate the best available approximation to the observed fields along either the Pioneer 10 or the Pioneer 11 trajectory, we would suggest the use of the (3I, 2E) coefficients of Table II for the corresponding spacecraft. The residuals are larger for the Pioneer 11 data than for those of Pioneer 10. This does not mean that the latter are superior. The Pioneer 11 data cover a much larger range of radial distance, of longitude, and of latitude than do the Pioneer 10 data. It is harder to make a good fit over the extended range, but the coefficients derived are likely to give a better representation of the field in regions where neither spacecraft made measurements. If one uses Pioneer 11 data only in the range $2.85 < r < 8\ R_J$, which is the range in which the Pioneer 10 data were taken, the residuals are much more like those for Pioneer 10. In the case of Pioneer 10, one gets a signifi-

TABLE II
Spherical Harmonic Coefficients Derived From Pioneer 11 and Pioneer 10 Vector Helium Magnetometer Data Taken Inside 8 R_J

Spacecraft	P11	P11	P11	P11	P10	P10	P10	P10
Number of orders:								
Interior	3	3	3	2	3	3	3	2
Exterior	2	1	0	2	2	1	0	2
RMS Residual (γ)	45.6	62.8	124.2	150.4	18.5	21.7	33.4	24.0

SPHERICAL HARMONIC COEFFICIENTS:

Interior dipole (Gauss)

	P11	P11	P11	P11	P10	P10	P10	P10
g_1^0	+4.092	+4.080	+4.135	+4.151	+4.156	+4.090	+3.696	+4.143
g_1^1	−0.494	−0.497	−0.519	−0.539	−0.441	−0.431	−0.321	−0.497
h_1^1	+0.533	+0.539	+0.627	+0.611	+0.579	+0.593	+0.559	+0.582

Interior quadrupole (Gauss)

	P11	P11	P11	P11	P10	P10	P10	P10
g_2^0	−0.033	−0.043	−0.034	+0.053	+0.164	+0.161	+0.097	+0.198
g_2^1	−0.871	−0.892	−0.813	−0.658	−0.424	−0.840	−1.140	−0.664
h_2^1	−0.108	−0.105	−0.074	−0.052	−0.153	−0.043	+1.099	−0.310
g_2^2	+0.323	+0.318	+0.275	+0.188	+0.232	+0.277	+0.419	+0.318
h_2^2	−0.448	−0.449	−0.330	−0.258	−0.360	−0.374	−0.779	−0.181

Interior octupole (Gauss)

g_3^0	-0.113	-0.210	-0.562	$+0.873$	$+0.958$	$+0.857$
g_3^1	-0.718	-0.769	-0.695	$+0.594$	$+0.909$	-0.082
h_3^1	-0.072	-0.117	-0.169	$+0.318$	$+0.492$	$+0.349$
g_3^2	$+0.245$	$+0.280$	$+0.368$	-0.026	-0.169	$+1.050$
h_3^2	-0.186	-0.231	-0.258	-0.708	-0.109	$+0.840$
g_3^3	-0.172	-0.135	-0.142	-0.340	-0.344	-0.678
h_3^3	-0.066	-0.040	$+0.182$	-0.009	-0.149	-0.302

Exterior dipole (Gamma)

$\bar{g}_1^0$	$-193.$	$-135.$	$-255.$	$-129.$	$-121.$	$-123.$
$\bar{g}_1^1$	$+62.$	$+0.$	$+111.$	$-13.$	$-12.$	$-$
h_1^1	$+42.$	$-37.$	$+169.$	$+9.$	$+22.$	$+12.$

Exterior quadrupole (Gamma)

$\bar{g}_2^0$	$+4.9$		$+10.0$	$+1.0$		$+5.1$
$\bar{g}_2^1$	$+5.6$		$+19.1$	$+0.7$		-2.1
h_2^1	-9.0		-18.4	$+4.3$		$+10.4$
$\bar{g}_2^2$	-14.8		-27.8	$+1.2$		$+1.7$
h_2^2	-6.8		-16.6	$+2.9$		$+5.1$

cantly better residual if one includes at least the exterior dipole source. The roughly equal rms residuals for the (3I, 1E) and (2I, 2E) cases suggest that for Pioneer 10 the exterior quadrupole source is as important as the interior octupole source. We see from the Pioneer 11 data that if a good fit is desired nearer the planet's surface, the interior octupole, as should be expected, is more important than the exterior quadrupole and that retaining both is considerably more worthwhile than in the Pioneer 10 case.

The rms fieldstrength is $7000\,\gamma$ for the Pioneer 11 data used in these fits and is $3600\,\gamma$ for the Pioneer 10 data. Thus the better cases in Table II give very good ratios of rms residuals to rms fieldstrengths ($< 1\%$). This gives us high confidence in the validity of both the data and the spherical harmonic analyses.

To obtain the coefficients in Table II requires the solution of n simultaneous equations in n unknowns, where n ranges from 23 to 15. It is important to consider how well conditioned these equations are, i.e., how much the coefficients solved for change when the input data are given small, random changes. This can be studied (Lawson and Hanson 1974) by a critical value analysis or, less precisely but quite adequately in non-critical cases, by determining the condition number for each solution. The larger the condition number, the less accurately the unknowns are determined. Some sophistication must be used in interpreting condition numbers because they are changed significantly when a in Eq. (8) is changed or when, in the usual process for efficient matrix inversion, the matrix is pre-conditioned by multiplying its rows and columns by suitably chosen constants. The values of the condition numbers that we get for the various Pioneer 11 cases shown in Table II are 706 for the (3I, 2E) case, 189 for the (3I, 1E) case, 172 for the (3I, 0E) case, and 88 for the (2I, 2E) case. In our judgment, these condition numbers support our confidence that the coefficients provide a reasonable representation of the field the accuracy of which can be roughly estimated from the rms residuals and the variation of the coefficients from case to case.

Table I lists the offset dipoles deduced by the use of Eqs. (12)–(14) for the cases given in Table II. Table I also gives the offset dipoles D_4 and D_2 fitted directly to the data by the use of Eqs. (4)–(6). These offset dipoles can be regarded as a convenient way of summarizing the main characteristics of the various models.

It is encouraging to see that most of the cases give essentially the same vector magnetic moment and that the Pioneer 11 results give reasonably consistent positions of the dipole. From this, and the comparison of the coefficients in Table II, we feel that the interior dipole coefficients are determined with high accuracy, the interior quadrupole coefficients are reasonably accurate, and the interior octupole coefficients show enough scatter from case to case to suggest that they are not determined with much accuracy. It is very likely that if higher orders were used in the fit, the residual would be reduced somewhat but it is unlikely that the coefficients would be accurate enough to be used with any confidence in predicting fields at points not on the trajectory. The exterior quadrupole coefficients appear to be equally unreliable, based on the large scatter from case to case. Since they are very small, this does not matter much unless one wishes to estimate the field for values of r greater than 8 R_J. The exterior dipole terms, in particular the $\bar{g}_1^0$ term, appear to be determined to within a factor of about 30% and should probably be used in fitting the field.

If the residuals are evaluated at points between 8 R_J and 12 R_J, that is, points outside the range used for determining the coefficients, they are reasonably small in all cases except the (31, 2E) and (21, 2E) cases. Whenever exterior quadrupole coefficients are used for values of r significantly larger than those used in the determination of the coefficients, the residuals go up rapidly. This is to be expected since the exterior quadrupole coefficients depend sensitively on just what data are used in the fitting process and are multiplied by spherical harmonics that increase rapidly with r. These very same factors apply to the octupole terms for fields closer to the surface than periapsis.

Discussion of the Pioneer Results

To compare the Pioneer models with the description of Jupiter's field derived from radio astronomy, it is best to deal separately with the observations at decimetric and decametric wavelengths. Inferences based on the decimetric observations are appropriate to a radial distance of approximately 2 R_J, the region from which the most intense synchrotron radiation is received at Earth. On the other hand, the decametric emissions are considered to originate in regions near the planet, such as at ionospheric levels at the base of the Io flux tube.

The various decimetric models are reasonably consistent (Carr and Gulkis 1969; Mead 1974)[2] and agree well with the Pioneer results. The tilt angle between the dipole and Jupiter's rotation axis is $10° \pm 1°$. The longitude of the pole in the northern hemisphere has a System III value of $230° \pm 5°$ at the epoch of Pioneer 11. The polarity of the field is southward at the equator, i.e., opposite to the polarity of the earth's magnetic dipole.

The Pioneer observations have led to significant improvements in our knowledge of the fieldstrength and the offset of Jupiter's dipole. Previous radio astronomy estimates of the fieldstrength at a given location were considered uncertain by at least a factor of 3. Several estimates of the offset had been derived from decimetric observations which indicated that the offset was ≤ 0.1 R_J (McCulloch and Komesaroff 1973; Berge 1974). The offset based on the Pioneer data appears consistent with inferences drawn from the

---

[2] See pp. 677 and 693.

decimetric data, especially in view of the relatively large uncertainties associated with the radio astronomy results.

There is evidence in the decimetric polarization data of a possible magnetic anomaly near a System III longitude of 220° (Conway and Stannard
1972). An analysis of the Pioneer 10 data in terms of two offset dipoles, a 12
parameter fit, has been reported by Smith *et al.* (1974*b*). One dipole represents the main field source while the other could simulate a local anomaly
such as might arise near or above the surface of Jupiter. The longitude of the
second dipole in this model has been found to be 210°, in good agreement
with the radio astronomy observations. It will be interesting to see if essentially the same model will also fit the Pioneer 11 observations.

In order to compare the Pioneer results with inferences based on decametric emissions, which are thought to originate near the surface, it is necessary to extrapolate the Pioneer observations to low altitudes. One such
extrapolation is shown in Fig. 4 and is based on the 15-coefficient spherical
harmonic analysis (3I, 2E). It leads to a maximum fieldstrength in the northern hemisphere of 14.4 Gauss and to a value of 10.8 Gauss in the southern
hemisphere.

It is customary to identify the frequency of the decametric emissions with
the electron cyclotron frequency at the source. A relatively sharp cutoff in
the maximum frequency of the emission has persistently been observed in

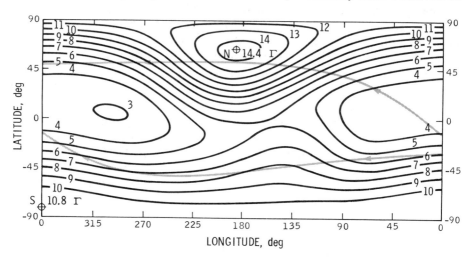

Fig. 4.   Equal-intensity contours of Jupiter's surface field. Contours along which the field magnitude is constant are shown. They are based on an extrapolation of the Pioneer 11 observations to the oblate spheroidal surface using the model derived from the 23 coefficient (3I, 2E)
spherical harmonic analysis. The Jovigraphic latitude and System III longitude are indicated
at regular intervals along the ordinate and abscissa. Crosses identify the maximum fieldstrengths in the two hemispheres. The grey curve shows the projection onto the surface of the
Pioneer 11 trajectory while the spacecraft was within 8 R_J of Jupiter. It provides some information about the extent of the latitude, longitude coverage while mapping the planet's field.

the vicinity of 40 MHz. This cutoff is usually interpreted as a maximum field-strength of 14 Gauss along the locus of the source region. The close correlation between the observation at Earth of decametric radio bursts and the phase angle of Io has been interpreted to mean that the source is located at the foot of the magnetic flux tube passing through or near Io. Using a method previously employed by Mead (1974), the approximate locus of the Io flux tube has been determined using Pioneer model D_4. Along this locus, the maximum surface field corresponding to Fig. 4 is found to be 14 Gauss, a value which agrees quite well with that inferred from the decametric data and models.

The Pioneer observations may be compared with current views regarding the origin of planetary magnetic fields (Gubbins 1974). Planetary fields are thought to be the consequence of a self-sustaining dynamo operating in the fluid, electrically conducting interior of the planet. Motions in the core which lead to the generation of the planetary field are considered to be the effect of a sufficiently vigorous convection inside the rotating fluid interior. The enormous mass of Jupiter implies a sufficiently large internal pressure to cause a transition of hydrogen, the most abundant element, from a molecular to a metallic phase (Hubbard and Smoluchowski 1973). The latter appears to have adequate electrical conductivity to support the currents required by dynamo theory. In addition, the internal generation of heat, evidence for which is provided by the well-known radiation imbalance, is expected to lead to fluid convection in the core, a condition which is conducive to the development and maintenance of a planetary dynamo.

The magnetic field in the core is undoubtedly very large as well as very irregular. However, the rapid decrease with distance of the successively higher-order magnetic moments makes it likely that only the lowest-order terms will make a significant contribution to the field at the surface. A quantitive measure of the degree of the irregularity of the planetary field is provided by the ratio of the higher-order multipole moments to the dipole moments. The Pioneer 11 Vector Helium Magnetometer results yield quadrupole and octupole moments that are 20 and 15%, respectively, of the dipole moment. Since the corresponding values at Earth are 13 and 9%, only $\simeq 0.6$ times smaller, Jupiter's field appears to be only slightly more irregular than that of Earth.

The existence of an offset, or alternatively a finite quadrupole moment, is thus attributable to the irregular nature of the fields and motions inside the core. The earth's magnetic dipole has a similar offset of 0.06 R_E. Offset dipoles have also been reported in magnetic stars.

The near correspondence between the magnetic axes of Jupiter and Earth and their rotation axes is presumably fundamental. Apparently the rotation of the planet, which is rapid compared to internal motions, lends an axial symmetry to the fluid motions driving the dynamo. Various arguments can be made to account for the strength of Jupiter's field as compared to the earth's field, e.g., that Jupiter is ten times larger and rotates twice as rapidly. However, evidence is accumulating that naturally-occurring dynamos are commonly, and perhaps fundamentally, oscillatory (Rikitake 1966). The oscillations give rise to periodic variations in fieldstrength and to reversals in polarity over geological time intervals. If this view is correct, neither the strength nor polarity of a planetary field is fundamental, but their values will depend on the phase of the cycle during which the field is observed.

THE MIDDLE MAGNETOSPHERE

As noted in the introduction, it is possible observationally to distinguish two regions in which plasma currents strongly perturb the dipole field. The distinction can be seen in the behavior of both the field direction and its magnitude.

Perhaps the single parameter that serves this purpose best is the field latitude (Fig. 5). In the inner magnetosphere, there is clear evidence of 10-hour variations associated with the rotating dipole. At 30 R_J, and at low spacecraft latitudes (Pioneers 10 and 11 inbound and Pioneer 10 outbound), the field latitude tends to adopt a more or less radial orientation ($\delta \to 0$) and this is a characteristic feature of the middle magnetosphere. At higher latitudes (Pioneer 11 outbound), the field also tends to be radial. In the outer magnetosphere, beyond 50–60 R_J, the latitude tends toward -45 to $-90°$, implying a southward directed field. The field direction is also much more variable but as a result of what appear to be random fluctuations. These conditions hold both near the equator and at higher latitudes.

The two regions can also be seen in the field longitude and magnitude. In the middle magnetosphere, the longitude (Fig. 6) shows that the field tends to spiral, especially at low magnetic latitudes whereas at higher latitudes the field tends to follow the dipole direction more closely. Frequent reversals of the longitude angle by 180° are seen, many of them occurring with a 10-hour period. In the middle magnetosphere, the field magnitude (Fig. 7) tends to be relatively constant ($\simeq 10\gamma$) especially near the equator but with occasional dips to low values of $\simeq 1\gamma$. These field decreases are correlated with the changes in the longitude angle by 180°.

In the outer magnetosphere the field longitude is very irregular. It is also poorly defined, principally because of the southward field orientation, and small fluctuations can cause large apparent changes in the field direction. The field magnitude is also much more irregular and periodic 10-hour variations, if they exist, are difficult to discern.

The properties of the middle magnetosphere are dominated by a thin current sheet. The most obvious evidence for this current sheet are quasi-periodic crossings in which abrupt changes in longitude by 180° are accompanied by a relatively large reduction in field magnitude. An example of a current sheet crossing is shown in Fig. 8. This crossing was seen in Pioneer

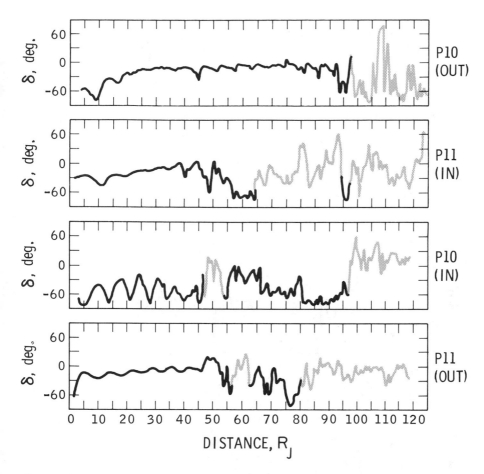

Fig. 5.   Field latitude inside Jupiter's magnetosphere. The latitude of the field is shown as a function of radial distance vector in *RJ* coordinates (**R**, the planetocentric radius vector to Pioneer, defines the *X* axis and **J**, Jupiter's rotation axis, defines the prime meridian or *XZ* plane). Pioneer 10 and 11 data, both inbound and outbound are shown as indicated in the right-hand column. The data have been ordered in terms of local time with the pre-dawn data in the top panel and with the pre-noon data in the bottom panel. Data obtained inside the magnetosphere are black and data in a grey tone were obtained in the magnetosheath or interplanetary space.

10 data outbound at a local time of 5^h 30^m. A dayside crossing at 9^h 30^m local time observed with Pioneer 11 is shown in Fig. 9 and exhibits essentially the same features. The combined Pioneer 10 and Pioneer 11 observations of the current sheet in both the predawn and morning regions of the magnetosphere are consistent with the current flowing around Jupiter in the form of a ring. The observed variation in field direction and magnitude is consistent with the properties of other well-known plasma current layers such as

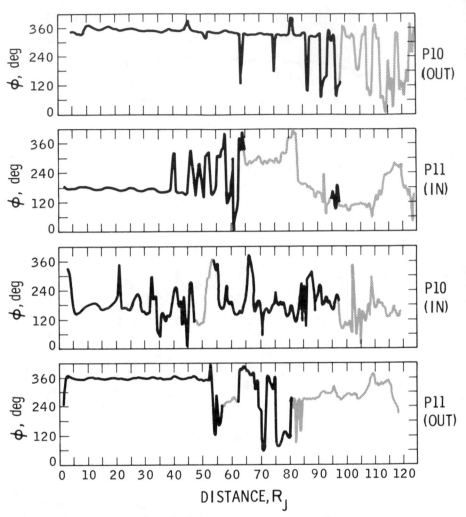

Fig. 6. Field longitude inside the magnetosphere. The field longitude in *RJ* coordinates is
shown as a function of radial distance. The format of this figure is the same as that of Fig. 5.

have been observed in the earth's magnetotail or at interplanetary sector
boundaries.

The sheet current flows principally in an azimuthal direction. The magnitude of the oppositely-directed fields on the two sides of the current layer
implies a typical linear current density of 10^{-2} amp m^{-1}. The thickness of
the current sheet, estimated from the duration of the crossings to be $\simeq 1\ R_J$,
may be used to convert this value to an average current density per unit area
of 2×10^{-10} amp m^{-2}. If the current sheet is assumed to extend typically
between 30 and 60 R_J, a total current of 6×10^7 amp is implied. A current
of this magnitude would be expected to produce a quasi-uniform northward

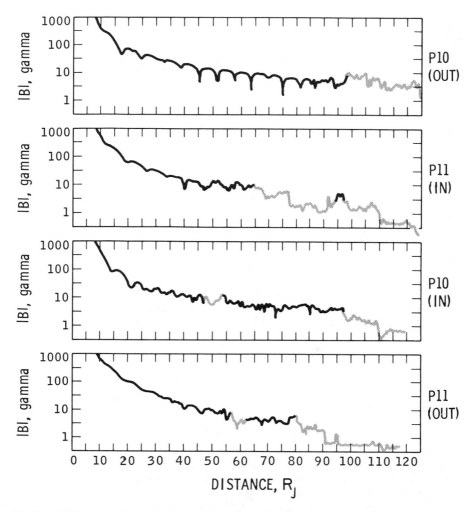

Fig. 7.   Field magnitude inside the magnetosphere. The format used here is the same as that
described in the caption to Fig. 5.

field in the inner magnetosphere of $\simeq 10\,\gamma$. Thus, the current sheet does not
significantly perturb the dipole field in the inner magnetosphere. The quasi-
uniform field of $100\,\gamma$, inferred from the spherical harmonic analysis to originate external to $10\,R_J$, is presumably caused by a ring current that is nearer
Jupiter and distributed over a greater range of latitude than the current sheet.

The origin of the current sheet is presumed to be centrifugal force acting
on relatively dense, low-energy plasma trapped in the magnetic field. Because of the large scale of Jupiter's magnetosphere and its rapid rate of
rotation, centrifugal force will become increasingly larger than other stresses,
such as that associated with the magnetic field gradient, with increasing

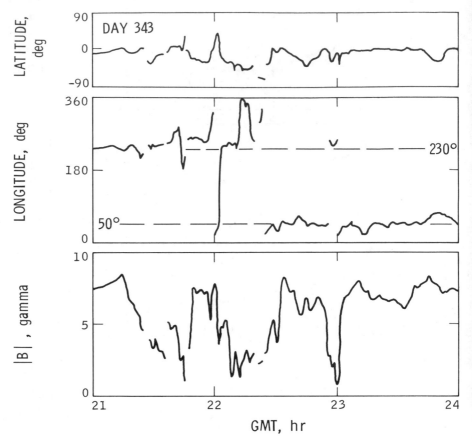

Fig. 8.   Pioneer 10 current sheet crossing. One-minute averages of the spherical components of the field are shown in *SJ* coordinates (the primary axis, *X* is parallel to **S**, the direction from Jupiter to the sun, and Jupiter's rotation axis, **J** defines the *XZ* plane). This crossing of the current sheet, which occurred in the outbound data at a large Jovigraphic distance, and at a local time of $5^h\,30^m$ shows the reduction in field magnitude, the reversal in field longitude and the tendency for the field to point southward inside the current sheet.

distance. From an equivalent point of view, the trapped plasma is being whirled rapidly around Jupiter, at speeds comparable to the solar wind speed, and centrifugal force then stretches the lines of force radially outward, much like rubber bands.

Evidence that centrifugal force causes the current sheet is provided by the orientation of the current layer at large Jovigraphic distances. The perturbation field, $\mathbf{P} = \mathbf{B} - \mathbf{D}$, i.e., the difference between the observed field **B** and the field at the point of observation corresponding to the dipole field **D**, is produced by the current sheet. The perturbation field has been shown to be relatively invariant with respect to the rotation axis of Jupiter **J** (Smith *et al.* 1974*b*). This result has been interpreted to indicate that, at large dis-

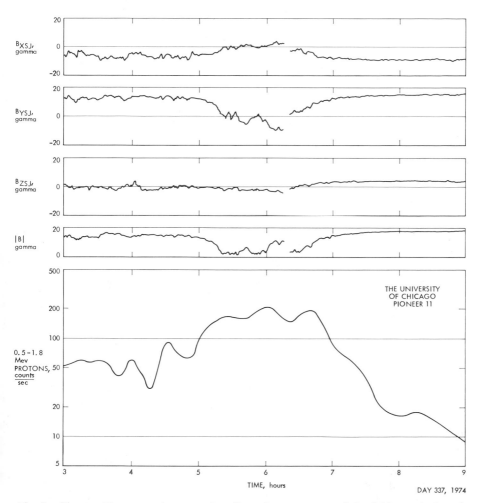

Fig. 9. Pioneer 11 current sheet crossing. One-minute averages of the field magnitude and three components are shown also in *SJ* coordinates. The lower panel contains the count rate of trapped protons being measured simultaneously by the University of Chicago Experiment. These data were obtained at $\simeq 40\ R_J$ and at a local time of $\simeq 9^h\ 30^m$. The spacecraft penetrated the current layer shortly after 5^h and then returned to a location below the current sheet by 7^h. The behavior of the X and Y components is equivalent to a change in longitude of $\simeq 180°$. There is a very good correlation between the particle flux and the field magnitude.

tances from Jupiter, the current sheet tends to be parallel to the rotational equator although not necessarily lying in it (Fig. 10). This configuration would lead to a radial orientation of the field adjacent to the current sheet at large radial distances as observed for the field latitude angle. Inside the current sheet, the latitude angle exhibits a southward orientation, implying that the continuous component of the field is principally normal to the current sheet.

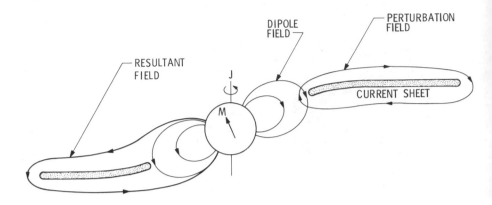

Fig. 10. Model of the Jovian current system. This model, which is based on the Pioneer observations, shows the tilted dipole **M** and Jupiter's rotation axis *J*. The two components of the magnetospheric field associated with the dipole and the thin current sheet are shown in the right-hand side. The resultant field, obtained by a vectorial addition of the dipole and ring current fields, is shown on the left-hand side where it is seen that the fieldlines are effectively stretched out parallel to the equator at large distances. The current sheet curves over to join the magnetic dipole equator nearer the planet, causing the current sheet to be higher on one side than on the other.

The average longitude angle of the field adjacent to the current sheet shows that the current flow is not purely azimuthal but includes a radial component. The presence of both radial and azimuthal current components causes the magnetic field to spiral, rather than to continue to lie in meridian planes, as the radial distance increases. The general tendency for the field to spiral is obvious in Fig. 6. The Pioneer 10 data outbound and Pioneer 11 data inbound give the clearest indication that the field does not lie in the meridian plane (i.e., within 10° of $\phi = 180°$ or 360°). The fields adjacent to the current sheet in these intervals are consistent with this general orientation of the field as can be seen by referring to Fig. 8. Thus, although the field longitude changes by 180° as the current sheet is crossed, the fields on adjacent sides are not 180° and 360°, but are 150° and 330°.

The spiraling is apparently associated with the current sheet itself. At higher latitudes there is evidence that the observed field does not spiral but has approximately the orientation of the dipole field. The inbound data on Pioneer 10 (Fig. 6) show a region in which the field longitude tends to follow the dipole variation (with a slight delay) between 180° and 360° which corresponds to the observations being made at higher magnetic latitudes. However, between 0° and the longitude corresponding to the spiral direction, there are few if any points giving rise to a "negative clipping" of the periodic longitude variations. During these times the spacecraft was apparently near the current sheet whose field then exceeded the dipole field and essentially determined the field direction. The absence of significant spiraling away from the current sheet also explains why no spiral is evident in the outbound Pioneer 11 data (Fig. 6) which were acquired at a latitude of $\simeq 30°$.

The absence of spiraling at higher latitudes is only one aspect of the tendency for the field above and below the current sheet to be more dipole-like. Thus, the outbound Pioneer 11 data in the middle magnetosphere show the field (Fig. 7) to be only slightly larger than the dipole field (typically much less than a factor of 2 stronger). A study of the radial component shows that it obeys an inverse cube (r^{-3}) radial dependence reasonably well, again indicating the strong influence of the dipole field at moderately high latitudes. Finally, the latitude angle (Fig. 5) varies with a 10-hour period that is correlated with the magnetic latitude of Pioneer (although a phase lag is present). Calculations of the perturbation field in this region show that, although the direction is still invariant with regard to Jupiter's rotation axis, it is not principally radial but contains a significant northerly component [$\psi = \cos^{-1}$ (**P** · **J**) $\simeq 30°$], as expected for a ring current. Thus, although the field due to the current sheet is present at higher latitudes, it does not dominate the dipole field as it does near the equator.

Current sheet crossings are presumably caused by a motion of the current layer across the spacecraft, either as a result of the rotation of the planetary dipole, a latitudinal motion inherent to the current sheet, or some combination of both. The typical rapidity of the crossings is evidence that the current sheet is very thin as compared to the scale of the magnetosphere. A thickness of 1 to 2 R_J has been estimated on the basis of the time required for the longitude to change 180° and an estimated latitudinal velocity (which is presumed to be on the order of 30 km sec^{-1}, i.e., large compared to the velocity of the spacecraft).

It has been proposed that the current sheet, although roughly parallel to the equator at large distances, does not lie in the equator but curves over to join the dipole magnetic equator nearer the planet (Smith *et al.* 1974*b*). The current sheet would then have the shape of a warped disc that would appear to wobble up and down as it rotates giving rise to periodic current sheet crossings. It might be supposed that the disc would remain relatively rigid so that actual motions of the current layer in the rotating system would not be needed to explain the observed current sheet crossings. However, this model is difficult to reconcile with the repeated current sheet crossings in the Pioneer 10 data outbound. The latitude of the spacecraft, which being $\simeq 10°$ is essentially the same as the dipole tilt and the bending of the current sheet below the magnetic equator as it becomes parallel to the Jovigraphic equator, should preclude any crossings at large distances. One way out of this dilemma is to assume that the current sheet is not actually rigid but either "flaps" up and down or is subject to some form of "fluting" so that local bulges or ripples form along the outer extremity and then cross the spacecraft.

There is a close correspondence between the intensity of the energetic particles trapped in Jupiter's middle magnetosphere and the location of the current sheet. An example of this correlation is shown in Fig. 9 which contains the University of Chicago measurements of energetic protons at the time of one of the Pioneer 11 current sheet crossings (J. Simpson, personal communication). There are numerous examples of this kind in the Pioneer 10 and 11 particle data. In the middle magnetosphere, as far as the mirroring radiation particles are concerned, the current sheet essentially defines the magnetic equator.

The correspondence between the particle fluxes and the magnetic longitude or latitude of the spacecraft is more obscure. The phase relation between the particle maxima, the current sheet crossings and the longitude of the spacecraft in the dipole field are not well understood. A study of these relations by Northrop *et al.* (1974) led to the suggestion that the fieldlines are slipping in Jupiter's ionosphere and are thereby introducing large longitude shifts between the dipole and the field in the middle and outer magnetosphere. An alternative hypothesis, proposed by Chenette *et al.* (1974) on the basis of their energetic particle observations, is that the particle maxima are invariant in time rather than longitude. This hypothesis could mean that Jupiter's magnetosphere is either pulsating or undergoing other large-scale time variations.

THE OUTER MAGNETOSPHERE

The magnetic fields adjacent to the magnetopause are shown in Fig. 11. For this magnetopause crossing, which was the second of the outermost crossings in the Pioneer 11 data at 94 R_J, the field magnitude inside the magnetosphere is only 5 γ, a value that is smaller than usual, as anticipated for the most distant magnetopause crossings. The transition from inside (earlier) to outside (later) takes place in only a minute or so, implying a relatively thin boundary. A decrease in fieldstrength to nearly zero is seen, a characteristic feature frequently observed in traversals of the earth's magnetopause current layer. The field decrease near $7^h 10^m$ UT is suggestive of a close approach to the magnetopause but without an actual crossing. Inside the magnetosphere, the field is predominantly southward, a characteristic feature of the outer magnetosphere generally. The direction of the normal to the magnetopause has been determined from the usual variance analysis (Sonnerup and Cahill 1968) and, to a good approximation, is directed radially outward as anticipated for a blunt magnetosphere.

The persistent southward component in the outer magnetosphere is presumably a consequence of the observations being made beyond the outermost edge of the eastward-flowing current sheet as well as of the effect of magnetopause currents resulting from the confinement of the magnetosphere by the shocked solar wind. In addition, the planetary field contributes to the resultant southward field. If it is assumed that the impinging solar wind causes an approximate doubling of the field interior to the magnetopause, then for a 10 γ field, the dipole and current sheet would contribute a total of

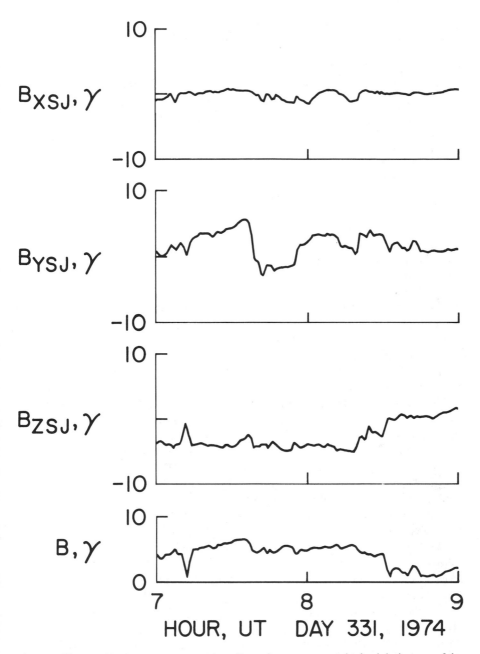

Fig. 11. Pioneer 11 magnetopause crossing. One-minute averages obtained during one of the inbound magnetopause crossings are shown in *SJ* coordinates (defined in the caption to Fig. 8). At 7^h (Earth Receipt Time in this figure) Pioneer 11 was inside the magnetosphere. Apparently the magnetopause approached the spacecraft at $\simeq 7^h\ 10^m$, then retreated and finally moved outward across Pioneer at $8^h\ 30^m$. The crossing took only a minute or so and was characterized by a jump in field magnitude and by a southward field inside the magnetosphere.

5γ. At 60 R_J the equatorial dipole field is $\simeq 2\gamma$, so that the current sheet would account for 3γ. Closer to the middle magnetosphere the ring current and dipole field contributions would be expected to be larger, and the contribution of magnetopause currents smaller, than these values.

When the magnetosphere is not very inflated and the magnetopause is at distances < 50–$60\ R_J$, the field magnitude just inside the magnetopause often appears large enough to withstand the inward pressure of the post-shock solar wind. For example, a fieldstrength inside the magnetopause of 10γ corresponds to a pressure $B^2/8\pi$ of 4×10^{-10} dynes cm^{-2}. The latter is approximately the dynamic pressure of the average solar wind at Jupiter of 5×10^{-10} dynes cm^{-2} (assuming $V = 360$ km sec^{-1}, $N = 0.2$ cm^{-3}). The presence of the magnetosheath magnetic field just outside the magnetopause, which is frequently 1/3 to 1/5 of the magnetospheric field just inside the magnetopause, could imply that low-energy magnetospheric plasma also contributes to the pressure balance. There are times when particle pressure inside the magnetopause must make a significant contribution. Certainly, a wide variety of plasma betas have been inferred for the outer magnetosphere (Smith *et al.* 1974a; Wolfe *et al.* 1974). There may also be a significant contribution in the form of dynamic pressure associated with the motion of the magnetopause which the Pioneer observations have shown to be frequently or continuously in motion (Wolfe *et al.* 1974).

Throughout the outer magnetosphere no periodic field changes are evident in the magnetometer data (Figs. 5, 6, 7). The persistent southward orientation means that the field longitude angle is susceptible to small irregularities in the field which give rise to large deflections in ϕ. Thus, a periodic variation associated with the dipole rotation would be difficult to detect in the longitude angle. Periodic variations would be best seen in the behavior of the field latitude. However, if periodic changes are present in δ, they are also obscured by the irregular fluctuations in field direction that are a characteristic feature of the outer magnetosphere. The clearest evidence that the field varies with the rotation period of Jupiter is provided by the particle measurements. The trapped radiation customarily found in this region has been observed to vary regularly with a 10-hour period.

There is no evidence that the field in the outer magnetosphere tends to follow a spiral. It might be supposed that the persistent lagging of the outer-most fieldlines observed in the middle magnetosphere would ultimately lead to a "wrapping up" of the field between the current sheet and the magneto-pause. Such a model has, for example, been proposed by Piddington (1969). However, the strong tendency for the field in the outer magnetosphere to be directed north-south rather than predominantly east-west shows that this does not occur. Furthermore, the fieldlines passing through the outer magnetosphere are presumably the outer extremity of those observed at higher latitudes above the current sheet which do not appear to spiral.

COMPARISON WITH MAGNETOSPHERIC MODELS

Most models of Jupiter's magnetosphere can be divided into two major types. In the "closed" magnetosphere (e.g., Brice and Ioannidis 1970; Hill *et al.* 1974) both the high-energy particle radiation and the lower-energy plasma are trapped in the planetary magnetic field, however badly it may be distorted. The other model consists of an "open" magnetosphere (e.g., Michel and Sturrock 1974; Eviatar and Ershkovich 1976; Kennel and Coroniti 1974; H. Alfvén, personal communication) containing regions in which convecting plasma is transporting magnetic fields that are open (the fieldlines are not simply and continuously connected between one planetary hemisphere and the other). Details of these basic models can be considered to show how each might lead to an interpretation of the Pioneer observations. Although it may not be possible at this time to choose unequivocally between the models, the observations establish definite constraints which each of them must satisfy.

Closed Magnetosphere

In the closed magnetosphere model (Fig. 12) the current sheet arises as the result of significant centrifugal stresses exerted on trapped low-energy plasma by the large scale and rapid rotation of Jupiter. This outward stress gives rise to an azimuthal current which, in turn, causes strong radial fields. This view is supported by the observation that the perturbation vector **P** in the vicinity of the current sheet is essentially invariant with respect to Jupiter's rotation axis and, at large radial distances near the equator, tends to be perpendicular to **J**. The existence of a north-south field component within the current sheet (approximate value 1γ), as observed, is critical because it is essentially this component that causes the plasma to co-rotate.

The current sheet gives rise to strong fields such that at large distances near the equator the field tends to be more or less parallel to the Jovigraphic equator. The field will have oppositely directed polarities above and below the current layer. Passage of such a thin sheet over the spacecraft then causes a reduction in fieldstrength to nearly zero, i.e., the value of the normal component, with a subsequent recovery and a reversal in field longitude by 180° if the spacecraft emerges on the opposite side.

Since the trapped plasma essentially co-rotates with the planetary field, except for a relatively slow differential drift of protons and electrons that gives rise to the current, the current sheet would be expected to form a ring with Jupiter at its center. The latter is supported by the Pioneer 10 observation of the current sheet just behind the dawn terminator (5[h] local time) and by Pioneer 11 observations of the current sheet in the dayside hemisphere at 9[h] local time. There is also convincing evidence in the inbound, dayside Pioneer 10 data that the current sheet was present, although not as well developed as when it was observed while Pioneer 10 was outbound.

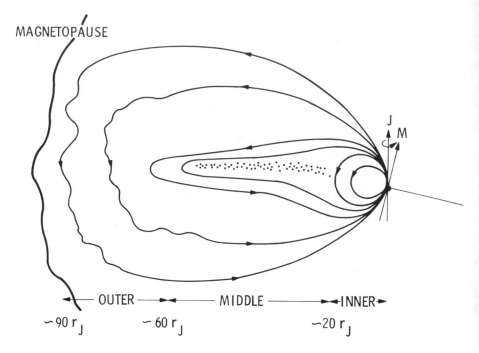

Fig. 12.   A closed magnetosphere model. The approximate locations of the inner, middle, and outer magnetosphere are shown. Dipole fieldlines near Jupiter, stretched-out fieldlines near the plasma sheet (dotted) and closed fieldlines at higher latitude passing through the outer magnetosphere are shown. The model is based on inferences drawn from the Pioneer observations.

The current sheet changes the topology of the equatorial field from being dipole-like to being strongly stretched out. This stretching of fieldlines radially outward will displace the lines of force of the planetary field from the inner to the outer magnetosphere and help explain the southward directed fields between the outer edge of the current sheet and the magnetopause. At higher latitude in the middle magnetosphere, the intensity and direction of the current sheet field do not perturb the dipole field nearly as strongly as nearer the equator, and more dipole-like fields are to be expected.

Their southward orientation implies that the fieldlines in the outer magnetosphere are closed and might be co-rotating with Jupiter. The observation of a dipole-like field at relatively high latitudes above the current sheet, and the expectation that these fields pass through the outer magnetosphere, suggest that the outer magnetosphere will co-rotate. In order to avoid a large current in the outer magnetosphere associated with the very rapid co-rotation velocities and stresses, it would then appear necessary to assume a relatively low plasma density in the outer magnetosphere. If a significant plasma pressure is needed to help the magnetic field hold off the shocked solar wind, the further assumption would be necessary that the outer magnetospheric plasma, although having a low density, is very hot.

On the other hand, it is possible that the field in the outer magnetosphere is not co-rotating. For example, the fieldlines might be "cut" in the ionosphere and the flux tubes then need not co-rotate with the planet. The issue of avoiding large currents in the outer magnetosphere associated with centrifugal stresses then does not arise. According to particle measurements made inbound on both Pioneer 10 and Pioneer 11, however, the trapped energetic particles in the outer magnetosphere are co-rotating with Jupiter (Trainor *et al.* 1974; Van Allen *et al.* 1975). Thus, the particle observations imply that the field is co-rotating.

One of the principal difficulties associated with the closed magnetosphere model may be its inability to account for the observed spiraling of the fields adjacent to the current sheet. This observation implies that the current is not purely azimuthal but has a significant radial component. In terms of trapped plasma and the general approach which attributes the observed current to body stresses inside the magnetosphere, this observation appears to require a significant azimuthal stress. One obvious possibility is drag exerted by the solar wind as it flows around the magnetosphere, which must then be transmitted through the outer magnetosphere to the current sheet.

Open Magnetosphere

Most outflow models assume that, beyond some radial distance such as $30\ R_J$, the planetary field will not be able to make the plasma co-rotate nor prevent it from being spun outward and escaping (Fig. 13). Outward convection is presumed to set in when the co-rotation velocity becomes approximately equal to the Alfvén speed. This condition is equivalent to the magnetic field stress being unable to compensate for the centrifugal acceleration of the co-rotating plasma which then flows radially outward. The model bears numerous similarities to the basic model of the solar wind or, perhaps even more, to recently proposed models of the interaction between the solar wind and the interstellar medium which leads to the development of the heliosphere (Parker 1963; Axford 1972).

The outflow model can accommodate spiral fields quite naturally. By analogy with the solar wind, the spiral angle, α, is given by $\tan^{-1}\ (V_c/V_R)$ where the co-rotation velocity $V_c = \Omega_J r$, Ω_J is the angular rate of rotation of Jupiter and V_R is the radial outflow or convection velocity. As a consequence of the observed spiral angle, e.g., $\alpha \simeq 30°$ at $80\ R_J$, the convection velocity would be relatively large, i.e., $V_R \simeq 2V_c \simeq 10^3\,\mathrm{km\ sec^{-1}}$. Such high velocities would probably exceed both the Alfvén velocity and the magnetosonic wave velocity, and imply that the outflow would be supersonic.

The outflow model is also expected to lead to a current sheet. As the oppositely-directed fields in the northern and southern hemispheres are drawn outward, a neutral sheet should form near the equator. The intensity of the

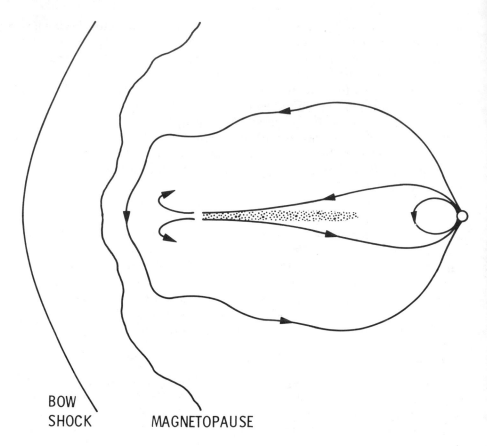

Fig. 13.   An equatorial outflow model. The dipole field is shown near the planet. The fieldlines
are shown as being open at greater distances as a consequence of plasma flowing radially
outward. Since the Pioneer observations imply closed, co-rotating southward-directed fields
in the outer magnetosphere, the outflow is indicated as being deflected inside the magnetosphere. The subsequent flow (curved arrows) is unspecified but could be around the middle
magnetosphere and down the magnetotail.

current in this sheet would simply be determined by the magnitude of the
fields adjacent to it. The direction of the current would be principally azimuthal and, if the flow is axisymmetric, the current would also close around
Jupiter to form a ring.

In order for the model to be consistent with the observations, the plasma
would have to be deflected before it reaches the outer magnetosphere. The
southward directed fields in the outer dayside magnetosphere represent a
barrier through which the outflowing plasma could not penetrate. The plasma would have to exit in some other region of the magnetosphere, perhaps
somewhere along the flank or more probably down the tail. If the flow is
supersonic, as speculated above, the plasma would presumably first have to

pass through a shock and become thermalized so that it could be deflected inside and along the magnetopause. Advocates of the outflow model often predict the existence of some form of shock or transition region inside the magnetosphere, separating basically radial from basically azimuthal flow.

The outflow model appears to be contradicted by several aspects of the observations. One problem associated with the model is the presence of a significant north-south field component passing through the current layer. The usual interpretation of such a component when observed inside a current sheet dominated by convection is that it results from fieldline merging. However, the sense of the component in this instance, which is southward, appears to be opposite that anticipated for an outflowing plasma. Another problem for the model is the observation that the trapped particle flux is a maximum within, or adjacent to, the current sheet. It is difficult to see how energetic particles could be trapped in a region containing outflowing plasma and open fields.

The most obvious choice of a boundary in the Pioneer data that might be interpreted as a shock inside the magnetosphere appears to be the innermost "magnetopause crossing" observed on both Pioneers inbound and outbound at 50 to 60 R_J (Wolfe et al. 1974; Mihalov et al. 1975; Smith et al. 1975a; Smith et al. 1975b). Since the plasma just outside this boundary should flow around the middle magnetosphere, and might convect any embedded magnetic fields along with it, the outflowing plasma in this region could have the appearance of magnetosheath plasma. This interpretation is an alternative to the one frequently proposed, namely that time variations in the dimensions of the Jovian magnetosphere, presumably associated with changes in the dynamic pressure of the solar wind, cause occasional re-entries of the spacecraft into the magnetosheath.

However, the supposition that this innermost boundary is not the magnetopause faces several difficulties that appear insurmountable. First, there is an extended region inside this boundary in which the field is southward (Fig. 5). Second, the trapped energetic particles in this region have been observed to be co-rotating (Van Allen et al. 1975). Third, the plasma analyzer measurements indicate that the flow outside the boundary is inward, not outward (Wolfe et al. 1974). All these observations appear contrary to what would be expected for a transition region corresponding to outward flow.

Another open magnetosphere model has been proposed in which the fieldlines are open at latitudes above and below the current sheet, rather than within or adjacent to it (Fig. 14). Such a model has been developed by Goertz et al. (1976), following a suggestion by Van Allen, in order to explain the very strong periodic modulation of the energetic particle fluxes observed in the outbound Pioneer 10 data. The field within, and adjacent to, the current sheet is assumed to be closed and can then contain trapped particles. At higher latitudes, however, the fieldlines are open and the particle fluxes drop to essentially interplanetary levels. The distended closed dipole fieldlines pre-

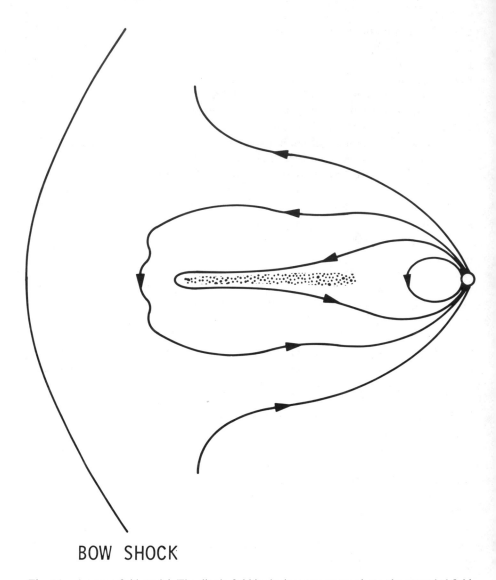

BOW SHOCK

Fig. 14.   An open field model. The dipole field in the inner magnetosphere, the extended fieldlines and plasma sheet or current sheet (dotted) in the middle magnetosphere and the closed
fieldlines at mid-latitudes and passing through the outer magnetosphere are shown. The fieldlines at higher latitudes are open and are presumably being carried into the magnetotail. In
this and Figs. 12 and 13, the direction to the sun is not specified so that the view may be
considered to represent either the noon or dawn meridian if so desired.

sumably account for the southward fields observed between the outer edge
of the current sheet and the magnetopause. So far, the model has only been
proposed for the pre-dawn sector of the magnetosphere but it might be extended successfully to the dayside magnetosphere.

In regions where the fieldlines are open, a southward field component would not be expected and the field should be principally radial. There are indeed data intervals, observed principally from Pioneer 10 but also from Pioneer 11, when the observed field is nearly radial. (In RJ coordinates, $\delta = 0$ corresponds to a radial orientation, the distinction between radial and equatorial only becoming obvious at a relatively high latitude.) One test of a radial field is that the radial component should decrease by r^{-2}. The radial component measured on Pioneer 10 outbound between 20 and 80 R_J decreases approximately as $r^{-1.7}$. At 10° latitude, however, it is difficult to distinguish unambiguously between a radial and an equatorial field on this basis. Thus, this result can only be considered consistent with the model and should not be taken to be a conclusive discriminator. The Pioneer 11 data outbound at higher latitude follow an r^{-3} dependence much more closely than r^{-2}, as noted above, implying that the fieldlines are dipolar in the dayside hemisphere up to latitudes of $\simeq 30°$. It is possible of course that the field is open at even higher latitudes.

Once the fieldlines become radial, it is possible to estimate whether or not the presumably closed fields at lower latitude could account for the magnetic flux observed in the outer magnetosphere. Within a given longitude interval $\delta\phi$, the southward flux through the outer magnetosphere would be approximately $B_0 (r_3^2 - r_2^2) \delta\phi / 2$, where r_2 and r_3 are the inner and the outer radial limits of the outer magnetosphere. In the middle magnetosphere the magnetic flux at radial distance r_1, passing radially outward through the solid angle determined by $\delta\phi$ and δ_M, the latitude of the last closed fieldline, is $B_R \delta\phi \sin \delta_M r_1^2$. Hence, conservation of magnetic flux implies that $\sin \delta_M = B_0 (r_3^2 - r_2^2)/2B_R r_1^2$. Taking typical values of $r_1 = 45$, $r_2 = 60$, $r_3 = 80 R_J$ and $B_0 = B_R = 10 \gamma$, implies $\delta_M = 45°$. However, the region containing southward fields in the outbound Pioneer 10 data is smaller, lying between 80 and 95 R_J. Taking $r_1 = 45 R_J$ yields $\delta_M = 12°$, a value that is consistent with open fields at much lower latitudes as predicted by Goertz et al. (1976).

These estimates lead to significantly different latitudes at which the fieldlines might be open in the dayside and pre-dawn magnetosphere. The shape of the magnetosphere may possibly influence the intensity of the current sheet and affect the latitude at which the field becomes open. The distance to the magnetopause is greater near dawn than near noon and centrifugal force is then potentially much more effective near dawn. If this conjecture is correct, the magnetopause could be more disc-like on the flank than near local noon.

What causes the open fieldlines is as yet unspecified. In the Goertz et al. (1976) model, a mathematical representation of the field observed by Pioneer 10 outbound, ignoring magnetopause currents, simply yields open fieldlines when extrapolated to higher latitudes. It is uncertain whether the fieldlines are being convected into the magnetotail of Jupiter by plasma entering the magnetosphere, e.g., through some type of cusp as at Earth, or are being convected outward. The opening of fieldlines at higher latitudes, however, does not appear readily compatible with plasma flow caused by centrifugal force. The most likely region for outflow to occur is near the equator where the field is weakest and the centrifugal force is greatest.

Acknowledgements. Our co-investigators are D. S. Colburn, P. J. Coleman, Jr., P. Dyal and C. P. Sonett. We are grateful to Allan Frandsen for his assistance in reducing and displaying the data. E.J.S. acknowledges helpful discussions with H. Alfvén regarding the importance of the observed spiraling of the current sheet. This chapter presents the results of one phase of research carried out at the Jet Propulsion Laboratory, California Institute of Technology, under contract NAS7-100, sponsored by the National Aeronautics and Space Administration. The contribution by L.D. was partially supported by NASA Grant NGR-05-002-160.

DISCUSSION

A. C. Riddle: At a special meeting on May 21, 1975, in Tucson, Arizona, of 25 radio astronomers and other interested persons (see Riddle and Warwick 1976), a resolution was made as follows:

"We have concluded that the provisional rotational period adopted for Jupiter's System III (1957.0) longitude measure is inadequate for current use. Consequently, we propose the adoption of a new System III measure for which the sidereal rotation rate of Jupiter is $870°536$ per Ephemeris Day. Further, we propose that the epoch shall be 1965 Jan. 1, $0^h\ 0^m\ 35°73$ ET, that the longitude at epoch of the central meridian, as observed from Earth, shall be $217°956$ and that the system be called System III (1965)."

J. G. Melville: From the Jovian decimetric radio observations, Conway and Stannard (1972) have proposed the existence of a magnetic field anomaly near the surface of Jupiter and at a System III (1957) longitude of 220°. In order to see if such a feature could be detected by the Pioneer 10 magnetic field measurements we fit the data to a dual dipole model. The main dipole represents the large-scale features of the magnetic field, while the second dipole is intended to approximate the proposed field anomaly. These results, reported earlier by Melville and Jones (1974) and by Smith *et al.* (1974*b*), are listed below.

Main dipole	*Second dipole*
$M = +4.02$ Gauss R_J^3	$M = +0.066$ Gauss R_J^3
$C_x = -0.200\ R_J$	$C_x = -0.8221\ R_J$
$C_y = -0.034\ R_J$	$C_y = +0.4605\ R_J$
$C_z = +0.068\ R_J$	$C_z = -0.1184\ R_J$
Tilt $= 9°92$	Tilt $= 77°5$
Long. III (1957) $= 226°4$	Long. III (1957) $= 218°5$

rms residual $= 89.3\ \gamma$

Note: x, y, z is a right-handed cartesian coordinate system with z parallel to Jupiter's spin axis and x at zero longitude in System III (1957).

The fit is better than the best single offset dipole fit which has an rms residual of $102.5\,\gamma$. A more realistic model must include the field produced by magnetospheric currents. In the region of this study ($r < 7.5\,R_J$) we approximate the field due to these currents by a constant field. When a dual dipole plus constant field is fit to the data a significant drop in the rms residual is obtained as seen in the parameters listed below.

Main dipole	Second dipole	Constant Field
$M = +4.37$ Gauss R_J^3	$M = +0.082$ Gauss R_J^3	
$C_x = -0.084\ R_J$	$C_x = -0.744\ R_J$	$B_x = -16.0\gamma$
$C_y = +0.102\ R_J$	$C_y = +0.282\ R_J$	$B_y = +31.6\gamma$
$C_z = +0.049\ R_J$	$C_z = +0.187\ R_J$	$B_z = +185.1\gamma$
Tilt $= 9°68$	Tilt $= 77°6$	
Long. III (1957) $= 225°2$	Long. III (1957) $= 206°7$	
	rms residual $= 29.6\gamma$	

Here the position of the second dipole is at $0.82\ R_J$ with a System III (1957) longitude of $200°8$.

There are several interesting differences between this 15 parameter fit and the simpler D_2 Model (Smith *et al.* 1974*b*). The tilt and System III longitude of the main dipole differ slightly from the D_2 values. Also a contour map of the surface magnetic fieldstrength shows a local region of field enhancement near the second dipole. One sees that the position of the second dipole is approximately that suggested by Conway and Stannard for the proposed field anomaly. Finally, the large constant field component suggests large currents in the inner magnetosphere.

REFERENCES

Axford, W. I. 1972. The interaction of the solar wind with the interstellar medium. *Solar Wind*. (C. P. Sonett, P. J. Coleman, Jr., and J. M. Wilcox, eds.), pp. 609–658. NASA SP-308. Washington D.C.: U.S. Government Printing Office.

Bartels, J. 1936. Eccentric dipole approximating the earth's magnetic field. *Terr. Magn.* 41: 225–250.

Berge, G. L. 1974. Position and Stokes parameters of integrated 21 cm radio emission of Jupiter and their variation with epoch and central meridian longitude. *Astrophys. J.* 191:775–784.

Bevington, P. R. 1969. *Data reduction and error analysis for the physical sciences*. New York: McGraw Hill.

Brice, N. M., and Ioannidis, G. A. 1970. The magnetospheres of Jupiter and Earth. *Icarus* 13:173–183.

Carr, T. D., and Gulkis, S. 1969. The magnetosphere of Jupiter. *Ann. Rev. Astron. Astrophys.* 7:577–618.

Chapman, S., and Bartels, J. 1940. *Geomagnetism*. pp. 639–668. London: Oxford University Press.

Chenette, D. L.; Conlon, T. F.; and Simpson, J. A. 1974. Bursts of relativistic electrons from Jupiter observed in interplanetary space with the time variation of the planetary rotation period. *J. Geophys. Res.* 79:3551–3558.

Conway, R. G., and Stannard, D. 1972. Non-dipole terms in the magnetic fields of Jupiter and earth. *Nature* 239:142–143.

Eviatar, A., and Ershkovich, A. I. 1976. The Jovian magnetopause and outer magnetosphere. In draft.

Goertz, C. K.; Jones, D. E.; Randall, B. A.; Smith, E. J. and Thomsen, M. F. 1976. Evidence for open field lines in Jupiter's magnetosphere. Dept. of Physics and Astronomy preprint, U. of Iowa; *J. G. R.* (special Jupiter issue). In press.

Gubbins, D. 1974. Theories of geomagnetic and solar dynamos. *Rev. Geophys. Space Phys.* 12:137–154.

Hill, T. W.; Dessler, A. J.; and Michel, F. C. 1974. Configuration of the Jovian magnetosphere. *Geophys. Res. Lett.* 1:3–6.

Hubbard, W. B., and Smoluchowski, R. 1973. Structure of Jupiter and Saturn. *Space Sci. Rev.* 14:599–662.

Kennel, C. F., and Coroniti, F. V. 1974. Is Jupiter's magnetosphere like a pulsar's or Earth's? *The magnetospheres of the earth and Jupiter.* (V. Formisano, ed.) pp. 451–477. Dordrecht-Holland: D. Reidel Publ. Co.

Lawson, C. L., and Hanson, R. J. 1974. *Solving least squares problems.* New York: Prentice Hall.

McCulloch, P. M., and Komesaroff, M. M. 1973. Location of the Jovian magnetic dipole. *Icarus* 19:83–86.

Mead, G. D. 1974. Magnetic coordinates for the Pioneer 10 Jupiter encounter. *J. Geophys. Res.* 79:3514–3521.

Melville, J. G., and Jones, D. E. 1974. A study of single and dual dipole magnetic field models for Jupiter: Pioneer. Paper presented at the spring 1974 meeting of the Utah Academy of Sciences, Arts, and Letters.

Michel, F. C., and Sturrock, P. A. 1974. Centrifugal instability of the Jovian magnetosphere and its interaction with the solar wind. *Planet. Space Sci.* 22:1501–1510.

Mihalov, J. D.; Collard, H. R.; McKibben, D. D.; Wolfe, J. H.; and Intriligator, D. S. 1975. Pioneer 11 encounter: preliminary results from the Ames Research Center plasma analyzer experiment. *Science* 188:448–451.

Northrop, T. G.; Goertz, C. K.; and Thomsen, M. F. 1974. The magnetosphere of Jupiter as observed with Pioneer 10. 2. Nonrigid rotation of the magnetodisc. *J. Geophys. Res.* 79: 3579–3582.

Parker, E. N. 1963. *Interplanetary dynamical processes.* New York: Inter-Science Publishers.

Piddington, J. H. 1969. *Cosmic electrodynamics.* New York: J. Wiley and Sons.

Riddle, A. C., and Warwick, J. W. 1976. Redefinition of System III longitude. *Icarus* (special Jupiter issue). In press.

Rikitake, T. 1966. *Electromagnetism and the earth's interior.* New York: American Elsevier Publ. Co.

Smith, E. J.; Connor, B. V.; and Foster, Jr., G. T. 1975a. Measuring the magnetic fields of Jupiter and the outer solar system. *IEEE Trans. on Magnetics.* MAG-11:962–980.

Smith, E. J.; Davis, Jr., L.; Jones, D. E.; Colburn, D. S.; Coleman, Jr., P. J.; Dyal, P.; and Sonett, C. P. 1974a. Magnetic field of Jupiter and its interaction with the solar wind. *Science* 183:305–306.

Smith, E. J.; Davis, Jr., L.; Jones, D. E.; Coleman, Jr., P. J.; Colburn, D. S.; Dyal, P.; Sonett, C. P.; and Frandsen, A. M. A. 1974b. The planetary magnetic field and magnetosphere of Jupiter: Pioneer 10. *J. Geophys. Res.* 79:3501–3513.

Smith, E. J.; Davis, Jr., L.; Jones, D. E.; Coleman, Jr., P. J.; Colburn, D. S.; Dyal, P.; Sonett, C. P. 1975b. Jupiter's magnetic field, magnetosphere, and interaction with the solar wind: Pioneer 11. *Science* 188:451–455.

Sonnerup, B. U. O. and Cahill, Jr., L. J. 1968. Magnetopause structure and attitude from Explorer 12 observations. *J. Geophys. Res.* 72:171–183.

Trainor, J. H.; McDonald, F. B.; Teegarden, B. J.; Webber, W. R.; and Roelof, E. C. 1974. Energetic particles in the Jovian magnetosphere. *J. Geophys. Res.* 79:3600–3613.

Van Allen, J. A.; Randall, B. A.; Baker, D. N.; Goertz, C. K.; Sentman, D. D.; Thomsen, M. F.; and Flindt, H. R. 1975. Pioneer 11 observations of energetic particles in the Jovian magnetosphere. *Science* 188:459–462.

Wolfe, J. H.; Mihalov, J. D.; Collard, H. R.; Intriligator, D. S.; McKibbin, D. D.; and Frank, L. A. 1974. Pioneer 10 observations of the solar wind interaction with Jupiter. *J. Geophys. Res.* 79:3489–3500.

RESULTS FROM THE
GSFC FLUXGATE MAGNETOMETER ON PIONEER 11

M. H. ACUÑA and N. F. NESS
NASA Goddard Space Flight Center

The main magnetic field of Jupiter has been measured by the Goddard Space Flight Center (GSFC) Fluxgate Magnetometer on Pioneer 11; the analysis reveals it to be rather complex. In a centered spherical harmonic representation with maximum order $n = 3$ (designated GSFC Model O_4) the dipole term, with opposite polarity to Earth's, has a magnitude of 4.28 Gauss $R_J{}^3$, and it is tilted by 9°.6 towards the System III longitude of 232° (epoch 1974.9). However, the quadrupole and octupole moments are significant, namely 24% and 21% of the dipole moment, and this leads to a deviation of the planetary magnetic field at distances $< 3\ R_J$ from a simple offset tilted dipole. The model shows a north polar fieldstrength of 14 Gauss, while in the south it is 10.4 Gauss. In the northern hemisphere the "footprint" of the Io-associated flux tube passes directly over the polar region. Derived L-shell parameters for the radiation belts predict enhanced absorption effects due to the satellites Amalthea and Io as a result of the field distortion. Warping of the charged-particle magnetic equator from a plane is also predicted.

A high-field, triaxial fluxgate magnetometer (FGM) was placed on Pioneer 11 by GSFC to measure the strong planetary magnetic field. A brief note based upon real time, quick-look data obtained during encounter on 3–4 December 1975 reported the discovery of a distorted, main magnetic field of the planet with significant quadrupole and octupole contributions (Acuña and Ness 1975b). A subsequent analysis of the data based upon a more comprehensive data tape, improved information on spacecraft attitude and A-D converter performance, and comparisons with charged particle data has allowed a more accurate determination of these higher-order moments and their consequences for the motion of charged particles and the absorption effects associated with Io and Amalthea.

[830]

It is the purpose of this chapter to elaborate on the FGM results and their implications for the study of trapped particles, planetary radio emissions and planetary interiors.

INSTRUMENTATION

The FGM provides instantaneous triaxial vector measurements of the 3 components of the magnetic field using a 10-bit precision A-D converter. The maximum measurable field is 10 Gauss along each orthogonal axis and the quantization step size is $\pm 600\gamma$ for field intensities less than 2 Gauss. Measurements are made once every 36 seconds in an inertial reference frame synchronized with the rotation of the spacecraft (revolution period of the spacecraft is 12 sec).

A redundant measurement of the magnetic field component parallel to the spin axis of the spacecraft was obtained using two orthogonally mounted biaxial sensors. The analog signal from the redundant Z-axis of the second sensor was digitized by the 6-bit A-D converter of the Pioneer spacecraft, and its value was transmitted at the same rate as that of the measurements of the primary vector.

Although no sensitivity calibrations of the magnetometer were performed in flight, no drift of zero levels of the systems was noted from pre- and post-encounter measurements in the weak interplanetary field. Since the output of each axis is biased to one half of the A-D converter's dynamic range, this provided direct verification of the calibration of the A-D converters.

The redundancy provided by the dual Z-axis measurements was used to monitor the sensitivity calibration of each sensor versus the other and thus give a measure of the internal consistency of the data. The maximum field measured along the Z-axis was approximately 1 Gauss which represents only 5 quantization steps of the spacecraft's A-D converter. Thus, in the absence of additional information, and based on the nominal performance of this converter (± 1 count), it was possible to verify the sensitivity calibration to within $\pm 20\%$. Our preliminary report based on quick-look data (Acuña and Ness 1975b) is consistent with this verification of the sensitivity calibration, but subsequent comparisons with directions of charged particle anisotropy observed near closest approach (Van Allen, and Fillius $et\ al.$ personal communications) indicated the possibility of a slight drift of the sensitivity calibration in the X-Z sensor. Recently three calibration voltages associated with the spacecraft's A-D converter, accurately measured at three selected transition points (6%, 50%, 81%), have become available. They allow a more precise intercomparison of the Z-axis data, revealing a 10% sensitivity increase in the X-Z sensor. This small but significant effect is illustrated in Fig. 1 where the redundant Z-axis data have been plotted versus the main Z-axis data. The straight lines represent the original and corrected calibrations for the X-Z sensor.

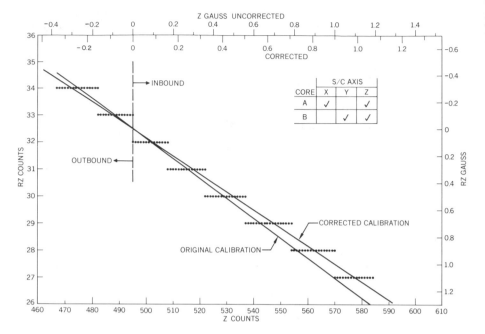

Fig. 1.   Comparison of z-axis component magnetic field data from the A core (Z) using the 10 bit A-D converter of the instrument with the redundant B core (RZ) using the 6 bit A-D converter of the Pioneer 11 spacecraft. Solid curves are preflight and corrected calibrations.

This correction yields excellent agreement with charged particle data obtained by other experimenters from the Pioneer spacecraft (Van Allen *et al.* 1975;[1] Fillius *et al.* 1975[2]) and good agreement with magnetic field data (Smith *et al.* 1975).[3] A complete description of the instrument has been published elsewhere (Acuña and Ness 1975*a*).

MEASUREMENTS AND ANALYSES

The triaxial measurements of the magnetic field are corrected for phase and amplitude distortion by the spinning spacecraft and then rotated to a Jupiter-centered spherical coordinate system. A total of 685 vector measurements were obtained during the close encounter period, from 01^h 20^m to 09^h 26^m on 3 December 1974, corresponding to distances less than 6 R_J. Due to an unfavorable bit-rate allocation for this instrument during occultation, no measurements were obtained during that period when the spacecraft was operated in a memory storage mode.

[1]See pp. 954 and 38.
[2]See p. 925.
[3]See p. 788.

The data have been analyzed in terms of a Schmidt normalized spherical harmonic expansion fitted to the observations in a least-squares sense. In view of the quantization step size, it is reasonable to neglect external harmonic terms in such a representation for distances less than 10 R_J. Such terms are associated with the highly distorted magnetosphere of Jupiter observed at distances greater than 20–25 R_J (Smith *et al.* 1975).

The observations between 1.7–6.0 R_J have been fitted in the least-squares sense so as to minimize the vector residual to harmonic expansions of the following form:

$$\mathbf{B} = -\nabla V$$

$$V = a \sum_{n=1}^{n\ max} \sum_{m=0}^{n} \left(\frac{a}{r}\right)^{n+1} (g_n{}^m \cos m\phi + h_n{}^m \sin m\phi)\, P_n{}^m (\cos \theta).$$

(1)

The inclusion of quadrupole and octupole terms leads to a significant reduction in the root mean square (rms) deviation of the residuals. An offset tilted dipole can be obtained by using 6 of the 8 coefficients in a quadrupole expansion.

Table I summarizes results for the dipole moment in all of the GSFC expansions thus far taken. Note the variation in the characteristics of the dipole term as the order of the expansion increases. This variation is an expected phenomenon, due to the cross-coupling of harmonic coefficients, in situations like this in which the data are not "complete" in a mathematical sense so that a set of mutually dependent coefficients occurs. The comparison of the measured magnetic field with that predicted from the spherical harmonic representations is shown in Fig. 2 for the O_4 model. The overall agreement with the observations is excellent in all three components of the field.

It is essential in any such representation to conduct an error analysis to determine the statistical stability of the various sets of coefficients. The use of the rms parameter, while appropriate, is a measure that is not complete. In the general problem of solving a set of linear equations, such as those obtained in the minimization of the vector residual, it is also useful to consider the condition number of the matrix to be inverted, regardless of the specific method finally employed to affect a solution. The condition number is related to the ratio of the magnitude of the largest to the smallest eigenvectors of the matrix to be inverted. The smaller the number the more stable the estimates of the unknowns become. Table I shows that the condition number for the O_4 model is quite modest, being only 50, while it would increase to 340 for a hexadecapole model. Inclusion of external terms, such as those by Smith *et al.* (1975), increase the condition number to very high values so that there may exist considerable uncertainty in some of the coefficients derived in this way. Neither the uncertainties of the trajectory nor the size of the quantization step adversely affect the determination of the set of coefficients for O_4.

TABLE I
Summary

	Dipole	Offset Tilted Dipole	Quadrupole	Octupole
Moment (Gauss $R_J{}^3$)	4.36	4.35	4.36	4.28
Tilt	8.41°	9.52°	9.83°	9.6°
Longitude (West Long. III)	246°	236°	235°	232°
Vector rms (Gauss)	0.021	0.0113	0.0106	0.0085
Condition number	—	—	—	50
Offset (R_J)				
X_0		− 0.0618		
Y_0	—	+ 0.024	—	—
Z_0		− 0.015		
$\dfrac{\text{rms residual}}{\text{rms field}^{\text{a}}}$	5.97%	3.2%	3.01%	2.4%

[a] rms field = 0.35172 Gauss

Extrapolation of the magnetic field model to the surface of the planet is illustrated in Fig. 3. The magnitude of the field of Jupiter for the O_4 model is shown along with other parameters related to Pioneer 11 and the satellite Io. An important feature of our model is the hemispherical and azimuthal asymmetry in both the surface field and in the field topology which results from the significant values of the quadrupole and octupole moments. Table II summarizes the magnitude and phase of the higher-order moments for the quadrupole and O_4 models. The table also includes a computation of the normalized magnitude of the higher-order moments relative to the dipole term. It is seen that in the O_4 model the quadrupole is approximately 24% of the dipole term while the octupole is approximately 21%. These values exceed those of the earth, which are respectively 14 and 9%.

It is appropriate at this point to consider comparison of this GSFC O_4 magnetic field model with those derived from other instruments on Pioneers 10 and 11. Smith *et al.* (1974) proposed, on the basis of their Pioneer 10 Helium Vector Magnetometer measurements, that the planetary field was well represented by an offset tilted dipole with moment 4.0 Gauss $R_J{}^3$ tilted at 10°.6 towards a System III longitude $\lambda_{\text{III}} = 222°$ in December 1973. On the

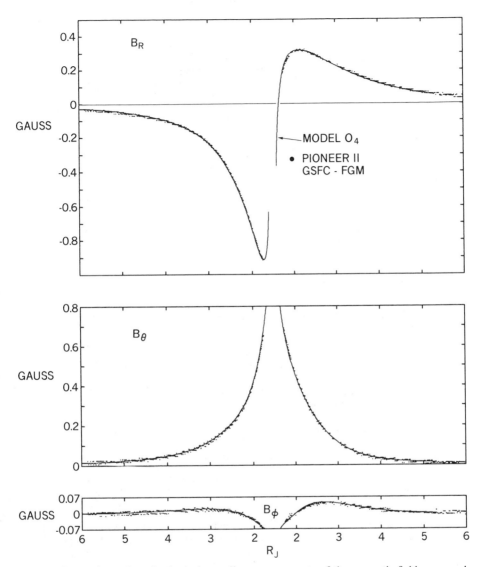

Fig. 2. Comparison plot of spherical coordinate components of the magnetic field measured on Pioneer 11 with the GSFC O_4 model.

basis of charged particle measurements from Pioneer 10, Van Allen *et al.* (1974) derived a centered dipole tilted $9°5$ toward $\lambda_{III} = 230°$ which was reported to better represent the results as measured by radiation belt characteristics. The dipole term in the GSFC O_4 model (tilt = $9°6$ and $\lambda_{III} = 232°$) shows good agreement with the Van Allen *et al.* model but less satisfactory agreement with the Smith *et al.* (1974) model. (Note that λ_{III} is computed for 1973.9.)

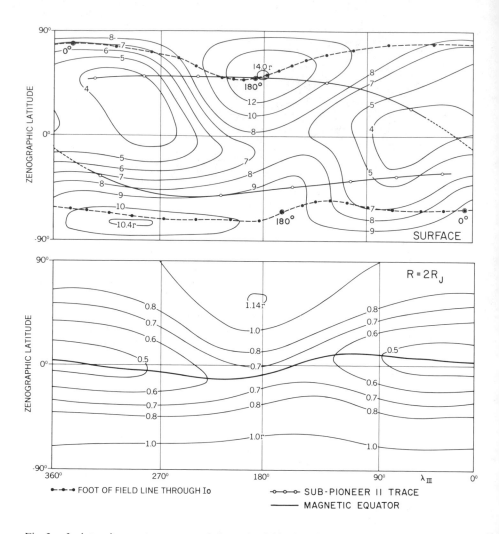

Fig. 3.   Isointensity contour maps of the main field of Jupiter at the surface of the planet (assuming 1/15.4 flattening) and at 2 R_J for the O_4 model. The trace of the Pioneer 11 trajectory is shown in the upper panel, as is the trace of the footprint of the flux tube associated with Io.

More recently Smith *et al.* (1975) have presented a spherical harmonic analysis in which they report a dipole term possessing a tilt angle of 9°.9 and a longitude of 227° while the quadrupole and octupole moments are 20% and 15% of the dipole moment, respectively. The GSFC O_4 model is in good agreement with these results, the minor differences being probably due to different procedures in mathematical analysis of the data, the number of data points included, and the radial distances considered.

TABLE II
Summary of Higher-Order Moments
(+ East Longitude III)

n	m	Quadrupole $C_n{}^m$	$\phi_n{}^m$	Octupole (O_4) $C_n{}^m$	$\phi_n{}^m$
1	0	+ 4.298		+ 4.218	
1	1	+ 0.745	+ 125°	+ 0.715	+ 128°
1	—	+ 4.36	(100%)[a]	+ 4.28	(100%)[a]
2	0	− 0.233		− 0.203	
2	1	+ 0.527	+ 172°	+ 0.872	+ 182°
2	2	+ 0.246	− 26°	+ 0.521	− 25°
2	—	+ 0.627	(14.4%)	+ 1.036	(24.2%)
		(= M_n)			
3	0			− 0.233	
3	1			+ 0.585	− 128°
3	2			+ 0.515	+ 5°
3	3			+ 0.374	+ 47°
3	—			+ 0.895	(20.9%)

[a]By definition.

$$\phi_n{}^m = \frac{1}{m} \arctan \left(\frac{h_n{}^m}{g_n{}^m}\right)$$

$$C_n{}^m = [\,(g_n{}^m)^2 + (h_n{}^m)^2\,]^{\frac{1}{2}}$$

$$M_n{}^2 = \sum_{m=0}^{n} (g_n{}^m)^2 + (h_n{}^m)^2$$

IMPLICATIONS: CHARGED PARTICLES

The motion of charged particles in a planetary magnetic field has been studied extensively only in the case of Earth (see reviews by Schulz 1974 and West 1975). In these studies, it has been found necessary to use a high-order harmonic representation of the terrestrial field even though the quadrupole and octupole moments are only 14 and 9% of the dipole moment and higher multipoles are significantly less. This is because the guiding center for the cyclotron motion of the charged particles, as it moves from northern to southern mirror points, drifts irregularly in longitude. While on Earth the electrons drift eastward and the protons westward, the drift motion of charged particles at Jupiter is opposite because the dipole moment of Jupiter is opposite to that of the earth.

In initially describing the relative position of the spacecraft trajectory with respect to the planetary magnetic field, it is convenient to utilize the

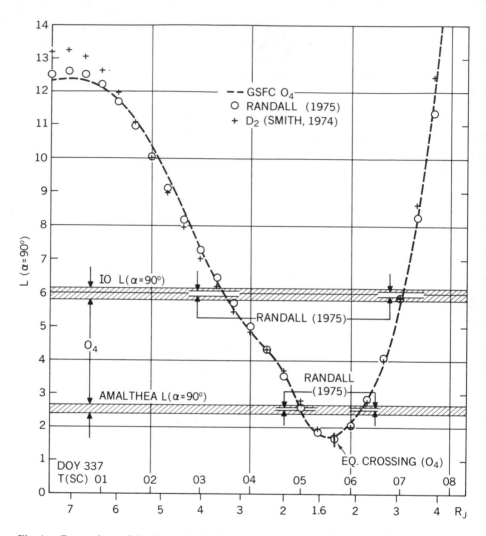

Fig. 4.   Comparison of the derived L shell parameter for the O_4, D_2 and Randall models along the Pioneer 11 trajectory. The L values are computed for charged particles which mirror at the spacecraft, i.e., a local pitch angle of 90°. Indicated regions of absorption effects due to Io and Amalthea are identified for the O_4 and Randall models.

classical L-shell parameter that measures (in an appropriately normalized sense) the equatorial distance of the fieldline about which a charged particle moves. For charged particles, the equatorial point along the fieldline is defined to be where the field intensity is a minimum.

We have calculated the corresponding L values derived from the GSFC O_4 model along the Pioneer 11 spacecraft trajectory as shown in Fig. 4. Those charged particles which would impact Io and Amalthea in their complicated cyclotron, bounce and drift motion, can be expected to be absorbed

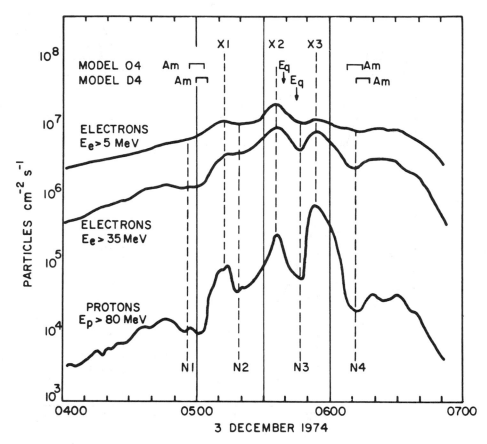

Fig. 5.   Comparison of electron and proton fluxes measured by the experiment of the University of California at San Diego (Fillius *et al.* 1975) near closest approach to Jupiter by Pioneer 11. The figure illustrates a multiple peak structure, and the predicted times of absorption due to Amalthea and the crossing of the charged particle equator (EQ).

by these satellites (Mead and Hess 1973; Hess *et al.* 1974). We have also computed the corresponding L shells swept out by Io and Amalthea for the O_4 model; these are included in Fig. 4. For Io, the O_4 L-values swept out are 6.13 to 5.79, while for Amalthea they are 2.66 to 2.34. The predictions of magnetic field models for satellite sweeping have met with varied success according to the charged particle instruments on Pioneer 11 (Fillius *et al.* 1975; Simpson *et al.* 1975; Trainor *et al.* 1975 and Van Allen *et al.* 1975). Part of the difficulty at Io may be associated with the demonstrated existence of a source of charged particles observed near this satellite (McIlwain and Fillius 1975).

    In order to fully study the problems of satellite sweeping, a more complete treatment is needed for the distributions of the drift shells of the particles and their pitch angles and for energy-angle characteristics of the indi-

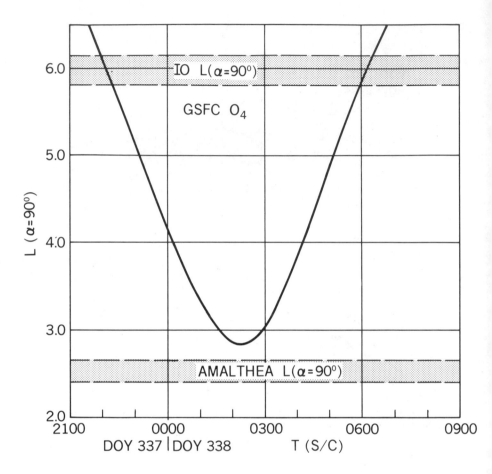

Fig. 6.   Same as Fig. 5 but for Pioneer 10 trajectory. Note that Pioneer 10 came to within 0.18 R_J of Amalthea's geometrical sweeping region.

vidual detectors. Here only the simplest view has been taken, namely that of geometrical sweeping of particles, and without proper consideration of the differences of L with pitch angle (at the satellites and at Pioneer 11).

The situation at Amalthea is interesting because of the complex nature of the radiation belt structure observed there by Pioneer 11 (Fillius *et al.* 1975). Figure 5 presents a comparison of expected absorption effects associated with Amalthea as predicted by two models of the planetary field: D_4 (Smith *et al.* 1975), and O_4 with the observations by Pioneer 11. It is seen that the O_4 model provides a better correspondence of the observed characteristics of minima and maxima of the multiple peak structure of the radiation belt close to the planet. This provides a completely independent but powerful test which endorses the physical validity of the O_4 model. However, a completely satisfactory explanation of the observed structure is still lacking since effects of particle void diffusion have been neglected. It should be noted that with larger multipole moments the complexity of charged particle motion and drift shells will be enhanced over that familiar from terrestrial studies which show such other important phenomena as longitudinal variations in the loss cones.

It has been noted on Pioneer 10 (Fillius and McIlwain 1974; Simpson *et al.* 1974; Trainor *et al.* 1974), that near the periapsis of the trajectory there was a reduction in some of the detector count rates which could not be explained by the then available D_2 model of the Jovian magnetic field. With the observations of the broader range of sweeping associated with Amalthea, in the more complex magnetic field model O_3, it was then suggested (Fillius *et al.* 1975) that the strange behavior observed on Pioneer 10 probably was associated with the sweeping effects of Amalthea. Figure 6 shows the L-shell parameter of the Pioneer 10 trajectory which indicates that Pioneer 10 periapsis was indeed within $\sim 0.18\ R_J$ from the simple geometric sweeping region associated with Amalthea when the O_4 model is used.

The interpretation of satellite sweeping effects using only the L-shell parameter is a rather more simplified approximation than justified in the case of Jupiter. The complete study should include a consideration of the mirror point distribution of the particles and, in addition, the position of the spacecraft should be given in B, L space in such a way that associated with each L shell value there exists a corresponding specific value for the field intensity. The high-order moments of the Jovian magnetic field can be expected to contribute to some splitting of the L shells so that the satellite sweeping regions may be more extended than the simple geometric absorption considered thus far (see review by Roederer 1972 and references cited therein).

The hemispherical and azimuthal asymmetries of the O_4 field model provide a quantitative confirmation of the speculated source mechanism and locations of periodic escape of particles from the radiation belts of Jupiter into interplanetary space as discussed by Hill *et al.* (1974) and Vasyliunas (1975).

IMPLICATIONS FOR PLANETARY RADIO EMISSIONS

Prior to direct measurements of the Jovian magnetic field, groundbased radio astronomy contributed substantially to our understanding of the magnetic field of Jupiter (see review by Carr and Gulkis 1969,[4] and references contained therein). Not only the approximate field magnitude at the equator of 1 to 10 Gauss was predicted but also the polarity sense of the magnetic field was established. The most precisely determined parameter came from the modulation of the electric field polarization characteristics of decimetric radio emissions observed from electrons radiating in the synchrotron mode

---

[4]See pp. 621, 693, and 1146.

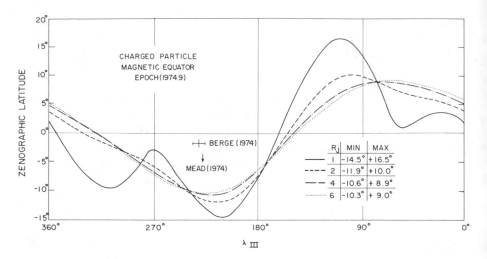

Fig. 7.   Latitude versus longitude plot of minimum *B* intensity along a fieldline at various zeno-
centric distances.

close to the planet, primarily between 1.5 to 4 R_J. These observations provided a good estimate of the tilt and longitude of the equivalent dipole magnetic axis of the planet Jupiter (Komesaroff *et al.* 1970; Berge 1974), as well
as established its rotation rate relative to coordinate systems I and II, defined
on the basis of atmospheric observations. The variation of polarization and
its intensity, however, could not be accurately matched to that of any simple
dipole and it was suggested that non-dipolar contributions were responsible
(Conway and Stannard 1972). Figure 7 shows the shape of the charged particle equator determined from O_4 as observed at different radial distances
and illustrates clearly that this, the principal region of emissions, deviates
significantly from a simple flat planar geometry.

The latitude variation of the charged particle equator is by no means a
simple sinusoid, although at all distances from 1 to 6 R_J the curves show a
minimum latitude near $\lambda_{III} = 225°$. This value determines the longitude of the
equivalent dipole and Fig. 7 includes the extrapolated values from Berge
(1974) and Mead (1974). The variations in the peak-to-peak excursions of
these curves also suggest a reasonable explanation for the various differences
in derived magnetic field characteristics associated with separate studies of
the decimetric emission at different wavelengths. Since the principal contributions to the different wavelength regions will originate from different regions in the radiation belts, their differing geometrical characteristics will be
reflected accordingly into the radiated emissions by their respective positions.

The modulation of the decametric radio emission by Io has been an enigmatic phenomenon studied in recent years. Based upon the octupole model
O_4, Fig. 8 shows the position of the Io-associated flux tube as seen from the
north zenographic pole. The fieldlines which thread Io, as the satellite moves

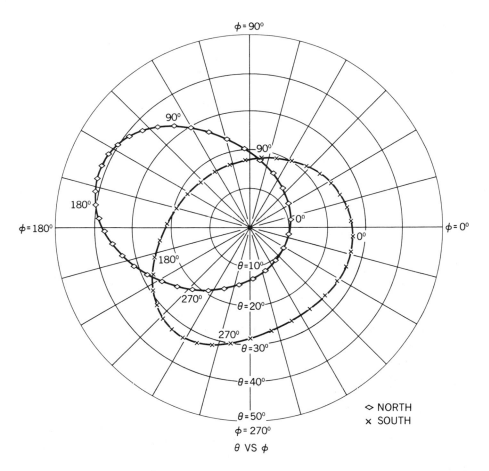

Fig. 8.   Intersection of field lines threading Io with the surface of Jupiter according to the GSFC O_4 model. The zenographic longitude of Io is indicated alongside the individual north-south footprints. This polar projection has longitudes in System III (Epoch 1974.9), but positive eastward.

in zenographic longitude, are identified as Io's position varies from 0 to 360°. A comparison of this figure with Fig. 3 shows that the northern footprint of Io passes directly over the polar regions of the planetary field. Thus it seems reasonable to speculate that the magnetic field geometry and corresponding Io flux tube support the idea that the decametric emission is associated with particle precipitation into the auroral regions of the planet.

Further study is necessary to elaborate more fully on the implications of the magnetic field measurements by the GSFC FGM with respect to decimetric and decametric radio emissions. Combined with definitive measurements of the distribution functions of charged particles, it should be possible to affect comparisons of the groundbased observations with those predicted from *in situ* models currently under development.

Some attempts to reconcile the Io footprint and flux tube geometry with decametric observations have been made (J. K. Alexander and R. A. Smith personal communication).[5] The limited data of Pioneer 11 do not permit determination of higher-order multipoles than those in O_4 and it is felt that the field geometry within 0.5 R_J of the surface and at the surface of Jupiter may not be sufficiently accurately determined to yield completely satisfactory results in any of these studies.

IMPLICATIONS: PLANETARY INTERIORS

The most plausible source mechanism of planetary magnetism is thought to be a dynamo operating in the interior and most probably driven by a convection process (Busse 1975; Gubbins 1974; Soward and Roberts 1975). The significant higher-order moments can be interpreted as representing two different phenomena associated with the planetary field. The first is the displacement of the equivalent dipole center from the center of the planet, which yields quadrupole and higher-order terms when the field is described by a *centered* harmonic expansion. This mathematical artifact can be removed by an expansion centered on the displaced dipole position, the magnitude and direction of the displacement determined from the centered quadrupole terms so as to reduce to zero the three terms g_{20}, g_{21} and h_{21}: The remaining quadrupole terms g_{22} and h_{22} have to do with "warping" of the magnetic equator. When this is done, the remaining quadrupole and octupole moments give an indication of the proximity of the source region generating the magnetic field to the surface of the planet. A preliminary calculation based on a spherical harmonic expansion of order $n = 3$ (internal terms up to octupole) yields offset quadrupole and octupole moments which are 6.2 and 19% of the dipole moment, respectively. The corresponding values for Earth are 6.25 and 9.9%, respectively (Hilton and Schulz 1973). On the basis of this preliminary calculation, no definitive statements can be made about the relative size of the source regions at Earth and Jupiter. Since the octupole may not be accurately determined, due to the limited data imposed by the Pioneer 11 trajectory, it is only possible to speculate that these source regions are similar in relative size; the main difference between the two planets is that in the case of Jupiter there is apparently a slightly larger equivalent dipole displacement from the center.

Further studies comparing Earth and Jupiter's planetary fields will be exploited to provide more precise insight on the mechanisms of planetary magnetic dynamos. The solar dynamo appears to be related to a more dynamic process and thus comparison with multipolar representations of the solar field are sensitive to the specific solar period of interest.

---

[5]See p. 1146.

SUMMARY

The main magnetic field of Jupiter, as measured by the GSFC FGM on Pioneer 11, is found to be more complex than indicated by the results of the Pioneer 10 Helium Vector Magnetometer as reported. At distances less than 3 R_J, the magnetic field is observed to increase more rapidly than an inverse-cubed distance law associated with any simple dipole model. Contributions from higher-order multipoles are significant, with the quadrupole and octupole being 24 and 21% of the dipole moment, respectively.

Because the data are not uniformly distributed over a surface enclosing the planet, cross coupling exists in the determination of the sets of coefficients depending upon the maximum order of the multipole considered. For the GSFC octupole model O_4 the dipole term is found to have a moment of 4.28 ± 0.15 Gauss R_J^3 tilted $9°6 \pm 0°3$ towards System III longitude of $232 \pm 3°$. Considerable hemispherical and azimuthal asymmetry of the magnetic field is found with maximum field intensities of 14 Gauss in the northern polar regions and 10.4 Gauss in the southern polar region.

The deviation of the main planetary magnetic field from a simple dipole leads to distortion of the L shells of the charged particles and to warping of the magnetic equator. Enhanced absorption effects associated with Io and Amalthea are predicted as a result, and the O_4 model provides better agreement with the observed absorption effects at Amalthea as reflected in the multiple peak structure of the radiation belts close to the planet. These effects are consistent with the conclusions derived from many years of decimetric and decametric radio observations regarding characteristics of the planetary field. The magnitude of the higher-order moments described in an offset dipole coordinate system suggest that the relative sizes of the source regions generating the field at Jupiter and Earth are of comparable magnitude.

Acknowledgements. We appreciate the outstanding efforts and the prompt processing and analysis of these data by F. W. Ottens and R. F. Thompson of the NASA Goddard Space Flight Center. Important discussions of these results with our colleagues, in particular J. K. Alexander, S. M. Gulkis, J. Warwick, J. Van Allen, R. W. Fillius and J. Roederer are also appreciated. The L-shell values are provided as the initial result of a collaborative effort with J. Roederer of the University of Denver in order to study the motion of charged particles in the magnetosphere of Jupiter.

REFERENCES

Acuña, M. H., and Ness, N. F. 1975a. The Pioneer XI high field fluxgate magnetometer. *Space Sci. Inst.* 1:177–188.

———. 1975b. Jupiter's main magnetic field measured by Pioneer 11. *Nature* 253:327–328.

Berge, G. L. 1974. The position and Stokes parameters of the integrated 21-cm radio emission of Jupiter and their variation with epoch and central meridian longitude. *Astrophys. J.* 191: 775–784.

Busse, F. H. 1975. A model of the geodynamo. *Geophys. J. Roy. Astron. Soc.* In press.

Carr, T. D., and Gulkis, S. 1969. The magnetosphere of Jupiter. *Ann. Rev. Astron. Astrophys.* 7:577–618.

Conway, R. G., and Stannard, D. 1972. Non-dipole terms in the magnetic fields of Jupiter and the earth. *Nature* 239:142–143.

Fillius, R. W., and McIlwain, C. E. 1974. Measurements of the Jovian radiation belts. *J. Geophys. Res.* 79:3589–3599.

Fillius, R. W.; McIlwain, C. E.; and Mogro-Campero, A. 1975. Radiation belts of Jupiter: a second look. *Science* 188:465–467.

Gubbins, D. 1974. Theories of the geomagnetic and solar dynamos. *Rev. Geophys. Space Phys.* 12:137–154.

Hess, W. N.; Birmingham, T. J.; and Mead, G. D. 1974. Absorption of trapped particles by Jupiter's moons. *J. Geophys. Res.* 79:2877–2880.

Hill, T. W.; Carbary, J. F.; and Dessler, A. J. 1974. Periodic escape of relativistic electrons from the Jovian magnetosphere. *Geophys. Res. Lett.* 1:333–336.

Hilton, H. H., and Schulz, M. 1973. Geomagnetic potential in offset dipole coordinates. *J. Geophys. Res.* 78:2324–2330.

Komesaroff, M. M.; Morris, D.; and Roberts, J. A. 1970. Circular polarization of Jupiter's decimetric emission and the Jovian magnetic field strength. *Astrophys. Lett.* 7:31–36.

McIlwain, C. E., and Fillius, R. W. 1975. Differential spectra and phase space densities of trapped electrons at Jupiter. *J. Geophys. Res.* 80:1341–1345.

Mead, G. D. 1974. Pioneer 10 mission: Jupiter encounter. *J. Geophys. Res.* 79:3514–3521.

Mead, G. D., and Hess, W. N. 1973. Jupiter's radiation belts and the sweeping effect of its satellites. *J. Geophys. Res.* 78:2793–2811.

Roederer, J. G. 1972. Geomagnetic field distortions and their effects on radiation belt particles. *Rev. Geophys. Space Phys.* 10:599–630.

Schulz, M. 1974. Geomagnetically trapped radiation. SAMSO Report TR-74-264 (N75-26965).

Simpson, J. A.; Hamilton, D. C.; Lentz, G. A.; McKibben, R. B.; Perkins, M.; Pyle, K. R.; Tuzzolino, A. J.; and O'Gallagher, J. J. 1975. Jupiter revisited: first results from the University of Chicago charged particle experiment on Pioneer 11. *Science* 188:455–459.

Simpson, J. A.; Hamilton, D. C.; McKibben, R. B.; Mogro-Campero, A.; Pyle, K. R.; and Tuzzolino, A. J. 1974. The protons and electrons trapped in the Jovian dipole magnetic field region and their interaction with Io. *J. Geophys. Res.* 79:3522–3544.

Smith, E. J.; Davis, Jr., L.; Jones, D. E.; Coleman, Jr., P. J.; Colburn, D. S.; Dyal, P.; and Sonett, C. P. 1975. Jupiter's magnetic field, magnetosphere and interaction with the solar wind: Pioneer 11. *Science* 188:451–455.

Smith, E. J.; Davis, Jr., L.; Jones, D. E.; Coleman, Jr., P. J.; Colburn, D. S.; Dyal, P.; Sonett, C. P.; and Frandsen, A. M. A. 1974. The planetary magnetic field and magnetosphere of Jupiter: Pioneer 10. *J. Geophys. Res.* 79:3501–3513.

Soward, A. M., and Roberts, P. H. 1975. Recent developments in dynamo theory. *Ann. Rev. Fluid Mech.* 7. In press.

Trainor, J. H.; McDonald, F. B.; Stillwell, D. E.; Teegarden, B. J.; and Webber, W. J. 1975. Jovian protons and electrons: Pioneer 11. *Science* 188:462–465.

Trainor, J. H.; McDonald, F. B.; Teegarden, B. J.; Webber, W. J.; and Roelof, E. C. 1974. Energetic particles in the Jovian magnetosphere. *J. Geophys. Res.* 79:3600–3613.

Van Allen, J. A.; Baker, D. N.; Randall, B. A.; and Sentman, D. D. 1974. The magnetosphere of Jupiter as observed with Pioneer 10. I. Instrument and principal findings. *J. Geophys. Res.* 79:3559–3577.

Van Allen, J. A.; Randall, B. A.; Baker, D. N.; Goertz, C. K.; Sentman, D. D.; Thomsen, M. F.; and Flindt, H. R. 1975. Pioneer 11 observations of energetic particles in the Jovian magnetosphere. *Science* 188:459–462.

Vasyliunas, V. M. 1975. Modulation of Jovian interplanetary electrons and the longitude variation of decametric emissions. *Geophys. Res. Lett.* 2:87–88.

West, Jr., H. I. 1975. Advances in magnetospheric physics 1971–1974. *Rev. Geophys. Space
Phys.* 13. In press.

RESULTS OF THE PLASMA ANALYZER EXPERIMENT ON PIONEERS 10 AND 11

D. S. INTRILIGATOR
University of Southern California

and

J. H. WOLFE
NASA Ames Research Center

The Ames Research Center Plasma Analyzer Experiments on Pioneers 10 and 11 have determined that the characteristics of the solar wind interaction with the Jovian magnetosphere are basically similar to those observed for the solar wind interaction at Earth and differ mainly in terms of the scale size of the interaction. The Jovian magnetosheath flow field and the calculated normals to the Jovian magnetosphere indicate that the Jovian magnetosphere is extremely thick and blunt in shape. The size of the Jovian magnetosphere in the sunward (dayside) direction can change by as much as a factor of two in response to relatively minor changes in the solar wind dynamic pressure. The outer dayside Jovian magnetosphere is inflated with a high-beta ($\beta \simeq 1; \beta \equiv 4\pi N\kappa T B^{-2}$) thermal plasma.

In this chapter we briefly summarize some of the Pioneer 10 and Pioneer 11 plasma ion and electron observations of the solar wind interaction with the Jovian particles and field environment. Many of the general features of this interaction at Jupiter are similar to those observed for this interaction at Earth. The reader is referred to Wolfe and Intriligator (1970) for a detailed review of the solar wind interaction with the earth's environment. In this chapter we discuss some of our Pioneer 10 and 11 observations of the Jovian magnetosheath and magnetosphere.

I. INSTRUMENT DESCRIPTION

The Ames Research Center Plasma Analyzer Experiments on Pioneer 10 and Pioneer 11 consist of dual 90° quadrispherical electrostatic analyzers, multiple charged particle detectors, and attendant electronics. This analyzer system is capable of determining the incident plasma distribution parameters

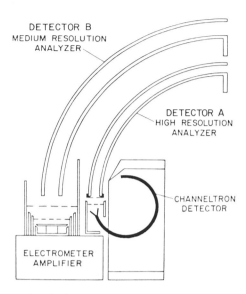

Fig. 1.   Central, cross-sectional schematic of the analyzer and detector portions of the Ames
Research Center Plasma Analyzer Experiment, dual 90° quadrispherical electrostatic analyzers, on Pioneers 10 and 11.

over the energy range of 100–18,000 eV for protons and approximately 1–
500 eV for electrons. A central, cross-sectional drawing of the analyzer and
detector portions of the experiment is shown in Fig. 1. The A detector, or
high-resolution quadrispherical analyzer (the inner analyzer system shown
in Fig. 1), has an analyzer constant of 9 (charged particle acceptance energy
per unit charge divided by the analyzer plate potential) with a mean radius
of the analyzer plate of 9 cm, and 0.5 cm separation. The high-resolution
analyzer is used for ion analysis only and utilizes 26 Bendix type CEM 4012
Channeltrons, operated in the pulse-counting mode, for ion detection. The
Channeltron detectors are arranged in a semicircle at the base of the analyzer
plates and cover the angular range of ±51° with respect to the entrance
aperture normal. The Channeltrons have an angular separation of approximately 3° near the central portion of the analyzer and approximately 8°
separation at the extremes of the analyzer. The Channeltron bias voltage
can be changed in two sections (left and right halves) by ground command
in eight discrete steps over the range of 2600–4400 V. Analysis of flight data
has shown that 51 of the 52 Channeltrons on both Pioneer 10 and Pioneer 11
have operated flawlessly since launch and no appreciable degradation has
been observed prior to, during, or subsequent to the Jupiter encounter.
    The B detector, or medium-resolution analyzer (the outer analyzer sys-
tem in Fig. 1), has a 12 cm mean radius and 1 cm plate separation, giving an
analyzer constant of 6. The medium-resolution analyzer is used for both ion

and electron detection and utilizes five flat-surface current collectors and electrometer amplifiers. Each of the three central current collectors has a 15° view width and covers an angular view range of $\pm 22°5$ with respect to the entrance aperture normal. The two outside collectors have an angular width of $47°5$ each and are located at $\pm 46°25$ with respect to the center of the analyzer.

Since detector *A* and detector *B* operate independently, a complete cross-check between the two analyzers is possible. The combined analyzer system covers the dynamic range for charged particle fluxes from approximately 1×10^2 to 3×10^9 cm^{-2} sec^{-1} and is capable of resolving proton temperatures down to at least 2×10^3 °K. Both analyzers on Pioneer 10 and Pioneer 11 were calibrated prior to launch in the Plasma Ion Calibration Facility of the Ames Research Center. These prelaunch calibrations are utilized in a least-squares fit to the flight data for a variety of possible distribution models in order to determine the plasma ion distribution parameters.

Although there are a number of possible operating modes for the experiment, the principal mode utilized during the encounter phase of the Pioneer 10 and 11 missions is one in which the energy per unit charge acceptance analyzer potential is stepped every one-half revolution of the spacecraft, and all current collectors and Channeltrons are read out at the roll angle of the spacecraft recording the highest flux value of the plasma protons.

The Plasma Analyzer Experiment is situated on the Pioneer 10 and 11 spacecraft such that the entrance apertures view back toward the earth (and therefore the sun) through a wide slit in the back of the spacecraft high-gain antenna reflector as shown in Fig. 2. The entrance aperture normals are oriented parallel to the spacecraft spin axis, thus allowing a complete angular scan of the earthward hemisphere every half spacecraft revolution. The edges of the antenna reflector limit instrument viewing to $\pm 73°$ with respect to the spacecraft spin axis.

II. MAGNETOSHEATH OBSERVATIONS

Observations obtained by the ARC Plasma Analyzer Experiments (Wolfe *et al.* 1974*a,b*; Intriligator and Wolfe 1974; Mihalov *et al.* 1975) on Pioneer 10 and Pioneer 11 during the flybys of Jupiter in 1973 and 1974 have provided our first observations of the solar wind interaction with Jupiter. The Pioneer 10 and 11 trajectories (as viewed from the north) referenced to Jupiter and projected into a plane parallel to the ecliptic and passing through the center of Jupiter are shown in Fig. 3. The trajectories referenced to Jupiter as seen from Earth are shown in Fig. 4. The Pioneer 10 trajectory was inclined $\sim 14°$ to the Jovian equator. The Pioneer 11 trajectory was inclined $\sim 50°$.

The first unambiguous indication of the interaction of the solar wind with the Jovian magnetic field was observed on Pioneer 10 on November 26, 1973, at $\sim 19^h 46^m$ UT spacecraft time. The telemetry signals were actually

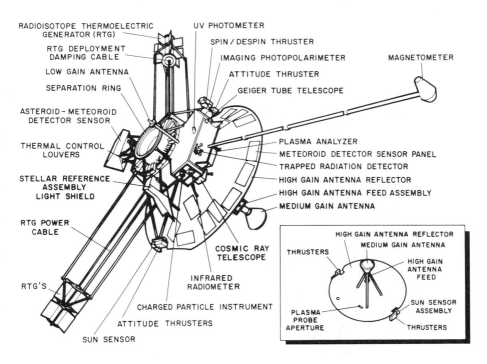

Fig. 2. Pioneer 10 and 11 spacecraft. Insert indicates Plasma Analyzer Experiment entrance aperture in high gain antenna reflector.

received at ~ 20^h 31^m UT on Earth, ground-received time (GRT), corresponding to a one-way radio propagation time of ~ 45 minutes. (Note that unless it is specifically indicated as GRT, spacecraft time will be used throughout this chapter.) At this time the Pioneer 10 spacecraft was inbound toward Jupiter at a Jovicentric radial distance of 108.9 R_J ($R_J = 71,372$ km). The solar wind ion spectrum shown in Fig. 5 was taken in the interplanetary medium (the spectrum on the left) at 19^h 05^m UT GRT on November 26, 1973 (day 330), about 1 hour and 25 minutes before the Jovian bow shock crossing and the other ion spectrum (on the right) was taken in the Jovian magnetosheath at 4^h 51^m UT GRT on November 27, 1973, about 8 hours and 20 minutes after the shock crossing. In this figure the counts are the digitized output from the plasma analyzer per ion energy channel for the detector recording the peak. The ion fluxes are digitized to 9-bit accuracy (0–511). Although the ion characteristics in the magnetosheath were quite variable, the spectrum shown in Fig. 5 is considered to be typical. The ragged appearance of this spectrum is most likely due to fluctuations in the magnetosheath ion characteristics during the period required to obtain the spectrum and is therefore considered to be an artifact in the data, caused by sample aliasing. The observation of this drastic change in the ion spectral characteristics (Fig. 5) is interpreted as the encounter of the Pioneer 10 spacecraft

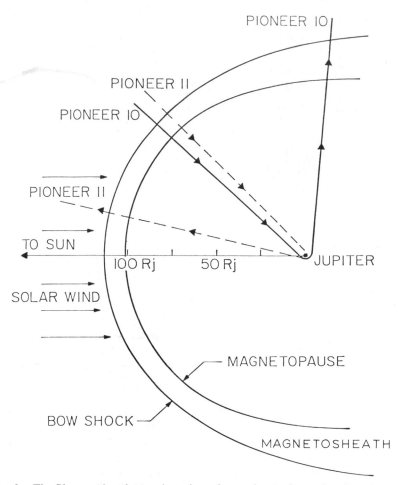

Fig. 3.   The Pioneer 10 and 11 trajectories referenced to Jupiter and projected onto a plane parallel to the ecliptic plane, with a schematic shape and with the location of the Jovian bow shock and magnetopause drawn in.

with a detached bow shock wave standing off from Jupiter's magnetosphere and in many respects is quite similar to the case at Earth.

For the interplanetary ion spectrum shown in Fig. 5 the proton peak is seen near 1 keV, and the doubly charged helium peak near 2 keV. This interplanetary spectrum corresponds to a solar wind convective speed of ∼ 441 km sec^{-1}, a proton number density of 0.12 cm^{-3}, and an isotropic proton temperature of $6.1 \times 10^4\,°K$. It should be noted that this solar wind speed and number density correspond to an anomalously low solar wind dynamic pressure (by about a factor of 4), compared with that normally observed by this experiment in the interplanetary medium near 5 A.U. (An anomalously

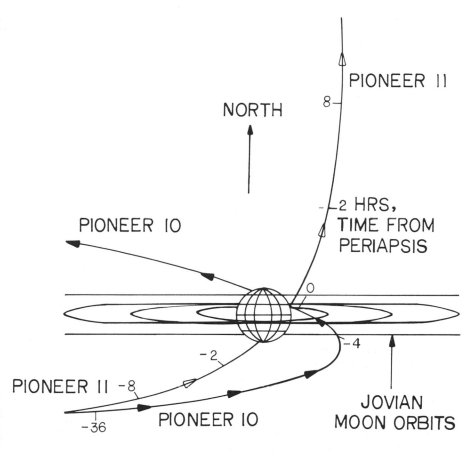

Fig. 4. The Pioneer 10 and 11 trajectories referenced to Jupiter as seen from Earth.

low solar wind dynamic pressure was also observed by Pioneer 11 just upstream from the Jovian bow shock.) The ion distribution parameters for this first magnetosheath traversal were mostly obtained from Detector A, the high-resolution analyzer. The large flow angle ($\sim 40°$) in the magnetosheath plasma flow direction with respect to the spacecraft spin axis and the high plasma temperature and attendant low density precluded obtaining reliable measurements from the medium-resolution analyzer. This large deflection in flow direction, from approximately anti-sunward to a large angle with respect to the spin axis, was observed as the spacecraft crossed the bow shock and, with the exception of a 220 minute period commencing at $\sim 5^h$ 00^m UT on November 27, 1973, persisted throughout the entire Pioneer 10 magnetosheath traversal. During the exceptional 220 minute period, for example between $8^h 00^m$ and $9^h 00^m$ UT on November 27, 1973, the flow directions were both toward the center of the plasma analyzer acceptance angle,

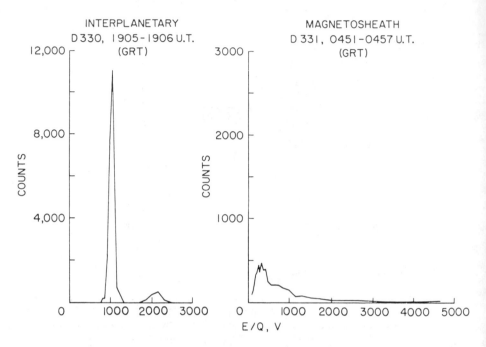

Fig. 5.   Comparison of the plasma ion spectra taken upstream and downstream from Jupiter's bow shock for the inbound portion of the Pioneer 10 flyby of Jupiter in 1973.

and in addition, the plasma currents were enhanced. Average distribution parameters of the magnetosheath with a bulk speed of 273 km sec^{-1}, a proton number density of 0.62 cm^{-3}, and a proton temperature of 3.5×10^5 °K were calculated for this time. The distribution parameters for the magnetosheath spectrum of Fig. 5 are similar; here the bulk velocity is ~ 191 km sec^{-1}, and the isotropic temperature is $\sim 2 \times 10^5$ °K. The magnetosheath flow field characteristics are discussed in more detail below.

The first Jupiter bow shock crossing by Pioneer 11 was observed at 3^h 39^m on November 26, 1974 at a planet-centered distance of 109.7 R_J. At this time the solar wind bulk speed (spacecraft frame of reference) decreased from 480 to 328 km sec^{-1} with a concurrent shift of the flow direction of $\sim 50°$ and a proton isotropic temperature increase from $\sim 2 \times 10^4$ to 5×10^5 °K. The transition time for the proton bulk speed is longer than 10 minutes in this case. The time for changes in the proton bulk speed at these shock crossings generally seems to be longer than the times for changes in the flow direction and temperature. The proton number density increased by at least a factor of 2.5 across the shock transition from an upstream value of 0.06 cm^{-3}. The anomalously low dynamic pressure of the upstream solar wind and the plasma and magnetic field characteristics on both sides of the bow shock indicate that the initial Pioneer 10 and 11 bow shock crossings are consistent with the expansion of the Jovian magnetosphere and its associated bow shock overtaking the spacecraft. Smith *et al.* (1974*a,b*, 1975)[1] have identified these shocks as perpendicular shocks. Intriligator and Smith (Smith *et al.* 1975), using simultaneous plasma and magnetic data, have estimated the speed of these shocks by employing the relation

$$V_{SHOCK} = \frac{(B_2 V_2 - B_1 V_1)}{(B_2 - B_1)} \tag{1}$$

where B is the fieldstrength and V is the solar wind speed; the subscript "2" refers to downstream and "1" to upstream. The Pioneer 11 bow shock speed is then ~ 100 km sec^{-1}. Mihalov *et al.* (1976) have studied the shock jump conditions in detail and found that the bow shocks encountered on Pioneers 10 and 11 have a high Alfvén Mach number similar to that seen at the earth (Wolfe and Intriligator 1970).

Half-hour averages of the proton bulk velocities, number densities, and isotropic temperatures corresponding to the first crossing of Jupiter's bow shock and magnetosheath on Pioneers 10 and 11 are shown in Fig. 6. The deflection on the Jovian magnetosheath was so large that the peak of the proton distribution was seen in the farthest out Channeltron. As a result, although one can compute reliable proton velocities and temperatures in this region, one can only compute lower limits for the proton number densities except for the few hours (indicated above) when the flow was not as highly deflected. The large deflection in the Jovian magnetosheath is much larger than the analogous case in the earth's magnetosheath and implies a very blunt shape for the Jovian magnetosphere.

The inbound magnetopause crossings are identified by the disappearance of detectable flowing plasma. The first inbound magnetopause crossing observed on Pioneer 10 was at $\sim 19^h 53^m$ UT on November 27, 1973 at a distance of 96.36 R_J. The first inbound magnetopause crossing observed on Pioneer 11 was at $\sim 2^h 45^m$ UT on November 27, 1974 at a distance of 97.3 R_J. During the Pioneer 10 and 11 inbound and outbound trajectories multiple crossings were observed of the Jovian bow shock and magnetopause. These Pioneer 10 and 11 events are summarized in Tables I and II, respectively. Tables I and II indicate that during the inbound trajectories, Jupiter's magnetopause was observed to be as far out as 97 R_J and as close as 46 R_J from the planet. This change of a factor of 2 in size indicates that the Jovian magnetosphere is extremely responsive in size to relatively minor changes in the solar wind pressure. These observations are also indicative of the inflated spongy character of Jupiter's outer magnetosphere.

Intriligator (1975*a,b*) and Intriligator and Wolfe (1974) have summarized some of the Pioneer 10 plasma electron observations in the Jovian magnetosheath and the outer magnetosphere. Figure 7 shows two examples of plasma

---

[1]See p. 823.

electron spectra obtained in the Jovian magnetosheath. In this figure the
electron counts are the digitized output from the plasma analyzer per energy
channel for the collector recording the peak. The electron fluxes are digitized
to 9-bit accuracy (0–511) covering (logarithmically) the dynamic range from
approximately 10^{-14} to 10^{-9} amperes cm^{-2}. [The paper by Intriligator and
Wolfe (1974) has a more detailed discussion of the electron measurements
and the uncertainties involved.] The magnetosheath spectrum on November 27 in Fig. 7 was obtained during the first extended magnetosheath traversal on Pioneer 10. The important feature in this spectrum is the presence
of the enhanced high-energy tail of the spectrum between ~ 50 eV and 200
eV. The existence of this high-energy tail indicates that there is nonthermal
heating of the solar wind electrons in the magnetosheath. This heating is
similar to that observed for the case of the earth's bow shock (Wolfe and
Intriligator 1970). The ions, on the other hand, although heated across the
Jovian bow shock, appear to have very little nonthermal component (Mihalov
et al. 1976). The magnetosheath spectrum on December 1 in Fig. 7 was obtained during the second extended inbound magnetosheath traversal on
Pioneer 10. The spacecraft reentered the Jovian magnetosheath at a radial
distance of 54 R_J and the magnetosheath ion flow field persisted for approximately eleven hours and was abruptly terminated at a radial distance of 46.5
R_J. The electron spectrum on December 1 shown in Fig. 7 is typical of the
electron spectra obtained during the second magnetosheath traversal. The
greatly enhanced high-energy tail in this spectrum, as compared to the
magnetosheath spectrum on November 27, indicates that the electron temperatures were significantly higher during the December 1, 1973 magnetosheath traversal than during the earlier traversal. This increase in electron
temperature is qualitatively consistent with what one would expect for a
contraction of the Jovian magnetosphere due to an increase in the solar
wind dynamic pressure as has been observed for the case of the earth's bow
shock (Spreiter and Alksne 1969). Moreover, the electron temperature increase and the simultaneous changes in the plasma ions and the magnetic
field during the second magnetosheath traversal all imply the importance of
compressional effects due to an increase in the external solar wind dynamic
pressure rather than a reorientation of a thin magnetodisk due to only a
directional change in the solar wind. A quantitative analysis of the contraction of the magnetosphere must await more detailed knowledge of the topology of the Jovian magnetosphere which is impossible to ascertain from only
two flybys.

Figures 8 and 9 summarize the bow shock crossings (the circles) and the
magnetopause crossings (the squares) observed on Pioneer 11 both inbound
and outbound. The upper graph in each of these figures gives a projection
on Jupiter's equatorial plane as viewed from the north. The lower graph
gives an orthogonal projection on a plane that contains the Jupiter-Sun direction. In Fig. 8 the arrows indicate the vector velocity (both magnitude and

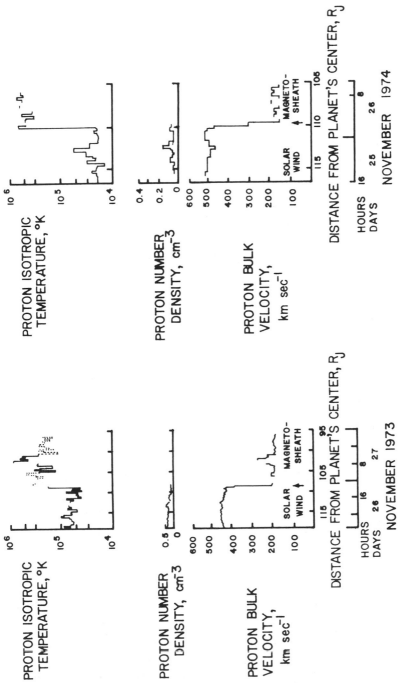

Fig. 6. Half-hour averages of the plasma proton parameters measured just upstream and downstream of the first inbound crossing of the Jovian bow shock on Pioneer 10 (data on the left) in 1973, and on Pioneer 11 (data on the right) in 1974. The times of the bow shock crossings are indicated with arrows. The temperatures indicated with dashes in the Pioneer 10 data are approximate values derived from scans through the velocity distribution that do not include the peak. The Pioneer 11 data are preliminary and the spacecraft velocity has not been removed from the speed measurements.

TABLE I
Magnetosheath Boundaries Observed by the Pioneer 10 Ames Research Center Plasma Analyzer Experiment; S, Shock; M, Magnetopause.

Boundary observation	Spacecraft time		Distance (R_J)
	1973 date	UT	
---------------------Inbound---------------------			
S	26 Nov	$19^h\ 46^m\ \pm 2^m$	108.90
M	27 Nov	$19\ 53\ \pm 2$	96.36
M	1 Dec	$02\ 33\ \pm 6$	54.32
M	1 Dec	$13^h\ 35^m_.7 \pm 2^m_.2$	46.50
---------------------Outbound---------------------			
M	10 Dec	$11^h\ 53^m_.4 \pm 0^m_.5$	97.92
M	12 Dec	$09\ 43.2 \pm 0.5$	121.52
M	12 Dec	$09\ 58.2 \pm 0.5$	121.66
S	12 Dec	$14\ 53.2 \pm 1.5$	124.14
S	12 Dec	$19\ 50.7 \pm 6$	126.64
M	13 Dec	$01\ 58.1 \pm 0.5$	129.73
M	14 Dec	$18\ 50\ \pm 1$	150.08
S	18 Dec	$03\ 28\ \pm 1$	188.87
S	20 Dec	$21\ 45.1 \pm 0.5$	220.54
S	20 Dec	$22\ 33.9 \pm 0.5$	221.41
S	21 Dec	$02\ 12\ \pm 2$	223.13
S	21 Dec	$06\ 43\ \pm 2$	225.27
S	21 Dec	$10\ 27\ \pm 2.8$	227.04
S	21 Dec	$11\ 58\ \pm 2$	227.76
S	21 Dec	$18\ 48.5 \pm 9.5$	230.99
S	21 Dec	$19\ 29.2 \pm 2.9$	231.31
S	22 Dec	$06\ 05\ \pm 10$	236.31
S	22 Dec	$17\ 57.6 \pm 1.8$	241.44
S	22 Dec	$18\ 05\ \pm 1$	241.97
S	22 Dec	$18\ 11.8 \pm 1.5$	242.02
S	22 Dec	$18\ 15.7 \pm 0.1$	242.05
S	22 Dec	$19^h\ 28^m_.0 \pm 2^m_.7$	242.62

Note: Magnetosheath-type bursts at $02^h\ 25^m$ and $13^h\ 45^m$, December 1, and $09^h\ 47^m_.7$, December 12. Greatly reduced magnetosheath plasma flux at $13^h\ 24^m_.8 \pm 0^m_.5$, December 1.

direction) of the magnetosheath flow field. Using the plasma parameters to locate the position of the magnetopause, the (vector) normal to the magneto-pause was calculated using the magnetic field parameters (graciously provided by E. J. Smith). These normals to the magnetopause are indicated by the arrows in Fig. 9. The magnetosheath flow field indicated in Fig. 8 and the

TABLE II

Magnetosheath Boundaries Observed
by the Pioneer 11 Ames Research Center
Plasma Analyzer Experiment;
S, Shock; M, Magnetopause.

Boundary observation	Spacecraft time		Distance (R_J)
	1974 date	UT	
---------------------Inbound----------------------			
S	26 Nov	$03^h \ 39^m.3 \pm 0^m.9$	109.7
M	27 Nov	02 46 $+ 2$	97.3
M	27 Nov	07 52 ± 26	94.5 ± 0.2
S	27 Nov	13 06.1 ± 0.1	91.6
S	28 Nov	14 35.5 ± 0.3	77.5
M	29 Nov	$13^h \ 18^m.7 \pm 0^m.8$	64.5
--------------------- Outbound ---------------------			
M	6 Dec	$08^h \ 06^m.5 \pm 1^m.0$	56.6
M	6 Dec	18 28 ± 2.8	62.7
M	8 Dec	00 36 ± 3.5	80.0
S	8 Dec	20 14.5 ± 0.8	90.8
S	9 Dec	02 56.3 ± 1.2	94.5
S	9 Dec	$03^h \ 46^m.0 \pm 0^m.1$	95.0

normals to the magnetopause indicated in Fig. 9 each emphasize the extremely thick, blunt shape of the Jovian magnetosphere.

III. MAGNETOSPHERIC OBSERVATIONS

At $20^h \ 38^m$ UT on November 27, 1973 Pioneer 10 crossed the Jovian magnetopause at a radial distance of 96 R_J and entered the outer Jovian magnetosphere, where the electron component of a thermal plasma was measured at an energy of a few eV (Intriligator and Wolfe 1974; Intriligator 1975b). Figure 10 shows an example of an electron spectrum measured in the outer magnetosphere. This spectrum was obtained at ~ 96 R_J on November 27, 1973 after the first extended magnetosheath traversal. The peak evident here near 4 eV was consistently observed throughout the entire outer ($\geq 15 \ R_J$) Jovian magnetosphere. We cannot conclude the existence of a thermal plasma in the inner magnetosphere where the background due to energetic charged particles precludes their observation. If in the Jovian magnetosphere there had been a co-rotating plasma it would not have been included in the field of view of the plasma analyzer experiment since, as indicated in Fig. 2, the spacecraft antenna limits the field of view of the experiment.

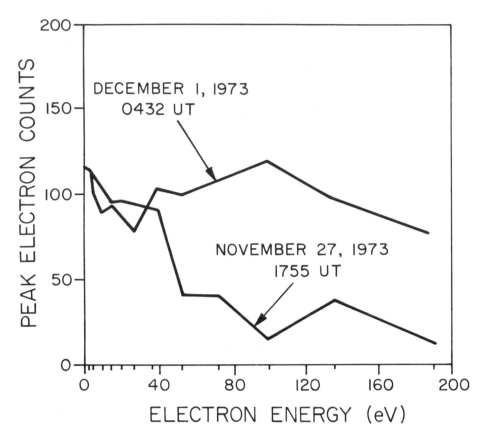

Fig. 7.   Electron spectra taken in the Jovian magnetosheath. The vertical axis indicates the
electron counts, the digitized output from the plasma analyzer per energy channel for the
collector recording the peak (see text). Electron fluxes are digitized to nine-bit accuracy
(0–511) covering (logarithmically) the dynamic range from approximately 10^{-14} to 10^{-9} amperes of the energy of the electrons in volts. The small vertical lines on the horizontal axis
indicate the locations of the individual electron energy channels.

If one assumes that the magnetopause boundary is a tangential discontinuity then the pressure balance across it can be calculated. For the
inbound case the thermal component of the magnetospheric electrons was
consistently observed near 4 eV and associated with a temperature of $\sim 5 \times$
10^{4} °K. Assuming $T_e \sim T_i$, the magnetic field values reported by Smith *et al.*
(1974*a*) and the magnetosheath ion parameters (where they could be determined) reported by Wolfe *et al.* (1974*b*) imply that the dayside magnetosphere has a plasma beta near unity corresponding to a number density of a
few particles cm^{-3}. It is cautioned that the above value of beta and the
corresponding number density is considered to be an upper limit since the
possible magnetospheric pressure contribution from the observed nonthermal
plasma electrons and unobservable energetic electrons between 500 eV and
~ 50 keV has not been accounted for.

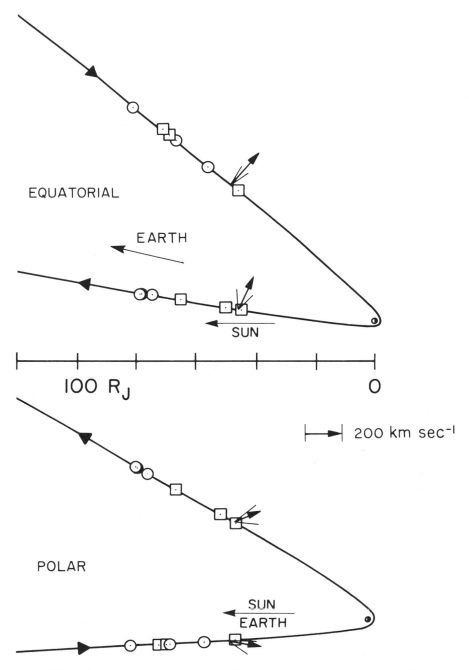

EQUATORIAL

EARTH

SUN

100 R$_J$

0

200 km sec^{-1}

POLAR

SUN

EARTH

Fig. 8. The (vector) velocity magnetosheath flow field observed on Pioneer 11. The upper graph gives a projection of the Pioneer 11 trajectory on Jupiter's equatorial plane, viewed from the north. The lower graph gives an orthogonal projection of the trajectory on a plane that contains the Jupiter-Sun direction. Locations of the bow shock crossings (circles) and magnetopause crossings (squares) are also shown.

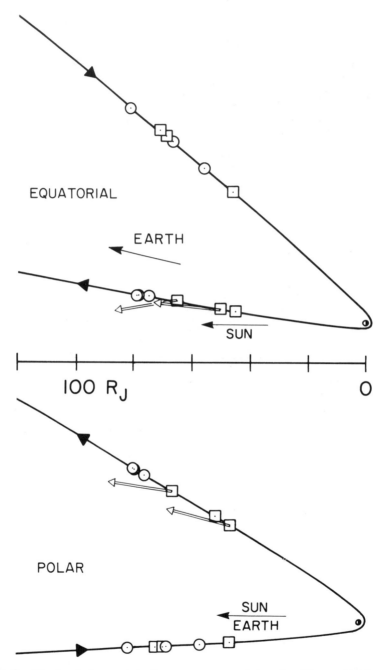

EQUATORIAL

EARTH

SUN

100 R$_J$ 0

POLAR

SUN
EARTH

Fig. 9.   The (vector) normals to the magnetopause observed on Pioneer 11. The upper graph
gives a projection of the Pioneer 11 trajectory on Jupiter's equatorial plane, viewed from the
north. The lower graph gives an orthogonal projection of the trajectory on a plane that contains the Jupiter-Sun direction. Locations of the bow shock crossings (circles) and magnetopause crossings (squares) are also shown.

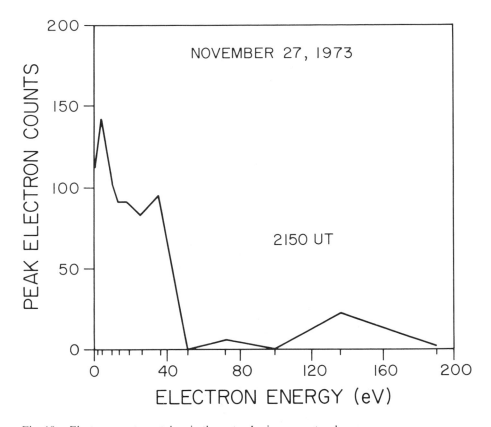

Fig. 10.   Electron spectrum taken in the outer Jovian magnetosphere.

Table III gives estimates of magnetospheric plasma properties obtained for Pioneer 10 from two different methods (Wolfe *et al.* 1974*b*). The first assumes pressure balance across the magnetopause, expressed by

$$n_1 k(T_{e1} + T_{i1}) + \frac{B_1^2}{2\mu_0} = n_2 k(T_{e2} + T_{i2}) + \frac{B_2^2}{2\mu_0}. \tag{2}$$

The second method uses the aerodynamic analogy [cf. Spreiter *et al.* 1966] and is described below. In the above pressure balance equation the subscripts 1 and 2 refer to magnetosheath and magnetosphere parameters, respectively; n is the plasma ion number density; T_e and T_i are the electron and ion temperatures, respectively; B is the magnetic field magnitude; k is the Boltzmann constant; and μ_0 is the magnetic permeability of free space, equal to $4\pi \times 10^{-7}$ H m^{-1}; $T_e \sim T_i$ is assumed; and $T_{i2} \sim 5 \times 10^4\,°$K ($\sim 4$ eV electrons) is assumed on the basis of the measured magnetospheric electron energy spectra.

The magnetosheath values for the December 13, 1973 crossing are less reliable than the other Pioneer 10 values owing to the divergence of the

TABLE III

Estimated Plasma Properties of the Jovian Magnetosphere on the Assumption of Pressure Balance Across the Magnetopause

Date 1973	Time (hours, UT)	Magnetosheath			Magnetosphere		
		Thermal Pressure (dyn cm^{-2} $\times 10^{-11}$)	Magnetic Energy Density (ergs cm^{-3} $\times 10^{-11}$)	Calculated Thermal Pressure (dyn cm^{-2} $\times 10^{-11}$)	Calculated Beta	Calculated Ion Number Density, cm^{-3}	Magnetic Energy Density (ergs cm^{-3} $\times 10^{-11}$)
Dec. 10	11^h 53^m	12	44	12	0.28	8.4	44
Dec. 13	01^h 58^m	0.4[a]	2.9	0.4	0.13 / 0.2[b]	0.27 / 0.43[b]	2.9
Dec. 14	18^h 50^m	7.1	4.1	7.6	2.1 / 2.8[b]	5.5 / 7.3[b]	3.6

[a]Measured value that is not too reliable.
[b]Estimated by using aerodynamic analogy.

plasma flow direction near the outer limit of angular acceptance detector B, the medium-resolution detector. The December 10 crossing appeared to occur at a time of extreme conditions. Also, the observed magnetic field profile across the magnetopause appears as if a wide layer were crossed (E. J. Smith, personal communication 1974). The results in Table III obtained with the aerodynamic analogy use Pioneer 10 free-stream plasma parameters, however, and ignore the fields outside this "layer" since they are obtained using the values closest in time to the indicated magnetopause crossings. The time delay between the magnetospheric and the free-stream measurements by Pioneer 10 makes this method less reliable owing to neglect of possible time variations in the external free-stream conditions. The assumed condition is

$$Kmn_1{}^*v^{*2}\cos^2\theta + n_1{}^*K(T_{e1}{}^* + T_{i1}{}^*) + \frac{B_1{}^{*2}}{2\mu_0} = n_2k(T_{e2} + T_{i2}) + \frac{B_2{}^2}{2\mu_0} \quad (3)$$

where m is the proton mass, v is the bulk velocity, the asterisks denote free-stream quantities, K is taken as unity, and θ is the angle between the magnetopause normal and the free-stream plasma flow direction. Angle θ is obtained from calculations made for the case at Earth given by Spreiter et al. (1966). The values of θ were tested against calculated values obtained by using the magnetic field measured across the magnetopause (E. J. Smith, personal communication, 1974) and by assuming that the magnetopause is a tangential discontinuity; the earth analogy values were much larger than the calculated values. This result also implies a magnetopause body shape more blunt than that of the earth.

It is important to discuss the validity of the magnetospheric observations obtained during the Jupiter flyby of Pioneer 10. It is recognized that the measurement of low-energy charged particles (of a few eV) is exceedingly difficult and subject to error primarily due to spacecraft potential effects. This is particularly true for the case of Jupiter's magnetosphere where the spacecraft is subjected to high intensities of energetic charged particles (Fillius and McIlwain 1974;[2] Simpson et al. 1974;[3] Trainor et al. 1974a,b;[4] Van Allen et al. 1974[5]). In general, as is the case at Earth, spacecraft charge build-up produces an unknown perturbing effect on these low-energy electron measurements which is different for each spacecraft. At least in the outer portion of the Jovian magnetosphere, however, the consistency of the 4 eV electron peak argues in favor of the dominance of the spacecraft potential by the ambient thermal plasma and the photoelectrons since if the spacecraft potential were changing one would expect the 4 eV peak to be affected. Note that at 5 A.U. the flux of photoelectrons from the spacecraft is below the instrument threshold. In the inner Jovian magnetosphere, however,

[2]See pp. 896 and 1193.   [4]See p. 961.
[3]See p. 738.   [5]See p. 928; also see p. 891.

where thermal electron measurements are obscured by high background, the possibility of spacecraft charge build-up cannot be excluded. It is argued, nevertheless, that large charge build-up probably did not occur on Pioneer 10 since the effects to various spacecraft systems caused by arcing were not observed when the spacecraft passed into the Jovian shadow. This problem is complicated but one can speculate that if there had been a large charge build-up on the spacecraft before it entered the shadow of the planet then in the absence of the photoelectrons, during solar occultation, arcing might have occurred if there were no thermal plasma present to neutralize the charge build-up. In the presence of a thermal plasma, however, the spacecraft cannot be charged to a potential much higher than the local plasma potential. Thus, this may imply the existence of a thermal plasma in the inner Jovian magnetosphere. It is also tantalizing to speculate that in addition to the solar wind, the thermal plasma observed in the outer Jovian magnetosphere may have its origin in the inner Jovian magnetosphere and perhaps the ionosphere. It should also be noted that during the inbound portion of the flyby trajectory of Pioneer 11 there was some evidence of the temporary effects to spacecraft systems of arcing due to spacecraft charge build-up, but this occurred only near Europa.

An extensive investigation of the plasma distribution in the inner magnetosphere has been made by Frank *et al.* (1976) (see Fig. 11) using data from the Plasma Analyzer Experiment on Pioneer 10. The difficulties encountered in subtracting the background due to energetic charged particles, however, lead to large uncertainties with regard to the absolute densities. Spacecraft charging could also lead to uncertainties; however, there is no strong evidence for this effect for the Pioneer 10 flyby. The evidence for spacecraft charging on Pioneer 11, in contrast to the lack of definite evidence on Pioneer 10, might possibly be due to the much higher magnetic latitude of the Pioneer 11 trajectory and the different position of Europa itself for the Pioneer 11 flyby as compared to its position during the Pioneer 10 flyby. However, despite these uncertainties Fig. 11 illustrates a magnetospheric plasma distribution model which fits the data.

The observations of the electron component of the thermal plasma and the inferred ion component are consistent with this thermal plasma being the primary controlling factor causing the inflation of Jupiter's outer magnetosphere. It is interesting to note that the Pioneer 10 observations made in the daylight hemisphere for Jupiter's magnetosphere are more reminiscent of the case in the earth's magnetotail. That is, in the outer ($\geq 15\ R_J$) Jovian dayside hemisphere the magnetic fieldlines are stretched out away from the planet (Smith *et al.* 1974*a*) and there is a high-beta plasma so that in this respect it is similar to the earth's magnetotail (Wolfe and Intriligator 1970). It is also important to recognize that these thermal electrons were observed everywhere in the outer Jovian magnetosphere and not simply confined to the disk-like configuration observed for the energetic ($\gtrsim 5$ MeV) electrons

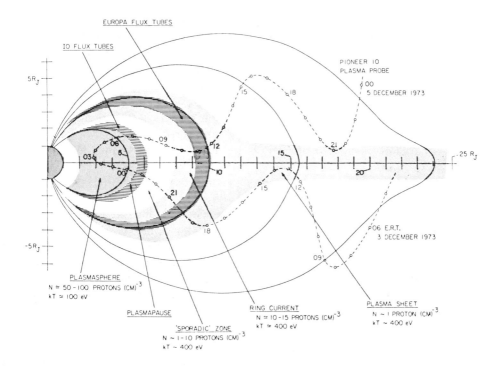

Fig. 11.   Summary of major plasma features deep within the Jovian magnetosphere as viewed in a magnetic meridional plane.

(Fillius and McIlwain 1974; Simpson *et al.* 1974; Trainor *et al.* 1974*a,b*; Van Allen *et al.* 1974). The energetic electrons observed in the outer magnetosphere appear to be modulated by the planet's period of rotation (Fillius and McIlwain 1974; Simpson *et al.* 1974; Trainor *et al.* 1974*a,b*; Van Allen *et al.* 1974). Trainor *et al.* (1974*a*) reported, however, that the ∼ 10 hour periodicity in the outer magnetosphere was not nearly as significant for the lower-energy electrons ($\lesssim 5$ MeV) or for the protons. Since the plasma electrons do not seem to be associated with the periodic behavior observed for the energetic ($\gtrsim 5$ MeV) electrons this also supports arguments in favor of a thick magnetosphere as opposed to a "magneto-disk" implied by observation of the Pioneer 10 energetic electrons alone.

IV.   CONCLUSIONS

The plasma electron and ion observations by the Ames Research Center Plasma Analyzer Experiments on Pioneers 10 and 11 are completely consistent and indicate that the Jovian magnetosphere is very blunt in shape and extremely responsive in size to changes in the solar wind pressure. The observations show that the interaction of Jupiter's magnetic field with the solar wind is similar in many ways to that at Earth, but that the scale size is over 100 times larger. Jupiter is found to have a detached standing bow shock wave of high Alfvén Mach number. Like the earth, Jupiter has a prominent magnetopause that deflects the magnetosheath plasma and excludes its direct entry into the Jovian magnetosphere. Unlike that of Earth, the sunward hemisphere of Jupiter's outer magnetosphere is found to be highly inflated with thermal plasma and a high-beta ($\beta \simeq 1$) region that is highly responsive to changes in dynamic pressure of the solar wind. The shape of Jupiter's magnetosphere is found to be extremely thick and blunt.

Acknowledgements. This work was supported by NASA Ames Research Center and the University of Southern California. We are indebted to C. F. Hall and the Pioneer Project Office for their outstanding management and operations throughout the Pioneer 10 and 11 missions; to TRW Systems Group for their building of the spacecraft; and to Time Zero Laboratories for construction of the Ames Research Center Plasma Analyzer Experiments. We also thank our colleagues H. R. Collard, D. D. McKibbin, and J. D. Mihalov at NASA Ames Research Center and L. A. Frank at the University of Iowa for their contributions to the success of this experiment.

REFERENCES

Fillius, R. W., and McIlwain, C. E. 1974. Radiation belts of Jupiter. *Science* 183:314–315.

Frank, L. A.; Ackerson, K. L.; Wolfe, J. H.; and Mihalov, J. D. 1976. Observations of plasma in the Jovian magnetosphere. *J. Geophys. Res.* 81:457–468.

Intriligator, D. S. 1975a. Pioneer 10 observations of the solar wind and its interaction with Jupiter: plasma electron results. *The magnetospheres of the earth and Jupiter* (V. Formisano, ed.), pp. 297–300. Dordrecht, Holland: D. Reidel Publ. Co.

———. 1975b. Pioneer 10 observations of the Jovian magnetosphere: plasma electron results. *The magnetospheres of the earth and Jupiter* (V. Formisano, ed.), pp. 313–316. Dordrecht, Holland: D. Reidel Publ. Co.

Intriligator, D. S., and Wolfe, J. H. 1974. Initial observations of plasma electrons from the Pioneer 10 flyby of Jupiter. *Geophys. Res. Lett.* 1:281–284.

Mihalov, J. D.; Collard, H. R.; McKibbin, D. D.; Wolfe, J. H.; and Intriligator, D. S. 1975. Pioneer 11 encounter: preliminary results from the Ames Research Center plasma analyzer experiment. *Science* 188:448–451.

Mihalov, J. D.; Wolfe, J. H.; and Frank, L. A. 1976. Survey for non-maxwellian plasma in the Jovian magnetosheath. *J. G. R.* (special Jupiter issue).

Simpson, J. A.; Hamilton, D.; Lentz, G.; McKibben, R. B.; Mogro-Campero, A.; Perkins, M.; Pyle, K. R.; Tuzzolino, A. J.; and O'Gallagher, J. J. 1974. Protons and electrons in Jupiter's magnetic field: results from the University of Chicago experiment on Pioneer 10. *Science* 183:306–309.

Smith, E. J.; Davis, Jr., L.; Jones, D. E.; Colburn, D. S.; Coleman, Jr., P. J.; Dyal, P.; and Sonett, C. P. 1974a. Magnetic field of Jupiter and its interaction with the solar wind. *Science* 183:305–306.

Smith, E. J.; Davis, Jr., L.; Jones, D. E.; Colburn, D. S.; Coleman, Jr., P. J.; Dyal, P.; Sonett, C. P.; and Frandsen, A. M. A. 1974b. The planetary magnetic field and magnetosphere of Jupiter: Pioneer 10. *J. Geophys. Res.* 79:3501–3513.

Smith, E. J.; Davis, Jr., L.; Jones, D. E.; Coleman, Jr., P. J.; Colburn, D. S.; Dyal, P.; and Sonett, C. P. 1975. Jupiter's magnetic field, magnetosphere, and interaction with the solar wind: Pioneer 11. *Science* 188:451-455.

Spreiter, J. R., and Alksne, A. Y. 1969. Plasma flow around the magnetosphere. *Magnetospheric physics* (D. J. Williams and G. D. Mead, eds.), Vol. 7, pp. 11-50. Washington, D. C.: American Geophysical Union.

Spreiter, J. R.; Summers, A. L.; and Alksne, A. Y. 1966. Hydromagnetic flow around the magnetosphere. *Planet. Space Sci.* 14:223-253.

Trainor, J. H.; McDonald, F. B.; Teegarden, J. H.; Webber, W. R.; and Roelof, E. C. 1974*a*. Energetic particles in the Jovian magnetosphere. *J. Geophys. Res.* 79:3600-3613.

Trainor, J. H.; Teegarden, B. J.; Stilwell, D. E.; McDonald, F. B.; Roelof, E. C.; and Webber, W. R. 1974*b*. Energetic particle population in the Jovian magnetosphere. A preliminary note. *Science* 183:311-313.

Van Allen, J. A.; Baker, D. N.; Randall, B. A.; Thomsen, M. F.; Sentman, D. D.; and Flindt, H. R. 1974. Energetic electrons in the magnetosphere of Jupiter. *Science* 183:309-311.

Wolfe, J. H.; Collard, H. R.; Mihalov, J. D.; and Intriligator, D. S. 1974*a*. Preliminary Pioneer 10 encounter results from the Ames Research Center plasma analyzer experiment. *Science* 183:303-304.

Wolfe, J. H., and Intriligator, D. S. 1970. The solar wind interaction with the geomagnetic field. *Space Sci. Rev.* 10:511-596.

Wolfe, J. H.; Mihalov, J. D.; Collard, H. R.; McKibbin, D. D.; Frank, L. A.; and Intriligator, D. S. 1974*b*. Pioneer 10 observations of the solar wind interaction with Jupiter. *J. Geophys. Res.* 79:3489-3500.

PLASMA PHYSICS AND WAVE-PARTICLE
INTERACTIONS AT JUPITER

F. L. SCARF
Space Sciences Department of TRW Systems

The dynamics of the inner magnetosphere of Jupiter is conceptually similar to that of the earth, despite the vast difference in size and in energy of the trapped particles. However, the satellites provide important sources of plasma and field-aligned currents, and significantly affect the trapped particle populations. It appears that in this region the trapped electron fluxes are near the stable limit set by whistler mode wave-particle interactions; however, the proton fluxes may be controlled by other instabilities. The whistler waves cause pitch-angle diffusion, and the precipitating electrons should significantly affect the ionospheric properties and influence ionosphere-magnetosphere coupling. It is expected that enhanced precipitation and wave-induced anomalous conductivity develop along fieldlines threading the inner satellites, and strong plasma wave-radiation field coupling must account for the intense levels of decametric emissions. The outer magnetosphere of Jupiter is dominated by the high-β ($\beta = 4\pi N\kappa T B^{-2}$) plasma spun out by centrifugal forces. Here wave-particle interactions can provide local acceleration, they can affect particle diffusion, and they can lead to fieldline merging. If Jupiter's magnetosphere flows radially outward to form a planetary wind, wave-particle interactions should lead to a second collisionless shock within the magnetopause. Other relevant wave-particle interactions involve whistlers generated by atmospheric lightning, and electron plasma oscillations associated with suprathermal particles in the upstream solar wind and magnetosphere; local measurement of the wave frequency for the electron plasma oscillations provides a sheath-independent determination of the total plasma density.

Knowledge about Jupiter's magnetosphere first became available during the early days of exploration of the earth's Van Allen belts, when remote measurements of decimeter emissions from Jupiter were made (Drake and Hvatum 1959) and interpreted in terms of synchrotron radiation from an extremely energetic electron population trapped in a magnetic field considerably stronger than the earth's field. In a very general sense, these facts already suggested that Jupiter was immersed in the streaming solar-wind plasma; since the extrapolated wind pressure at 5.2 A.U. is only 1/27 of the value at 1 A.U., this conclusion implied that the Jupiter magnetosphere must be enormous in comparison with that of the earth (Carr and Gulkis 1969; Scarf 1969). At this stage of investigation it was also generally concluded that the intense decametric radio bursts from Jupiter (Burke and Franklin 1955) were probably associated with local phenomena that developed at relatively low altitudes, perhaps in the ionosphere where the local fieldstrength is higher than in the heart of the radiation belt. With the discovery of the Io modulation effect on the decametric emissions (Bigg 1964), the concept of strong ionosphere-magnetosphere coupling at Jupiter gained acceptance and it was clear that for Io, and perhaps for other inner satellites, strong satellite-magnetosphere interactions were operative (see Warwick 1970).

Once the existence of a large dipolar magnetic field at Jupiter was established, a number of theoreticians noted that the strong field and high planetary rotation rate would lead to properties very different from those found at Earth. It was observed that centrifugal forces would dominate the plasma configuration beyond a few Jupiter radii (Ellis 1965; Melrose 1967; Gledhill 1967; Piddington 1967; Brice and Ioannidis 1970), and well before the Pioneer 10 and 11 encounters these theoretical speculations suggested that Jupiter's magnetosphere has characteristics similar to those used in constructing models of rapidly rotating stellar objects, such as pulsars. In fact, even without any local spacecraft measurements an Earth-based observer could now conclude that the plasma environment of Jupiter has much in common with various astrophysical bodies; charged particles are clearly accelerated to relativistic energies in the magnetosphere, and cosmic rays from Jupiter (Pizella and Venditti 1973) as well as radio emissions are detectable from Earth with intensity modulations correlated with Jupiter's rotation rate.

Major advances in understanding the overall configuration of Jupiter's magnetosphere stem from the work of Brice and his associates, who first developed quantitative models of the plasma density distribution for a rapidly rotating magnetized planet with sufficiently strong gravity to preclude escape of cold ionospheric plasma into the magnetosphere (Brice and Ioannidis 1970; Ioannidis and Brice 1971; Brice and McDonough 1973). While this research on the photo-electron "slingshot" model was being carried out it became evident that dynamical phenomena in the earth's magnetosphere are controlled to a major extent by the development of local plasma instabilities and wave-particle interactions. These ideas were first brought together in 1971, when detailed theoretical models were presented at the Jupiter Radiation Belt Workshop at the Jet Propulsion Laboratory (Brice 1972; Thorne and Coroniti 1972; Kennel 1972). It was assumed: (a) that Jupiter has a strong planetary dipole field (of order 1–10 gauss at the surface); (b) that the thermal plasma distribution is given by the Brice-Ioannidis slingshot distribution of exospheric photoelectrons; (c) that magnetosheath electrons and protons are injected with high values of $\mu = \kappa T_\perp / B$ across a porous high-β outer boundary resembling a plasmapause, rather than a magnetopause ($\beta = 4\pi N\kappa T B^{-2}$); and (d) that these high-μ particles convect and diffuse inward (conserving μ) to attain very high energies in the inner belt of Jupiter. It was assumed that the resultant trapped particle fluxes are limited by local gyroresonant ion cyclotron or whistler mode plasma instabilities (as on Earth) and by synchrotron radiation losses.

Based on knowledge derived from observations in the earth's magnetosphere, it was clear even before the Pioneer 10 and 11 encounters that several other microscopic plasma instabilities had great significance at Jupiter. Current driven instabilities involving ion-acoustic wave modes are especially important because these waves act on both ions and electrons and they provide an efficient mechanism for plasma heating and energy dissipation. This type of instability leads to thermalization at the bow shock, a possible process for dissipation of merging fieldlines, and an effective finite electrical conductivity that plays an important non-linear role in regulating the auroral field-aligned currents flowing between the ionosphere and magnetosphere.

At Jupiter, it is certain that all of the corresponding phenomena are important, with a number of additional applications. For instance, the satellite motions through the magnetospheric plasma must drive strong field-aligned currents that close in the ionosphere, and auroral-type phenomena are likely to be involved in the modulation of the decametric emissions. In another area, a second collisionless shock, with very novel current-driven instabilities, should develop within the plasmapause if the Jupiter thermal plasma does flow radially outward, resembling a miniature solar wind (Michel and Sturrock 1974; Kennel and Coroniti 1975).

Wave-wave interactions are also of major importance at Earth and Jupiter. It has long been known that the decametric radiation levels are much too high to be explained in terms of incoherent radiation from any reasonable particle population. Similar conclusions are now advanced for the terrestrial analog, the very intense kilometric noise bursts that appear to originate one to two Earth radii above discrete auroral arcs (Gurnett 1974). In both applications it is concluded that local plasma waves organize the radiating particles and greatly enhance the coupling to the radiation field.

In the following sections of this chapter the wave mode classification is discussed briefly using near-Earth observations, and these ideas on wave-particle interactions at Jupiter are re-examined on the basis of post-encounter information from Pioneers 10 and 11.

I. BACKGROUND: WAVE MODES AND PLASMA PARAMETERS AT EARTH

Plasma wave modes of importance in magnetospheric physics are related to the electron and ion gyrofrequencies, $f_c^\pm = eB/2\pi m_\pm c$, and to the electron and ion plasma frequencies, $f_p^\pm = (4\pi N e^2/m_\pm)^{\frac{1}{2}}/2\pi$, where N is the plasma density and B is the local magnetic fieldstrength. As an example of the frequency range for typical upstream solar-wind conditions near the earth, these expressions give $f_p^- = 20$ kHz, $f_p^+ = 470$ Hz, $f_c^- = 140$ Hz, and $f_c^+ = 0.076$ Hz, with $N = 5$ cm^{-3}, $B = 5$ gamma. These modes can also be described as generalized electromagnetic waves (having both E and B wave components), or as electrostatic oscillations (similar to compressional sound waves, with space charge variations that produce only electric field wave components). For instance, the whistler mode ($f \leq f_c^-$) is electromagnetic, but ion sound waves ($f \lesssim f_p^+$) and electron plasma oscillations ($f \approx f_p^-$) are electrostatic.

Figure 1 shows some examples of natural plasma waves observed near Earth using the broadband or waveform channels of the plasma wave instruments on the OGO-5 (Orbiting Geophysical Observatory) and IMP-6 (Interplanetary Monitoring Platform) spacecraft. As shown in the top panels of the figure, the electron plasma frequency is significant in two distinct ways. Suprathermal electrons in the region upstream from the bow shock lose energy by radiating narrowband electrostatic waves (upper left) at $f = f_p^-$ (Fredricks et al. 1971). Secondly, f_p^- is a well known critical frequency for electromagnetic waves in the sense that waves with $f > f_p^-$ can be transmitted through a plasma, while waves with $f < f_p^-$ cannot propagate. The cutoff of outer magnetosphere electromagnetic noise at the plasma frequency is shown in the upper right panel of Fig. 1. Measurement of this cutoff frequency provides an excellent technique for determining the density of the plasma; in fact, since this technique is based on analysis of electromagnetic wave modes having wavelengths in the kilometer range, the presence of the spacecraft and its plasma sheath is unimportant, and observation of the f_p^--cutoff provides an absolute and sheath-independent evaluation of the total plasma density (Gurnett and Frank 1974).

At the collisionless bow shock (left center), currents flow and the two-stream instability generally produces intense electrostatic turbulence in the ion acoustic mode, with $f \lesssim f_p^+$, and $V(phase) = \omega/k \approx \sqrt{\kappa T_-/m_+}$ (Fredricks et al. 1970). For this type of interaction, the wave-particle scattering can be well described in terms of an anomalous electrical conductivity (Sagdeev 1965), $\sigma = j/E(DC) = Ne^2/m\nu_{(eff)}$, where

$$\nu_{(eff)} \approx (2\pi^3)^{\frac{1}{2}} f_p^- \left[\epsilon_0 E^2(wave)/2N\kappa T_-\right]. \tag{1}$$

The resistive heating associated with this interaction is very effective in producing the shock dissipation and particle acceleration; for a typical bow shock at Earth $\nu_{(eff)} \approx 1$–10 collisions per second, while the conventional coulomb or binary collision rate in the solar wind at 1 A.U. is on the order of one collision per day.

The wave-particle interactions represented by the data in the remaining panels of Fig. 1 are extremely important at Earth because the various plasma wave modes with $f \approx f_c^-$ provide stable trapping limits for relatively ener-

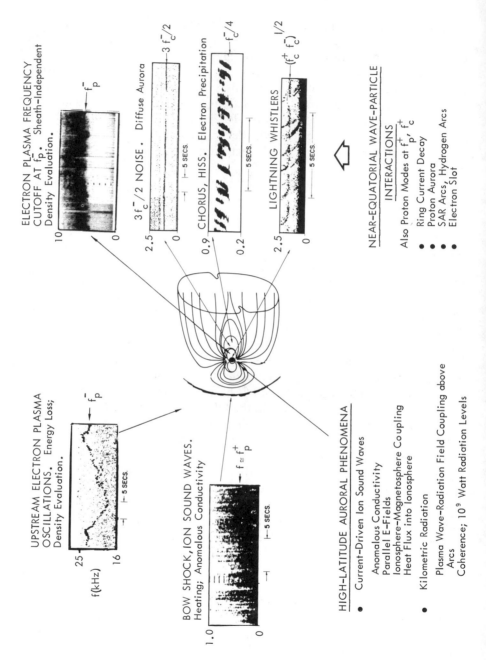

Fig. 1.   Summary of major wave-particle interaction phenomena detected near Earth. Broadband telemetry observations from OGO-5 and IMP-6 are used to make up these frequency-time diagrams. The arrows show typical locations in the upstream region, at the bow shock, and in the near-equatorial region of the magnetosphere. Detailed explanations and discussions of the high-latitude interactions and wave-particle interactions in the proton mode appear in the text.

getic magnetospheric electrons, as well as pitch angle diffusion for lower energy electrons, leading to the diffuse aurora and precipitation heat flux into the ionosphere and upper atmosphere.

Dungey (1963) and Cornwall (1964) first suggested that natural electromagnetic whistler mode waves (chorus, hiss, lightning-whistler signals) can lead to precipitation of trapped electrons. Shortly thereafter Kennel and Petschek (1966) proposed their classic self-consistent theory. It was based on use of the natural gyroresonance plasma instability for wave growth (amplification associated with the loss cone of trapped particles or with $T_\perp >$ $T_\parallel$ pitch angle distributions) and on the concept of turbulent pitch angle diffusion arising from the local wave-particle interactions. For the whistler mode, electrons with resonant energy,

$$E_R \simeq \frac{B^2}{8\pi N} \frac{f_c^-}{f} \left(1 - \frac{f}{f_c^-}\right)^3 \tag{2}$$

amplify waves with frequency $f (< f_c^-)$ and the effective pitch-angle diffusion coefficient, D_α, is then roughly given by

$$D_\alpha = (2\pi f_c^-) \, [nE(wave)/cB]^2 \tag{3}$$

where n is the index of refraction, and E is the amplitude of the electric component of the wave (Kennel and Petschek 1966).

This early plasma instability theory predicted overall stable trapping limits and precipitation patterns that agreed quite well with available observations, and the pioneering effort has since stimulated an enormous amount of more detailed analytical activity. The most convincing verifications of the Kennel-Petschek ideas on radiation belt structure involve analytical treatments based on use of generally measured spectral characteristics of wave turbulence. In a series of papers, Lyons et al. (1972) and Lyons and Thorne (1973) assumed a distribution of oblique waves with amplitude and spectral shape that match many measurements of extremely low frequency (ELF) hiss within the plasmapause, and they examined the effects of these waves on the trapped particles in detail. These authors derived energy-dependent and L-dependent pitch-angle distribution lifetimes and equilibrium distribution functions that agree very well with the complex profiles measured on OGO-5 and Explorer 45 within the plasmasphere.

A parallel theory of ion cyclotron turbulence has also been developed and applied to explain ring-current decay and proton precipitation in the earth's magnetosphere. Cornwall et al. (1970) first noted that the sudden decrease in resonant energy at the plasmapause should lead to wave growth, greatly enhanced pitch angle scattering and rapid decay of ring currents. In subsequent reports associated theories were developed to relate the presumed ion cyclotron turbulence at the plasmapause to stable auroral red (SAR) arc generation and formation of an electron slot (see the review by Coroniti 1973). These predictions are in good general agreement with some energetic particle observations from Explorer 45 in the sense that during the recovery phase of geomagnetic storms, the inner edge of the proton ring-current decays just within the plasmapause in a manner consistent with expectations for a moderate ion-cyclotron instability that does not yield strong diffusion. However, it is evident that many strong events of proton precipitation are not connected with plasmapause phenomena on Earth, and it has been suggested that electrostatic instabilities with $f \approx f_p^+$ act to limit the proton population.

The most intense waves detected in the magnetosphere with frequencies related to the local electron gyrofrequency f_c^- are the $(n + \frac{1}{2})f_c^-$ emissions first discussed by Kennel et al. (1970). Subsequent analyses by Fredricks and Scarf (1973), Scarf et al. (1973a), and Gurnett and Shaw (1973) suggested that these strong emissions are related to substorms, that the most common observation of this mode is at $f \approx 3f_c^-/2$ (see the broadband E-field dynamic spectrum on the right side of Fig. 1), and that the mode is electrostatic.

Wave amplitudes higher than 10 millivolts per meter are measured during substorms, and the associated pitch-angle diffusion coefficient for this mode is roughly

$$\bar{D}_\alpha(electrostatic) \simeq 10^{-2} \, (eE/p)^2/f \qquad (4)$$

where E is the wave amplitude, f is the frequency, and p is the particle momentum (Scarf et al. 1973a). A definitive study of the effects of $(n + \frac{1}{2})f_c^-$ emissions on magnetospheric particle distributions has been carried out recently by Lyons (1974) who demonstrated that the commonly observed turbulence amplitude of 1–10 mV m^{-1} is sufficient to put electrons with energies of several kilovolts in strong diffusion, and that 100 mV m^{-1} waves would even lead to strong diffusion for 100 keV electrons. It now seems certain that the $(n + \frac{1}{2})f_c^-$ modes play an important role in providing the source for precipitation of high-latitude electrons, especially during geomagnetically active periods, and this interaction is regarded as the ultimate origin of the diffuse aurora.

As indicated in the lower left part of Fig. 1, several wave-particle interactions play significant roles in the earth's auroral regions. Important aspects of magnetosphere-ionosphere coupling are intimately associated with the current systems in the auroral and polar ionosphere, and with the closure of these auroral current systems to produce divergenceless current patterns. Of particular interest are the field-aligned current systems above the peak of the F2 layer and out into the deep magnetosphere. Sufficiently strong currents in plasmas always produce plasma waves, and the wave spectrum then provides a set of equivalent scattering centers with which the current-carrying particles may "collide." Thus, even in a classically collisionless plasma, an effective collision frequency is produced, which in turn is used to define a microscopic transport coefficient such as the "anomalous" resistivity previously discussed in connection with the earth's bow shock [Eq. (1)].

Wave-produced anomalous resistivity associated with magnetospheric current systems has as a consequence the possible development of a potential drop along a flux tube, or in other words, a field-aligned electric field. Kindel and Kennel (1971) have carried out detailed calculations on this topic for plasma conditions characteristic of the topside auroral ionosphere. They argued that field-aligned currents in the auroral arcs on the poleward edge of the oval are sufficiently intense to destabilize the electrostatic ion cyclotron and ion acoustic modes in the topside ionosphere, and that the resultant plasma turbulence creates an anomalous resistance to the current flow. Therefore a self-consistent potential drop appears along the fieldline to drive the magnetospherically imposed currents through the anomalous resistor. These parallel electric fields can accelerate the current-carrying electrons into the ionosphere. Coroniti and Kennel (1972) have suggested that some of the observed auroral arc electron precipitation fluxes can be runaway electrons from the anomalous resistance region.

Anomalous field-aligned resistance is also of importance in the auroral oval region because a field-aligned current into the ionosphere at the equatorward edge of the oval is required to feed the poleward Hall current, and this current can also be unstable in the topside ionosphere. Coroniti and Kennel propose that the anomalous resistance forces the earth's ionosphere to polarize into a Cowling electrojet configuration, which creates the intense auroral electrojet currents that are magnetically detected during substorms. The runaways from this anomalous resistor of the equatorward oval are field-aligned protons of several keV that can precipitate into the ionosphere and produce the observed hydrogen $H\beta$ arcs. The precipitating electrons also provide a significant heat flux (of the order 1–10 ergs cm^{-2} sec^{-1}) into the ionosphere and upper atmosphere of the earth.

It is thus generally recognized that the structure of the earth's high-latitude ionosphere is directly governed by the magnetospheric configuration and microscopic processes involving waves and parallel electric fields that develop well above the ionosphere itself. Evans (1974) shows that details of some auroral electron energy spectra can be explained by assuming a field-aligned potential difference of several hundred volts to several kilovolts extending from a lower boundary (at an altitude near 2000–2500 km) to the hot magnetospheric plasma. In related areas Scarf et al. (1973b) described the local detection of field-aligned currents and enhanced electrostatic wave turbulence at relatively high altitudes on closed L shells on the night side of the earth. In addition Fredricks et al. (1973), Fredricks and Russell (1973), and Scarf et al. (1975) have discussed correlations of plasma wave data with magnetometer signatures indicating that unstable field-aligned current systems flow on the boundary of the dayside polar cusp.

Striking non-linear interactions in the earth's high-latitude plasma environment also involve wave-wave coupling. Specifically, this mechanism apparently leads to the remarkably intense kilometer-wave radiation from the regions above auroral arcs. Gurnett (1974) shows that at peak intensity the total power radiated is about 10^9 W. This is approximately one percent of the maximum energy dissipated by the precipitating auroral particles. It appears highly unlikely that any incoherent radiation mechanism can account for the tremendous power radiated above the auroral arcs, and it may be necessary to invoke mode coupling in which local plasma waves organize the radiating particles to provide a high degree of coherence.

In conclusion, it may be stated that in almost all regions of the earth's magnetosphere the collisionless plasma distribution functions are significantly non-Maxwellian and they are either continuously unstable with respect to generation of plasma waves, or readily perturbed into unstable states so that enhanced wave levels develop during geomagnetic storms and substorms. The plasma waves act back on the particle distributions to produce pitch-angle scattering, spatial diffusion, acceleration, or heating. The plasma turbulence also provides an important mechanism for energy dissipation in collisionless shocks and regions of field annihilation; it can lead to anomalous resistivity effects that allow macroscopic parallel DC electric fields to be maintained. It appears that plasma waves also play an important role in providing the coherence that leads to intense terrestrial electromagnetic radiation from the magnetospheric plasma above auroral arcs.

II. THE PLASMA ENVIRONMENT OF JUPITER

As noted above, the basic questions concerning the overall dynamics of Jupiter's magnetosphere and its interaction with the solar wind, the origin of the energetic trapped particles and radio emissions, the satellite interactions and ionosphere-magnetosphere coupling all involve plasma physics and phenomena of wave-particle interactions. In order to discuss any of these fundamental phenomena in a quantitative manner, it is necessary to determine how the plasma density, temperature, velocity and imbedded magnetic fieldstrength vary throughout the magnetosphere. In fact, Pioneer 10 and 11 did provide extremely important and definitive information on the configuration of Jupiter's magnetic field. However, these spacecraft did not carry comprehensive magnetospheric plasma probes, so there are still important unanswered questions concerning the distribution of plasma in Jupiter's magnetosphere.

Figure 2 contains an attempt to represent the present state of knowledge of the plasma environment around Jupiter. The central panel shows various evaluations of the plasma density profile, based on theoretical studies and interpretations of some relatively indirect observations from Pioneer 10. The theoretical curves are derived from exospheric models similar to the one originally proposed by Brice and Ioannidis (1970), who noted that Jupiter's enormous gravity would not allow cool ionospheric plasma to escape beyond the synchronous orbit ($r \approx 2R_J$); however, these authors also observed that ionospheric photoelectrons with $\kappa T_e \approx 5$–10 eV could readily populate the Jovian magnetosphere, and that centrifugal forces would then tend to provide

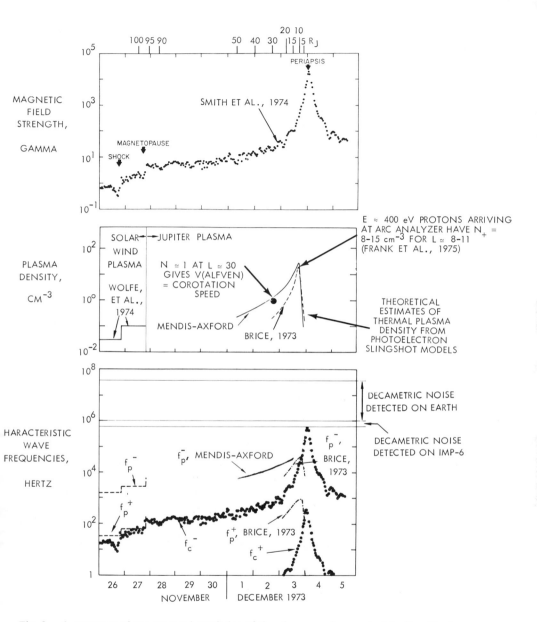

Fig. 2. A summary of our present knowledge of the plasma environment of Jupiter. The top panel has hourly averages of the field magnitude from Pioneer 10. Pioneer 11 data now provide general verification and additional observations into 1.6 R_J. However, to a large extent, present concepts of the plasma density profile are based on the theories of Brice, Axford and Mendis. In addition, as indicated in the central panel, Wolfe et al. (1974) used the solar wind analyzer on Pioneer 10 to derive plasma densities in the upstream wind and magnetosheath; Frank et al. (1975) used data from the same instrument to evaluate $E \simeq 0.1$–0.4 keV proton densities within $L \simeq 13$. The corresponding characteristic cyclotron and plasma frequencies are plotted in the bottom panel.

a concentration initially increasing with radial distance and with proximity to the spin equator. The complete photoelectron "slingshot" plasma distribution only has N increasing with L to a maximum near $L \simeq 10$–12; at this point several processes (interchange instabilities, recombination, etc.) serve to limit the plasma density, leading to the steady decrease shown in Fig. 2.

The curve labeled "Brice, 1973" in the central panel of Fig. 2 is a modified version of the original Brice-Ioannidis profile, based on use of improved estimates of photoelectron yields (Brice, personal communication, 1973). The plasma density model developed by Mendis and Axford (1974) also takes into account gravitational and rotational effects, as well as recombination and flux tube interchanges in the outer magnetosphere ($L > 10$); this model assumes that in the $2 < L < 10$ region there is diffusive equilibrium along fieldlines, and the resulting density profile is similar to that given by the 1973 Brice model.

There is presently little direct information on the validity of these models, but some general confirmation comes from analysis of the very large magnetic field distortions detected in the outer magnetosphere or magnetodisk. The top panel of Fig. 2 shows hourly average values of the magnetic field measured during the inbound passage of Pioneer 10 (Smith *et al.* 1974). This field magnitude profile can be well fitted by the expression

$$B(L) = 4 \text{ gauss}\left[\frac{1}{L^3} + \frac{1}{(30)^2 L}\right] \tag{5}$$

and it has been suggested (Hill *et al.* 1974) that the basic "break" from a dipolar to a non-dipolar radial variation at $L = 30$ can be explained if the co-rotation speed at $L = 30$ is about equal to the local Alfvén speed, $V_A = B(4\pi N m_+)^{\frac{1}{2}}$. This theoretical speculation on the origin of the strong field distortion in the distant magnetodisk leads to an estimate of $N \simeq 1$ cm^{-3} at $L = 30$, and the point is marked by the isolated dot in the plasma density panel of Fig. 2. Scarf (1975) also has observed that the currents, needed to give co-rotation of the Brice-Mendis-Axford thermal plasma, produce negligible field distortion in the entire inner magnetosphere; however, the field perturbations do rise rapidly with increasing L. These specific photoelectron slingshot distributions lead to $\Delta B \simeq B$ near $L = 30$, and the self-consistent magnetic field becomes sufficiently distorted so that strict co-rotation is no longer possible, in general agreement with Pioneer observations.

The solar-wind plasma probe on Pioneer 10 provided additional local information on low-energy protons within the Jovian magnetosphere; but for a number of reasons, results such as these are very difficult to extract from the data, and several interpretations are possible. For instance, Frank *et al.* (1975) describe the instrumental problems on Pioneer 10, and they note that solar-wind plasma analyzers lack the sensitivity and energy range required for comprehensive surveys of the hot quasi-isotropic plasmas found within planetary magnetospheres. Moreover, very energetic electrons penetrate into the instrumentation and these particles generate substantial background currents that vary in response to the particle anisotropy and the uneven shielding as the spacecraft spins. Thus, the effective in-flight sensitivity is relatively poor, and Frank *et al.* (1975) show that proton densities lower than about 8 cm^{-3} are masked by the penetrating particle background for $r \lesssim 15\ R_J$. Nevertheless, in the region $L \simeq 8-11$ (inbound and outbound) Frank *et al.* report proton densities ranging between 8 and 15 cm^{-3}, with a broad peak near $L \approx 9-10$, in good agreement with the predictions of the Brice-Mendis-Axford density models (see Fig. 2). There are, however, two significant differences between predictions of these models and the Pioneer 10 observations: (1) In the $L \simeq 8-11$ region the protons arrive at the plasma analyzer with energies in the range of 400 eV, rather than the 5–10 eV range predicted by the Brice-Axford-Mendis calculations; (2) within $L \simeq 6-7$, Frank *et al.* report on the detection of a relatively dense (N_+ up to 60 cm^{-3}) proton population with $\kappa T_+ \simeq 100$ eV, while the slingshot models predict a steeply declining density of cool protons in this inner region.

There are many possible explanations for these differences, because there are many possible sources of low energy plasma at Jupiter, other than the ionospheric photoelectron source considered in the Brice-Mendis-Axford models. Table I contains a list of very plausible additional sources of nonthermal low-energy plasma at Jupiter. It is clear that reasonable magnetosphere-ionosphere interactions, satellite-magnetosphere interactions, or local wave-particle interactions can lead to the appearance of non-thermal plasma throughout the inner magnetosphere.

It is also clear that interactions between spacecraft and plasma can account for some local observations of non-ambient plasma characteristics. In Earth orbit, photoelectron emission and secondary electron emission from the spacecraft surfaces are both generally significant, and plasma probe measurements always involve ambient distributions plus populations of charged particles that are emitted from or trapped near the spacecraft. At Jupiter, these complexities may also involve sputtering from spacecraft surfaces. Moreover, a plasma probe on a spacecraft always measures the spectrum of ions or electrons arriving at the detector, but these charged particles are locally accelerated or retarded by sheath fields. The energy shifts associated with the plasma sheath can be very important; De Forest (1972) has shown that ATS-5 (Applications Technology Satellite) charged up to 10 kV (negative) when the electron temperature at synchronous Earth orbit (6.6 R_E) reached 10–20 kV during substorms.

The possible relevance of this charging problem at Jupiter has been discussed by Scarf (1973, 1975) and by Mendis and Axford (1974). Much higher spacecraft charging levels are expected at Jupiter, although the potential problems depend on the details of the spacecraft electrical and mechanical configuration. Real knowledge of the plasma characteristics at Jupiter will only be obtained if a program of *in situ* and sheath-independent measure-

TABLE I
Possible Origin of Non-Thermal Low Energy Plasma at Jupiter

1. Secondary electron emission from satellite atmospheres, planetary atmospheres (spacecraft surfaces).

2. Escape and ionization (photo, charge exchange, impact) of satellite atmospheres.

3. Sputtering on the satellite (spacecraft) surfaces.

4. Acceleration of polar wind plasma.

5. Local acceleration associated with satellite $\mathbf{V} \times \mathbf{B}$ electric fields, satellite differential charging (sun-shadow asymmetries, satellite wake effects).

6. Resistive acceleration: current-driven waves interacting with particles in the presence of parallel electric fields (auroral analogy, anomalous resistivity).

7. Cyclotron acceleration by resonant wave-particle interactions.

8. Heating by landau damping of waves (sar arc analogy).

9. Betatron acceleration.

10. Particle heating in field-merging regions.

ments is carried out. For example, the measurement of the plasma frequency cutoff, if such waves are present, does give a sheath-independent determination of $N(total)$ (see upper panels in Fig. 1), since this technique utilizes oscillations of such long wavelength that the characteristics are unaffected by the presence of the spacecraft or its sheath.

In summary, at present there is a reasonable body of evidence suggesting that for $L \gtrsim 8$, the ambient plasma density at Jupiter can be described by the Brice-Mendis-Axford curves shown in the central panel of Fig. 2. Although there are persistent questions about the origins, thermal characteristics and latitude variations of the magnetospheric plasma, these density and B-field profiles can be used to compute the characteristic frequencies of plasma waves that are important at Jupiter. These curves are shown in the lower panel of Fig. 2, and it can be seen that the critical local wave measurements are in a range extending well below 10–20 kHz, for $L \gtrsim 10$.

III. PLASMA INSTABILITIES AND TRAPPED PARTICLE FLUXES

In their presentation at the JPL Radiation Belt Workshop, Thorne and Coroniti (1972) discussed a self-consistent Jupiter radiation belt model in which the cyclotron resonance instabilities limit the trapped fluxes. Thorne and Coroniti used the 1971 cold density model of Ioannidis and Brice [the 1973 Brice model has $N(L)$ reduced by a factor of five from this earlier version]. They assumed that the solar wind provides a source for the radiation belts, and they applied the concept of inward radial diffusion with conservation of magnetic moment. The basic conclusions from these initial calculations were that electromagnetic plasma instabilities would limit the trapped electron fluxes in a "filter region" within $L \simeq 18$–20, to values independent of the strength of the injection source. These authors predicted a maximum integral flux, $J(\text{max}, L)$, varying as $L^{-4.5}$, with $J \simeq 10^8$ cm^{-2} sec^{-1} for $L \simeq 5$. However, the workshop "nominal" and "upper limit" models correspond to significantly lower electron fluxes, as indicated in the bottom panel of Fig. 3.

In the post-Workshop period, Brice (personal communication, 1973) re-evaluated the expected thermal plasma density and, as noted above, he proposed a reduction in $N(L)$ by a factor of five from the original model. The top panel in Fig. 3 shows an expanded plot of this Brice model of 1973, along with the density profile of Mendis and Axford (1974). These theoretical density estimates for the Jovian magnetosphere provide the framework for the most comprehensive quantitative evaluations of the resonant particle energies [Eq. (2)], the pitch-angle diffusion coefficient [Eq. (3)], and the stable trapping limit.

As an example, the lower panel in Fig. 3 contains a specific prediction made before Pioneer 10 encounter by Coroniti (1974) of the trapped flux of energetic ($E \gtrsim 1$ MeV) electrons. The self-consistent model that yields this stable trapping limit assumes that: (a) electrons are transported inward from the solar wind or (more likely) from the Jovian magnetic tail by radial diffusion driven by fluctuating ionospheric dynamo fields; (b) in the outer zone ($7 < L < 20$), electron fluxes are limited to the stably trapped level by whistler mode precipitation loss, as computed using the Brice 1973 density function; (c) the Galilean satellites sweep out all electrons that have access to the satellites.

The measured points represented by dots in Fig. 3 are the flux values from the Trapped Radiation Detector of the University of California at San Diego (UCSD) on Pioneer 10 that were used in formulating the Radiation Workshop Model derived at the Ames Research Center in 1974 (Fillius personal communication).[1] It can be seen that there is really excellent agreement between Coroniti's pre-encounter theory and the actual observations.

After the electron data of Pioneer 10 were widely disseminated, Coroniti (1975) reviewed the theoretical situation in more detail. He observed that in the inner zone ($3 < L < 6$) the measured electron fluxes approximately follow the $L^{-4.5}$ scaling expected for loss-free radial diffusion. In the region $10 < L < 20$, the observed fluxes are very close to those predicted by the whistler mode stable-trapping theory, and there is evidence that plasma turbulence and wave-particle interactions cause electron losses in this region. Moreover, Coroniti (1975) noted that the precipitating electrons having $E \gtrsim 100$ keV will carry a heat flux of 0.1 to 1 erg cm^{-2} sec^{-1} to Jupi-

[1]See p. 896.

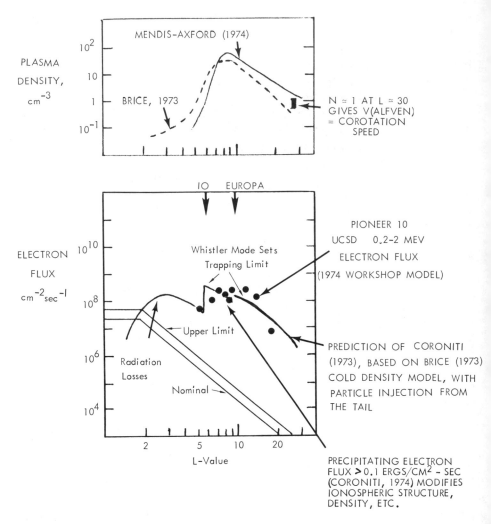

Fig. 3.   The density distributions of cold plasma (in the upper panel) predict that the instability of the whistler mode will provide significant pitch-angle scattering for relativistic electrons beyond about $L = 6$, where $B^2/8\pi N$ is sufficiently low. The associated stable trapping limit (as computed by Coroniti before the Pioneer 10 encounter) is plotted here along with some data points from Pioneer 10 (Fillius, personal communication) and other pre-encounter expectations based on the *ad hoc* "nominal" and "upper limit" models.

ter's upper atmosphere, and this precipitation flux is comparable with solar ultraviolet as an ionizing source.

The Pioneer 10 measurements also indicate that absorption losses of trapped particles at Io and Europa are less severe than some people anticipated. The radiation belts at these L shells are, however, sufficiently intense that significant energy fluxes to their surfaces, and/or to Io's ionosphere, are expected. It is unclear whether the satellites can also energize particles. This point will be discussed further in Sec. IV.

Despite the impressive agreement indicated in Fig. 3, it must be recognized that there are significant problems that arise in reconciling theory and observation. Moreover, the instability theories applied so far to Jupiter have been neither complete nor fully self-consistent. The simplest theories involve whistler mode waves propagating parallel to the magnetic field; with this approximation only the first-order cyclotron resonance is important, and the important wave-particle interactions occur on separate L shells near the equator. However, whistler mode waves actually propagate over a range of angles with respect to the B-field, and higher-order resonances are effective in producing wave-particle interactions. As shown by Lyons, *et al.* (1972), the actual pitch-angle diffusion of electrons in the earth's radiation belt involves oblique whistler mode wave propagation, all cyclotron harmonic resonances and the Landau resonance, and wave-particle interactions that occur at all geomagnetic latitudes.

The left hand side of Fig. 4 shows some predictions of the more complete calculations of Lyons *et al.* (1972). The theoretical pitch-angle distribution for $E \simeq 1$ MeV, $L = 2.5$ depends on the cold plasma density (N), the whistler mode center frequency ($f_m = \omega_m/2\pi$), turbulence bandwidth ($\delta f = \delta\omega/2\pi$) and wave propagation angle, Θ_W. In fact, the details of the pitch-angle distribution depend primarily on the wave characteristics, rather than on the plasma parameters such as $N(cold)$.

The central panel in Fig. 4 contains a detailed comparison between observation and this more complete theory. Lyons *et al.* (1972) have used *measured* characteristics of whistler mode plasma-wave turbulence to compute the equatorial pitch-angle distributions given by the solid curves; the data points represent differential flux measurements from the Livermore electron-proton spectrometer on OGO-5, and it can be seen that the predicted "hat" near 90° is actually observed.

A number of reports by the Pioneer 10 and 11 investigators contain discussions of the whistler mode plasma instability. Recently, Van Allen *et al.* (1975) considered the compatibility of the observed angular distributions of $E_e > 21$ MeV electrons for $L \lesssim 7$ with the predictions of the Kennel-Petschek (1966) whistler interaction theory, and the right hand panel in Fig. 4 shows the relevant Pioneer 10 and 11 data points on an omnidirectional flux versus B_0/B plot.

The Pioneer 10 and 11 experimenters are not to be blamed for the smooth curve drawn through these points in Fig. 4, but it is clear that the observations presently available from Jupiter are again suggestive of a hat-shaped equatorial pitch-angle distribution similar to that shown on the left side of the figure.

However, even if attention is focused only on the electrons and on the whistler instability, many fundamental questions still remain. For instance,

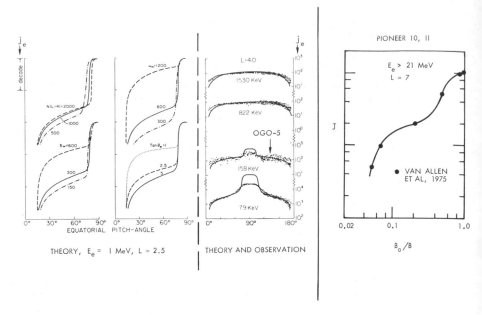

Fig. 4.   Equatorial pitch-angle distributions associated with electron interactions with whistler mode waves: The "hat-shaped" distribution is predicted by the full theory of Lyons *et al.* (1972), while the calculations for $E_e \simeq 1$ MeV, $L = 2.5$ show that the detailed shape of the distribution depends on the wave characteristics (center frequency ω, bandwidth $\delta\omega$, propagation angle Θ_W) rather than on the cold density N. The central panel contains a comparison of electron data from OGO-5 and theoretical predictions, based on measured whistler mode wave characteristics. The Pioneer 10 and 11 observations are not strictly comparable in this form but they are suggestive of a hat-shaped distribution based on pitch-angle scattering by whistler mode waves.

McIlwain and Fillius (1975) and Baker and Van Allen (1975) have constructed differential energy spectra and phase space densities of trapped electrons at Jupiter, and both groups find a continuously decreasing trend in the phase space density as Pioneer 10 moved from about $L \simeq 12$ to periapsis. However, if the Brice-Axford-Mendis density distribution is correct, then the instability should not be operative at low L shells, such as $L \lesssim 5$. Moreover, with this distribution, the plasma density falls off rapidly with distance above or below the Jovian spin equator, and whistler mode waves are strongly confined to the equatorial region (Liemohn 1973; Melander and Liemohn 1975). Of course, if the thermal plasma is not given by the Brice-Axford-Mendis models, these conclusions do not apply. Since all of the phenomena (wave growth and propagation, wave-particle interactions, precipitation and associated ionospheric contribution to magnetospheric plasma, etc.) are all strongly coupled, it will undoubtedly be a formidable task to achieve complete understanding of the dynamical phenomena affecting the electrons in the inner belt of Jupiter. In fact, since electrostatic waves (such as the $3 f_c^-/2$ emissions) can also produce strong precipitation events, careful and comprehensive measurements of the thermal plasma, the wave characteristics, and the trapped electron angular and energy distributions are needed to resolve all of these questions. Indeed, Sentman and Van Allen (1975) have noted that some Pioneer 10 data bring into question the entire picture of inward radial diffusion for electrons at Jupiter; they show that the "normal" loss-cone distribution is found only in the central core of the magnetosphere ($r \lesssim 12\ R_J$), while for $r \simeq 12\text{--}25\ R_J$, the electron distributions are bi-directional with peaks near zero and 180° pitch angles.

The corresponding problems of wave-particle interactions involving protons are also of considerable interest, and the initial (pre-encounter) calculations of stable trapping limits were actually performed to evaluate the possible proton hazard at Jupiter. Nevertheless, the instability picture involving proton mode waves at Jupiter is still relatively unclear, even after the Pioneer 10 and 11 encounters. It does appear that the measured proton flux in the inner belt is well below the stable trapping limits associated with ion cyclotron waves (Simpson *et al.* 1974; Coroniti 1975), but the proton energies for which the calculations were performed may not have been adequately measured during these initial encounters. In addition, electrostatic instabilities may be operative in this region with relatively high-β ($\beta = 4\pi N\kappa TB^{-2}$) plasma, and electrostatic wave-particle interactions may limit the fluxes.

IV. CURRENT-DRIVEN INSTABILITIES, AURORAL ANALOGS, AND SATELLITE INTERACTIONS

In the earth's magnetosphere, it is believed that the field-aligned currents flowing in and out of the auroral oval map into the boundaries of the polar cleft (or cusp), and that the boundaries of the plasma sheet map into the geomagnetic tail. Some of the complex microscopic processes associated with these current systems are schematically described in the lower drawing of Fig. 5. In fact, the basic phenomena that develop in these high-latitude regions of a rotating planet with a sun-oriented magnetosphere are extremely nonlinear, and the planetary ionosphere and atmosphere are strongly coupled to the magnetosphere by the field-aligned current systems. Convection in the ionosphere is driven by magnetospheric phenomena; ionospheric fieldlines "slide over" the fieldlines below the ionosphere which are held in place by the highly conducting planet, while neutral atmospheric winds are driven by ion-neutral collision effects.

Kennel and Coroniti (1975) have discussed these phenomena for Jupiter in the light of Pioneer 10 observations, and Northrop *et al.* (1974) numerically analyze oscillations in maximum count rates observed by Pioneer 10 for energetic particles in terms of fieldline "slippage" in the Jovian ionosphere. This study yields an estimated electric field of several volts per meter in the polar ionosphere of Jupiter; Northrop *et al.* (1974) speculate

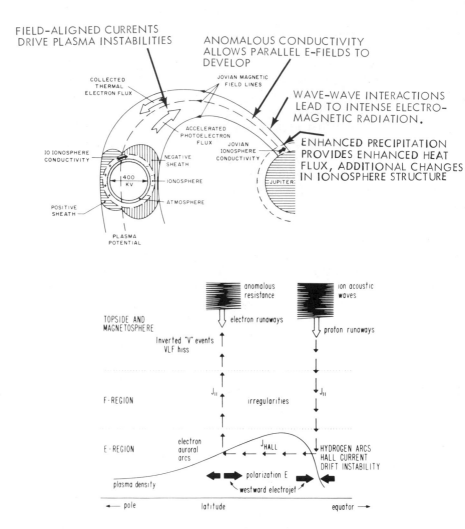

Fig. 5. The lower drawing summarizes present concepts of wave-particle interaction phenomena associated with current-driven instabilities in the region of the earth's auroral oval
(Kindel and Kennel 1971; Coroniti and Kennel 1973). Similar processes must develop in
Jupiter's high-latitude region. In addition, the Jovian satellites should drive analogous current
systems that close in the ionosphere, as indicated in the upper drawing (based on a specific
model of Shawhan *et al.* 1975), and current-driven instability phenomena should be operative
on fieldlines threading the satellites.

that this potential drop is a measure of the maximum energy available to
particles from a reconnection process.

Local wave-particle interactions must affect the dynamics of these high-
latitude phenomena in a number of significant ways. Currents in plasmas
involve electron-proton drifts, two-stream instabilities, and wave-particle

interactions that lead to anomalous conductivity [Eq. (1)]. In these high-latitude regions, the parallel electric fields that develop when the wave-particle interactions give rise to sufficiently modified conductivity can produce local particle acceleration, similar to the situation thought to occur above discrete auroral arcs at Earth (Evans 1974). Beam-plasma inter-actions that involve turbulence of microscopic plasma waves also lead to strong heat conduction fluxes into the ionosphere (Fontheim 1975). At Jupiter, such high-latitude beams may be associated with current systems in the magnetosphere, or with precipitating particles, but in any case, significant ionospheric perturbations should be induced by magnetospheric beams. Finally, the electrostatic plasma waves associated with the current-driven instabilities serve to organize the phases of radiating charged particles, and this coherence effect of plasma waves leads to an enormous increase in the electromagnetic radiation from the high-latitude region at Earth (Gurnett 1974) or at Jupiter (Scarf 1974).

All of these high-latitude phenomena should be operative at the Jupiter analog of the auroral oval extended into space. This oval may well involve the boundary of the magnetodisk as a generalized plasma sheet. However, at present so little is known about the overall configuration of Jupiter's magnetosphere and the presence or absence of merging, convection, etc., that it is not really fruitful to pursue this analogy in detail.

On the other hand, it may be stated with reasonable certainty that the fieldlines threading the inner satellites of Jupiter map out ionospheric ovals in which all of these auroral-type processes do occur. Specifically, there are $V \times B$ electric fields across the surfaces of these satellites (Goldreich and Lynden-Bell 1969; Gurnett 1972; Shawhan et al. 1973) or across their iono-spheres (Shawhan et al. 1975), and differential charging can also lead to large DC potential differences (Scarf 1973; Mendis and Axford 1974). These potential differences must give rise to field-aligned current systems that close in the ionosphere, as described in the example of Io at the top of Fig. 5. The labels attached here indicate how local phenomena involving plasma waves along the Io flux tube, for instance, can affect ionospheric structure and radio emissions.

Plasma wave phenomena should also play an important role in coupling any charged particles emitted from the satellite surfaces or atmospheres to the Jupiter plasma. There are many potential mechanisms in which the inner satellites act as magnetospheric sources of energetic electrons, heavy ions, etc. (see the review by Shawhan et al. 1975), but in general these satellite-associated particles will represent suprathermal bumps or drifting beams in the overall magnetospheric plasma. Wave-particle interactions should generally develop to provide thermalization or "pickup" of these satellite-associated particles.

A final possible satellite effect may have to do with detection of lightning from Jupiter's atmosphere. Bar-Nun (1974) argues that thunderstorms should be very frequent at Jupiter, with a rate of Earth-like lightning strokes 10^4 times as great as on Earth. At Earth, these lightning signals readily propagate into the magnetosphere in the whistler mode, but Melander and Liemohn (1975) show that for the density distribution of Mendis and Axford (1974) only proton whistlers will propagate to the Jovian equator. However, if the inner satellites are strong sources of plasma of moderate energy [such as the 100 eV to 400 eV protons discussed by Frank et al. (1975)], this plasma will readily spread all along the relevant fieldlines and "open up" the whistler mode window to the ionosphere. Thus, it is conceivable that the fieldlines threading the satellites will have a whistler mode connection to the upper atmosphere so that lightning will be readily detectable here.

V.   MAGNETODISK PHENOMENA

The present theories of Jupiter's outer magnetosphere fall into three classes.

1. *Radial outflow models.* Spurred on by the pulsar analog, various authors (Michel and Sturrock 1974; Kennel and Coroniti 1975) have assumed that a magnetodisk would be generated by a miniature solar wind, driven by particles flowing out from Jupiter's ionosphere or generated elsewhere in the $L < 20$ region, and spun up by the planet's rapid rotation. Here the plasma flow in the disk would be radially outward, and the magnetic field would be wrapped into a "garden-hose" configuration.

2. *Earth-like convection models.* It has also been suggested that re-connection of the solar-wind magnetic field with Jupiter's field would drive an internal convection pattern similar to that of the earth (Kennel and Coroniti 1975). The extended disk in this model would be due to the fact that the flow has a greater energy density than that of the magnetic field ($\beta \gg 1$). Here, the flow would be, at times, directed in the antisolar direction and its speed would exceed the local Alfvén speed.

3. *Quasi-static models.* Here it is assumed that particles from Jupiter's atmosphere or inner magnetosphere fill *closed* fieldlines and do not flow rapidly outward. Their inertia pulls the fieldlines out into a disk. There are two possibilities for the expected flow speed. If Jupiter communicates its rotation directly to the fieldlines, a flow with the local co-rotation speed and direction would be expected. On the other hand, if Jupiter's atmosphere cannot support complete co-rotation, the flow speed would be less than (perhaps much less than) the co-rotation speed of the solid body (Scarf 1975).

There are many ways in which wave-particle interactions are important in the outer magnetosphere of Jupiter, no matter which of these models is correct. For instance, with the radial outflow model a second collisionless shock forms within the magnetopause, and the flow is super-Alfvénic and/or supersonic, so that Alfvén waves or acoustic oscillations are readily generated.

For the re-connection model, the Jupiter problem also presents some extremely unusual features since the planet rotates very rapidly, with respect to the speed of the solar wind past the magnetosphere. As noted by Kennel and Coroniti (1975), five hours after an interplanetary fieldline connects at the nose of Jupiter's magnetosphere, the interplanetary "parcel" would be over the polar cap, but the ionospheric foot of the fieldline would have rotated to local midnight. This "twisting up" can lead to the formation of an X-type null configuration, and it is known that current-driven plasma instabilities do play a major role in dissipating energy at such a null (Bratenahl and Yeates 1970).

For any of these outer-magnetosphere models of Jupiter, it now seems obvious that the radiation belt particles do not come directly from the solar wind into a simple dipolar or tail magnetic field, as previous theory assumed; if the electrons simply diffuse in from large radial distances, then loss-cone distributions will be expected well beyond $L = 12$, in disagreement with the observations of Sentman and Van Allen (1975). Clearly, these particles must traverse, or be generated in, Jupiter's outer magnetosphere, and any acceleration process operative in this high-β plasma region must involve stochastic wave-particle interactions. However, in the high-β plasma of the outer magnetosphere of Jupiter, there are such a large number of theoretical candidates for sources of collisionless acceleration that further progress in this area must be deferred until *in situ* plasma wave measurements are made.

Acknowledgments. I thank F. V. Coroniti, W. Fillius, L. A. Frank, D. A. Gurnett and J. A. Van Allen for very helpful discussions of their recent observations and analyses. The preparation of this chapter was carried out under the auspices of the Independent Research and Development Program at TRW Systems.

DISCUSSION

T. R. McDonough: Might the characteristics of the 100 eV proton fluxes reported by Frank *et al.* (1975) have been affected by spacecraft charging phenomena?

F. L. Scarf: There is no way to determine this with certainty, based on information from the Pioneer instruments. It has been shown that the Pioneer 10 charging could have been severe if the only ambient sources of low energy plasma were associated with photoemission from the sunlit portions of the spacecraft and from the Jovian ionosphere (Scarf 1975). In this case, 10 eV protons might arrive at the analyzer with an effective energy of several hundred electron volts, with respect to spacecraft ground. However, if the protons described by Frank *et al.* are, in fact, real, then Pioneer would not charge significantly and the reported spectral characteristics would be correct. An unambiguous determination of the true situation requires a plasma density evaluation that is sheath-independent (such as that derived from a wave measurement) or more elaborate measurements of the complete electron and proton distribution functions.

    A. Eviatar: Did you find any evidence in the behavior of the spacecraft systems that would indicate extreme charging?

    F. L. Scarf: On Pioneer 10 and 11 a substantial number of anomalies were detected near $L \simeq 12-13$, or just in the region where current-balance considerations would suggest a sheath reversal. On Pioneer 10 these anomalies included spurious commands for the photopolarimeter, changes in the level of the spacecraft receiver and commutator anomalies for the Trapped Radiation Detector of the University of California. On Pioneer 11, spacecraft heaters and the conscan mode were spontaneously turned on at the same magnetic L shells. These anomalies are similar to those frequently detected in Earth orbit when substorm plasma injections lead to rapid variations in spacecraft potential.

REFERENCES

Baker, D. N., and Van Allen, J. A. 1975. Energetic electrons in the Jovian magnetosphere. Univ. of Iowa Tech. Rept. 75–13.

Bar-Nun, A. 1975. Thunderstorms on Jupiter. *Icarus* 24:86–94.

Bigg, E. K. 1964. Influence of the satellite Io on Jupiter's decametric emission. *Nature* 203: 1008–1010.

Bratenahl, A., and Yeates, C. M. 1970. Experimental study of magnetic flux transfer at the hyperbolic neutral point. *Phys. Fluids* 13:2696–2709.

Brice, N. 1972. Energetic protons in Jupiter's radiation belts. *Proc. Jupiter Rad. Belt Workshop* (A. J. Beck, ed.) pp. 283–302. JPL Tech. Memo. 33–543.

Brice, N. M., and Ioannidis, G. A. 1970. The magnetospheres of Jupiter and earth. *Icarus* 13: 173–183.

Brice, N., and McDonough, T. R. 1973. Jupiter's radiation belts. *Icarus* 18:206–219.

Burke, B. F., and Franklin, K. L. 1955. Observations of a variable radio source associated with the planet Jupiter. *J. Geophys. Res.* 60:213–217.

Carr, T. D., and Gulkis, S. 1969. The magnetosphere of Jupiter. *Ann. Rev. Astron. Astrophys.* 7:577–618.

Cornwall, J. M. 1964. Scattering of energetic trapped electrons by very low-frequency waves. *J. Geophys. Res.* 69:1251–1258.

Cornwall, J. M.; Coroniti, F. V.; and Thorne, R. M. 1970. Turbulent loss of ring current protons. *J. Geophys. Res.* 75:4699–4709.

Coroniti, F. V. 1973. The ring current and magnetic storms. *Radio Sci.* 8:1007–1011.

———. 1974. Energetic electrons in Jupiter's magnetosphere. *Astrophys. J. Suppl.* 27:261–282.

———. 1975. Denouement of Jovian radiation belt theory. *Magnetospheres of earth and Jupiter.* (V. Formisano, ed.) pp. 391–410. Dordrecht, Holland: D. Reidel Publ. Co.

Coroniti, F. V., and Kennel, C. F. 1972. Can the ionosphere regulate magnetospheric convection? *J. Geophys. Res.* 78:2837–2851.

De Forest, S. E. 1972. Spacecraft charging at synchronous orbit. *J. Geophys. Res.* 77:651–659.

Drake, F. D., and Hvatum, S. 1959. Non-thermal microwave radiation from Jupiter. *Astron. J.* 64:329–330.

Dungey, J. W. 1963. Loss of Van Allen electrons due to whistlers, *Planet. Space Sci.* 11: 591–595.

Ellis, G. R. A. 1965. The decametric radio emissions of Jupiter. *Radio Sci.* 69D:1513–1530.

Evans, D. S. 1974. Precipitating electron fluxes formed by a magnetic field-aligned potential difference. *J. Geophys. Res.* 79:2853–2858.

Fontheim, E. G. 1975. Beam-plasma interactions as a heat source in the magnetosphere. *Geophys. Res. Lett.* 2:150–153.

Frank, L. A.; Ackerson, K. L.; Wolfe, J. H.; and Mihalov, J. D. 1975. Observation of plasmas in the Jovian magnetosphere. Univ. of Iowa Res. Rept. 75–5; *J. Geophys. Res.* In press.

Fredricks, R. W.; Crook, G. M.; Kennel, C. F.; Green, I. M.; Scarf, F. L.; Coleman, P. J., Jr.; and Russell, C. T. 1970. OGO-5 observations of electrostatic turbulence in bow shock magnetic structures. *J. Geophys. Res.* 75:3751–3768.

Fredricks, R. W., and Russell, C. T. 1973. Ion cyclotron waves observed in the polar cusp. *J. Geophys. Res.* 78:2917–2925.

Fredricks, R. W., and Scarf, F. L. 1973. Recent studies of magnetospheric electric field emissions above the electron gyrofrequency. *J. Geophys. Res.* 78:310–314.

Fredricks, R. W.; Scarf, F. L.; and Frank, L. A. 1971. Nonthermal electrons and high frequency waves in the upstream solar wind. 2: Analysis and Interpretation. *J. Geophys. Res.* 76:6691–6699.

Fredricks, R. W.; Scarf, F. L.; and Russell, C. T. 1973. Field-aligned currents, plasma waves, and anomalous resistivity in the disturbed polar cusp. *J. Geophys. Res.* 78:2133–2165.

Gledhill, J. A. 1967. Magnetosphere of Jupiter. *Nature* 214:155–156.

Goldreich, P., and Lynden-Bell, D. 1969. Io, a Jovian unipolar conductor. *Astrophys. J.* 156: 59–78.

Gurnett, D. A. 1972. Sheath effects and related charged-particle acceleration by Jupiter's satellite Io. *Astrophys. J.* 175:525–533.

———. 1974. The earth as a radio source: Terrestrial kilometric radiation. *J. Geophys. Res.* 79:4227–4238.

Gurnett, D. A., and Frank, L. A. 1974. Thermal and suprathermal plasma densities in the outer magnetosphere. *J. Geophys. Res.* 79:2355–2361.

Gurnett, D. A., and Shaw, R. R. 1973. Electromagnetic radiation trapped in the magnetosphere above the plasma frequency. *J. Geophys. Res.* 78:8136–8149.

Hill, T. W.; Dessler, A. J.; and Michel, F. L. 1974. Configuration of the Jovian magnetosphere. *Geophys. Res. Lett.* 1:3–6.

Ioannidis, G., and Brice, N. M. 1971. Plasma densities in the Jovian magnetosphere: Plasma slingshot or Maxwell demon? *Icarus* 14:360–373.

Kennel, C. F. 1972. Stably trapped proton limits for Jupiter. *Proc. Jupiter Rad. Belt Workshop* (A. J. Beck, ed.) pp. 347–361. JPL Tech. Memo. 33–543.

Kennel, C. F., and Coroniti, F. V. 1975. Is Jupiter's magnetosphere like a pulsar's or earth's? *Magnetospheres of earth and Jupiter.* (V. Formisano, ed.) pp. 451–477. Dordrecht, Holland: D. Reidel Publ. Co.

Kennel, C. F., and Petschek, H. E. 1966. Limit on stably trapped particle fluxes. *J. Geophys. Res.* 71:1–28.

Kennel, C. F.; Scarf, F. L.; Fredricks, R. W.; McGehee, J. H.; and Coroniti, F. V. 1970. Electric field observations in the magnetosphere. *J. Geophys. Res.* 75:6136–6152.

Kindel, J. M., and Kennel, C. F. 1971. Topside current instabilities. *J. Geophys. Res.* 76:3055–3078.

Llemohn, H. B. 1973. Wave propagation in the magnetosphere of Jupiter. *Astrophys. Space Sci.* 20:417–429.

Lyons, L. R. 1974. Electron diffusion driven by magnetospheric electrostatic waves. *J. Geophys. Res.* 79:575–580.

Lyons, L. R., and Thorne, R. M. 1973. Equilibrium structure of radiation belt electrons. *J. Geophys. Res.* 78:2142–2149.

Lyons, L. R.; Thorne, R. M.; and Kennel, C. F. 1972. Pitch-angle diffusion of radiation belt electrons within the plasmasphere. *J. Geophys. Res.* 77:3455–3474.

McIlwain, C. E., and Fillius. 1975. Differential spectra and phase space densities of trapped electrons at Jupiter. *J. Geophys. Res.* 80:1341–1345.

Melander, B., and Liemohn, H. 1975. CMA propagation diagrams for the Jovian magnetosphere. *Icarus.* In press.

Melrose, D. B. 1967. Rotational effects on the distribution of thermal plasma in the magnetosphere of Jupiter. *Planet. Space Sci.* 15:381–393.

Mendis, D. A., and Axford, W. I. 1974. Satellites and magnetospheres of the outer planets. *Rev. Earth Planet. Sci.* 2:419–474.

Michel, F. L., and Sturrock, P. A. 1974. Centrifugal instability of the Jovian magnetosphere and its interaction with the solar wind. *Planet. Space Sci.* 22:1501–1510.

Northrop, T. G.; Goertz, C. K.; and Thomsen, M. F. 1974. The magnetosphere of Jupiter as observed with Pioneer 10. 2. Nonrigid rotation of the magnetodisc. *J. Geophys. Res.* 79: 3579–3587.

Piddington, J. H. 1967. *Jupiter's magnetosphere.* Univ. of Iowa Tech. Rept. 67–63.

Pizzella, G., and Venditti, G. 1973. Evidence for emissions of cosmic rays by Jupiter. Paper presented at *13th Internat. Cosmic Ray Conf.,* Aug. 1973, Denver, Colorado. Proceedings publ. by Univ. of Denver, 1975, pp. 1129–1134.

Sagdeev, R. Z. 1965. Ohm's law resulting from instabilities. *Proc. Symp. Applied Math.* 18: 281–286.

Scarf, F. L. 1969. Characteristics of the solar wind near the orbit of Jupiter. *Planet. Space Sci.* 17:595–608.

————. 1973. Plasma physics phenomena in the outer planet magnetospheres. *Amer. Inst. Aeronautics and Astronautics* Paper No. 73–566.

————. 1974. A new model for the high frequency decametric radiation from Jupiter. *J. Geophys. Res.* 79:3835–3839.

————. 1975. The magnetospheres of Jupiter and Saturn. *Magnetospheres of earth and Jupiter* (V. Formisano, ed.) pp. 433–449. Dordrecht, Holland: D. Reidel Publ. Co.

Scarf, F. L.; Fredricks, R. W.; Kennel, C. F.; and Coroniti, F. V. 1973a. Satellite studies of magnetospheric substorms on August 15, 1968. *J. Geophys. Res.* 78:3119–3130.

Scarf, F. L.; Fredricks, R. W.; Russell, C. T.; Kivelson, M.; Neugebauer, M.; and Chappell, C. R. 1973b. Observation of a current-driven plasma instability at the outer zone plasma sheet boundary. *J. Geophys. Res.* 78:2150–2164.

Scarf, F. L.; Fredricks, R. W.; Russell, C. T.; Neugebauer, M.; Kivelson, M.; and Chappell, C. R. 1975. Current-driven plasma instabilities at high latitudes. *J. Geophys. Res.* 80:2030–2040.

Sentman, D. D., and Van Allen, J. A. 1975. Angular distributions of electrons of energy $E_e >$ 0.06 MeV in the Jovian magnetosphere. *J. Geophys. Res.* In press.

Shawhan, S. D.; Goertz, C. K.; Hubbard, R. F.; Gurnett, D. A.; and Joyce, G. 1975. Io-accelerated electrons and ions. *Magnetospheres of earth and Jupiter.* (V. Formisano, ed.) pp. 375–389. Dordrecht, Holland: D. Reidel Publ. Co.

Shawhan, S. D.; Hubbard, R. F.; Joyce, G.; and Gurnett, D. A. 1973. Sheath acceleration of photoelectrons by Jupiter's moon Io. *Photon and particle interactions with surfaces in space* (R. Grard, ed.) pp. 405–413. Dordrecht, Holland: D. Reidel Publ. Co.

Simpson, J. A.; Hamilton, D. C.; McKibben, R. B.; Mogro-Campero, A.; Pyle, K. R.; and Tuzzolino, A. J. 1974. The protons and electrons trapped in the Jovian dipole magnetic field region and their interaction with Io, *J. Geophys. Res.* 79:3522–3544.

Smith, E. J.; Davis, L., Jr.; Jones, D. E.; Coleman, P. J., Jr.; Colburn, D. S.; Dyal, P.; Sonett, C. P.; and Frandsen, A. M. A. 1974. The planetary magnetic field and magnetosphere of Jupiter: Pioneer 10. *J. Geophys. Res.* 79:3501–3513.

Thorne, R. M., and Coroniti, F. V. 1972. A self-consistent model for Jupiter's radiation belts. *Proc. Jupiter Rad. Belt Workshop* (A. J. Beck, ed.) pp. 363–402. JPL Technical Memorandum 33–543.

Van Allen, J. A.; Randall, B. A.; Baker, D. N.; Goertz, C. K.; Sentman, D. D.; Thomsen, M. F.; and Flindt, H. R. 1975. Pioneer 11 observations of energetic particles in the Jovian magnetosphere. *Science* 188:459–462.

Warwick, J. W. 1970. *Particles and fields near Jupiter.* NASA-CR 1685. Washington, D.C.: Government Printing Office.

Wolfe, J. H.; Mihalov, J. D.; Collard, H. R.; McKibbin, D. D.; Frank, L. A.; and Intriligator, D. S. 1974. Pioneer 10 observations of the solar wind interaction with Jupiter. *J. Geophys. Res.* 3489–3500.

THE TRAPPED RADIATION BELTS OF JUPITER

W. FILLIUS
University of California at San Diego

We review the data and initial analyses from the University of California, San Diego (UCSD) instruments on Pioneers 10 and 11. The Pioneer measurements are the first ever made in the Jovian magnetosphere, and, as they are still too fresh and too copious to be interpreted completely, we encounter unanswered and new problems in our discussion. Energetic electrons of Jovian origin are found in interplanetary space. Although we now know that the Jovian magnetosphere is larger than had been expected and is inflated by distributed currents, there is uncertainty regarding its configuration and the dynamics of energetic electrons contained in the outer region. The five innermost satellites of Jupiter orbit within the radiation belts and affect the intensities and angular distributions of the radiation by both absorbing and injecting particles. In the inner region, radial and pitch angle diffusion are dominant processes as they are at Earth. Near the Pioneer 11 periapsis there are multiple peaks in the proton and electron intensities that have not been explained. We also present the Pioneer 10 and Pioneer 11 flyby trajectories in several coordinate systems which may be useful for studying the behavior of the trapped radiation.

This chapter reviews the preliminary results from the Trapped Radiation Detector instrument of the University of California at San Diego. This is one of several charged particle instruments on board Pioneers 10 and 11, and some of our findings overlap those of other experiment teams. We shall refer to their work when we feel it augments ours, but for a complete view of their findings, the reader is referred to the companion pieces in this volume by Van Allen, by Simpson and McKibben, and by McDonald and Trainor.[1] The Pioneer data are too recent, copious, and undigested to allow a definitive review at this time. Our chapter will not look backward, then, but for-

------

[1]See pp. 928, 738, and 961.

ward, and as we present some of the major features of the Jovian radiation belts, we shall emphasize the questions which they pose to our understanding.

I. THE UCSD INSTRUMENT AND THE FLYBY TRAJECTORY

Tables I and II summarize the capabilities of the UCSD Trapped Radiation Detectors during the Pioneers 10 and 11 flybys. The instrument package contains five sensors of three different kinds, consumes 2.9 watts, and weighs 1.77 kg. The instrumental characteristics are described in more detail by Fillius and McIlwain (1974b).

The encounter trajectories have been described by Hall (1974, 1975) and by Mead (1974). We complete this section by showing the flyby trajectories in coordinate systems helpful for studies of the radiation belts (Figs. 1–5). In Fig. 1 the trajectories are projected onto Jupiter's equatorial plane to show the local time coverage, which is predominantly in the morning quadrant. The coverage in magnetic latitude is shown in Fig. 2. The plane of this figure is defined to contain Jupiter's internal dipole moment vector and the spacecraft (or moon). Thus the magnetic equator always intersects this plane in a horizontal line through the origin, and magnetic latitude appears in the usual way. Because the dipole is tilted by $10°6$ with respect to the spin axis, a fixed observer would see the magnetic meridian plane wobble back and forth with the planetary rotation period. This wobble is what causes the spacecraft and satellite loci to oscillate in latitude. The fixed observer would also see the meridian plane swing as if hinged on the dipole vector and follow the spacecraft (or satellite) in local time. It is because of this swing that the spacecraft appears not to go around the planet. The dipole magnetic field-line through Io is shown for illustration.

The gyrocenter of a trapped particle travels along a magnetic fieldline, bouncing back and forth between mirror points. It is hard to trace this motion in Fig. 2 because of the varying curvature of the fieldlines. Therefore Fig. 3 shows the trajectories in magnetic dipole coordinates, where lines of force are straight. The vertical axis is related to latitude by $B / B_{eq} = \sqrt{4 - 3 \cos^2 \lambda} / \cos^6 \lambda$, where λ is the magnetic latitude, and B_{eq} is the fieldstrength at the magnetic equator. The shaded areas represent L shells where sweeping of trapped radiation can be accomplished by the satellites (see Sec. IV.A).

In addition to bouncing in latitude, particles also drift in longitude. Longitude drift is driven by electric fields and by the gradient and curvature of the magnetic field. In the idealized case, where there are no electric fields parallel to the magnetic field, electric field drift causes particles to circle the planet at just the planetary rotation frequency. In this case we refer to the magnetosphere as "co-rotating" and, by viewing the longitudinal motion of the particles in a coordinate system fixed to the planet, we can forget about this component of the drift. Figures 4 and 5 show the paths of the Pioneer spacecraft and the five innermost satellites in such a system. Magnetic field

TABLE I

Characteristics of the UCSD Trapped Radiation Detector on Pioneer 10 at Jupiter Encounter, December 1973

Designation	Channels	Discrimination Levels	Particle Sensitivity	Geometric Factor
C	C1	31 photoelectrons	$>$ 6 MeV electrons	11.5 cm^2 sr
	C2	65 photoelectrons	$>$ 9 MeV electrons	4.5 cm^2 sr
Cerenkov counter	C3	135 photoelectrons	$>$13 MeV electrons	0.5 cm^2 sr
E	E1	0.089 MeV	$>$0.16 MeV electrons	1.3 $\times 10^{-2}$ cm^2 sr
	E2	0.19 MeV	$>$0.255 MeV electrons	1.04 $\times 10^{-2}$ cm^2 sr
Electron scatter counter	E3	0.40 MeV	$>$0.460 MeV electrons	5.7 $\times 10^{-3}$ cm^2 sr
M	M1	0.40 MeV	$>$35 MeV electrons	0.038 cm^2
Minimum ionizing	M2	0.85 MeV	background	
Particle counter	M3	1.77 MeV	$>$80 MeV protons	

TABLE II

Preliminary Characteristics of the UCSD Trapped Radiation Detector
on Pioneer 11 at Jupiter Encounter, December 1974

Designation	Channels	Discrimination Levels	Particle Sensitivity	Geometric Factor
C	C1	31 photoelectrons	$\gtrsim$ 5 MeV electrons	~ 13.5 cm^2 sr
	C2	65 photoelectrons	$\gtrsim$ 8 MeV electrons	~ 5.9 cm^2 sr
Cerenkov counter	C3	135 photoelectrons	$\gtrsim$ 12 MeV electrons	~ 1.0 cm^2 sr
	CDC	10^{-13}–10^{-5} amp	$\gtrsim$ 1 MeV electrons	~35 cm^2 sr
E	E1	0.089 MeV	> 0.16 MeV electrons	1.3 × 10^{-2} cm^2 sr
	E2	0.19 MeV	> 0.255 MeV electrons	1.04 × 10^{-2} cm^2 sr
Electron scatter counter	E3	0.40 MeV	> 0.460 MeV electrons	5.7 × 10^{-3} cm^2 sr
M	M1	0.40 MeV	>35 MeV electrons	0.038 cm^2
Minimum ionizing	M2	0.85 MeV	background	
Particle counter	M3	1.77 MeV	>80 MeV protons	
SP scintillator	SPDC	10^{-14}–10^{-5} amp	>150 keV protons	7.4 × 10^{-23} amp ev^{-1} cm^2 sec sr(p)
			> 10 keV electrons	7.4 × 10^{-23} amp ev^{-1} cm^2 sec sr(e)
SE scintillator	SEDC	10^{-14}–10^{-5} amp	>150 keV protons	2 × 10^{-24} amp ev^{-1} cm^2 sec sr(p)
			> 10 keV electrons	1.4 × 10^{-23} amp ev^{-1} cm^2 sec sr(e)

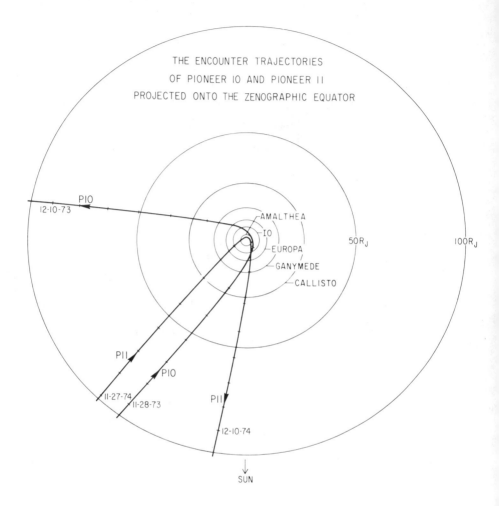

Fig. 1.   Projection of the Pioneer 10 and 11 trajectories on the equatorial plane of Jupiter. The dates mark the first and last magnetopause crossings for each flyby.

drift will still occur, and particles of different signs will go in the directions shown. Although the satellites have prograde orbits as seen from Earth, they move in the retrograde sense in this coordinate system. In this figure one can visualize the periodic motion of the satellites through the trapped radiation and determine the longitudinal relationship between satellites and spacecraft when the spacecraft crossed a satellite's L shell.

The dipole magnetic field representations shown here are quantitatively accurate only inside $\sim 10\ R_J$. However, we have extended the figures beyond this limit of validity in order to give a qualitative picture of the region farther out.

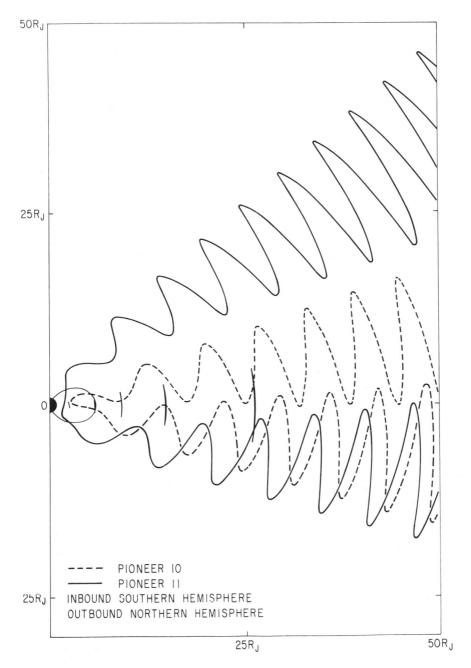

Fig. 2. The orbits of Jupiter's five innermost satellites and the trajectories of Pioneers 10 and 11 projected on a magnetic meridian plane. The outbound leg of the Pioneer 11 flyby was at high magnetic latitude, whereas the other three legs were all near the magnetic equator. Although the magnetic field is not well represented by a dipole beyond 5 to 15 R_J, we show the projection farther out for illustrative purposes.

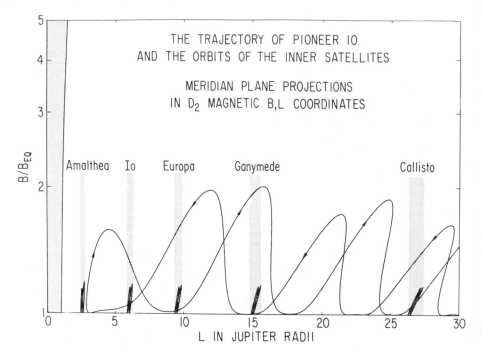

Fig. 3a.   Meridian plane projection in magnetic dipole coordinates. The equator lies along the abscissa, and the southern hemisphere is reflected into the upper quadrant. Magnetic lines of force are straight vertical lines from the equator to the planet's surface, which arches upward from the left side of the figure. The ordinate is the value of the scalar magnetic field normalized to the value at the equator.

II.   INTERPLANETARY ELECTRONS OF JOVIAN ORIGIN

Perhaps the first novelty with which Jupiter greeted the incoming Pioneer spacecraft was bursts of energetic electrons of Jovian origin. Figure 6 shows a time profile of Jovian electrons recorded by the UCSD Cerenkov detector on Pioneer 10 between April, 1973 and encounter with Jupiter in December of that year. Since the counter, which responds to particles with a velocity $> 0.75c$, records both cosmic ray nucleons and relativistic electrons, two channels were used in order to separate components by solving two simultaneous linear equations. The data are one-day averages, and the apparent negative counting rates are merely the result of systematic and statistical errors in the separation procedure. This profile resembles those shown by Chenette *et al.* (1974) and by Teegarden *et al.* (1974). Note that electron bursts appear as far away from Jupiter as 1 A.U. Re-analysis of data taken at Earth orbit by earlier Interplanetary Monitoring Platform

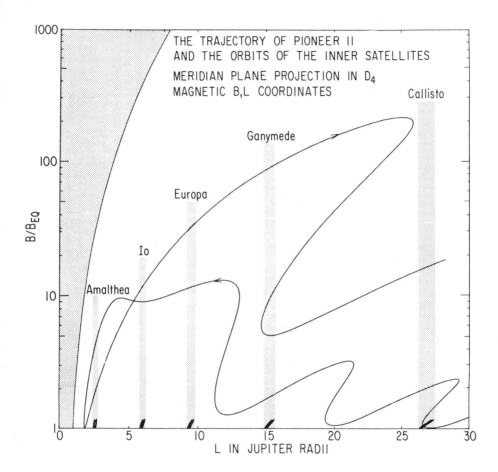

Fig. 3*b*. Same as for Fig. 3*a* but for Pioneer 11.

(IMP) spacecraft revealed that Jovian electrons are detectable as far away as 4 A.U. (Teegarden *et al.* 1974).

Figure 7 shows a series of bursts recorded by the Cerenkov counter just before Pioneer 10 reached the Jovian magnetosphere. The anisotropy dials show that the flux tends to be higher when the detector faces west in solar-ecliptic coordinates. This is the direction in which the magnetic field spiral leads away from the sun, and the electrons are flowing inward from Jupiter. One of the most significant features of these bursts is that peaks tend to reoccur at ten-hour intervals. As the rotation period of Jupiter is ~ 10 hr, this periodicity is a dramatic signature of the origin of these electrons. (More precisely the rotation period of Jupiter is 9 hours, 55 minutes, 29.711 seconds

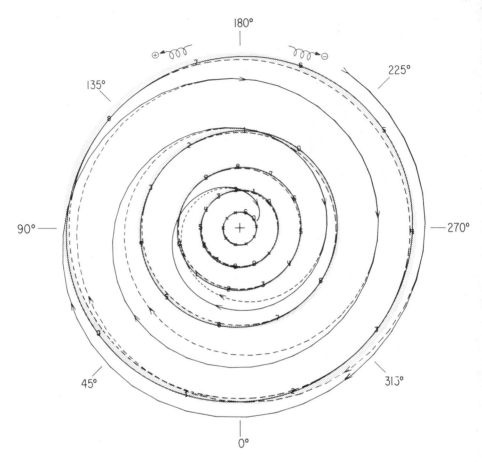

Fig. 4. The spacecraft trajectory for Pioneer 10, inbound (left) and outbound (right) and the five satellite orbits in a D_2 magnetic coordinate system fixed to Jupiter and rotating with the planet at the System III (1957.0) rotation rate. The track of each object is represented twice. The dashed lines represent the radial distance from the planet and the solid lines show the L values. Where the lines coincide an object is on the magnetic equator and where they are farthest apart it is the maximum latitude. The L shells traversed by each of the satellites are shaded. Plasma and low-energy trapped particles co-rotate with Jupiter and so remain fixed in this coordinate system, but high-energy particles drift in circles, electrons westward and protons eastward. The zero on each satellite's track indicates the position of the satellite when the spacecraft crossed its L shell, and the other figures indicate where the satellite was a given number of hours earlier. The large amount of information available in these figures makes them useful, but because it takes some concentration to interpret them, we call these figures vertigo diagrams.

according to the System III (1965.0) convention. In our discussions we round it off to ten hours.)

The mode of escape of these particles and its implications for the stability and structure of the Jovian magnetosphere are questions open to in-

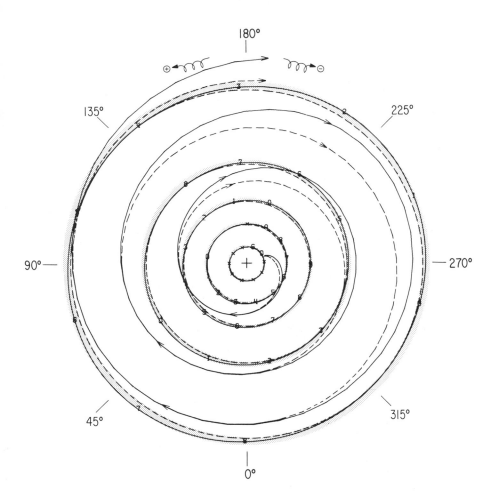

vestigation. So are the characteristics of their radial and longitudinal propagation, and particularly how they retain their cohesion and periodicity. Studies of their directionality, energy spectra, and frequency spectra should shed light on these questions.

Any new ideas generated by this study on Jovian particle propagation in the heliosphere will have a direct effect on the models of propagation of solar and galactic cosmic rays. One of the advantages of using Jovian electrons as test particles in the heliosphere is that Pioneers 10 and 11 have directly sampled the regions of their origin (the Jovian magnetosphere), whereas in the case of solar cosmic rays we do not have *in situ* measurements of these particles at their origin, and in the case of galactic cosmic rays we have not yet sampled their fluxes in interstellar space.

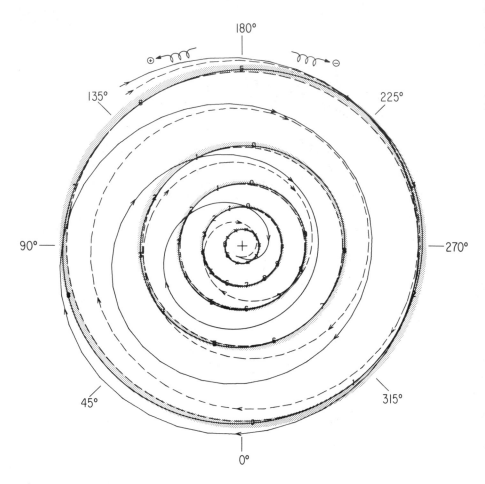

Fig. 5.   Same as for Fig. 4 but for Pioneer 11 and D₄ jovimagnetic coordinates.

III.   THE CONFIGURATION OF THE MAGNETOSPHERE

Energetic particles are trapped within the magnetosphere. Figure 8 for Pioneer 10 and Fig. 9 for Pioneer 11 show the entire radiation belts in profile. One easily distinguishes the inner magnetosphere, $R \lesssim 20 \, R_J$, and the magnetopause crossings, where there are abrupt steps between interplanetary and trapped flux levels. There are at least three magnetopause crossings on each inbound and outbound pass. The position of the magnetopause is evidently variable, moving inward and outward rapidly in response to changes in the solar wind. This and other evidence have been cited (Mihalov et al. 1975; Wolfe et al. 1974)[2] to picture the magnetosphere as a blunt, spongy region.

---

[2]See p. 856.

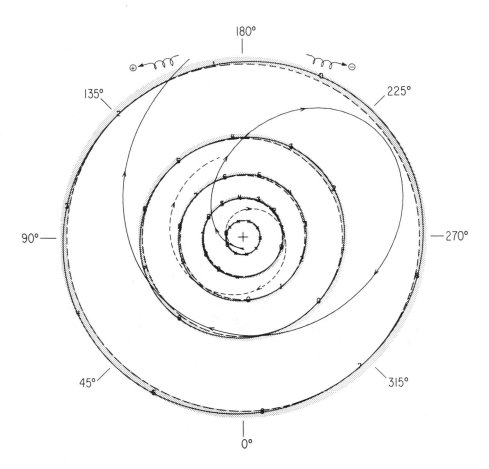

Large-scale fluctuations with a 10-hr period dominate the trapped electron fluxes outside 20 R_J. The first explanation of this periodicity held that the radiation belts were confined to a thin disc near the magnetic equator. The modulation was attributed to wobbling of the disc at the planetary rotation rate (Fillius and McIlwain 1974a; Simpson et al. 1974a; Trainor et al. 1974b; Van Allen et al. 1974). This explanation was compelling until the outbound half of the Pioneer 11 encounter. Whereas all previous data were acquired in and near the disc at low latitudes, the Pioneer 11 outbound leg was at a latitude well outside the disc. Thus it was a surprise to see the 10-hr periodicity continue as before, with intensity peaks higher even than during the inbound leg.

There is not a consensus of opinion as to why this pass does not follow the predictions of the original model. Van Allen et al. (1975) concluded that the disc is blunted in the subsolar direction (the outbound direction of Pioneer 11). Fillius et al. (1975a) suggested that, if local time was not the cause, the latitude profile might be bifurcated, with peaks at the equator and

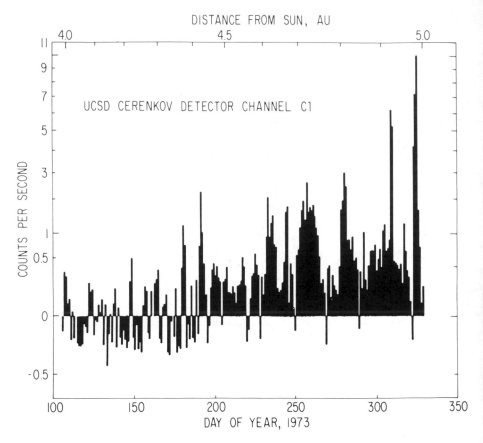

Fig. 6.   Electrons from the Jovian magnetosphere found in interplanetary space before the Pioneer 10 encounter. The data points are one-day averages, and the counting rate from cosmic ray nucleons has been subtracted out. The contribution from cosmic ray nucleons has been eliminated by a linear subtraction procedure. Negative counting rates are caused by the errors and statistical uncertainties in this procedure.

at high latitude, too. Such a profile is a possible configuration of the analytical model magnetosphere since published by Barish and Smith (1975). Simpson *et al.* (1975) argued that, although there may be some modulation due to a disc-like structure of the radiation belts, the primary cause of the 10-hr periodicity is time-dependent. The maxima and minima occur as determined by time, and they appear nearly simultaneously throughout the sunward side of the radiation belts regardless of latitude.

Much attention has been given to the phase of the variations. Figure 9*b* includes tic marks placed at the phase expected in the disc model and the time model. The reader may see for himself that neither model makes a good

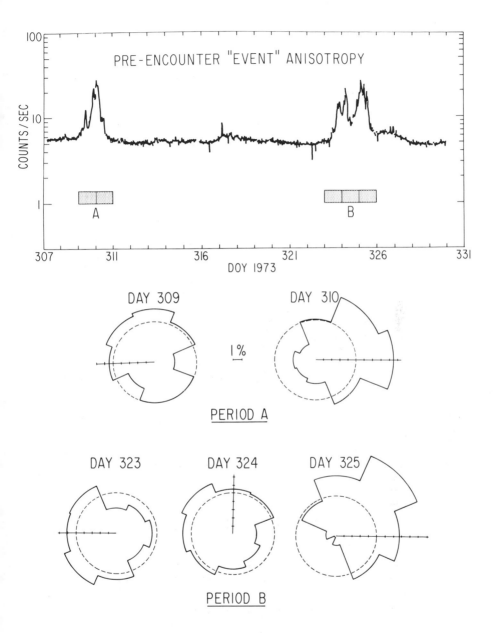

Fig. 7. Bursts of Jovian electrons that occurred in interplanetary space several days before encounter. No subtraction has been performed in this figure to eliminate the nucleonic cosmic ray counting rate of about 5 sec^{-1}. The cogwheels represent the directionality of the counting rates in the spacecraft equatorial plane, which is normal to the spacecraft-Earth vector. On days 310 and 325 there was a marked flow away from Jupiter.

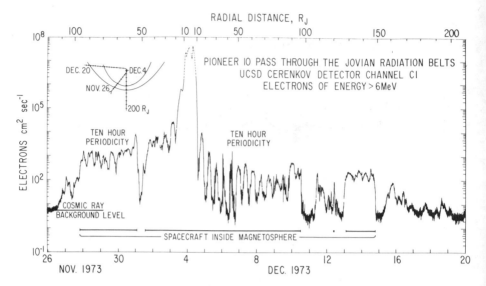

Fig. 8. Jupiter's radiation belts, end to end. It took about three weeks for each of the Pioneer spacecraft to fly through this enormous region of space. The radiation intensity in the inner magnetosphere is four orders of magnitude higher than in the surroundings. The outer magnetosphere is characterized by periodic fluctuations at the planetary rotation rate. Because of variations in the size of the outer magnetosphere, the spacecraft crossed and recrossed the magnetospheric boundary, or magnetopause, several times inbound and outbound.

fit. Qualifications may be added to either model. However, it seems that the preliminary appraisal of the data is inconclusive.

More thorough studies can be expected to shed light on the configuration of the magnetosphere. In addition to the phase of the particle peaks, the magnetometer data are certainly important. The magnetic field is discussed by E. J. Smith et al.[3] in this volume. Combined studies of the magnetic field and particle distributions are being undertaken.

Still more information is contained in the particle angular distributions. One significant result is the east-west anisotropy of the low-energy protons which indicates that these particles co-rotate with the planet. (Trainor et al. 1974a; Van Allen et al. 1975.) Near the inner magnetosphere both spacecraft encountered intense field-aligned fluxes of highly energetic particles. Figure 10 shows an example that persisted for several hours. Although these events are evidently of high significance, their interpretation is ambiguous (Fillius and McIlwain 1974b). Present studies are in an undigested state.

This region differs from the earth's radiation belts both in the larger dimensions and in the influence of the rapid rotation of the planet. Basic theoretical understanding is lacking for such features as the effect of a distributed

---

[3]See p. 788.

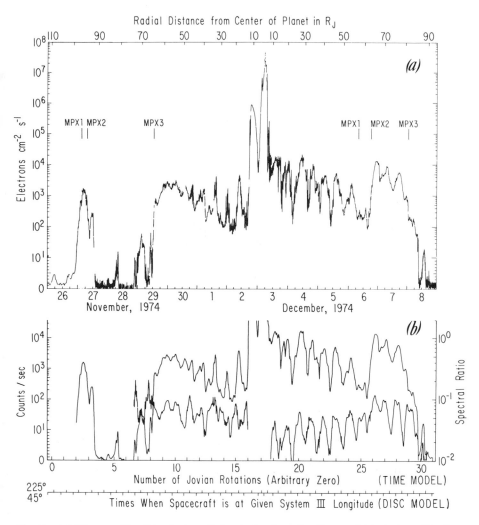

Fig. 9.   (a) Time profile of Jupiter's radiation belts along the Pioneer 11 trajectory. The flux is of electrons of energy E > 5 MeV. The labels MPX-1, MPX-2, and MPX-3 mark the times when the spacecraft entered or left the magnetosphere. (b) Running one-hour averages of data, filtered to show the ten-hour periodicity more clearly. The middle trace is the flux of > 5 MeV electrons, the same as in (a) above. The bottom trace is the ratio of two channels with energy thresholds above and below 5 MeV. Higher ratios indicate harder spectra.

ring current, the possibility of non-co-rotation or slippage of the magnetic field, and particle convection and acceleration in the possibly turbulent field. Clarification of these features can be expected to improve our understanding of such diverse subjects as pulsars and laboratory plasmas as well.

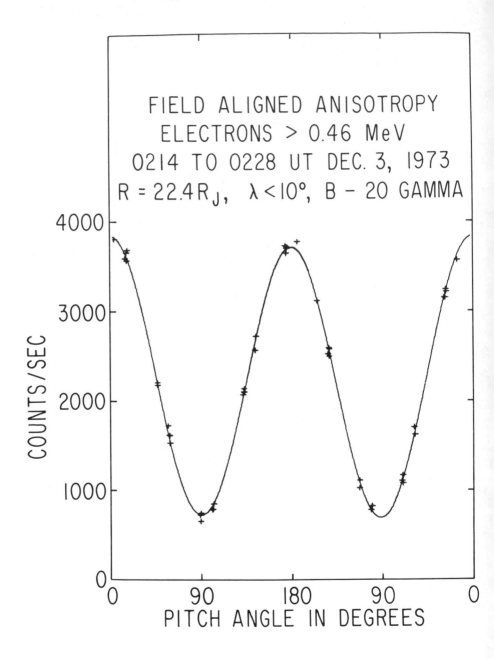

Fig. 10.   The pitch angle distribution of ~ 1/2 MeV electrons during an episode of a field-aligned, or "dumbbell," angular distribution. This episode occurred while the inbound Pioneer 10 spacecraft was very near the equator at an *L* value of about 22 R_J.

IV. INTERACTIONS BETWEEN THE TRAPPED
RADIATION BELTS AND JUPITER'S SATELLITES

A unique feature at Jupiter is the presence of satellites within the radiation belts. Besides being useful probes of trapped particle behavior, they generate extraordinary particle and electrodynamic effects which give them intrinsic interest. Before the Pioneer mission, the Io-controlled decametric emission had already broadcast evidence that these satellites could have remarkable electrodynamic effects. The recent observation of sodium emission lines from a halo surrounding Io (Brown and Yung)[4] and the discovery by Pioneer 10 of a partial torus of hydrogen emission in Io's orbit (Judge *et al.*)[5] heighten the interest. The energetic particle detectors on Pioneers 10 and 11 add to the list of satellite-associated features, the absorption of energetic trapped radiation and the injection or acceleration of energetic particles. These are the best observations until 1979 when the Mariner spacecraft flying to Jupiter will make a near encounter with Io and take a much closer look.

A.   *Absorption of Trapped Radiation*

All particle experiments on Pioneers 10 and 11 observed absorption at the three innermost satellites, Amalthea, Io, and Europa. Figures 11 and 12 illustrate the absorption features in several channels of the UCSD instrument. The flux or counting rate is plotted versus time, and the labels indicate the times when the spacecraft crossed the particle drift shells occupied by each satellite. Absorption at Ganymede is difficult to verify because the latitude excursions of the spacecraft caused rapid changes near this L shell, and there seem to be other variations which should be accounted for before a definite association can be made.

Mogro-Campero and Fillius (1975) and Mogro-Campero[6] constructed a statistical model for the rate of absorption of trapped particles at a satellite, and then, exploiting this effect as a probe of trapped particle dynamics, derived diffusion coefficients for the radial motion of the particles in the Jovian magnetosphere. Ignoring complications from variations in magnetic latitude of the satellites, it is an easy model to deal with. The key simplification is a statistical approximation to the probability that a particle may be absorbed by a satellite. Noting that diffusive excursions in the motion of particles tends to redistribute them randomly, complete randomization is assumed throughout a sweeping region of width ΔL in less time than the recurrence period, P, of the satellite in a frame of reference rotating at the particles' drift rate. If ΔL is greater than the satellite's diameter d, the probability per unit time of a

[4]See p. 1102.
[5]See p. 1079.
[6]See p. 1196.

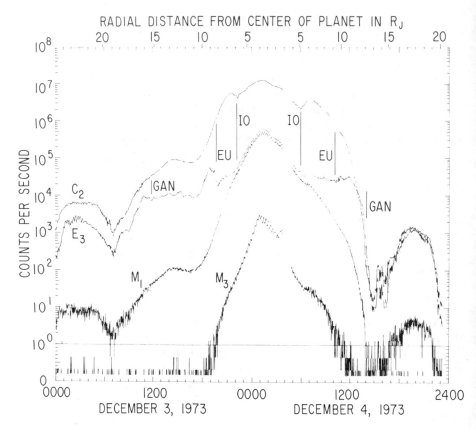

Fig. 11.   Profile of Jupiter's inner radiation belt taken by Pioneer 10. Channel C2 counts electrons of energy $E > 9$ MeV; E3, electrons of $E > 0.43$ MeV; M1, electrons of $E > 35$ MeV; and M3, protons of $E > 80$ MeV. There are clear absorption features at the positions of Io and Europa, and a questionable one at Ganymede. The dumbbell pitch angle distribution shown in the last figure occurred during the subsidiary peak at the left of this figure.

particle being met by the satellite and absorbed is given by $d/(P\Delta L)$. Thus the rate of change of the particle density, τ, is given by

$$\frac{d\tau}{dt} = -\tau d/(P\Delta L) .\qquad (1)$$

Particles diffusing inward are absorbed at this rate for as long as they remain inside the sweeping region. This time is ΔL divided by the diffusion velocity, $D n$, where D is the diffusion coefficient and n is the average slope $\frac{\partial}{\partial L}(\ln \tau)$.

The fraction of particles that diffuse through the region without being absorbed is given by integrating Eq. (1) from $t = 0$ to $t = \Delta L/(D n)$. Note that the arbitrary width ΔL cancels out of the result:

$$\tau/\tau_0 = \exp\left(-d/PDn\right) .\qquad (2)$$

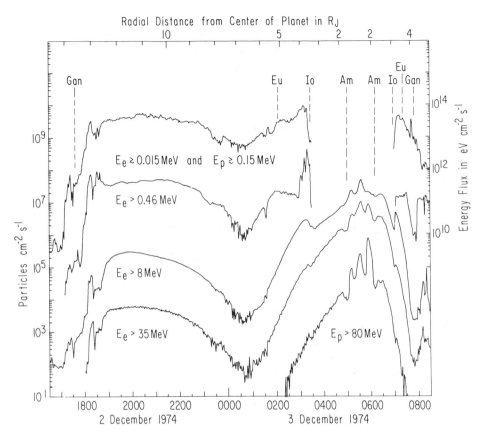

Fig. 12.   Pioneer 11 profile of Jupiter's inner radiation belt. The uppermost trace shows the combined energy flux for electrons and protons above the lower threshold and below an upper limit of ~0.1 MeV for electrons and several MeV for protons. The right-hand scale refers to the uppermost trace only; all other profiles should be measured against the left-hand scale. Pioneer 11 passed the orbit of Amalthea in addition to the other satellites, and absorption features were seen for this satellite as well.

With observations of τ/τ_0 Eq. (2) can be inverted to solve for D:

$$D = -d/[Pn \ln(\tau/\tau_0)] \,. \tag{3}$$

The statistical assumption may not be accurate for all types and energies of particles. However, the more the positions of the particles are randomized, the better the assumption will be. Randomization will be enhanced by instabilities in the wakes of the satellites in the manner discussed by R. A. Smith[7] and by Huba and Wu (1976). Particles with substantial drift rates ($E \gtrsim 1$ MeV) will approach the satellite at a different longitude each time around, and this introduces another dimension of variability. Because the detectors have wide energy ranges, their responses cover a spectrum of

---

[7]See p. 1181.

different conditions, which may be appropriately represented by a statistical ensemble. Furthermore, the statistical assumption can be checked for self-consistency. We calculate the rms displacement of a set of particles in a time P, following the derivation given by Reif (1965):

$$< \Delta L^2 > = 4DP. \tag{4}$$

Using a satellite's period for P, from Eq. (4) we can obtain $\sqrt{< \Delta L^2 >}$, or the width of the region over which the particles' positions are effectively randomized. For most values of D obtained by Mogro-Campero and Fillius, $\sqrt{< \Delta L^2 >} \gg d$, as required by the statistical approach.

Deterministic models have been used to derive values for the diffusion coefficient by Simpson et al. (1974b) and by Thomsen and Goertz (1975). Their values have not differed substantially from the results of the statistical model, but with suitable refinement the deterministic approach may offer the ability to look further into the interaction between particles of satellites and to account for some of the detail in the observations of trapped radiation.

B. Effect on Angular Distributions

The angular distribution of the trapped radiation is also affected by the satellites. As predicted by Mead and Hess (1973) the probability of absorption depends upon a particle's equatorial pitch angle. This is simply a geometrical effect depending upon whether or not the particle's bounce motion carries it across the satellite's orbital path. The vertigo diagrams (Figs. 4 and 5) demonstrate how the magnetic latitude of the satellites varies with planetary longitude, and illustrate the zones where a particle which mirrors close to the equator can slip under the orbital path of a satellite and escape absorption.

The UCSD Trapped Radiation Detector observed changes in the pitch angle distributions which are consistent with this model of selective absorption. Traces 2 and 3 in Fig. 13 show how the pancake-shaped angular distribution of 9 MeV electrons sharpens as the particles diffuse past Io. Evidently particles with small equatorial pitch angles are absorbed preferentially. A more complete interpretation of Fig. 13 includes pitch angle diffusion to reduce the sharpness of the angular distribution inside Io. This interpretation is discussed at greater length by Fillius et al. (1976). As exploited in that paper, the satellite serves again as a probe of trapped particle behavior. Further insights may be expected from analysis of these absorption features.

C. Injection of Energetic Radiation

Besides being a sink of particles, Io is also a source. Injection of particles is predicted by a sheath model for Io's interaction with the magnetosphere (Shawhan 1976). During the Pioneer 10 flyby, local peaks were detected near the magnetic L shell of Io. Fillius and McIlwain (1974b) suggested that these were caused by a local source of particles, and McIlwain and Fillius (1975) demonstrated that the phase space density went through a local maximum

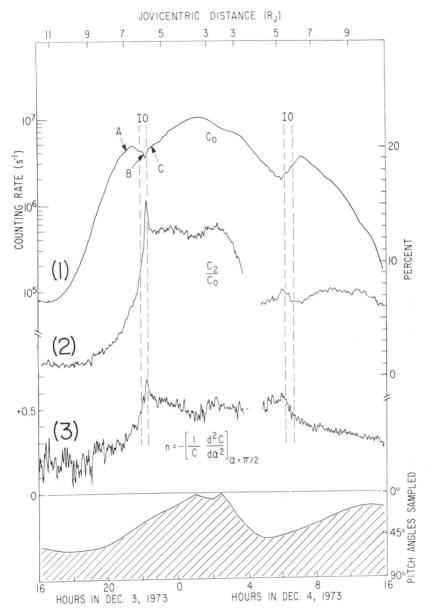

Fig. 13.   Angular distribution data for the UCSD Cerenkov counter on Pioneer 10. The top trace is the spin-averaged count rate for electrons of $E > 9$ MeV, and shows the Io absorption features. The second trace shows the modulation amplitude at twice the spin frequency. In the region of Io the higher amplitude indicates a sharper pancake angular distribution, but because the detector is in the spacecraft equatorial plane, it samples the pitch angle distribution obliquely. The third trace shows the index, n, that would result from representing the pitch angle distribution near $\alpha = \pi/2$ in the form $\sin^n \alpha$. Again, higher values indicate a sharper pancake, and we see that the pitch angle distribution is sharper when Pioneer 10 crosses Io's L shell outbound as well.

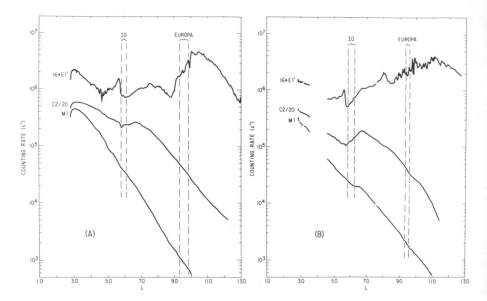

Fig. 14.   Counting rates in the inner region plotted versus the magnetic parameter, L, calculated using D_2 magnetic coordinates. Channel E1 counts electrons of energy $E > 0.16$ MeV; C2 counts electrons of $E > 9$ MeV; and M1, electrons of $E > 35$ MeV. Note that the E1 and C2 counting rates are multiplied by scaling factors in order to share the same vertical axis. (A) Pioneer 10 inbound; (B) Pioneer 10 outbound.

here, proving the existence of a source. The peaks appear in Fig. 14 just inside the inner edge of Io's sweeping region. They occurred only for Channels E1, E2, and E3 for electron energies $E_e > 150$, 250, and 460 keV. The C and M detectors ($E_e > 6$ MeV and 35 MeV) did not record such a feature.

As shown in Figures 4a and 4b, Pioneer 10 crossed the L shell of Io 2 hours (inbound) and $5\frac{1}{2}$ hours (outbound) after Io had passed the same longitude. Thus it was never near the instantaneous flux tube occupied by Io. Pioneer 11, on its inbound pass, was fortunate to come very close to the Io flux tube (see Fig. 5a). The exact miss distance depends upon the magnetic field model because there was a large latitude difference between spacecraft and satellite, and differences in magnetic declination are critical to the connection. The miss distance could be quoted as the actual distance from the spacecraft to the nearest point on the surface of the flux tube, but it is more useful to project the satellite and spacecraft positions to the magnetic equator along lines of force and measure the miss distance in the equatorial plane. Projected in this manner the miss distance from the center of Io was $\sim 13,000$ km (7 R_{Io}) in D_4 coordinates, and ~ 7000 km (4 R_{Io}) in the D_2 system. (The D_2 and D_4 magnetic models are defined by E. J. Smith et al.[8])

As Pioneer 11 passed the Io flux tube, all channels of Detector E jumped

---

[8]See pp. 788 and 39.

to the highest counting rates recorded in either flyby. This peak can be seen in Fig. 12, and in more detail in Fig. 15. Neither of the higher-energy channels recorded such a spike, and the particle energies therefore were below several MeV. Certainly a local source is needed to explain such an impulsive event.

Figure 15 also shows the relative coordinates of the spacecraft and Io as projected to the equator using the D_2 system. It is similar for D_4. Pioneer 11 apparently passed to the west of the flux tube, and the electron spike ended abruptly just as the spacecraft crossed the L shell of Io. As co-rotating magnetospheric plasma goes from west to east faster than Io's orbital motion, the trajectory is on the upstream side of the satellite in this projection. This statement should be treated with caution, because the projection could look different with a more complex magnetic field model. Also, Io itself could cause local disturbances which perturb the field. However, if the D_2 projection is correct, particles upstream of Io should cause no more surprise than the other phenomena associated with this remarkable satellite.

One of the virtues of the sheath acceleration model is that it makes specific predictions which can be tested. The electron power content, energy spectra, angular distributions, and spatial extent are all important quantities to be determined. This information will be better known after more sophisticated analyses have been performed on the data, but in the preliminary inspection there are differences between predictions and observations. The most significant of these differences are that (1) the pitch angle distribution is peaked perpendicular to the fieldline rather than parallel; (2) there seem to be electrons with energies exceeding the maximum available sheath potential; and (3) the peak found by Pioneer 11 was outside the Io-sweeping region whereas those found by Pioneer 10 were inside. Although this is disappointing, the sheath model does achieve a major success in predicting an electron source of about the right energy, and it seems more likely that the sheath model needs to be embellished than that it should be dropped.

V. THE BEHAVIOR OF ELECTRONS IN THE INNER JOVIAN MAGNETOSPHERE

The inner radiation belts of Jupiter consist of stably trapped particles in a dipole-like magnetic field, and the methods of radiation-belt theory developed for Earth are applicable with minor modifications. The Pioneer 10 flyby demonstrated that, as in the case of Earth, the inner Jovian radiation belt is populated by particles which have diffused inward toward the planet (McIlwain and Fillius 1975). The value of the radial diffusion coefficient has been estimated from a study of the absorption of these particles by the satellites (Sec. IV). Evidence for pitch angle diffusion has been found in the angular distributions and in an analysis of the radial profile of the trapped electron density (Fillius et al. 1976).

These effects are easiest to see after the data have been converted to

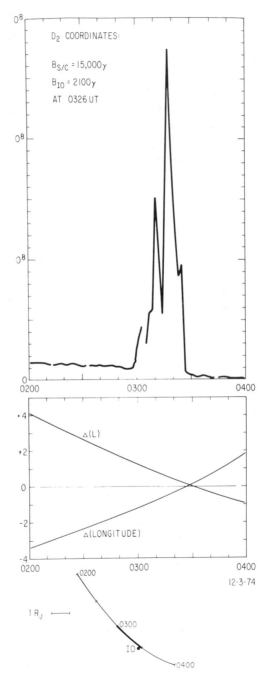

Fig. 15. Detail of the Pioneer 11 flyby of the Io flux tube, showing the local spike of energetic electrons. The relative magnetic coordinates of Io and Pioneer 11 were calculated using the D_2 field model. The relationship looks the same for D_4, but it might be different in a higher order model.

density, τ, in phase space, which we express in units of $(eV\ sec)^{-3}$. We relate τ to the radiation intensity by $\tau = 900\,j/(p\,c)^2$ where $p\,c$ is the particle momentum times the velocity of light expressed jointly in MeV, and j is the flux of particles $cm^{-2}\ sec^{-1}\ sr^{-1}\ MeV^{-1}$. As a consequence of Liouville's Theorem, the phase space density should be constant along a dynamical trajectory. It follows that, in the absence of sources or sinks, τ should be constant everywhere, and it also follows that in time-stationary circumstances a local maximum is a sure manifestation of a source. Trajectories that intersect an absorbing surface, such as the planet or a satellite, become vacated forward of the point of intersection. The image of the absorber is projected forward as a cluster of vacant trajectories — a forbidden cone in velocity space. Although Liouville's Theorem assures us that the cones always retain their identities, the orbits of trapped particles are so complex that forbidden and allowed cones become microscopically intermingled in a very complicated way. No realizable detector can resolve the individual allowed and forbidden cones; all that can be measured is a macroscopic average. A sink, local or not, dilutes the volume of phase space that is occupied at full density with an inextricable volume that is vacant, and a detector senses this dilution as a reduction in the apparent phase space density. A theoretical basis for dealing with this mixture has been worked out (Birmingham *et al.* 1967; Birmingham *et al.* 1974) and a diffusion-like equation is obtained for the ensemble-averaged phase space density. It is this, the macroscopic or ensemble-averaged phase space density, that we refer to in this chapter as τ.

Figure 16 shows near-equatorial profiles of τ versus L for electrons over a range of values of the first adiabatic invariant. These are from the Pioneer 10 inbound pass, and the method used to obtain τ from the data is described by McIlwain and Fillius (1975). It is immediately apparent that the major source is on the left, there is a sink on the right, and the net diffusive flow is toward the planet. At low energies, a subsidiary maximum occurs just inside the orbit of Io at $L = 5.9$. This corresponds to the peak seen in Fig. 14, but since it appears in the phase space density, Liouville's Theorem proves that it is caused by a local source.

If Fig. 16 is a solution of the radial transport equation, one can demonstrate that losses must take place throughout the region $3 \leqslant L \leqslant 10$. Lumping sources and sinks into one term, S, the radial equation is

$$\frac{\partial \tau}{\partial t} = L^2 \frac{\partial}{\partial L}\left(\frac{D}{L^2}\frac{\partial \tau}{\partial L}\right) + S \tag{5}$$

where D is the radial diffusion coefficient. Inside $10\ R_J$ most profiles in Fig. 16 can be described by a power law with slope of 4. Thus we will write as a solution of Eq. (5), $\tau(L) = \tau_1 L^{n_\tau}$ with $n_\tau \cong 4$. Mogro-Campero[9] has represented the diffusion coefficient as a power law in L, namely $D(L) = D_1 L^{n_D}$, with $n_D \cong 4$. Anticipating that it is a sink, we treat the source/sink term as

[9]See p. 1203.

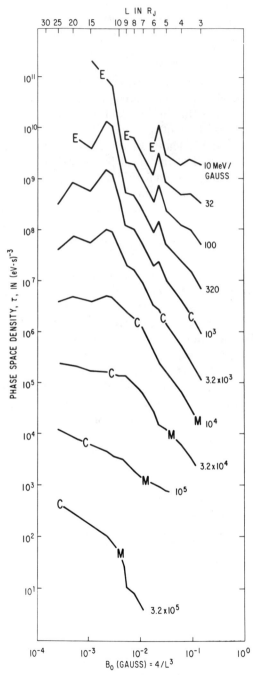

Fig. 16. Phase space density τ versus L at several values of the first adiabatic invariant. The second invariant is approximately zero. The physical processes acting on the particles are more apparent after the data have been converted to density in phase space.

an exponential in time, and write $S = \dfrac{d\tau}{dt} = \dfrac{d}{dt}\left[\tau_0 \exp\left(\dfrac{-t}{T}\right)\right] = -\dfrac{\tau}{T}$. Finally,

assuming that Fig. 16 represents a steady state solution, $\dfrac{\partial \tau}{\partial t} = 0$. Substituting these representations into Eq. (5) and differentiating, we get

$$\frac{1}{T} = \frac{D(L)}{L^2}\,(n_D + n_\tau - 3)\,n_\tau\,. \tag{6}$$

Note that a lack of distributed losses would correspond to $T = \infty$, and in that case the equation would balance only if $n_D + n_\tau - 3 = 0$. As it stands, the equation will not balance without distributed losses, and their lifetime is given by

$$T \cong \frac{1}{20}\frac{L^2}{D(L)}\,. \tag{7}$$

At $L = 5$, $T \simeq 1$ yr.

The identity of the loss mechanism has been discussed by Fillius et al. (1976) and they conclude that the particles are lost from the equator by pitch angle diffusion into the planetary loss cone. The particle trajectories are thus vacated in the atmosphere, which is accessed along the line of force after pitch angle scattering takes place at the equator.

VI. MULTIPLE PEAKS NEAR PIONEER 11 PERIAPSIS

Data from Pioneer 11 exhibit multiple peaks in the particle flux profiles near the closest approach to the planet. These are shown in Fig. 17 and their positions are listed in Table III. Note that the same features appear in both electron and proton profiles. Two of the minima, N1 and N4, are reasonably attributed to sweeping by Amalthea. There is some imprecision in matching L values, but this is presumably caused by uncertainties in the present magnetic field models. The remaining minima (and the maxima which complement them) are unexplained.

The reason for this multiple structure is a mystery, and it is one of the new and challenging problems of the Jovian radiation belts. Hypotheses that are specific to only one particle species cannot account for the fact that protons and electrons both exhibit the same features. Thus latitude-dependent synchrotron radiation losses, or regions of critical wave-particle interactions, seem to be incomplete explanations. One may speculate that our proton detector is really responding to electrons. The reader is referred to our earlier paper (Fillius and McIlwain 1974b) for an account of the particle identification and background elimination procedures, but he will probably derive greater satisfaction from the fact that the University of Chicago fission detector independently recorded similar proton features (Simpson et al. 1975). We regard the experimental evidence as convincing.

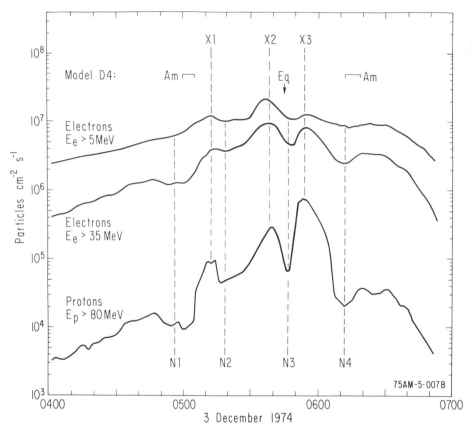

Fig. 17.   Electron and proton fluxes measured near the closest approach of Pioneer 11 to Jupiter (1.6 R_J from the center of the planet at 5^h 23^m). The multiple peak structure is un-explained, and could be accounted for by magnetic field anomalies, a dust ring, or some other cause.

   It remains, then, to explain the unexpected features with a mechanism that operates on both protons and electrons. The simplest is absorption by a solid body. Indeed, we have attributed minima N1 and N4 to the sweeping effects of Amalthea. If the magnetic field is so distorted that the particle drift shells become rippled, the spacecraft could have passed through these features again. Alternatively, in a distorted field some of the drift shells might dip down to the planet's surface and be emptied. Detailed calculations of particle drift shells will be necessary to test these hypotheses. Furthermore, the results of the calculations will only be as good as the magnetic field models, and the magnetometer experimenters have cautioned us against extrapolating to the planet's surface. However, it may still be possible to test these ideas.

TABLE III
Zenocentric Coordinates
for Particle Features in Figure 17

Feature in Fig. 17	(R_J)	Zenocentric Coordinates	
		Latitude (deg)	Longitude III (deg)
N1	1.78	-39.5	312
X1	1.62	-24.6	342
N2	1.60	-18.2	351
X2	1.66	$+3.9$	22
N3	1.74	$+12.0$	33
X3	1.82	$+18.3$	43
N4	2.11	$+31.0$	67

A more radical hypothesis has been raised by Acuña and Ness (1976). They suggest that, if the drift shells are not sufficiently complex, the possibility should be considered of another, hitherto undetected, satellite inside the orbit of Amalthea. It would not have to be a single mass, for a ring of smaller particles could do the job as well. Indeed, since minima N2 and N3 are inside the Roche limit, a particle ring is more likely. Considering the similarities between Jupiter and Saturn, there seems to be no *a priori* reason against a dust ring near Jupiter, too. However, there are obvious questions which need to be investigated. Why has there been no optical detection of such a ring? What would be the gravitational effect on the other satellites and on the Pioneer spacecraft? Would one expect the Pioneer Meteoroid Detectors[10] to detect the ring when they passed near it? This exciting hypothesis has obvious problems, and it will take some time to sort out all of the possibilities and ramifications.

Acknowledgement. We thank C. McIlwain, A. Mogro-Campero, and W-H. Ip for their collaboration on the work reviewed here. This research was supported by NASA Contracts NAS 2-5602, NAS 2-6552, and NASA Grant NASA-NGL 05-005-007.

---

[10]See p. 1062.

REFERENCES

Acuña, M. H., and Ness, N. F. 1976. The complex main magnetic field of Jupiter. *J. Geophys. Res.* In press.

Barish, F. D., and Smith, R. A. 1975. An analytical model of the Jovian magnetosphere. *Geophys. Res. Lett.* 2:269–272.

Birmingham, T.; Hess, W.; Northrup, T.; Baxter, R.; and Lojko, M. 1974. The electron diffusion coefficient in Jupiter's magnetosphere. *J. Geophys. Res.* 79:87–97.

Birmingham, T. J.; Northrup, T. G.; and Fälthammar, C. G. 1967. Charged particle diffusion by violation of the third adiabatic invariant. *Phys. Fluids* 10:2389–2398.

Chenette, D. L.; Conlon, T. F.; and Simpson, J. A. 1974. Bursts of relativistic electrons from Jupiter observed in interplanetary space with the time variation of the planetary rotation period. *J. Geophys. Res.* 79:3551–3558.

Fillius, R. W., and McIlwain, C. E. 1974a. Radiation belts of Jupiter. *Science* 183:314–315.

————. 1974b. Measurements of the Jovian radiation belts. *J. Geophys. Res.* 79:3589–3599.

Fillius, R. W.; McIlwain, C. E.; and Mogro-Campero, A. 1975. Radiation belts of Jupiter: a second look. *Science* 188:465–467.

Fillius, R. W.; McIlwain, C.; Mogro-Campero, A.; and Steinberg, G. 1976. Evidence that pitch angle scattering is an important loss mechanism for energetic electrons in the inner radiation belt of Jupiter. *Geophys. Res. Lett.* 3:33–36.

Hall, C. F. 1974. Pioneer 10. *Science* 183:301–302.

————. 1975. Pioneer 10 and 11. *Science* 188:445–446.

Huba, J. D., and Wu, C. S. 1976. On the sweeping of energetic particles by Io. In draft.

McIlwain, C. E., and Fillius, R. W. 1975. Differential spectra and phase space densities of trapped electrons at Jupiter. *J. Geophys. Res.* 80:1341–1345.

Mead, D. 1974. Magnetic coordinates for the Pioneer 10 Jupiter encounter. *J. Geophys. Res.* 79:3514–3521.

Mead, G. D., and Hess, W. N. 1973. Jupiter's radiation belts and the sweeping effect of its satellites. *J. Geophys. Res.* 78:2793–2811.

Mihalov, J. D.; Collard, H. R.; McKibbin, D. D.; Wolfe, J. H.; and Intriligator, D. S. 1975. Pioneer 11 encounter: preliminary results from the Ames Research Center plasma analyzer experiment. *Science* 188:448–451.

Mogro-Campero, A., and Fillius, W. 1975. The absorption of trapped particles by the inner satellites of Jupiter and the radial diffusion coefficient of particle transport. *J. Geophys. Res.* In press.

Reif, F. 1965. *Fundamentals of statistical and thermal physics.* New York: McGraw-Hill, Inc. pp. 486–488.

Shawhan, S. D. 1976. Io sheath-accelerated electrons and ions. *Icarus* (special Jupiter issue). In press.

Simpson, J. A.; Hamilton, D.; Lentz, G.; McKibben, R. B.; Mogro-Campero, A.; Perkins, M.; Pyle, K. R.; Tuzzolino, A. J.; and O'Gallagher, J. J. 1974a. Protons and electrons in Jupiter's magnetic field: results from the University of Chicago experiment on Pioneer 10. *Science* 183:306–309.

Simpson, J. A.; Hamilton, D. C.; Lentz, G. A.; McKibben, R. B.; Perkins, M.; Pyle, K. R.; Tuzzolino, A. J.; and O'Gallagher, J. J. 1975. Jupiter revisited: first results from the University of Chicago charged particle experiment on Pioneer 11. *Science* 188:455–458.

Simpson, J. A.; Hamilton, D. C.; McKibben, R. B.; Mogro-Campero, A.; Pyle, K. R.; and Tuzzolino, A. J. 1974b. The protons and electrons trapped in the Jovian dipole magnetic field region and their interaction with Io. *J. Geophys. Res.* 79:3522–3544.

Teegarden, B. J.; McDonald, F. B.; Trainor, J. H.; Webber, W. R.; and Roelof, E. C. 1974. Interplanetary MeV electrons of Jovian origin. *J. Geophys. Res.* 79:3615–3622.

Trainor, J. H.; McDonald, F. B.; Teegarden, B. J.; Webber, W. R.; and Roelof, E. C. 1974a. Energetic particles in the Jovian magnetosphere. *J. Geophys. Res.* 79:3600–3613.

Trainor, J. H.; Teegarden, B. J.; Stilwell, D. E.; McDonald, F. B.; Roelof, E. C.; and Webber, W. R. 1974b. Energetic particle population in the Jovian magnetosphere: a preliminary note. *Science* 183:311–313.

Thomsen, M. F., and Goertz, C. K. 1975. Satellite sweep-up effects at Jupiter. *Trans. Am. Geophys. Union* 56:428–429.

Van Allen, J. A.; Baker, D. N.; Randall, B. A.; Thomsen, M. F.; Sentman, D. D.; and Flindt, H. R. 1974. Energetic electrons in the magnetosphere of Jupiter. *Science* 183:309–311.

Van Allen, J. A.; Randall, B. A.; Baker, D. N.; Goertz, C. K.; Sentman, D. D.; Thomsen, M. F.; and Flindt, H. R. 1975. Pioneer 11 observations of energetic particles in the Jovian magnetosphere. *Science* 188:459–462.

Wolfe, J. H.; Mihalov, J. D.; Collard, H. R.; McKibbin, D. D.; Frank, L. A.; and Intriligator, D. S. 1974. Pioneer 10 observations of the solar wind interaction with Jupiter. *J. Geophys. Res.* 79:3489–3500.

HIGH-ENERGY PARTICLES IN THE
JOVIAN MAGNETOSPHERE

J. A. VAN ALLEN
The University of Iowa

A digest is given of the principal observational findings of the University of Iowa experiments on Pioneers 10 and 11 during their passages through the magnetosphere of Jupiter in November–December 1973 and November–December 1974, respectively. Data and phenomenological interpretations are presented on absolute energy spectra, angular distributions, and positional distributions of electrons in several energy ranges for energies $E_e > 40$ keV and of protons in the energy range $0.61 < E_p < 3.41$ MeV. Some provisional suggestions are made on the physical dynamics of the Jovian magnetosphere and on their differences from those of the earth's magnetosphere.

In 1955 Burke and Franklin (1955) reported the first persuasive evidence that Jupiter is a non-thermal source of radio noise. Their early observations of sporadic bursts of noise were made principally at a frequency $f = 22.2$ MHz (decametric wavelengths). During the subsequent twenty years, a large body of knowledge has been developed on the properties of the Jupiter system as a non-thermal (as well as a thermal) radio emitter. Valuable reviews of the original work have been given by various authors as the subject has developed, most recently by Carr and Gulkis (1969) and by Warwick (1970).[1] The presently known phenomena fall into two qualitatively different classes.

In the decametric range ($f < 40$ MHz):

(a) The emissions are sporadic.

(b) The angular diameter of the source, if phase-incoherent, is less than 0.1 arcsecond, i.e., less than 1/400th of that of the planet (Dulk 1970).

(c) The probability of occurrence of bursts is synchronous with the rotation of the planet and in fact defines a planet-fixed longitude system (System III) with high precision.

(d) The probability of occurrence of bursts is further modulated in a pattern related to the orbital phase of the satellite Io relative to the planet-earth line.

---

[1] See pp. 621, 693 and 1146.

The decametric emissions are attributed to instabilities in plasma distributions within the external magnetic field of the planet. The upper limit of the spectrum of decametric bursts ($f \approx 40$ MHz) has been interpreted as the electron gyro-frequency at the source. It corresponds to a magnetic field strength of 14 gauss, probably in the ionosphere of the planet.

Non-thermal emissions in the decimetric range ($f \gtrsim 200$ MHz) are of a quite different nature:

(a) The source is a toroidal region encircling the planet with the central plane of the toroid tilted about 10° to the equatorial plane.
(b) The major diameter of the toroid is about two times the diameter of the planet.
(c) The radiation is strongly plane polarized with the electric vector parallel to the central plane of the toroid.
(d) The flux density S diminishes by only a factor of ~ 2 as the wavelength increases from 5 to 180 cm.
(e) The emission is essentially time-independent in intensity and spectral form, though there is some evidence for a small fractional variation in intensity over periods of several years.

Following the original suggestion of Drake and Hvatum (1959), the decimetric emission is interpreted as synchrotron radiation by relativistic electrons ($E_e \gtrsim 0.5$ MeV) trapped in the external dipolar magnetic field of the planet, as in the Van Allen radiation belts of the earth.

In adopting the synchrotron interpretation, many authors have proposed models of the energy spectra and distribution of relativistic electrons in the Jovian magnetosphere. The radio data do not fix the various parameters of such models uniquely but do confine their values to a finite volume in parameter space. No useful information is obtained on energetic protons.

The non-thermal radio emissions of Jupiter distinguish it from all of the other planets, except the earth and possibly Saturn (within present knowledge), and have provided a special motivation for the *in situ* investigation of Jupiter.

I. IN SITU OBSERVATION OF THE MAGNETOSPHERE OF JUPITER

Particle-detecting instruments on the two NASA Ames Research Center spacecraft Pioneers 10 and 11 have provided a large body of definitive information on the energy spectra, angular distributions, and positional distributions of energetic electrons and protons in the Jovian magnetosphere. Detailed results from the two flyby missions have been published previously in the 25 January 1974 issue of *Science*, in the 1 September 1974 issue of the *Journal of Geophysical Research*, and in the 2 May 1975 issue of *Science*. This chapter does not attempt to give a comprehensive review of all these results. Rather, it emphasizes some of the principal findings of the University of Iowa experiments on Pioneers 10 and 11 and offers some interpretative suggestions.

The hyperbolic encounter trajectory of Pioneer 10 with Jupiter was prograde in a plane inclined $13°.8$ to the planet's equatorial plane and passed through periapsis at a radial distance of 2.85 R_J (1 $R_J = 71,372$ km, the adopted value of the equatorial radius of the planet) at $3^h 12^m$ ERT [earth received time (UT) of the telemetered data] on 4 December 1973. The encounter trajectory of Pioneer 11 was retrograde in a plane inclined at $51°.8$ and passed through periapsis at a radial distance of 1.60 R_J at $6^h 3^m$ ERT on 3 December 1974.

Our instrument has been described in detail by Baker (1973), Van Allen et al. (1974a), and Van Allen et al. (1975). Briefly, we measure the absolute intensities, energy spectra, and angular distributions of electrons ($E_e > 0.040$ MeV) in a number of energy ranges and, on Pioneer 11, the absolute intensities and angular distributions of protons $0.61 < E_p < 3.41$ MeV, all on a point-by-point basis along the encounter trajectories. Positional data and attitude data for the rotating spacecraft are supplied by the Ames Research Center and the Jet Propulsion Laboratory as are all of the raw telemetry data.

II. PIONEER 10 OBSERVATIONS

Survey of Observations

Figure 1 shows an ecliptic-plane projection of the encounter trajectory of Pioneer 10 relative to the planet and Fig. 2 shows the trajectory as projected on the (non-uniformly) rotating magnetic meridian plane through the spacecraft. The adopted magnetic field model as specified in Fig. 2 is one derived from our own particle measurements as sketched in a later section.

Figures 3, 4, 5, and 6 show absolute intensities of electrons in several energy ranges as a function of ERT during the course of the encounter. The most noteworthy features of a general nature are the following:

(a) The physical dimensions of the magnetosphere are on the order of twice as great as had been expected from scaling the earth's magnetosphere to the radio astronomical value of the magnetic moment of the planet and to the anticipated solar wind conditions there. On the sunward side, the presence of energetic particles was first detected at 109 R_J (7.78×10^6 km) and on the dawn side it was observed out to over 150 R_J (10.7×10^6 km). Thus assuming that the dimension on the dusk side is also 150 R_J, the magnetosphere subtends an angle of 2° as viewed from the earth at opposition.

(b) There is a remarkably high spectral intensity of relativistic electrons in the outer magnetosphere, even in the magnetosheath. Such energetic electrons are incompatible with the expectations for a thermalized solar wind.

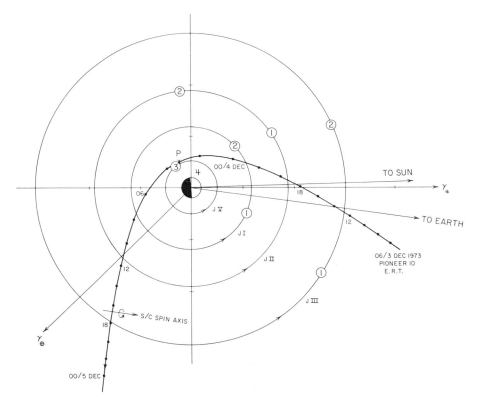

Fig. 1. Projection on the ecliptic plane of the hyperbolic encounter trajectory of Pioneer 10 with Jupiter and the orbits of the four inner satellites. The numbers *1* and *2* on the orbits of Io (J I), Europa (J II), and Ganymede (J III) show the positions of each of these satellites at the times that the spacecraft crossed the *L* shell of that satellite, inbound and outbound, respectively. The number *3* shows the position of Amalthea (J V) at the time that the space-craft was at periapsis, marked *P*; $\gamma_{2\iota}$ designates Jupiter's vernal equinox, and $\gamma_{\oplus}$ designates earth's vernal equinox. Note that the spacecraft's spin axis is parallel to the planet-Earth line throughout the encounter (Van Allen *et al.* 1974*a*).

(c) On 1 December at a radial distance of about 50 R_J, there was a 6-hour duration drop of intensities to magnetosheath values. This result, when taken together with the magnetic and plasma data, corresponds to a transient compression of the outer boundary of the magnetosphere by a factor of two in radial dimension, thus signifying the delicacy of the pressure balance in the outer magnetosphere.

(d) The omnidirectional intensity of electrons of energy $E_e > 0.060$ MeV has a maximum value of 4×10^8 cm^{-2} sec^{-1} in the inner magneto-sphere. The energy spectrum is relatively flat out to ≈ 10 MeV and then falls off more steeply toward higher energies.

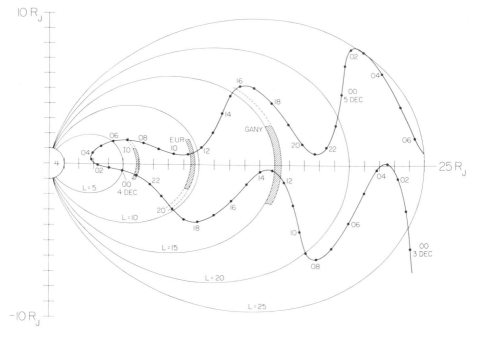

Fig. 2. The time-labeled trace of Pioneer 10's encounter trajectory in magnetic polar coordinates (magnetic meridian plane projection) for the Jovian centered dipole: tilt = 9°.5;
λ_{III} (1957) = 230°. The cross hatching shows the regions that bound the orbits of Io, Europa,
and Ganymede in this coordinate system. The time is earth received time (ERT) (Van
Allen *et al.* 1974*a*).

   (e) In our lower-energy channels there is a signature in the intensity —
time profiles associated with passage through the magnetic shells of the
satellites Ganymede, Europa, and Io (cf. Fig. 2). The general nature of the
intensity versus radial distance curve strongly suggests that each of the satellites inhibits (by physical sweeping) the inward diffusion of particles and
greatly reduces the intensity that would have existed inward of their magnetic shells if they were not present.[2] These effects are progressively less
important at higher electron energies. There is an apparent peak in the intensity of lower-energy electrons interior to the Io shell on the inbound pass,
suggesting local injection.
   (f) Most notably on the outbound portion of the trajectory (Fig. 5) but
also evident on the inbound portion (particularly for higher-energy electrons)
(Fig. 3), there is a persistent well-ordered 10-hour modulation of intensity.
This is identified with the rotational period of the planet.
   (g) Throughout the encounter, even within the inner magnetosphere
(Fig. 4) there are marked temporal variations of intensity especially at the

_____
[2]See p. 1203.

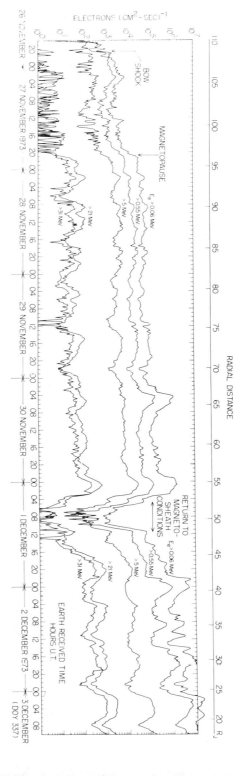

Fig. 3. Absolute omnidirectional intensities of electrons in five integral energy ranges as a function of time during Pioneer 10's inbound traversal of the outer magnetosphere (Baker and Van Allen 1976a).

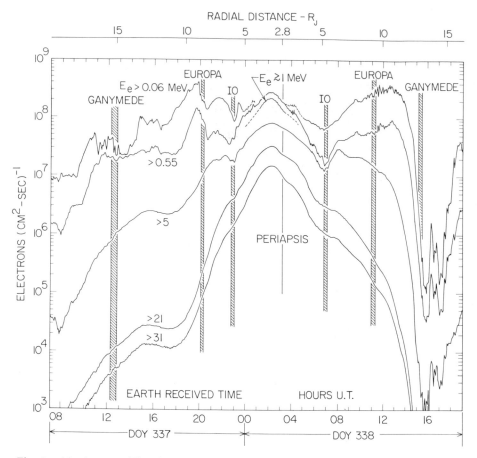

Fig. 4.   Absolute omnidirectional intensities of electrons in six integral energy ranges as a
function of time during Pioneer 10's traversal of the inner magnetosphere. The geometrical
sweeping regions of the three inner Galilean satellites are shown by vertical shaded bars
(Baker and Van Allen 1976*b*).

lower energies. These variations are taken to signify dynamic, unstable
conditions.

The Magnetodisc

The magnetometer data (Smith *et al.* 1974*a*)[3] show that inside a radial
distance $r \approx 12\ R_J$ the magnetic field is approximately dipolar whereas outside of 20 R_J it is strongly distended. The region $12 < r < 20\ R_J$ is one of
transition. Taking these results together with our particle data we have proposed (Van Allen *et al.* 1974*b*) the disc model of Jupiter's magnetosphere
shown in Fig. 7 with magnetic field lines near the magnetic equator being

[3]See p. 788.

closed, and hence capable of trapping charged particles out to $r \gtrsim 100\ R_J$. The model corresponds most particularly to data on the outbound portion of the trajectory (Fig. 5). In the interest of presenting a first-order schematic model we suggested originally that the central plane of the thin magnetodisc was coincident with the magnetic equatorial plane and therefore inclined at $9°5$ to the rotational equatorial plane of the planet. The corresponding wobbling of the disc as the planet rotates would reproduce the observations (cf. Fig. 2 as extended to greater radii). A second feature of the original suggestion was that the magnetodisc had approximate axial symmetry about the magnetic dipolar axis of the planet. This second feature was already known to be only a crude approximation as is evident from comparing Fig. 3 (sunward side) and Fig. 5 (dawn side). On the basis of these Pioneer 10 data and more decisively on the basis of high-latitude data from Pioneer 11 to be considered below, it is now clear that the magnetodisc is much blunted (i.e., extends to much higher latitudes) on the sunward side of the magnetosphere. Restricting consideration to the dawnward form of the magnetodisc, Smith *et al.* (1974*b*) have contested the first feature of our model, its rigid coincidence with the magnetic equatorial plane. On the basis of a semi-intuitive interpretation of their magnetic field data, they suggest that the magnetodisc is bent away progressively from the magnetic equatorial plane toward parallelism with the rotational equatorial plane as r increases. However, Goertz (1976) has demonstrated that the bent (or flapping) magnetodisc model of Smith *et al.* is inconsistent with both simple geometric and detailed physical considerations (cf. Hill *et al.* 1974). Goertz finds that the best-fit model is one more nearly resembling the rigid disc model and he further shows that the magnetic field observations themselves are consistent with the rigid model even though they were invoked by the magnetometer experiments to support the bent disc model.

Orientation of the Magnetic Dipolar Moment of the Planet

The decimetric radio astronomical observations (see above) have yielded values for the tilt of the dipole, its (lack of) eccentricity, and the System III longitudes of its poles. The decametric observations have yielded an approximate value of the absolute magnetic moment, and both decimetric and decametric evidence has shown that the magnetic north pole is in the northern hemisphere of the planet, corresponding to the opposite polarity relationship between magnetic moment and rotational angular momentum that exists for the earth.

The conventional expectation is that direct magnetic vector measurements will give superior determinations of all of the above quantities as well as those corresponding to higher order moments. This is doubtless true, given sufficient instrumental accuracy *and* a sufficiently comprehensive, three-dimensional network of observations. However, point-by-point measurements along a single flyby trajectory have substantial deficiencies in

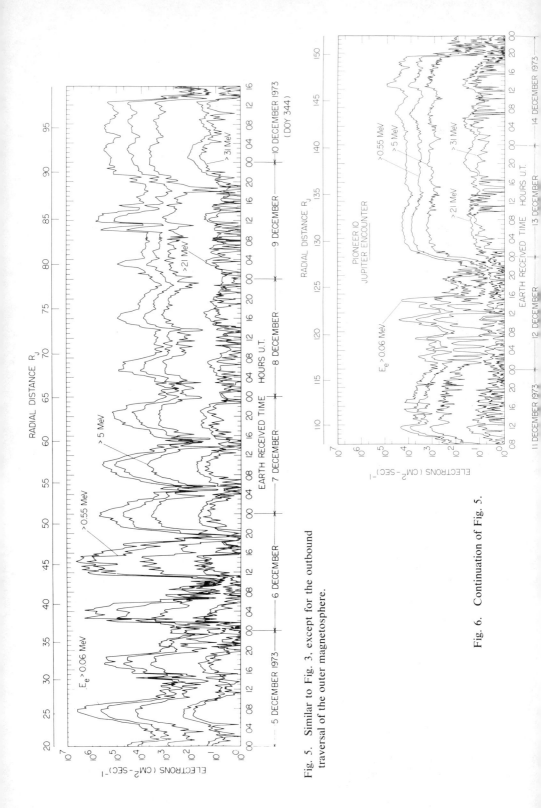

Fig. 5. Similar to Fig. 3. except for the outbound traversal of the outer magnetosphere.

Fig. 6. Continuation of Fig. 5.

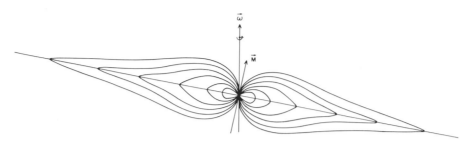

Fig. 7. A heuristic sketch of the magnetodisc model of the outer magnetosphere of Jupiter, showing topology of the magnetic field and the region of trapped energetic particles. The outer tip of the sketch is at a radial distance of 100 to 150 R_J. The planet's rotational axis is denoted by ω and its magnetic axis by **M** (Van Allen *et al.* 1974*b*). See text for qualifying discussion.

meeting the second criterion. We show that the orientation of the magnetic dipolar moment can be determined by particle measurements alone, without use of magnetic field data. This method has, in fact, a certain potential superiority in that the particle data automatically correspond to a longitudinally averaged situation rather than to a point-by-point one. The technique is to adopt the tilt of the dipole axis and the longitude of its north pole as free parameters, and then to find the values of these two parameters that yield the best closure of inbound and outbound data (corrected to the equator) on energetic electron intensities as a function of magnetic shell parameter. The System III longitude of the spacecraft increases from 231° at $r = 4\ R_J$ inbound to 264° at $r = 4\ R_J$ outbound. Because of this small difference in longitude, we considered it futile to introduce three additional parameters to characterize an offset, i.e., we assumed a centered dipole.

The process is illustrated by the succession of Figs. 8, 9, and 10 wherein latitude corrections have been made by assuming various simple pitch-angle dependences as shown. Our best values are 9°.5 for the tilt and 230° for the System III (1957.0) longitude of the north pole. The closure of data except for $L < 4$ is relatively insensitive to the assumed angular distribution. It deteriorates perceptively for tilts different from the above value by $\pm$ 0°.5 and for longitudes different by $\pm$ 3°. There is no dipolar model which will produce closure for $L \gtrsim 12$, unless one invokes a very special *ad hoc* dependence of the angular distribution. We carried out our analysis at a time when the tilt determined by the magnetometer (Smith *et al.* 1974*a*) was 14°.7, a value that we found unacceptable (Fig. 8). A fuller analysis of the magnetometer data later yielded a tilt of 10°.6 (Smith *et al.* 1974*b*).

Energy Spectra and Phase Space Densities

Exemplary energy spectral data for electrons, taken from an extensive study by Baker and Van Allen (1976*a*), are shown in Fig. 11. The adopted

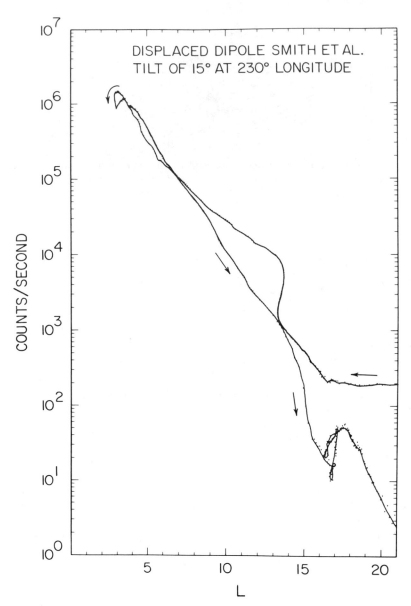

Fig. 8.   Inbound and outbound counting rates of Detector C ($E_e > 21$ MeV) on Pioneer 10 as a function of L for the Smith *et al.* (1974*a*) model of the Jovian magnetic field. All rates are corrected to magnetic equatorial values, assuming a pitch-angle (α) distribution of uni-directional intensity $j \propto \sin^4 \alpha$ (Van Allen *et al.* 1974*a*).

spectral form is

$$\frac{\mathrm{d}J}{\mathrm{d}E} = k\,E^{-1.5}\,(1 + E/H)^{-n} \tag{1}$$

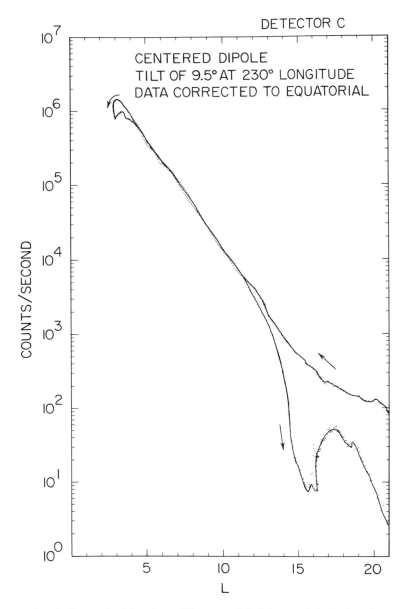

Fig. 9.   A plot similar to Fig. 8 but for a different model of the magnetic field as shown.

where J is the omnidirectional intensity in $cm^{-2}\ sec^{-1}$, E is the kinetic energy in MeV, and H and n are fitting parameters.

Energy spectra are valuable for certain purposes, but one obtains a more useful formulation for diffusional and other dynamical analyses by converting the energy spectra to phase space density as a function of magnetic mo-

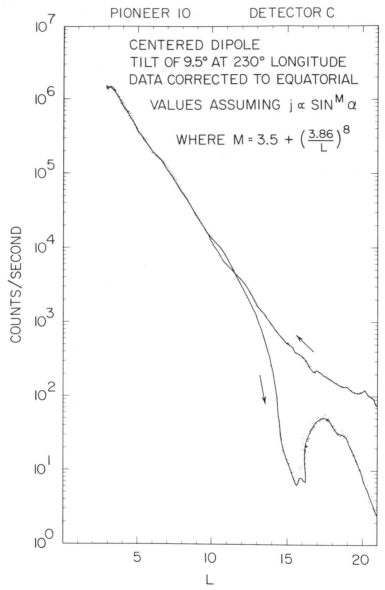

Fig. 10.   A final version of Fig. 9, assuming a simple L dependence of m in the angular distribution $j \propto \sin^m \alpha$.

ment $f(\mu)$ (cf. McIlwain and Fillius 1975). Such a formulation is shown in Fig. 12. The values of all three $f(\mu)$ diminish decisively and strongly toward lesser r for $r < 12\ R_J$, thus making it virtually certain that the predominant source of inward diffusing electrons is outside of that radius. $f(\mu_1)$ and $f(\mu_2)$ have absolute maxima at considerably greater r but vary indecisively for

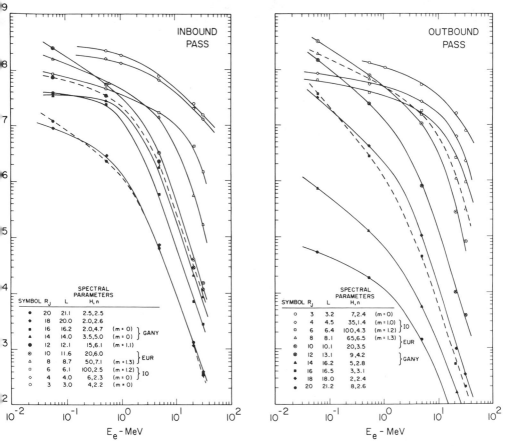

Fig. 11. Sample electron energy spectra within the central magnetosphere ($r < 20\ R_J$). The magnetic shell parameters L and magnetic latitudes λ are based on the centered dipole model with a tilt of $9°5$ toward System III (1957.0) longitude of 230°. Integral intensity points are connected by smooth curves and approximate spectral parameters are given (Baker and Van Allen 1976a).

$r > 12\ R_J$; $f(\mu_3)$ appears to diminish monotonically toward lesser r for $r < 60\ R_J$, though it is less well determined than $f(\mu_1)$ and $f(\mu_2)$.

Angular Distributions

A detailed study of the Pioneer 10 data on angular distributions of electrons has been made by Sentman and Van Allen (1976). Angular distributions having a maximum at pitch angle $\alpha = 90°$ and minima at 0° and 180° are called "pancake" distributions and those having a minimum at 90° and maxima at 0° and 180° are called "dumbbell" distributions. The following summary of observations is quoted from the above cited paper:

> "The run of observations suggests the following summary of the spatial distribution of electron $E_e > 0.06$ MeV pitch-angle distributions for the regions of the Jovian magnetosphere surveyed during the Pioneer 10 encounter.

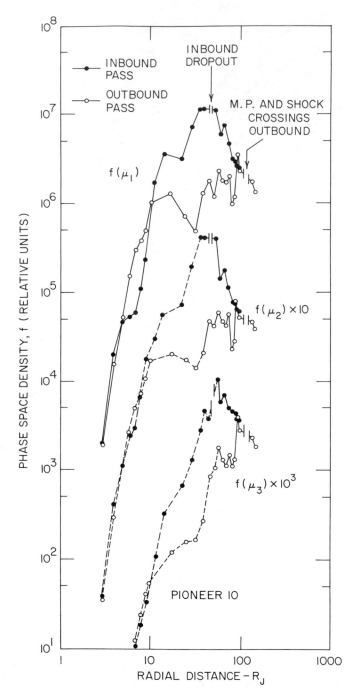

Fig. 12.   Radial dependence of phase space densities $f(\mu)$ of near-equatorial electrons having a constant value of the first adiabatic invariant μ; μ_1, μ_2, and μ_3 are equal to 1.3×10^3, 1.7×10^4, and 5.9×10^5 MeV Gauss^{-1}, respectively.

"(1) The inner core hard trapping region $R \lesssim 12 \, R_J$ displays pitch angle distributions that are pancake in character for all measured low-energy electrons. These distributions are of the same type as those deduced previously for higher energy electrons ($E_e > 21$ MeV) in this region [Van Allen et al. 1974a, 1974b; Simpson et al. 1974; Fillius and McIlwain 1974].

"(2) The region between 12 and 25 R_J shows strong field-aligned bidirectional anisotropies. The maximum strength of these anisotropies occurs at approximately $20 \pm 3 \, R_J$. Electrons of the lowest energies observed show the largest degree of anisotropy, while electrons with energies greater than about 5 MeV show very little of the anisotropic behavior prevalent at lower energies.

"(3) The nature of electron pitch-angle distributions in the outer magnetosphere $R \gtrsim 25 \, R_J$ is dependent on local time and on effective magnetic latitude. Near the noon meridian at distances between ~ 25 and 40 R_J weak anisotropies show periodic behavior corresponding to the 10-hour rotational period of the planet. The nature of the periodicities at this local time is such that dumbbells occur on or near the magnetic equator, while weak pancakes appear at off-equatorial latitudes. The distributions on the noonside of Jupiter at distances of 60–100 R_J show weak dumbbells (bidirectional). The occurrence and persistence of these anisotropies seem to show a correlation with the degree of disorderliness in the magnetic field, such as is displayed during DOY's 333 and 334. Two brief instances of unidirectional field-aligned streaming suggest the existence of dynamical processes in this region. Near the dawn meridian at distances greater than 25 R_J the particle distributions for each of the detected energies are predominantly isotropic. Exceptions to this isotropy occurred briefly for the lowest energy electrons at distances of about 40 R_J."

The corresponding interpretative remarks of Sentman and Van Allen are as follows:

"The pancake distributions inside of 12 R_J are considered to be roughly analogous to those in the strong trapping regions of the terrestrial radiation belts. Presumably these distributions result from inward radial diffusion with approximate conservation of the first and second adiabatic invariants μ and J and from the inevitable existence of loss cones near $\alpha = 0°$ and 180°.

"In quasi-trapping regions of the terrestrial magnetosphere, dumbbell distributions prevail [Haskell 1969]. It is tempting to apply analogous considerations to the distributions in the Jovian magnetodisc, though we do not pretend to offer any full explanation of the observations.

"The following further remarks are of a qualitative and descriptive nature.

"(a) Isotropic and dumbbell distributions suggest quasi-trapping or durable trapping with a wide distribution of mirror points. The suggestion is stronger for dumbbell distributions. In neither case is it rigorous, because open field lines having kinks and many back-scattering centers can masquerade as closed field lines. An analysis of the magnetic field topology in the magnetodisc by the Euler potential technique does, however, strongly indicate that magnetic field lines near the central plane of the magnetodisc are indeed closed, whereas the relatively un-populated lines above and below the magnetodisc are open [Goertz et al. 1976].

"(b) On the upper and lower faces of the magnetodisc (defined as the distant magnetospheric region populated by electrons $E_e > 0.06$ MeV), there is presumably a transition region of time-variable openness and closedness. Particles having pitch angles near 0° and 180° will leak out preferentially during transient openness, thus tending to produce pancake distributions, as observed. An illuminating example of such an effect occurred on DOY 331 immediately inside the sunward magneto-pause . . . Another occurred on DOY 335 immediately following the particle intensity 'dropout' on that day.

"(c) Further local time and real-time effects on the angular distributions may be expected during the cyclic diurnal expansion and compression of the outer magnetosphere as well as during transient fluctuations induced by variable solar wind pressure.

"(d) The possibility that the strong dumbbell distributions in the radial range $12-25$ R_J might be a natural result of injection of low energy electrons at the magnetopause and subsequent inward diffusion in the much distended magnetodisc field has been investigated by a numerical experiment. For this purpose there was constructed a magnetic field model approximating the main features of the distended field observed by Smith et al. [1974b] on the dawn side of Jupiter. Particles with various pitch angles were injected at 50 R_J on the magnetic equator and imagined to diffuse inward under the conservation of the adiabatic invariants μ and J. The equatorial pitch angles of the particles were then numerically evaluated as a function of radial distance from Jupiter. The calculations showed that all particle pitch angles drifted toward 90° as the particles diffused toward the undistorted dipole field, even when the particles were in the severely distended region of the model. Therefore, pitch angle distributions comprising such particles should show a progressive increase at $\alpha = 90°$ and depletion at the ends of the distribution as inward diffusion proceeds. In particular, iso-tropic injection at 50 R_J should result in pancake distributions every-where interior to this point if μ and J are conserved and if large scale electric fields are absent. Thus such a simple set of assumptions regarding particle diffusion conditions does *not* account for the observed dumbbell distributions at 20 R_J."

III. PIONEER 11 OBSERVATIONS

Survey of Observations

Figures 13 and 14 show the encounter trajectory of Pioneer 11 and Figs. 15, 16, and 17 show the time dependence of absolute intensities of electrons in several energy ranges and of protons in a chosen energy range. These figures are comparable to those given earlier for Pioneer 10.

The Pioneer 11 observations are generally confirmatory of those with Pioneer 10 but because of improvements in our instrument and because of the quite different geometrical nature of the encounter trajectory, important new results were obtained (Van Allen *et al.* 1975).

Magnetodisc

On the inbound trajectory the first crossing of the bow shock was identified by the magnetometer (Smith *et al.* 1975)[4] and plasma analyzer (Mihalov *et al.* 1975)[5] experimenters at $4^h\ 20^m$ ERT on 26 November 1974 at $r = 109.7\ R_J$. Thereafter there occurred two magnetopause crossings, then two bow shock crossings, and a "final" durable crossing of the magnetopause at about 65 R_J on 29 November (Fig. 15). Again we note there is a 10-hour modulation similar to that observed by Pioneer 10 on its inbound trajectory, at a similar local time. Absolute particle intensities are of comparable magnitudes for both cases but details of intensity versus r curves are quite different.

The observations of the outer magnetosphere by Pioneer 11 on the outbound portion of its trajectory are generally similar to those inbound despite the much higher latitude outbound (Fig. 14). It is noted that both inbound and outbound trajectories of Pioneer 11 (Fig. 13) are on the sunward side of the planet. Thus, there are three generally similar runs of data (Pioneer 10 inbound and Pioneer 11 inbound and outbound) in the outer magnetosphere and one that is quite different (Pioneer 10 outbound). The three similar cases are all on the sunward side of the planet whereas the different case (Figs. 5 and 6) is on the dawn side. If this limited body of results is taken to be characteristic of the situation, then the magnetodisc, which is thin on the dawn side, is much blunted and extends to much higher latitudes on the sunward side but still exhibits the 10-hour modulation. Although nothing is yet known about the configuration of the magnetodisc on the dusk and anti-sunward sides of the planet, it already seems reasonably certain that the magnetodisc undergoes diurnal pumping to a quite important degree. The significance of this pumping process in energizing particles and in inducing their diffusion has not yet been adequately considered.

Distribution of Quite Energetic Electrons in the Inner Magnetosphere

In Fig. 18, we show a family of iso-counting rate contours in the magnetic meridian plane as derived from the counting rate versus time data from detec-

[4]See p. 788.   [5]See p. 855.

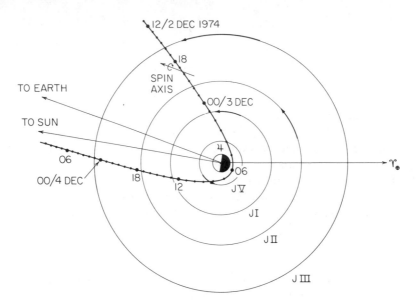

Fig. 13. Projection on the ecliptic plane of the hyperbolic encounter trajectory of Pioneer 11 with Jupiter and the orbits of the four inner satellites. The heavy arc with an arrow on each satellite orbit represents the motion of that satellite between inbound and outbound crossings of its *L* shell by the spacecraft.

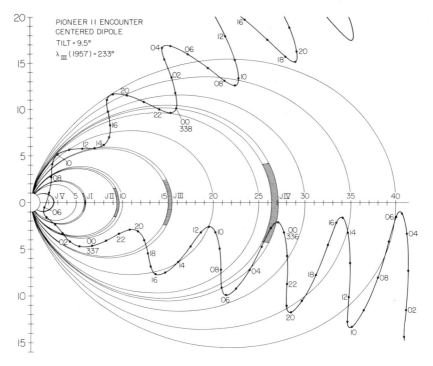

Fig. 14. A trajectory diagram for Pioneer 11, similar to Fig. 2. Note that the longitude of the magnetic pole has been updated from 1973 to 1974 by an increase of 3° (Van Allen *et al.* 1975).

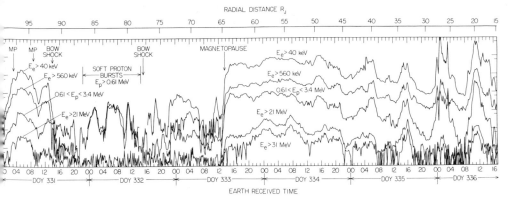

Fig. 15. Absolute omnidirectional intensities of electrons in four integral energy ranges and of protons in one differential energy range as a function of time during Pioneer 11's inbound traversal of the outer magnetosphere. "DOY" means day of year; e.g., DOY 331 is 27 November (of 1974).

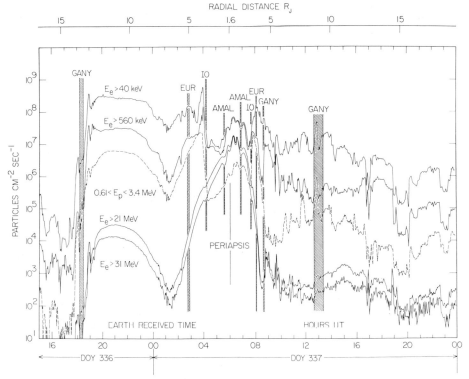

Fig. 16 Absolute omnidirectional intensities of electrons in four integral energy ranges and of protons in one differential energy range as a function of time during Pioneer 11's traversal of the inner magnetosphere. The geometrical sweeping regions of the four inner satellites are shown by vertical shaded bars.

[947]

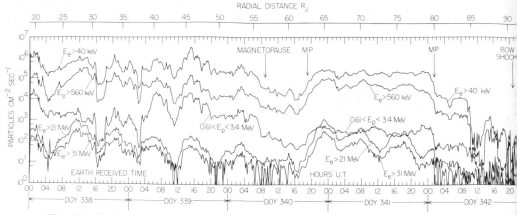

Fig. 17.   Similar to Fig. 15 except for outbound traversal of the outer magnetosphere.

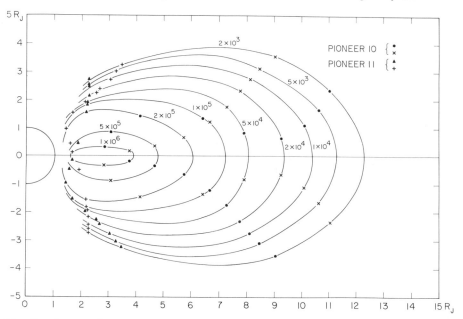

Fig. 18.   Iso-counting rate contours (counts per second) for energetic electron detector C. Omnidirectional intensities of electrons $E_e > 21$ MeV are found by multiplying the counting rate by 23. This figure shows combined observations from the Pioneer 10 and Pioneer 11 missions based on the use of a centered dipole model with a tilt of 9°5 toward System III (1957.0) longitudes of 230° and 233°, respectively. Circles and triangles are observed points for Pioneer 10 and Pioneer 11, respectively; x's and crosses are corresponding reflections in the magnetic equatorial plane (Van Allen *et al.* 1975).

tor C on both Pioneer 10 and Pioneer 11. The contours refer to electrons $E_e > 21$ MeV. In both cases, the counting rate was a relatively smooth, and apparently stable, function of position. Hence this distribution can be taken, provisionally at least, to represent the time-independent situation and hence

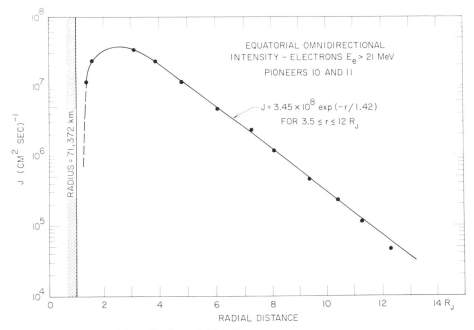

Fig. 19.   An equatorial profile through Fig. 18.

to provide basic data for calculating the synchrotron emission throughout the region of interest. The equatorial profile of the contours in Fig. 18 is shown in Fig. 19. The maximum in the omnidirectional intensity is $J = 3.8 \times 10^7$ cm^{-2} sec^{-1} at $r = 2.5\ R_J$, and the dependence of J on r is a falling exponential in the range $3.5 < r < 12\ R_J$ with an e-folding length $1.42\ R_J$. Analyses of this type are underway for both lower and higher energy electrons. The latitude dependence of omnidirectional intensity J along magnetic shells having a constant value of the McIlwain parameter L is shown as a function of the parameter (B_0/B) in Fig. 20. In a dipolar field at constant L,

$$\frac{B_0}{B} = \frac{\cos^6 \lambda}{[4 - 3 \cos^2\lambda]^{\frac{1}{2}}} \qquad (2)$$

where B_0 is the magnetic field strength at the equator and B is its value at latitude λ. We have previously suggested (Van Allen et al. 1975) that these curves correspond to the expectation of pitch-angle scattering by the resonant interaction between whistler-mode noise and high-energy electrons in the manner first discussed for the terrestrial magnetosphere by Kennel and Petschek (1966).

Angular Distribution of Particles

Our Pioneer 11 observations of the angular distributions of electrons in various energy ranges generally confirm those with Pioneer 10, as quoted

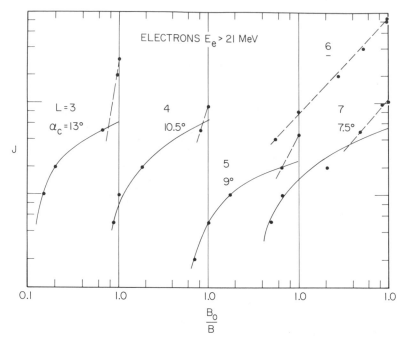

Fig. 20.   Dependence of the omnidirectional intensity of electrons $E_e > 21$ MeV on (B_0/B) for several values of L as derived from Fig. 18. The black circles are the observed points; solid lines are according to the Kennel-Petschek theory of whistler-mode, pitch-angle scattering. α_c is the loss cone angle at the equator as found by fitting the theoretical curve to the observations. The vertical scale is arbitrary and different for each set of data (Van Allen *et al.* 1975).

above (Sentman and Van Allen 1976). Significant new observations of particle streaming will be referred to in a later section.

By virtue of our solid-state detector on Pioneer 11, but not on Pioneer 10, we were able to make an important new class of angular distribution measurements throughout the encounter period, except during occultation of the spacecraft by the planet. (Stored data were obtained during this period of 42 minutes but at too low a bit rate to yield angular distributions.) In the inner magnetosphere the angular distributions of protons $0.61 < E_p < 3.41$ MeV are usually of "pancake" form with a maximum intensity at pitch angle $\alpha = 90°$ and minima at $\alpha = 0°$ and $180°$. In some cases, however, there is a secondary minimum at $\alpha = 90°$ and two maxima symmetrically positioned at greater and lesser values of α ("butterfly" distributions). Examples of both types are given in Fig. 21.

Such angular distributions are valuable for many different purposes. One of these is the checking of magnetic field models. The axis of the uniformly rotating spacecraft (11.89 sec period) is pointed continuously at the earth to

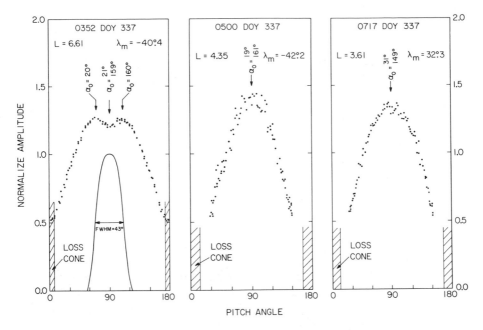

Fig. 21.   Some sample pitch-angle distributions of protons $0.61 < E_p < 3.41$ MeV in the inner magnetosphere (Pioneer 11). The instrumental profile is shown in the left-hand panel. α_0 is the equatorial pitch angle and λ_M is the magnetic latitude (Sentman and Van Allen 1976).

an accuracy of better than 1°. By means of sun- and star-sensors the orientation of reference axes fixed in the spacecraft is known as a function of time. Hence, detector counting rates as a function of time can be referenced to an inertial coordinate system. We have analyzed our data as a function of the roll angle ϕ, measured from the ascending node of the spacecraft's equatorial plane on the ecliptic, by making a second-order Fourier analysis of counting rates

$$f(\phi) = M \left[1 + K \cos 2 \left(\phi - \Delta \right) \right]. \qquad (3)$$

Values of K and Δ as a function of time for a 6-hour period around periapsis are shown in Fig. 22. The computed values of Δ have an estimated accuracy of $\pm 1°$. Shown as smooth curves are the acute magnetic cone angle θ_B and the magnetic clock angle ϕ_B plus 90°, kindly supplied by Smith *et al.* of the magnetometer team. The acute magnetic cone angle (0°–180°) is the angle between the rotational axis of the spacecraft and the local magnetic vector. The magnetic clock angle is the angle between the ascending node of the spacecraft's equatorial plane on the ecliptic and the projection of the magnetic vector on that equatorial plane. For a pancake distribution one expects that $\phi_B + 90° = \Delta$, under the physical assumption that the particle distribution has axial symmetry around the local magnetic vector. The agreement is

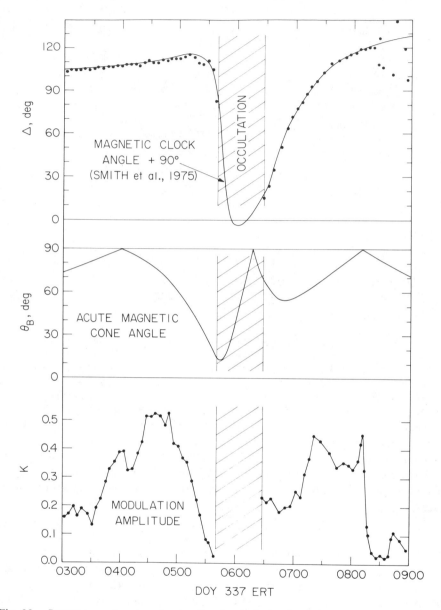

Fig. 22. Parameters of a second-order Fourier analysis of angular distributions of protons $0.61 < E_p < 3.41$ MeV in the inner magnetosphere (black circles) and magnetic cone angle θ_B and magnetic clock ϕ_B plus 90° (smooth curves in upper two panels). The detailed values of θ_B and ϕ_B from magnetometer data were kindly supplied by E. J. Smith.

seen to be excellent, except for $K < 0.1$; such low values of K occur when θ_B goes toward zero (a degenerate case for finding Δ) and/or when the angular distribution itself approaches isotropy (e.g., after 0820 ERT in Fig. 22).

Satellite Effects

The presence of the three inner Galilean satellites — Ganymede, Europa and Io — and Amalthea within the relatively well-ordered magnetosphere of Jupiter provides a wealth of new phenomena and provides bases for diagnosing the dynamics of the magnetospheric processes. Such a situation does not exist in the earth's magnetosphere because the moon's orbit lies outside of the well-ordered magnetosphere.

The Jovian satellites induce plasma instabilities and contribute to the emission of decametric radio noise; they absorb particles that are diffusing across their orbits and thus have an important influence on the intensities of particles in the inner magnetosphere; they may act as sources for the acceleration of particles and injection of such particles in the vicinity of their orbits; and they emit gas into the magnetospheric system.

All of these phenomena are under study. A recent paper by Thomsen and Goertz (1975) suggests the sophistication that present observations permit in going beyond rudimentary inferences on the diffusion coefficients of charged particles as a function of particle species, L and energy.

There are two prominent satellite effects that are immediately obvious in our Pioneer 11 data. The first is the marked peaks in intensity of electrons $E_e > 40$ keV and > 560 keV that occur just outside of the crossing of Io's magnetic shell on the inbound trajectory (Fig. 16). These peaks may correspond to the theoretically expected injection of energetic electrons from the plasma sheath around Io (Gurnett 1972; Shawhan *et al.* 1975). The observed situation is less clear on the outbound trajectory. There are also marked dips in the intensity of the lower energy electrons at the Io shell both inbound and outbound, and lesser effects on the intensity of higher energy electrons.

The second prominent effect is the great reduction of the intensity of protons $0.61 < E_p < 3.41$ MeV interior to the magnetic shell of Io (Fig. 16). Inbound, the intensity decreases discontinuously by a factor of 100, and outbound (at a different latitude) it increases discontinuously by a factor of 30. This phenomenon is being analyzed in greater detail than can be reported fully at present. As one example, the radial positions of the two discontinuities and their relationships to the magnetic shell of Io, considered as an inert physical absorber of particles, provide a valuable test of magnetic field models. Such a test of three current magnetic models is shown in Fig. 23. The observed spin-averaged proton intensities (without correction for angular distributions or for differing latitude) have been replotted against L, as calculated separately for each case. A true magnetic model should result in (a) coincidence of the inbound and outbound discontinuities and (b), the "proper" positioning of the satellite shell relative to the discontinuities. We do not yet know the best theoretical basis for application of the second criterion. The spherical harmonic model of Smith *et al.* (1975) and the Randall centered dipole model both produce excellent inbound-outbound closure of the particle data, especially if one supposes that a proper latitude correction will normalize the magnitudes of the two discontinuities. The Acuña-Ness 03 model (1975)[6] produces inferior closure at Io, but superior representation of the lesser but perhaps significant Amalthea effect.

Co-rotation

For the preliminary examination of co-rotation effects we have made a first-order Fourier analysis of the angular distributions by the formula

$$f(\phi) = M' \left[1 + K' \cos (\phi - \Delta') \right].$$ (4)

On the inbound trajectory and for protons $0.61 < E_p < 3.41$ MeV, co-rotation is clearly evident from the magnetopause at 65 R_J in to at least 30 R_J (Fig. 24). The phase angle Δ' is near 180° and K' is approximately a linear function of radial distance r with a value 0.32 at $r = 65$ R_J. These results correspond to rotation at the planet's angular velocity of an isotropic angular distribution of protons $E_p \sim 1$ MeV in the rotating frame of reference; the spectral index $\gamma \approx 1.8$ in a differential energy spectrum of power law form, the index being approximately independent of r over the range $65 > r > 30$ R_J. This result generally confirms Pioneer 10 observations by Trainor et al. (1974).[7] Our larger values of K' correspond to our lower energy range, and our lesser values of γ are plausibly consistent with the trend of their energy spectra toward lower energies.

The co-rotation of the outer magnetosphere is apparently established as one of its fundamental properties.

Magnetic Field-Aligned Streaming

In a recently made analysis of Pioneer 11 data Sentman et al. (1975) find that there is net streaming of both electrons $E_e > 40$ keV and $E_e > 560$ keV and protons $0.61 < E_p < 3.41$ MeV *away* from the planet along high-latitude field lines. This result is compatible with the recent suggestion of Nishida (1975) that energetic particles undergo trans-L shell diffusion at low altitudes without significant change of energy. The observed outward streaming coupled with his hypothesis and other considerations provides a plausible explanation for: the remarkable pitch-angle distributions near the equator in the range $12 < L < 30$ as discussed in a previous section, the presence of particles of $\approx$ MeV energy at the outer edge of the magnetosphere, and hence, via conventional inward diffusion processes, of those having magnetic moments of several hundred MeV per gauss in the inner magnetosphere. The recirculation of energetic particles emerges as an important dynamical feature of the Jovian magnetosphere. It is also suggested that emission of energetic particles into interplanetary space occurs at the polar caps rather than at the equator.

---

[6]See p. 832.
[7]See p. 964.

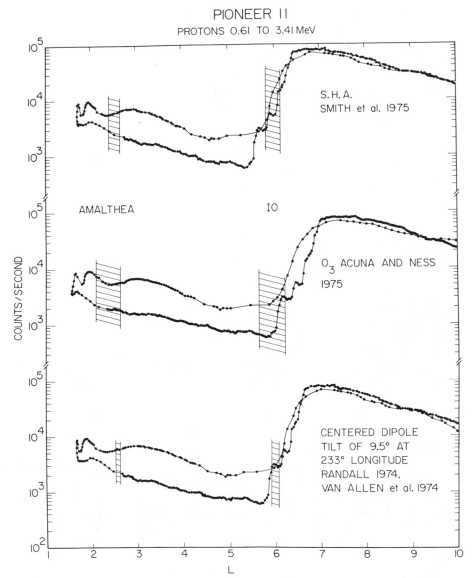

Fig. 23. Comparison of the inbound-outbound Io sweep-out effect for three magnetic models (courtesy B. A. Randall).

IV. REMARKS ON ORIGIN OF ENERGETIC PARTICLES IN THE JOVIAN MAGNETOSPHERE

The foregoing summaries of Pioneer 10 and 11 observations produce several puzzles in interpretation and lead to the impression that the dynamics of the magnetosphere of Jupiter differ in important ways from those of the earth's.

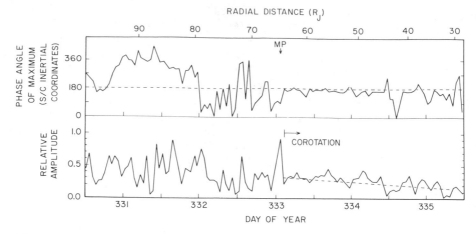

Fig. 24.   Parameters of a first-order Fourier analysis of the angular distributions of protons 0.61 $< E_p <$ 3.41 MeV in the outer magnetosphere (Sentman and Van Allen 1976).

1. The presence of substantial intensities of electrons having energies greater than 20 MeV in the magnetosheath and outer magnetosphere of Jupiter is incompatible with simple thermalization of the solar wind, which can result in energies of only about one keV.

2. The radial dependence of the angular distribution of electrons, as summarized in Sec. I, is altogether incompatible with their inward diffusion from the magnetopause if both first and second adiabatic invariants are conserved.

3. As shown in Fig. 12, the phase space densities of electrons $f(\mu_1)$ and $f(\mu_2)$ (those dominating the electron population in the outer magnetosphere) have their most decisive and unequivocal declines inside of $\sim$ 12 R_J, whereas for greater radii the phase space densities are more-or-less indecisively dependent on radius. Hence losses of particles during inward diffusion are mostly confined to $r <$ 12 R_J. Unless another good reason for this behavior can be found, the data of Fig. 12 permit the possibility of injection (i.e., an effective source) anywhere in the range between 12 R_J and the magnetopause.

4. A relatively well-ordered and time-stationary distribution of high-energy electrons exists only for $r <$ 12 R_J (Figs. 10, 18, and 19 and the discussion relative thereto). For greater radii the distribution is chaotic and time-dependent.

5. The inner satellites (J III, J II, J I, and J V) produce a fluctuating and complex structure in the distribution of the lower energy electrons, even within the inner magnetosphere. This is presumed to be the consequence of plasma instabilities and perhaps accelerative processes associated with their movement through the ambient plasma of the inner magnetosphere *and* of the coupling of the magnetic lines of force to a turbulent ionosphere (Brice and McDonough 1973).

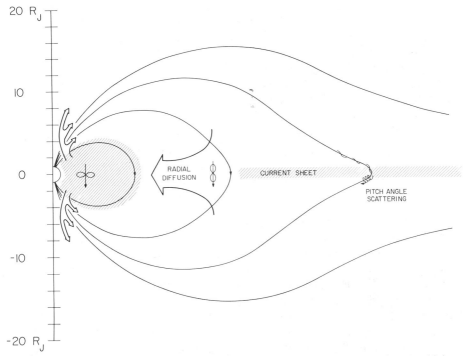

Fig. 25.   A schematic magnetic meridian cross section of the Jovian magnetosphere combining suggestions of Nishida (1975) and our Pioneer 11 observations on particle streaming. The crucial new features of this model are trans-L shell diffusion from low to high latitudes near the planet and pitch-angle scattering in the neutral sheet within the equatorial ring current sheet.

6. Our discovery of net field-aligned streaming away from the planet at high latitudes of both electrons and protons supports the suggestion of Nishida that cross-L diffusion of particles at low altitudes with little or no change in energy is an important process and further corresponds to the recirculation of energetic particles (Fig. 25).

7. The co-rotation of energetic particles persists out to the magnetopause (Fig. 24) whereas for the earth this situation terminates at the outer boundary of the plasmasphere ($\approx 4\ R_E$), far inside of the magnetopause.

The foregoing summary suggests, heuristically, that the capture of solar wind particles may be a relatively minor feature of the dynamics of the Jovian magnetosphere and that the internal acceleration of internally available particles may be the dominant feature. The problem of a magnetized conducting sphere rotating within a stationary plasma has been discussed in an illuminating manner by Hones and Bergeson (1965). The more general problem of an arbitrarily shaped conducting magnet moving and rotating in an arbitrary manner within a plasma has been considered by Birmingham and Northrop (1968). The latter authors conclude:

" . . . the principal result of this paper, that the bounce-averaged rate of energy change of trapped particles is of $0(\epsilon^3)$, small in the adiabatic parameter [$\epsilon = m/e$, the ratio of mass to charge of the particles as discussed by Northrop, 1963] is a theorem, proved for the idealized model in which $E_{||} = 0$ [the electric field parallel to the magnetic vector **B**] and **B** arises solely from and moves rigidly with a conducting magnet. If the entire magnetic field surrounding the earth were due to the earth's internal currents (and $E_{||} = 0$), the wobbling field caused by the earth's rotation would not affect trapped particle energies. The existence in reality, however, of small finite parallel resistivity and concomitant $E_{||}$ and the presence of nonrigidly convected magnetic field sources, e.g., ring, magnetopause, and tail currents, pushes us into . . . predictions of much more rapid energy changes."

My impression from these papers is that the axially asymmetric and non-rotating situation caused by the solar wind flow past Jupiter is essential to the development of its magnetosphere, whether or not the solar wind particles themselves are captured into the system.

On the basis of demonstrated co-rotation of the outer magnetosphere of Jupiter and of the fact that the co-rotation velocity at $100\,R_J$, say, is ≈ 1200 km sec^{-1}, about three times the typical solar wind velocity, Gold (1976) has argued that the rotational energy of the planet and not the solar wind is the dominant source of energy for driving magnetospheric processes. He further points out that the existence of the solar wind is nonetheless essential to establishing the physical conditions under which transfer of rotational energy to charged particles is possible. Comparable suggestions on rather different but kindred bases have been made by Scarf.[8]

Acknowledgements. This paper is based primarily on the work of the author and his students and colleagues at the University of Iowa — B. A. Randall, D. N. Baker, D. D. Sentman, M. F. Thomsen, C. K. Goertz, and H. R. Flindt. Their spirited and devoted collaboration is gratefully acknowledged.

REFERENCES

Acuña, M. H., and Ness, N. F. 1975. Jupiter's main magnetic field measured by Pioneer 11. *Nature* 253:327–328.

Baker, D. N. 1973. University of Iowa Pioneer 10 instrument calibration, University of Iowa Research Report 73–26.

Baker, D. N., and Van Allen, J. A. 1976a. Energetic electrons in the Jovian magnetosphere. *J. Geophys. Res.* In press.

——. 1976b. Revised Pioneer 10 absolute electron intensities in the inner Jovian magnetosphere. Submitted to *J. Geophys. Res.*

[8]See p. 890.

Birmingham, T. J., and Northrop, T. G. 1968. Charged particle energization by an arbitrarily moving magnet. *J. Geophys. Res.* 73:83–86.

Brice, N., and McDonough, T. R. 1973. Jupiter's radiation belts. *Icarus* 18:206–219.

Burke, B. F., and Franklin, K. L. 1955. Observations of a variable radio source associated with the planet Jupiter. *J. Geophys. Res.* 60:213–217.

Carr, T. D., and Gulkis, S. 1969. The magnetosphere of Jupiter. *Annual review of astronomy and astrophysics.* (L. Goldberg, D. Layzer, and J. G. Phillips, eds.) vol. 7, pp. 577–618. Palo Alto, California: Annual Reviews, Inc.

Drake, F. D., and Hvatum, S. 1959. Non-thermal microwave radiation from Jupiter. *Astron. J.* 64:329–330.

Dulk, G. A. 1970. Characteristics of Jupiter's decametric radio source measured with arc-second resolution. *Astrophys. J.* 159:671–684.

Fillius, R. W., and McIlwain, C. E. 1974. Measurements of the Jovian radiation belts. *J. Geophys. Res.* 79:3589–3599.

Goertz, C. K. 1976. The current sheet in Jupiter's magnetosphere. Submitted to *J. Geophys. Res.*

Goertz, C. K.; Jones, D. E.; Randall, B. A.; Smith, E. J.; and Thomsen, M. F. 1976. Evidence for open field lines in Jupiter's magnetosphere. *Icarus* (special Jupiter issue). In press.

Gold, T. 1976. The magnetosphere of Jupiter. *Icarus* (special Jupiter issue). In press.

Gurnett, D. A. 1972. Sheath effects and related charged-particle acceleration by Jupiter's satellite Io. *Astrophys. J.* 175:525–533.

Haskell, G. P. 1969. Anisotropic fluxes of energetic particles in the outer magnetosphere. *J. Geophys. Res.* 74:1740–1748.

Hill, T. W.; Dessler, A. J.; and Michel, F. C. 1974. Configuration of the Jovian magnetosphere. *Geophys. Res. Lett.* 1:3–6.

Hones, Jr., E. W., and Bergeson, J. E. 1965. Electric field generated by a rotating magnetized sphere. *J. Geophys. Res.* 70:4951–4958.

Kennel, C. F., and Petschek, H. E. 1966. Limit on stably trapped particle fluxes. *J. Geophys. Res.* 71:1–28.

McIlwain, C. E., and Fillius, R. W. 1975. Differential spectra and phase space densities of trapped electrons at Jupiter. *J. Geophys. Res.* 80:1341–1345.

Mihalov, J. D.; Collard, H. R.; McKibbin, D. D.; Wolfe, J. H.; and Intriligator, D. S. 1975. Pioneer 11 encounter: preliminary results from the Ames Research Center plasma analyzer experiment. *Science* 188:448–451.

Nishida, A. 1975. Outward diffusion of energetic particles from the Jovian radiation belt. Preprint, Inst. of Space and Aero. Science, University of Tokyo.

Northrop, T. G. 1963. *The adiabatic motion of charged particles.* New York: Interscience Publishers.

Sentman, D. D., and Van Allen, J. A. 1976. Angular distributions of electrons of energy $E_e >$ 0.06 MeV in the Jovian magnetosphere. *J. Geophys. Res.* In press.

Sentman, D. D.; Van Allen, J. A.; and Goertz, C. K. 1975. Recirculation of energetic particles in Jupiter's magnetosphere. *Geophys. Res. Lett.* 2:465–468.

Shawhan, S. D.; Goertz, C. K.; Hubbard, R. F.; Gurnett, D. A.; and Joyce, G. 1975. Io-accelerated electrons and ions. *The magnetospheres of the earth and Jupiter.* (V. Formisano, ed.) pp. 375–389. Dordrecht-Holland: D. Reidel Publishing Co.

Simpson, J. A.; Hamilton, D. C.; McKibben, R. B.; Mogro-Campero, A.; Pyle, K. R.; and Tuzzolino, A. J. 1974. The protons and electrons trapped in the Jovian dipole magnetic field region and their interaction with Io. *J. Geophys. Res.* 79:3522–3544.

Smith, E. J.; Davis, Jr., L.; Jones, D. E.; Coleman, Jr., P. J.; Colburn, D. S.; Dyal, P.; and Sonett, C. P. 1975. Jupiter's magnetic field, magnetosphere, and interaction with the solar wind: Pioneer 11. *Science* 188:451–454.

Smith, E. J.; Davis, Jr., L.; Jones, D. E.; Colburn, D. S.; Coleman, Jr., P. J.; Dyal, P.; and Sonett, C. P. 1974a. Magnetic field of Jupiter and its interaction with the solar wind. *Science* 183:305–306.

Smith, E. J.; Davis, Jr., L.; Jones, D. E.; Coleman, Jr., P. J.; Colburn, D. S.; Dyal, P.; Sonett, C. P.; and Frandsen, A. M. A. 1974b. The planetary magnetic field and magnetosphere of Jupiter: Pioneer 10. *J. Geophys. Res.* 79:3501–3513.

Thomsen, M. F., and Goertz, C. K. 1975. Satellite sweep-up effects at Jupiter (abstract SM 63). *EOS Trans. Amer. Geophys. Union* 56:428–429.

Trainor, J. H.; McDonald, F. B.; Teegarden, B. J.; Webber, W. R.; and Roelof, E. C. 1974. Energetic particles in the Jovian magnetosphere. *J. Geophys. Res.* 79:3600–3613.

Van Allen, J. A.; Baker, D. N.; Randall, B. A.; and Sentman, D. D. 1974a. The magnetosphere of Jupiter as observed with Pioneer 10. Part 1. Instrument and principal findings. *J. Geophys. Res.* 79:3559–3577.

Van Allen, J. A.; Baker, D. N.; Randall, B. A.; Thomsen, M. F.; Sentman, D. D.; and Flindt, H. R. 1974b. Energetic electrons in the magnetosphere of Jupiter. *Science* 183:309–311.

Van Allen, J. A.; Randall, B. A.; Baker, D. N.; Goertz, C. K.; Sentman, D. D.; Thomsen, M. F.; and Flindt, H. R. 1975. Pioneer 11 observations of energetic particles in the Jovian magnetosphere. *Science* 188:459–462.

Warwick, J. W. 1970. *Particles and fields near Jupiter,* NASA CR-1685, National Aeronautics and Space Administration, Washington, D.C.; U.S. Government Printing Office.

OBSERVATIONS OF ENERGETIC
JOVIAN ELECTRONS AND PROTONS

F. B. McDONALD

and

J. H. TRAINOR
NASA Goddard Space Flight Center

This chapter is a summary of the particle observations in the Jovian magnetospher. by the Goddard/University of New Hampshire experiment on Pioneers 10 and 11. The data suggest that stable trapping does not exist in the outer magnetosphere. The proton energy spectra are of the form E^{-4} in the outer region and become more complex in the inner region. Large first-order anisotropies are observed in the angular distribution of the proton component. These represent the combined effect of a co-rotation anisotropy and particle flow along the magnetic fieldlines. It is found that particle absorption by several of the Jovian satellites is an important effect. In particular, Io dominates the proton component of the inner Jovian magnetosphere. This effect clearly establishes that radial diffusion is the dominant acceleration process in this region. Electron increases were observed in interplanetary space on Pioneer 10 as it approached within 1 A.U. of Jupiter. These discrete bursts or increases were typically several hundred times the normal quiet-time electron flux, and became much more frequent as Pioneer 10 approached Jupiter, resulting in the quasi-continuous presence of large fluxes of these electrons in interplanetary space. In view of the origin of these electrons at Jupiter, and the similarity of these increases to quiet-time electron increases previously observed at Earth, the temporal presence of the quiet-time increases at Earth is examined. It is found that these increases have a 13-month periodicity indicating a Jovian origin for the events near the earth as well. The spectra of electrons observed in Jupiter's magnetosphere, on Pioneer 10 in interplanetary space near Jupiter, for the quiet-time increases near the earth. and for the ambient electron spectrum are all remarkably similar. These two lines of evidence suggest the possibility that Jupiter could be the source of most of the ambient electrons at low energies in the heliosphere.

The passage of Pioneers 10 and 11 by Jupiter in December 1973 and December 1974, respectively, have revealed a very large, complex and dynamic magnetosphere. In addition, measurements made enroute to the planet suggest that Jupiter is the dominant source of 0.2–30 MeV electrons in the heliosphere (Chenette *et al.* 1974; Teegarden *et al.* 1974). These Jovian electrons are frequently observed at Earth in the form of quiet-time electron increases. In this chapter we present a summary of the experimental observations of Jovian energetic particles made by the Goddard / University of New Hampshire cosmic ray experiments on Pioneers 10 and 11. The experiments on both spacecraft were essentially identical and consisted of a set of three complimentary solid-state telescopes (Stilwell *et al.* 1975). These telescopes (Fig. 1) covered an extended range in energy and charge spectra and intensity (Table I). Two of the detector systems — the high-energy telescope (HET) and the low-energy telescope I (LET I) — were designed primarily for galactic cosmic ray studies. With their moderate geometric factors and negligible side and back shielding, they saturate at particle fluxes above $\sim 10^5$ particles cm^{-2} sec^{-1} sr^{-1}. This limited their usefulness to the outer Jovian magnetosphere ($\gtrsim 25\ R_J$ for HET and $\gtrsim 15\ R_J$ for LET I). However, in this outer region they provided precise particle identification and energy spectra. In particular, the multi-parameter HET detector was of great value in studying MeV electrons.

The LET II telescope (Fig. 1 and Table I) was specifically designed to study trapped particles in the Jovian magnetosphere. The heavy shielding and small geometric factor (0.015 cm^2 sr) of the LET-II permit measurements up to flux levels of 4×10^6 particles cm^{-2} sec^{-1} sr^{-1} without need for correction. The combination of aluminum and lead shielding will stop electrons $\lesssim 25$ MeV and protons $\lesssim 140$ MeV. Electrons and protons are identified by a two-parameter analysis technique. The front detector S_1 of thickness 50 μm has an electronic threshold such that the energy loss of an electron penetrating to S_2 will be below this threshold, while any protons penetrating and stopping in S_2 will have an energy loss above threshold. The very thin S_1 detector, in anti-coincidence with S_2, measures protons from 0.2 to 2.1 MeV in four energy ranges, with negligible electron contamination, ordinarily. The upper range never suffers electron contamination, even from pulse pileup (Trainor *et al.* 1974*a*). Angular distributions are measured for selected rates in all three telescopes by dividing each spacecraft rotation into 8 angular sections. The field of view of all three detector systems was normal to the spacecraft spin axis.

During both Jovian encounters the onset of saturation in the regions of high particle flux are abrupt and well defined. Negligible corrections are necessary prior to entering or leaving this saturation point. Inside 10 R_J substantial corrections are necessary to the LET II data for Pioneer 10. These are well understood and have been discussed in detail by Trainor *et al.* (1974*a*). Corrections are necessary for the LET II data on Pioneer 11 only for a narrow region near periapsis (Trainor *et al.* 1975).

The magnetospheric observations are a strong function of the spacecraft trajectory with respect to the planet. The Pioneer 10 and 11 encounter trajectories as viewed in Jovian local time are shown in Fig. 2. Pioneer 10

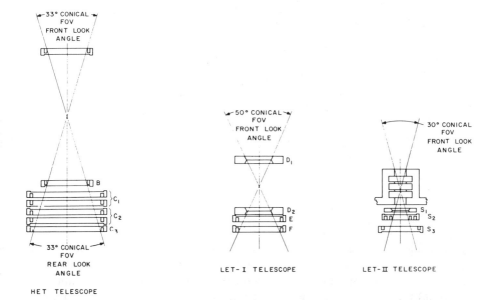

Fig. 1.   Schematic drawing of the Goddard / University of New Hampshire telescopes on Pioneers 10 and 11.

TABLE I
Detector Characteristics

Detector	Energy / Particle Range
HET	2.1–8.0 MeV electrons 20–500 MeV/Nuc protons and alphas 40–120 MeV/Nuc medium nuclei
LET-I	0.4–3 MeV/Nuc protons (single parameter analysis) 3–21 MeV/Nuc protons and alphas 6–40 MeV/Nuc medium nuclei
LET-II	0.05–2.1 MeV electrons (electron-proton separation above 0.12 MeV) 0.2–21 MeV protons

was on a prograde trajectory which approached Jupiter from a direction of some 30° west of the sun and circled the planet in a counterclockwise direction and exited toward the dawn meridian. Pioneer 11 approached Jupiter in the morning quadrant at an angle of some 40° west of the sun, circled the planet in a clockwise or retrograde direction and exited at high latitudes toward the direction of the sun. The Pioneer 10 trajectory was inclined ~ 14°

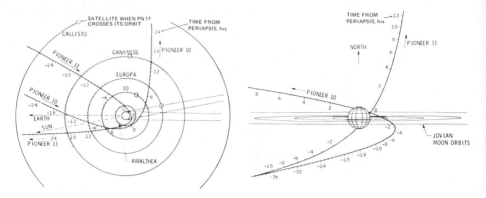

Fig. 2.   The Pioneer 10 and 11 trajectories referenced to Jupiter and projected on a plane parallel to that of the ecliptic (after Hall 1975).

to the Jovian equator whereas that of Pioneer 11 was inclined $\sim 50°$. From Fig. 2 it would be expected that the data from the inbound Pioneer 11 trajectory should be intermediate between that observed on the inbound and outbound Pioneer 10 trajectories. In both cases the observations are mainly on the front side of the Jovian magnetosphere with no data from the tail region.

I.   OVERVIEW OF THE JOVIAN MAGNETOSPHERE

Figures 3, 4, 5, and 6 provide an overview of the Pioneer 10 and 11 encounter for particle energies near the upper and lower ranges of the instrument (Trainor *et al.* 1974a, 1975). On Pioneer 10 a steady increase in the flux of > 0.5 MeV protons was observed $\sim 5\ R_J$ before crossing the bow shock at $\sim 109\ R_J$. On Pioneer 11 protons of this energy were present for days in advance of crossing into the magnetosphere. This is not unexpected since the outer Jovian magnetosphere is filled with unstably trapped protons and electrons. Inside the magnetosheath, it is convenient to divide the Jovian magnetosphere into three regions.

1.   The outer Jovian magnetosphere. This region extends from the bow-shock crossing ($108.9\ R_J$ for Pioneer 10 and $109.7\ R_J$ for Pioneer 11) to $\sim 40\ R_J$.
2.   A transition region from ~ 40 to $25\ R_J$.
3.   The inner Jovian magnetosphere inside $25\ R_J$.

The definition of the boundaries is somewhat arbitrary since, except for the bow-shock crossing, there is a gradual transition from one region to the next.

The outer Jovian magnetosphere is a region of quasi-trapping and diffusion. Electrons and protons both show remarkably constant energy spectra

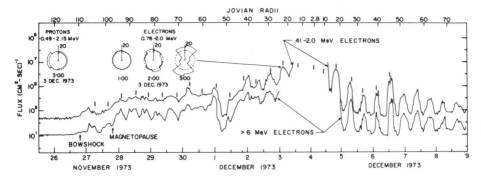

Fig. 3.   Electron time histories during the Jovian encounter of Pioneer 10. Angular distributions are given for selected time periods. Tick marks show points when Pioneer 10 was predicted to be closest to the magnetic equator, based upon a rigidly rotating field due to a tilted dipole. In all cases spacecraft time is used (after Trainor *et al.* 1974*a*).

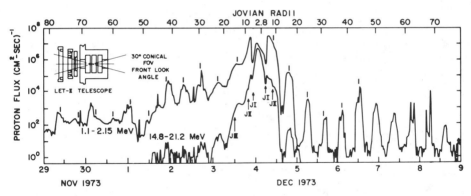

Fig. 4.   Proton time histories in two different energy intervals from Pioneer 10. Times of crossing the inner Galilean satellite orbits are shown, respectively, as J I, J II, and J III (after Trainor *et al.* 1974*b*).

of the form $E^{-\gamma}$ with $\gamma_e = 1.5$–2.0 and $\gamma_p \simeq 4$. This near constancy of γ suggests that almost no acceleration occurs in the region. The MeV electron component displays a rather well-defined 10-hr periodicity. This is not unexpected since the normal magnetic latitude of the spacecraft will vary with the 10-hr rotation period of the planet. This periodicity is not nearly as well defined for lower-energy electrons or for the protons. The changes in the proton and electron fluxes are frequently uncorrelated. There are often rapid changes in the angular distributions.

In the transition region between 40 and 20 R_J the angular distribution becomes better defined and is dominated by a combination of co-rotation anisotropy and flow along the fieldlines apparently into the upper atmosphere of the planet. The proton energy spectra begin to change from a single power

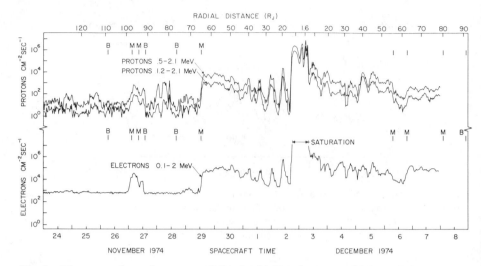

Fig. 5.   Fluxes are shown for protons (0.5 to 2.1 MeV and 1.2 to 2.1 MeV) and electrons (0.1 to 2 MeV) for the period 24 November to 8 December 1974 from Pioneer 11. The locations of crossing of the bow shock (*B*) and magnetopause (*M*) are noted (Wolfe *et al.* 1974). The relatively high background in this electron measurement from the LET-II telescope amounts to ~ 2 count sec^{-1} and is due to gamma rays from the radioisotope power supply (after Trainor *et al.* 1975).

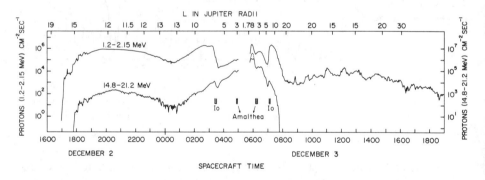

Fig. 6.   Flux profiles of 1.2 to 2.15 MeV and 14.8 to 21.2 MeV protons measured in the inner, core region of the Jovian magnetosphere by Pioneer 11. The expected locations of crossing of Amalthea's and Io's orbits are shown (after Trainor *et al.* 1975).

law to one best fitted by the sum of two power laws, suggesting that some acceleration is taking place. The 10-hr periodicities for the electrons and protons are in phase with each other and are well defined for the proton component. There is only a gradual increase in the intensity of the 10-hr peak intensities displayed by the MeV electron component between the bow shock and 30 R_J. The peak proton fluxes are nearly constant in the outer region.

The inner magnetosphere region begins at ~ 20–$25\ R_J$. This is the approximate interval where the Jovian magnetic field should rigidly rotate with the planet (Trainor *et al.* 1974*a*). In this region there is a very rapid increase in both the 1–2 MeV and 15–21 MeV proton flux (Figs. 4 and 6). This rapid increase is interpreted as being the onset of stable and durable trapping. Both the co-rotation and field-aligned anisotropies decrease strongly. As we shall discuss later, the inner Jovian satellites, especially Io, play an important role in absorbing the nucleon component. On Pioneer 11 there is a sharp peak observed close to perijove. This structure is not understood.

The behavior of the outbound trajectory of Pioneer 10 was very different from that observed inbound. The dominant feature is the 10-hr periodicity which produces peak-to-valley ratios of as much as 10^5 for a 20° excursion in latitude (Figs. 3 and 4). The peak fluxes are near the predicted magnetic equator. The data from the inbound trajectory of Pioneer 11 appears to be intermediate between that observed for the inbound and outbound Pioneer 10 passes. The high-latitude data from the outbound pass of Pioneer 11 (Figs. 5 and 6) are similar to or greater than that observed on Pioneer 10 for 1.2–2.2 MeV protons near the equator. This result is unexpected since on the basis of a "magnetic disc" model, the intensity would decrease very rapidly at large distances from the equatorial current sheet. Note in Fig. 5 that the flux of 14–21 MeV protons does decrease very rapidly. In the following sections these experimental observations are discussed in greater detail.

II. THE NUCLEON COMPONENT

Energy Spectra and Charge Compositions

As we have emphasized previously, one of the characteristics of these observations is the accuracy with which the energy spectra and charge composition can be determined. The electron spectra will be discussed in Sec. III, which deals with Jovian electrons in interplanetary space. A set of 7 typical proton spectra are shown in Fig. 7 (Trainor *et al.* 1974*a*). The data from the magnetosheath region and from the Jovian radiation belts in to $\sim 40\ R_J$ seem to be well fitted by a simple power law with an exponent of 4 but varying from 4.2 to 3.0 for brief periods. We interpret this lack of systematic change as indicating that very little acceleration occurs in the outer magnetosphere.

Beginning with the measurements inside 41 R_J the spectra are better fitted by the sum of two power laws rather than by a simple power law (Fig. 7*d*). In Fig. 7*f* one can see that the LET I telescope, with its much larger geometrical factor and lack of shielding, has now become saturated and its apparent count rates have fallen. Figure 7*g*, for comparison, shows the spectra taken in the first flux minimum outbound in the region 12.6–13.5 R_J. It is well fitted by a power law of exponent 3.5. The Pioneer 11 measurements, which were generally at higher latitudes, could be represented by a single power law with the index slowly changing from 4 to 3.

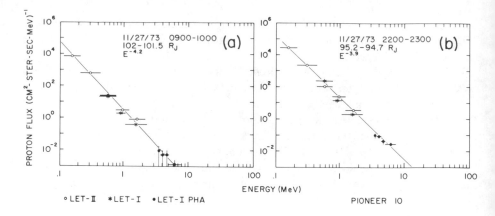

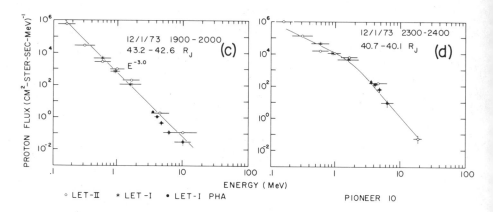

Fig. 7.   Proton differential energy spectra measured by the LET I detector (asterisks) and by LET II (open circles) from Pioneer 10 at three locations along the trajectory. The LET I pulse-height analyzer data is indicated by the solid circles (after Trainor *et al.* 1974*a*).

Composition studies using the LET I telescope were carried out in the outer part of the inner region as well as over the complete outer Jovian magnetosphere. Only helium could be identified. Even in this case the measured fluxes were so small that it was necessary to average data over fairly long periods of time to obtain reasonable statistics. Figure 8 shows the ratio of alpha and proton intensities as a function of Jovian radius for the inbound pass of Pioneer 10. A general decreasing trend in the ratio is apparent with one exception, namely the point between 40 and 55 R_J. This point, however, occurs at the time when Pioneer 10 re-entered the magnetosheath (Wolfe *et al.* 1974)[1]. The abnormally high value of the point suggests that conditions

---

[1]See p. 852.

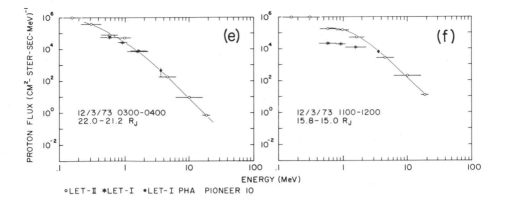

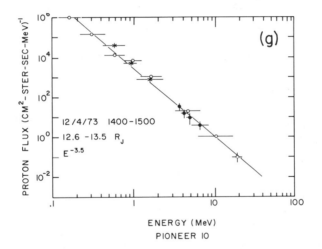

near the boundary of the magnetosphere prevailed during this time and the point is therefore quite consistent with the observations of Wolfe *et al.* The ratio varies between 6×10^{-3} and 6×10^{-4} over the region measured and is to be compared with the solar wind alpha-proton ratio, which varies over the range 0.01 to 0.1 (Wolfe *et al.* 1966; Gosling *et al.* 1967). Thus the alpha-proton ratio in the outer part of the Jovian magnetosphere is closest to the solar wind values.

The alpha-proton ratio in the earth's magnetosphere shows little *L* dependence but apparently has a very strong latitude dependence (D. J. Williams, personal communication, 1974). The value of the ratio at Earth spans the range 10^{-2} to 10^{-4}. With the limited statistical accuracy of our data it is difficult to establish whether or not a strong latitude dependence exists. The Earth observations, however, were made in regions where particles are stably trapped, and this situation is almost certainly not the case for many of the measurements presented here.

Proton and Electron Angular Distributions

It is expected that the measured particle anisotropies will be of fundamental importance in understanding the Jovian magnetosphere, and further
that these may be very different from those observed in the earth's magnetosphere. For example, the great extent of the Jovian magnetosphere combined
with the relatively short rotation period means that co-rotation anisotropies
should be an important aspect of the particle angular distributions. Furthermore, the weak non-dipole Jovian magnetic field in the 50–25 R_J region
($\sim 20\gamma$) (Smith *et al.* 1974)[2] suggest that anisotropies produced by particle
intensity gradients could be significant. The data from the LET II detectors
(Fig. 1) were found to be most useful for this study. The detector look direction is normal to the spin axis and the data for a given rate is summed over
five spacecraft rotations. The 30° opening angle of the LET II detector is
sufficiently less than the 45° sector, and no deconvolution of the data is
required.

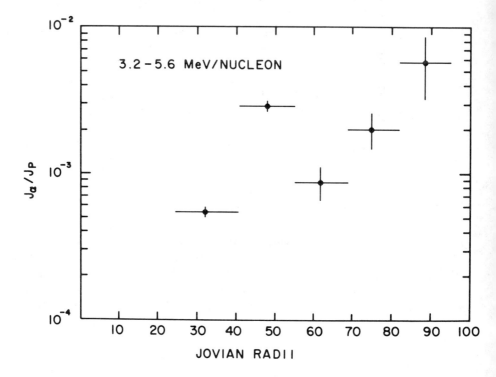

Fig. 8. Alpha-proton ratio as a function of Jovian radius. The ratio is taken at equal intervals
of energy per nucleon (after Trainor *et al.* 1974*a*).

---

[2]See p. 808.

The data from representative periods between 40 and 35 R_J is shown in Fig. 9 for the inbound pass of Pioneer 10 for 0.49–2.15 MeV protons and 0.78–1.0 MeV electrons. It is apparent that there is a large first-order anisotropy of the type and direction expected for either a gradient or co-rotation effect. The electron distributions are essentially isotropic.

The proton distributions from 25 to 19 R_J (Fig. 10) show a well-defined first-order anisotropy. The electrons at the beginning and end of this period are almost completely isotropic. However, there is a well-defined double-ended anisotropy as the spacecraft approaches the normal location of the magnetic equator. Similar behavior was briefly observed during other periods. These dumbbell distributions appear to be field-aligned and are not understood at the present time.

A harmonic analysis of the form

$$J(\Theta) = A_0 + A_1 \sin(\Theta - \Theta_1) + A_2 \sin 2(\Theta - \Theta_2) \qquad (1)$$

where Θ is the angle between the detector look direction and the normal to the ecliptic plane and is contained in the plane normal to the spacecraft spin axis, was performed on the 3600-sec averages. The resulting values for A_1/A_0 and A_2/A_0 for the 1.2 to 2.15 MeV interval are shown in Figs. 11 and 12 (Trainor et al. 1974a). The ratio A_1/A_0 is a measure of the first-order anisotropy. The expected co-rotation anisotropy is given by

$$\xi = \frac{f(v_1) - f(v_2)}{f(v_1) + f(v_2)}$$

$$v_1 = V_r + v, \quad v_2 = V_r - v \qquad (2)$$

where $f(v) \propto [J(E)]/E$ is the nonrelativistic particle distribution function, v is the particle velocity, and $V_r = \omega r$ is the rotation velocity.

For power law spectra this equation is of the form

$$\xi = \frac{E_2^{\gamma+1} - E_1^{\gamma+1}}{E_2^{\gamma+1} + E_1^{\gamma+1}}. \qquad (3)$$

Physically, this effect arises from the motion of the magnetic field lines with respect to the spacecraft. The particle distribution is expected to be symmetric with respect to the fieldline. When the detector is looking in the direction of motion of B, a particle of velocity V in the B frame of reference will have a velocity $V - V_r$ as seen by the detector. When the detector is looking in a direction opposite to the motion of the fieldline, then the particle will have a velocity $V + V_r$. The net effect is to produce a first-order or Compton-Getting type anisotropy in the frame of reference of the spacecraft. For the 1.2 to 2.2 MeV region $E \simeq 1.4$ MeV, $v = 1.6 \times 10^7$ m sec^{-1}, and $V_r = 12.6N \times 10^3$, where N is the distance to the observing point measured in units of R_J (7.137 $\times 10^7$ m). At 25 R_J, $V_r = 3 \times 10^5$ m sec^{-1}, $V_r/v = 0.019$, and $\xi = 18\%$ (at 50 R_J this increases to 37%). The expected co-rotation anisotropy for $\gamma = 4$ is

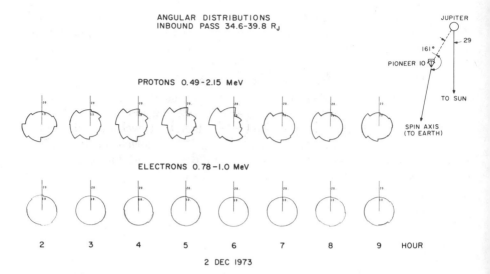

Fig. 9.   Polar plots of angular distributions of protons and electrons between 35 and 40 R_J for Pioneer 10. The top of the figure represents the direction towards the north ecliptic pole. The normal out of the page is in the direction of the spacecraft spin axis (after Trainor *et al.* 1974a).

shown as a dashed line in Fig. 11. The agreement with the innermost values of A_1/A_0 is good. The large peaks between 25 and 50 R_J in Fig. 11 are much larger than would be produced by co-rotation effects. A gradient anisotropy would be on the order of $(R_g/J)(dJ/dR)$, with R_g the gyro-radius of the protons. This would be in the same direction as the co-rotation anisotropy. However, it requires a $\sim 100\%$ increase per R_J, and the large increases in A_1/A_0 are associated with the minima in the counting rates. Gradient anisotropies do not appear to cause important effects.

In Fig. 13 A_1/A_0, Θ_1, α and the corresponding averaged detector counting rate are plotted as a function of time for the inbound pass of Pioneer 10; α is the direction between the projection of the magnetic field into the plane of the spacecraft and the direction of the first-order anisotropy. Small values of α represent current flow toward the planetary equator and large values (i.e., 90–$180°$) represent a net particle flux toward the planet. The dashed lines intersect the A_1 peaks. It is clear that large values of A_1/A_0 are associated with flux minima and values of $\alpha \sim 180°$, and therefore that particles are flowing down the fieldline to the atmosphere. Thus, this region of the magnetosphere is continuously leaking substantial fluxes into the atmosphere. This is further evidence that particles in the outer and transition regions are not stably trapped. The maximum in A_2/A_0 appears to be located (Fig. 12) close to the equatorial region.

In the inner region, the protons approach isotropy (Fig. 14) until the orbit of Io. Immediately within Io's orbit there is a complex "butterfly" distri-

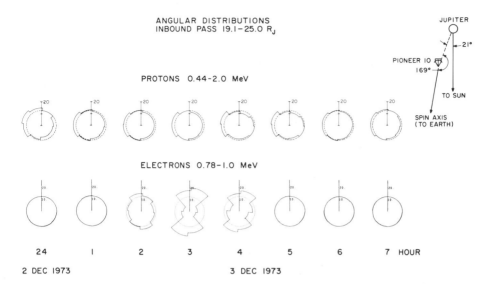

Fig. 10.   Polar plots of electron and proton data between 19 and 25 R_J on the inbound Pioneer 10 pass. The orientation of the plots is the same as that of Fig. 9 (after Trainor *et al.* 1974*a*).

bution and then a strong second harmonic component develops that is symptomatic of stable trapping. On Pioneer 11 the high-latitude data (Fig. 14) in the inner region is that which is expected of a location where many particles are mirroring. In addition there is a well-defined loss cone.

The Inner Region and the Effects of the Jovian Satellites

It was observed on both Pioneer 10 and 11 that Io is able to almost completely remove the lower-energy (1.2 to 2.1 MeV) protons (Trainor *et al.* 1974*a*, 1975; Van Allen *et al.* 1975[3]). Inbound on Pioneer 11, more than 99% of the protons in this energy interval were removed in the L region which Io occupies (Fig. 14*a*). This drop by a factor of ~ 100 compares with a factor of ~ 60 noted from Pioneer 10 data; but Pioneer 11 was at magnetic latitudes above 40°, and the larger effect there is in agreement with predictions (Mead and Hess 1973; Birmingham *et al.* 1974; Hess *et al.* 1974).

Figure 14 shows the count rate data for 1.2 to 2.1 MeV protons for Pioneers 10 and 11 as a function of L, and angular distributions measured on the spacecraft at the same times are shown at the indicated L locations. The magnetic latitude is also indicated periodically, as well as the predicted regions occupied by Io. The fit of the data for the removal of particles at Io is fair, but the fit is lacking to the extent that on both Pioneer 10 and Pioneer 11 the count rates were dropping appreciably, at times substantially after the inner-

---

[3]See pp. 929 and 1203.

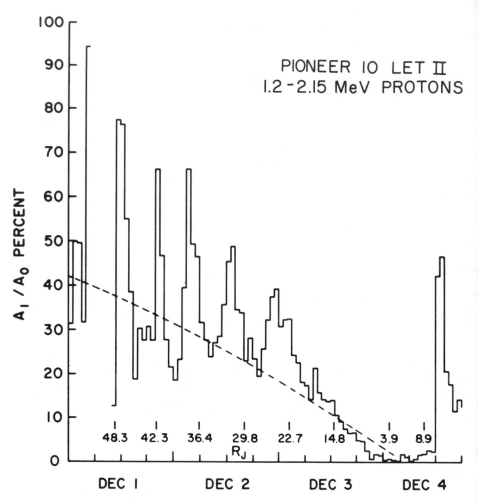

Fig. 11. Magnitude of the first harmonic in the angular distribution of 1.2–2.15 MeV protons as a function of time and Jovian radius for the inbound pass of Pioneer 10. The dashed line indicates the magnitude of the first-order anisotropy expected from co-rotation (after Trainor *et al.* 1974a).

most *L* shell predicted to be swept by Io. The indication that the flux inside Io is less than that encountered outside establishes the fact that radial diffusion is the dominant acceleration process for the nuclear component in the inner region. As previously mentioned, the origin of the sharp proton spike at perigee is not understood.

No substantial removal of protons while passing through Europa's orbit is apparent from the Pioneer 11 data. On Pioneer 10 there is a well-defined decrease at Europa. These contrary data may well be an indication of strong azimuthal or wake effects.

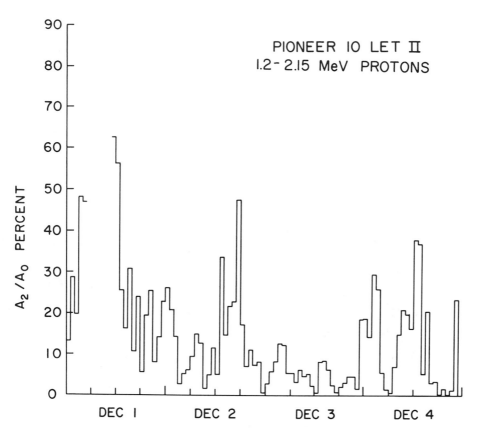

Fig. 12. Amplitude of the second harmonic of the 1.2–2.15 MeV proton distributions (after Trainor *et al.* 1974*a*).

III. THE ELECTRON COMPONENT

As we have previously discussed, the energetic electron component shows a well-defined 10-hr periodicity in the outer and transition regions. High counting rates prevented any meaningful electron study in the inner region. Figure 15 shows the differential electron spectra measured near the magnetic equator by the HET and LET II telescopes on the inbound and outbound passes. The spectra in this energy range are remarkably hard and similar over the region outside $\sim 24\ R_J$ where we have measurements, and they are very similar to the spectra of electrons leaking from the Jovian magnetosphere, measured on Pioneer 10 many months before encounter. This, of course, is quite different from the behavior in the earth's radiation belts. We have found no obvious correlation between magnetic latitude and the spectral shape in the 0.12 to 8.0 MeV region. The proton component in the

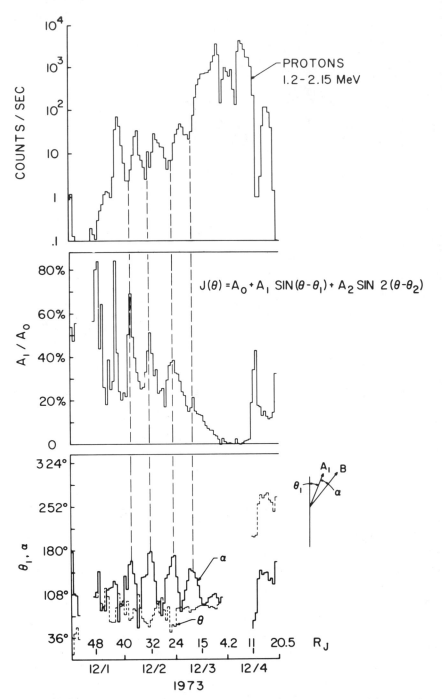

Fig. 13. Plot of the 1.2–2.15 MeV proton counting rate, A_1/A_0, Θ_1 and α, for the inbound pass of Pioneer 10. Θ_1 is the direction of the first-order harmonic and α is the angle between A_1 and the projection of B in the plane of the spacecraft.

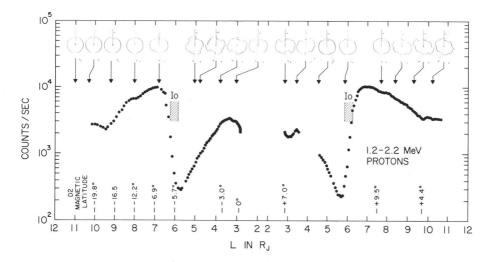

Fig. 14a. Count rate data for the 1.2 to 2.1 MeV protons are shown for Pioneer 10 versus L calculated from the D_2 model of Jupiter's magnetic field (Chenette *et al.* 1974). The predicted regions of L to be swept by Io and Europa are shown. Angular distributions shown are summed in the experiment data system in 45° sectors of spin. The detector has a full field view of 30°. The projection of the magnetic field vector on the sector plane is shown (after Trainor *et al.* 1975).

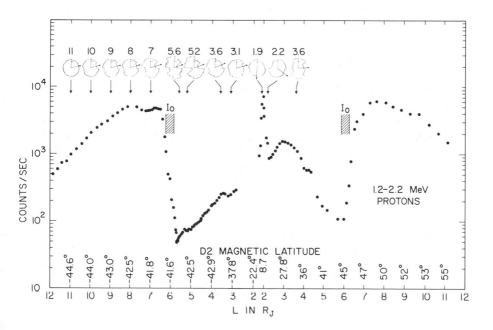

Fig. 14b. As for 14a but for Pioneer 11.

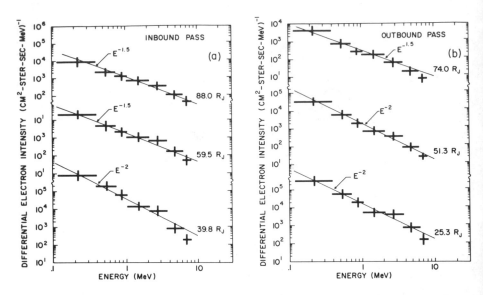

Fig. 15.   Electron energy spectra at different locations in and near the Jovian magnetosphere.

outer magnetosphere does not appear to be stably trapped; this should also apply to the electrons in this region. It would be expected that some of these would diffuse into interplanetary space; and, in fact, these Jovian electrons are observed in interplanetary space in very large numbers. The injection process is more complex than that of simple diffusion from the outer magnetosphere.

As Pioneer 10 approached Jupiter both the University of Chicago (Chenette *et al.* (1974)[4] and the Goddard-University of New Hampshire experiments (Teegarden *et al.* 1974) observed low-energy (~ 0.2 to 8 MeV) electron increases at > 1 A.U. from the planet. These discrete bursts or increases were typically several hundred times the normal quiet-time electron flux and became much more frequent as the spacecraft approached Jupiter (Fig. 16). Close to Jupiter, but well outside its magnetosphere, there is the quasi-continuous presence of large fluxes of these electrons.

These observations suggested that Jovian electrons should be observable at 1 A.U. Previously it had been reported that the 3 to 12 MeV electron component detected by the Interplanetary Monitoring Platform (IMP) series (McDonald *et al.* 1972) frequently identified positive increases of these electrons that could not be associated with discrete solar events. These quiet-time increases represented a factor of 3 to 5 increase in intensity and lasted from 5 to 12 days. They displayed a remarkable anti-correlation with low-energy proton events, and their amplitude was observed to generally dimin-

---

[4]See p. 755.

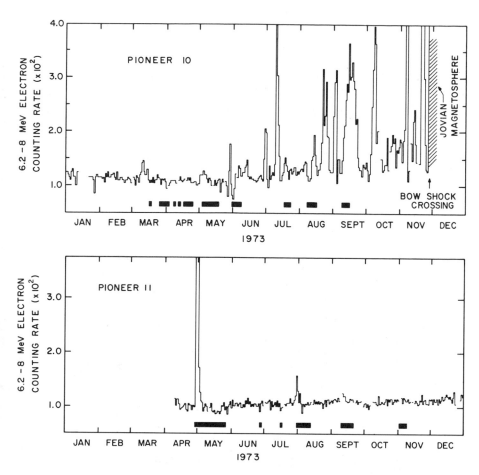

Fig. 16. The daily average counting rates for 6.2 to 8 MeV electrons on Pioneer 10 and Pioneer 11. The black rectangles indicate the larger solar cosmic-ray events. Most of these are of the low-energy co-rotating type (after Teegarden *et al.* 1974).

ish toward solar maximum. Re-examination of the IMP data revealed that these increases have a 13-month periodicity (Fig. 17) indicating a Jovian origin for the quiet-time events observed near Earth (Teegarden *et al.* 1974).

In Fig. 18 the time of the year that the earth crosses Jupiter's fieldline is plotted for each year from 1964 to 1974. The vertical bars in the plot give the duration of the periods when the quiet-time increases were present. With the exception of the 1964 bar, all the periods fall close to or contain the predicted time of crossing Jupiter's fieldline. This is a further indication that Jovian electrons are being detected at 1 A.U. Furthermore, the spectra for electrons observed in Jupiter's outer magnetosphere, for those in interplanetary space near Jupiter, for the quiet-time increases near the earth, and for

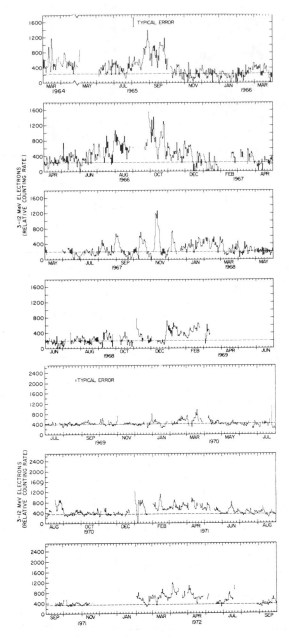

Fig. 17.   Results from Interplanetary Monitoring Platform IMP 3, IMP 4 (McDonald *et al.* 1972), and IMP 5 (M. A. I. Van Hollebeke, personal communication 1974); 3 to 12 MeV data plotted in 13-month intervals. The dashed line is a convenience for identifying the electron increases. There still may be major solar contributions in some periods such as 28 July to 10 August 1966, and 20 September to 10 October 1966. The 13-month periodicity is clearly defined, and the amplitude decreases over solar maximum (1969). A background subtraction has not been made for the data after July 1969 (after Teegarden *et al.* 1974).

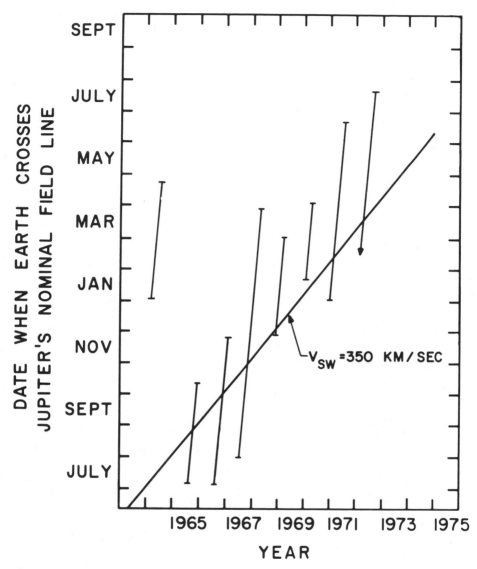

Fig. 18.   The vertical bars give the duration of the periods when quiet-time increases were present during the various epochs in Fig. 5; arrow indicates when the length of the bar is uncertain due to a data gap. The diagonal line represents the time of the year that an idealized spiral interplanetary magnetic field line would connect the earth and Jupiter assuming a constant plasma velocity of 350 km sec⁻¹ (after Teegarden et al. 1974).

the ambient electron spectrum are all remarkably similar. These lines of evidence suggest that Jupiter is the source of most of the low-energy electrons observed at 1 A.U.

If we examine the entire energy spectrum of electrons (Fig. 19), it is seen that the entire low-energy spectrum from 0.2 to 40 MeV is remarkably dif-

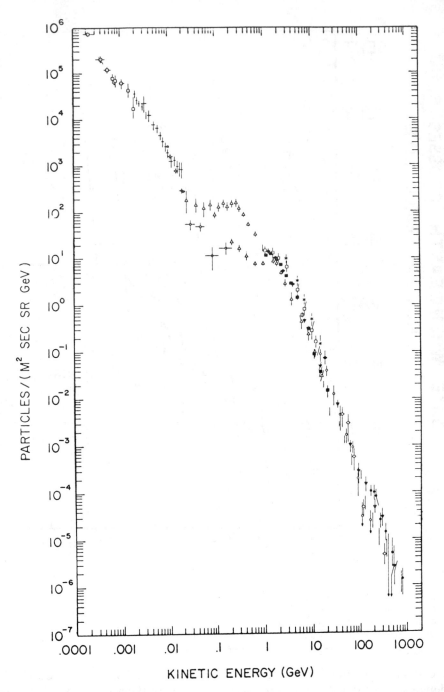

Fig. 19.   Data on electron energy spectra from 0.2 MeV to 10^3 GeV. This represents the combined efforts of many different groups. The two data sets between 50 MeV and 1 GeV show typical solar cycle variations due to interplanetary modulation. The increase below 40 MeV may be largely of Jovian origin.

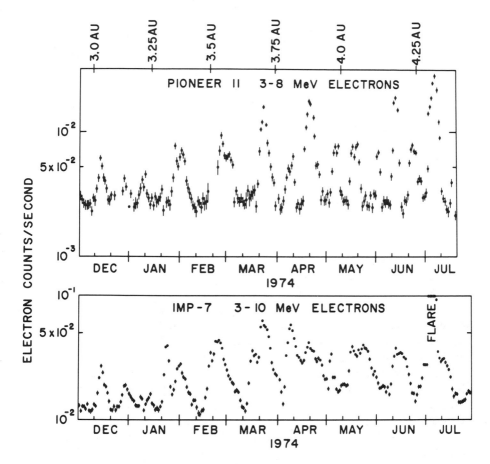

Fig. 20.   Simultaneous observation of MeV electrons on Pioneer 11 and IMP 7. The sensitivity of Pioneer 11 has been extended over that of Fig. 16 by the use of pulse-height information.

ferent from the relatively flat differential spectra measured from ~ 40 MeV to ~ 1 GeV. Above 1 GeV the measurements approach a power law of $\sim E^{-3}$. Since the spectral shape of the low-energy component is consistent with that of knock-on electrons produced in interstellar space by higher-energy nucleons, it has been generally assumed that they were interstellar secondaries. However, this model requires that the total solar modulation at low energies be less than a factor of 5. The total solar modulation is the important question, and this can be estimated by comparing simultaneous Pioneer and IMP data.

In Fig. 20 daily averages have been plotted for an 8-month period extending from December 1973 to July 1974. The sensitivity of the Pioneer data has been increased by using the pulse-height information. It is readily seen that Jovian electrons were present for the entire 8-month period. The mid-December 1973 increase occurred essentially simultaneously at the two

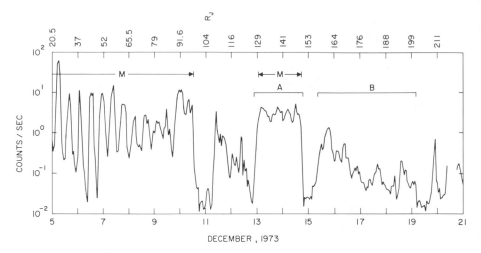

Fig. 21.   Time history of 5–7 MeV electrons for the outbound pass of Pioneer 10. The regions marked *M* represent periods inside the Jovian magnetosphere. *A* and *B* refer to similar periods shown in Fig. 20 for Pioneer 11.

spacecraft. Using the known geometric factors and background response a gradient of $\sim 150\%$ per A.U. was found, with the flux increasing outbound. Furthermore, it is seen that the events have become larger on Pioneer 11. It appears possible that the 0.2 to 40 MeV electrons observed near Earth are mainly of Jovian origin. In addition, it is expected that Jovian protons will be released at the same time. Their energy spectra are steep ($\Phi \propto E^{-4}$), and in the Jovian outer magnetosphere they are generally $\sim 1\%$ of the electron flux at ~ 1 MeV. The flux expected away from Jupiter is thus small and difficult, if not impossible, to separate from the solar protons.

At the time of the December 1973 electron increase, Pioneer 10 was still on the fringes of the dawn side of the Jovian magnetosphere at a distance of some 125–200 R_J from the planet. The Pioneer 10 magnetospheric data for this period are shown in Fig. 21. The periods when the spacecraft is within the magnetosheath region (Wolfe *et al.* 1974) are indicated by the intervals labeled *M*. The time of the Pioneer 11 electron increase (Fig. 22) is indicated by two intervals labeled *A* and *B*. The Jovian data for Region *A* represents an expansion of the magnetosphere past the spacecraft, and the peak-to-valley ratio of the 10-hr modulation is greatly reduced. In Region *B* the spacecraft is outside the magnetosphere and sees a series of discrete electron injections. Pioneer 11 may see a superposition of these increases.

This ejection of particles into interplanetary space appears to be a general characteristic of planetary magnetospheres—Jupiter being the standout because of the large size of the leakage. It has been reported (Simpson *et al.* 1974) that Mercury also injects particles into the interplanetary medium, and

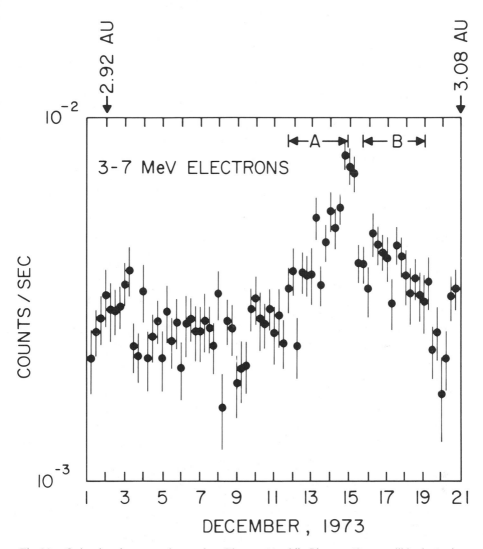

Fig. 22.   Quiet-time increase observed on Pioneer 11 while Pioneer 10 was still in the Jovian magnetosphere. Period *A* represents the interval to peak intensity and *B* represents the remainder of the event.

such activity has long been associated with the earth's own magnetosphere. In fact, an experiment of the Applied Physics Laboratory on IMP 7 and IMP 8 shows that when the 0.3 to 1 MeV proton flux is at the lowest level, essentially all the nucleon flux is coming from the earth's magnetosphere (Krimigis *et al.* 1975).

DISCUSSION

T. Gehrels: Pioneer 10 and Pioneer 11 flew by Jupiter at solar minimum. Would you not expect MJS 77 to see a drastically changed magnetosphere at Jupiter, about the time of solar maximum?

J. H. Trainor: We are far from certain of the details and topography of the outer magnetosphere of Jupiter, but it does appear that the magnetosphere on the sunward side beyond 40 or 50 R_J is very "soft" and readily compressed. Since the occurrence of interplanetary disturbances will be much higher during solar maximum one could expect conditions to be much more dynamic and disturbed in the outer magnetosphere at that time.

The propagation of these effects to the inner magnetosphere is uncertain. Arguments can be made for believing that the flux of energetic particles trapped in the inner regions would increase at solar maximum, and other plausible arguments can be made which lead to the opposite conclusion. One does not know enough yet to allow reasonable predictions.

REFERENCES

Birmingham, T.; Hess, W.; Northrop, T.; Baxter, R.; and Lojko, M. 1974. The electron diffusion coefficient in Jupiter's magnetosphere. *J. Geophys. Res.* 79:87–97.

Chenette, D. L.; Conlon, T. F.; and Simpson, J. A. 1974. Bursts of relativistic electrons from Jupiter observed in interplanetary space with the time variation of the planetary rotation period. *J. Geophys. Res.* 79:3551–3558.

Gosling, J. T.; Asbridge, J. R.; Bame, S. J.; Hundhausen, A. J.; and Strong, I. B. 1967. Measurements of the interplanetary solar wind during the large geomagnetic storm of April 17–18, 1965. *J. Geophys. Res.* 72:1813–1821.

Hall, C. F. 1975. Pioneer 10 and Pioneer 11. *Science* 188:445–446.

Hess, W. N.; Birmingham, T. J.; and Mead, G. D. 1974. Absorption of trapped particles by Jupiter's moons. *J. Geophys. Res.* 79:2877–2880.

Krimigis, S. M.; Kohl, J. W.; and Armstrong, T. P. 1975. The magnetospheric contribution to the quiet-time low-energy nucleon spectrum in the vicinity of earth. *Geophys. Res. Lett.* 2: 457–460.

McDonald, F. B.; Cline, T. L.; and Simnett, G. M. 1972. Multifarious temporal variations of low-energy relativistic cosmic-ray electrons. *J. Geophys. Res.* 77:2213–2231.

Mead, G. D., and Hess, W. N. 1973. Jupiter's radiation belts and the sweeping effect of its satellites. *J. Geophys. Res.* 78:2793–2811.

Simpson, J. A.; Eraker, J. H.; Lamport, J. E.; and Walpole, P. H. 1974. Electrons and protons accelerated in Mercury's magnetic field. *Science* 185:160–166.

Smith, E. J.; Davis, Jr., L.; Jones, D. E.; Coleman, Jr., P. J.; Colburn, D. S.; Dyal, P.; Sonett, C. P.; and Frandsen, A. M. A. 1974. The planetary magnetic field and magnetosphere of Jupiter: Pioneer 10. *J. Geophys. Res.* 79:3501–3513.

Stilwell, D.; Joyce, R. M.; Trainor, J. H.; White, Jr., H. P.; Streeter, G.; and Bernstein, J. 1975. Pioneer 10/11 and Helios A/B cosmic ray instruments. *IEEE Trans. Nucl. Sci.* 22:570–574.

Teegarden, B. J.; McDonald, F. B.; Trainor, J. H.; Webber, W. R.; and Roelof, E. 1974. Interplanetary MeV electrons of Jovian origin. *J. Geophys. Res.* 79:3615–3622.

Trainor, J. H.; McDonald, F. B.; Stilwell, D. E.; Teegarden, B. J.; and Webber, W. R. 1975. Jovian protons and electrons: Pioneer 11. *Science* 188:462–465.

Trainor, J. H.; McDonald, F. B.; Teegarden, B. J.; Webber, W. R.; and Roelof, E. C. 1974*a*. Energetic particles in the Jovian magnetosphere. *J. Geophys. Res.* 79:3600–3613.

Trainor, J. H.; Teegarden, B. J.; Stilwell, D. E.; and McDonald, F. B. 1974*b*. Energetic particle population in the Jovian magnetosphere: a preliminary note. *Science* 183:311–313.

Van Allen, J. A.; Randall, B. A.; Baker, D. N.; Goertz, C. K.; Sentman, D. D.; Thomsen, M. F.; and Flindt, H. R. 1975. Pioneer 11 observations of energetic particles in the Jovian magnetosphere. *Science* 188:459–462.

Wolfe, J. H.; Mihalov, J. D.; Collard, H. R.; McKibben, D. D.; Frank, L. A.; and Intriligator, D. S. 1974. Pioneer 10 observations of the solar wind interaction with Jupiter. *J. Geophys. Res.* 79:3489–3500.

Wolfe, J. H.; Silva, R. W.; McKibben, D. D.; and Mason, R. H. 1966. The compositional, anisotropic, and nonradial flow characteristics of the solar wind. *J. Geophys. Res.* 71:3329–3335.

PART V

Satellites

THE JOVIAN SATELLITES

D. MORRISON
University of Hawaii

and

J. A. BURNS
Cornell University

We discuss the physical and, to a lesser extent, the dynamical properties of the satellites, with primary emphasis upon the Galilean satellites. Each Galilean satellite is described separately, and its surface properties, orbit, and probable internal composition and evolutionary history are considered. We intercompare the Galilean satellites, noting the division into an inner pair characterized by high density, almost certain differentiation, and depletion of volatiles, and an outer pair, composed primarily of water, that have poorly understood histories. However, the surface properties are not well correlated with the bulk properties: Io and Europa, in spite of their apparent low content of volatiles, have bright surfaces of ices or evaporites of former water solutions, while Ganymede and particularly Callisto have lower albedo surfaces apparently composed largely of silicate or carbonaceous material. All of the Galilean satellites have rotations that are tidally locked, and all are presumably influenced by the Jovian magnetospheric particles with which they interact. Such external processes, rather than internal ones, may be responsible for most of their photometric and spectrophotometric properties. Only Io is known to have an atmosphere, and this atmosphere and the extended gas torus to which it gives rise are fundamentally affected by the magnetosphere. We also review the few physical data that are available for Amalthea and the outer satellites. We suggest that the density of Amalthea may be found by observing its tidal distortion.

The satellites of Jupiter have had important effects on our concept of the universe and our understanding of some basic physical laws. For example, when Galileo discovered the four largest Jovian satellites in 1610, the Earth was shown not to be the only center of motion in the universe, and this new perspective helped persuade many people of the failings of the geocentric cosmology. Light was first found to travel at a finite speed in 1675 by Olaf Römer from Cassini's timings of eclipses of the Galilean satellites. Such fundamental contributions to science by the Jovian satellites are to be ex-

[991]

pected in the future too: the upgrading of the Arecibo radio telescope may permit general relativity theories to be tested through precise radar probing of the orbits of the Galilean satellites, and it is likely that the Jovian satellites tell better of the primordial conditions in Jupiter's vicinity than does Jupiter itself. In fact, a true understanding of the origin and the evolution of the Jovian satellite family may be crucial for an understanding of the origin of the solar system.

Recent physical studies are dramatically demonstrating that the Galilean satellites are four different worlds as deserving of detailed study as the four terrestrial planets which they resemble in size. We now see them as falling into two groups with different bulk composition, reminiscent of the division of the planets into terrestrial and Jovian classes. The surfaces as well as the interiors are of different composition for each satellite, implying different geological histories. All of these objects, but most particularly Io with its atmosphere and associated gaseous ring around Jupiter, are apparently influenced by energetic particles in the Jovian magnetosphere. Most of what we know about the physical properties of these satellites is new, and their study is one of the fastest moving fields in contemporary solar system research. At present this work is almost entirely groundbased, but before the end of the nineteen seventies we anticipate a substantial increase in knowledge as a result of the reconnaissance of the Jovian satellite system by the two Mariner Jupiter/Saturn (MJS) spacecraft.

Most studies of the Jovian satellites, through 1971 and 1973 respectively, are discussed by Newburn and Gulkis (1973) in their review paper on the outer solar system or by Morrison and Cruikshank (1974) in their comprehensive review on the physical properties of the satellites. In 1974, the IAU colloquium, "Planetary Satellites," considered the Jovian satellites in some detail; the review papers from it are edited by Burns (1976*b*), and most of the contributed papers have appeared in special issues of *Icarus* (vol. *24*, no. 4 and vol. *25* no. 3) and *Celestial Mechanics* (vol. *12,* no. 1). Because of these recent sources of detailed information, this review of the Jovian satellites will emphasize results published or observations made since the review paper by Morrison and Cruikshank (1974), including those presented at the 1974 IAU colloquium. It will first discuss the satellites as a whole, listing their orbital and physical properties and some general physical processes that act on them, and then go on to consider the satellites individually.

I. GENERAL INFORMATION

The four major satellites of Jupiter — the Galilean satellites — are moon-sized objects which are easily seen with a small telescope and which would be visible even with the naked eye were it not for their nearness to the brilliant Jupiter. Excepting the moon, they were the first planetary satellites to be discovered. Simon Marius, who said he found them at the same time as

Galileo did, named them Io, Europa, Ganymede and Callisto, after lovers of Jupiter; they are also known as J 1, J 2, J 3 and J 4, in order of increasing distance from their parent.

Nearly three centuries elapsed before the next Jovian satellite, J 5 or Amalthea, was discovered visually by Barnard in 1892; it is the innermost and is named after the nymph who nursed Jupiter. The eight outer satellites were found photographically, with the most recent being Kowal's discovery of Leda (J 13) (Kowal *et al.* 1975) in 1974 with the 122-cm Schmidt Telescope on Palomar Mountain. The members of the outer group are designated by numerals corresponding to the order of their discovery and have recently been named by the IAU after lovers of Jupiter. The small satellites in prograde orbits have names ending in *a*, and those in retrograde orbits have names ending in *e*. More historical details are given in Porter (1960) and in the popular book by Wetterer (1971).

Table I lists the orbital and physical properties of the known Jovian satellites in order of distance from the planet. The five innermost satellites — also called the *regular satellites* — move prograde on nearly circular, equatorial orbits. The orbits of these satellites are measurably affected by all the classical perturbations: the orbits of Io, and especially Amalthea, are strongly perturbed by the large oblateness of the rapidly spinning Jupiter; the solar attraction produces significant perturbations of Callisto's orbit; and the Galilean satellites are massive enough to induce substantial mutual perturbations as indicated by the puzzling Laplace resonance, which precisely relates the mean motions of the inner three. The Laplace resonance, unlike most satellite resonances (Greenberg 1973, 1976,[1]), is probably not produced by tidal friction (Burns 1976*a*; Sinclair 1975); its existence may indicate that gas drag, or even electromagnetic drag, on the satellites was once more substantial than that present today. The regularity of the orbits of the inner satellites is usually ascribed to their formation in a dense proto-planetary nebula, as well as to subsequent internal energy loss in the satellites themselves, which generally circularizes orbits (Goldreich 1963). However, the pertinence of the recently proposed, and probable, liquid interiors to this circularization process has never been investigated.

Although Amalthea, by far the smallest of the regular satellites, may have a distinct origin, it is probable that all of the regular satellites originated dynamically in the same manner from the circumplanetary swarm of proto-Jupiter (Safronov and Ruskol 1976; Cameron and Pollack p. 78). The observed decrease in mass density of the regular satellites with distance from Jupiter (see Table I) is thought to have been caused by higher temperatures in the central planetary nebula at the time of origin (Pollack and Reynolds 1974; Cameron and Pollack, p. 81). The differing physical properties of the surfaces of the Galilean satellites (e.g., albedos) to be considered in

---

[1]See p. 128.

TABLE I[a]

ORBITAL DATA

PHYSICAL DATA

	Semimajor Axis 10³km	(a/R)	Period (days)	e	i	$\bar{V}$[b]	$V(1,0)$[c]	p_V	R (km)	$M(10^{23}\text{g})$	Inverse Mass	$\rho(\text{g cm}^{-3})$
5 Amalthea	181.3	(2.55)	0.489	0.003	0°4	13.0	+6.3		120 ± 30		$\sim 10^{-8}$	
1 Io	421.6	(5.95)	1.769	0.000	0.0	5.0	−1.68	0.63	1820 ± 10	891	$(4.696 \pm 0.06) \times 10^{-5}$	3.52 ± 0.07
2 Europa	670.9	(9.47)	3.551	0.000	0.5	5.3	−1.41	0.64	1525 ± 25	487	$(2.565 \pm 0.06) \times 10^{-5}$	3.28 ± 0.18
3 Ganymede	1070.	(15.1)	7.155	0.001	0.2	4.6	−2.09	0.43	2635 ± 25	1490	$(7.845 \pm 0.08) \times 10^{-5}$	1.95 ± 0.06
4 Callisto	1880	(26.6)	16.689	0.01	0.2	5.6	−1.05	0.17	2500^{+75}_{-150}	1074	$(5.65 \pm 0.10) \times 10^{-5}$	$1.63^{+0.34}_{-0.20}$
13 Leda	11110	(156)	240	0.146	26.7	20	+13.3[d]		(1 → 7)		$\sim 2 \times 10^{-14}$	
6 Himalia	11470	(161)	250.6	0.158	27.6	14.8	+8.0	0.03	(85 ± 10)		$\sim 4 \times 10^{-9}$	
10 Lysithea	11710	(164)	260	0.130	29.0	18.4	+11.7[d]		(3 → 16)		$\sim 2 \times 10^{-12}$	
7 Elara	11740	(165)	260.1	0.207	24.8	16.4	+9.3	0.03	(40 ± 10)		$\sim 4 \times 10^{-10}$	
12 Ananke	20700	(291)	617	0.17	147.	18.9	+12.2[d]		(3 → 14)		$\sim 5 \times 10^{-12}$	
11 Carme	22350	(314)	692	0.21	164.	18.0	+11.3[d]		(4 → 20)		$\sim 10^{-11}$	
8 Pasiphae	23300	(327)	735	0.38	145.	17.7	+11.0[d]		(4 → 23)		$\sim 10^{-11}$	
9 Sinope	23700	(333)	758	0.28	153.	18.3	+11.6[d]		(3 → 18)		$\sim 5 \times 10^{-12}$	

[a]The data are taken primarily from the summary by Morrison et al. (1976). Masses are based on Pioneer 10 values (Anderson et al. 1974) for the Galilean satellites and for all outer satellites on estimated radii, tabulated by Morrison et al. (1976), and an assumed density of 3.0 g cm⁻³. The magnitudes of the outer satellites are from Andersson (1974). The orbits of the outer satellites vary considerably. Proper eccentricity and proper inclinations are given for the Galilean satellites.

[b]Mean opposition visual magnitude.

[c]Visual magnitude at zero phase reduced to 1 A.U. distance from both Earth and sun; it does not include the opposition surge exhibited by the Galilean satellites.

[d]Based on a mean color index of +0.8.

detail in the following sections are probably a result of not only their individual internal compositions but also the particular radiation environments in which the satellites find themselves today (cf. E. J. Smith *et al.* p. 788; Van Allen p. 928; Mogro-Campero p. 1190; and below).

The sizes given in Table I for the Galilean satellites are found either from optical measurements, stellar occultations, or mutual occultations (cf. Morrison and Cruikshank 1974). Their masses are derived from the perturbations on the Pioneer 10 trajectory (Anderson *et al.* 1974;[2]); these masses are somewhat different from the classical values, particularly for Io.

The outer satellites — or the *irregular satellites* as they are often called — are located at relative distances from Jupiter similar to those of the terrestrial planets from the sun. They are, according to the orbital elements displayed in Table I, confined to two groups. The inner group, consisting of Himalia (J 6), Elara (J 7), Lysithea (J 10), and Leda (J 13), have similar prograde orbits of moderate inclination (25° to 30°) and moderate eccentricity (0.15 to 0.20) with semimajor axes of 1.1 to 1.2 × 10⁷ km. The outer group — Pasiphae (J 8), Sinope (J 9), Carme (J 11), and Ananke (J 12) — have orbits similar in eccentricity (0.17 to 0.38) and inclinations (18° to 35°) but different in direction (retrograde) and semimajor axis (2.1 to 2.4 × 10⁷ km). Both groups lie well beyond the Jovian magnetosphere; they are located about half way to the point at which they would be definitely under the control of the solar gravitational field (Kuiper 1951). Their orbits are very substantially affected by solar perturbations (Aksnes 1976).

For the outer satellites listed in Table I, lower bounds for the radii were calculated on the assumption that the satellite has a geometric albedo of 1.0 and upper bounds by the smaller of either (a) a diameter of 1/3 arcsec, which would be detectable visually, or (b) an albedo of 0.03. The mass estimates listed in Table I were obtained with a radius about two or three times the lower bound, on the assumption that the outer satellites are probably dark, like most asteroids (Chapman *et al.* 1975). Similarly, a density of 3 g cm⁻³ has been assumed for this tabulation. We caution the reader that these estimated masses are only order-of-magnitude guesses.

Before proceeding to discuss each satellite individually, we mention a few phenomena that influence the behavior of all the satellites: (1) stability of ices; (2) tidal effects that tend to synchronize rotation with orbital revolutions; (3) correlations of color and albedo with orbital phase; (4) possible interactions with the radiation and meteoroid environment; and (5) a suggested secular brightening of all the satellites.

(1) Water ice is marginally stable at Jupiter's distance. The maximum surface temperatures of the satellites are in the range 135°K to 165°K, just where the vapor pressure of water ice is strongly dependent upon temperature (Watson *et al.* 1963; Lebofsky 1975). The lifetime of a layer of ice on a planetary surface depends very critically on the temperature history, and

[2]See p. 113.

thus on the albedo and location on the surface. In addition, the escape velocity and possible presence of an atmosphere are important for controlling the migration of condensates from one surface location to another. At the mean temperatures of the Jovian satellites, less than one meter of ice will be lost over the age of the solar system, but as temperatures rise above 150°K, as they do each rotation near the equators of the satellites, many meters of ice could be lost in a few million years, either by migration to cooler regions or by escape from a water-vapor atmosphere. Although two of the satellites, Europa and Ganymede, have been known since the early 1970's to have substantial quantities of exposed water ice on their surfaces, the observed distribution and amount of ice are not well understood in terms of current theoretical arguments on stability. Ices of methane and ammonia, probably common in the systems of Saturn and Uranus, are clearly not stable in the Jovian system.

(2) Tides produced by Jupiter on a satellite will synchronize its rotation rate with its orbital revolution rate in a time proportional to a^6/R^2, where R is the satellite radius and a is its orbital semimajor axis. Using this result, it is easy to show that the five inner Jovian satellites should be tidally locked, while the rotations of the outer eight have been essentially unaffected by tides (Peale 1976). Thus, when one observes a regular satellite at a given orbital angle θ from superior conjunction, one is looking at a particular face of the satellite. Tidal despinning may have provided a moderate internal heat source for the Galilean satellites early in their lifetimes.

(3) One of the most obvious regularities of the Galilean satellites is the progression of photometric properties from Io outwards to Callisto. First, there is the decrease in average albedo with increasing distance from Jupiter (see Table I). Second are the changes between satellites in the photometric differences (both albedo and color) of leading and trailing hemispheres. Each satellite has one hemisphere that is redder and 10% to 30% darker than the other; the darker hemisphere is oriented toward the direction of orbital motion for Callisto and away from it for the inner three. As shown by Johnson (1971), Blanco and Catalano (1974), Veverka (1976a), Morrison and Morrison (1976), and Millis and Thompson (1975), there is a monotonic shift in the orientation of the dark hemisphere with respect to the satellite orbital velocity vector from Io to Callisto.

(4) The above mentioned photometric regularity also occurs in the satellite system of Saturn. It naturally suggests that the surface properties are being influenced by some external agent, such as differential accretion of meteoritic material or differential interaction with the Jovian magnetospheric particles. Uneven coloration may occur through direct impacts with interplanetary meteoroids: these impacts will be concentrated on the leading hemisphere because of the satellite's orbital velocity. The focusing and the impact energy will increase for satellites closer to Jupiter because of their higher orbital velocities. The effect of impacts on albedos depends on the

nature of the satellite surface and whether the debris is asteroidal or cometary. The boundary between the leading and trailing hemispheres will not be
distinct because of the satellite's own gravitational attraction and because of
the distribution of ejection velocities from the impact. Some ejecta will escape the satellite only to return later after circling the planet for a short time
on a nearby orbit. Other ejecta will remain near the impact site, while an
intermediate set of ejection velocities will be sufficient to distribute debris
as much as halfway around the satellite. Small particles — either captured
material or collisional debris — that find themselves within the Jovian sphere
of influence will gradually be dragged into Jupiter by the Poynting-Robertson
effect which withdraws energy from the particle's orbit because the re-
emitted thermal radiation produces a reaction force on the particle (cf. Burns
1976a). Their velocities as they move past the satellite may be small and
they may be gently deposited on satellite surfaces without producing impact
craters. These processes might be significant solely for Callisto if its characteristic capture collision time is considerably shorter than the time for the
Poynting-Robertson drift, so that the material cannot reach the inner satellites. Radiation pressure on small particles, however, causes large oscillations in eccentricity which probably spread debris periodically past Callisto
(Burns 1976a). The particles may develop an electric charge which will considerably complicate their dynamics (Mendis and Axford 1974) in the Jovian
magnetic field.

Although charged particle impacts may also substantially affect surface
features, laboratory studies of the effects of such impacts, particularly into
ice, are not available. The present Jovian magnetosphere (E. J. Smith et al.
p. 790) encloses all the Galilean satellites, but only the inner two, or perhaps
three, are especially susceptible to charged particle impacts (Van Allen p.
932). On the basis of present data, only Io and Amalthea are known to absorb
the particles effectively (Mogro-Campero p. 1190). Radiation belt particles
have three distinct gross motions: the particles spiral around their magnetic
fieldlines but are loosely tied to them, bouncing back and forth between mirror points in the northern and southern hemispheres of Jupiter's magnetic
field; the particles drift around Jupiter under the action of the curvature and
gradient of the magnetic field, going in directions determined by the sign of
their charge; and the electrons and protons diffuse radially inward from some
unknown external source. The bounce motion of some of the particles should
increase the probability of their collision with polar regions of the satellites,
perhaps accounting for the unusual polar coloring of Io and Europa. The
strength of the interaction between the particles and the satellites is dependent on energy and species, because the drift velocities, cyclotron radii and
radial diffusion coefficient depend upon particle mass and velocity. The interaction also depends on the satellite. In the most obvious way, the satellite
size is important; its electric or magnetic properties (whether the satellite is
a conductor or an insulator or has its own magnetic field) determine the size

of the target (Burns 1968); and its orbital position affects the relative drift velocity. The magnetic field motion and the particle drifts cause a substantial relative velocity in the direction of the satellite orbital motion. The resulting impacts may account for the albedo and color variations with orbital phase angle on the inner satellites (Mendis and Axford 1974). We also note that co-rotation of the particles with Jupiter produces substantial velocities with respect to the satellites, and in addition, particles near Io (and possibly Europa) are accelerated by plasma-sheath effects (Gurnett 1972; Shawhan *et al.* 1975; Shawhan 1976). Particle impacts with the satellites themselves can also contribute to energizing the radiation field (E. J. Smith *et al.* p. 788; Shawhan 1976).

(5) Photoelectric observations of the Galilean satellites have been made for half a century. In general they indicate photometric stability, considering the inherent uncertainties in comparing results obtained by different authors on different photometric systems. Nevertheless, there is a fairly large scatter for the derived magnitude of Io, from which Blanco and Catalano (1975) have recently suggested quasi-periodic variability by as much as 0.2 mag. However, we note that in their compilation the greatest excursions from the mean occur for those observations most difficult to transform to a common photometric system. Also, transformation problems are likely to be larger for Io than for the other Galilean satellites because of its very red color. More remarkable is the recent finding by Millis and Thompson (1975) and Jones and Morrison (1976) that *all* the Galilean satellites appear to have brightened by 0.02 to 0.05 mag between 1973 and 1974. We also mention in this context that Titan has been brightening by about 0.02 mag/yr for the past three years (Andersson 1976; Lockwood 1975*a,b*; Noland *et al.* 1974; Jones and Morrison 1975). The suggestion of such large secular changes in satellite brightnesses is certainly puzzling and clearly deserving of further study.

II. CALLISTO

Callisto, as the outermost of the Galilean satellites, has been least affected by the thermal and particle radiation of Jupiter, and it is therefore likely to be more representative of unperturbed conditions at 5 A.U. from the sun than the other Galilean satellites.

The bulk composition of a body such as a satellite is characterized by its mean density, and hence by its diameter and mass. The mass of Callisto (1.074×10^{26} g) is fairly well determined from Pioneer 10 and 11 data (Anderson *et al.* 1974; Null, personal communication 1975). The only values for the diameter, however, are based on visual measurements. From an analysis of a variety of such observations, giving highest weight to his own double-image micrometer data, Dollfus (1970) derives a diameter of 5000 km. This result is in excellent agreement with the filar micrometer measurements made by Barnard in the 1890's (5040 km) but one must note that both of these procedures have tended to overestimate the sizes of the other three Galilean satellites by a few percent (Morrison and Cruikshank 1974). One modern observation of possible relevance is the broadband infrared brightness temperature as reviewed by Morrison (1976), who used Callisto as a calibrator for the radiometric technique of deriving asteroid radii and albedos (cf. Morrison 1973; Jones and Morrison 1974). This calibration, adjusted to give the correct mean value for all the standard objects, yields a diameter of 4700 km for Callisto. However, while these results suggest that Dollfus' diameter is more likely to be too large than too small, we do not feel that a justification yet exists to suggest a more reliable alternative. For this chapter, we adopt $D = 5000 \, (+ 150, - 300)$ km for Callisto, a value that yields a bulk density of $1.6 \, (+ 0.3, - 0.2)$ g cm^{-3}. This density is among the lowest measured for any satellite (Morrison et al. 1976); note, however, that it is not significantly lower than that of Ganymede (see Table I) with present uncertaintties in mass and diameter.

A density of between 1.5 and 2.0 g cm^{-3} implies the presence of a substantial amount of water, according to the condensation sequences suggested by Lewis (1972a,b) and others. Indeed, these models suggest that water will be the dominant constituent by mass.

It is impossible to determine from present data whether Callisto differentiated as first suggested by Lewis (1971), with the silicate and other heavy material sinking toward the center, surrounded by a thick mantle of liquid or solid water. Consolmagno and Lewis (p. 1035) have computed the thermal histories of the Galilean satellites in some detail; they find that the compositions implied by densities of 2.0 and greater inevitably lead to nearly complete melting with consequent destruction of the primitive crusts, but if Callisto has a density of only 1.6, its original crust may have survived to the present. Thus it seems highly probable that the interior of this satellite is differentiated, but it is at least possible that Callisto (alone among the Galilean satellites) still retains a crust of the primitive, undifferentiated material deposited in the last stages of accretion.

The composition and structure of the surface of a satellite can be studied by a variety of photometric, polarimetric, spectroscopic, and radiometric techniques. For Callisto, diagnostic data on the surface composition are sparse, although it is possible to suggest that several important materials are *not* dominant. The mean visual geometric albedo (defined by linear extrapolation of the phase law for $\alpha > 6°$ to $\alpha = 0°$) of 0.17 (Morrison and Morrison 1976) is clearly too low for any relatively pure frost or ice and too high for a pure carbonaceous chondritic material. These are important constraints, since water ice is probably the primary bulk constituent of Callisto and is present on the surfaces of Ganymede and Europa, while carbonaceous material predominates on the surfaces of asteroids in the outer part of the belt

(Chapman *et al.* 1975) and is generally expected to be an important product of condensation in this part of the solar nebula (Lewis 1972*a,b*).

Infrared spectrophotometry (1.0 to 2.5 μm) by Pilcher *et al.* (1972) and Fink *et al.* (1973) have revealed no spectral features, and together with broadband infrared photometry suggests that the albedo of Callisto is approximately constant from 1 to 3 μm. There is perhaps a hint of absorptions due to water ice, but the upper limit for the fraction of the surface (as seen from Earth) that can be covered with this material is 25% (Pilcher *et al.* 1972). Spectrophotometry from 0.3 to 1.1 μm (e.g., Johnson 1971; Wamsteker 1972), as illustrated for all the Galilean satellites in Fig. 1, reveals no features for Callisto other than the general decline in albedo shortward of 0.5 μm shared by all these satellites. The ultraviolet absorber responsible for the low ultraviolet albedos has not been identified (cf. Johnson and Pilcher 1976). None of the curves in Fig. 1 shows any absorption feature near 1.0 μm that might indicate the presence of iron-bearing silicates. The flatness of the spectra also rules out lunar type materials rich in glasses as well as many of the mineralogical assemblages present in meteorites (Gaffey

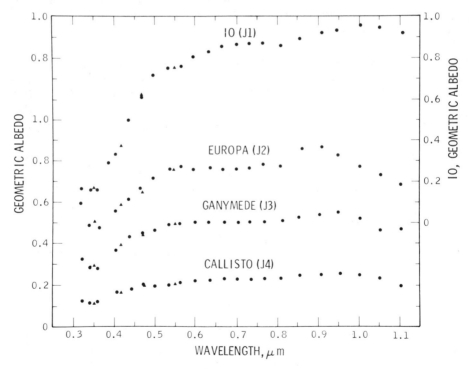

Fig. 1.   The spectral geometric albedos (0.3–1.1 μm) of the leading sides of the Galilean satellites. The curves are scaled to the mean albedo of each individual satellite (Morrison and Morrison 1976). Note the decrease in reflectance in the ultraviolet common to all the satellites (from Johnson and Pilcher 1976).

1974) and asteroids (Chapman *et al.* 1975). We note, however, that to exclude these materials because of the absence of certain spectral features, one
implicitly must assume that a substantial fraction of the observed light is
reflected from the rocky material. If as much as 15 or 20% of the surface is
covered with bright material, then virtually all of the light we see is reflected
from this high-albedo substance, and the dark material that covers most of
the surface does not noticeably influence the spectra. Such dark material
could be either carbonaceous (with a flat spectral reflectivity) or very red in
color as is the case for the Trojan asteroids (McCord and Chapman 1974).
On the basis of present spectrophotometric data, the best we can conclude is
that Callisto probably has a surface composed primarily of some unidentified
silicaceous rock, without a great deal of glassy or carbonaceous material.

Broadband photometry and polarimetry, particularly when obtained as
a function of both rotational (θ) and solar (α) phase angles, contribute to our
knowledge of the distribution of photometric properties over the surface.
A broadband photometric study in blue light carried out in 1926 and 1927
by Stebbins and Jacobsen at Lick Observatory first demonstrated that Callisto is in synchronous rotation, that its leading side is darker than its trailing
side, and that it has a remarkably large photometric phase effect; that is, its
magnitude changes very rapidly with solar phase angle, especially at small
angles ($\alpha \leq 4°$). Further, they showed that Callisto's photometric behavior
cannot be represented satisfactorily by a single phase law, but rather that
the leading and trailing faces have quite different opposition effects. Although
a number of subsequent studies provided much additional information on
color variations of Callisto, only very recently has this 50-year-old study
been improved upon in terms of defining the dependence of visual brightness
on orbital and solar phase.

The most recent V-magnitude lightcurves are illustrated in Fig. 2. We
have included in this figure (and in the similar illustrations given for the other
three Galilean satellites) the y-band data (transformed to V magnitudes) obtained at Mauna Kea Observatory by Morrison *et al.* (1974) in 1973 and by
Jones and Morrison (1976) in 1974 with the V observations obtained the
same two years at Lowell Observatory by Millis and Thompson (1975).
These graphs represent an attempt to separate the light variations into functions of θ and α. In the upper curves, dependence on α has been removed by
assuming the phase laws shown in the lower curve—one for $0 \leq \theta < 180°$,
one for $180° \leq \theta < 360°$. In the lower curves the mean dependence of V
on θ (for $\alpha \simeq 6°$), shown in the upper curves, has been removed in order to
generate V as a function of α only, again for the leading and trailing faces.
As noted by Veverka (1976*a*), the assumption that the dependence is separable in functions of α and θ is an arbitrary way to reduce data in which the
magnitude is in fact an unknown function of two variables, but in the absence
of a very large number of observations, it will suffice as a first approximation.
We stress, however, that the upper (rotation) curves for Callisto apply only

at $\alpha > 4°$; near opposition the effect of differing phase functions on leading and trailing faces is to reduce the lightcurve amplitude nearly to zero.

The large phase effect ($dV/d\alpha = 0.026$ mag deg^{-1}) on Callisto suggests a dark, rough, porous surface material (cf. Veverka 1976a)— in other words, a regolith, at least on a microscopic scale. The rapid increase in brightness at very small α (the opposition surge) also indicates a porous material, and Veverka (1970) has shown that the difference in opposition surge between the two hemispheres is consistent with a ratio of the degree of porosity, leading/trailing, greater than 1.6. The phase function can also be used to set some limits on the phase integral q defined as the ratio of the Bond albedo and the geometric albedo for a given wavelength. Veverka (1970, 1976a) summarizes the technique, which is due to Stumpff (1948), for obtaining an upper limit to q from the phase coefficient. For Callisto, this maximum value of the phase integral is 0.6, which is essentially the lunar value. In most studies of Callisto (e.g., Morrison 1976) $q = 0.6$ has been assumed.

The lightcurve in the upper panel of Fig. 2 tells us only that the leading face is some 15% darker, on the average, than the trailing face. (All other satellites for which reliable photometry exists show the opposite orientation, with the lighter hemisphere leading; Iapetus is the exception.) The curve is not simple, and presumably it is the result of averages over many small regions of differing brightness. Note that the satellite brightens more rapidly than it fades, a property common to all the Galilean satellites (Millis and Thompson 1975; Morrison and Morrison 1976). The photometric asymmetry is closely aligned with the orbital motion. The reasons for this effect, and for the differences in phase of photometric maximum from one satellite of Jupiter to another, have been discussed in the previous section, but remain mysterious.

Callisto varies in color as well as albedo as it rotates (cf. Morrison et al. 1974; Johnson and Pilcher 1976; Millis and Thompson 1975), in the sense that the darker parts are redder. As a consequence, the amplitude of the lightcurve is about twice as great at 0.35 μm as at 0.55 μm. Presumably this behavior indicates that the higher-albedo material is more neutral in color while the darker material is redder. If so, it suggests that the dark material is not carbonaceous.

Observations of linear polarization as a function of α and θ are, like photometry, indicators of the albedo and roughness of the surface. Dollfus (1975), and Gradie and Zellner (1973) have shown that the leading (darker) side has a greater (in absolute value) minimum polarization and a larger angle of polarization minimum than does the trailing side. Dollfus has argued that these differences are so great that they indicate a surface of smooth rock on the trailing hemisphere, while Veverka (1976b) feels that the data are compatible with a regolith on both hemispheres and that the differences are due to differences in particle size or degree of compaction between the darker and lighter faces of the satellite. This controversy is at present unsolved.

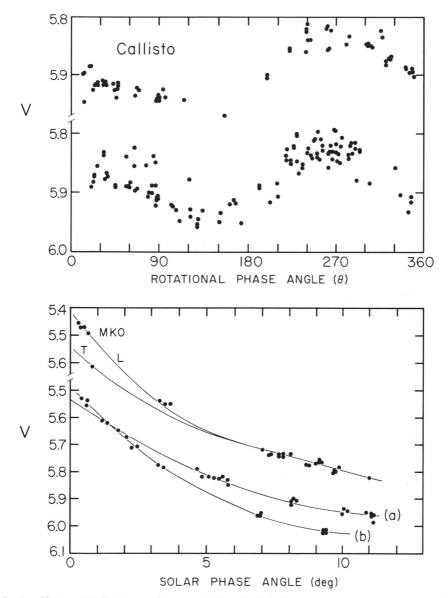

Fig. 2. Photometric lightcurves in V (~ 0.55 μm) of Callisto, showing the dependence of brightness on both solar phase angle (α) and orbital position angle (θ). At superior geocentric conjunction, $\theta \equiv 0$. Observations from two different observatories made during 1973 and 1974 are illustrated. Rotation curve a is from Mauna Kea Observatory (Morrison *et al.* 1974; Jones and Morrison 1976) and curve b is from Lowell Observatory (Millis and Thompson 1975); both are reduced to $\alpha - 6°$ based on the phase curves in the lower part of the figure. Curves c and d are from Mauna Kea, and e and f are from Lowell. Note the differences between the leading and trailing hemispheres of this satellite. The numerical coefficients of the fitted curves are given by Jones and Morrison (1976) and Millis and Thompson (1975).

Disk-averaged temperatures for Callisto have been measured in the infrared and at microwave frequencies at wavelengths from 8 μm to 2.8 cm (reviewed by Morrison 1976; see also Ulich and Conklin 1975; Berge and Muhleman 1975; and Rieke, personal communication 1975). All values are consistent with simple, grey-body emission from a satellite with the albedo and diameter given in Table I. The thermal response of the surface of Callisto to changing insolation has also been measured (Morrison and Cruikshank 1973; Morrison and Hansen 1976) and indicates a material of very low thermal conductivity, again suggesting a porous, fragmented surface.

Virtually nothing is known about the surface markings of Callisto. Also unexplored is the possibility of an atmosphere, although Fink *et al.* (1973) obtained upper limits of 0.5 cm-atm for CH_4 and NH_3 on all four Galilean satellites. The only other limit on an atmosphere is provided by the low value of the thermal conductivity, which suggests that the surface pressure must be less than 1 mbar (Morrison and Cruikshank 1973).

In summary, Callisto is a world of the same size as Mercury but probably composed primarily of water, either liquid or solid. Orbiting almost two million kilometers from Jupiter, it is little affected by the energetic particles or the thermal radiation of the planet. The surface is largely composed of dark, rocky minerals, with perhaps some water ice present. It is remarkable primarily for the surprising photometric and polarimetric differences between hemispheres and for the fact that it is one of the few satellites that is darker on its leading rather than its trailing hemisphere. The surface apparently consists of a regolith, rough and porous on scales of one centimeter or less. The surface mineralogy, large-scale roughness, distribution of albedo features, and nature of any atmosphere are still unknown and will probably require observations from the MJS flybys before they can be understood.

III.   GANYMEDE

Ganymede is the largest Galilean satellite and the second largest satellite in the solar system, with a size intermediate between that of Mercury and Mars. Its diameter has been determined from a stellar occultation (Carlson *et al.* 1973) to within 10 or 20 kilometers, and its mass was measured by Pioneer 10 (see Table I), so that the bulk density of 1.95 g cm^{-3} is known to within a few percent. The main uncertainty in this quantity arises from the possibility of a substantial atmosphere on this satellite. If the surface pressure is greater than $\sim 10^{-3}$ mbar, the occultation diameter will refer to some point in the atmosphere rather than to the solid surface, and the density quoted above will be a lower limit to the density of the solid body. However, as discussed below, the existence of an atmosphere of this magnitude is doubtful, so we will adopt the occultation diameter without adjustment, noting however the possibility that the true diameter could be somewhat smaller.

With its bulk density of about 2 g cm^{-3}, Ganymede must be similar to Callisto in being largely composed of water. It is, in addition, more likely

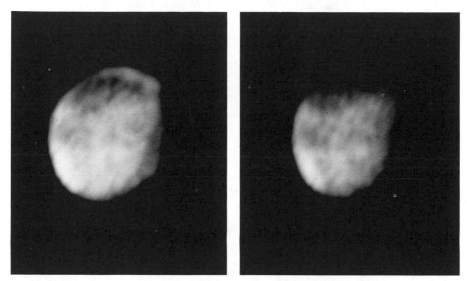

Fig. 3.   Ganymede as seen by the imaging photopolarimeter on Pioneer 10. On the left is the blue image while the red is on the right. North is up and the rising limb is to the left (from Gehrels 1976).

than Callisto to be fully differentiated, due both to its greater proximity to Jupiter and to the apparently somewhat higher content of radionuclides as suggested by its higher density. Interior models (Lewis 1971, 1974; Consolmagno and Lewis p. 1042) suggest that the satellite has a rocky core, surrounded by a mantle of liquid water and topped by a thin (100 to 200 km) ice crust. About 15% by mass is in the core, which is composed primarily of silicates and metallic oxides. Only small amounts of CH_4 and NH_3, both in solution in the water, are expected.

Figure 3 is a Pioneer image of Ganymede (Gehrels 1976) and is the best view we have of the surface features on any satellite of Jupiter. The surface is clearly made up of regions of different albedo with sizes down to the resolution limit of about 400 kilometers. No polar cap is seen nor is there any other obvious correlation of albedo with latitude, nor any indication of a strongly bimodal distribution of albedo. However, since the north pole is tipped $\sim 20°$ away from the viewer, a small north polar cap may be present. The brightest resolved areas have about twice the albedo of the darkest areas; if, as seems likely, the typical size of albedo features is below the resolution limit of Fig. 3, the true range of contrast on the satellite is greater. It is impossible to recognize topographical or morphological features at this resolution, and there seems to be no way to determine whether the albedo contrasts are the result of variations in surface mineralogy or are due to volatiles on the surface.

On a larger spatial scale, we can be confident of the stability of the albedo features on Ganymede, since the lightcurves have not changed over the past

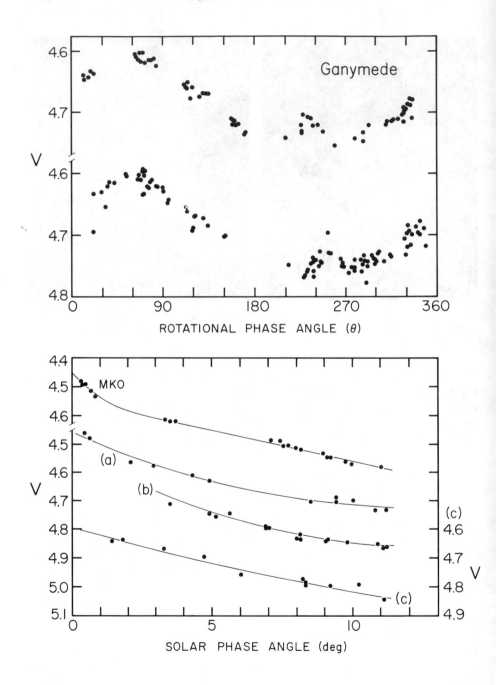

Fig. 4.   Photoelectric lightcurves in V of Ganymede. Rotation curve a and phase curve c are from Mauna Kea (Jones and Morrison 1976) and curves b and d are from Lowell (Millis and Thompson 1975). (See caption of Fig. 2 for more details.)

half century. Figure 4 illustrates a lightcurve and phase curve for this satellite based on 1973 and 1974 observations. The upper curve is similar in amplitude and shape to that of Callisto, except that it is shifted by nearly half a cycle in phase, with the brighter hemisphere facing generally forward, the maximum brightness occuring at $\theta \simeq 60°$. The color variations of Ganymede are also slightly smaller than those of Callisto, with a peak-to-peak amplitude of only ~ 0.01 mag in the $(b$-$y)$ index and with a total amplitude at 0.35 μm that is less than twice as great as at 0.55 μm (Morrison et al. 1974; Millis and Thompson 1975).

The dependence of brightness on solar phase angle shown in the lower panel of Fig. 4 is very different from that of Callisto. The slope of the linear part of the curve, i.e., the phase coefficient, is large $(0.022 \pm 0.002$ mag deg^{-1} in V), but the opposition surge at small phase is quite modest. This behavior suggests a surface less porous than that on Callisto (cf. Veverka 1976a) but it is not strongly diagnostic of the nature of the surface, particularly when we consider that these photometric parameters must refer to some ill-defined average over the properties of surface materials with a wide variety of albedos and compositions.

Radiometric observations of Ganymede at wavelengths from 8 μm to 2.8 cm are generally consistent with expectations for a body in equilibrium with the insolation (Morrison 1975). The only possible recent exceptions are a 3-mm brightness temperature of $129 \pm 23°$K (Ulich and Conklin 1975) and a temperature at 2.8 cm of $55 \pm 14°$K (Pauliny-Toth et al. 1974). Infrared eclipse radiometry by Hansen (1973) and Morrison and Cruikshank (1973) agree in assigning a very low thermal inertia ($\sim 10^4$ erg cm^{-2} sec$^{\frac{1}{2}}$ °K^{-1}) to the uppermost layers and suggesting a transition to a region of higher thermal conductivity a centimeter or so below the surface. Radiometric measurements have also been used (Morrison 1976) to estimate a phase integral of ~ 1.0 for this satellite, while Veverka (1970, 1976a), using the Stumpff approximation, suggests from the phase coefficient an upper limit to q of 1.0. Visible polarimetry of Ganymede is consistent with the presence of low-opacity surface materials such as water frost (Dollfus 1975; Veverka 1976b).

A clear identification of the chemical composition of the higher-albedo material on Ganymede was provided by 1- to 3-μm spectra obtained by Pilcher et al. (1972) and Fink et al. (1973), who demonstrated the presence of water-ice features, although they differed as to the fraction of the surface (as seen from Earth) covered. In a more recent analysis of the same data, Kieffer and Smythe (1974) confirm that there is no indication of any other species in these infrared spectra. Both the geometric albedo in the visible of only 0.43 (Morrison and Morrison 1976) and the structure apparent in the Pioneer 10 picture, of course, argue for the presence of some darker material in addition to the ice, and Pilcher et al. (1972) have concluded that the band depths in their infrared spectra were also consistent with a less than total coverage of the surface by frost. Spectrophotometry between 0.3 and 1.1 μm unfortunately does not reveal diagnostic features, although Ganymede shows the drop in albedo shortward of 0.5 μm characteristic of all the Galilean satellites (see Fig. 1). Following the analysis by Pilcher *et al.*, it has generally been assumed that the surface of Ganymede consists of roughly equal parts of some unidentified silicate material and water ice. With such a two-component model the visible lightcurve of the satellite would be understood as resulting from differing fractions of ice covering the surface on the leading and trailing hemispheres. One would then expect differences between the infrared spectra of the two sides of Ganymede, as are shown in Fig. 5.

It is of interest to inquire to what degree the ice and rock are mixed on the surface. The image shown in Fig. 3 suggests that the distribution is clumpy on a scale of hundreds of kilometers. Further information is available from the appearance of temperature-dependent features in the ice spectra. Fink and Larson (1975) find that these features indicate that the ice is at a temperature substantially lower than the average for the satellite as measured radiometrically, and they conclude that the ice maintains a temperature appropriate to its high albedo and not to the average albedo. We note that Kieffer and Smythe (1974) do not confirm this low temperature, but Fink and Larson argue that the problem may arise from the comparison by Kieffer and Smythe of laboratory and satellite spectra of substantially different resolution. The color temperature in the 8- to 25-μm region also is best matched with two components of different albedo, each at its own equilibrium temperature (Rieke, personal communication 1975). It follows that the ice and rock are not mixed on a scale of centimeters, the distance appropriate for thermal contact during one diurnal period of the satellite. The mean subsolar temperature on Ganymede is $160 - 165°K$ and, according to standard vapor pressure curves, substantial quantities of water ice would have evaporated in regions where such temperatures are reached. In this case we might expect the water ice to migrate from the equatorial regions to the poles, yet no polar caps are seen in Fig. 3. On the other hand, if the ice always is limited to a maximum temperature appropriate to its own high albedo rather than to the average albedo, then it is stable even near the equator, and perhaps polar caps should not be expected.

It is clear that the surfaces of the Galilean satellites will occasionally be cratered by the impact of large meteorites, and the stability of topographic features produced by such impacts, or by possible endogenic processes, has been investigated by Johnson and McGetchin (1973). They conclude that if the strength of the crustal material is that appropriate for water ice (as is possibly the case for all of the Galilean satellites, but particularly so for Ganymede and Europa), then at the known temperatures of these objects large-scale topography will be destroyed by flow on time scales of only millions of years. Thus we expect no large topographic features on Ganymede. We would interpret MJS images of a heavily cratered surface on any of the Galilean satellites as evidence that the crust is silicate in composition and that the satellite has not fully differentiated.

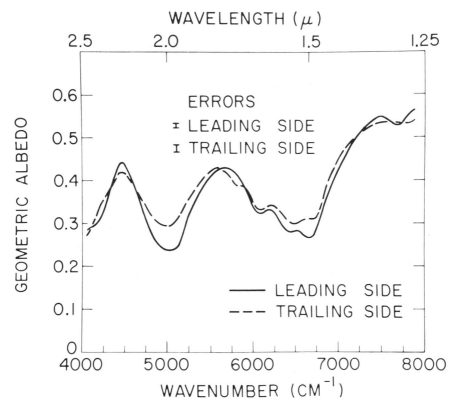

Fig. 5. Geometric albedos for the leading and trailing faces of Ganymede scaled to the same value at 5625 cm⁻¹. Each error bar is the average of the standard deviations for all wavelengths (from Pilcher *et al.* 1972).

The reality and nature of the atmosphere of Ganymede are in some dispute. The spectroscopic upper limits on CH_4 and NH_3 set by Fink *et al.* (1973) correspond to partial pressures $\sim 10^{-3}$ mbar. However, a positive detection of an atmosphere with surface pressure $> 10^{-3}$ mbar has been suggested by Carlson *et al.* (1973) on the basis of their measured occultation lightcurves. These authors concluded that the star dimmed gradually rather than instantaneously when occulted, although the signal-to-noise level in the photometry was such as to permit other interpretations. Recently, a mechanism for the maintenance of a substantial atmosphere on Ganymede has been suggested by Yung and McElroy (1975). Noting that, at the mean subsolar temperature calculated for this satellite, a substantial partial pressure of water vapor is expected, they showed that the water would dissociate and that sufficient amounts of the resulting oxygen would be retained to establish an equilibrium oxygen and water-vapor atmosphere with a surface pressure of about 10^{-3} mbar. Again, however, this conclusion rests on the assumption that the ice temperature is equal to the mean local temperature, and if the ice is as cold as indicated by Fink and Larson (1975), the mechanism proposed by Yung and McElroy will not work. We conclude that the existence of a substantial atmosphere on Ganymede is very much an open question that may be answered in the future, either by groundbased spectroscopy or by MJS observations.

Ganymede appears to be a weak absorber of charged particles. On the outbound pass of Pioneer 10 and the inbound pass of Pioneer 11, the instrument of the University of Iowa (Van Allen p. 000) detected a sharp dropoff in particle flux at all energies outside the orbit of Ganymede. This effect may be caused by Ganymede or be due to the high magnetic latitudes of the spacecraft or be due to a moving outer boundary to the radiation belts. Pioneer 11 did nevertheless detect transient, highly anisotropic bursts of MeV protons near Ganymede's orbit, presumably locally accelerated or injected nearby. Since similar bursts are seen elsewhere in the magnetosphere, it is not certain whether Ganymede is involved in this phenomenon. There is a suggestion of a possible detection of a hydrogen torus near Ganymede's orbit (Judge, personal communication 1975).

Ganymede recently became the first natural satellite other than the moon to be detected by radar. Goldstein and Morris (1975) obtained a return from the satellite at a wavelength of 12.6 cm on 6 nights near the 1974 opposition. The scattering efficiency is 12%, at least as great as that of any of the terrestrial planets. Ganymede also appears substantially rougher at this wavelength than the terrestrial planets.

In summary, Ganymede is a planet-sized satellite probably composed primarily of water, likely in the liquid state, with a rock core and a water-ice crust. The surface consists of roughly equal exposures of ice and rock material of unknown mineralogy, probably in the form of a loose regolith maintained by meteoroid impacts and volatilization of the frost. The ice and rock may be not mixed but rather be distributed in discrete patches, with the ice concentrated slightly on the leading hemisphere. There may be an atmosphere with surface pressure between 10^{-3} and one mbar composed primarily of gases other than NH_3 or CH_4, but the arguments for its presence are not compelling. Because of the structural weakness of the ice crust, only the most recently formed large-scale topographic features are expected to be visible.

IV. EUROPA

Europa is the smallest of the Galilean satellites, being even slightly smaller than the moon. The visually determined diameter (Dollfus 1970) is 3100 km, and recent analyses of photometric observations of occultations of this satellite by Io are in good agreement with this value. Here we will adopt the diameter of 3050 km obtained in the detailed analysis of the occultations by Aksnes and Franklin (1975b). Our best estimate of the uncertainty is ± 50 km. The bulk density then is 3.2 ± 0.2 g cm^{-3}, similar to that of Io and the moon but substantially larger than the densities of Ganymede and Callisto. This density suggests that Europa formed under circumstances in which water did not condense; it is presumably composed primarily of silicates and metal oxides, with only a modest amount of volatiles. Io has a similar density, so that the Galilean satellites divide into two clear groups according to content of water. Pollack and Reynolds (1974) have shown that this difference is to be expected if Jupiter had a substantial luminosity at the time of formation of the satellites, so that the temperature inside the orbit of Ganymede was too high for the condensation of water. A high temperature in this region of the pre-satellite nebula might also be produced by the increased opacity near the planet of the thickening cloud around the proto-Jupiter (Cameron 1976). Since the luminosity history of Jupiter early in its life can be calculated, time scales of a few million years for satellite formation can be deduced, as illustrated in Fig. 6, taken from the discussion by Fanale *et al.* (1976). The formation of the Galilean satellites is discussed more extensively in this volume by Cameron and Pollack (p. 78).

Europa is the only one of the Galilean satellites for which the composition of the surface material is well determined. Both the very high geometric albedo in V of 0.64 (Morrison and Morrison 1976) and the shallow negative branch of the polarization-phase curve (Dollfus 1975; Veverka 1976*b*) suggest that most of the satellite is covered by ice or frost; and the infrared spectroscopy obtained in 1972 has confirmed that this material is water frost. Pilcher *et al.* (1972) have concluded that virtually the whole surface is covered by this material. Kieffer and Smythe (1974), in a recent re-analysis of the data, derive an upper limit of 28% to the exposure of any other material on the surface. The radiometric temperatures of Europa are quite low (brightness temperature $\sim 130°K$ at 10 μm and $\sim 120°K$ at 20 μm), consistent with a high-albedo frost surface (Morrison 1976). Fink and Larson (1975) deduce a spectroscopic temperature for the ice of $95 \pm 10°K$, a value that seems surprisingly low in comparison with the radiometric values, if the surface is almost wholly ice covered.

Fanale *et al.* (1976) and Consolmagno and Lewis (p. 1045) have discussed the internal evolution of Europa on the assumption that it is composed primarily of carbonaceous chondritic minerals. They expect that, early in its history, internal temperatures would rise to the point where extensive outgassing of volatiles would take place, ultimately exceeding 700°K several hundred kilometers below the surface. Outgassing of most of the interior water would create an ice crust about 75 km thick. Depending on the rate of generation of heat in the interior today, the lower few tens of kilometers of this crust might now be melted. The arguments given by Johnson and McGetchin (1973) suggest that there will be little topographic structure.

Although spectroscopically Europa appears to be covered with ice, short-wavelength photometry and spectrophotometry show major differences in color and albedo on the surface. Figure 7 illustrates the dependence of V magnitude on θ and α. The phase coefficient and opposition effect (lower panel) are small, consistent with the presence of multiple scattering in a high-albedo, porous material (Veverka 1973, 1976*a*). The amplitude of the light-

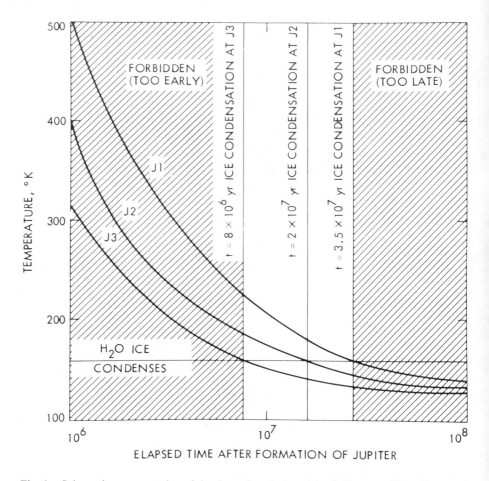

Fig. 6. Schematic representation of the thermal evolution of the Galilean satellites. Illustrated are the surface temperatures of Io, Europa, and Ganymede as functions of time (from Fanale *et al.* 1976).

curve, however, is larger than for any of the other Galilean satellites, with the brighter hemisphere facing approximately forward ($\theta_{max} \simeq 75°$). The color curves (Johnson 1971; Morrison *et al.* 1974) for this satellite are somewhat anomalous, in that the rotational variation at 0.47 μm is smaller than at 0.55 μm. At shorter wavelengths the amplitude increases rapidly, however, so that at 0.35 μm it is more than twice as great as in the visual curve illustrated in Fig. 7. Further, since the curves in different colors vary in shape as well as in amplitude, they cannot be interpreted strictly in terms of a two-component model, with one material dark and red and the other bright and neutral.

The series of occultations by Io observed photoelectrically in 1973 have been used in an attempt to derive additional information on the distribution of albedo and color on the surface of Europa (Murphy and Aksnes 1973;

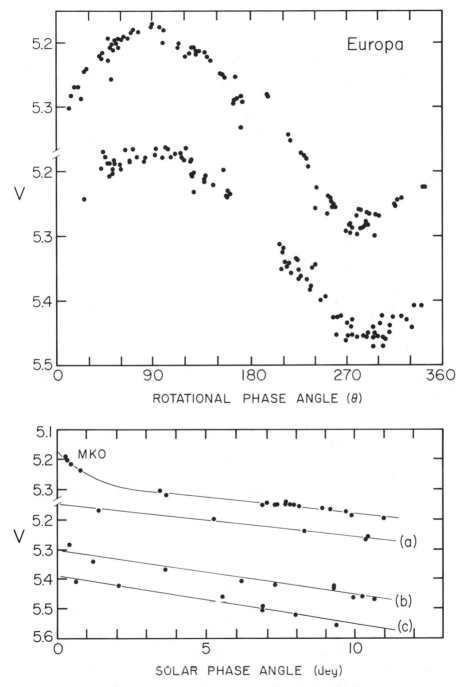

Fig. 7.   Photoelectric lightcurves in *V* of Europa. Rotation curve *a* and phase curve *c* are from Mauna Kea (Jones and Morrison 1976) and curves *b* and *d* are from Lowell (Millis and Thompson 1975). (See caption of Fig. 2 for more details.)

Vermilion *et al.* 1974; Aksnes and Franklin 1975*a*; Greene *et al.* 1975). These authors have suggested, on the basis of analyses in which it was assumed that there were no errors in the predicted latitudes of the satellites, that Europa is brighter and less red toward the north pole than at the equator. The presence of bright polar caps on this satellite is also indicated by the visual observations reported by Murray (1975). However, a recent analysis of the occultation data by Aksnes and Franklin (1975*b*) suggests that latitude errors may be substantial and that a model in which the satellite has a uniform albedo while adjustment of latitude terms is allowed provides a better match to the lightcurves than do the earlier interpretations. We therefore conclude that, while Europa may well have polar caps, the existing data do not require them, and indeed the only evidence for their existence comes from visual observations.

Europa lies outside most of the intense Jovian radiation belts. Perhaps for this reason, or because of the increase in the diffusion coefficient with increasing L or simply due to its small size, Europa appears not to absorb a significant fraction of the radiation belt particles near its orbit. Some minor decreases in most particle fluxes were found by Pioneer 10 on crossing Europa's L shell but very little by Pioneer 11, possibly because of the latter's high-magnetic-latitude trajectory (Fillius p. 915; McDonald and Trainor p. 966; Simpson and McKibben p. 744; Van Allen p. 947). Strong absorptions of electrons from 0.16 to 1 MeV and of 15 to 20 MeV protons were recorded by Pioneer 10 but not by Pioneer 11; instruments on both spacecraft saw considerable structure in the particle fluxes near Europa's L shell.

In summary, Europa has the size and density of the moon, but unlike our satellite most of its surface is covered with water ice. Since its interior composition may also be different, the agreement in density may be largely coincidental. The surface ice probably was outgassed from the interior. This ice, particularly on the trailing side and away from the poles, is contaminated by a darker and redder material of unknown nature, or else the ice itself is darkened and reddened in these regions by interaction with the Jovian magnetospheric particles. Little atmosphere or surface topography is expected.

V. IO

Io is one of the most intriguing objects in the solar system. Several other review chapters in this volume discuss Io, particularly its atmosphere (Brown and Yung p. 1102), its hydrogen torus (Judge *et al.* p. 1077), and its interaction with the Jovian radiation environment (Mogro-Campero p. 1190); we will therefore devote less space here to Io than it really deserves, because other authors explore it in more detail in their chapters. Our main purpose will be to compare Io with the other Jovian satellites.

Io is the innermost of the Galilean satellites and has therefore been subject to the greatest influence of Jupiter. It seems clear that Jupiter has dominated this satellite and is responsible for many of its unique properties. In particular, the high level of thermal radiation from Jupiter at the time of its formation may have eliminated water ice and other materials of high volatility from its composition (Pollack and Reynolds 1974; Fanale *et al.* 1976; Cameron and Pollack p. 000). The formation scenarios discussed by these authors suggest that Io accreted during a span of five million years, beginning within a few million years of the time Jupiter collapsed into approximately its present size (cf. Bodenheimer 1974, 1976) and that the main chemical constituents of the satellite were partially hydrated silicates with a bulk density about 3.0 g cm^{-3}.

The diameter of Io has been measured to within a few kilometers from photometry of the 1971 occultation of β Scorpii (Taylor 1972; O'Leary and van Flandern 1972) and the mass is known from Pioneer 10 data; the resulting density is 3.5 ± 0.1 g cm^{-3}, the best-determined value for any Galilean satellite. Fanale *et al.* (1976) calculate that if the satellite formed with carbonaceous chondritic composition, the temperature would have risen to above $1000°K$ throughout most of the satellite in less than a billion years and might be as high as $3000°K$ at the center today. This heating would have caused extensive degassing of chemically bound water and its subsequent migration to the surface. The interior would eventually become molten and fully differentiated, and the water percolating to the surface would carry with it large quantities of soluble salts such as sodium sulfate, epsomite, gypsum and bloedite. Thus sodium, sulfur, calcium, etc. could be concentrated in the crust, while the final bulk density would agree with the measured value for Io.

It is a well-known fact, however, that infrared spectra of Io fail to show any sign of the prominent ice bands visible on Europa and Ganymede. Figure 8 illustrates a recent spectrum of very high quality in the 1- to 3-μm region obtained by Fink *et al.* (1975). The spectrum of this satellite is remarkable not only for the absence of water features but for its exceedingly high albedo throughout the infrared. One of the important problems of satellite research today is the identification of this surface material.

Photometry and spectrophotometry of Io in the visual part of the spectrum do not provide much diagnostic information on composition. As is shown in Fig. 1, this satellite exhibits the usual drop in albedo shortward of 0.5 μm, but it is more pronounced than that of the other Galilean satellites, and there is some indication of weak absorption features near 0.55 and 0.80 μm. There may also be a feature near 1.1 μm, but to date no very satisfactory intercomparison has been made between the spectrophotometry shown in Fig. 1 and that at longer wavelengths, so the behavior of the albedo from about 1.1 to 1.3 μm is not well defined. The lightcurves and phase curves in V are illustrated in Fig. 9. The amplitude of the rotational curve is smaller than for Europa. Both the phase coefficient and the magnitude of the opposition surge are larger, a result that indicates a difference in surface materials or texture between these two satellites, in spite of their similar albedos. The color variations on Io (Morrison *et al.* 1974; Millis and Thompson 1975)

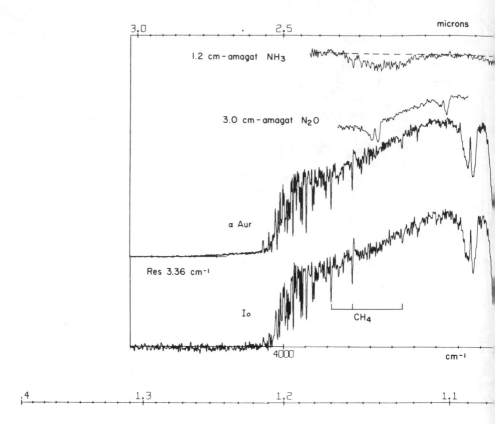

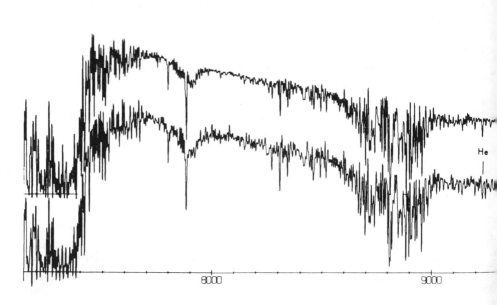

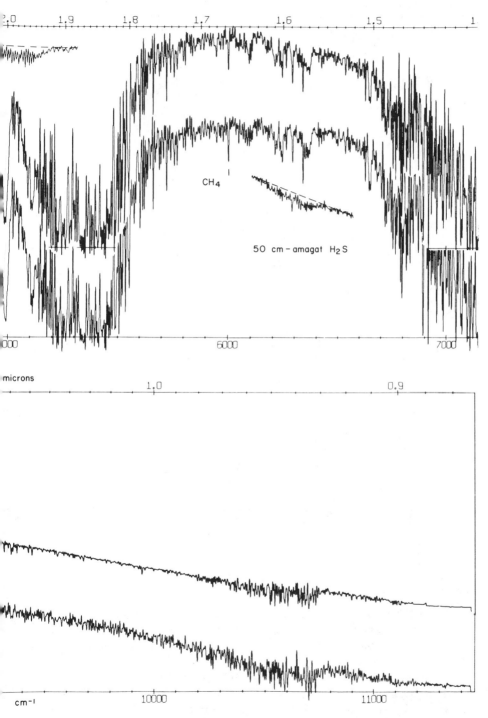

Fig. 8. Infrared spectrum of Io obtained with a Fourier transform spectrometer at an effective spectral resolution of 3.36 cm⁻¹ (from Fink *et al.* 1976).

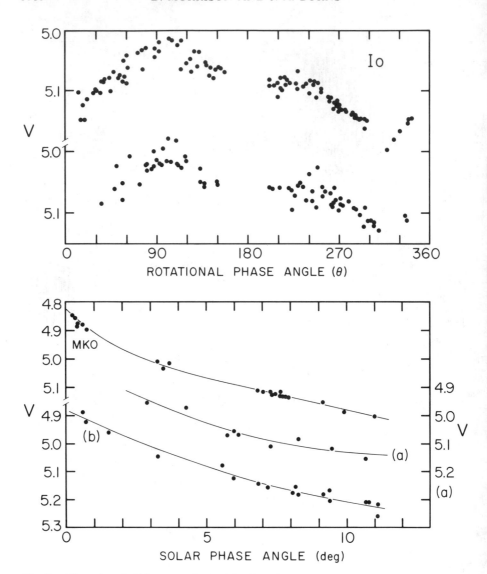

Fig. 9.   Photoelectric light curves in V of Io. Rotation curve a and phase curve c are from Mauna Kea (Jones and Morrison 1976) and curves b and d are from Lowell (Millis and Thompson 1975). (See caption of Fig. 2 for more details.)

are greater than on any other satellite, with a total amplitude at 0.35 μm of more than a magnitude. The color effects are centered near $\theta = 280°$, at which longitude there seems to be a concentration of dark, very red material, much different from that which covers most of the surface.

Additional information on the distribution of albedo and color on Io is provided by some very fine groundbased images, both photographic (Minton 1973) and visual (Murray 1975), obtained while the satellite transited Jupiter. These show dark, reddish polar caps, as illustrated in Fig. 10. It is tempt-

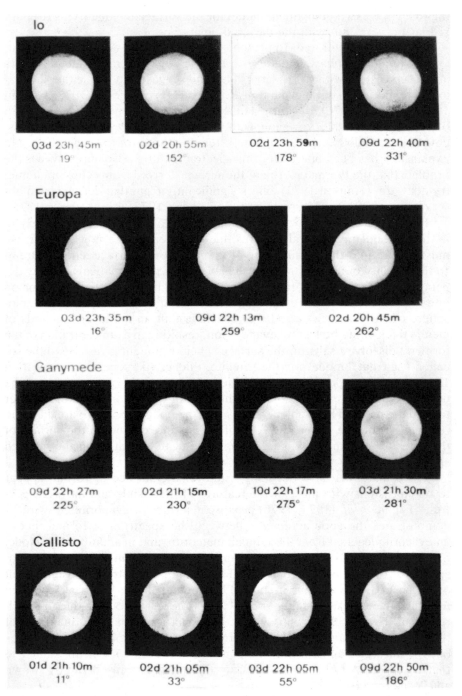

Fig. 10.   Sketches of the Galilean satellites based on observations at the Pic du Midi 108-cm telescope (from Murray 1975).

ing to invoke a two-component model for the surface by identifying the dark red spot at $\theta = 280°$ with the dark red polar caps.

In the thermal infrared, Io presents other mysteries. Although most indications point to a phase integral $q \geqslant 1.0$, the 8- to 25-μm broadband colors are not consistent with such a large value and suggest that Io is either anomalously bright near 10 μm or faint near 20 μm (Morrison et al. 1972; Hansen 1973; Morrison 1976). Recently, Hansen (1975) and Armstrong et al. (1976), observing in the 10-μm region with higher spectral resolution, have discovered anomalously high brightness temperatures for Io near 8 μm. No explanation has been offered for this spectral feature. Also unresolved is the problem that the thermal inertia of the surface derived from eclipse radiometry at 10 μm (Hansen 1973) differs significantly from that derived from observations at 20 μm (Morrison and Cruikshank 1973) (cf. Morrison 1976).

If large quantities of water were outgassed to the surface of Io early in its lifetime but no spectral features due to water ice are now present, we must inquire into the possible fate of the ice. At the surface temperature of Io the stability of ice over long periods of time is problematical and may depend rather critically on energy sources in the Jovian particle belts or on interactions between the Jovian magnetosphere and water vapor in the atmosphere of Io. Fanale et al. (1976) conclude that an ice layer hundreds of meters thick could be lost by evaporation, resulting in a concentration of the formerly dissolved salts on the surface. Their argument gives rise to the so-called "salt flat" model for the satellite. Pilcher (personal communication 1975) doubts that such large quantities of water could be removed, but he does conclude that evaporation from the equator might concentrate the water toward the poles. The problem here is that the polar caps on Io are dark and red, not bright like water ice. Also, if ice is still exposed on the surface of the satellite, one must explain the absence of its signature in the infrared spectra.

Three materials have been suggested for the surface of Io that appear to be consistent with most of the features of the visible and infrared spectrum. Fanale et al. (1974, 1976) propose a surface of evaporites. Figure 11 demonstrates the good agreement between the spectrum of Io and that of an evaporite leached from the Orgueil meteorite, and in addition their model has the advantage of providing a source for the sodium seen in the atmosphere of Io (to be discussed below). Figure 11 also illustrates the agreement of the observed spectrum with that of elemental sulfur, a material suggested for the surface by Wamsteker et al. (1974). The third model is due to McElroy et al. (1974, 1975) who suggest ammonia ice with large quantities of dissolved sodium and calcium. The presence of the dissolved metals washes out the infrared bands of the ammonia and produces a spectrum similar to that observed. All of these models are discussed in more detail by Brown and Yung (p. 1102).

Of these materials, the origin of both sulfur and the evaporite salts can

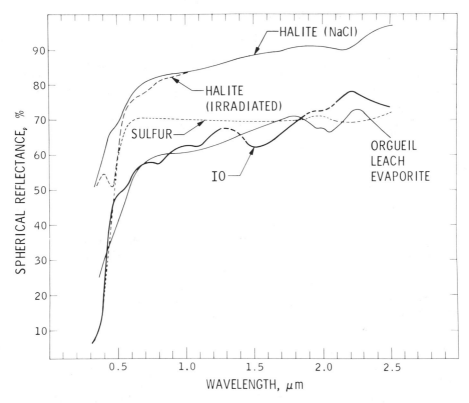

Fig. 11.   Comparison between the spectrum of Io and that of an evaporite leached from the
Orgueil meteorite. Also shown are spectra of selensulfur, natural halite and the halite sample
after irradiation by protons (from Fanale *et al.* 1976).

be understood in terms of outgassing of the vapors of aqueous solutions from
the interior, although the problem of removing the primordial "oceans" of
ice by evaporation is not resolved. Ammonia ice in large quantities is less
likely cosmochemically.

The question of the existence of an atmosphere on Io has been discussed
repeatedly during the past decade (e.g., Sinton 1973), primarily in the context of the transient post-eclipse brightening of the satellite seen photometrically by some observers. Recent attempts to establish the reality of this
effect (Fallon and Murphy 1971; Franz and Millis 1971; O'Leary and Miner
1973; Cruikshank and Murphy 1973; Franz and Millis 1974; Millis *et al.*
1974; Frey 1975) are inconclusive, and, even if the effect is real, it may be
unrelated to the atmosphere. The first definite evidence for an atmosphere
was obtained in 1973 by Brown (1974), who detected the Na D-lines in
emission in the spectrum of Io when the satellite was near elongation and
hence had the maximum Doppler shift with respect to the deep Na D-

absorption lines in the solar spectrum. We will discuss briefly the sodium atmosphere and extended emission region below.

The 1971 occultation of β Scorpii by Io yielded an upper limit to the refractivity of a possible atmosphere corresponding to maximum partial pressure of CH_4 and N_2 of $\sim 1 \times 10^{-7}$ bar (Smith and Smith 1972; Bartholdi and Owen 1972). A somewhat more stringent upper limit of $\sim 1 \times 10^{-8}$ bar for the surface pressure was derived by Kliore et al. (1974, 1975) from the Pioneer 10 occultation by Io. All of these observations suggest a very tenuous atmosphere, so much so that the exosphere probably extends to the surface and particles leaving the surface move in approximately ballistic paths with little interaction. Present data do not provide a value for the density of the atmosphere of Io, nor do they exclude the possibility that one or more of the other Galilean satellites has an atmosphere as substantial as that on Io. Both the presence of unusual materials on the surface and the intensive bombardment of the satellite by high-energy protons and electrons single out Io, and suggest possible mechanisms for the generation of its atmosphere as well as for rendering even a very tenuous atmosphere observable.

The Pioneer 10 occultation experiment (Kliore et al. 1974, 1975) discovered the existence of an ionosphere on Io extending, on the day side, from near the surface to an altitude of at least 700 km, with a peak electron density of 6×10^4 cm^{-3} at 100-km altitude. The nightside electron density was nearly an order of magnitude smaller. The ultraviolet photometer on the same spacecraft observed a hydrogen torus in the orbit of Io with greater density in the trailing portion (Judge and Carlson 1974; Carlson and Judge 1974, 1975), although the interpretation of the observations will be altered if part of the emission is produced near Amalthea as well as in the orbit of Io (Judge et al. p. 1068). Continuing observations of the Na D-emission from Io have indicated that the sodium cloud is also extended in the orbit of the satellite, apparently forming a partial torus around Jupiter (Trafton et al. 1974; Bergstralh et al. 1975; Brown et al. 1975; Macy and Trafton 1975; Münch and Bergstralh 1975). Analyses of these observations (in the papers indicated above) as well as those by Brown and Chaffee (1974), McElroy et al. (1974), Matson et al. (1974), Fanale et al. (1974, 1976), and Carlson et al. (1975), indicate that the cloud consists of an inner, optically thick region that might properly be considered part of the atmosphere of Io and an extended, optically thin emission region extending up to 60 arcsec from the satellite in the orbital plane. The temperature in the inner cloud has been estimated at from 500° to 3000°K, and the flux of sodium atoms escaping from Io at $\sim 10^7$ cm^{-2} sec^{-1}. The emission itself is due to resonant scattering of sunlight. The atmosphere of Io is more extensively discussed in this volume by Brown and Yung (p. 1140).

Pioneer measurements showed that Io interacts strongly with energetic protons and electrons in the Jovian radiation belt. MeV protons are almost entirely absorbed by the satellite (McDonald et al. p. 973; Van Allen p. 953). At higher energies, absorption of protons is still considerable; tens of percent of those diffusing past Io are removed in the interval 14.8 to 21.2 MeV (McDonald *et al.*, p. 966). A similar fraction of electrons above 40 keV is absorbed, while almost no depletion takes place at energies above 20 MeV (Van Allen p. 956; Fillius p. 914). Since the drift velocity of these very high-energy electrons relative to Io is approximately zero, the electrons can diffuse past Io without striking it.

Matson *et al.* (1974) have proposed that sputtering, following the impacts of energetic particles into Io's surface, is the source of the sodium seen concentrated at the orbital distance of Io and in its orbital plane (Brown and Yung p. 1138). If their hypothesis is tenable, other elements, abundant on Io's surface according to usual cosmochemical distributions, should also be sputtered and therefore be present in the atmosphere. Several searches (Trafton 1976; Wehinger 1976) have been made for calcium, aluminum, potassium and other elements without success. [In September 1975, however, Trafton (1975) did succeed in detecting potassium.] Upper limits for several plausible gases derived by Fink *et al.* (1976) from the infrared spectrum illustrated in Fig. 8 are: NH_3, 0.12 cm atm; CH_4, 0.12 cm atm; N_2O, 0.4 cm atm; and H_2S, 24 cm atm. At this time, the question of whether the sputtering mechanism can yield the large amount of sodium observed is being considered by several groups, but no quantitative model yet seems entirely satisfactory (cf. Nash *et al.* 1975; Brown and Yung p. 1138).

In the mid 1960's it was discovered that the probability of receiving sporadic but intense bursts of decametric radiation from Jupiter depends strongly on the orbital position of Io (Carr and Desch p. 697). The probability is sharply enhanced when Io is at orbital phase angles near 90° and 225°. Several models developed to account for Io's influence are reviewed by R. A. Smith (p. 1146). The models of Goertz and of Goldreich and Lynden-Bell require the presence of a conducting body, and the conductivity of the observed ionosphere appears to be sufficient. In view of these models, it is interesting to note that electrons are found to be energized near Io: Pioneer 10 observed a concentration of soft electrons just inside Io's *L* shell, and Pioneer 11 measured the flux of electrons of energy > 0.46 MeV to jump by an order of magnitude when the spacecraft was within 6000 km of the presumed location of Io's flux tube. The energy lost in decametric radiation, if it comes from Io's orbital energy, may measurably influence Io's orbital evolution.

In summary, Io is a Mercury-sized object with lunar density. Perhaps as a result of interactions with the planet, Io displays many remarkable properties. Its spectrophotometric surface properties are unlike those of any other object in the solar system, and the composition of its surface remains one of the most important problems in current satellite research. Io is the only one of the Jovian satellites known to have an atmosphere and ionosphere, and in addition it apparently gives rise to toroidal clouds of sodium and hydrogen surrounding Jupiter. The mechanisms for the production and maintenance of this extended atmosphere also are not understood. More data are available on Io than on any of the other Galilean satellites, and it is being vigorously investigated by many workers; yet, at the time of this writing many fundamental properties of Io are still a mystery.

VI. AMALTHEA

Amalthea (J 5), because of its small relative size and its proximity to Jupiter, is probably unlike the other regular satellites in several important respects. Unfortunately, it is an extremely difficult object to observe due to scattered light from nearby Jupiter, and consequently little is known about it. However, it should not be a much more difficult object to observe than Saturn's satellite Mimas, which has recently been investigated both by photographic (Koutchmy and Lamy 1975) and area-scanning photometry (Franz 1975). Early visual observations of Amalthea, coupled with more recent photographic results, suggest that the satellite's orbit has substantial eccentricity and inclination (Sudbury 1969), although the exact values are not well known (cf. Greenberg p. 127 for more details). A better knowledge of its orbital characteristics would be useful in arriving at more precise values of the Jovian gravitational coefficients.

The surface and interior of Amalthea may be unique in the Jovian system. If the formation of the Galilean satellites has been affected by early thermal radiation from Jupiter as proposed by Pollack and Reynolds (1974) and Cameron and Pollack (p. 81), the heating of Amalthea at the time of its origin should have been even more severe. Temperatures as high as $1500°K$ are likely, and hence Amalthea may be largely refractory in composition, much like Mercury. Its exact composition depends significantly on the temperature and hence on the opacity of the nebula surrounding proto-Jupiter: a high-opacity nebula produces temperatures near $1500°K$ while $500°K$ is more probable for a low-opacity nebula (Pollack, personal communication 1975). Refractory materials generally contain a high percentage of radionuclides, which will heat the interior further, counteracting to some extent the effect of Amalthea's high ratio of surface area to mass. Amalthea's small size may have been caused by this heating or by the fact that Jupiter, since it had not finished its gravitational collapse, was scarcely inside Amalthea's orbit at the time of the satellite's origin.

The closeness of Amalthea to Jupiter means that strong tides act on Amalthea, elongating it towards Jupiter and squashing it normal to its orbital plane. The magnitude of Amalthea's distortion provides an indirect approach to finding its mean density; the density probably cannot be determined otherwise since its mass is so small. A homogeneous satellite without strength, having a mean radius R_s and density ρ_s, will be distorted into a triaxial ellipsoid of semi-axes $[R_s (1 + 35d/12), R_s (1 - 10d/12), R_s (1 - 25d/12)]$ pointing toward the planet, normal to the satellite orbital plane and in the orbital plane, respectively; d equals $(\rho_J/\rho_s) (R_J/a)^3$, where the subscript J refers to Jupiter and a is the satellite semimajor axis (Jeffreys 1970). Thus, if Amalthea responds like a fluid over the age of the solar system, which is more likely for ice than for silicates, the tidal bulge might be detectable by accurate photometry or spacecraft imaging. The cross sections seen from Earth at elongation and conjunction differ because of the tidal distortion by ten to thirty percent, depending on the density. The maximum difference in diameters, which would be visible only by a spacecraft, is about 30 km for a high-density model and nearly 100 km for a low-density model. Since MJS resolution on Amalthea is expected to be about 10 km, detection of the tidal distortion of a silicate Amalthea by spacecraft imaging could be somewhat marginal, particularly when surface roughness is accounted for. However, the resolution should be easily sufficient to distinguish a silicate or refractory composition from one of water ice; it should therefore test the Pollack and Reynolds (1974) scenario for the origin of the inner satellites. A similar tidal distortion has been proposed by Harris and Soter (1976) to account for the non-spherical shape of Phobos. Tidal distortion should also be observable on the icy Mimas, for which it is likely to be more than ten percent.

While the V magnitude usually quoted for Amalthea is 13.0 (Harris 1961), it is quite possible that the satellite is fainter and that it is brightest at eastern elongation (Van Biesbroeck 1946), as are the other inner regular satellites. There are no photoelectric or photographic determinations of its brightness. Recently, however, Rieke (1975) has succeeded in measuring Amalthea in the thermal infrared, at wavelengths from 8 to 25 μm. He fits the spectral dependence of these thermal measurements to deduce a color temperature of 155 $\pm$ 15°K. On the assumption that Amalthea radiates with unit emissivity, the temperature and the measured infrared fluxes yield a diameter of 240 $\pm$ 60 km. If the visual magnitude at the time of the observations, all made in 1974 near eastern elongation, is taken to be 13, then Rieke obtains a geometric albedo of $\sim$ 0.1, very much lower than that of Io. Although this albedo has substantial uncertainties due to both the photometry and the radius, it does indicate a rocky rather than icy surface, consistent with the high vapor pressure of ice at the temperature of the satellite. The observed temperature of Amalthea is consistent with a dark object in equilibrium with the insolation (Rieke 1975) and does not indicate a substantial heating by the numerous radiation belt particles which are known to interact with Amalthea, as described below.

Amalthea, moving deep within the Jovian radiation belts, was found by Pioneers 10 and 11 to absorb both keV and MeV electrons (Simpson and McKibben p. 745) and MeV protons (Mogro-Campero p. 1195; McDonald and Trainor p. 966), particularly at higher energies. Particle absorption by Amalthea significantly decreased trapped particle fluxes at all energies (Mogro-Campero p. 1195). Two minima in the particle fluxes that cannot be related to any known satellite have been suggested by Ness (personal communication 1975) as the signatures of an undiscovered satellite at 1.7 R_J. Amalthea may even have its own hydrogen torus. A reasonable alternative to the usual interpretation of the Pioneer ultraviolet photometer data suggests that the observed asymmetry is due to Amalthea rather than to a real asymmetry in the Io cloud (Judge *et al.* p. 1093).

VII. THE OUTER SATELLITES

Very few observations have been made of the outer satellites beyond the photographs used for their discoveries and orbit computations. A search for outer Jovian satellites has recently been initiated at Hale Observatories and has already resulted in the discovery of Leda (Kowal p. 000). Kowal (personal communication 1975) expected to find more than just one satellite, presuming the size distribution of other satellites to be similar to that of asteroids, and he still believes several more will be found before 1980. A recent determination of the magnitudes of several of the outer Jovian satellites (cf. Table I) by Andersson (1974) shows some differences from the classical values. However, with the exception of the 1.1-magnitude discrepancy for Pasiphae (J 8), the two sets of results are surprisingly consistent, indicating that the satellites are probably not very irregular objects such as small captured asteroids or collisional debris might be expected to be. Photoelectric measurements of Himalia (J 6) during two nights by Andersson (1974) suggest magnitude variations of about 0.3 mag but no obvious periodicity with time such as might be produced by rotation of an irregular body; hence, the photometric period and amplitude of even this brightest of the outer satellites remain unknown.

With the exception of Himalia, and perhaps Elara (J 7), the outer satellites are difficult objects to observe photoelectrically because of their faintness and their poorly defined ephemerides (Aksnes 1976). Hence virtually no color information is available for most of them. This ignorance of color is particularly unfortunate since Andersson's photoelectric data on Himalia (see Fig. 12) indicate a color ($B - V = 0.68$, $U - B = 0.46$) that makes this object unique in the solar system. It is worth noting that the two Trojans plotted in Fig. 12 lie on the opposite side of the color diagram from Himalia.

The first attempts to detect thermal radiation from Himalia and Elara in the infrared have failed, indicating that their albedos are greater than 0.03 (Cruikshank 1975). It is important that such observations be attempted again with greater sensitivity, since the Trojan asteroids, which have been considered a possible source for the outer satellites (Kuiper 1956), are known from radiometric measurements to be very dark (Cruikshank 1975). One might presume that, given the same initial composition, the surfaces of the outer Jovian satellites and the Trojan asteroids should be similar because of their common sizes and distances from the Sun. However, some important process, such as interaction with the high-energy proton flux of Jovian origin found by Pioneer 11 in nearby interplanetary space (Simpson *et al.* 1975), might act on one species and not the other.

The comparatively small sizes of the outer satellites, their irregular orbits, and their great distances from Jupiter speak of an origin different from that

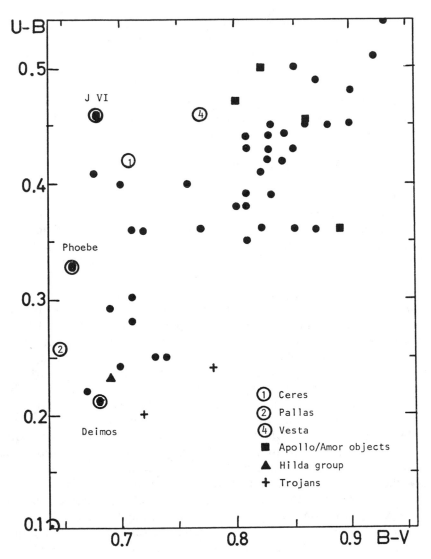

Fig. 12. Two-color diagram for some asteroids and small satellites. Note the position of
Himalia (J 6) in the upper left-hand corner and the two Trojan asteroids in the central bottom
(from Andersson 1974).

of the inner regular satellites (cf. Cameron and Pollack p. 78; Greenberg p.
123). The most commonly accepted view is that they were captured from the
nearby asteroid belt (Kuiper 1951), perhaps shortly after the solar system
originated, when the space density of asteroids in the vicinity of Jupiter was
considerably higher than it is today (Kaula and Bigeleisen 1975; Weidenschilling 1975). However, as long as energy dissipation does not intervene,
capture is a reversible process and thus a flow of planetoids to and from some
storage place, perhaps the triangular Lagrangian points where the Trojan

asteroids are located, is to be expected (Kuiper 1951). Furthermore, computer calculations (Everhart 1973) illustrate that captures, besides being temporary, are extremely uncommon occurrences. More realistic numerical modeling would be a useful addition to our understanding of the origin of the outer satellites.

The existence of the inner and outer satellite groups can be qualitatively understood in any capture scenario. Retrograde orbits are to be expected for the outer satellites, since they have been shown to be more stable than direct orbits semi-analytically by Moulton (1914) and numerically by Hunter (1967), Hénon (1970), and Chebotarev and Bozhkova (1964). The satellite clusters are each located about halfway to the point at which circular equatorial orbits obviously become unstable (Kuiper 1956). Their location may then be determined simply by a playoff between the depth into Jupiter's potential well at which capture is probable versus the distance from which escape of a satellite on an elliptic orbit is easy. On the other hand, the clustering may not indicate multiple captures but simply result from fragmentation taking place after, or during, a single capture.

In work frequently quoted, Bailey (1971, 1972) has argued that capture may occur solely through the inner Lagrangian point, and only at either Jupiter's perihelion or aphelion; perihelion captures in Bailey's model lead to direct orbits which lie very near the actual inner irregular satellite cluster whereas captures at aphelion produce retrograde orbits with semimajor axes approximately those of the outer satellite cluster. Bailey's results are, however, in considerable dispute; numerical calculations (Hunter 1967; Everhart 1973) have illustrated many captures and escapes at points other than the inner Lagrangian point. Errors in the calculation have been pointed out by Heppenheimer (1975), who has proposed his own capture criterion. Davis (1974) has called attention to the fact that the orbital elements of Jupiter that produce Bailey's striking numerical agreement occur only occasionally because the Jovian orbit varies under the action of the other planets.

The similarity of the orbital elements of both clusters of irregular satellites suggested to Bronshten (1968) and Colombo and Franklin (1971) that the members of both groups originated from the same event. They propose that a collision occurred between objects inside Jupiter's sphere of influence, eliminating excess orbital energy relative to Jupiter, and was followed by capture. This hypothesis is discussed in more detail by Greenberg (p. 125). A photometric survey of the outer satellites to look for similarities in their surface properties would be a valuable test of this class of capture theories.

One should not too easily assume capture as a proven scenario for the origin of the outer satellites. The apparent lack of large brightness variations for any outer satellite and the unusual color of Himalia mean that these satellites are probably not simply captured asteroids, but could be objects that have had a special origin or that have been acted upon by some process not affecting the Trojan asteroids. The size distribution of the bodies does not match that of either collisional fragments or asteroids; each of these types of fragmented bodies would contain many more small ones. (See Gehrels' comment following paper by Greenberg p. 130.) More observations are needed; they will be difficult for most of these tiny bodies. However, measurements of Himalia and Elara, which are substantially larger than the others, are not impossible and are strongly encouraged.

VIII. CONCLUSION

To date, practically all studies of the Jovian satellites have been made by groundbased optical astronomy. However, the early 1970's have shown that other techniques will become increasingly important in the future. Pioneers 10 and 11 in 1973–74 greatly improved our knowledge of the masses of the satellites, gave a first *in situ* glimpse of the atmosphere of Io, and imaged several satellites directly. They also improved our understanding of the nature of the Jovian magnetosphere and particle environment in which the inner satellites are immersed. In 1974 radar signals were first returned from Ganymede, and in the next year or two extensive surface mapping and orbital studies of the Galilean satellites should become possible with the upgraded Arecibo radar. At the end of the 1970's two MJS spacecraft are planned to fly by the satellites with imaging capabilities that will for the first time provide detailed spatial resolution. All of the Galilean satellites will be studied with resolution, coverage, and phase angle similar to the best Mariner 6 and 7 far-encounter images of Mars, and for Io and Ganymede there will be substantial coverage at low sun angles and much higher resolution, comparable to Mariner 10 coverage of Mercury. Present mission profiles call for 8% coverage of Io at 1-km resolution and 1% coverage of Ganymede at 1.3-km resolution near their terminators. More extensive and detailed imaging, comparable to that obtained by Mariner 9 on Mars, will await the launch of a Mariner orbiter, which will provide repeated close flybys of each of the Galilean satellites. Only through the use of these more sophisticated techniques will we be able to unravel many of the mysteries that the satellites present to us today.

REFERENCES

Aksnes, K. 1976. Properties of satellite orbits: ephemerides, dynamical constants and satellite phenomena. *Planetary satellites.* (J. A. Burns, ed.) Tucson, Arizona: University of Arizona Press.

Aksnes, K., and Franklin, F. A. 1975a. Results of 1973 occultations of Europa by Io. *Astron. J.* 80:56–63.

——. 1975b. DeSitter's theory "melts" Europa's polar cap. *Nature* 258:503–505.

Anderson, J. D.; Null, G. W.; and Wong, S. K. 1974. Gravitational parameters of the Jupiter system from the Doppler tracking of Pioneer 10. *Science* 183:322–323.

Andersson, L. E. 1974. *A photometric study of Pluto and satellites of the outer planets.* Ph.D. dissertation. Indiana University, Bloomington, Indiana.

——. 1976. The variability of Titan: 1896–1974. *Planetary satellites.* (J. A. Burns, ed.) Tucson, Arizona: University of Arizona Press.

Armstrong, K. R.; Minton, R. B.; Rieke, G. H.; and Low, F. J., 1976. Jupiter at 5 microns. *Icarus* (special Jupiter issue). In press.

Bailey, J. M. 1971. Origin of the outer satellites of Jupiter. *J. Geophys. Res.* 76:7827–7832.

———. 1972. Studies on planetary satellites. Satellite capture in the three-body elliptical problem. *Astron. J.* 77:177–182.

Bartholdi, P., and Owen, F. 1972. The occultation of Beta Scorpii by Jupiter and Io. II. Io. *Astron. J.* 77:60–65.

Berge, G. L., and Muhleman, D. O. 1975. Callisto: disk temperature at 3.71 centimeter wavelength. *Science* 187:441–443.

Bergstralh, J. T.; Matson, D. L.; and Johnson, T. V. 1975. Sodium D-line emission from Io: synoptic observations from Table Mountain Observatory. *Astrophys. J.* 195:L131–L135.

Blanco, C., and Catalano, S. 1974. On the photometric variations of the Saturn and Jupiter satellites. *Astron. Astrophys.* 33:105–111.

———. 1975. Secular variations of Io and Titan. *Icarus* 25:585–587.

Bodenheimer, P. 1974. Calculations of the early evolution of Jupiter. *Icarus* 23:319–325.

———. 1976. Contraction models for the evolution of Jupiter. *Icarus* (special Jupiter issue). In press.

Bronshten, V. A. 1968. On the origin of the irregular satellites of Jupiter. (In Russian.) *Astron. Vestnik* 2:No. 1, 29-36.

Brown, R. A. 1974. Optical line emission from Io. *Exploration of the planetary system.* (A. Woszczyk and C. Iwaniszewski, eds.) pp. 527–531. Dordrecht, Holland: D. Reidel Publ. Co.

Brown, R. A., and Chaffee, Jr., F. H. 1974. High resolution spectra of sodium emission from Io. *Astrophys. J.* 187:L125–L126.

Brown, R. A.; Goody, R. M.; Murcray, F. J.; and Chaffee, F. H. 1975. Further studies of line emission from Io. *Astrophys. J.* 200:L49–L53.

Burns, J. A. 1968. Jupiter's decametric radio emission and the radiation belts of its Galilean satellites. *Science* 159:971–972.

———. 1976a. Orbital evolution. *Planetary satellites.* (J. A. Burns, ed.) Tucson, Arizona: University of Arizona Press.

Burns, J. A., ed. 1976b. *Planetary satellites.* Tucson, Arizona: University of Arizona Press.

Cameron, A. G. W. 1976. Formation of the outer planets and satellites. *Planetary satellites.* (J. A. Burns, ed.) Tucson, Arizona: University of Arizona Press.

Carlson, R. W.; Bhattacharyya, J. C.; Smith, B. A.; Johnson, T. V.; Hidayat, B.; Smith, S. A.; Taylor, G. E.; O'Leary, B. T.; and Brinkmann, R. T. 1973. An atmosphere on Ganymede from its occultation of SAO 186800 on 7 June 1972. *Science* 182:53–55.

Carlson, R. W., and Judge, D. L. 1974. Pioneer 10 ultraviolet photometer observations at Jupiter encounter. *J. Geophys. Res.* 79:3623–3633.

———. 1975. Pioneer 10 ultraviolet photometer observations of the Jovian hydrogen torus: the angular distribution. *Icarus* 24:395–399.

Carlson, R. W.; Matson, D. L.; and Johnson, T. V. 1975. Electron input ionization of Io's sodium emission cloud. *Geophys. Res. Lett.* 2:469–472.

Chapman, C. R.; Morrison, D.; and Zellner, B. 1975. Surface properties of asteroids: a synthesis of asteroid spectrophotometry, radiometry, and polarimetry. *Icarus* 25:104–130.

Chebotarev, G. A., and Bozhkova, A. 1964. *Bull. Inst. Theor. Astron. Leningrad* 9:No. 6.

Colombo, G., and Franklin, F. A. 1971. On the formation of the outer satellite groups of Jupiter. *Icarus* 15:186–191.

Cruikshank, D. P. 1975. Radiometric studies of Trojan asteroids and Jovian satellites 6 and 7. *Bull. Am. Astron. Soc.* 7:377–378.

Cruikshank, D. P., and Murphy, R. 1973. The post-eclipse brightening of Io. *Icarus* 20:7–17.

Davis, D. R. 1974. Secular changes in Jovian eccentricity: effect on the size of capture orbits. *J. Geophys. Res.* 79:4442–4443.

Dollfus, A. 1970. Diamètres des planètes et satellites. *Surfaces and interiors of planets and satellites.* (A. Dollfus, ed.) pp. 46–139. New York: Academic Press.

———. 1975. Optical polarimetry of the Galilean satellites of Jupiter. *Icarus* 25:416–431.

Everhart, E. 1973. Horseshoe and Trojan orbits associated with Jupiter and Saturn. *Astron. J.* 78:316–328.

Fallon, F. W., and Murphy, R. E. 1971. Absence of post-eclipse brightening of Io and Europa in 1970. *Icarus* 15:492–496.

Fanale, F. P.; Johnson, T. V.; and Matson, D. L. 1974. Io: a surface evaporite deposit? *Science* 186:922–924.

――――. 1976. Io's surface and the histories of the Galilean satellites. *Planetary satellites.* (J. A. Burns, ed.) Tucson, Arizona: University of Arizona Press.

Fink, U.; Dekkers, N. H.; and Larson, H. P. 1973. Infrared spectra of the Galilean satellites of Jupiter. *Astrophys. J.* 179:L155–L159.

Fink, U., and Larson, H. P. 1975. Temperature dependence of the water-ice spectrum between 1 and 4 microns: applications to Europa, Ganymede and Saturn's rings. *Icarus* 24:411–420.

Fink, U.; Larson, H. P.; and Gautier, T. N. 1976. New upper limits for atmospheric constituents on Io. *Icarus* 27. In press.

Franz, O. G. 1975. A photoelectric color and magnitude of Mimas. *Bull. Am. Astron. Soc.* 7:388.

Franz, O. G., and Millis, R. L. 1971. A search for an anomalous brightening of Io after eclipse. *Icarus* 14:13–15.

――――. 1974. A search for posteclipse brightening of Io in 1973. II. *Icarus* 23:431–436.

Frey, H. 1975. Posteclipse brightening and non-brightening of Io. *Icarus* 25:439–446.

Gaffey, M. J. 1974. *A systematic study of the spectral reflectivity characteristics of the meteorite classes with applications to the interpretation of asteroid spectra for mineralogical and petrological information.* Ph.D. dissertation. Massachusetts Institute of Technology, Cambridge, Mass.

Gehrels, T. 1976. Picture of Ganymede. *Planetary satellites.* (J. A. Burns, ed.) Tucson, Arizona: University of Arizona Press.

Goldreich, P. 1963. On the eccentricity of satellite orbits in the solar system. *Mon. Not. Roy. Astron. Soc.* 126:257–268.

Goldstein, R. M., and Morris, G. A. 1975. Ganymede: observations by radar. *Science* 188:1211–1212.

Gradie, J., and Zellner, B. 1973. A polarimetric survey of the Galilean satellites. *Bull. Am. Astron. Soc.* 5:404.

Greenberg, R. J. 1973. Evolution of satellite resonance by tidal dissipation. *Astron. J.* 78:338–346.

――――. 1976. Orbit-orbit resonances among natural satellites. *Planetary satellites.* (J. A. Burns, ed.) Tucson, Arizona: University of Arizona Press.

Greene, T. F.; Vermilion, J. R.; Shorthill, R. W.; and Clark, R. N. 1975. The spectral reflectivity of selected areas on Europa. *Icarus* 25:405–415.

Gurnett, D. A. 1972. Sheath effects and related charged-particle acceleration by Jupiter's satellite Io. *Astrophys. J.* 175, 525–533.

Hansen, O. L. 1973. Ten micron eclipse observations of Io, Europa, and Ganymede. *Icarus* 18:237–246.

――――. 1975. Infrared albedos and rotation curves of the Galilean satellites. *Icarus* 26:24–29.

Harris, A., and Soter, S. 1976. The tidal distortion of Phobos. Preprint.

Harris, D. L. 1961. Photometry and colorimetry of planets and satellites. *Planets and satellites.* (G. P. Kuiper and B. M. Middlehurst, eds.) pp. 272–342. Chicago, Illinois: University of Chicago Press.

Hénon, M. 1970. Numerical exploration of the restricted problem. VI. Hill's case: non periodic orbits. *Astron. Astrophys.* 9:24–36.

Heppenheimer, T. A. 1975. On the presumed capture origin of Jupiter's outer satellites. *Icarus* 24:172–180.

Hunter, R. B. 1967. Motion of satellites and asteroids under the influence of Jupiter and the sun. I. Stable and unstable satellite orbits. *Mon. Not. R. Astron. Soc.* 136:245–265.

Jeffreys, H. 1970. *The earth* (5th ed.). p. 208. Cambridge, England: Cambridge University Press.

Johnson, T. V. 1971. Galilean satellites: Narrowband photometry 0.30 to 1.10 microns. *Icarus* 14:94–111.

Johnson, T. V., and McGetchin, T. R. 1973. Topography on satellite surfaces and the shape of asteroids. *Icarus* 18:612–620.

Johnson, T. V., and Pilcher, C. B. 1976. Satellite spectrophotometry and surface composition. *Planetary satellites.* (J. A. Burns, ed.) Tucson, Arizona: University of Arizona Press.

Jones, T. J., and Morrison, D. 1974. A recalibration of the radiometric/photometric method of determining asteroid sizes. *Astron. J.* 79:892–895.

———. 1975. The secular brightening of Titan. *Bull. Amer. Astron. Soc.* 7:384.

———. 1976. Six-color photometry of the Galilean satellites. In draft.

Judge, D. L., and Carlson, R. W. 1974. Pioneer 10 observations of the ultraviolet glow in the vicinity of Jupiter. *Science* 183:317–318.

Kaula, W. M., and Bigeleisen, P. 1975. Early scattering by Jupiter and its collision effects in the terrestrial zone. *Icarus* 25:18–33.

Kieffer, H. H., and Smythe, W. D. 1974. Frost spectra: comparison with Jupiter's satellites. *Icarus* 21:506–512.

Kliore, A.; Cain, D. L.; Fjeldbo, G.; Seidel, B. L.; and Rasool, S. I. 1974. Preliminary results of the atmospheres of Io and Jupiter from Pioneer 10 S-band occultation experiment. *Science* 183:323–324.

Kliore, A. J.; Fjeldbo, G.; Seidel, B. L.; Sweetnam, D. N.; Sesplaukis, T. T.; Woiceshyn, P. M.; and Rasool, S. I. 1975. Atmosphere of Io from Pioneer 10 radio occultation measurements. *Icarus* 24:407–410.

Koutchmy, S., and Lamy, P. L. 1975. Study of the inner satellites of Saturn by photographic photometry. *Icarus* 25:459–465.

Kowal, C. T.; Aksnes, K.; Marsden, B. G.; and Roemer, E. 1975. Thirteenth satellite of Jupiter. *Astron. J.* 80:460–464.

Kuiper, G. P. 1951. On the origin of the irregular satellites. *Proc. Natl. Acad. Sci.* 37:717–721.

———. 1956. On the origin of the satellites and the Trojans. *Vistas in Astronomy,* vol. 2 (A. Beer, ed.) pp. 1631–1666. Oxford, England: Pergamon Press.

Lebofsky, L. A. 1975. Stability of frosts in the solar system. *Icarus* 25:205–217.

Lewis, J. S. 1971. Satellites of the outer planets: their physical and chemical nature. *Icarus* 15:174–185.

———. 1972a. Metal/silicate fractionation in the solar system. *Earth Planet. Sci. Lett.* 15: 286–290.

———. 1972b. Low temperature condensation from the solar nebula. *Icarus* 16:241–252.

———. 1974. The chemistry of the solar system. *Scientific American* 230:50–65.

Lockwood, G. W. 1975a. The secular and orbital brightness variations of Titan, 1972–1974. *Astrophys. J.* 195:L137–L139.

———. 1975b. Planetary brightness changes: evidence for solar variability. *Science* 190: 560–562.

Macy, W. W., and Trafton, L. M. 1975. Io's sodium emission cloud. *Icarus* 25:432–438.

Matson, D. L.; Johnson, T. V.; and Fanale, F. P. 1974. Sodium D-line emission from Io: sputtering and resonant scattering hypothesis. *Astrophys. J.* 192:L43–L46.

McCord, T., and Chapman, C. R. 1974. Spectrophotometry of 624 Hektor and other asteroids. *Bull. Am. Astron. Soc.* 6:373.

McElroy, M. B., and Yung, Y. L. 1975. The atmosphere and ionosphere of Io. *Astrophys. J.* 196:227–250.

McElroy, M. B.; Yung, Y. L.; and Brown, R. A. 1974. Sodium emission from Io: implications. *Astrophys. J.* 187:L127–L130.

Mendis, D. A., and Axford, W. I. 1974. Satellites and magnetospheres of the outer planets. *Ann. Rev. Earth Planet. Sci.* 2:419–474.

Millis, R. L., and Thompson, D. T. 1975. UBV photometry of the Galilean satellites. *Icarus* 26:408–419.

Millis, R. L.; Thompson, D. T.; Harris, B. J.; Birch, P.; and Sefton, R. 1974. A search for post eclipse brightening of Io in 1973. I. *Icarus* 23:425–430.

Minton, R. B. 1973. The polar caps of Io. *Comm. Lunar Planet. Lab.* 10:35–39.

Morrison, D. 1973. Determination of radii of satellites and asteroids from radiometry and photometry. *Icarus* 19:1–14.

——. 1976. Radiometry of satellites and the rings of Saturn. *Planetary satellites.* (J. A. Burns, ed.) Tucson, Arizona: University of Arizona Press.

Morrison, D., and Cruikshank, D. P. 1973. Thermal properties of the Galilean satellites. *Icarus* 18:224–236.

——. 1974. Physical properties of the natural satellites. *Space Sci. Rev.* 15:641–739.

Morrison, D.; Cruikshank, D. P.; and Burns, J. A. 1976. Introducing the satellites. *Planetary satellites.* (J. A. Burns, ed.) Tucson, Arizona: University of Arizona Press.

Morrison, D.; Cruikshank, D. P.; and Murphy, R. E. 1972. Temperatures of Titan and the Galilean satellites at 20 μ. *Astrophys. J.* 173:L143–L146.

Morrison, D.; and Hansen, O. L. 1976. Radiometry of two eclipses of Callisto. In preparation.

Morrison, D.; and Morrison, N. D. 1976. Photometry of the Galilean satellites. *Planetary satellites.* (J. A. Burns, ed.) Tucson, Arizona: University of Arizona Press.

Morrison, D.; Morrison, N. D.; and Lazarewicz, A. 1974. Four-color photometry of the Galilean satellites. *Icarus* 23:399–416.

Moulton, F. R. 1914. On the stability of direct and retrograde satellites. *Mon. Not. Roy. Astron. Soc.* 75:40–57.

Münch, G., and Bergstralh, J. 1975. Sodium D-line emission from Io: spatial brightness distributions from multi-slit spectra. *Bull. Amer. Astron. Soc.* 7:386.

Murphy, R. E., and Aksnes, K. 1973. Polar cap on Europa. *Nature* 244:559–560.

Murray, J. B. 1975. New observations of surface markings on Jupiter's satellites. *Icarus* 25:397–404.

Nash, D. B.; Matson, D. L.; Johnson, T. V.; and Fanale, F. P. 1975. Na-D line emission from rock specimens by proton bombardment: implications for emissions from Jupiter's satellite Io. *J. Geophys. Res.* 80:1875–1879.

Newburn, R. L., and Gulkis, S. 1973. A survey of the outer planets Jupiter, Saturn, Uranus, Neptune, Pluto and their satellites. *Space Sci. Rev.* 14:179–271.

Noland, M.; Veverka, J.; and Goguen, J. 1974. New evidence for the variability of Titan. *Astrophys. J.* 194:L157–L158.

O'Leary, B. T., and van Flandern, T. C. 1972. Io's triaxial figure. *Icarus* 17:209–215.

O'Leary, B., and Miner, E. 1973. Another possible post-eclipse brightening of Io. *Icarus* 20:18–20.

Pauliny-Toth, I. K.; Witzel, A.; and Gorgolewski, S. 1974. The brightness temperatures of Ganymede and Callisto at 2.8 cm wavelength. *Astron. Astrophys.* 34:129–132.

Peale, S. 1976. Rotation histories of the natural satellites. *Planetary satellites.* (J. A. Burns, ed.) Tucson, Arizona: University of Arizona Press.

Pilcher, C. B.; Ridgway, S. T.; and McCord, T. B. 1972. Galilean satellites: identification of water frost. *Science* 178:1087–1089.

Pollack, J. B., and Reynolds, R. T. 1974. Implications of Jupiter's early contraction history for the composition of the Galilean satellites. *Icarus* 21:248–253.

Porter, J. G. 1960. The satellites of the planets. *J. Brit. Astron. Assn.* 70:33–59.

Rieke, G. H. 1975. The temperature of Amalthea. *Icarus* 25:333–334.

Safronov, V. S., and Ruskol, E. L. 1976. Accumulation of satellites. *Planetary satellites.* (J. A. Burns, ed.) Tucson, Arizona: University of Arizona Press.

Shawhan, S. D. 1976. Io accelerated electrons and ions. *J.G.R.* (special Jupiter issue). In press.

Shawhan, S. D.; Goertz, C. K.; Hubbard, R. F.; Gurnett, D. A.; and Joyce, G. 1975. Io-accelerated electrons and ions. *Proceedings of the Magnetospheres of Earth and Jupiter,* Dordrecht, Holland: D. Reidel Publ. Co. In press.

Simpson, J. A.; Hamilton, D. C.; Lentz, G. A.; McKibben, R. B.; Perkins, M.; Pyle, K. R.; Tuzzolion, A. J.; and O'Galagher, J. J. 1975. Jupiter revisited: first results from the University of Chicago charged particle experiment on Pioneer 11. *Science* 188:455–459.

Sinclair, A. T. 1975. The orbital resonance amongst the Galilean satellites of Jupiter. *Mon. Not. Roy. Astron. Soc.* 171:59–72.

Sinton, W. M. 1973. Does Io have an ammonia atmosphere? *Icarus* 20:284–296.

Smith, B. A., and Smith, S. A. 1972. Upper limits for an atmosphere on Io. *Icarus* 17:218–222.

Stumpff, K. 1948. Concerning the albedos of planets and the photometric determination of the diameters of asteroids. (In German.) *Astron. Nachr.* 276:108–128.

Sudbury, P. V. 1969. The motion of Jupiter's fifth satellite. *Icarus* 10:116–143.

Taylor, G. E. 1972. The determination of the diameter of Io from its occultation of β Scorpii C on May 14, 1971. *Icarus* 17:202–208.

Trafton, L. 1975. Detection of a potassium cloud near Io. *Nature* 258:690–692.

——. 1976. A search for emission features in Io's extended cloud. *Icarus* (special Jupiter issue). In press.

Trafton, L.; Parkinson, T.; and Macy, W. 1974. The spatial extent of sodium emission around Io. *Astrophys. J.* 190:L85–L89.

Ulich, B. L., and Conklin, E. K. 1975. Brightness temperature measurements at 3-mm wavelength. *Bull. Am. Astron. Soc.* 7:391.

Van Biesbroeck, G. 1946. The fifth satellite of Jupiter. *Astron. J.* 52:114.

Vermilion, J. R.; Clark, R. N.; Greene, T. F.; Seamans, J. F.; and Yantis, W. F. 1974. Low resolution map of Europa from four occultations by Io. *Icarus* 23:89–96.

Veverka, J. 1970. *Photometric and polarimetric studies of minor planets and satellites.* Ph.D. dissertation. Harvard University, Cambridge, Mass.

——. 1973. The photometric properties of natural snow and of snow-covered planets. *Icarus* 20:304–310.

——. 1976a. Photometry of satellite surfaces. *Planetary satellites.* (J. A. Burns, ed.) Tucson, Arizona: University of Arizona Press.

——. 1976b. Polarimetry of satellite surfaces. *Planetary satellites.* (J. A. Burns, ed.) Tucson, Arizona: University of Arizona Press.

Wamsteker, W. 1972. Narrow-band photometry of the Galilean satellites. *Comm. Lunar Planet. Lab.* 9:171 and 10:70.

Wamsteker, W.; Kroes, R. L.; and Fountain, J. A. 1974. On the surface composition of Io. *Icarus* 23:417–424.

Watson, K.; Murray, B. C.; and Brown, H. 1963. The stability of volatiles in the solar system. *Icarus* 1:317–327.

Wehinger, P. 1976. Mapping of the sodium emission associated with Io and Jupiter. *Icarus* (special Jupiter issue). In press.

Weidenschilling, S. J. 1975. Mass loss from the region of Mars and the asteroid belt. *Icarus* 26:361–366.

Wetterer, M. K. 1971. *The Moons of Jupiter.* New York: Simon and Schuster.

Yung, Y. L., and McElroy, M. B. 1975. Ganymede: possibility of an oxygen atmosphere. *Bull. Am. Astron. Soc.* 7:387.

STRUCTURAL AND THERMAL MODELS OF ICY GALILEAN SATELLITES

G. J. CONSOLMAGNO

and

J. S. LEWIS
Massachusetts Institute of Technology

Thermal history models are presented for a suite of possible initial structures. Complete melting and differentiation of the ice component of Europa and Ganymede due to internal heat sources are predicted. A thick crust of an undifferentiated mixture of silicates and ice is possible for Callisto.

That the larger icy satellites of the outer solar system should melt due to internal heat sources was first proposed by Lewis (1971) on the basis of cosmochemical composition models and steady-state heat flow approximations. A detailed description of possible cosmochemistries has been presented by Consolmagno and Lewis (1976), and detailed thermal models, based on computer simulation of time dependent generation and transport of heat, are given by Consolmagno (1975). This chapter will present the parts of the work that are relevant to Galilean satellites Europa, Ganymede, and Callisto. We presume that Io is essentially ice-free.

Details of the modeling process, including assumptions made, and a complete description of the computer programs used, has been given by Consolmagno (1975). Briefly, we have confined ourselves to mixtures of water ice and rocky material only. The nature and evolution of the rocky material has been for the most part ignored and for simplicity, we have assumed it to be ferromagnesium silicates with the thermal properties of serpentine. We have run the models for two densities of the rocky material, 2.5 and 3.7 g cm^{-3}. Presumably, the lower figure applies when hydrous silicates are present in the rocky material. The initial bodies are assumed to be homogeneous mixtures of silicates and water ice of the appropriate phase at a temperature of 100°K, with the silicate having a chondritic abundance of uranium, thorium,

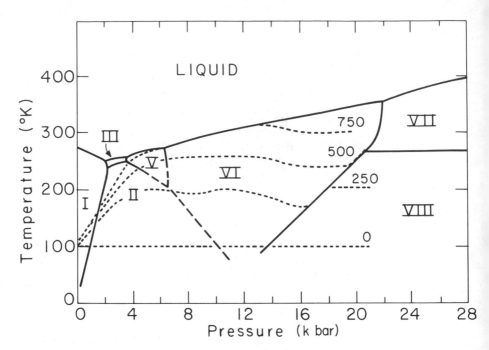

Fig. 1. Phase diagram of water ice. Dotted lines show thermal profiles over 250 M yr intervals for Callisto's homogeneous low-density silicate model.

and potassium. Heats of phase change for the various high-pressure phases of ice are considered (see Fig. 1). As ice with silicates melts, the silicates are allowed to fall towards the planet's center through melted regions. As they fall they lose potential energy; this energy is added to the heat budget of the region through which they fall. Liquid regions are held at a constant temperature, with the heat flux from below transported to the top of the region by convection.

Transport of heat by solid-state creep convection of the ice is not considered. The viscosity of the high-pressure phases of ice is not known, but may possibly be low enough that such creep may be important.

A partial-differences scheme similar to that used by Toksöz et al. (1972) solves the equation of heat conduction. Conductivity and heat capacity varies with temperature and radius, and heat production varies with radius and time. Time intervals are of 1 million years, and the temperature is found for 26 radially symmetric layers within the body.

In our model the thicknesses of these layers are allowed to change with time. As the body warms up, the ice will change phase and may eventually melt, and this water may be displaced by the infalling silicates; thus the densities of the various parts of the planet will be changing at various times. Therefore, each layer will expand or contract; and it will move in or out as the volume of the layers beneath it changes.

In general, the thickness of these layers varies from 75 to 150 km in our Ganymede and Callisto models, with proportionately smaller layers in the Europa model. This is a serious limit to our model, as the thickness of the crust cannot be determined more accurately. However, we can use the extent of melting in the uppermost region to guess an approximate size of the crust. When our model predicts that this top layer is completely melted, we can only say that the crust (which must exist, since the surface temperature is only 100°K) is very thin, and perhaps easily broken.

The number of layers (and hence the accuracy of our estimate as to their size) is controlled by the criteria for convergence of our partial-differences method and by the practical considerations of computer time, which grows as the cube of the number of layers. For each minute of computer time spent on a model with 100-km thick layers, over an hour would be needed for one with layers 25 km thick.

In addition to the homogeneous models, a heterogeneously accreted model of Callisto is also considered, with pure ice covering a rocky core.

In all models, the density of water was taken as 1 g cm^{-3}. This simplifies certain calculations and it is a good estimate for the smaller icy bodies of Saturn (which were modeled at the same time) but is certainly too low for satellites like Ganymede and Callisto. Giving water in Ganymede a density of 1.2 will make the final radius of our model equal to the observed radius.

With this description, we now turn to the models.

CALLISTO

The size and density of Callisto are not quite as well known as those of the other Galilean satellites.[1] Its radius is 2500 ± 150 km, and its mass roughly $(91 \pm 10) \times 10^{24}$g, giving a density ranging anywhere from 1 to 2 g cm^{-3}. However, using the figures of 2500 km and 91×10^{24}g, we find a density of 1.4 g cm^{-3}. Despite the fact that Callisto has the lowest density of the Galilean satellites, the surface of Callisto appears to be much darker than the others, having a spherical Bond albedo of only 0.2.

Rather than trying to fit our models to these uncertain numbers, we have instead found the silicate-to-ice ratio from cosmic abundances, assuming the abundance of iron, magnesium, and silicon are applied to the silicates, and any oxygen left after the formation of the silicates goes towards the formation of water, following the scheme of Lewis (1971). This gives a mixture of 60% by weight for water ice and 40% for silicates.

Since Callisto represents the type of satellite predicted by cosmic abundances of water ice and rock, it has received the most thorough modeling. We present here three extremes of the models: a homogeneously accreted body with silicates of density 2.5; one with silicates of density 3.7; and a

[1]See p. 998.

heterogeneous accretion model, with silicates of density 3.7 forming the core, and various phases of pure water ice overlying it.

Model 1: Homogeneous Accretion with Silicate Density 2.5

The first model starts, as mentioned before, at a uniform temperature of 100°K. At the center is an ice (VIII) and silicate region, extending to 1300 km; next is an ice (VI) and silicate region, extending to 1700 km; then an ice (II) and silicate region, to 2300 km; and finally an ice (I) and silicate crust, giving the planet an initial radius of 2400 km.

Figure 1 gives a detailed picture of how the ice warms up and changes phase. By 600 million years (M yr) after formation, the first melting occurs, at the ice (V)-ice (VI)-water triple point. The melting region grows, mostly towards the center, until at 1250 M yr the planet develops the structure seen in Fig. 2. By 3000 M yr, the undifferentiated crust has shrunk to 150 km in thickness, with another 100 km of partly melted and differentiated ice below it. But the crust gets no thinner, as the heat sources in the silicates decay, and the partly melted region refreezes.

At present, the crust of Callisto by this model will be 250 km thick, and range in composition from primitive material at the surface to partly differentiated material at the base of the crust. Beneath this crust will be a mantle of liquid water 900 km thick, overlying a 1400-km silicate core. Finally, it should be noted that the temperatures in the core would reach 1000°K by 2000 M yr; by this time, one would expect dehydration of the silicates to begin, resulting in some moderation of temperatures within the core.

Model 2: Homogeneous Accretion with Silicate Density 3.7

With the density of the silicates set upwards to 3.7, the overall picture is similar. Because of the more dense material the crustal radius is less by 100 km, making cooling of the interior slightly more efficient so that the planet is not quite so differentiated after 1250 M yr (Fig. 3); however, the greater density of the silicates virtually doubles the heat generated by gravitational infall (since a larger fraction of the total mass is involved), and the smaller volume fraction occupied by the silicates lowers the conductivity of the crust, so that by 3000 M yr the crust is a bit thinner than in the lower density silicate case. But, it begins to refreeze by the present time, and a thick (250 km) crust results, the upper part of which remains undifferentiated.

Model 3: Heterogeneous Accretion with Silicate Density 3.7

In this model we start with a "dry" core of silicates 1350 km in radius, covered by a 100-km thick region of pure ice (VIII), a 200-km thick region of ice (VI), a 700-km region of ice (II), and a 60-km crust of ice (I). The heat flux from the core slowly melts the overlying ice layers: first melting does not occur until 700 M yr, and by 1500 M yr the water region is only 250 km

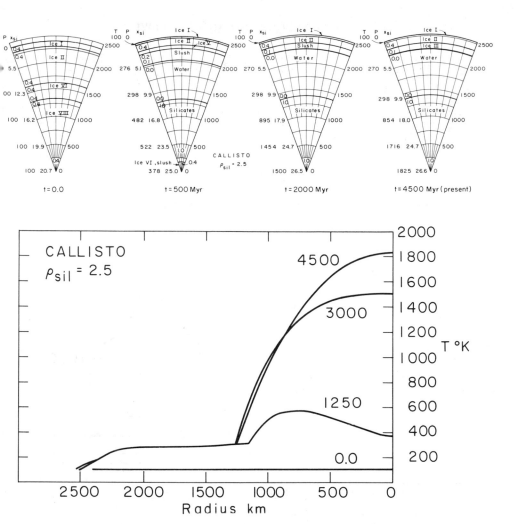

Fig. 2. Pie sections and temperature versus radius curves for Callisto, assuming 40% by weight for silicates of density 2.5 g cm⁻³, 60% by weight for water ice. Temperature in degrees Kelvin, pressure in kilobars. The weight fraction of silicates is listed down the left-hand side of the pie sections and the radius in kilometers, on the right. The sections labeled "ice I," "ice II," etc., represent those regions where the indicated ice phase is stable. Note, however, the weight percent of silicates which also exist in those regions.

thick, with another 100 km above it melting. By 3000 M yr the water region has nearly tripled in size and continues to grow to the present. The present structure is roughly in steady state, with a 100-km crust of ice (I), 200-km region of ice (II), and an 850-km liquid mantle overlying the silicate core (Fig. 4).

Central temperatures and pressures reach very high values: the model predicts a temperature of nearly 3000°K, at a pressure of 45 kbar. Naturally, silicates will undergo considerable phase changes at these conditions, and we

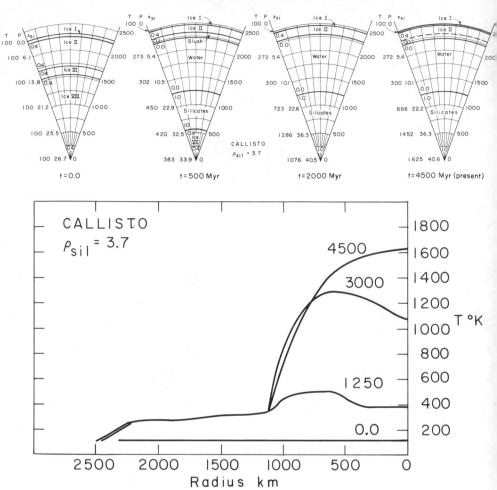

Fig. 3. Pie sections and temperature versus radius curves for Callisto, assuming 40% by weight for silicates of density 3.7 g cm^{-3}, 60% by weight for water ice. See caption to Fig. 2 for further explanation.

should expect a substantial differentiation and evolution of silicate phases. The silicate phase changes might, on the one hand, absorb considerable amounts of heat which would otherwise contribute to the heat flux being convected to melt the crust. On the other hand, the differentiation of the silicates may serve to enrich the upper layers of the silicate core in uranium and thorium, thus in effect increasing the local heat flux. Such reactions in the core would start early in the object's history, at about the same time that melting of ice becomes important, and would probably have a more profound effect on the ultimate internal structure of such a body than the simple melting of ice modeled here.

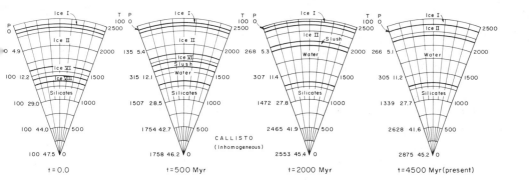

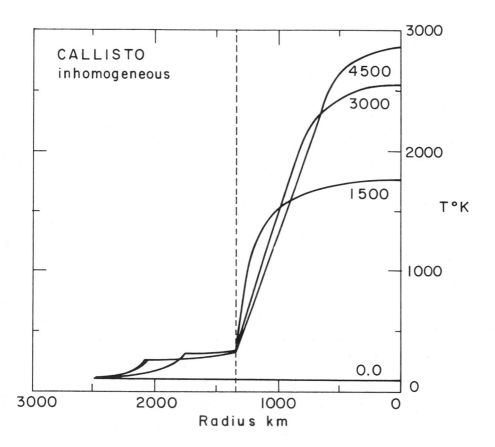

Fig. 4. Pie sections and temperature versus radius curves for an inhomogeneously accreted Callisto. Here sections labeled "ice I," "ice II," etc., represent regions of pure ice of the indicated phase. Again, the temperature is indicated in degrees Kelvin, pressure in kilobars, radius in kilometers, and time in million years from formation.

GANYMEDE

Ganymede has a radius of 2635 km, known to within 10 km by stellar occultation, and a mass of 154×10^{24}g, known as in Europa's case by orbital characteristics. This gives a mean observed density of 2.0 (Morrison and Cruikshank 1974). The surface of Ganymede is also known to be icy, like Europa's. The surface albedo is high (0.4), and ice has been observed in its spectrum (Pilcher *et al.* 1972).

Since Ganymede's density is well known, we can use it as a constraint on the amounts of ice and rocky material present. Assuming a density of 1.2 g cm^{-3} for the ices, we use the formula:

$$x_{sil} = \left(\frac{\rho_{sil}\, \rho_{ice}}{\rho_{total}} - \rho_{sil} \right) / (\rho_{ice} - \rho_{sil}) \qquad (1)$$

to determine the weight percent of ices and silicates. We find that the composition for Ganymede can range from 20% water and 80% silicates of density 2.5, to 40% water and 60% silicates of density 3.7.

Low-density Silicate Model

Starting with the 2.5 g cm^{-3} silicate, we again set up our satellite at an initial temperature of 100°K, as we did for Callisto. We have at the center an ice (VIII) and silicate core extending to 2100 km, overlying ice (VI) and silicate, and ice (II) and silicate mantles, and a thin ice (I) and silicate crust (Fig. 5). After 250 M yr substantial melting has occurred near the surface, resulting in differentiation from a radius of 2250 km to the surface. Under the water and silicates comes a layer of ice (VI) and silicates, and a core of ice (VII) and silicates. By 2000 M yr the planet is completely differentiated and remains so to the present. We would expect only a thin icy crust on the surface.

The central temperatures are in excess of 1000°K in the silicate core as early as 1000 M yr after formation, and so we can expect that any hydrous silicates will have dehydrated by this time. Such a process would require heat, resulting in a slightly cooler core than illustrated. After dehydration the planet with more dense silicates would probably resemble more the second model of Ganymede, using silicates of density 3.7.

High-density Silicate Model

This model starts out with twice as much ice, and thus needs more time to evolve. However, by 500 M yr it has reached the point illustrated in Fig. 4, with a 100-km crust of undifferentiated "dirty" ice, a 300-km region of slush (where the ice is melting), a thin mantle of water underlain by a region of silicates, then high-pressure ices (VI) and (VII) with silicates. By 2000 M yr this model also has completely differentiated, but with a thicker crust than the 2.5 g cm^{-3} silicate model. Since this ice has been totally melted and differentiated, it thus would look "clean." This structure would continue to the present (Fig. 6).

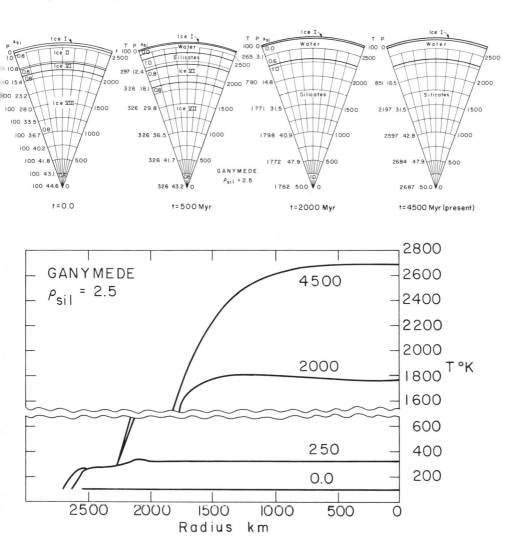

Fig. 5.   Pie sections and temperature versus radius curves for Ganymede, assuming 80% by weight for silicates of density 2.5 g cm^{-3}, 20% by weight for water ice. See caption to Fig. 2 for further explanation.

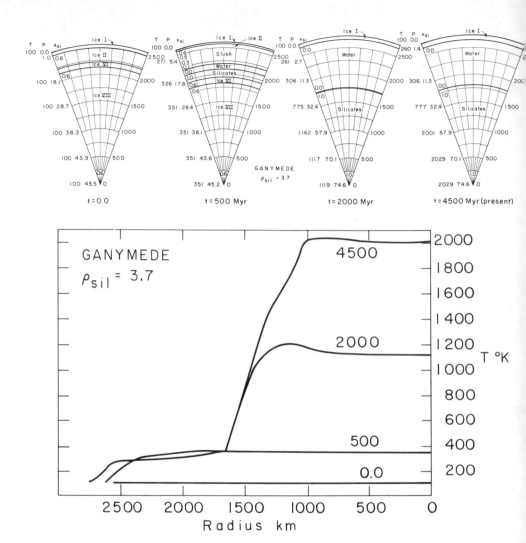

Fig. 6. Pie sections and temperature versus radius curves for Ganymede, assuming 60% by weight for silicates of density 3.7 g cm⁻³, 40% by weight for water ice. See caption to Fig. 2 for further explanation.

EUROPA

Europa has a mass of $48 \pm 1 \times 10^{24}$g, as determined by the characteristics of its orbit, and a radius of 1550 ± 150 km, as determined by groundbased observations, giving a total density of 3.1 ± 1 g cm⁻³ (Morrison and Cruikshank 1974). Within this uncertain range, it is possible that little or no ice need be present. The albedo of Europa is 0.6, suggestive of an icy surface.

Pilcher *et al.* (1972) have reported observing characteristic absorption bands of ice in the reflectance spectrum, which would indicate that ice makes up at least a small part of the satellite's chemistry.

Model

Since Europa's density is greater than 2.5, we assume only the 3.7 density material for the silicates. Using the same formula as for Ganymede [see Eq. (1)], a density of 3.1 would result from a mixture of 10% by weight of water ice and 90% by weight of silicates. The structure is illustrated in Fig. 7.

We begin with a core of silicates and ice (VIII) extending to a radius of 1200 km, covered with mantles of silicates and ice (VI), and silicates and ice (II), with silicates and ice (I) as a crust on the surface. After 250 M yr, substantial melting has already taken place, resulting in differentiation of the water and silicates. Melting has proceeded almost to the surface, and a crust of pure ice (I) now exists. Due to the high pressures in the center, however, the melting point of ice (VII) has not yet been reached there and so this region remains undifferentiated. By 2000 M yr the entire satellite has been melted and differentiated. The high heat flux from the core, carried by the convecting water, has again completely melted the crust — any layer of ice on the surface will be too thin to be detected by our model grid. Temperatures in the center of the core reach nearly 2000°K.

After 4500 M yr, a structure similar to what we expect for the present may exist. A thin crust of ice covers a convecting region of water, which is cooling off the upper layers of the silicate core. Heat production in the core has dropped as well, as the radioactive nuclides decay. However, the center is still effectively isolated from the surface and continues to heat up reaching temperatures of 2800°K. This central temperature is not to be strictly believed, since we have ignored the chemistry and melting behavior of the rocky material, which will be significant long before these high temperatures are achieved. This chemistry was described briefly in discussions (see above) of the heterogeneous model of Callisto.

COMPARISON OF MODELS WITH OBSERVED FEATURES

The only presently observable intrinsic features of the Galilean satellites are their densities and their surface properties. The densities show a striking relation to distance from Jupiter, with the most rock-like (or perhaps most ice-depleted) satellite Io being the closest, and the satellite most typical of cosmic abundances of ice and silicate (the least ice-depleted) being Callisto. This description is matched in our models of the outer three satellites. A reason for this simple decrease in density has been proposed by Pollack *et al.* (1974) who describe the Galilean system as a small-scale solar system, with the composition of satellite-forming materials being determined by the local pressure and temperature in a proto-Jupiter nebula.

The surface features are more difficult to explain. The darkest satellite is (by density arguments) the iciest, while the brightest (Io) should have no ice at all. Ice features are seen in the spectra of Europa and Ganymede, but only hinted at in the spectrum of Callisto, which should be the iciest of all. Again, polarimetric and thermometric results for Europa and Ganymede are consistent with ice and frost on their surfaces, but are more difficult to explain for Callisto.

To explain these differences, Fanale *et al.* (1976) have proposed that the surface of Io is covered with salts leached from its interior, Europa is covered with a thin crust of ice, Ganymede has a crust of differentiated ice, and Callisto's crust is an undifferentiated ice-plus-rock mixture. Our thermal models lend support to this idea.

In our models, we predict Callisto will always have a thick undifferentiated crust, while even the topmost iteration layers of our Ganymede and Europa models are predicted to be melted and differentiated. In reality, since the surface temperature is presumably always below the melting point of ice, there will always be an unmelted crust too thin to be seen by our model. Judging from the predicted heat flow through the mantle, this crust will be roughly 30 km thick, thus much more easily punctured, destroyed, or otherwise altered than the 200 km crust on Callisto.

Altering the Primitive Crust

Any of a variety of events may have occurred which could have thoroughly melted the primitive crust: a superabundance of radioactive nuclides, rapid accretion, a period of increased solar heating, electromagnetic heating from an increased solar wind, or tidal effects. In addition, the uncertainty in the radius of Callisto means it may well have a density as great as that of Ganymede. By our model, this would mean an enrichment of radionuclide-bearing rocky material, and thus a thermal history similar to Ganymede's.

There is the question of the stability of such a crust; ice plus silicates would presumably be denser than pure water. However, depending on the

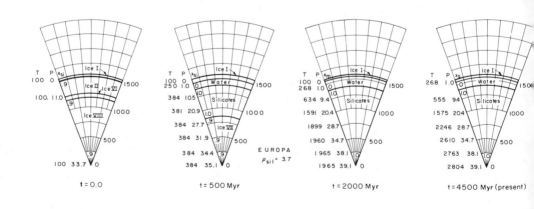

amount and density of the silicate phase, such a difference may not be too extreme, since water under high pressure approaches the density of ice (I) plus silicates.

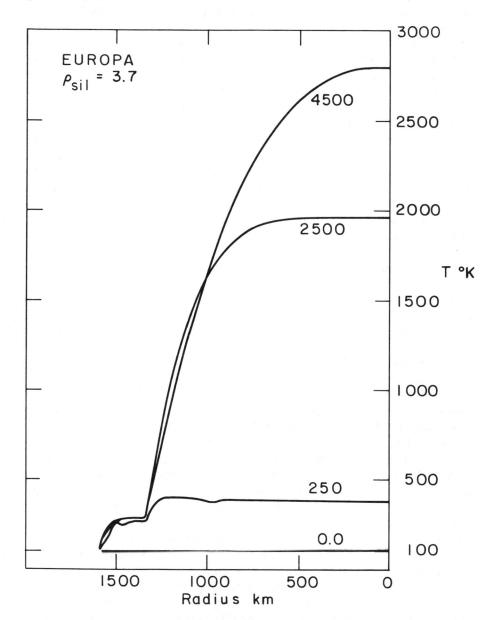

Fig. 7. Pie sections and temperature versus radius curves for Europa, assuming 90% by weight for silicates of density 3.7 g cm⁻³, 10% by weight for water ice. See caption to Fig. 2 for further explanation.

There is the possibility that impacts may cause the crust to break, and this coupled with the above mentioned density instability would lead to a catastrophic destruction of the crust. The efficiency of this mechanism depends on the nature of the crust and the impacting material. If the impacting material originated from the same region of space as the satellites themselves, it will likely have low density (being at least part ice itself) and would likely approach the satellites at a low relative velocity. Thus, especially large objects would be needed to break through a thick crust. While the flux of such objects is not well known, it is likely to be very low after the first 1000 M yr, while crusts thin enough to be broken will not have been formed until that time. On the other hand, if the important source for colliding material is cometary in origin, then its composition cannot be presumed to be only ice, and its velocity is likely to be very high, especially due to the acceleration caused by Jupiter's mass. Thus impacts, though not necessarily common, will be of high energy. Whether they would be able to break through a 200 km thick crust is still uncertain, however.

A more quantitative treatment of this subject would be of interest. In order to predict how impacts would affect the surfaces of these objects, we would need to know not only the chemical nature, velocities, and flux of impacting objects, but also the strength of an ice and rock crust at $100°K$; none of these factors are well known at present. The best guess at present is that impacts might puncture the thin (~ 30 km) crusts predicted for Europa and Ganymede. It may be possible that impacts similar to those postulated for the formation of lunar mare basins would occur to puncture Callisto's crust.

Our models predict considerable thermal expansion, and this may produce significant cracks in the crust, leading to upwelling of the less-dense liquid material underneath and eventually to catastrophic overturn of the crustal layers. But the thermal expansion appears to be on a slow enough time scale that plastic flow of the ice should heal such cracks as they develop.

An undifferentiated crust is not necessary to explain the low albedo of Callisto. If the surface were evenly dusted by a uniform layer of an especially dark material, such as a carbonaceous chondrite material, the water absorption features would be obscured and the darkness of the surface could be explained (Johnson and Fanale 1973). Since the observed albedo of Callisto, though lower than expected for ice, is still fairly high, we might expect such an effect to be important. If the Poynting-Robertson effect due to the brightness of Jupiter would allow small dark particles to drift inwards towards the planet, then such particles would first encounter Callisto, the outermost of the Galilean satellites, and provide darkening material. However, even in this case a thick, stable crust (although not necessarily a dirty, undifferentiated one) would be desirable to allow this dust to accumulate over a long period of time.

If Europa and Ganymede were formed of homogeneous mixtures of ice and silicates with chondritic abundances of radioactive nuclides, while solid-state creep is unimportant, they will almost completely melt, leaving thin, easily altered primitive crusts. One cannot say with equal assurance that the crust of Callisto will not melt, only that internal heat sources alone may not be sufficient to melt it. However, it seems plausible that a thick crust, as predicted for Callisto, will survive.

CONCLUSIONS

The densities of Callisto, Ganymede, and possibly Europa indicate that they are made of a mixture of ice and rocky material. If the rocky material contains a chondritic abundance of radionuclides, then these bodies will be substantially melted and differentiated.

A Callisto-sized object, with cosmic abundances of water and rocky materials in a homogeneous mixture will begin to melt within 750 M yr, be substantially melted by 1200 M yr after formation, and persist to the present with a 200-km thick rock-and-ice crust, a liquid water mantle about 1000 km thick, and a 1200-km rocky core.

Ganymede, with a higher silicate/ice ratio to match its observed density, will begin to melt by 500 M yr and perhaps as early as 250 M yr (depending upon the nature and amount of silicates present); by 2000 M yr virtually all of it will have melted; and at present it will have a crust of ice less than 100 km thick, a liquid water mantle 400–800 km thick and a large rocky core 1800–2200 km in diameter.

Europa, even if it started with a composition of as much as 10% ice (which seems to be a maximum considering its high density) would be substantially melted and differentiated within 500 M yr. Internal temperatures will certainly be high enough to devolve water from any hydrous minerals present (which is probably a more likely source for surface ice) by 2500 M yr. A Europa with 10% water content would have a 70-km ice crust at present, a 100-km water mantle, and a rocky core 1400 km in radius.

Internal heat sources seem to be sufficient for Europa and Ganymede to completely melt at some time in their history (at least to within 30 km of the surface). This is not true of Callisto, and may be a clue to understanding why the surface of Callisto appears to be much darker than that of Europa or Ganymede. However, the still uncertain density for Callisto (with resulting uncertainty in the amount of rocky material present), the undefined nature of the rocky material, and the large number of other, unmodeled events which could have affected the surfaces of all these bodies, makes such a suggestion speculative.

Inhomogeneously accreted bodies, consisting of rocky cores covered by rock-free ice, will melt more slowly than homogeneous rock-and-ice mixture bodies. However, by the present time such bodies would have icy crusts and water mantles comparable to those of the homogeneously accreted bodies. Their crusts, of course, would always be clean, rock-free ice.

Regardless of the model of accretion, any Galilean-sized object with at least a cosmic ratio of rocky to icy material will show a significant internal structure, and the process of differentiating the interior will likely be extensive enough to affect the surfaces of these satellites.

Acknowledgements. We thank R. G. Prinn, J. E. Hart and R. G. Donnelly for invaluable advice. The generous help of S. S. Barshay, P. Briggs, M. B. Fegley, Jr., M. Gaffey, C. Peterson, D. N. Skibo and S. J. Weidenschilling has been most encouraging. This research was supported by NASA Grant NGL-22-009-521.

DISCUSSION

T. R. McDonough: You mentioned magnetic effects due to a super solar wind. What are these effects?

G. J. Consolmagno: This was the theory of Sonett *et al.* (1975) that the sun's early magnetic field, carried by the solar wind, might heat up bodies in the solar system by induction.

W. Kaula: For Sonett's heating-by-electromagnetic-induction to work on silicate bodies, their electrical conductivity must be raised by first heating them to a few $100°K$ by some other mechanism. Is there a similar "threshold" effect to heat the icy parts of Galilean satellites by this process, and, if so, what?

G. J. Consolmagno: The electrical conductivity of pure ice measured in the laboratory is low. However, the conductivity of the material making up the Galilean satellites, containing ice and a rocky component of unknown particle size, and whose composition may include anything from electrolytic salts to unoxydized metal, is unknown. Heating by electromagnetic induction is merely mentioned as one of a number of possible heating events which we have ignored.

W. H. McCrea: The surface of Europa is apparently very patchy; is your model compatible with this?

G. J. Consolmagno: The simple theory which has for Europa and Ganymede "clean," differentiated ice crusts, while Callisto has a "dirty," undifferentiated crust, could not explain this. However, if the surfaces are dusted with some darker material, then we can explain this patchiness. Europa and Ganymede, with thin icy crusts as we predict, would be more easily punctured by an impact; liquid water could then flow from the mantle onto the surface forming a flat, clean plain in a manner somewhat similar to the way we think the maria may have been formed on the moon. In that case, we would expect the surface of Europa to be patchy. Callisto, on the other hand, could not maintain a thin crust even if some outside event completely melted it, so any dust that it collects is more likely to remain uncovered, hence its darker color. Still, we should not constrain the top 200 kilometers of these satellites by observations of the top 2 microns. Our models cannot pretend to give a detailed picture of the surfaces.

J. B. Pollack: There are two corollaries that follow from the early high-luminosity phase of Jupiter as applied to the composition of the Galilean satellites. First, the temperatures in the nebula in which they formed barely become low enough to permit water-ice condensation. Hence the temperatures probably never become low enough to permit condensation of ammonia or methane clathrates. Second, the early high-luminosity phase inhibited water-ice condensation in the early phases of the formation of Ganymede and Callisto. As a result, just after their formation these satellites may have had a strong chemical zonation with the rocky material concentrated toward the center and water ice toward the surface.

REFERENCES

Consolmagno, G. J. 1975. Thermal history models of icy satellites. Master's thesis, Massachusetts Inst. of Technology, Cambridge, Mass.

Consolmagno, G. J., and Lewis, J. S. 1976. Preliminary thermal history models of icy satellites. *Planetary satellites.* (J. A. Burns, ed.) Tucson, Arizona: University of Arizona Press.

Fanale, F. P.; Johnson, T. V.; and Matson, D. L. 1976. Io's surface and the evolution of the Galilean satellites. *Planetary satellites.* (J. A. Burns, ed.) Tucson, Arizona: University of Arizona Press.

Johnson, T. V., and Fanale, F. P. 1973. Optical properties of carbonaceous chondrites, and their relationship to asteroids. *J. Geophys. Res.* 78:8507–8518.

Lewis, J. S. 1971. Satellites of the outer planets: their physical and chemical nature. *Icarus* 15:174–185.

Morrison, D., and Cruikshank, D. P. 1974. Physical properties of the natural satellites. *Space Sci. Rev.* 15:641–739.

Pilcher, C. B.; Ridgeway, S. T.; and McCord, T. B. 1972. Galilean satellites: identification of water frost. *Science* 178:1087–1089.

Pollack, J. B., and Reynolds, R. T. 1974. Implications of Jupiter's early contraction history for the Galilean satellites. *Icarus* 21:248–253.

Sonett, C. P.; Colburn, D. S.; and Schwartz, K. 1975. Formation of the lunar crust: an electrical source of heating. *Icarus* 24:231–255.

Toksöz, M. N.; Solomon, S. C.; Minear, J. W.; and Johnston, D. H. 1972. The thermal evolution of the moon. *Moon* 4:190–213.

THE JOVIAN METEOROID ENVIRONMENT

D. H. HUMES
Langley Research Center

Meteoroid fluxes in interplanetary space at 5 A.U. and near Jupiter have been measured with the penetration detectors on Pioneer 10 and Pioneer 11. The data suggest that the size distribution of meteoroids at 5 A.U. is the same as that at 1 A.U. Furthermore, the high flux observed near Jupiter appears due to meteoroids in orbit about the sun being gravitationally focused toward the planet. There is no indication of a large population of particles in orbit around Jupiter. Calculations show that the mass influx of meteoroids on Jupiter is 170 times that on the earth, but this does not have a significant effect on the energy balance of Jupiter. The high speed with which meteoroids enter the Jovian atmosphere causes them to be brighter than those on Earth so that the number of visible meteors is 5800 times the number in the earth's atmosphere.

Before the Pioneer 10 mission, it was presumed that Jupiter must attract many more meteoroids than the earth does because it is so massive and near the asteroid belt. The data obtained with the penetration detectors on Pioneers 10 and 11 have shown that, indeed, the meteoroid population near Jupiter is greater. It was discovered, however, that the asteroid belt is not responsible for this high meteoroid population (Humes *et al.* 1974). The population of small meteoroids is not greater in the asteroid belt than it is in adjacent regions of space, the concentration being essentially constant between 1 A.U. and 5 A.U.

In this chapter the size distribution of meteoroids at 5 A.U. is calculated from the ratio of the penetration fluxes in the Pioneer 10 and Pioneer 11 detectors at 5 A.U., using the interplanetary data and not the encounter data because the former can be compared directly. The spatial density of meteoroids in interplanetary space at 5 A.U. is then calculated assuming different distributions of orbits. I shall demonstrate that the data obtained on Pioneers 10 and 11 during their encounters with Jupiter can be explained by gravitational focusing. A model of the Jovian meteoroid environment is derived and the total mass and energy added to Jupiter are calculated. The meteors created when meteoroids enter the Jovian atmosphere also will be considered. The altitude at which the meteoroid material is deposited in the Jovian atmosphere is calculated, and the number of visible meteors occurring in the Jovian atmosphere is estimated.

I. DATA

The meteoroid penetration detectors on Pioneer 10 and 11 are pressured cells, each having an exposed area of 2.45×10^{-3} m^2 (O'Neal 1974). When a meteoroid penetrates a cell wall, the gas escapes and the loss of pressure causes the signal that a penetration has occurred. Each cell can detect only one meteoroid so the sensitive area of the experiment decreases with each penetration. The cell walls are 25 μm thick stainless steel for the Pioneer 10 detectors which make the detectors sensitive to meteoroids larger than 2×10^{-9}g at typical impact velocities (Humes *et al.* 1974). The cell walls are 50 μm thick stainless steel on Pioneer 11 making the detectors sensitive to particles larger than 1.6×10^{-8}g. There are 234 cells on each spacecraft. The experiments on each spacecraft were fabricated as two essentially independent instruments to increase reliability. The Channel (0) instruments on each spacecraft process data from 108 cells, while Channel (1) instruments handle the data from the other 126 cells.

The Channel (1) instrument on Pioneer 10 began to malfunction 6 days after the spacecraft was launched. The Channel (1) data from Pioneer 11 is still being evaluated. Consequently, only the Channel (0) data from Pioneers 10 and 11 are considered in this chapter.

The time histories of penetrations in Channel (0) are shown in Fig. 1 (Humes *et al.* 1975). 54 cells were penetrated on Pioneer 10 while the spacecraft was between the earth and Jupiter, and 20 cells penetrated on Pioneer 11 in the same region of space.

There was a sharp increase in the number of meteoroid penetrations as Pioneer 10 passed near Jupiter (Kinard *et al.* 1974). There were only two penetrations on Pioneer 11. In addition, a spurious count was registered soon after the second penetration, while the spacecraft was near periapsis and the radiation was high. This has raised some question as to whether the two indicated penetrations were real or whether they were counts induced by radiation (Humes *et al.* 1975). It is clear, however, that no more than two meteoroid penetrations occurred during encounter. The counter, of the number of cells that have been penetrated, is inhibited for approximately 83 minutes after each event so that multiple counts will not be registered. The inhibit time is significant when compared to the time between penetrations on Pioneer 10 during the Jupiter encounter. Some cells could have been penetrated while the counter was inhibited and those penetrations would not have been detected. In the analysis of the Pioneer 10 data, the experiment was considered to be turned off for the 83 minutes following each recorded

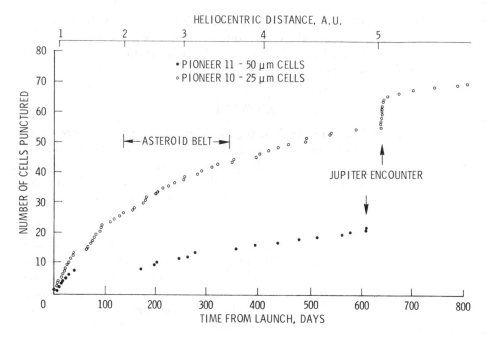

Fig. 1. Time histories of meteoroid penetrations detected on the Channel (0) instruments on Pioneer 10 and Pioneer 11.

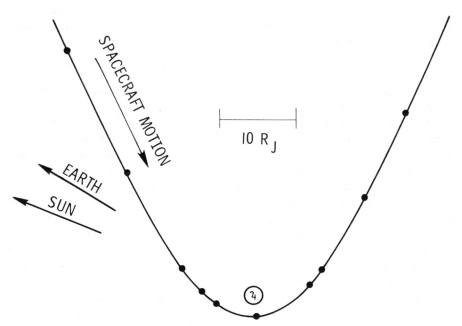

Fig. 2. Sketch of Pioneer 10 trajectory near Jupiter showing positions of spacecraft when meteoroid penetrations were detected. The view is from the north.

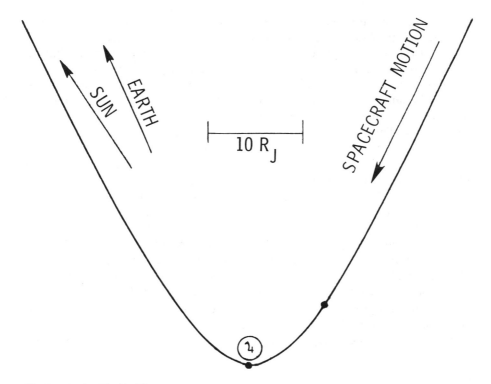

Fig. 3.   As for Fig. 2; Pioneer 11.

penetration. A small error is introduced in the calculated flux if a few un-detected penetrations occurred, namely from not knowing the remaining sensitive area. The attitude of the spacecraft and the direction of Jupiter's motion are calculated from the direction to the earth and the sun in Figs. 2 and 3. The spin axis of the spacecraft, which is the axis of symmetry of the high-gain antenna, is always pointed toward the earth and the penetration detectors are mounted on the back of the dish.

The total number of meteoroid penetrations recorded on Pioneers 10 and 11 is small and they occurred over a wide range of space so that subtle variations in the concentration of meteoroids cannot be detected. Furthermore, there is a considerable uncertainty associated with the mean values of the fluxes and spatial densities of meteoroids calculated from the Pioneer 10 and 11 data.

II. SIZE DISTRIBUTION

If meteoroids in solar orbit are being focused toward Jupiter, the size distribution of the particles which strike the planet will be the same as the size distribution of the particles at 5 A.U. outside the sphere of influence of

Jupiter's gravitational field. The size distribution at 5 A.U. is easily calculated from interplanetary data of Pioneers 10 and 11. The transfer trajectories
between Earth and Jupiter were similar for the two spacecraft so that the
meteoroid penetration data can be compared directly. Assuming that there is
no difference in the orbital distributions of meteoroids with masses between
2×10^{-9}g and 1.6×10^{-8}g, the ratio of the penetration rates on the two
spacecraft is equal to the ratio of the cumulative number of meteoroids of
these sizes. This is not true of the encounter data because the trajectories
were very different and many factors will affect the penetration rates.

The penetration flux for the two detectors is shown in Fig. 4 as a function
of the heliocentric range of the spacecraft. The penetration flux is equal to
N_P/At, where N_P is the number of penetrations that occurred in a certain
region of space, A is the sensitive area of the detectors, and t is the time spent
in the region. The horizonal bars show the region of space considered in calculating the average penetration flux indicated by the data points. The regions for Pioneers 10 and 11 were chosen to facilitate comparison and to include a significant number of meteoroid penetrations (6 to 13). Large regions
where no meteoroid penetrations occurred are identified by the gaps. The
uncertainty in the results, because the sample of meteoroid penetrations is
so small, is illustrated by the 90% confidence limits, i.e., the vertical bars of
each point. Only the interplanetary data were used to obtain the flux near 5

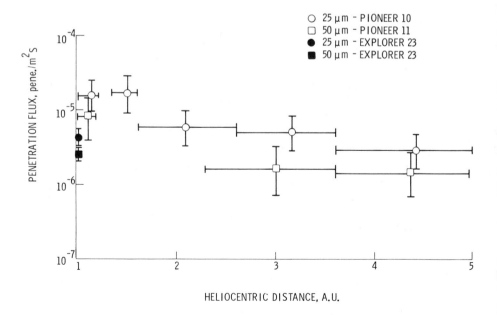

Fig. 4.   Penetration flux in 25 μm and 50 μm stainless steel between 1 and 5 A.U. The uncertainty in the data is shown by the vertical bars which are 90%-confidence limits.

A.U. The mean flux values for the 25 and 50 μm detectors differed by a factor of 1.97 at 5 A.U. and by 1.91 at 1 A.U.; the size distribution of the meteoroids at 5 A.U. is similar to that near 1 A.U. Penetration fluxes in 25 and 50 μm stainless steel detectors on Explorer 23 (O'Neal 1968) are also shown in Fig. 4; the difference is a factor of 1.71. On Explorer 16, which also is an Earth-orbiting spacecraft but with beryllium copper detectors, the fluxes in the 25 and 50 μm detectors differed by a factor of 1.95 (Hastings 1964). The penetration fluxes on Explorer 23 are lower than those on Pioneer 10 and 11 because the detectors were randomly oriented with respect to the meteoroid environment on Explorer 23 while they were placed in the most favorable place to receive meteoroid impacts on Pioneers 10 and 11 (Humes *et al.* 1974).

The meteoroid mass distribution is often fitted to a power law expression

$$N = k_1 \, m^\alpha \tag{1}$$

where N is the cumulative number of meteoroids of mass m and greater, k_1 is a constant, and α is the mass distribution index. The value of α at 5 A.U. is found by the Pioneer 10 and 11 data to be -0.34; this is for meteoroids with masses near 10^{-8}g and is not expected to reflect the size distribution of much smaller or larger meteoroids.

It has been observed that for meteoroids near the earth the value of α is not constant over a wide range of masses; Alvarez (1968) gives

$$\alpha = -1.728 - 0.1694 \, (\log_{10} m), \quad m < 1.26 \times 10^{-2}g \tag{2}$$

$$\alpha = -1.408, \qquad\qquad\qquad m > 1.26 \times 10^{-2}g \tag{3}$$

where m is the mass of the meteoroid in grams.

Assuming that the size distribution of meteoroids at 5 A.U. is the same as that for meteoroids near the earth, the expressions given by Alvarez are used to calculate the size distribution of Jovian meteoroids for the entire mass range.

III.   SPATIAL DENSITY

The spatial density of meteoroids with masses $\geq 2 \times 10^{-9}$g and $\geq 1.6 \times 10^{-8}$g in interplanetary space at 5 A.U. is calculated from the penetration fluxes measured at that distance, by assuming the distributions of mass and orbital parameters. The orbital parameters are needed for the relative speed between spacecraft and meteoroids, and also to determine the direction from which they approach the spacecraft, and consequently the *effective* area of the detectors exposed to the meteoroids. The mass distribution is needed to normalize the data to 2×10^{-9}g and 1.6×10^{-8}g when the assumed meteoroid orbits indicate lower impact speeds at 5 A.U. than at 1 A.U. and consequently give the detectors a different threshold sensitivity.

The threshold sensitivity of the detectors, i.e., the minimum mass particle that can penetrate the cell wall, is according to Frost (1970)

$$m = \frac{l^{2.841}}{k_2^{2.841}\,\rho^{0.473}\,v^{2.486}} \tag{4}$$

where l is the thickness of the detector wall, k_2 is a constant associated with the detector material, ρ is the mass density of the meteoroid, and v is the impact speed of the particle. Equation (4) is for particles that strike the detector normal to the surface. It is assumed that Eq. (4) is valid for impacts at oblique angles when the normal component of velocity is substituted for the impact speed.

The cumulative spatial density for 2×10^{-9}g and 1.6×10^{-8}g meteoroids in interplanetary space at 5 A.U. was calculated with

$$S = \frac{\phi_P}{\bar{v}\eta}\, 10^{\alpha[\log_{10}(\bar{v}/v_m)^{2.486}]} \tag{5}$$

where ϕ_P is the penetration flux measured by the detector, $\bar{v}$ is the average relative speed between meteoroids and spacecraft, η is the ratio of the effective area of the detectors to the actual area, α is the meteoroid mass distribution index, and v_m is the reference speed for which the sensitivity of the detectors was established. This is the simple formula for calculating the spatial density from a flux, $S = \phi_P/\bar{v}$, with two correction factors, namely η to correct for the variation in projected area of the detectors with direction and for the portion of the detectors obstructed by other spacecraft components, and $10^{\alpha[\log_{10}(\bar{v}/v_m)^{2.486}]}$ to normalize the data to the nominal mass sensitivity selected for the detectors. The effective area of the detectors is defined as the projected area in the direction of the meteoroid flux not obstructed by other parts of the spacecraft. Values of effective area are given by Humes et al. (1974) as a function of the meteoroid approach angle. Assuming that the mass density is 0.5 g cm^{-3} and that the most probable impact angle is $26° \pm 5°$, the value of v_m is 15.7 km sec^{-1} when Eq. (5) is being used to calculate the spatial density of meteoroids larger than 2×10^{-9}g from the Pioneer 10 data, and 15.0 km sec^{-1} for $> 1.6 \times 10^{-8}$g from Pioneer 11. The mass density of 0.5 g cm^{-3} is suggested by Kessler (1970) as typical of cometary meteoroids. There is no evidence in the Pioneer 10 and 11 data of a significant population of small asteroidal meteoroids (Humes et al. 1974).

To calculate the spatial density with Eq. (5) it is necessary to know the speed and direction with which the meteoroids approach the spacecraft, i.e., the orbits of the meteoroids must be known. Unfortunately, these we do not know and we therefore consider two distributions of meteoroid orbits. The first has all meteoroids in direct circular orbits and the second is that of Southworth and Sekanina (1973) which is based on the observation of radio

TABLE I

Calculated Cumulative Spatial Density
of Meteoroids at 5 A.U.

Assumed Distribution of Meteoroid Orbits	Cumulative Spatial Density, m^{-3}	
	2×10^{-9}g	1.6×10^{-8}g
Direct circular	6.9×10^{-10}	3.4×10^{-10}
Southworth-Sekanina	5.7×10^{-10}	2.8×10^{-10}

meteors, at 1 A.U., while they formulated a model for the orbital distribution throughout the solar system. It is their distribution of meteoroids that cross 5 A.U. that has been used in this analysis. The calculated cumulative spatial density for 2×10^{-9}g and 1.6×10^{-8}g meteoroids at 5 A.U. is given in Table I for the two assumed distributions of orbits.

The spatial density of smaller and larger meteoroids at 5 A.U. is not known. But, if it is assumed that the size distribution at 5 A.U. is the same as that given by Alvarez (1968) for meteoroids near the earth, the cumulative spatial density at 5 A.U. becomes, per m^3,

$$\log_{10}S_c = -17.872 - 1.728 \log_{10}m - 0.0847 (\log_{10}m)^2,$$
$$m < 1.26 \times 10^{-2}\text{g} \quad (6)$$

$$\log_{10}S_c = -17.570 - 1.408 \log_{10}m,$$
$$m > 1.26 \times 10^{-2}\text{g} \quad (7)$$

or

$$\log_{10}S_{ss} = -17.787 - 1.728 \log_{10}m - 0.0847 (\log_{10}m)^2,$$
$$m < 1.26 \times 10^{-2}\text{g} \quad (8)$$

$$\log_{10}S_{ss} = -17.485 - 1.408 \log_{10}m,$$
$$m > 1.26 \times 10^{-2}\text{g} \quad (9)$$

where S_c is for direct circular orbits, S_{ss} is for the Southworth-Sekinina distribution, and m is the meteoroid mass in grams.

The logarithmic terms were chosen so that the derivate of the logarithm of spatial density with respect to the logarithm of mass, which is the mass distribution index, α, was equal to that given by Alvarez (1968) (see Sec. II). The constants were calculated from the Pioneer 10 and 11 data (Table I). The results are shown in Fig. 5.

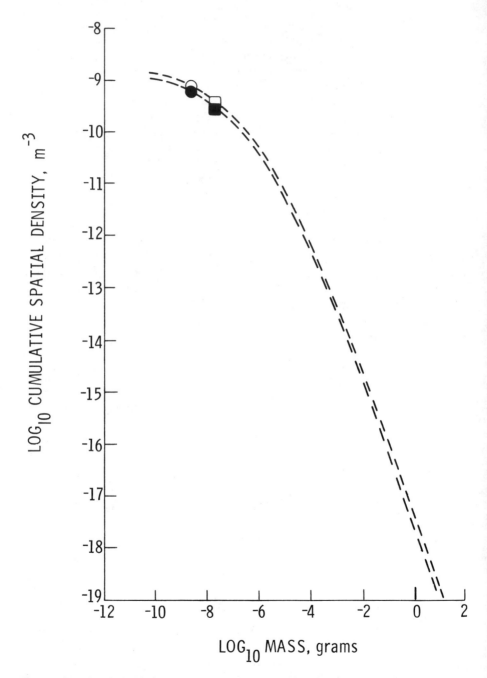

Fig. 5. Calculated cumulative spatial density of meteoroids at 5 A.U. From Pioneer 10 data: the ○ assume circular orbits, the ● assume Southworth-Sekanina orbit distribution. From Pioneer 11 data: the □ assume circular orbits, the ■ assume Southworth-Sekanina orbit distribution.

IV. GRAVITATIONAL FOCUSING

When a stream of meteoroids approaches a planet, they are deflected from their solar orbits and focused towards the planet. The resulting flux was predicted by Öpik (1951) to be

$$\phi = S\, u \left(1 + \frac{v_e^2}{u^2}\right) \tag{10}$$

where S is the spatial density, u is the unaccelerated speed of the meteoroids relative to the planet, and v_e is the escape velocity from the planet. The flux predicted by Eq. (10) is for any arbitrary surface, not necessarily a solid surface, and the flux at any distance from the planet is found from the escape velocity at that distance.

The penetration flux on a spacecraft as a function of distance r from the planet is

$$\phi_P(r) = S_P(r)\, u \left(1 + \frac{v_e^2(r)}{u^2}\right) \eta(r)\, f \tag{11}$$

where $S_P(r)$ is the cumulative spatial density in interplanetary space of meteoroids with a mass greater than the threshold mass of the detector — and it is a function of the distance from the planet because the threshold mass depends on the impact speed; u is the unaccelerated speed of the meteoroids relative to the planet, $v_e(r)$ is the escape velocity, $\eta(r)$ is the ratio of the effective-to-actual areas of the spacecraft component being considered, and f converts the flux on the planet to that on a moving spacecraft. Furthermore, f is assumed to be equal to $[v_e^2(r) + v_{sc}^2(r)]^{\frac{1}{2}} / v_e(r)$ where $v_{sc}(r)$ is the velocity of the spacecraft relative to the planet.

The theoretical penetration flux for Pioneer 10 during the encounter with Jupiter was calculated using Eq. (11) and the results are shown in Fig. 6. The impact speed of the meteoroids on the spacecraft was assumed to be $[v_e^2(r) + v_{sc}^2(r)]^{\frac{1}{2}}$. Two values of u were considered, namely 0.6 km sec^{-1}, which is the average relative speed between Jupiter and meteoroids that are in direct circular orbits, and 2.8 km sec^{-1} which is the average weighted relative speed with the Southworth-Sekanina orbital distribution. The penetration flux measured on Pioneer 10 (Humes et al. 1974) is shown for comparison. The theory and measurements are in agreement within the accuracy of the measurements.

The Pioneer 11 data obtained during the encounter with Jupiter are also consistent with gravitational focusing theory. The penetration flux has not been calculated because there were, at most, only two meteoroid penetrations. The approach leg was similar to that of Pioneer 10, in that the spacecraft was travelling in the direction toward which the meteoroid detectors were pointed. If the meteoroid flux around Jupiter is omnidirectional, the Pioneer 11 detectors would be expected to receive about half the number of Pioneer 10 penetrations during the approach to periapsis because of the size

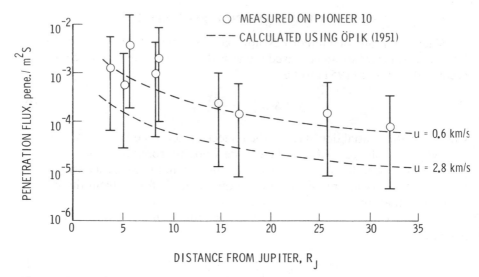

Fig. 6.   Penetration flux for 25 μm stainless steel near Jupiter. The uncertainty in the flux measured on Pioneer 10 is shown by 90%-confidence limits.

distribution of the meteoroids. On the other hand, during the receding leg of Pioneer 11, the spacecraft was traveling in the opposite direction from which the meteoroid detectors were pointed. In this case, the effective exposed area of the experiment is very small and no meteoroid penetrations would be expected.

It is concluded, therefore, that the increase in the penetration fluxes observed when Pioneers 10 and 11 passed near Jupiter was the result of gravitational focusing of meteoroids having solar orbits. There is no reason to believe that there is a large population of meteoroids in orbits around Jupiter.

V.   JOVIAN METEOROID ENVIRONMENT

The gravitational focusing theory can be used to calculate the flux of meteoroids into the atmosphere of Jupiter. Assuming that meteoroids at 5 A.U. are in circular orbits so that u is 0.6 km sec^{-1} and using 60 km sec^{-1} for the escape velocity from Jupiter, the flux of meteoroids on Jupiter is

$$\phi(m) = S(m) \, (6 \times 10^6) \tag{12}$$

where $\phi(m)$ is the cumulative flux of meteoroids of mass m and greater in m^{-2} sec^{-1}, and $S(m)$ is the cumulative spatial density of meteoroids of mass m and greater in interplanetary space at 5 A.U. in m^{-3}. Accepting the size distribution of Alvarez discussed in Sec. II, the flux of meteoroids on Jupiter is, in m^{-2} sec^{-1},

$$\phi(m) = 10^{[-11.009 - 1.728 \, \log_{10} m - 0.0847 \, (\log_{10} m)^2]}, \quad m < 1.26 \times 10^{-2} \text{g} \tag{13}$$

$$\phi(m) = 10^{[-10.707 - 1.408 \log_{10} m]}, \qquad\qquad m > 1.26 \times 10^{-2} \text{g} \quad (14)$$

where m is the mass of the meteoroids in grams. The results are plotted in Fig. 7.

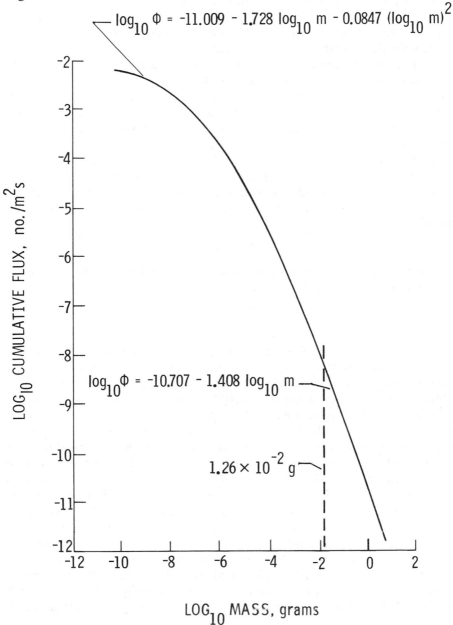

Fig. 7.   Calculated cumulative flux of meteoroids into the Jovian atmosphere.

The total mass influx from the meteoroid environment described by Eqs. (13) and (14) is 3.38×10^{-13} g cm^{-2} sec^{-1}. This leads to a mass accretion on Jupiter of 2.16×10^8 g sec^{-1}. A similar calculation for the total mass influx on Earth, based on Pioneer 10 and 11 data obtained near 1 A.U., showed that the influx on Jupiter is 170 times the influx on Earth. The Pioneer 10 and 11 data obtained near 1 A.U. are consistent with the Explorer 23 meteoroid data obtained in orbit around the earth when meteoroids are assumed to have the orbital distributions of Southworth and Sekanina (1973).

The kinetic energy is absorbed by the Jovian atmosphere, most of it being converted to heat while a small amount is converted to light while creating Jovian meteors. The energy flux, assuming that all meteoroids enter the Jovian atmosphere at the escape velocity of 60 km sec^{-1}, is 6 ergs cm^{-2} sec^{-1}. This is, of course, a small amount compared to Jupiter's excess radiation of approximately 5×10^4 ergs cm^{-2} sec^{-1} (Ingersoll et al.).[1]

The depth to which the meteoroids penetrate was calculated using the equations of meteor physics (Lebedinec and Šuškova 1968) and a model of the Jovian atmosphere. The density of the Jovian atmosphere was assumed to be, in g cm^{-3},

$$\rho_a = (3.25 \times 10^{-4})\, e^{-0.0535\,H} \tag{15}$$

where H is the altitude above a reference surface in km. The pressure in the atmosphere was assumed to be, in atmospheres,

$$P_a = 2830\, \rho_a^{0.968}. \tag{16}$$

The reference surface is where the pressure is equal to one Earth atmosphere. This is near the altitude at which the visible clouds are observed on Jupiter. Equations (15) and (16) are empirical fits to the nominal Jovian atmosphere presented by Divine (1971). The calculation of the heating and ablation and the subsequent destruction of the meteoroid was performed with the numerical technique used by Humes and Bess (1972) to study Martian meteors. The depth to which meteoroids can penetrate the Jovian atmosphere is shown in Fig. 8 where the atmospheric density at the altitude at which the meteoroids expire is plotted as a function of the initial meteoroid mass. It can be seen that the residue of the meteoroids, i.e., the dust and gases, is deposited high in the Jovian atmosphere far above the visible clouds.

Jupiter's large gravitational field and the composition of its atmosphere tend to make Jovian meteors (the optical phenomena produced when meteoroids ablate and decelerate in the atmosphere) different from those on Earth. Meteoroids enter the Jovian atmosphere at 60 km sec^{-1} compared to about 25 km sec^{-1} on Earth, which tends to make Jovian meteors brighter. However, the Jovian atmosphere is composed primarily of hydrogen which is much lighter than the abundant constituents of the earth's atmosphere,

---

[1]See p. 202.

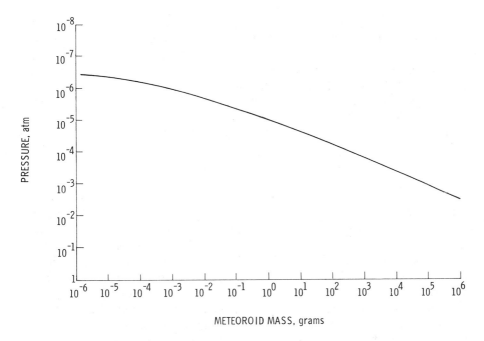

Fig. 8. Calculated pressure at the altitude in the Jovian atmosphere where incoming meteoroids expire.

and this tends to make Jovian meteors dimmer than those on Earth. The net effect of these two conditions was calculated using the meteor simulation technique previously mentioned. It was assumed that the brightness of a meteor is proportional to the rate of loss of the kinetic energy of the ablated atoms and that the proportionality factor is a linear function of the speed (Jacchia 1948), so that

$$I \propto \frac{dm}{dt} v^3 \qquad (17)$$

where I is the luminous power of a meteoroid which is traveling at speed v and ablating at the rate dm/dt. It was further assumed that the proportionality factor is independent of the composition of the atmosphere through which the meteoroid is passing. The meteors visible to the unaided eye in the earth's atmosphere are caused by meteoroids with masses $\gtrsim 10^{-3}$g. The calculations showed that visible meteors will be produced in the Jovian atmosphere by meteoroids with $\frac{1}{15}$ the mass of those in the earth's atmosphere, assuming they are observed at the same range. At Jupiter, the number of meteoroids which produce visible meteors is therefore 34 times the number of those with masses in excess of 10^{-3}g, because the mass distribution index is -1.3 for the mass range around 10^{-3}g. Thus there are 5800 times as many visible meteors in the Jovian atmosphere as for the earth, due to the combined effects of a higher influx of meteoroids and the greater entry speeds.

On clear nights on the earth, meteors visible to the unaided eye occur, on the average, at 10 minute intervals (McKinley 1961). The range of the meteors is about 100 km from the observer. In the Jovian atmosphere visible meteors are calculated to occur at about 0.1 sec intervals, when the range is 100 km. This suggests that a spacecraft observing the dark side of Jupiter would be an excellent laboratory from which to study meteors.

DISCUSSION

A. F. Cook: Figure 6 exhibits data points which reflect single impacts and intervals of time associated with them by splitting the intervals to the preceding and following impacts. Accordingly, the 90% confidence limits are themselves poorly defined and Öpik's value of the dispersion in orbital velocities of meteoroids fits the observations. It is to be preferred since it corresponds to an origin from comets in low-inclination direct orbits of moderate eccentricity near and between the orbits of Jupiter and Saturn.

The sensitivity of the television systems of the Mariner Jupiter Saturn 1977 spacecraft is such that only very bright fireballs will be seen on the dark side of Jupiter. I have used the near Earth number versus mass distribution function to predict that fireballs should be seen at a reasonable rate. Observation of these fireballs should give us another point on the number versus mass curve and even failure to observe any would set an important upper limit.

REFERENCES

Alvarez, J. M. 1968. Preliminary meteoroid-penetration model. *The Explorer XXIII micrometeoroid satellite.* (R. L. O'Neal, ed.) NASA TN D-4284, pp. 85–89.
Divine, N. 1971. *The Planet Jupiter (1970).* NASA SP-8069. Washington, D.C.; Government Printing Office, pp. 60–63.
Frost, V. C. 1970. *Protection against meteoroids.* NASA SP-8042. Washington, D.C.; Government Printing Office, p. 30.
Hastings, E. C. ed. 1964. *The Explorer 16 micrometeoroid satellite, suppl. 3, preliminary results for the period May 27, 1963, through July 22, 1963.* NASA TM X-949, p. 7.
Humes, D. H.; Alvarez, J. M.; Kinard, W. H.; and O'Neal, R. L. 1975. Pioneer 11 meteoroid detection experiment: preliminary results. *Science* 188:473–474.
Humes, D. H.; Alvarez, J. M.; O'Neal, R. L.; and Kinard, W. H. 1974. The interplanetary and near-Jupiter meteoroid environments. *J. Geophys. Res.* 79:3677–3684.
Humes, D. H., and Bess, T. D. 1972. *Meteoroid impacts on Mars and the secondary particle environment.* NASA TN-6951, pp. 6–9.
Jacchia, L. G. 1948. *Ballistics of the upper atmosphere.* Harvard College Observatory and Center of Analysis of Massachusetts Institute of Technology, Technical Report Number Two, p. 9.
Kessler, D. J. 1970. *Meteoroid environment model—1970 (interplanetary and planetary).* NASA SP-8038. Washington, D.C.; Government Printing Office, p. 34.

Kinard, W. H.; O'Neal, R. L.; Alvarez, J. M.; and Humes, D. H. 1974. Interplanetary and near-Jupiter meteoroid environments: preliminary results from the meteoroid detection experiment. *Science* 183:321–322.

Lebedinec, V. N., and Šuškova, V. B. 1968. Evaporation and deceleration of small meteoroids. *Physics and dynamics of meteors*. (Ľ. Kresák and P. M. Millman, eds.) New York: Springer-Verlag, pp. 193–204.

McKinley, D. W. R. 1961. *Meteor science and engineering*. New York: McGraw-Hill Book Company, Inc., p. 114.

O'Neal, R. L., ed. 1968. *The Explorer 23 micrometeoroid satellite — description and results for the period November 6, 1964, through November 5, 1965*. NASA TN D-4284, p. 15.

O'Neal, R. L., ed. 1974. *Description of the meteoroid detection experiment flown on the Pioneer 10 and 11 Jupiter flyby missions*. NASA TN D-7691, pp. 13–78.

Öpik, E. J. 1951. Collision probabilities with the planets and the distribution of interplanetary matter. *Proc. Roy. Irish Acad.* 54:165–199.

Southworth, R. B., and Sekanina, Z. 1973. *Physical and dynamical studies of meteors*. NASA CR-2316. Washington, D.C.: Government Printing Office, pp. 15–35.

PIONEER 10 AND 11 ULTRAVIOLET PHOTOMETER OBSERVATIONS OF THE JOVIAN SATELLITES

D. L. JUDGE, R. W. CARLSON, F. M. WU and U. G. HARTMANN
University of Southern California

The Pioneer 10 and 11 Jupiter probes have provided several opportunities for observation of the Jovian satellites J I through J V. From these data a tenuous atmosphere of hydrogen for Io has been identified and, in the present work, an additional short-wavelength ($\lambda < 800$ Å) emission associated with an extended cloud centered on Io is reported and interpreted as arising from the radiative decay of excited atomic ions. Characteristic x-rays produced by electron bombardment of Io's surface may also contribute to the signal. Emission features associated with Amalthea (J V) and Europa (J II) are also observed. Signals apparently associated with J V occur in the long-wavelength channel while emissions were observed in the short-wavelength channel during J II observations. The data of the long-wavelength channel are interpreted as arising from atomic hydrogen Lyman-α emission in all cases. The source species for the short-wavelength emissions cannot at this time be unambiguously determined, but the wavelength range of the signals is well established.

The present observations of the Jovian satellites, J I through J V, have proved to be extremely interesting, providing a number of new and unexpected results. At the time the Pioneer 10 and 11 missions were planned, none of the satellites of Jupiter were known to have an atmosphere, either bound or unbound. The earth-based observation of a sodium cloud associated with Io, and that of a possible tenuous atmosphere at Ganymede have both occurred since the Pioneer 10 launch in 1972.[1] [For a review of earlier work the reader is referred to articles by Newburn and Gulkis (1973), Morrison and Cruikshank (1974), and Carlson (1975).] The sodium observation has been of particular interest and has given rise to intensive efforts to understand the complex interaction of this satellite with the Jovian magnetosphere. Io

---

[1]See pp. 1103 and 1004.

is relatively unique among the thirteen known Jovian satellites, with several observations to set it apart from its companions: (a) the decametric radio modulation, (b) the now known Na and H clouds, (c) the strong variation in color from pole to equator, (d) Io's significant sweeping and injection of charged particles from and into Jupiter's magnetosphere, and finally, (e) the most recent observation reported here of extended, short-wavelength emission with $\lambda < 800$ Å. Some of the possible interpretations of this observation will be discussed in later sections.

In addition to the sodium and hydrogen observations at Io, and the extended cloud of these gases in orbit about Jupiter, there is further evidence that ultraviolet emissions are associated with another Galilean satellite, Europa (J II), and with the innermost satellite, Amalthea (J V). Prior to the groundbased observations of sodium at Io, the only suggested Jovian satellite atmospheres were those of Ganymede and Io, the former tentatively identified by stellar occultation measurements while post-eclipse brightening suggested an atmosphere on the latter. Since stellar occultations are rare and only give limited information about the actual atmospheric constituents, direct photometric observation can be highly rewarding.

The wavelength region investigated in the present study has been limited to observations shortward of ~ 1400 Å, a region for which the earth's atmosphere is opaque. The ultraviolet photometer which obtained the data to be discussed covered two broad spectral regions shown in Fig. 1. The short-wavelength band is sensitive only to emissions shortward of ~ 800 Å, and includes the 584 Å resonance line of helium. The region of sensitivity of the long-wavelength channel includes the hydrogen resonance line at 1216 Å. Although detection of emissions other than those from neutral hydrogen and helium are certainly possible, the instrument design was based on the expected dominance of these two gases in the Jovian and interplanetary glow. The broadband nature of the instrument response, however, requires careful interpretation of emissions from less well-characterized sources, such as the Jovian satellites. In fact, as discussed in subsequent sections, the demands of self-consistency require that signals observed in the "helium" or short-wavelength channel be ascribed to a source other than helium atoms for the present satellite observations.

Photometer data presented in the following sections are left in terms of count rates, unless a reasonable estimate of source geometry could be determined. Conversion to an equivalent apparent brightness requires detailed information about the spatial extent of the source being viewed, as well as convolution with the angular response of the instrument and the viewing geometry appropriate to the particular observation.

The next two sections present the details of the encounter viewing geometry for the photometer and the equations necessary to convert count rates to brightness. Subsequent and main sections will discuss the Io observations in each of the two channels, followed by their respective interpreta-

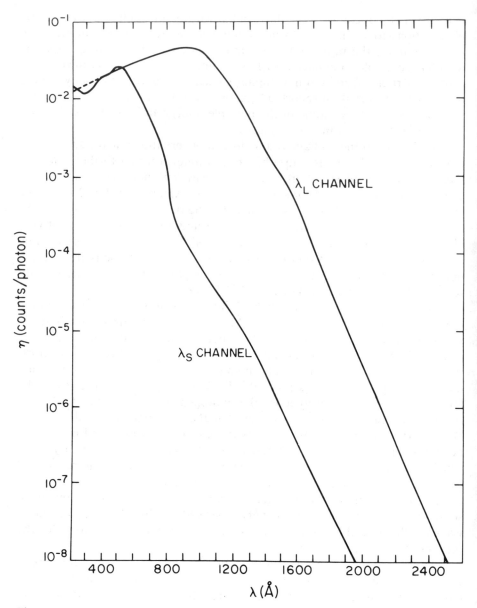

Fig. 1. Spectral sensitivity of the two-channel ultraviolet photometer of the University of
Southern California on board both Pioneers 10 and 11.

tions. Next, the Amalthea and Europa observations will be presented with a
discussion of plausible source mechanisms for these heretofore unreported
emissions. Finally, two short sections on Ganymede and Callisto complete
the satellite observational survey.

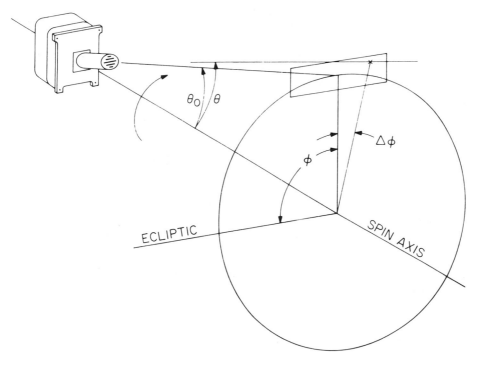

Fig. 2.   The optical axis of the photometer is at an angle $\theta_0 \sim 20°$ to the spin axis of the space-craft. The angle ϕ is the clock angle relative to the ecliptic. The instantaneous field of view is nominally $1° \times 10°$.

I. VIEWING GEOMETRY

The field of view of the photometer is limited by a mechanical collimator whose optical axis is oriented at an angle $\theta_0 = 20°$ to the spacecraft spin axis (Fig. 2). The scanning motion that results from spacecraft rotation sweeps the field of view over the surface of a $40°$ cone having its vertex at the space-craft and oriented along the spin axis. During both Pioneer 10 and 11 en-counters the spin axis was parallel to the spacecraft-Earth line and nearly in the plane of the Jovian satellite orbits; the curve traced by the intersection of the instrument's line-of-sight with the orbital plane of the satellites is a hyperbolic section as sketched in Fig. 3. As the photometer clock angle ϕ in-creases from zero through $90°$ to $180°$, the projection of the field of view onto the plane of the satellite orbits follows the hyperbola from one asymptote through its vertex to the other asymptote. At the time of the observations, prior to periapsis, the spacecraft was below the orbital plane and all satellite emission features therefore occur with clock angles $0 < \phi < 180°$. Viewed from Jupiter's north pole, looking down onto the plane of the satellites, Figs.

USC/XUV VIEWING GEOMETRY

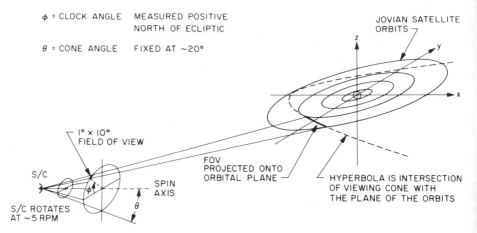

φ = CLOCK ANGLE   MEASURED POSITIVE
                  NORTH OF ECLIPTIC

θ = CONE ANGLE   FIXED AT ~20°

JOVIAN SATELLITE ORBITS

1° × 10° FIELD OF VIEW

S/C

SPIN AXIS

FOV PROJECTED ONTO ORBITAL PLANE

HYPERBOLA IS INTERSECTION OF VIEWING CONE WITH THE PLANE OF THE ORBITS

S/C ROTATES AT ~5 RPM

Fig. 3.   The field of view projected onto the satellite orbital plane traces a hyperbola as the spacecraft rotates from $\phi = 0°$ to $\phi = 180°$.

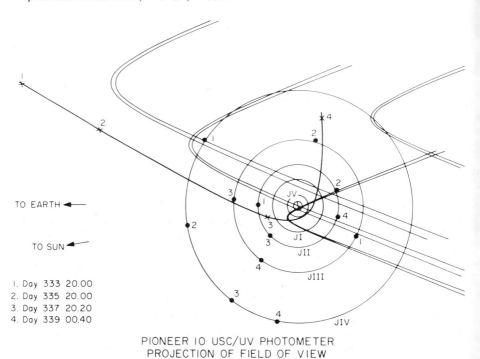

TO EARTH ◄———

TO SUN ◄———

1. Day 333 20.00
2. Day 335 20.00
3. Day 337 20.20
4. Day 339 00.40

PIONEER 10 USC/UV PHOTOMETER
PROJECTION OF FIELD OF VIEW
(IN THE PLANE OF THE SATELLITE ORBITS)

Fig. 4a.   Viewing hyperbolas in the satellite plane as seen from above Jupiter's north pole for Pioneer 10. The width of the hyperbolas represents the 1° cone angle width over which the photometer is sensitive. Four different spacecraft positions are shown and the satellite orbits are labeled by each satellite's position at that time.

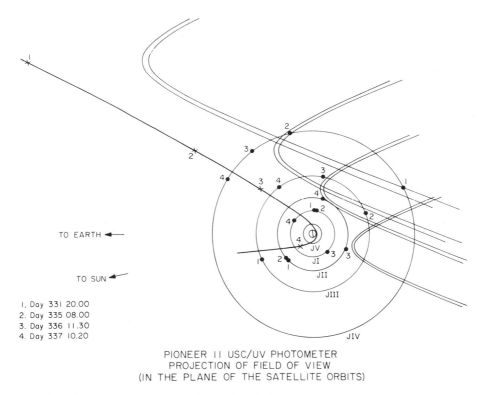

TO EARTH ◄───

TO SUN ◄──

1. Day 331 20.00
2. Day 335 08.00
3. Day 336 11.30
4. Day 337 10.20

PIONEER 11 USC/UV PHOTOMETER
PROJECTION OF FIELD OF VIEW
(IN THE PLANE OF THE SATELLITE ORBITS)

Fig. 4b.   The same sequence as for Fig. 4a but for Pioneer 11.

4a and 4b show a computer drawn progression of encounter viewing over several days for Pioneers 10 and 11, respectively. The width of the hyperbolas as shown reflects the 1° cone angle over which the photometer is sensitive.

Satellite viewing opportunities are indicated on the time axes of the data plots (Figs. 5, 6 and 7) and inset figures are included to show the orbits and positions of the satellites in relation to the field of view at that time, as well as the shadow of Jupiter. Fiducial marks along the orbits represent one-hour intervals, and the horizontal line points toward Earth. The section of the field of view shown in each case is centered near $\phi \sim 15°$ with $\Delta\phi \sim \pm 5°$.

II.   RELATION OF COUNT RATE
TO APPARENT SURFACE BRIGHTNESS

The rate at which photons are received by the photometer due to a small source element with apparent surface brightness $4\pi I$ photons sec^{-1} cm^{-2} and projected area Δa is

$$S = \frac{A}{4\pi r^2} \times 4\pi I \times \Delta a \text{ photons sec}^{-1} \qquad (1)$$

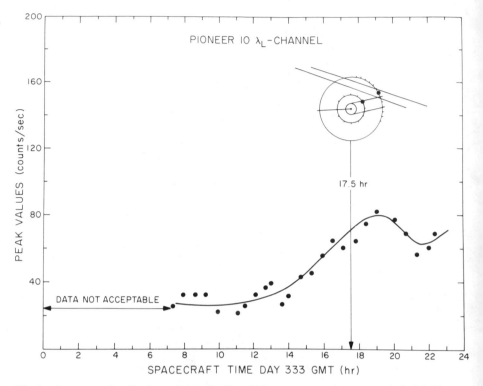

Fig. 5.   Long-wavelength channel data for Day 333 was averaged over a period of 0.5 hr to obtain each point shown. The inset shows the positions of Amalthea and Io at 17^h 50^m when Io is centered in the field of view with clock angle $\phi \sim 10°$.

where r is the distance between the spacecraft and the source of emission: A is the effective area of the acceptance aperture, and is a function of source position (θ, ϕ) (Carlson and Judge 1974). For a source at the center of the field of view (θ_0, ϕ_0), A is just the geometrical area of the aperture which is 1.66 cm².

The response of the instrument to a finite extended source, such as the disc of Jupiter or an Io cloud is obtained by integrating $A(\theta, \phi)$ over the field of the object. For a source of strength $4\pi I(\theta, \phi)$ extending from θ_1 to θ_2 in cone angle and ϕ_1 to ϕ_2 in clock angle, the photon rate in photons per sec for the photometer at (θ_0, ϕ_0) is

$$S = \frac{\sin \theta_0}{4\pi} \int_{\theta_1}^{\theta_2} d\theta \int_{\phi_1}^{\phi_2} d\phi \, 4\pi I(\theta, \phi) \, A(\theta, \phi - \phi_0). \qquad (2)$$

Here, we have used the fact that $da = r^2 \sin \theta_0 \, d\theta \, d\phi$. The apparent surface brightness of a source with uniform emission, in photons sec⁻¹ cm⁻², can then be written as

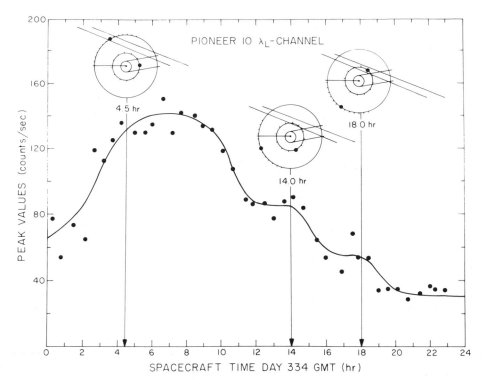

Fig. 6.   Long-wavelength channel data for Day 334 was averaged over a period of 0.5 hr to obtain each point shown. The insets show the positions of Amalthea and Io for the times indicated. The structure at $14^h\,00^m$ and $18^h\,00^m$ is interpreted as emissions associated with Amalthea. The uncertainty in the data may be more than the point scatter would indicate.

$$4\pi I \simeq S / \left[\frac{\sin \theta_0}{4\pi} \int_{\theta_1}^{\theta_2} A(\theta, 0) \times (\phi_2 - \phi_1)\, d\theta \right].$$ (3)

Here we have neglected the variation of $A(\theta, \phi - \phi_0)$ in ϕ. This is justified since we know $\Delta\phi\ (= \phi_2 - \phi_1)$ for the cases considered is always much smaller than the photometer resolution in clock angle ($\sim 25°$). For example, for a typical Pioneer 10 observation a source with a vertical dimension of $1\ R_J$ (perpendicular to the orbital plane) will subtend only $\Delta\phi \sim 2°$ at Day 333 and $\Delta\phi \sim 2.5°$ at Day 334; thus, using $A(\theta, 0)$ rather than $A(\theta, \phi)$ will introduce little error for most sources. For an extended source centered in the field of view, with $\theta_2 - \theta_1$ larger than $1°$, such as the Io cloud, the above formula can be further approximated by

$$4\pi I \approx 6.3 \times 10^4 \times S / \Delta\phi \text{ photons sec}^{-1}\text{ cm}^{-2}.$$ (4)

Here, we have evaluated the integral in Eq. (3) and used $\theta_0 = 20°$; $\Delta\phi$ is ex-

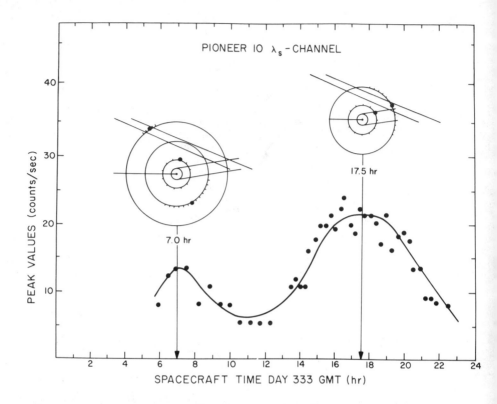

Fig. 7.   Short-wavelength channel data for Day 333 was averaged over a period of 1 hr to obtain each point shown. The inset at $17^h\ 50^m$ shows the position of Amalthea and Io at that time. The inset at $7^h\ 00^m$ includes the position of Europa as well. The small peak at $7^h\ 00^m$ could be characteristic x-rays from Europa's disc, whereas the signal at $17^h\ 50^m$ is interpreted as due to radiative decay of excited atomic ions centered on Io.

pressed in degrees. (To obtain $4\pi I$ in Rayleighs, divide by 10^6.) For very small sources, such as the satellite discs, angular variations can be neglected altogether and the conversion of S to brightness becomes a function of area ratios:

$$4\pi I = \frac{4S}{1.66}\left(\frac{r}{R}\right)^2 \text{ photons sec}^{-1}\text{ cm}^{-2}. \qquad (5)$$

Here R is the radius of the disc, and r is the distance from the spacecraft to the satellite.

Finally, we note that the photon flux S is related to the count rate C of the instrument by

$$S = C/\eta_\lambda \qquad (6)$$

where η_λ is the spectral response of the photometer at wavelength λ. The value of η_λ as a function of λ for both the long-wavelength and short-wavelength channels is shown in Fig. 1. In particular, the sensitivity of the short-wavelength channel at $\lambda = 584$ Å is $1/59$ counts per photon and that of the long-wavelength channel at $\lambda = 1216$ Å is $1/88$ counts per photon.

III. LONG-WAVELENGTH CHANNEL OBSERVATIONS OF IO

The long-wavelength channel data from Pioneer 10 (Fig. 6) have been discussed at length previously (Carlson and Judge 1974, 1975; Judge and Carlson 1974). These data have been interpreted as resonantly scattered Lyman-α from atomic hydrogen in orbit about Jupiter, Io being the "source" of the particles. The current extended analysis of the data confirms the earlier results and, in addition, suggests that temporal variations in the cloud density cannot be ruled out from the data. The viewing opportunities for Io and the measured peak count-rates are summarized in Table I.

The long-wavelength channel (λ_L) data as a function of time for Days 333 and 334, 1973, with the viewing geometry, are given in Figs. 5 and 6. The data were averaged for either 30 or 60 minutes, depending on data quality, throughout the observation period. Since the spacecraft was spinning with a period of about 12 seconds it was possible to extract signals buried in the high-energy particle background noise by utilizing *a priori* knowledge of the location of Io. For each spin, at the appropriate clock angle (rotation position), a feature is expected corresponding to emission from Io. Thus, by both temporal and spatial filtering, enhancement of the signal-to-noise ratio was achieved. Recognition of a true emission peak was based on both its width in clock angle (minimum FWHM of 25° for a point source), and its continued presence at the appropriate clock angle for a time at least as long as that during which the satellite itself occupied the field of view.

The data of Day 333 clearly show that emissions are detected at least 3 hours before Io enters the field of view, that the peak emission occurs at 19^h 00^m, ~ 1.5 hr after Io is centered in the field of view, and that signals are present thereafter through the end of the day. The dip following the peak emission corresponds to a time when the trailing portion of the cloud being viewed is shadowed by Jupiter and from these data both the excitation mechanism (resonant scattering of solar Lyman-α) and the spatial extent of the cloud have been inferred (Carlson and Judge 1974). Because of the effect of Jupiter's shadow, neither the true peak counting rate nor the exact extent of the cloud asymmetry can be unequivocally determined from the data of Day 333. For the Day 334 viewing opportunity, however, the situation is not complicated by shadow effects and the peak counting rate of ~ 140 counts sec^{-1} occurs ~ 2.5 hr after Io is centered in the field of view and emissions attributable to Io continue for perhaps 12 hr thereafter. (The "lumps" at 14^h and 18^h are possibly due to Amalthea.) For an orbital rate of $\sim 8°5$ per hour, the peak emission trails Io by $\sim 20°$ and the total extent of the cloud is esti-

TABLE I

Viewing Opportunities of Io (JI) (Field of View Centered on JI)

GMT (Day/hr)	S/C-satellite distance (R_J)	Comments	λ_L-channel signal[a] Peak (counts/sec)	λ_L-channel signal[a] Comments[b]	λ_S-channel signal[a] Peak (counts/sec)	λ_S-channel signal[a] Comments[b]
Pioneer 10						
333/17.5	75	Part of Io cloud in Jupiter's shadow	65	—	20	—
334/4.5	61	—	130	—	<20	limited by instrument resolution
335/7.0	56	Jupiter in the field of view; Io behind Jupiter	<350	limited by instrument resolution	11	possible Jupiter emission
336/4.5	32	Jupiter starts to leave the field of view	172	part of the signal is due to Jupiter emission	<70	limited by instrument resolution
336/21.0	28		No peak	extreme background variation	<80	limited by instrument resolution
337/12.0	18		<200	limited by instrument resolution	<250	limited by instrument resolution

[a] The averaging time is 0.5 hr for the λ_L-channel and 1 hr for the λ_S-channel, except on Day 336, for which the λ_S-channel data were averaged for 0.5 hr.

[b] Resolution of the instrument = background/64.

mated to be $\sim 120°$. The peak counting rate of Day 333 is only about one half or less of that observed on Day 334 despite the fact that the distance from the spacecraft to Io changed only from ~ 75 to ~ 61 R_J. Even though the effect of the shadow cannot be accurately assessed without more information on cloud size and shape than can be inferred from present data, the relative counting rates for the two days cannot rule out changes in the cloud parameters of perhaps 20% over this time interval. These changes could be either true temporal variations, a coupling of the emissions to the orbital phase of Io, or coupling to the magnetic field of Jupiter.

IV. INTERPRETATION OF THE λ_L OBSERVATIONS OF IO

The emission data of the long-wavelength channel previously reported have been interpreted as coming from an atomic hydrogen cloud in orbit about Jupiter with the following characteristics (Carlson and Judge 1974, 1975):

column density	3×10^{12} cm^{-2}.
cloud angular extent	$\sim 1/3$ of Io's orbit in the orbital plane.
cloud thickness	~ 1 R_J (assumed).
emission intensity	300 Rayleighs of Lyman-α.
total number of hydrogen atoms	$\sim 2 \times 10^{33}$ atoms.
loss process	charge exchange with magnetospheric protons.
possible production processes	(1) proton flux neutralized at the surface or in a tenuous atmosphere (McDonough 1975).
	(2) photodissociation of atmospheric NH$_3$ (McElroy and Yung 1975).
	(3) sputtering of hydrogen-bearing compounds at the surface of Io.
	(4) electron impact dissociation of a hydrogen-bearing molecular atmosphere (Johnson, *et al.* 1976).
cloud temperature	$\sim 3000°$K (may not be in thermal equilibrium).

The present improved data are in good agreement with the earlier results. However, the large signal, and the apparently increased asymmetry evident on Day 334 relative to Day 333 (possibly a shadow effect), suggest that the high temperature of the cloud, inferred from its asymmetry, may not be required in order to explain the data. If a sporadic source were invoked, a much lower average atmospheric temperature could be accommodated, in agreement with the Pioneer 10 radio occultation data which suggest a temperature of 400°K (although one must hasten to note that the derived ionospheric temperatures are model dependent). Since it is known that the Jovian magnetosphere (at least the outer regions) is highly dynamic and that the magnetospheric particles interact strongly with Io, as well as with several other inner satellites, a time-dependent source, and hence atmosphere, would be consistent with other Pioneer observations. On the other hand, if the processes of the hydrogen and sodium atom source are related, the temporal constancy of the sodium cloud (Bergstralh *et al.* 1975) would suggest a constant hydrogen source.

The extent to which a current system of the type suggested by Shawhan *et al.* (1975) and Goldreich and Lynden-Bell (1969) connecting Jupiter and Io could account for either a variable or steady-state source of atmospheric particles is not yet clear. It should be noted, however, that the magnitude of any such current system is at least not always as great as suggested by the above authors. The Pioneer 11 spacecraft flew within 6000 km of the flux tube connecting Jupiter with Io. According to E. J. Smith (personal communication), the helium vector magnetometer did not observe a field change either in magnitude of direction during this time. Taking their quantization of $500\,\gamma$ (high field value) as an upper limit to the field due to a current system of the Io flux tube, it is found that a current as low as 10^5 A would have produced a measurable perturbation. This maximum acceptable current is two orders of magnitude below the value of Shawhan *et al.,* and one order below that of Goldreich and Lynden-Bell.

V.   SHORT-WAVELENGTH CHANNEL OBSERVATIONS OF IO

Obvious but unexpected signals were observed in the short-wavelength channel of the Pioneer 10 photometer on Day 333 beginning at $\sim 14^h\,00^m$, when the field of view crossed into the orbit of Io but before Io itself entered the view. Prior to this time, the signal-to-noise ratio was less than one, the noise after filtering being about 10 counts $\sec^{-1}$ on Day 333. The signals reached a maximum of ~ 20 counts $\sec^{-1}$ as Io itself came into the field of view and continued, but with diminishing value, as Io left the field of view. At $\sim 23^h\,00^m$ the signal count rate decreased to the noise level and no features were observed thereafter.

An example of the data as a function of clock angle is given in Fig. 8, and clearly indicates an emission feature at the 10° clock angle of Io. The extent of the emitting cloud is difficult to estimate. However, only ~ 2 hr are required for Io itself to cross the field of view while the emission feature persists for more than 6 hr (see Fig. 7). It is obvious that the short-wavelength channel emission signals extend beyond the solid surface of Io, and into a region at least 2 to 3 R_J distant from Io (approximately $\pm 20°$).

If one assumes that the vertical dimension of the cloud is 1 R_J, as was adopted for the hydrogen cloud, then the observed count rate of 20 counts

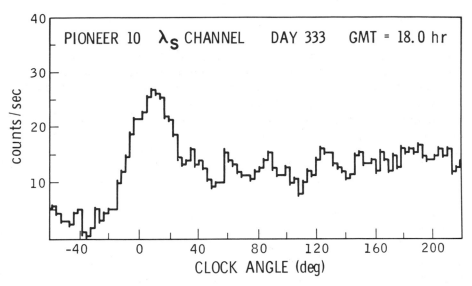

Fig. 8.   The λ_S signal versus clock angle after averaging over a period of 1 hr. The feature at $\phi \sim 10°$ is interpreted as radiative decay of excited atomic ions in an extended region of space centered on Io.

sec^{-1} leads to a surface brightness of ~ 40 Rayleighs. On the other hand, if the emission feature comes mostly from a source the size of Io, a brightness of ~ 25 kiloRayleighs (kR) is required using the detection efficiency at 584Å.

In order to limit the possible sources of the emission we note that the short-wavelength channel is sensitive to radiation at wavelengths $\lesssim 800$ Å. In the observation of Jupiter, there is no problem identifying the source of the emission feature in this channel since a consistent model of Jupiter's atmosphere would suggest only He as a significant scattering source in this wavelength range (Carlson and Judge 1974). However, both the constituents and excitation mechanisms appropriate to Io's atmosphere remain somewhat uncertain. Thus, positive identification of the short-wavelength emission is difficult to establish. In the following sections some possible sources for producing the emission features with $\lambda < 800$ Å at Io are discussed.

VI.   INTERPRETATION OF THE λ_S OBSERVATIONS OF IO

A number of species are possible sources for the short-wavelength radiation. The most likely candidates are neutral helium, excited ions, bremsstrahlung and characteristic x-rays. Each is considered as a source of either extended emissions or restricted to the disc of Io, and is discussed below.

A. Extended Cloud Emission

Helium. The first extended short-wavelength emission to be considered is the possibility of helium resonance emission, since the He I resonance line (584 Å) is a prominent feature in the solar ultraviolet spectrum. This solar emission, incident on a helium atmosphere, will be resonantly scattered with an excitation rate (*g*-value)² at Io of 6×10^{-7} sec^{-1} atom^{-1}.

Helium atoms can of course also be excited by particle impact. The excitation rates for particle excitation, accurate to an order of magnitude, are summarized in Table II. The electron and proton fluxes and cross-section data from Carlson *et al.* (1975) are also listed in Table II. From this table we see that the particle impact excitation rates for producing excited helium are less than the excitation rate by resonance scattering. Thus, if it is assumed that resonance scattering dominates the observed emission at Io, the presently observed brightness of 40 R (assuming a source 1 R_J in vertical extent) requires a He column density of $4\pi I/g = 7 \times 10^{13}$ atoms/cm^{-2}. Since helium escapes so rapidly from Io, the production of a helium torus would be inevitable (if Io was a source of helium). We therefore rule out a localized atmosphere around Io (i.e., within a few scale heights) as the source of observed emissions. It would also be impossible for such an atmosphere to produce the observed intensity by resonance scattering, since even if the atmosphere were thick enough to scatter the entire solar line, an intensity of only ~ 250 R would be produced, which is to be compared with 25 kR if from the highly localized atmosphere.

Trafton (1976) has placed an upper limit to the density of metastable 2^3S helium atoms near Io of 4×10^8 cm^{-2}. The relative population of metastable and ground-state helium atoms can be related by examining the mechanisms for production and loss of triplet helium in the near Jupiter environment. The most efficient means of populating the triplet levels is electron excitation, and there does exist thermal plasma at the orbit of Io with sufficient energy (~ 100 eV) for excitation (Frank *et al.* 1976). In the absence of other destruction mechanisms, electron collisions would produce a population of excited He characteristic of the electron temperature. However, other processes can destroy metastable helium; these include photoionization, two-photon decay, and electron ionization. The destruction rates for these processes are respectively: $\sim 6 \times 10^{-5}$, $\sim 2 \times 10^{-5}$ (McElroy 1965), and $\sim 6 \times 10^{-6}$ sec^{-1} (estimated using Gryzinski's cross sections and the plasma measurements of Frank *et al.* 1976). The cross section for producing triplet helium atoms can be estimated from the experiments of St. John *et al.* (1964) and Jobe and St. John (1967). We find a total triplet excitation cross section at 100 eV of $\sim 1.5 \times 10^{-18}$ cm² which gives a production rate of 3×10^{-8} per second per He atom. The ratio of neutral to metastable atoms is then

---

²$g = F_\lambda \lambda^2 (\pi e^2/m_e c^2) f = (8.83 \times 10^{-13}$ cm$) \times (\pi F_\lambda) \lambda^2 f$. For He 584 Å, $\pi F_\lambda = 7.4 \times 10^8$ sec^{-1} cm^{-2} Å^{-1} at the orbit of Jupiter (Maloy *et al.* 1976) and $f = 0.28$.

TABLE II

The Helium Excitation Rate Produced by Particle Impact

Particle	Flux (sec^{-1} cm^{-2})	Cross Section (cm^2)	Excitation Rate (atom^{-1} sec^{-1})	Reference for Cross Section
Electron	$\sim 10^{10}$	$\leqslant 10^{-16}$	$< 10^{-6}$	Altshular (1952)
Proton	$\sim 10^9$	$\leqslant 10^{-17}$	$< 10^{-8}$	Bell (1961)
H atom	$\sim 10^{7}$ [a]	$\leqslant 10^{-18}$	$< 10^{-11}$	Bates and Crothers (1967)

[a]High-energy H flux = column density of H/lifetime of low-energy H $\sim 10^{12}/10^5 =$ 10^7 sec^{-1} cm^{-2}.

$(6 \times 10^{-5} + 2 \times 10^{-5} + 6 \times 10^{-6})/3 \times 10^{-8} = 3 \times 10^3$. This results in an upper limit to helium atoms of $\sim 1.2 \times 10^{12}$ cm^{-2} based on Trafton's (1976) measurements, which argues that the emissions observed by Pioneer 10 are not due to helium.

Consideration of possible production and loss mechanisms for helium can also provide an estimate of the amount of helium in the Io cloud. For the earth, radioactive decay of uranium and thorium and their subsequent decay products provides the atmospheric helium. Such a source sufficient to supply the required column density at Io would be surprising indeed, unless Io is unusually enriched in radioactive elements, and is not considered here.

A second potential source of helium is the neutralization of magnetospheric alpha particles at Io. The total α-particle flux in the Jovian magnetosphere has not been determined to great precision, but order of magnitude limits on both the source flux and subsequent loss of helium atoms are discussed below.

Pioneer 10 data indicate that the ratio of the flux of α-particles to that of protons is a highly variable function of particle energy and distance from Jupiter. Trainor *et al.* (1974) find $6 \times 10^{-4} < J_\alpha/J_p < 6 \times 10^{-5}$ for 3–6 MeV per nucleon particles between 30 R_J and 90 R_J. Simpson *et al.* (1974) find values for the ratio at 1 MeV per nucleon as large as 2×10^{-2} for some distances in the 30–90 R_J range. Giving slightly more weight to the lower-energy data where fluxes are higher, a reasonable upper limit might be $J_\alpha/J_p \lesssim 10^{-3}$. With low-energy proton fluxes possibly as high as 10^9 cm^{-2} sec^{-1} (Frank *et al.* 1976) at the orbit of Io, the estimated upper bound to the α-particle flux would be $J_\alpha \lesssim 10^6$ cm^{-2} sec^{-1}, although it must be recognized that the initial sources of the energetic magnetospheric particles and the thermal plasma could be quite different, exhibiting quite different particle abundances.

If the α-particles swept up by Io's disc are neutralized and leave the satellite as ground-state helium atoms via Jeans escape, their mean expansion velocity $<v> = \sqrt{2kT/3m_\alpha} \simeq 0.46\,\text{km sec}^{-1}$. To explain the observed orbital extent of $\pm\Delta\theta \lesssim 20°$ requires a lifetime $\tau_\alpha \lesssim 1/3R\Delta\theta <v>_\alpha$ which is $\sim 1.1 \times 10^5$ sec (Carlson and Judge 1974) implying a total column density, in the extended cloud, of $n \sim (R_{Io}/R_J)^2 J_\alpha \tau_\alpha \sim 7.4 \times 10^7\,\text{cm}^{-2}$. The observed signal, however, requires a column density of $n \sim 7 \times 10^{13}\,\text{cm}^2$, several orders of magnitude higher than that estimated from an optimistic interpretation of available particle flux data. In addition, to limit a possible helium cloud to within 20° of Io requires a highly efficient loss process. The most probable process is likely to be electron impact ionization which has a maximum cross section of $\sigma \lesssim 4 \times 10^{-17}\,\text{cm}^2$ (Lotz 1967) peaked near 0.1 keV. This would require a low-energy electron flux $J_{el} \sim 1/\tau\sigma \gtrsim 2.3 \times 10^{11}\,\text{cm}^{-2}\,\text{sec}^{-1}$ at 0.1 keV, and is not consistent with the measured value which is estimated to be $< 10^{10}\,\text{cm}^{-2}\,\text{sec}^{-1}$.

Therefore, in addition to Trafton's measurements, there are several theoretical reasons why atomic helium is probably not the source of the signal observed in the short-wavelength channel: (a) the most reasonable estimate of the integral alpha-particle flux at the orbit of Io is insufficient to explain the observed signal intensity, and (b) even if adequate amounts of helium were somehow produced by other mechanisms, there appears to be no loss process that is efficient enough to limit the extent of such a cloud to the observed 40°. Having shown that atomic helium is not a likely source of the observed emissions, we next consider atomic ions since substantial amounts of sodium are known to exist in the vicinity of Io (Trafton *et al.* 1974; Brown 1974).

Sodium and Other Atoms. Another potential source of the observed short-wavelength emission is the Na II resonance lines near 400 Å. These emissions can be produced by both particle ionization-excitation and resonant scattering of solar radiation by the ion. We consider first the loss mechanisms for Na atoms, which produce Na^+ ions in the extended atmosphere, in addition to the excitation processes leading to emission of the Na II resonance lines.

It has been estimated by Macy and Trafton (1975) that the sodium atoms are lost from the extended Io cloud at a rate of $6 \times 10^{-6}\,\text{sec}^{-1}$. It has been argued by Carlson *et al.* (1975) that the primary Na loss mechanism is through electron ionization with a rate of $\sim 5 \times 10^{-6}\,\text{sec}^{-1}$, in good agreement with the observed sodium cloud geometry. However, the charge exchange of sodium with magnetospheric protons also contributes to the loss rate of sodium in the Io cloud. Charge transfer between H^+ and Na was not included by Macy and Trafton (1975) and Carlson *et al.* (1975) because of the large internal energy defect (8.5 eV) between the ground states of hydrogen and sodium, leading to a low cross section at energies of the thermal plasma. But, as pointed out by Grüebler *et al.* (1970), the energy defect between the first excited state $(n = 2)$ of hydrogen and the ground state of Na is only about

1.7 eV, and the condition for near resonance is satisfied. Thus, the contribution of the $n = 2$ state of hydrogen may result in significant charge transfer between H$^+$ and Na. The cross section for this charge transfer is $\sim 10^{-14}$ cm^2 in the energy range from 1 to 10 keV (Grüebler et al. 1970). This is quite large compared to the cross section for electron ionization of sodium, $\sim 4 \times 10^{-16}$ cm^2, for energies from 10 to 100 eV.

To estimate the importance of this charge transfer process we note that the proton flux at Io's orbit in the 0.108–4.8 keV region can, according to Frank et al., be represented by a thermal distribution with a temperature of ~ 100 eV, but with perhaps an additional high-energy tail, with a spectral index of ~ 1.3. The integral flux greater than 1 keV is then found to be $\sim 10^8$ cm^{-2} sec^{-1}, while the mean cross section for charge transfer in the 1–10 keV range is $\sim 7 \times 10^{-15}$ cm^2, and results in a sodium loss rate due to proton impact of $\sim 7 \times 10^{-7}$ sec^{-1}. The relative importance of this process clearly depends on the actual value of the proton flux adopted.

The Na loss process may occur through simultaneous ionization and excitation. The emission rate of Na II lines through such loss is considered below. The excited states of Na$^+$, such as 2p^5(^{2}p$^0_{\frac{3}{2}}$)3s and 2p^5(^{2}p$^0_{\frac{1}{2}}$)3d are about 30 to 40 eV above the 2p^6 ground state. The wavelengths for the transitions from these excited states to the ground state are between 300 and 400 Å, and would thus be detected in the short-wavelength channel. For electrons, the cross section for the ionization excitation process is $\sim 2 \times 10^{-17}$ cm^2, with a maximum at $E = 200$ eV (Omidvar et al. 1972). For a flux density of 10^{10} electrons cm^{-2} sec^{-1}, the excitation rate is then 2×10^{-7} sec^{-1} per Na atom. Using an average Na column density of 10^{10}–10^{11} cm^{-2} in the vicinity of Io leads to an estimated emission rate of 0.002–0.02 R for the Na II lines. This intensity is over two orders of magnitude less than the observed emission. The cross section for ionization excitation of Na through H$^+$ impact has not been measured. However, this ionization excitation cross section can be estimated from the model of Gryzinski (1965) for charge transfer, giving a value $\sim 10^{-17}$ cm^2, with a maximum at 40 keV. Assuming the H$^+$ integral flux for energies greater than 10 keV to be 10^7 cm^{-2} sec^{-1}, an excitation rate of 10^{-10} sec^{-1} results, and is much smaller than that for electron ionization excitation.

Once created in an extended atmosphere, ground state Na II may be excited by electrons or protons, or may resonantly scatter incident solar radiation; these effects are shown below to be negligible.

The electron and proton excitation cross sections of Na$^+$ have not been measured. An upper limit to the electron excitation cross section is found by adopting the ionization cross section of Na II, $\sigma \sim 10^{-17}$ cm^2 at $E \sim 100$ eV (Lotz 1967). An electron flux on the order of 10^{10} cm^{-2} sec^{-1} then yields an excitation rate of $\sim 10^{-7}$ sec^{-1}, which is about the same order of magnitude as that for ionization excitation of sodium. Since the proton ionization cross section only becomes significant at about 1 MeV, the proton ionization excitation rate will be on the order of 10^3 smaller than that for electron excitation ionization. The large difference in rates results from the fact that the proton and electron ionization cross section peaks occur at greatly different energies and hence fluxes. The ratio of electron to proton flux at their peak cross sections yields the factor of 10^3. If we assume proton ionization excitation to have the same efficiency as that for electrons, proton excitation will clearly be negligible.

Next we consider Na II resonance scattering at 372 Å and 300 Å. These lines have not been isolated in the solar spectrum; an upper limit to the flux might be placed at $\sim 10^7 \; \text{sec}^{-1} \; \text{cm}^{-2} \; \text{Å}^{-1}$. For $\lambda \sim 372$ Å, this implies a resonance scattering rate of $\sim 10^{-8} \; \text{sec}^{-1}$, even smaller than that for electron excitation of Na^+. Unless the Na^+ density is significantly greater than neutral Na in the vicinity of Io, these relatively low excitation rates argue against particle and photon impact on Na^+ in producing the observed emissions. The limited spatial extent of the short-wavelength emissions strengthens this argument. Newly created ions will be swept up by the co-rotating Jovian magnetic field and form a cylindrically symmetric distribution since the rotation period is probably much less than the lifetime. Such a distribution was not observed in the emission profile.

The limited orbital extent of the emissions ($\pm 20°$) is consistent with the limited orbital extent of the observed neutral Na emissions (Brown and Yung;[3] Trafton *et al.* 1974), and the strongest emission mechanisms discussed above are associated with the observed neutral sodium cloud. However, it appears that radiation from sodium ions alone cannot account for Io's short-wavelength emission. There exists the possibility that other elements (e.g., K, Ne, N) are present with sodium in the Io cloud. These elements can also contribute to the observed short-wavelength emission through simultaneous ionization and excitation by energetic particles. For example, for atomic nitrogen the ionization cross section of the 2s shell ($\sim 10^{-16} \; \text{cm}^2$ at $E \sim 100$ eV) by electron impact is comparable to the cross section for the 2p shell (Omidvar *et al.* 1972; Lotz 1967). Removal of the 2s electron leaves the ion in the excited 2s $2p^3$ $3S^0$ state, which then decays emitting 645 Å radiation. Using $10^{10} \; \text{cm}^{-2} \; \text{sec}^{-1}$ as the electron flux and $10^{-16} \; \text{cm}^2$ as the cross section for excitation ionization leads to a reaction rate of $10^{-6} \; \text{sec}^{-1}$. Thus a nitrogen abundance of 10^{13} atom cm^{-2} is required to account for the λ_S-channel intensity measurement of 10 Rayleighs.

B.   *Surface Emission*

Although it is apparent that the λ_S-channel emissions appear to come from an extended region around Io, the total peak signal may contain a substantial contribution from the disc of Io itself. We consider below three

---

[3]See p. 1103.

processes: (a) emission from excited Na^+ produced at the surface of Io by sputtering, (b) characteristic x-rays produced by energetic particle bombardment of the elements in Io's surface layer, and (c) bremsstrahlung from high-energy particles impacting the surface. The third will be shown to be negligible, whereas the other two may contribute to the observed signal.

Surface Sputtering. The primary source of sodium atoms and ions in the cloud of Io is thought to be sodium-rich evaporite salts, such as sulfates (Fanale *et al.* 1974) on the surface of Io which release sodium through sputtering by magnetospheric particles (Matson *et al.* 1974).

Recently Nash *et al.* 1975 measured the sodium-D line emission produced by proton bombardment of silicate and halite rock samples. The sputtering yield (number of sputtered sodium atoms in the 3p state per proton) was found to be 10^{-3}. A total sputtering yield of 10^{-2} for proton bombardment (including all the ejected atoms and ions) has been reported by Carter and Colligon (1968). The ratio of neutral atoms (including all states of excitation) to ions in the sputtered ejecta is unknown. However, it has been observed that the majority of particles sputtered from alkali halides under ion bombardment are ions (Campbell and Cooper 1972; Batanou 1963), in contrast to metals where the sputtered particles are primarily neutral atoms. Thus, it seems reasonable to assume that the proton sputtering yield for ions might be as high as 10^{-2} and that some of the ions are in excited states.

Since the low-energy proton flux at Io is about 10^9 protons cm^{-2} sec^{-1}, under the assumptions discussed above, the production of excited sodium ions by proton sputtering will be less than but perhaps comparable to 10^7 ions cm^{-2} sec^{-1}. The flux of particles which produce sputtered atoms and ions could include heavier magnetospheric nuclei and re-impacting ions, in which case the emission rate could be greater. If the emission comes only from the decay of sputtered ions, the brightness of Io's disc would be 10 R. This estimate ignores secondary processes since it is noted that Na^+ ions which are sputtered from NaCl crystals have an energy of only 1–10 eV (Miyagawa 1973; Stuart *et al.* 1969). For the Na-D lines emission, however, Nash *et al.* (1975) observed that a 3.0 keV proton flux of density 1.7×10^{13} cm^{-2} sec^{-1} produced an intensity of 2.6×10^{10} photons cm^{-2} sec^{-1}, while a 5.0 keV flux of 2.8×10^{13} cm^{-2} sec^{-1} only increased the intensity to 3.5×10^{10} cm^{-2} sec^{-1}. If it is assumed that the emission efficiency (photons/proton) is dependent on the incident power density P according to a power law, $\epsilon \propto P^\gamma$ ($\gamma \sim 0.3$), then a 1 MeV proton flux of density 10^7 cm^{-2} sec^{-1} could produce a Na-D line emission of several kiloRayleights (Nash *et al.* 1975). If the same power law holds for the Na^+ emission lines, a brightness as large as 10 kR could result. Such an extrapolation to high energies and to Na^+ is a rather gross assumption, however, and the 10 kR value should be considered as an upper limit.

Characteristic X-Rays. A second possible source of short-wavelength radiation from the disc is the characteristic radiation. Here the energy loss by a bombarding particle is primarily through excitation of inner shell electrons of the surface atoms. The radiation which results when the atoms return to the unexcited state is the characteristic radiation or valence-band emission, depending upon the initial energy level of the electron involved in the radiation. The emission spectra of possible elements in Io's surface layer, such as Na, Mg and Al, are well known (Arakawa and Williams 1973; Rooke 1963; Hansen and Arakawa 1972). The wavelengths of the *L*-shell band emission are 200–400 Å, and are thus within the sensitivity range of our short-wavelength channel.

The cross section for inner shell ionization due to bombarding electrons has been derived by Gryzinski (1965) and is given by

$$\sigma_i = \frac{n\sigma_0}{U_i^2}\, g_i(E_e/U_i). \tag{7}$$

Here $\sigma_0 = \pi e^4 = 6.56 \times 10^{-14}$ eV^2 cm^2, U_i is the ionization energy of inner shell electrons, n is the number of electrons in the shell, and E_e is the energy of the incident electron. The function $g_i(E_e/U_i)$ contains the energy dependence of the cross section and has a maximum value of about 0.2 for $E_e/U_i \sim 2$ to 10. For the 2p electrons of the sodium atom, $U_i \simeq 50$ eV, and the *L*-shell ($n = 8$) ionization will be produced primarily by 0.1–1 keV electrons. The *L*-shell ionization cross section in this energy range is then 10^{-17} to 10^{-16} cm^2.

The relationship between the x-ray production cross section σ_x and the ionization cross section σ_i is $\sigma_x = \bar{\omega}\sigma_i$. Here $\bar{\omega}$ is the average radiation yield for the appropriate atomic shell (i.e., the fraction of the shell vacancies which are filled radiatively). For the *L*-shell of an atom with atomic number < 30, the value of $\bar{\omega}$ depends on the structure of the solid and is taken[4] to be between 10^{-2}–10^{-4}. Thus, the electron cross section for producing *L*-shell radiation is $\sigma_x \sim 10^{-21}$–10^{-18} cm^2.

The number of x-rays created by each incoming electron, per cubic centimeter, is

$$\phi = \sigma_x(\text{cm}^2) \times \left(\frac{\text{number of atoms}}{\text{cm}^3}\right)$$
$$= \sigma_x\left(\frac{N_0\rho}{A}\right) \tag{8}$$

where N_0 is Avogadro's number, A is atomic mass and ρ is the density of the

---

[4]$\bar{\omega}$ for the *L*-shell of atoms with $Z < 26$ has not been measured or calculated. The quoted range of values adopted here was estimated from the existing values of $\bar{\omega}$ for other atoms. See, for example, Chen and Crasemann (1971).

material. Taking into account the self absorption of the x-rays as they escape from the material, the x-ray yield per electron is equal to

$$Y = \int_0^{R_0/\rho} \phi e^{-\mu\rho x} dx$$

$$= \frac{\phi}{\mu\rho}(1 - e^{-\mu R_0})$$

(9)

where μ is the mass absorption coefficient for x-rays, and R_0 is the range of an incident electron in gm cm^{-2}. The photon absorption cross sections of simple molecules, in the 400 Å spectral region, are on the order of $\sigma \sim 10^{-17}$ cm^2 (see, for example, Lee et al. 1973). The mass absorption coefficient at these wavelengths for an atom with $A \sim 20$ gm/mole is then $\mu = (N_0/A)\sigma = 3 \times 10^5$ cm^2 gm^{-1}. For a range, R_0, on the order of 10^{-5} gm cm^{-2} (Katz and Penfold 1952) the term $\exp(-\mu R_0) \sim \exp(-3)$ is much less than 1. The yield of photons per electron can thus be simply written as

$$Y \sim \phi_x/\mu\rho$$

(10)

$$\sim \sigma_x/\sigma.$$

For σ_x between 10^{-21}–10^{-18} cm^2, the x-ray yield Y is 10^{-4} to 10^{-1} photons per electron.

Assuming a flux density of 10^{10} cm^{-2} sec^{-1} for low-energy electrons, the characteristic radiation intensity is 10^{-3}–10^0 kR, and thus would not contribute significantly to the signal of the λ_S channel. However, we note that the above estimate did not consider the presence of the "secondary" electrons inside the material, which can cause additional x-radiation. Lacking accurate cross section data, the possibility that characteristic emissions also contribute to the signals detected by the short-wavelength channel on Pioneer 10 cannot be ruled out.

No essential difference exists between the cross section for ionization by protons and by electrons, if they have the same velocity. Since the mass of a proton is a factor of 1836 that of an electron, L-shell ionization of sodium requires MeV protons. The available proton flux in this range is small, on the order of 10^7 cm^{-2} sec^{-1}, and the characteristic radiation due to proton excitation is thus expected to be much less significant than that due to electron excitation.

Bremsstrahlung. The third potential source of short-wavelength radiation from the disc, bremsstrahlung, is a continuous x-ray emission spectrum resulting from the inelastic collision of electrons with nuclei. The spectral distribution of bremsstrahlung for material with atomic number Z bombarded by an electron with energy E_i is

$$dI = 2KZ(E_i - E_\lambda) \, dE_\lambda \text{ for } 0 < E_\lambda < E_i$$

(11)

where I is the bremsstrahlung energy, E_λ is the photon energy and K is a constant. For $E_i < 2.5$ MeV, K is $(0.7 \pm 0.2) \times 10^{-3}$ MeV^{-1} (Evans 1955).

The bremsstrahlung spectrum extends over all energies from 0 to E_i. However, the detection range of the λ_S channel is limited, as we have noted, to radiation with $\lambda < 800$ Å ($E_\lambda \sim 1.5 \times 10^{-5}$ MeV). Moreover, for radiation with $\lambda < 1$ Å ($E_\lambda \sim 1.2 \times 10^{-3}$ MeV), there is no angular resolution in the ultraviolet photometer since the instrument collimator is transparent to such radiation; hence, no peak signal could be detected. Thus, the total detectable bremsstrahlung photon flux is

$$n = \phi(E_i) \int_{1.5 \times 10^{-5}}^{E_m} \frac{2KZ(E_i - E_\lambda)}{E_\lambda} \, dE_\lambda$$

$$= 2KZ \, \phi(E_i) \left[E_i \, \ln\frac{E_m}{1.5 \times 10^{-5}} + (E_m - 1.5 \times 10^{-5}) \right]$$

(12)

where $\phi(E_i)$ is the electron flux density at energy E_i and E_m is the smaller of E_i and 1.2×10^{-3} MeV; both E_i and E_m are expressed in MeV.

To estimate the bremsstrahlung contribution we calculate the signal to be expected from both keV and MeV electrons. We adopt an electron flux of 10^9 cm^{-2} sec^{-1} at 1 keV and 10^8 cm^{-2} sec^{-1} at 1 MeV. Using $Z \sim 10$ and $K = 0.7 \times 10^{-3}$ MeV^{-1}, the detectable bremsstrahlung flux from Io's disc is 4×10^4 photons cm^{-2} sec^{-1} for keV electrons and it is $\sim 9 \times 10^6$ photons cm^{-2} sec^{-1} for MeV electrons. These values are several orders of magnitude less than the intensity inferred from the λ_S-channel data. Thus, the bremsstrahlung radiation cannot contribute significantly to the signal observed in that channel. It should be noted that the above calculation does not include the effects of self-absorption; including this will further reduce the bremsstrahlung flux. While protons also give rise to bremsstrahlung, the production cross section is only $(1/1836)^2$ that of electrons and is accordingly insignificant.

Summarizing our interpretations of the Io short-wavelength emission, we note separate contributions from the area surrounding the satellite and from its surface. The most plausible interpretation of the data suggests that the observed emissions arise from (1) decay of excited Na$^+$ and other ions produced by particle impact ionization-excitation of the neutrals in the extended cloud, and (2) the decay of excited Na$^+$, produced by sputtering at the disc, as well as possible soft x-ray band emissions. The resolution of the photometer in both clock angle and wavelength does not allow a quantitative estimate of the various contributions to the total signal, but our data strongly suggest that the above mentioned mechanisms are those most likely to be consistent with the known properties of this most spectacular of the Galilean satellites.

VII. AMALTHEA

Little is known about Amalthea (J V), partially because of its small size, $R_{JV} \sim 120$ km (Rieke 1975), and because its low relative brightness (visual magnitude 13) and its closeness to Jupiter ($2.2\ R_J$) make observation difficult. In the absence of accurate mass and radius data it is not possible to determine its density. However, the orbital characteristics would imply that J V belongs to the Galilean satellite group. Magnetospheric particle impact on Amalthea may be much more intense than that for the Galilean satellites, and one might expect to observe ultraviolet emissions due to such an interaction. The present ultraviolet observations of Amalthea represent the first such measurements and are therefore of considerable interest.

The viewing opportunity for Amalthea extended from approximately 12^h 00^m on Day 334 to $19^h\ 00^m$ of Day 336. During this period Amalthea entered the field of view 9 times. The observational results are summarized in Table III. Most of these data, however, are not useful either because the background count rate was too great or because Jupiter was also in the field of view. The experimental data are presented below followed by an interpretation of the measurements.

A. Long-Wavelength Channel Observations

The long-wavelength data shown in Fig. 6 indicate that a weak signal is present at the end of Day 334. The signal count rate is ~ 40 counts sec^{-1} and well above the noise level. The source of the signal has been assumed to be the cloud of hydrogen trailing Io (Carlson and Judge 1975). However, according to the viewing geometry, these signals could also be due to a hydrogen cloud of Amalthea. In support of this interpretation it may be noted in Fig. 6 that the signals are not monotonically decreasing with increasing orbital angle measured from Io, as expected if Io were the only source of the observed signal. An increasing signal is observed at $14^h\ 00^m$ of Day 334. This coincides with the fact that the extent of the field of view inside Amalthea's orbit increases significantly at this time. It is thus tentatively suggested that a portion of the apparent asymmetry of the observational data (with respect to Io's orbital position) is caused by emission from a hydrogen cloud at the orbit of Amalthea.

It should be noted, however, that at this time the spacecraft is less than $7\ R_J$ below the orbital plane, while it is still far away from Jupiter (about 60 R_J). A sphere with a diameter of $1\ R_J$ at Io's orbital position would subtend about 2°5 in clock angle. On the other hand, two points in the orbital plane with a separation of Io's orbital radius ($\sim 6\ R_J$) also only subtend $\sim 2°$ in clock angle. Thus, for this geometry it is not possible to resolve (in clock angle) two sources with a separation of less than Io's orbital radius, and Amalthea's contribution to the total signal cannot be unambiguously isolated from that

TABLE III

Viewing Opportunities of Amalthea (JV) (Field of View Centered on JV)

GMT (Day/hr)	S/C-satellite distance (R_J)	Comments	λ_L-channel signal[a] Peak (counts/sec)	λ_L-channel signal[a] Comments[b]	λ_S-channel signal[a] Peak (counts/sec)	λ_S-channel signal[a] Comments[b]
Pioneer 10						
334/18.0	59	Amalthea only in view	50	possible emission from extended (100°) Io cloud	20	limited by background variation
334/19.0	57	Amalthea only in view	40	possible emission from extended (100°) Io cloud	<20	limited by instrument resolution
335/4.5	54	Jupiter entering the field of view	<250	limited by resolution of instrument	10	possible Jupiter emission
335/9.0	47	Jupiter in the field of view	<250	limited by resolution of instrument	13	partly due to Jupiter emission
335/15.5	48	J V signal blocked by Jupiter	<100	limited by background variation	<20	limited by instrument resolution
335/22.0	39	Jupiter in the field of view	183	primarily Jupiter emission	<60	limited by instrument resolution
336/2.5	41	Jupiter in the field of view	172	primarily Jupiter emission	<100	limited by instrument resolution
336/11.0	31	possible contribution from trailing Io cloud	No peak	extreme background variation	<90	limited by background variation
336/13.0	32	possible contribution from trailing Io cloud	No peak	extreme background variation	<90	limited by background variation

[a] The averaging time is 0.5 hr for the λ_L-channel and 1 hr for the λ_S-channel, except on Day 336, for which the λ_S-channel data were averaged for 0.5 hr.

of the distant tail of Io's cloud. The viewing geometry in the orbital plane is given in Fig. 6.

It is difficult to establish the extent of an Amalthea cloud from the present data. However, it can be estimated from the expression (Carlson and Judge 1974)

$$\Delta\theta \simeq 3 <v> \tau/R_A. \qquad (13)$$

Here $\Delta\theta$ is the angular extent of the cloud measured with respect to Amalthea. $<v>$ and τ are the mean velocity and mean life-time of the atoms, respectively. R_A is the orbital radius of Amalthea. With a 100 eV proton flux of 2×10^9 cm^{-2} sec^{-1} at Amalthea's orbit (Frank et $al.$ 1976) and assuming that the H atoms are lost through proton charge exchange with a cross section of $\simeq 4 \times 10^{-15}$ cm^2, τ is $\simeq 1 \times 10^5$ sec. The temperature of Amalthea is 150°K (Rieke 1975) and the corresponding Jeans mean expansion velocity is $<v>$ ~ 0.9 km sec^{-1}. Hence, we get $\Delta\theta \sim 1.9$ radian.

The total angular extent of the cloud is thus $\simeq 3.8$ rad (218°). Accordingly, the Amalthea cloud is probably not a complete torus. Using 40 counts sec^{-1} as the optical emission from the Amalthea cloud and assuming the vertical dimension of the cloud to be 1 R_J (as for Io), the surface brightness would be about 100 Rayleighs, for a Lyman-α source.

During the viewing opportunity on Day 335, the background count rate was quite high, and no emission feature was evident. Later during the day and early the next day, signals were present, but these were mostly due to Jupiter. After 11^h 00^m of Day 336, because of background noise, no useful data were obtained.

B. Short-Wavelength Channel Observations

A signal with a count rate of about 20 counts sec^{-1} appeared on Day 334 at 18^h 00^m as Amalthea first entered the field of view. However, the signals were spread over a clock angle of 80° and only persisted for ~ 1 hr. Since the signal was present for such a short time and was spread over such a large angular extent, it is difficult to account for the observations in terms of optical emissions; most likely they must be attributed to interference from a unidirectional high-energy particle flux. For such a case, assuming a 25° optical emission[5] superimposed on the wide background signal, the optical emission is estimated to be less than ~ 10 counts sec^{-1}. For the viewing opportunities of Day 335 and later, the short wavelength channel data are either very noisy or contaminated by emission contributions from Jupiter. Assuming 10 counts sec^{-1} to be the maximum optical emission from Amalthea's disc, its maximum brightness near 400 Å is estimated at 2500 kR.

---

[5]An optical emission from a point source will extend over $\sim 25°$ in clock angle.

VIII.   EUROPA

The density of Europa is nearly the same as that of Io (~ 3 gm cm^{-3}). Unlike Io, its spectral reflectance clearly indicates that H_2O covers most of the satellite's surface, which Fanale *et al.* (1974) suggest is composed mainly of hydrated salt and ice. The lack of evidence for sodium in the visual observational data may simply indicate that either there is less sodium available at the surface to be sputtered or that the presence of H_2O at the surface makes the sputtering process less efficient. The possibility of meteoritic material supplying the sodium seems unlikely, although it cannot be ruled out. Ganymede apparently has more meteoroid impacts than Europa, yet there is no direct observational evidence in either the ultraviolet or visible spectral region for sodium as a surface or atmospheric constituent on that satellite (Morrison and Cruikshank 1974).

Europa, like Io, apparently interacts strongly with the magnetospheric particles. This is evident from the sweeping of high-energy particles at Europa's orbit, as at the orbits of Amalthea and Io.

A.   *Short-Wavelength Channel Observations of Europa*

Referring to Fig. 8, the data show a clear emission signal of ~ 13 counts sec^{-1} peaked at $7^h\ 00^m$ hours of Day 333. At this time Io is almost directly on the other side of Jupiter from Europa and is not likely to contribute to the observed signal. Emissions are present for about 2 hours, which is the time required for the satellite to pass through the field of view. Assuming a source of Europa's size ($R = 1550$ km), the apparent brightness is about 14 kilo-Rayleighs (at $\lambda \sim 400$ Å). Other Europa viewing opportunities are either contaminated by signals from other objects in the field of view or limited by noise, permitting only upper limits to be determined, as given in Table IV.

B.   *Interpretation of Europa's Short-Wavelength Channel Observations*

The proton density in the vicinity of Europa is 10 cm^{-3} with energy 400 eV (Frank *et al.* 1976) and the proton flux is $\sim 3 \times 10^8$ cm^{-2} sec^{-1}.

Proton bombardment of Europa's surface would be expected to sputter surface atoms into the surrounding space and thereby create an atmosphere. Since there is no evidence for an atmosphere or ionosphere at Europa, the sputtering process at the surface may be insignificant, perhaps due to the water frost thought to be covering much of the surface.

Europa's surface is also bombarded by an intense electron flux. This bombardment will cause characteristic x-ray radiation from the surface elements. The possible elements in Europa's surface, such as Na, Mg, K, and Ca have valence band x-ray emissions with wavelengths between 200 and 700 Å, and within the sensitivity of the short-wavelength channel. Assuming a surface composition and electron flux similar to that at Io and using the same rather uncertain cross section data, a perhaps conservative estimate of the x-ray contribution to the observed brightness of Europa is $\lesssim 1$ kR. Further consideration of this and other processes is required to establish the origin of the observed short-wavelength emissions.

IX. GANYMEDE

The mean density of Ganymede is about 2 gm cm^{-3} and it probably has a small rocky core and a liquid mantle covered with ices of water, ammonia and methane. The surface is apparently reasonably uniform as observed in eclipse radiometry, although large-scale variations in its albedo are evident. The infrared spectrum implies water ice at the surface, an observation consistent with the other available optical data.

It has also been suggested through stellar occultation measurements that Ganymede possesses a tenuous atmosphere (Carlson *et al.* 1973). The constituents have not been determined nor has the stability of such an atmosphere. An atmosphere with surface pressure greater than 10^{-3} mb was observed during the above stellar occultation. Since no infrared emission features associated with the atmosphere have been observed, it is perhaps composed of constituents such as N_2, O_2, Ne, or A, which have no dipole moment.

Particle impact at Ganymede is less significant than at the orbits of the inner satellites. This is perhaps important in explaining the Pioneer 10 and 11 data which indicate that no ultraviolet emissions are associated with Ganymede. Table V contains a summary of our findings and, using the Pioneer 11 data from Day 335, the upper limits of apparent brightness for the disc are estimated to be ~ 4 kR for $\lambda \sim 1200$ Å and ~ 2 kR for $\lambda \sim 400$ Å.

X. CALLISTO

The photometric observations of Callisto indicate that its surface may consist of high reflectance silicates. Callisto's low density of 1.4 gm cm^{-3} would be caused by a relatively minor amount of heavy elements, implying little radioactive heating; because Callisto was formed at a relatively large distance from Jupiter, tidal heating may also be small. Further, there is no strong evidence for water ice at the surface as inferred from infrared observations.

The charged particle flux at Callisto is greatly reduced from that at the inner satellites, and is probably not important in determining the production and/or loss of an atmosphere.

No ultraviolet emission signals were observed from Callisto. Using the Pioneer 11 data at Day 335 given in Table VI, the upper limit of apparent brightness for Callisto's disc is estimated to be 9 kR for $\lambda \sim 1200$ Å and 7 kR for $\lambda \sim 400$ Å.

TABLE IV

Viewing Opportunities of Europa (JII) (Field of View Centered on JII)

GMT (Day/hr)	S/C-satellite distance (R_J)	Comments	λ_L-channel signal[a]		λ_S-channel signal[a]	
			Peak (counts/sec)	Comments[b]	Peak (counts/sec)	Comments[b]
Pioneer 10						
332/9.5	85	—	<230	limited by instrument resolution	<11	limited by instrument resolution
333/7.0	70	angular separation between J II and J I is 130°	<40	limited by instrument resolution	13	—
335/11.0	57	Jupiter in the field, Europa at the back of Jupiter	<300	limited by instrument resolution	12	Jupiter emission contribution
337/8.0	9	—	<150	limited by instrument resolution	<250	limited by instrument resolution
Pioneer 11						
336/23.0	12	—		No data due to high charged particle and/or radiation background	<150	limited by instrument resolution
337/1.0	10	—		No data due to high charged particle and/or radiation background	<150	limited by instrument resolution

[a] The averaging time is 0.5 hr for the λ_L-channel and 1 hr for the λ_S-channel.

[b] Resolution of the instrument = background/64.

TABLE V

Viewing Opportunities of Ganymede (J III) (Field of View Centered on J III)

GMT (Day/hr)	S/C-satellite distance (R_J)	Comments	λ_L-channel signal[a] Peak (counts/sec)	λ_L-channel signal[a] Comments[b]	λ_S-channel signal[a] Peak (counts/sec)	λ_S-channel signal[a] Comments[b]
Pioneer 10						
334/6.0	77	View is within 10° of Io	120	probably all due to Io cloud emission	<20	limited by instrument resolution
336/17.0	22	—	No peak	extreme background variation	<30	limited by instrument resolution
Pioneer 11						
335/12 5	42	—	<45	limited by instrument resolution	<10	limited by background variation
336/9.0	20	—	<64	limited by instrument resolution	<60	limited by instrument resolution

[a]The averaging time is 0.5 hr for the λ_L-channel and 1 hr for the λ_S-channel, except on Pioneer 10 Day 336, for which the λ_S-channel data were averaged for 0.5 hr.

[b]Resolution of the instrument = background/64.

TABLE VI

Viewing Opportunities of Callisto (JIV) (Field of View Centered on JIV)

GMT (Day/hr)	S/C-satellite distance (R_J)	Comments	λ_L-channel signal[a] Peak (counts/sec)	λ_L-channel signal[a] Comments[b]	λ_S-channel signal[a] Peak (counts/sec)	λ_S-channel signal[a] Comments[b]
Pioneer 10						
333/21.0	44	Io cloud in the field of view	60	Io cloud emission contribution	10	Io cloud emission contribution
Pioneer 11						
331/19.0	107	—	<90	limited by background variation	<90	limited by background variation
335/6.0	28	—	<60	limited by instrument resolution	<80	limited by instrument resolution

[a]The averaging time is 0.5 hr for the λ_L-channel and 1 hr for the λ_S-channel.

[b]Resolution of the instrument = background/64.

XI. SUMMARY

The ultraviolet photometers on Pioneers 10 and 11 have established the three innermost of five Jovian satellites as sources of optical emissions with $\lambda < 1400$ Å. Amalthea's apparent hydrogen cloud and possible but uncertain short-wavelength emissions are the least well established of the present results, and further observations are highly desirable. Io is most certainly accompanied by a cloud of atomic hydrogen that resonantly scatters the solar Lyman-α at 1216 Å. It is also a source of shorter wavelength emissions, $\lambda < 800$ Å, that could result from the radiative decay of excited Na^+ and other atomic ions in its near environment. Soft x-ray band emissions of elements in Io's surface layer and radiative decay of sputtered ions from Io's surface also contribute to the detected short-wavelength signals. Additional laboratory data are required to establish whether or not these sources are significant, however. Signals in this lower wavelength range are detected from the surface of Europa as well. Although the origin of Europa's emissions remain somewhat uncertain, their existence underscores the common characteristic shared by all three of the inner satellites: the observed emissions and clouds are associated with a significant sweeping up of energetic particles in the Jovian magnetosphere.

Acknowledgement. This work was supported by Contract NAS2-6558 with the NASA Ames Research Center.

REFERENCES

Altshular, S. 1952. Excitation cross section for helium atoms. *Phys. Rev.* 87:992–994.

Arakawa, E. T., and Williams, M. W. 1973. Satellites in the x-ray emission spectra of Li, Be, and Na. *Phys. Rev.* 138:4075–4078.

Batanou, G. M. 1963. Emission of charged particles from alkali halide crystals bombarded with potassium ions. *Soviet Phys.-Solid State* 4:1306–1312.

Bates, D. R., and Crothers, D. S. F. 1967. Inelastic collisions between hydrogen and helium atoms. *Proc. Phys. Soc.* 90:73–80.

Bell, R. J. 1961. Excitation of helium to the $(1s, 2p)^1P$ and $(1s, 3p)^1P$ states by proton and alpha particle impact. *Proc. Phys. Soc.* (London) 78:903–911.

Bergstralh, J. T.; Matson, D. L.; and Johnson, T. V. 1975. Sodium D-line emission from Io: synoptic observations from Table Mountain Observatory. *Astrophys. J.* 195:L131–L135.

Brown, R. A. 1974. Optical line emission from Io. *Exploration of the planetary system.* (A. Woszczyk and C. Iwaniszewska, eds.) pp. 527–531. Dordrecht-Holland: D. Reidel Publ. Co.

Campbell, III, A. B., and Cooper, C. B. 1972. Mass spectrometric study of sputtering of KB by low-energy Ar^+ and Xe^+ ions. *J. Appl. Phys.* 43:863–866.

Carlson, R. W. 1975. Atmospheres of outer planet satellites. *A.I.A.A. Progress in astronautics and aeronautics.* (E. Greenstaed, D. Intriligator and M. Dryer, eds.) Cambridge, Mass.: Massachusetts Institute of Technology Press. In press.

Carlson, R. W.; Bhattacharyya, J. C.; Smith, B. A.; Johnson, T. V.; Hidayat, B.; Smith, S. A.; Taylor, G. E.; O'Leary, B. T.; and Brinkmann, R. T. 1973. An atmosphere on Ganymede from its occultation of SAO 186800 on 7 June 1972. *Science* 182:53–55.

Carlson, R. W., and Judge, D. L. 1974. Pioneer 10 ultraviolet photometer observations at Jupiter encounter. *J. Geophys. Res.* 79:3623–3633.

———. 1975. Pioneer 10 ultraviolet photometer observations of the Jovian hydrogen torus: the angular distribution. *Icarus* 24:395–399.

Carlson, R. W.; Matson, D. L.; and Johnson, T. V. 1975. Electron impact ionization of Io's sodium emission cloud. *J. Geophys. Res. Lett.* 2:469.

Carter, G., and Colligon, J. S. 1968. *Ion bombardment of solids.* New York: American Elsevier.

Chen, M. H., and Crasemann, B. 1971. Theoretical L_2- and L_3-subshell fluorescence yields and L_2-L_3 X Coster-Kronig transition probabilities. *Phys. Rev.* A4, 1–7.

Evans, R. D. 1955. *The atomic nucleus.* New York: McGraw-Hill.

Fanale, F. P.; Johnson, T. V.; and Matson, D. C. 1974. Io: a surface evaporite deposit? *Science* 186:922–924.

Frank, L. A.; Ackerson, K. L.; Wolfe, J. H.; and Mihalov, J. D. 1976. Observations of plasma in the Jovian magnetosphere. *J. Geophys. Res.* 81:457–468.

Goldreich, P., and Lynden-Bell, D. 1969. Io, a Jovian unipolar inductor. *Astrophys. J.* 156: 59–78.

Grüebler, W.; Schmelzbach, P. A.; König, V.; and Marmier, P. 1970. Charge exchange collisions between hydrogen ions and alkali vapour in the energy range of 1 to 20 keV. *Helv. Phys. Acta* 43:254–271.

Gryzinski, M. 1965. Classical theory of atomic collisions. I. Theory of inelastic collisions. *Phys. Rev.* 138:A336–358.

Hansen, W. F., and Arakawa, E. T. 1972. High energy x-ray satellites of $L_{2,3}$ emission bands of Na, Mg, Al, and Si. *Z. Physik* 251:271–288.

Jobe, J. D., and St. John, R. M. 1967. Absolute measurements of the 2^1P and 2^3P electron excitation cross sections of helium atoms. *Phys. Rev.* 164:117–121.

Johnson, T. V.; Matson, D. L.; and Carlson, R. W. 1976. The atmosphere of Io: new limits based on magnetospheric plasma models. *Geophys. Res. Lett.* In preparation.

Judge, D. L., and Carlson, R. W. 1974. Pioneer 10 observations of the ultraviolet glow in the vicinity of Jupiter. *Science* 183:317–318.

Katz, L., and Penfold, S. 1952. Range-energy relations of electrons and the determination of beta-ray end point energies by absorption. *Rev. Mod. Phys.* 24:28–44.

Lee, L. C.; Carlson, R. W.; Judge, D. L.; and Ogawa, M. 1973. The absorption cross sections of N_2, O_2, CO, NO, CO_2, N_2O, CH_4, C_2H_4, C_2H_6 and C_4H_{10} from 180–700 Å. *J. Quant. Spectrosc. Radiat. Transfer* 13:1023–1031.

Lotz, W. 1967. Electron-impact ionization cross sections and ionization rate coefficients for atoms and ions. *Astrophys. J. Suppl.* 14:207–268.

Macy, W., and Trafton, L. 1975. A model for Io's atmosphere and sodium cloud. *Astrophys. J.* 200:510–519.

Maloy, J. O.; Carlson, R. W.; Hartmann, U. G.; and Judge, D. L. 1976. Rocket measurements of the profile and intensity of the solar HeI λ584 Å resonance line. *J. Geophys. Res.* In preparation.

Matson, D. L.; Johnson, T. V.; and Fanale, F. P. 1974. Sodium D-line emission from Io: sputtering and resonant scattering hypothesis. *Astrophys. J.* 192:L43–L46.

McDonough, T. R. 1975. A theory of the Jovian hydrogen torus. *Icarus* 24:400–406.

McElroy, M. B. 1965. Excitation of atmospheric helium. *Planet. Space Sci.* 13:403–433.

McElroy, M. B., and Yung, Y. L. 1975. The atmosphere and ionosphere of Io. *Astrophys. J.* 196:227–250.

Miyagawa, S. 1973. Energy spectrum of Na^+ ions sputtered from NaCl. *J. Appl. Phys.* 44: 5617–5618.

Morrison, D., and Cruikshank, D. P. 1974. Physical properties of the natural satellites. *Space Sci. Rev.* 15:641–739.

Nash, D. B.; Matson, D. L.; Johnson, T. V.; and Fanale, F. P. 1975. Production of Na-D line emission from rock specimens by proton bombardment: implications for Io emissions. *J. Geophys. Res.* 80:1875–1879.

Newburn, R. L., and Gulkis, S. 1973. A survey of the outer planets, Jupiter, Saturn, Uranus, Neptune, and Pluto, and their satellites. *Space Sci. Rev.* 14:179–271.

Omidvar, K.; Kyle, H. L.; and Sullivan, E. C. 1972. Ionization of multielectron atoms by fast charged particles. *Phys. Rev.* A5:1174–1187.

Rieke, G. H. 1975. The temperature of Amalthea. *Icarus* 25:333–334.

Rooke, G. A. 1963. Plasmon satellites of soft x-ray emission spectra. *Phys. Letters* 3:234–236.

Shawhan, S. D.; Goertz, C. K.; Hubbard, R. F.; Gurnett, D. A.; and Joyce, G. 1975. Io-accelerated electrons and ions. *The magnetospheres of Earth and Jupiter.* (V. Formisano, ed.) pp. 375–389. Dordrecht-Holland: D. Reidel Publ. Co.

Simpson, J. A.; Hamilton, D. C.; McKibben, R. B.; Mogro-Campero, A.; Pyle, K. R.; and Tuzzolino, A. J. 1974. The protons and electrons trapped in the Jovian dipole magnetic field region and their interaction with Io. *J. Geophys. Res.* 79:3522–3544.

St. John, R. M.; Frank, L. M.; and Lin, C. C. 1964. Absolute electron excitation cross section of helium. *Phys. Rev.* 134:A888–A897.

Stuart, R. V.; Wehner, G. K.; and Anderson, G. S. 1969. Energy distribution of atoms sputtered from polycrystalline metals. *J. Appl. Phys.* 40:803–812.

Trafton, L. 1976. A search for emission features in Io's extended cloud. *Icarus* (special Jupiter issue). In press.

Trafton, L.; Parkinson, T.; and Macy, W. 1974. The spatial extent of sodium emission around Io. *Astrophys. J.* 190:L85–L89.

Trainor, J. H.; McDonald, F. B.; Teegarden, B. J.; Webber, W. R.; and Roelof, E. C. 1974. Energetic particles in the Jovian magnetosphere. *J. Geophys. Res.* 79:3600–3613.

IO, ITS ATMOSPHERE AND OPTICAL EMISSIONS

R. A. BROWN and Y. L. YUNG
Harvard University

Io is surrounded by a halo of atoms which radiate in emission lines. This was discovered by groundbased observations in 1973, a year before Pioneer 10 passed through the Jovian system. Earlier optical and radio observers had reported other anomalies associated with Io, and the Pioneer spacecraft discovered an ultraviolet emission cloud around the satellite and a substantial ionosphere. A new field of planetary research is dedicated to integrating these phenomena into a model of Io and of the Jovian environment with which it strongly interacts.

Sodium dominates the optical emission cloud around Io. The production rate is very large and has not yet been satisfactorily explained. Sputtering by charged particle bombardment may play an important role. The primary excitation mechanism is the resonant scattering of sunlight. While the sodium cloud is brightest near Io, it extends around the entire orbit and is present at low levels in the general Jovian environment. The ionospheric electron density on Io is comparable to that on Mars, a surprising result since the solar flux is much weaker. Here, again, Io's charged particle environment may play an important part. The observed electron profiles plus the constraints provided by the emission cloud are consistent with at least two models of Io's neutral atmosphere.

The four bright satellites of Jupiter have a unique perennial interest and historical importance. Their discovery by Galileo in 1610 was among the very first achievements of the telescope. Early observers intensively chronicled Io's rapid synodic motion—and this was to a very practical purpose, since it established an accurate and accessible timepiece. By comparison with local solar time, terrestrial longitude could be measured. In the late seventeenth century Römer found an annual phase lag in "Io time" and first determined the finite speed of light.

Modern studies of the Galilean satellites focus on their physical properties. Interest has been sparked by the view that the Jovian system may be a textbook in microcosm of the solar system. The observational record, especially for Io, is replete with startling discoveries. Among them are a possible post-eclipse brightening (Binder and Cruikshank 1964) and a modulation of

Jovian decametric radiation (Bigg 1964). Recently we have learned that Io is a source of atomic resonance line emission and that it possesses a substantial ionosphere and therefore an atmosphere. These phenomena are related, and our current understanding of them and their relationship is reviewed here.

The chapter is divided into two parts. The first is primarily observational, and its focus is on measurements of the line emission from Io and its environment. The second part is primarily theoretical, and it deals with current models of Io's atmosphere. No single existing theory satisfactorily treats all the processes which are thought to be important; however, a variety of models for Io's surface and environment have been developed in the context of recent observational data (McElroy and Yung 1975; Matson *et al.* 1974; Macy and Trafton 1975).

In Sec. I the sodium D-line measurements are documented and set in their observational context. Section II reviews Pioneer 10's observation of an ultraviolet emitting region associated with Io, which is probably a cloud of atomic hydrogen. Section III reports on the current state of the search for other line emissions. Section IV discusses future observational work.

In Sec. V we discuss the ionosphere and conclude that the atmosphere may be cooler than the observed sodium and that the exobase probably stands well above Io's surface. In Sec. VI we discuss possible sources of hydrogen; its escape into the surrounding space is not a problem. In Sec. VII we treat the alternative sources of sodium; its escape *is* a problem. Having motivated the observational constraints on the neutral atmosphere, in Sec. VIII we show that two different models are consistent with the observations.

I. SODIUM LINE EMISSION FROM IO

The observation of sodium D-line emission associated with Io was first made by Brown in 1973 (see Brown 1974). A Doppler shift analysis of a high-resolution spectrum positively identified Io as the source. Only Io of the four Galilean satellites showed D-line emission to the limit of the instrumental sensitivity. The strong emission varied with large amplitude on a short time scale, and was possibly correlated with Io's orbital position.

The sodium phenomenon has been intensively observed during the 1973 and 1974 oppositions of Jupiter, but it is still rather poorly understood. Beyond its existence, the only unambiguous facts are that the emission also comes from the neighboring space around Io (Trafton *et al.* 1974) and that the brightness of the sodium nearest Io is definitely correlated with the satellite's orbital phase angle measured from superior heliocentric conjunction (Bergstralh *et al.* 1975). The observational challenge, somewhat simplified, is to record the characteristics of a spatial continuum which is a spectral point-source in the presence of a spectral continuum which is a spatial point-source. The former (the sodium cloud) and the latter (Io) also undergo temporal variations with regard to observational perspective and brightness. Therefore, it is important for this chapter to set the sodium measurements in their observational context. The first three subsections lay that ground-work. In A we discuss the sunlight incident on Io and reflected from its surface. In B we characterize and label the spatial regimes relevant to the observational problem. In C we review some fundamental physics of the sodium atom which relates directly to the D-line observations.

In D the dominant excitation mechanism for the D-line emission is es-tablished by reviewing the observations and conclusions of Bergstralh *et al.* (1975). That mechanism is solar resonant scattering, and its recognition is important for interpreting the high resolution spectra and photometric studies discussed in E. In F we discuss low-intensity sodium emission recorded in the Jovian environment remote from Io.

A. Incident and Reflected Solar Light Near the D-Lines

The solar radiation incident on the Jupiter system has direct importance to the optical observations of Io's sodium in two ways. First, this light is resonantly scattered by free atoms of sodium. The relevant quantity for studying this phenomenon is the amount of solar radiation available for scat-tering, that is, falling within the D_1 and D_2 absorption coefficient profiles for a certain Doppler shift of the sodium with respect to the sun. Second, the solar light reflected from Io's surface serves to calibrate measurements of the sodium emission photometrically. For that procedure the important quan-tity is the flux at Earth from Io in the vicinity of the D-lines. We call this quantity πF_I (.59), and it refers to the continuum flux from Io at 0.59 μm between Fraunhofer absorption lines.

The solar spectrum near the D-lines at $\lambda_{D_2} = 5889.95$ Å and $\lambda_{D_1} = 5895.92$ Å is marked by deep and broad absorptions due to sodium in the sun. A high-resolution, photometric spectrum of the light from the whole solar disc has been obtained and reduced by C. D. Slaughter of Kitt Peak National Observatory. It has been published by Parkinson (1975) and is reproduced in Fig. 1 and Table I. The Doppler shift due to the relative motion with respect to the sun determines the points on the D_1 and D_2 solar absorp-tion lines which can excite sodium in Io's environment. The range of Doppler shifts associated with Io's orbital motion (17.3 km sec^{-1}) is about $\pm$ 0.34 Å.

Referring to the observational geometry shown schematically in Fig. 2, the Io-Sun Doppler shift is proportional to sin (θ_{SHC}) except for a small correction due to Jupiter-Sun motion. The tabulated values, γ_{D_1} and γ_{D_2}, are fractions of $\pi F_\odot$ (.59), the incident solar flux in the nearby continuum. The ratio $\gamma_{D_2}/\gamma_{D_1}$ remains remarkably steady at about 0.85 over the range sampled by Io.

$\pi F_\odot$ (.59) can be obtained by a wavelength interpolation in a table of mean solar disc intensities (Allen 1973), $F_\odot'$ (λ), which refers to the con-tinuum between lines

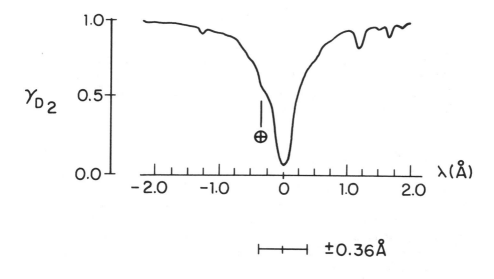

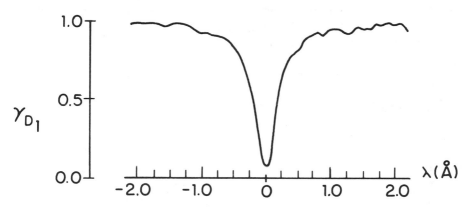

Fig. 1. The solar spectrum in integrated light in the regions of the sodium D_2 and D_1 Fraunhofer lines (obtained by C. D. Slaughter). The horizontal bar shows the region tabulated in Table I and indicates the regions of interest for the Io problem. In the D_2 spectrum there are several telluric H_2O lines, the largest of which is marked and none of which have been removed (From Parkinson 1975).

$$\pi F_\odot (.59) = \pi F_{\odot}' (.59) \left(\frac{R}{r}\right)^2$$

$$= 5.9 \times 10^{13} \frac{1}{r^2} \text{ photon cm}^{-2} \text{ Å}^{-1} \text{ sec}^{-1} \tag{1}$$

where R is the solar radius and r is the Jupiter-Sun distance measured in A.U. At "mean opposition," $r = a \equiv 5.203$ A.U., and

$$\pi F_\odot^{m.o.} (.59) = 2.02 \times 10^{12} \text{ photon cm}^{-2} \text{ Å}^{-1} \text{ sec}^{-1}. \tag{2}$$

$\pi F_I (.59)$ is derived from measurements of the flux, $\pi F_I (V)$, from Io in the V-spectral pass-band which have been recently reviewed by Morrison and Morrison (1976). The V-filter is 90 Å wide, and the effective wavelength for Io's color is about 0.55 μm. Io's flux is expressed in terms of the absolute calibration of the V-magnitude system:

$$\pi F_I (V) = \pi F_0 (V)^{-0.4 \, V_I} \tag{3}$$

where $\pi F_o (V) = 3.72 \times 10^{-9}$ erg cm^{-2} sec^{-1} Å^{-1} (Allen 1973). V_I is Io's V-magnitude, which varies with all four independent parameters defining the observational geometry. Dependence on r and Δ is simply inverse-square. A strongly backscattering phase function with an "opposition surge" in brightness is embodied in the α-dependence: Io is 50% brighter at opposition than at maximum solar phase angle (referred to the same Earth-Io separation). A periodic variation with orbital phase angle θ_{SGC} has an amplitude of 8% and implies synchronous rotation of the satellite. Figure 3a shows V_I (α), which is referred to "mean opposition" distances, $r = a$ and $\Delta = a - 1$. Shown in Fig. 3b is the nominal variation of Io's magnitude with rotational phase, $\delta V_I (\theta_{SGC})$. These data are taken from the Morrisons' paper, and the result is that $\pi F_I (V)$ can be computed with a few percent uncertainty for the time of a particular observation:

$$V_I = V_I (\alpha) + \delta V_I (\theta_{SGC}) + 5 \log \left(\frac{r\Delta}{a(a-1)} \right). \tag{4}$$

Using $\pi F_I (V)$, $\pi F_I (.59)$ can be determined by accounting for Io's color.

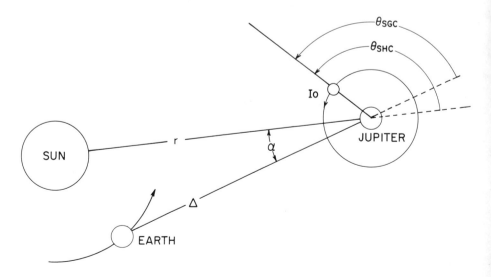

Fig. 2.   The geometry of groundbased observations of Io.

TABLE I

The solar spectrum in integrated light near the D lines. The height of the spectrum is given as a fraction of the continuum height.[a]

$\Delta\lambda$(mÅ)	γ_{D_2}	γ_{D_1}	$\Delta\lambda$(mÅ)	γ_{D_2}	γ_{D_1}	$\Delta\lambda$(mÅ)	γ_{D_2}	γ_{D_1}
360	.6298	.7315	100	.1898	.2409	− 110	.2266	.2828
50	.6195	.7256	90	.1591	.1986	− 20	.2617	.3239
40	.6095	.7176	80	.1312	.1601	− 30	.2951	.3616
30	.5995	.7091	70	.1069	.1276	− 40	.3252	.3948
20	.5893	.6995	60	.0872	.1022	− 50	.3513	.4233
10	.5786	.6889	50	.0727	.0841	− 60	.3736	.4478
300	.5672	.6777	40	.0629	.0719	− 70	.3929	.4691
90	.5552	.6661	30	.0566	.0641	− 80	.4101	.4883
80	.5426	.6542	20	.0527	.0592	− 90	.4256	.5061
70	.5296	.6417	10	.0503	.0564	− 200	.4394	.5229
60	.5170	.6281	5	.0495	.0557	− 10	.4517	.5389
50	.5051	.6139	0	.0490	.0555	− 20	.4633	.5541
40	.4942	.5991	− 5	.0489	.0558	− 30	.4747	.5687
30	.4836	.5840	− 10	.0491	.0566	− 40	.4858	.5832
20	.4725	.5685	− 20	.0510	.0597	− 50	.4963	.5977
10	.4602	.5524	− 30	.0549	.0650	− 60	.5060	.6119
200	.4466	.5353	− 40	.0612	.0733	− 70	.5155	.6259
90	.4317	.5169	− 50	.0706	.0858	− 80	.5246	.6396
80	.4155	.4974	− 60	.0843	.1041	− 90	.5320	.6531
70	.3975	.4763	− 70	.1035	.1291	− 300	.5357	.6662
60	.3768	.4532	− 80	.1283	.1609	− 10	.5360	.6789
50	.3524	.4274	− 90	.1582	.1986	− 20	.5361	.6910
40	.3237	.3979	− 100	.1915	.2401	− 30	.5410	.7025
30	.2913	.3641				− 40	.5538	.7136
20	.2870	.3259				− 50	.5729	.7241
110	.2226	.2842				− 360	.5940	.7339

[a]These data are shown graphically in Figure 1 and are from Parkinson (1975).

The spectral reflectivity studies of Johnson and McCord (1970) show Io to be 6% brighter at 0.59 μm than at 0.55 μm:

$$\pi F_I (.59) = 1.06 \; \pi F_I (V) .$$

As an illustration, we consider the case of a "mean" opposition and neglect the effect of orbital phase.

$$V_I^{m.o.} = 4.84 . \qquad (5)$$

Then

$$\pi F_I^{m.o.} (.59) = 12.9 \text{ photon cm}^{-2} \text{ Å}^{-1} \text{ sec}^{-1} . \qquad (6)$$

Given Io's size, $R_I = 1830$ km, and the incident solar flux, the value of $\pi F_I (.59)$ is extraordinarily high for all observing geometries. This is illustrated by the geometric albedo

$$p(.59) = \frac{\pi F_I{}^{m.o.} (.59)}{\dfrac{1}{\pi} \cdot \pi F_\odot{}^{m.o.} (.59) \cdot \dfrac{\pi R_I{}^2}{(a-1)^2}} = 0.75 \ . \tag{7}$$

Another relevant dramatization of Io's brightness is the comparison of πF_I (.59) to the flux πF_ℓ which would be observed from a white Lambert sphere exactly replacing Io in size and position. Ignoring the orbital phase modulation, the ratio πF_I (.59)$/\pi F_\ell$ varies from 1.1 to 0.7 as the solar phase angle varies from 0° to the maximum of 12°. We shall discuss below in D the peculiar significance of this effect for sodium observations; it makes sodium in Io's atmosphere difficult to observe from Earth.

B. The Spatial Domains

Io is not a point source; its diameter subtends 1.2 arc-seconds at opposition and 0.8 arc-seconds at conjunction with the sun. Under typical observing conditions, "seeing" further spreads Io's radiation into a peaked distribution two or more arc-seconds full-width at half-maximum ($FWHM$). This distribution is Io's continuum surface brightness, and its integral over solid angle is the continuum flux. To the extent that the surface brightness is approximated by a Gaussian distribution of angular size σ arcsec $FWHM$, the central intensity is

$$J_0 \ (\lambda) = 3.75 \times 10^{10} \left(\frac{\pi \ F_I \ (\lambda)}{\sigma^2} \right) . \tag{8}$$

This result can be expressed as an apparent emission rate[1] for the center of the seeing distribution if πF_I (λ) has the units photon cm^{-2} sec^{-1} Å^{-1}

$$E_0 \ (\lambda) = 0.472 \left(\frac{\pi F_I(\lambda)}{\sigma^2} \right) \ \mathrm{MR} \ \text{Å}^{-1} . \tag{9}$$

Continuing the illustration in the previous section, at opposition the apparent emission rate in the center of a Gaussian seeing disc 2 arcsec $FWHM$ is 1.5 MR Å^{-1}.

For discussion purposes, the sodium in Io's vicinity can be partitioned into three regions. Region A is spatially coincident with Io's visible disc and corresponds to sodium in Io's bound atmosphere. Region B has direct association with Io and extends $5-15$ arcsec from it; this is the sodium cloud first identified by Trafton. Extremely faint D-line emission has been reported extending many arcmin away from Io, originating in remote parts of Io's orbit and beyond, by Wehinger and Wyckoff (1974), Mekler and Eviatar (1974) and Mekler et al. (1975). This is Region C.

---

[1]Apparent emission rate $(E) = 4\pi \times 10^{-6} J$, where J is the measured surface brightness in photons sec^{-1} cm^{-2} ster^{-1} (Appendix II of Chamberlain 1961). E is measured in Rayleighs, and kR and MR are 10^3 and 10^6 Rayleighs respectively. The qualifier "apparent" is used because E is equal to a column emission rate only for an optically thin cloud. The term is also commonly used in optically thick situations.

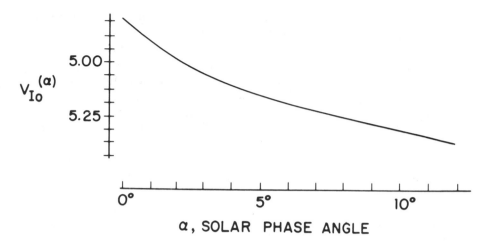

Fig. 3a. The dependence of Io's magnitude on the solar phase angle (Morrison and Morrison 1976).

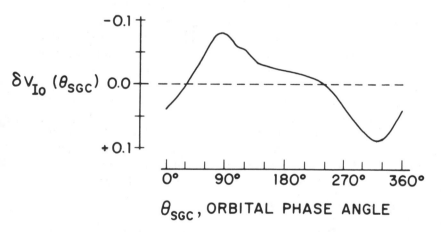

Fig. 3b. The dependence of Io's magnitude on the orbital phase angle (Morrison and Morrison 1976).

    The observability of these regions is determined by both their brightness and their contrast with the background. For Region A that background is the bright disc of Io, and the contrast is poor if a resonant scattering mechanism alone excites the sodium in Io's atmosphere. This point is discussed below in D. For Regions B and C the backgrounds are the telluric sodium nightglow and the halos of scattered continuum radiation from Jupiter and its satellites. The former is substantially fainter even than Region C, and it is observationally distinguishable from D-line radiation originating in the Jovian system: it exhibits no Doppler shift and is spatially uniform over the telescopic field. The scattered continuum radiation is a serious impediment to observations at spectral resolution lower than $\sim$ 10,000. As discussed in Sec. IV. this has prevented imaging the sodium cloud with conventional techniques. The apparent emission rate of Region B is tens of kR. Absolute photometry of Region C has not been reported, but a comparison of the detections with the 5577 Å line in the telluric airglow implies an apparent emission rate less than one kR. The distinction between Regions B and C is drawn on the basis of existing observations which do not manifest the transition from one region to the other. The demonstration that the distinction is artificial, or the definition of the boundary between Regions B and C, awaits further study.

Since the bulk of the existing sodium observations have been made with astronomical slit spectrographs, some specific remarks are in order about the reduction of such spectra. A usual procedure is to record a spectrum of Io itself and to use the instrumental response to Io's reflected solar continuum to calibrate sodium emission spectra. If the entire disc of Io passes through the entrance aperture, the recorded spectral continuum near the D-lines will be the instrumental response to πF_I (.59). Since spectral and spatial resolution are not independent for these instruments, however, some fraction of Io's continuum image is lost on the slit jaws during observations which press for higher spectral resolution. In such cases, the distribution of Io's continuum surface brightness must be modeled in its projection on the slit, for only by knowing the actual number of photons incident on the detector is it possible to calibrate the system's response. This point would be academic if the emission came only from Region A, since the effects of seeing and truncation by the slit jaws would affect continuum and emission identically. But this is not the case for Io, and the analyses of spectra taken with Io in the spectrograph slit depend on hypotheses about the spatial distributions of the continuum and emission components (see e.g., Brown et $al.$ 1975).

C.   Atomic Physics of the Sodium D-Lines

The sodium D-lines[2] are transitions between the ground state $(3s^2S_{\frac{1}{2}})$ and the first excited state which is split by spin-orbit coupling into two fine structure components $(3p^2P_{\frac{1}{2}}, 3p^2P_{\frac{3}{2}})$. The positions and oscillator strengths of the lines are:

$$\begin{aligned}
\lambda_{D_1} &= 5895.92 \text{ Å} & \nu_{D_1} &= 16956.183 \text{ cm}^{-1} & f_{D_1} &= 0.327 \\
\lambda_{D_2} &= 5889.95 \text{ Å} & \nu_{D_2} &= 16973.379 \text{ cm}^{-1} & f_{D_2} &= 0.655 \quad (10)
\end{aligned}$$

---

[2]For a review of the physics of the sodium atom with special emphasis on observations of the telluric sodium airglow, see Chamberlain (1961).

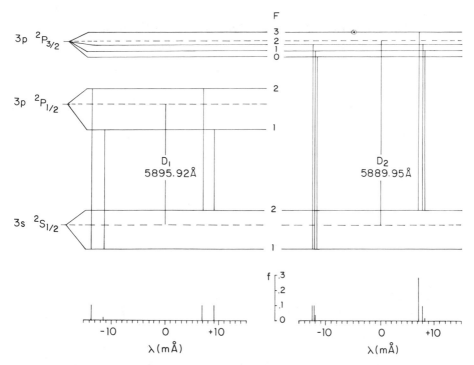

Fig. 4. A schematic representation of the energy levels in atomic sodium between which the D-lines originate. The wavelengths of the centroids λ_{D_1} and λ_{D_2} are given. Allowed transitions are indicated by lines connecting the appropriate sublevels. For each component the oscillator strength position and the wavelength position relative to the centroid is shown below.

The only stable isotope of sodium is $_{11}Na^{23}$ with nuclear spin $I = \frac{3}{2}$. Hyperfine structure results from the coupling between the nuclear spin $\mathbf{I}$ and the total electronic angular momentum $\mathbf{J}$ to produce $\mathbf{F} = \mathbf{I} + \mathbf{J}$. The resultant F-manifolds and the allowed transitions ($\Delta F = \pm 1, 0$) are illustrated in Fig. 4. Table II gives the oscillator strengths of these hyperfine transitions and their wavelength positions with respect to the D_1 and D_2 centroids, λ_{D_1} and λ_{D_2}.

For each hyperfine transition, the absorption coefficient of an individual atom at rest is distributed over wavelength in a Lorentz line shape with full-width at half-maximum $\Delta\nu_L = 3.10^{-4}$ cm^{-1} or $\Delta\lambda_L = 0.1$ m Å. The integral of this distribution is

$$\alpha(F_s, F_p) = \frac{\pi e^2}{mc} f(F_s, F_p) \tag{11}$$

which has units of cm^2 sec^{-1}.

TABLE II[a]

*The Oscillator Strengths and the Wavelength Positions
of the D_1 and D_2 Hyperfine Components*

$^2P_{\frac{1}{2}}$

	D_1	$F = 1$	$F = 2$
$^2S_{\frac{1}{2}}$	$F = 1$	0.020 − 11.4 mÅ	0.102 − 13.6 mÅ
	$F = 2$	0.102 + 9.1 mÅ	0.102 + 6.9 mÅ

$^2P_{\frac{3}{2}}$

	D_2	$F = 0$	$F = 1$	$F = 2$	$F = 3$
$^2S_{\frac{1}{2}}$	$F = 1$	0.041 − 12.0 mÅ	0.102 − 12.2 mÅ	0.102 − 12.6 mÅ	——
	$F = 2$	——	0.020 + 8.3 mÅ	0.102 + 7.9 mÅ	0.287 + 7.2 mÅ

[a]These transitions are shown schematically in Fig. 4.

For an ensemble of atoms in thermal motion, the absorption coefficient per atom is the convolution of the Lorentzian with a Doppler (Gaussian) profile with *FWHM*

$$\Delta\nu_d = 2.5 \times 10^{-3} \sqrt{T} \text{ cm}^{-1}, \tag{12}$$

or

$$\Delta\lambda_d = 7.16 \times 10^{-7} \lambda \sqrt{\frac{T}{23}} \approx 0.88 \sqrt{T} \tag{13}$$

where T is the temperature in °K and the unit is 10^{-3} Ångstrom. For naturally occurring temperatures, then, the absorption coefficient per atom of the hyperfine line (F_s, F_p) is effectively distributed in the Doppler profile

$$\alpha_\nu(F_s, F_p) = \left(\frac{\pi e^2}{mc}\right) f(F_s, F_p) \frac{2(\ln 2)^{\frac{1}{2}} c}{\sqrt{\pi}\, \Delta\nu_D} \exp\left[-\left(\frac{\nu - \nu(F_s, F_p)}{\Delta\nu_d}\right)^2 \ln 2\right] \tag{14}$$

or, written as a function of wavelength,

$$\alpha_\lambda(F_s, F_p) = \left(\frac{\pi e^2}{mc}\right) f(F_s, F_p) \frac{2(\ln 2)^{\frac{1}{2}}}{\sqrt{\pi}\, \Delta\lambda_D} \exp\left[-\left(\frac{\lambda - \lambda(F_s, F_p)}{\Delta\lambda_D}\right)^2 \ln 2\right]. \tag{15}$$

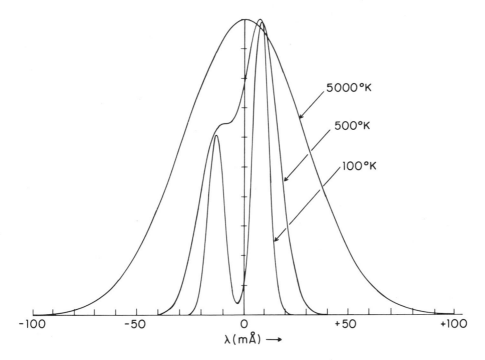

Fig. 5. The theoretical D_1 absorption coefficient as a function of wavelength relative to the centroid λ_{D_1} for sodium kinetic temperatures 100, 500 and 5000°K. The profiles have been normalized to the same maximum height.

The absorption coefficient profiles per atom are the sums of $\alpha_\lambda(F_s, F_p)$ or $\alpha_\nu(F_s, F_p)$ over the allowed transitions, assuming the states F_s are populated according to their multiplicity:

$$\alpha_\nu = \sum_{F_s, F_p} \alpha_\nu(F_s, F_p) \tag{16}$$

and likewise for α_λ.

The shape of the D_1 absorption coefficient as a function of wavelength is shown in Fig. 5 for 100, 500 and 5000°K. Only for temperatures higher than a few thousand degrees are the D-lines adequately described by a single component of strength f_{D_1} or f_{D_2} distributed in a Gaussian $\Delta\lambda_d$ $FWHM$ centered on λ_{D_1} or λ_{D_2}.

Non-thermal mechanisms may produce a velocity dispersion in the atoms sampled by a particular observation of Io's sodium cloud (Matson *et al.* 1974; Trafton and Macy 1975). If thermodynamic equilibrium does not apply, the absorption coefficients cannot be characterized by a single temperature, and the fundamental quantity is the detailed distribution of atomic velocities.

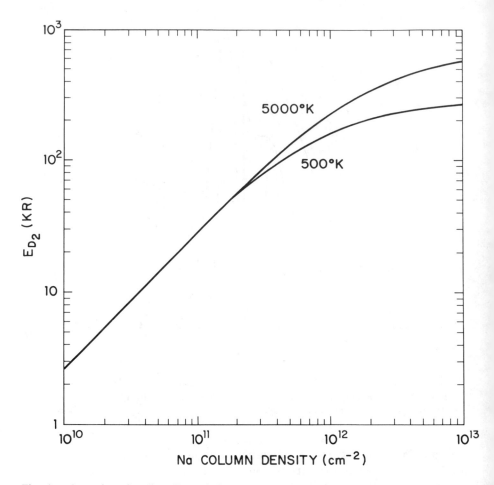

Fig. 6a.   Intensity of sodium D_2 emission versus sodium column abundance for 500 and 5000°K. The continuum solar flux is taken to be 2.02×10^{12} photons cm^{-2} sec^{-1} Å^{-1}, and $\gamma_{D_2} = 0.61$.

Combining the work above with that in Subsections A and B, we may set the physical scale of the interaction of sodium with solar flux in Io's environment. Anticipating an observational result discussed below in E, we consider a sodium cloud with column abundance N atoms cm^{-2} which is sufficiently "hot" to ignore hyperfine structure. The optical depth in the center of the D_1 line is then

$$\tau_{D_1} = \alpha_{\nu_{D_1}} N = 1.1 \times 10^{-10}\, T^{-\frac{1}{2}} N \,, \qquad (17)$$

while τ_{D_2} is twice as great. The column abundance corresponding to unit optical depth at 5000°K is 6.6×10^{11} atoms cm^{-2}.

If Io's sodium cloud is optically thin, its apparent emission rate due to the resonant scattering of solar flux in a D-line may be estimated as follows.

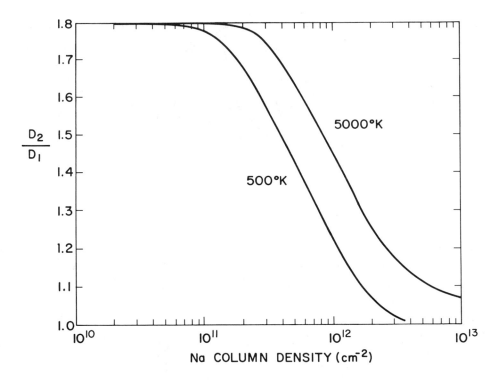

Fig. 6b.   Ratio of emergent D_2 to D_1 apparent emission rates for 500°K and 5000°K. γ_{D_2} and γ_{D_1} are taken to be 0.61 and 0.72 respectively.

[Here the reader is again referred to Chamberlain (1961).] For plane-parallel geometry and zero solar phase angle

$$E = gN \qquad (18)$$

where the g-factor is given by

$$g \equiv \left[\gamma \cdot \pi F_\odot \, (.59) \cdot \frac{\lambda^2}{c} \right] \frac{\pi e^2}{mc} f. \qquad (19)$$

For the D_1 line

$$E_{D_1} \equiv 0.20 \, \gamma_{D_1} \, N \,. \qquad (20)$$

As an example, take $\gamma_{D_1} = 0.5$ and $N = 10^{11}$ atoms cm^{-2}, then

$$E_{D_1} \equiv 10 \, \text{kR} \,. \qquad (21)$$

As the column abundance of sodium increases, a detailed radiative transfer calculation is required to compute E as a function of temperature and optical thickness. The gross effect is that the D_2/D_1 ratio is reduced and the emission line profiles are broadened due to saturation in the line core. These

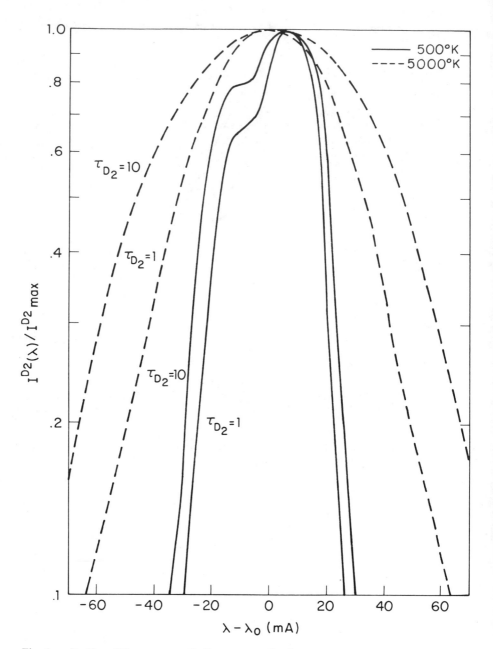

Fig. 6c.   Profiles of the emergent D_2 line co-normalized to the same peak intensity.

effects are illustrated in Fig. 6 which shows E_{D_2}, D_2/D_1 and the D_2 emission line profile versus column abundance for a sodium cloud. These results are taken from McElroy and Yung (1975) and include modifications for hyperfine structure and the Rayleigh phase function in some components of the D_2 line, suggested by Macy (personal communication).

D. The D-Line Excitation Mechanism

The early D-line observations of Brown (1974) included some very high emission fluxes from Io: several Ångstroms equivalent width for $D_1 + D_2$ when a spectrometer of large spatial aperture was used. The corresponding apparent emission rates were tens of MR if the emission came from Region A and impossibly high to explain by solar resonant scattering. On this basis, the first theory of the D-line excitation, proposed by McElroy et al. (1974), suggested an auroral mechanism involving a resonant transfer of energy to the sodium atoms from vibrationally excited N_2. Such a mechanism had been suggested by Hunten (1965) to account for D-line emission in terrestrial aurorae. In addition to explaining the high intensities, this model was attractive in that it was specific to the D-lines (no other emissions had been discovered) and could explain the broadened lines seen by Brown and Chaffee (1974), which are discussed below.

Figure 7 shows a spectrum from Trafton et al. (1974) illustrating the discovery that the sodium emission extends into space beyond Io. The circular aperture of the spectrometer was 4.7 arc-seconds in diameter, and it was set on the sky next to Io as shown. The recorded apparent emission rates were $E_{D_1} = 7.6 \pm 0.5$ kR and $E_{D_2} = 14.6 \pm 0.5$ kR, with $D_2/D_1 = 1.9 \pm 0.2$. This is Region B by the definitions in Sec. I.B, and its recognition led directly to the suggestion that resonant scattering of sunlight is indeed the emission mechanism (Trafton et al. 1974; Matson et al. 1974). The work presented in Sec. I.C shows that these intensities and their ratio are consistent with sunlight being scattered from an optically thin sodium layer. The anomalously high emissions from Io seen by Brown (1974) can be understood if this emission covered an area within $10-15$ arcsec of Io.

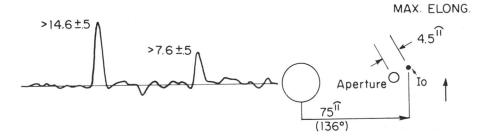

Fig. 7. A D-line spectrum of Region B from Trafton et al. (1974). The apparent emission rates in kR for the D_1 and D_2 lines are shown, and $D_2/D_1 = 1.9 \pm 0.2$. The observing geometry is shown schematically on the right. θ_{SGC} is shown in parenthesis. The elongation point is indicated with an arrow.

Bergstralh *et al.* (1975) measured the brightness of Io's sodium emission almost nightly in July and August of 1974. The measurements were all taken with the same instrumental arrangement and were reduced identically. The 3 × 8 arcsec spectrograph aperture was centered on Io, recording all Io's continuum flux plus emission from a definite portion of the surrounding space. The consistency of the observational technique clearly revealed the time signature of the resonant scattering mechanism: the variation of the emission in direct proportion to the amount of sunlight available for scattering at the appropriate Doppler-shifted D-line wavelengths (see Fig. 1). Figure 8 shows the observations and their fit to the predictions of the scattering theory. The departures of the cloud intensity from simple direct proportion to the insolation are possibly attributable to an asymmetrical sodium distribution and the varying observational perspective, which produce varying optical thickness along the line of sight as defined by the projection of the spectrograph aperture on the sky.

Evidence against any residual auroral mechanism is the absence of anomalies which could be attributed to an emission core from Region *A*. An upper limit of ~ 130 kR is set by Bergstralh *et al.* (1974) to such constant additional source. This result is supported by the absence of $\gtrsim 2$ kR emission during or near solar eclipse on Io; null findings have been obtained by a number of observers including Macy and Trafton (1975).

If the *only* D-line excitation mechanism is solar resonant scattering, then an important conclusion can be drawn: sodium in Region *A* is virtually unobservable from Earth. This was first pointed out by Brown *et al.* (1975). The reason for this curiosity is that there is very low contrast between Io's bright surface and sodium atoms in the atmosphere of Io. Those atoms scatter solar photons nearly isotropically [see the discussion by Parkinson (1975)], so that an optically thick sodium layer close around Io in Region *A* will have nearly the phase function of a Lambert sphere in scattering D-line photons. But Io's surface itself returns to us approximately as much reflected solar light as a Lambert sphere in that whole wavelength range (see Sec. I.A). Since the terrestrial observer cannot distinguish between a photon scattered from Io's surface and one scattered by a sodium atom in Region *A*, it is difficult to detect that sodium by looking for excess brightness in the D-lines.

E. *Physical Measurements of the Region* **B** *Sodium Cloud*

The D_1 and D_2 line emissions from Io are sufficiently broad to be resolved with high resolution astronomical spectrographs. This was first reported by Brown and Chaffee (1974). Spectra at higher resolution have been reported and analyzed by Macy and Trafton (1975) and Brown *et al.* (1975). Accounting for instrumental broadening, the recorded lines measure about $60 - 80$ m Å *FWHM*. All of the spectra were recorded with Io in the spectrograph slit and the continuum was subtracted to isolate the emission fea-

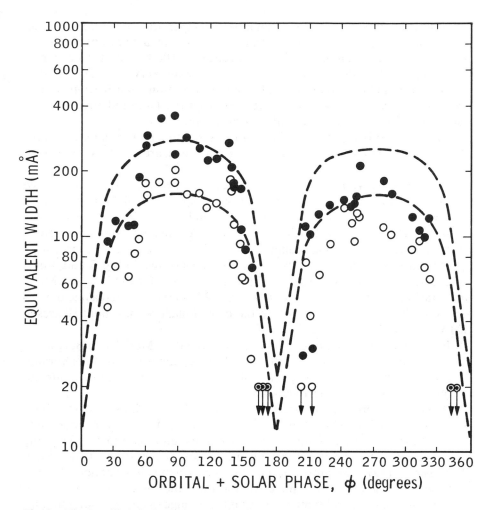

Fig. 8.   The D_2 (filled circles) and D_1 (open circles) emission line equivalent widths observed by Bergstralh *et al.* (1975). The data are plotted versus orbital plus solar phase angle, which is θ_{SHC}, the angle from superior heliocentric conjunction. The dashed curves are the solar excitation values scaled to match the 90° emission peak for both D_2 and D_1.

tures. Nevertheless, from the conclusion drawn in D above, the emission lines probably originate in Region *B*, but the light has been mixed into Region *A* observationally by blurring effects collectively called "seeing."

A simple measurement of the recorded line width does not translate directly into a kinetic temperature for the sodium. Several effects contributing to the linewidth must be accounted for: instrumental broadening, radiative transfer effects, variation of Doppler shifts and intensities during the observation, and hyperfine structure of sodium. Removing these effects, Brown *et al.* (1975), using a single-component model, report a residual breadth which translates to a sodium kinetic temperature of $5200 \pm 2300°K$. Macy and Trafton (1975) have interpreted their similar linewidth measurement to imply a much lower temperature, $500 - 1000°K$, on the basis of a two-component scheme. They envision an optically very thick kernel close around Io in addition to an optically thin contribution from Region B. The broad line from the cooler source results from the large radiative transfer correction associated with the thick component.

Bulk motion is a potential source of systematic error in determinations of the sodium temperature from Doppler-shift measurements as mentioned in Sec. I.C.

The few observations of the sodium emission in Region B which have combined photometric calibration with good spatial resolution are consistent with a partial torus of sodium leading and trailing Io in its orbit. This elongated cloud is optically thin to within a few arcsec from Io, where it reaches at least unit optical depth. The published observations are, first, spectra taken at localized points near Io, such as shown in Fig. 7 and otherwise described in Trafton et al. (1974), Trafton and Macy (1975) and Macy and Trafton (1975). Second, Brown et al. (1975) have reported a one dimensional profile of Region B parallel to the orbital plane within about 10 arcsec of Io. This is shown in Fig. 9.

Preliminary two-dimensional images described by Münch and Bergstralh (1975) are also consistent with the partial torus geometry. The uncalibrated long-slit spectra revealing Region C show that the torus may be complete at low emission levels; these data are discussed below.

A recorded apparent emission rate can be translated into a column abundance along the line of sight on the basis of Eq. (18). However the g-value must be known in order to do this with accuracy, and that requires knowing what velocity characterizes the bulk motion of that portion of the cloud. Brown et al. (1975) assumed, in the absence of better information, that the value of g for the optically-connected cloud was the same as for Io. Sodium distributed along Io's orbit must surely have a range of sodium-Sun Doppler shifts and therefore experience a range of g-values. The only published observations pertinent to this topic are the 22 October 1973 spectra of Trafton et al. (1974) and a Region C spectrum of Mekler and Eviatar (1974). The latter is discussed below. The former reports the same sodium-Earth Doppler shift for rather widely separated spots in Region B. Trafton and Macy (1975) have recently reported asymmetrical D-line profiles in spectra taken at very high resolution which may point to differential bulk motion in the Region B sodium cloud on the scale of a few arcsec.

The question of the D_2/D_1 ratio in Region B differing from the value ($D_2/D_1 \approx 1.8$) expected for a solar resonant scattering mechanism has been rather widely discussed in the literature. There are simply not enough high-quality measurements of that ratio to indicate that there is a problem.

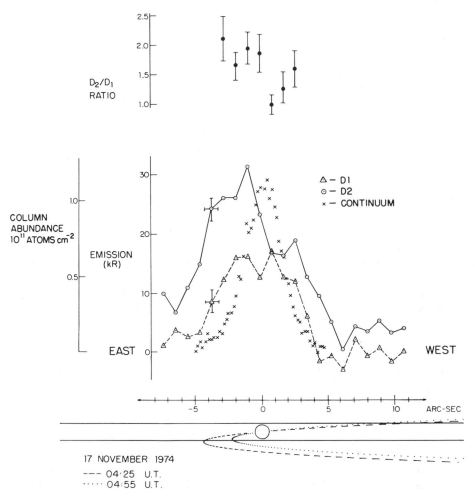

Fig. 9. A one-dimensional profile of the Region *B* sodium emission. At bottom, the geometric disc of Io is shown fixed in the projected slit. The orbit, which drifted with respect to the slit, is sketched for the beginning and end of the observation. At center, the D_1 and D_2 emission profiles are plotted on the same abscissa. Each point is both an average over one arc-second of height along the spectrograph slit and an integration over 0.64 Å along the dispersion. The continuum profile locates Io's image and is scaled arbitrarily. The sodium column abundance scale is valid for the D_2 emission. At the top, the D_2/D_1 ratio is plotted over the range in which it is well determined. (From Brown *et al.* 1975).

F.   Sodium in Region C

Detection of D-line emission remote from Io has been reported by Mekler and Eviatar (1974), Wehinger and Wyckoff (1974), Mekler *et al.* (1975) and Wehinger *et al.* (1976). All these observations were made with the same instrument, namely the fast and efficient low-dispersion image tube spectrograph at the Wise Observatory of Tel-Aviv University. Little quantitative work has been reported, but visual inspections of the plates have led the investigators to describe a highly flattened distribution of sodium emission concentrated in the orbital plane of the satellites. In that plane it is discernable to about 4 (Mekler *et al.* 1975) or 10 (Wehinger and Wyckoff 1974) arcmin from Jupiter. It is visible $\frac{1}{2}$ to 1 arcmin above and below the orbital plane, and even over the poles of Jupiter (Wehinger *et al.* 1976). Mekler *et al.* (1975) refer to an inner limit to the distribution about 1 arcmin from Jupiter.

To dramatize the extent of Region C, we note that the elongations of Io's orbit are a little more than 2 arcmin from Jupiter. During the past two years the jovicentric declination of the earth has been less than about $1°7$, which means that the semi-minor axis of Io's apparent orbit was less than about 4 arcsec for the period of these observations.

Wehinger *et al.* (1976) report photometric comparison of their Region C observations with the 5577 Å telluric airglow line due to atomic oxygen. They find D-line intensities between 100 R and 2 kR in the orbital plane.

Mekler and Eviatar (1974) report that on one occasion they observed a strong tilt of the D emission line with respect to the telluric airglow lines on the same spectrum. The sodium line extended about 2 arcmin along the projected slit, which was oriented at 25° with respect to the orbital plane. The Doppler-shift difference between the two ends of the emission line translates to a velocity difference of 200 km sec^{-1}. Wehinger *et al.* (1976) report that they have not observed this tilting effect.

The photometric reduction of image tube spectra of the Region C emission must be done with great care. Visual inspection of the Mekler *et al.*, (1975) spectra show strong contamination by non-linear response in the photographic emulsion. The emission lines are superimposed on Jupiter's or a satellite's scattered continuum which varies over the height of the slit and biases the response to the emission lines to different parts of the film's characteristic curve. The result is seen as non-uniformity over the height of the slit of the 5577 Å airglow line. It is not clear to what extent this effect has influenced the qualitative conclusions drawn by the observers from these plates.

II. ULTRAVIOLET EMISSION OBSERVED FROM PIONEER 10

The "H" channel of the two-color ultraviolet photometer aboard Pioneer 10 recorded emission originating in the vicinity of Io. These measurements have been described by Judge and Carlson (1974) and Carlson and Judge (1974, 1975).[3] We summarize here the rather sparse observational knowledge

---

[3]See p. 1079.

about the source, its spatial distribution and the mechanism which excites the ultraviolet emission.

The investigators tentatively identified the emission as Lyman α resonantly scattered by atomic hydrogen. The transition has a large oscillator strength, the solar flux is high at that wavelength (1216 Å), and hydrogen is an abundant element. However, the identification cannot be considered definitive. Figure 10 shows the spectral responses of the two channels. The fact that the emission was seen only in the "H" channel simply constrains the wavelength to be greater than 800 Å. However, increasingly higher emission rates are required than those derived on the basis of the Lyman α identification if the actual wavelength is longer than 1216 Å.

The ultraviolet emission comes from an elongated section of Io's orbit approximately centered on the satellite. This is shown in Fig. 11, which is from Carlson and Judge (1975).

An upper limit to the extent of the emission normal to Io's orbital plane was established by the emission's disappearance in solar eclipse by Jupiter: it is less than one Jupiter diameter thick. This observation also implies that the excitation mechanism is solar resonance scattering.

If the emission is Lyman α, the apparent emission rate of the hydrogen cloud is ~ 300 R. The radial column abundance is calculated to be $\sim 3 \times 10^{12}$ cm^{-2}, which corresponds approximately to unit optical thickness in Lyman α. The implied total population of the cloud is $\sim 2 \times 10^{33}$ atoms (Carlson and Judge 1974).

III.   OTHER OPTICAL LINE EMISSIONS

No optical emissions from Io or its vicinity have yet been positively identified other than those discussed in Secs. I and II, though there have been intensive but not exhaustive searches.

Mekler and Eviatar (1974) reported the possible detection of emission by neutral calcium at 4227 Å. Using the Wise Observatory's image tube spectrograph, a continuum spectrum of Io (Io in the spectrograph slit) and a spectrum of the moon were taken. They compared the central intensities of two Fraunhofer lines at 4227 Å (CaI) and 4236 Å (FeI), and found the calcium feature for Io to be 10% less deep than for the moon. Assuming the iron line to be constant, the observers ascribed this difference to calcium emission from Io's Region A. This procedure is doubtful due to the substantially different strengths of the two Fraunhofer features which were compared. Two possible sources of systematic error not considered by Mekler and Eviatar are the consequences of non-linear detection and the ring effect, most recently discussed by Barmore (1975).

The same calcium resonance line has been sought by other observers using higher-precision techniques, but it has not been detected (Macy and

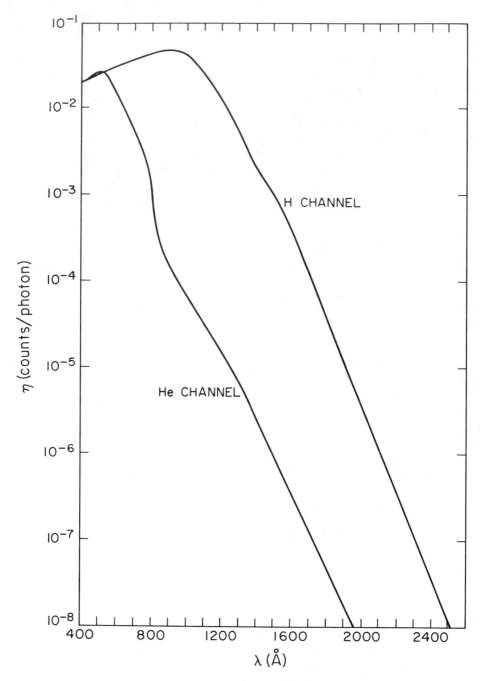

Fig. 10. Spectral response of the UV ultraviolet photometer on Pioneer 10. The quantum efficiency in photoelectrons per photon is shown for the two channels as a function of wavelength and includes photoelectric response, filter transmission and collection efficiency. (From Carlson and Judge 1974.)

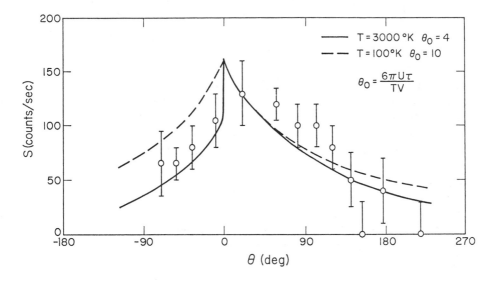

Fig. 11.   The observed H Lyman-α intensity as a function of the orbital angle with respect to Io. Positive values of θ are for the trailing portion of the cloud. The figure is taken from Carlson and Judge (1975), and the curves are for their models of the hydrogen cloud formation.

Trafton 1975; Brown *et al.* 1975). The upper limits to 4227 Å emission quoted by these authors is about 5 kR referred to the size of Io's disc.

A possible detection of helium emission from Io at 10,830 Å was reported by Cruikshank *et al.* (1974). They have since announced that the observation was spurious and due to an instrumental problem that is now understood. Macy and Trafton (1975) have reported that 0.5 kR of 10,830 Å emission from the spatially uniform telluric airglow was observed when they attempted confirmation of the earlier report.

Other negative results have been reported. From Trafton *et al.* (1974) with no quoted upper limit: 6160.7 Å (NaI), 8194.8 Å (NaI), 6707.8 Å (LiI), 7699.0 Å (KI), 5183.7 Å (MgI) and 6300.2 Å (OI). Macy and Trafton (1975) report the following upper limits: 5 kR of 6707.8 Å (LiI), 5 kR of 7699.0 Å (KI) and 0.17 kR of 4571 Å (MgI). These measurements were made with Io in the spectrograph slit.

Recently Trafton (1976) has reported observations of Region *B* in the spectral range 3100 Å to 8700 Å. No emission feature other than the D-lines was found. Upper limits are 0.4 – 0.6 kR in the red, 0.6 – 1.6 kR in the green and blue, and up to 5 kR in the violet.

The Region *C* spectra of Wehinger *et al.* (1976) were examined for emission features other than the sodium D-lines and telluric airglow lines. The detection limit at 5900 Å was below 100 R. No new features were discovered between 3500 Å and 9000 Å.

IV. FUTURE OBSERVATIONAL WORK ON IO'S LINE EMISSIONS

Observations during the 1975–76 apparition of Jupiter will probably result in a clearer description of the spatial distribution and motions of Io's
sodium cloud. Io's orbital period is about 42 hours and we view its orbit from
the side, so a series of two-dimensional D-line images are required to visualize the cloud's morphology. The instrumental difficulties inherent in this
task are formidable. Jupiter and a halo of its light scattered in the optics
dominate the field of view, and in the immediate vicinity of Io the skirts of
Io's seeing disc also contribute a continuum background. To achieve good
contrast between the narrow emission lines and this background, conventional techniques require high spectral resolution. Continuing an example
given in Sec. I.B, consider Io with opposition magnitude and 2 arcsec seeing. Suppose we wish to image a portion of Region B where the continuum
apparent emission rate of Io is down from its central value by a factor $q =
E/E_0$ while $E_0 = 1.5$ MR Å^{-1}. Using a filter with bandpass $\Delta\lambda$ the ratio of
D-line to continuum signal is

$$\frac{S}{N} = \frac{E_{\mathrm{D}}}{\Delta\lambda \, q \, E_0}.$$ (22)

Taking $q \simeq 10^{-2}$ and $E_{\mathrm{D}} = 10$ kR, the requirement on the spectral resolution for $S/N = 10$ is

$$\Delta\lambda \lesssim 0.1 \text{ Å}$$ (23)

which is rather severe.

Imaging techniques at high spectral resolution will provide important
information about the sodium's motion by measuring the Doppler shift. An
innovative imaging technique is being developed by Mertz (1975) for studying Io's sodium cloud. With an interferometer he creates fringes across the
telescopic field in D-line radiation; the background continuum radiation
produces no fringes. The distribution of sodium is present as AC information
on the detector and is recovered by a heterodyne technique. A similar technique has been described by Blamont and Courtès (1955).

The search for other optical emissions from Io's environment will be
continued. To discover such emissions from Region B, near Io, is an easier
observational task than looking for features on the continuum spectrum of
Io. This point follows from the discussion in Sec. I.D of contrast with Io's
continuum radiation. Io's albedo rises steeply from the ultraviolet to near infrared, with Lambert-sphere-equivalence achieved near 0.6 μm. The Region
A contribution of an isotropic resonant scatterer is in the sense of an emission line for shorter wavelengths and an absorption feature for longer wavelengths, but the contrast is low. However, if the constituent is also in Region
B, then its resonance line may have good observational contrast against the
dark sky for spectra taken with Io just outside the spectrograph slit.

V. IO'S IONOSPHERE

The only existing evidence of an atmosphere on Io is the two electron density profiles in Fig. 12, deduced by Kliore *et al.* (1975) from the S-band occultation experiment on Pioneer 10. The electron density distribution on the dayside shows a peak concentration of 6×10^4 cm^{-3} at an altitude of 100 km and an extent of 750 km with a nearly constant scale height of 200 km. On the nightside the peak electron density is 1×10^4 cm^{-3} and the ionosphere appears to cut off abruptly above 200 km. These peak electron densities are comparable to those observed on Mars and Venus (Kliore *et al.* 1965; Kliore *et al.* 1967), even though the solar flux at Io is much weaker. This sets important constraints on the rates of production and loss of ions. The contrast between the day and night profiles suggests that ions are removed in a time shorter than an Io day (42 hours).

In this section we review four existing models for Io's ionosphere. The basic assumptions and predictions of each are summarized in Table III. As a group they can be criticized for being parochial; they are based on our experience with the atmospheres of the terrestrial planets and ignore, or do not fully account for, the interaction between Io and the Jovian magnetosphere. Detailed plasma interaction models are currently being studied by F. M. Neubauer,[4] Shawhan (1975) and the Harvard group under McElroy. We conclude this section with a summary of what we have learned from studying the ionosphere of Io.

A. The Webster et al. Model

The Webster *et al.* model (1972) was motivated by the decametric radiation theory of Goldreich and Lynden-Bell (1969). Proposed before the Pioneer 10 encounter, it is inadequate in a number of ways, but it conveys the spirit of later photochemical models. The bulk atmosphere is assumed to consist largely of methane with a surface number density of 10^{13} molecules cm^{-3}. The primary process of interest is photoionization of CH_4 by ultraviolet sunlight:

$$CH_4 + h\nu \rightarrow CH_4^+ + e \qquad (24)$$

followed by dissociative recombination

$$CH_4^+ + e \rightarrow CH_3 + H . \qquad (25)$$

The electron number density $[n_e]$ is simply given by

$$[n_e] = \{J[CH_4]/\alpha\}^{\frac{1}{2}} \qquad (26)$$

where $[CH_4]$, J and α respectively denote CH_4 number density (cm^{-3}), photoionization rate (sec^{-1}) and dissociative recombination coefficient (cm^3

---

[4]Oral presentation at the Io Week Workshop, Feb. 22, 1975, Harvard University.

TABLE III
Summary of Io Model Ionospheres

ASSUMPTIONS	Webster et al.	McElroy & Yung (A)	McElroy & Yung (B)	Whitten et al.
Atmospheric Composition (cm^{-3})	CH_4 (1×10^{13})	NH_3, N_2, Na (10^{11})[a], (10^9), (3×10^6)	NH_3 (10^{11})	Ne (4×10^9)
Atmospheric Temperature	100°K	500°K	500°K	160°K
Major Ion	CH_4^+	Na^+	NH_3^+	Ne^+
Electron Source	$CH_4 + h\nu \rightarrow CH_4^+ + e$	$Na + h\nu \rightarrow Na^+ + e$	$NH_3 + p \rightarrow NH_3^+ + p + e$ $NH_3 + e \rightarrow NH_3^+ + e + e$	$Ne + h\nu \rightarrow Ne^+ + e$
Electron Loss	$CH_4^+ + e \rightarrow CH_3 + H$	$Na^+ + e \rightarrow Na + h\nu$ Diffusion to surface	$NH_3^+ + e \rightarrow NH_2 + H$	Diffusion to surface
PREDICTIONS				
Peak Electron Density	7×10^3 cm^{-3}	6×10^4 cm^{-3}	6×10^4 cm^{-3}	1×10^5 cm^{-3}
Altitude of Peak	200 km	100 km	100 km	50 km
Scale Height	58 km	200 km	200 km	200 km
Time Constant	10^3 sec[b]	one Io day	not estimated	one Io day

[a] at night $[NH_3] = 10^9$ cm^{-3} [b] estimated by present authors

sec^{-1}). The model successfully predicts the existence of an ionosphere. The computed $[n_e]$ profile is illustrated in Fig. 13. By varying the surface number density of CH_4 and the atmospheric temperature[5] as parameters we can adjust the model to agree with the Pioneer 10 observations in many essential aspects. But the small peak electron density is an intrinsic feature of this model, indeed of any other model in which the dominant ion is a molecular ion produced by photoionization and removed by dissociative recombination (cf. Model 1 of McElroy and Yung 1975).

It is interesting to note that the first model of an atmosphere on Io suggests that Io would be a large source of hydrogen. Photolysis of CH_4 by sunlight readily leads to production and escape of hydrogen at a rate $\sim 10^{10}$ atoms cm^{-2} sec^{-1}:

$$CH_4 + h\nu \rightarrow CH_3 + H. \qquad (27)$$

Webster *et al.* did not pursue the consequences of methane dissociation.

B.   The McElroy and Yung Model (A)

The column abundance of sodium atoms in Region *B* has been observed to be $\sim 10^{11}$ cm^{-2} (Brown *et al.* 1975). The amount of sodium in Region *A* cannot be estimated directly from observations for reasons discussed in Sec. I.D. But, if Io is the source of sodium atoms in the cloud, the atmosphere probably contains more than 10^{12} sodium atoms cm^{-2}.

The possibility of an ionosphere with Na$^+$ as the major ion was first recognized by McElroy *et al.* (1974) prior to the Pioneer 10 encounter. The idea was pursued and developed by McElroy and Yung (1975). They proposed an ammonia atmosphere with sodium as a minor component. Because of its low ionization potential (5.14 eV), sodium atoms are readily ionized by absorption of ultraviolet photons shortward of 2412 Å:

$$Na + h\nu \rightarrow Na^+ + e. \qquad (28)$$

The Na$^+$ ions will be removed via radiative recombination:

$$Na^+ + e \rightarrow Na + h\nu \qquad (29)$$

and by molecular diffusion to the surface. Note the crucial difference between an atomic ion like Na$^+$ and a molecular ion like CH_4; the radiative recombination coefficient for an atomic ion is typically about 10^5 times slower than the dissociative recombination coefficient for a molecular ion. Consequently the presence of even a small concentration of sodium atoms in the atmosphere could lead to the formation of large numbers of sodium ions. The model includes details of molecular diffusion, diurnal variation, and bulk vertical motion due to condensation and evaporation of ammonia. The results for electron number densities are illustrated in Fig. 14.

[5]In an internally consistent model, the temperature must be calculated from the heating rate based on atmospheric composition and is not a free parameter.

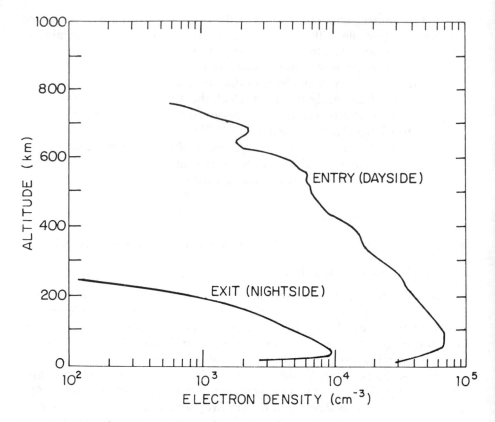

Fig. 12.   Entry (dayside) and exit (nightside) electron density profiles as measured on Io by
Pioneer 10. (Kliore *et al.* 1975.)

C.   The McElroy and Yung Model (B)

A model involving charged energetic particles in the Jovian radiation belt as the ionization source was explored by McElroy and Yung (1975). The major atmospheric gas is assumed to be NH_3. NH_3^+ ions are primarily produced by proton or electron impact:

$$NH_3 + p \rightarrow NH_3^+ + e + p$$
$$NH_3 + e \rightarrow NH_e^+ + e + e. \tag{30}$$

The loss mechanism is dissociative recombination:

$$NH_3^+ + e \rightarrow NH_2 + H. \tag{31}$$

The cross-section of a 10 Mev proton or a 10 kev electron is about 10^{-18} cm^2 and varies as $\log E/E$, where E is the energy. The calculation shows that it is possible to account for the principal features of the observed ionosphere

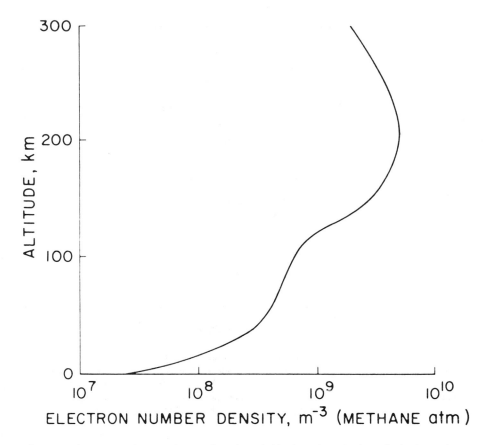

Fig. 13. Electron number density as a function of altitude at the subsolar point in the methane atmosphere assumed for Io. (Webster *et al.* 1972.)

with a flux of 3×10^7 cm^{-2} sec^{-1} of 10 Mev protons impinging on an atmosphere with a surface number density equal to 3×10^{11} cm^{-3}. The results are shown in Fig. 15. To explain the diurnal difference, it must be postulated that the energetic particles are entering Io's atmosphere asymmetrically, a view consistent with the implications of the electromagnetic interactions in the vicinity of Io (Shawhan *et al.* 1974; Shawhan 1976).

D. The Whitten et al. Model

Recently Whitten *et al.* (1975) proposed a neon ionosphere. Io is assumed to have a tenuous neutral atmosphere composed of neon with surface density about 4×10^9 cm^{-3}. Photoionization is the main source of Ne$^+$. Loss of ions is due to diffusion to the surface. The main feature of the model, as the authors pointed out, is that the neutral atmosphere is relatively cold ($\sim 160°$K) and tenuous. The neon model does not treat Io's hydrogen source,

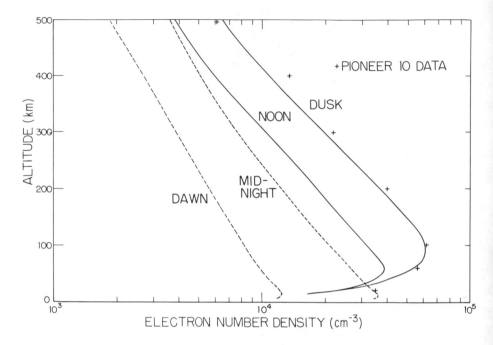

Fig. 14.   Electron density profiles for Io's ionosphere. Na^+ is the major ion. NH_3 is the major neutral species. In the daytime the surface number density of Na and NH_3 is taken to be 3.7×10^6 cm^{-3} and 5.0×10^{10} cm^{-3} respectively. At night NH_3 is reduced by a factor of 5 and a vertical downward motion of 5 m sec^{-1} is imposed on the atmosphere. The calculation is carried out for an isothermal atmosphere at 500°K. (McElroy and Yung 1975).

and it ignores a potentially important reaction with sodium. Charge transfer by neon ions to sodium atoms is probably fast,

$$Ne^+ + Na \rightarrow Ne + Na^+ . \tag{32}$$

The chemical lifetime for Ne^+ is given by

$$\tau = \frac{1}{k \cdot [Na]} \tag{33}$$

where k is the charge-transfer rate coefficient and $[Na]$ is the number density of sodium atoms. If we assume $k \approx 10^{-9}$ cm^3 sec^{-1} and $[Na] \approx 10^5$ cm^{-3}, then $\tau \approx 10^4$ sec, a value which is comparable to the diffusion time constant and suggests that Na^+ probably cannot be ignored. This reaction may be important.

E.   Tentative Conclusions about Io's Neutral Atmosphere

From the study of a variety of models of Io's ionosphere, we are able to suggest tentative answers to two important questions about Io's neutral at-

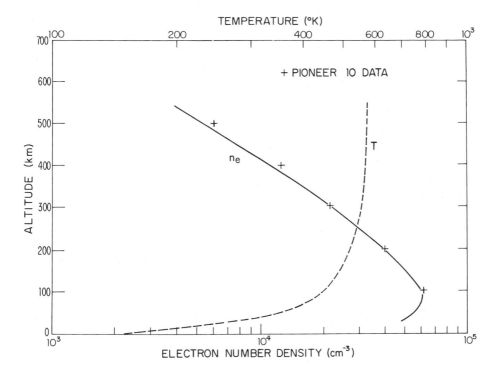

Fig. 15.   Electron number density and atmospheric temperature profiles computed with the McElroy and Yung Model *B*. (McElroy and Yung 1975.)

mosphere: how hot was it at the time of the encounter and how dense is it? For ambipolar diffusive equilibrium, the plasma scale height is given by

$$H = \frac{k}{m_0 g} \left[\frac{T_e + T_i}{A_i} \right]$$ (34)

where m_0 is the atomic mass unit, k is the Boltzmann constant, g is Io's gravitational acceleration, A_i is the mass number of ion species, and T_e and T_i are the electron and ion temperatures respectively (cf. Bauer 1973). Assuming $T_e = T_i$ and using $g = 178$ cm sec^{-2} (Anderson *et al.* 1974), the scale height observation implies

$$\frac{T_i}{A_i} = 21.5 .$$ (35)

We must emphasize that the expression is derived on the assumption of diffusive equilibrium for the ions. Inclusion of a solar-wind type interaction can seriously modify the interpretation (Cloutier *et al.* 1969). For protons the implied temperature is 21.5°K, for Na$^+$ it is 500°K. In thermodynamic equilibrium, T_i equals T_n (the temperature of the neutral atmosphere), but in general T_i is higher than T_n. The inequality

$$T_n < T_i = 21.5\, A_i \tag{36}$$

implies, for reasonable values of A_i, that the neutral atmosphere did not exceed 1000°K at the time of Pioneer 10 encounter.

The inference of a lower limit to the density of the neutral atmosphere from the detected ionosphere is model-dependent. In the Whitten *et al.* Model, the absolute lower limit is $\sim 4 \times 10^9$ cm^{-3}. Below this value loss of ions by diffusion will be too fast and the computed ionospheric peak will be too small to account for the observations.

In the McElroy and Yung Model A, the constraint of the ionosphere is

$$N_0 \cdot [\text{Na}]_0 \approx 10^{16}\ \text{cm}^{-6} \tag{37}$$

where N_0 is the neutral number density and $[\text{Na}]_0$ is the sodium number density at the surface. The flux of neutral Na atoms to the surface is given by

$$\phi = - D \left\{ \frac{d[\text{Na}]}{dz} + \frac{[\text{Na}]}{H} \right\} \approx - D\, \frac{[\text{Na}]_0}{H}\ \text{cm}^{-2}\ \text{sec}^{-1} \tag{38}$$

where D is the diffusion coefficient and H is the scale height. Combining Eqs. (37) and (38),

$$\phi \times N_0^2 \approx 3 \times 10^{27}\ \text{cm}^{-8}\ \text{sec}^{-1}. \tag{39}$$

Thus, if we require $\phi < 10^7$ cm^{-2} sec^{-1}, then $N_0 \gtrsim 1.7 \times 10^{10}$ cm^{-3}.

In McElroy and Yung Model B the relationship between N_0 and the flux of energetic particles f_0 is given by

$$N_0 f_0 \approx 3 \times 10^{18}\ \text{cm}^{-5}\ \text{sec}^{-1}. \tag{40}$$

The Pioneer 10 measurements impose an upper limit of about 10^9 cm^{-2} on f_0, and hence $N_0 > 3 \times 10^9$ cm^{-3} for this model.

Since McElroy and Yung Models A and B only when combined touch on all the observational aspects—the sodium and hydrogen as well as the ionosphere—we consider that a best estimate to the neutral density is

$$N_0 \approx 10^{10}\ \text{cm}^{-3} \tag{41}$$

a value which is about two orders of magnitude lower than the upper limit deduced from the observations of the β-Scorpii occultation (Taylor *et al.* 1971; Smith and Smith 1972).

VI. IO AS A SOURCE OF HYDROGEN

As discussed in Sec. II, the Pioneer 10 ultraviolet observations suggest that Io is surrounded by an extensive hydrogen cloud containing $\sim 10^{33}$ hydrogen atoms. To account for the cloud's geometry Carlson and Judge

(1974) and McElroy and Yung (1975) independently estimated a lifetime of $\sim 10^5$ sec for the hydrogen atom. This lifetime is considerably shorter than the photoionization lifetime of $\sim 10^8$ sec and is probably due to charge transfer of hydrogen atoms with low-energy protons in the Jovian magnetosphere. The calculations suggest the existence of a low-energy proton flux $\sim 10^9$ cm^{-2} sec^{-1}, a result that has recently been supported by the detailed analysis of plasma observations by Pioneer 10 in the neighborhood of Io's flux tube (Frank *et al.* 1975). If Io were the only source of hydrogen, a mean flux of $\sim 10^{11}$ atoms cm^{-2} sec^{-1} must be supplied from the surface of Io to maintain the observed cloud in the steady state. But we have no observational evidence that the hydrogen cloud is in a steady state, so the actual demand on the hydrogen source could be much lower. We will next consider the implications for the hydrogen source in three subsections. In A we examine photolysis of NH$_3$ as a source of H, and in B we discuss the possibility of cycling hydrogen with magnetospheric protons. In C possible escape mechanisms for hydrogen are discussed.

A. Photolysis of NH$_3$ as the Source of Hydrogen

If the hydrogen cloud were assumed to be in a steady state with the supply coming solely from Io, the required flux would be 10^{11} atoms cm^{-2} sec^{-1} at Io's surface. Such a large escape rate can best be appreciated by noting that it would represent loss of over 10 km of surface material to space in the age of the solar system. It is reasonable to hypothesize that the surface of Io contains a volatile material rich in hydrogen. Evaporation of this material and its subsequent photolysis can provide a large source of atomic hydrogen. Ammonia frost is an obvious candidate, but arguing against its presence is the lack of ammonia absorption features in the infrared spectrum of Io (Fink *et al.* 1973; Pilcher *et al.* 1972). McElroy *et al.* (1974) have thoroughly discussed the various aspects of the ammonia hypothesis. The most important reason for assuming an ammonia atmosphere is the need to supply a huge escape rate of hydrogen. The most serious problem associated with this model is the escape of nitrogen, the terminal product of photodissociation of ammonia. Note that the nature of the problem is not changed if we assume a methane atmosphere. Photolysis of a hydrogen-bearing molecule releases the hydrogen atoms and leaves the heavy "residue" in the atmosphere. The rate at which the heavy component would accumulate must essentially equal the rate of hydrogen escape and would result in an atmosphere in excess of the observed upper limit in less than 10 years. To circumvent the problem of a rapid build-up of the end-products of photolysis, McElroy *et al.* (1974) postulated escape of a large amount of gases during periods of intense heating in the upper atmosphere initiated by the precipitation of energetic particles from the radiation belt.

B. Proton Exchange

Fanale *et al.* (1975, 1976) have made the most complete study of the surface of Io based on its photometric properties, and they also considered cosmochemical composition and thermal history. They concluded that Io's properties could best be explained by postulating that it is largely covered by evaporite salts, rich in sodium and sulphur. They considered the presence of large amounts of frost unlikely. In view of this model it is reasonable to ask whether there is an alternative to ammonia ice as a source for the hydrogen cloud observed by Pioneer 10.

Protons in the extensive magnetosphere of Jupiter obviously represent a potential hydrogen source. Frank *et al.* (1975) have analyzed the directional, differential spectrum for proton densities $\frac{\partial J}{\partial E}$ (cm²-sec-ster-eV)⁻¹ at positions inside the Io flux tubes. The proton spectrum is approximated well by a Maxwellian distribution with proton number density $N_p = 60$ cm⁻³ and mean energy $\epsilon_p = 95$ eV. The flux of protons impinging on Io is on the order of

$$N_p \sqrt{\frac{2\epsilon_p}{M_p}} \approx 10^9 \text{ cm}^{-2} \text{ sec}^{-1} \tag{42}$$

where M_p is the proton mass.

This falls about two orders of magnitude short of the required source, so a large number of protons outside the observed energy range must be postulated to make this a viable mechanism. McDonough (1975) proposed an interaction between Io and a low-energy component of the magnetospheric plasma. The latter is assumed to originate from Jupiter and could provide a source of hydrogen atoms to Io. But again it is not clear that this mechanism can provide enough hydrogen to explain the observations.

In view of the uncertainties of observations and models we do not regard this deficiency as serious, especially since we have no observational evidence for the stability of the hydrogen cloud. The magnetospheric protons could be thermalized and accumulated in the atmosphere and blown off sporadically when the atmosphere is heated by an electromagnetic interaction.

C. Hydrogen Escape

Escape of a light species like hydrogen from Io is an efficient process. A hydrogen atom with the thermal kinetic energy associated with temperatures in excess of 190°K can escape Io's gravity. The escape rate is given by the Jeans formula

$$\phi = \frac{n_c v_0}{2(\pi)^{\frac{1}{2}}} (1 + \lambda)e^{-\lambda} \tag{43}$$

where n_c is the hydrogen number density at the exobase, $v_0 = (2kT/m)^{\frac{1}{2}}$ and $\lambda = GM_I/rkT$ (Chamberlain 1965). All the current models on the dynamics of the hydrogen cloud (Carlson and Judge 1974; Smyth personal communication 1975) are based on Jeans' theory, modified somewhat to include effects of the three-body gravitational interaction and the possibility of non-thermal velocity distribution for the escaping atoms. A crucial assumption in deriving the Jeans result is that the atmosphere is in hydrostatic equilibrium:

$$\frac{\mathrm{d}}{\mathrm{d}r}\,(nkT) + \frac{GM_I mn}{r^2} = 0 \tag{44}$$

where n = number density of gas, and other symbols have their usual meanings. Parker (1958) has shown that for thermal energy large compared with gravitational energy the hydrostatic equation must be replaced by a set of hydrodynamic equations

$$mnv\,\frac{\mathrm{d}v}{\mathrm{d}r} + \frac{\mathrm{d}}{\mathrm{d}r}\,(nkT) + \frac{GM_I mn}{r^2} = 0 \tag{45}$$

$$\frac{\mathrm{d}}{\mathrm{d}r}\,(r^2 nv) = 0\,. \tag{46}$$

Physically, Eq. (45) is the equation of motion obtained by adding an inertial term to Eq. (44), and Eq. (46) therefore is the equation of continuity. When the thermal energy per particle exceeds its gravitational potential energy, that is,

$$\frac{3}{2}\,kT > \frac{GM_I m}{r} \tag{47}$$

the gas is free to expand rapidly or "blow-off" (Öpik 1963). For Io the critical temperature for the stability of a hydrogen-dominated exosphere is 216°K (Gross 1974), so from our understanding of the ionosphere the temperature in Io's upper atmosphere probably exceeds the blow-off temperature for hydrogen.

For hydrogen alone, whether the escape occurs by Jeans escape or by blow-off makes little difference. The limiting factor is probably in the ultimate supply of hydrogen atoms and diffusion through the atmosphere (Hunten 1973). Perhaps the most appealing feature of the blow-off mechanism is that it may not be as selective as Jeans escape in discriminating against heavy atoms like sodium (Gross 1972). We will return to this point when we discuss the escape of sodium atoms.

VII.   IO AS A SOURCE OF SODIUM

The observations of sodium emission in Region B reveal that the cloud is remarkably stable (Bergstralh et al 1975) and that solar resonant scattering is probably the excitation mechanism. This suggests that if Io were the source of sodium the required escape rate would be of order 10^7 atoms cm^{-2}

sec^{-1} (McElroy and Yung 1975; Macy and Trafton 1975). This deduction is based on an estimated population of 10^{30} sodium atoms and a photoionization lifetime of 10^6 sec (Hunten 1954). Comparison with Sec. VI shows that the estimated sodium source is four orders of magnitude smaller than the hydrogen source; the hydrogen atoms are approximately 10^3 times more numerous in Region B, and they are depleted about 10 times faster.

In the following subsections A and B we discuss two possible sources for sodium in the atmosphere of Io, namely sputtering from the surface and a meteoritic source. In Sec. C we shall discuss escape processes which can move this sodium into Region B where it is observed. In Sec. D we comment briefly on the significance of the absence of other metallic emission lines.

A.   Sputtering

It is difficult to release sodium from the surface since metallic sodium as well as most common sodium compounds have low vapor pressures at Io's surface temperature ($\sim 140°K$). Matson et al. (1974) proposed a sputtering mechanism involving the energetic particles in the Jovian magnetosphere. To date, this mechanism is the only plausible means we know for ejecting sodium from the surface into the atmosphere, but there are a number of difficulties, most of which have been discussed by Matson et al. (1974). The first potential problem is the presence of an atmosphere which may attenuate the influx of energetic particles. If the surface density of the neutral atmosphere is 10^{10} cm^{-3} and its scale height is 100 km, it will only be thin for interactions having cross-sections smaller than about 10^{-17} cm^2. Since that is approximately the cross-section for a one-MeV proton collision with a molecule, high energy protons probably reach the surface without substantial attenuation. The second problem is to have an adequate number of incident particles. In the keV – MeV range, the measured proton flux falls short of that required by more than an order of magnitude. It is conceivable that an "avalanche" process discussed by Matson et al. (1974) could remedy this deficiency, and we expect detailed models to carry this idea further.

B.   Meteoritic Source

A number of papers (Morrison and Cruikshank 1974; Mekler et al. 1975; Sill 1974) have suggested meteoritic material as the source of Io's sodium. This idea is largely motivated by our understanding of the terrestrial sodium emission. Hunten and Wallace (1967) and Donahue and Meier (1967) first suggested that the origin of Earth's sodium atoms could be connected to the presence of dust. Possible mechanisms for releasing the sodium atoms include meteoric ablation (Gadsden 1968) and sublimation of dust particles (Fiocco and Visconti 1973; Fiocco et al. 1974).

Measurements of micrometeoroids in the Jovian environment by Pioneer 10 and 11 (Humes et al. 1974)[6] indicated that the mass flux of meteoroids

---

[6]See p. 1064.

intercepted by Io could be two orders of magnitude higher than that measured for Earth ($\sim 2 \times 10^{-16}$ g cm^{-2} sec^{-1}). Assuming a sodium concentration of 0.66% by weight, a value typical of chondritic meteorites (Junge *et al.* 1962), we estimate the net sodium source to be $\sim 3 \times 10^6$ atoms cm^{-2} sec^{-1}. The sodium atoms can be released from the meteorites by evaporation at the moment of impact or by sputtering after the material has settled down on the surface. The first mechanism faces the difficulty of explaining why sodium has not been observed on any one of the other Galilean satellites. The second mechanism would suggest that the sodium content in the Jovian meteorites must be higher by at least a factor of 10 than that typical of chondritic meteorites. These difficulties must be answered by any future theory that seriously proposes a meteoritic origin of Io's sodium.

C. Escape of Sodium

While the atmosphere could be thin to the incident energetic particles, the discussion in Sec. V.E indicates that it probably appears thick to the sputtered atoms, which typically have an energy of a few eV and a cross-section of $\sim 10^{-15}$ cm^2. The sodium atoms are probably thermalized in the atmosphere. The meteoritic source also yields sodium only to the bound atmosphere, so an escape mechanism must be formulated. Jeans escape would require an exospheric temperature of order 10,000°K, which could be achieved sporadically (as described in Sec. VI.A). Matson *et al.* (1974) considered a solar wind-type interaction between the upper atmosphere and the magnetospheric protons. Since the sodium problem is probably related to that of the hydrogen, we can gain some insight by examining Io's hydrogen escape. Gross (1972) has shown that hydrogen in a moderately hot thermosphere is not stable against a blow-off (see Sec. VI.C). If a hydrogen blow-off occurs with sodium as a minor component, it is possible that sodium will be dragged along.

D. Implications of Undetected Emissions

The absence of other metallic emission lines is puzzling. Considerations of cosmic abundance, oscillator strength and solar radiation lead us to expect that a number of optical resonance lines, notably Ca (4227 Å), Al (3962 Å), K (7665 Å), could have intensities comparable to those of the sodium D-lines. But as discussed in Sec. III, no other metallic lines have yet been identified, despite considerable effort devoted to the searches.

It is of interest to note a similar circumstance for the comets. The visible spectrum of Comet Mkros shows an extensive sodium emission, but no other metallic lines have been detected (Greenstein and Arpigny 1962). Greenstein and Arpigny considered their absence significant and drew attention to the difference in dissociation potentials of the oxides and hydrides of metals. In general, the dissociation potential of the hydrides is low, e.g., 2.2 eV for NaH, < 2.5 eV for MgH, < 3.1 eV for AlH, and < 1.7 eV for CaH. The oxides have higher dissociation potentials, e.g., 3.7 eV for MgO, < 3.7 eV for AlO, and 5.9 eV for CaO. Therefore, if NaH were the parent molecule for atomic Na, it could easily be dissociated. But if CaO were the parent molecule for atomic Ca, it would not be dissociated as easily.

We may note the truly unique nature of sodium on Io. Not only is sodium the only metallic line detected so far, but Io is also the only one of the four Galilean satellites to possess this emission. It is difficult to think that a meteoritic source would favor a particular satellite or that sputtering by MeV protons would discriminate against other metallic atoms. It appears that the absence of other metallic emissions from Io and the absence of sodium emission from the other Galilean satellites could be used to provide useful constraints on the chemical nature of the surfaces of the Galilean satellites or on the mechanism for releasing the metallic atoms from their bound state.

VIII. SUMMARY

The current state of our knowledge of Io's atmosphere is summarized schematically in Fig. 16 which includes all the observational constraints. The quantities enclosed between brackets are hypothesized "missing links," advanced theoretically to complete the picture. We distinguish two types of models, characterized by the source for hydrogen. In the Type 1 Model, hydrogen is supplied by dissociation of ammonia. Proton charge exchange is the primary source of hydrogen in the Type 2 Model. Both models have the common feature that meteoritic impact or sputtering of the surface provides a source for sodium.

The main problem of the Type 1 Model is the need to free nitrogen at about the same rate as hydrogen. Thermal escape would require an exospheric temperature of order 10,000°K, a value in conflict with the implication in the ionosphere data of a temperature less than 1000°K. We are forced to conclude that the escape process is sporadic and was not operative during the Pioneer 10 encounter. On the other hand, the remarkable temporal stability of the sodium cloud implies that any sporadic process must occur with sufficient frequency so that the amount of sodium ejected each time is small compared with the total contents of Region B. This demand is not difficult to meet since the sodium cloud has a photoionization lifetime of about two weeks.

The main problem with the Type 2 Model is that the observed flux of charged particles is insufficient to supply the observed hydrogen cloud maintained in the steady state. The discrepancy is not considered serious since we have no observational evidence for the stability of the hydrogen cloud, and an abundant supply of low-energy protons below the threshold of the Pioneer detectors cannot be ruled out. This model opens up the attractive possibility of a relatively cold upper atmosphere as implied by the ionosphere measurement. Escape of hydrogen can be accomplished by Jeans

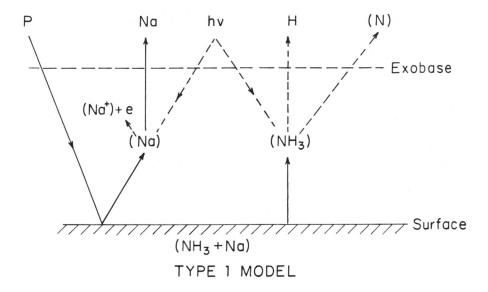

TYPE 1 MODEL

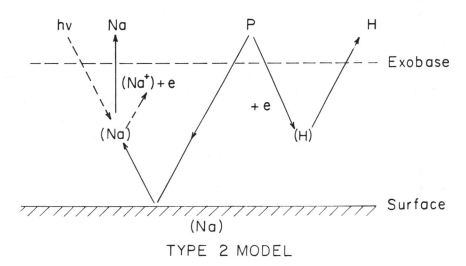

TYPE 2 MODEL

Fig. 16.   Schematic diagram of Type 1 and Type 2 Models of Io's atmosphere.

escape of a blow off process. The escape rate of sodium is many orders of magnitude lower than that of hydrogen. A blow-off will result in the loss of the entire exosphere, dragging away any minor species like sodium.

We note that within the context of present observational constraints, Type 1 Model is sporadic. The sporadic nature is crucially introduced to resolve the conflict between the high exospheric temperature (10,000°K) needed to free nitrogen (to a lesser extent, sodium) and the observed ionospheric temperature ($\sim$ 500°K). The Type 2 Model probably is sporadic too. The observed proton flux is not high enough to maintain the hydrogen cloud in a steady state. Escape of sodium depends on the sporadic blow-off of hydrogen from the upper atmosphere.

Acknowledgements. We are pleased to acknowledge many helpful discussions with our colleagues R. Goody, M. McElroy, F. Murcray, and W. Smyth. We thank R. Carlson, A. Eviatar, D. Hunten, D. Judge, D. Matson, M. Rosen, L. Trafton, and P. Wehinger for critical comments on the manuscript. This research was funded by the Atmospheric Sciences Section of the National Science Foundation under grant DES72-01472 A03 to Harvard University. One of us (Yuk L. Yung) acknowledges support by Kitt Peak National Observatory under NASA Contract NAS7-100.

DISCUSSION

D. M. Hunten: Meteoroid evaporation is usually considered to be the source of mesospheric sodium on Earth. Also, we should not be too literal about relative abundances. Several of the atoms one would expect to see are not found in terrestrial twilight, and we do not understand the selectivity. There seems to be something special about sodium that keeps it in the atomic form so that it is readily detected.

R. W. Carlson: The probable loss mechanism for the sodium atoms is ionization by electrons in the thermal plasma. Frank *et al.* have analyzed data from the Plasma Analyzer Experiment on Pioneer 10 and find densities of $\sim$ 50 cm^{-3} at the orbit of Io, with an ion temperature of $\sim$ 100 eV. Assuming that the electron and ion temperatures are the same, and using the ionization cross section of Na, they find a lifetime of $\sim$ 10^5 sec, roughly an order of magnitude shorter than the photoionization lifetime.

L. Trafton: Io's cloud sometimes exhibits sodium profiles with a high degree of asymmetry. These suggest that macroscopic motion is important. This implies that temperatures of about 5000°K, that you derive from the line widths, may be excessive. Also, a large column abundance of sodium can cause the line to appear broad at low temperatures because of saturation.

D. L. Matson: The key question about Io's atmosphere is its pressure. This quantity is poorly known because the currently available values are model dependent. The variety of models which have been considered do not include all of the physical processes which are known to occur at Io. For example, the field and forces due to the plasma sheaths have not yet been considered.

REFERENCES

Allen, C. W. 1973. *Astrophysical Quantities.* London: The Athlone Press.

Anderson, J. D.; Null, G. W.; and Wong, S. K. 1974. Gravity results from Pioneer 10 Doppler data. *J. Geophys. Res.* 79:3661–3664.

Barmore, F. E. 1975. The filling-in of Fraunhofer lines in the day sky. *J. Atmos. Sci.* 32:1489–1493.

Bauer, S. J. 1973. *Physics of planetary ionospheres.* New York: Springer Verlag.

Bergstralh, J. T.; Matson, D. L.; and Johnson, T. V. 1975. Sodium D-line emission from Io: synoptic observations from Table Mountain Observatory. *Astrophys. J.* 195:L131–L135.

Bigg, E. K. 1964. Influence of the Satellite Io on Jupiter's decametric emission. *Nature* 203:1008–1010.

Binder, A. B., and Cruikshank, D. P. 1964. Evidence for an atmosphere on Io. *Icarus* 3:299–305.

Blamont, J. E., and Courtès, G. 1955. Noveau procédé d'étude photométrique des émissions monochromatiques du ciel nocturne. *Ann. Géophys.* 11:252–254.

Brown, R. A. 1974. Optical line emission from Io. *Exploration of the planetary system.* (A. Woszczyk and C. Iwaniszewska, eds.) pp. 527–531. Dordrecht, Holland: D. Reidel Publ. Co.

Brown, R. A., and Chaffee, F. H. 1974. High-resolution spectra of sodium emission from Io. *Astrophys. J.* 187:L125–L126.

Brown, R. A.; Goody, R. M.; Murcray, F. J.; and Chaffee, F. H. 1975. Further studies of line emission from Io. *Astrophys. J.* 200:L49–L53.

Carlson, R. W., and Judge, D. L. 1974. Pioneer 10 ultraviolet photometer observations at Jupiter encounter. *J. Geophys. Res.* 79:3623–3633.

———. 1975. Pioneer 10 ultraviolet photometer observations of the Jovian hydrogen torus: the angular distribution. *Icarus* 24:395–399.

Chamberlain, J. W. 1961. *Physics of the aurora and airglow.* New York: Academic Press.

———. 1965. Planetary coronae and atmospheric evaporation. *Planet. Space Sci.* 11:901–960.

Cloutier, P. A.; McElroy, M. G.; and Michel, F. C. 1969. Modifications of the Martian ionosphere by the solar wind. *J. Geophys. Res.* 74:6215–6228.

Cruikshank, D. P.; Pilcher, C. B.; and Sinton, W. M. 1974. Io: reported helium emission. *IAU Circular* Nos. 2693 and 2722.

Donahue, T. M., and Meier, R. R. 1967. Distribution of sodium in the daytime upper atmosphere as measured by a rocket experiment. *J. Geophys. Res.* 72:2803–2829.

Fanale, F. P.; Johnson, T. V.; and Matson, D. L. 1976. Io's surface and the histories of the Galilean satellites. *Planetary Satellites.* (J. Burns, ed.) Tucson: Univ. of Arizona Press.

Fanale, F. P.; Nash, D. B.; Johnson, T. V.; and Matson, D. L. 1975. The surface of Io: a progress report. *Bull. Amer. Astron. Soc.* 7:386.

Fink, U.; Dekkers, N. H.; and Larson, H. P. 1973. Infrared spectra of the Galilean satellites of Jupiter. *Astrophys. J.* 179:L155–L159.

Fiocco, G.; Fua, E.; and Visconti, G. 1974. Origin of the upper atmospheric Na from sublimating dust: a model. *Ann. Géophys.* 30:517–528.

Fiocco, G., and Visconti, G. 1973. On the seasonal variation of upper atmospheric sodium. *J. Atm. Terr. Phys.* 35:165–171.

Frank, L. A.; Ackerson, K. L.; Wolfe, J. H.; and Mihalov, J. D. 1975. Observations of plasmas in the Jovian magnetosphere. *J. Geophys. Res.* 81:457–468.

Gadsden, M. 1968. Sodium in the upper atmosphere: meteoric origin. *J. Atm. Terr. Phys.* 30:151–161.

Goldreich, P., and Lynden-Bell, D. 1969. Io, a Jovian unipolar inductor. *Astrophys. J.* 156:59–78.

Greenstein, J. L., and Arpigny, C. 1962. The visual region of the spectrum of comet Mkros (1957d) at high resolution. *Astrophys. J.* 135:892–905.

Gross, S. H. 1972. On the exospheric temperature of hydrogen dominated planetary atmospheres. *J. Atmos. Sci.* 29:214–218.

———. 1974. The atmospheres of Titan and the Galilean satellites. *J. Atmos. Sci.* 31:1413–1420.

Humes, D. H.; Alvarez, J. M.; O'Neal, R. L.; and Kinard, W. H. 1974. The interplanetary and near-Jupiter meteoroid environments. *J. Geophys. Res.* 79:3677–3684.

Hunten, D. M. 1954. A study of sodium in twilight. I. theory. *J. Atmos. Terr. Phys.* 5:44–56.

———. 1965. Excitation of the sodium emission in aurora. *J. Atm. Terr. Phys.* 27:583–586.

———. 1973. The escape of H_2 from Titan. *J. Atmos. Sci.* 30:726–732.

Hunten, D. M., and Wallace, L. 1967. Rocket measurement of the sodium dayglow. *J. Geophys. Res.* 72:69–79.

Johnson, T. V., and McCord, T. B. 1970. Galilean satellites: the spectral reflectivity 0.30-1.10 micron. *Icarus* 13:37–42.

Judge, D. L., and Carlson, R. W. 1974. Pioneer 10 observations of the ultraviolet glow in the vicinity of Jupiter. *Science* 183:317–318.

Junge, C. E.; Oldenberg, O.; and Wasson, J. T. 1962. On the origin of the sodium present in the upper atmosphere. *J. Geophys. Res.* 67:1027–1039.

Kliore, A.; Cain, D. L.; Levy, G. S.; Eshleman, V. R.; Fjeldbo, G.; and Drake, F. D. 1965. Occultation experiment: results of first direct measurement of Mars's atmosphere and ionosphere. *Science* 149:1243–1248.

Kliore, A.; Fjeldbo, G.; Seidel, B. L.; Sweetnam, D. N.; Sesplaukis, T. T.; and Woiceshyn, P. M. 1975. Atmosphere of Io from Pioneer 10 radio occultation measurements. *Icarus* 24:407–410.

Kliore, A.; Levy, G. S.; Cain, D. L.; Fjeldbo, G.; and Rasool, S. I. 1967. Atmosphere and ionosphere of Venus from Mariner V S-Band radio occultation measurement. *Science* 158:1683–1688.

Macy, W., and Trafton, L. 1975. A model for Io's atmosphere and sodium cloud. *Astrophys. J.* 200:510–519.

Matson, D. L.; Johnson, T. V.; and Fanale, F. P. 1974. Sodium D-line emission from Io: sputtering and resonant scattering hypothesis. *Astrophys. J.* 192:L43–L46.

McDonough, T. R. 1975. A theory of the Jovian hydrogen torus. *Icarus* 24:400–406.

McElroy, M. B., and Yung, Y. L. 1975. The atmosphere and ionosphere of Io. *Astrophys. J.* 196:227–250.

McElroy, M. B.; Yung, Y. L.; and Brown, R. A. 1974. Sodium emission from Io: implications. *Astrophys. J.* 187:L127–L130.

Mekler, Yu., and Eviatar, A. 1974. Spectroscopic observations of Io. *Astrophys. J.* 193:L151–L152.

Mekler, Yu.; Eviatar, A.; and Coroniti, F. V. 1975. Sodium in the Jovian atmosphere. *Astrophys. Space Sci.* In draft.

Mertz, L. 1975. Heterographic techniques for astronomy. Paper delivered at OSA Meeting, "Imaging in Astronomy." Cambridge, Massachusetts, June 18–21, 1975.

Morrison, D., and Cruikshank, D. P. 1974. Physical properties of the natural satellites. *Space Sci. Rev.* 15:641–739.

Morrison, D., and Morrison, N. D. 1976. Photometry of the Galilean satellites. *Planetary Satellites*. (J. Burns, ed.) Tucson: Univ. of Arizona Press.

Münch, G., and Bergstralh, J. T. 1975. Sodium D-line emission from Io: spatial brightness distribution from multislit spectra. *Bull. Am. Astron. Soc.* 7:386.

Öpik, E. J. 1963. Selective escape of gases. *Geophys. J.* 7:490–509.

Parker, E. N. 1958. Dynamics of the interplanetary gas and magnetic fields. *Astrophys. J.* 128:664–675.

Parkinson, T. D. 1975. Excitation of the sodium D-line emission observed in the vicinity of Io. *J. Atmos. Sci.* 32:630–633.

Pilcher, C. B.; Ridgway, S. T.; and McCord, T. B. 1972. Galilean satellites: identification of water frost. *Science* 178:1087–1089.

Shawhan, S. D. 1976. Io sheath-accelerated electrons and ions. *Icarus* (special Jupiter issue). In press.

Shawhan, S. D.; Goertz, C. K.; Hubbard, R. F.; Gurnett, D. A.; and Joyce, G. 1974. Io: accelerated electrons and ions. Paper presented at Neil Brice Memorial Symposium, *The magnetosphere of the earth and Jupiter,* May, 1974, Frascati, Italy.

Sill, G. T. 1974. Sources of sodium in the atmosphere of Io. *Bull. Am. Astron. Soc.* 6:384.

Smith, B. A., and Smith, S. A. 1972. Upper limits for an atmosphere on Io. *Icarus* 17:218–222.

Taylor, G. E.; O'Leary, B.; Van Flandern, T. C.; Bartholdi, P.; Owen, F.; Hubbard, W. B.; Smith, B. A.; Smith, S. A.; Fallon, F. W.; Devinney, E. J.; and Oliver, J. 1971. The occultation of Beta Scorpii C by Io on May 14, 1971. *Nature* 234:405–406.

Trafton, L. 1976. A search for emission features in Io's extended cloud. *Icarus* (special Jupiter issue). In press.

Trafton, L., and Macy, W. 1975. The geometry of Io's sodium cloud. *Bull. Am. Astron. Soc.* 7:386.

Trafton, L.; Parkinson, T.; and Macy, W. 1974. The spatial extent of sodium emission around Io. *Astrophys. J.* 190:L85–L89.

Webster, D. L.; Alksne, A. Y.; and Whitten, R. C. 1972. Does Io's ionosphere influence Jupiter's radio bursts? *Astrophys. J.* 174:685–696.

Wehinger, P., and Wyckoff, S. 1974. Jupiter I. *IAU Circular No. 2701.*

Wehinger, P. A.; Wyckoff, S.; and Frohlich, A. 1976. Mapping of the sodium emission associated with Io and Jupiter. *Icarus* (special Jupiter issue). In press.

Whitten, R. C.; Reynolds, R. T.; and Michelson, P. F. 1975. The ionosphere and atmosphere of Io. *Geophys. Res. Lett.* 2:49–51.

MODELS OF JOVIAN DECAMETRIC RADIATION

R. A. SMITH
NASA Goddard Space Flight Center

We present a critical review of theoretical models of Jovian decametric radiation, with particular emphasis on the Io-modulated emission. The problem is divided into three broad aspects: the mechanism coupling Io's orbital motion to the inner exosphere, the consequent instability mechanism by which electromagnetic waves are amplified, and the subsequent propagation of the waves in the source region and the Jovian plasmasphere. At present there exists no comprehensive theory that treats all of these aspects quantitatively within a single framework. Acceleration of particles by plasma sheaths near Io appears to be a promising explanation for the coupling mechanism, while most of the properties of the emission may be explained in the context of cyclotron instability of a highly anisotropic distribution of streaming particles. The present state of the theory is evaluated, and some suggested approaches for future work are discussed.

The nature and origin of the Jovian decametric radiation (DAM) have prompted much theoretical effort, especially since the remarkable modulating effect of Io was demonstrated by Bigg (1964). The problem occupies an unusual niche in the gallery of radio-astronomical phenomena. As is true of emission from quasars, pulsars, and the sun, DAM presents a rich phenomenology. Unlike the cases of quasars and the sun, however, which exhibit detailed line spectra at optical wavelengths, radio observations [of both DAM and the weaker decimetric emission (DIM)] provided the only information about the plasma environment of Jupiter before the recent probes of the Jovian magnetosphere by Pioneers 10 and 11. Thus, for nearly twenty years the physical characteristics of the Jovian magnetosphere were inferred solely from radio astronomy. In this respect the problem is similar to that of pulsars. Like the sun, however, Jupiter is close enough that a detailed observational morphology of its radio emissions may be constructed. This unusual combination of circumstances, together with the extensive development of understanding of terrestrial magnetospheric processes, has allowed for wide variability in the mixture of inductive and deductive arguments in various individual attempts to explain DAM.

[1146]

Despite the great amount of theoretical effort that has so far been expended, we have not yet arrived at a comprehensive, convincing understanding of *DAM*. Nevertheless, a review of the theoretical work to date is both pertinent and timely. The results from the Pioneer 10 and 11 missions have excited widespread interest in Jovian physics in general, and much of the conceptual framework in which we evaluate these results has been developed in the course of attempts to elucidate *DAM*. We anticipate that the relations between the problems of understanding *DAM* in particular and the Jovian magnetosphere in general will be more widely appreciated in the future. Furthermore, although we do not view any of the attempts heretofore to explain various facets of *DAM* as wholly successful, we believe that in the paradigmatic nature of science a successful theory will emerge from the gradual modification and consolidation of previous work, rather than from radically new hypotheses. In this spirit, we attempt in this chapter to present a critical framework for the evaluation of the theoretical efforts to date, in the hope that such a framework will prove useful in both motivating and evaluating future work. The bulk of our discussion concerns the Io-modulated *DAM*; we discuss the Io-independent emission briefly near the end (Sec. IV-A).

A current review of *DAM* observations appears in this volume (by Carr and Desch[1]). Previous reviews by Warwick (1964, 1967, 1970) and by Carr and Gulkis (1969)[2] deal with observations of both *DAM* and *DIM* and the inferences about the Jovian magnetosphere drawn from these observations. In addition, Warwick (1967) presents a critical review of some of the theoretical work up to 1967.

Our approach in this chapter is to separate the theoretical problem into various more-or-less distinct facets, and to consider these facets both in themselves and in their structural relationships to each other. Most of the theoretical works to date deal principally with only one or two of the different elements of the problem, and so we discuss each work in the most appropriate contexts. We assume that the reader is familiar with the observations as they are described in the above-mentioned reviews, and so our references to observational results will usually be superficial and will seldom be referenced; the article by Carr and Desch in this volume will generally provide adequate observational background for our discussion.

Some preliminary semantic definitions may be in order. As we have already done above, we shall sometimes distinguish between the phenomenology and morphology of *DAM*. By "phenomenology" we mean those aspects that are directly observed, such as intensity, time scales, bandwidth, polarization, and such features of the dynamic spectra as frequency drifts and modulation lanes. In contrast, we mean by "morphology" the organization of the data into such features as rotational profiles of occurrence prob

---

[1]See p. 693.   [2]See also p. 621.

ability and polarization, the similarity of the dynamic spectra of different events observed under the same geometrical conditions, the declination effect, the induction by Bigg of Io-modulation, the concomitant recognition of an Io-independent component, and so forth. Moreover, we do not distinguish between the familiar decametric emissions and the hectometric emissions observed from satellites by Brown (1974a,b) and by Desch and Carr (1974). Except for the component of hectometric emissions that was designated by Brown (1974b) as "mid-frequency" (MF) emission, the hectometric emission appears to be merely a continuation of the decametric spectrum to longer wavelengths, and we shall include the hectometric range in the traditional designation DAM. We note, however, that at this writing (mid-1975) the observational statistics are insufficient to determine whether either the "normal" or the MF components of hectometric emission are modulated by any of the Jovian satellites.

Before plunging into the literature, we discuss some theoretical preliminaries.

The high intensities, limited bandwidths, short time scales, and sporadic nature of individual DAM bursts indicate that they originate in stimulated emission from one or more collective micro-instabilities. The remarkable repeatability of the dynamic spectra observed under similar geometrical conditions, and the complex beaming pattern manifested in the source morphology, imply that the structure of the ambient medium is stable in the long-term sense and determines both the local conditions for plasma instability and the subsequent propagation of the radiation. Although this statement seems indisputable in the general sense, it must nevertheless be taken with some qualification. From detailed study of certain repeatable features of dynamic spectra, Warwick (1963) showed that these features always appeared for particular corresponding central meridian longitudes (CML), within a range of $\pm 9°$. Warwick interpreted these results as implying that the emission was beamed into a cone of 9°. Alexander (1975), however, compared several pairs of identical-looking dynamic spectra observed one Jovian year ($\simeq 11.9$ yr) apart. Although the members of each pair of spectra commonly shared quite detailed features, Alexander concluded from a study of the rotation rates inferred from the different pairs that the beaming pattern might be subject to longitude shifts over a range of up to $\pm 10°$. Another point is so obvious that it is not often made explicitly, but we feel that perhaps for that very reason its implications are sometimes lost sight of in the literature. This point is that the occurrence probability distribution in the (CML, γ_{Io}) plane [γ_{Io} is the phase of Io from superior geocentric conjunction (SGC)] is conventionally interpreted as indicative of a beaming pattern. The occurrence probability over most of the plane, however, is substantially less than unity. This fact alone indicates that there must be some variable influence on the modulation mechanism or the waveguide structure of the plasma medium, or both.

Plasma instabilities require a source of free energy. Large-amplitude magnetohydrodynamic (*MHD*) waves or shocks may provide free energy under some circumstances. In micro-instabilities (by which we mean instabilities that are predicted only in kinetic-theoretical analyses and cannot be predicted from fluid theory) the free energy resides in some nonthermal feature of the particle distribution function $F(\mathbf{r}, \mathbf{v}, t)$. Such a nonthermal feature may be manifested in either the configuration-space or velocity-space dependence of F. For example, density gradients in a magnetized plasma create drift currents, which may drive a variety of instabilities. Velocity-space features such as beams of high-velocity particles, gaps in the distribution in some velocity range, or anisotropy in either pitch angles or Larmor phases, may result in instability.

The repeatability of *DAM* spectra suggests that the instability occurs for waves of frequency at or related to one of the characteristic frequencies of the plasma, as is commonly the case. In the high-frequency regime, these are the electron plasma frequency f_{pe}, the electron gyrofrequency f_{ce}, and the upper hybrid frequency $f_{UH} = (f_{pe}^2 + f_{ce}^2)^{\frac{1}{2}}$. At Jupiter, the *DAM* frequency range may be identified with these characteristic frequencies only near the planet: roughly speaking, at altitudes of $\lesssim 1\ R_J$.

Thus, the central problem in the theory of *DAM* is to elucidate the *coupling mechanism* whereby the orbital motion of Io provides a free-energy source in the inner plasmasphere. The existence and nature of an instability are determined by the form in which the free energy appears. The coupling mechanism and the subsequent instability mechanism by which the radio emission is amplified are, however, viewed here as two logically distinct elements of the problem. Roughly speaking, we might expect that the coupling mechanism primarily affects the *DAM* morphology, while the nature of the instability is reflected in the phenomenology. This statement is intended simply as a general guide, however, and not as an organizing postulate.

The third element of the problem is the propagation of the radiation in the Jovian plasmasphere. Propagation effects are expected to be manifested in both the phenomenology and morphology. As we shall see below, for example, certain instability mechanisms that have been proposed actually amplify waves that are trapped in the Jovian plasmasphere. Therefore, some mode-conversion process is necessary in order for the radiation to propagate into free space. In this case the polarization of the escaping wave, which is generally the datum by which observers identify the base mode, may be determined by the escape mechanism rather than by the instability. The escape mechanism may also determine the initial wave vector of the escaping ray, thus determining the ray path in the plasma medium and the beaming pattern.

Provided that the boundaries between the different aspects of the problem are not regarded as impenetrable, our formulation is a useful one which enables us to construct a general conceptual framework within which to consider particular theories.

I. THE COUPLING MECHANISM

The question of the coupling mechanism has been addressed, to varying degrees, by many authors. Smith and Wu (1974) divided the majority of the proposed mechanisms into three broad categories:

A. *Wave models,* in which the motion of a conducting Io through the magnetoplasma may generate whistlers or *MHD* waves that propagate to the Jovian ionosphere;

B. *Acceleration models,* in which the electric field induced in the frame of Io by the satellite's motion through the ambient magnetic field is invoked to accelerate particles to energies from tens to hundreds of keV. These particles then precipitate along the Io-threaded flux tube (*IFT*);

C. The *sweeping model,* in which energetic particles trapped on the *IFT* are removed from the trapping volume by impingement on Io.

To our knowledge, the above categories embrace all the cases in the literature in which a coupling mechanism has been examined quantitatively. This does not mean that their corresponding instability mechanisms have been considered quantitatively; as we shall see in Sec. II, no quantitative model of an instability mechanism that might result from *MHD*-wave coupling has yet been presented in the literature. Whether or not such mechanisms may exist, the generation of *MHD*-wave disturbances by Io may have important consequences which are discussed below.

In addition to the coupling models described above, other more speculative hypotheses have been advanced. Duncan (1970) suggested that the source morphology was determined by a combination of the geocentric beaming pattern and heliocentric effects. He constructed a model of the Jovian magnetosphere in which the subsolar point of the magnetopause was at 5 R_J, and conjectured that the strong confinement of Io modulation to orbital phases near $\gamma_{Io} \simeq 90°$ *SGC* and 240° *SGC* implied that Io crossed the magnetopause at about these phases. Duncan further speculated that these magnetopause crossings would precipitate trapped particles into the Jovian ionosphere, although he did not indicate why or how this might occur. To evaluate Duncan's hypothesis, however, it suffices to note that the Pioneer 10 and 11 probes have revealed the Jovian magnetosphere to be extended to characteristic dimensions of 50–100 R_J in the dayside hemisphere.

Another phenomenological suggestion in a spirit similar to that of Duncan's was made by Conseil *et al.* (1971, 1972). These authors inferred a correlation between the velocity of the solar wind at Jupiter (extrapolated from satellite measurements at 1 A.U.) and the phase of Io at the onset of *DAM* storms. They suggested that the correlation might be attributed to a bulge in the plasmapause, similar to that of the terrestrial plasmapause, the location of which varied in response to solar wind ram pressure. They speculated that Io might trigger *DAM* storms upon passing through the plasmapause boundary.

The statistical significance of the correlation asserted by Conseil *et al.* is unclear; it has been questioned by Carr and Desch.[3] Furthermore, a Jovian plasmapause would be unlikely to contain a bulge analogous to that in the terrestrial case. It is believed that the terrestrial plasmapause occurs at the boundary between closed and open streamlines arising from the combined flow fields of co-rotation and convection. In the Jovian magnetosphere, co-rotation would be expected to dominate convection out to large distances, because of the rapid rotation rate of Jupiter. A comparison of the flow pattern to be expected at Jupiter and that in the terrestrial magnetosphere is given by Brice and Ioannidis (1970). Although this picture may be subject to considerable modification in light of the unexpectedly large extension of the Jovian magnetosphere and the dominance of its outer structure by plasma outflow, it should not be greatly changed in the inner region. A true plasmapause at Jupiter would probably be determined by the breakdown of co-rotation.

Frank *et al.* (1975), however, inferred the existence of a plasmapause at the orbit of Io, where the density of protons of energies 100 eV–4.80 keV was observed to increase by a factor of about 3 to 4 on the inbound pass of Pioneer 10. As may be seen in Fig. 1, however, this density decreases again inside Io's orbit, before resuming a general increasing trend as the radial distance decreases. These observations could also be explained by injection of these particles by Io, rather than as a plasmapause structure.

Whether or not the structure of the thermal-plasma density near Io is indicative of a plasmapause, it is clearly complicated. The basic idea behind the suggestion of Conseil *et al.* was that the stimulation of *DAM* by Io might depend on the local plasma density. This density may, in fact, be an important parameter in the *MHD*-wave model of Goertz and Deift (1973; cf. also Deift and Goertz 1973; Goertz 1973*a,b*) and the recent acceleration models of Shawhan *et al.* (1974) and of Smith and Goertz (1975).

We now consider the quantitative categories (A)–(C). As was mentioned above, in most of the works discussed in this review the major focus is on only one aspect of the total problem; in the majority of the papers we consider in this section it is the coupling mechanism. In most of these papers, however, the authors had some hypothesis in mind concerning the subsequent generation or propagation of the electromagnetic waves. In some other works, such as that of Wu (1973), two or more aspects of the problem are treated integrally. Therefore, in order not to fragment our exposition unduly, we shall occasionally anticipate some points of the later discussion in order to place the immediate topic in context.

A. Wave Models

Ellis (1965) first speculated that the coupling might be due to either *MHD* or whistler (which he called "electromagnetic") waves generated by

---

[3]See p. 728.

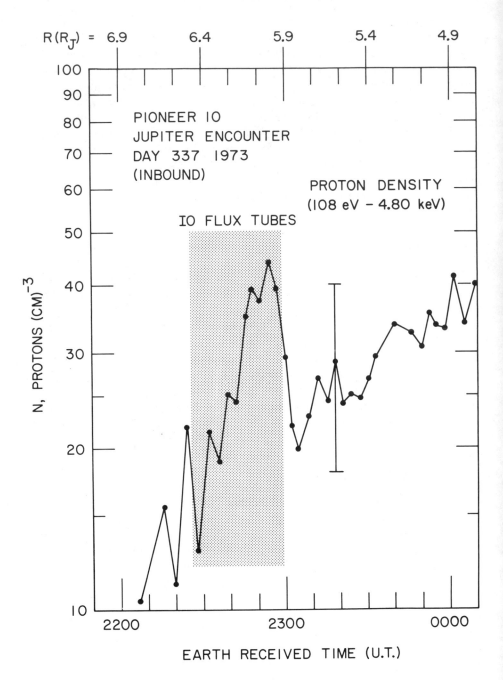

Fig. 1. The density of protons in the energy range 108 eV − 4.80 keV near Io, as measured by Frank *et al.* (1975). These authors fit their proton density measurements throughout the inner plasmasphere to Maxwellian distributions of characteristic temperature ≃ 100 eV and densities between 50 and 100 cm⁻³.

Io. He did not investigate how such waves might be excited, except to note that the relative velocity of Io normal to the co-rotating magnetic field was greater than the phase velocities of these modes, all of which have resonances (wave number $k \rightarrow \infty$) at angles $\theta < \pi/2$ with respect to the field. Ellis, therefore, assumed that these low-frequency modes might be generated by Cerenkov radiation from Io. This is by no means obvious, however, because in order to be a source for the waves, Io must either have an appropriate current distribution on its surface or cause such a current distribution in the ambient plasma by perturbing its flow.

Ellis hypothesized that the Io-generated waves would interact with preexisting particle streams or "bunches" to stimulate the electromagnetic radiation observed as DAM, in analogy to the well-known phenomenon of stimulated very low frequency (VLF) emissions in the terrestrial magnetosphere. Thus, the whistler waves are the important ones in this context; this mechanism was also considered by Chang (1963), before Io-modulation had been recognized. It is implicit in this scheme that the resulting DAM waves travel along field lines, and that the source region would therefore be the CML.

McCulloch (1971) developed Ellis' ideas for the coupling more quantitatively. He too assumed whistlers to be generated by Io, and did ray tracing in a model magnetosphere. By varying the parameters of the model, and by assuming that the coupling region where the whistlers interacted with particle streams was at magnetic latitudes between 75° and 80°, McCulloch obtained a distribution of coupling regions in the (CML, γ_{Io}) plane, for which the relative intensities of the propagated whistlers corresponded well with the distribution of occurrence probability of DAM (cf. especially his Figs. 7, 9, and 10). In particular, McCulloch's results indicated that the time required for the propagation of the whistlers from Io to the Jovian ionosphere was of the order of 33 min. This is the time required for Io to move 15° in orbital phase relative to the co-rotating medium; in this way, McCulloch accounted for the asymmetry of the Io phase, for peak occurrence probability, about the Earth-Jupiter line (Fig. 2).

The parameters which McCulloch found necessary to describe his model magnetosphere, in order to obtain good agreement between the coupling mechanism and observational morphology, are not in accord with the Pioneer observations. For example, McCulloch used an ionospheric electron temperature T_e of 2000°K, whereas Kliore et al. (1975) have inferred $T_e \simeq$ 750°K. Furthermore, McCulloch states (1971, p. 1302) that he uses a density model for the thermal plasma of the form derived by Melrose (1967). But he also states that the "most satisfactory models, however, were those for which the electrons were concentrated in a disk about the magnetic equator, and the plasma frequency and gyrofrequency at Io were nearly equal" (McCulloch 1971, p. 1303). The latter condition $f_{ce} \simeq f_{pe}$ is actually in accord with observations: for a dipole moment of 4G R_J^3 (Smith et al.

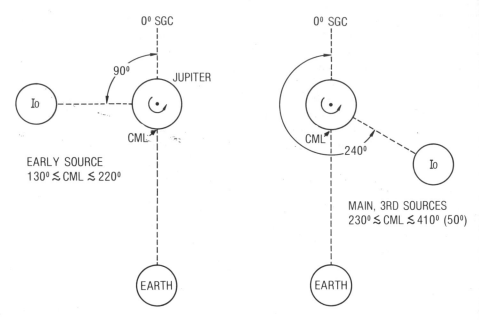

Fig. 2. The geometrical configuration of the Earth-Jupiter-Io system for peak occurrence probability in the various sources. Note the asymmetry in Io phase with respect to the Earth-Jupiter line. The ranges of the central meridian longitude (*CML*) for the sources are in System III (1957.0) coordinates, at the Pioneer 11 epoch (1974.9). *SGC* is the superior geocentric conjunction.

1975), the field strength $B_0 \simeq 0.02$ G at Io and $f_{ce} = 56$ kHz, while for a plasma density at Io of $N_0 \simeq 30$ cm^{-3} (Frank *et al.* 1975), $f_{pe} \simeq 55$ kHz. The feature of a plasma disk in the equatorial plane, however, is not contained in Melrose's density model, in which the density decreases monotonically with latitude along a field line.[4] Therefore, the possibility exists that McCulloch's work may be internally inconsistent. Finally, McCulloch's ray tracing calculations assumed a centered, tilted dipole field of equatorial field strength 15 G; it is not clear in context (cf. p. 1306) whether *ad hoc* anomalies, such as were postulated by Ellis and McCulloch (1963) were incorporated or not. As has been shown by Acuña and Ness (1975) and Smith *et al.* (1975),[5] however, the field near the surface is strongly influenced by higher-order moments.

Apart from questions about the associated instability mechanism and consequent propagation effects, these considerations indicate that, at the very least, coupling by whistlers must be re-examined in the context of the plasma environment deduced from the Pioneer observations. This comment does not apply uniquely to the work of McCulloch, however.

---

[4]The plasma disk *is* contained in density models developed by Gledhill (1967) and Goertz (1973*a*, 1975).

[5]See p. 834 and p. 791, respectively.

Other authors have concentrated on low-frequency ($f \ll f_{pi}$) *MHD* waves to couple Io to the *DAM* source regions. Marshall and Libby (1967) suggested that such waves might induce spin transitions in free radicals in the Jovian ionosphere, presumably because of the presence of off-diagonal terms in the pressure tensor, in analogy to the interaction of phonons with free radicals in crystals (Warwick 1967).

The first quantitative estimates of *MHD*-wave coupling were made by Warwick (1967), who considered Io to be perfectly conducting and, therefore, argued that it would push aside the ambient field. He estimated that for "resonant" waves of phase velocity $V_{ph} \simeq V_{Io}$ (where $V_{Io} = 56$ km sec^{-1} is the speed of Io relative to the co-rotating plasma) and wavelength $2\ R_{Io}$, waves of amplitude $B_w = (V_A / \pi V_{Io})\ B_0$ might be produced, where V_A is the Alfvén velocity near Io. At the current estimate of $V_A \simeq 700$ km sec^{-1}, this would give $B_w \simeq 4\ B_0$. From other considerations, involving large but finite conductivity for Io, Warwick (1970) estimated $B_w \simeq 2.6\ B_0$. Waves of this magnitude are inherently nonlinear, and would be expected to steepen into shocks as they propagated toward Jupiter. Warwick (1970) estimated that $\sim 4 \times 10^{16}$ watts would thus be radiated from Io. He did not attempt to partition this energy among the various *MHD* modes (shear Alfvén, compressional Alfvén, and magnetosonic). Instead, he noted that if the *MHD* radiation from Io were isotropic, about 3×10^9 W would be intercepted by an area of 1.6×10^5 km^2 (the presumed size of the *DAM* source region) at the surface of Jupiter.

The estimates by Warwick (1970) were made partly in the context of a critique of the paper by Goldreich and Lynden-Bell (1969), which is discussed below. These latter authors assumed that the electric field induced across Io by its motion relative to the Jovian magnetic field would drive a current system that closed in the planetary ionosphere. In contrast, Warwick asserted that the current system would be closed within Io itself. This idea was also adopted by Schmahl (1970), who considered Io to have a permanent dipole moment $\mathbf{M}_{Io}$ induced by the DC average of its ambient magnetic field, and thus to be a point current source

$$\mathbf{j}_s(\mathbf{r},\ t) = c\mathbf{M}_{Io} \times \nabla\ \delta\ (\mathbf{r} - \mathbf{V}_{Io}\ t)\ . \tag{1}$$

Schmahl calculated that 8×10^9 W would be delivered to the Jovian ionosphere in the shear-Alfvén mode, which propagates essentially along the *IFT*.

In a contrasting approach, Goertz and Deift (1973) assumed that currents leak from Io into the ambient plasma. They attempted to calculate the steady flow pattern of the ambient medium around Io. The resulting distortions of the ambient magnetic field are to be understood in the sense of *MHD* wave fields. Some particularly striking features of their model are: (a) the perturbations of the field ahead of Io propagate to the Jovian ionosphere as solitons and are reflected as solitons, leading to a stretching of the field roughly analogous to the stretching of a string; (b) the existence of an

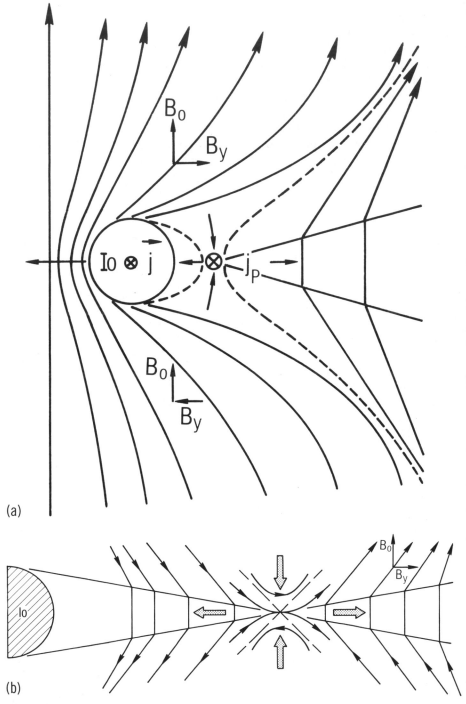

Fig. 3. (a) The magnetic field topology near Io and (b) detail of the x-type neutral point reconnection region behind Io (after Goertz and Deift 1973).

x-type neutral point behind Io (Fig. 3). The x-type configuration is asserted to occur when the field lines leading Io are stretched too far; reconnection then takes place behind the satellite, leading to the propagation of Alfvén waves up the *IFT*. In effect, Io "plucks" the magnetic field lines, like plucking a violin string. The angular limit of distortion before reconnection occurs is about 15°, accounting for the Io-phase asymmetry. Goertz and Deift estimate the maximum power radiated in Alfvén waves to be $\geq 2.7 \times 10^{15}$ W.

In Goertz and Deift's model, the wavelength of the Alfvén waves generated by reconnection is taken to be 6 R_{Io} (i.e., $k \simeq R_{Io}^{-1}$). The resulting frequency of the waves is between $1-3$ Hz, depending on the plasma density at Io. This density is determined from a magnetospheric model developed by Goertz (1973a). Deift and Goertz (1973) consider the propagation of these waves along the *IFT*. They find a "transmission coefficient" for the waves into the Jovian ionosphere: the depth to which the Alfvén waves penetrate—and, therefore, the gyrofrequency at this depth—is a function of the wave frequency. Goertz (1973b) equated the gyrofrequency at the local altitude of penetration, as a function of longitude, to the maximum *DAM* frequency at that longitude, showing good agreement with observation (Goertz and Haschick 1972).

Goertz and Deift attempted to consider the three-dimensional nature of the *MHD* flow around Io. It is not clear, however, that they did so consistently. They assumed the conductivity of the ambient medium to be a scalar, whereas in a low-β plasma ($\beta \equiv 8\pi N K_B T_e / B^{-2}$) the tensor properties of the conductivity are significant. Furthermore, even within the context of their model, the field lines which obey the geometry of Fig. 3 are limited to those lines near the stagnation point of the flow; in principle, these lines degenerate to a set of measure zero, and it is then unclear to what extent their analyzed flow is dominated by the actual three-dimensional flow. In particular, the efficiency of flux reconnection at x-type neutral points may be drastically lower in three-dimensional flow than in two-dimensional flow. Furthermore, the back-flow towards Io associated with the reconnection is not taken into account.

In defense of these authors, it must be noted that the three-dimensional *MHD* flow past a spherical body is an intractable problem, particularly in a low-β plasma. Nevertheless, the spirit of our comments is that a two-dimensional solution may be greatly misleading. In the common example, a tight-rope walker confined to the vertical plane could never fall off the tight-rope.

In addition, Deift and Goertz consider incompressible plane waves propagating along the *IFT*. It is not clear whether, in the situation of wave generation by flux reconnection, compressible modes or finite wave packets (Goldstein *et al.* 1974) might be more appropriate.

Finally, we remark that the entire discussion by Goertz and Deift (1973), Deift and Goertz (1973), and Goertz (1973b) is heavily dependent for its quantitative significance upon the magnetospheric model of Goertz (1973a).

None of the wave-coupling models discussed include a quantitative discussion of a consequent instability mechanism. In respect to the instability mechanism, however, the power available in the coupling mechanism is of vital importance because the efficiency of conversion to electromagnetic waves is necessarily low. We shall take 10^8 W to be a nominal power requirement for electromagnetic waves in DAM bursts (Warwick 1970). Therefore, the power in the coupling mechanism must be $\gg 10^8$ W.

McCulloch (1971) estimates that 2×10^9 W may be transferred from Io via whistlers. Taken by itself this figure would indicate a conversion efficiency of 0.05, which is incredibly high. The presumed interaction mechanism, however, is the phase organization of trapped particles (cf. Sec. II), and therefore the particles may be more important in the energy balance than the whistlers.

Because no instability mechanism has been demonstrated for coupling by low-frequency MHD waves, we take the viewpoint that the conversion efficiency must be compatible with the power available in the MHD waves. [This viewpoint is consistent with the suggestion by Goertz (1973b) that the DAM radiation may originate from particles accelerated in the MHD wavefronts.] In this light, Schmahl's estimate of $\simeq 10^{10}$ W in shear-Alfvén waves requires a very high conversion efficiency of 10^{-2}. On the other hand, the maximum-power estimates of 2.7×10^{15} W by Goertz and Deift and 4×10^{16} W by Warwick (1970) are much too high to agree with the recent determination by Goldstein (1975) that the secular acceleration of Io is limited to a value

$$\frac{\dot{p}}{p} < 3 \times 10^{-18} \text{ sec}^{-1} \qquad (2)$$

where p is the orbital period of Io. Because angular momentum is conserved in the system of Jupiter and its satellites, this finding limits the power dissipated by Io to $\sim 10^{14}$ W. Although such a power is consistent with conversion efficiencies for DAM bursts of $\gtrsim 10^{-6}$, it also indicates that the coupling mechanisms considered by Warwick (1970) and by Goertz and Deift cannot work at the maximum powers estimated by their authors.

It is certain, however, that Io must create some MHD disturbance in the magnetospheric plasma. Whether or not these waves are responsible for the coupling to DAM, they may have other significant effects. One such effect could be to provide local ionospheric heating near the base of the IFT. A crude calculation (cf. Appendix) yields an estimate that the waves might heat the ionosphere to temperatures of the order of 10^4 to 10^5 °K. Because the plasma temperature may, under some conditions, be an important parameter in the contexts of instability mechanisms and wave propagation, such heating may have significant consequences for these aspects of the theory.

B. Acceleration Models

The starting point for acceleration models is that, owing to the motion of Io with velocity $|\mathbf{V}_I| = 56$ km sec^{-1} relative to the co-rotating plasma, an induced electric field $\mathbf{E}_{ind} = \mathbf{V}_I \times \mathbf{B}_I/c$ appears in Io's frame, where $\mathbf{B}_I$ is the ambient magnetic field at Io. If Io is not a good conductor, then this field would be attenuated only by volume polarization throughout its interior. Piddington and Drake (1968), however, suggested that Io might be highly conducting and so its interior would be screened from the induced field by surface polarization charges. As a result, they asserted that the induced electromagnetic field of 400 kV could be mapped onto the field lines of the *IFT*, which are assumed to be equipotentials. The existence of the potential is communicated along the field line at the Alfvén speed $V_A = B_0/(4\pi\rho_m)^{\frac{1}{2}}$, where B_0 is the magnetic field strength and ρ_m the mass density of the plasma.[6] Provided that the conductivity σ is large enough so that the field does not diffuse through Io in a time shorter than the propagation time τ_A of Alfvén waves to the Jovian ionosphere, a DC circuit which closes in the Jovian ionosphere will be established. The resistive diffusion time τ_m, over a scale length R_{Io}, is given by

$$\tau_m = 4\pi \, \sigma \, \frac{R_{Io}^2}{c^2}. \tag{3}$$

The precise value of τ_A is not crucial in the present context. For purposes of estimating a constraint on σ, we may set $\tau_A \simeq 60$ sec; this is the time required for Io to move a distance equal to its own diameter. Then for $\tau_m > 60$ sec, we require $\sigma > 1.5 \times 10^5$ sec$^{-1} = 1.7 \times 10^{-5}$ mho m^{-1}.

Goldreich and Lynden-Bell assumed sufficient conductivity for Io that the DC circuit model would be valid. They asserted that this requires $\tau_m > 2\tau_A$ as we have defined τ_A. In actuality, the DC circuit will be established for $\tau_m \sim \tau_A$, because only a one-way trip for an Alfvén wave is necessary to establish the potential in the ionosphere. (One may turn on a light switch and get light very quickly; no electron in the wires need traverse the entire circuit.) Goldreich and Lynden-Bell, however, actually required more than the existence of a circuit which closes in the Jovian ionosphere. They argued that if $\tau_m \gg 2\tau_A$, the entire *IFT* will be frozen to Io, only slipping in the ionosphere in order to follow rigidly Io's orbital motion. The importance of this assumption to Goldreich and Lynden-Bell was that it enabled them to argue that the field vanishes within the flux tube and that the current is carried in thin sheets on the boundary of the *IFT*. From this argument, and from some assumptions regarding the thickness of these sheets and the plas-

---

[6]Strictly speaking, it is communicated at the phase velocity ω/k of an Alfvén wave, where $\omega/k = V_{ph} = c/(1 + c^2/V_A^2)^{\frac{1}{2}}$. For $V_A \ll c$, $V_{ph} \to V_A$; in the limit $V_A \to \infty$, $V_{ph} \to c$. The opposite limit, $V_A \to 0$, cannot be recovered from the Alfvén-wave dispersion relation.

ma density, they inferred a requisite energy for the current-carrying particles of a few keV.

Goldreich and Lynden-Bell assumed that there would be some potential drop along each field line. They then solved the current circuit equations to find the shape of the potential distribution across the IFT. Their solution has the correct form if one assumes that the current flow may be proportional to the voltage (Smith and Goertz 1975). For the potential to drop along the field line, however, requires some parallel resistivity. Goldreich and Lynden-Bell did not discuss the origin of this resistivity, and so their solution for the potential was left in terms of an unspecified constant V_0.

The models by Piddington and Drake and by Goldreich and Lynden-Bell are oversimplified in their assumptions regarding the MHD interaction of Io with the ambient plasma. Gurnett (1972) considered the effect of the formation of potential sheaths around Io. An object placed in a plasma attains a surface charge because the flux of thermal electrons hitting its surface is greater than the flux of ions. As a result, there develops a negative surface potential which is shielded from the ambient plasma by a Debye sheath. The equilibrium state is one in which the current to the surface is zero. If the surface emits a photoelectron current larger than the current it collects from the thermal flux, the potential at the surface becomes positive and a photoelectron sheath of the opposite polarity develops. Gurnett showed that in the case of a sunlit Io, part of which emits photoelectrons and part of which is in darkness, both types of sheaths occur. Their respective current densities J_D (for the Debye sheath) and J_p (for the photoelectron sheath) and areas A_D and A_p are related by the current balance condition:

$$J_p A_p = J_D A_D .$$ (4)

Gurnett's model was considered in more detail by Hubbard et al. (1974), and was extended to account for the existence of an ionosphere at Io by Shawhan et al. (1974). The principal features of the latter modification are: (a) the conductivity required to close the current system is provided by the ionosphere of Io rather than by the solid satellite; (b) the sheaths are assumed to form at some ill-defined height at the "top" of the ionosphere. The electron density profile of Io's ionosphere deduced by Kliore et al. (1974) shows a precipitous drop at an altitude of about 750 km, where the scale height changes abruptly from ~ 220 km to ~ 36 km. Shawhan et al. interpreted this as the sheath region. Adding this altitude to the radius of Io to give an "effective" radius for the conducting system, the total induced electromagnetic field becomes 570 kV. The sheath model also assumes a current system that closes in the Jovian ionosphere; the model is schematically depicted in Fig. 4.

Gurnett (1972) considered an idealized steady-state model of the current system (Fig. 5) and concluded that if the height-integrated Pedersen conductivity Σ_p of the Jovian ionosphere is lower than $\simeq 5$ mho, the IFT would

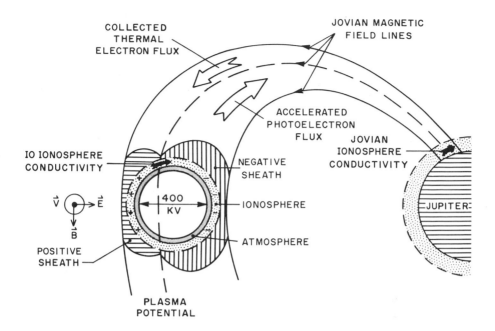

Fig. 4.   Schematic of the Io sheath model (after Shawhan *et al.* 1974).

move with Io and the bulk of the potential drop in the external circuit would occur in the Jovian ionosphere, rather than in the sheaths. Conversely, Gurnett asserted that if $\Sigma_\rho \gtrsim 5$ mho, the *IFT* slips by Io and substantial voltage drops occur across the sheaths. After noting these conclusions, Hubbard *et al.* (1974) assumed that the major potential drops occurred across the sheaths [cf. their Eq. (6)]. This assumption was also made by Shawhan *et al.* (1974) and by Shawhan (1976). All of these authors assumed that particles traversing the sheaths would gain potential energy equal to the sheath potential.

Gurnett estimated J_D and J_p independently, as did Hubbard *et al.* Similarly, Hubbard *et al.* estimated the sheath thicknesses as independent parameters, and similar estimates were made by Shawhan *et al.* These latter authors, however, recognized the sheaths to be potential double layers, which have been investigated by Block (1972), Carlqvist (1972), Knorr and Goertz (1974), and Goertz and Joyce (1975). Such layers form when electrons and ions in a plasma with $T_e \lesssim T_i$ drift relative to one another with a velocity $U \gtrsim V_e$, where V_e is the electron thermal velocity. A potential ϕ_0 then develops over a scale length L. Smith and Goertz (1975) noted that the development of the double layer was subject to the Buneman instability (Buneman 1958), which produces acoustic turbulence in the frequency range $\omega_i \ll \omega \ll \omega_e$. The convection of the turbulence at the group velocity, which

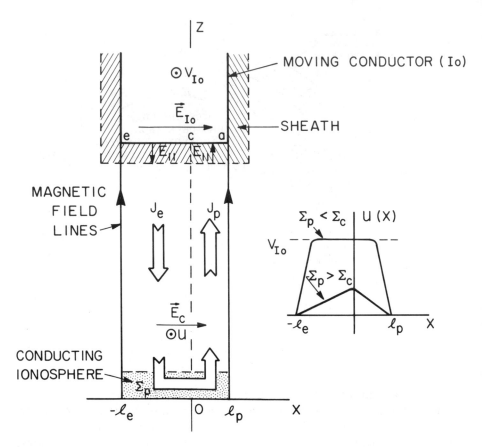

Fig. 5. The idealized circuit analyzed by Gurnett (1972). The inset shows the plasma flow velocity in the context of Gurnett's analysis. Σ_ρ is the height-integrated Pedersen conductivity of the Jovian ionosphere. For Σ_c less than a critical value $\Sigma_c \simeq 5$ mho, the electric field in the ionosphere is large and the flux tube tends to co-rotate with Io. For $\Sigma_\rho > \Sigma_c$, the voltage drop across the ionosphere is small and the plasma slips by Io.

is given by $V_{g,B} \simeq 0.4\ U$ for the most unstable waves, leads to a natural scale length L for the sheaths, given by

$$L \simeq \frac{V_{g,B}}{\gamma_B} \tag{5}$$

where $\gamma_B \simeq \omega_e / 18$ is the maximum growth rate of the Buneman instability. Smith and Goertz also showed that the limiting drift velocity U was related to the sheath potential ϕ_s by

$$\frac{1}{2}\ m\ U^2 \lesssim \frac{1}{2}\ e\ \phi_s\ .$$

By considering a steady-state current system similar to that analyzed by

Gurnett (1972), Smith and Goertz showed that the maximum potential drop across the double layers is comparable to the potential induced across Io.

We stress that all of these analyses regarding the sheath model assume the current circuit to be closed in the Jovian ionosphere. The self-consistent system of drift currents in the ionosphere of Io, the sheath regions, and the ambient plasma into which Io is penetrating (and which is therefore magneto-hydrodynamically perturbed) is not analyzed, and it has not been demonstrated conclusively that the current system does not close locally near Io. We note, however, that the current system cannot close in a region which moves with Io, because in such a region $\nabla \times \mathbf{j} = 0$.

Related to this point is the question of whether the conductivity of Io's ionosphere is itself sufficient to carry the current without causing a large drop of the emf. Shawhan *et al.* (1974) estimated the conductance of Io's ionosphere to be 260 mho, a value sufficient by more than an order of magnitude. The atmospheric model of McElroy and Yung (1975) is also compatible with a conductance of this order. This model assumes N_2 and NH_3 to be the primary atmospheric constituents. Whitten *et al.* (1975) developed a model in which the primary constituent is neon. In this model, the ionospheric structure inferred by Kliore *et al.* (1974) is obtained with a much smaller neutral density than in McElroy and Yung's model, and the ionospheric conductivity is correspondingly lower, perhaps becoming marginal with regard to the requirements of the sheath model (Shawhan 1976).

C. The Sweeping Model

The *MHD*-wave and acceleration models discussed above all entail the assumption that Io is a good conductor. In contrast, Wu (1973) assumed that the conductivity of Io was sufficiently low that the Jovian magnetic field would diffuse readily through it. In that case, particles trapped on the field lines threading Io would impinge on the satellite and be "swept" from the magnetosphere. Sweeping had been previously discussed by Mead and Hess (1973) in the context of its effects on the structure of the Jovian radiation belts.

Energetic particles trapped in a dipole magnetic field execute a complete bounce, from a given mirror point to the conjugate mirror point and back again, in the time

$$\tau_b \simeq (1.3 - 0.56 \sin \alpha_0) \, 4L \, \frac{R_J}{v} \tag{6}$$

where α_0 is the equatorial pitch angle, L is the conventional magnetic shell parameter, and v is the speed of the particle. For particles at $L \simeq 6$, mirroring at gyrofrequencies corresponding to the *DAM* frequency range, $\alpha_0 \lesssim 0.1$ radian, and so $\tau_b \simeq 30 \, R_J/v$. In the frame of reference co-rotating with Jupiter, Io moves a distance equal to its own diameter in a time $\tau_s \simeq 60$ sec. Therefore, particles that execute a half-bounce in a time shorter than τ_s will

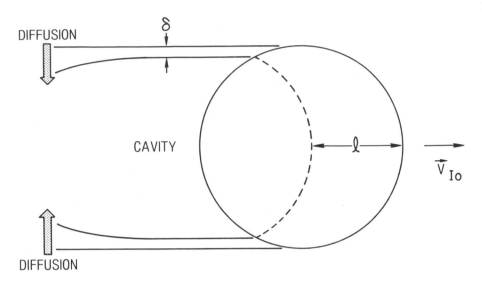

Fig. 6.   Cross section of the energetic-particle cavity in Io's orbital plane.

impinge upon Io and be removed from the field lines described by their guiding center motions. The threshold velocity v_c above which particles will be swept in this way is determined by equating $\tau_b(v_c)/2 = \tau_s$. For $L = 6$, $v_c \simeq 1.8 \times 10^4$ km sec^{-1}, corresponding to threshold energies of approximately 0.8 keV for electrons and 1.6 MeV for protons. Particles above the appropriate threshold are effectively swept from the Io flux tube.

The instability mechanism due to the sweeping effect depends on the energetic protons. In Sec. II we shall consider the form of the proton distribution function; here it suffices to note that it includes as a parameter a characteristic energy E_p.

The sweeping of energetic particles by Io results in a cavity in the density distribution of such particles. In the orbital plane of Io, the cavity has a cross section similar to a wake and remains for some time after the passage of Io; diffusion across the cavity boundary tends to replace the energetic particles on some time scale τ_D, that is long compared to the growth time of the electromagnetic instability. The cavity cross section in Io's orbital plane is illustrated in Fig. 6. The density of particles with energies above the sweeping threshold falls to zero across the distance δ. The bow portion of the cavity where the sweeping is incomplete has a characteristic dimension ℓ given by

$$\ell = \frac{\tau_b(E_p)}{2} V_{Io} \; ; \tag{7}$$

thus the bow of the completely swept portion has a cross section which is a semicircle displaced from the bow of Io by the distance ℓ. The dimension δ of the lateral boundary layer is given by the intersection of this profile with the trailing limb of Io:

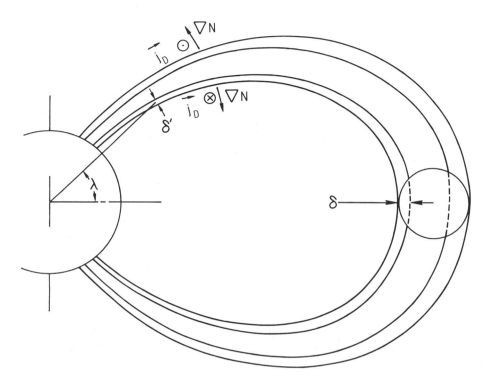

Fig. 7.   Mapping of the density-gradient layer along the Io flux tube. Also shown are the directions of the density gradients and drift currents at the inner and outer walls of the cavity.

$$\delta = R_{Io} \left[1 - \cos \sin^{-1} \left(\frac{\ell}{2 R_{Io}} \right) \right]. \tag{8}$$

Equation (8) is valid as long as it gives a value for δ greater than the characteristic proton Larmor radius $R_p = (E_p / M\Omega_p^2)^{\frac{1}{3}}$, where $\Omega_p = 2\pi f_{cp}$ is the angular proton gyrofrequency; R_p is the physical lower bound on δ. The equatorial boundary thickness δ maps to values $\delta'(\lambda) < \delta$ at magnetic latitudes $|\lambda| > 0$ (Fig. 7):

$$\delta'(\lambda) = \delta - \int_0^\lambda \frac{d\delta}{d\lambda'} \, d\lambda' . \tag{9}$$

Assuming $R_p(0)/\delta \ll 1$, we have $R_p(\lambda)/\delta' < 1$ also, and the boundary thickness at higher latitudes is given approximately by the mapping for a centered collinear dipole:

$$\frac{\delta'}{\delta} - \frac{\cos^3 \lambda}{(4 - 3 \cos^2 \lambda)^{\frac{1}{2}}} . \tag{10}$$

Owing to the density gradient of energetic protons across the walls of the *IFT*, drift currents appear. These currents drive instabilities at the upper and lower hybrid resonances; these instabilities are discussed in later sections.

D. *Evidence From Pioneers 10 and 11*

The trajectories of Pioneers 10 and 11 did not pass through the instantaneous *IFT*, and so the evidence from these probes regarding the coupling mechanisms discussed above is indirect and inferential. Inbound, Pioneer 11 passed about 6000 km from the instantaneous *IFT* as defined by the D_2 field model of Smith *et al.* (1975) at about $03^h\ 00^m$ on 3 December 1974 (Fillius *et al.* 1975). The spacecraft was then at approximately $-40°$ magnetic latitude, and so its distance from the *IFT* was about twenty times the radius of the flux tube itself at that latitude. No distortions of the magnetic field near this region have been reported in the literature, indicating that the *MHD* perturbations arising from Io's penetration of the magnetospheric plasma are either unresolvable or are localized at this latitude to distances much less than 6000 km from the *IFT*. Thus, it is unlikely that Io warps its flux tube by 15°, as suggested by Goertz and Deift. No direct evidence about the actual *MHD* nature of the Io-magnetosphere interaction can be obtained from trajectories such as those of Pioneers 10 and 11, however.

The particle data from both Pioneer 10 and 11 show that sweeping occurs and is effective. In this regard, it must be remembered that sweeping of the *IFT* is followed by diffusion of the energetic particles back into the swept region. The Pioneer 10 data were often interpreted to indicate that sweeping is much less effective than predicted by Mead and Hess; typical reductions in the particle fluxes at the *IFT* observed on Pioneer 10 were by factors of about two to five, and the sweeping effect seemed generally to be confined to lower-energy particles. Mead and Hess, however, considered only the slow processes of electric and magnetic diffusion, as they are understood to operate in the terrestrial magnetosphere, to balance the sweeping effects of the various Jovian satellites. They were led to predict large decreases, by several orders of magnitude, of the trapped particle fluxes at the orbits of each of the moons. Smith (1973), however, showed that the sweeping effect leads not only to the high-frequency electromagnetic instability described by Wu, but also to a low-frequency instability which produces strong diffusion across the boundaries of the *IFT*. (To the extent that sweeping is effective at other satellites, the same mechanisms operate there also.) Moreover, the diffusion coefficient is energy-dependent, increasing with the particle energy. Thus, the relative mildness of the sweeping effect as observed in the Pioneer 10 data may be attributed to the fact that the spacecraft crossed the orbit of Io several hours behind Io, and the density gradient had been eliminated in the case of higher-energy particles and considerably reduced at lower energies. Further evidence of this interpretation may be found in the Pioneer 11 data. The trajectory of Pioneer 11 was such that the spacecraft passed the Io orbit much closer to Io, both inbound and outbound, than did Pioneer 10; sweeping was found by all the particle experiments to be quite severe.

There is clear evidence for acceleration of electrons at Io, but its interpretation in terms of existing models is ambiguous. On both the inbound and outbound crossings of Pioneer 10, Fillius and McIlwain (1974) observed enhancements, by a factor of about two, in the electron flux at energies $E >$ 0.16 MeV. On the inbound crossing of Pioneer 11, Fillius et al. (1975) observed a dramatic increase – by a factor of thirty – in the electron flux at $E > 0.46$ MeV. These injections can be readily understood within the context of acceleration in double layers.

II. THE INSTABILITY MECHANISM

There is some variability in the literature in the nomenclature used for the various normal modes of plane waves in plasma. Therefore, we shall first define the nomenclature to be used here. We shall not be concerned with low-frequency modes for which the ion motion is important. In the high-frequency regime there are two normal modes in a uniform magnetized plasma. These are designated as the ordinary (o) and extraordinary (x) modes. The refractive indices $n_{o,x}$ ($\equiv ck/\omega$) of these modes are given by the well-known Appleton-Hartree formula:

$$n_{o,x}^2 = 1 - \frac{2\omega_e^2(1 - \omega_e^2/\omega^2)}{2(\omega^2 - \omega_e^2) - \Omega_e^2 \sin^2\theta \pm \Omega_e \Delta} \tag{11}$$

where $\omega_e = 2\pi f_{p,e}$, $\Omega_e = 2\pi f_{c,e}$, θ is the angle between the wave vector $\mathbf{k}$ and the ambient magnetic field $\mathbf{B}_0$, and $\Delta \equiv [\Omega_e^2 \sin^4\theta + 4(\omega^2 - \omega_e^2)\cos^2\theta]^{\frac{1}{2}}$. The upper sign refers to the o-mode, and the lower sign to the x-mode. If variations in the magnetic field or plasma density occur on a length scale much greater than k^{-1}, Eq. (11) may be taken to describe the local propagation of the waves.

The ratio $\epsilon \equiv \omega_e^2/\Omega_e^2$ is an important parameter in determining the nature of the normal-mode propagation. Throughout the region of the Jovian plasmasphere of interest for DAM, $\epsilon \ll 1$. This is an unusual parameter regime in both laboratory and astrophysical plasmas; the only other known region in the solar system in which $\epsilon \ll 1$ is the high-latitude region above the terrestrial plasmasphere. Interestingly enough, this is the source region of the terrestrial kilometric radiation (Gurnett 1974; Kaiser and Stone 1975).

In Fig. 8 we show the topology of the dispersion curves ω versus k and the refractive indices n^2 versus ω of the o- and x-modes for propagation at large angles to the magnetic field. The x-mode is seen to consist of two branches, the slow branch (over most of which $n_x^2 > 1$) and the fast branch ($n_x^2 < 1$). Both modes exhibit cutoffs, occurring at ω_e for the o-mode and $\omega_x = \Omega_e \{ [1 + 2\epsilon \pm (1 + 4\epsilon)^{\frac{1}{2}}]/2 \}^{\frac{1}{2}}$ for the x-mode, where the upper (lower) sign refers to the fast (slow) branch. For the fast x-branch, the cutoff frequency is approximately given by $\omega_x = \Omega_e(1 + \epsilon)$. In addition, the slow x-branch exhibits a resonance at $\omega = \Omega_e(1 + \epsilon/2 \sin^2\theta)$.

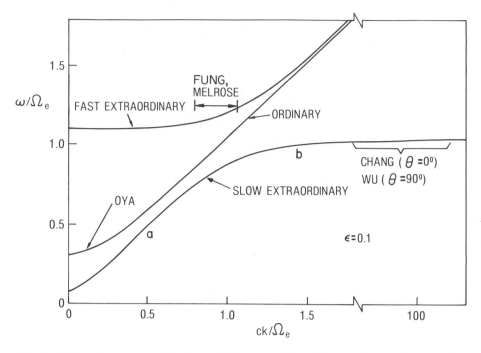

Fig. 8. Dispersion curves and indices of refraction of the ordinary and extraordinary modes for $\epsilon \ll 1$. The approximate regions of instability found by various authors are indicated.

At $\theta = 90°$ the o-mode is linearly polarized and the x-mode is right elliptically polarized, with $\mathbf{E} \perp \mathbf{B}_0$. As $\theta \to 0$, the x- and o-modes go over continuously to right and left circular polarizations, respectively. The portion a-b of the slow branch is the whistler regime.

The x-mode cannot propagate in the "stop band" between the slow-branch resonance and the fast-branch cutoff. The axes of Fig. 8 are normalized to Ω_e; for a fixed wave frequency, increasing ω/Ω_e generally corresponds to moving towards lower magnetic fields. Therefore, we see that waves on the slow x-branch may not propagate from the Jovian plasmasphere into interplanetary space. Strictly speaking, the stop band exists only in the cold-plasma limit; thermal effects remove the resonance and lead to a non-zero minimum in the group velocity of the slow branch (Aubry *et al.* 1970).[7] The finite temperature of the plasma also implies a collision frequency ν_c, however, and waves with extremely slow group velocity V_g are attenuated by collisions over a scale length V_g/ν_c. If the group velocity is sufficiently small, or the spatial region corresponding to the stop band (the "stop zone") is sufficiently extensive, the attenuation in traversing the stop zone may still

---

[7]Other thermal effects on the dispersion relations are negligible for our purposes, provided $V_e^2/c^2 \ll \epsilon$. Although $\epsilon \ll 1$, the low temperature of the Jovian plasma ensures the validity of this inequality.

be large. Thus, as a general rule the existence of the stop band must be taken into account in assessing the propagation of waves. In particular, instability mechanisms in which the waves are amplified on the slow branch also imply the necessity of some mode-coupling process leading to the escape of waves in either the o-mode or the fast x-mode.

The cold-plasma approximation may be deduced from the more rigorous kinetic theory in the asymptotic limit of low temperatures, under the conditions

$$\frac{k_\perp V_e}{\Omega_e} \to 0 \tag{12}$$

$$\left| \frac{\omega - \ell\, \Omega_e}{k_{||} V_e} \right| \gg 1 \quad (\ell = 0, 1, 2, \ldots) \tag{13}$$

where $k_\perp = k \sin \theta$, $k_{||} = k \cos \theta$. Expression (12) is the condition that finite-Larmor-radius effects may be neglected. Expression (13) is the condition that the thermal electrons do not exhibit Landau and cyclotron resonance with the wave as they spiral along field lines. Such resonances may quickly damp the wave.

Inequalities (12) and (13) arise from the inclusion of thermal effects in the plasma dielectric tensor. In the cold-plasma limit the dielectric is Hermitian, giving refractive indices which are either purely real (oscillatory solutions) or purely imaginary (exponentially damped solutions). Thermal effects modify somewhat the Hermitian part of the dielectric, which determines the normal-mode frequencies. As noted above, however, these modifications are of order V_e^2/c^2 for the electromagnetic modes, and are generally negligible. The thermal effects also introduce a non-Hermitian part to the dielectric. It is this part which implies damping of the normal modes; the damping is weak under the conditions (12) and (13).

In the same manner, the inclusion of a population of energetic particles also modifies the plasma dielectric, in general introducing a frequency shift into the real part of the refractive index and, in addition, an imaginary part to the frequency, which may then be written

$$\omega(\mathbf{k}) = \omega_r(\mathbf{k}) + i\gamma(\mathbf{k}, \omega_r). \tag{14}$$

The usual approach adopted in linear instability analysis is to consider an energetic particle population of low density relative to the thermal background plasma. Then the real frequency ω_r is governed by the cold-plasma dispersion relation, while the growth or damping of the waves is determined by the energetic particle distribution. The dispersion relation is then of the form

$$D(\mathbf{k}, \omega) = C(\mathbf{k}, \omega) + \sum_s g_s(\mathbf{k}, \omega) \tag{15}$$

where $C(\mathbf{k}, \omega)$ is the cold-plasma dispersion relation and $g_s(\mathbf{k}, \omega)$ represents the contribution of energetic particles of species s. In principle, the thermal damping effects may be included in $\Sigma_s\, g_s$, but this is not usually done; in particular, it is not done in the theories we shall discuss here. Therefore, the tendency for instability must be compared *a posteriori* with conditions (12) and (13), to determine that the instability is stronger than the competing damping by thermal particles.

All of the instability mechanisms we shall consider depend on an anisotropic distribution of a small energetic-particle population. Instabilities may exist in various parameter regimes in both ω_r and in the parameters that describe the anisotropy, e.g., $T_\perp / T_\parallel$. The nature of the instability may be closely related to the nature of the coupling, and some coupling mechanisms are compatible with only certain types of instability mechanisms. To some extent, individual coupling mechanisms have generally been proposed with some particular type of consequent instability mechanism in mind, even when the coupling and instability have not both been investigated by the same author. Conversely, in some cases a proposed instability mechanism includes the tacit assumption that there exists an unspecified coupling mechanism which produces either the required anisotropy in the energetic-particle distribution or the conditions necessary for an existing marginally stable anisotropy to be driven unstable to electromagnetic wave amplification.

It is sometimes asserted in the literature that the *DAM* emission mechanism must be cyclotron emission. There is no compelling evidence, either observational or theoretical, that this is the case. Goldreich and Lynden-Bell argue for coherent cyclotron emission primarily on the basis of the sharp beaming pattern of the emission, and also on the high intensity. Such beaming, however, is an inherent feature of all of the mechanisms discussed below. In addition, there is the complicating factor that, if amplification occurs on the slow x-branch, the beaming pattern may be closely related to or determined by the mode-coupling process by which the *DAM* escapes from Jupiter. Thus, cyclotron emission is only one of the possibilities for the instability mechanism.

Before discussing individual instability mechanisms, it should be noted that a complete understanding of *DAM* may involve more than one mechanism. There are at least two distinct types of Io-modulated *DAM*, the decasecond L and millisecond S bursts. They have quite distinct time scales — from about 0.5 to 3 sec for L-bursts and 1 to 10 milliseconds for S-bursts — which do not merge continuously into one another. Moreover, the S-bursts exhibit true frequency drifts, which are always negative. In contrast, L-bursts exhibit frequency drift only in their overall envelope; this is probably a geometric effect of the beaming pattern. Although both L- and S-bursts share the same general morphology of Io-modulation, these facts imply that they may be caused by separate instability mechanisms.

It is also unclear whether the Io-modulated and Io-independent L-bursts imply different emission mechanisms. With the recent recognition of an Io-independent component of the early source region (Carr and Desch;[8] Bozyan and Douglas 1976) one might be tempted to hypothesize that the Io-modulated *DAM* is simply an enhancement of a weak continuous emission. It would then seem necessary, however, to shift from the common interpretation that the morphology of the Io-modulation implies a beaming pattern, and instead to suppose that the Io-modulation is indeed only possible for certain geometrical configurations relative to the observer. This latter hypothesis is physically untenable. Moreover, the declination effect in *CML* is related to the Io-independent *DAM* (Goertz 1971) while the modulated component exhibits a declination effect in Io phase (Conseil 1972; Lecacheux 1974; Thieman *et al.* 1975; Bozyan and Douglas 1976). Thus, it seems probable that the Io-independent and Io-modulated components are due to distinct mechanisms, although these arguments do not conclusively rule out the possibility that they may be closely related.

Thus, in assessing the viability of a given instability mechanism, one should attempt to do so in the several contexts of Io-independent *L*-bursts and Io-modulated *L*- and *S*-bursts. In the first of these contexts, the lack of a demonstrable coupling mechanism is irrelevant.

Finally, we note that the mechanisms we shall discuss have generally been analyzed only in the linear regime. Linear instability analysis may be quite misleading, however, because it provides no information on the saturation level of the instability. In general, a complete stability analysis must include consideration of the nonlinear regime.

A. The Sweeping Model

Of the instability mechanisms we discuss, the sweeping model of Wu (1973) is unique in that the instability is driven by anisotropy in configuration space rather than in velocity space. As was mentioned in Sec. I, the density gradients $\nabla N_{e,i}$ of trapped energetic particles across the boundary of the *IFT* produce drift currents in the cavity "wall"; these currents are perpendicular to both $\mathbf{B}_0$ and N. Wu considered the motion of each energetic particle species transverse to $\mathbf{B}_0$ to be characterized locally by a mean energy $E_{\perp,s}$. The resulting drift velocity of the s-th species is then given by

$$\langle v_{x,s} \rangle = -\frac{q_s}{|q_s|} \frac{1}{\Omega_s} \left(\frac{\mathrm{d}\ell n\, N_s}{\mathrm{d}y} \right) \frac{E_{\perp,s}}{m_s} \equiv v_{D,s} \qquad (16)$$

where q_s is the charge, the density gradient is taken to be positive in the y-direction, and the drift motion is in the $+x$ direction for electrons and the $-x$ direction for protons.

Wu considered instability at the upper hybrid frequency ω_{UH} and for very short wavelengths, $ck/\Omega_e \gg 1$. Under these conditions the energetic electrons contribute only a negligible shift in the real frequency, and the

---

[8]See p. 713.

instability is produced by the non-resonant contribution of the protons. Taking an isotropic Gaussian for the local distribution of the energetic protons, Wu found the growth rate of waves with $\mathbf{k} \perp \mathbf{B}_0$ to be given by

$$\gamma = \frac{1}{(8\pi)^{\frac{1}{4}}} \frac{1}{(kR_p)^{\frac{3}{2}}} \left(\frac{N_p}{N_e} \frac{M}{m} \epsilon\right)^{\frac{1}{2}} \left(\frac{k_x v_D}{\Omega_e} - 1\right)^{\frac{1}{2}} \omega_e \tag{17}$$

where subscripts p and e denote energetic protons and thermal electrons, respectively, $R_p = (E_{\perp,p}/M)^{\frac{1}{2}}/\Omega_p$, and k_x is the x-component of the wave vector. Maximum growth occurs at

$$\frac{k_x v_D}{\Omega_e} \simeq \frac{v_D}{c} \frac{ck}{\Omega_e} = \frac{3}{2}. \tag{18}$$

Taking $v_D = E_\perp/M\Omega_p\delta'$, where δ' is the thickness of the density-gradient layer (see Sec. I-C), one finds v_D in the range $1-4 \times 10^3$ km sec^{-1} for 5 MeV $\lesssim E_\perp < 10$ MeV. Thus, the wavenumber at maximum growth is on the order of $cK/\Omega_e \simeq 100$. Wu did not address the question of how the waves penetrated the stop band.

Wu's analysis was done for $\mathbf{k}$ strictly perpendicular to $\mathbf{B}_0$. Because the analysis was a local one, it also requires a finite component of k in the y-direction, so that

$$|k_y| \left|\frac{\partial \ell n \, N_p}{\partial y}\right|^{-1} \gg 1. \tag{19}$$

Although it is clear from Eq. (17) that γ increases as k_x/k increases, the inequality (19) is not sufficient to allow a precise determination of the azimuthal width of the beaming pattern; at best, one can say only that it is a pencil beam. In the direction parallel to $\mathbf{B}_0$, the width of the pencil beam is sharply delimited by Eq. (13), which requires

$$\left|\frac{k_\parallel}{k}\right| \ll \frac{\epsilon}{2} \frac{1}{kR_e} = \frac{\epsilon}{2} \frac{c}{V_e} \left(\frac{\Omega_e}{ck}\right). \tag{20}$$

The current-driven instability also contains the feature of a low-frequency cutoff, because an essential requirement for the instability to occur is that, locally, $\epsilon > 2m/M \simeq 10^{-3}$. In general, ϵ decreases with increasing altitude, at least until the point of maximum zenopotential height (Melrose 1967; Smith 1973; Goertz 1975).

The power available in this mechanism was estimated by Wu (1973) to be about 10^9 W, adequate for the *DAM* requirement. Smith (1973) estimated the power to be $10^9 \eta$ W, where $\eta \ll 1$ is an efficiency factor. Smith also pointed out that the estimated requirement of $\simeq 10^8$ W depends on an assumed beaming into a solid angle of approximately 0.1 steradian; from the above discussion this solid angle might be much too large in the context of the sweeping instability. Both of these estimates assumed growth rates of order $\gamma \sim 10^{-2}$ sec^{-1}. Because the growth rate depends on the energetic-proton density, the precise value of the estimated power may have little significance. The ability of the mechanism to meet the power requirement of Io-modulated *DAM*, however, seems marginal.

Comparison of the predictions and requirements of the sweeping model with Pioneer observations tends to cast doubt on the viability of the mechanism for Io-modulated *DAM*. As was noted above, sweeping seems to occur effectively enough to establish the necessary density gradient. Because the mechanism relies on trapped protons, however, and the higher-order moments of the magnetic field make the surface field highest in the northern hemisphere, the instability is limited at any longitude to the region between the topside Jovian ionosphere in the southern hemisphere and the conjugate mirror point. Because the growth rate of the instability is proportional to N_e, the emission is essentially limited to the southern hemisphere. In order to span the range of *DAM* frequencies, therefore, the peak ionospheric electron density, N_{e0}, must be sufficient to raise the upper hybrid frequency in the ionosphere up to about 40 MHz. The maximum gyrofrequency at the foot of the *IFT* is about 27 MHz, giving the requirement that $N_{e0} \gtrsim 10^7$ cm^{-3}. The maximum density inferred by Fjeldbo *et al.* (1975), however, is on the order of 3×10^5 cm^{-3}. Thus, unless this determination is too low by a factor of 30, the sweeping mechanism is essentially limited to frequencies below 30 MHz (see Fig. 9).

Other aspects of the sweeping mechanism will be discussed below. In particular, it may be quite relevant to the question of the "Io-independent" *DAM*.

B. Maser-Like Mechanisms

A number of authors have considered wave amplification by electron distributions which are anisotropic in velocity space. Such mechanisms are often called "maser-like," because the anisotropic distribution is analogous to an inverted level population. The most commonly analyzed situation is that in which the mean perpendicular energy $T_\perp$ is greater than the mean parallel energy $T_{||}$.

Chang (1963) discussed the amplification of *x*-mode waves by an anisotropic distribution of relativistic electrons. Although he referred to these waves as whistlers, his analysis was in fact valid for the entire *x*-mode. He only considered the cases of propagation parallel and anti-parallel to $\mathbf{B}_0$, however. Chang found two parameter regimes in which amplification could occur. Of these, the only case relevant to the Jovian exosphere, in which $\epsilon \ll 1$, is the frequency regime

$$\omega_e^2 \gg \Omega_e \omega - \omega^2 > 0 , \quad \Omega_e/\omega \simeq 1 . \tag{21}$$

Because $\epsilon \ll 1$, however, inequality (21) is equivalent to the condition

$$\frac{\omega}{\Omega_e} = 1 - \xi , \quad \xi \ll \epsilon . \tag{22}$$

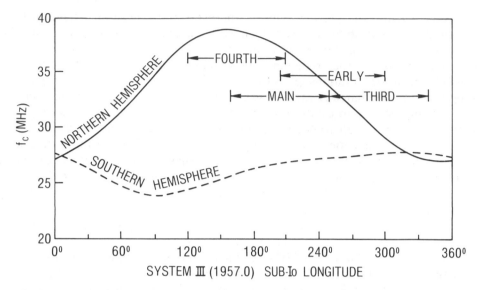

Fig. 9.   The gyrofrequency at the feet of the Io flux tube, in the magnetic field model of Smith *et al.* (1975). The approximate sub-Io longitude ranges of the various sources are indicated (Epoch 1974.9).

Thus, as in the case of the sweeping model, Chang's mechanism requires $ck/\Omega_e \gg 1$ (see Fig. 8). For parallel propagation, however, these waves are strongly damped by the thermal electrons because, for $\xi \ll \epsilon$ and $\epsilon \ll 1$, we have

$$\left| \frac{\omega - \Omega_e}{k_\| V_e} \right| \ll \epsilon \, \frac{c}{V_e} \frac{\Omega_e}{ck} \ll 1 \tag{23}$$

in violation of inequality (13).

In an analysis complementary to that of Chang (1963), Goldstein and Eviatar (1972) discussed amplification at $\theta = 90°$ on the slow x-branch by a loss-cone distribution of trapped electrons. Goldstein and Eviatar (1972), however, explicitly recognized the necessity for the waves to penetrate the stop zone. They argued that the amplification would be sufficiently strong so that the radiation would simply tunnel through the stop band, suffering exponential attenuation but emerging into the outer region of the exosphere with observable intensity. They asserted that this mechanism could be responsible for the Io-independent *DAM*. As we shall discuss below, the stop zone is probably sufficiently thin that exponential attenuation of waves incident upon it is negligible. Assuming this to be so, the instability analysis of Goldstein and Eviatar might account for Io-independent *DAM* by emission from a loss-cone distribution near marginal stability.

Hirshfield and Bekefi (1963) considered the amplification of x-mode waves at Ω_e, with $\theta = 90°$. In their model the distribution function was assumed to be isotropic; the maser-like effects were due to the dependence of the gyrofrequency on the relativistic electron mass. This dependence leads to coherent interaction between the particles and the wave field through phase synchronization, such as occurs in synchrotron accelerators (Kuckes and Sudan 1971).

Although it was apparently not recognized by Hirshfield and Bekefi, the unstable waves at the gyrofrequency are below the stop band. They also predicted amplification at higher harmonics of the gyrofrequency, but at unobservable levels. Moreover, their work was done before the discovery of the Io-modulation, and so the distribution function used in their analysis was based on the assumption of direct access to the Jovian magnetosphere of streams of low-energy particles having a solar origin.

Goldreich and Lynden-Bell (1969) also postulated coherent cyclotron amplification for the emission mechanism, but recognized the problem of traversing the stop zone. They discussed the instability as seen in the frame of the streaming particles; in this frame their results are essentially similar to those of Hirshfield and Bekefi. Goldreich and Lynden-Bell asserted that in Jupiter's frame the waves would be Doppler-shifted above the stop band and so would escape the magnetosphere. This conclusion was erroneous, however, because a wave with phase velocity smaller than c in one frame will not have phase velocity greater than c in any other frame. Thus, the Doppler-shifted waves postulated by Goldreich and Lynden-Bell would not be normal modes in the magnetospheric plasma, and would be heavily damped away from the source region.

All of the mechanisms so far discussed provide for amplification on the slow x-branch. With the exception of the sweeping-model instability, which is non-resonant and is carried by protons, these mechanisms all invoke resonant-electron effects. These are possible only on the slow x-branch, where the phase velocities of the waves may be exceeded by particle velocities. For the o-mode and the fast x-branch, however, the phase velocities are greater than c, and resonant interaction is impossible. There are, nonetheless, maser-like mechanisms to amplify these waves; they are sometimes called subluminous instabilities. Such mechanisms were hypothesized by Ellis (1962, 1963, 1965) and by Ellis and McCulloch (1963), who assumed the existence of electron "bunches" precipitating into the Jovian ionosphere. The mechanism was investigated quantitatively in a series of papers by Fung (1966a,b,c). The first of these papers was concerned with terrestrial VLF emission and so neglected terms of order Ω_e^2/ω_e^2, which are large in the Jovian exosphere. Fung (1966b) extended the analysis to include these terms. Both of these analyses assumed precipitating electron distributions of the form

$$F_e(p) \sim \delta(p_{||} - p_{||0})\, \delta(p_\perp - p_{\perp 0}) \qquad (24)$$

and allowed for arbitrary angles of wave propagation. In his third paper Fung extended the formalism to include finite temperatures $T_\perp$ and $T_{||}$ in the precipitating stream.

In this mechanism the emission is Doppler-shifted to a normal mode above the stop band by the streaming velocity of the particles. Radiation at the fundamental is confined to a finite cone, in the forward direction, of wave-vector directions about the magnetic field, while emission at higher harmonics may occur in all directions but is much weaker.

Melrose (1973, 1975) also considered subluminous instability from a streaming anisotropic distribution, which he assumed to be a shifted bi-Maxwellian:

$$F_e(\beta, \alpha) = \frac{1}{(2\pi)^{\frac{3}{2}}\beta_{\perp0}^2\,\beta_{\|0}}\;\exp\left[\frac{-\beta^2\sin^2\alpha}{\beta_{\perp0}^2} - \frac{(\beta\cos\alpha - \beta_s)^2}{\beta_{\|0}^2}\right]. \qquad (25)$$

Here $\beta = v/c$, α is the particle pitch angle, and $\beta_s = \langle\beta\cos\alpha\rangle$ is the normalized streaming velocity. The Doppler-shifted frequency of the s-th harmonic, radiated in the direction θ by a particle of pitch angle α and Lorentz factor γ, is

$$\omega = \frac{s\Omega_e}{\gamma(1 - n_x\,\beta\cos\alpha\cos\theta)}. \qquad (26)$$

At the fundamental, the condition that the radiated wave be Doppler-shifted above the stop band is then

$$n_x\,\beta_s\cos\theta > \epsilon \qquad (27)$$

which is not excessively stringent in view of the smallness of ϵ.

Melrose finds a maximum growth rate on the order of

$$\gamma_{max} \simeq \frac{\eta\epsilon}{n_x\,\beta_{\|0}}\,\Omega_e \qquad (28)$$

where $\eta \equiv N_s/N_e$ is the ratio of the stream density to that of the background plasma. The bandwidth of the growing waves is on the order of

$$\frac{\Delta\omega}{\Omega_e} \simeq n_x\,\beta_{\|0} \ll 1 \qquad (29)$$

which is in qualitative accord with observations.

Melrose finds a threshold condition on the degree of anisotropy necessary in order for amplification to occur; the condition is

$$\beta_{\perp0}^2 > \beta_{\|0}. \qquad (30)$$

Although this inequality may seem stringent, it is actually fairly easy to meet under certain models for the acceleration of the particles. For instance, the accelerated spectrum in the sheath model is essentially a delta-function in energy at any given point at Io; the particular injection energy depends on the spatial structure of the sheath potential (Smith and Goertz 1975). Because the energy is in general much larger than the temperature of the particles before acceleration, the accelerated distribution will map at high latitudes essentially to a distribution of the form (24) assumed by Fung. The parallel motion may be equated locally to Melrose's β_s, while his $\beta_{\perp 0}$ may be equated to the characteristic transverse velocity. In the limit of this model, inequality (30) is virtually satisfied. This condition appears to be an artifact of the assumed form (25) for the distribution function; Fung (1966b) found instability for distributions of the form (24) in which $\beta_\perp^2 < \beta_0$.

Melrose (1975) used quasilinear arguments to estimate the saturation energy density W_w of the amplified waves as

$$\frac{W_w}{W_s} = \frac{2}{\gamma_{max}} \frac{\beta_s \, c \, \beta_{\perp 0}^2}{L_B(\beta_s^2 + \beta_{\perp 0}^2)} \tag{31}$$

where L_B is the gradient scale length of the magnetic field and $W_s = N_s \times mc^2(\beta_{\perp 0}^2 + \beta_s^2)/2$ is the energy density in the stream. He evaluated this for two different sets of beam parameters, obtaining results which bracket an observational estimate of 4×10^{-12} erg cm^{-3} required. This latter estimate is made using a very narrow beaming cone of solid angle 0.01 steradian. In general, however, the energetic requirement of DAM appears to be well met within the context of Melrose's calculations.

The maximum growth rates predicted by Melrose's theory are on the order of $10^2 \lesssim \gamma \lesssim 10^3$ sec^{-1}. Thus the mechanism may be most relevant to L-bursts, rather than to S-bursts, which are already fully developed on this time scale.

Finally, we note that Melrose did not quantitatively address the question of the beaming pattern of his mechanism. As we shall discuss in Sec. III, this is an important topic for comparison of the predictions of the mechanism with the observational phenomenology.

We have discussed only mechanisms that have been treated quantitatively in the literature. Other suggestions have been advanced by Warwick (1961, 1963), Zheleznyakov (1966), and Marshall and Libby (1967). For a discussion of these works, the reader is referred to the review by Warwick (1967).

III. WAVE PROPAGATION IN THE JOVIAN PLASMASPHERE

Compared with the aspects of coupling and instability mechanisms, propagation of the DAM waves in the Jovian plasmasphere has received relatively little attention. It is an important topic, however, which must be considered as an integral part of a complete theory of DAM. The beaming pattern implied by the DAM morphology is undoubtedly due to a combination of the inherent beaming pattern of the instability mechanism and the subsequent propagation of the waves. Conversely, the phenomenology may provide the basis for inferences regarding the plasma medium and the source regions.

As an example of the latter approach, some authors have considered the polarization. Warwick and Dulk (1964) argued that the highly elliptical polarization of decametric waves upon emergence from the Jovian magnetosphere indicates that the base mode of the emission is the extraordinary mode, an interpretation which has generally been accepted in the literature. The sense of polarization of this mode is right (left)-handed when the wave vector has a component parallel (antiparallel) to B_0. Although the limiting polarization depends upon the total ray path, the former case would generally be expected to occur for emission from the northern hemisphere, and the latter for emission from the southern hemisphere. At high frequencies (> 18 MHz), DAM is predominantly right-polarized, while at lower frequencies the degree of left-polarization increases as the frequency decreases. The axial vector exhibits a quasi-sinusoidal rotation profile, which is conventionally interpreted as indicating that at a given longitude one or the other hemisphere dominates the beaming into a given direction. This reasoning should be considered with caution, however, for within storms that are rich in time structures there may exist considerable polarization diversity in the frequency-time plane.

Notwithstanding the argument of Warwick and Dulk about the base mode of the emission, Warwick (1970) pointed out that the radiation incident upon the terrestrial ionosphere is not in a base mode. Nearly all of the Faraday rotation observed, however, is attributable to propagation through the terrestrial ionosphere; Warwick estimated that less than 360° of Faraday rotation occurs in the Jovian magnetosphere. Warwick argued that this fact placed a small upper limit on the column density $\int N_e \, ds$ through which the radiation propagates at Jupiter, and that this limit on the column density was incompatible with density models such as those of Gledhill (1967) and Goertz (1973a, 1975), in which there is a disk-shaped concentration of plasma in the equatorial plane.[9] Goertz (1974) challenged this argument; he showed that, if the radiation is generated entirely in a base mode, then during propagation through a spatially varying medium it will couple to the other mode just sufficiently to preserve its polarization, and so Faraday rotation could not occur.

As we have discussed in Sec. II, several proposed instability mechanisms actually amplify waves on the slow x-branch; such waves cannot escape directly to free space. The general topology of the stop zones has been discussed by Liemohn (1973) and by Gulkis and Carr (1966); these latter authors suggested that the dimension of the stop zone might be longitudinally asymmetric, so that the DAM beaming pattern—as manifested in the source morphology—might be governed by the location of the thinnest stop zones. They noted, however, that the mechanism by which waves could penetrate the stop zones was unclear.

---

[9]Although the density models of Goertz and Gledhill are topologically similar, they are derived from different considerations and therefore differ by orders of magnitude in the actual densities predicted at a given radial distance in the plasma disk.

Oya (1974) attempted to account for the escape of waves through the stop band by considering mode conversion of extraordinary waves propagating inward towards the planet. The conversion occurs by the reflection of the o-mode at the plasma level $\omega_e = \omega$. There are a number of difficulties with the geometry inherent in Oya's model. It suffices to note, however, that if the peak ionospheric electron density is 3×10^5 as reported by Fjeldbo *et al.*, then the corresponding plasma frequency is about 5 MHz, and the critical condition is therefore not met at DAM frequencies.

The importance of the stop zones, however, is greatly dependent on the temperature and peak density of the Jovian ionosphere. Smith (1975) showed that for $N_0 = 3 \times 10^5$ (Fjeldbo *et al.* 1975) and $T_0 = 750°$ (Kliore *et al.* 1975), the attenuation of waves during traversal of the evanescent region is negligible for frequencies $\lesssim 20$ MHz. This fact has profound consequences for the explanation of Io-independent DAM, which we discuss in Sec. IV.

As we have seen in Sec. II, theories in which instability occurs on the slow x-branch, such as those of Goldstein and Eviatar (1972) and of Wu (1973), amplify waves of very low group velocity $V_g \ll c$. Smith *et al.* (1972) and Wu *et al.* (1973) attempted to account for the Io-phase asymmetry (referred to in Sec. I) in terms of the low group velocity of the waves before their escape from the Jovian magnetosphere. In addition, these authors suggested that the sense of frequency drift of the modulation lanes — negative in the main and third sources and positive in the early source — was caused by the vectorial combination of a low group velocity in the medium relative to the observer. This latter suggestion stemmed from an interpretation of the modulation lanes in terms of drifts as a modulated structure in the emission itself. Smith (1973) showed that this interpretation could not be correct if the modulation structure in the emission were impressed upon the IFT; instead, a longitudinal modulation is required within the framework of the group-delay hypothesis.

Smith (1973) also found that refractive effects upon upper-hybrid waves were strong, tending to make the waves propagate into the stop zone even when the wave vector is initially perpendicular to the normal to the stop zone surface. In view of Smith's later results concerning the thickness of the stop zones, it is unlikely that group-delay effects play any significant role in either the phenomenology or morphology of DAM.

One caveat to be made about this last statement is that it implicitly assumes geometrical optics to be valid everywhere. All of the instability mechanisms discussed in Sec. II imply either subsequent propagation through a thin evanescent region or, as in the case of Melrose's mechanism, that the radiation is initially beamed toward the planet and subsequently reflected toward the observer. All studies of wave propagation to date have been in the context of geometrical optics, in which the medium is slowly varying and the propagation at any point is governed by the Appleton-Hartree equation with the local plasma parameters. The condition for the validity of geometrical optics, which is essentially a Wentzel-Kramers-Brillouin approach, is

$$\frac{c}{\omega n^2} \, |\nabla n| \ll 1 \, . \tag{32}$$

Near resonance and reflection, inequality (32) breaks down. Nevertheless, application of geometrical optics in the case of reflection, for example, usually yields good results for the ray trajectory even when the solution has been carried through a thin region where WKB analysis is invalid (Ginzburg 1970). Strictly speaking, however, the breakdown of inequality (32) in some region implies that a full-wave solution must be carried out in that region and matched to the WKB solution in regions where geometrical optics is valid. It is possible that such an analysis would explain some of the fine structure observed in decametric spectra.

IV. SUMMARY AND DISCUSSION

For the most part the discussion above has treated coupling, instability, and propagation separately, although occasional references have been made to the relations between them. In this section we attempt to draw the different aspects together in evaluating the present state of the theory. We shall also indicate some of the major unsolved problems, and suggest some possible approaches for future work.

A. Io-Independent DAM

As we have already remarked, there are not one but three distinct decametric emissions: Io-independent L-bursts and Io-modulated L- and S-bursts.

Since the discovery of the Io-modulation only one paper, that of Goldstein and Eviatar (1972), has been addressed specifically to the question of an emission mechanism that cannot be related to a coupling mechanism. This mechanism appears to be viable for the Io-independent DAM, especially in light of the indications that tunneling of the radiation through the evanescent regions does not lead to severe attenuation. The radiation may emanate from particles trapped on L-shells close to but inside the orbit of Io, and so may be unaffected by any magnetospheric perturbations local to Io. The waves are inherently beamed sharply perpendicular to the magnetic field, in accord with the declination effect. The chief shortcoming of the work of Goldstein and Eviatar is that they did not estimate the level at which the instability would saturate due to pitch-angle scattering by the radiation.

Another theory which may account for the Io-independent DAM is the sweeping model of Wu (1973). Although this model was originally proposed to explain Io-modulated emission, there are several reasons why it may be more relevant to the unmodulated radiation. Among these are the sharp beaming pattern perpendicular to the magnetic field, and the fact that, as discussed in Sec. II, the instability is limited to frequencies below 30 MHz within the context of the density inferred by Fjeldbo et al. (1975). In addition, the growth rates evaluated by Wu and by Smith (1973) are rather low, and cannot account for the time scales of Io-modulated L-bursts.

Although sweeping is certainly a coupling mechanism, the real question is whether its effects are localized close to Io. Smith (1973) showed that in addition to the upper-hybrid instability, the drift currents induced by sweeping also drive a long-wavelength instability at the lower-hybrid frequency $\omega_{UH} = \Omega_i [1 + \epsilon(M/m)/(1 + \epsilon)]^{\frac{1}{2}}$. Smith evaluated the spatial diffusion coefficient (Dupree 1967)

$$D_{\perp} = max \left[\frac{\gamma(k)}{k^2} \right] \tag{33}$$

and showed that it depended on both the thermal plasma density and the local gyrofrequency, in addition to its dependence upon the energetic-proton distribution. In general, the diffusion is slower at high altitudes, so that the density gradient is maintained longer at low frequencies. Smith suggested that the Io-independent DAM is in fact due to sweeping, but is poorly correlated to the position of Io because of the slow diffusion at low frequencies, and thus appears statistically independent of Io. Another diffusion mechanism has recently been investigated by Huba and Wu (1975).

Wu (1973) and Smith (1973) neglected the longitudinal drift motion of the protons in the dynamics of the sweeping process. This drift is oppositely directed to Io's orbital motion relative to the co-rotating plasma. As shown by Thomsen and Goertz (1975), the swept cavity drifts in the same way. Thus, the effective sweeping time is reduced from 13 hours to about 4 hours, and it is likely that the cavity persists along the entire Io orbit. Although the diffusion tends to reduce the magnitude of the density gradient during this period, the gradient is never completely eliminated, and so the drift-current-driven instability is always occurring.

A possible impediment to the sweeping model as the cause of unmodulated DAM might exist if in fact Io causes local heating of the ionosphere by MHD waves, as suggested in Sec. I. Such heating would drive thermal plasma up into the IFT, increasing the scale height H of the thermal-plasma density gradient. If $H \gg \epsilon R_J$, then the thickness of the stop zone becomes equal to H, so that x-mode waves could not penetrate the stop zone. We might expect this enhanced density "filament" to cool quickly by thermal conduction to the ionosphere, however, and so be localized close to Io.

B. Io-Modulated L-Bursts

A complete theory for Io-modulated DAM would synthesize all three aspects of the problem. Starting from a particular demonstrated coupling mechanism, the theory would predict the form of the resulting free-energy source, for example the velocity-space distribution of accelerated particles. It would then incorporate this distribution into an instability analysis, demonstrating that the instability could account for the various aspects of *DAM* phenomenology such as time scales, bandwidth, polarization, and spectral power. The inherent beaming pattern of the radiation in the source region would be evaluated, and its subsequent propagation in and emergence from the Jovian plasmasphere would be compared with *DAM* morphology. In addition, the theory should contain some natural explanation for the probabilistic nature of the emission.

Needless to say, such a comprehensive theory has not yet been attained. Nevertheless, some promising work has been done on different aspects of the problem. Smith and Goertz (1975) have given the sheath model of acceleration a firm theoretical basis. The emission mechanism of Melrose is compatible with the sheath coupling model, and seems able to account for most of the necessary features of the phenomenology. It yet remains, however, to synthesize these two models, and to investigate the propagation and reflection of the resulting radiation.

Continued investigations of other mechanisms for both coupling and instability might also yield viable new models. For example, Palmadesso *et al.* (1975) have recently proposed a parametric instability mechanism for the terrestrial kilometric radiation. This theory seems attractive, and might also have application in the parameter regime of *DAM*.

There can be no question that propagation effects contribute significantly to the *DAM* morphology, and probably also to the phenomenology. Any instability mechanism has an inherent beaming pattern in the source region. The structures of the magnetic field and the exospheric density distribution combine to modify this pattern into the directivity pattern that is conventionally inferred from the morphology. A complete theory of *DAM* must include a study of the "antenna pattern" of the Jovian plasmasphere to show that it is compatible with this morphology. Such a study would serve two purposes. First, to the extent that it is dependent upon a model of the exospheric structure, it would help to elucidate that structure. Second, the emerging radiation pattern depends upon the beaming pattern of the instability in the source region. Thus, study of the propagation may place boundary conditions upon the instability mechanism which allow one to discriminate between likely mechanisms. Because the instability mechanism is, in general, also dependent upon the ambient plasma regime, we might expect that the synthetic study of coupling, instability, and wave propagation will lead to a convergent picture.

C. Three Major Problems

None of the models we have discussed appear able to account for the Io-modulated *S*-bursts. Ellis (1965) interpreted them as being due to streams of $30-60$ keV electrons moving outward. It is unclear why such streams should always be accelerated outward, or why they should occur primarily in the early-source region. It is also unclear why they should give rise to a different instability than that responsible for the L-bursts. As noted in Sec. II, the well-separated time scales of the L- and S-bursts, and the time frequency drifts of the S-bursts, indicate that different mechanisms produce the decasecond and millisecond emissions. For the moment the S-bursts remain enigmatic.

A second major question is the explanation of the asymmetry in Io phase for emission viewed from east and west of the Earth-Jupiter line. This may be a local-time propagation effect; if so, its explanation would have important implications about the structure of the Jovian plasmasphere.

Finally, one of the most intriguing problems regarding DAM is the explanation of the modulation lanes observed in both L- and S-bursts. Riihimaa (1971, 1974) has documented an extensive morphology for these features. The modulation lanes probably contain a great deal of information about the mechanism and propagation of DAM.

EPILOGUE

We began this chapter with the observation that a comprehensive understanding of DAM has yet to be attained. Nevertheless, many authors have contributed a great deal of insight to our perspective of the problem. The conceptual framework for the theory is largely in place, and we may reasonably expect much progress in the future.

Acknowledgements. I am grateful to J. K. Alexander, C. K. Goertz, and J. W. Warwick for many stimulating and informative discussions. I have also benefited from conversations with T. D. Carr, M. D. Desch, and A. F. Kuckes. It is a pleasure to thank J. A. Van Allen, S. D. Shawhan, C. K. Goertz, and the staff of the Department of Physics and Astronomy, University of Iowa, for their gracious hospitality and technical assistance during the completion of this work.

APPENDIX: HEATING OF THE JOVIAN IONOSPHERE BY MAGNETOHYDRODYNAMIC WAVES

In Sec. I-A we mentioned the possibility that MHD waves generated by Io might provide local heating of the Jovian ionosphere. This comment was based on the following rough but indicative calculation, due to Wu and Smith (unpublished).

Assume the waves steepen into shocks during their propagation along the IFT. The typical dimension of a collisionless shock is c/ω_i, where $\omega_i = 2\pi f_{pi}$ is the angular ion plasma frequency. Then for a wave amplitude of δB, the current density j is given by

$$j = \frac{c}{4\pi} |\nabla \times \mathbf{B}| \simeq \frac{\omega_i \delta B}{4\pi}. \tag{34}$$

The Joule heating rate of the plasma is then

$$NK_B \frac{\partial T}{\partial t} = \frac{j^2}{\sigma} = \frac{\omega_i^2 \delta B^2}{16\pi^2 \sigma}. \tag{35}$$

The conductivity σ can be either anomalous or the usual Coulomb conductivity

$$\sigma_c \simeq \frac{K_B T^{\frac{3}{2}}}{m_e^{\frac{1}{2}} e^2 \, \ell n \, \wedge} \tag{36}$$

where $\wedge \equiv (K_B T)^{\frac{3}{2}} / (4\pi N)^{\frac{1}{2}} \, e^3$. Use of σ_c will lead to an upper bound for the heating. Substituting (35) for (36), we obtain

$$\frac{\partial}{\partial t} T^{\frac{5}{2}} = \frac{5}{32} \frac{m_e^{\frac{1}{2}} \, e^2 \, \ell n \, \wedge \, \omega_i^2 \, \delta B^2}{\pi^2 \, N \, K_B^{\frac{5}{2}}}. \tag{37}$$

Because the temperature dependence on the right hand side of Eq. (37) is logarithmic, we may obtain an approximate solution by taking $\ell n \wedge$ to be constant; we assume $\ell n \wedge \simeq 20$. Assuming that all the heating occurs in the ionosphere, in a time interval τ, we find

$$T(\tau) \simeq T_0 \left[1 + \frac{1.7 \times 10^{13}}{T_0^{\frac{5}{2}}} \delta B^2 \tau \right]^{\frac{2}{5}}. \tag{38}$$

Taking $T_0 \simeq 750°K$ (Kliore *et al.* 1975) and $\tau \simeq 60$ sec (the time for which the *IFT* threads Io), (38) becomes

$$T(\tau = 60 \text{ sec}) \simeq 10^6 \, \delta B^{\frac{4}{5}}. \tag{39}$$

We may estimate δB in terms of the amplitude δB_0 of the perturbation by Io. For a rough estimate, we consider a generalized Poynting theorem. Let U be the total energy density (electric, magnetic, and kinetic) in the wave, and let $\mathbf{V}_g$ denote the group velocity. The generalized Poynting theorem is:

$$\frac{\partial U}{\partial t} + \mathbf{V}_g \cdot \nabla U + U \nabla \cdot \mathbf{V}_g = 0. \tag{40}$$

We assume that a *WKB* solution is possible. (Strictly speaking, this is not consistent with our estimate of $|\nabla \times \mathbf{B}|$ from the shock condition; we discuss this below.) The Alfvén wave dispersion relation is

$$\omega = \frac{\mathbf{k} \cdot \mathbf{V}_A}{(1 + V_A^2/c^2)^{\frac{1}{2}}} \tag{41}$$

and therefore, $\mathbf{V}_g = \alpha \mathbf{B}$ where $\alpha \equiv [4\pi N m_p (1 + V_A^2/c^2)]^{-\frac{1}{2}}$ and therefore

$$\nabla \cdot \mathbf{V}_g = B(s) \frac{\partial}{\partial s} \alpha(s) \equiv f(s) \tag{42}$$

where $\partial \alpha / \partial s$ is the directional derivative along $\mathbf{B}$. Assuming a solution of the form

$$U = W \exp - \int_0^s dx \, \frac{f(x)}{V_g(x)} \qquad (43)$$

Eq. (40) transforms to

$$\frac{\partial W}{\partial t} + V_g \frac{\partial W}{\partial s} = 0 \qquad (44)$$

the solution of which is simply

$$W(s, t) = W\left[s', t - \int_{s'}^s \frac{dx}{V_g(x)} \right]. \qquad (45)$$

The boundary condition on the solution is

$$W(0, t) = U_{Io}(t). \qquad (46)$$

Furthermore, the integrand in (43) is simply

$$\frac{f(s)}{V_g(s)} = \frac{\partial}{\partial s} \ell n \, \alpha \qquad (47)$$

so that

$$U(s) = U_{Io} \frac{\alpha_{Io}}{\alpha(s)} = U_{Io}\left[\frac{N(s)}{N_{Io}} \frac{1 + V_{AIo}^2/c^2}{1 + V_A^2(s)/c^2} \right]^{\frac{1}{2}}. \qquad (48)$$

In an Alfvén wave, energy is partitioned in the ratios

$$\frac{E^2}{8\pi} : \frac{B^2}{8\pi} : \frac{1}{2} \rho V^2 = \frac{V_A^2}{c^2} : 1 : 1. \qquad (49)$$

Thus, neglecting V_A^2/c^2 throughout, we may combine (39) and (48) to find

$$T(\tau) \simeq 10^6 \left(\frac{N_0}{N_{Io}} \right)^{\frac{2}{5}} (\delta B_{Io})^{\frac{4}{5}} \qquad (50)$$

where N_0 is the ionospheric electron density. Assuming $N_{Io} \simeq 30$ cm^{-3} and $N_0 \simeq 3 \times 10^5$ cm^{-3}, we have $T(\tau) \simeq 4 \times 10^7 (\delta B_{Io})^{\frac{4}{5}}$. The power radiated by Io in Alfvén waves is approximately

$$P \simeq V_{A,\, Io} \, \pi R_{Io}^2 \left(\frac{\delta B_{Io}^2}{8\pi} \right). \qquad (51)$$

Taking $V_{A,\, Io} \simeq 7 \times 10^3$ km sec^{-1}, we require $\delta B_{Io} \simeq 2 \times 10^{-5}$ G if $P = 10^{10}$ W, and 2×10^{-3} G if $P = 10^{14}$ W. (The former figure is Schmahl's estimate of the power, the latter is the upper limit for which the effect on Io's orbital period would not be observable.) The corresponding temperatures according to (50) are $7 \times 10^3 \,^{\circ}$K and $3 \times 10^5 \,^{\circ}$K.

The estimate (50) is quite crude, however, because the arguments leading to it were not entirely consistent. On the one hand, we assumed that the waves steepen into shocks in order to estimate the Joule heating current. On the other hand, we assumed that the wave propagation was compatible with a WKB solution in order to find the steepened amplitude. The effect of the first assumption is to overestimate the numerical coefficient in Eq. (50); the effect of the second is to underestimate δB. The value of the estimate should not be taken too seriously; rather, the calculation should be viewed in the spirit that the large discrepancy between the temperature as given by (50) and the nominal ionospheric temperature of 750°K indicates that transient local heating by Io may be significant.

REFERENCES

Acuña, M., and Ness, N. F. 1975. The complex magnetic field of Jupiter. NASA X-690-75-42. *J. Geophys. Res.* 80. In press.

Alexander, J. K. 1975. Note on the beaming of Jupiter's decameter-wave radiation and its effect on radio rotation period determinations. *Astrophys. J.* 195:227–233.

Aubry, M. P.; Bitoun, J.; and Graff, P. 1970. Propagation and group velocity in a warm magnetoplasma. *Radio Sci.* 5:635–645.

Bigg, E. L. 1964. Influence of the satellite Io on Jupiter's decametric emission. *Nature* 203: 1008–1009.

Block, L. P. 1972. Potential double layers in the ionosphere. *Cosmic Electrodynamics* 3: 349–376.

Bozyan, F. A., and Douglas, J. N. 1976. Directivity and stimulation of Jovian decametric radiation. *Icarus* (special Jupiter issue). In press.

Brice, N. M., and Ioannidis, G. A. 1970. The magnetospheres of Jupiter and earth. *Icarus* 13:173–183.

Brown, L. W. 1974a. Jupiter emission observed near 1 MHz. *Astrophys. J.* 192:547–550.

———. 1974b. Spectral behavior of Jupiter near 1 MHz. *Astrophys. J.* 194:L159–L162.

Buneman, O. 1958. Instability, turbulence, and conductivity in current-carrying plasma. *Phys. Rev. Lett.* 1:8–9.

Carlqvist, P. 1972. On the formation of double layers in plasmas. *Cosmic Electrodynamics* 3:377–388.

Carr, T. D., and Gulkis, S. 1969. The magnetosphere of Jupiter. *Ann. Rev. Astron. Astrophys.* 7:577–618.

Chang, D. B. 1963. Amplified whistlers as the source of Jupiter's sporadic decametric radio emission. *Astrophys. J.* 138:1231–1241.

Conseil, L. 1972. Contribution à l'étude du rayonnement de Jupiter sur ondes décamétriques. Thèse du doctorat 3è cycle, Université de Paris.

Conseil, L.; Leblanc, Y.; Antonini, G.; and Quemada, D. 1971. The effect of the solar wind velocity on the Jovian decametric emission. *Astrophys. Lett.* 8:133–137.

———. 1972. Study of a Jovian plasmasphere and the occurrence of Jupiter radio bursts. *Cosmic plasma physics.* (K. Schindler, ed.) pp. 27–35. New York: Plenum.

Deift, P. A., and Goertz, C. K. 1973. The propagation of Alfvén waves along Io's flux tube. *Planet. Space Sci.* 21:1417–1429.

Desch, M. D., and Carr, T. D. 1974. Decametric and hectometric observations of Jupiter from the RAE-1 satellite. *Astrophys. J.* 194:L57–L59.

Duncan, R. A. 1970. A theory of Jovian decametric emission. *Planet. Space Sci.* 18:217–228.

Dupree, T. H. 1967. Nonlinear theory of drift-wave turbulence and enhanced diffusion. *Phys. Fluids* 10:1049–1055.

Ellis, G. R. A. 1962. Cyclotron radiation from Jupiter. *Aust. J. Phys.* 15:344–353.

———. 1963. The radio emissions from Jupiter and the density of Jovian exosphere. *Aust. J. Phys.* 16:74–81.

———. 1965. The decametric radio emissions of Jupiter. *J. Res. NSB/USNC-URSI Radio Science* 69D:1513–1530.

Ellis, G. R. A., and McCulloch, P. M. 1963. The decametric radio emissions of Jupiter. *Aust. J. Phys.* 16:380–397.

Fillius, R. W., and McIlwain, C. E. 1974. Measurements of the Jovian radiation belts. *J. Geophys. Res.* 79:3589–3599.

Fillius, R. W.; McIlwain, C. E.; and Mogro-Campero, A. 1975. Radiation belts of Jupiter: a second look. *Science* 188:465–467.

Fjeldbo, G.; Kliore, A.; Seidel, D.; Sweetnam, D.; and Cain, D. 1975. The Pioneer 10 radio occultation measurements of the ionosphere of Jupiter. *Astron. Astrophys.* 39:91–96.

Frank, L. A.; Ackerson, K. L.; Wolfe, J. H.; and Mihalov, J. D. 1975. Observations of plasmas in the Jovian magnetosphere. Preprint 75–5. Dept. of Phys. and Astron., University of Iowa.

Fung, P. C. W. 1966a. Excitation of backward Doppler-shifted cyclotron radiation in a magnetoactive plasma by an electron stream. *Planet. Space Sci.* 14:335–346.

———. 1966b. Excitation of cyclotron radiation in the forward subluminous mode and its application to Jupiter's decametric emissions. *Planet. Space Sci.* 14:469–481.

———. 1966c. Excitation of cyclotron electromagnetic waves in a magnetoactive plasma by a stream of charged particles, including temperature effects of the stream. *Aust. J. Phys.* 19:489–499.

Ginzburg, V. L. 1970. *Propagation of electromagnetic waves in plasmas.* Oxford: Pergamon.

Gledhill, J. A. 1967. Magnetosphere of Jupiter. *Nature* 214:155–156.

Goertz, C. K. 1971. Variation of source A position of the Jovian decametric radiation. *Nature* 229:151–152.

———. 1973a. Jupiter's ionosphere and magnetosphere. *Planet. Space Sci.* 21:1389–1398.

———. 1973b. The Io-controlled decametric radiation. *Planet. Space Sci.* 21:1431–1445.

———. 1974. Polarization of Jovian decametric radiation. *Planet. Space Sci.* 72:1491–1500.

———. 1975. Plasma in the Jovian magnetosphere. Preprint 75–16. Dept. of Phys. and Astron., University of Iowa.

Goertz, C. K., and Deift, P. A. 1973. Io's interaction with the magnetosphere. *Planet. Space Sci.* 21:1399–1415.

Goertz, C. K., and Haschick, A. 1972. Variation of highest cutoff frequency of the Io-controlled Jovian decametric radiation. *Nature* 235:91–94.

Goertz, C. K., and Joyce, G. 1975. Numerical simulation of the plasma double layer. *Astrophys. Space Sci.* 32:165–173.

Goldreich, P., and Lynden-Bell, D. 1969. Io, a Jovian unipolar inductor. *Astrophys. J.* 156:59–78.

Goldstein, M. L., and Eviatar, A. 1972. The plasma physics of the Jovian decameter radiation. *Astrophys. J.* 175:275–283.

Goldstein, M. L.; Klimas, A. J.; and Barish, F. D. 1974. On the theory of large amplitude Alfvén waves. *Solar wind three.* (C. Russell, ed.) pp. 385–387. Los Angeles, California: Institute of Geophysics, University of California at Los Angeles.

Goldstein, Jr., S. J. 1975. On the secular change in the period of Io. *Astron. J.* 80:532–539.

Gulkis, S., and Carr, T. D. 1966. Asymmetrical stop zones in Jupiter's exosphere. *Nature* 210:1104–1105.

Gurnett, D. A. 1972. Sheath effects and related charged-particle acceleration by Jupiter's satellite Io. *Astrophys. J.* 175:525–533.

———. 1974. The earth as a radio source: terrestrial kilometric radiation. *J. Geophys. Res.* 79:4227–4238.

Hirshfield, J. L., and Bekefi, G. 1963. Decameter radiation from Jupiter. *Nature* 198:20–22.

Huba, J. D., and Wu, C. S. 1975. A local diffusion process associated with the sweeping of energetic particles by Io. In draft.

Hubbard, R. F.; Shawhan, S. D.; and Joyce, G. 1974. Io as an emitter of 100-keV electrons. *J. Geophys. Res.* 79:920–928.

Kaiser, M. L., and Stone, R. G. 1975. Earth as an intense planetary radio source: similarities to Jupiter and Saturn. *Science* 189:285–287.

Kliore, A.; Cain, D. L.; Fjeldbo, G.; Seidel, B. L.; and Rasool, S. I. 1974. Preliminary results on the atmospheres of Io and Jupiter from the Pioneer 10 S-band occultation experiment. *Science* 183:323–324.

Kliore, A.; Fjeldbo, G.; Seidel, B. L.; Sesplaukis, T. T.; Sweetnam, D. W.; and Woiceshyn, P. M. 1975. Atmosphere of Jupiter from the Pioneer 11 S-band occultation experiment: preliminary results. *Science* 188:474–476.

Knorr, G., and Goertz, C. K. 1974. Existence and stability of strong potential double layers. *Astrophys. Space Sci.* 31:209–223.

Kuckes, A. F., and Sudan, R. N. 1971. Coherent synchrotron deceleration and the emission of type II and type III solar radio bursts. *Solar Phys.* 17:194–211.

Lecacheux, A. 1974. Periodic variations of the position of Jovian decameter sources in longitude (System III) and phase of Io. *Astron. Astrophys.* 37:301–304.

Liemohn, H. D. 1973. Wave propagation in the magnetosphere of Jupiter. *Astrophys. Space Sci.* 20:417–429.

Marshall, L., and Libby, W. F. 1967. Stimulation of Jupiter's radio emission by Io. *Nature* 214:126–128.

McCulloch, P. M. 1971. Theory of Io's effect on Jupiter's decametric radio emissions. *Planet. Space Sci.* 19:1297–1312.

McElroy, M. B., and Yung, Y. L. 1975. The atmosphere and ionosphere of Io. *Astrophys. J.* 196:227–250.

Mead, G. D., and Hess, W. N. 1973. Jupiter's radiation belts and the sweeping effect of its satellites. *J. Geophys. Res.* 78:2793–2811.

Melrose, D. B. 1967. Rotational effects on the distribution of thermal plasma in the magnetosphere of Jupiter. *Planet. Space Sci.* 15:381–393.

———. 1973. Coherent gyromagnetic emission as a radiation mechanism. *Aust. J. Phys.* 26:229–247.

———. 1975. An interpretation of Jupiter's decametric radiation and the terrestrial kilometric radiation as direct amplified gyro-emission. Preprint, University of Colorado.

Oya, H. 1974. Origin of Jovian decameter wave emissions—conversion from the electron cyclotron plasma wave to the ordinary mode electromagnetic wave. *Planet. Space Sci.* 22:687–708.

Palmadesso, P.; Coffey, T. P.; Ossakow, S. L.; and Papadopoulos, K. 1975. A direct mechanism for the generation of terrestrial kilometric radiation. Preprint, Naval Research Laboratory, Washington, D. C.

Piddington, J. H., and Drake, J. F. 1968. Electrodynamic effects of Jupiter's satellite Io. *Nature* 217:935–937.

Riihimaa, J. J. 1971. Radio spectra of Jupiter. Report S-22, Dept. of Electrical Engineering, University of Oulu, Finland.

———. 1974. Modulation lanes in the dynamic spectra of Jupiter's decametric radio emission. *Ann. Acad. Sci. Fenn.* VI. A:1–39.

Schmahl, E. J. 1970. Io, an Alfvén-wave generator. Ph.D. dissertation. Dept. of Astro-Geophysics, University of Colorado, Boulder, Colorado.

Shawhan, S. D. 1976. Io sheath-accelerated electrons and ions. *Icarus* (special Jupiter issue). In press.

Shawhan, S. D.; Goertz, C. K.; Hubbard, R. F.; Gurnett, D. A.; and Joyce, G. 1974. Io-accelerated electrons and ions. *The magnetospheres of earth and Jupiter* (V. Formisano and C. Kennel, eds.) pp. 375–389. Dordrecht, Holland: D. Reidel Publ. Co.

Shawhan, S. D.; Gurnett, D. A.; Hubbard, R. F.; and Joyce, G. 1973. Io-accelerated electrons: predictions for Pioneer 10 and Pioneer 11. *Science* 182:1348–1350.

Smith, E. J.; Davis, Jr., L.; Jones, D. E.; Coleman, Jr., P. J.; Colburn, D. S.; Dyal, P.; and Sonett, C. P. 1975. Jupiter's magnetic field, magnetosphere, and interaction with the solar wind: Pioneer 11. *Science* 188:451–455.

Smith, R. A. 1973. On the Io-modulated Jovian decametric radiation. Ph.D. dissertation, Department of Phys. and Astron., University of Maryland, College Park, Maryland.

———. 1975. Stop zones in the Jovian plasmasphere. In draft.

Smith, R. A., and Goertz, C. K. 1975. On the modulation of the Jovian decametric radiation by Io. I. Acceleration of charged particles. Submitted to *Astrophys. J.*

Smith, R. A., and Wu, C. S. 1974. Implications of the Pioneer 10 measurements of the Jovian magnetic field for theories of Io-modulated decametric radiation. *Astrophys. J.* 190:L91–L95. (Addendum: *ibid.* 193:L101–L102.)

Smith, R. A.; Wu, C. S.; and Zmuidzinas, J. S. 1972. The geometry and dynamic spectra of Io-modulated Jovian decametric radio emissions. *Astrophys. J.* 177:L131–L136.

Thieman, J. R.; Smith, A. G.; and May, J. 1975. Motion of Jupiter's decametric sources in Io phase. *Astrophys. Lett.* 16:83–86.

Thomsen, M. F., and Goertz, C. K. 1975. Satellite sweep-up effects at Jupiter (abstract). *EOS Trans. Am. Geophys. Union* 56:428.

Warwick, J. W. 1961. Theory of Jupiter's decametric radio emission. *Ann. N. Y. Acad. Sci.* 95:39–60.

———. 1963. Dynamic spectra of Jupiter's decametric emission, 1961. *Astrophys. J.* 137:41–60.

———. 1964. Radio emission from Jupiter. *Ann. Rev. Astron. Astrophys.* 2:1–22.

———. 1967. Radiophysics of Jupiter. *Space Sci. Rev.* 6:841–891.

———. 1970. Particles and fields near Jupiter. NASA CR-1685. Washington, D. C.: U. S. Government Printing Office.

Warwick, J. W., and Dulk, G. A. 1964. Faraday rotation on decametric radio emissions from Jupiter. *Science* 145:380–383.

Whitten, R. C.; Reynolds, R. T.; and Michelson, P. F. 1975. The ionosphere and atmosphere of Io. *Geophys. Res. Lett.* 2:49–51.

Wu, C. S. 1973. Modulation of the Jovian decametric radio emissions by Io. *Astrophys. J.* 186:313–326.

Wu, C. S.; Smith, R. A.; and Zmuidzinas, J. S. 1973. Theory of decametric radio emissions from Jupiter. *Icarus* 18:192–205.

Zheleznyakov, V. V. 1966. The origin of Jovian radio emission. *Sov. Astron. AJ* 9:617–625.

ABSORPTION OF RADIATION BELT PARTICLES BY THE INNER SATELLITES OF JUPITER

A. MOGRO-CAMPERO
University of California, San Diego

The study of trapped particle absorption by the inner Jovian satellites is reviewed from the viewpoint of radiation belt physics. Both pre- and post-Pioneer work is discussed but the emphasis is on methods used to deduce radial diffusion coefficients of particle transport from particle data. The phenomenon of particle absorption as observed by experiments on Pioneers 10 and 11 is considered; absorption effects are found to depend on the satellite, and on particle energy and species. Approximate diffusion coefficients derived from the data are found to follow a steeper spatial dependence than previously expected. The assumptions and limitations of absorption analysis and diffusion coefficient estimation are pointed out.

In this paper we discuss the absorption of radiation belt particles by the satellites from the standpoint of radiation-belt physics. We are concerned with the passive role of satellites as absorbers of trapped radiation and with the implications relevant to trapped particle transport throughout the inner magnetosphere of Jupiter.

Since Pioneer 10 was the first spacecraft which traversed the Jovian magnetosphere, it is convenient to divide the development of the problem into pre- and post-Pioneer 10 phases, which we label Phase 1 and Phase 2. This division is obviously chronological, and in the following sense conceptual. During Phase 1 the central question can be summarized as: given a dynamic situation in which trapped particles undergo radial diffusion, what is the effect of satellite absorption on the particle intensity profiles for values of the diffusion coefficient assumed or derived independently? In Phase 2 the question becomes: given that trapped particle intensity profiles have been observed which indicate absorption by the satellites, what values can we deduce for the diffusion coefficient of radial transport?

We review the work done in Phases 1 and 2, and summarize the methods and results obtained in Phase 2. Values are then presented for the radial diffusion coefficient as a function of particle energy, drift frequency and L (throughout this chapter L is the dimensionless McIlwain shell parameter). These values are derived by using one of the methods presented here and all of the Pioneer 10 and 11 data.

The reader who wishes background information on radiation-belt physics may consult books by Hess (1968), Roederer (1970), or more specifically on trapped particle diffusion, by Schulz and Lanzerotti (1974).

I. PHASE 1

Before spacecraft measurements in the close environment of Venus and Mars were performed, it was thought possible that these planets might have significant radiation belts. In this context, the absorption effect of a planetary satellite was considered for the first time. Phobos, which orbits Mars at ~ 3 Martian radii from the center of the planet, was a natural candidate for study. Thus, Singer (1962) treated the sweeping effect of Phobos on the Martian radiation belt. He obtained expressions for the particle absorption lifetime in the case of a magnetic dipole field displaced by an arbitrary amount from the center of the planet in the equatorial plane, and he concluded that trapped particle intensity measurements in the region of Phobos could be used to deduce the particle injection rate and ultimately the source of the trapped radiation.

About a decade later, in July 1971, a group of scientists and engineers met for a Workshop at the Jet Propulsion Laboratory, Pasadena, California, in order to review the current state of Jupiter radiation belt knowledge and to recommend a best set of models for the determination of spacecraft design requirements. Two papers at that meeting (one by Hess and one by Mead) dealt with the effects of Jupiter's satellites on the trapped radiation; one of these (Mead 1972) can be found in the Workshop Proceedings. In Mead's paper, characteristic spatial and temporal parameters of trapped protons and of the inner satellites were evaluated and discussed with respect to particle absorption. The radiation belt models resulting from the Workshop (Divine 1972) did not include the effect of the satellites, since it was assumed that the diffusion of protons past the satellites occurs without interference. It was realized, however, that alternative models with a strong absorption effect by the Jovian satellites were also plausible (Davis 1972). A few months later, in an extension of the discussions of the Workshop, Jacques and Davis (1972) solved the trapped particle equations of transport for diffusion of particles by violation of the third adiabatic invariant, including losses due to satellite absorption and synchrotron radiation. They estimated values for diffusion coefficients due to (a) deformation of the magnetic field by the solar wind ($\propto L^{10}$); (b) randomly fluctuating electric fields ($\propto L^6$), and (c) interchange of flux tubes by ionospheric winds $[\propto L^2 (L-1)]$. The first two are processes thought to be important at Earth, and the third is based on a mechanism suggested at the Workshop by Brice (1972). Jacques and Davis concluded that only the third mechanism could explain the electrons required to produce the observed synchrotron radiation, and that in this case the effect of Io would be neither overwhelming nor negligible.

All studies of particle absorption by the satellites have assumed that trapped particles co-rotate with the planet. Mead and Hess (1973) made a thorough study of the absorption process for low-energy particles, i.e., those for which the drift period is close to the planetary rotation period. They pointed out that because of the tilt of the magnetic dipole field with respect to the planetary rotation axis, the absorption process would be pitch-angle dependent. Assuming violation of the third adiabatic invariant by deformation of the magnetic field due to the solar wind (resulting in a diffusion coefficient $D \propto L^{10}$), they concluded that the inner satellites would act as a barrier to the inward diffusion of particles, in agreement with the conclusion of Jacques and Davis (1972) for this type of diffusion coefficient.

Brice and McDonough (1973) estimated the strength of the diffusion coefficient due to electric field fluctuations produced by neutral winds in the ionosphere, following an earlier suggestion by Brice (1972). For slowly drifting particles, i.e., those whose drift period is close to the planetary rotation period, they found that the diffusion coefficient $D \propto L^3$. For fast drift particles, $D \propto L^5$ and $D \propto L^{3.5}$ for nonrelativistic and relativistic particles, respectively. Their values for D led them to conclude that no serious losses would occur due to particle absorption by the Jovian satellites, in agreement with the conclusion of Jacques and Davis (1972) for a similar type of diffusion coefficient.

Coroniti (1974) also considered radial diffusion driven by fluctuating electric fields originating from atmospheric neutral wind turbulence, and derived a diffusion coefficient $D \propto L^3$. He investigated plasma instabilities which could limit the intensity of stably trapped electron fluxes and absorption of particles by the satellites. His estimate for the value of the diffusion coefficient led him to conclude that the satellites would not act as major barriers to the passage of inward diffusing particles.

Hess *et al.* (1973, 1974) estimated sizable reductions in the trapped particle intensities (up to one order of magnitude, depending on equatorial pitch angle) near the orbits of Ganymede, Europa and Io due to particle absorption by these satellites. They used a diffusion coefficient $D = kL^a$, with the values of k and a (1.7×10^{-9} sec^{-1} and 2, respectively) deduced by Birmingham *et al.* (1974) by fitting the observed radial distribution of Jupiter's decimeter radio emission to a model of trapped electrons. Birmingham *et al.* noted that their diffusion coefficient had roughly the same radial dependence, but was considerably smaller in magnitude than the upper limit diffusion coefficients for field-line exchange driven by ionospheric winds suggested by Brice and McDonough (1973), and by Jacques and Davis (1972).

In summary, prior to the first flyby of Jupiter by Pioneer 10, many authors had considered the possible effect of absorption of trapped particles by the

Jovian satellites. It was generally felt that some of the inner Jovian satellites would produce observable effects on the trapped particle intensities, and that the diffusion coefficient of radial transport was probably of the type $D \propto L^a$, with $a \simeq 2$ to 3 for slow-drift particles.

II. PHASE 2

All charged particle experimenters on Pioneer 10 reported features in the trapped particle intensity profiles which could clearly be attributed to satellite absorption (Fillius and McIlwain 1974a,b; Simpson et al. 1974a,b; Trainor et al. 1974a,b; Van Allen et al. 1974a,b). Similar results have been reported recently for Pioneer 11 (Fillius et al. 1975a;[1] Simpson et al. 1975;[2] Trainor et al. 1975;[3] Van Allen et al. 1975[4]). Electron absorption is also evident when one considers the particle phase-space densities computed from the data (McIlwain and Fillius 1975; Baker and Van Allen 1975). Vesecky (1975) has collected the Pioneer 10 observations relevant to satellite absorption and compared them with theoretical expectations developed in Phase 1.

Since data are available for both inbound and outbound passes of Pioneers 10 and 11, it is possible to investigate whether absorption is a strong function of longitude. In spite of the wide range of longitudes involved (Table I), the large-scale features in the particle intensity profiles of a given detector which are attributable to satellite absorption are similar in the four Pioneer passes with the exception of the MeV protons at Europa as reported by Trainor et al. (1975). A possible local rapid diffusion process may be partly responsible for the longitudinal uniformity (Huba and Wu 1975).

A discernible energy and species dependence of particle absorption is clearly identifiable in the Pioneer data. For example, in Fig. 1 are shown the counting rate profiles of three electron detectors for the inbound and outbound passes of Pioneer 10. The average energies to which these detectors are sensitive are ~ 1, 10, and 25 MeV, so that the energy dependence of absorption at Io and Europa is evident. A dependence of particle absorption on spacecraft magnetic latitude should be observable due to the expected preferential absorption of small equatorial pitch-angle particles (Mead and Hess 1973). The Pioneer 10 trajectory is suitable for the observation of this effect at Io (from Table I we see that the magnetic latitude of Pioneer 10 was $-6°$ and $14°$ for the inbound and outbound passes, respectively). Figure 1 is especially tuned to observe this effect, since the particle counting rates shown are those measured perpendicular to the local magnetic field vector. The pitch-angle dependence of absorption is qualitatively confirmed by the middle trace in Fig. 1, where a larger absorption effect is evident in the case of the higher-latitude outbound pass. An analysis of particle anisotropies in

---

[1]See also p. 913.   [3]See also p. 964.
[2]See also p. 743.   [4]See also p. 929.

TABLE I

Pioneer-Satellite Longitude Intervals and
Pioneer Magnetic Latitude at Times When the
Spacecraft Traversed the L Shells of a Satellite

Pioneer 10 or 11 and Satellite	Inbound or Outbound	Longitude Interval[a] (deg)	Spacecraft Magnetic Latitude[b] (deg)
P 10 – Europa	I	334	− 19
	O	126	+ 4
P 11 – Europa	I	76	− 43
	O	204	+ 53
P 10 – Io	I	51	− 6
	O	157	+ 14
P 11 – Io	I	(a few)	− 41
	O	163	+ 45
P 11 – Amalthea	I	129	− 34
	O	30	+ 23

[a]This longitude interval is relevant for particle absorption, and is defined as the longitude angle (0° to 360°) from the spacecraft to the satellite in the direction of motion of the satellite in the frame co-rotating with Jupiter. The accuracy of these values is a few degrees. This has been emphasized in the table by the entry for P 11 – Io, I. A study of the Pioneer 11 proximity to the Io flux tube requires a more accurate description.
[b]Magnetic latitudes are those given by the D2 magnetic field model of Smith *et al.* (1974). The magnetic latitude of the satellites is bounded by ± 11°.

the vicinity of Io's orbit (e.g., see Fig. 2) was also found to be consistent with the concept that particles with small equatorial pitch angles are absorbed preferentially (Simpson *et al.* 1974*b*; Trainor *et al.* 1974*a*; Fillius *et al.* 1975*b*). In the case of electrons > 0.06 MeV, however, Sentman and Van Allen (1975) do not observe clearly identifiable effects on the angular distributions associated with passage through the magnetic shells of Ganymede, Europa, and Io. Angular effects for electrons may be more difficult to observe because of pitch-angle scattering (Fillius *et al.* 1975*b*).

We have seen that the absorption effect is observed to depend on particle species and energy. This fact seemed at first puzzling from the viewpoint of the theoretical framework developed in Phase 1 (Hess *et al.* 1974). However, the slow drift approximation which is prevalent in these models is not valid for many of the particle energies. Simpson *et al.* (1974*b*) have studied particle absorption by Io and showed that the impact probability is depen-

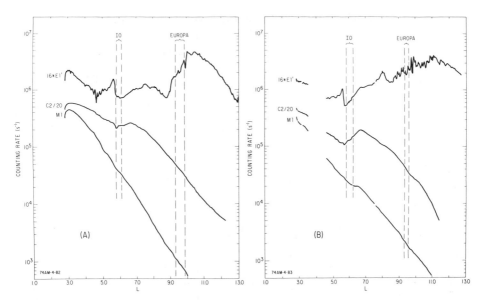

Fig. 1.   The energy dependence of particle absorption at Io and Europa. Counting rates of three electron detectors of the University of California, San Diego experiment for (A) the inbound trajectory, and (B) the outbound trajectory of Pioneer 10. These are the counting rates perpendicular to the local magnetic field vector. The average energy of response varies with location in the Jovian magnetosphere. Approximate values are 1, 10, and 25 MeV from top to bottom (from Mogro-Campero and Fillius 1976).

dent on energy and species. Furthermore, they have obtained a probable value for the radial diffusion coefficient of $\sim$ 1 MeV protons at Io. Mogro-Campero *et al.* (1975) have discussed particle absorption by the inner satellites and they showed that the particle-satellite collision time, the radial diffusion coefficient, and the resulting absorption probability are expected to be dependent on species and energy. They also showed that the energy dependence of absorption exhibited by their electron detectors was in qualitative agreement with the expected energy dependence. Mogro-Campero and Fillius (1976) discussed several approaches to obtain estimates of the radial diffusion coefficient. They have derived an expression for the sweeping time at a given satellite, and by using Pioneer 10 electron data they obtained values for the diffusion coefficient at Europa and Io in the energy range of $\sim$ 0.7 to 14 MeV. They concluded that the diffusion coefficient is a function of energy and L. Based on this formulation and on Pioneer 11 data, Fillius *et al.* (1975a) have recently reported values for the diffusion coefficient of $\sim$ 100 MeV protons and $\sim$ 90 MeV electrons at Amalthea, providing a stronger basis for contentions of energy and spatial dependences of the diffusion coefficient.

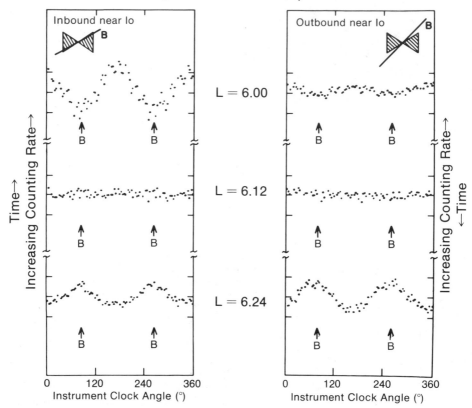

Fig. 2.   The change in the low-energy (0.5 – 1.8 MeV) proton anisotropy as the orbit of Io is crossed, both inbound and outbound, as observed with the University of Chicago experiment on Pioneer 10. The diagrams near the top of the figure show the relationship between the magnetic field direction and the detector acceptance band (from Simpson *et al.* 1974*b*).

III.   METHODS USED IN PHASE 2

The Sweeping Time

The sweeping time is defined as the time required for the satellite to absorb a given fraction of the trapped particles within its sweeping region. The following considerations apply:

1.   It is always assumed that the trapped radiation co-rotates with the planet.
2.   Because of our ignorance of the electric and magnetic field configuration in the immediate vicinity of the satellites, it is assumed that trapped-particle motion is determined exclusively by the undistorted planetary

magnetic field (the D2 dipole model of Smith *et al.* 1974 is used throughout this paper). Thus, if a trapped particle trajectory intersects a satellite, the particle is removed from the trapped particle population. The electron/proton energy in MeV at which the gyroradius equals the satellite radius is 680/220 at Amalthea, 970/410 at Io, and 220/25 at Europa. These energies are all above those measured on the Pioneer missions where absorption effects have been noticed. It is therefore not possible for these particles to avoid impact with the satellite when their center of gyration lies within the satellite.

3. Particles may escape absorption by leapfrogging the satellite in their longitudinal drift motion during half a bounce period. This is not possible for electrons of the energies in consideration ($0.1 \lesssim E \lesssim 100$ MeV) for the four innermost Jovian satellites (Mogro-Campero and Fillius 1976), but it is an important point to consider e.g., in the case of protons with $\lesssim 7$ MeV at Europa. For convenience at this time, we restrict our attention to cases where this longitude skipping mechanism is unimportant.

4. The extent of the sweeping region ΔL, is a function of equatorial pitch angle. This is equivalent to the preferential absorption of small pitch-angle particles mentioned previously. This concept, as well as the geometry of particle absorption, can be illustrated by considering satellite trajectories in magnetic field coordinates. Such a trajectory for Io is shown in Fig. 3. It is clear from this figure that particle absorption is more likely for small equatorial pitch-angle particles, since their magnetic latitude coverage is wider than that of the satellite.

The time required for removal by Io of 90% of the electrons and protons at $L = 6.19$ has been evaluated by computer simulation as a function of energy by Simpson *et al.* (1974*b*), and is shown in Fig. 4. At low energies the smooth behavior corresponds to complete absorption within one drift period. At higher energies the fine structure is produced by particles whose drift velocity is in some simple ratio to the velocity of Io in the co-rotating frame. The effects of radial diffusion were not considered.

When reviewing the Pioneer data we have concluded that longitudinal uniformity of the particle intensity profiles in the vicinity of the satellites is probably a good approximation. Processes contributing to this uniformity are the combined effects of the particle's drift motion, random radial displacements characteristic of their radial diffusive motion, and possible localized turbulent diffusion (Huba and Wu 1975).

By postulating that longitudinal uniformity would be achieved in time scales comparable or shorter than the satellite orbital period in the co-rotating frame, Mogro-Campero and Fillius (1976) derived the following expression for the sweeping time at a satellite:

$$T_s = - g \, t_c \, \ell n \, y \tag{1}$$

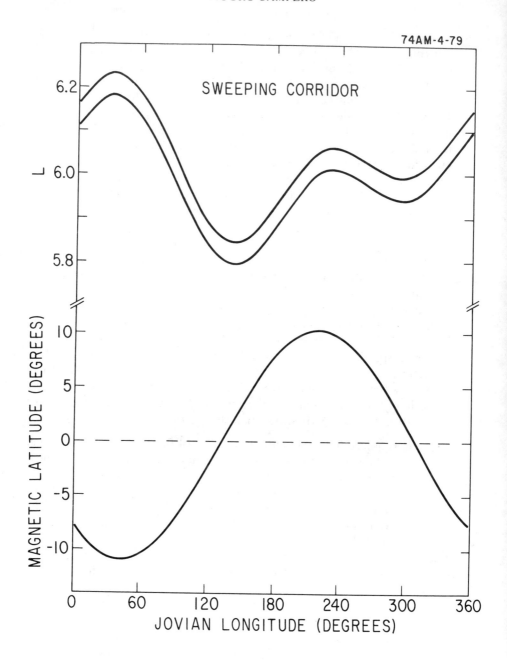

Fig. 3.   Io's trajectory in the magnetic field coordinates L and magnetic latitude versus Jovian System III longitude (Mead 1974). The width of the band in L corresponds to Io's diameter and determines the sweeping corridor for particles with gyroradii $\ll$ Io's radius. Io passes through a given longitude in the sweeping corridor once every orbital period in the frame of reference co-rotating with Jupiter (from Mogro-Campero and Fillius 1976).

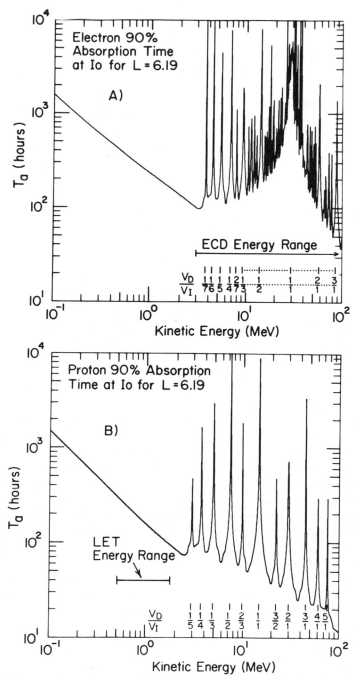

Fig. 4. The time required for removal by Io of 90% of the (A) electrons and (B) protons initially present in a uniformly populated drift shell at $L = 6.19$ in the absence of radial diffusion, computed as a function of energy. The ratio of particle drift velocity to Io's orbital velocity in the co-rotating frame is labeled V_D/V_I (from Simpson et al. 1974b).

where g is a geometrical factor which depends on the satellite and the particle equatorial pitch angle, y is the fraction of particles surviving absorption (to be determined from the data), and $t_c = |f_s \mp f_p|^{-1}$ with f_s = satellite orbital frequency and f_p = particle drift frequency ($-/+$ refers to electrons/protons). $T_s \propto t_c$ had been suggested earlier by Mogro-Campero *et al.* (1975). Although it has always been assumed that the relevant satellite diameter for particle absorption is the physical diameter d of the satellite, it is illustrative to discuss the dependence of the geometrical factor g on d. This is useful in considering for example the result of changing the effective satellite diameter by internally or externally generated magnetic fields. In the simplest case of trapped particles with equatorial pitch angles $< 67°$, $g \propto d^{-1}$ (Mogro-Campero and Fillius 1976).

The sweeping time given by Eq. (1) is shown in Fig. 5 as a function of particle energy at Io. In comparing the energy dependence of the solutions in Figs. 4 and 5, we notice that at the higher energies ($E \gtrsim 3$ MeV), the smooth energy dependence in Fig. 5 approximately follows the envelope of minima in Fig. 4. In order to compare absolute values we have evaluated Eq. (1) at $L = 6.19$ and for $y = 0.1$, and we find that in this high-energy region Eq. (1) gives values of a factor of ~ 3 higher than the envelope of minima in Fig. 4. At the lower energies Fig. 4 is relatively energy independent and a comparison of absolute values results in a maximum discrepancy at 0.1 MeV, where Eq. (1) lies a factor of ~ 6 lower than the value in Fig. 4.

In Fig. 5, the strong energy dependence of the electron sweeping time at ~ 30 MeV is a consequence of the inefficient particle absorption which occurs with zero relative velocity between the drifting particles and the satellite. The other "resonant" energies for electrons are 14.5 MeV at Amalthea, 21 MeV at Europa, and 14 MeV at Ganymede. The strong energy dependence makes it difficult to estimate sweeping times for particles of energy near the resonant energies. The fact that the observations are consistent with inefficient absorption at these energies (e.g., see the higher-energy electrons in Fig. 1) is an indirect confirmation of the assumption of trapped particle co-rotation. It is interesting to note that if no co-rotation is assumed, the situation is reversed in the sense that protons and not electrons will exhibit the resonance phenomenon. For example, ~ 5 MeV protons at Io, and ~ 40 MeV protons at Amalthea would diffuse past these satellites with ease. Protons of ~ 1 to 2 MeV and those of ~ 18 MeV are observed to be significantly absorbed at Io, but no measurements exist at ~ 5 MeV so that this possibility cannot be excluded. On the other hand, if particles do not co-rotate, the observed inefficient absorption of high-energy electrons discussed above would remain unexplained.

The Radial Diffusion Coefficient

The equation describing the radial diffusion of particles by violation of the third adiabatic invariant is (Schulz and Lanzerotti 1974):

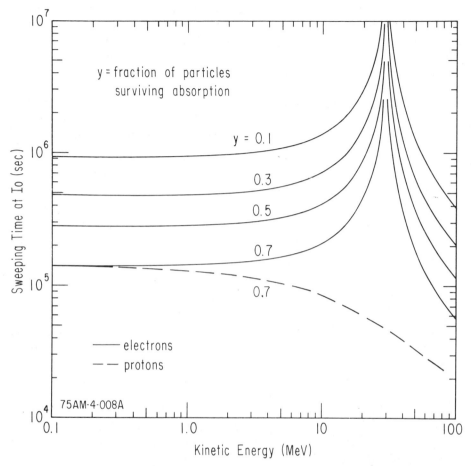

Fig. 5. The sweeping time at Io as given by Eq. (1) in the text, as a function of particle energy. Selected values of the fraction of particles surviving absorption are indicated. The value used for the geometrical factor g in Eq. (1) corresponds to particles which mirror at magnetic latitudes $\gtrsim 11°$.

$$\frac{\partial f}{\partial t} = L^2 \frac{\partial}{\partial L}\left(\frac{D}{L^2}\frac{\partial f}{\partial L}\right) \tag{2}$$

where f is the phase space density of particles such that $f\, d^3x\, d^3p$ is the number of particles in the spatial volume element d^3x and in the momentum volume element d^3p; it must be evaluated at constant first and second invariants. The relationship between the differential particle flux j (cm^{-2} sec^{-1} ster^{-1} MeV^{-1}) measured perpendicular to the magnetic field vector and $f = j/p^2$. Since L is the dimensionless shell parameter, the dimension of the diffusion coefficient D is inverse time.

Particle absorption in the satellite sweeping region ΔL can be treated by adding a loss or sink term $S = - f / T_\ell$ to the right-hand side of Eq. (2). The time T_ℓ is the exponential decay time of particles in the absence of diffusive flow as can be readily seen from Eq. (2) with sink term but no diffusion term. From Eq. (1) we see that the particle population may in fact be considered to decay exponentially with time, so that $T_\ell = g \, t_c$.

Since the objective is to obtain values for the diffusion coefficient, Eq. (2) with sink term must be solved for D. For $D \propto L^a$ and steady state conditions ($\partial f / \partial t = 0$), Mogro-Campero and Fillius (1976) obtain

$$D = \frac{f / T_\ell}{\left[\dfrac{(a-2)}{L} \dfrac{\partial f}{\partial L} + \dfrac{\partial^2 f}{\partial L^2} \right]} \tag{3}$$

valid in the sweeping region ΔL. The parameter a can be obtained by solving Eq. (2) in the steady state in an L-region with no satellite absorption, resulting in

$$a = 2 - L \frac{\partial^2 f / \partial L^2}{\partial f / \partial L} . \tag{4}$$

In order to evaluate these expressions we require a substantial set of differential energy measurements covering a wide range of spatial locations and energies. Similar approaches have been profitable in the case of Earth's radiation belt (Schulz and Lanzerotti 1974), but other methods may be more appropriate for the analysis of the first stage of Jovian radiation belt exploration. An estimate of the diffusion coefficient can be obtained by solving the diffusion Eq. (2) with sink term for the phase space density f, and comparing this with f determined by the observations. The value of D for the best fit can then be chosen. Mogro-Campero and Fillius (1976) have used this method for a particular case where they could determine f. The following expressions derived from the diffusion equation by different approximations have been used in considering Jovian radiation belt data [Mogro-Campero and Fillius (1976); and similar expressions in a paper by Simpson *et al.* (1974*b*)]:

$$D \simeq \frac{(\Delta L)^2}{4 T_s} \tag{5}$$

$$D \simeq - \frac{\Delta L}{T_s \dfrac{\partial}{\partial L} (\ln f)} \tag{6}$$

where T_s is the sweeping time and ΔL the sweeping region, both defined above. Another approximate expression can be obtained by solving Eq. (27) of Mead and Hess (1973) for D. In the notation of this paper one obtains

$$D \simeq \frac{(\Delta L)^2}{T_\ell \left(\cosh^{-1} \frac{1}{y} \right)^2}. \tag{7}$$

If T_s from Eq. (1) is used in (5), Eqs. (5) and (7) give results which differ at most by a factor of 2 in the range $0.001 < y < 0.9$ (by definition y must be in the range $0 \leqslant y \leqslant 1$).

IV. VALUES OF THE RADIAL DIFFUSION COEFFICIENT

Values of the diffusion coefficient based on the approximate formulas (5) and (6) have been evaluated at the orbits of satellites: for ~ 1 MeV protons at Io (Simpson et al. 1974b), for ~ 0.8 MeV and ~ 12 MeV electrons at Io and Europa (Mogro-Campero and Fillius 1976), and for ~ 100 MeV protons and ~ 90 MeV electrons at Amalthea (Fillius et al. 1975a). Mogro-Campero and Fillius (1976) find evidence for an energy and spatial dependence of the diffusion coefficient.

Any given experiment is limited in its coverage of energy and species. It is also difficult to compare diffusion coefficients deduced by different experimenters since their methods are not identical. Furthermore, estimates of diffusion coefficients have not yet been made by all Pioneer experimenter teams. It therefore seemed of interest to compute diffusion coefficients using all of the Pioneer 10 and 11 data reported so far, by one method. This has been done by using Eq. (5), with T_s given by Eq. (1). It must be cautioned that Eq. (5) is a crude approximation, but since most of the Pioneer data has not been reduced to produce particle phase space densities, better approximations such as that of Eq. (6), and especially of Eqs. (3) and (4) are not yet possible for the whole set of data.

The diffusion coefficients as a function of particle kinetic energy are shown in Figs. 6 and 7. An overall positive slope is apparent. A more reasonable parameter is probably the particle-drift frequency, since radial diffusion is caused by violation of the third adiabatic invariant in a time comparable to the particle-drift period in an inertial frame. Therefore, particles with the same drift frequency at a given L should have the same diffusion coefficient. The relationship between particle species, energy, L value, and drift frequency is illustrated in Fig. 8. The diffusion coefficients as a function of particle-drift frequency are shown in Figs. 9 and 10. The frequency dependence shows a minimum near the co-rotation frequency and is an indirect measure of the power spectrum of the electric and magnetic field irregularities which are responsible for violating the particles' third adiabatic invariant (e.g., see Schulz and Lanzerotti 1974). Since the diffusion coefficient depends on drift frequency, the L dependence is displayed in three frequency regimes in Fig. 11.

From the results in Fig. 11 we see that it is not possible to identify a simple spatial dependence for the diffusion coefficient. There is uncertainty

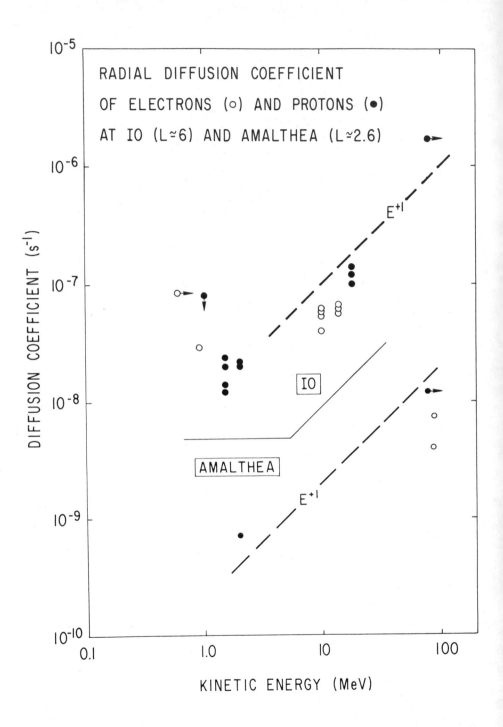

Fig. 6.   Diffusion coefficients as a function of particle kinetic energy for protons and electrons at Io and Amalthea. The dashed line is meant to serve as a reference slope.

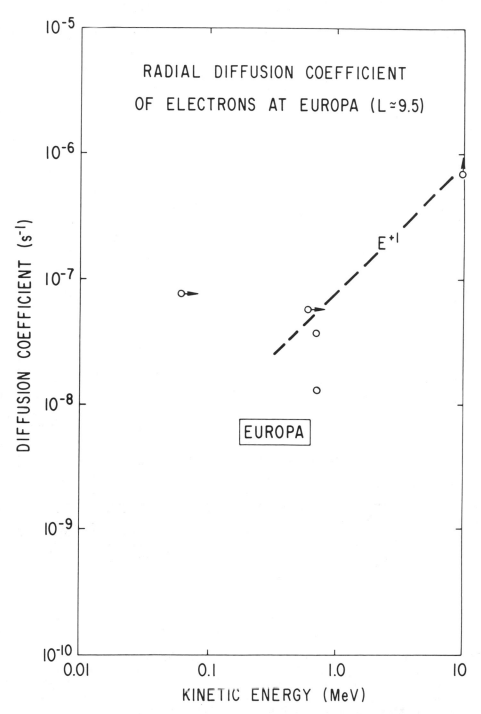

Fig. 7. Same as Fig. 6 but for Europa.

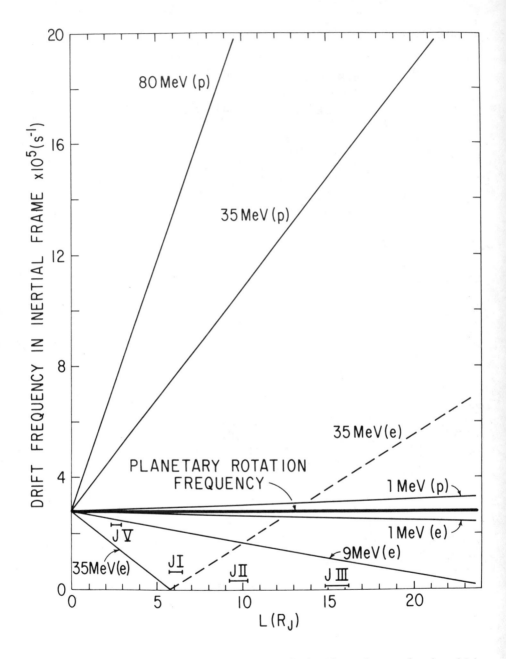

Fig. 8. Drift frequencies of protons and electrons of selected energies as a function of L in an inertial frame. The ranges of L values traversed by the satellites are indicated at their appropriate orbital frequencies (J V is Amalthea, J I is Io, J II is Europa, and J III is Ganymede). The dashed line corresponds to negative frequencies i.e., opposite to the direction of planetary rotation (from Mogro-Campero *et al.* 1975).

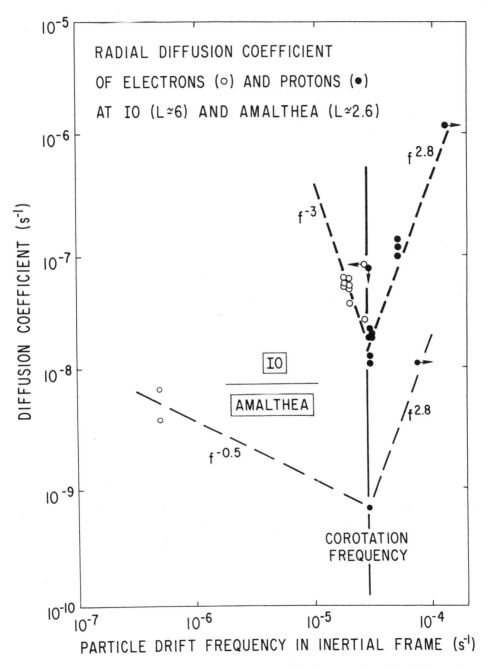

Fig. 9. Diffusion coefficients as a function of particle drift frequency in an inertial frame for protons and electrons at Io and Amalthea. A minimum seems to occur at the planetary rotation frequency (co-rotation frequency).

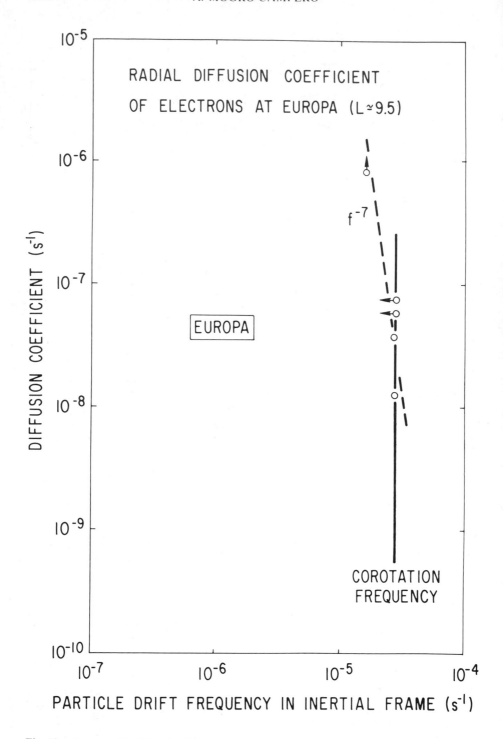

Fig. 10.   Same as Fig. 9 but for Europa.

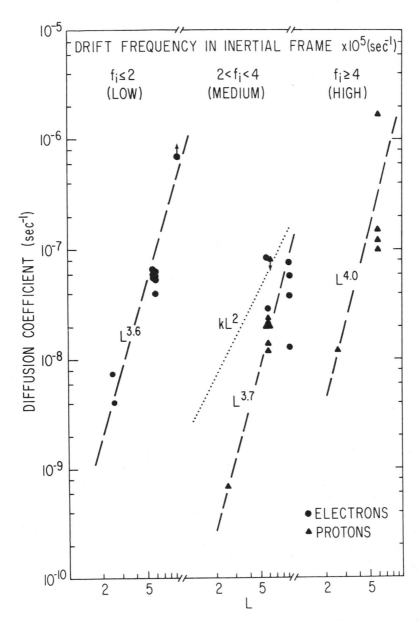

Fig. 11. The diffusion coefficient as a function of L based on Pioneer 10 and 11 trapped particle data in the vicinity of the orbits of Amalthea ($L \simeq 2.6$), Io ($L \simeq 6$) and Europa ($L \simeq$
9.5). The approximations used are described in the text. The diffusion coefficients have been
classified into low-, medium-, and high-frequency regimes, with arbitrary boundaries. The
planetary rotation frequency (the low-energy limit of the particle-drift frequency in the inertial frame) lies approximately in the middle of the medium-frequency regime. The dashed
lines are rough visual fits drawn to illustrate the steepness of spatial dependences. The dotted
line with $k \simeq 2 \times 10^{-9}$ sec^{-1} was deduced from radio observations by Birmingham $et\ al.$
(1974) before the Pioneer particle data became available.

particularly at medium frequencies in going from Io to Europa. An overall simplifying statement would be to deduce from these results that if a power law must be chosen, the exponent $a \simeq 4$ is a reasonable value. The urge to fit the spatial dependence into a power law comes from our experience at Earth and from theoretical considerations (see Sec. I), where $D \propto L^a$ is a reasonable approximation with $a = 10$, 6 or ~ 3 depending on whether the mechanism responsible for the violation of the third adiabatic invariant is due to, respectively,

(a) deformation of the magnetic field by the solar wind,
(b) randomly fluctuating electric fields, or
(c) field-line exchange driven by ionospheric winds.

We also have $D = kL^2$, with $k = 2 \times 10^{-9}$ sec^{-1} given by Birmingham et al. (1974) from their fits to radio observations at $L \lesssim 4$ (this result has been included in Fig. 6 as a dotted line). The value $a \simeq 4$ is higher than the favorite pre-Pioneer values of $a \simeq 2$ or 3.

V. SUMMARY AND COMMENTS

In Sec. I we have reviewed the work done on the absorption of trapped particles by the Jovian satellites in the pre-Pioneer 10 era. The emphasis at that stage was on determining the relative importance of such absorption, and it was generally felt to be a relevant process in the inner magnetosphere. Much of the conceptual framework needed to study the absorption process and its consequences for the trapped particle distributions was developed at this time. The power law exponent $a \simeq 2$ to 3 in $D \propto L^a$ was favored.

In Sec. II we have reviewed the work performed after Jupiter encounter data became available from Pioneer 10. Absorption effects at the inner satellites were observed by charged particle detectors on Pioneers 10 and 11. Particle angular distributions are consistent with expected absorption effects, and the absorption effects seem to have little or no longitude dependence. An energy and species dependence is clearly observable in the data, and progress has been made in accounting for this effect.

In Sec. III, the methods used in the analysis of particle data have been reviewed. A major uncertainty in considerations of the absorption process is the electric and magnetic field configuration in the vicinity of the satellites. These fields may change the absorption cross section which is presently taken to be the geometrical one. Methods are available for computing the time required for a satellite to absorb a given fraction of the trapped particles within its sweeping region. It is possible to calculate the diffusion coefficient of particle radial motion by using parameters obtained in considering the absorption process and by neglecting other terms (such as pitch-angle diffusion) in the diffusion equation. The data currently available probably are too unsophisticated to use the more powerful theoretical methods so that approximate expressions for the diffusion coefficient have been derived and used.

In Sec. IV we have referenced current values of the diffusion coefficient and presented a first attempt at a unified approach based on the complete set of available particle data. For self-consistency in the absorption analysis, diffusion coefficients derived from the data should be inserted into the diffusion equation which can then be used to predict the particle intensity profiles. The spatial dependence of the diffusion coefficient derived from particle observations appears steeper ($a \simeq 4$ in $D \propto L^a$) than the pre-Pioneer expectations. It should be noted, however, that the early studies were concerned mostly with the region around $L = 2$, where the synchrotron radio source is important.

Substantial improvement in absorption analysis is probably not to be expected from more refined magnetic field models, except possibly near Amalthea (Fillius *et al.* 1975*a*). For example, the maximum sweeping regions ΔL for the D2 model of Smith *et al.* (1974) are 0.28, 0.40, and 0.55 for Amalthea, Io and Europa. The corresponding values for the O_3 magnetic field model of Acuña and Ness (1975)[5] are 0.44, 0.57, and 0.63.

Acknowledgements. I thank R. W. Fillius and C. E. McIlwain for helpful comments and discussions. This work was supported in part by NASA Contract NAS 2-6552 and by NASA Grant NGL 05-005-007.

DISCUSSION

T. R. McDonough: The L^3 dependence of the diffusion coefficient which we got in the Brice theory was dependent on the model of the fluctuating winds in the upper atmosphere which we assumed. The real behavior of these winds may give a different L-dependence.

A. Mogro-Campero: I agree. I would like to mention, in that connection, that the L^2 or L^3 dependence in the pre-Pioneer literature was derived to fit the radio data, which is mainly around $L = 2$. The Pioneer data, however, cover larger values of L, and the dependence of the diffusion coefficient on L might become less steep as one approaches the planet.

C. K. Goertz: I would like to point out that the actual problem of deriving a diffusion coefficient from the observed absorption of particles by the satellites is much more complicated.

1. The satellites absorb the particles every 15 or 20 hours and thus the problem becomes strongly time dependent.
2. The solution of the diffusion equation depends on the initial and boundary conditions which are time dependent. Every time the satellite passes a particular L shell it interacts with the holes created previously.

---

[5]See, however, p. 38.

3.   The holes created by the satellites drift in azimuth. I do not believe that azimuthal symmetry is a valid assumption.

A. Mogro-Campero: I think it is clear to all who have worked on this problem that its treatment can be made more complicated. What is not clear is whether the additional complications provide a better description of the physical phenomena. Furthermore, in obtaining numerical values for diffusion coefficients one must take into account the nature of the observations available and our current knowledge of Jovian magnetospheric parameters.

The effect of the holes created by the satellites and their drift in time has been discussed by Simpson *et al.* (1974*b*). Justifications for the assumptions made by different authors are given in their papers. The relative merit of the approach you have outlined will be judged when it is presented in the literature.

D. Harris: Could these data be used to compute the net charge on Io?

A. Mogro-Campero: The net charge on Io, due to its absorption of electrons and protons of the energies which have been measured, can be estimated. The contribution to the net charge from the probably more numerous lower-energy particles, whose fluxes have not been measured, is unknown, so that the computed net charge would be of little value.

REFERENCES

Acuña, M. H., and Ness, N. F. 1975. Jupiter's main magnetic field measured by Pioneer 11. *Nature* 253:327–328, and personal communication.

Baker, D. N., and Van Allen, J. A. 1975. Energetic electrons in the Jovian magnetosphere. *J. Geophys. Res.* In press.

Birmingham, T.; Hess, W.; Northrop, T.; Baxter, R.; and Lojko, M. 1974. The electron diffusion coefficient in Jupiter's magnetosphere. *J. Geophys. Res.* 79:87–97.

Brice, N. 1972. Energetic protons in Jupiter's radiation belts. *Proceedings of the Jupiter radiation belt workshop.* (A. J. Beck, ed.) Tech. Mem. 33–543, pp. 283–302. Pasadena, California: Jet Propulsion Laboratory.

Brice, N., and McDonough, T. R. 1973. Jupiter's radiation belts. *Icarus* 18:206–219.

Coroniti, F. V. 1974. Energetic electrons in Jupiter's magnetosphere. *Ap. J. Suppl.* 27:261–281.

Davis, Jr., L. 1972. Comments on models of the Jovian radiation belts. *Proceedings of the Jupiter radiation belt workshop.* (A. J. Beck, ed.) Tech. Mem. 33–543, pp. 517–525. Pasadena, California: Jet Propulsion Laboratory.

Divine, N. 1972. Post-workshop models of Jupiter's radiation belts. *Proceedings of the Jupiter radiation belt workshop.* (A. J. Beck, ed.) Tech. Mem. 33–543, pp. 527–542. Pasadena, California: Jet Propulsion Laboratory.

Fillius, R. W., and McIlwain, C. E. 1974*a*. Radiation belts of Jupiter. *Science* 183:314–315.
——— . 1974*b*. Measurements of the Jovian radiation belts. *J. Geophys. Res.* 79:3589–3599.

Fillius, R. W.; McIlwain, C. E.; and Mogro-Campero, A. 1975*a*. Radiation belts of Jupiter: a second look. *Science* 188:465–467.

Fillius, W.; McIlwain, C. E.; Mogro-Campero, A.; and Steinberg, G. 1975*b*. Pitch angle scattering as an important loss mechanism for energetic electrons in the inner radiation belt of Jupiter. Submitted to *Geophys. Res. Lett.*

Hess, W. N. 1968. *The radiation belt and magnetosphere.* Waltham, Mass.: Blaisdell Pub. Co.

Hess, W. N.; Birmingham, T. J.; and Mead, G. D. 1973. Jupiter's radiation belts: can Pioneer 10 survive? *Science* 182:1021–1022.

———. 1974. Absorption of trapped particles by Jupiter's moons. *J. Geophys. Res.* 79:2877–2880.

Huba, J. D., and Wu, C. S. 1975. A local diffusion process associated with the sweeping of energetic particles by Io. *University of Maryland preprint.* College Park, Maryland.

Jacques, S. A., and Davis, Jr., L. 1972. Diffusion models for Jupiter's radiation belt. *California Inst. Technology Report.* Pasadena, California.

McIlwain, C. E., and Fillius, R. W. 1975. Differential spectra and phase space densities of trapped electrons at Jupiter. *J. Geophys. Res.* 80:1341–1345.

Mead, G. D. 1972. The effect of Jupiter's satellites on the diffusion of protons. *Proceedings of the Jupiter radiation belt workshop.* (A. J. Beck, ed.) Tech. Mem. 33–543, pp. 271–276. Pasadena, California: Jet Propulsion Laboratory.

———. 1974. Magnetic coordinates for the Pioneer 10 Jupiter encounter. *J. Geophys. Res.* 79:3514–3521.

Mead, G. D.; and Hess, W. N. 1973. Jupiter's radiation belts and the sweeping effect of its satellites. *J. Geophys. Res.* 78:2793–2811.

Mogro-Campero, A., and Fillius, R. W. 1976. The absorption of trapped particles by the inner satellites of Jupiter and the radial diffusion coefficient of particle transport. *J. Geophys. Res.* In press.

Mogro-Campero, A.; Fillius, R. W.; and McIlwain, C. E. 1975. Electrons and protons in Jupiter's radiation belts. *Space Research.* Vol. XV, Berlin: Akademie Verlag. In press.

Roederer, J. G. 1970. *Dynamics of geomagnetically trapped radiation.* Berlin: Springer-Verlag.

Schulz, M., and Lanzerotti, L. J. 1974. *Particle diffusion in the radiation belts.* Berlin: Springer-Verlag.

Sentman, D. D., and Van Allen, J. A. 1975. Angular distributions of electrons of energy $E_e >$ 0.06 MeV in the Jovian magnetosphere. *J. Geophys. Res.* In press.

Simpson, J. A.; Hamilton, D.; Lentz, G.; McKibben, R. B.; Mogro-Campero, A.; Perkins, M.; Pyle, K. R.; Tuzzolino, A. J.; and O'Gallagher, J. J. 1974a. Protons and electrons in Jupiter's magnetic field: results from the University of Chicago experiment on Pioneer 10. *Science* 183:306–309.

Simpson, J. A.; Hamilton, D. C.; Lentz, G. A.; McKibben, R. B.; Perkins, M.; Pyle, K. R.; Tuzzolino, A. J.; and O'Gallagher, J. J. 1975. Jupiter revisited: first results from the University of Chicago charged particle experiment on Pioneer 11. *Science* 188:455–459.

Simpson, J. A.; Hamilton, D. C.; McKibben, R. B.; Mogro-Campero, A.; Pyle, K. R.; and Tuzzolino, A. J. 1974b. The protons and electrons trapped in the Jovian dipole magnetic field region and their interaction with Io. *J. Geophys. Res.* 79:3522–3544.

Singer, S. F. 1962. Radiation belts of Venus and of Mars (with consideration of sweeping effect of Phobos). *Space Age Astronomy.* (A. J. Deutsch and W. B. Klemperer, eds.) pp. 444–461. New York: Academic Press.

Smith, E. J.; Davis, Jr., L.; Jones, D. E.; Coleman, Jr., P. J.; Colburn, D. S.; Dyal, P.; Sonett, C. P.; and Frandsen, A. M. A. 1974. The planetary magnetic field and magnetosphere of Jupiter: Pioneer 10. *J. Geophys. Res.* 79:3501–3513.

Trainor, J. H.; McDonald, F. B.; Stilwell, D. E.; Teegarden, B. J.; and Webber, W. R. 1975. Jovian protons and electrons: Pioneer 11. *Science* 188:462–465.

Trainor, J. H.; McDonald, F. B.; Teegarden, B. J.; Webber, W. R.; and Roelof, E. C. 1974a. Energetic particles in the Jovian magnetosphere. *J. Geophys. Res.* 79:3600–3614.

Trainor, J. H.; Teegarden, B. J.; Stilwell, D. E.; McDonald, F. B.; Roelof, E. C.; and Webber, W. R. 1974b. Energetic particle population in the Jovian magnetosphere: a preliminary note. *Science* 183:311–313.

Van Allen, J. A.; Baker, D. N.; Randall, B. A.; and Sentman, D. D. 1974a. The magnetosphere of Jupiter as observed with Pioneer 10. 1. Instrument and principal findings. *J. Geophys. Res.* 79:3559–3578.

Van Allen, J. A.; Baker, D. N.; Randall, B. A.; Thomsen, M. F.; Sentman, D. D.; and Flindt,
    H. R. 1974*b*. Energetic electrons in the magnetosphere of Jupiter. *Science* 183:309–311.
Van Allen, J. A.; Randall, B. A.; Baker, D. N.; Goertz, C. K.; Sentman, D. D.; Thomsen,
    M. F.; and Flindt, H. R. 1975. Pioneer 11 observations of energetic particles in the Jovian
    magnetosphere. *Science* 188:459–462.
Vesecky, J. F. 1975. Features of Jupiter's trapped particle environment associated with Jovian
    satellites—Pioneer 10 results and X-ray observations. *Geophys. J. Roy. Astron. Soc.* 41:
    331–346.

Glossary, List of Contributors, and Index

GLOSSARY[1]

albedo
geometric albedo: ratio of planet brightness at zero phase angle to the brightness of a perfectly diffusing disk with the same position and apparent size as the planet; Bond albedo: fraction of the total incident light reflected by a spherical body.

Alfvén waves
waves moving perpendicularly through a magnetic field.

aliasing
in a discrete Fourier transform, the overlapping of replicas of the basic transform, usually due to undersampling.

amagat
a unit of density; 2.687×10^{19} molecules per cm^3 under standard conditions.

Å
Ångstrom $= 10^{-8}$ cm.

arcsec
second of arc.

A.U.
Astronomical Unit $= 1.496 \times 10^{13}$ cm $\simeq 500$ light seconds.

B
magnetic field vector.

bit
binary digit (0 or 1).

Boussinesq equations
hydrodynamic equations pertaining to the onset of convection in a fluid by allowing for the variations of density only insofar as buoyancy forces are concerned.

c
velocity of light.

cm
centimeter.

CML
central meridian longitude.

Cerenkov radiation
visible (and more energetic) radiation caused by an electromagnetic shock wave arising from charged particles moving with velocities greater than the speed of light in the medium.

---

[1]In addition to *Astrophysical Quantities* by C. W. Allen (London: Athlone Press, 1973) we used *Glossary of Astronomy and Astrophysics* by J. Hopkins (Chicago: University of Chicago Press, 1976).

chondrite	a stony meteorite usually characterized by the presence of chondrules, which are small spherical grains, usually composed of iron, aluminum, or magnesium silicates. Carbonaceous chondrites are characterized by the presence of carbon compounds, while their Type I contains no chondrules.
Coriolis effect	the acceleration which a body in motion shows when observed in a rotating frame.
cosmogony	the study of the origin of celestial systems, especially the solar system.
cosmology	the study of the origin, structure, and evolution of the universe.
DAM	decametric radiation.
DIM	decimetric radiation.
differentiation (in a planet)	a process whereby the primordial substances are separated, the heavier elements sinking to the center and the lighter ones rising to the surface.
DOY	day of year (e.g., 3 Dec. 1973 and also 3 Dec. 1974 are DOY 337).
E layer	the part of Earth's ionosphere (~ 150 km) where the temperature gradient reverses and starts to rise; it reflects "shortwave" radio waves.
eV	electron volt $= 1.60 \times 10^{-12}$ ergs.
ERT	Earth received time.
ET	Ephemeris time (nearly Universal time, but corrected for changes in Earth rotation).
$FWHM$	full width of a spectral line at half-maximum intensity.
G star	star of spectral type G, for example the sun.
Galilean satellites	the four largest satellites of Jupiter: Io (J I), Europa (J II), Ganymede (J III), and Callisto (J IV).
γ	$1 \gamma = 10^{-5}$ Gauss.
IAU	International Astronomical Union.
IFT	Io-threaded flux tube; a tube of magnetic fieldlines connecting Io and Jupiter.
Jy	Jansky; 1 Jy $= 10^{-26}$ W m^{-2} Hz^{-1}.
keV	kilo electron volt.

°K	degrees Kelvin.
L	the McIlwain parameter, defined to be $L = R/\cos^2\lambda$ in a dipole field where R is radial distance in R_J and λ is the magnetic latitude.
MHD	magnetohydrodynamics.
mag	astronomical magnitude proportional to $-2.5 \log_{10}I$ where I is the intensity.
metallic hydrogen	a hypothetical form of hydrogen in which the molecules have been forced by extremely high pressures to assume the lattice structure typical of metals.
mb or mbar	millibar $= 10^{-3}$ bar $= 10^3$ dyn cm$^{-2} = 0.987 \times 10^{-3}$ atm.
Mbar	megabar $= 10^6$ bar.
MeV	million electron volt.
MJS	Mariner Jupiter/Saturn missions.
μm	1 μm = 1 micrometer = 1 micron $= 10^{-4}$ cm.
nm	nanometer $= 10^{-7}$ cm.
oblateness	ratio of the difference between the equatorial and polar radii to the equatorial radius.
obliquity	the angle between a planet's axis of rotation and the pole of its orbit.
Pa	pascal; 1 Pa $= 10^{-5}$ bar.
perijove	the point in the orbit of a spacecraft or satellite where it is closest to Jupiter's center of mass.
phase angle	solar phase angle: the angle subtended at the (center of the) planet by the directions to sun and observer.
plume	a rising column of gas over a maintained source of heat.
Rayleigh	1 R $= 10^6$ photons emitted in all directions per cm^2 vertical column per second.
R_J	Jovian radius $\simeq$ 71,400 km.
Ry	Rydberg; 1 Ry = 13.5978 eV.
scale height	the height at which a given parameter changes by a factor e.
sec	second of time.
SGC	superior geocentric conjunction.

System I Jovian rotation for latitudes $-10°$ to $+10°$, sidereal period 9^h 50^m $30\overset{s}{.}003$. Longitude was $47\overset{°}{.}31$ at 1897 July 14 Greenwich Mean Noon.

System II Jovian rotation for latitudes $-90°$ to $-10°$ and $+10°$ to $+90°$, sidereal period 9^h 55^m $40\overset{s}{.}632$. Longitude was $96\overset{°}{.}58$ at 1897 July 14 Greenwich Mean Noon.

System III Jovian radio period, sidereal rotation rate of $870\overset{°}{.}536$ per Ephemeris Day; sidereal period 9^h 55^m $29\overset{s}{.}71$. CML was $217\overset{°}{.}956$ at 1965.0.

T Tauri stars eruptive variable subgiant stars associated with interstellar matter and believed to be in the process of gravitational contraction.

terminator the line of sunrise or sunset on a planet or satellite.

Trojans Trojan asteroids occur in two (namely the ones preceding and following Jupiter in its orbit) of the five Lagrangian Points, which are points in the orbital plane of two massive particles in circular orbits around the common center of gravity where a third particle of negligible mass can remain in equilibrium.

UBV the photometric system described in *Basic Astronomical Data* (K. Aa. Strand, ed.; University of Chicago Press, 1963).

UT Universal time, which is the same as Greenwich mean time, counted from 0 hours beginning at Greenwich mean midnight.

Van Allen belts two doughnut-shaped belts of energetic charged particles trapped in Earth's magnetic field (~ 3000 km, and 18,000–20,000 km above the surface).

Voigt profile profile of a spectral line allowing for the effects of Doppler broadening combined with a Lorentz (damping) profile.

WKB Wentzel-Kramers-Brillouin approximation to Schrödinger's equation.

zodiacal light a faint glow that extends away from the sun in the ecliptic plane; it is caused by scattering of sunlight by interplanetary particles.

LIST OF CONTRIBUTORS TO THIS BOOK

M. H. Acuña, Goddard Space Flight Ctr., Greenbelt, Md.
J. K. Alexander, Goddard Space Flight Ctr., Greenbelt, Md.
L. W. Alvarez, Univ. of Calif., Berkeley, Calif.
J. D. Anderson, Jet Propulsion Lab., Pasadena, Calif.
K. R. Armstrong, Univ. of Ariz., Tucson, Ariz.
S. K. Atreya, Univ. of Mich., Ann Arbor, Mich.
W. I. Axford, Max Planck Inst. für Aeronomie, Lindau, W. Germany
A. L. Baker, Univ. of Ariz., Tucson, Ariz.
W. A. Baum, Lowell Obs., Flagstaff, Ariz.
D. B. Beard, Univ. of Kans., Lawrence, Kans.
R. Beebe, New Mex. State Univ., Las Cruces, N. Mex.
R. Beer, Jet Propulsion Lab., Pasadena, Calif.
G. L. Berge, Calif. Inst. of Technology, Pasadena, Calif.
J. T. Bergstralh, Jet Propulsion Lab., Pasadena, Calif.
E. C. Beshore, Univ. of Ariz., Tucson, Ariz.
T. Birmingham, Goddard Space Flight Ctr., Greenbelt, Md.
C. Blenman, Univ. of Ariz., Tucson, Ariz.
P. Bodenheimer, Lick Obs., Santa Cruz, Calif.
F. A. Bozyan, Univ. of Texas, Austin, Texas
R. A. Brown, Harvard Univ., Cambridge, Mass.
W. E. Brunk, NASA Headquarters, Washington, D.C.
T. Burke, NASA Headquarters, Washington, D.C.
J. A. Burns, Cornell Univ., Ithaca, N.Y.
F. H. Busse, Univ. Fridericiana Karlsruhe, Karlsruhe, W. Germany
A. G. W. Cameron, Harvard College Obs., Cambridge, Mass.
R. W. Carlson, Univ. of Southern Calif., Los Angeles, Calif.
T. D. Carr, Univ. of Florida, Gainesville, Fla.
C. R. Chapman, Planetary Sciences Inst., Tucson, Ariz.
D. L. Coffeen, Goddard Inst. for Space Studies, New York, N.Y.
M. Combes, Obs. de Paris, Meudon, France
G. J. Consolmagno, Univ. of Ariz., Tucson, Ariz.
A. F. Cook, Harvard Univ., Cambridge, Mass.
F. V. Coroniti, Univ. of Calif., Los Angeles, Calif.
D. P. Cruikshank, Univ. of Hawaii, Honolulu, Hawaii

P. C. Crump, Mauna Kea Obs., Hilo, Hawaii
R. E. Danielson, Princeton Univ., Princeton, N.J.
L. Davis, Jr., Calif. Inst. of Technology, Pasadena, Calif.
C. de Bergh, Obs. de Paris, Meudon, France
M. D. Desch, Univ. of Florida, Gainesville, Fla.
A. J. Dessler, Rice Univ., Houston, Texas
J. R. Dickel, Univ. of Ill., Urbana, Ill.
T. M. Donahue, Univ. of Mich., Ann Arbor, Mich.
J. N. Douglas, Univ. of Texas, Austin, Texas
M. Dryer, Nat. Oceanic and Atmospheric Admin., Boulder, Colo.
B. J. Duncan, Marshall Space Flight Ctr., Huntsville, Ala.
J. L. Elliot, Cornell Univ., Ithaca, N.Y.
J. Elston, Univ. of Ariz., Tucson, Ariz.
Th. Encrenaz, Obs. de Paris, Meudon, France
A. I. Ershkovich, Tel-Aviv Univ., Ramat-Aviv, Israel
V. R. Eshleman, Stanford Univ., Stanford, Calif.
A. Eviatar, Tel-Aviv Univ., Ramat-Aviv, Israel
F. Fanale, Jet Propulsion Lab., Pasadena, Calif.
C. B. Farmer, Jet Propulsion Lab., Pasadena, Calif.
T. D. Faÿ, Marshall Space Flight Ctr., Huntsville, Ala.
R. F. Fellows, NASA Headquarters, Washington, D.C.
W. Fillius, Univ. of Calif., La Jolla, Calif.
U. Fink, Univ. of Ariz., Tucson, Ariz.
G. Fjeldbo, Jet Propulsion Lab., Pasadena, Calif.
J. W. Fountain, Univ. of Ariz., Tucson, Ariz.
L. A. Frank, Univ. of Iowa, Iowa City, Iowa
A. Frohlich, Tel-Aviv Univ., Ramat-Aviv, Israel
D. Gautier, Obs. de Paris, Meudon, France
T. Gehrels, Univ. of Ariz., Tucson, Ariz.
L. J. Gleeson, Monash Univ., Clayton, Victoria, Australia
C. K. Goertz, Univ. of Iowa, Iowa City, Iowa
T. Gold, Cornell Univ., Ithaca, N.Y.
J. C. Golson, Univ. of Ariz., Tucson, Ariz.
J. C. Gradie, Univ. of Ariz., Tucson, Ariz.
R. J. Greenberg, Univ. of Ariz., Tucson, Ariz.
A. Grossman, Iowa State Univ., Ames, Iowa
S. Gulkis, Jet Propulsion Lab., Pasadena, Calif.
H. Gunawardene, Sri Lanka Astron. Assoc., Colombo, Sri Lanka
C. F. Hall, Ames Research Ctr., Moffett Field, Calif.
R. Hanel, Goddard Space Flight Ctr., Greenbelt, Md.
M. S. Hanner, Max Planck Inst. für Astronomie, Heidelberg, W. Germany
J. E. Hansen, Goddard Inst. for Space Studies, New York, N.Y.
D. Harris, Univ. of Ariz., Tucson, Ariz.
U. G. Hartmann, Univ. of Southern Calif., Los Angeles, Calif.
R. Hide, Meteorological Office, Bracknell, Berks., England
T. W. Hill, Rice Univ., Houston, Texas
J. D. Huba, Univ. of Maryland, College Park, Md.
W. B. Hubbard, Univ. of Ariz., Tucson, Ariz.
D. H. Humes, Langley Research Ctr., Hampton, Va.

G. E. Hunt, Meteorological Office, Bracknell, Berks., England
D. M. Hunten, Kitt Peak Nat. Obs., Tucson, Ariz.
J. L. Inge, Lowell Obs., Flagstaff, Ariz.
A. P. Ingersoll, Calif. Inst. of Technology, Pasadena, Calif.
D. S. Intriligator, Univ. of Southern Calif., Los Angeles, Calif.
W. H. Ip, Univ. of Calif., La Jolla, Calif.
W. M. Irvine, Univ. of Mass., Amherst, Mass.
D. Jackson, Univ. of Kans., Lawrence, Kans.
J. R. Jokipii, Univ. of Ariz., Tucson, Ariz.
A. D. Jones, III, Jet Propulsion Lab., Pasadena, Calif.
D. E. Jones, Brigham Young Univ., Provo, Utah
D. L. Judge, Univ. of Southern Calif., Los Angeles, Calif.
W. M. Kaula, Univ. of Calif., Los Angeles, Calif.
Y. Kawata, Goddard Inst. for Space Studies, New York, N.Y.
R. M. Kellerman, Royal Obs., Edinburgh, Scotland
C. Kennel, Univ. of Calif., Los Angeles, Calif.
M. J. Klein, Jet Propulsion Lab., Pasadena, Calif.
A. Kliore, Jet Propulsion Lab., Pasadena, Calif.
M. J. Komesaroff, CSIRO, Sydney, Australia
C. T. Kowal, Calif. Inst. of Technology, Pasadena, Calif.
S. M. Krimigis, Johns Hopkins Univ., Silver Spring, Md.
T. D. Kunkle, Univ. of Hawaii, Honolulu, Hawaii
H. H. Lane, National Science Foundation, Washington, D.C.
H. P. Larson, Univ. of Ariz., Tucson, Ariz.
J. Lecacheux, Obs. de Paris, Meudon, France
J. S. Lewis, Mass. Inst. of Technology, Cambridge, Mass.
H. B. Liemohn, Battelle N.W., Richmond, Wash.
F. J. Low, Univ. of Ariz., Tucson, Ariz.
R. Lüst, Max Planck Inst., München, W. Germany
J. P. Maillard, Obs. de Paris, Meudon, France
S. Marinus, Univ. of Ariz., Tucson, Ariz.
H. Mark, Ames Research Ctr., Moffett Field, Calif.
T. Z. Martin, Univ. of Hawaii, Honolulu, Hawaii
D. L. Matson, Jet Propulsion Lab., Pasadena, Calif.
M. S. Matthews, Univ. of Ariz., Tucson, Ariz.
T. Maxworthy, Univ. of Southern Calif., Los Angeles, Calif.
W. H. McCrea, 37 Houndean Rise, Lewes, Sussex, England
P. M. McCulloch, Univ. of Tasmania, Hobart, Tasmania
F. B. McDonald, Goddard Space Flight Ctr., Greenbelt, Md.
T. R. McDonough, Cornell Univ., Ithaca, N.Y.
C. E. McIlwain, Univ. of Calif., La Jolla, Calif.
R. B. McKibben, Univ. of Chicago, Chicago, Ill.
G. D. Mead, Goddard Space Flight Ctr., Greenbelt, Md.
B. Melander, Univ. of Wash., Seattle, Wash.
J. G. Melville, Brigham Young Univ., Provo, Utah
D. A. Mendis, Univ. of Calif., La Jolla, Calif.
G. Michel, Lab. Aimé Cotton, Orsay, France
J. D. Mihalov, Ames Research Ctr., Moffett Field, Calif.
R. B. Minton, Univ. of Ariz., Tucson, Ariz.

A. Mogro-Campero, General Electric Res. and Dev. Ctr., Schenectady, N.Y.
D. Morrison, Univ. of Hawaii, Honolulu, Hawaii
D. O. Muhleman, Calif. Inst. of Technology, Pasadena, Calif.
E. A. Müller, Obs. de Genève, Genève, Switzerland
G. Münch, Calif. Inst. of Technology, Pasadena, Calif.
R. E. Murphy, Maryland Acad. of Sciences, Baltimore, Md.
N. F. Ness, Goddard Space Flight Ctr., Greenbelt, Md.
G. Neugebauer, Calif. Inst. of Technology, Pasadena, Calif.
R. A. Norden, Univ. of Ariz., Tucson, Ariz.
G. W. Null, Jet Propulsion Lab., Pasadena, Calif.
G. S. Orton, Calif. Inst. of Technology, Pasadena, Calif.
T. C. Owen, State Univ. of N.Y., Stony Brook, N.Y.
D. A. Peck, Univ. of Ariz., Tucson, Ariz.
C. B. Pilcher, Univ. of Hawaii, Honolulu, Hawaii
M. Podolak, Princeton Univ., Princeton, N.J.
J. B. Pollack, Ames Research Ctr., Moffett Field, Calif.
C. Ponnamperuma, Univ. of Maryland, College Park, Md.
A. Prakash, Cornell Univ., Ithaca, N.Y.
R. G. Prinn, Mass. Inst. of Technology, Cambridge, Mass.
V. Radhakrishnan, Raman Research Inst., Bangalore, India
B. A. Randall, Univ. of Iowa, Iowa City, Iowa
D. M. Rank, Univ. of Calif., Berkeley, Calif.
L. G. Redekopp, Univ. of Southern Calif., Los Angeles, Calif.
E. J. Reese, New Mex. State Univ., Las Cruces, N. Mex.
A. C. Riddle, Univ. of Colo., Boulder, Colo.
S. T. Ridgway, Kitt Peak Nat. Obs., Tucson, Ariz.
G. H. Rieke, Univ. of Ariz., Tucson, Ariz.
S. K. Runcorn, The University, Newcastle-upon-Tyne, England
C. Sagan, Cornell Univ., Ithaca, N.Y.
E. E. Salpeter, Cornell Univ., Ithaca, N.Y.
F. L. Scarf, TRW Systems, Redondo Beach, Calif.
B. L. Seidel, Jet Propulsion Lab., Pasadena, Calif.
S. D. Shawhan, Univ. of Iowa, Iowa City, Iowa
G. T. Sill, Univ. of Ariz., Tucson, Ariz.
J. A. Simpson, Univ. of Chicago, Chicago, Ill.
W. M. Sinton, Univ. of Hawaii, Honolulu, Hawaii
W. L. Slattery, Univ. of Ariz., Tucson, Ariz.
B. A. Smith, Univ. of Ariz., Tucson, Ariz.
E. J. Smith, Jet Propulsion Lab., Pasadena, Calif.
R. A. Smith, Goddard Space Flight Ctr., Greenbelt, Md.
R. Smoluchowski, Princeton Univ., Princeton, N.J.
C. P. Sonett, Univ. of Ariz., Tucson, Ariz.
D. Stannard, Univ. of Manchester, Macclesfield, Ches., England
D. J. Stevenson, Cornell Univ., Ithaca, N.Y.
P. H. Stone, Mass. Inst. of Technology, Cambridge, Mass.
W. B. Streett, U.S. Military Acad., West Point, N.Y.
D. F. Strobel, Naval Research Lab., Washington, D.C.

D. Sweetnam, Jet Propulsion Lab., Pasadena, Calif.
F. W. Taylor, Jet Propulsion Lab., Pasadena, Calif.
B. J. Teegarden, Goddard Space Flight Ctr., Greenbelt, Md.
V. G. Teifel, Astrofizicheskii Inst., Alma-Ata, U.S.S.R.
M. F. Thomsen, Univ. of Iowa, Iowa City, Iowa
M. G. Tomasko, Univ. of Ariz., Tucson, Ariz.
M. Townsend, Univ. of Ariz., Tucson, Ariz.
L. M. Trafton, Univ. of Texas, Austin, Texas
J. H. Trainor, Goddard Space Flight Ctr., Greenbelt, Md.
V. P. Trubitsyn, Inst. Fiziki Zemli, Moscow, U.S.S.R.
G. L. Tyler, Stanford Univ., Stanford, Calif.
J. A. Van Allen, Univ. of Iowa, Iowa City, Iowa
L. Vapillon, Obs. de Paris, Meudon, France
J. E. Veverka, Cornell Univ., Ithaca, N.Y.
L. V. Wallace, Kitt Peak Nat. Obs., Tucson, Ariz.
W. Wamsteker, European Southern Obs., Santiago, Chile
J. W. Warwick, Univ. of Colo., Boulder, Colo.
L. H. Wasserman, Lowell Obs., Flagstaff, Ariz.
A. B. Weaver, Univ. of Ariz., Tucson, Ariz.
P. A. Wehinger, Tel-Aviv Univ., Ramat-Aviv, Israel
J. Westphal, Calif. Inst. of Technology, Pasadena, Calif.
G. P. Williams, Princeton Univ., Princeton, N.J.
P. M. Woiceshyn, Jet Propulsion Lab., Pasadena, Calif.
J. H. Wolfe, Ames Research Ctr., Moffett Field, Calif.
R. D. Wolstencroft, Univ. of Hawaii, Honolulu, Hawaii
S. K. Wong, Jet Propulsion Lab., Pasadena, Calif.
R. Woo, Jet Propulsion Lab., Pasadena, Calif.
C. S. Wu, Univ. of Maryland, College Park, Md.
F. M. Wu, Univ. of Southern Calif., Los Angeles, Calif.
S. Wyckoff, Tel-Aviv Univ., Ramat-Aviv, Israel
F. C. Yang, Jet Propulsion Lab., Pasadena, Calif.
A. Young, Texas A & M Univ., College Station, Texas
Y. L. Yung, Harvard Univ., Cambridge, Mass.
V. Zappala, Osservatorio Astron. di Torino, Pino Torinese, Italy
Y. Zeau, Obs. de Paris, Meudon, France
V. N. Zharkov, Inst. Fiziki Zemli, Moscow, U.S.S.R.

INDEX

The names of authors in this volume are listed in *italic* type in the index, and the page numbers of their contributions are in **boldface** type.